Habitats and Biota
of the Gulf of Mexico:
Before the Deepwater
Horizon Oil Spill

Volume 1: Water Quality, Sediments, Sediment Contaminants, Oil and Gas Seeps, Coastal Habitats, Offshore Plankton and Benthos, and Shellfish

Habitats and Biota of the Gulf of Mexico: Before the Deepwater Horizon Oil Spill

Volume 1: Water Quality, Sediments, Sediment Contaminants, Oil and Gas Seeps, Coastal Habitats, Offshore Plankton and Benthos, and Shellfish

Edited by

C. Herb Ward

Rice University, Houston, TX, USA

Authors

Mark R. Byrnes

Richard A. Davis, Jr.

Mahlon C. Kennicutt II

Ronald T. Kneib

Irving A. Mendelssohn

Gilbert T. Rowe

John W. Tunnell, Jr.

Barry A. Vittor

C. Herb Ward

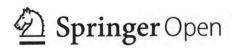 Springer Open

Editor
C. Herb Ward
Department of Civil
 and Environmental Engineering
Rice University
Houston, TX, USA
wardch@rice.edu

ISBN 978-1-4939-8053-6   ISBN 978-1-4939-3447-8   (eBook)
DOI 10.1007/978-1-4939-3447-8

Cover Design: Map data from National Oceanic and Atmospheric Administration National Centers for Environmental Information ETOPO1 Global Relief Model and ESRI USA Base Layer. Map Design by Christopher Dunn, Ramboll Environ, Portland, ME.

Printed on acid-free paper

This Springer imprint is published by Springer Nature
The registered company is Springer Science+Business Media LLC
The registered company address is: 233 Spring Street, New York, NY 10013, U.S.A.

Preface

The Deepwater Horizon accident and oil spill in the Gulf of Mexico from the Macondo well began on April 20, 2010. Oil flowed into the Gulf for 87 days until the well was capped on July 15, 2010, and declared sealed on September 19, 2010. The United States (USA) Government initially estimated that a total oil discharge into the Gulf of 4.9 million barrels (210 million U.S. gallons) resulted from the spill; however, the estimate was challenged in litigation, reduced to 3.19 million barrels by a trial court, and remains in dispute. A massive cleanup, restoration, and research program followed and continues to the present, mostly funded by BP Exploration & Production Inc. (BP).

The Deepwater Horizon accident and oil spill quickly polarized factions of both the government regulatory and scientific communities, which resulted in a continuing barrage of conflicting opinions and reports in the media and at scientific meetings. In the aftermath of the oil spill, it quickly became apparent that much of the differences in opinion being expressed about biological and ecological effects were based on individual perceptions of the status and health of the Gulf of Mexico before the spill. Because of the very large differences between the Deepwater Horizon oil spill and the next largest oil spill in the Gulf (Ixtoc l), few comparisons of pre-spill conditions and post-spill effects could be made.

BP funded cooperative research with government agencies on the effects of the Gulf oil spill and external competitively awarded independent research through their $500 million Gulf of Mexico Research Initiative (GoMRI) program. However, little of the research addressed the status and ecological health of the Gulf of Mexico *before* the Deepwater Horizon accident to serve as baseline to help assess post-spill effects.

Perhaps because of my 30-year background as the founding Editor in Chief of *Environmental Toxicology and Chemistry*, in teaching oil spill cleanup courses in the 1980s, in editing the *The Offshore Ecology Investigation* volume, and my work on tar ball formation from oil spilled in the Gulf, BP asked me to identify potential authors with appropriate expertise to research and write baseline white papers on the status and ecosystem health of the Gulf of Mexico before the Deepwater Horizon accident. Dozens of potential authors were identified and vetted for conflicts. Those selected as authors of white papers were given complete freedom to research and write their papers. I worked with the authors much in the mode of a journal editor to help them develop advanced drafts of their papers suitable for external peer review. As editor I researched and selected the peer reviewers for each paper and worked with the authors to address peer reviewer comments, which at times required preparation of additional text, figures, and tables. Author coordination meetings were held at the James A. Baker III Institute for Public Policy at Rice University.

After most of the white papers had been written, edited, and vetted by peers, BP proposed to publish them as a SpringerOpen two-volume series under the Creative Commons License for noncommercial use to promote wide distribution and free access.

   In organizing and editing this two-volume series on baseline conditions in the Gulf of Mexico before the Deepwater Horizon oil spill, I have been assisted by Diana Freeman and Mary Cormier at Rice University, Alexa Wenning, Michael Bock, Laura Leighton, Jonathan Ipock[a], and Richard Wenning at Ramboll Environ and Catherine Vogel who prepared the text and figures for preparation of page proofs by Springer. All involved in writing and editing this book series have been compensated for their time and efforts.

C. Herb Ward, Series Editor

A.J. Foyt Family Chair of Engineering, Professor of Civil and Environmental Engineering, and Professor of Ecology and Evolutionary Biology Emeritus and Scholar in Environmental Science and Technology Policy, Baker Institute for Public Policy, Rice University, Houston, TX.

---

[a]The late Jonathan "Jon" Ipock (1986-2015) tragically died too young. While working with Ramboll Environ, Inc., he tirelessly obtained documents, compiled data and references, and prepared maps and graphs for Chapter 7 (Offshore Plankton and Benthos of the Gulf of Mexico), Chapter 9 (Fish Resources of the Gulf of Mexico), and Chapter 11 (Sea Turtles of the Gulf of Mexico). During his short career Jon worked at two environmental consulting firms for more than eight years, first as a volunteer student intern, then as an associate ecologist. Jon's thirst for ecology was endless; he eagerly learned all he could and was one of ecology's rising stars.

About the Editor

C. Herb Ward

C. Herb Ward is Professor Emeritus at Rice University. He held the A. J. Foyt Family Chair of Engineering and was Professor of Civil and Environmental Engineering in the George R. Brown School of Engineering and Professor of Ecology and Evolutionary Biology in the Weiss School of Natural Sciences. He is now a Scholar in Environmental Science and Technology Policy in the James A. Baker III Institute for Public Policy at Rice University. He received his B.S. (1955) in Biology and Agricultural Science from New Mexico College of Agriculture and Mechanical Arts; his M.S. (1958) and Ph.D. (1960) in Microbial Diseases, Physiology, and Genetics of Plants from Cornell University; and the M.P.H. (1978) in Environmental Health from the University of Texas School of Public Health. He is a Registered Professional Engineer in Texas and a Board-Certified Environmental Engineer by the American Academy of Environmental Engineers. He was the founding Chair of the Department of Environmental Science and Engineering, Chair of the Department of Civil and Environmental Engineering, and the inaugural Director of the Energy and Environmental Systems Institute at Rice University. He also served as Director of the U.S. Environmental Protection Agency (USEPA)-sponsored National Center for Ground Water Research and the U.S. Department of Defense (DoD)-sponsored Advanced Applied (Environmental) Technology Development Facility. Dr. Ward was a member of the USEPA Science Advisory Board and served as Chair of the Scientific Advisory Board of the DoD Strategic Environmental Research and Development Program (SERDP). He is the founding Editor in Chief of the scientific journal *Environmental Toxicology and Chemistry* and led the development of the journal of *Industrial Microbiology and Biotechnology*. He is a Fellow in the American Academy of Microbiology (AAM), Society of Industrial Microbiology and Biotechnology (SIMB), and the Society of Environmental Toxicology and Chemistry (SETAC). Dr. Ward received the Pohland Medal for Outstanding Contributions to Bridging Environmental Research, Education, and Practice and the Brown and Caldwell Lifetime Achievement Award for Environmental Remediation in 2006, the Water Environment Federation McKee Medal for Achievement in Groundwater Restoration in 2007, and the SIMB Charles Thom Award for Bioremediation Research in 2011 and was recognized as a Distinguished Alumnus by New Mexico State University in 2013. He was a coauthor of the 2011 AAM report, *Microbes and Oil Spills*.

About the Authors

Mark R. Byrnes

Mark R. Byrnes is Principal Coastal Oceanographer at Applied Coastal Research and Engineering, Cape Cod, Massachusetts, a company he cofounded in 1998 that specializes in the analysis and modeling of coastal and estuarine physical processes. He received a B.A. in earth science from Millersville University in 1978 and a Ph.D. in geological oceanography from Old Dominion University in 1988. For the past 26 years, Dr. Byrnes has been a Principal Investigator on more than 90 wetland, estuarine, and nearshore process studies as (1) Research Scientist at the U.S. Army Engineer Research and Development Center and Coastal and Hydraulics Laboratory (formerly the Coastal Engineering Research Center); (2) Coastal Geology Section Chief for the Louisiana Geological Survey; (3) Research Professor at the Coastal Studies Institute, Louisiana State University; and (4) Consulting Oceanographer at Applied Coastal. Dr. Byrnes' primary research interests focus on regional sediment transport processes controlling coastal change and geomorphic evolution of estuarine and nearshore depositional systems, including detailed sediment budget evaluations, wetland loss delineation and classification, and wetland/shoreline restoration strategies. Much of his research has been conducted throughout the northern Gulf of Mexico; however, he has completed a number of regional sediment transport studies along the northwest U.S. coast at the mouth of the Columbia River and Grays Harbor and along the eastern seaboard from Florida to Massachusetts. He has authored more than 100 publications on coastal and estuarine physical processes.

Richard A. Davis, Jr.

Richard A. Davis is Distinguished University Professor Emeritus at the University of South Florida (USF) and Visiting Professor/Research Associate at the Harte Research Institute for Gulf of Mexico Studies at Texas A&M University-Corpus Christi. He received his B.S. (1959) from Beloit College, his M.A. (1961) from the University of Texas at Austin, and his Ph.D. (1964) from the University of Illinois-Champaign/Urbana, all in geology, and was a Postdoctoral Fellow at the University of Wisconsin-Madison (1964–1965). Dr. Davis was Assistant and Associate Professor of Geology at Western Michigan University (1965–1973), with a sabbatical year at the University of Texas, Marine Science Institute. He moved to the USF as Professor and Chair of the Department of Geology (1973–1982), Associate Dean (1982–1984), and acting Dean (spring 1984) and was promoted to Distinguished University Professor in 1988. He was the first Director of the USF Environmental Science and Policy Program (1994–1997) and retired from full-time employment in 2000.

Dr. Davis has been a Senior Fulbright Scholar and a visiting Professor at Duke University, University of North Carolina-Chapel Hill, University of Melbourne, University of Waikato (New Zealand), University of Sydney, Copenhagen University, University of Huelva (Spain), and University of Utrecht. He was a visiting Scholar at the Senckenberg Institute in Germany multiple times and a visiting Lecturer at the Universities of Tongii, Nanjing, and Peking in China. He has supervised 57 master's theses and 3 doctoral dissertations and served on multiple dissertation committees at other domestic and foreign universities. He was President of the Southeastern Geological Society, Councilor for Mineralogy, and Secretary-Treasurer of the Society for Sedimentology (SEPM), received the Shepard Medal in Marine Geology from SEPM, and is a Fellow of the Geological Society of America. Dr. Davis specializes in coastal geology with emphasis on beaches, tidal inlets, and barrier islands. He has published more than 150 peer-reviewed papers, many field trip guidebooks, and authored or edited 20 books ranging from introductory textbooks to research monographs. He was Associate Editor of the *Journal of Sedimentary Petrology* from 1984 to 1989.

Mahlon C. Kennicutt II

Mahlon C. Kennicutt is a founding Member and former Director (1998–2004) of the Geochemical and Environmental Research Group (GERG) and is Professor Emeritus of Oceanography at Texas A&M University (TAMU). He received his B.S. degree in chemistry from Union College (1974) and a Ph.D. in oceanography (1980) from TAMU. At GERG he was involved in more than $100 million of research funding, spent more than 575 days at sea, mentored 21 M.S. and Ph.D. graduate students, published over 130 scientific articles and 9 chapters in books, and participated in submersible cruises on the Johnson Sea-Link, the Diaphus, the U.S. Navy NR-1, and Pisces II submarines. In 2004, Dr. Kennicutt was named Director of Sustainable Development in the Office of the Vice President for Research at TAMU and continued to lead the Sustainable Coastal Margins Program created in 2000. In the Oceanography Department he taught oceanography, polar science, and science and policy. His research interests include environmental chemistry, organic geochemistry, the fate and effects of pollutants, environmental monitoring, ecosystem health, Antarctic environmental issues, and sustainability science.

Dr. Kennicutt first went to Antarctic as a graduate student in 1977, which marked the beginning of more than 22 years of research on the impact of humans on Antarctica. He served as the U.S. Delegate to the Scientific Committee on Antarctic Research (SCAR) for 14 years and was a SCAR Vice President from 2004 to 2008 and President from 2008 to 2012. He was an ex officio Member of the National Academies Polar Research Board for 14 years and a Science Advisor to the U.S. State Department Antarctic Treaty Delegation for 7 years and attended ten Antarctic Treaty Consultative Meetings. He has served on numerous U.S. National Academies' committees including on the effects of oil and gas exploration on the North Slope of Alaska. He is currently a Trustee and Chair of the International Science Panel of the New Zealand Antarctic Research Institute. Dr. Kennicutt has been named a National Associate of the U.S. National Academy of Sciences for life and was awarded the U.S. Antarctic Service Medal. An Antarctic geographic feature was officially named Kennicutt Point in 2006.

Ronald T. Kneib

Ronald T. Kneib is currently a Private Ecological Consultant and Sole Proprietor of RTK Consulting Services based in Hillsboro, New Mexico. He is also Senior Research Scientist Emeritus with the University of Georgia Marine Institute on Sapelo Island where he lived and conducted research in tidal marshes for 30 years (1980–2010). Dr. Kneib earned his M.S. (1976) and Ph.D. (1980) in ecology at the University of North Carolina-Chapel Hill. His most recent research links change in landscape structure with ecological processes across a range of spatial scales and includes the study of variation in functional genomics of fish as related to spatial characteristics of tidal marsh landscapes. He has authored or coauthored over 90 peer-reviewed publications, been awarded over 30 major research grants, and been an invited plenary speaker at numerous international conferences and workshops. These contributions focus on temporal and spatial variation in ecological interactions and production dynamics of fish and inverte-brate assemblages in salt-marsh and mangrove-dominated wetlands worldwide. He has served on the editorial boards of the international journals *Marine Ecology Progress Series*, *Wetlands*, and *Endangered Species Research*.

Dr. Kneib is certified as a Senior Ecologist by the Ecological Society of America's (ESA's) Board of Professional Certification and has been involved in ecological consulting since 1994. He has assisted private industry and government agencies with major restoration projects in tidal wetlands on the Atlantic, Pacific, and Gulf coasts of the United States and is also a Member of the Ecological Society of America's Rapid Response Team. In that capacity, he has represented the ESA membership in Washington, DC, by contributing to activities aimed at informing environmental policy associated with application of the Clean Water Act to tributaries and associated wetlands. In Georgia, he served on the scientific advisory board of the Center for a Sustainable Coast and was appointed by Commissioners of the Georgia Department of Natural Resources to multiple 3-year terms on the Georgia Coastal Advisory Council. He frequently provides expert witness testimony in court cases involving coastal development under Georgia's Coastal Marshlands Protection Act and the Erosion & Sedimentation Act.

Irving A. Mendelssohn

Irving A. Mendelssohn is Professor Emeritus in the Department of Oceanography and Coastal Sciences at Louisiana State University (LSU). He received his B.A. in biology from Wilkes College (1969), M.S. in marine sciences from the College of William and Mary (1973), and Ph.D. in botany from North Carolina State University (1978). He has taught courses and conducted research in coastal and wetland ecology at LSU since 1977. Professor Mendelssohn's primary research interests are the influence of environmental constraints and disturbance on coastal vegetation distribution and productivity and the management and restoration of coastal and wetland ecosystems. He has conducted wetland research throughout the United States, including the northern Gulf of Mexico, the Chesapeake Bay, the southeastern U.S. barrier strand, the Florida Everglades, and San Francisco Bay and internationally in Canada, Mexico, Portugal, Belize, Denmark, the Netherlands, China, New Zealand, and Australia. Dr. Mendelssohn has served as an Associate Editor for the journals *Wetlands* and *Estuaries* and the European botanical journal *Flora*. His honors include the Society for Wetland Scientists' Merit Award, the LSU Rainmaker Designation for outstanding teaching and research, Fellow of the Society of Wetland Scientists, honorary Doctorate from the University of Aarhus in Denmark for his collaborative research, and honorary Professor at Qingdao Institute of Marine Geology in China. He has published more than 130 scientific papers and mentored over 30 graduate students. Irv Mendelssohn is cofounder and cotrustee, with his wife, Karen McKee, U.S. Geological Survey Emeritus Scientist, of *The Wetland Foundation*™, a nonprofit private foundation devoted to fostering wetland education and research and providing financial support and guidance to students of wetland science.

Gilbert T. Rowe

Gilbert T. Rowe has more than 40 years of experience in deep-ocean ecological studies, as reflected in his 140 plus peer-refereed publications, his numerous shipboard campaigns, and his supervision of graduate students who have gone on to prestigious careers. After gaining degrees at Texas A&M University (TAMU) (B.S., 1964; M.S., 1966) and Duke (Ph.D., 1968), he spent 10 years at the Woods Hole Oceanographic Institute (1968–1979) and 8 years at the Brookhaven National Laboratory on Long Island (1979–1987). He returned to Texas as Head of the Oceanography Department at TAMU-College Station (1987–1993). In 2003, he transferred to TAMU-Galveston as Head of Marine Biology (2003–2008). A Regents Professor in the TAMU System, he is also an elected Fellow of the American Association for the Advancement of Science and a Fulbright Scholar in Chile. He has been the Chairperson of the Marine Biology Interdisciplinary (Graduate) Degree Program in the Texas A&M System. Between 2000 and 2005, Dr. Rowe was the Program Manager of the Deep Gulf of Mexico Benthos (DGoMB) program, a 5.3 million dollar effort involving more than a dozen ocean scientists from around the world. This research resulted in numerous publications in peer-refereed journals, including an issue of *Deep-Sea Research II* dedicated to the deep Gulf of Mexico. The DGoMB program is now providing an extensive database on which to base future exploration and exploitation of the deep Gulf of Mexico. His current research is focused on comprehensive assessment of the biological cycling of carbon in the Arabian (Persian) Gulf and mercury cycling in the Arabian Gulf with support from the Qatar National Research Fund. Rowe's body of work has made significant strides in fundamental understanding of animal zonation patterns, total biomass distributions, sediment community respiration, benthic-pelagic coupling, and carbon cycling. The food web models developed by Rowe are contributing to an evaluation of deep-ocean ecosystem services and a better understanding of the complexities of deep-ocean sustainability.

John W. Tunnell, Jr.

John W. Tunnell is Associate Director and Endowed Chair of Biodiversity and Conservation Science, Harte Research Institute for Gulf of Mexico Studies (HRI), and Professor Emeritus, Fulbright Scholar, and Regents Professor, Department of Life Sciences, College of Science and Engineering, Texas A&M University-Corpus Christi. He received his B.S. (1967) and M.S. (1969) degrees in biology from Texas A&I University (now Texas A&M University-Kingsville) and his Ph.D. in biology (1974) from Texas A&M University; he also studied in Colorado, California, and Florida. He started his career at Texas A&M University at Corpus Christi (now TAMU-CC) in 1974. At TAMU-CC he is Founder and former Director of the Center for Coastal Studies (1984–2009), and he assisted in the development of the Harte Research Institute (2001–2005) and its building (2003–2005). Dr. Tunnell is a Marine Ecologist and Biologist focusing primarily on coastal and coral reef ecosystems. He has extensive expertise working on coastal ecology of Texas, seashells of the Gulf of Mexico, the environmental impacts of marine oil spills, and coral reef ecology in Mexico. He has studied and published on vertebrate fossils from the seabed, sponges, brachiopods, seabirds, and Gulf of Mexico biodiversity.

Dr. Tunnell has received numerous awards, most notably a Fulbright Scholar Award (1985–1986), Regents Professor Award (1998), TAMU-CC Alumni Distinguished Professor Award (2003), Gulf Guardian Award, Bi-National Category (2006 and 2008), and the TAMU-CC Excellence in Scholarly Activity Award (2006–2007). Through spring 2015, he has published 107 peer-reviewed manuscripts and 66 technical reports and 6 books and received 154 research grants and contracts worth more than $10 million. His most notable books are *The Laguna Madre of Texas and Tamaulipas* (2001), *Coral Reefs of the Southern Gulf of Mexico* (2007), and *Encyclopedia of Texas Seashells* (2010), all published by Texas A&M University Press. He has advised or co-advised 70 M.S. students, 6 Ph.D. students, and 4 postdoctoral research associates. Dr. Tunnell is also Editor of two book series for Texas A&M University-Press.

Barry A. Vittor

Barry A. Vittor received his doctoral degree in marine ecology from the University of Oregon in 1971 and was then appointed Assistant Professor of Marine Science at the University of Alabama. He was promoted to Associate Professor in 1976. In 1977 he left the University to form a private environmental consulting firm (Barry A. Vittor & Associates, Inc.). His research in the Gulf of Mexico has focused primarily on two areas: coastal wetlands and benthic community ecology. Research methods used in his investigations include aerial photogrammetry, remote sensing, field surveys and mapping, and laboratory analysis of biological collections. His wetland work has included mapping and characterization of salt marshes and freshwater wetlands and creation and restoration of these extremely productive habitats for public and private sector clients. He is recognized as a leading expert in wetland delineation and compensatory mitigation and in federal and state regulations pertaining to wetland resources.

Dr. Vittor has performed benthic community assessments in most areas of the Gulf of Mexico, from the Florida Keys to Brownsville, Texas. These studies have involved taxonomic analysis of macroinfaunal organisms collected in estuarine, nearshore, and deepwater habitats. His taxonomy laboratory developed a taxonomic guide to the polychaetes of the northern Gulf for the Department of Interior to standardize the systematics of this group of benthic taxa among several large-scale marine ecological studies sponsored by the Bureau of Land Management. He has also conducted studies in deep-sea environments of the northeastern Gulf, including studies of natural seeps and an investigation of the effects of oil drilling and platform operations on benthic macroinfauna and meiofauna. He has written numerous reports on benthic communities in coastal habitats, including multi-year monitoring studies of areas subjected to anthropogenic alterations by navigation improvements, petroleum production, and industrial discharges. Dr. Vittor has served on numerous scientific panels that have addressed coastal environmental issues such as dredging impacts, impacts of oil and gas drilling and production, contaminated sediment assessment strategies, and coastal zone management.

External Peer Reviewers

Charles M. Adams
Food and Resource Economics
University of Florida
Gainesville, FL, USA

John B. Anderson
Sedimentology and Earth Science
Rice University
Houston, TX, USA

Susan S. Bell
Marine and Restoration Ecology
University of South Florida
Tampa, FL, USA

William F. Font
Fish Ecological Parasitology
Southeastern Louisiana University
Hammond, LA, USA

Mark A. Fraker
Marine Mammal Ecology
TerraMar Environmental Research LLC
Ashland, OR, USA

Jonathon H. Grabowski
Ecology and Fisheries Biology
Northeastern University
Boston, MA, USA

Frank R. Moore
Bird Migration and Ecology
University of Southern Mississippi
Hattiesburg, MS, USA

Pamela T. Plotkin
Sea Turtle Behavioral Ecology
Texas A&M University
College Station, TX, USA

Steve W. Ross
Marine Fish Ecology
University of North Carolina
Wilmington, NC, USA

Roger Sassen
Marine Geochemistry
Texas A&M University
College Station, TX, USA

Greg W. Stunz
Marine Biology and Fisheries
Texas A&M University
Corpus Christi, TX, USA

John H. Trefry
Chemical Oceanography
Florida Institute of Technology
Melbourne, FL, USA

Edward S. Van Vleet
Chemical Oceanography
University of South Florida
St. Petersburg, FL, USA

Contents

VOLUME 1

VOLUME 2

List of Figures

List of Tables

CHAPTER 1

HABITATS AND BIOTA OF THE GULF OF MEXICO: AN OVERVIEW

C. Herb Ward[1] and John W. Tunnell, Jr.[2]

[1]Rice University, Houston, TX 77005, USA; [2]Texas A&M University-Corpus Christi, Corpus Christi, TX 78412, USA
wardch@rice.edu; wes.tunnell@tamucc.edu

1.1 INTRODUCTION AND OVERVIEW OF CHAPTER TOPICS

The Gulf of Mexico is the ninth largest body of water in the world, and it is recognized as 1 of 64 Large Marine Ecosystems by the U.S. National Oceanic and Atmospheric Administration (NOAA) (Kumpf et al. 1999). Economically and ecologically the Gulf is one of the most productive and important bodies of water (Tunnell 2009; Fautin et al. 2010; NOS/NOAA 2011; Yoskowitz et al. 2013), occupying a surface area of more than 1.5 million square kilometers (km^2) (579,153 square miles [mi^2]), a maximum east–west dimension of 1,573 km (977 mi), and 900 km (559 mi) from north to south between the Mississippi Delta and Yucatán Peninsula. The shoreline, which extends counterclockwise from Cape Sable, Florida, to Cabo Catoche, Quintana Roo, Mexico, is approximately 5,696 km (3,539 mi) long, and it includes another 380 km (236 mi) of Gulf shoreline in Cuba from Cabo San Antonio in the west to Havana in the east (Tunnell 2009; Fautin et al. 2010).

The Gulf of Mexico basin resembles a bowl with a shallow rim around the edges. The shallow continental shelves, generally less than 200 meters (m) (656 feet [ft]), are narrow and terrigenous in the west, moderately broad and terrigenous in the north, and wide carbonate platforms in the east, adjacent to the Florida and Yucatán peninsulas. Approximately 32 % of the Gulf is continental shelf, 41 % is continental slope (200–3,000 m/656–9,843 ft), and 24 % is abyssal plain (more than 3,000 m/9,843 ft). The deepest area (more than 3,800 m/12,467 ft) occurs within the Sigsbee Deep (Darnell and Defenbaugh 1990; Tunnell 2009; Darnell 2015).

Warm, tropical waters enter the Gulf of Mexico from the Caribbean Sea between the Yucatán Peninsula and Cuba via the Yucatán Straits, where it forms the primary Gulf current—the Loop Current. Large eddies occasionally spin off this large current system and move westward (Sturges and Lugo-Fernandez 2005). After penetrating northward into the Gulf, the Loop Current loops eastward and then southward, then exits the Gulf via the Florida Straits between Florida and Cuba, where it forms one of the world's strongest and most important currents—the Gulf Stream.

As a large receiving basin, the Gulf of Mexico receives extensive watershed drainage from five countries (Canada, Cuba, Guatemala, Mexico, and the United States [U.S.]), including over two-thirds of the continental United States. The Mississippi River dominates the drainage systems in the north, and the Grijalva-Usumacinta River System dominates in the south. Thirty-three major river outlets and 207 bays, estuaries, and lagoons are found along the Gulf coastline (Kumpf et al. 1999).

C.H. Ward (ed.), *Habitats and Biota of the Gulf of Mexico: Before the Deepwater Horizon Oil Spill*,
DOI 10.1007/978-1-4939-3447-8_1

Biologically, shallow waters in the northern Gulf are warm temperate (Carolinian Province); those in the south are tropical (Caribbean Province) (Briggs 1974; Fautin et al. 2010). Oyster reefs and salt marshes are the dominant habitat type in the northern Gulf. Low-salinity estuaries and shallow-water seagrass beds are common in clearer, more saline bays. In the tropical southern Gulf of Mexico, mangrove swamps line bay and lagoon shorelines with oyster reefs, some salt marshes, and seagrasses distributed in similar salinity conditions as the northern Gulf. Along the western Gulf coastline, uniquely wedged between two wet regions, the Laguna Madre of Texas and Tamaulipas exist as the most famous of only five hypersaline lagoons in the world (Tunnell and Judd 2002). This highly productive lagoon has extensive clay dunes, wind-tidal flats, and shallow seagrass beds in a semiarid region. Offshore, coral reefs are common in the Florida Keys, Cuba, and the southern Gulf off the state of Veracruz and on the Campeche Bank (Tunnell et al. 2007). The Flower Garden Banks south of the Texas-Louisiana border represent the only coral reefs in the northern Gulf, but numerous other topographic highs or hard bottoms are found on the normally flat, soft substratum of the continental shelves of the northern Gulf (Rezak and Edwards 1972; Rezak et al. 1985; Ritchie and Keller 2008; Ritchie and Kiene 2012). Unique, recently discovered, and highly diverse habitats in deeper Gulf waters include chemosynthetic communities and communities of deepwater corals (*Lophelia* reefs) (CSA International Inc. 2007; Brooks et al. 2008; Cordes et al. 2008).

The purpose of this book series is to summarize the state of knowledge of the Gulf of Mexico environment, as well as the status and trends of its biota and habitats, before the Deepwater Horizon oil spill. Few books have ever attempted to cover the entire Gulf or most of it (Galtsoff 1954; Gore 1992; Kumpf et al. 1999), although one was released in 2015 (Darnell 2015). Alternatively, some books have covered one particular topic or discipline of the entire Gulf, as in the list below:

- Economy—Cato (2009)
- History—Weddle (1985)
- Ecosystem-based management—Day and Yanez-Arancibia (2013)
- Geology/geological oceanography—Rezak and Henry (1972); Buster and Holmes 2011
- Physical oceanography—Capurro and Reid (1972), Sturges and Lugo-Fernandez (2005)
- Biology—Pequegnat and Chace (1970)
- Shore ecology—Britton and Morton (1989)
- Biodiversity—Felder and Camp (2009)
- Beaches—Davis (2014)
- Sea-level change—Davis (2011)
- Fishes—McEachran and Fechhelm (1998, 2005)
- Marine mammals—Würsig et al. (2000)

Other books have focused on multiple topics within a particular region, such as Caso et al. (2004) and Withers and Nipper (2009) which both focus on the southern Gulf of Mexico.

Thirteen white papers on selected topics of the Gulf of Mexico were commissioned by BP and appear as Chapters 2 through 14 in these volumes. The chapters focus on baseline knowledge of the Gulf of Mexico before the Deepwater Horizon accident on April 20, 2010.

Chapters on water quality, sediment contaminants, and natural oil and gas seepage help define the physical and chemical settings for the diversity of coastal and marine habitats in the Gulf of Mexico. Plankton and benthos systems are analyzed to illustrate energy capture, trophic levels, and food webs. Chapters on status and population trends of shellfish and finfish are followed by an economic analysis of Gulf recreational and commercial fisheries. The final chapters on sea turtles, resident and migratory birds, and marine mammals, as well as fish and other animal diseases, explore threatened and endangered species issues through analysis of the historical and current status of selected indicator species.

All chapters have been written by recognized experts in the subjects covered, and most of the authors have lengthy careers in Gulf of Mexico research and are now known as distinguished, regents, or emeritus professors. Each chapter is well illustrated and referenced. Some chapters have appendices with additional supporting and reference material. Author biographies are provided in the front matter of this volume.

This introductory chapter provides a brief overview of the environmental assessment chapters that follow, stressing key points in each and ending with a conclusions section for all chapters. The reader is referred to the individual chapters for further detail on each topic covered.

1.2 WATER QUALITY IN THE GULF OF MEXICO (CHAPTER 2)

Water quality is a measure of a water body's suitability for ecosystems and/or human use. Water quality is a vital characteristic that determines how societies and humans use and value aquatic environments and other associated natural resources. Coastal and offshore environments are some of the greatest natural assets in the United States, and much of their value is critically dependent upon good water quality. Coastal, shelf, and deepwater environments are subject to numerous processes, interactions, influences, and stresses, which in turn determine the quality of the water they contain. The determinants, current status, and historical trends in water quality in the northern Gulf of Mexico are reviewed in Chapter 2, which is authored by Mahlon C. Kennicutt II, Professor Emeritus of Chemical Oceanography at Texas A&M University and long-time oceanographic researcher in the Gulf of Mexico.

The information reviewed in Chapter 2 was drawn from periodic summaries of national coastal condition reports prepared by various federal, state, and local agencies and programs. These summaries were reviewed, but the underlying primary data that provided the basis for the reports was not reanalyzed. The assessments involved were produced by a large number of expert government and academic scientists based on a vast amount of data and information from primary sources and the peer-reviewed literature. Within this context, the synthesized data comes from hundreds of sources including national program reports; water quality reporting at the federal, state, and local levels; locally organized monitoring programs; and the published literature. The reports and data collection programs are primarily from the 1990s to the mid-2000s, and they often utilize differing metrics, indicators, and methodologies for assessing and rating water quality. These region-wide assessments in this time period of approximately 20 years are the most relevant and up-to-date means of defining the present day status and trends in water quality in the northern Gulf of Mexico.

Good water quality is a concept derived from a suite of characteristics, so there is no single definition. Two key determinants of water quality in the Gulf of Mexico are physiographic setting and human activities. Several important measures of water quality include water clarity, degree of eutrophication (excessive aquatic plant growth caused by nutrient enrichment), and chemical (petroleum and nonpetroleum pollutants) and biological (pathogens) contamination. Natural and anthropogenic effects on water quality are dynamic on many scales, and this leads to considerable variability in space and time. Impacts on water quality caused by multiple factors can be additive and/or synergistic. Thus, the cumulative effects of natural and anthropogenic influences and processes that ultimately determine overall water quality and the type and mix of components used to define water quality are highly site dependent.

The patterns, current status, and historical trends in water quality in the Gulf of Mexico are complex and variable in both space and time. Assessments performed over the past two decades lead to the conclusion that water quality in most estuaries and coastal environments of the northern Gulf of Mexico is highly influenced by human activities. One of the most prevalent causes of degraded coastal water quality in the northern Gulf is anthropogenic addition of nutrients such as nitrogen and phosphorous, which cause widespread coastal eutrophication. Multiple impacts from eutrophication include lower dissolved oxygen concentrations, increases in chlorophyll *a*, and diminished water clarity, all of which can lead to toxic/nuisance algal blooms and loss of submerged aquatic vegetation (Figure 1.1). Although variable over time,

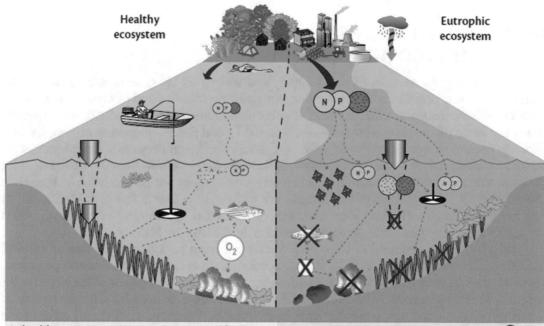

Figure 1.1. Comparison of a healthy system with no or low eutrophication to an unhealthy system exhibiting eutrophic symptoms (Figure 2.6 from Chapter 2 herein; modified from Bricker et al. 2007).

overall ecological conditions in Gulf estuaries were judged as "fair to poor," and assessments consistently concluded that water quality was "fair." At some locations, water quality appeared to be improving because of environmental regulations and controls; in other localities, conditions were continuing to deteriorate.

The current status and historical trends in water quality are highly site specific. Many Gulf of Mexico coastal environments exhibit high levels of eutrophication, and chlorophyll *a* concentrations were high, particularly along the west coast of Florida, in coastal Louisiana, and in lower coastal areas of Texas. Abundance of macroalgae and epiphytes were moderate to high in a number of locations, and low dissolved oxygen concentrations were routinely observed, particularly along coastal Florida and in the Mississippi River Plume. Loss of submerged aquatic vegetation was a consistent problem in many estuaries, and nuisance/toxic algal blooms were pervasive in many, especially in Florida, western Louisiana, and the lower Texas coast. The few improvements observed over time were attributed to better management of point and nonpoint sources of nutrients (e.g., wastewater outfalls and agricultural runoff). The intensity of human activities generally correlates with high eutrophication, although in many instances impairment of use was difficult to directly or solely relate to eutrophication or water quality. In comparing a 1999 assessment to a 2007 assessment, eutrophication conditions had worsened in one system and improved in another. A complete or comprehensive trend analysis was not possible because indicators were not always comparable. In one report, of the 38 Gulf of Mexico estuaries studied, 13 were predicted to develop worsening conditions in the future (Figure 1.2). Main factors expected to influence future trends in water quality were control and mitigation of urban runoff, wastewater treatment, industrial expansion, atmospheric deposition, animal operations, and agriculture activities. No estuaries had conditions that were expected to improve, and worsening conditions were predicted in all systems for which data were available. Trends in human population distributions, increasing development pressures, and human-associated activities were the primary factors used to predict if water quality will worsen in the future.

Direct measurements of chemical pollutants dissolved in marine waters are limited. While chemical contaminants can, and probably do, make limited contributions to water quality degradation, especially in coastal areas where concentrations are highest, impacts are masked by the overwhelmingly dominant factor of eutrophication that degrades water quality. The northwestern Gulf of Mexico has some of the highest average annual inputs of petroleum into North American marine waters, as a result of the high volume of tanker traffic, large numbers of oil and gas platforms, contaminated inflows from the Mississippi River, and the occurrence of natural oil and gas seeps. Indirect indications of possible impacts from chemical contaminants on water quality included the detection of contaminants in biological tissues and sediments. Elevated tissue concentrations of total polychlorinated biphenyls (PCBs), dichlorodiphenyltrichloroethane (DDT), dieldrin, mercury, cadmium, and toxaphene have been detected in fish tissue. However, contaminants can accumulate in biological tissues via pathways other than uptake from water. Fish consumption advisories due to mercury contamination were reported as common along the northern Gulf of Mexico, and beaches have been routinely closed or under advisories due to elevated levels of bacteria.

Once outside the influence of coastal processes, however, water quality in the Gulf of Mexico is good and has been good for a long time. Exceptions to this are hypoxic zones on the continental shelf caused by nutrient enriched waters flowing from the Mississippi River, waters just above natural oil and gas seeps, as well as localized and ephemeral effects on water quality due to the discharge of produced waters around oil and gas platforms. However, outer continental shelf, slope, and abyssal Gulf of Mexico waters remain mostly unimpaired by

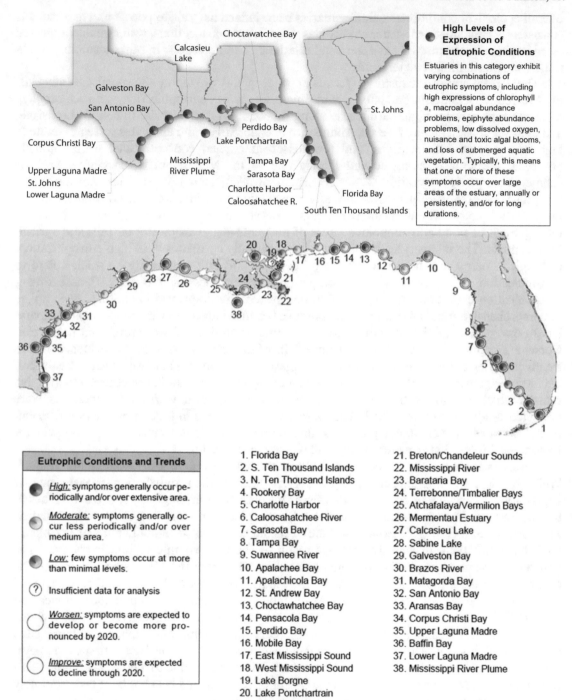

Figure 1.2. Level of expression of eutrophic conditions and future trends (Figure 2.8 from Chapter 2 herein; modified from Bricker et al. 1999).

human activities, principally because of the low levels of pollutant discharges and the large volume and mixing rates of receiving waters. Coastal Gulf of Mexico water quality is highly influenced by humans, and this will continue to be true for the foreseeable future. For the most part, future trends in water quality will be dependent on the decisions made by the populations that live, recreate, and work along the northern Gulf of Mexico coast in regard to controlling and/or mitigating those factors that degrade water quality.

Some major conclusions resulting from an extensive review of the available literature and applicable databases on water quality in the Gulf of Mexico include the following:

- Patterns and trends in Gulf of Mexico water quality are highly variable in space and through time. Water quality in Gulf of Mexico coastal environments is highly influenced by human activities, and the primary cause of degraded water quality is excess nutrients. Water quality rapidly improves with distance offshore. More than 60 % of assessed estuaries were either threatened or impaired for human use and/or aquatic life over the time period of this review—from the 1990s to the mid-2000s.

- Eutrophication has produced low dissolved oxygen and increased chlorophyll *a* concentrations, diminished water clarity, and other secondary effects including toxic/nuisance algal blooms and loss of submerged aquatic vegetation. Degraded coastal water quality was also indicated by contaminants in biological tissues and sediments, fish consumption advisories, and beach closing/advisories due to bacterial contamination.

- Gulf of Mexico continental shelf/slope and abyssal water quality was and continues to be good. Exceptions are hypoxic zones on the continental shelf, waters just above natural oil and gas seeps, and ephemeral effects due to produced water discharges during petroleum extraction. Along the northwest/central Gulf of Mexico continental shelf, the seasonal occurrence of waters with low concentrations of oxygen is geographically widespread. These "dead zones" are highly seasonal, and it has been suggested they result from water column stratification driven by weather coupled with Mississippi River outflow that delivers excess nutrients (mostly from agricultural lands) to the offshore region. It has been suggested that anthropogenic changes to the Mississippi River drainage basin and its discharges have increased the frequency and intensity of hypoxic events.

1.3 SEDIMENTS OF THE GULF OF MEXICO (CHAPTER 3)

The Gulf of Mexico is a Mediterranean-type sea with limited fetch and low tidal ranges (microtidal) throughout. The basin is somewhat like a miniature ocean in that it contains all of the main bathymetric provinces of an ocean along with a complicated coastal zone with many estuaries, barrier islands, and other features. Sediments of the Gulf of Mexico basin are the focus or emphasis of this chapter, and discussions are restricted primarily to surface sediments and only to Holocene sediments where subsurface materials are included. Richard A. Davis, who has studied the sediments and coastal geology of the Gulf of Mexico for more than 45 years, is author of this chapter. He is Distinguished Research Professor Emeritus of Coastal Geology and Sedimentology at the University of South Florida, as well as Visiting Research Associate at the Harte Research Institute for Gulf of Mexico Studies at Texas A&M University-Corpus Christi.

A broad spectrum of depositional environments exists in the Gulf of Mexico from the coast to deep water (Figure 1.3). These sediments and the processes that distribute them vary greatly over these diverse environments. The primary mechanisms that move the sediments are waves, tides, currents, and gravity. With the exception of gravity, weather is a significant influence on all of these processes. Topography also can be a factor in how these processes distribute sediments. Most sediment has its origin from the adjacent land, primarily via fluvial (stream and river) transport. Some sediment is also produced in situ through chemical or biogenic processes. Direct precipitation of calcium carbonate and evaporite minerals takes place

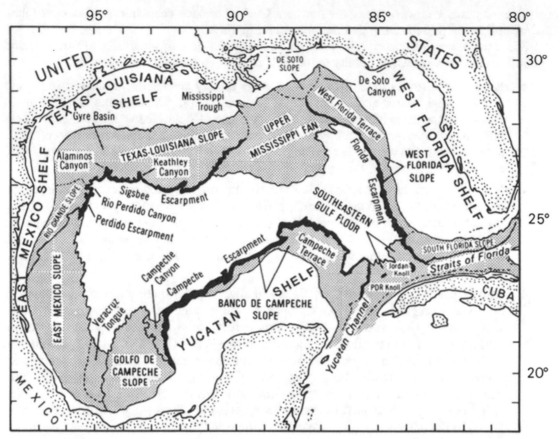

Figure 1.3. Physiographic map showing the major provinces of the Gulf of Mexico (Figure 3.2 from Chapter 3 herein; from Martin and Bouma 1978; AAPG© [1978], reprinted by permission of the AAPG whose permission is required for further use).

primarily on the Florida and Yucatán platforms, or continental shelves, but also in some coastal lagoons. Skeletal debris may comprise most of the modern bottom sediment on these two platforms, but this debris is also widely distributed in a range of varying amounts throughout all Gulf environments.

Deep Gulf environments tend to be dominated by mud in a combination of terrigenous and biogenic sediments. Biogenic components are planktonic and include coccoliths, foraminiferans, diatoms, and radiolarians. Deep terrigenous sediments are delivered by both gravity-driven processes along the continental slope and through the water column. Sedimentary gravity processes dominate the continental slope, where the sediments come to rest as deposits called turbidites. This is also a region of significant topographic relief with the steepest slopes in the Gulf of Mexico. The continental shelf is well known, and it also shows considerable variety. The carbonate platforms are shelf provinces dominated by biogenic debris and low rates of sedimentation. The remainder of the shelf is dominated by terrigenous mud and sand with varying amounts of biogenic debris. The rate of accumulation and the volume of modern sediment on the continental shelf range widely. In general, areas bounded by rivers receive the greatest volume of sediment at the highest rates of delivery. The Mississippi Delta is the extreme of this generality in that it covers almost the entire continental shelf. Some shelf areas, such as those bordering southern Texas and northern Mexico, have received little modern sediment.

Coastal sedimentary environments display the widest variety of sediments and sedimentary processes. Beaches, tidal inlets, tidal flats, wetlands, and estuaries include the full spectrum of sediments, but terrigenous sediments dominate them. The outer barrier island/inlet complexes are comprised of sediments that combine those recently delivered to the coast with the older sediments that have been reworked by rising sea level over the past 18,000 years. Sand dominates this depositional system with waves, tides, and currents being the main processes that deliver and maintain the sediments. Landward of the barrier system, sedimentary environments are much lower in physical energy. In these areas, mud is the dominant sediment texture and biogenic sediment is relatively abundant. Tidal range on the Gulf is low, meaning that tidal currents are minimal except in the inlets. Estuaries tend to be shallow, thus waves can be important in modifying sediment distribution. Sediment delivery is dominated by fluvial discharge so climate and seasons are other important factors.

In summary, the nature and distributions of sediments in the Gulf of Mexico are similar to the ocean basins. There are basically two primary provinces: terrigenous sediments carried from land to the northern and western portions of the basin and carbonate sediments that originate on the Florida and Yucatán platforms. Changes in sea level over the past several thousand years have had a major influence on sediment distributions.

The coastal systems of the Gulf of Mexico contain the most complicated sediment distributions and are dominated by terrigenous sediments. Coastal sediments are composed of mud and sand with biogenic organic debris; sand is dominating in the barrier-inlets systems and mud is the largest sediment component in the estuaries and lagoons.

1.4   SEDIMENT CONTAMINANTS OF THE GULF OF MEXICO (CHAPTER 4)

Sediments are vital components of the health of aquatic environments, but the presence of elevated concentrations of contaminants can degrade sediment quality, thereby adversely affecting organisms and ecosystems and possibly human health. The most widely found chemicals in the sediments of the northern Gulf of Mexico that have the highest likelihood of causing detrimental biological effects include polycyclic aromatic hydrocarbons (PAHs), pesticides, PCBs, and the following metals: lead, mercury, arsenic, cadmium, silver, nickel, tin, chromium, zinc, and copper (Pb, Hg, As, Cd, Ag, Ni, Sn, Cr, Zn, and Cu, respectively). The potential for harmful effects by these chemicals is multifold and includes issues related to toxicological and physicochemical properties, widespread use and release by humans, bioavailability, accumulation in sediments and lipid-rich biological tissues, and persistence in the environment. Contaminants have been released into the Gulf of Mexico for many years, and this continues today by a wide range of human activities, which are most highly concentrated in coastal areas. Accidental or intentional releases of contaminants can be traced to population centers and urban-associated discharges; agricultural practices; industrial, military, and transportation activities; and the exploration for and the production of oil and gas. The sources, current status, and historical trends of sediment contaminants in the northern Gulf of Mexico have been reviewed and summarized in this chapter by Mahlon C. Kennicutt II, Professor Emeritus of Chemical Oceanography at Texas A&M University and long-time Gulf of Mexico oceanographic researcher.

Polycyclic aromatic hydrocarbons and some of the metals have natural, as well as human-related sources. A certain amount of these chemicals ultimately end up in coastal, and to a much lesser extent, offshore sediments. Releases or inputs of these chemicals into the environment are spatially and temporally variable in both composition and concentration. Sediments are integrators of these inputs, as well as the breakdown and removal processes. The mixture of contaminants and their concentrations found in sediments at any given locale are often unique and variable over small spatial scales. Nationally sponsored regional assessments of the Gulf of Mexico are available with detailed information from the mid-1980s to the present, and these have been used extensively in this chapter. In the 1980s, the NOAA National Status and Trends Program observed that the highest concentrations of contaminants in sediments were located close to population centers. In the 1990s, the U.S. Environmental Protection Agency (USEPA) Environmental Monitoring and Assessment Program and Regional Environmental Monitoring and Assessment Program concluded that although measurable concentrations of contaminants were present in almost all estuaries of the northern Gulf of Mexico, less than 25 % of the estuarine area had contaminant concentrations that exceeded concentrations suspected of causing biological effects. USEPA's first National Coastal Condition Report (USEPA 2001) for the 1990s (1990–1997) concluded that overall coastal conditions were fair to poor with 51 % of the estuaries of the northern Gulf of Mexico in good ecological condition showing few signs of degradation due to contamination. Sediment quality at the remaining locations was judged to be poor, and contaminant concentrations exceeded concentrations suspected of causing biological effects at many locations. Most exceedances (exceeding set standards) were for pesticides and metals, while PCB and PAH exceedances occurred at <1 % of the locations. Enrichments in these sediment contaminant concentrations were directly attributed to humans. For the year 2000, the National Coastal Condition Report II (USEPA 2004) concluded that the overall condition of the northern Gulf of Mexico coast was fair. Effects range median (ERM) exceedances occurred mainly in Texas and Mobile Bay, and no exceedances were observed along the Florida Gulf Coast. Pesticides and metals exceeded concentrations suspected of causing biological effects at some locations, but only a few PCB and PAH exceedances were observed. In the National Coastal Condition Report III (USEPA 2008) covering 2001 and 2002, the sediment contaminant index was rated as fair and poor for 1 % and 2 % of coastal area, respectively, indicating that about 97 % of coastal areas had fewer than five chemicals that exceeded sediment concentrations suspected of causing biological effects. Elevated concentrations of pesticides and metals, and occasionally PCB and PAH, were observed in sediments, but only a few of them exceeded concentrations suspected of causing biological effects. In the National Coastal Condition Report IV (USEPA 2012), which covers 2003–2006, the sediment contaminants indicator was rated good, with 2 % and about 3 % of coastal area rated as fair and poor, respectively, indicating about 95 % of the coastal area had fewer than five chemicals that exceeded concentrations suspected of causing biological effects. Elevated concentrations of both metals and pesticides, as well as occasionally PCB and PAH, in sediments were observed, but few of the concentrations exceeded biological effects values. Table 1.1 is a summary of these four reports. Finer scale monitoring in selected bays revealed steep gradients in contaminant concentrations near the shore in close proximity to population centers and industrial complexes. The highest concentrations of contaminants in most coastal sediments were generally restricted to "hot spots" of limited spatial extent associated with unique contaminant sources; however, a few bays contained extensive areas of contaminated sediments.

Contaminant concentrations in sediments quickly decrease with distance offshore. Petroleum hydrocarbons found in continental shelf and slope sediments are almost exclusively due to natural oil and gas seepage. Few releases of petroleum in the offshore region that are attributable to humans reach the underlying sediments. The one exception to this is the discharge of petroleum and metal-contaminated drilling muds and cuttings from offshore oil and gas exploratory platforms. Deposits of contaminated sediments from these discharges are generally restricted to within a few hundred meters of the discharge point, and they usually occur as thin veneers less than a few meters thick, which become diluted with uncontaminated sediments with time due to the action of currents. Considering the immense area of sea bottom in the offshore region, these localized, contaminated sediment deposits are expected to have limited and local-only impact. Contaminant concentrations in these offshore areas are low, and PCBs and pesticides are generally absent. Contaminated sediments close to platforms measured

Table 1.1. Summary of Results from USEPA National Coastal Condition Reports (NCCR) I, II, III, and IV for Percent of Coastal Area Exceeding ERL (effects range low) and ERM (effects range median) Values of Chemicals in Sediments (Table 4.2 from Chapter 4 herein; from USEPA 2001, 2004, 2008, 2012)

NCCR (years of data collection)	Pesticides	Metals	PCB	PAH
I (1990–1997)	43 %	37 %	<1 %	<1 %
II (2000)	12–14 % with one pesticide or PCB exceedances	28 %	12–14 % with one pesticide or PCB exceedances	Rare
				<1 %
	<14 %		≪14 %	
III (2001–2002)	97 % of coastal with <5 ERL exceedances	97 % of coastal with <5 ERL exceedances	97 % of coastal with <5 ERL exceedances	97 % of coastal with <5 ERL exceedances
	<3 %	<3 %	≪3 %	<3 %
IV (2003–2006) [Note: 1 % of ERM exceedances were for silver in a Florida Bay]	95 % of coastal with <5 ERL exceedances	95 % of coastal with <5 ERL exceedances	95 % of coastal with <5 ERL exceedances	95 % of coastal with <5 ERL exceedances
	<5 %	<5 %	≪5 %	≪5 %
	<1 % exceed ERM	<1 % exceed ERM	<1 % exceed ERM	<1 % exceed ERM
	<2 % with <5 ERL exceedances	<2 % with <5 ERL exceedances	<2 % with <5 ERL exceedances	<2 % with <5 ERL exceedances

over a period of years were similar with a few exceptions, for example increases in Pb concentrations and microbial degradation of petroleum. This is most likely due to the low energy setting and slower rates of removal processes. It can be reasonably expected that offshore areas will remain relatively uncontaminated by chemicals attributable to humans for the foreseeable future.

Chemical contaminants in sediments continue to threaten environment quality in the coastal regions of the northern Gulf of Mexico, but sediment contamination is much less extensive in offshore regions. Elevated concentrations of pesticides and metals in coastal areas are of most concern; however, the mixtures of chemicals and their concentrations can be highly variable in both time and space. In coastal areas, pesticides and metals account for most exceedances of concentrations suspected of causing biological effects, but these exceedances appear to be decreasing with time. Nationally sponsored, region-wide assessments suggest a decrease in contamination of coastal sediments in the northern Gulf of Mexico, but there is a high degree of spatial and temporal variability from location to location. Use of some chemicals has been banned in the United States and/or decreased over time; for example, certain pesticides and sediment concentrations are expected to continue to decline. Continued reductions in emissions and discharges, as well as remediation of contaminated sites, can be expected to accelerate improvements in sediment contaminant levels, thereby reducing the role of sediment contaminants in degrading environmental quality in the northern Gulf of Mexico.

Summary findings on contaminants in Gulf of Mexico sediments resulting from extensive review of relevant literature and government synthesis reports include the following:

- Contaminant concentrations and distributions in Gulf of Mexico sediments are spatially and temporally heterogeneous over small scales due to variations in inputs, sediment deposition and accumulation rates, susceptibility to and rates of removal, chemical form, physicochemical properties, and the physical settings of receiving waters.

- Contaminants are found widely in Gulf of Mexico coastal sediments and coastal estuaries. Coastal sediments were judged to be in good to poor condition with concentrations of metals and pesticides in more than 40 % and concentrations of PAHs and PCBs in less than 1 % of coastal sediments exceeding levels suspected of causing biological effects. Within bay systems, steep gradients in contaminant concentrations were observed near population centers, agricultural activities, and industrial complexes. Contaminant concentrations decrease with distance offshore, since these regions are remote, with few exceptions, from most contaminant inputs. Natural petroleum seepage is the major source of hydrocarbons in northern-central Gulf of Mexico continental shelf/slope sediments.

- In general, levels of pesticides and contaminant metals appear to have decreased with time in coastal sediments in response to water pollution control regulations.

1.5 OIL AND GAS SEEPS IN THE GULF OF MEXICO (CHAPTER 5)

Hydrocarbon seepage is a prevalent, natural worldwide phenomenon that has occurred for millions of years, and it is especially widespread in the deepwater region of the Gulf of Mexico. As one of the most prolific oil and gas basins in the world, the Gulf of Mexico has abundant deep-seated supplies of oil and gas to migrate to the surface. The deepwater region of the Gulf of Mexico is an archetype for oil and gas seepage, and most of our worldwide knowledge of petroleum seeps is based on studies of this region. The essential geological conditions for seepage are met in many areas of the deepwater region of the Gulf of Mexico, including multiple, deeply buried mature source rocks and migration pathways to the surface. The northern Gulf of Mexico basin has been a depocenter for massive amounts of sediments over geologic time, and salt tectonics are prevalent, setting boundaries on the geographic patterns of petroleum seepage. Gulf of Mexico seeps are highly variable in composition and volume and include gases, volatiles, liquids, pitch, asphalt, tars, water, brines, and fluidized sediments.

These seeps occur on land and beneath the ocean, and they are biogenic, thermogenic, or mixed in origin. Oil and gas seeps are well known and widespread in the Gulf of Mexico region, but they are most prevalent in deeper water areas. These seeps release considerable amounts of oil and gas to the environment each year, and they are estimated to account for about 95 % of oil annually discharged to Gulf of Mexico waters. Biogenic gas seeps have a microbial metabolic origin, and microbial methane is pervasive in recent marine sediments throughout the world's oceans, including the Gulf of Mexico. Thermogenic hydrocarbons, on the other hand, rise to the surface from more deeply buried source rock horizons or accumulations. As a prolific petroleum basin, vast amounts of oil and gas have been generated beneath the deepwater of the Gulf Mexico, giving rise to widespread thermogenic seeps, which often comingled with wider spread biogenic gas seepage.

Geological processes control the location and intensity of thermogenic oil and gas seeps. The essential geological conditions that lead to thermogenic petroleum seepage include source rocks and migration pathways to the surface. Deeply buried source rocks underlie the deep waters of the Gulf of Mexico, and salt tectonics has created extensive fractures and faults in these subsurface strata. Buoyant hydrocarbons migrate along geological layers crossing strata via these fractures and faults. The distribution and interactions of these two phenomena control

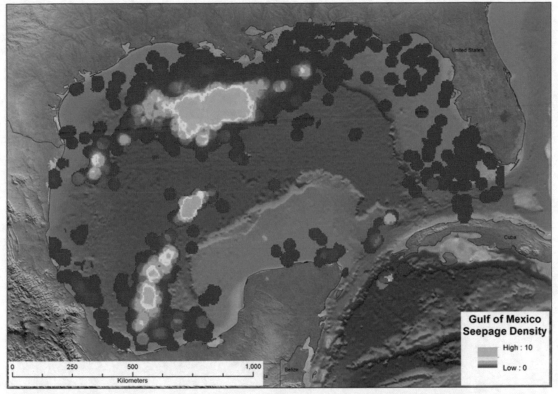

Figure 1.4. Oil and gas seepage in the Gulf of Mexico (determined from analysis of synthetic aperture radar, graphic provided by CGG's NPA Satellite Mapping, used with permission) (Figure 5.42 from Chapter 5 herein).

the amounts and spatial patterns of thermogenic oil and gas seepage with most seeps occurring in the northwestern and central deepwater region of the Gulf of Mexico (Figure 1.4). These seeps are dynamic on various timescales and can be ephemeral or persist for years. All seep compositions, whether on the land or in the sea, reflect the source of gases and liquids and postseepage alteration, or weathering, processes. Some seeps are 100 % methane, while others contain a range of petroleum hydrocarbons. Seeps interact with the surrounding environment and can range from unaltered to severely altered, mostly by microbial degradation. Seep gases can be free, adsorbed to mineral or organic surfaces, and/or entrapped in mineral inclusions.

Phenomena commonly associated with marine petroleum seeps include sea-surface slicks, water-column bubble streams and plumes, elevated hydrocarbon concentrations in sediments, seafloor mounds and pockmarks, and precipitation of authigenic minerals (Figure 1.5). The seepage of oil and gas into marine sediments initiates a complex biogeochemical cycle, and seafloor acoustic properties are altered in areas of seepage due to the presence of gases and fluids, lithification, disruption of sediment layers, and gas hydrate formation and decomposition. Gas hydrates can occur in sediments in water depths below about 500 m (1,640 ft), an upper temperature and pressure boundary for stability.

A unique ecology has evolved in association with oil and gas seeps based on chemosynthesis and symbioses. Assemblages of microbial species mediate the geological and biogeochemical processes that are essential for supporting what are commonly referred to as cold-seep communities. Cold-seep, chemosynthetic communities are common at macroseeps across the northern Gulf of Mexico continental slope, on the abyssal plain, and in the southern reaches of

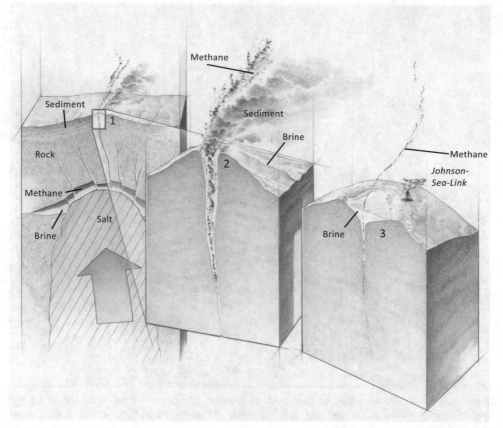

Figure 1.5. Schematic diagram of a typical marine seep location and associated features: (1) rising pillars of salt (diapirs) fracture the overlying strata creating migration pathways from deep-seated reservoirs to the near-surface; (2) the efflux of gases and fluidscan disrupt and mix with overlying sediments creating seabed mounds and/or craters that are often associated with gas and/or liquid plumes in the overlying water column; and (3) seeping brines that are denser than sea water can accumulate in the depression forming a sea-bottom lake of high salinity water (Figure 5.1 from Chapter 5 herein. MacDonald and Fisher 1996; Bruce Morser/National Geographic Creative, used with permission). *Johnson-Sea-Link* refers to a scientific research submersible (http://oceanexplorer.noaa.gov/technology/subs/sealink/sealink.html).

the Gulf of Mexico. At these locations, bacteria oxidize hydrocarbons to carbon dioxide or bicarbonate ions, which favor the formation of hard ground substrate in otherwise mostly muddy environments. Other bacteria reduce sulfate ions to hydrogen sulfide, an essential nutrient for many of the free-living and symbiotic bacteria. Common and distinctive macrofauna at these chemosynthetic community sites include tubeworms, mussels, and clams.

The prevalence, persistence, number, and volume of oil and gas seeps in the Gulf of Mexico have created a spectrum of biological, chemical, and physical characteristics that are typical of seep sites. The fluxes of crude oil, gas, and brine seepage vary over time, and therefore, cold-seep community assemblages evolve and can die when seepage abates or ceases. Thermogenic oil and gas seeps and biogenic gas seeps are pervasive and intrinsic features of the Gulf of Mexico, and thermogenic seeps will persist as long as oil and gas continue to migrate to the seafloor. Petroleum seepage in the Gulf of Mexico has occurred for millions of years and is widespread and active today.

1.6   COASTAL HABITATS OF THE GULF OF MEXICO (CHAPTER 6)

Vegetated coastal and marine habitats of the Gulf of Mexico provide a wealth of ecosystem services, such as food, employment, recreation, and natural system maintenance and regulation to the three countries bordering the Gulf: the United States, Mexico, and Cuba. The economic, ecologic, and aesthetic values of these habitats benefit human well-being as illustrated by the desire of humans to live on or near the coast. Ironically, the attraction of coastal shorelines and their varied habitats to people, along with associated demands on the exploitation of natural resources, have led to environmental pressures and degradation, or loss of many vegetated coastal and marine habitats in the Gulf of Mexico. Nevertheless, coastal habitats of the Gulf continue to represent vital components of the Gulf of Mexico ecosystem. This chapter has been written by a team of experts to cover the wide variety of topics included. Irving A. Mendelssohn is Professor Emeritus of Oceanography and Coastal Sciences at Louisiana State University and a coastal ecologist who has studied Louisiana coastal marshes for several decades; Mark R. Byrnes is a coastal oceanographer who specializes in analysis and modeling of coastal and estuarine processes at Applied Coastal Research and Engineering in Mashpee, Massachusetts; Ronald T. Kneib is a population and community ecologist and sole proprietor of RTK Consulting Services based in Hillsboro, New Mexico; and Barry A. Vittor is a wetlands and benthic community ecologist and owner of Barry A. Vittor & Associates in Mobile, Alabama.

The coastal habitats chapter reviews the physical and biological processes that control habitat formation, change, and ecological structure and function. The goal was to provide baseline information by which resource managers and decision makers can better understand and manage these important natural resources. Emphasis has been given to those vegetated marine habitats that occur immediately adjacent to the Gulf of Mexico, including barrier islands and beaches, salt marshes and mangroves, seagrasses, and reed marshes at the mouth of the Mississippi River. Also included are intertidal flats and subtidal soft bottom habitats because of their close spatial association with many of the dominant vegetated habitats.

Diverse coastal depositional systems evolved along the 6,077 km (3,776 mi) land–water interface in response to various patterns in upland drainage; groundwater supply; sediment availability; wind, wave, and current processes; relative sea-level rise; and physiographic characteristics of margin deposits. However, three depositional environments dominate: (1) carbonate deposits in Mexican States of Campeche (east of Laguna de Términos), Yucatán, and Quintana Roo, as well as the northwestern coast of Cuba and the southwestern coast of Florida; (2) terrigenous sediment in the northern Gulf of Mexico; and (3) terrigenous fluvial input along the Tamaulipas, Veracruz, and Tabasco coasts of Mexico, resulting in a mixture of fine-grained terrigenous clastics and carbonate sediment.

Vegetated marine habitats dominate these depositional shorelines, and although qualitatively similar throughout the Gulf, they vary in relative importance depending upon their location. Regional climate, geology, and riverine influence are key drivers of geographical habitat differences. Mangrove habitat is more prevalent in the Southern Gulf of Mexico Ecoregion, as well as the South Florida/Bahamian Atlantic and Greater Antilles Ecoregions, compared with the Northern Gulf of Mexico Ecoregion, where salt marshes dominate (Figure 1.6). Seagrasses occur throughout much of the Gulf, but areal extent is lower in the northern Gulf due to reduced water clarity and salinity associated with major riverine discharges of the Mississippi/Atchafalaya drainage basins. Arid environments resulting from low precipitation and high evapotranspiration in southern Texas-northwestern Mexico and the northern Yucatán generate hypersaline conditions and sedimentary habitats where rooted vegetation is stunted, absent, or replaced by algal assemblages. Such conditions stand in

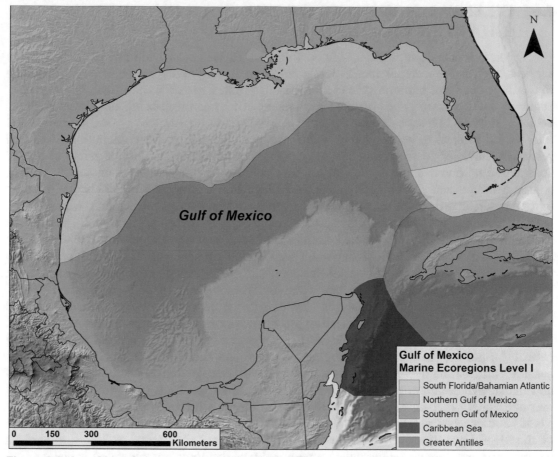

Figure 1.6. Level I marine ecoregions of the Gulf of Mexico (Figure 6.2 from Chapter 6 herein; data from Spalding et al. 2007; Wilkinson et al. 2009; and basemap from CEC 2007; French and Schenk 2005).

contrast to much of the rest of the Gulf of Mexico, where high precipitation and lush vegetated marine habitats occur. Barrier islands and beaches, as well as intertidal flats and subtidal soft bottoms, occur throughout much of the Gulf.

Vegetated habitats throughout the Gulf of Mexico play a key role in providing organic matter essential for the trophic support of coastal faunal assemblages, refugia from predation, and nursery grounds for highly valued fisheries species. For example, macroinvertebrates that live near or on the bottom (epifauna) and within the substrate (infauna) provide an important trophic base for secondary consumers. Macroinvertebrates are distributed primarily on the basis of sediment texture and quality, and vegetative cover type. Most of the numerically dominant epifaunal and infaunal taxa are found throughout the Gulf of Mexico, while others exhibit more limited geographic distributions. Species that are adapted to finer and organic-rich sediments characterize the Mississippi Estuarine and Texas Estuarine Ecoregions, while some species in the Eastern Gulf Neritic Ecoregion and South Florida/Bahamian Atlantic Ecoregion are associated primarily with biogenic sediments on the West Florida Shelf and Campeche Banks in the Southern Gulf of Mexico Ecoregion (Figure 1.7).

Coastal habitat epifauna and infauna, which play an important role in the trophic dynamics of the Gulf of Mexico ecosystems, exhibit a wide range of feeding strategies and are critical to the conversion of vegetative detritus available to higher trophic levels. Few of these taxa are

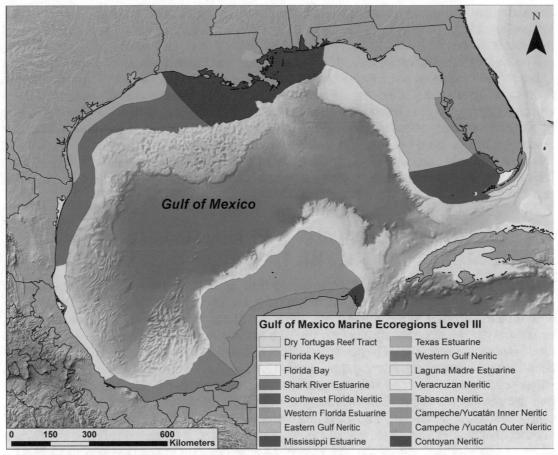

Figure 1.7. Level III marine ecoregions for the Gulf of Mexico (Figure 6.3 from Chapter 6 herein; data from Wilkinson et al. 2009 and basemap from CEC 2007; French and Schenk, 2005).

migratory as juveniles and adults, but their life histories often include a planktonic larval stage; consequently dispersal and recruitment is limited to the early life stages. Nekton, in contrast, are characterized by their mobility, and so their assemblages in the region's coastal habitats are a subset of the fishes, natant crustaceans, cephalopods, marine reptiles, and marine mammals found along the beaches, bays, lagoons, and tidal channels of the Gulf of Mexico. It is difficult to describe a characteristic nekton assemblage for individual marine habitats because the habitat of many nekton species includes multiple types of coastal wetlands; species richness and abundance are often greatest at the boundaries (i.e., edges) between subtidal (e.g., embayments) and intertidal (e.g., salt marshes) wetland habitats.

Overall, nekton assemblages connect vegetated marine habitats across the coastal landscape of the Gulf by functioning to facilitate significant energy transformations and production transfers among coastal wetland habitats and from estuaries to nearshore coastal marine environments via either diel, tidal and ontogenetic migrations (e.g., penaeid shrimps, gulf menhaden), or size-structured predator–prey interactions.

In summary, coastal habitats have experienced the greatest temporal changes in areas most susceptible to relative sea-level rise, tropical cyclones, and human disturbances. Consequently, the deltaic coast of Louisiana has the most substantial land and habitat changes in the Gulf of Mexico. Conversely, the more stable coasts of the Yucatán Peninsula, Cuba, and southwestern Florida show the least amount of change. Human disturbances are evident in areas of significant industrial activity and tourism. Human impacts are in large part tied to periodic and chronic stressors and disturbances associated with urban, agricultural, and industrial activities. Draining and filling of wetlands for human habitation, agricultural development, and industrial expansion have dramatically impacted coastal habitats throughout the Gulf. Also, overfishing and related activities have threatened important commercial fisheries in some areas of the Gulf. Other stressors such as nutrient enrichment and resulting eutrophication and hypoxia, altered hydrology from multiple causes, invasive species, and chemical pollutants including those associated with energy extraction and production have challenged the health and sustainability of vegetated marine habitats. In addition, natural disturbances driven by hurricanes, underlying geology, and floods and drought are exacerbated by human impacts. Information provided in this review should facilitate effective management and restoration of coastal habitats in the Gulf of Mexico as environmental change continues to alter their structure and function and reshape their associated biotic assemblages.

1.7   OFFSHORE PLANKTON AND BENTHOS OF THE GULF OF MEXICO (CHAPTER 7)

The plankton and benthos of the offshore Gulf of Mexico are reviewed in this chapter because of their importance as food sources for all major groups of larger organisms of economic importance to recreational or commercial fisheries (large invertebrates and finfish), or for the charismatic megafauna (mammals, birds, turtles) that are generally not subject to direct human consumption. The health and status of these groups, which can be defined by their abundance, biomass, diversity, and productivity, regulates the diversity and biomass of the larger organisms in the food web that consume them. In turn, the terminal elements of a food web are not sustainable if their food supplies fail or if their food sources are altered significantly. Finfish, commercially important invertebrates, turtles, birds, and mammals, are covered in other chapters. Gilbert T. Rowe, the author of this chapter, is Regents Professor and former Chair of the Marine Biology Interdisciplinary Degree Program in the Department of Marine Biology at Texas A&M University at Galveston.

This chapter addresses communities or assemblages of organisms in a variety of habitats. These assemblages of organisms can each be defined by their quantitative abundances and biomasses, as well as their biodiversity within volumes of water or sea-surface areas. In addition, and where useful and available, the dominant organisms of these assemblages are listed by their common and scientific names, but comprehensive species lists for these assemblages are not provided, although references in the literature cited contain such lists. The Gulf of Mexico offshore is divided into salient habitats that contain their own suites of organisms. These habitats include (1) continental shelves; (2) deep continental margins and adjacent abyssal plain; (3) methane seeps; and (4) live (hard) bottoms, partitioned according to water depths (e.g., hermatypic coral reefs in the Mexican Exclusive Economic Zone (EEZ), coral banks on salt diapirs [e.g., Flower Gardens Banks National Sanctuary off Texas], Alabama Pinnacles, Florida Middle Ground, Viosca Knolls, and Florida Lithoherms). In addition, some important exceptional habitats within these broader habitats are highlighted (continental shelf hypoxia off Louisiana, large submarine canyons [Mississippi, DeSoto, Campeche], deep iron stone sediments, and asphaltine outcroppings).

Several functional groups of organisms are reviewed: (1) phytoplankton, separated into nearshore (neritic) and open ocean assemblages; (2) zooplankton, also separated into neritic and offshore populations, with somewhat more extensive coverage of the ichthyoplankton because of its potential importance to fisheries; and (3) benthos, divided by size into the microbiota, meiofauna, macrofauna, megafauna, and demersal (near-bottom dwelling) fishes. In each case,

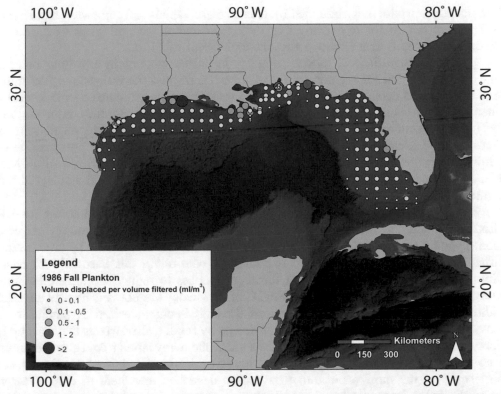

Figure 1.8. Zooplankton displacement volume in SEAMAP samples from fall sampling in the upper 200 m (656 ft) (larger than 330 μm mesh net) (data from the Southeast Area Monitoring and Assessment Program [SEAMAP], http://www.gsmfc.org/seamap.php) (Figure 7.9 from Chapter 7 herein).

brief explanations are given about what biological processes or environmental characteristics of a particular habitat control the distributions of the organisms being summarized.

Several significant generalizations can be concluded, as summarized in the baseline information referred to above. In general, low productivity and biomass of many of the larger habitats indicate that the Gulf is oligotrophic (low plant nutrients) when compared to similar habitats at higher latitudes or to continental margins characterized by tropical or equatorial upwelling. This overall generalization is based on geographically widespread assessments of phytoplankton, zooplankton, and benthic biomass. The offshore plankton and benthos are characterized by exceptional geographic variation in biomass, productivity and diversity that are controlled by physical processes and regional geology. A narrow band of highly productive habitats hug the coast around the entire circumference of the Gulf of Mexico. These biologically rich zones are fertilized regionally by rivers and, to a limited extend, upwelling. This fertilization leads to stressful seasonal hypoxia in a limited area on the continental shelf off Louisiana west of the Mississippi River. The rich nearshore productivity is in stark contrast to offshore habitats that by and large are characterized by low biomass and low productivity because the source water of the open Gulf is the nutrient-depleted Caribbean Sea (Figure 1.8). This offshore water enters the Gulf of Mexico from the Caribbean via the Yucatán Strait. This Caribbean water forms the Loop Current that curls to the right, flows back down the west coast of Florida, and then exits into the North Atlantic via the strait between Cuba and Florida. The Loop Current spins off warm eddies that create a patchwork of warm water bodies of low surface plankton productivity bounded by intermediate habitats of somewhat higher productivity. This mottling of the upper layers of the open ocean affects all levels of the food web hundreds of kilometers from shore across the entire Gulf of Mexico.

Deep benthos, regardless of its size category, declines exponentially as a function of depth and the delivery of detrital organic matter to the seafloor, and the well-established statistical regressions of these declines tend to be below similar biomass estimates on other worldwide continental margins. Likewise, the benthic biomass going down across the continental margin of the northern Gulf appears to be higher than that across the continental margin of the southern Gulf of Mexico. The deep zooplankton and the benthos species composition of the Gulf fall into depth-related zones along the continental margin of the northern Gulf of Mexico. That is, all groups of organisms appear to be zoned into discrete depth intervals, but there is substantial overlap in species composition between zones.

Several important exceptions to the oligotrophic conditions mentioned above are evident. The Louisiana continental shelf west of the Mississippi River Delta is annually subjected to seasonal hypoxia because of excessive nutrient (primarily nitrate) inflow by river water and stratification caused by fresh water. Containing or controlling this harmful and recurring condition is problematic, but improving farming practices to reduce the nitrate loading and diverting the freshwater before it reaches the Gulf of Mexico are possible helpful alternatives. In addition, much of the continental slope of the Gulf is characterized by patches of large chemosynthetic benthic organisms that are sustained by fossil hydrocarbons that seep up to the seafloor from deep deposits within the sediments. While many similar cold-seep communities have now been discovered on continental margins worldwide, the Gulf of Mexico appears to support some of the most prolific that have been described anywhere to date. Clearly, the majority of what is known today about the species composition and the chemistry, as well as physiological modes of existence, of such communities is based on studies conducted in the Gulf of Mexico.

Another exceptional habitat type with high diversity and biomass are several large submarine canyons, which are presumed to support high regional biomass by accumulating or focusing organic detritus. Likewise, such habitats provide physical complexity that enhances species richness. In addition, hard bottoms, sometimes referred to as *live* bottoms, are scattered intermittently across the entire Gulf of Mexico continental margin. These are inherently more difficult to evaluate because quantitative evaluations have to consider their three-dimensional aspects in most cases. The hard bottom also makes sampling difficult for traditional gear. Numerous sessile, large benthic organisms, both plants and animals, are attached to the seafloor in these habitats and provide a diverse physical environment that provides niches for a long list of inhabitants, from small cryptic invertebrates to large finfishes. While diversity and species lists have been compiled for these habitats utilizing cameras and direct observations with self-contained underwater breathing apparatus (SCUBA) in shallow water and submersibles and remotely operated underwater vehicles (ROV) in deeper water, quantifying biomass and rates of processes remains extremely difficult, if not impossible. Therefore, comparisons between such habitats are relative. Shallow topographic highs, or banks, on the continental shelf contain hermatypic corals that depend on light because the corals contain photosynthetic zooxanthellae (microalgae) within their tissues. Many of these banks are important to recreational fisheries, as are the many habitats formed by offshore oil and gas platforms (artificial reefs). Such complex structures are also fascinating destinations for SCUBA divers. An important example of this situation is the Flower Garden Banks natural reefs, which are surrounded by oil and gas platforms (Figure 1.9). At greater depths, such as the Alabama Pinnacles, hard bottoms on seafloor prominences have long provided popular fishing spots, but

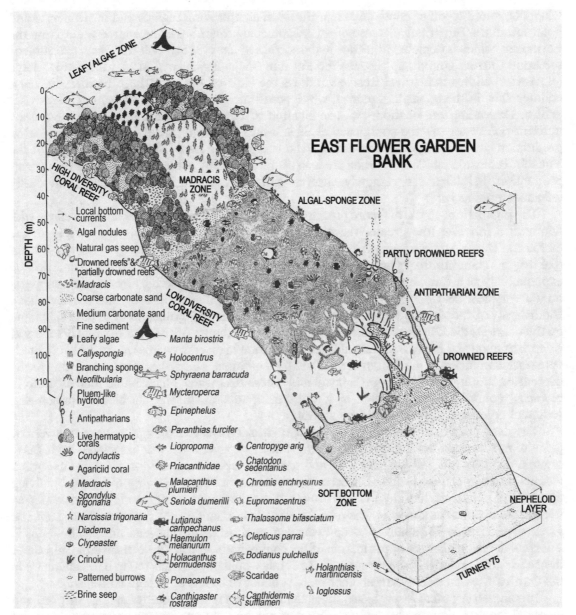

Figure 1.9. Diagram of faunal and floral zonation down the side of the East Flower Garden Bank coral reef on top of a salt diapir on the outer continental shelf off Texas. Note the salt pond and stream on the lower boundary and the bubbles appearing intermittently across the entire depth interval. Copied from Rezak et al. (1985) and based on Bright et al. (1984) (republished from Rezak et al. (1985) with permission of John Wiley and Sons Inc.; permission conveyed through Copyright Clearance Center, Inc.) (Figure 7.54 from Chapter 7 herein).

these are too deep for recreational SCUBA. Little is known about what lives on the steep, unexplored escarpments surrounding the deep Gulf of Mexico central basin.

This chapter on the plankton and benthos of the Gulf of Mexico demonstrates that the principal ecosystem components, at the lower end of the food web (phytoplankton, zooplankton, mid-water fishes, and seafloor organisms) in most habitats are characteristic of an oligotrophic ecosystem. That is, the biota is relatively low in numbers of organisms and biomass when compared to other continental margins such as upwelling regions and in temperate and polar latitudes. The principal cause of this oligotrophic condition is the source water from the Caribbean, which is depleted of nitrate in the surface to about 125 m (410 ft). The penetration of the Loop Current coming up through the Yucatán Channel spins off large, warm anticyclonic (clockwise) eddies that travel westward across the Gulf of Mexico. These features induce a counter flow in the opposite direction, which sometimes includes cyclonic (counterclockwise) eddies. Depending on location, this combination of complicated surface currents can draw nutrient-rich water off the continental shelf and into deep Gulf water, where phytoplankton production can be marginally enhanced offshore. Upwelling zones along the west coast of the Yucatán Peninsula and West Florida are also characterized by some intensification of primary production. Satellites can remotely observe most of these offshore regions of modestly enhanced productivity.

The populations in the offshore plankton represent a near-surface fauna that declines with depth as a *biocline* (the greater the distance from the surface, the more depauperate the biomass). This biocline occurs in the top 100–200 m (328–656 ft), and by a depth of 1 km (0.6 mi), the standing stocks are very limited. All size groups of multicellular organisms decline exponentially as a function of depth and distance from land, so that the abyssal plain supports only a very few seafloor organisms (fishes; zooplankton; mega-, macro-, and meiobenthos). Biodiversity of the macrobenthos follows a different pattern as a function of depth, depending on the taxon studied. In general, there is a mid-depth maximum of the macrofaunal diversity at a depth of about 1.2 km (0.8 mi). In addition, a zonation in diversity across a physical gradient is apparent with increasing depth in macrofauna, megafauna, and fishes, most likely due to the decreasing amount of food sources available. These deepwater oligotrophic (depauperate in biomass) conditions are reflected in low sediment mixing, as well as biodegradation and sediment community biomass and respiration.

The deep continental margin of the Gulf of Mexico has exceptionally complex layers of pelagic and terrigenous sediments overlying thick salt layers that are associated with fossil organic deposits (oil and gas). This oil and gas seeps up to the seafloor where it supports a distinctive and peculiar fauna. The seep-supported assemblages are very old, possibly living upwards of centuries, based on in situ growth rate experiments. Authigenic carbonate deposited at old seep areas provide substrate for deep-living, cold-water corals such as *Lophelia pertusa*, which provide habitat for deep-living demersal fishes, crustaceans, and echinoderms in a narrow depth band at the upper margin of the continental slope in the northeastern Gulf. Since the open Gulf is relatively oligotrophic, these corals would not be expected to be as abundant in the Gulf of Mexico as they are in other more productive basins or at high latitudes.

In summary, potential problems in sustaining the offshore biota (plankton, nekton, and benthos) include climate change, turbidity currents and slumps, eutrophication, oil and gas industry accidents, hypoxia, overfishing, trawling the bottom, and hurricanes. The luxuriant growths associated with topographic highs (reefs and banks) are potentially threatened by all of the above. The establishment of areas such as the Flower Gardens Banks National Sanctuary offers some protection from directly intrusive activities, but it does not provide protection from climate-induced changes that are more global. The thousands of oil and gas industry platforms in the Gulf of Mexico seem to have had a positive effect on biodiversity and fishing, but there is no uniform acceptance of these relationships. Removal of platforms on the other hand is thought to be a threat to thriving recreational fishermen and charter boat operators.

1.8 SHELLFISH OF THE GULF OF MEXICO (CHAPTER 8)

Shellfish species are highly regarded as seafood delicacies of great value. In the Gulf of Mexico, four of the five top species by value and poundage of landings are shellfish species, and therefore, great attention has been focused on their biology and fisheries. Gulf-wide, there are at least 49 officially recognized shellfish species among the three surrounding Gulf countries of the United States, Mexico, and Cuba. Of these 49 species, 28 are mollusks, 18 are crustaceans, and 3 are echinoderms. The greatest diversity of shellfish species is found in the tropical waters of the southern Gulf of Mexico, but the largest abundances and values are found in the temperate northern Gulf. Regarding the three countries surrounding the Gulf of Mexico, 16 shellfish species are taken within U.S. waters, 46 from Mexico, and 6 from Cuba. The main purpose of this chapter is to summarize the status and trends of the five major shellfish species in the northern Gulf of Mexico. The author of this chapter is John W. Tunnell, Jr. who has studied the biology and ecology of Gulf of Mexico marine life for almost 50 years. He is Associate Director and Endowed Chair of Biodiversity and Conservation Science at the Harte Research Institute for Gulf of Mexico Studies, as well as Professor Emeritus, Regent's Professor, and Fulbright Scholar, at Texas A&M University-Corpus Christi.

The waters and species of the Gulf do not recognize political boundaries. Many species range much wider than just the northern Gulf, and since the Gulf of Mexico is recognized as a large marine ecosystem, an overview of all Gulf shellfish species is provided first for better understanding of the species and their aquatic habitats. Within this chapter, shellfish species are broken into three separate categories: (1) major (5 species); (2) moderate, but important (6 species); and (3) minor (38 species) (Table 1.2). Although the moderate and minor species are briefly covered Gulf-wide, the major focus is on the northern Gulf species of brown, pink, and white shrimp, Eastern oyster, and blue crab.

Table 1.2. Relative Size and Importance of Gulf of Mexico Shellfish Fisheries (Table 8.3 from Chapter 8 herein)

Species	Country
Major fishery	
1. Eastern oyster	USA, MX
2. Brown shrimp	USA, MX
3. Pink shrimp	USA, MX
4. White shrimp	USA, MX
5. Blue crab	USA, MX
Moderate but important fishery	
1. Queen Conch	USA, MX, CU
2. Yucatán Octopus	MX
3. Mangrove Oyster	MX, CU
4. Atlantic Seabob	USA, MX
5. Spiny Lobster	USA, MX, CU
6. Florida Stone Crab	USA
Minor fishery	
1. Milk Conch	MX
2. West Indian Fighting Conch	MX

(continued)

Table 1.2. (continued)

Species	Country
3. Banded Tulip	MX
4. True Tulip	MX
5. Horse Conch	MX
6. Knobbed Welk	MX
7. Crown Conch	MX
8. West Indian Chank	MX
9. Squids (three species)	USA, MX
10. Common Octopus	MX
11. Transverse Ark	MX
12. Southern Ribbed Mussel	MX
13. American Horse Mussel	MX
14. Stiff Pen Shell	MX
15. Bay Scallop	USA, MX
16. Tiger Lucine	MX
17. Carolina Marsh Clam	MX
18. Florida Cross-barred Venus	MX
19. Southern Quahog	MX
20. Atlantic Rangia	MX
21. Brown Rangia	MX
22. Rock Shrimp	USA, MX
23. Royal Red Shrimp	USA
24. Spotted Lobster	MX
25. Swimming Crabs (six species)	MX
26. Gulf Stone Crab	USA, MX
27. Cuban Stone Crab	USA, MX, CU
28. Blue Land Crab	MX, CU
29. Sea Cucumbers (three species)	MX

USA United States, *MX* Mexico, *CU* Cuba

The biology and ecology of each species is presented, as well as its current status and historical trends over the past several decades. All species are known to vary widely or fluctuate in population levels in accordance with varying environmental conditions from year to year. In addition to these natural fluctuations, shrimp harvests also have been affected by exogenous factors, such as rising fuel costs, market competition from imported shrimp, and fleet damage from hurricanes. Overall, the shrimp populations—the most valuable of all Gulf shellfish species—seem to be flourishing, while the shrimp fishery is in decline due to these and other factors (Figure 1.10).

Oysters show the same annual environmental fluctuations, but the fishery appears to be fairly stable overall, except for hurricane damage in some places and a decadal decline in stock assessment in Louisiana (Figure 1.11). The biggest concern with oysters is the continued loss of oyster reef habitat. The blue crab fishery is quite variable from state to state with Louisiana

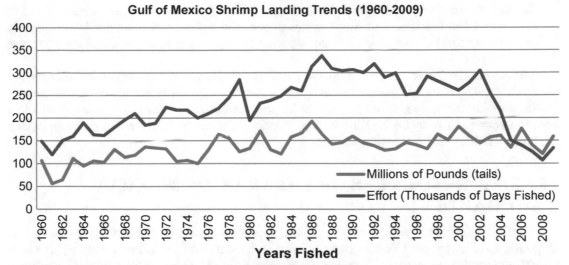

Figure 1.10. Fishery-dependent total Gulf of Mexico (U.S.) shrimp landing trends from 1960 to 2009 using NOAA Fisheries fishery-dependent data (Figure 8.17 from Chapter 8 herein; data from NOAA Fisheries).

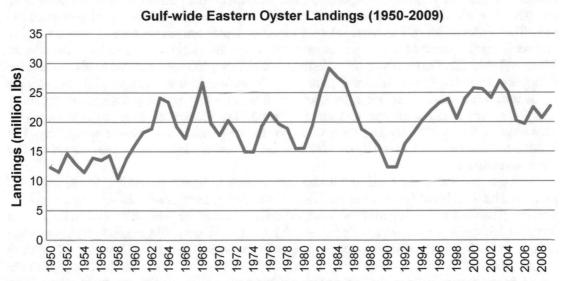

Figure 1.11. Gulf-wide Eastern oyster landings (pounds of meats) from 1950 to 2009 in the northern Gulf of Mexico (Figure 8.37 from Chapter 8 herein; data from NOAA Fisheries).

showing a continued growth; Louisiana has had the largest fishery over the past two decades. Texas shows a decrease in not only the fishery but also in the species populations statewide during the same timeframe. Gulf-wide there is agreement that healthy bays and estuaries lead to more productive fisheries, thus, conservation of some habitats and the restoration of others is needed.

In summary, four of the top five species by value and poundage of landings in the Gulf of Mexico are shellfish species (brown and white shrimp, Eastern oyster, and blue crab), but there are 49 species (28 mollusks, 18 crustaceans, and 3 echinoderms) currently taken as commercial shellfish species in the Gulf of Mexico.

Population trends of shellfish in the Gulf vary widely from year to year, primarily due to environmental fluctuations, but some landings are also influenced annually by exogenous factors (such as market competition from imported shrimp, rising fuel costs, and fleet damage due to hurricanes).

Shrimp populations are flourishing, but the shrimp fishery, the most valuable fishery in the Gulf of Mexico, is in decline due to exogenous factors, especially cheap, imported shrimp. Oyster populations appear fairly stable, but landings in Louisiana have been low for almost a decade compared to the 1990s. Blue crab populations fluctuate widely with Louisiana having the largest fishery and increasing catches.

1.9 FISH RESOURCES OF THE GULF OF MEXICO (CHAPTER 9)*

The Gulf of Mexico, with its unique oceanographic and hydrographic conditions, as well as its geological setting, provides a great diversity of habitats and therefore a dynamic ichthyofaunal community with more than 1,443 finfish species, 51 shark species, and 42 ray/skate species. This chapter evaluates and summarizes the Gulf of Mexico ichthyofaunal community and shark/ray complex, as well as population dynamics of selected key fish species of commercial and recreational importance. General distribution patterns and life history processes of fishes are evaluated, importance and contributions of fishes to the Gulf ecosystem and fisheries are described, and factors contributing to their spatiotemporal dynamics are identified. Fifteen fish species were selected for an in-depth analysis because of their ecological and economic importance and representativeness of the diversity of fish species in the Gulf of Mexico (Table 1.3). This analysis includes their life history processes, trophic levels, population dynamics, habitats, and fisheries (Figure 1.12). In addition, four groups of shark species: (1) coastal large shark complex, (2) coastal small shark complex, (3) pelagic shark complex, and (4) prohibited shark groups, and some important ray species are included in the analysis. This chapter was written by Dr. Yong Chen, Professor of Fisheries Science in the School of Marine Sciences at the University of Maine in Orono and a widely recognized expert on fish stock assessment.

Fish species within the Gulf of Mexico vary greatly in their distribution, life history, and preferred habitat. Most fish species generally use estuaries and inshore shallow waters as their nursery grounds for feeding and for refuge in order to avoid large predators when they are in larval and juvenile stages. Many finfish species spawn offshore, but currents transport their pelagic larvae into inshore shallow waters and estuaries where they spend their early life history stages. Many of the highly migratory finfish and shark species move into the estuaries in the spring to spawn in inshore shallow waters, so their young can utilize the highly productive inshore habitats for feeding and refuge. Water temperature, level of salinity, food availability, life history stage, and avoidance of predators are five of the most important habitat factors influencing the spatiotemporal distribution, recruitment dynamics, and movement of most fish species in the Gulf of Mexico. There are great diversities in the spatiotemporal distribution of different fish species in the Gulf, with some species being ubiquitous, because they are tolerant of large environmental gradients or variations, and other species being more restricted in their distributions because they require more specific types of habitat or narrow ranges of environmental parameters.

---

*Refer Chapter 9–14 are in Volume 2.

Table 1.3. Key Finfish Species of High Commercial and/or Recreational Importance in the Gulf of Mexico, Listed by Habitat (Table 9.3 from Chapter 9 herein)

Habitat	Finfish Species
Benthic	Rock hind grouper (*Epinephelus adscensionis*), Yellowfin grouper (*Mycteroperca venenosa*), Scamp grouper (*Mycteroperca phenax*), Red hind (*Epinephelus guttatus*), Atlantic goliath grouper (*Epinephelus itajara*), Nassau grouper (*Epinephelus striatus*), **Red grouper (*Epinephelus morio*)**, Gag grouper (*Mycteroperca microlepis*), Yellowedge grouper (*Hyporthodus flavolimbatus*), Mutton snapper (*Lutjanus analis*), Blackfin snapper (*Lutjanus buccanella*), **Red snapper (*Lutjanus campechanus*)**, Lane snapper (*Lutjanus synagris*), Silk snapper (*Lutjanus vivanus*), Yellowtail snapper (*Ocyurus chrysurus*), Vermillion snapper (*Rhomboplites aurorubens*), **Tilefish (*Lopholatilus chamaeleonticeps*)**, Blueline snapper (*Lutjanus kasmira*), Golden snapper (*Lutjanus inermis*), **Red drum (*Sciaenops ocellatus*)**, Black drum (*Pogonias cromis*), Bluefish (*Pomatomus saltatrix*), Common snook (*Centropomus undecimalis*), Crevalle jack (*Caranx hippos*), Spotted seatrout (*Cynoscion nebulosus*), and **Striped mullet (*Mugil cephalus*)**
Pelagic and highly migratory	Skipjack (*Katsuwonus pelamis*), Albacore (*Thunnus alalunga*), Bigeye (*Thunnus obesus*), **Atlantic bluefin tuna (*Thunnus thynnus*)**, Yellowfin tuna (*Thunnus albacores*), Small tunas, **Atlantic blue marlin (*Makaira nigricans*)**, White marlin (*Tetrapturus albidus*), **Atlantic sailfish (*Istiophorus albicans*)**, and **Atlantic swordfish (*Xiphias gladius*)**
Pelagic	**Dolphinfish (*Coryphaena hippurus*)**, Spanish mackerel (*Scomberomorus maculatus*), Cobia (*Rachycentron canadum*), Atlantic thread herring (*Opisthonema oglinum*), **King mackerel (*Scomberomorus cavalla*)**, Spanish sardine (*Sardinella aurita*), **Menhaden (*Brevoortia* spp.)**, and **Greater amberjack (*Seriola dumerili*)**

Species *highlighted* were selected for evaluation

Finfish and shark species support important commercial and recreational fisheries, and these are two of the most important industries in the Gulf of Mexico. Gulf fisheries are some of the most productive in the world. Overall, approximately 25 % of U.S. commercial fish landings and 40 % of recreational harvest occur in the Gulf of Mexico. However, a wide variety of long-term anthropogenic and natural stressors, such as rapid coastal development with subsequent degraded water quality and habitat loss, heavy fishing pressure, a large quantity of bycatch in shrimp fisheries, climate change, and natural disasters have negatively impacted the Gulf of Mexico ecosystem and its fishery species. The Gulf receives about 50 % of all U.S. watershed discharge, and there are over 3,100 point source outfalls in the northern Gulf. Pesticides and fertilizers (nutrients) used in the watersheds of the states bordering the Gulf exceed those used in any of the other coastal zones in the United States. During a 1997–2000 assessment, 59 % of the estuarine areas of the Gulf, which are essential nursery and spawning grounds for many finfish and shark species, were considered impaired or threatened. A 2007 study suggested that 78 km^2 (30 mi^2) of coastal wetlands were being lost annually, and that 20–100 % of the seagrass had been destroyed in some areas of the Gulf of Mexico. High fishing mortality in the Gulf, as a result of target fishery and bycatch, reduces stock reproductive potential and impairs the ability of fish stocks to recover from low fish stock abundance. Many fish stocks of high commercial and recreational importance in the Gulf of Mexico were found to be overfished (population level too low) and/or in a state of overfishing (fishing mortality too high) in the 1990s and

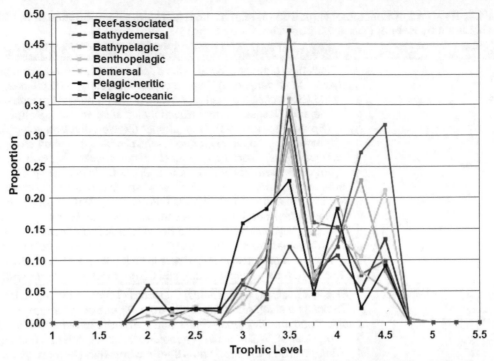

Figure 1.12. **The distribution of trophic levels for fish, shark, and ray species of different habitats in the Gulf of Mexico. Trophic level measures the number of steps the fish, shark, or ray is from the start of the food chain: 1 = primary producers that make their own food, such as plants and algae; 2 = primary consumers, such as herbivores consuming primary producers; 3 = secondary consumers, such as carnivores eating herbivores; 4 = tertiary consumers, such as carnivores eating other carnivores; and 5 = apex predators that are at the top of the food chain with no predators (data from FishBase 2013) (Figure 9.2 from Chapter 9 herein).**

2000s. These long-term anthropogenic and natural stressors have reduced resilience and robustness of the ichthyofaunal community in the Gulf with respect to human and natural perturbations. Management regulations recently adopted in the fisheries industry, to limit fishing efforts and bycatch in the shrimp fishery, appear to have worked for some finfish species by reducing the number of overfished fish populations and the frequency of occurrence of overfishing in the Gulf of Mexico. Summary findings include the following:

- No formal stock assessments were done for the vast majority of fish species in the Gulf of Mexico immediately prior to the Deepwater Horizon oil spill.

- Of the 15 finfish species evaluated in this chapter, 5 species were being overfished and/or were in the status of overfishing in 2010, including red snapper, red grouper (some local subpopulations), Atlantic bluefin tuna, Atlantic blue marlin, and greater amberjack.

- Of 39 shark species included in the shark Fisheries Management Plan in the Gulf of Mexico, 19 species have been listed as commercially and recreationally prohibited species because of very low population biomass and poor stock conditions.

- Finfish species evaluated in this study that were determined not overfished in the Gulf of Mexico immediately before the Deepwater Horizon oil spill included menhaden, Atlantic swordfish, Atlantic sailfish, red drum, striped mullet, tilefish, king mackerel, Gulf flounder, and dolphinfish.

1.10 COMMERCIAL AND RECREATIONAL FISHERIES OF THE GULF OF MEXICO (CHAPTER 10)

Given its diversity of species, the Gulf of Mexico offers opportunities to both commercial and recreational fishermen. The objective of this chapter is to provide a systematic examination of the commercial and recreational fishing sectors of the Gulf of Mexico, focusing on a variety of topics. The coauthors of this chapter are Walter R. Keithly and Kenneth J. Roberts of Louisiana State University (LSU). Keithly is Associate Professor in the Center for Natural Resource Economics and Policy, Department of Agricultural Economics, and Roberts is Associate Vice Chancellor Emeritus of the LSU Ag Center.

Commercial fisheries are generally described and reported by either landings in weight or value in dollars. Aggregate finfish and shellfish landings attributed to the U.S. Gulf states fluctuate, but the ranking of the states does not change much from year to year (Figure 1.13). Louisiana ranks first due to landings in the five major species (menhaden, brown and white shrimp, blue crab, and oysters).

When examined at the state level, the dockside value of all landings is mostly concentrated in Louisiana and Texas, with shares of 43 % and 26 %, respectively (Figure 1.14). Economic impacts include sales, income, and value added, originating from both landings and imports (Figure 1.15).

With respect to the commercial sector, some of the topics considered in this chapter include trends in production of various species, the value of production associated with these various species, the impact of imports on dockside prices, and processing. Overall, long-term landings of most key commercial species (menhaden, shrimp, blue crab, and oyster) appear to be stable, and recognized changes, where noted, appear to be tied to regulations to manage fish stocks. This is particularly true with respect to finfish stocks. Of all the commercial fisheries examined, the shrimp fishery faces the greatest obstacles in terms of long-term viability. The increasing volume of imports has led to a significant decline in the price that shrimpers receive for the harvested product, and in turn, a reduction in profitability. This reduction has led to a substantial downsizing of the shrimp fishing industry with current effort in the fishery

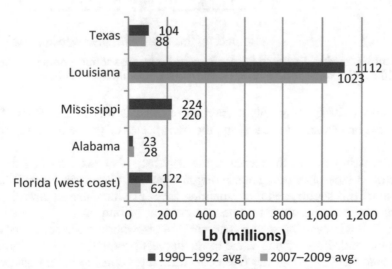

Figure 1.13. Average annual landings by state, 1990–1992 and 2007–2009 (1 lb is equal to 0.454 kg) (Figure 10.6 from Chapter 10 herein; data source from personal communication with National Marine Fisheries Service (NMFS), Fisheries Statistics Division).

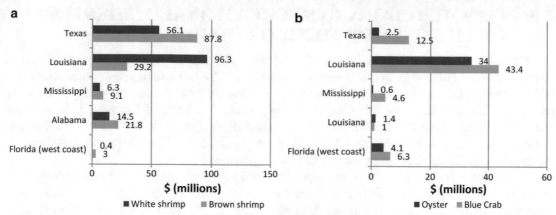

Figure 1.14. Value of commercial landings by state and species (shrimp, *left panel*; oysters and blue crab, *right panel*), 2007–2009 average (Figure 10.9 from Chapter 10 herein; data source from personal communication with NMFS, Fisheries Statistics Division).

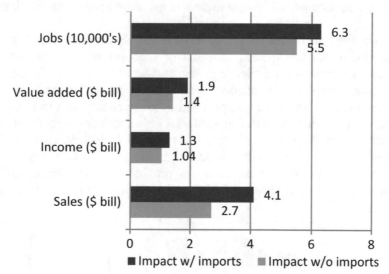

Figure 1.15. Gulf of Mexico commercial seafood industry economic impact, 2009 (Figure 10.10 from Chapter 10 herein; data source, U.S. Department of Commerce 2011).

(measured in days fished) being only a fraction of what it was in the 1990s. This statement applies to both the brown and white shrimp, the two species of prime relevance in the northern Gulf of Mexico.

Like the harvesting sector, the increasing import base also has impacted the Gulf shrimp-processing sector. A steadily eroding marketing margin and, presumably, profit has culminated in consolidation of this sector, and the remaining firms are increasing output in an attempt to counterbalance the declining marketing margin per unit of output.

Direct jobs in the harvesting sector generate jobs elsewhere in the economy via companies that supply inputs and those adding value to the harvest product, which is ultimately, in turn, used by the consumer. In four of the five U.S. Gulf states considered in this analysis (Florida was excluded because the west coast data could not be differentiated from the east coast), seafood industry jobs averaged 92,000 annually from 2007 to 2009. However, the four-state employment fell from 109,000 in 2007 to 63,000 in 2009. Income impacts for the four states equaled $2.1 billion in 2009, and that represented a decline when compared to 2007.

Regarding the recreational sector, topics considered in this chapter include expenditures and impact, angler participation, trips, and catch and harvest. The analysis was based almost exclusively on Marine Recreational Information Program statistics, the new name for Marine Recreational Fisheries Statistics Survey (MRIP/MRFSS), the most continuous and long-term monitoring program on recreational fishing patterns available. Texas opted out of this program and, therefore, is largely excluded from this report with the exception of expenditures and impacts. At the top end in terms of economic impacts, about 42,000 jobs were generated in Florida in response to recreational fishing activities, with an associated $2.4 billion in income. On the bottom end, about 3,200 jobs were generated in Mississippi, with an associated income of $162 million. Louisiana was in the middle of these numbers, with the generation of almost 20,000 jobs and almost $1.0 billion in additional income. Table 1.4 shows the economic impact associated with Gulf of Mexico angling activities from 2006 to 2009.

Overall, marine recreational fishing participation in three of the four states increased significantly from the mid-1990s, with Mississippi being the sole exception. While fishing participation increased substantially, much of the growth occurred prior to the mid-2000s. It is likely that the combination of high fuel prices in recent years, along with the downturn in the economy, negatively influenced both participation and the number of trips.

While the MRFSS/MRIP represents the primary data source for tracking participation over time, state-issued marine fishing license sales also can be used to track changes, but this is subject to a number of caveats. A comparison between MRFSS/MRIP participation estimates and license sales for both Louisiana and Mississippi was prepared to determine whether license sales track with MRFSS/MRIP estimates in a reasonable manner. Disturbingly, some significant differences were noted with the MRFSS/MRIP estimates, which exceeded license sales by a large margin. While there are explanations for these observed differences (for example, a license is not required for saltwater fishing in Louisiana for those under the age of 16), the differences are large enough to justify further examination of the MRFSS/MRIP participation data.

The number of Gulf angler trips (excluding Texas) increased from about 17 million annually during the decade of the 1990s to 23 million annually during the 2000s, with a sharp increase in the number of angler trips beginning in 2000. The explanation for this sharp increase in the number of angler trips is open to speculation, but it does coincide with a sharp increase in the number of nonresident participants in Florida. Florida accounted for approximately 70 % of total Gulf trips during the analysis period, and about one-half of those trips were in inland waters. Louisiana accounted for another 17 % of the total, and about 85 % of the Louisiana-based trips were taken in inland waters.

Given that the vast majority of Louisiana's fishing activities take place in inshore waters, it comes as no surprise that targeting behavior and catch are also largely associated with those species utilizing inshore habitat, and the two primary species include red drum and spotted seatrout. Fully 50 % of all Louisiana-based angling trips target spotted seatrout, and with the catch averaging about 20 million fish per year, Louisiana accounts for about 60 % of the Gulf's total spotted seatrout catch, in terms of numbers of fish. Similarly, Louisiana accounts for about 80 % of the U.S. Gulf of Mexico red drum harvest, in terms of pounds.

While there is considerable red drum and spotted seatrout catch in Florida waters, the state can also lay claim to a large offshore fishery component, where reef fish are generally the target.

In summary, given its diversity of species, the Gulf of Mexico offers ample opportunities to both commercial and recreational fishermen. Both of these sectors generate considerable

Table 1.4. Economic Impacts Associated with Gulf of Mexico Angling Activities, 2006–2009 (Table 10.5 from Chapter 10 herein; Data Source from U.S. Department of Commerce 2011)

Location	Jobs	Sales ($1000 s)	Value Added ($1000 s)	Income ($1000 s)
2006				
Florida (West Coast)	75,257	7,823,752	4,235,087	NA
Alabama	6,572	630,181	325,523	NA
Mississippi	3,731	490,501	189,450	NA
Louisiana	26,612	2,382,034	1,199,333	NA
Texas	34,175	4,197,011	2,154,891	NA
Total	**146,347**	**15,523,479**	**8,104,284**	**NA**
2007				
Florida (West Coast)	65,799	6,829,434	3,704,818	NA
Alabama	6,759	654,353	337,493	NA
Mississippi	4,707	616,930	239,021	NA
Louisiana	27,446	2,453,392	1,234,449	NA
Texas	23,382	3,004,862	1,514,791	NA
Total[a]	**128,093**	**13,558,971**	**7,030,572**	**NA**
2008				
Florida (West Coast)	54,589	5,650,068	3,075,710	NA
Alabama	4,719	455,093	235,481	NA
Mississippi	2,930	382,778	148,837	NA
Louisiana	25,590	2,297,078	1,156,796	NA
Texas	25,544	3,288,135	1,656,545	NA
Total[a]	**113,372**	**12,073,152**	**6,273,369**	**NA**
2009				
Florida (West Coast)	42,314	4,369,022	1,532,821	2,385,738
Alabama	4,924	474,746	155,663	245,437
Mississippi	3,188	417,080	105,472	162,099
Louisiana	19,688	1,774,692	578,767	894,123
Texas	22,127	2,846,858	910,011	1,434,733
Total[a]	**92,241**	**9,900,398**	**3,282,734**	**5,122,130**

[a]The "total" figures should be considered a minimum since they do not account for any trade among individual Gulf States (estimated by authors)

Note: NA not available. *Source*: U.S. Department of Commerce (various issues) (available at: http://www.st.nmfs.noaa.gov/st5/publication/fisheries_economics_2009.html)

economic impacts locally, within each of the Gulf States, and throughout the entire nation. In general, commercial landings of most primary species appear to be stable and cases of instability, where observed, tend to be tied to regulations created to manage fish stocks. However, the largest component of the commercial fishing sector—the shrimp fishery—is confronted with obstacles to long-run viability, with the primary obstacle being increasing imports. Increasing imports have led to a decline in dockside price and a concomitant downsizing of the industry.

In the recreational sector, the number of Gulf angler trips (excluding Texas) increased from an estimated 17 million annually during the decade of the 1990s to 23 million annually during the most recent decade. About 70 % of total Gulf recreational trips were based in Florida (west coast); Louisiana accounted for 17 % of total recreational fishing activity. An estimated 92 thousand jobs (including Texas) were generated as a result of Gulf recreational fishing in 2009, with generated income totaling about $5.1 billion.

1.11   SEA TURTLES OF THE GULF OF MEXICO (CHAPTER 11)

The Gulf of Mexico provides important sea turtle nesting habitat, oceanic habitat for juvenile sea turtle growth and development, critical foraging habitat for juvenile and adult sea turtles, and important mating and inter-nesting habitat for adults. Five species of sea turtles are found in the Gulf of Mexico, including the Kemp's ridley, loggerhead, green, leatherback, and hawksbill. Available nesting, distribution, abundance, and habitat use information is summarized in this chapter to characterize the distribution and abundance of sea turtles in the Gulf prior to the Deepwater Horizon event. Life history information is also summarized for each species of sea turtle, and Gulf-of-Mexico-specific data are presented, when available (Figure 1.16). Roldán A. Valverde, Dyson Endowed Professor at Southeastern Louisiana University and Kym Rouse Holzwart, formerly a Certified Senior Ecologist with ENVIRON International Corporation (now Ramboll Environ, Inc.), have written this chapter. Ms. Holzwart is now an Environmental Scientist with the Hillsborough County Conservation and Environmental Lands Management Department in Central Florida.

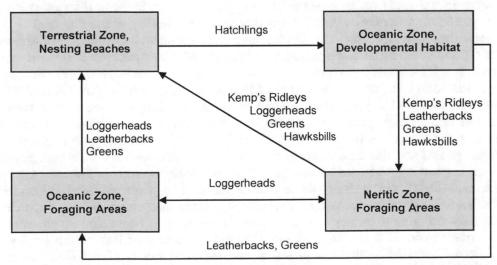

Figure 1.16. Generalized life cycles of sea turtle species that occur in the Gulf of Mexico (Figure 11.1 from Chapter 11 herein).

Available data indicate that the current populations of the five turtle species that inhabit the Gulf of Mexico are well under historical levels. All of these populations were heavily exploited in the Gulf and in the Caribbean for centuries after Europeans came to the New World. Sea turtle fisheries data from the Gulf of Mexico began to be collected in the late 1800s. These data clearly show a steep rise in the exploitation of sea turtles in the Gulf, with a subsequent collapse in catch by the early 1900s. The collapse in the Gulf was so pronounced that markets in the Gulf began to be supplied by sea turtles caught in other regions, until the Endangered Species Act and the Convention for International Trade on Endangered Species were set into place to stop sea turtle fisheries in 1973. Since then, some recovery of the populations has been documented. However, current pressures of fisheries-associated bycatch mortality, mainly in longline and shrimping fisheries, along with pollution and habitat destruction, have significantly hampered the recovery of these populations back to historic levels. Many anthropogenic and natural threats still affect Gulf sea turtles, and this information is summarized in Table 1.5, where impacts are quantified, where possible, using bycatch, stranding, and other threats data.

The Kemp's ridley has made a remarkable recovery from the brink of extinction in the Gulf of Mexico since conservation efforts focused on stressors affecting all life stages. The number of Kemp's ridleys in the Gulf has increased dramatically in recent years, and the population trajectory is promising.

Subpopulations from peninsular Florida, the northern Gulf, the Dry Tortugas, and the Greater Caribbean of the northwest Atlantic Ocean loggerhead population occur in the Gulf of Mexico during some portion of their life cycle. Annual loggerhead nesting on peninsular Florida beaches increased from 1979 to 2000 but then declined from 2001 to 2009. More data are needed to determine the long-term trends of the northern Gulf and Greater Caribbean loggerhead subpopulations. High cumulative threats and significant mortalities to oceanic and neritic juveniles, as well as adults, of the northwest Atlantic Ocean loggerhead population currently result from bycatch in multiple fisheries. The significant overlap between the northwest Atlantic Ocean loggerhead population range and the coastal and oceanic areas where fisheries occur, results in the death of thousands of loggerheads each year.

Loggerheads are the most abundant sea turtle in the western Gulf of Mexico; the majority of loggerheads that occur there are neritic juveniles. In addition, large juveniles have been associated with hard substrates, such as reefs and oil production areas (Figure 1.17), and appear to use these areas for resting. Core areas within the loggerhead's range in the Gulf include several oil and gas platforms that may be visited frequently on a daily, weekly, or monthly basis.

Despite being greatly depleted in the past, green turtle populations in the Gulf of Mexico are increasing, and green turtle nesting along the Mexican Gulf Coast has increased in recent years and remains relatively stable. In addition, nesting at major rookeries in the wider region, such as Tortuguero, Costa Rica, and the east coast of Florida, including Archie Carr National Wildlife Refuge, has increased significantly since the 1970s. While fibropapilloma tumors have been reported in all sea turtle species, the frequency of these tumors is much higher in green turtles when compared to that for the other species, and this disease remains a threat to green turtles. Green sea turtles are also dependent on healthy seagrass meadows for their foraging areas. Although impacts to green turtles resulting from incidental bycatch in fisheries are not as significant as those for loggerheads, some green turtles die each year from fisheries interactions.

The available data for the pelagic leatherback verifies that leatherbacks use the Gulf of Mexico as a foraging area, that they are often found in areas containing an abundance of jellyfish (their main food source), and that they are less abundant than Kemp's ridleys and loggerheads. Determining the current status and historical trends of the Gulf leatherback population is challenging because of their extensive migrations, large foraging areas, and significant data gaps. However, increased leatherback nesting in Florida may indicate that the leatherback population in the Gulf of Mexico, Caribbean, and northwest Atlantic Ocean area is

Table 1.5. Summary of Anthropogenic and Natural Threats Affecting the Various Ecosystems Used by Sea Turtle Populations in the Gulf of Mexico (Table 11.6 from Chapter 11 herein; from NMFS and USFWS 2008; Bolten et al. 2011; NMFS, USFWS, SEMARNAT 2011)

Threat	Terrestrial Zone[a]	Neritic Zone[a]	Oceanic Zone[a]
Incidental capture in commercial and recreational fisheries			
Trawls		X	X
Gill nets		X	X
Dredges		X	X
Pelagic and bottom long lines		X	X
Seines		X	
Pound nets and weirs		X	
Pots and traps		X	
Hook and line		X	X
Illegal harvest			
Eggs	X		
Juveniles		X	
Adults	X	X	
Nesting beach alterations			
Cleaning	X		
Human presence	X		
Driving on beach (cars and off-road vehicles)	X		
Artificial lighting	X	X	
Construction	X		
Nourishment and restoration	X	X	
Sand mining	X	X	
Armoring and shoreline stabilization (drift fences, groins, jetties)	X		
Other anthropogenic impacts			
Channel dredging and bridge building		X	
Boat strikes		X	X
Oil and gas exploration (including seismic activity), development, and production	X	X	X
Stormwater runoff		X	X
Oil and chemical pollution and toxins	X	X	X
Algal Blooms, including Red Tides		X	

(continued)

Table 1.5. (continued)

Threat	Terrestrial Zone[a]	Neritic Zone[a]	Oceanic Zone[a]
Hypoxia		X	
Marine debris ingestion and entanglement	X	X	X
Military activities and noise pollution	X	X	X
Industrial and power plant intake, impingement, and entrainment		X	
Dams and water diversion		X	
Sea level rise due to climate change	X		
Temperature change due to climate change	X	X	X
Trophic changes due to fishing and benthic habitat alteration		X	X
Natural impacts			
Predation	X	X	X
Beach erosion and vegetation alteration	X		
Habitat modification by invasive species	X	X	
Pathogens and disease	X	X	X
Hurricanes and severe storms	X	X	
Droughts		X	
Cold-stunning		X	

[a]Terrestrial zone = nesting beach where females excavate nests and lay eggs, where embryos develop; Neritic zone = inshore marine environment from the surface to the sea floor, including bays, sounds, and estuaries, as well as the continental shelf, where water depths do not exceed 200 m (656 ft); and Oceanic zone = open ocean environment from the surface to the sea floor where water depths are greater than 200 m (656 ft)

stable or increasing. Large numbers of leatherback sea turtles are captured each year in the Gulf as bycatch in pelagic longline fisheries.

Hawksbills are the rarest of the five species of sea turtle that occur in the Gulf of Mexico, and their current abundance is only a fraction of historical levels because millions were killed for tortoiseshell (jewelry, combs, brushes, buttons, etc.) during the past 100 years. Significant threats to hawksbills include destruction of nesting habitat, their dependence on coral reefs (one of the world's most endangered ecosystems) for food and shelter, and the continued trade in hawksbill products. Impacts from bycatch in Gulf fisheries to hawksbill sea turtles are minimal.

In summary, because sea turtles are difficult to study and since some species have been studied more than others, there are significant gaps in the data available by species, as well as by life stage. However, despite the data gaps and limitations associated with selected data sets

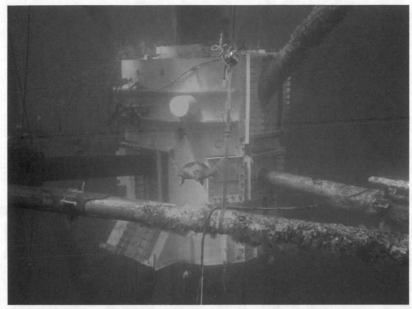

Figure 1.17. Loggerhead sea turtle swimming under an oil and gas platform (photograph courtesy of Ed Elfert, Chevron Corporation, photographer unknown) (Figure 11.23 from Chapter 11 herein).

(nesting and stranding data), characterizing the life history, distribution, and abundance, as well as summarizing impacts for the five species, was possible qualitatively, and sometimes quantitatively. In addition to revealing important data gaps, this summary highlights the variability of sea turtle data and the importance of long-term datasets. Changes in sea turtle populations can be detected, especially on nesting beaches; however, determining the causes of the change is extremely difficult because adequate baseline data are not available, multiple anthropogenic and natural threats affect all life stages of sea turtles, threats affect life stages and species differentially, multiple threat effects may be synergistic, and impacts may not be detected for many years. As new threats emerge and attempts are made to quantify the impacts of various threats in order to develop sea turtle conservation plans and solutions, these issues will continue to be challenges in the future.

1.12 AVIAN RESOURCES OF THE GULF OF MEXICO (CHAPTER 12)

The Gulf of Mexico is a complex mosaic of many habitat types, influenced by political, economic, social, and biological factors, as well as global climate change, sea-level rise and land subsidence, tides, storms, and hurricanes. The Gulf of Mexico ecosystem is a matrix of tropical, subtropical, and temperate habitats, which include different landmasses and different land margin interfaces. Large peninsulas (Florida, Yucatán), large islands (Cuba), barrier islands, offshore islands or keys, barrier beaches, sandy and gravel beaches, open water, mangroves, saltmarshes, and brackish marshes intergrade with freshwater marshes, swamps, and more upland habitats. Joanna Burger, who is Distinguished Professor of Biology at Rutgers University, is the author of the avian resources chapter. She has written more than 20 books and published 500 peer-reviewed journal articles, many of which are on birds.

The Gulf of Mexico is one of the most important regions in the Western Hemisphere for birds. Birds from North America funnel over or around the Gulf during their migratory flights,

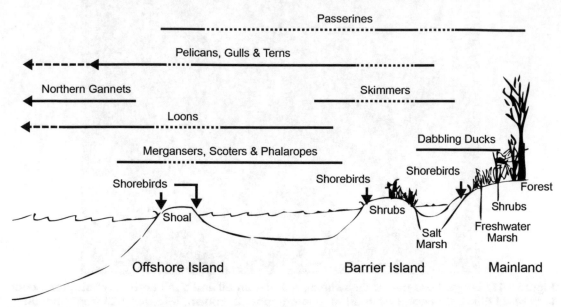

Figure 1.18. Schematic of spatial gradient for birds wintering in the Gulf of Mexico, from open water (pelagic zone) to upland habitats. Solid line indicates normal habitat use, dotted line indicates area not usually used, and dashed line means frequency is less (Figure 12.7 in Chapter 12 herein). © J. Burger.

birds from both north and south come to winter along Gulf shorelines or on the open water, and many species of birds breed in the Gulf. Thus, the coastal areas around the Gulf of Mexico serve as a hotspot of avian diversity (Figure 1.18).

Habitat availability and suitability are important distinguishing criteria within the Gulf of Mexico. Habitat availability is whether habitat is present and available that meets the needs of the species or species groups, such as open sandy beaches for shorebirds to feed, salt marshes for clapper rail and seaside sparrow to breed and forage, isolated islands with suitable vegetation for brown pelicans, terns, skimmers, herons, and egrets to nest, and bare sandy beaches for snowy plover to breed and forage. Habitat suitability, on the other hand, refers to whether the habitat will actually meet the needs of birds with respect to providing adequate places to forage, roost, breed, and migrate free from predators, human disturbance, high tides and storm tides, and other weather-related events. Available habitat must meet the species requirements in terms of vegetation, elevation, and physiognomy, while habitat suitability relates to whether the habitat is usable in terms of predator isolation and freedom from human disturbance. Factors that affect suitability often relate to exposure to the elements (storms, tides, winds, hurricanes, floods, and over the long term, sea-level rise), exposure to predators and people, the degree of competition from conspecific and interspecific interactions, presence of pollutants, and physical disruptions. In short, the habitat has to allow survival and reproduction. In many cases, suitable avian resource habitat is only available on islands or cays isolated from the mainland.

Habitat loss is a major factor affecting bird populations in the Gulf of Mexico and affects all birds, whether residents, migrants, or wintering species. Also, it influences all aspects of their daily lives from breeding and nesting to foraging and having sufficient safe places to roost. Loss of habitat is most severe at the land-sea margin, and it is most severe where anthropogenic activities occur, where the land is modified and is no longer suitable, or where land is completely developed.

Pollutants have affected behavior and populations of birds in the Gulf of Mexico, although this has been to a far lesser degree than habitat loss and modification. The use of DDT in the 1950s and 1960s had a great effect on fish-eating birds, such as osprey, wading birds, and brown pelicans, which declined dramatically. Pelicans were especially hard hit; they were largely extirpated as a successful breeding bird from some regions of the Gulf. In addition, mercury has affected behavior and reproduction in both resident birds (great egrets and other fish-eating birds), and migrants (common loon). Oil, on the other hand, can cause immediate mortality and chronic injury, but it has not been demonstrated to permanently affect any populations of birds in the Gulf. Plastics and fishing lines also cause mortality in the Gulf, particularly in foraging seabirds, but the long-term effects are unclear.

Understanding avian assemblages that use the Gulf of Mexico entails examining several different factors: migrant versus resident, solitary versus colonial nesting, ground versus tree nesting, method of foraging, and location of foraging. The 15 indicator species examined in the avian resources chapter illustrate all of these different lifestyles and behavioral patterns (Table 1.6). Obviously, nesting on the ground exposes nests, eggs, and chicks to ground predators, tidal flooding, and human disturbance, while nesting in trees exposes birds to aerial predators but usually protects them from mammalian predators. Nesting on low islands usually prevents mammalian predators from surviving, because high tides or severe storms wash them away, but nesting there also exposes the birds to flooding from high tides and storms during the breeding season. In addition, the indicator species illustrate different life strategies: some delay breeding, some have small clutch size, others have long parental care, and still others have long lifespans such as common loon and royal tern. Some species (e.g., mottled duck and clapper rail) breed when they are only 1 year old, but they have large clutches and short lifespans. These factors generally determine how fast a species can recover from any negative event or stressor, whether natural or manmade.

The selected indicator species illustrate and are representative of the range of population trends: some are increasing, others are decreasing, and in some, the variation from year to year is so great that it is difficult to ascertain trends. In other species, site fidelity to a specific colony location is so low that it is nearly impossible to census them accurately, and often their populations fluctuate wildly from year to year, depending upon water levels. Nonetheless, for the 15 indicator species, Christmas Bird Count data indicate clear declines over the past 45 years for certain species (mottled duck, black skimmer, and seaside sparrow), and clear increases for others (brown pelican, great egret, and laughing gulls, although data from the last 15–20 years indicate that laughing gull is now declining).

Overall declines seem to be due primarily to habitat loss, coupled with human disturbance and other disruptions to beach, saltmarsh, and coastal environments. Dramatic increases are often the result of laws and regulations (endangered species laws, cessation of the use of pesticides, as with brown pelican and osprey), to specific management practices (whooping crane, piping plover), to habitat creation (brown pelican), inadvertent management (dredge spoil islands for snowy plover and other beach nesting species), and possibly to global warming (more northern movement of southern species, such as roseate spoonbill).

The avian communities and resources of the Gulf of Mexico are varied and diverse, largely because of the diversity of habitats, the richness of the marine-land interface, the presence of a gradient from temperate to tropical, and the geography of the Gulf, which places it as the funnel point for Nearctic-neotropical migrants. Fluctuations in the avian community occur because of short-term and long-term stressors that render habitat either suitable or unsuitable. Habitat loss and destruction in the Gulf, which is continuing at an alarming rate, due to both

Table 1.6. Summary of Rationale for Selection of Indicator Species[a] (Table 12.7 from Chapter 12 herein)

Species	Endangered and Threatened	Largely a Gulf Species	Resident	Migrant	Colonial	Solitary	Open Ocean	Mud Flat	Beach or Sand	Sand, Light Vegetation	Marsh
Common Loon				X		X	X				
Brown Pelican		X	X		X		X				
Great Egret			X		X			X			X
Reddish Egret		X	X		X			X			X
Roseate Spoonbill		X	X		X			X			X
Mottled Duck		X	X			X					X
Osprey			X			X					
Whooping Crane	X	X		X	X		X				X
Clapper Rail			X			X					X
Snowy Plover		X	X			X			X		
Piping Plover	X			X		X			X		
Laughing Gull			X		X		X	X	X	X	X
Royal Tern			X				X		X	X	
Black Skimmer			X				X		X	X	
Seaside Sparrow			X			X					X

[a]The last five columns are habitat categories

natural and anthropogenic causes, will result in changes to the bird communities. Protection and management can only counter these losses and changes, and this requires monitoring to assess the overall health of avian communities. Finally, the needs and requirements of the avian communities must be viewed within the context of the human communities that also thrive along the Gulf Coast, and management, protection, and conservation of birds must be designed with the human dimension in mind. The following are important conclusions resulting from this analysis of the avian resources of the Gulf of Mexico:

- The Gulf of Mexico (and environs) is one of the most important places for birds in the Western Hemisphere because it has species whose major ranges are in both North and South America, and hosts a wide range of migrants. Nearly 400 species have been reported from the Gulf.

- Approximately 31 % of the 395 species found in the Gulf have been recorded in all areas of the Gulf.

- The high diversity in birds in the Gulf of Mexico is due to the Gulf's diversity of habitats, richness of marine-land interface, a gradient from tropical to temperate, and the geography of the Gulf which places it as the funnel point for Nearctic-neotropical migrants.

- Most birds that use saltwater to brackish ecosystems are seabirds, herons and egrets, shorebirds, waterfowl, gulls, terns, and specialized marsh species such as clapper rail and seaside sparrow. Assessment of 15 indicator species for the Gulf shows that mottle ducks, black skimmer, and clapper rail have declined over the last 45 years, while brown pelican, great egret, and osprey have increased. Declines seem to be related to habitat loss, coupled with human disturbance and other disruptions.

- Higher species diversity of birds is found in the southern Gulf of Mexico than in the northern coast.

- A higher percentage of some colonial species nesting in North America do so in Louisiana and Texas rather than elsewhere along the Gulf.

- Habitat loss is the primary threat facing birds in the Gulf of Mexico, due to both natural and anthropogenic causes, and it is occurring at an ever-increasing rate. One of the greatest impacts on avian populations in the Gulf of Mexico is habitat loss (either because it is less available, or because what is available is no longer suitable), followed by human disturbance.

- Populations of birds in the Gulf have varied greatly over the past 50 years; some have increased and some have declined.

1.13   MARINE MAMMALS OF THE GULF OF MEXICO (CHAPTER 13)

The Gulf of Mexico has a rich marine mammal fauna with approximately 22 species that occur commonly within this semitropical area (Table 1.7). One is the vegetarian sirenian, the West Indian manatee, which occurs mainly in Florida, but with some individuals migrating into Alabama, Mississippi, and Louisiana as well. All of the rest are cetaceans, which are members of the whale and dolphin clades, and there are no porpoises, sea lions, fur seals, or true seals in the Gulf. Bernd Würsig, who is a Regents Professor in the Departments of Marine Biology and Wildlife and Fisheries Sciences at Texas A&M University, is the author of this chapter. He has written a book on the marine mammals of the Gulf of Mexico, as well as several other books on marine mammals of the world and published numerous peer-reviewed papers on marine mammals during his long and distinguished career.

Table 1.7. Potential Marine Mammal Species in the Gulf of Mexico (Table 13.1 from Chapter 13 herein; from Würsig et al. 2000)

Species	Main Reasons for Former/Present Listing
North Atlantic right whale, *Eubalaena glacialis*	1 Stranding, one sighting of 2; reports of former hunting
Blue whale, *Balaenoptera musculus*	2 Strandings
Fin whale, *Balaenoptera physalus*	5 Strandings and rare sightings
Sei whale, *Balaenoptera borealis*	5 Strandings
Humpback whale, *Megaptera novaeangliae*	Occasional strandings and rare sightings
Minke whale, *Balaenoptera acutorostrata*	Occasional strandings; and rare sightings, Florida Keys
Bryde's whale, *Balaenoptera edeni*	**Strandings and quite common sightings**
Sperm whale, *Physeter macrocephalus*	**Common sightings**
Pygmy sperm whale, *Kogia breviceps*	**Common sightings**
Dwarf sperm whale, *Kogia sima*	**Common sightings**
Cuvier's beaked whale, *Ziphius cavirostris*	**Multiple strandings and occasional sightings**
Blainville's beaked whale, *Mesoplodon densirostris*	**4 Strandings and occasional sightings**
Sowerby's beaked whale, *Mesoplodon bidens*	1 Stranding
Gervais' beaked whale, *Mesoplodon europaeus*	**Multiple strandings and occasional sightings**
Killer whale, *Orcinus orca*	**Common sightings**
Short-finned pilot whale, *Globicephala macrorhynchus*	**Common sightings**
Long-finned pilot whale, *Globicephala melas*	Inferred but with no confirmed records
False killer whale, *Pseudorca crassidens*	**Medium common sightings**
Pygmy killer whale, *Feresa attenuata*	**Medium common sightings**
Melon-headed whale, *Peponocephala electra*	**Common sightings**
Rough-toothed dolphin, *Steno bredanensis*	**Common sightings**
Risso's dolphin, *Grampus griseus*	**Common sightings**
Common bottlenose dolphin, *Tursiops truncatus*	**Common sightings**
Pantropical spotted dolphin, *Stenella attenuata*	**Common sightings**
Atlantic spotted dolphin, *Stenella frontalis*	**Common sightings**
Spinner dolphin, *Stenella longirostris*	**Common sightings**
Clymene dolphin, *Stenella clymene*	**Common sightings**
Short-beaked common dolphin, *Delphinus delphis*	Inferred due to former misidentifications
Long-beaked common dolphins, *Delphinus capensis*	Inferred but with no evidence
Fraser's dolphin, *Lagenodelphis hosei*	**Occasional sightings**
West Indian manatee, *Trichechus manatus*	**Common sightings**

Those in **bold** are the 21 species presented in Chapter 13 that occur commonly within the Gulf

The most ubiquitous and best-known cetacean in the Gulf is clearly the common bottlenose dolphin, which occurs in coastal bays and estuaries, as well as nearshore and deeper waters. There are also upper continental shelf Atlantic spotted dolphins, the deepwater fish and squid eaters, such as the so-called *blackfish* and beaked whales, and members of the tropical genus *Stenella*, including Clymene and spinner dolphins that prefer lower continental shelf and deep waters of the Gulf. Numerically, the most common cetacean is the pantropical spotted dolphin, but the one with most biomass is the sperm whale, which is common in mid-depth waters off Louisiana and the shelf break off Texas. Bryde's whale is the only common baleen whale in the Gulf, and it inhabits upper and mid-slope waters, typically in the eastern Gulf of Mexico. All 22 species covered in Chapter 13 have descriptive information about them (size, color, shape, etc.) and range and distribution with a map, habitat, and field photo.

Recorded knowledge of marine mammals of the Gulf began with commercial whaling of sperm and pilot whales, as well as Risso's dolphins, in the 1700s and 1800s, but it progressed to natural history observations and one of the first volunteer stranding organizations, the Texas Marine Mammal Stranding Network, in the 1970s. In the 1980s and beyond, there have been considerable ship and aerial survey efforts to describe marine mammal populations of the Gulf, with the most intensive work accomplished in the 1990s, linking species, habitats utilized, and oceanographic parameters, under the auspices of the large multidisciplinary, U.S. government-funded project termed GulfCet.

While manatees generally use riverine and shallow oceanic waters for food and safety, the various species of cetaceans utilize all habitats of the Gulf. The GulfCet studies determined that sperm whales and smaller toothed whales are generally associated with the more productive cold-core upwelling gyres and eddies than the warm-core rings that break off from the Loop Current that comes from the south, out of the Caribbean. This fact gives the cetacean fauna of the northern Gulf a most-dynamic and ever-changing spatial dimension that needs to be viewed and considered in light of monthly to yearly changes of physical and biological oceanography.

The sperm whale and West Indian manatee are listed as endangered in the United States, but sperm whales are doing reasonably well worldwide, and there is no reason to believe that the Gulf of Mexico population is in imminent peril. The manatee numbers are in the low thousands of animals off Florida, subject to mortality largely due to periodic cold spells and recreational boat collisions, but hope exists as conservation, management, and public awareness efforts improve. Major anthropogenic threats exist for all marine mammals, but they do not appear to be as intensive in the Gulf of Mexico as in several other ocean basins. These threats include prey depletion, incidental mortality and injury due to fisheries, intentional and direct takes, vessel strikes, disturbance, acoustic (noise) pollution, chemical contamination, ingestion of solid debris, natural oil seeps, and aspects of ecosystem change.

1.14 DISEASES AND MORTALITIES OF FISHES AND OTHER ANIMALS IN THE GULF OF MEXICO (CHAPTER 14)

It is presumed that the health of animals in the Gulf of Mexico would follow along with the health of the Gulf ecosystem. Although there is no widespread monitoring program to measure the health of multiple Gulf species or the ecosystem, episodes of fish kills, infections, and abnormalities in marine species have been documented in the Gulf of Mexico for decades. Acute, mass mortalities have attracted the most attention, but when such an event occurs, attempts are usually made to ascribe a single cause for them. However, elevated mortalities are usually due to a convergence of factors, with interacting hosts, agents, and environmental conditions producing a "perfect storm." Such interacting factors are always present to some degree, but bringing them all together at once seems to be rare. Some microbial agents, parasite infections, and environmental conditions occur in large cycles of multiple years, or even decades, but whether this results from some underlying periodicity or from random co-occurrence of contributing factors is not clear.

The laboratory of Robin Overstreet and William Hawkins, both Professors Emeritus, at the Gulf Coast Research Laboratory of the University of Southern Mississippi in Ocean Springs has been one of the leading facilities for tracking parasites and diseases in coastal and marine species in the northern Gulf of Mexico for over four decades. Their detailed research program and broad study of taxonomy, systematics, development and life histories, diagnoses and management of diseases, ecology, pathogenesis and host–parasite relationships, as well as public health studies, provide the foundation for this overview of diseases and mortalities of coastal and marine species in the Gulf of Mexico, with an emphasis on fishes.

Physical and chemical factors generally trigger large-scale mortalities. Eutrophication occurs throughout the Gulf where high nutrient input occurs, and low oxygen levels associated with eutrophication produce a major stress leading to fish mortality, but it also leads to disease and parasite-caused mortality. Red tides have a major influence on the health of fishes and other animals from the West Coast of Florida and occasionally elsewhere in the Gulf. Mass mortalities from sudden cold spells, which occur primarily inshore where it is hard for some animals to escape, are more disastrous in South Texas and South Florida, because species there are not as well acclimated to tolerate rapid temperature changes as they are in higher latitudes of the Gulf. Likewise, excessive heat, hypersalinity, sulfate reduction, sediments, and drilling fluids all have been implicated in mortality events, but they produce more localized effects. Hurricanes can occur anywhere in the Gulf, but resulting fish kills depend on the geography of the areas the hurricanes pass through and impacts to the environment. As with most catastrophic events, the presence and absence of specific parasites can provide a good indication of environmental health and its restoration.

Few diseases cause mass mortality. When investigated, the cause of such events usually involves one or more stresses, with an interaction between the host, disease agent, and the environment. Most diseases involving infectious agents are usually shown to be highly restricted to certain geographic areas or to certain species. The most obvious infectious disease and mass mortality event in the Gulf of Mexico came from a catfish die-off occurring in 1996 that eventually spread from Texas to Florida and was caused by, either directly or indirectly, a virus. It is not known whether that virus becomes intermittently introduced or if it always occurs in the habitat in low numbers until some threshold is surpassed, triggering a pandemic. Some event, such as reproductive activity in the catfish, may have served as the stressor, but no catastrophic event coincided with the mortality. What seems to be the same agent infects fishes in the southern Gulf of Mexico, South America, Africa, and India.

Parasites often cause disease conditions and mortalities in hosts, usually intermediate hosts, as a part of the parasitic strategy to complete its life history. However, these effects tend to be ongoing at a low level without harm to an overall population or to the ecosystem. In cases where mass mortality occurs, changes in anthropogenic or natural environmental conditions are usually involved. Major stress can affect resistance of a host to disease organisms, especially bacterial or protozoal agents. Diseases caused by a few species seem to serve as a means of host population control. Parasites, even when not harming their hosts, can be extremely useful as bioindicators in providing information about stock assessment, biological activities of hosts such as migration and feeding, restoration of habitats, and habitat and ecosystem health.

Neoplasms, some virally induced, have seldom been observed or reported in Gulf of Mexico fishes, although their occurrence has likely been underestimated, but elsewhere, neoplasms have served as good indicators of various contaminants, particularly sediment-bound PAHs. Consequently, more attention to documenting them is warranted. Developmental abnormalities and histopathological alterations, which have been seen in many Gulf species, can indicate levels of stress from a variety of environmental factors.

Regarding vertebrates other than fish, data on disease conditions are uneven. The best-known condition in sea turtles is fibropapillomatosis, and it appears to have multiple causes. Bird mortality events are sometimes ascribed to bacterial, fungal, and viral infections, but the effects of these agents can be exacerbated by environmental conditions that reduce energy and deplete needed resources. Brevitoxins and morbillivirus have been implicated in periodic marine mammal mortalities, but the cause of others is unclear, and most data are based on skewed samples from strandings.

Concerning invertebrates, diseases of penaeid shrimps and the blue crab have been well documented, but the effect of these diseases on host populations in the Gulf remains unclear. In the eastern oyster, the protozoan disease known as *dermo* has received a great deal of research attention. Researchers know that its impact on oyster populations varies widely according to salinity, temperature, genotype of the infectious agent, and perhaps interaction with specific contaminants, but its variation and severity from location to location in the Gulf has not been adequately explained. Other agents and fouling agents affect oysters also, but their impacts and interactions are less well studied. Loss of corals by bleaching and disease has had a major influence on tropical and subtropical Gulf communities, because along with their loss, there has been a loss of the associated fishes and invertebrates in the coral community.

Although almost 100 images of a wide variety of diseases, parasitic infections, and other causative agents are shown in Chapter 14 on various Gulf of Mexico invertebrates and vertebrates, only a few examples are shown here to demonstrate that variety (Figures. 1.19, 1.20, and 1.21).

To better understand diseases and mortalities in the Gulf of Mexico, there is a need for monitoring both diseases and mortalities; conducting more long-term, broad-scaled field work; acquiring more expertise; and developing more critical tools for evaluating health of the animals and health of the ecosystem.

Figure 1.19. Southern flounder, *Paralichthys lethostigma*, exhibiting relatively common bacterial lesion on blind side of specimen from Pascagoula estuary, Mississippi, 1987 (Figure 14.9 from Chapter 14 herein).

Figure 1.20. A few of the many pouch lice, *Piagetiella peralis*, infesting the gular pouch of an American white pelican (Figure 14.74 from Chapter 14 herein).

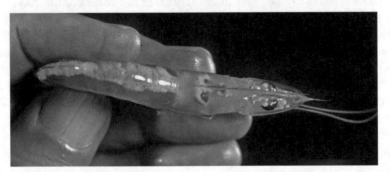

Figure 1.21. White shrimp, *Litopenaeus setiferus*, with the microsporidian *Agmasoma penaei* in the cephalothorax and along the dorsum, superficially appearing like developing gonads (Figure 14.79 from Chapter 14 herein).

1.15   CONCLUSIONS

The major conclusions of this collection of chapters are included here as an overview of the Gulf of Mexico environment, as well as the current status and historical trends of species and habitats prior to the Deepwater Horizon oil spill:

1. *Water quality*: Patterns and trends in water quality are highly variable in space and through time in the Gulf of Mexico, and coastal environments are highly influenced by human activities where the primary cause of degraded water quality is excess nutrients. Water quality rapidly improves with distance offshore. More than 60 % of assessed estuaries were either threatened or impaired for human use and/or aquatic life over the time period of this review that spans the 1990s to the mid-2000s.

2. *Water quality*: Eutrophication has produced low dissolved oxygen and increased chlorophyll *a* concentrations, diminished water clarity, and other secondary effects including toxic/nuisance algal blooms and loss of submerged aquatic vegetation. Degraded coastal water quality was also indicated by contaminants in biological tissues

and sediments, fish consumption advisories, and beach closing/advisories due to bacterial contamination.

3. *Water quality*: Water quality of Gulf of Mexico continental shelf/slope and abyssal waters was and continues to be good. Exceptions are hypoxic zones on the continental shelf, waters just above natural oil and gas seeps, and ephemeral effects due to produced water discharges during petroleum extraction. Along the northwest/central Gulf of Mexico continental shelf, the seasonal occurrence of waters with low concentrations of oxygen is geographically widespread. These "dead zones" are highly seasonal, and it has been suggested they result from water column stratification driven by weather coupled with Mississippi River outflow that delivers excess nutrients (mostly from agricultural lands) to the offshore region.

4. *Sediments*: Sediment nature and distributions in the Gulf of Mexico are similar to ocean basins. There are basically two primary provinces: terrigenous sediments carried from land to the northern and western portions of the basin, and carbonate sediments that originate on the Florida and Yucatán platforms. Sea-level changes over the past several thousand years have had a major influence on sediment distributions.

5. *Sediments*: Sediments in coastal systems of the Gulf of Mexico have the most complicated distributions, and are dominated by sediments of terrestrial origin. Coastal sediments are composed of mud and sand with biogenic organic debris: sand is dominating in the barrier-inlets systems and mud is the largest sediment component in the estuaries and lagoons. Deep Gulf environments tend to be dominated by mud in a combination of terrigenous and biogenic (coccoliths, diatoms, foraminiferans, and radiolarians) sediments.

6. *Sediment contaminants*: Concentrations and distributions of sediment contaminants in the Gulf of Mexico are spatially and temporally heterogeneous over small scales due to variations in inputs, sediment deposition and accumulation rates, susceptibility to and rates of removal, chemical form, and physicochemical properties and the physical settings of receiving waters.

7. *Sediment contaminants*: Sediment contaminants are found widely in Gulf of Mexico coastal bays and estuaries. Coastal sediments were judged to be in good to poor condition with concentrations of metals and pesticides in more than 40 % and concentrations of PAHs and PCBs in less than 1 % of coastal sediments exceeding levels suspected of causing biological effects. Within bay systems, steep gradients in contaminant concentrations were observed near population centers, agricultural activities, and industrial complexes. Contaminant concentrations decrease with distance offshore, since these regions are remote, with few exceptions, from most contaminant inputs. Natural petroleum seepage is the major source of hydrocarbons in northern-central Gulf of Mexico continental shelf/slope sediments.

8. *Sediment contaminants*: In general, levels of pesticides and contaminant metals appear to have decreased with time in coastal sediments in response to water pollution control regulations.

9. *Oil and gas seeps*: Hydrocarbon seepage is a prevalent, natural worldwide phenomenon that has occurred for millions of years, and it is especially widespread in the deepwater region of the Gulf of Mexico, which is an archetype for oil and gas seepage and where most worldwide studies and knowledge of petroleum seeps are based. Gulf of Mexico seeps are highly variable in composition and volume and include gases, volatiles, liquids, pitch, asphalt, tars, water, brines, and fluidized sediments.

10. *Oil and gas seeps*: Hydrocarbon seeps occur on land and beneath the ocean, and they are biogenic, thermogenic, or mixed in origin. These seeps release considerable amounts of oil and gas to the environment each year, estimated at about 95 % of oil annually discharged to Gulf of Mexico waters. Cold-seep, chemosynthetic communities are common at macroseeps across the northern Gulf of Mexico continental slope, on the abyssal plain, and in the southern reaches of the Gulf of Mexico.

11. *Coastal habitats*: Vegetated coastal and marine habitats of the Gulf of Mexico provide a wealth of ecosystem services, such as food, employment, recreation, and natural system maintenance and regulation to the three countries bordering the Gulf. Salt marshes dominate vegetated shorelines in the northern Gulf, and mangroves dominate in the tropical south.

12. *Coastal habitats*: Coastal vegetated habitats have experienced the greatest temporal changes in areas most susceptible to relative sea-level rise, tropical cyclones, and human disturbances. Consequently, the deltaic coast of Louisiana has the most substantial land and habitat changes in the Gulf of Mexico. Conversely, the more stable coasts of the Yucatán Peninsula, Cuba, and southwestern Florida show the least amount of change. Human disturbances are evident in areas of significant industrial activity and tourism. Human impacts are in large part tied to periodic and chronic stressors and disturbances associated with urban, agricultural, and industrial activities. Draining and filling of wetlands for human habitation, agricultural development, and industrial expansion have dramatically impacted coastal habitats throughout the Gulf.

13. *Coastal habitats*: Nutrient enrichment and resulting eutrophication and hypoxia, altered hydrology from multiple causes, invasive species, and chemical pollutants including those associated with energy extraction and production have challenged the health and sustainability of vegetated marine habitats. In addition, natural disturbances driven by hurricanes, underlying geology, and floods and drought are exacerbated by human impacts.

14. *Offshore biota*: Offshore plankton and benthos of the Gulf of Mexico at the lower end of the food web (phytoplankton, zooplankton, mid-water fishes, and seafloor organisms) in most habitats are characteristic of an oligotrophic ecosystem. That is, the biota is relatively low in numbers of organisms and biomass when compared to other continental margins such as upwelling regions and in temperate and polar latitudes. The principal cause of this oligotrophic condition is the source water from the Caribbean, which is depleted of nitrate in the surface to about 125 m (410 ft).

15. *Offshore biota*: Offshore plankton populations represent a near-surface fauna that declines with depth as a biocline (the greater the distance from the surface, the more depauperate the biomass). All size groups of multicellular organisms decline exponentially as a function of depth and distance from land, so that the abyssal plain supports only a very few seafloor organisms (fishes; zooplankton; mega-, macro-, and meiobenthos). Biodiversity of the macrobenthos follows a different pattern as a function of depth, depending on the taxon studied. In general there is a mid-depth maximum of macrofaunal diversity at about 1.2 km (0.75 mi) in depth. In addition, a decreasing zonation in diversity across a physical gradient is apparent with increasing depth in macrofauna, megafauna, and fishes, most likely due to the decreasing amount of food sources available.

16. *Offshore biota*: Other distinctive offshore biota assemblages or habitats include chemosynthetic benthic fauna associated with and sustained by fossil hydrocarbon seeps in the northwestern Gulf, deep-living, cold-water corals, such as *Lophelia pertusa*, which

provide distinctive habitat for demersal species in a narrow depth band at the upper margin of the continental slope in the northeastern Gulf, and lastly, high diversity hard-bottom areas which are spread across the continental margin as topographic highs (reefs and banks) or low-relief live bottoms.

17. *Offshore biota*: Potential problems in sustaining the offshore biota (plankton, nekton, and benthos) include climate change, turbidity currents and slumps, eutrophication, oil and gas industry accidents, hypoxia, overfishing, trawling the bottom, and hurricanes.

18. *Shellfish*: Shellfish include four of the top five commercial species by value and poundage of landings in the Gulf of Mexico. These include brown and white shrimp, Eastern oyster, and blue crab, but there are 49 species total (28 mollusks, 18 crustaceans, and 3 echinoderms) that are currently taken as commercial shellfish species in the Gulf.

19. *Shellfish*: Population trends of shellfish in the Gulf vary widely from year to year, primarily due to fluctuations in environmental conditions (such as temperature, salinity, etc.), but some landings are also influenced annually by exogenous factors (such as market competition from imported shrimp, rising fuel costs, and fleet damage due to hurricanes).

20. *Shellfish*: Shrimp populations are flourishing, but the shrimp fishery—the most valuable fishery in the Gulf—is in decline due to exogenous factors, especially cheap, imported shrimp. Oyster populations appear fairly stable, but landings in Louisiana have been low for almost a decade compared to the 1990s. Blue crab populations fluctuate widely due to varying environmental conditions with Louisiana having the largest fishery and increasing catches.

21. *Fish resources*: Fish resources from the Gulf of Mexico total 1,536 species and include 1,443 finfish, 51 sharks, and 42 rays/skates.

22. *Fish resources*: Gulf fisheries are some of the most productive in the world with approximately 25 % of commercial fish landings and 40 % of recreational harvest in the United States coming from the Gulf.

23. *Fish resources*: A wide variety of long-term anthropogenic and natural stressors, such as coastal development with subsequent degraded water quality and habitat loss, heavy fishing pressure, a large quantity of bycatch in shrimp fisheries, climate change, and natural disasters have negatively impacted the Gulf of Mexico ecosystem and its fishery species.

24. *Fish resources*: Of the 15 finfish species evaluated, 5 species were being overfished and/or were in the status of overfishing in 2010, including red snapper, red grouper (some local subpopulations), Atlantic bluefin tuna, Atlantic blue marlin, and greater amberjack.

25. *Fish resources*: Of 39 shark species included in the shark Fisheries Management Plan in the Gulf of Mexico, 19 species have been listed as commercially and recreationally prohibited species because of very low population biomass and poor stock conditions.

26. *Fish resources*: Finfish species evaluated that were determined not overfished in the Gulf of Mexico immediately before the Deepwater Horizon oil spill included menhaden, Atlantic swordfish, Atlantic sailfish, red drum, striped mullet, tilefish, king mackerel, Gulf flounder, and dolphinfish.

27. *Commercial and recreational fisheries*: Commercial fisheries are generally described and reported by either landings in weight or value in dollars. Aggregate finfish and shellfish landings attributed to the U.S. Gulf states fluctuate, but the ranking of the states does not change much from year to year. Louisiana ranks first in landings in the

five major Gulf species (menhaden, brown and white shrimp, blue crab, and oysters). Dockside value of all landings is mostly concentrated in Louisiana and Texas, with shares of 43 % and 26 %, respectively.

28. *Commercial and recreational fisheries*: In the recreational fisheries sector, the number of Gulf angler trips (excluding Texas) increased from an estimated 17 million annually during the decade of the 1990s to 23 million annually during the most recent decade. About 70 % of total Gulf recreational trips were based in Florida (west coast); Louisiana accounted for 17 % of total recreational fishing activity. An estimated 92,000 jobs (including Texas) were generated as a result of Gulf recreational fishing in 2009, with generated income totaling about $5.1 billion. Spotted seatrout and red drum are the popular inshore species for recreational fishermen, and reef fish (snapper and grouper) are the most popular offshore species.

29. *Sea turtles*: Five species of sea turtles are found in the Gulf of Mexico, including the Kemp's ridley, loggerhead, green, leatherback, and hawksbill.

30. *Sea turtles*: The Gulf of Mexico provides important sea turtle nesting habitat, oceanic habitat for juvenile sea turtle growth and development, critical foraging habitat for juvenile and adult sea turtles, and important mating and inter-nesting habitat for adults.

31. *Sea turtles*: All sea turtle populations were heavily exploited in the Gulf and Caribbean for centuries after the Europeans arrived in the New World, and even though all species have been protected since 1973 and some recovery has occurred, none of the populations have returned to historic levels. Current pressures from fisheries-associated bycatch mortality, mainly in longline and shrimping fisheries, along with pollution and habitat destruction, have significantly hampered recovery.

32. *Sea turtles*: The Kemp's ridley is the most endangered sea turtle species in the world, and it only nests in the Gulf of Mexico. It has made a remarkable recovery from the brink of extinction in the Gulf since conservation efforts focused on stressors affecting all life stages. The number of Kemp's ridleys in the Gulf has increased dramatically in recent years, and the population trajectory is promising.

33. *Avian resources*: The Gulf of Mexico is one of the most important places for birds in the Western Hemisphere because it has species whose major ranges are in both North and South America, and hosts a wide range of migrants. Nearly 400 species have been reported from the Gulf.

34. *Avian resources*: The high diversity of birds in the Gulf of Mexico is due to the Gulf's diversity of habitats, richness of marine-land interface, a gradient from tropical to temperate, and the geography of the Gulf which places it as the funnel point for Nearctic-neotropical migrants.

35. *Avian resources*: Most birds that use saltwater to brackish ecosystems are seabirds, herons and egrets, shorebirds, waterfowl, gulls, terns, and specialized marsh species such as clapper rail and seaside sparrow. Assessment of 15 indicator species for the Gulf shows that mottle ducks, black skimmer, and clapper rail have declined over the last 45 years, while brown pelican, great egret, and osprey have increased.

36. *Avian resources*: A higher species diversity of birds is found in the southern Gulf of Mexico, compared to the northern coast.

37. *Avian resources*: A higher percentage of some colonial species nesting in North America do so in Louisiana and Texas rather than elsewhere along the Gulf.

38. *Avian resources*: Habitat loss is the primary threat facing birds in the Gulf of Mexico, due to both natural and anthropogenic causes, and it is occurring at an ever-increasing-rate.

39. *Avian resources*: Populations of birds in the Gulf have varied greatly over the past 50 years; some have increased and some have declined.

40. *Marine mammals*: While 31 species of marine mammals have been listed for the Gulf of Mexico, only 28 are confirmed and 22 species occur commonly. These 22 species include one sirenian, the West Indian manatee, and 21 cetaceans (whales and dolphins); there are no porpoises, sea lions, fur seals, or true seals in the Gulf.

41. *Marine mammals*: The most ubiquitous and best-known cetacean in the Gulf is the common bottlenose dolphin, which occurs in coastal bays and estuaries, as well as nearshore and deeper waters. Numerically, the most common cetacean is the pantropical spotted dolphin, but the one with most biomass is the sperm whale, which is common in mid-depth waters off Louisiana and the shelf break off Texas.

42. *Marine mammals*: The sperm whale and West Indian manatee are listed as endangered in the United States, but sperm whales are doing reasonably well worldwide, and there is no reason to believe that the Gulf of Mexico population is in imminent peril. The manatee numbers are in the low thousands of animals off Florida, subject to mortality largely due to periodic cold spells and recreational boat collisions, but hope exists as conservation, management, and public awareness efforts improve.

43. *Marine mammals*: Major anthropogenic threats exist for all marine mammals, but they do not appear to be as intensive in the Gulf of Mexico as in several other ocean basins. These threats include prey depletion, incidental mortality and injury due to fisheries, intentional and direct takes, vessel strikes, disturbance, acoustic (noise) pollution, chemical contamination, ingestion of solid debris, natural oil seeps, and aspects of ecosystem change.

44. *Diseases and mortalities*: There is no widespread monitoring program to measure the health of Gulf species or the Gulf ecosystem, but episodes of fish kills, infections, and abnormalities in marine species have been documented in the Gulf of Mexico for decades. Eutrophication and associated low oxygen have led to fish mortality in certain areas, and red tides, severe cold, excessive heat and hypersalinity have caused localized mass mortalities. A virus caused a massive, widespread die-off of catfish from Texas to Florida in 1996.

45. *Diseases and mortalities*: Parasites and various diseases have caused stress and mortality in selected invertebrates, fish, turtles, birds, and mammals of the Gulf of Mexico for decades, but there is no widespread metric or system to track the health of these animals in the Gulf of Mexico ecosystem.

REFERENCES

Bolten AB, Crowder LB, Dodd MG, MacPherson SL, Musick JA, Schroeder BA, Witherington BE, Long KJ, Snover ML (2011) Quantifying multiple threats to endangered species: An example from loggerhead sea turtles. Front Ecol Environ 9:295–301

Bricker SB, Clement CG, Pirhalla DE, Orlando SP, Farrow DRG (1999) National estuarine eutrophication assessment: Effects of nutrient enrichment in the nation's estuaries. NOAA (National Oceanic and Atmospheric Administration), National Ocean Service, Special

Projects Office and the National Centers for Coastal Ocean Science, Silver Spring, MD, USA, 328 p

Bricker SB, Longstaff B, Dennison W, Jones A, Boicourt K, Wicks C, Woerner J (2007) Effects of nutrient enrichment in the nation's estuaries: A decade of change. NOAA coastal ocean program decision analysis series 26. National Centers for Coastal Ocean Science, Silver Spring, MD, USA, 328 p

Briggs JC (1974) Marine zoogeography. McGraw Hill, New York, NY, USA, 475 p

Bright TJ, Kraemer GP, Minnery GA, Viada ST (1984) Hermatypes of the Flower Garden Banks, Northwestern Gulf of Mexico: A comparison to other western Atlantic reefs. Bull Mar Sci 34:461–476

Britton JC, Morton B (1989) Shore ecology of the Gulf of Mexico. University of Texas Press, Austin, TX, USA, 387 p

Brooks J, Fisher C, Roberts H, Bernard B, MacDonald I, Carney R, Joye S, Cordes E, Wolff G, Goehring E (2008) Investigations of chemosynthetic communities on the lower continental slope of the Gulf of Mexico, interim report 1. OCS study MMS 2008–2009. U.S. Department of the Interior, Minerals Management Service, Gulf of Mexico OCS Regional Office, New Orleans, LA, USA, 332 p

Buster NA, Holmes CW (eds) (2011) Gulf of Mexico origin, waters, and biota, vol 3, Geology. Texas A&M University Press, College Station, TX, USA, 446 p

Capurro LRA, Reid JL (eds) (1972) Texas A&M University oceanographic studies, vol 2, Contributions on the physical oceanography of the Gulf of Mexico. Gulf Publishing Company, Houston, TX, USA, 288 p

Caso M, Pisanty I, Ezcurra E (eds) (2004) Diagnostico Ambiental del Golfo de Mexico, vols 1 and 2. Instituto Nacional de Ecologia, Mexico DF, 1108 p

Cato JC (ed) (2009) Gulf of Mexico origin, waters, and biota, vol 2, Ocean and coastal economy. Texas A&M University Press, College Station, TX, USA, 110 p

CEC (Commission for Environmental Cooperation) (2007) North America Elevation 1-Kilometer Resolution Map, 3rd edn. Collaborators include U.S. Department of the Interior, USGS, National Atlas of the U.S. CEC, Montréal, Québec, Canada. http://www.cec.org/Page.asp? PageID¼924&ContentID¼2841&SiteNodeID¼497&BL_ExpandID¼. Accessed 19 Jan 2015

Cordes EE, McGinley MP, Podowski EL, Becker EL, Lessard-Pilon S, Viada S, Fisher CR (2008) Coral communities of the deep Gulf of Mexico. Deep-Sea Res 55:777–787

CSA International Inc (2007) Characterization of northern Gulf of Mexico deepwater hard bottom communities with emphasis on *Lophelia* corals. OCS MMS 2007-044. U.S. Department of the Interior, Minerals Management Service, Gulf of Mexico OCS Regional Office, New Orleans, LA, USA, 169 p

Darnell RM (2015) The American Sea—a natural history of the Gulf of Mexico. Texas A&M University Press, College Station, TX, USA, 554 p

Darnell RM, Defenbaugh RE (1990) Gulf of Mexico. Environmental overview and history of environmental research. Am Zool 30:3–6

Davis RA Jr (2011) Sea-level change in the Gulf of Mexico. Texas A&M University Press, College Station, TX, USA, 172 p

Davis RA Jr (2014) Beaches of the Gulf Coast. Texas A&M University Press, College Station, TX, USA, 244 p

Day JW, Yanez-Arancibia A (2013) Gulf of Mexico origin, waters, and biota, vol 4, Ecosystem-based management. Texas A&M University Press, College Station, TX, USA, 446 p

Fautin D, Dalton P, Incze LS, Leong JC, Pautzke C, Rosenberg A, Sandifer P, Sedberry G, Tunnell JW Jr, Abbott I, Brainard RE, Broudeur M, Eldredge LG, Feldman M, Moretzsohn F, Vroom PS, Wainstein M, Wolff N (2010) An overview of marine biodiversity in United States waters. PLoS One 5(8), e11914. doi:10.1371/journal.pone.0011914

Felder DL, Camp DK (2009) Gulf of Mexico origin, waters, and biota, vol 1, Biodiversity. Texas A&M University Press, College Station, TX, USA, 1393 p

FishBase (2013) http://fishbase.org/. Accessed 31 May 2013

French CD, Schenk CJ (2005) Shaded relief image of the Gulf of Mexico (shadedrelief.jpg). USGS, Central Energy Resources Team, Reston, VA, USA. http://pubs.usgs.gov/of/1997/ofr-97-470/OF97-470L/graphic/data.htm. Accessed Jan 2015

Galtsoff PS (1954) Gulf of Mexico. Its origin, waters and marine life: Fishery bulletin of the Fish and Wildlife Service, fishery bulletin 89. U.S. Government Printing Office, Washington, DC, USA, 604 p

Gore RH (1992) The Gulf of Mexico: A treasury of resources in the American Mediterranean. Pineapple Press, Sarasota, FL, USA, 384 p

Kumpf H, Steidinger K, Sherman K (eds) (1999) The Gulf of Mexico large marine ecosystem: Assessment, sustainability, and management. Blackwell Science, Inc., Malden, MA, USA, 704 p

MacDonald IR, Fisher C (1996) Life without light. National Geographic, October:86–97

Martin RG, Bouma AH (1978) Physiography of the Gulf of Mexico. In: Bouma AH, Moore GT, Coleman JM (eds) Framework, facies, and oil-trapping characteristics of the upper continental margin. Am Assoc Petrol Geol Stud Geol 7:3–19.

McEachran JD, Fechhelm JD (1998) Fishes of the Gulf of Mexico, vol 1. University of Texas Press, Austin, TX, USA, 1112 p

McEachran JD, Fechhelm JD (2005) Fishes of the Gulf of Mexico, vol 2. University of Texas Press, Austin, TX, USA, 1004 p

NMFS (National Marine Fisheries Service) USFWS (United States Fish and Wildlife Service) (2008) Recovery plan for the Northwest Atlantic population of the Loggerhead Sea Turtle (Caretta caretta): second revision. Office of Protected Resources, National Marine Fisheries Service, Silver Spring, MD, USA, 325 p

NMFS, USFWS, SEMARNAT (Secretary of Environment and Natural Resources, Mexico) (2011) Bi-national recovery plan for the Kemp's Ridley Sea Turtle (Lepidochelys kempii), second revision. National Marine Fisheries Service, Silver Spring, MD, USA, 174 p

NOS (National Ocean Service), NOAA (2011) The Gulf of Mexico at a glance: A second glance. U.S. Department of Commerce, Washington, DC, USA, 51 p

Pequegnat WE, Chace FA (eds) (1970) Texas A&M University oceanographic studies, vol 1, Contributions on the biology of the Gulf of Mexico. Gulf Publishing Company, Houston, TX, USA, 270 p

Rezak R, Edwards GS (1972) Carbonate sediments of the Gulf of Mexico. In: Rezak R, Henry VJ (eds) Contributions on the geological and geophysical oceanography of the Gulf of Mexico, Texas A&M University Oceanography Studies 3. Gulf Publishing Company, Houston, TX, USA, pp 263–280

Rezak R, Henry VJ (1972) Contributions on the geological and geophysical oceanography of the Gulf of Mexico, vol 3, Texas A&M University oceanography studies. Gulf Publishing Company, Houston, TX, USA, 303 p

Rezak R, Bright TJ, McGrail DW (1985) Reefs and banks of the Northwestern Gulf of Mexico. John Wiley & Sons, New York, NY, USA, 259 p

Ritchie KB, Keller BD (eds) (2008) A scientific forum on the Gulf of Mexico: The islands in the stream concept. Marine sanctuaries conservation series NMSP-08-04. U.S. Department of Commerce, National Oceanic and Atmospheric Administration, National Marine Sanctuary Program, Silver Spring, MD, USA, 108 p

Ritchie KB, Kiene WE (eds) (2012) Beyond the horizon: A forum to discuss a potential network of special ocean places to strengthen the ecology, economy and culture of the Gulf of Mexico. In: Proceedings of the forum, Mote Marine Laboratory, Sarasota, FL, USA, May 11–13, 2011, 138 p

Spalding MD, Fox HE, Allen GR, Davidson N, Ferdaña ZA, Finlayson M, Halpern BS, Jorge MA, Lombana A, Lourie SA, Martin KD, McManus E, Molnar J, Recchia CA, Robertson J (2007) Marine ecoregions of the world: A bioregionalization of coastal and shelf areas. BioScience 57:573–583

Sturges W, Lugo-Fernandez A (eds) (2005) Circulation in the Gulf of Mexico: Observations and models. American Geophysical Union, Washington, DC, USA, 347 p

Tunnell JW Jr (2009) The Gulf of Mexico. In: Earle SA, Glover LK (eds) Ocean: An illustrated atlas. National Geographic Society, Washington, DC, USA, pp 136–137

Tunnell JW Jr, Judd FW (eds) (2002) The Laguna Madre of Texas and Tamaulipas. Texas A&M University Press, College Station, TX, USA, 346 p

Tunnell JW Jr, Chavez EA, Withers K (eds) (2007) Coral reefs of the Southern Gulf of Mexico. Texas A&M University Press, College Station, TX, USA, 194 p

U.S. Department of Commerce (2011) Fisheries economics of the United States, 2009: Economics and sociocultural status and trends series. U.S. Dept. Commerce, NOAA Technical Memorandum NMFS-F/SPO-118. https://www.st.nmfs.noaa.gov/st5/publication/index.html. Accessed 12 Feb 2013

USEPA (U.S. Environmental Protection Agency) (2001) National coastal condition report. EPA/620-R-01-005. Office of Research and Development and Office of Water, Washington, DC, USA

USEPA (2004) National coastal condition report II. EPA/620-R 03-002. Office of Research and Development and Office of Water, Washington, DC, USA

USEPA (2008) National coastal condition report III. EPA/842-R-08-002. Office of Research and Development and Office of Water, Washington, DC, USA

USEPA (2012) National coastal condition report IV. EPA/842-R-10-003. Office of Research and Development and Office of Water, Washington, DC, USA

Weddle RS (1985) Spanish Sea: The Gulf of Mexico in North American Discovery, 1500–1685. Texas A&M University Press, College Station, TX, USA, 457 p

Whelan J (2004) When seafloor meets ocean, the chemistry is amazing. Originally published online February 13, 2004: in print vol 42, no 2, April 2004. http://www.whoi.edu/oceanus/viewArticle.do?id=2441. Accessed 14 Sept 2014

Wilkinson T, Wiken E, Bezaury-Creel J, Hourigan T, Agardy T, Herrmann H, Janishevski L, Madden C, Morgan L, Padilla M (2009) Marine ecoregions of North America. Commission for Environmental Cooperation, Montreal, Canada, 200 p

Withers K, Nipper M (eds) (English translation) (2009) Environmental analysis of the Gulf of Mexico, vols 1 and 2. Harte Research Institute for Gulf of Mexico Studies, special publication no. 1. Available as e-book only on HRI website, 710 p

Würsig B, Jefferson TA, Schmidly DJ (2000) The marine mammals of the Gulf of Mexico. Texas A&M University Press, College Station, TX, USA, 232 p

Yoskowitz D, Leon C, Gibeaut J, Lupher B, Lopez M, Santos C, Sutton G, McKinney L (2013) Gulf 360: State of the Gulf of Mexico. Harte Research Institute for Gulf of Mexico Studies, Texas A&M University-Corpus Christi, Corpus Christi, TX, USA, 52 p

CHAPTER 2

WATER QUALITY OF THE GULF OF MEXICO

Mahlon C. Kennicutt II[1]

[1]Texas A&M University, College Station, TX 77843, USA
mckennicutt@gmail.com

2.1 INTRODUCTION

Water quality is a vital characteristic in determining how societies and humans use and value aquatic environments and associated natural resources. Coastal and offshore environments are some of the greatest assets of the United States, and much of their value is critically dependent on the quality of the water they contain (Pew Oceans Commission 2003; U.S. Commission on Ocean Policy 2004). The Gulf of Mexico accounts for approximately 13.5 percent (%) of the U.S. coastline. A considerable portion of the economies of the states that border the Gulf of Mexico—Texas, Louisiana, Mississippi, Alabama, and Florida—are dependent on resources and services provided by the maritime environment. Water quality is a derived concept that is usually assessed based on a water body's suitability for ecosystems and/or human use (USGS 2001). Coastal, shelf, and deep water environments are subject to a variety of processes, interactions, influences, and stresses that determine the quality of the water they contain.

In this chapter, the determinants of, the status of, and the trends in water quality in the Gulf of Mexico are reviewed. This review draws on periodic summaries of national coastal conditions by various federal, state, and local agencies and programs. These summaries are reviewed but the underlying primary data are not reanalyzed. The assessments involved were produced by a large number of expert government personnel and academicians based on a vast amount of data and information from primary sources and peer-reviewed literature. These assessments are based on comparable information that strengthen conclusions and allow for comparisons over time. The synthesized data comes from hundreds of sources including national program reports; water quality reporting at the federal, state, and local levels; locally organized monitoring programs; and published literature. These reports and data collection programs span the 1990s to the mid-2000s and often use differing metrics, indicators, and methods for assessing and rating water quality. The time period considered was based on the date that approaches to assessing water quality were adopted region wide and the date of the most recent, complete assessment. The approximate 20-year time period is also most relevant to defining the present day status and trends in water quality in the northern Gulf of Mexico. Data collected pre-1990 is unlikely to reveal significant additional insights and is difficult to integrate with later assessments due to inconsistencies in the methods and approaches used. The end date of the period of time considered was based on the latest, fully vetted national assessment (USEPA 2012). National assessment reports lag data collection by several years due to the process involved. In addition, assessment and rating tools have evolved over time within programs. While standard approaches were often used, caution was taken when comparing data and assessments across many years and multiple programs, though trends in water quality can be discerned. For ease of reference, the methods used to assess and rate water quality are summarized in Appendix A for most of the reports and monitoring programs included in this summary.

© The Author(s) 2017

C.H. Ward (ed.), *Habitats and Biota of the Gulf of Mexico: Before the Deepwater Horizon Oil Spill*,
DOI 10.1007/978-1-4939-3447-8_2

2.2 DETERMINANTS AND MEASURES OF WATER QUALITY

Good water quality is a concept that is derived from a suite of characteristics, and therefore has no single definition. Important determinants of water quality in the Gulf of Mexico are physiographic setting and human activities. Measures of water quality include water clarity, degree of eutrophication, and chemical (petroleum and non-petroleum pollutants) and biological (pathogens) contamination. Natural and anthropogenic effects on water quality are dynamic on many scales leading to considerable variability in space and over time. Impacts on water quality by multiple factors can be additive and/or synergistic. The cumulative effect of natural and anthropogenic influences and processes ultimately determines water quality. The type and mix of components used to define water quality is highly site dependent. It is useful to assess water quality by also considering other indicators of environmental condition such as sediment quality, ecosystem health, and sediment, organismal, and beach contamination. In this chapter these other aspects are only considered in the context of conclusions about water quality and are more comprehensively treated within the national assessments.

2.2.1 Physiographic Setting

The geology, morphology, and oceanographic setting of the Gulf of Mexico are first order determinants of water quality. This review restricts itself to the northern Gulf of Mexico stretching from the southern tip of the Florida Keys to the Texas/Mexico border. Runoff from nearly two-thirds of the continental United States empties into the Gulf of Mexico, primarily via the Mississippi River system and its tributaries (NOAA 1985; USEPA 2006). The geomorphology of the Gulf of Mexico coastal region is characterized by flat coastal plains with adjacent marine environments that are subject to high rates of sediment deposition. A major feature of the Gulf of Mexico is estuaries that have formed large deltas at river mouths reflective of high-energy inflows into lower energy offshore environments. Suspended sediment carried by runoff is deposited in shallow coastal waters and redistributed by nearshore currents often forming sand bars and enclosing shallow, saline lagoons that are most common along the Texas coast. The inlets to these lagoons are often narrow and limit the exchange of water with the open Gulf of Mexico. These restrictions of inflow cause lagoon circulation to be primarily wind driven (NOAA 1985; USEPA 2006). Tidal range and influence in shallow coastal plain estuaries of the Gulf Coast is small varying between 0.3 meters (m) (1 foot [ft]) in Louisiana and Texas to 1.1 m (3.6 ft) in Florida (NOAA 1985). Hurricanes are common from June to late November and can have a dramatic effect on water quality by increasing freshwater inflow due to precipitation and saltwater intrusions due to storm surge. Annual rainfall varies from an average of 1.2 m (3.9 ft) in western Florida to 1.4 m (4.6 ft) in Alabama, Mississippi, and Louisiana to 0.6 m (2.0 ft) in south Texas (NOAA 1985; USEPA 2006). The Gulf Coast includes feeding, spawning, and nursery habitats for fish, wildlife, and plant species that support submerged aquatic vegetation communities that stabilize shorelines from erosion, reduce non-point source loadings, improve water clarity, and provide wildlife habitat. Water quality can be influenced by a wide variety of natural processes including atmospheric transport and deposition, erosion of solids and sediments, runoff, and exchanges between surface water and groundwater.

Most estuarine systems in the Gulf of Mexico are located in low-lying watersheds. The Gulf of Mexico region includes the Mississippi River basin as well as small coastal watersheds in Florida (Figure 2.1a). The watershed area to estuarine area ratio exerts a significant influence on water quality, especially in areas adjacent to dense populations of humans. This ratio can be used as an indicator of the influence of watershed-based inputs on the estuary. Estuaries in the

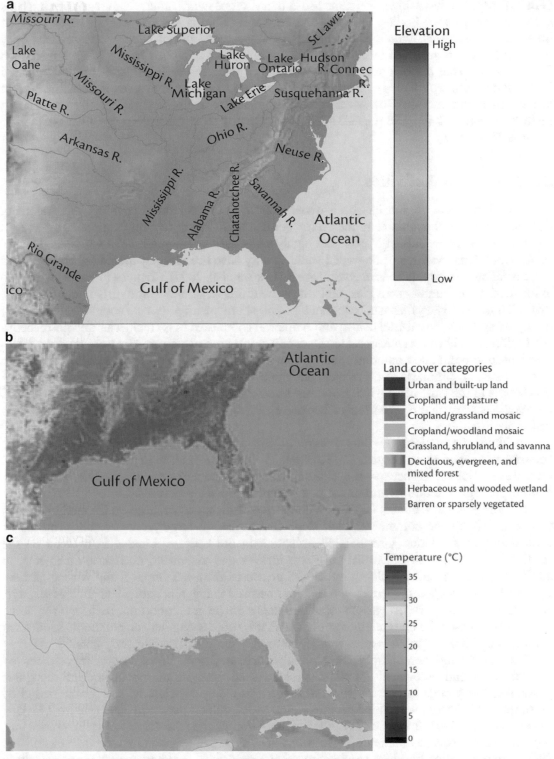

Figure 2.1. (a) Elevation and major rivers of the Gulf of Mexico; (b) land cover categories along the Gulf of Mexico; and (c) sea surface temperature (°C—degree Celsius) in the Gulf of Mexico (modified from Bricker et al. 2007).

Gulf of Mexico have high watershed-to-estuary ratios with input from large watersheds entering small water bodies. Rainfall amounts and patterns also influence the delivery of nutrients to estuaries. Watersheds located in the western Gulf of Mexico are relatively dry with land cover dominated by grassland, shrub land, and savanna (Figure 2.1b). The eastern Gulf of Mexico has a subtropical climate with higher annual rainfall and land covers dominated by croplands and woodlands. Climate along the coast is modulated by ocean temperatures that are warm along the Gulf of Mexico. Annual mean temperatures reflect this modulating influence (Figure 2.1c). The present average number of frost days along the Gulf of Mexico coast is 12 per year.

2.2.2 Human Activities

The Gulf Coast region has been under pressure due to human development for many decades. Studies conclude that the water quality of the majority of estuaries and coastal environments of the Gulf of Mexico are highly influenced by human-related activities (Bricker et al. 2007). Observations of degraded water quality have been largely attributed to dense and increasing human populations in coastal areas (Bricker et al. 2007). Changes in water quality are associated with human activities such as agriculture; residential and urban development; diversion of waterways; coastal construction and shoreline alterations; recreational activities; transport systems; fossil fuel usage; and industrial complexes (e.g., refineries and petrochemical facilities). These activities create the conditions that cause eutrophication, nutrient introductions, and point and non-point source pollutant releases. It is beyond the scope of this paper to exhaustively summarize land usages, the scope and history of human activities, and population trends in the Gulf of Mexico. However, a select set of snapshots are provided as a view of the types of pressures on the Gulf of Mexico that influence the status and trends observed in water quality.

In 2006, the National Estuary Program (NEP) identified major environmental concerns focused in coastal areas (Figure 2.2) (USEPA 2006). Some environmental concerns affect all estuaries and others affect specific locations due to unique climactic, hydrologic, geologic, or geomorphologic conditions and/or the mix of anthropogenic pressures.

In most instances, human influences diminish with distance offshore, so the quality of deep waters overlying the continental shelf/slope and abyss are largely outside the influence of coastal human activities. One notable exception in the Gulf of Mexico is hypoxia on the continental shelf linked to Mississippi River inflows and associated nutrient enrichments. In addition, offshore water quality is subject to pressures from offshore oil and gas exploration and production, shipping, recreational and commercial fishing, and natural oil and gas seepage. Atmospheric transport of various contaminants can be an important pathway for some pollutants to enter the marine environment, and this process can deliver pollutants significant distances offshore in some instances (e.g., mercury from coal-fired power plants).

Population and demographics are closely correlated with the stressors experienced by coastal areas and associated water resources. As an example of increasing anthropogenic pressures, the population of the 48 coastal counties along the Gulf of Mexico increased by more than 133 % from 4.9 million people in 1960 to 11.3 million people in 2000 (Figure 2.3) (U.S. Census Bureau 1991, 2001; USEPA 2006). Population density for these coastal counties was 746 persons per square kilometer (persons/km^2) (1,933 persons per square mile (persons/mi^2) in 2000 with population densities varying from 251 persons/km^2 (651 persons/mi^2) for the Galveston Bay complex to 20 persons/km^2 (53 persons/mi^2) for the Coastal Bend Bays region (U.S. Census Bureau 2001; USEPA 2006).

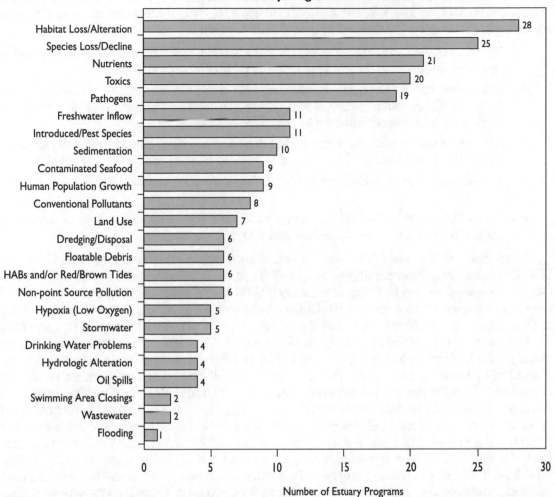

Figure 2.2. Environmental concerns for U.S. estuaries; numbers indicate how many of 28 national estuaries of significance are experiencing a particular concern; HAB—Harmful Algal Blooms (modified from USEPA 2006).

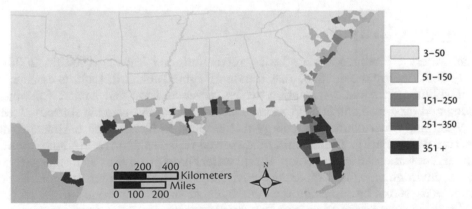

Figure 2.3. Population along the Gulf of Mexico coast in 2003; numerical units are in thousands of persons/mi^2 (modified from Bricker et al. 2007).

The Gulf of Mexico is a focus for commerce and supports considerable and varied recreational activities. In 1999, the Gulf of Mexico Program summarized the major effects humans were having on the Gulf of Mexico (USEPA 1999):

- Texas, Louisiana, and Alabama ranked first, second, and fourth in the nation in 1995 in terms of discharging the greatest amounts of toxic chemicals.

- More than half of the oyster-producing areas along the northern Gulf of Mexico are permanently or conditionally closed. These closure areas are growing as a result of increasing human and domestic animal populations.

- Diversions and consumptive water use for human activities have significantly changed the quantity and timing of freshwater inflows to Gulf of Mexico coastal habitats.

- Louisiana is losing coastal wetlands at the rate of approximately 65 km^2 (25 mi^2) per year.

- Up to 18,000 km^2 (7,000 mi^2) of oxygen deficient (hypoxic) bottom waters have been observed offshore of the Louisiana and upper Texas coasts.

Land use within Gulf of Mexico estuarine watersheds was summarized by the U.S. Environmental Protection Agency (USEPA) in 1999 (USEPA 1999). Gulf of Mexico estuaries were estimated to be approximately 30,000 km^2 (11,600 mi^2) representing 42 % of the total estuarine surface area of the United States excluding Alaska. The Mississippi River drainage area was estimated to be more than 4 million km^2 (1.5 million mi^2), which is more than 55 % of the total area of the conterminous United States. The Gulf of Mexico was receiving an average of 27,473 cubic meters per second (m^3/s) of freshwater inflow daily which was more than 50 % of the daily average for the continental United States. Land use within a watershed determines the materials carried by runoff into adjacent coastal areas. In classifying the land-use categories of the five Gulf States (not the entire watershed), forest and agriculture occupied approximately 58 % of the land area. Forests provide filtration for sediment and nutrients from runoff, stabilize shorelines, and reduce erosion. In the Gulf of Mexico many forests are distant from the shore and are being rapidly replaced by urban and agricultural expansions. Agricultural land included pasture and cropland. Pastureland included grassy areas to raise and feed livestock, and cropland was cultivated for various food products. Other land uses located close to the coastline included wetland habitats (17 %) and urban areas (5 %). While the mix of activities varies with time and place, this snapshot provides an overview of the types of activities that are and will continue to be important for water quality along the northern Gulf of Mexico coastal region.

2.2.3 Water Clarity

Clear waters are valued for aesthetics, recreation, and drinking (USEPA 2008). Water clarity is quantified by the depth of penetration of light (Table 2.1). Light is essential for the health of submerged aquatic vegetation, which serves as food and habitat for other biota. Suspended and dissolved solids that can have natural and anthropogenic sources affect water clarity. Wind and other sources of energy that suspend sediments and particulate matter in water affect water clarity. The amount of dissolved organics and the productivity of phytoplankton affect water clarity and color. Turbid waters have positive as well as negative effects on marine environments. In high-energy environments, turbid waters support healthy and productive ecosystems by supplying the materials that sustain estuarine substrates (i.e., sediments), by being a source of food, and by providing protection for estuarine organisms from predators. In contrast, turbid waters also harm coastal ecosystems by burying benthic

Table 2.1. Criteria for Assessing Water Clarity as an Indicator of Water Quality in Coastal Gulf of Mexico Environments (modified from USEPA 2008)

Area	Good	Fair	Poor
Sites in coastal waters with naturally high turbidity	>10 % light at 1 m	5–10 % light at 1 m	<5 % light at 1 m
Sites in coastal waters with naturally normal turbidity	>20 % light at 1 m	10–20 % light at 1 m	<10 % light at 1 m
Sites in coastal waters that support submerged aquatic vegetation	>40 % light at 1 m	20–40 % light at 1 m	<20 % light at 1 m
Regional assessments criteria of condition as good, fair or poor	Less than 10 % of the coastal area is in poor condition, and more than 50 % of the coastal area is in good condition	10–25 % of the coastal area is in poor condition, and more than 50 % of the coastal area is in combined fair and poor condition	More than 25 % of the coastal area is in poor condition

communities, inhibiting filter feeders, and/or blocking light needed by photosynthetic vegetation. Within an estuary, water clarity can be highly variable over short distances and through time due to tides, storm events, mixing by winds, and changes in incident light. Water clarity is highly variable; it is usually measured based on a ratio of observed clarity in comparison to a reference condition.

One measure of water clarity—turbidity—measures the amount of light that passes through the water over a given distance. Suspended materials include soil inorganic (e.g., clay, silt, and sand) and organic (e.g., bacteria, algae, plankton, and zooplankton) particles. Suspended particles vary in size and affect water clarity and color. Suspended solids/sediments come from non-point sources (e.g., stormwater runoff, stream erosion, agricultural runoff, urban runoff, and leaching of soils) and point sources (i.e., construction projects and industrial or sewage treatment plant discharges). Total suspended solids (TSS) are defined as that material indefinitely suspended in solution but retained on a sieve size of two micrometers (2 µm). Settleable solids refer to material that does not remain suspended or dissolved when water is motionless. Settleable solids may include large particulate matter or insoluble particles. The total inorganic and organic substances dissolved in water are called total dissolved solids (TDS). Dissolved solids are usually defined as material that passes through a sieve size of 2 µm (APHA 1992). TDS is normally only an indicator of water quality for freshwater because saltwater contains dissolved ions that are included in measurements of TDS. The sources of TDS are similar to those for suspended solids. Chemicals commonly dissolved in water include calcium, phosphates, nitrates, sodium, potassium, and chloride. These chemicals are found in various types of runoff from land surfaces and occur as cations, anions, molecules, and/or aggregates. Contaminants that can partially occur in a dissolved state include hydrocarbons, metals, and persistent organic pollutants. Naturally occurring TDS are formed during the weathering of rocks and soils. Processes that affect turbidity in estuaries include resuspension, deposition, and advection of sediment. Tide-dominated estuaries are naturally turbid because strong tidal currents tend to resuspend sediments. Tidal currents can mobilize fine sediments, and turbidity can vary considerably during daily tidal cycles. Trapping and flocculation of sediment at the salinity discontinuity (mixing zone) between freshwater and seawater can cause a turbidity maximum. Criteria have been developed to assess water clarity in the coastal Gulf of Mexico based on light penetration (Table 2.1).

2.2.4 Eutrophication

In 1999, the National Oceanic and Atmospheric Administration (NOAA) reported the results of a national estuarine eutrophication survey recognizing the persistent and pervasive nature of this environmental problem in the nation's coastal regions:

> One of the most prominent barometers of coastal environmental stress is estuarine water quality, particularly with respect to the inputs of nutrients. Coastal and estuarine waters are now among the most heavily fertilized environments in the world. Nutrient sources include point (e.g., wastewater treatment plants) and non-point (e.g., agriculture, lawns, and gardens) discharges. These inputs are known to have direct effects on water quality. For example, in extreme conditions, excess nutrients can stimulate excessive algal blooms that can lead to increased metabolism and turbidity, decreased dissolved oxygen, and changes in community structure and condition described by ecologists as eutrophication. Indirect effects can include impacts to commercial fisheries, recreation, and even public health. (Bricker et al. 1999)

Assessments of eutrophication are based on several of the most utilized measures of water quality: dissolved inorganic nitrogen (DIN), dissolved inorganic phosphorus (DIP), dissolved oxygen, and chlorophyll *a* concentrations (Figure 2.4). Water clarity is also affected by eutrophication, but water clarity is treated separately above in Section 2.2.3.

Nutrients are essential elements that support biological productivity in coastal waters and sustain healthy and functioning ecosystems. Nutrients of particular concern for water quality are those that contain nitrogen and phosphorus. Nutrients from various sources can increase estuarine concentrations above background levels, increasing rates of organic matter synthesis. These nutrient additions can lead to eutrophication and degraded water quality (Figure 2.5).

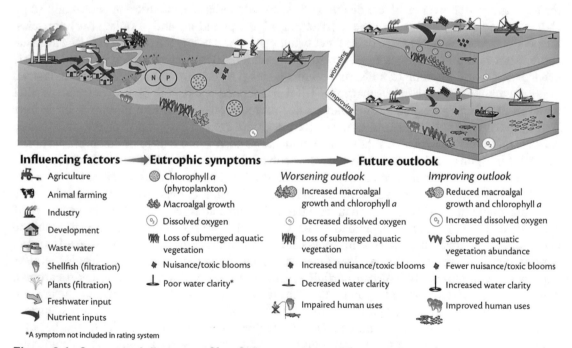

Figure 2.4. Conceptual diagrams of key features, major nutrient sources, and resulting symptoms related to eutrophication in the Gulf of Mexico (modified from Bricker et al. 2007).

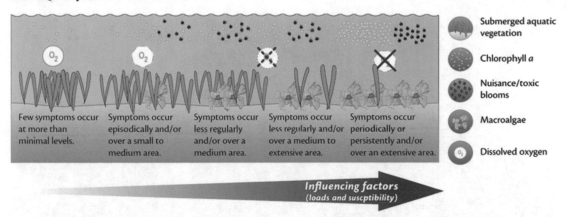

Figure 2.5. Relationship between eutrophication condition, associated trophic symptoms, and influencing factors—nitrogen loads and susceptibility (modified from Bricker et al. 2007).

Excess plant production increases chlorophyll concentrations, decreases water clarity and lowers concentrations of dissolved oxygen due to aerobic decomposition of organic matter. If nutrients are present at concentrations less than needed, the growth and reproduction of organisms is limited. Nutrient additions to aquatic systems occur naturally due to geological weathering and ocean upwelling. In coastal areas, human population growth has increased nutrient inputs many times their natural levels accelerating eutrophication (Figure 2.5). Nutrient increases can threaten biota and lead to impairments of aesthetics, health, fishing opportunities and success, tourism, and real estate values (Figure 2.6).

Nitrogen is usually the primary limiting nutrient for growth of algae in marine waters (Pedersen and Borum 1996). Nitrogen can be found in several different forms in aquatic systems including ammonia (NH_3^+), total nitrogen, nitrites (NO_2-), and most commonly nitrate (NO_3-). Phosphorus in aquatic systems occurs as organic phosphate and inorganic phosphate. Plants use inorganic phosphorus while animals can use either organic or inorganic phosphate to form tissues. Organic and inorganic phosphorus can be dissolved in water or can occur as particulates (e.g., attached to eroded soil). Animals meet their organic phosphorus nutritional requirements by consuming aquatic plants, other animals, and/or decomposing plant and animal detritus. Plants and animals excrete wastes containing both nitrogen and phosphorus.

Nitrogen and phosphorus are released upon the death of an organism by a process termed *remineralization*. Remineralization occurs when bacteria convert organic matter to particulate or dissolved inorganic nitrogen and phosphorus. Inorganic nitrogen and phosphorus in sediments can be resuspended into the water column by bottom dwelling organisms, human activity, diffusion, and/or currents and winds. Remineralized nutrients reenter the food web, once again beginning the cycle. Excess nitrogen and phosphorus are released to aquatic environments by agriculture practices (e.g., application of chemical fertilizer, manure, and organic matter); residential and urban development (e.g., lawn fertilizer, pet wastes, and failing septic systems); and wastewater discharges (e.g., untreated or treated wastewater and sewage) (Figure 2.6). One of the largest inputs of excess nitrogen is the Mississippi River system that delivers excess fertilizer from the heartland of the United States to the Gulf of Mexico (Figure 2.7).

The amount of oxygen dissolved in water is a basic measure of water quality. Organisms in aquatic environments need oxygen to support aerobic respiration. Low oxygen concentrations can reduce aquatic biomass and diversity. Oxygen enters water by diffusion from the

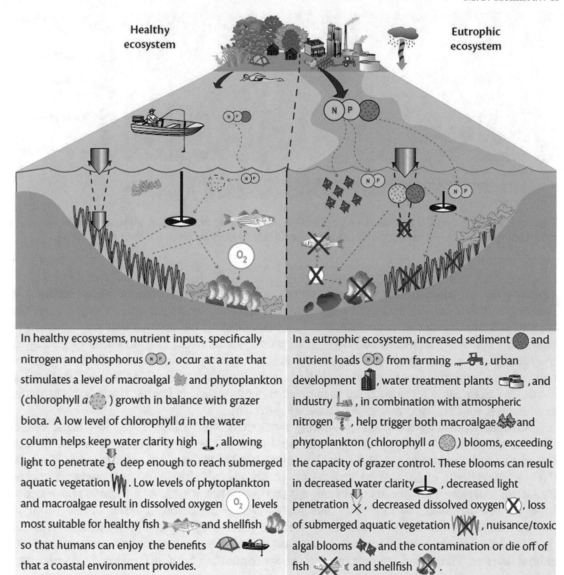

Figure 2.6. Comparison of a healthy system with no or low eutrophication to an unhealthy system exhibiting eutrophic symptoms (modified from Bricker et al. 2007).

atmosphere, by turbulent mixing with the atmosphere, and by release during photosynthesis. Dissolved oxygen is removed from water by diffusion into the overlying atmosphere if concentrations exceed solubility, respiration, and aerobic decomposition (remineralization) of organic matter. Water with less than 1 milligram per liter (mg/L) is anoxic (lethal), and water with less than 5 mg/L of oxygen is suboxic (stressful to most organisms); and water with more than 7 mg/L of oxygen is considered desirable for aquatic life (Table 2.2).

Chlorophyll *a* concentrations are another basic measure of water quality. Chlorophyll *a* indicates the amount of algae (or phytoplankton) growing in a water body. High concentrations of chlorophyll *a* indicate the potential for overproduction of algae resulting in degraded water quality (Table 2.2).

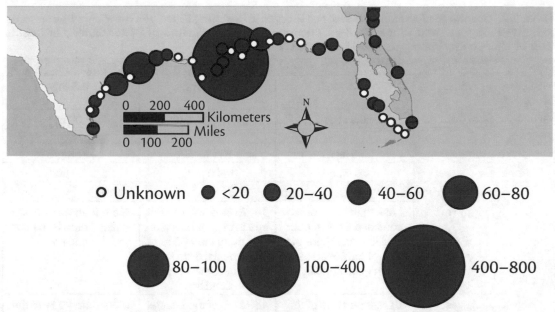

Figure 2.7. Nitrogen loads ($\times 10^6$ tons per year) for the Gulf of Mexico. High nitrogen loads correspond with high agricultural activity and the Mississippi River outflow (modified from Bricker et al. 2007).

2.2.5 Chemical Contaminants

Most marine environments are subject to a complex and time-variant mixture of factors that collectively degrade water quality. While contaminant chemicals have the potential to affect water quality, it is usually difficult to unambiguously ascribe degraded water quality to contamination alone (the major exception being excess nutrient releases which can be considered chemical contaminants). There are a few scenarios where chemical contaminants may be the primary cause of degraded water quality such as a major oil spill or locations associated with the manufacture of chemicals (e.g., pesticide manufacturing operations). However, chemical contaminants can, and do, contribute to the degradation of water quality with follow-on effects on associated organisms and ecosystems.

The chemicals that are most often the focus of environmental concern because of known toxicological properties and their wide usage by humans include aromatic hydrocarbons, metals, and persistent organic pollutants. In this review, these chemicals are collectively referred to as *contaminants* (excluding nutrients which are separately considered above). Some contaminants have natural as well as anthropogenic origins. In this review, contaminants are categorized into two major classes: petroleum and non-petroleum (although some non-petroleum chemicals are synthesized from petroleum), and their effects on water quality are separately considered. Non-petroleum contaminants are further subdivided into organic and inorganic contaminants. Each category of chemical contaminants has different sources, environmental fates and effects, toxicities, and potentials to degrade water quality.

Petroleum, including products refined from petroleum, contains a complex mixture of potentially toxic compounds. The class of compounds that accounts for most of the toxicity of petroleum is polycyclic aromatic hydrocarbons (PAHs) (NRC 2003). PAH concentrations are often used as an indicator of petroleum contamination, but other measures, such as oil and grease gravimetrically (by weight) determined as total extractable hydrocarbons or gas

Table 2.2. Criteria for Assessing Dissolved Inorganic Nitrogen, Dissolved Inorganic Phosphorus, Dissolved Oxygen, and Chlorophyll *a* Concentrations as Indicators of Water Quality in Coastal Gulf of Mexico Environments (modified from USEPA 2008)

Indicator	Good	Fair	Poor
Dissolved inorganic nitrogen	<0.1 mg/L	0.1–0.5 mg/L	>0.5 mg/L
Dissolved inorganic phosphorus	<0.01 mg/L	0.01–0.05 mg/L	>0.05 mg/L
Dissolved oxygen	>5 mg/L	2–5 mg/L	<2 mg/L
Chlorophyll *a*	<0.5 µg/L	5–20 µg/L	>20 µg/L
Regional assessment criteria as good, fair, or poor			
Dissolved inorganic nitrogen	Less than 10 % of the coastal area is in poor condition, and more than 50 % of the coastal area is in good condition	10–25 % of the coastal area is in poor condition, and more than 50 % of the coastal area is in combined fair and poor condition	More than 25 % of the coastal area is in poor condition
Dissolved inorganic phosphorus	Less than 10 % of the coastal area is in poor condition, and more than 50 % of the coastal area is in good condition	10–25 % of the coastal area is in poor condition, and more than 50 % of the coastal area is in combined fair and poor condition	More than 25 % of the coastal area is in poor condition.
Dissolved oxygen	Less than 5 % of the coastal area is in poor condition, and more than 50 % of the coastal area is in good condition	5–15 % of the coastal area is in poor condition, and more than 50 % of the coastal area is in combined fair and poor condition	More than 15 % of the coastal area is in poor condition
Chlorophyll *a*	Less than 10 % of the coastal area is in poor condition, and more than 50 % of the coastal area is in good condition	10–20 % of the coastal area is in poor condition, and more than 50 % of the coastal area is in combined fair and poor condition	More than 20 % of the coastal area is in poor condition

mg/L milligram(s) per liter (parts per million (ppm)), *µg/L* microgram(s) per liter (parts per billion (ppb))

chromatographically resolved compounds determined by flame ionization or mass spectroscopy detection, are also used. These methods quantify different portions of petroleum and are subject to different interferences (including the measurement of non-petroleum materials), so results are usually difficult to compare. PAHs are complex mixtures of sometimes hundreds of compounds. Petroleum is released to the environment by intentional and/or unintentional discharges and spills and as byproducts of petroleum usage by humans (NRC 2003). Petroleum is also released to the environment by natural processes such as oil and gas seepage.

Non-petroleum, organic contaminants include polychlorinated biphenyls (PCBs), chlorinated pesticides, and other synthetic chemicals. These chemicals usually, but not always, contain halogens—particularly chlorine and bromine—accounting in part for their toxicity. These chemicals are widely used by humans for various purposes and are ubiquitous in marine environments (see Appendix B for descriptions of common organic contaminants). Non-petroleum, inorganic contaminants include various metals. The most common metals of environmental concern include lead, mercury, arsenic, cadmium, silver, nickel, selenium, chromium, zinc, and copper. These metals are known to have toxicological properties. Organo-metallic compounds are also included in this subcategory of compounds including tributyltin (used in antifouling paints) and methylmercury (a microbial metabolic derivative of mercury). Metals are released to the environment by human activities including vehicle emissions, industrial processes, improper use or disposal of metallic products, and pesticides (see Appendix C for descriptions of common metal contaminants). Many metals also occur naturally in crustal rocks and minerals. Beyond the contaminants mentioned above, there are also a series of other human-derived chemicals that have the potential to cause environmental degradation including improper disposal of unused pharmaceuticals, household chemicals, and personal hygiene products; fire retardants (brominated compounds); and endocrine-disrupting or mimicking compounds. However, most monitoring programs rarely systematically measure these chemicals in the waters of the northern Gulf of Mexico, so the extent and impact of these chemicals remains largely unknown.

In this chapter, the potential for petroleum contamination to degrade water quality is partially inferred from annual mass loadings of petroleum to the northern Gulf Mexico. Estimates of the inputs of petroleum to the Gulf of Mexico are summarized by the National Research Council's (NRC's) *Oil in the Sea III: Inputs, Fates, and Effects* report (NRC 2003). This report is the most recent comprehensive compilation of petroleum inputs to the northern Gulf of Mexico and is based on data from the 1990s (NRC 2003). The 9-year averages provided are representative of longer-term trends in the region. However, the absolute amounts associated with various sources are expected to vary with time. Mass loading estimates cannot be used to infer petroleum concentrations in environmental matrices such as water but do provide some insight into the origins, geographic distribution, and magnitude of petroleum inputs within limits (Tables 2.3 and 2.4). Within these limitations, mass loadings of petroleum are qualitatively compared and contrasted with observed spatial and temporal patterns in water quality in the northern Gulf of Mexico to determine if any relationship exists.

Due to low solubility in water, contaminant concentrations in water are usually low, challenging even the most sensitive analytical methods. Therefore, most water quality monitoring programs do not routinely measure the concentrations of contaminants in water (except nutrients). However, contaminants may contribute to degraded water quality even though ambient water concentrations are low. Because of their hydrophobic properties, contaminants preferentially accumulate in biological tissues and sediments. Over time organisms exposed to low levels of contaminants in water will continue to accumulate contaminants because contaminant solubility in lipid-rich biological tissues far exceeds their solubility in water. Biological tissue contaminant concentrations can potentially indicate the presence of contaminants in water that may not be detectable by direct analysis. However, there are complications in inferring that contaminants are present in water by their presence in biological tissues. Contaminants can accumulate in organismal tissues via pathways other than uptake from contaminated water. Some organisms ingest contaminated sediments. Other organisms consume contaminated dietary foodstuffs. Some organisms remove contaminants from their systems through depuration and excretion. In many organisms, physiological processes that can detoxify contaminant chemicals are quite advanced, while other organisms have little innate ability to detoxify contaminants. Higher trophic level organisms such as fish consume contaminated organisms, and the levels of contaminants increase by a process termed *biomagnification*. Larger and larger organisms consume greater and greater biomass to support their higher metabolic demands. The organisms themselves, as well as their living foodstuffs, may have

Table 2.3. Concepts Related to the Behavior of Petroleum in the Environment Important to Interpreting Mass Loadings of Petroleum to Marine Environments (modified from NRC 2003)[a]

Process/factor	Definitions/concepts/description/importance
Weathering	A series of changes in physical and chemical properties: • Weathering processes occur at very different rates • Weathering rates are not consistent and are usually highest immediately after release • Weathering processes and rates at which they occur depend more on the type of oil than on environmental conditions • Most weathering processes are highly temperature dependent
The size of the release and the impact on organisms	Loading rates, in units of mass per unit time, are useful for comparing the relative importance of various types of loadings and describing the spatial distribution of loadings • Petroleum is a complex group of mixtures, and each group may contain widely varying relative amounts of hundreds (or more) compounds • Many of the compounds are apparently benign. Many others, such as some types of PAH, are known to cause toxic effects in some marine organisms • Predicting the environmental response to a specific release of a known quantity of a refined petroleum product (which contains far fewer compounds than crude oil) requires site-specific information about the nature of the receiving water body • Loading and impact are distinct and it is not possible to directly assess environmental damage from petroleum hydrocarbon mass loading rates • Effects tend to reflect the amount of toxic hydrocarbon compounds reaching a marine organism and the differing susceptibility of various organisms, populations, and ecosystems to the effects of these hydrocarbons • Effects tend to reflect the amount of toxic hydrocarbon compounds reaching a marine organism and the differing susceptibility of various organisms, populations, and ecosystems to the effects of these hydrocarbons • Ecotoxicological responses are driven by the dose of petroleum hydrocarbons available to an organism, not the amount of petroleum released into the environment • Dose is rarely directly proportional to the amount released because of the complex environmental processes acting on the released petroleum • The type of petroleum released and the susceptibility of the target organisms must both be considered • It is often difficult to reach consensus on the magnitude and duration of environmental effects

(continued)

Table 2.3 (continued)

Process/factor	Definitions/concepts/description/importance
Bioavailability	The amount of petroleum made available to an organism through various environmental processes (whether for ingestion or absorption) is referred to as being biologically available, or simply, bioavailable • The release of equal amounts of the same substance at different times or locations may have dramatically different environmental impacts • Bioavailability can describe the net result of physical, chemical, and biological processes that moderate the transport of hydrocarbon compounds from their release points to the target organisms • Processes acting on petroleum as it moves from the release point to the marine organism can alter the chemical composition of the petroleum mixture, which in turn likely alters the toxicity by selectively enriching or depleting toxic components • Physical weathering processes may encapsulate some or all of the petroleum in forms that are less available to organisms such as tar balls • Physiological and behavioral processes moderate the movement of petroleum from the surrounding environment into marine organisms • Individual petroleum components pass into organisms at different rates depending on their physical and chemical properties • Organisms respond to hydrocarbons in their surroundings and moderate or accentuate exposure • Once the hydrocarbons are in the organisms, there is a wide variation in the types and magnitudes of physiological responses. Many organisms metabolize and excrete hydrocarbons creating more toxic intermediates

[a]These concepts apply most directly to spilled petroleum; however, the general principles apply to all petroleum once released to the environment "regardless of source

Table 2.4. Processes that Move Petroleum Hydrocarbons Away from the Point of Origin (modified from NRC 2003)[a]

Input type	Persistence	Evaporation	Emulsification	Dissolution	Oxidation	Horizontal transport or movement	Vertical transport or movement	Sedimentation	Shoreline stranding	Tar balls
Seeps	Years	H	M	M	M	H	M	M	H	H
Spills										
• Gasoline	Days	H	NR	M	L	L	L	NR	NR	NR
• Light distillates	Days	M	L/L	H	L	M	H	L	L	NR
• Crudes	Months	M	M	M	M	M	M	M	H	M
• Heavy distillates	Years	L	M	L	L	H	L	H	H	H
Produced water	Days	M	NR	M	M	L	L	L	L	NR
Vessel operation	Months	M	L	M	L	M	L	L	L	M
Two-stroke engines	Days	H	NR	M	L	L	L	L/NR	NR	NR
Atmospheric	Days	H	NR	M	M	H	NR/NR	L	NR	NR
Land-based	U	M	L	L	L	M	M	M	NR	U

Note: H high; *M* moderate; *L* low; *NR* not relevant; *U* unknown

[a]Each input is ranked using a scale of high, medium, and low that indicates the relative importance of each process. The table is intended only to convey variability and is based on many assumptions providing a general idea of the relative importance of these processes. The importance of a particular process will depend on the details of the spill event or release. These concepts apply most directly to spilled petroleum. "The chemical and physical character of crude oils or refined products greatly influences how these compounds behave in the environment as well as the degree and duration of the environmental effects of their release" (NRC 2003)

migrated from distant locations or roamed over great distances. Mobile marine organisms can range over quite large distances and tissue contaminant concentrations reflect what may be a complex history of dosages, exposures, and excretions. All of these factors are highly variable from one species to the next. Contaminant concentrations in organism tissues are the end product of these complex physiological processes and interactions with the environments they live in, confounding the attribution of tissue contaminant sources to specific water bodies.

While recognizing the limitations on interpretations of the data, contaminant concentrations in biological tissues and sediments can provide a qualitative indication that contaminants may be contributing to degraded water quality. A comprehensive review of contaminants in biological tissues in the northern Gulf of Mexico is beyond the scope of this review; however, limited considerations of data on fish consumption advisories are used to identify which chemicals are of greatest concern. The geographic distribution of advisories can pinpoint contaminant hot spots and be compared with the distribution of degraded water quality to identify co-occurrences, but cause and effect is difficult to infer for the reasons identified. Sentinel, sessile organisms, such as filter-feeding bivalves (oysters and mussels), filter and accumulate particles from large volumes of water acting as time integrators of exposure to contaminants in water. Contaminant concentrations in the tissues of these organisms are good indicators of local contamination and can be used to infer possible contaminant-related degraded water quality. Oyster tissue contaminant distributions for the northern Gulf of Mexico are reported and reviewed elsewhere (Kimbrough et al. 2008). The distribution and types of contaminants in sediments can also be used to infer possible contaminant-related degraded water quality. Distributions and origins of the common contaminants in sediments of the northern Gulf of Mexico are reported elsewhere (this volume, Chapter 4).

2.2.6 Water Quality Impairment and Biological Contaminants

Water quality impairment assessments synthesize diverse sets of information to describe the overall condition of marine waters. These assessments indicate the status of water quality and are used to inform the public about risks associated with various uses of marine waters. Assessments of the presence of biological contamination (pathogens) in waters and assessments of chemical contaminants in organisms consumed by the public can provide indications of possible water quality degradation. This chapter reviews the methods used to detect and report the presence of biological contaminants and the translation of these and other data into assessments of how well waters are supporting designated uses, including the criteria for beach closings. These summaries are from documents referenced in the assessments used in this review, and it should be noted that guidance criteria are under continuous review and may have been revised subsequent to the issuing of these assessment reports.

States report water quality assessment information and water quality impairments under Sections 305(b) and 303(d) of the Clean Water Act. These assessments compare field data to state water quality standards (USEPA 2001). Water quality standards include narrative and numerical criteria that are used to judge if water bodies are capable of supporting specific, designated uses without undue risk to public health. These criteria set specific goals that need to be met to prevent degradation of water quality. The criteria are used to evaluate whether the designated uses of water bodies are supported as follows:

- Fully supporting: These waters meet applicable water quality standards, both criteria and designated use.

- Threatened: These waters currently meet water quality standards, but states are concerned they may degrade in the future.

- Partially supporting: These waters meet quality standards most of the time, but exhibit occasional exceedances.

- Not supporting: These waters do not meet water quality standards.

The data is then integrated and compared to established criteria to ascertain if designated uses can be supported with acceptable risk to public health. Categories of water use include aquatic life support; drinking water supply; recreation activities such as swimming, fishing, and boating; and fish and shellfish consumption by humans (USEPA 2001). A water body classified as partially supporting or not supporting its usages is considered impaired. Each state monitors water quality parameters differently, so generalities about condition are often difficult to make based on these data alone. States also issue consumption advisories to inform the public of elevated concentrations of chemical contaminants detected in local fish and shellfish tissues.

Public health may be at risk due to polluted bathing beaches. USEPA established the Beaches Environmental Assessment, Closure, and Health (BEACH) Program and the Program Tracking, Advisories, Water Quality Standards, and Nutrients (PRAWN) to better define the extent of beach contamination in the United States (USEPA 2001, 2008). A few states have comprehensive beach monitoring programs while others have only limited or no beach monitoring programs, making comprehensive assessments of the problem in a region like the northern Gulf of Mexico difficult. However, beach water contamination, particularly by pathogens, is considered to be a persistent problem based on the number of beach closings and swimming advisories issued each year (USEPA 2003a, b). The integration of these data into assessments of impairment provide an indication of water quality issues and assist in identifying possible causative agents that may require regulatory action.

Pathogens can have detrimental effects on water quality. Biological contaminants are introduced to receiving waters by a variety of processes. Fecal bacteria indicate the possible presence of pathogens in water and the risk of humans contracting diseases from the ingestion of contaminated surface water or raw shellfish (USEPA 2003b). Contact with contaminated water can lead to ear or skin infections, and inhalation of pathogen-contaminated water can cause respiratory diseases. These infections and diseases are due to exposure to bacteria, viruses, protozoans, fungi, and/or parasites that live in the gastrointestinal tract of humans and the feces of warm-blooded animals (USEPA 2003b). Concentrations of fecal bacteria, including fecal coliforms, enterococci, and *Escherichia coli* in water are used to indicate fecal contamination (USEPA 2003b). Enterococci and *E. coli* have been shown to correlate with outbreaks of disease, and USEPA recommends them as indicators of biological contamination (USEPA 2003b). Sources of pathogenic organisms include malfunctioning septic systems, overboard discharges of untreated sewage from boats, sewer overflows, improperly stored/ used animal manure, pet wastes, and improperly working waste treatment facilities. *E. coli* counts often increase after storm events such as heavy thundershowers or continuous rain. USEPA recommends various bacteriological assay methods to detect indicator pathogens. USEPA bacteriological criterion for restricting bathing in recreational marine water, based on no less than five samples equally spaced over a 30-day period, is that the geometric mean of the enterococci densities should not exceed 35 per 100 milliliters (mL) of water. Because states often adopt their own methodologies and criteria for assessing biological contamination of waters and issuing advisories, comparisons across monitoring programs should be made with caution. For this review, a limited number of the reports of beach closings and the reasons for these closings are provided as an indication of degraded water quality; however, it is not an exhaustive treatment of all available data for the northern Gulf of Mexico which is reviewed elsewhere.

2.3 COASTAL WATER QUALITY

Based on the importance of coastal resources, a coordinated effort to monitor their condition has been in place since the early 1990s in the United States (Bricker et al. 1999; USEPA 2001, 2004, 2008, 2012). One of the first comprehensive, national assessments of estuarine eutrophication was NOAA's *National Estuarine Eutrophication Assessment* in 1999 (Bricker et al. 1999). This was followed in subsequent years by National Coastal Condition Reports that "...describe and summarize the ecological and environmental conditions in U.S. coastal water and highlight exemplary...programs that assess coastal ecological and water quality conditions." The USEPA Office of Wetlands, Oceans and Watersheds' Coastal Programs created these reports to provide a "comprehensive picture of the health of the nation's coastal waters." The reports are based on data collected from a variety of sources coordinated by USEPA, NOAA, the U.S. Geological Survey (USGS), U.S. Fisheries and Wildlife Service (USFWS), and coastal states. One aspect of these national assessments is a region-by-region consideration of water quality. To describe water quality in the Gulf of Mexico, the regional trends in these reports are summarized as well as a discussion of site-specific monitoring results. The reviews of regional assessments are followed by summaries of a series of state-of-the-bay reports that highlight water quality on a finer spatial scale. These are summaries and not a reanalysis of primary, underlying data.

2.3.1 NOAA's Estuarine Eutrophication Assessment (1999)

In 1999, NOAA's National Estuarine Eutrophication Assessment provided the first comprehensive assessment of water quality in the northern Gulf of Mexico (Bricker et al. 1999). The assessment was based primarily on the results of a national survey conducted by NOAA from 1992 to 1997 supplemented by information on nutrient inputs, population projections, and land use from a variety of sources. This assessment catalyzed future USEPA National Coastal Condition Reports. The assessment was conducted at a workshop of experts that participated in a nationwide survey. The report is described as presenting "... the results of a comprehensive National Assessment to address the problem of estuarine eutrophication. The assessment includes evaluations of eutrophic conditions, human influence, impaired estuarine uses, future conditions, data gaps and research needs, and recommendations for a national strategy to respond to the problem..." (Bricker et al. 1999). Eutrophication "...refers to a process in which the addition of nutrients to water bodies stimulates algal growth. In recent decades, human activities have greatly accelerated nutrient inputs, causing the excessive growth of algae and leading to degraded water quality and associated impairments of estuarine resources for human (and ecological) use..." (Bricker et al. 1999).

The report provided regional assessments including the northern Gulf of Mexico. The assessment concluded that "...the expression of high eutrophic conditions is extensive, and human influence is substantial, in the Gulf of Mexico region. Although there is a great diversity of estuary types, common characteristics, such as low tidal flushing, warm water, and long algal growing seasons, create conditions that make many of the region's estuaries susceptible to eutrophic problems. The most significant symptoms in the overall expression of eutrophic conditions are low dissolved oxygen and loss of submerged aquatic vegetation. Impaired resource uses are evident in many, but not all, of the affected systems. Conditions are expected to worsen in more than half of the estuaries by 2020..." (see Figure 2.8).

Of the 38 Gulf of Mexico estuaries and the Mississippi River Plume, 20 estuaries exhibited high levels of at least one of the symptoms of eutrophication. Chlorophyll *a* concentrations were high in 12 estuaries mainly on the coasts of western Florida, Louisiana, and lower Texas.

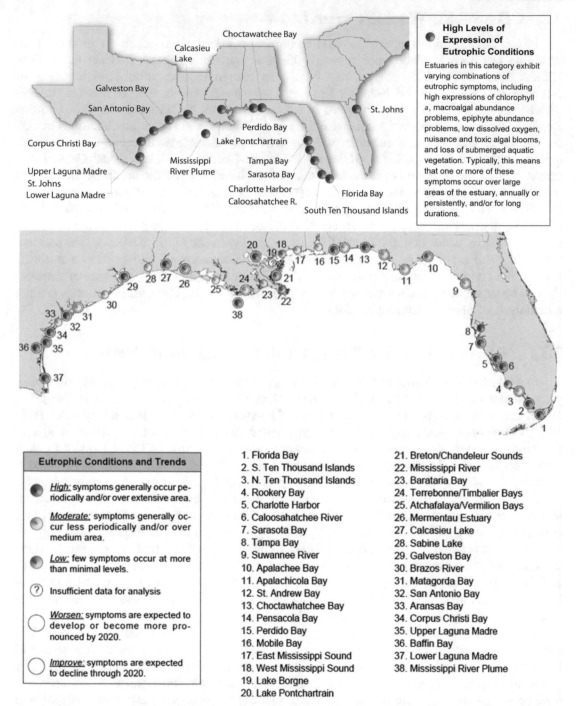

Figure 2.8. Level of expression of eutrophic conditions in the Gulf of Mexico and future trends (modified from Bricker et al. 1999).

Epiphytes were moderate to high in eight estuaries. Macroalgal abundance was moderate to high in seven estuaries. Low dissolved oxygen concentrations were observed in four estuaries along the Florida coast and in the Mississippi River Plume. Submerged aquatic vegetation loss was observed in 28 estuaries, and eight were considered to have high levels of loss along the Florida, western Louisiana, and the lower Texas coasts.

High eutrophic conditions were expressed as loss of submerged aquatic vegetation, increased turbidity associated with high concentrations of chlorophyll *a*, and low levels of dissolved oxygen. Moderate to high levels of nuisance/toxic algal blooms and epiphyte abundance were observed as well. It was noted that conditions seemed to be improving due to better management of point and non-point nutrient sources at some locations. The authors concluded that the Gulf of Mexico was well studied and the data synthesis robust (Bricker et al. 1999). It was also concluded that human influence was high in more than half of the estuaries studied and that this was linked with high expressions of eutrophication. Those areas considered to be most influenced by humans included the Mississippi River Plume, Lake Pontchartrain, Upper and Lower Laguna Madre, and Baffin Bay. Estuaries with lower levels of human influence were Rookery Bay, the Suwannee River, Apalachee Bay, and Breton/Chandeleur Sounds (Figure 2.8) (Bricker et al. 1999).

The factors that had greatest influence on expressions of eutrophication in the Gulf of Mexico were low tidal energy, low flushing rates with increased nutrient inputs, and low dissolved oxygen levels generally due to warm waters and long growing seasons. Nitrogen inputs were considered moderate. Bricker et al. (1999) conclude that impaired uses were difficult to define as being directly related to eutrophication but results suggest that the most impaired uses were recreational and commercial fishing, shellfishing, and loss of submerged aquatic vegetation. Of the 38 estuaries, 23 were predicted to develop worsening conditions during the following 20 years, and six estuaries were judged to be at high risk of worsening eutrophication in the future including the Mississippi River Plume, Lake Pontchartrain, Corpus Christi Bay, Upper and Lower Laguna Madre, and Baffin Bay. Three estuaries were judged to have the potential to decrease eutrophic symptoms in the future, including Florida Bay, Breton and Chandeleur Sounds, and Mermentau Estuary.

2.3.2 USEPA's National Coastal Condition Reports I (2001) and II (2004)

The need for regular assessments of coastal conditions to identify problem areas and judge long-term trends to inform management and regulatory decisions was highlighted by the NOAA eutrophication survey (Bricker et al. 1999). The first National Coastal Condition Report was issued in 2001 based on information collected from 1990 to 1997 (USEPA 2001) and the second was issued in 2004 based on monitoring data collected from 1997 to 2000 (USEPA 2004). These reports concluded that the overall condition of Gulf of Mexico coastal waters was fair to poor (Figure 2.9).

The USEPA Environmental Monitoring and Assessment Program (EMAP) collected environmental stressor and response data from 1991 to 1995 at 500 locations from Florida Bay, Florida, to Laguna Madre, Texas. The conclusions of EMAP were similar to those of NOAA (USEPA 1999; Bricker et al. 1999), that is, eutrophication was one of the most critical problems facing northern Gulf of Mexico ecosystems. EMAP concluded that excess nitrogen enters Gulf of Mexico estuaries via fertilizer runoff from agricultural and residential land, animal manure, and atmospheric deposition. In addition, the region has the highest number of wastewater treatment plants and the most land devoted to agriculture with the most applied fertilizer in the United States. Many Gulf of Mexico estuaries showed evidence of pre-eutrophic or eutrophic conditions. Four indicators of nutrient enrichment were used to assess the overall nutrient status of estuaries: the NOAA Estuarine Eutrophication Survey (Bricker et al. 1999), state 305 (b) assessment of nitrogen level, state 305(b) assessment of chlorophyll levels, and the Rabalais et al. (1992) evaluation of nutrient increases. Nutrient problems ranged from minimal in

a
Overall
Gulf
Good Fair Poor
Water Clarity
Dissolved Oxygen
Coastal Wetlands
Eutrophic Condition
Sediment
Benthos
Fish Tissue
b
Gulf
Overall
Score (2.4)
Good Fair Poor
Water Quality Index (3)
Sediment Quality Index (3)
Benthic Index (2)
Coastal Habitat Index (1)
Fish Tissue Index (3)

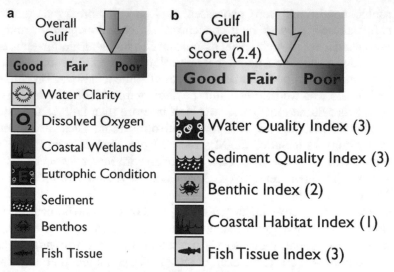

Figure 2.9. (a) Overall condition of Gulf of Mexico coastal resources was rated fair to poor in 2001 (modified from USEPA 2001) and (b) 2004 (modified from USEPA 2004).

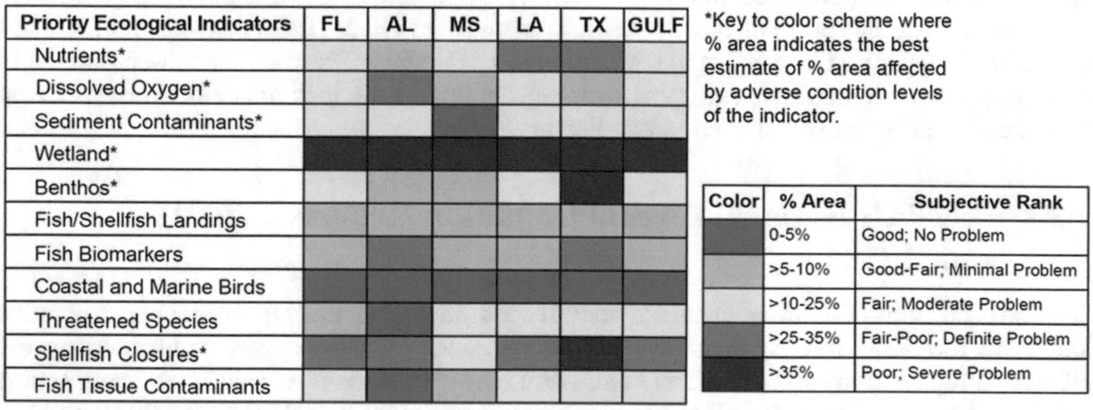

Priority Ecological Indicators
FL
AL
MS
LA
TX
GULF
Nutrients*
Dissolved Oxygen*
Sediment Contaminants*
Wetland*
Benthos*
Fish/Shellfish Landings
Fish Biomarkers
Coastal and Marine Birds
Threatened Species
Shellfish Closures*
Fish Tissue Contaminants
*Key to color scheme where % area indicates the best estimate of % area affected by adverse condition levels of the indicator.
Color
% Area
Subjective Rank
0-5%
Good; No Problem
>5-10%
Good-Fair; Minimal Problem
>10-25%
Fair; Moderate Problem
>25-35%
Fair-Poor; Definite Problem
>35%
Poor; Severe Problem

Figure 2.10. Estimates of the status of ecological conditions along the northern Gulf of Mexico (modified from USEPA 1999).

Alabama to definite problems in Louisiana and Texas with overall moderate problems throughout the northern Gulf of Mexico. Low dissolved oxygen concentrations in estuaries were attributed to stratification, metabolism, seasonal storm events, and depth/tide regimes. Low dissolved oxygen was often exacerbated by anthropogenic nutrient enrichment, habitat modifications, and channelization. Using EMAP and NOAA data and the Rabalais et al. (1992) assessment of oxygen depletion, Gulf of Mexico estuaries were ranked as fair overall with most estuaries east of the Mississippi River exhibiting persistent low dissolved oxygen. A USEPA report card representing the best estimate of ecological condition was produced (Figure 2.10). For the overall Gulf of Mexico, 8 of 11 indicators were ranked as fair to poor. Estuaries on the Florida coast had fewer problems than other Gulf States. Alabama coasts rated good to fair for most of the indicators with problems indicated by low dissolved oxygen concentrations. Mississippi rated good to fair for all indicators except wetland loss. Louisiana and Texas estuaries exhibited problems associated with excess nutrients. Estuaries in the northern Gulf of Mexico had significant but variable environmental problems. The report

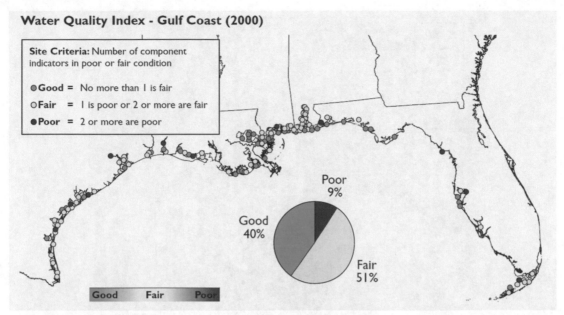

Figure 2.11. Water quality index data for northern Gulf of Mexico estuaries from 1996–2000 (modified from USEPA 2004).

concluded that there had been some improvement in the condition of estuaries since the Clean Water Act was passed, as indicated by the relatively moderate problems with water quality indicators such as nutrients and dissolved oxygen (USEPA 1999).

The assessment process was revised and the indices used to determine coastal condition were redefined; direct comparisons with previous assessments should be made with caution. From 1996 to 2000, Gulf of Mexico estuaries ranked poor for eutrophic condition with 38 % of the estuarine area having a high expression of eutrophication (Bricker et al. 1999). Estuaries with poor water quality conditions were found in all five states but the contributing factors were different. The water quality index used in 2004 (based on five indicators: nitrogen, phosphorus, chlorophyll *a*, water clarity, and dissolved oxygen) showed that 40 % of the estuaries rated good, 51 % fair, and 9 % poor (Figure 2.11).

Water clarity in Gulf Coast estuaries was judged to be fair in the 2001 assessment (USEPA 2001). Water clarity was estimated by the penetration of light through the water column. For 22 % of the waters in Gulf of Mexico estuaries, less than 10 % of surface light penetrated to a depth of 1 m (3.3 ft) (Figure 2.12a). In the 2004 assessment, Texas and Louisiana estuaries had poor water clarity (Figure 2.12b) (USEPA 2004) while overall water clarity in Gulf of Mexico estuaries was again judged to be fair. In the 2001 assessment, dissolved oxygen conditions in Gulf of Mexico estuaries were generally good except in a few highly eutrophic regions. EMAP estimates for Gulf of Mexico estuaries concluded that about 4 % of the bottom waters in Gulf of Mexico estuaries had hypoxic conditions or low dissolved oxygen concentrations (less than 2 parts per million [ppm]) on a continuing basis in the late summer (Figure 2.13a).

Affected areas included Chandeleur and Breton Sounds in Louisiana, some shoreline regions of Lake Pontchartrain, northern Florida Bay, and smaller estuaries associated with Galveston Bay, Mobile Bay, Mississippi Sound, and the Florida panhandle. In the 2004 assessment, dissolved oxygen conditions in northern Gulf of Mexico estuaries were assessed to be good. Less than 1 % of the bottom waters exhibited hypoxia (less than 2 mg/L dissolved oxygen) in the late summer (Figure 2.13b). Affected areas included Mobile Bay, Alabama,

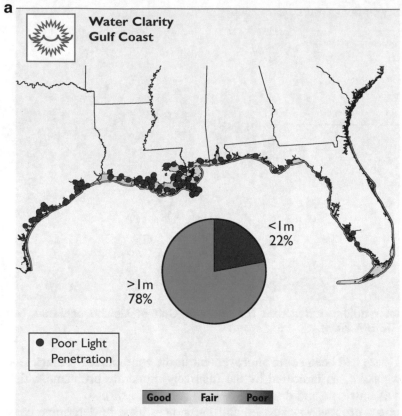

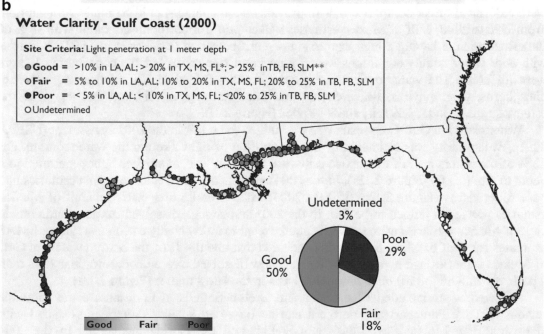

Figure 2.12. (a) Light penetration and locations for sites with less than 10 % light penetration (modified from USEPA 2001) and (b) water clarity for Gulf of Mexico estuaries (*FL = Florida estuaries except Tampa Bay [TB] and Florida Bay [FB], **SLM = Southern Laguna Madre (modified from USEPA 2004).

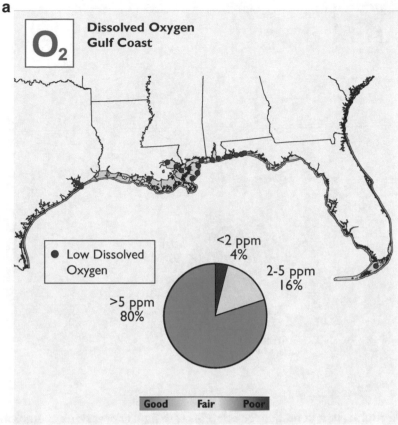

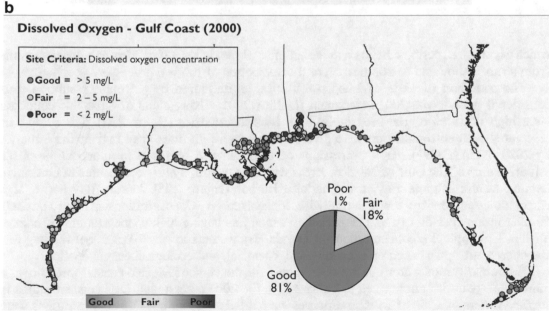

Figure 2.13. Dissolved oxygen concentrations for Gulf of Mexico estuaries: (a) sites with less than 2 ppm in the 2001 assessment (modified from USEPA 2001) and (b) dissolved oxygen criteria from 1996 to 2000 (modified from USEPA 2004).

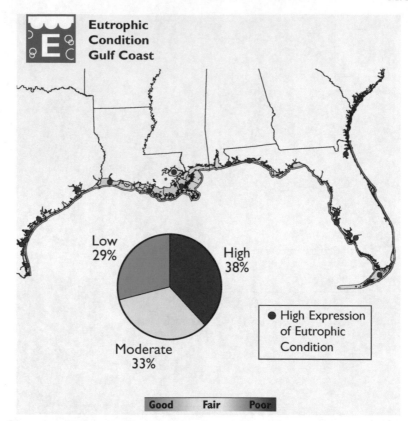

Figure 2.14. Eutrophication condition for estuaries with high expressions of eutrophication (modified from USEPA 2001).

which experiences periodic hypoxia in the summer. Hypoxia in Gulf of Mexico estuaries results from stratification and eutrophication or a combination of the two processes.

The condition of Gulf of Mexico estuaries, as measured by eutrophic condition, was considered poor in the 2001 assessment (USEPA 2001). Expressions of eutrophic condition were high in 38 % of the area in Gulf of Mexico estuaries (Figure 2.14). The symptoms associated with eutrophication were predicted to increase in more than half of the estuaries by 2020 (NOAA 1997). High expressions of chlorophyll *a* occurred in about 30 % of the estuarine area of the Gulf of Mexico. Areas with high chlorophyll *a* were located in Louisiana, Laguna Madre, Tampa Bay, and Charlotte Harbor (Figure 2.15). Florida Bay had a high eutrophic condition but low chlorophyll *a* concentrations. Concentrations of approximately 50 micrograms per liter (µg/L) classified an estuary as having high concentrations of chlorophyll *a*. Chlorophyll *a* concentrations in Florida Bay were as low as 20 µg/L but the bay was considered eutrophic based on other physical, chemical, and ecological characteristics.

A comparison of water quality assessments for the Gulf of Mexico coastal waters over a number of years is summarized in Table 2.5. In the 2004 assessment, DIN concentrations in surface waters of Gulf of Mexico estuaries were rated as good, but DIP concentrations were rated as fair (Figure 2.16a, b (USEPA 2004). High concentrations of DIN (greater than 0.5 mg/L) occurred in 2 % of the estuarine area (Figure 2.16a). Florida Bay sites were rated poor if DIN exceeded 0.1 mg/L or if DIP exceeded 0.01 mg/L based on lower expected nutrient concentrations in tropical and subtropical waters. The Houston Ship Channel, Texas and the Back Bay of Biloxi, Mississippi, exhibited high concentrations of nitrogen and phosphorus. The

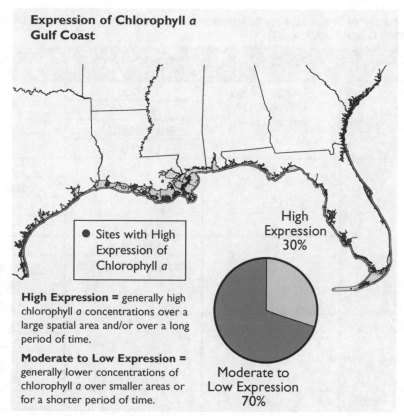

Figure 2.15. Chlorophyll *a* concentrations in Gulf of Mexico estuaries and those locations with high expression of chlorophyll *a* (modified from USEPA 2001).

Perdido River in Alabama was hypoxic and exhibited high chlorophyll *a* concentrations. DIN concentrations above 0.5 mg/L were observed in the Houston Ship Channel, Texas; Calcasieu River, Louisiana; and Back Bay of Biloxi, Mississippi. In Gulf of Mexico coastal waters elevated DIN concentrations were not expected during the summer because freshwater input is usually lower and dissolved nutrients are rapidly taken up by phytoplankton. Elevated DIP concentrations (greater than 0.05 mg/L) occurred in 11 % of Gulf of Mexico estuaries (Figure 2.16b). Tampa Bay and Charlotte Harbor, Florida, had high DIP concentrations because of the natural occurrence of phosphate rocks and anthropogenic sources in their watersheds. Coastal chlorophyll *a* concentrations in Gulf of Mexico estuaries were rated good. Eight percent of the estuarine area in the Gulf Coast region had high concentrations of chlorophyll *a* (Figure 2.16c).

2.3.3 USEPA National Estuarine Condition (2006)

In 2006, a report was issued presenting monitoring data that provided a perspective on the condition of U.S. NEP estuaries (USEPA 2006). The data were collected by the National Coastal Assessment (NCA) group and individual NEPs and their local partners.

The overall condition of NEP estuaries in the Gulf of Mexico for 1997–2003 was rated as fair based on four indices of estuarine condition (Figures 2.17 and 2.18). The assessment was based on data collected from 221 sites sampled in Gulf of Mexico estuaries during the summers of 2000, 2001, and 2002. The region's water quality index was rated as fair

Table 2.5. Summary of Water Quality Assessments for Gulf of Mexico Coastal Waters (modified from NOAA 1997; USEPA 2001, 2004)

Water quality indicator	NOAA 1997	USEPA	
		2001	2004
Eutrophication condition	Poor 38 % high expression of eutrophication	Poor	NA[a]
		High 38 %	
		Moderate 33 %	
		Low 29 %	
Water quality index	NA	NA	Overall fair
			Good 40 %
			Fair 51 %
			Poor 9 %
Chlorophyll a concentrations	NA	High 30 %	Overall fair
		Moderate to low 70 %	Good 51 %
			Fair 38 %
			Poor 8 %
DIN concentrations	NA	–	Overall good
			Good 89 %
			Fair 9 %
			Poor 2 %
DIP concentrations	NA	–	Overall fair
			Good 58 %
			Fair 31 %
			Poor 11 %
Water clarity	NA	Overall fair	Overall Fair
		Good 78 % (>1 m)	Good 50 %
		–	Fair 18 %
		Poor 22 % (<1 m)	Poor 29 %
Dissolved oxygen	NA	Overall good	Overall good
		Good 80 %	Good 81 %
		Fair 16 %	Fair 18 %
		Poor 4 % hypoxic	Poor 1 %

[a]NA not applicable

(Figures 2.18 and 2.20). A summary of the percentage of estuarine area rated good, fair, poor, or missing for each water quality parameter is presented in Figure 2.19.

Estuarine water quality was rated as 21 % good, 65 % fair, and 13 % poor. The Gulf of Mexico region was rated overall as good for DIN concentrations with 88 % good, 8 % fair and 3 % poor. Elevated DIN concentrations were not expected to occur during the summer in Gulf of Mexico waters because freshwater input is lower and nutrients are rapidly taken up by phytoplankton (Figure 2.20). The estuaries studied were rated fair for DIP concentrations with 22 % rated poor. Gulf Coast estuaries were rated fair overall for chlorophyll a concentrations.

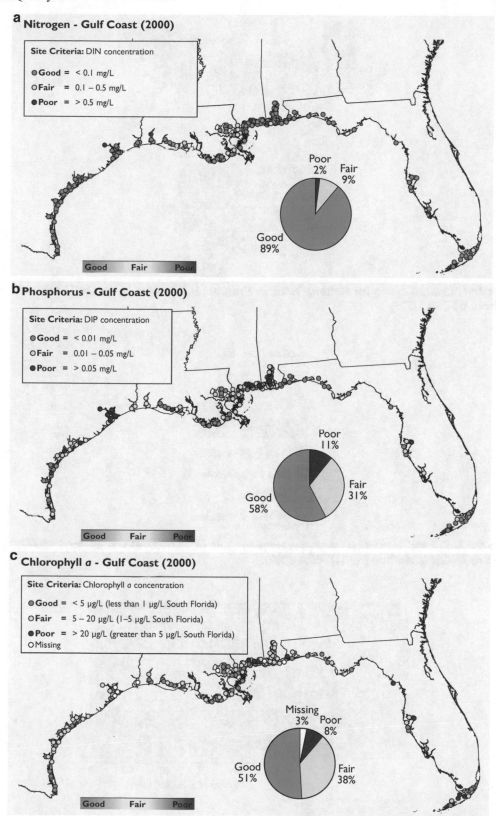

Figure 2.16. (a) Dissolved inorganic nitrogen (DIN) concentrations, (b) dissolved inorganic phosphorus (DIP) concentrations, and (c) chlorophyll *a* concentrations for Gulf of Mexico estuaries in 2000 (modified from USEPA 2004).

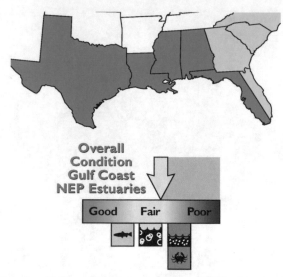

Figure 2.17. Overall rating for National Estuary Program (NEP) sites in the Gulf of Mexico (modified from USEPA 2006).

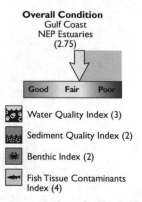

Figure 2.18. Overall condition of representative Gulf of Mexico estuaries for 2000–2003 was judged to be fair (modified from USEPA 2006).

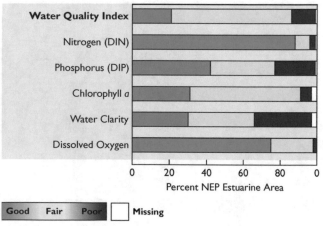

Figure 2.19. Percentage of representative Gulf of Mexico estuaries achieving each rating for individual components of the water quality index for 2000–2003 (modified from USEPA 2006).

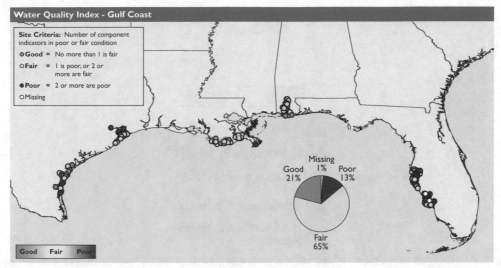

Figure 2.20. Water quality index for representative Gulf of Mexico estuaries for 2000–2003 (modified from USEPA 2006).

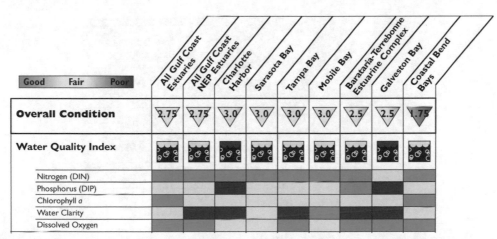

Figure 2.21. Comparison of overall condition and water quality index for Gulf of Mexico estuaries for 2000–2003 (modified from USEPA 2006).

Chlorophyll *a* conditions were rated as 6 % poor, 60 % fair, and 31 % good. Overall water clarity in Gulf of Mexico estuaries was rated as poor with 31 % poor, 36 % fair, and 30 % good. Gulf of Mexico estuaries were rated as good overall for dissolved oxygen concentrations with 2 % poor, 23 % fair, and 75 % good. Survey results of Gulf of Mexico estuaries allowed for a comparison of sites across the region. All Gulf Coast estuaries were rated as fair for overall condition from 2000 to 2003 (Figure 2.21).

2.3.4 USEPA's National Coastal Condition Report III (2008)

In 2008, the third National Coastal Condition Report was issued based on data collected between 2001 and 2002 (USEPA 2008). The overall condition of the coastal waters of the Gulf of Mexico region was rated as fair to poor and water quality was rated as fair (Figures 2.22 and 2.24). The assessment was based on data collected from 487 locations in Florida, Alabama,

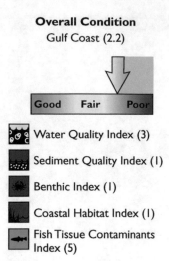

Figure 2.22. Overall condition of Gulf of Mexico coastal waters for 2001–2002 was rated fair to poor (modified from USEPA 2008).

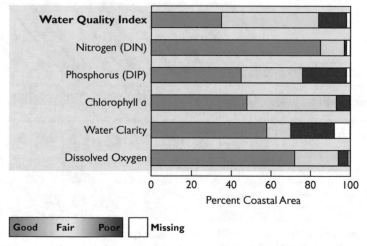

Figure 2.23. Percentage of coastal area achieving each ranking for the water quality index and components of the indicator in the Gulf of Mexico for 2001–2002 (modified from USEPA 2008).

Mississippi, Louisiana, and Texas. Water quality condition was rated as 14 % poor and 49 % fair (Figure 2.23). The water quality index was based on five indicators DIN, DIP, chlorophyll *a*, water clarity, and dissolved oxygen (Figure 2.24).

Estuaries with poor water quality conditions were found in all five Gulf States but the reason differed among states. At locations in Texas, Louisiana, and Mississippi, poor water clarity and high DIP concentrations contributed to poor water quality ratings. Poor conditions at locations in several Texas bays were due to high chlorophyll *a* concentrations. Only three locations in Louisiana had high concentrations of both DIN and DIP. Many locations rated poor or fair for individual components of the indicator, but were rated fair by the overall water quality index. For comparison, NOAA's Estuarine Eutrophication Survey rated the Gulf Coast as poor for eutrophic condition with 38 % of the coastal area exhibiting high expressions of eutrophication (Bricker et al. 1999).

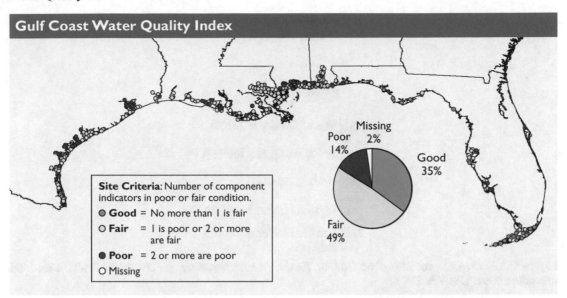

Figure 2.24. Water quality index for Gulf of Mexico coastal waters for 2001–2002 (modified from USEPA 2008).

The northern Gulf of Mexico region was rated as good for DIN concentrations but fair for DIP concentrations from 2001 to 2002. Different criteria for DIN and DIP concentrations were applied in Florida Bay because coastal Florida was considered a tropical estuary. DIN concentrations were rated poor in 1 % of Gulf of Mexico coastal areas including three sites in Louisiana's East Bay, Atchafalaya Bay, and the Intracoastal Waterway between Houma and New Orleans, Louisiana. DIP concentrations were rated poor in 22 % of Gulf of Mexico coastal areas with locations in Tampa Bay and Charlotte Harbor, Florida, highest in DIP due to the occurrence of natural geological formations of exposed phosphate rock in the watersheds and anthropogenic DIP. Gulf of Mexico estuaries were rated fair overall for chlorophyll *a* concentrations from 2001 to 2002 with 7 % poor and 45 % fair. High concentrations of chlorophyll *a* occurred in the coastal areas of all five Gulf States. Water clarity in the northern Gulf of Mexico region was rated fair from 2001 to 2002 with 22 % rated as poor. Lower-than-expected water clarity was observed throughout the northern Gulf of Mexico with poor conditions concentrated in Mississippi, the Coastal Bend region of Texas, and Louisiana. The criteria used to assign water clarity ratings varied across Gulf of Mexico coastal waters based on natural variations in turbidity levels, regional expectations for light penetration related to submerged aquatic vegetation distributions, and local water body management goals. Gulf of Mexico estuaries were rated as fair overall for dissolved oxygen concentrations with 5 % rated as poor. Hypoxia in Gulf of Mexico coastal waters generally resulted from stratification, eutrophication, or a combination of these two conditions. Mobile Bay, Alabama, has regularly experienced hypoxic events during the summer since colonial times, most likely due to natural events (May 1973).

2.3.5 USEPA's National Coastal Condition Report IV (2012)

In 2012, the fourth National Coastal Condition Report was issued based on data collected between 2003 and 2006 (USEPA 2012). The overall condition and water quality of the coastal waters of the Gulf of Mexico region were rated as fair (Figures 2.25 and 2.27). The assessment was based on data collected from 879 locations in Florida, Alabama, Mississippi, Louisiana, and

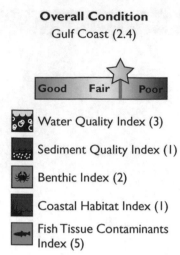

Overall Condition
Gulf Coast (2.4)

Water Quality Index (3)

Sediment Quality Index (1)

Benthic Index (2)

Coastal Habitat Index (1)

Fish Tissue Contaminants
Index (5)

Figure 2.25. Overall condition of Gulf of Mexico coastal waters for 2003–2006 was rated fair (modified from USEPA 2012).

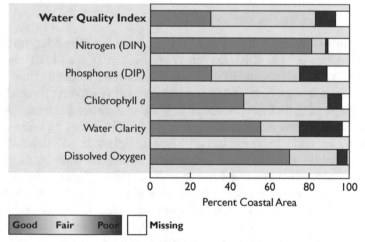

Figure 2.26. Percentage of coastal area achieving each ranking for the water quality index and components of the indicator in the Gulf of Mexico for 2003–2006 (modified from USEPA 2012).

Texas. Water quality condition was rated as 10 % poor and 53 % fair (Figure 2.26). Due to hurricanes Katrina and Rita, Alabama and Louisiana did not collect data in 2005. As before, the water quality index was based on DIN, DIP, chlorophyll *a*, water clarity, and dissolved oxygen. Poor water quality conditions were found across the region but the reason differed among states. Poor water clarity, high DIP concentrations, and high chlorophyll *a* concentrations contributed to poor water quality ratings. Three sites in Louisiana had high concentrations of DIN and DIP. A lower percentage of Gulf of Mexico coastal areas rated good for the water quality index than the component indicators as indications of poor or fair conditions did not always coincide. The NOAA Estuarine Eutrophication Survey in 1999 rated the Gulf Coast poor for eutrophic condition with approximately 38 % of the coastal area exhibiting high expressions of eutrophication (Bricker et al. 1999). The northern Gulf of Mexico was rated good for DIN concentrations and fair for DIP concentrations. Criteria for DIN and DIP concentrations in Florida Bay differed from other areas because it is considered to be a tropical estuary. DIN

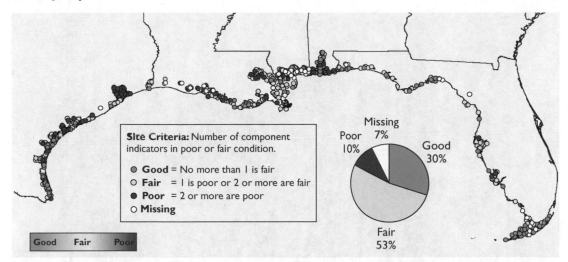

Figure 2.27. Water quality index for Gulf of Mexico coastal waters for 2003–2006 (modified from USEPA 2012).

concentrations were poor in 1 % of the coastal area at several sites in Louisiana and Texas from 2003 to 2004. DIP concentrations were rated poor for 14 % of the coastal area including sites in Tampa Bay and Charlotte Harbor, Florida, due to naturally occurring phosphate rock in the watersheds and anthropogenic sources of DIP. The region was rated fair for chlorophyll *a* concentrations with high concentrations of chlorophyll *a* occurring in all five Gulf Coast States. Water clarity in the Gulf of Mexico region was rated fair with 21 % of the coastal area rated poor. Poor water clarity conditions were observed most frequently in Texas and Louisiana. The region was rated good for dissolved oxygen concentrations with less than 5 % (4.8 %) of the coastal area rated poor. Hypoxia generally resulted from stratification, eutrophication, or a combination of these two conditions. Mobile Bay, Alabama, experiences regular hypoxic events during the summer. These occurrences have been known since colonial times and are believed to be natural events (May 1973) (Figure 2.27).

2.3.6 State of the Bays

In the previous sections, water quality was summarized on a regional basis for the northern Gulf of Mexico highlighting sites with specific water quality issues. In this summary, individual bays and estuarine complexes in the Gulf of Mexico are considered to provide a finer spatial scale view of water quality in Gulf of Mexico estuaries. These summaries draw on information produced as part of the NEP. The NEP was established under Section 320 of the 1987 Clean Water Act Amendments as a USEPA effort to protect and restore the water quality and ecological integrity of major U.S. estuaries. At the time of assessment, there were 28 estuaries designated of national significance, and six of them were located in the northern Gulf of Mexico.

2.3.6.1 Texas Bays

Water quality for bays in the state of Texas is summarized based on monitoring data collected in Galveston Bay and the Coastal Bend Bays and Estuaries (CBBE) complex (Figure 2.28). Both bays host an NEP.

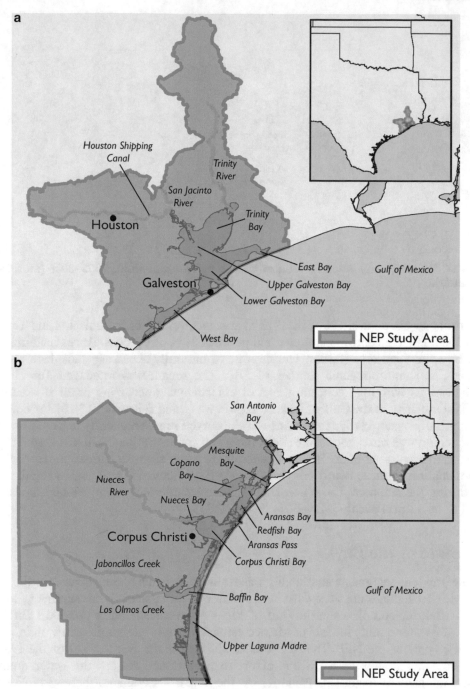

Figure 2.28. Map of National Estuary Program Study Areas (a) Galveston Bay Estuary Complex and (b) Coastal Bend Bays and Estuaries, Texas (modified from USEPA 2006).

Galveston Bay is a subtropical estuary located on the southeastern shore of the upper Texas Gulf Coast. The bay is composed of five major sub-bays: Trinity, Upper Galveston, Lower Galveston, East, and West bays. The combined area of the five sub-bays was estimated to be 1,554 km^2 (600 mi^2) surrounded by 1,885 km (1,171 mi) of shoreline (GBEP 2005). The estuary receives inflow from the Trinity and San Jacinto rivers and is bordered by low-lying wetlands, two barrier islands, and a peninsula. The waters of Galveston Bay were considered well mixed and shallow averaging 2.1 m (7 ft) and shallower in places due to oyster reefs (GBEP 2005). The bay volume has increased over the last 50 years due to natural and anthropogenic subsidence, sea level rise and dredging (Lester and Gonzalez 2003). Major habitats in the bay include estuarine and freshwater marsh, mudflats, sea grass beds, oyster reefs, and open water. The watershed includes a variety of habitats ranging from open prairies and coastal wetlands to riparian hardwoods and pine-dominant forests. These habitats support numerous plant, fish, and wildlife species. Galveston Bay is extensively used for recreational and commercial activities. Potential human impacts are large due to the surrounding populations. Galveston Bay is one of the largest sources of seafood for Texas and a major national oyster-producing estuary. The oysters, crabs, shrimp, and finfish harvested from Galveston Bay were estimated to be worth approximately $19 million per year (Lester and Gonzalez 2003). At the time, one-third of the Texas commercial fishing income and more than one-half of the state's recreational fishing expenditures came from Galveston Bay (GBEP 2005). The Port of Houston was the second largest port in the United States in tonnage and the eighth largest port in the world in 2002 (Lester and Gonzalez 2003). Along with the port cities of Texas City and Galveston, the Port of Houston supports petrochemical industries that were the largest in the nation and the second largest in the world in 2006 (Port of Houston Authority 2006). These industries produced one-half of the nation's chemicals and represented one-third of the nation's petroleum refining capacity. Extending back from the river mouths, the Galveston Bay watershed covered 85,469 km^2 (33,000 mi^2) at the time including the metropolitan areas of Houston-Galveston and Dallas-Fort Worth, home to nearly half of the population of Texas in 2005 (GBEP 2005). Galveston Bay environmental concerns include wetland loss and habitat degradation, point and non-point source pollution, and chemical and refined product spills from barges and industry (Lester and Gonzalez 2003). Non-point source pollution in Galveston Bay includes runoff from thousands of gas stations, residential lawns, failing septic systems, driveways, parking lots, industries, farms, and other sources. Accidental spills and the deliberate dumping of oil and other contaminants harm the habitat and living resources of Galveston Bay. Galveston Bay was also subject to introductions of aquatic and terrestrial exotic nuisance species, contaminated runoff from urbanized areas, and the diversion of fresh water inflows. Some sediment in the Houston Ship Channel exceeded levels of concern for a number of hazardous chemicals including PCBs, DDT (dichlorodiphenyltrichloroethane), dioxin, and metals in 2006.

The Coastal Bend Bays and Estuaries (CBBE) complex include three of the seven estuaries along the Texas coast. The northerly portion of the CBBE Program (CBBEP) includes San Antonio, Mesquite, Redfish, Copano, and Aransas Bays. The middle portion includes Nueces Bay and Corpus Christi Bay, the largest of the bays, and discharges into the Gulf of Mexico at Aransas Pass. The most southerly portion includes Upper Laguna Madre and Baffin Bay. The area was estimated to include 121 km (75 mi) of Texas coastline and 1,334 km^2 (515 mi^2) of water (CBBEP 2005). The area included barrier islands, tidal marshes, sea grass meadows, open bays, oyster, and serpulid worm reefs, wind tidal flats, and freshwater marshes. The CBBEP supports recreational, commercial, industrial, and residential uses including sport boat fishing, bird watching, and windsurfing. The commercial fishing industry annually harvested, on average, more than eight million pounds of finfish, shrimp, and crab (Tunnell et al. 1996). The area was estimated to contain 40 % of the state's total sea grass acreage, nursery areas for fish and shellfish, and habitats for other wildlife including birds, sea mammals, and marine turtles (CBBEP 1998). Corpus Christi Bay was the nation's fifth largest port and included the third largest refinery and petrochemical complex in the United States in 2005 (CBBEP 2005). The region's population was 550,000 in 1995 and was projected to be nearly one million by 2050

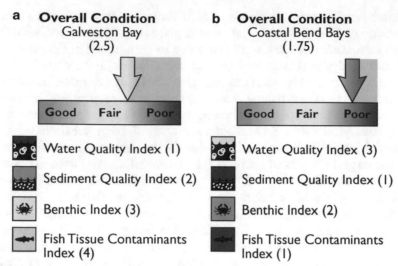

Figure 2.29. Overall condition of (a) Galveston Bay and (b) Coastal Bend Bays in 2000 (modified from USEPA 2006).

(CBBEP 1998). Freshwater was in short supply in semiarid southern Texas due to many competing demands. Residential and business water use in this region was expected to increase by 50 % by 2050 and industrial demand was expected to double (CBBEP 1998). Freshwater is vital to the human population and is closely tied to the health of coastal ecosystems.

In the 2006 assessment, the overall condition of Galveston Bay and the Coastal Bend was rated as fair to poor, respectively (Figure 2.29). The water quality index was rated poor for Galveston Bay and fair for the Coastal Bend (Figures 2.29 and 2.31). In NOAA's Estuarine Eutrophication Survey in 1997, Galveston Bay was listed as having medium chlorophyll *a* concentrations and medium-to-low DIN and DIP concentrations with elevated concentrations occurring in tidal freshwater areas (NOAA 1997). In 2006, Galveston Bay was rated fair for DIN concentrations and poor for DIP concentrations. Thirteen percent of the estuarine area was rated poor for DIN concentrations, and 68 % of the estuarine area was rated poor for DIP concentrations (Figure 2.30). Galveston Bay was rated fair overall for chlorophyll *a* concentrations with 4 % poor, 71 % fair, and 13 % good with data unavailable for 12 % of the estuarine area. Water clarity in Galveston Bay was rated poor overall because 28 % of the estuarine area was rated poor. Water clarity for turbid estuaries was rated poor if light penetration at 1 m (3.3 ft) was less than 10 % of surface illumination. Dissolved oxygen conditions in Galveston Bay were rated as good overall with 71 % good and 29 % fair (Figure 2.31).

In NOAA's Estuarine Eutrophication Survey in 1997, the Coastal Bend was listed as having medium to hyper-eutrophic chlorophyll *a* levels and low to high DIN and DIP concentrations with elevated concentrations occurring in tidal freshwater areas (NOAA 1997). In 2006, the Coastal Bend was rated good overall for DIN concentrations with 99 % of the estuarine area rated as good (Figure 2.30) and was rated fair overall for DIP concentrations with 4 % as poor, 46 % fair, and 50 % good (Figure 2.30). Chlorophyll *a* concentrations in the Coastal Bend Bays were rated good overall with 5 % rated as poor, 40 % fair, and 55 % good. Water clarity in the Coastal Bend was rated fair overall because 16 % of the estuarine area was rated poor. In Corpus Christi and Aransas bays, water clarity was rated poor if light penetration at 1 m (3.3 ft) was less than 10 % of surface illumination. Dissolved oxygen concentrations in the Coastal Bend were rated as good overall with 70 % good and 30 % fair.

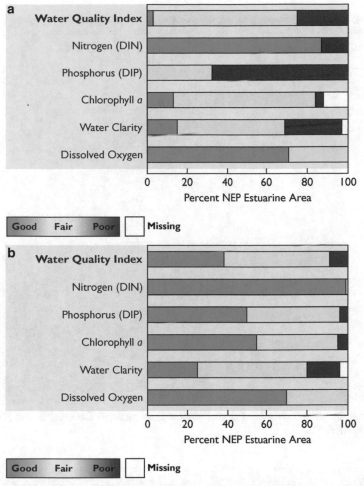

Figure 2.30. Percentage of estuarine area achieving each rating for water quality index and its components (a) Galveston Bay and (b) Coastal Bend Bays (modified from USEPA 2006).

2.3.6.2 Louisiana Bays

Water quality for bays in the state of Louisiana is summarized based on monitoring data collected in the Barataria-Terrebonne Estuary (Figure 2.32). The Barataria-Terrebonne Estuary hosts an NEP. The Barataria-Terrebonne estuary is located between the Mississippi and Atchafalaya rivers in southern Louisiana and covers approximately 16,800 km² (6,500 mi²) (Caffey and Breaux 2000). Bayou Lafourche separates the area into two basins: Barataria Basin to the east and Terrebonne Basin to the west. The mixing of saltwater and freshwater begins offshore where water, sediment, nutrients, and pollutants from the Mississippi River comingle with the salty water of the Gulf of Mexico. Industrial and municipal effluents enter the Mississippi River between Baton Rouge and New Orleans and contribute to nutrient and contaminant loads in the estuary system. Several natural and man-made waterways transect the estuary system including the Gulf Intracoastal Waterway and the Barataria Waterway. Open water and wetlands were the predominant land-use classifications in the region, and it had been increasing in area since 1956. More than three-quarters of the area (approximately 12,900 km² or 5,000 mi²) was classified as open water or wetlands with approximately 4,050 km² (1,562 mi²) used for urban and agricultural activities (Moore and Rivers 1996).

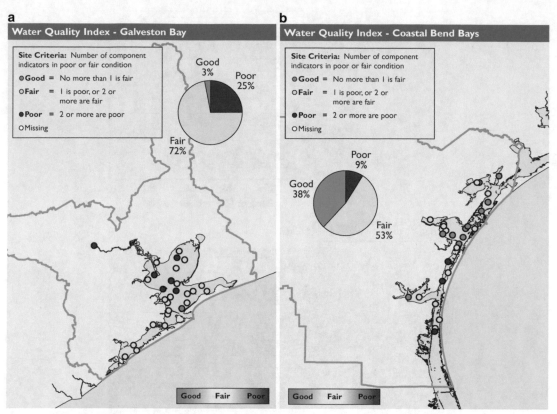

Figure 2.31. Water quality index for (a) Galveston Bay and (b) Coastal Bend Bays in 2000–2001 (modified from USEPA 2006).

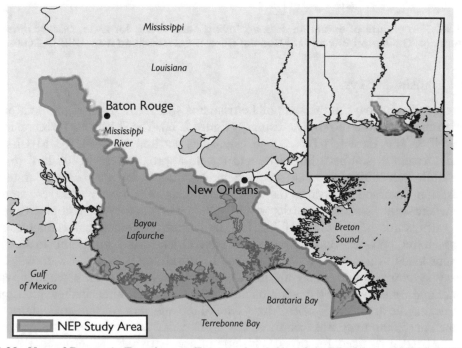

Figure 2.32. Map of Barataria-Terrebonne Estuary, Louisiana (modified from USEPA 2006).

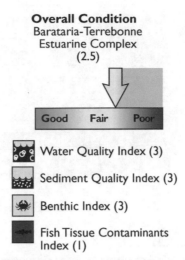

Figure 2.33. Overall condition of Barataria-Terrebonne estuarine area (modified from USEPA 2006).

The issues affecting the area include habitat loss, hydrological modification, reduced sediment flows (reduction in sediment inputs), eutrophication, pathogen contamination from untreated sewage and stormwater discharges, toxic substances, and declines in living resources (Battelle 2003). Sediment loss (depletion) in conjunction with the subsidence (sinking) of marshes was considered the most significant problem in the Barataria-Terrebonne Estuarine Complex at the time. The construction of levees to control flooding diminished freshwater inflow and sediments reaching the estuaries. Sea level rise, erosion, canal dredging, and the construction of navigation and oil-exploration channels contributed to wetland loss. Hydrological modifications had created paths for high salinity waters to intrude inland impacting freshwater plants causing animals to adapt or relocate. At the time, about 38.8 km^2 (15 mi^2) of wetlands were being lost each year and 0.0002 km^2 (0.05 acres) of the coastal wetlands was turning to open water every 15 minutes (min) (BTNEP 2002). The loss of habitat adversely affects the health of fish and wildlife populations and stymies economic development.

The overall condition of the Barataria-Terrebonne Estuarine Complex was rated fair based on four indices of estuarine condition, and water quality was also rated as fair (Figure 2.33). Figure 2.34 summarizes the percentage of estuarine area rated as good, fair, poor, or missing for each parameter considered. This assessment was based on data from 25 locations sampled in 2000 and 2001.

Based on survey results, the water quality index for the Barataria-Terrebonne Estuarine Complex was rated fair (Figure 2.35). In NOAA's Estuarine Eutrophication Survey in 1997, Barataria Bay was listed as having high to hyper-eutrophic chlorophyll *a* concentrations and high DIN and DIP concentrations (NOAA 1997). In the same report, the Terrebonne and Timbalier bays were listed as having high chlorophyll *a* and DIP concentrations and moderate DIN concentrations. In the 2006 report, DIN and DIP concentrations in the estuarine area were rated as good overall. For both component indicators, 4 % were rated poor, 16 % fair, and 80 % good. Chlorophyll *a* concentrations in the Barataria-Terrebonne Estuarine Complex were rated fair overall with 4 % of the estuarine area rated poor, 64 % fair, and 32 % good. Water clarity was rated poor overall with 52 % of the estuarine area rated poor, 20 % fair, and 28 % good. Dissolved oxygen conditions in the estuarine area were rated good overall with none of the estuarine area rated poor, 4 % fair, and 96 % good. Eutrophic conditions and nutrient levels in the Barataria-Terrebonne Estuarine Complex were monitored at a series of 15 locations; all were classified as having medium or high nutrient conditions under NOAA guidelines. During

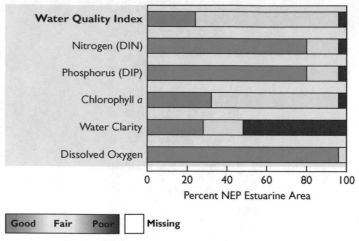

Figure 2.34. Percentage of Barataria-Terrebonne estuarine area achieving each rating for each component indicator of the water quality index (modified from USEPA 2006).

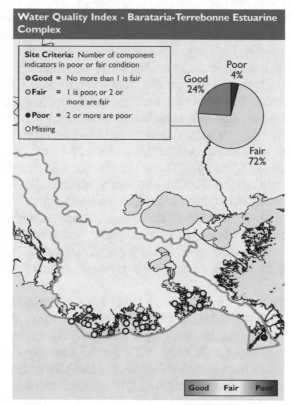

Figure 2.35. Water quality index for Barataria-Terrebonne Estuarine Complex, 2000–2001 (modified from USEPA 2006).

the 20 years before the assessment, measurements of chlorophyll *a* levels provided evidence of eutrophication with many locations exhibiting an increase in chlorophyll *a* concentrations over time (Rabalais et al. 1995). Hypoxic events were being induced by inflows of wastewater treatment plant effluent and agricultural runoff. Nearshore bottom water dissolved oxygen concentrations varied from 4 to 8 mg/L, and indications of persistent hypoxia from mid-May to mid-September were observed (Rabalais et al. 1995). Hypoxic conditions occurred in poorly flushed areas, deeper channels, and areas receiving organic loading from sewage or other wastewater outfalls. Pathogens from sewage pollution were associated with illnesses in humans who swam in contaminated waters or consumed contaminated oysters. Fecal coliform came from poorly functioning septic systems, pastureland runoff, and animal waste. Copper, lead, arsenic, chromium, and cadmium concentrations declined in concentration since the 1980s, whereas mercury levels remained fairly constant. Although contamination was fairly widespread, the areas of most concern were on the periphery such as Oyster Bayou and Tiger Pass. Toxics were detected in fish and crustaceans of the Barataria-Terrebonne Estuarine Complex including pesticides, metals, volatile organic compounds (VOCs), and PCBs (Rabalais et al. 1995).

2.3.6.3 Mississippi and Alabama Bays

Water quality for bays in Mississippi and Alabama is summarized based on monitoring data for Mobile Bay (Figure 2.36). Mobile Bay hosts an NEP. Mobile Bay is a submerged river valley at the transition between the coastal zone of the Mobile Bay watershed and the Gulf of Mexico. The Mobile Bay watershed covered approximately 115,500 km^2 (44,600 mi^2) including two-thirds of Alabama and portions of Mississippi, Georgia, and Tennessee at the time of assessment (NOAA 1985; Mobile Bay NEP 2002). At that time, it was the fourth largest watershed by flow volume in the United States and the sixth largest river system in area (Mobile Bay NEP 2002). The surface waters of Mobile Bay were estimated to cover approximately 1,060 km^2 (409 mi^2) with an average depth of approximately 3 m (10 ft) (NOAA 1985; Mobile Bay NEP 2002). Freshwater flows into the bay through several rivers (e.g., the Mobile-Tensaw, Blakely, Apalachee, Dog, Deer, Fowl, and Fish rivers). The bay's primary opening to the Gulf of Mexico is the Main Pass, located between Dauphin Island and the Fort Morgan Peninsula. Covering approximately 749 km^2 (289 mi^2) of marsh, swamp and forested wetlands, the Mobile-Tensaw River Delta was the largest intact delta in the United States at the time of assessment (Wallace 1994; Auburn University 2004). The bay basin includes barrier islands,

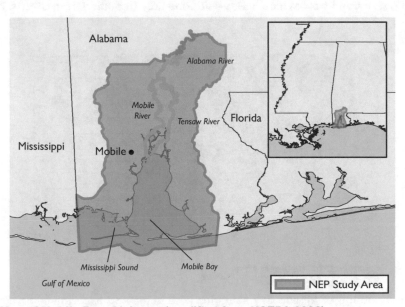

Figure 2.36. Map of Mobile Bay, Alabama (modified from USEPA 2006).

tidal marshes, cypress swamps, bottomland hardwoods, and oyster reefs. Portions of Mobile Bay support commercial fisheries, industry, tourism and recreation, and coastal development. It was estimated that 4.85 million metric tons (5.35 million tons) of sediment annually entered the estuary with 33 % deposited in the Mobile-Tensaw Delta, 52 % in the bay, and 15 % flowing out into the Gulf of Mexico (Mobile Bay NEP 2002). Mobile Bay's salinity regime is complex and highly variable because winds and tides affect the inflow of salty Gulf of Mexico waters into the bay. Salinity varied with depth in the bay and in the major river channels (Braun and Neugarten 2005).

The overall condition of Mobile Bay was rated as fair based on four indices of estuarine condition, and water quality was rated as fair (Figures 2.37 and 2.39). The assessment of the estuarine status rated each parameter in the water quality as good, fair, poor, or missing (Figure 2.38). The water quality index for Mobile Bay was rated as fair based on data collected at 66 locations (Figure 2.39). In NOAA's 1997 Estuarine Eutrophication Survey, Mobile Bay

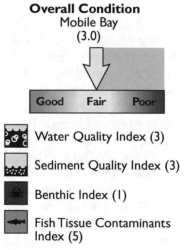

Figure 2.37. Overall condition of Mobile Bay estuarine area (modified from USEPA 2006).

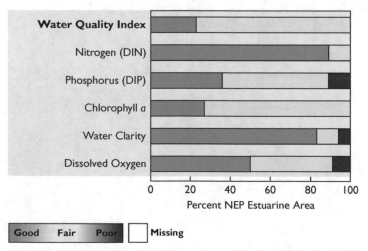

Figure 2.38. Percentage of Mobile Bay estuarine area achieving each indicator of water quality (modified from USEPA 2006).

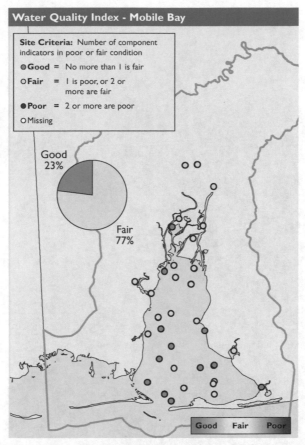

Figure 2.39. Water quality index for Mobile Bay, 2000–2001 (modified from USEPA 2006).

was listed as having medium levels of chlorophyll *a* and medium-to-low DIN and DIP concentrations (NOAA 1997).

DIN and DIP concentrations in Mobile Bay were rated good and fair overall, respectively. Concentrations of DIN were rated as good in 89 % of the estuarine area and fair in the remaining 11 %. Within the estuarine area, 11 % was rated poor for DIP concentrations, 53 % fair, and 36 % good. Chlorophyll *a* concentrations were rated as fair overall. No poor chlorophyll *a* conditions occurred with 73 % rated as fair and the remaining 27 % rated good. Water clarity in Mobile Bay was rated good overall. Mobile Bay experiences high river flow which causes naturally turbid water. Water clarity was rated as poor in 6 % of the estuarine area, 11 % fair, and 83 % good. Dissolved oxygen conditions in Mobile Bay were rated as fair overall with 9 % rated poor, 41 % fair, and 50 % good.

2.3.6.4 Florida Bays

Water quality for bays in the state of Florida is summarized based on monitoring data for Tampa and Sarasota bays (Figure 2.40). Both bays host NEPs.

At the time, Tampa Bay was Florida's largest open water estuary spanning approximately 1,036 km^2 (400 mi^2) and draining approximately 5,957 km^2 (2,300 mi^2) of land (Figure 2.40a) (TBEP 2003). The watershed includes the upper reaches of the Hillsborough River, east to the headwaters of the Alafia River, and south to the headwaters of the Manatee River. Freshwater enters the bay from the Lake Tarpon Canal and the Hillsborough, Palm, Alafia, Little Manatee,

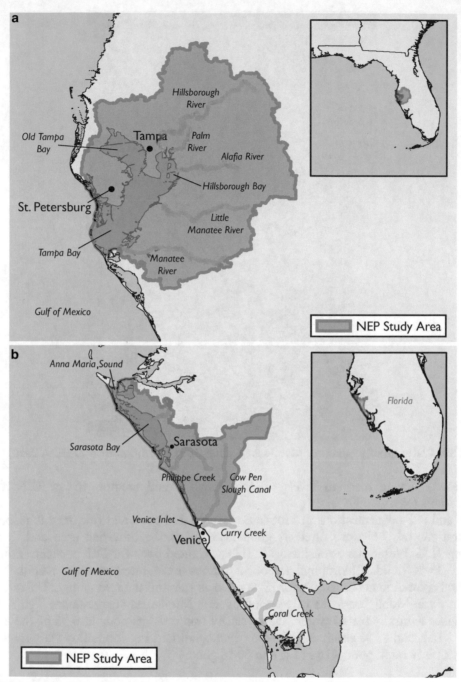

Figure 2.40. Maps of (a) Tampa Bay and (b) Sarasota Bay (modified from USEPA 2006).

and Manatee rivers. The Intracoastal Waterway empties into the bay via Boca Ciega Bay and into the Gulf of Mexico via the Southwest Channel and Passage Key Inlet. Sarasota Bay, located on the southwestern coast of Florida, covers approximately 135 km^2 (52 mi^2) of surface water area and is a small, subtropical estuary (Figure 2.40b). The bay's watershed includes Manatee and Sarasota counties and covers approximately 389 km^2 (150 mi^2) of land. The bay extends from Venice Inlet to Anna Maria Island including the barrier islands and the mainland

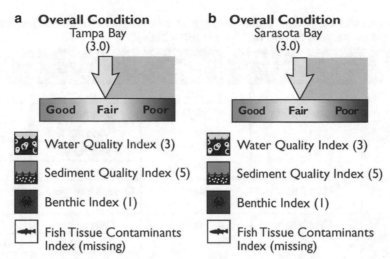

a Overall Condition
Tampa Bay
(3.0)

Good   Fair   Poor

Water Quality Index (3)

Sediment Quality Index (5)

Benthic Index (1)

Fish Tissue Contaminants
Index (missing)

b Overall Condition
Sarasota Bay
(3.0)

Good   Fair   Poor

Water Quality Index (3)

Sediment Quality Index (5)

Benthic Index (1)

Fish Tissue Contaminants
Index (missing)

Figure 2.41. Overall condition of (a) Tampa Bay and (b) Sarasota Bay in 2000 (modified from USEPA 2006).

east to Interstate 75 (SJRWMD 2002). Sarasota Bay was classified as an Outstanding Florida Water Body and an Estuary of National Significance in 1987 (SBNEP 2000; FDEP 2005). Sarasota Bay is the largest and deepest bay between Tampa Bay and Charlotte Harbor. The bay is flushed by passes (Big Sarasota, New, and Longboat) making its waters much clearer than those of smaller bays to the south (Roberts, Little Sarasota, and Blackburn bays) (Florida Center for Community Design and Research 2004). Over the years, Sarasota Bay's water quality has improved due to the provision of more freshwater from the surrounding watershed. Most of the bay's estuarine areas are designated as recreational-use waters for fishing and swimming. Sarasota Bay's watershed is highly urbanized.

The overall condition and the water quality index for Tampa Bay and Sarasota Bay were rated fair in 2000 (Figures 2.41 and 2.43; Table 2.6). A summary of the percentage of estuarine area of each bay rated good, fair, poor, or missing for each parameter of the water quality index is provided in Figure 2.42. This assessment was based on data collected in 2000 from 25 to 20 locations sampled in Tampa Bay and Sarasota Bay, respectively (Figure 2.43).

Comparing NOAA's Estuarine Eutrophication Survey (NOAA 1997) and results from the 2000 survey (USEPA 2006) some improvements were noted. Nitrogen was a major pollutant of concern for Florida's bays. In Sarasota Bay nitrogen was being transported to the bay by base flow, wastewater, stormwater, and atmospheric deposition (SBNEP 2000). Atmospheric deposition of total nitrogen to the surface of Tampa Bay accounted for about one-quarter of the nitrogen loading (about 707 metric tons or 780 tons per year) (Poor et al. 2001). This did not include deposition of nitrogen in the watershed washed into the estuary by stormwater. When both direct and indirect pathways were considered, more than 50 % of the total nitrogen loading to Tampa Bay originated from atmospheric sources, while only 15 % of total nitrogen loading was derived from atmospheric deposition in Sarasota Bay (Poe et al. 2005) (Figure 2.44).

In Sarasota Bay, human activities such as management of waste and the operation of automobiles and watercraft contributed a much larger fraction of nitrogen and other contaminants that degrade water quality than did base flow and atmospheric sources (Figure 2.44). Increased development had resulted in excess nitrogen pollution and stormwater runoff into Sarasota Bay. Stormwater and suspended matter were transported into Sarasota Bay by tributaries resulting in the poorest water quality. Overall water quality monitoring data showed improvements in Tampa and Sarasota Bay. In Tampa Bay, estimates showed that nitrogen

Table 2.6. Comparison of Water Quality Indicators between 1997 and 2000 in Tampa Bay and Sarasota Bay (modified from NOAA 1997 and USEPA 2006)

Water quality indicator	NOAA 1997[a]		USEPA 2006 (percentages of area)	
	Tampa Bay	Sarasota Bay	Tampa Bay	Sarasota Bay
Water quality index	NA	NA	Fair	Fair
Chlorophyll *a* concentrations	Med./V. High	High	Overall fair	Overall fair
			Good 32 %	Good 20 %
			Fair 52 %	Fair 75 %
			Poor 16 %	Poor 5 %
DIN concentrations	Med./High	Med.	Overall low (good)	Overall low (100 % good)
DIP concentrations	Med./High	High	Overall fair	Overall fair
			Good 16 %	Good 75 %
			Fair 72 %	Fair 10 %
			Poor 12 %	Poor 15 %
Water clarity	NA	NA	Overall poor	Overall fair
			Good 36 %	Good 15 %
			Fair 36 %	Fair 65 %
			Poor 28 %	Poor 10 %
Dissolved oxygen	NA	NA	Overall good	Overall fair
			Good 88 %	Good 80 %
			Fair 12 %	Fair 15 %
			Poor 0 %	Poor 5 %

[a]*NA* not applicable, *Med.* medium, *V. High* very high

loading for 1995–2003 was higher than for 1985–1994 mostly due to rains and runoff associated with an El Niño event in 1997–1998 (Poe et al. 2005). In Sarasota Bay, data for 1968–1991 indicates that nutrient and chlorophyll *a* levels were decreasing in the bay. Data for 1980–2002 suggests that DIN and chlorophyll *a* concentrations had declined over the long term in Sarasota Bay. Inorganic phosphorus levels also declined, but increases were noted in some years (Dixon 2003). In general, trends across Sarasota Bay are the same, though there were differences in the magnitude of the changes depending on location within the bay, especially areas receiving water from tributaries. Occasionally elevated levels of bacteria in Tampa Bay waters were detected most likely due to septic system malfunctions and stormwater runoff during rainfall events. Bacteria levels were seen as a potential public health concern for recreational swimming and boating activities. In 2000, a survey showed that the human health risk from bacterial contamination was low throughout Tampa Bay with only 2 of 22 locations exceeding guidelines for human health (Rose et al. 2001).

2.3.7 Coastal Water Quality and Petroleum

Coastal regions are the locations where most chemical contaminants are used and released to the environment, and nearshore environments are also the sites of delivery of land-derived chemical inputs via river- and precipitation-associated runoff and atmospheric deposition. Therefore, if chemical contaminants play a significant role in degrading water quality, they

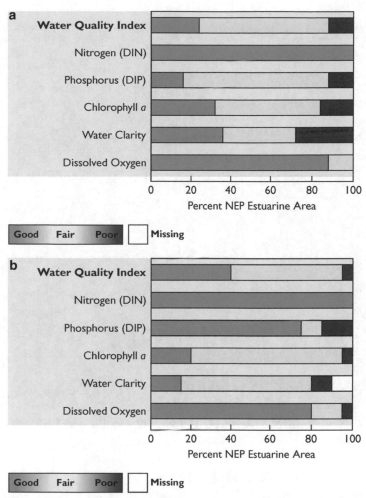

Figure 2.42. Percentage of estuarine area achieving each rating for the water quality index and its components (a) Tampa Bay and (b) Sarasota Bay (modified from USEPA 2006).

are most likely to be detectable in coastal water bodies, with an exception being the immediate effects of large volume oil releases in offshore regions (e.g., spills). As discussed, few water quality assessment studies directly measure chemical concentrations in water due to the low concentrations, so other approaches must be used to assess the role of chemical contaminants in degrading water quality. Two approaches to assessments were described in the introduction. One approach considers the mass loadings of contaminants to receiving water bodies, and the second considers the detection of contaminants in lipid-rich organismal tissues that preferentially accumulate, and in some instances magnify, chemical contamination. As described in the introduction, chemical contaminants can be classified as petroleum or non-petroleum with the latter category subdivided into organic and inorganic non-petroleum contaminants (for detailed descriptions of contaminants in these categories see the introduction and Appendices B and C). As noted, these categories of chemical contaminants have different sources, environmental fates, and toxicities and thus different potentials for affecting water quality. The most comprehensive analysis of annual mass loadings of contaminants to the northern Gulf of Mexico is available for petroleum. The NRC's *Oil in the Sea III*: *Inputs, Fates, and Effects* report (NRC 2003) summarizes annual mass loadings in the coastal northern Gulf of Mexico

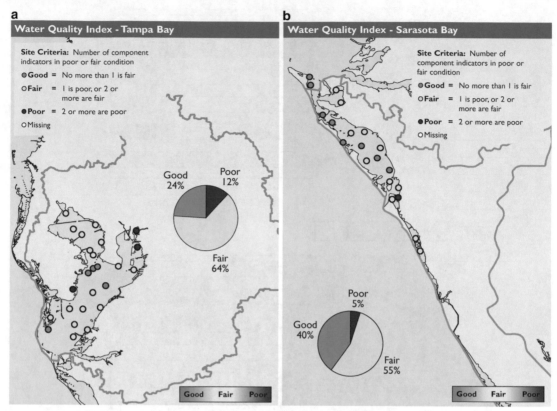

Figure 2.43. Water quality index data for (a) Tampa Bay and (b) Sarasota Bay in 2000 (modified from USEPA 2006).

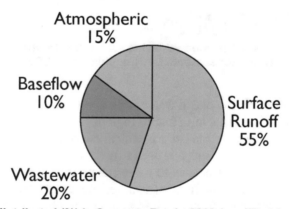

Figure 2.44. Nitrogen distributed (%) in Sarasota Bay in 2000 (modified from USEPA 2006).

for petroleum for 1990–1999. While the report was issued several years ago and the data is from the 1990s, these 9-year average mass loadings are indicative of longer-term trends regarding the role of petroleum contamination in degrading water quality. While the absolute amounts associated with specific releases will vary with time, the NRC report estimates are within the time frame of assessments of coastal water quality conditions along the northern Gulf of Mexico covered in this review. Therefore, the trends identified in coastal water quality can be compared and contrasted, at least qualitatively, with the trends discerned from petroleum mass

Table 2.7. Average Annual Mass Loadings of Petroleum (tonnes) to the Coastal Gulf of Mexico from 1990 to 1999 (1 tonne = 1 metric ton(ne) = 1.102 U.S. short tons) (modified from NRC 2003)

Zone (coastal)	North Central/ Northeastern	North Central/ Northwestern	South Central/ Southwestern
Sum seeps[a]	*na*	*na*	*na*
Platforms	Trace[b]	90	nd[c]
Atmospheric	Trace	trace	nd[c]
Produce	Trace	590	Trace
Sum extraction	*Trace*	*680*	*Trace[c]*
Pipelines	Trace	890	Trace
Tank vessel	140	770	80
Coastal facilities	10	740	nd[d]
Atmospheric	Trace	Trace	Trace
Sum transportation	*160*	*2,400*	*90*
Land-based	1,600	11,000	1,600
Recreational vessels	770	770	nd[e]
Vessels > 100 gigatonne (spills)	30	100	Trace
Vessels > 100 gigatonne (op discharge)	Trace	Trace	Trace
Vessels < 100 gigatonne (op discharge)	Trace	Trace	Trace
Atmospheric	60	90	100
Aircraft[f]	na	na	na
Sum consumption	*2,500*	*12,000*	*1,700*

[a]No known seeps in these regions
[b]Estimated loads of less than 10 tonnes per year reported as "trace"
[c]Lack of precise locations for platforms in this zone precluded determining whether spills or other releases occurred less than 3 mi from shore, thus all values for this zone reported as "offshore"
[d]No information on the existence of coastal facilities was available for this region
[e]Populations of recreational vessels were not available for these regions
[f]Purposeful jettisoning of fuel not allowed within 3 mi of land

loadings. The NRC report also assesses petroleum inputs to other North American coastal waters, providing useful comparisons with Gulf of Mexico estimates. Much of the oil and gas production in North America is located in the Gulf of Mexico, so conclusions about petroleum contamination in North American marine environments are largely applicable to the Gulf of Mexico. The following assessment is constrained by the limitations to this approach discussed in the introduction (e.g., mass loadings reflect the intensity and location of petroleum usage but do not directly indicate biological or ecological impact or ambient water concentrations). This review provides comprehensive information about the sources, geographic distributions, and magnitude of petroleum contamination of the northern coastal Gulf of Mexico for completeness. Mass loadings of average annual petroleum inputs to the coastal Gulf of Mexico for 1990–1999 are summarized in Table 2.7 (NRC 2003).

The other categories of chemical contaminants also have the potential to impact water quality. However, there are no summaries of mass loadings for these contaminants similar to those provided by the NRC (2003) report for petroleum. In order to assess the potential impact of these other contaminants on water quality, the second approach described in the introduction—using data on the presence of contaminants in biological tissues—is employed.

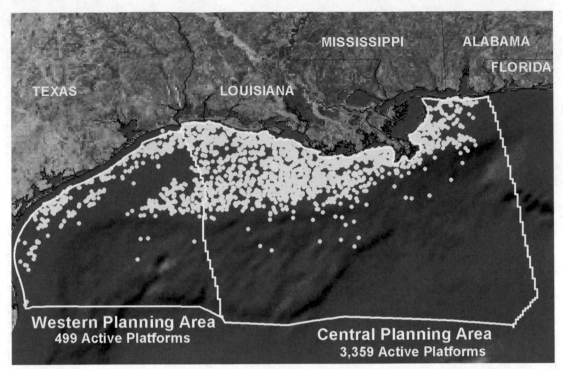

Figure 2.45. Map of the 3,858 oil and gas platforms in the Gulf of Mexico in 2006. The size of the dots used to note platform locations is highly exaggerated and the density of platforms is low (from NOAA 2012).

This qualitative indication of the role of contaminants in degrading of coastal water quality is considered in Section 2.3.8. The detection of petroleum in biological tissue is also reported in the national coastal assessments.

The Gulf of Mexico is one of the most prolific oil and gas provinces in the world and has been the site of oil and gas exploration and extraction activities for many decades. In 2006, there were nearly 4,000 oil and gas platforms in the northern Gulf of Mexico, mostly offshore of Louisiana and Texas (Figures 2.45 and 2.46). In recent years, new oil and gas exploration and production in the Gulf of Mexico has been concentrated on the continental shelf/slope and deeper water regions, but there is a long history of these activities in coastal waters and adjacent onshore areas (Figures 2.45 and 2.46). Activities associated with the transportation and consumption of petroleum are widespread in the Gulf of Mexico as well (Figure 2.47). Large petrochemical and refining complexes are located along the Texas coast making the Gulf of Mexico a major destination for seaborne and pipeline transportation of petroleum and refined products (NRC 2003). The widespread extraction, transportation, and consumption of petroleum in the northern Gulf of Mexico have resulted in chronic releases of petroleum to the environment for many years. In addition, major river systems, including the Mississippi River, deliver petroleum contaminants via runoff from the land. Adding to these anthropogenic sources of petroleum, the Gulf of Mexico is also the location of extensive natural oil and gas seepage (Figure 2.48). Once released to the environment, by whichever pathway, petroleum poses a range of environmental threats including the potential to degrade water quality. Beyond the more directly observable physical impacts, the toxicity of compounds that make up petroleum can affect organisms from the cellular to the population level (NRC 2003). Compounds that occur in petroleum, such as PAHs, are also known human carcinogens. Once weathered and

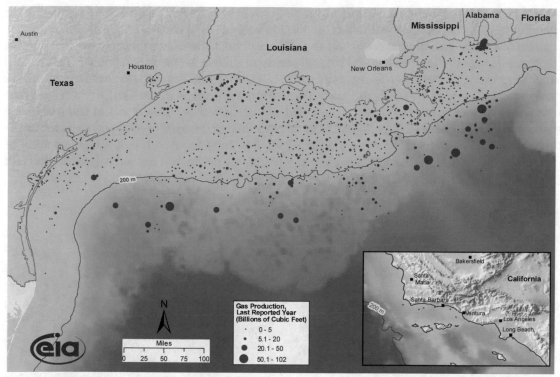

Figure 2.46. Offshore gas production in the Gulf of Mexico (from Energy Information Administration 2009).

mixed with particulate matter, oil in the environment often forms tar balls that float or, if dense enough, can sink to the sea floor. Floating tar balls are found throughout the Gulf of Mexico and can have direct effects on organisms due to uptake in diets or by adherence to surfaces of organisms. In general, tar balls are not expected to be a major factor in degrading water quality, but they are widely detected in marine environments, and the Gulf of Mexico is no exception.

2.3.7.1 Natural Oil and Gas Seeps

The seepage of oil and gas in marine environments is a natural phenomenon that occurs when oil and gas from deep subsea reservoirs migrate to surface seafloor sediments and into the overlying water column. Natural seepage of oil into the marine environment is the largest source of petroleum to the marine environment (NRC 2003). Annual releases due to oil and gas seeps are estimated to exceed 160,000 tonnes (176,000 tons) in North America alone, accounting for over 60 % of the petroleum entering marine waters (Figure 2.49). Almost all deeply buried petroleum reservoirs naturally leak to some extent, and marine environments overlying prolific oil and gas provinces, such as the northern Gulf of Mexico, are chronically subjected to natural oil and gas seepage. The effects of oil and gas seepage are generally restricted to closely associated sediments and benthic organisms and the formation of oil slicks at the air/sea interface. However, seeping oil and gas transits through the water column and aerobic microbial oxidation of hydrocarbons consumes oxygen. Gaseous and low molecular weight hydrocarbons dissolve in seawater based on their solubility, the temperature and salinity of the water, and the time in contact with water. The water column directly above oil and gas seeps can exhibit lowered oxygen concentrations due to aerobic microbial degradation of petroleum. In general, due to the well-mixed nature of marine waters these effects are restricted to a few meters or less up into the water column above the sediment/water interface. Hydrocarbon gases (e.g., methane, ethane, propane and butane) are more soluble in water than liquid hydrocarbons and more buoyant and often form plumes that can persist into the water column meters above seep locations and even reach the sea surface. Petroleum seeps in the Gulf of Mexico occur mostly in deeper water offshore regions and are discussed in more detail in the section on offshore water quality (Figure 2.48). In the coastal Gulf of Mexico few oil and gas seeps have been observed so natural oil seepage in this region is considered to be a negligible source of petroleum contamination, suggesting that this source of petroleum has an insignificant effect on coastal water quality (Table 2.7).

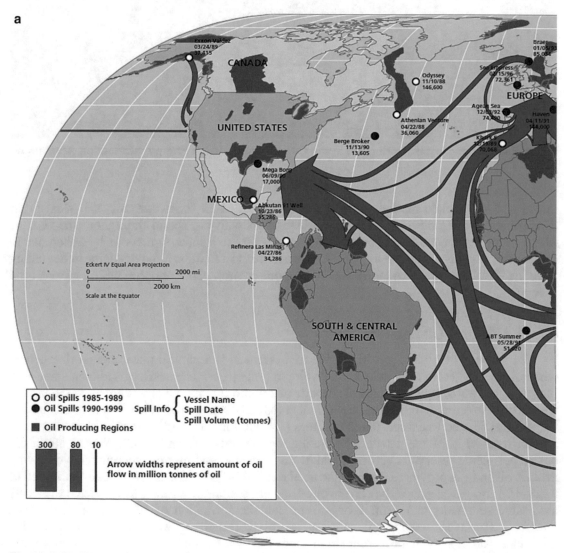

Figure 2.47. Worldwide seaborne flow of oil in 2000 in millions of tonnes (modified from NRC 2003).

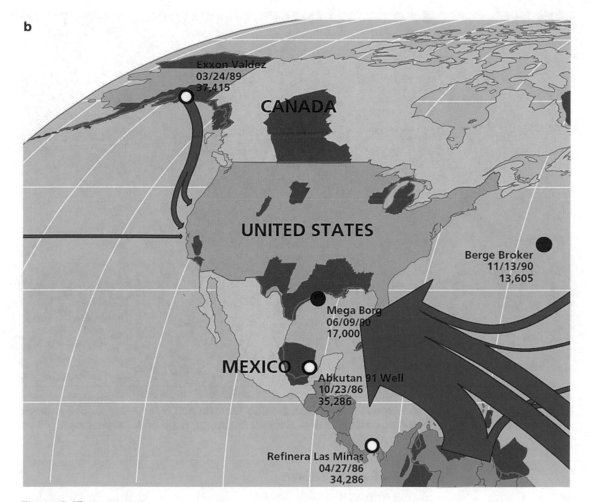

Figure 2.47. (continued)

2.3.7.2 Extraction of Petroleum

Extraction of oil and gas is a source of spills and other releases to the marine environments (NRC 2003). Extraction activities release petroleum and refined products to the surrounding water from platforms by discharging produced waters and by atmospheric releases and deposition (Figure 2.49) (NRC 2003). The nature and size of these releases are highly variable from site to site. Activities associated with oil and gas exploration or production introduced on average approximately 3,000 tonnes (3,307 tons) of petroleum to North American waters each year for the 1990–1999 time period, and annual totals for the coastal Gulf of Mexico were estimated at 680 tonnes (750 tons), almost all in the northwestern region (Table 2.7; Figures 2.49 and 2.50). Inputs from platforms can occur as spills or as chronic releases. For comparison, it was estimated that the IXTOC-I blowout released 476,000 tonnes (524,700 tons) of petroleum to the Gulf of Mexico over approximately 9 months in 1979 (NRC 2003). For the 1990–1999 time period, an estimated 150 tonnes (165 tons) of petroleum per year was accidentally spilled from platforms in North American waters (NRC 2003). The use of chemical dispersants on oil spills can materially change the behavior of oil in seawater.

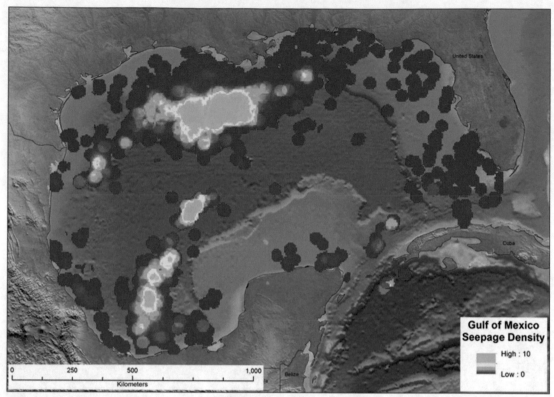

Figure 2.48. Oil and gas seepage in the Gulf of Mexico (determined from analysis of synthetic aperture radar, graphic provided by CGG's NPA Satellite Mapping, used with permission).

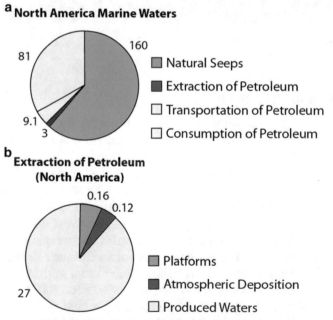

Figure 2.49. Average annual releases of petroleum hydrocarbons in thousands of tonnes (1 tonne = 1 metric ton(ne) = 1.102 U.S. short tons) to North American waters from (a) natural seeps and extraction, transportation, and consumption activities and (b) petroleum extraction from 1990 to 1999 (modified from NRC 2003).

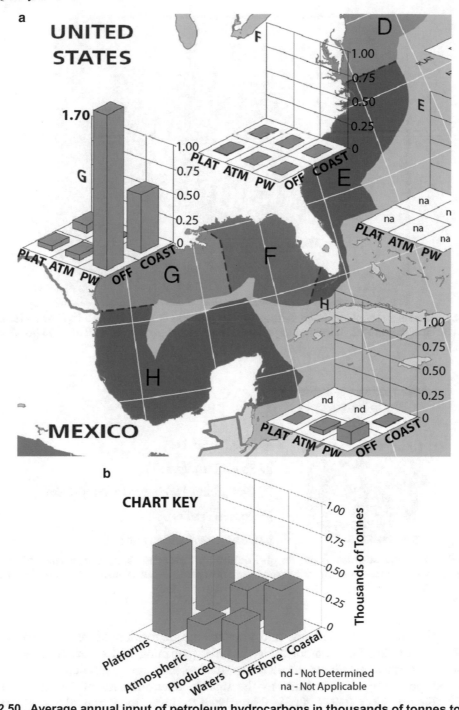

Figure 2.50. Average annual input of petroleum hydrocarbons in thousands of tonnes to the Gulf of Mexico from petroleum extraction for 1990–1999 (modified from NRC 2003).

2.3.7.3 Transportation of Petroleum

The transportation of petroleum releases varying amounts of petroleum from major spills to small regular operational releases. Petroleum hydrocarbon discharges into marine waters by transportation activities include pipeline spills, tank vessel spills, discharges from cargo

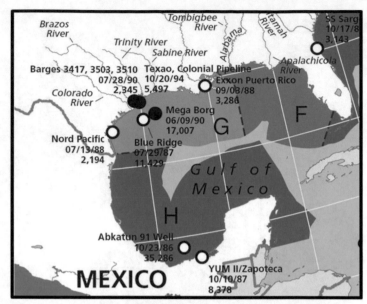

Figure 2.51. Distribution of selected vessel oil spills in the Gulf of Mexico in tonnes (*solid black dots* indicate spills included in the average annual mass loadings from 1990 to 1999 (modified from NRC 2003).

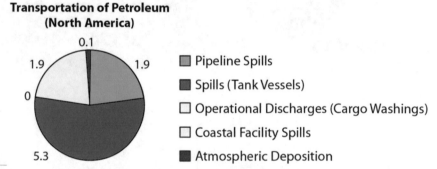

Figure 2.52. Average annual input of petroleum hydrocarbons in thousands of tonnes to North American marine environments from the transport of petroleum for 1990–1999 (modified from NRC 2003).

washings, spills at coastal facilities, and atmospheric deposition of releases from tankers (Figure 2.51) (NRC 2003). Transportation, including refining and distribution activities, of petroleum or refined products resulted in the release, on average, of 9,100 tonnes (10,031 tons) per year of petroleum to the marine environments of North America for 1990–1999 (Figure 2.52) (NRC 2003). From 1990 to 1999, total annual mass loading of petroleum from transportation activities for the coastal northwestern and northeastern Gulf Mexico were 2,400 tonnes (2,646 tons) and 160 tonnes (176 tons), respectively (Table 2.7). In the northwestern Gulf of Mexico the releases from pipelines, tank vessels, and coastal facilities were similar in magnitude, whereas in the northeastern Gulf of Mexico releases came almost exclusively from tank vessels (Table 2.7). Atmospheric deposition was considered negligible in both regions during this time period. Pipeline spills can occur as petroleum is transported from the source to refineries and from refineries to the consumer (NRC 2003). Tank vessels are allowed discharges of contaminated water related to cargo and propulsion machinery whereas non-tankers are only allowed machinery-related discharges (NRC 2003). Operational discharges from cargo washings are illegal in North American coastal waters (NRC 2003). Discharges of oil in ballast and tank washing from oil tankers are prohibited within 92.6 km (50 nautical miles) of the coast (NRC 2003). Discharges from coastal facilities include episodic spills as well as chronic releases (NRC 2003).

Releases due to the transportation of petroleum were approximately 9 % of the total petroleum input to the marine environments of North America during this time period. Most transportation-related releases of petroleum occurred in the western Gulf of Mexico where the majority of offshore platforms, pipelines, coastal oil refineries and chemical plants, and major ports are located (Figure 2.53). A major source of petroleum released to the Gulf of Mexico during the extraction process is the intentional discharge of produced waters (Figure 2.49b). Over 90 % (2,700 tonnes; 2,976 tons) of petroleum released during extraction activities during 1990–1999 was accounted for by produced water discharges which release low but continuous amounts of dissolved components and dispersed crude oil to the marine environment. Discharges of produced water have the potential to impact water quality across the northern Gulf of Mexico given the large number and density of petroleum platforms offshore Louisiana and Texas (Figure 2.45). The potential for impact from discharged waters is greatest in coastal or inland areas where flushing rates are low and petroleum tends to accumulate over time. Shallow water areas with restricted flow and dispersion (low flushing rates), water with a high concentration of suspended particulates, and fine-grained anaerobic sediments are especially vulnerable to water quality issues (Boesch and Rabalais 1989a, b; St. Pé KM 1990). In the Gulf of Mexico, coastal oil production occurs only in Louisiana and Texas. In the late 1990s the discharge of produced water in coastal waters was prohibited so this input has been greatly reduced since then (Boesch and Rabalais 1989a, b; St. Pé KM 1990; Rabalais et al. 1991).

Spills of petroleum associated with platforms accounted for approximately 5 % of the total inputs from extraction activities totaling 2.2–2.5 tonnes (2.4–2.8 tons) and 81 tonnes (89 tons) per year for 1990–1999 in the northeastern and northwestern coastal Gulf of Mexico, respectively, reflecting the low intensity of coastal oil and gas production in the northeastern Gulf of Mexico (NRC 2003). Again, these discharges were prohibited in the late 1990s.

2.3.7.4   Consumption of Petroleum

Once petroleum has been extracted, transported to refineries, and refined, it is delivered to the consumer. The major sources of petroleum releases related to consumption include land-based sources (river discharge and runoff), two-stroke vessel discharges, non-tank vessel spills, operational discharges, atmospheric deposition, and aircraft dumping (Figure 2.54). Consumption-related releases of petroleum are generally individually small; however, the ubiquity and number of releases collectively contribute the majority of anthropogenic petroleum to marine environments (Figure 2.54) (NRC 2003). On average, approximately 84,000 tonnes (92,594 tons) per year of petroleum were released to marine waters of North America for 1990–1999 (NRC 2003). Releases associated with the consumption of petroleum were approximately 70 % of the petroleum released from anthropogenic sources to North American waters during this time period. The majority of the consumption of petroleum occurs on land so together, river and waste and stormwater runoff are the largest sources of petroleum to coastal environments. Another important input of petroleum in coastal areas is leakage from two-stroke engines. Land runoff and two-stroke engines accounted for approximately 75 % of the petroleum introduced to North American waters by petroleum

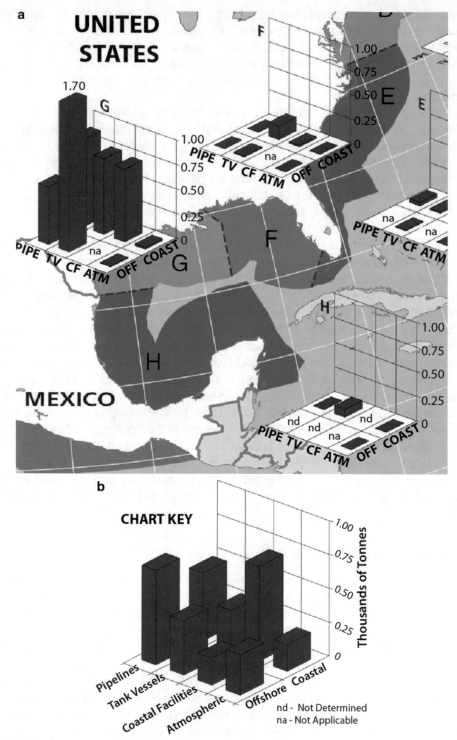

Figure 2.53. Average annual input of petroleum hydrocarbons in thousands of tonnes to the Gulf of Mexico from petroleum transportation from 1990 to 1999; (modified from NRC 2003).

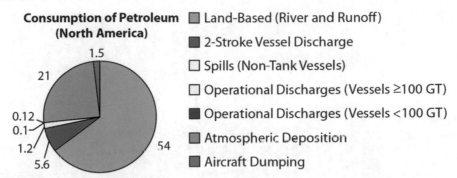

Figure 2.54. Average annual input of petroleum hydrocarbons in thousands of tonnes to North American marine environments from the consumption of petroleum from 1990 to 1999 (modified from NRC 2003).

consumption from 1990 to 1999. These activities are almost exclusively restricted to coastal waters. In the coastal Gulf of Mexico for 1990–1999, annual mass loadings of petroleum from activities associated with consumption were concentrated in the northwestern region and mostly associated with land-based sources (Figure 2.55). For the 1990–1999 time period, land-based sources contributed 12,000 tonnes (13,228 tons) and 1,600 tonnes (1,763 tons) of petroleum annually in the northwestern and northeastern coastal Gulf of Mexico, respectively. The next largest coastal source of petroleum was recreational vessels, which contributed 770 tonnes of petroleum annually to the northeastern and 770 tonnes (849 tons) to the northwestern Gulf of Mexico from 1990 to 1999. All other consumption-related inputs contributed less than 300 tonnes (331 tons) annually to the coastal Gulf of Mexico region for 1990–1999.

2.3.7.5   Spatial Variability of Petroleum Contamination

In summary, coastal northern Gulf of Mexico environments are subject to highly variable mixes of petroleum inputs that differ substantially for the northeastern and northwestern regions (Figure 2.56). For coastal waters, land-based sources of petroleum related to consumption activities are ubiquitous and dominate inputs across the northern Gulf of Mexico. For the 1990–1999 time period, the northwestern Gulf of Mexico received only 21 % of the total input from land-based sources in North America despite the large number of refineries in the region and riverine inflows from the Mississippi River (NRC 2003). However geographic distributions, admixtures of sources, and the magnitude of annual petroleum loadings do reflect the large petroleum industry located in the northwestern Gulf of Mexico that includes all phases of exploration, production and transportation. Transportation-related petroleum mass loadings in the northwestern Gulf of Mexico were about 15–25 times greater than in the northeastern Gulf of Mexico during the 1990s reflecting this concentration of industry (NRC 2003). As noted previously, petroleum contamination is rarely identified as the primary cause of degradation of coastal water quality, except in specific cases such as major oil spills. This is expected, as degraded water quality along the northern Gulf of Mexico has been largely attributed to excess nutrient loadings. Degraded coastal water quality and petroleum contamination in coastal regions are associated with human population patterns as both are predominantly anthropogenic in origin. The ubiquitous presence of petroleum contamination in the northern Gulf of Mexico would be expected to be at least a minor contributor to degraded water quality but these effects are masked by other more dominant factors such as nutrient enrichments.

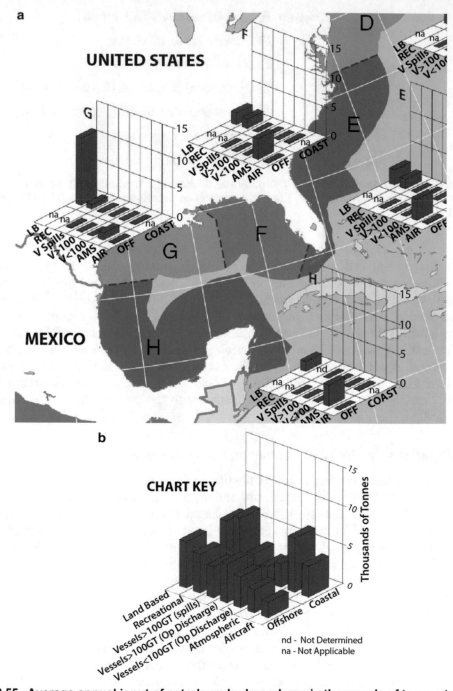

Figure 2.55. Average annual input of petroleum hydrocarbons in thousands of tonnes to the Gulf of Mexico from petroleum consumption for 1990–1999 (modified from NRC 2003).

2.3.8 Coastal Water Quality and Utilization of Water

Water quality is based on the suitability of a body of water for certain uses by ecosystems and/or humans and can be assessed based on how well human expectations are being met in terms of the services provided by a body of water. As described in the introduction, an integration of multiple indicators can be used to assess the impairment of valued activities.

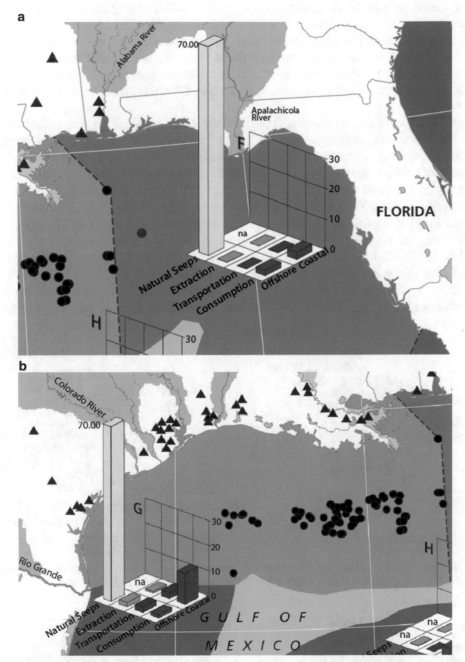

Figure 2.56. Average annual input of petroleum hydrocarbons in thousands of tonnes to the coastal Gulf of Mexico for 1990–1999 (*yellow* = natural seeps, *green* = extraction, *purple* = transportation, and *red* = consumption) (modified from NRC 2003).

Chemical and biological contaminants in water can contribute to impairment by causing acute and/or chronic human health effects, but unambiguous links to degraded water quality are often tenuous. Humans may be exposed to waterborne toxins or pathogens due to consumption of fish and shellfish and/or directly via contact with water. Impacts on ecosystem and human use provide insight into potential issues that might have an origin in water quality. Assessments of impairment also provide an indirect, qualitative assessment of the role of chemical and biological contaminants in degrading water quality within the limitations discussed in the introduction. The following assessments are presented as examples, but an exhaustive review of all information related to water impairment, beach closures and fish consumption reports is beyond the scope of this review as explicit links to water quality are difficult to discern. These examples also provide a qualitative indication of which contaminants may be responsible for impairments and identify hot spots of contamination for comparison with other indicators of water quality.

Based on 5 years of monitoring from 1991 to 1995, 51 % of northern Gulf of Mexico estuaries were assessed as unimpaired, 27 % impaired for human use, and 37 % impaired for aquatic life (percentages add to more than 100 % as estuaries can be impaired for both human and aquatic life use) (Figure 2.57a). For 1996–2000, the overall condition of northern coastal Gulf of Mexico estuaries was rated as fair with 35 % of the estuarine areas assessed as impaired for aquatic life use and 14 % impaired for human use (Figure 2.57b). Of the assessed estuaries, 20 % were in good ecological condition with no evidence of degradation. Of estuarine areas assessed along the northern Gulf of Mexico, 39 % were considered threatened. Gulf States assessed 48 % (18,845 km^2 [7,276 mi^2] of 39,668 km^2 [15,316 mi^2]) of the Gulf Coast estuaries for 1998 Clean Water Act Section 305(b) reports (Figures 2.58 and 2.59). In these reports it was not possible to distinguish between Atlantic Coast and Gulf of Mexico listings, so 305 (b) assessment information for Florida was included in 2001 Gulf of Mexico summaries. Of

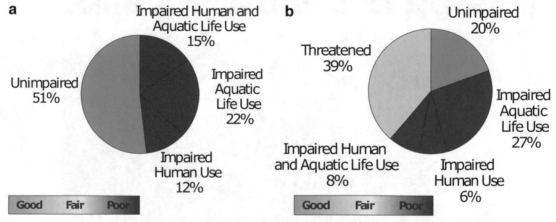

Figure 2.57. Gulf Coast estuarine condition estimates ±6 % based on 5 years of sampling, (a) for years 1991–1995 and (b) for years 1996–2000 (modified from USEPA 2001, 2004).

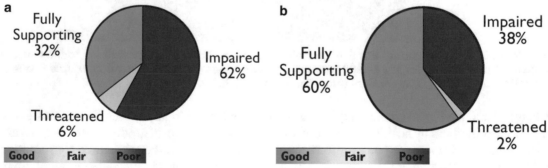

Figure 2.58. Water quality assessments in 1998 for northern Gulf of Mexico (a) estuaries and (b) shore lines (modified from USEPA 2001).

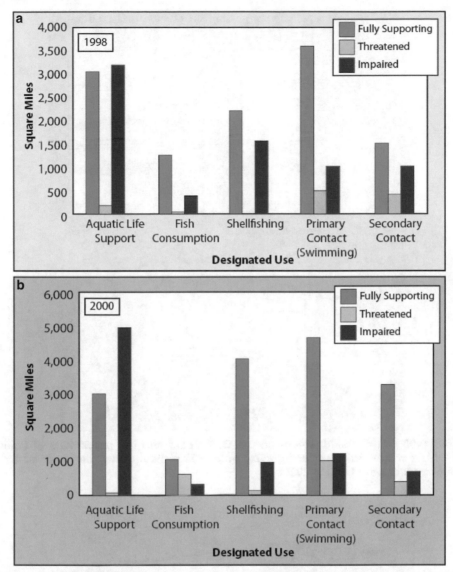

Figure 2.59. Individual use support for assessed estuaries in the Gulf Coast (a) 1998 (modified from USEPA 2001) and (b) 2000 (modified from USEPA 2004).

the assessed estuarine waters, 32 % fully supported their designated uses and 6 % were considered under threat for one or more uses (Figure 2.58a). Some form of contamination or habitat degradation impaired the remaining 62 % of the estuarine waters assessed. Individual use support for estuaries in 1998 and 2000 is shown in Figure 2.59. Of 16,195 coastal shoreline km (10,063 coastal shoreline mile), 296 km (184 mi) or 0.02 % were assessed in 2001. Of the shoreline miles assessed, 60 % fully supported the designated uses, 2 % were considered threatened for one or more uses, and 38 % were impaired by some form of contamination or habitat degradation (Figure 2.58b). In 2001, there were 233 waters in the Gulf of Mexico listed as impaired under Section 303(d) of the Clean Water Act. The percentage of listed waters impaired by major pollutant category is summarized in Figure 2.60. Of 41,069 km^2 (15,857 mi^2) of Gulf of Mexico estuaries 71 % (29,057 km^2 [11,219 mi^2]) were assessed for 2000 Clean Water Act 305(b) reports, which were generally based on data collected in the late 1990s (Figure 2.61).

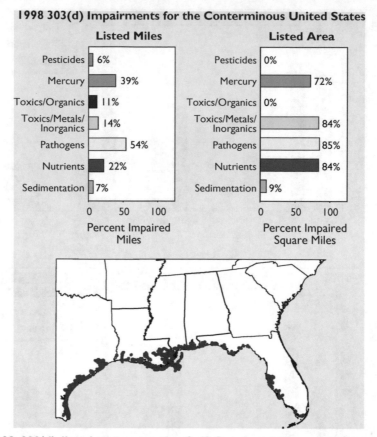

Figure 2.60. 1998 303(d) listed waters on the Gulf Coast and the percentage of listed waters impaired by the major pollutant categories. *Note*: **303(d) listing may be impaired by multiple pollutants (modified from USEPA 2001).**

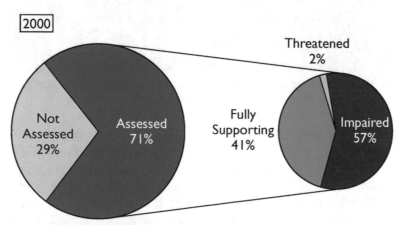

Figure 2.61. Water quality in assessed Gulf Coast estuaries in 2000 (modified from USEPA 2004).

As in 2001, it was not possible to distinguish between Atlantic and Gulf of Mexico listings; therefore, 305(b) assessment information for Florida was included in Gulf of Mexico summaries. Of the assessed estuarine waters along the northern Gulf of Mexico, 41 % fully support the designated uses and 2 % were considered threatened for one or more uses. Some form of pollution or habitat degradation impaired the remaining 57 % of assessed estuarine waters on the Gulf Coast.

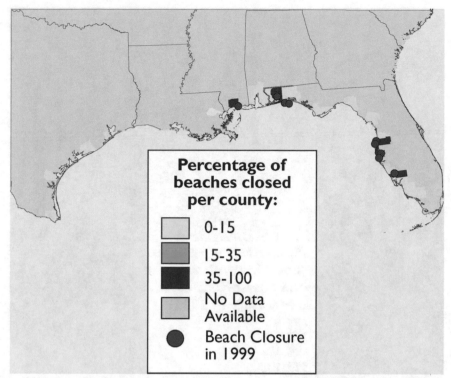

Figure 2.62. Locations of beaches for which information was available. Of the beaches submitting information, 13 % were closed at least once in 1999 (modified from USEPA 2001).

Information on monitoring and beach closures was reported to USEPA in 1999 by all Gulf States, except Louisiana (USEPA 2001). In total, 85 beaches reported with 85 % of respondents located in Florida. Of these 85 beaches, 79 % (67 beaches) had a water quality monitoring program. In Florida, 81 % of the beaches reported that monitoring was conducted in 1999 covering approximately 97 km (60 mi) of beach coastline. Ten beaches (14 % of those reporting) along Florida's coast reported closing at least once in 1999 (Figure 2.59). The primary reason for beach closures was elevated bacteria levels due to stormwater and other runoff. In Mississippi, only one coastal beach responded to USEPA's survey. The beach reported monitoring of 64 km (40 mi) of beach coastline that was partially closed twice in 1999. One beach in Louisiana on the south shore of Lake Pontchartrain was closed throughout 1998 due to elevated bacterial levels from sanitary sewer overflows and pipe breaks. In 2002, of the 176 coastal beaches in the Gulf of Mexico that reported information to USEPA, 37 % (65 beaches) were closed or under an advisory for some period of time. Florida's west coast had the most beaches with advisories or closures (Figure 2.62). Mississippi did not participate in the 2002 survey. Advisory and closure percentages for each county within each state are summarized in Figure 2.63.

Most advisories and closings at coastal beaches along the northern coastal Gulf of Mexico were due to elevated bacteria levels (Figures 2.64 and 2.66). Stormwater runoff, other unknown sources, and wildlife were frequently identified as sources of waterborne bacteria that resulted in advisories or closings. Unknown sources accounted for 36 % of the responses (Figure 2.65). In Florida, 39 % (52 of 134) of beaches reported an advisory or closing at least once during 2002. The primary reasons for public beach notifications were preemptive actions due to rainfall events or the detection of elevated bacteria levels from unknown sources, stormwater and other runoff, wildlife, boat discharges, septic systems, and publically owned treatment works

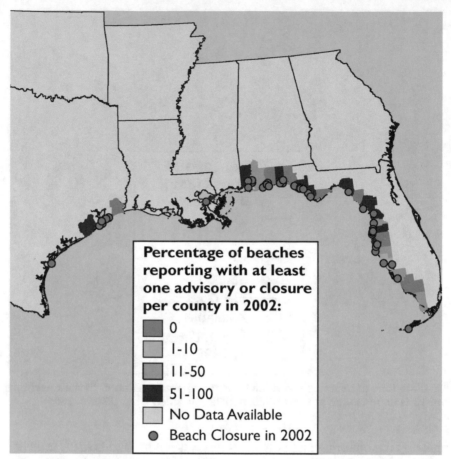

Figure 2.63. Percentage of Gulf Coast beaches with advisories or closures by county in 2003 (modified from USEPA 2004).

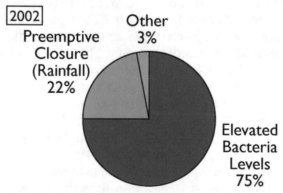

Figure 2.64. Reasons for beach advisories or closures on the Gulf Coast (modified from USEPA 2004).

(POTW) discharges. In Alabama, 4 of 11 responding beaches (36 %) reported advisories or closures during 2002 from elevated bacterial levels due to stormwater runoff, unknown sources, wildlife, and sewer line blockage or pipe breakage. In Louisiana, one beach on the south shore of Lake Pontchartrain reported being affected by a year-long advisory or closure during 2002 due to elevated bacterial levels from POTWs, sewer line blockage or pipe breakage,

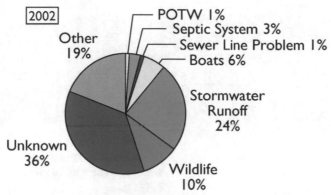

Figure 2.65. Sources of beach contamination on the Gulf Coast (modified from USEPA 2004).

and stormwater runoff. In Texas, 8 of 30 responding beaches reported advisories or closures during 2002 due to elevated bacteria levels from unknown sources, stormwater runoff, wildlife, septic systems, boat discharges, sanitary sewer overflows, and sewer line blockage or pipe breakage. Of the 619 coastal beaches in the northern Gulf of Mexico that reported to USEPA, 23 % (144 beaches) were closed or under an advisory in 2003. Florida's west coast had the most beaches with advisories or closures. Louisiana did not respond to the survey (USEPA 2006) (Figure 2.66).

Water quality can also be reflected in the number and type of fish consumption advisories. However, as indicated, a comprehensive review of seafood advisories in the northern Gulf of Mexico is beyond the scope of this review. Contaminants in fish and other seafood can be caused by a variety of sources other than direct uptake from water, but the levels of contaminants in fish tissues provide an indication of potential degraded water quality due to contaminants. A 3-year snapshot is provided as an example to illustrate the extent of the problems causing most concern in the northern Gulf of Mexico. In 2000, 2001, and 2003, there were 14, 13, and 14 fish consumption advisories in effect for the estuarine and marine waters of the Gulf of Mexico, respectively (Figure 2.67) (USEPA 2001, 2004, 2008). Most advisories (10, 12, and 2 in 2000, 2001, and 2003, respectively) were issued for mercury, and all Gulf States had one statewide coastal advisory in effect for mercury in king mackerel all 3 years. As a result of the statewide advisories, 100 % of the coastal miles of the northern Gulf of Mexico were under advisory for all 3 years and 64, 27, and 27 % of the estuarine square miles were under advisory in 2000, 2001, and 2003, respectively. Advisories placed on specific water bodies included additional pollutants and fish species. For example, in 2000, Bayou d'Inde in Louisiana was under an advisory for all fish and shellfish due to contamination by PCBs, mercury, hexachlorobenzene, and hexachlorobutadiene. Florida had four additional mercury advisories, in addition to the statewide coastal advisory. In Texas, the Houston Ship Channel was under advisory for catfish and blue crabs due to contamination by dioxins/furans (2000 and 2001). Most advisories (12) were issued for mercury, and each Gulf State had a statewide coastal advisory in effect for mercury in king mackerel. As a result of the statewide advisories, 100 % of the coastal miles in the Gulf of Mexico and 23 % of the estuarine square miles were under advisory in 2002 (Figure 2.67). In 2001, Florida had eight mercury advisories in effect for a variety of fish in addition to the statewide coastal advisory. In 2003, the Houston Ship Channel was under advisory for all fish species because of contamination by chlorinated pesticides and PCBs. Potential dioxin contamination in catfish and blue crabs resulted in additional advisories for the Houston Ship Channel.

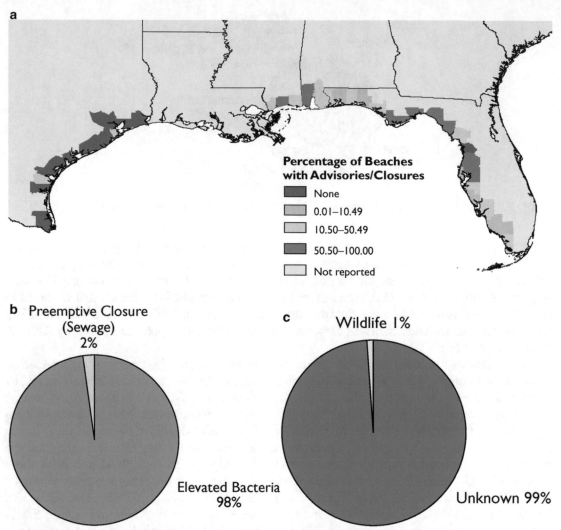

Figure 2.66. (a) Percentage of monitored beaches with advisories or closures by county for the Gulf Coast region; (b) reasons for beach advisories or closures for the Gulf Coast region; and (c) sources of beach contamination resulting in beach advisories or closures for the Gulf Coast region (modified from USEPA 2008).

Integrated assessments, beach closings, seafood consumption advisories, and contaminant levels in selected species show that degraded environmental conditions have impaired many northern Gulf of Mexico estuaries, shorelines, and beaches in regard to the services they provide to ecosystems and humans. Coastal environments are exposed to a wide range of influences that can degrade environmental quality. It is the cumulative effect of these factors that leads to impairment, making it difficult to ascribe degradation to a single causative factor such as water quality. However, degraded water due to chemical and biological contaminants is implicated as at least a contributor to degraded environments at numerous locations across the northern Gulf of Mexico. Human health has been demonstrated to be at risk due to consumption of seafood and exposure to contaminated waters that are contaminated by chemicals and pathogens. Upwards of 60 % of assessed estuaries were impaired for use by ecosystems and/or humans while many others were considered threatened. Locations of impairment are often closely associated with high concentrations of human populations (urban areas) along the coast that are also associated with human activities that introduce excess nutrients and contaminants

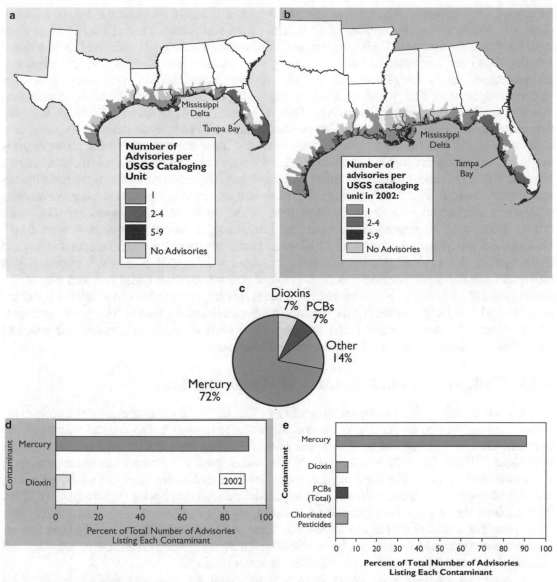

Figure 2.67. (a) Number of fish advisories active in 2000; (b) number of fish advisories active in 2002; (c) percentage of estuarine and coastal marine advisories issued for each contaminant on the Gulf Coast; (d) percentage of estuarine and coastal marine advisories issued for mercury and dioxin on the Gulf Coast in 2002; and (e) percentage of estuarine and coastal marine advisories issued for each contaminant on the Gulf Coast (modified from USEPA 2001, 2004, 2008).

to coastal environments. The reasons for impairment are highly variable and location dependent, and locations can be impaired due to more than one factor. The inflows of large river systems are also associated with impairment. Contaminant-related impairment at individual locations has been attributed to the presence of pesticides, mercury, other organic contaminants and pathogens. In the early 2000s, many advisories were issued due to the presence of mercury in certain species of fish; mercury is by far the most ubiquitous metal chemical contaminant detected in fish tissues along the northern Gulf of Mexico coast. At specific locations in highly urbanized and industrial estuaries, the concentrations of PCBs, chlorinated pesticides, and dioxins/furans in fish tissues resulted in the issuance of consumption advisories. However, it is unclear if these occurrences are caused by degraded water quality since chemical contaminants accumulate in biological tissues via other pathways (e.g., ingestion of contaminated sediments and dietary foods). For beach closing, this is almost exclusively associated with waterborne pathogens discharged into coastal waters from a variety of sources suggesting that water quality itself may be degraded. As indicated previously, a comprehensive review of beach closings, consumption advisories, and biological tissue contaminant concentrations is beyond the scope of this review, but the examples provided give insight into which chemical and biological contaminants in addition to petroleum are of environmental concern across the northern Gulf of Mexico. No comprehensive mass loading summaries are available for other organic and inorganic contaminants that are of environmental concern. However, extensive quantitative surveys of contaminated sediments and sentinel organism (oyster and mussels) contaminant burdens are available and reviewed elsewhere.

2.3.9 Temporal Trends in Coastal Water Quality

A question when considering water quality and its causes is whether conditions are getting better, getting worse, or staying the same. Since water quality in the coastal waters of the northern Gulf of Mexico has been assessed since 1991 these data can be used to detect trends over time (USEPA 2001, 2004, 2008). Only two water quality indicators were comparable in these two time frames: dissolved oxygen concentrations and water clarity. Year-by-year data showed no significant trend with time in the percent of area rated poor (Figure 2.68) (USEPA 2008). When the two time periods were compared, significantly more of the coastal area was rated poor for water clarity in the 2000–2002 time period than in the 1991–1994 time period. Longer-term temporal trends can be masked by interannual variations due to weather and climate that cause large short-term variations in water quality.

A second opportunity to assess long-term temporal changes was availed by NOAA's updating of the 1999 report on eutrophication in 2007 (Bricker et al. 1999, 2007). The updated assessment in 2007 identified eutrophication status and change since the 1999 report, tracked management progress, and identified potential solutions to eutrophication problems. These assessments gave insight into water quality trends over a 10-year period. Trends in eutrophication were assessed by examining influencing factors, eutrophic symptoms, overall eutrophic condition and future outlooks. The results were combined into an overall rating. As described previously, factors that influence eutrophication include nitrogen loading and the estuary's susceptibility to excess nutrients based on dilution and flushing rates. Overall eutrophic condition was based on an assessment of five indicators: chlorophyll *a* concentrations, macroalgae biomass, dissolved oxygen concentrations, submerged aquatic vegetation gain/loss, and nuisance/toxic blooms. Eutrophic condition was determined by evaluating the occurrence, spatial coverage, and frequency of these symptoms. In the 1999 report, the future outlook for eutrophic condition in the year 2020 was predicted based on expected changes in nutrient loads and an estuary's susceptibility to these loadings (Figure 2.69). The completeness and reliability of the assessment was a function of the availability and quality of data.

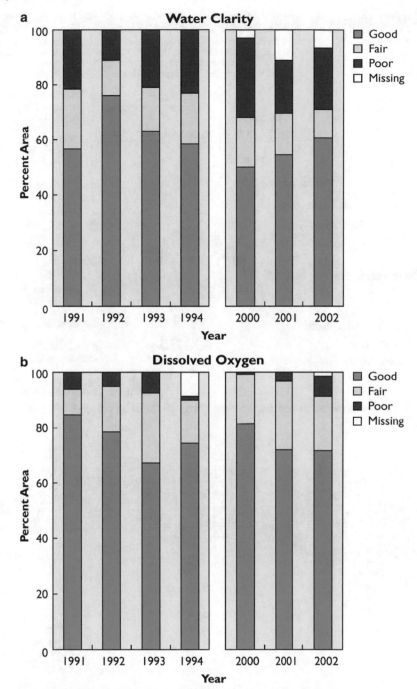

Figure 2.68. Percent of area of northern Gulf of Mexico waters rated as good, fair, poor, or missing for (a) water clarity and (b) dissolved oxygen concentrations measured over two time periods, 1991–1994 and 2000–2002 (modified from USEPA 2008).

The 1999 assessment concluded that Gulf of Mexico estuaries were mostly large, shallow, and poorly flushed leading to predictions of worsening eutrophication conditions. The estuaries tended to have large watersheds by area that support low to moderate human populations. Factors influencing eutrophication were high for a majority of assessed estuaries (Figure 2.70). A small proportion of estuaries had high or moderately high overall eutrophic condition in 2007

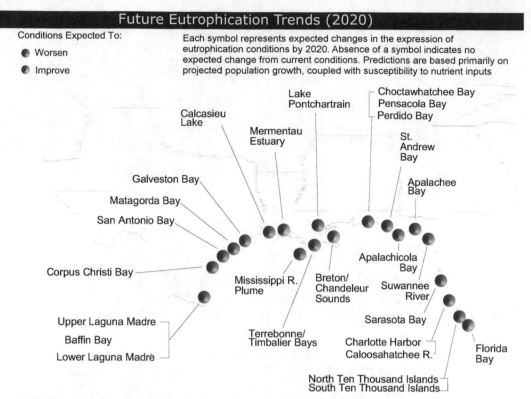

Figure 2.69. Expected trends in eutrophication through 2020 predicted in 1999 (modified from Bricker et al. 1999).

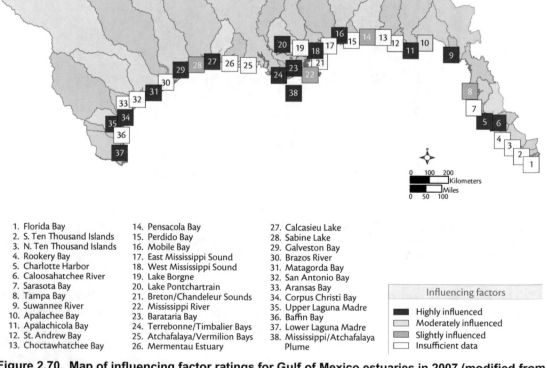

1. Florida Bay	14. Pensacola Bay	27. Calcasieu Lake
2. S. Ten Thousand Islands	15. Perdido Bay	28. Sabine Lake
3. N. Ten Thousand Islands	16. Mobile Bay	29. Galveston Bay
4. Rookery Bay	17. East Mississippi Sound	30. Brazos River
5. Charlotte Harbor	18. West Mississippi Sound	31. Matagorda Bay
6. Caloosahatchee River	19. Lake Borgne	32. San Antonio Bay
7. Sarasota Bay	20. Lake Pontchartrain	33. Aransas Bay
8. Tampa Bay	21. Breton/Chandeleur Sounds	34. Corpus Christi Bay
9. Suwannee River	22. Mississippi River	35. Upper Laguna Madre
10. Apalachee Bay	23. Barataria Bay	36. Baffin Bay
11. Apalachicola Bay	24. Terrebonne/Timbalier Bays	37. Lower Laguna Madre
12. St. Andrew Bay	25. Atchafalaya/Vermilion Bays	38. Mississippi/Atchafalaya Plume
13. Choctawhatchee Bay	26. Mermentau Estuary	

Figure 2.70. Map of influencing factor ratings for Gulf of Mexico estuaries in 2007 (modified from Bricker et al. 2007).

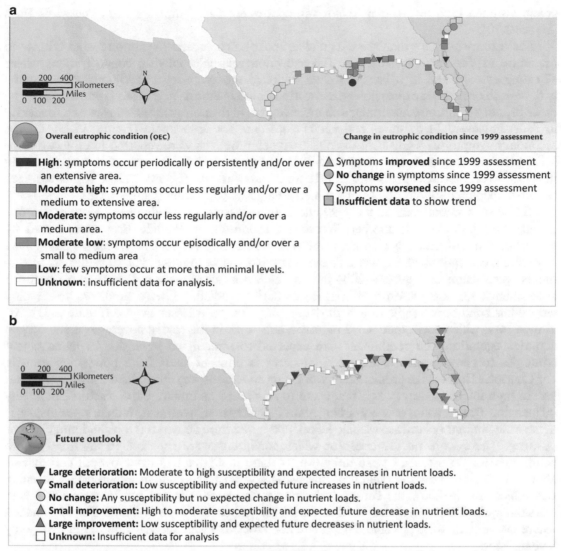

a

0 200 400
Kilometers
0 100 200
Miles

N

Overall eutrophic condition (OEC)

Change in eutrophic condition since 1999 assessment

■ **High**: symptoms occur periodically or persistently and/or over an extensive area.

■ **Moderate high**: symptoms occur less regularly and/or over a medium to extensive area.

■ **Moderate**: symptoms occur less regularly and/or over a medium area.

■ **Moderate low**: symptoms occur episodically and/or over a small to medium area

■ **Low**: few symptoms occur at more than minimal levels.

□ **Unknown**: insufficient data for analysis.

△ Symptoms **improved** since 1999 assessment
○ **No change** in symptoms since 1999 assessment
▽ Symptoms **worsened** since 1999 assessment
■ **Insufficient data** to show trend

b

0 200 400
Kilometers
0 100 200
Miles

N

Future outlook

▼ **Large deterioration**: Moderate to high susceptibility and expected increases in nutrient loads.

▽ **Small deterioration**: Low susceptibility and expected future increases in nutrient loads.

○ **No change**: Any susceptibility but no expected change in nutrient loads.

△ **Small improvement**: High to moderate susceptibility and expected future decrease in nutrient loads.

▲ **Large improvement**: Low susceptibility and expected future decreases in nutrient loads.

□ **Unknown**: Insufficient data for analysis

Figure 2.71. (a) Overall eutrophication condition and (b) future outlook for eutrophication conditions for the Gulf of Mexico estuaries (modified from Bricker et al. 2007).

(Figure 2.71). Gulf of Mexico estuaries were characterized as having high and often worsening chlorophyll *a* symptoms. Watershed nitrogen inputs were determined to be high in over 80 % of the estuarine systems assessed in the northern Gulf of Mexico. However, nitrogen loading data was limited, with no information available for about half of the estuaries. Nitrogen loadings were considered low for only two of the 38 estuaries—Tampa Bay and Pensacola Bay. Not unexpectedly, the Mississippi River had the largest nutrient load of all U.S. rivers at the time. Nutrient load estimates for the Mississippi River were used to calculate influencing factor ratings for both the Mississippi River and Mississippi/Atchafalaya Plume. Most estuaries in the northern Gulf of Mexico have shallow water depths and small tidal ranges that suggest low dilution and flushing rates. As a consequence, most estuaries were judged to have a moderate to high susceptibility to nutrient loading (Figures 2.70 and 2.71). The combination of effects of high nitrogen loads and moderate or high susceptibility to nutrients results in most estuaries being assigned high influencing factor ratings (except for Tampa Bay and Pensacola Bay) (Figure 2.70).

For estuaries where data were available, most eutrophication symptoms showed low to moderate expressions (Figure 2.71). The exception was chlorophyll *a* concentrations where 17 estuaries exhibited high level and five exhibited moderate level conditions. The systems with high chlorophyll *a* expression were mostly located in Florida and Texas (Figure 2.72a). The other primary symptom, macroalgae abundance, was high in only three estuaries and moderate in four; however, 24 estuaries had insufficient data for assessment (Figure 2.72b). Of the secondary symptoms, significant dissolved oxygen problems were reported in only two estuaries (Perdido Bay and the Mississippi Plume, Figure 2.72c). Five estuaries had moderate nuisance/toxic bloom expressions and 11 were rated as low (Figure 2.72d). All 11 assessed estuaries exhibited low-level loss of submerged aquatic vegetation (Figure 2.72e).

Based on comparisons of the 1999 and 2007 assessments, conditions were worse in one estuary and improved in another. Worsening conditions in Perdido Bay were caused by decreases in dissolved oxygen concentrations (Figure 2.73). In Mobile Bay, improved dissolved oxygen concentrations and fewer nuisance/toxic blooms were noted. For 16 estuaries, assessments were made in 1999 and 2004 but the indicators used were not comparable between assessments. Of the 38 Gulf of Mexico estuaries studied, 13 were predicted to develop worsening conditions, eight to a high degree and five to a lesser degree (Figure 2.73). For Tampa Bay, which had experienced regrowth and gains in the spatial coverage of submerged aquatic vegetation, the conditions were expected to remain the same due to management strategies to compensate for expected increases in nutrient loads from population growth. For Charlotte Harbor, the prediction of worsening conditions was due to land use changes from low to high intensity usage (e.g., rangeland to row crops or urban). Other factors potentially influencing future changes were urban runoff, wastewater treatment, industry, atmospheric deposition, animal operations (Sabine Lake), and agriculture activities (crops and rangeland or pasture). There were no estuaries for which conditions were expected to improve. Future conditions for 23 estuaries were unknown, making it difficult to draw overall conclusions about the region; however, many of the estuaries were expected to experience worsening eutrophication. In 2007, the future outlook was the same as it was in the early 1990s with worsening conditions predicted in all estuaries for which data were available. For 10 estuaries where evaluations were possible, 1999 predictions for 2020 were already realized in 2007, only 8 years later.

Galveston Bay water quality was monitored for a number of years at a finer spatial scale than the assessments described above to detect trends with time (USEPA 2006) (Figure 2.74). Indicators for monitoring water quality conditions in the estuary included dissolved oxygen, nitrogen (e.g., nitrate, nitrite, ammonia), total phosphorus, and chlorophyll *a* concentrations; TSS/turbidity; salinity; water temperature; pH; pathogens (e.g., Enterococci, fecal coliform); biochemical oxygen demand (BOD); and total organic carbon (TOC). Declines in annual average ammonia levels were observed in several areas of Galveston Bay with the most dramatic decline in the Houston Ship Channel. For the most part, annual average concentrations were below screening levels. Nitrate-nitrite concentrations were highest in the Houston Ship Channel which demonstrated an increasing trend from about 0 mg/L in 1969 to 1.75 mg/L in 2001. The Intracoastal Waterway East exhibited a significant decline in nitrate-nitrite, and the Trinity River had a significant decline in phosphorus (since 1969). None of the five sub-bays of Galveston Bay showed trends exceeding the estuarine screening levels for nutrients (Lester and Gonzalez 2003). Annual average concentrations of chlorophyll *a* had declined across all Galveston Bay sub-bays and tributaries since 1969, with the largest decreasing trend in chlorophyll *a* concentrations found in the Houston Ship Channel, San Jacinto River, and

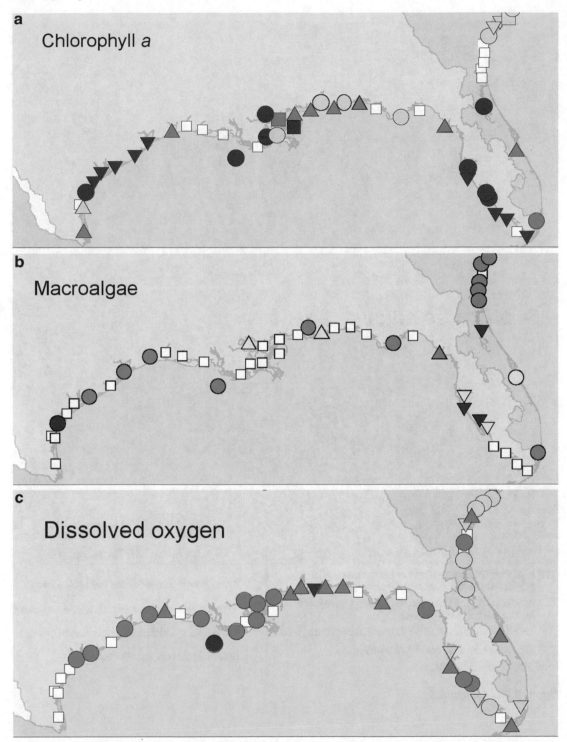

Figure 2.72. Expression of eutrophication symptoms: (a) chlorophyll *a*, (b) macroalgae, (c) dissolved oxygen, (d) nuisance/toxic algal blooms, and (e) submerged aquatic vegetation for Gulf of Mexico estuaries in 2007 (modified from Bricker et al. 2007).

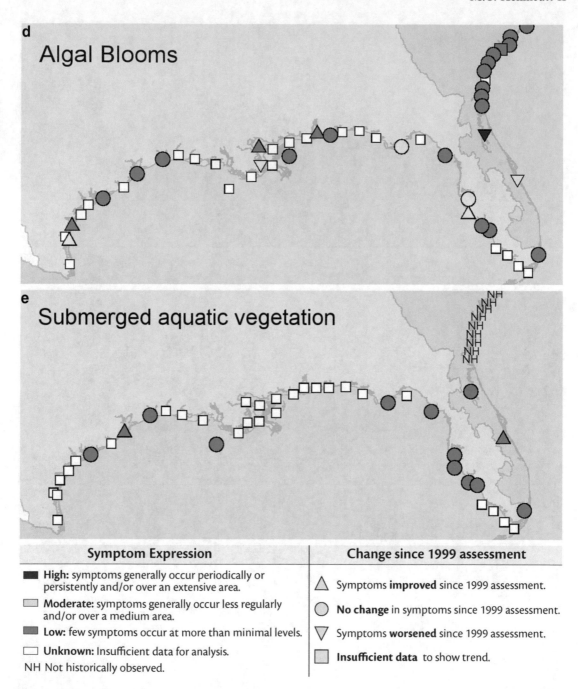

Figure 2.72. (continued)

Figure 2.73. Gulf of Mexico future outlook in 2004 and compared to the 1999 future outlook (modified from Bricker et al. 2007; *SAV* submerged aquatic vegetation).

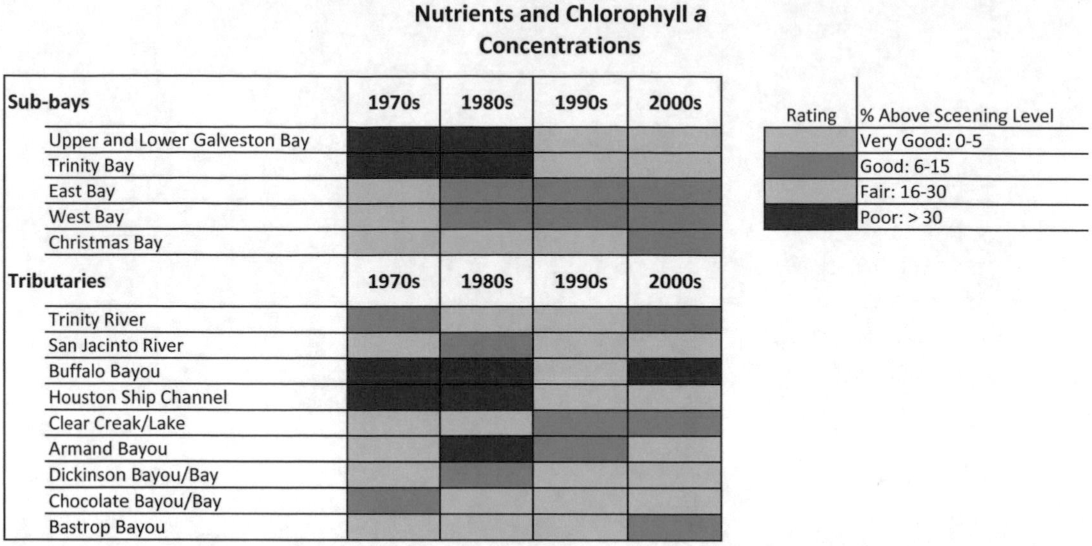

**Nutrients and Chlorophyll *a*
Concentrations**

Sub-bays	1970s	1980s	1990s	2000s
Upper and Lower Galveston Bay				
Trinity Bay				
East Bay				
West Bay				
Christmas Bay				

Rating	% Above Sceening Level
	Very Good: 0-5
	Good: 6-15
	Fair: 16-30
	Poor: > 30

Tributaries	1970s	1980s	1990s	2000s
Trinity River				
San Jacinto River				
Buffalo Bayou				
Houston Ship Channel				
Clear Creak/Lake				
Armand Bayou				
Dickinson Bayou/Bay				
Chocolate Bayou/Bay				
Bastrop Bayou				

Figure 2.74. Texas Commission for Environmental Quality (TCEQ) water quality ratings for Galveston Bay nutrients and chlorophyll *a* concentrations (modified from Lester and Gonzalez 2005).

Texas City Ship Channel. Monthly average concentrations of chlorophyll *a* did not show a trend in any of the five sub-bays in Galveston Bay. Survey data collected in 2000 and 2001 for the West Bay region averages were similar to previous Texas Commission for Environmental Quality (TCEQ) data, but chlorophyll *a* concentrations were slightly higher (Lester and Gonzalez 2003). Sub-bays were rated as moderate to good for the period 1990–2003, as compared to poor ratings for 2000–2001, though rating criteria varied among studies (Lester and Gonzalez 2005). Nutrients in Galveston Bay proper remained fairly constant during the year; however, nutrient concentrations in Galveston Bay tributaries were highest in the summer months. Overall, water quality was seen as improving in Galveston Bay since the 1970s (Lester and Gonzalez 2005). TSS showed declines in annual average concentrations across all sub-bays and tributaries of the Galveston Bay system, with the exception of Upper Galveston Bay, Lower Galveston Bay, and Cedar Bayou (Lester and Gonzalez 2003). Galveston Bay is naturally turbid because of its shallow depth and fine sediments. However, dredging activities, commercial fisheries, and natural and man-made erosion enhance natural turbidity.

Pathogens monitored in Galveston Bay included Enterococci, *E. coli*, and fecal coliform. According to the 2005 Galveston Bay Indicators Project, the areas of Galveston Bay with the greatest number of TCEQ criteria-level exceedances for fecal coliform bacteria were Buffalo Bayou, the Houston Ship Channel, Clear Creek, and Dickinson Bayou (Figure 2.75). A decline in fecal coliform was found in the East Intracoastal Waterway area but the other four major subareas of the bay did not show a trend in fecal coliform counts. The areas with the highest concentrations of Enterococci were the Houston Ship Channel, East Intracoastal Waterway, San Jacinto River, and Trinity Bay, whereas areas with the lowest concentrations were Galveston Channel, Texas City Channel, Christmas Bay, Bastrop Bayou Complex, Dickinson Bayou/ Dickinson Bay, and East Bay (Lester and Gonzalez 2003). In Galveston Bay, sediments, metals, and organic contaminants appeared to follow the same general spatial distribution, as do most other water quality parameters. Elevated concentrations of contaminants occurred in regions of runoff, freshwater inflow, and waste discharges, and lower, relatively uniform concentrations occur in the open bay. The upper Houston Ship Channel was generally the location of maximum concentrations of contaminants (Lester and Gonzalez 2005).

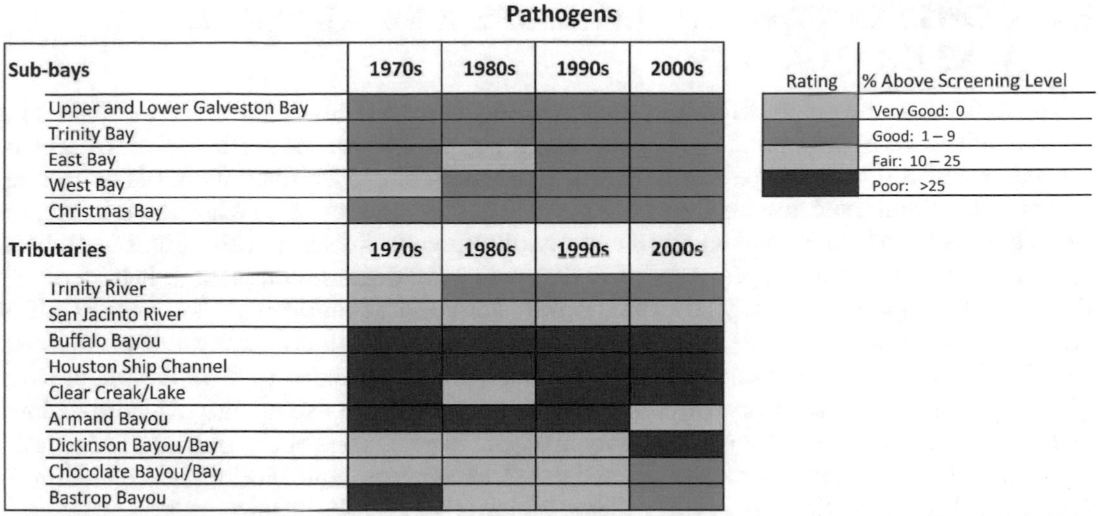

Pathogens

Sub-bays	1970s	1980s	1990s	2000s
Upper and Lower Galveston Bay				
Trinity Bay				
East Bay				
West Bay				
Christmas Bay				

Tributaries	1970s	1980s	1990s	2000s
Trinity River				
San Jacinto River				
Buffalo Bayou				
Houston Ship Channel				
Clear Creak/Lake				
Armand Bayou				
Dickinson Bayou/Bay				
Chocolate Bayou/Bay				
Bastrop Bayou				

Rating	% Above Screening Level
	Very Good: 0
	Good: 1 – 9
	Fair: 10 – 25
	Poor: >25

Figure 2.75. Texas Commission for Environmental Quality (TCEQ) water quality ratings for Galveston Bay pathogens (modified from Lester and Gonzalez 2005).

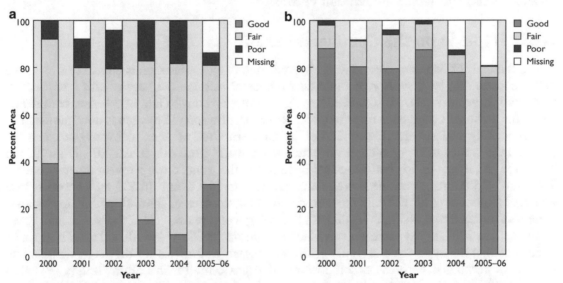

Figure 2.76. Percent area of Gulf Coast coastal waters in good, fair, poor, or missing categories for (a) water quality index and (b) DIN measured from 2000 to 2006 (modified form USEPA 2012).

Finally, 7 years of monitoring data (2000–2006) from Gulf Coast coastal waters was used to investigate temporal changes in water quality (National Coastal Conditions Reports II, III, and IV) (USEPA 2004, 2008, 2012). Interannual variation was evaluated by comparing annual estimates of percent area in poor condition for each indicator, and the associated standard error and trends in the percent area in poor condition for each indicator were evaluated using the Mann-Kendall test (USEPA 2012). The water quality index and its component indicators showed no significant linear trend over time in the percent area rated in poor condition (Figure 2.76).

2.4 CONTINENTAL SHELF/SLOPE AND ABYSSAL WATER QUALITY

In contrast to the record of monitoring programs in coastal environments, data concerning water quality on the continental shelf/slope and the abyssal deep of the northern Gulf of Mexico are sparse with a few notable exceptions. This is primarily due to the majority of offshore areas being remote from most human activities known to affect water quality. While these influences are often concentrated in coastal areas and rapidly lessen in intensity with distance offshore, human activities and natural processes have the potential to degrade continental shelf/slope and abyssal water quality. For many years the northwestern/central continental shelf of the Gulf of Mexico has been experiencing intermittent hypoxic events, commonly known as *dead zones*, associated with nutrient enrichment delivered to the Gulf of Mexico by the Mississippi River system. Atmospheric deposition of pollutants from the coast can extend into offshore regions. The most widespread anthropogenic activity in the offshore regions of the Gulf of Mexico is the exploration for, and the extraction of, oil and gas. A large percentage of oil and gas platforms in the Gulf of Mexico are located in the offshore regions (Figure 2.45). Transportation activities in the offshore area include commercial ship traffic both transiting and supplying platforms, a maze of petroleum pipelines to offshore facilities, commercial fishing fleets, and recreational boating. The offshore regions of the Gulf of Mexico are also the locations of most of the natural oil and gas seepage in the northern Gulf of Mexico.

2.4.1 Hypoxia on the Continental Shelf

In the Gulf of Mexico, coastal water hypoxia due to eutrophication is generally a localized occurrence within bays with vulnerable environmental settings (i.e., areas with low flushing rates and large inflows). However, along the northwest/central Gulf of Mexico continental shelf, the seasonal occurrence of waters with low concentrations of oxygen is now known to be geographically widespread (Figure 2.77). The northern Gulf of Mexico hypoxic zone is the second largest area of oxygen-depleted waters in the world (Rabalais et al. 2002). From 1985 to 1992, the areal extent of bottom-water hypoxia in the zone during midsummer averaged 7,770 km^2 (3,000 mi^2), and the average area doubled to 16,835 km^2 (6,500 mi^2) between 1993 and 1997 (Rabalais et al. 1999). In the summer of 2000, the area of the Gulf of Mexico hypoxic zone was reduced to 4,403 km^2 (1,700 mi^2) following a severe drought in the Mississippi River watershed. In 2002, the hypoxic zone had increased in size to 22,015 km^2 (8,500 mi^2). It has been suggested that the hypoxic zone results from water column stratification driven by weather and river flow combined with the decomposition of organic matter in bottom waters (Rabalais et al. 2002). River-borne organic matter along with the nutrients needed for phytoplankton growth enter the Gulf of Mexico via Mississippi River system discharge. Annual variability in the area of the hypoxic zone has been related to the rate of outflow of the Mississippi and Atchafalaya rivers, which is controlled by precipitation patterns that influence riverine discharge rates. The record of algal production preserved in sediment cores from the hypoxic zone show that algal production during the first half of the twentieth century in the Gulf of Mexico shelf was significantly lower, suggesting that anthropogenic changes to the basin and its discharges have increased the frequency and intensity of hypoxic events (CENR 2000; USEPA 2004). Since 1980, the basin's annual riverine discharge to the Louisiana shelf was estimated to be approximately 1.8 million metric tons (2 million tons) of nitrogen/year. It has been estimated that total nitrate-nitrogen flux tripled from the 1960s and 1970s to the 1980s and 1990s. More than half of this flux comes from non-point sources from the drainage of agricultural lands north of the confluence of the Ohio and Mississippi Rivers (CENR 2000). Gulf of Mexico continental shelf ecosystems and fisheries are affected by the hypoxia, with

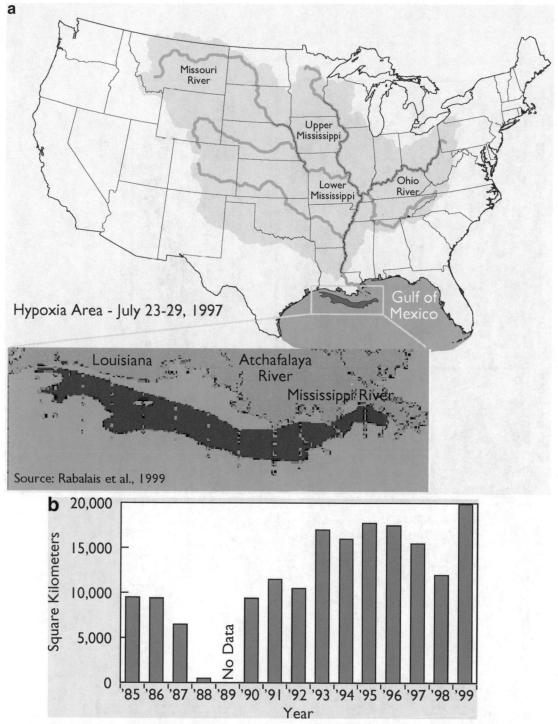

Figure 2.77. Hypoxic zone's (a) extent in 1997; (b) areal extent from 1985 to 1999; (c) spatial extent during July 1999, 2000, 2001; and (d) spatial extent of the Gulf Coast in July 2000, 2001, and 2002 (modified from USEPA 2001, 2004, 2008).

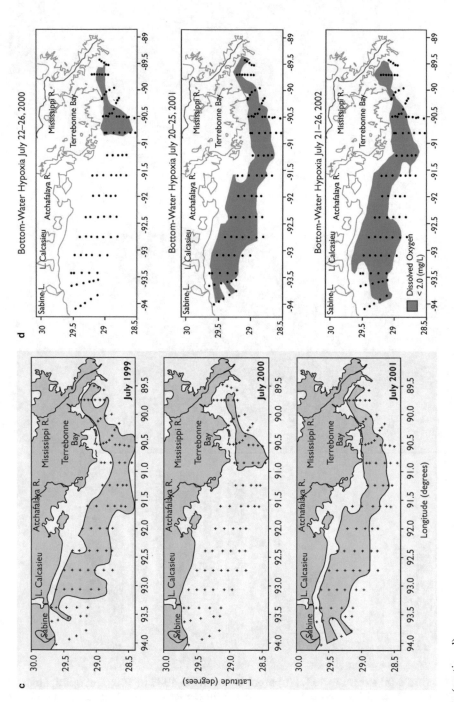

Figure 2.77. (continued)

mobile organisms trying to physically avoid the hypoxic zone. These hypoxic events are the most widespread example of degraded water quality in the offshore regions of the northern Gulf of Mexico.

2.4.2 Continental Shelf/Slope and Abyssal Water Quality and Contaminants

Contaminants have the potential to affect water quality in continental shelf/slope and abyssal environments but that potential is limited. In most instances, contaminants originate on land, in coastal estuaries, and/or are delivered to the coast in the inflows of river systems and runoff. In most instances contaminant concentrations tend to rapidly decrease with distance offshore. The major exception to this generality is petroleum contamination. In the continental shelf/slope and abyssal regions of the northern Gulf of Mexico, most petroleum contamination has been introduced by natural processes (i.e., oil and gas seepage). The vast majority of chemical contaminants, other than petroleum, are found in coastal areas where human activities are concentrated. However, contaminants can be introduced directly to the offshore by atmospheric deposition (e.g., mercury), disposal of drill muds and cuttings (e.g., petroleum and a suite of metals, mostly barium from drilling muds), discharge of produced waters (e.g., petroleum and trace amounts of metals), and the use and disposal of chemicals on oil and gas platforms and ships (e.g., local use of pesticides). On occasion, contaminants in coastal areas can persist and be transported to more distant offshore locations by ocean currents. Based on these considerations, expectations are that if contaminants other than petroleum are present in continental shelf/slope and abyssal waters, the concentrations in water would be exceedingly low and have little or no implications for offshore water quality. Other than the monitoring of contamination-associated discharges of drill cuttings and produced water at oil and gas platforms, few studies have measured chemical contaminants in offshore, northern Gulf of Mexico environments. On occasion, contaminants have been detected in sediments and biological tissues within a few hundred meters of oil and gas platforms. For petroleum contaminants the situation is quite different.

As previously noted, the most comprehensive and recent report on the sources and annual mass loadings of petroleum to U.S. marine environments is NRC's *Oil in the Sea III*: *Inputs, Fates, and Effects* report (NRC 2003). Those aspects of the NRC report relevant to understanding the impact of petroleum contamination on water quality have been provided in the introduction to this chapter and during consideration of petroleum contamination in coastal areas (Sections 2.2.5 and 2.3.7). The following assessments of petroleum in continental shelf/slope and abyssal environments are based on the NRC summary of data for 1990–1999. As before, the 9-year averages are considered representative of longer-term trends, and the loadings estimated in the NRC (2003) report for various sources of releases are expected to, and do, vary with time. The mass loadings of average annual petroleum inputs to the offshore Gulf of Mexico for 1990–1999 are summarized in Table 2.8 (NRC 2003). The conclusions reached in the following assessment of petroleum contamination in the continental shelf/slope and abyssal waters are subject to the limitations discussed in the introduction (e.g., mass loadings reflect the intensity and location of petroleum usage but do not directly indicate biological or ecological impact). Petroleum contamination has rarely been identified as a primary cause of the degradation of continental shelf/slope and abyssal water quality except in instances such as a major oil spill.

The Gulf of Mexico is prolific in oil and gas provinces and has been the site of exploration and extraction activities for many decades (Figures 2.45 and 2.46, Section 2.3.7). Current oil and gas exploration and production is concentrated in the deep water of the Gulf of

Table 2.8. Average Annual Mass Loadings of Petroleum (tonnes) to the Offshore Gulf of Mexico, 1990–1999 (modified from NRC 2003) (1 tonne = 1 metric ton(ne) = 1.102 U.S. short tons; 1 giga-tonne = 1 billion tonnes)

Zone (offshore)	North Central/ Northeastern	North Central/ Northwestern	South Central/ Southwestern
Sum seeps[a]	*70,000*	*70,000*	*na[a]*
Platforms	Trace[b]	50	61[c]
Atmospheric	Trace	60	40
Produced	Trace	1,700	130
Sum extraction	*Trace*	*1,800*	*231*
Pipelines	Trace	60	nd[d]
Tank vessel	10	1,500	nd[d]
Atmospheric	Trace	Trace	Trace
Sum transportation	*10*	*2,400*	*90*
Land-based[e]	na	na	na
Recreational vessels[f]	na	na	na
Vessels > 100 gigatonnes (spills)	70	120	Trace
Vessels > 100 gigatonnes (op discharge)	Trace	25	Trace
Vessels < 100 gigatonnes (op discharge)	Trace	Trace	Trace
Atmospheric	1,600	1,200	3,600
Aircraft[g]	80	80	20
Sum consumption	*1,800*	*1,400*	*3,600*

[a]No known seeps in these regions
[b]Estimated loads of less than 10 tonnes per year reported as "trace"
[c]Lack of precise locations for platforms in this zone precluded determining whether spills or other releases occurred less than 3 mi from shore, thus all values for this zone reported as "offshore"
[d]No information on the existence of coastal facilities was available for this region
[e]Land-based inputs are defined in this study as being limited to the coastal zone
[f]Recreational vessels are defined as being limited to operation with 3 mi of the coast
[g]Purposeful jettisoning of fuel not allowed within 3 mi of land

Mexico. Activities associated with the extraction, transportation, and consumption of petroleum have the potential to release petroleum to offshore water environments (Section 2.3.7.2 and Figures 2.49 and 2.50).

Petroleum inputs to the offshore Gulf of Mexico have a very different mix of sources and annual loadings when compared to coastal waters (Figure 2.78) (NRC 2003). In the offshore region, annual mass loadings of petroleum from natural oil and gas seeps were estimated to be 70,000 tonnes (77,162 tons) each for the northwestern and northeastern (almost all offshore Louisiana) offshore Gulf of Mexico in the 1990s (Table 2.8). Oil and gas seepage has been a feature of the Gulf of Mexico for thousands if not tens of thousands of years, so these estimates are not subject to the temporal fluctuations that are expected for anthropogenic releases of petroleum. The major uncertainties in petroleum loadings to the Gulf of Mexico are the accuracies of the methods used to make estimates. These estimates can have quite large uncertainties and vary depending on the estimation method. One single source contributed approximately 95 % of the petroleum input to the offshore northern Gulf of Mexico during the 1990s. Since most oil and gas platforms are located offshore Texas and Louisiana, releases related to extraction facilities were negligible in the northeastern Gulf of Mexico while 1,800

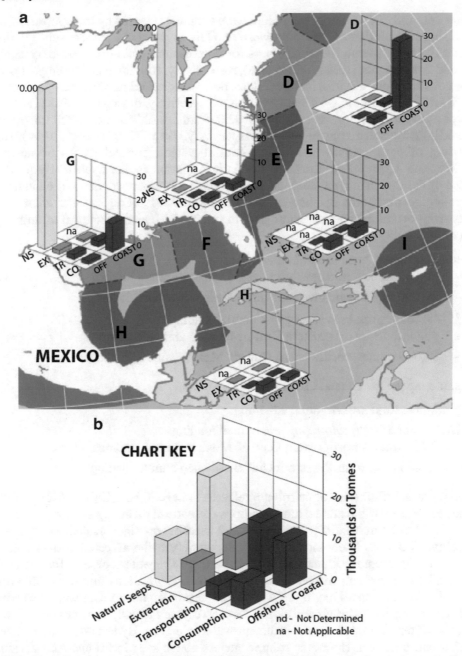

Figure 2.78. Variation in average annual input (thousands of tonnes) of petroleum to the marine environment in the Gulf of Mexico from 1990 to 1999 (*yellow* natural seeps, *green* extraction, *purple* transportation, *red* consumption) (modified from NRC 2003).

tonnes (1,984 tons) annually entered the northwestern Gulf of Mexico during the 1990s. Almost all of this petroleum release came from produced water discharges (Table 2.8). For comparison, the same inputs from extraction activities and produced water discharges were negligible amounts with 680 tonnes (750 tons) of petroleum released to northeastern and northwestern coastal waters combined during the same time period. Similarly, since most platforms and shore-based refineries and chemical complexes are in the northwestern Gulf of Mexico, 10 and

1,600 tonnes (11 and 1,764 tons) of petroleum were annually released by transportation activities to the offshore northeastern and northwestern Gulf of Mexico, respectively, in the 1990s (Table 2.8). For comparison, the same inputs for northeastern and northwestern coastal waters were 160 and 2,400 tonnes (176 and 2,646 tons), respectively. Annual mass loadings of petroleum related to consumption activities in the offshore northeastern and northwestern Gulf of Mexico were 1,800 and 1,400 tonnes (1,984 and 1,543 tons), respectively, during the 1990s (Table 2.8). For comparison, the same inputs were 2,500 and 12,000 tonnes (2,756 and 13,228 tons) for northeastern and northwestern coastal waters, respectively, from 1990 to 1999. This reflects the concentration of consumption activities in coastal waters, particularly in the northwestern Gulf of Mexico. A graphical summary of this information is displayed in Figure 2.78. The dominance of natural oil and gas seepage as a source of petroleum contamination in the Gulf of Mexico in general and in the offshore as compared to the coastal regions is evident.

Comparing overall petroleum loadings in the Gulf of Mexico as natural or anthropogenic annual loadings in the 1990s:

- 140,000 tonnes total *natural* annual loadings
  - 70,000 tonnes northeastern Gulf of Mexico annual loadings
  - 70,000 tonnes northwestern Gulf of Mexico annual loadings
- 25,400 tonnes total *anthropogenic* annual loadings
  - 4,400 tonnes northeastern Gulf of Mexico annual loadings
  - 21,000 tonnes northwestern Gulf of Mexico annual loadings

The same inputs for coastal Gulf of Mexico waters:

- Negligible total *natural* annual loadings
- 17,740 tonnes total *anthropogenic* annual loadings
  - 2,660 tonnes northeastern Gulf of Mexico annual loadings
  - 15,080 tonnes northwestern Gulf of Mexico annual loadings

Based on these summaries of petroleum releases to the offshore Gulf of Mexico during the 1990s, the magnitude of the annual mass loadings for natural oil and gas seepage suggests that this source of petroleum has the greatest potential to affect continental shelf/slope and abyssal water quality. The most likely indicator of water quality to be affected is dissolved oxygen concentrations. As seeping oil and gas transits through the water column, the water directly above oil and gas seeps can exhibit lowered oxygen concentrations due to aerobic microbial oxidation of hydrocarbons. Due to the well-mixed nature of the bottom waters overlying the Gulf of Mexico continental shelf and slope, these effects are usually restricted to a few meters or less of the water column above the sediment/water interface. Hydrocarbon gases often form plumes that can persist in the water column meters above seep locations. At individual seep sites, degradation of water quality appears to be spatially limited and ephemeral. In the offshore regions oxygen-rich deep waters from the Atlantic Ocean flow into the Gulf of Mexico from the Caribbean Sea with the major outflow being the Florida Straits (Jochens et al. 2005). The sources of dissolved oxygen in the upper waters (approximately 100–200 m [328–656 ft]) of the Gulf of Mexico are the atmosphere and photosynthesis, with wind and wave action controlling air-sea gas exchange. The depth to which photosynthesis occurs in the upper layers of the Gulf of Mexico depends on light penetration and nutrient concentrations. The source of dissolved oxygen in the deep waters of the Gulf of Mexico is the transport and mixing of oxygen-rich water from the Caribbean Sea delivered by currents via the Yucatán Channel. Deep oceanic circulation and the associated mixing are the only processes that replenish deepwater oxygen. The major sink for oxygen in the Gulf of Mexico, as in the world's oceans, is oxidation of organic matter. Organic matter consists of living organisms, detritus from living organisms (fecal pellets, secretions, dead organisms, etc.), continental detritus washed into the ocean via river runoff and in the Gulf of Mexico, petroleum. An oxygen minimum zone occurs in the Gulf of Mexico between 300 and 700 m (985 and 2,297 ft) due to the depletion of dissolved oxygen by processes occurring outside of the Gulf of Mexico and the decay of organic matter within Gulf of Mexico sediments and waters. The productivity of the Gulf of Mexico is not high enough to create extreme oxygen minimum zones as observed in other locations in the world's oceans. Other than the continental shelf hypoxia zones discussed above, dissolved oxygen concentrations indicate good water quality for continental shelf/slope and abyssal waters in the Gulf of Mexico. The impact of oil and gas seeps on dissolved oxygen concentrations was found to be negligible in the deep water of the Gulf of Mexico while localized effects might be measurable (Jochens et al. 2005). Although natural seepage of oil and gas into the Gulf of Mexico has been occurring for thousands of years, continental shelf/slope and abyssal water dissolved oxygen concentrations show no significant perturbations attributable to the presence of petroleum from natural seeps. As described above, localized low oxygen conditions have been reported in close proximity to the sediment/water interface at seep sites.

The other source of petroleum contamination to the offshore Gulf of Mexico that has the potential to affect continental shelf/slope and abyssal water quality is the massive volumes of discharged production waters from the many oil and gas platforms. The discharge of produced waters into the offshore waters of Louisiana and Texas is extensive (Figure 2.79). Estimates of produced waters discharged into outer continental shelf (OCS) waters of the northwestern Gulf were approximately 500×10^6 barrels per year (bbl/year) (21×10^9 gallons per year [gal/year]) with the majority of discharges occurring offshore of Louisiana (Rabalais et al. 1991). A more recent estimate (NRC 2003) indicated approximately 500×10^6 bbl/year (21×10^9 gal/year) for the OCS across the Gulf with an additional approximately 200×10^6 bbl/year (8.4×10^9 gal/year) for Louisiana territorial waters and approximately 4×10^6 bbl/year (167×10^6 gal/year) for Texas territorial waters for a total for the Gulf of Mexico of approximately 660×10^6 bbl/year (27.7×10^9 gal/year) of discharged produced waters. The offshore total volumes from the two estimates are similar. The amount of produced water generated increases as oil or gas fields are depleted and may be as high as 95 % of the product stream in older fields such as those offshore of Louisiana and Texas. A study directed at estimating the contribution of platform discharges on the hypoxic zone gives insight into the contribution of these point sources of pollution to the overall quality of continental shelf/slope and abyssal waters (Rabalais 2005). Organic carbon in produced waters has the potential to be degraded by aerobic microbes reducing dissolved oxygen concentrations. Nitrogen, mostly in the form of ammonium, has the potential to stimulate phytoplankton production some of which may be decomposed contributing to respiratory demand for oxygen. The amounts of organic carbon and ammonium (labile nitrogen) in produced water discharges were compared to those delivered by the Mississippi and Atchafalaya rivers. It was estimated that the contribution of carbon and nitrogen found in produced water discharges were minimal compared to riverine inputs (0.013 % of the total nitrogen delivered by the Mississippi River system, 0.008 % of the total DIN, and 0.002 % of the total ammonium at the time of the study). Petroleum discharged in production waters, measured as oil and grease, was minor compared to the Mississippi River input. The produced water contribution of organic carbon to the Gulf of Mexico hypoxic area was judged to be insignificant. Over the years discharges from platforms have been regulated and reduced, lowering the potential for degrading water quality even further. The USEPA Best Available Treatment Technology Economically Achievable for the National Pollutant Discharge Elimination System (NPDES) permit restricts the concentration of petroleum, measured as oil and grease, in produced water destined for ocean disposal to a monthly average of 29 mg/L (USEPA 1993). Produced waters must also meet toxicity criteria before discharge is allowed.

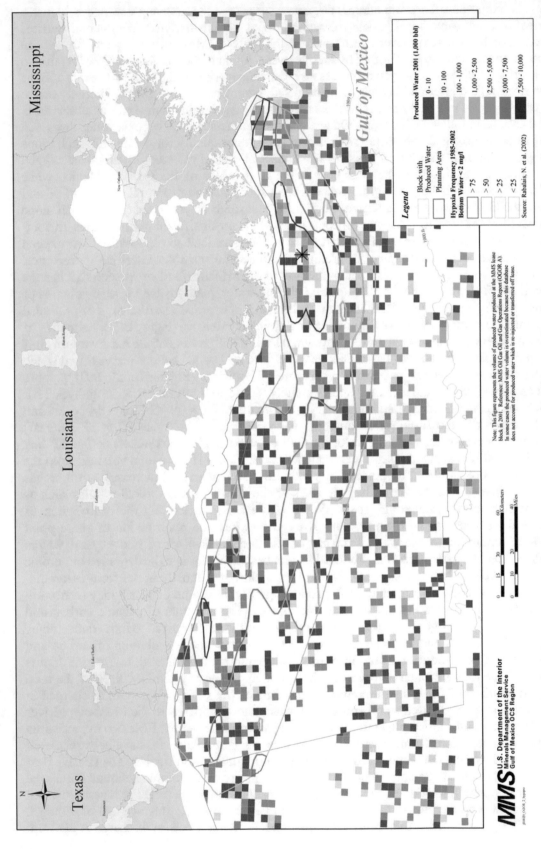

Figure 2.79. Distribution of produced water discharge volumes by OCS lease block for a portion of the Central and Western OCS superimposed on the frequency of midsummer, bottom-water hypoxia (modified from a map generated by the Minerals Management Service with input from Rabelais, 2005 for the distribution of hypoxia frequency).

The conclusion is that produced water discharges have minimal impact on water quality in the offshore regions of the Gulf of Mexico, and any effects that might be observed would be localized at discharge points and ephemeral.

As previously noted, non-petroleum contaminants also have the potential to degrade water quality; however, most monitoring programs only measure non-petroleum contaminants in sediments and biological tissues. Little information is available on the ambient concentrations of these chemicals in offshore waters though they are expected to be low. The concentrations of these chemicals in coastal waters are, in most instances, below the detection limits of standard analytical protocols, and it is reasonable to assume that concentrations in offshore water would be even lower. In several studies, contaminant concentrations in organism tissues collected close to offshore platforms not only contained no detectable petroleum, they also contained no detectable non-petroleum contaminants. Most non-petroleum contaminants result from chronic use of chemicals on land or the adjacent coastal areas. These influences are rapidly diminished seaward of source areas in coastal regions. The distance from the release point and the expected dilution with uncontaminated offshore waters further offshore suggest that non-petroleum contaminants do not degrade continental shelf/slope and abyssal water quality. However, some contaminants, such as those transported long distances by atmospheric (e.g., mercury) or oceanic processes and those contaminants that bioaccumulate and biomagnify, may be found in offshore marine organisms and sediments.

Continental shelf/slope and abyssal waters in the Gulf of Mexico are subjected to a variable mix of inputs that have the potential to degrade water quality. However, in most instances, no significant degradation of water quality has been observed, with one major exception—the input of nutrients from the Mississippi River system degrading water quality on the northwestern/central continental shelf. In offshore areas, natural oil and gas seeps are by far the dominant sources of petroleum loadings, but evidence is lacking that this has resulted in significant degradation of offshore water quality. The largest offshore source of anthropogenic petroleum contamination in the eastern Gulf of Mexico is spills from tank vessels. In the western Gulf of Mexico, it is produced water discharges, but the loading of petroleum from natural oil and gas seepage dwarfs these inputs. Petroleum inputs from activities associated with extraction, transportation, and consumption are chronic but low and widely geographically dispersed. These petroleum releases most often occur at the sea surface, which suggests that ambient water concentrations rapidly decrease due to dilution with uncontaminated waters. These factors account for a lack of observations of degraded water quality on the continental shelf/slope and abyss of the Gulf of Mexico. Similarly, waterborne biological contaminants (pathogens) are discharged almost exclusively in coastal areas (the exceptions being ship and platform sewage disposal). The viability of pathogens in seawater is limited, which reduces the possibility of long distance transport. Also, the effects of biological contaminants on water quality in deeper water regions of the northern Gulf of Mexico are expected to be negligible.

2.5 SUMMARY

The patterns and trends in water quality in the Gulf of Mexico are complex and variable in space and time. Assessments performed over more than two decades have concluded that water quality in a majority of estuaries and coastal environments along the northern Gulf of Mexico coast is highly influenced by human activities. One of the most prevalent causes of degraded water quality in the coastal areas of the Gulf of Mexico is excessive levels of anthropogenic nutrients that create widespread coastal eutrophication. Eutrophication lowers dissolved oxygen concentrations, increases chlorophyll *a* concentrations, diminishes water clarity, and can lead to toxic/nuisance algal blooms and loss of submerged aquatic vegetation. While variable over time, overall ecological conditions in Gulf estuaries have been judged as fair to poor, and assessments consistently have concluded that water quality is fair. In some locations, water quality appears to be improving due to environmental regulations and controls; at other sites, conditions have deteriorated. The status of and trends in water quality are highly site specific. Many Gulf of Mexico coastal environments exhibit high levels of eutrophication. Chlorophyll *a* concentrations are high, particularly along the coasts of western Florida, Louisiana, and lower Texas. Epiphytes (a variety of organisms that grow on other plants including submerged aquatic vegetation) and macroalgal abundances are moderate to high at a number of locations. Low dissolved oxygen concentrations have been routinely observed particularly along the Florida coast and in the Mississippi River Plume. The loss of submerged aquatic vegetation is a problem in many estuaries and nuisance/toxic algal blooms are pervasive in many estuaries especially in Florida, western Louisiana and the lower Texas coast. High levels of eutrophication have resulted in increased turbidity associated with high concentrations of chlorophyll *a*, low levels of dissolved oxygen, moderate to high levels of nuisance/toxic algal blooms and epiphyte abundances, and ultimately the loss of submerged aquatic vegetation. The few improvements observed over time are attributed to better management of point and non-point sources of nutrients. The intensity of human activities correlates with high eutrophication, though in many instances, impairment of use has been difficult to directly or solely relate to eutrophication or water quality. Comparing 1999 and 2007 assessments, eutrophication conditions worsened in one system and improved in another. A trend analysis was not possible because indicators were not always comparable. In one study, 13 of the 38 Gulf of Mexico estuaries studied were predicted to develop worsening conditions in the future. Factors expected to influence future trends in water quality were control and mitigation of urban runoff, wastewater treatment, industrial expansion, atmospheric deposition, animal operations, and agriculture activities. There were no estuaries where conditions were expected to improve and worsening conditions were predicted in all systems for which data were available (Bricker et al. 1999). Trends in human population distributions, accelerating development pressures, and human-associated activities were the main factors suggesting water quality will worsen in the future.

In regard to the effect of chemical pollutants on water quality, direct measurements of pollutants dissolved in marine waters are limited. While chemical contaminants can, and probably do, make limited contributions to degraded water quality, especially in coastal areas where concentrations are highest, these impacts are masked by the overwhelmingly dominant factor that degrades water quality—eutrophication. The northwestern Gulf of Mexico experiences some of the largest average annual inputs of petroleum to North American marine waters as a result of the high volumes of tanker traffic, the large numbers of oil and gas platforms, the contaminated inflows from the Mississippi River, and the occurrence of natural oil and gas seeps. Indirect indications of possible impacts of chemical contaminants on water quality include the detection of contaminants in biological tissues and sediments. Elevated tissue concentrations of total PCBs, DDT, dieldrin, mercury, cadmium, and toxaphene have been detected in fish tissue. However, contaminants accumulate in biological tissues via pathways other than uptake from water. Fish consumption advisories due to mercury contamination have been common along the northern Gulf of Mexico, and beaches have been routinely closed or under advisories due to elevated levels of bacteria. Once outside the influence of coastal processes, water quality is good and has been good for a long time in the Gulf of Mexico. Exceptions are hypoxic zones on the shelf, waters just above natural oil and gas seeps, and localized and ephemeral effects on water quality due to the discharge of produced waters. However, continental shelf/slope and abyssal Gulf of Mexico waters remain mostly unimpaired by human activities primarily due to the relatively low levels of pollutant discharges and the dilution due to the large volume and mixing rates of receiving waters. Coastal Gulf of Mexico water quality is highly influenced by humans and will continue to be for the foreseeable future.

In large part, future trends in water quality will be dependent on the decisions made by the populations that live, recreate, and work along the northern Gulf of Mexico coast in regard to controlling and/or mitigating those factors that degrade water quality.

REFERENCES

APHA (American Public Health Association) (1992) Standard methods for the examination of water and wastewater, 18th edn. American Public Health Association, Washington, DC, USA. 1,100 p

Auburn University (2004) Citizen guide to Alabama rivers, vol 5, Escatawpa, Mobile, and Tombigbee. Alabama Water Watch Program, Auburn University, Auburn, AL, USA. 15 p

Battelle (2003) Environmental indicators in the Barataria-Terrebonne Estuary System, technical manual: Indicators report development. Prepared by Battelle for the Barataria-Terrebonne National Estuary Program, Thibodaux, LA, USA

Boesch DF, Rabalais NN (eds) (1989a) Produced waters in sensitive coastal habitats: An analysis of impacts, Central Coastal Gulf of Mexico. OCS report/MMS 89-0031. U.S. Department of the Interior, MMS (Minerals Management Service), OCS Regional Office, New Orleans, LA, USA. 157 p

Boesch DF, Rabalais NN (eds) (1989b) Environmental impact of produced water discharges in Coastal Louisiana. Louisiana Division of the Mid-Continent Oil and Gas Association, Louisiana Universities Marine Consortium, Chauvin, LA, USA. 287 p

Braun D, Neugarten R (2005) Mobile-Tensaw River Delta, Alabama: Hydrological modifications impact study. Prepared for the Nature Conservancy, Arlington, VA, USA

Bricker SB, Clement CG, Pirhalla DE, Orlando SP, Farrow DRG (1999) National estuarine eutrophication assessment: Effects of nutrient enrichment in the nation's estuaries. National Oceanic and Atmospheric Administration, National Ocean Service, Special Projects Office and the National Centers for Coastal Ocean Science, Silver Spring, MD, USA. 328 p

Bricker SB, Longstaff B, Dennison W, Jones A, Boicourt K, Wicks C, Woerner J (2007) Effects of nutrient enrichment in the nation's estuaries: A decade of change. NOAA coastal ocean program decision analysis series 26. National Centers for Coastal Ocean Science, Silver Spring, MD, USA. 328 p

BTNEP (Barataria-Terrebonne National Estuary Program) (2002) Healthy estuary, healthy economy, healthy communities: Environmental indicators in the Barataria-Terrebonne Estuary System. BTNEP, Thibodaux, LA, USA. 31 p

Caffey RH, Breaux JB (2000) Portrait of an estuary: Functions and values of the Barataria-Terrebonne Estuary System. Louisiana Agricultural Center and Barataria-Terrebonne National Estuary Program, Thibodaux, LA, USA. 18 p

CBBEP (Coastal Bend Bays and Estuaries Program) (1998) Coastal bend bays plan to conserve and manage the coastal bend Bays of South Texas. Publication CCBNEP-1. Texas Natural Resource Commission, Austin, TX, USA. 80 p

CBBEP (2005) Online information. Coastal Bend Bays and Estuaries Program, Corpus Christi, TX, USA. http://www.cbbep.org. Accessed 18 Feb 2013

CENR (Committee on the Environment and Natural Resources) (2000) Integrated assessment of hypoxia in the Northern Gulf of Mexico. National Science and Technology Council Committee on Environment and Natural Resources, Washington, DC, USA. 58 p

Dixon LK (2003) Water quality trend analyses of the saline waters of Southwest Florida: 1960–2002. Mote Marine Laboratory technical report 919. Sarasota Bay National Estuary Program (SBNEP), Sarasota, FL, USA. 267 p

Energy Information Administration (2009) Map of gas production in offshore fields, lower 48 states. http://en.wikipedia.org/wiki/Offshore_oil_and_gas_in_the_US_Gulf_of_Mexico (original figure at http://www.eia.gov/oil_gas/rpd/offshore_gas.pdf). Accessed 21 June 2015

FDEP (Florida Department of Environmental Protection) (2005) Florida Department of Environmental Protection online information. Florida Department of Environmental Protection, Tallahassee, FL, USA. http://www.dep.state.fl.us. Accessed Aug 2005

Florida Center for Community Design and Research (2004) Sarasota County Water Atlas. University of South Florida online information. University of South Florida, School of Architecture and Community Design, Florida Center for Community Design and Research, Tampa, FL, USA. http://www.sarasota.wateratlas.usf.edu. Accessed 18 Feb 2013

GBEP (Galveston Bay Estuary Program) (2005) Galveston Bay Estuary Program online information. Galveston Bay Estuary Program, Houston, TX, USA. http://www.gbep.state.tx.us. Accessed 18 Feb 2013

Jochens AE, Bender LC, DiMarco SF, Morse JW, Kennicutt MC II, Howard MK, Nowlin WD Jr (2005) Understanding the processes that maintain the oxygen levels in the deep Gulf of Mexico: synthesis report. OCS study MMS 2005-032. U.S. Department of the Interior, Minerals Management Service, Gulf of Mexico OCS Region, New Orleans, LA, USA. 142 p

Kimbrough KL, Johnson WE, Lauenstein GG, Christensen JD, Apeti DA (2008) An assessment of two decades of contaminant monitoring in the Nation's Coastal Zone. NOAA technical memorandum NOS NCCOS 74. Silver Spring, MD, USA. 105 p

Lester LJ, Gonzalez L (2003) Status and trends database maintenance project final report. Prepared by Geotechnology Research Institute, Houston Advanced Research Center, for the Galveston Bay Estuary Program, Webster, TX, USA

Lester LJ, Gonzalez L (2005) Galveston bay indicators project final report. Galveston Bay Estuary Program, Webster, TX, USA. 150 p. http://www.galvbaydata.org/Portals/2/projects/reports/2005/GBIndicatorsFinalReport2005.pdf. Accessed 21 June 2015

May EB (1973) Extensive oxygen depletion in Mobile Bay, Alabama. Limnol Oceanogr 18:353–366

Mobile Bay NEP (National Estuary Program) (2002) Comprehensive conservation and management plan, vol I, A call to action. Mobile Bay National Estuary Program, Mobile, AL, USA

Moore DM, Rivers RD (eds) (1996) The estuary compact: A public promise to work together to save the Barataria and Terrebonne Basins. BTNEP, Thibodaux, LA, USA. 40 p

NOAA (National Oceanic and Atmospheric Association) (1985) National estuarine inventory—data atlas, Physical and hydrologic characteristics, vol 1. U.S. Department of Commerce, Rockville, MD, USA. 103 p

NOAA (1997) NOAA's estuarine eutrophication survey, Gulf of Mexico region, vol 4. Office of Ocean Resources Conservation and Assessment, Silver Spring, MD, USA. 77 p

NOAA (2012) Office of Ocean Exploration and Research. http://oceanexplorer.noaa.gov/explorations/06mexico/background/oil/media/platform_600.html. Accessed 21 June 2015

NRC (National Research Council) (2003) Oil in the sea III: Inputs, fates, and effects. National Academy Press, Washington, DC, USA. 278 p

Pedersen MF, Borum J (1996) Nutrient control of algal growth in estuarine waters. Nutrient limitation and the importance of nitrogen requirements and nitrogen storage among phytoplankton and species of macroalgae. Mar Ecol Prog Ser 142:261–272

Pew Oceans Commission (2003) America's living oceans, charting course for sea change. May. http://www.pewtrusts.org/~/media/Assets/2003/06/02/Full_Report.pdf?la=en. Accessed 18 Feb 2013

Poe A, Hackett K, Janicki S, Pribble R, Janicki A (2005) Estimates of total nitrogen, total phosphorus, total suspended solids, and biochemical oxygen demand loadings to Tampa Bay, Florida: 1999–2003. Technical report 06-04. Tampa Bay Estuary Program, St. Petersburg, FL, USA. 111 p

Poor N, Pribble R, Greening H (2001) Direct wet and dry deposition of ammonia, nitric acid, ammonium and nitrate to the Tampa Bay Estuary, FL, USA. Atmos Environ 35:3947–3955

Port of Houston Authority (2006) The Port of Houston Authority online information. Port of Houston Authority, Houston, TX, USA. http://www.portofhouston.com. Accessed 18 Feb 2013

Rabalais NN (2005) Relative contribution of produced water discharge in the development of hypoxia. OCS study MMS 2005-044. U.S. Department of the Interior, Minerals Management Service, Gulf of Mexico OCS Region, New Orleans, LA, USA. 56 p

Rabalais NN, McKee BA, Reed DJ, Means JC (1991) Fate and effects of nearshore discharges of OCS produced waters, vol II. Technical report. OCS study/MMS 91-0005. U.S. Department of the Interior, MMS, Gulf of Mexico OCS Regional Office, New Orleans, LA, USA. 337 p

Rabalais NN, McKee BA, Reed DJ, Means JC (1992) Fate and effects of produced water discharges in coastal Louisiana, Gulf of Mexico, USA. In: Ray JP, Engelhardt FR (eds) Produced water. Plenum Press, New York, NY, USA, pp 355–369

Rabalais NN, Dortch Q, Justic D, Kilgen MB, Klerks PL, Templet PH, Turner RE, Cole B, Duet D, Beacham M, Lentz S, Parsons M, Rabalais S, Robichaux R (1995) Status and trends of eutrophication, pathogen contamination, and toxic substances in the Barataria-Terrebonne Estuarine System. BTNEP Issue 22. BTNEP, Thibodaux, LA, USA. 265 p

Rabalais NN, Turner RE, Justiç D, Dortch Q, Wiseman WJ Jr (1999) Characterization of hypoxia: Topic 1 report for the integrated assessment on hypoxia in the Gulf of Mexico. NOAA coastal ocean program decision analysis series 15. NOAA Coastal Ocean Program, Silver Spring, MD, USA

Rabalais NN, Turner RE, Wiseman WJ Jr (2002) Gulf of Mexico hypoxia, a.k.a. "the dead zone". Annu Rev Ecol Syst 33:235–263

Rose JB, Paul JH, McLaughlin MR, Harwood VJ, Farrah S, Tamplin M, Lukasik G (2001) Healthy beaches Tampa Bay: Microbiological monitoring of water quality conditions and public health impacts. Technical report 03-01. Tampa Bay Estuary Program, St. Petersburg, FL, USA

SBNEP (Sarasota Bay National Estuary Program) (2000) Sarasota Bay 2000: A decade of progress. SBNEP, Sarasota, FL, USA. 61 p

SJRWMD (St. Johns River Water Management District) (2002) Profile gives snapshot of lagoon region. Indian River Lagoon Update Winter 2002

St. Pé KM (1990) An assessment of produced water impacts to low-energy, brackish water systems in Southeast Louisiana. Louisiana Department of Environmental Quality, Water Pollution Control Division, Baton Rouge, LA, USA. 199 p

TBEP (Tampa Bay Estuary Program) (2003) Baywide environmental monitoring report, 1998–2001. Technical publication 06-02. St. Petersburg, FL, USA. 19 chapters. http://www.tampabay.wateratlas.usf.edu/upload/documents/BaywideEnvironMonitorReport98_01.pdf. Accessed 21 June 2015

Tunnell JW, Dokken QR, Smith EH, Withers K (1996) Current status and historic trends of the estuarine living resources within the Corpus Christi Bay National Estuary Program Study

Area B, vol 1. CCBNEP-06A. Coastal Bend Bays and Estuaries Program, Corpus Christi, TX, USA. 581 p

U.S. Census Bureau (1991) 1990 census of population and housing: Population and housing unit counts, United States. 1990-CPH-2-1. U.S. Census Bureau, Washington, DC, USA. http://www.census.gov/population/www/cph-1.html. Accessed 8 Mar 2015

U.S. Census Bureau (2001) Your gateway to census 2000. U.S. Census Bureau online information. U.S. Department of Commerce, U.S. Census Bureau, Washington, DC, USA. http://www.census.gov/main/www/cen2000.html. Accessed 18 Feb 2013

U.S. Commission on Ocean Policy (2004) An ocean blueprint for the 21st century. U.S. Commission on Ocean Policy, Washington, DC, USA. http://www.opc.ca.gov/webmaster/ftp/pdf/docs/Documents_Page/Reports/U.S.%20Ocean%20Comm%20Report/FinalReport.pdf.   Accessed 18 Feb 2013

USEPA (U.S. Environmental Protection Agency) (1993) 40 CFR Part 435 [FRL-4537-1] RIN 2040-AA12. Oil and gas extraction point source category, offshore subcategory; effluent limitations guidelines and new source performance standards. Fed Reg 58(41):12454–12512

USEPA (1999) Ecological condition of estuaries in the Gulf of Mexico. EPA 620-R-98-004. National Health and Environmental Effects Research Laboratory, Gulf Ecology Division, Gulf Breeze, FL, USA. 71 p

USEPA (2001) National coastal condition report. EPA/620-R-01-005. Office of Research and Development and Office of Water, Washington, DC, USA. 232 p

USEPA (2003a) BEACH watch program: 2002 swimming season. Standards and Health Protection Division, Washington, DC, USA. http://yosemite.epa.gov/water/owrccatalog.nsf/9da204a4b4406ef885256ae0007a79c7/4bd032cc53003b0685256d4e00553b0e!OpenDocument

USEPA (2003b) Bacterial water quality standards for recreational waters (freshwater and marine waters) status report. EPA-823-R-03-008. Office of Water (4305T), Washington, DC, USA. 32 p

USEPA (2004) National coastal condition report II. EPA/620-R 03-002. Office of Research and Development and Office of Water, Washington, DC, USA. 286 p

USEPA (2006) National estuary program coastal condition report. EPA-842/B-06/001. Office of Water/Office of Research and Development. Washington, DC, USA. 445 p

USEPA (2008) National coastal condition report III. EPA/842-R-08-002. Office of Research and Development and Office of Water, Washington, DC, USA. 300 p

USEPA (2012) National coastal condition report IV. EPA/842-R-10-003. Office of Research and Development and Office of Water, Washington, DC, USA. 309 p

USGS (U.S. Geological Survey) (2001) Water quality in the nation's streams and aquifers overview of selected findings, 1991–2001. U.S. Geological Survey Circular 1265. http://pubs.usgs.gov/circ/2004/1265/#pdf. Accessed 21 June 2015

Wallace RK (1994) Mobile Bay and Alabama coastal waters fact sheet. Mississippi-Alabama Publication 94-017, Circular ANR-919. Alabama Cooperative Extension Service, Auburn, AL, USA

APPENDIX A

Table A.1. Summary of Methodologies for Judging Water Quality in Various Monitoring Programs in the Gulf of Mexico (most of these descriptions are taken verbatim from the reference indicated)

Information	Details
National estuarine eutrophication assessment: effects of nutrient enrichment in the nation's estuaries (Bricker et al. 1999)	
Data sources	The assessment was based primarily on the results of the National Estuarine Eutrophication Survey, conducted by NOAA from 1992 to 1997 supplemented by information on nutrient inputs, population projections, and land use drawn from a variety of sources (full report at http://ian.umces.edu/neea/pdfs/eutro_report.pdf, accessed June 21, 2015)
Methodology	A numerical scoring system was developed to integrate information on (1) primary symptoms: decreased light availability (chlorophyll *a* concentrations and problematic epiphytic and macroalgal growth), algal dominance (diatom/dinoflagellate ratios and benthic to pelagic dominance ratios), and increased organic matter decomposition (chlorophyll *a* concentrations and problematic macroalgal growth) and (2) secondary symptoms: loss of submerged aquatic vegetation (spatial coverage and trends), harmful algae (nuisance and toxic blooms), and low dissolved oxygen (anoxia, hypoxia, and stress) to determine the overall status of eutrophic symptoms in each estuary. This scoring system was implemented in three phases according to the methods described in detail the report

First, a single index value was computed from all primary symptoms. The scoring system gave equal weight to all three symptoms and considered the spatial and temporal characteristics of each. The scores for the three symptoms were then averaged, resulting in the highest values being assigned to estuaries having multiple primary symptoms that occur with great frequency, over large spatial areas of the estuary, and for extended periods of time. Likewise, the lowest scores indicate estuaries that exhibit few, if any, characteristics of the primary symptoms

Next, a single index value was computed from all secondary symptoms. The scoring system again gave equal weight to all symptoms and their spatial and temporal characteristics. The highest score of any of the three symptoms was then chosen as the overall secondary value for the estuary. This weights the secondary symptoms higher than the primary symptoms, because the secondary symptoms take longer to develop, thereby indicating a more chronic problem, and being more indicative of actual impacts to the estuary

Finally, the range of numeric scores assigned to primary and secondary symptoms was divided into categories of high, moderate, and low. Primary and secondary scores were then compared in a matrix so that overall categories could be assigned to the estuaries

Estuaries having high scores for both primary and secondary conditions were considered to have an overall "high" level of eutrophication. Likewise, estuaries with low primary and secondary values were assigned an overall "low" level of eutrophication. Scores were then assigned to the remaining estuaries based on interpretations of each estuary's combined values |
| *National Coastal Condition Report (NCCR I) (USEPA 2001)* | |
| Data sources | Coastal monitoring data from programs like EMAP and NOAA National Status and Trends (NS&T) Assessment and advisory data provided by states or other regulatory agencies and compiled in national databases (full report at http://water.epa.gov/type/oceb/assessmonitor/nccr/downloads.cfm, accessed June 21, 2015) |

(continued)

Table A.1 (continued)

Information	Details
Methodology	Overall condition for each coastal area was calculated by summing the scores for indicators and dividing by the number of indicators, where good = 5, fair = 3, and poor = 1. Characterizing coastal area (water quality indicators water clarity and dissolved oxygen) involves two value determinations. The first value is the definition of "poor" for an indicator. The definition of poor condition for each indicator is based on existing criteria, guidelines, and/or interpretation of scientific literature. The percent areas used for each indicator are value judgments and were largely determined by informally surveying environmental managers, resource experts, and the knowledgeable public
Water clarity	EMAP-Estuaries (EMAP-E) estimates water clarity by comparing the amount and type of light reaching the water surface to the light at a depth of 1 m. Water clarity is considered poor if less than 10 % of surface light reaches 1 m. The water clarity data were collected by the EMAP-E program unless otherwise noted. This measure is used to determine water quality as follows: good—less than 10 % of the coastal waters have poor light penetration, fair—10–25 % of the coastal waters have poor light penetration, and poor—more than 25 % of the coastal waters have poor light penetration
Dissolved oxygen	Dissolved oxygen (DO) is a fundamental requirement for all estuarine life. A threshold concentration of 4–5 ppm (five parts of oxygen per million parts of water) has been used by many states to set water quality standards. Concentrations below ~2 ppm are thought to be stressful to many estuarine organisms. These low levels most often occur in bottom waters and impact the organisms that live in the sediments. Low levels of oxygen (hypoxia) or lack of oxygen (anoxia) often accompany the onset of bacterial degradation, sometimes resulting in the presence of algal scums and noxious odors. In some estuaries, low levels of oxygen, at least periodically, are part of the natural ecology. Therefore, it is difficult to interpret whether the observed effects are natural or human induced. The DO data were collected under the EMAP-E program unless otherwise noted. This indicator is used to measure water quality as follows: good—less than 5 % of the coastal waters have less than 2 ppm DO, fair—5–15 % of the coastal waters have less than 2 ppm DO, and poor—more than 15 % of the coastal waters have less than 2 ppm DO
Eutrophication index	Eutrophication due to the accelerated input of nitrogen and phosphorus can promote a complex array of symptoms such as excessive growth of algae that may lead to other problems. For its National Estuarine Eutrophication Assessment, NOAA developed a system that evaluates several symptoms of eutrophication in an estuary to provide a single categorical value to represent the status of overall eutrophic condition for each estuary (Bricker et al. 1999). This value is the measure of eutrophic condition presented in this report. The primary symptoms examined for this value are chlorophyll a, macroalgal abundance, and epiphyte abundance. Secondary symptoms include loss of submerged aquatic vegetation, harmful algae, and low dissolved oxygen. This indicator is used to measure water quality as follows: good—less than 10 % of the coastal waters have symptoms indicating a high potential for eutrophication, fair—10–20 % of the coastal waters have symptoms indicating a high potential for eutrophication, and poor—more than 20 % of the coastal waters have symptoms indicating a high potential for eutrophication

(continued)

Table A.1 (continued)

Information	Details
Designated or desired uses	The following programs maintain databases repositories for information about how well coastal waters support their designated or desired uses. These uses are important factors in public perception of the condition of the coast and also say a lot about the condition of the coast as it relates to public health *Clean Water Act Section 305(b) and 303(d) Assessments*—States report water quality assessment information and water quality impairments under Sections 305(b) and 303(d) of the Clean Water Act. Water quality standards include narrative and numeric criteria that support specific designated uses and also specify goals to prevent degradation of good quality waters. Numeric criteria are used to evaluate whether the designated uses assigned to water bodies are supported. Data is consolidated into general categories. The most common designated uses are: aquatic life support; drinking water supply; recreation (such as swimming, fishing, and boating); and fish consumption. After comparing water quality data to the criteria set by water quality standards, waters are placed into the following categories: fully supporting—these waters meet applicable water quality standards, both criteria and designated use; threatened—these waters currently meet water quality standards, but states are concerned they may degrade in the near future; partially supporting—these waters meet water quality standards most of the time, but exhibit occasional exceedances; and not supporting—these waters do not meet water quality standards *Beach Closures*—There is growing concern about public health risks posed by polluted bathing beaches. Scientific evidence has documented a rise of infectious diseases caused by microbial organisms in recreational water. A primary goal of USEPA's Beaches Environmental Assessment, Closure, and Health (BEACH) Program, established in 1997, is to work to compile information on beach pollution to define the extent of the problem. A few states have comprehensive beach monitoring programs, many other states have only limited beach monitoring programs
National Coastal Condition Report (NCCR II) (USEPA 2004)	
Data sources	This report examined data sets from different agencies and areas of the country. Three types of data were presented in this report: coastal monitoring data from programs such as USEPA's EMAP and the NCA Program, NOAA's NS&T Program, and USFWS's National Wetlands Inventory (NWI); fisheries data for Large Marine Ecosystems (LMEs) from the National Marine Fisheries Service (NMFS), and assessment and advisory data provided by states or other regulatory agencies and compiled in national databases (full report at http://water.epa.gov/type/oceb/2005_index.cfm, accessed June 21, 2015)
Methodology	Five primary indices were created using data from national coastal programs: water quality index, sediment quality index, benthic index, coastal habitat index, and fish tissue contaminants index. These indices were selected because of the availability of relatively consistent data sets for these indicators. These indices do not address all characteristics of estuaries and coastal waters that are valued by society, but they do provide information on both ecological condition and human use of estuaries Characterizing coastal areas using each of the five indicators involved two steps. The first step was to assess condition at an individual site for each indicator. For each indicator, site condition rating criteria are determined based on existing criteria, guidelines, or the interpretation of

(*continued*)

Table A.1 (continued)

Information	Details
	scientific literature. The second step was to assign a regional rating for the indicator based on the condition of individual sites within the region. The regional criteria boundaries (i.e., percentages used to rate each regional condition indicator) were determined as a median of responses provided through a survey of environmental managers, resource experts, and the knowledgeable public. Evaluations for fish tissue contaminants were used to assess human use attainment The results of evaluations of estuarine condition were used to assess aquatic life use and human use attainment. If any of four indicators of condition—water quality condition, sediment quality, benthic condition, or habitat loss—received a poor rating at a given site, then the site was assessed as impaired for aquatic life use. Threatened aquatic life use was assessed as the overlap of fair conditions of these same indicators. A site was determined to be unimpaired for aquatic life use if all four indicators were rated good, or only one indicator was rated fair and no indicators were rated poor. Spatial areas were assigned a category of (1) impaired for aquatic life use only, (2) impaired for human use only, (3) impaired for both aquatic life use and human use, (4) threatened (for one or both uses), or (5) unimpaired (for both uses)
Water quality index	The water quality index consisted of five indicators: nitrogen, phosphorus, chlorophyll *a*, water clarity, and dissolved oxygen. The water quality index used in this report was intended to characterize acutely degraded water quality conditions. It did not consistently identify sites experiencing occasional or infrequent hypoxia, nutrient enrichment, or decreased water clarity. As a result, a rating of poor for the water quality index means that the site is likely to have consistently poor condition during the monitoring period. If a site is designated as fair or good, the site did not experience poor condition on the date sampled, but could be characterized by poor condition for short time periods. In order to assess the level of variability in the index at a specific site, increased or supplemental sampling is needed. DIN, DIP, chlorophyll *a*, water clarity, and dissolved oxygen were assessed for a given site (see below), the water quality index rating was calculated for the site based on these five indicators as: <u>good</u> if a maximum of one indicator is fair, and no indicators are poor; <u>fair</u> if one of the indicators is rated poor, or two or more indicators are rated fair; <u>poor</u> if two or more of the five indicators are rated poor; and <u>missing</u> if two components of the indicator are missing, and the available indicators do not suggest a fair or poor rating
Nutrients: nitrogen and phosphorus	DIN and DIP were determined chemically through the collection of filtered surface water at each site. DIN and DIP reference surface concentrations used to assess condition in this report were generally lower than those in the NOAA report because of the natural reduction in nutrient concentrations due to uptake by phytoplankton from spring to summer for the production of chlorophyll. Ratings for coastal monitoring sites in the Gulf of Mexico were for DIN concentrations: <u>good</u>—<0.1 mg/L, <u>fair</u>—0.1–0.5 mg/L and <u>poor</u>—>0.5 mg/L and for DIP concentrations: <u>good</u>—< 0.01 mg/L, <u>fair</u>—0.01–0.05 mg/L and <u>poor</u>—>0.05 mg/L. For regionals scores both DIN and DIP concentrations were: <u>good</u> if less than 10 % of the coastal area was in poor condition, and more than 50 % of the coastal area was in good condition; <u>fair</u> if 10–25 % of the coastal area was in poor condition, or more than 50 % of the coastal area was in combined poor and fair condition; and <u>poor</u> if more than 25 % of the coastal area was in poor condition

(*continued*)

Table A.1 (continued)

Information	Details
Chlorophyll *a*	Surface concentrations of chlorophyll *a* were determined from a filtered portion of water collected at each site and rating for coastal monitoring sites in the Gulf of Mexico were for chlorophyll *a* concentrations; good—<5 µg/L, fair—5–20 µg/L, and poor—>20 µg/L. For regionals scores Chlorophyll *a* concentrations were: good if less than 10 % of the coastal area was in poor condition, and more than 50 % of the coastal area was in good condition; fair if 10–20 % of the coastal area was in poor condition, or more than 50 % of the coastal area was in combined poor and fair condition; and poor if more than 20 % of the coastal area was in poor condition
Water clarity	Water clarity was estimated using specialized equipment that compared the amount and type of light reaching the water surface to the light at a depth of 1 m, as well as by using a Secchi disk. The water clarity indicator (WCI) was based on a ratio of observed clarity to reference conditions: WCI = (observed clarity at 1 m)/(reference clarity at 1 m). The reference conditions were determined by examining available data for the region. In the Gulf Coast conditions were set at 10 % of incident light available at a depth of 1 m for normally turbid locations, 5 % for naturally highly turbid conditions, and 20 % for regions with significant Submerged Aquatic Vegetation beds or active restoration programs. For individual sampling sites the WCI ratio is good if it is >2, fair if it is between 1 and 2, and poor if it is <1. For regional scores water clarity was: good if less than 10 % of the coastal area was in poor condition, and more than 50 % of the coastal area was in good condition; fair if 10 % to 25 % of the coastal area was in poor condition, or more than 50 % of the coastal area was in combined poor and fair condition; and poor if more than 25 % of the coastal area was in poor condition
Dissolved oxygen	Dissolved oxygen was measured as part of the survey. For individual sampling sites Dissolved oxygen was rated Good—>5 mg/L, Fair—2–5 mg/L and Poor—< 2 mg/L. For regional scores Dissolved oxygen concentrations were: good if less than 5 % of the coastal area was in poor condition, and more than 50 % of the coastal area was in good condition; fair if 5–15 % of the coastal area was in poor condition, or more than 50 % of the coastal area was in combined poor and fair condition; and poor if more than 15 % of the coastal area was in poor condition
Assessment and advisory data	Assessment and advisory data provided by states or other regulatory agencies was the third set of data used in this report to assess coastal condition. Several USEPA programs, including the Clean Water Act Section 305(b) Assessment Program, the National Listing of Fish and Wildlife Advisories (NLFWA) Program, and the Beaches Environmental Assessment, Closure, and Health (BEACH) Program, maintain databases that are repositories for information about how well coastal waters support their designated or desired uses. These uses are important factors in public perception of the condition of the coast and also address the condition of the coast as it relates to public health. The data for these programs were collected from multiple state agencies and data collection and reporting methods differed among states. Because of these inconsistencies, data generated by these programs are not included in the estimates of coastal condition
Designated or desired uses	Clean Water Act Section 305(b) Assessments and Beach Advisories and Closures data were utilized the same as in NCCR I (USEPA 2001)

(continued)

Table A.1 (continued)

Information	Details
National Estuary Program Coastal Condition Report (USEPA 2006)	
Data sources	The objective of this National Estuary Program Coastal Condition Report (NEP CCR) was to report on the condition of the nation's 28 NEP estuaries. The NEP CCR presented two major types of monitoring data for each NEP estuary: (1) data collected as part of USEPA's National Coastal Assessment (NCA) and (2) data collected by the individual NEPs or by the NEPs in partnership with interested stakeholders, including state environmental agencies, universities, or volunteer monitoring groups. Together, these data painted a picture of the overall condition of the coastal resources of the nation's NEP estuaries In addition to the NCA-based assessments, this report provided individual profiles of the 28 NEP estuaries that describe the indicators each NEP uses to address specific environmental concerns, including water and sediment quality, habitat quality, living resources, and environmental stressors, as appropriate. Each profile includes background information on the NEP estuary discussed, maps of the NEP study area, and data on the population pressures that affect the study area, including the total population (2000), population density (2000), and population growth rate (1960–2000) in NOAA-designated coastal counties that are within or transect the boundaries of the study area (Full Report at http://water.epa.gov/type/oceb/nep/upload/2007_05_09_oceans_nepccr_pdf_large_section1.pdf, accessed June 21, 2015)
Methodology	All of the methodologies, assessments, and ratings procedures for this report were the same as for NCCRII (USEPA 2004). The ratings in this report were based solely on NCA monitoring data and not the data collected by the individual NEPs. The NCA data were collected from 1997 through 2003 for four primary indices of estuarine condition (water quality index, sediment quality index, benthic index, and fish tissue contaminants index)
Effects of nutrient enrichment in the nation's estuaries: A decade of change (Bricker et al. 2007)	
Data sources	The evaluation included national data sets such as physical and hydrologic characteristics and nutrient loading
Methodology	This assessment evaluated the factors that influence water quality. Influencing factors that link a system's natural sensitivity to eutrophication and the nutrient loading and eutrophic symptoms actually observed illustrating the relationship between eutrophic conditions and use impairments. A system's eutrophic condition was assessed based on five water quality variables related to nutrient enrichment (chlorophyll *a*, macroalgal blooms, dissolved oxygen, loss of submerged vegetation and nuisance/toxic blooms) The data set included concentration or occurrence of problem conditions, and also characteristics such as duration, spatial coverage, frequency of occurrence of observed conditions, and data confidence. An increase in two of the primary symptoms indicates the first stage of water quality degradation associated with eutrophication. Epiphytes were omitted from this assessment due to the lack of a standard measure and data availability. Secondary symptoms are: low dissolved oxygen levels, loss of submerged aquatic vegetation, and occurrences of nuisance/toxic algal blooms. Nutrient concentrations were not used because they reflect the net biological, physical, and chemical processes such that even a severely degraded water body may exhibit low concentrations due to uptake by phytoplankton and macroalgae. Conversely, a relatively healthy system

(continued)

Table A.1 (continued)

Information	Details
	might have high nutrient concentrations due to low algal uptake as a result of light-limiting turbid waters, or may simply flush nutrients so quickly that phytoplankton do not have the opportunity to bloom extensively. For these reasons, nutrient concentrations may not be accurate indicators. In many estuaries, primary symptoms lead to more serious secondary symptoms, including low dissolved oxygen, loss of submerged aquatic vegetation (SAV), and nuisance/toxic blooms. In some cases, secondary symptoms can exist in the estuary without originating from primary symptoms. Such systems were consequently given a lower rating for nuisance/toxic blooms. Low ratings were also used because it is unclear whether offshore nuisance/toxic algal blooms grow and are maintained as a result of land-based nutrient sources (an increasing problem, regardless of bloom origin)
National Coastal Condition Report (NCCR III) (USEPA 2008)	
Data sources	NCCR III is based primarily on USEPA's National Coastal Assessment (NCA) data collected in 2001 and 2002. The NCA; NOAA's National Marine Fisheries Service (NMFS) and National Ocean Service; USFWS's National Wetlands Inventory (NWI); and USGS contributed most of the information presented in the report. Three types of data were presented in this report: Coastal Monitoring Data—Coastal monitoring data were obtained from programs such as USEPA's Environmental Monitoring and Assessment Program (EMAP) and NCA, NOAA's National Status & Trends (NS&T) Program, and FWS's NWI; Offshore Fisheries Data—These data are obtained from programs such as NOAA's Marine Monitoring and Assessment Program and Southeast Area Monitoring and Assessment Program. These data are used in this report to assess the condition of coastal fisheries in large marine ecosystems (LMEs); and Assessment and Advisory Data—These data are provided by states or other regulatory agencies and compiled in nationally maintained databases. These data provide information about designated-use support, which affects public perception of coastal condition as it relates to public health. The agencies contributing these data use different methodologies and criteria for assessment; therefore, the data cannot be used to make broad-based comparisons among the different coastal areas
Methodology	The data are used to rate indices and component indicators of coastal condition. The index scores are then used to calculate overall condition scores and ratings for the regions and the nation. The rating criteria for each index and component indicator in each region were determined based on existing criteria, guidelines, interviews with USEPA decision makers and other resource experts, and/or the interpretation of scientific literature. All of the methodologies, assessments, and ratings procedures for this report were the same as for NCCRII (USEPA 2004)
National Coastal Condition Report (NCCR IV) (USEPA 2012)	
Data sources	NCCR IV is based primarily on USEPA's NCA data collected between 2003 and 2006. The NCA, the NOAA's NMFS and NOS, and the USFWS's NWI contributed most of the information presented in this current report
Methodology	The data are used to rate indices and component indicators of coastal condition. The index scores are then used to calculate overall condition scores and ratings for the regions and the nation. The rating criteria for each index and component indicator in each region were determined based on existing criteria, guidelines, interviews with USEPA decision makers and other resource experts, and/or the interpretation of scientific literature. All of the methodologies, assessments, and ratings procedures for this report were the same as for NCCRIII (USEPA 2004)

APPENDIX B

Table B.1. Characteristics of Common Organic Contaminants in Marine Waters, Including Sources, Toxicity, and Fate in the Environment (modified from Kimbrough et al. 2008 and references therein)

Sources	Toxicity	Fate
Chlordanes: a group of organic pesticides called cyclodienes. It is a technical mixture whose principal components are alpha-chlordane, gamma-chlordane, heptachlor, and nonachlor)		
Chlordane, an insecticide, is a complex mixture of at least 50 compounds. It was used in the United States from 1948–1983 for agricultural and urban settings to control insect pests. It was also the predominant insecticide for the control of subterranean termites. Agricultural uses were banned in 1983, and all uses were banned by 1988	Exposure to chlordane can occur through eating crops from contaminated soil, fish and shellfish from contaminated waters, or breathing contaminated air. Chlordane can enter the body by being absorbed through the skin, inhalation, and ingestion. At high levels, chlordane can affect the nervous system, digestive system, brain, and liver, and is also carcinogenic. Chlordane is highly toxic to invertebrates and fish	Removal from both soil and water sources is primarily by volatilization and particle-bound runoff. In air, chlordane degrades as a result of photolysis and oxidation. Chlordane exists in the atmosphere primarily in the vapor-phase, but the particle-bound fraction is important for long-range transport. Chlordane binds to dissolved organic matter, further facilitating its transport in natural waters
DDT (dichlorodiphenyltrichloroethane)		
DDT was used worldwide as an insecticide for agricultural pests and mosquito control. Its use in the United States was banned in 1972, but it is still used in some countries today	Due to its environmental persistence and hydrophobic nature, DDT bioaccumulates in organisms. Many aquatic and terrestrial organisms are highly sensitive to DDT. As a result of DDT's toxic effect on wildlife, in particular birds, its usage was banned in the United States	DDT transforms to DDD and DDE, the latter being the predominant form found in the environment. Evaporation of DDT from soil followed by long distance transport results in its widespread global distribution. DDT and its transformation products are persistent and accumulate in the environment because they resist biodegradation. DDT that enters surface waters is subject to volatilization, adsorption to suspended particulates and sediment, and bioaccumulation. About half of the atmospheric DDT is adsorbed to particulates
Dieldrins		
Dieldrin is defined as the sum of two compounds, dieldrin and aldrin. Dieldrin and a related compound (aldrin) were widely used as insecticides in the 1960s for the control of termites around buildings and general crop protection from insects. In 1970, all uses of aldrin and dieldrin were canceled based on concern that	Exposure to aldrin and dieldrin occurs through ingestion of contaminated water and food products, including fish and shellfish, and through inhalation of indoor air in buildings treated with these insecticides. Aldrin is rapidly metabolized to dieldrin in the human body. Acute and long-term human exposures are associated	Aldrin is readily converted to dieldrin, while dieldrin is resistant to transformation. Dieldrin bioaccumulates and is magnified through aquatic food chains and has been detected in tissue of freshwater and saltwater fish, and marine mammals. Aldrin and dieldrin applied to soil are tightly bound, but may be transported to

(*continued*)

Table B.1 (continued)

Sources	Toxicity	Fate
they could cause severe aquatic environmental change and their potential as carcinogens. The cancellation was lifted in 1972 to allow limited use of aldrin and dieldrin, primarily for termite control. All uses of aldrin and dieldrin were again cancelled in 1989	with central nervous system intoxication. Aldrin and dieldrin are carcinogenic to animals and classified as likely human carcinogens	streams and rivers by soil erosion. Volatilization is the primary loss mechanism from soil. Dieldrin undergoes minor degradation to photodieldrin in marine environments
Polycyclic aromatic hydrocarbons (PAHs)		
Polycyclic aromatic hydrocarbons (PAHs) are found in creosote, soot, petroleum, coal, and tar. PAH can also have natural sources (e.g., forest fires, volcanoes) in addition to anthropogenic sources (automobiles emissions, home heating, coal-fired power plants). PAHs are formed from the fusing of benzene rings during the incomplete combustion of organic materials. They are also found in oil and coal. The main sources of PAHs to the environment are forest fires, coal-fired power plants, and automobile exhaust and local releases of oil	Made up of a suite of hundreds of compounds, PAHs exhibit a wide range of toxicities. Human exposure to PAHs can come as a result of being exposed to smoke from forest fires, automobile exhaust, home heating using wood, grilling and cigarettes. Toxic responses to PAHs in aquatic organisms include reproduction inhibition, mutations, liver abnormalities and mortality. Exposure to aquatic organisms can come as a result of oil spills, boat exhaust and urban runoff	The fate and transport of PAHs is variable and dependent on the physical properties of each individual compound. Most PAHs strongly associate with particles; larger PAH compounds (high molecular weight) associate to a higher degree with particles relative to smaller PAH compounds (low molecular weight). Smaller compounds predominate in petroleum products whereas larger compounds are associated with combustion
Polychlorinated biphenyl: there are 209 possible PCB (polychlorinated biphenyl) compounds, called "congeners" that were marketed as mixtures known as Aroclor)		
PCBs are synthetic organic chemicals composed of biphenyl substituted with varying numbers of chlorine atoms. They were manufactured between 1929 and 1977. PCB use was regulated in 1971, and new uses were banned in 1976. PCBs were used in electrical transformers, capacitors, lubricants and hydraulic fluids. Other uses included paints, adhesives, plasticizers and flame retardants. Manufacturing of PCBs for use as flame retardants and lubricants stopped in 1977. Currently, PCBs are predominately used in electrical applications and can still be found in transformers and electrical equipment	The main human exposure route for PCBs is through eating contaminated seafood and meats. PCBs are associated with skin ailments, neurological and immunological responses and at high doses can decrease motor skills and cause liver damage, and memory loss. Exposure of aquatic life to PCBs results in birth defects, lowered fecundity, cancer and death. PCBs are hazardous because they are toxic, degrade slowly and bioaccumulate	PCBs are persistent in the environment and associate with particles in aquatic systems as a result of their strong hydrophobic nature. They are long lived in the environment; improper disposal and leakage is responsible for environmental introduction

APPENDIX C

Table C.1. Characteristics of Common Metal Contaminants Including Origins, Toxicity, and Fate in the Environment (modified from Kimbrough et al. 2008 and references therein)

Contaminant	Origins	Toxicity	Fate
Arsenic (As)	Arsenic has natural and industrial sources. Products that contain arsenic include: preserved wood, semiconductors, pesticides, defoliants, pigments, antifouling paints, and veterinary medicines. In the recent past, as much as 90 % of arsenic was used for wood preservation. Atmospheric sources of arsenic include smelting, fossil fuel combustion, power generation, and pesticide application	Arsenic is toxic at high concentrations to fish, birds and plants. In animals and humans prolonged chronic exposure is linked to cancer. Inorganic arsenic, the most toxic form, represents approximately 10 % of total arsenic in bivalves. Less harmful organic forms, such as arsenobetaine, predominate in seafood.	Human activities have changed the natural biogeochemical cycle of arsenic leading to contamination of land, water and air. Arsenic in coastal and estuarine water occurs primarily from river runoff and atmospheric deposition. The major source of elevated levels of arsenic in the nation is natural crustal rock
Cadmium (Cd)	Cadmium occurs naturally in the earth's crust as complex oxides and sulfides in ores. Products that contain cadmium include batteries, color pigment, plastics and phosphate fertilizers. Industrial sources and uses include zinc, lead and copper production; electroplating and galvanizing; smelting; mining; fossil fuel burning; waste slag; and sewage sludge. Anthropogenic emissions, originate from a large number of diffuse sources	Cadmium is toxic to fish, salmonoid species and juveniles are especially sensitive, and chronic exposure can result in reduction of growth. Respiration and food represent the two major exposure pathways for humans to cadmium	Cadmium has both natural and non-point anthropogenic sources. Natural sources include river runoff from cadmium rich soils, leaching from bedrock, and upwelling from marine sediment deposits. Cadmium is transported by atmospheric processes as a result of fossil fuel burning, erosion, and biological activities. Land-based runoff and ocean upwelling are the main conveyors of cadmium into coastal environments. Elevated cadmium levels are primarily located in freshwater-dominated estuaries consistent with river transport of cadmium to coastal environments

(*continued*)

Table C.1 (continued)

Contaminant	Origins	Toxicity	Fate
Copper (Cu)	Copper is a naturally occurring ubiquitous element in the environment. Trace amounts of copper are an essential nutrient for plants and animals. Anthropogenic sources include mining, manufacturing, agriculture, sewage sludge, antifouling paint, fungicides, wood preservatives, and vehicle brake pads. The United States ranks third in the world for utilization and second in production. The USEPA phase-out of chromated copper arsenate (CCA) wood preservatives and the 1980s restrictions on tributyltin marine antifouling paint has stimulated a transition to copper-based wood preservatives and marine antifouling paint	Copper can be toxic to aquatic organisms; juvenile fishes and invertebrates are much more sensitive to copper than adults. Although copper is not highly toxic to humans, chronic effects of copper occur as a result of prolonged exposure to large doses and can cause damage to the digestive tract and eye irritation	The most common form of copper in water is Cu (II), it is mostly found bound to organic matter. Transport of copper to coastal and estuarine water occurs as a result of runoff and river transport. Atmospheric transport and deposition of particulate copper into surface waters may also be a significant source of copper to coastal waters
Lead (Pb)	Lead is a ubiquitous metal that occurs naturally in the earth's crust. Environmental levels of lead increased worldwide over the past century because of leaded gasoline use. Significant reductions in source and load resulted from regulation of lead in gasoline and lead based paints. High levels found in the environment are usually linked to anthropogenic activities such as manufacturing processes, paint and pigment, solder, ammunition, plumbing, incineration, and fossil fuel burning. In the communications	Lead has no biological use and is toxic to many organisms, including humans. Exposure of fish to elevated concentrations of lead results in neurological deformities and black fins in fish. Lead primarily affects the nervous system, which results in decreased mental performance and mental retardation in humans. Exposure to lead may also cause brain and kidney damage, and cancer	Loadings of lead into coastal waters are primarily linked with wastewater discharge, river runoff, atmospheric deposition, and natural weathering of rock. Lead can be found in air, soil, and surface water

(*continued*)

Table C.1 (continued)

Contaminant	Origins	Toxicity	Fate
	industry, lead is still used extensively as protective sheathing for underground and underwater cables, including transoceanic cable systems		
Mercury (Hg)	Mercury is a highly toxic, nonessential trace metal that occurs naturally. Elevated levels occur as a result of human activity. In the United States, coal-fired electric turbines, municipal and medical waste incinerators, mining, landfills, and sewage sludge are the primary emitters of mercury into the air	Mercury is a human neurotoxin that also affects the kidneys and developing fetuses. The most common human exposure route for mercury is the consumption of contaminated food. Children, pregnant women or women likely to become pregnant are advised to avoid consumption of swordfish, shark, king mackerel and tilefish and should limit consumption to fish and shellfish recommended by FDA and USEPA	In the environment, mercury may change forms between elemental, inorganic and organic. Natural sinks, such as sediment and soil, represent the largest source of mercury to the environment. Estimates suggest that wet and dry deposition accounts for 50–90 % of the mercury load to many estuaries, making atmospheric transport a significant source of mercury worldwide. Long-range atmospheric transport is responsible for the presence of mercury at or above background levels in surface waters in remote areas
Nickel (Ni)	Nickel is a naturally occurring, biologically essential trace element that is widely distributed in the environment. It exists in its alloy form and as a soluble element. Nickel is found in stainless steel, nickel-cadmium batteries, pigments, computers, wire, and coinage and is used for electroplating	Food is the major source of human exposure to nickel. Exposure to large doses of nickel can cause serious health effects, such as bronchitis, while long-term exposure can result in cancer. There is no evidence that nickel biomagnifies in the food chain	Nickel derived from weathering rocks and soil is transported to streams and rivers by runoff. It accumulates in sediment and becomes inert when it is incorporated into minerals. River and stream input of nickel are the largest sources for oceans and coastal waters. Atmospheric sources are usually not significant
Tin (Sn)	Tin sources in coastal water and soil include manufacturing and processing facilities. It also occurs in trace amounts in natural waters. Concentrations	Humans are exposed to elevated levels of tin by eating from tin-lined cans and by consuming contaminated seafood. Exposure to elevated levels of tin compounds	Tin enters coastal waters bound to particulates, and from riverine sources derived from soil and sediment erosion. Bio concentration factors for inorganic tin were

(continued)

Table C.1 (continued)

Contaminant	Origins	Toxicity	Fate
	in unpolluted waters and the atmosphere are often near analytical detection limits. Tin has not been mined in the United States since 1993	by humans leads to liver damage, kidney damage, and cancer	reported to be 1,900 and 3,000 for marine algae and fish. Inorganic tin can be transformed into organometallic forms by microbial methylation and is correlated with increasing organic content in sediment. Tin is regarded as being relatively immobile in the environment and is rarely detected in the atmosphere. It is mainly found in the atmosphere near industrial sources as particulates from combustion of fossil fuels and solid waste
Zinc (Zn)	As the fourth most widely used metal, zinc's anthropogenic sources far exceed its natural ones. The major industrial sources include electroplating, smelting and drainage from mining operations. The greatest use of zinc is as an anticorrosive coating for iron and steel products (sheet and strip steel, tube and pipe, and wire and wire rope). Canada is one of the largest producers and exporters of zinc. The United States is the largest customer for Canadian refined zinc, and the automobile industry is the largest user of galvanized steel	Zinc is an essential nutrient. Human exposure to high doses of zinc may cause anemia or damage to the pancreas and kidneys. However, zinc does not bioaccumulate in humans; therefore, toxic effects are uncommon and associated with excessively high doses. Fish exposed to low zinc concentrations can sequester it in some cases	Dissolved zinc occurs as the free hydrated ion and as dissolved complexes. Changes in water conditions (pH, redox potential, chemical speciation) can result in dissolution from or sorption to particles. In air, zinc is primarily found in the oxidized form bound to particles. Zinc precipitates as zinc sulfide in anaerobic or reducing environments, such as wetlands, and thus is less mobile, while remaining as the free ion at lower pHs. As a result of natural and anthropogenic activities, zinc is found in all environmental compartments (air, water, soil, and biota)
Butyltins	Tributyltin is used as an antifouling agent in marine paints applied to boat hulls. Slow release from the paint into the aquatic system retards organism attachment and increases ambient	Tributyltin is an extremely toxic biocide that is regulated as a result of its toxic effects (reproduction and endocrine disruption) on nontarget aquatic species. Organotin	Tributyltin is sparingly soluble in water and associates readily with suspended particles in the water column. Butyltins are persistent in the aquatic environment and accumulate in

(continued)

Table C.1 (continued)

Contaminant	Origins	Toxicity	Fate
	environmental levels. The United States partially banned the use of tributyltin in 1988 for use on boats less than 25 m in length, drastically limiting use on many recreational vessels	compounds are readily bio-accumulated by aquatic organisms from water but there is no evidence for biomagnification up the food chain. Sex changes have been shown to occur in gastropods exposed to elevated levels of tributyltin	sediment; therefore, they will continue to be a source of butyltin to the aquatic environment. Tributyltin transforms to dibutyltin and then to monobutyltin. Releases of organotins to the atmosphere are not significant due to their low vapor pressure and rapid photodegradation

CHAPTER 3

SEDIMENTS OF THE GULF OF MEXICO

Richard A. Davis, Jr.[1,2]

[1]University of South Florida, Tampa, FL 33620, USA; [2]Texas A&M University—Corpus Christi, Corpus Christi, TX 78412, USA
rdavis@usf.edu

3.1 INTRODUCTION

The Gulf of Mexico is a Mediterranean-type sea with limited fetch and low tidal ranges (microtidal) throughout. This basin is somewhat like a miniature ocean in that it contains all of the main bathymetric provinces of an ocean along with a complicated coastal zone (Figure 3.1). This chapter will consider the overall nature of the basin with emphasis on the sediments it contains. The discussion will be restricted to surface sediments and only to Holocene sediments where subsurface materials are included.

The Gulf of Mexico is a unique basin on the globe. It is located in the low, mid-latitudes and extends over multiple climatic zones. It includes regions where huge volumes of terrigenous sediments are delivered and others where terrigenous sediments are generally absent. The nature and distribution of sediments in the shallow Gulf margin have been controlled largely by the rise and fall of sea level during the waxing and waning of Quaternary glaciers. During that time, the shoreline migrated across virtually the entire continental margin, as we know it today. This has also had an influence on the sediments in the deep Gulf, from the continental slope to the abyssal plain.

The greatest terrigenous sediment supply is at the Mississippi Delta; next in volume is the Texas coast where numerous rivers cross the coastal plain regardless of the position of the shoreline. The northeast Gulf has also experienced a significant amount of terrigenous sediment influx. Similar sediment delivery along the coast of Mexico has occurred in the area south of Laguna Madre and north of Campeche Bay, but the sediment is different because of the extensive volcanic source rocks. There is virtually no sediment currently being delivered, nor has there been in the past, from the Florida Peninsula and the Yucatán Peninsula; both have been carbonate platforms throughout their existence. There is a veneer of terrigenous sediment on the Florida mainland, but the lack of well-developed drainage keeps it from being transported to the coast.

The deep Gulf environments are fairly similar to those of the world's oceans. The surface is rather flat with local relief of only a few meters. The sediments are a combination of fine terrigenous sediments and biogenic sediments contributed by various planktonic organisms. The terrigenous sediments are nearly all clay minerals that have come from the northern provinces of the Gulf States. The biogenic sediments are mostly foraminifera with some diatoms. The sediments are delivered to their sites of accumulation differently. Much of the terrigenous sediment comes to the abyssal plain via sediment gravity processes—especially turbidity currents. A small portion of the terrigenous sediment and all of the biogenic sediment settle through the water column.

The sediment on the continental slope typically is delivered in pulses or events by sediment gravity phenomena. Much of this occurred during low sea-level stages during the Quaternary. During these conditions, large streams that carried sediment extended across what is now the continental shelf, but was then the coastal plain. The mouths of these rivers were at, or near, the

© The Author(s) 2017
C.H. Ward (ed.), *Habitats and Biota of the Gulf of Mexico: Before the Deepwater Horizon Oil Spill*,
DOI 10.1007/978-1-4939-3447-8_3

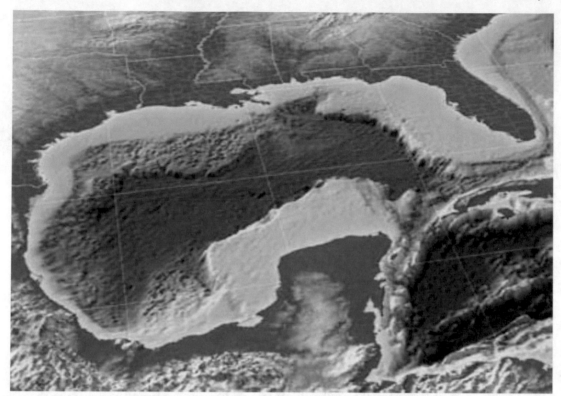

Figure 3.1. General topographic and bathymetric map of the Gulf of Mexico region (from Moretz-sohn et al. 2015).

slope-shelf break where sediment was discharged. Instead of developing deltas, as happens under present conditions, the sediment was transported down the relatively steep slope with some coming to rest on this surface and most making its way to the sediment fans and the deep abyssal environment.

The continental shelf is presently composed of a combination of modern sediments, delivered since sea level reached its present or near its present position, and sediments that were deposited during sea-level lowstands when this surface was accumulating mostly fluvial sediments in channels or floodplain deposits. Most of the inner shelf surface is now composed of modern sediments; much of the outer half of the shelf is relict sediments deposited in depositional environments different than those present.

Modern sediments delivered during present sea-level conditions currently dominate coastal environments. There are also sediments in these environments that are produced within the environment that they occupy as biogenic skeletal material. Because of the development and concentration of the population around modern coastal environments, these sediments tend to be polluted at some level.

In summary, the sediments of the Gulf of Mexico range widely in all respects. The following discussions will provide an introduction to their character and distribution.

3.2 BATHYMETRIC PROVINCES

The Gulf of Mexico (Figure 3.2) has a surface area of about 1.5 million square kilometers (km^2) (579,000 square miles [mi^2]), and 20 percent (%) of its area has a depth greater than 3,000 meters (m) (9,800 feet [ft]). The continental slope comprises 20 % of the Gulf, and the

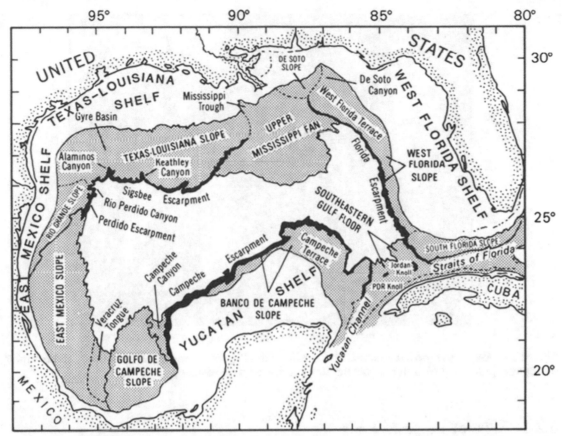

Figure 3.2. Physiographic map showing the major provinces of the Gulf of Mexico (from Martin and Bouma 1978). AAPG©[1978], reprinted by permission of the AAPG whose permission is required for further use.

continental shelf comprises 22 %. The coastal zone out to a depth of 20 m (65.6 ft) comprises 38 % of its area. Mean water depth of the Gulf is 1,615 m (5,299 ft), and the water volume of the Gulf is approximately 2.4 million cubic km (km³) (584,000 cubic mi [mi³]).

The shape of the basin is basically a simple cup with thick sediment sequences. It dates from Late Triassic time, about 150 million years ago. Sea level has experienced considerable change since that time as crustal plates have moved over the earth's surface. The time of highest sea level was during the Cretaceous Period, about 75 million years ago. The time of lowest sea level was only about 20,000 years ago when it was about 120–130 m (395–425 ft) below its present level (Salvador 1991). For purposes of this chapter, the Gulf will be divided into two provinces: terrigenous and carbonate (Uchupi 1975). These terms refer to the type of sediments that characterize each of the provinces (Figure 3.2). Terrigenous sediments are derived from land through river runoff, and carbonate sediments are precipitated in the Gulf generally as skeletal material (primarily from invertebrates) or as direct precipitates. These will be discussed in detail later in the chapter.

The physiography of the Gulf basin has been controlled by numerous geologic phenomena. They include (1) rifting, (2) subsidence, (3) development of carbonate platforms, (4) Gulf-wide changes in sea level, (5) formation of salt domes, (6) gravity slumping, and (7) sediment gravity flows (Bryant et al. 1991).

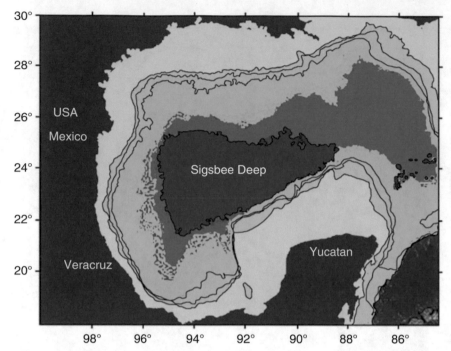

Figure 3.3. Map showing the location and extent of the Sigsbee Deep in the dark outlined area near the center (courtesy of Universidad Nacional Autonoma de Mexico).

3.2.1 Sigsbee Abyssal Plain

Much of the abyssal plain portion of the Gulf is called the Sigsbee Deep (Figure 3.3). It lies in the central portion of the basin and reaches a maximum depth of 3,750–4,384 m (12,300–14,383 ft) depending on which author you read (Turner 1999). This area is one of the flattest places on earth. Its surface slope is 1:10,000 and smooth (Bryant et al. 1991) except for the Sigsbee Knolls, which are diapiric salt domes that represent the only significant relief on this surface.

3.2.2 Mississippi Fan (Cone)

The Mississippi Fan (cone) is a deepwater feature (Figure 3.4) that extends from the outer continental margin off the mouth of the Mississippi River to the abyssal plain (Sigsbee Deep). It lies between the Mississippi Trough and the De Soto Canyon. This fan covers 300,000 km^2 (116,000 mi^2) (Twichell 2011). The Sigsbee Escarpment, the margin of the Jurassic Luann salt, is on the northwest edge of the fan. The Mississippi Canyon is the sediment's main pathway to the fan. On its southeastern edge, it grades into the Florida abyssal plain.

This huge accumulation of sediment (Figure 3.5) was developed primarily during Pleistocene time and is linked to the rise and fall of sea level resulting from expansion and contraction of ice sheets on the continents (Bryant et al. 1991). It has been calculated that during the Pleistocene lowstands of sea level, the rate of sediment delivery was about 13 times what it is now (Perlmutter 1985). Now sediment is being transported to the fan very slowly. When the course of the Mississippi River moved to the east late in the Quaternary, the Mississippi Canyon was removed as a major conduit of sediment to the fan.

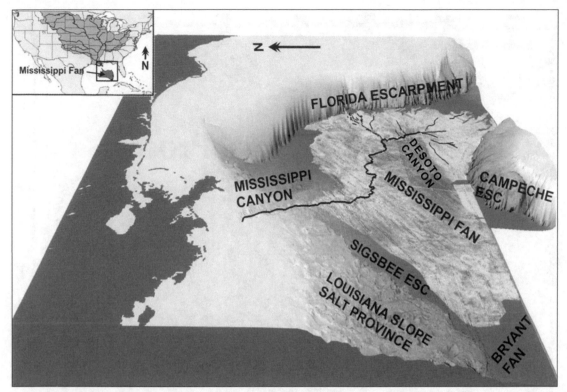

Figure 3.4. A three-dimensional schematic diagram of some of the major deepwater provinces in the Gulf of Mexico (from Twichell 2011).

3.2.3 Continental Slope

There is a range of morphologies on the continental slope. The aforementioned Florida Escarpment is very steep and has modest relief on the underlying carbonate strata of Mesozoic Age. The Yucatán Platform is similar. Much of the slope comprises small basins that are produced by the movement of salt, typically in diapiric fashion. The basins are 10–12 km (6.2–7.5 mi) in diameter with relief of 150–300 m (490–980 ft) (Bryant and Liu 2000). There are about 90 of these small basins.

The De Soto submarine canyon crosses the continental slope at the western end of the Florida Platform (Figure 3.4). Data show that this canyon is very old, and it has not received significant sediment accumulation since the latter part of the Cretaceous Period (Bryant et al. 1991). By contrast, the Mississippi Canyon (Figure 3.5) is one of the youngest such physiographic features in the Gulf of Mexico having been formed in the late Quaternary. It was also filled with sediment in a rather short time.

From the western edge of the Mississippi Canyon across the Texas slope, the surface of the continental slope is quite complex with abundant salt and shale diapiric structures that have relief of tens of meters (Figure 3.6). The Rio Grande slope is complicated by multiple structural ridge systems. Farther to the south, salt diapiric systems interact with these ridge systems and further complicate the bathymetry (Bryant et al. 1991). Proceeding into the Bay of Campeche, the nature of the slope becomes similar to that off the coasts of Texas and Louisiana with complex salt diapirs.

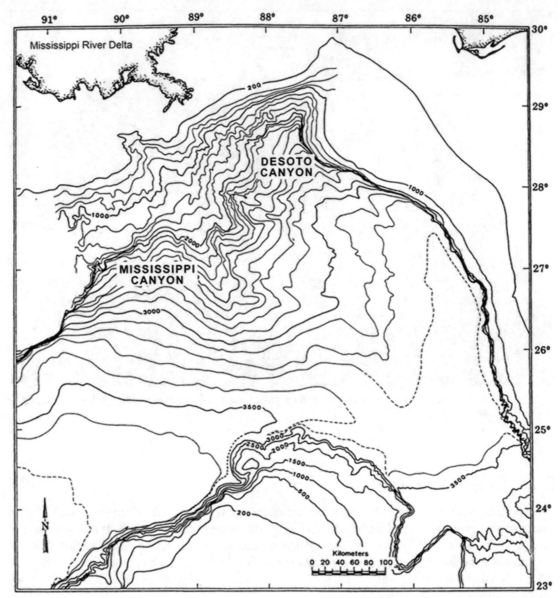

Figure 3.5. Bathymetric map of the Mississippi Fan. Reprinted from Bouma et al. 1985, Chapter 21- Mississippi Fan, Gulf of Mexico (AH Bouma, CE Stelting, JM Coleman), p. 144, with permission of Springer.

3.2.4 Continental Shelf

The continental shelf of the Gulf of Mexico displays a range of morphologies depending on the framework geology. The shelf adjacent to the Florida Peninsula is the widest and has the lowest gradient. This shelf is the submarine extension of the Florida Platform, a thick accumulation of limestone that extends back to the Jurassic Period. Its present expression is as a limestone surface with scattered carbonate and terrigenous sediment. It extends more than 500 km (300 mi) and is more than 150 km (93 mi) wide along the 75-m (245-ft) isobaths (Hine and Locker 2011). The maximum width is 240 km (149 mi). The gradient ranges from 0.2 to 4.0 m per km (m/km) (0.67–1.25 ft/mi) overall but is steeper (6–9 m/km) (20–30 ft/mi) in the

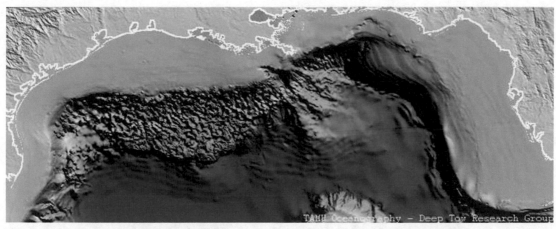

Figure 3.6. A schematic map of the continental slope of the northern Gulf of Mexico in the context of the continental shelf and the deep Gulf basin (courtesy of Texas A&M University, Deep Tow Research Group).

nearshore area. This shelf essentially terminates at De Soto Canyon on the north and at the Portales Terrace on the south. The Florida Escarpment (Figure 3.7) is at the edge of the shelf and is one of the steepest submarine slopes in the world at a slope of 45° (Bryant et al. 1991).

The shelf in the northeastern Gulf is narrower and steeper with numerous shelf ridges (Figs. 3.8 and 3.9) and relict deltas (Hine and Locker 2011). The fabric of this relief trends northwest–southeast. The shelf bathymetry in this part of the Gulf reflects the combination of the numerous cycles of sea-level change along with the dominance of fluvial influence. The De Soto Canyon (Figure 3.2) is the largest physiographic feature off the Florida panhandle. Its headland extends to within 25 km (15.5 mi) of the shoreline. The shelf widens to about 100 km (62 mi) on either side (Hine and Locker 2011).

There is essentially no continental shelf around the Mississippi Delta; the active delta extends across the shelf. The shelf to the west of the delta is rather similar along the entire Texas coast. It is wide, has low relief, and has a moderate gradient between that of the northeast Gulf and west Florida. There are many relict reefs and knolls on the Texas shelf that provide several meters of relief on an otherwise rather flat surface. The crests of salt domes that have protruded upward through younger Mesozoic and Cenozoic strata provide some relief on the east Texas shelf.

Moving to the west, the Louisiana and Texas shelf is broad and flat with a width that ranges from 32 to 90 km (20–56 mi) (Bryant et al. 1991). It is scattered with relict reefs and salt dome diapirs off Louisiana and east Texas. There are also numerous filled fluvial channels that developed during Quaternary lowstands of sea level.

A major change in the bathymetry of the Texas shelf occurs near the middle, what is commonly called the Coastal Bend. Here there is a gradual transition from shelf bathymetry to the slope in distinct contrast to the relative abrupt change in bathymetry between these two provinces on the remainder of the Texas shelf (Figure 3.10). This zone lies between the ancestral deltas of the Rio Grande River to the south and the Brazos–Colorado delta to the north.

Around the west Gulf shelf off Mexico there is a major narrowing of this province from about 80 km (50 mi) at the Rio Grande to less than 10 km (6.2 mi) off the volcanic province near Veracruz, Mexico. There is a marked widening of the shelf toward the Yucatán Peninsula. Salt diapirs are an influence west of the Campeche Bank (Bryant et al. 1991).

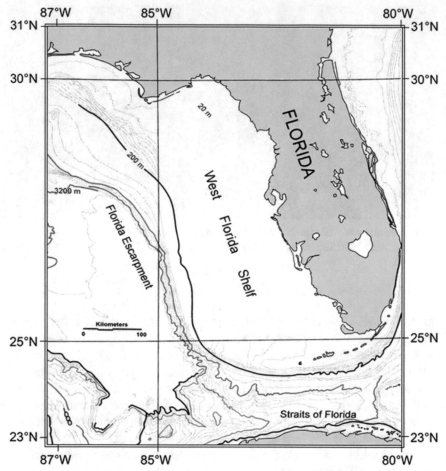

Figure 3.7. The Florida shelf and escarpment that forms the west margin of the Florida Platform (Hine et al. 2003b). Reprinted from Marine Geology, Vol 200, Hine AC et al., The west-central Florida inner shelf and coastal system: A geologic conceptual overview and introduction to the special issue, Figure 3, Copyright 2003, with permission from Elsevier.

The shelf along the Mexican mainland is similar to that in the northwest Gulf but is generally not as wide. It is gently sloping and has little relief on it. Like the Florida shelf, the slope on three sides of the Yucatán shelf is steep—up to 35°. Because the northern coast of Cuba is adjacent to the Florida Straits, it also has a narrow and relatively steep shelf.

3.2.4.1   Relict Sediment Cover

On most of the shelf, the relief of the relict sediment cover is limited to shore-parallel quartz sand sediment bodies that are interpreted as being relict barriers that were abandoned during rapid sea-level rise in the Pleistocene era. Most are late Pleistocene or Holocene in age, but some might be older. There are relict Quaternary reefs along various isobaths, particularly offshore of Texas. The relict cover is dominated by lowstand depositional environments such as fluvial channels, floodplains, and deltas (Anderson and Fillon 2004).

Another important component of the shelf is the presence of these relict barrier islands. They are distributed around the entire Gulf of Mexico but with various relationships to the modern sediment blanket. On the Florida shelf, these relict barriers are small and rest primarily on limestone bedrock (Figure 3.11).

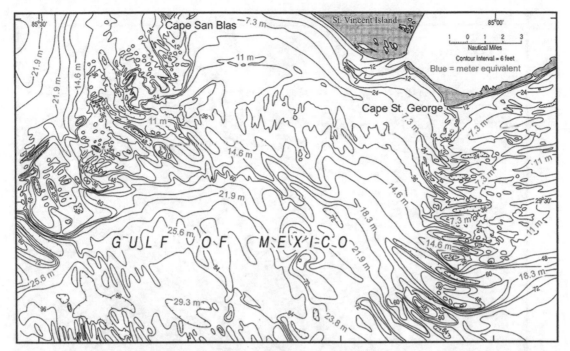

Figure 3.8. Bathymetric map of a portion of the continental shelf off the panhandle of Florida showing considerable relief of linear features that are sand bodies reworked from fluviodeltaic sediments of Quaternary sea-level lowstand deposition (from McBride et al. 2004: reprinted with permission from the Society for Sedimentary Geology).

Similar sand bodies are present off the panhandle coast of Florida and have been described in McKeown et al. (2004). Further west on the shelf, much larger relict sand bodies are present on the Louisiana shelf. The largest of these is the Ship Shoal (Figure 3.12). These relict sand bodies are the sites of possible nourishment sand for the Louisiana coast.

Similar sediment bodies are also present on the Texas coast. From east to west, they are the Sabine Bank, Heald Bank, and Freeport Rocks. All of these relict sediment bodies (Figure 3.13) represent coastal accumulations of sand and shell material that were deposited during slow-downs or stillstands of the Holocene sea-level rise (Rodriguez et al. 1999).

3.2.4.2   Modern Sediment Cover

On the west Florida shelf, the modern sediment cover displays essentially no relief on the shelf surface except for the above-described sand bodies. The surface of the shelf throughout most of its extent is the pre-Quaternary carbonate strata with sinkholes and karstic terrain—both widespread and abundant. The Yucatán shelf of Mexico is similar. The rest of the northern Gulf shelf is a mixture of relict and modern sediment surfaces. The continental shelf around the Yucatán Platform is in some ways similar to that in Florida. It is up to 240 km (150 mi) wide on the north but quite narrow on the east. The surface is scattered with karstic features and reefs (Logan et al. 1969).

The shelf in Cuba is unlike the rest of the Gulf of Mexico because it is a collision area. The Caribbean plate moved into this region pushing between the North American and South American plates. As a result, this margin is narrow with high relief and numerous structural components including faults.

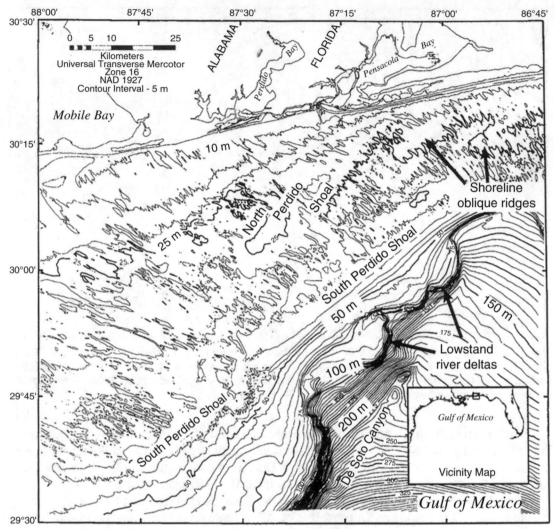

Figure 3.9. Bathymetric map off the northern Gulf Coast with considerable relief caused by northwest–southeast trend (from McBride et al. 2004: reprinted with permission from the Society for Sedimentary Geology).

3.2.5 Coastal Environments

3.2.5.1 Beach and Nearshore Zone (Barrier Islands)

Barrier islands and their contained beaches are extensive around the Gulf Coast. These barriers are young; some are only decades old, and the oldest is about 7,000 years old. Their size tends to be related to the abundance of sediment. They range from only a kilometer (0.6 mi) or so to 150 km (93 mi) in length. The relief may be up to 15 m (49 ft). Most of these barriers are wave dominated, but there are also mixed-energy barriers, most of which are on the Florida coast. The surf zone, just offshore of the beach, is commonly characterized by longshore bars and intervening troughs (Davis and FitzGerald 2003), the number of which is the result of bottom gradient and sediment availability. This zone is the most dynamic of the entire Gulf in that waves and currents are continuously present and ever changing.

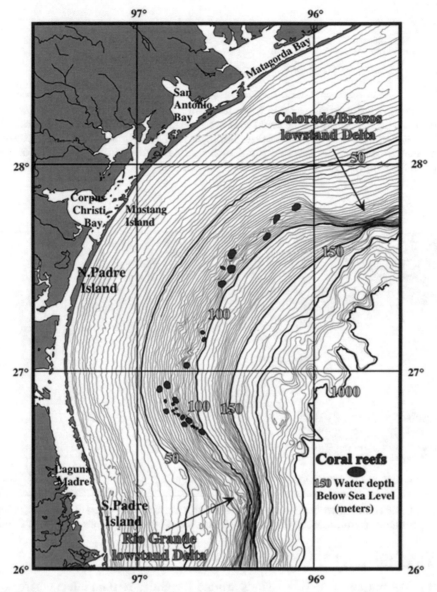

Figure 3.10. Bathymetric map of the shelf in the central and south Texas continental margin where there is a major change in the surface configuration. To the north and south, the bathymetry is as expected with a smooth shelf and a change to a steep slope. The central area is quite different with an embayment and multiple coral reefs (from Berryhill 1987). AAPG[©][1987], reprinted by permission of the AAPG whose permission is required for further use.

3.2.5.2 Dunes

Most of the Gulf Coast is surrounded by sand dunes. The dunes result primarily from onshore wind blowing over the dry beach. These dunes range widely in size depending on the availability of sediment. The largest tends to be on the panhandle of Florida, the Matagorda Peninsula, and Padre Island, Texas. In parts of the Gulf Coast, dunes are completely absent, primarily in the southwestern and Big Bend parts of Florida.

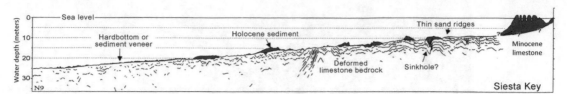

Figure 3.11. Interpreted seismic profile from the inner continental shelf of Florida showing the presence of small sediment bodies (*black*) that are interpreted as relict barrier island sand bodies (Locker et al. 2003). Reprinted from Marine Geology, Vol 200, Locker SD et al., Regional stratigraphic framework linking continental shelf and coastal sedimentary deposits of west-central Florida, Figure 3, Copyright 2003, with permission from Elsevier.

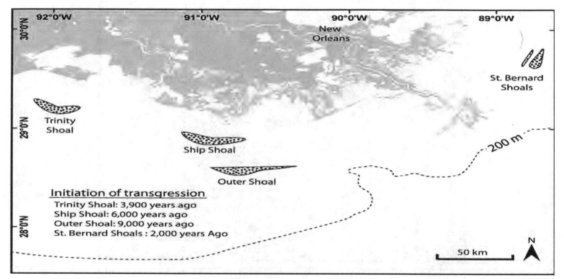

Figure 3.12. Major relict sand bodies on the Louisiana continental shelf. These huge sand bodies are potential nourishment material for the Louisiana coast (from Rogers and Kulp 2009).

3.2.5.3 Tidal Inlets

Breaks in the numerous barrier islands around the Gulf are tidal inlets where considerable tidal flux is transported during each tidal cycle. The volume of water transported through these inlets ranges among four orders of magnitude depending on the size of the estuaries (Davis 1988). The depth of the inlets ranges from only 1 m (3.3 ft) to more than 30 m (98 ft). Inlets tend to have large sediment accumulations at both the Gulf side (ebb-tidal deltas) and the landward side (flood-tidal deltas). The size of the inlet tends to be directly related to the tidal prism (water budget) that passes through the inlet during an individual tidal cycle. Inlets of the Gulf Coast range from tide-dominated through mixed-energy to wave-dominated. Small unstable inlets have closed over historical time, and hurricanes have generated new ones. In general, little sediment is passing from estuaries into the Gulf.

3.2.5.4 Wetlands

Coastal wetlands are widespread along the many coastal bays on the Gulf of Mexico. In the low latitudes—generally south of about 30°—wetlands are dominated by mangroves: red

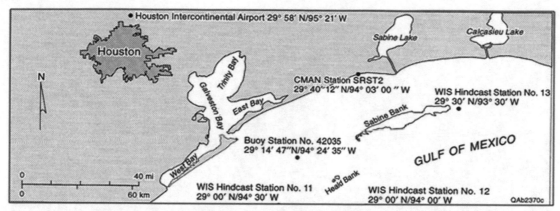

Figure 3.13. Locations and geography of Sabine and Heald Banks on the east Texas shelf. These are relict barrier islands similar to those off Louisiana (from Morton and Gibeaut 1995).

(*Rhizophora mangle*), black (*Avicennia germinans*), and/or white (*Laguncularia racemosa*). Some marsh grass can also be present. North of that latitude, the wetlands are dominated by salt marsh with cordgrass (*Spartina*) and rushes (*Juncus*) being the dominant vegetation. These wetlands extend over only a few tenths of a meter of elevation due to the small tidal range throughout the Gulf. The combination of dammed rivers, hurricanes, and sea-level rise has caused a tremendous reduction in the area of wetlands on the Gulf Coast, especially along the northern coast.

3.2.5.5   Estuaries

The Gulf of Mexico is surrounded by many estuaries; however, they are small and scarce on the Yucatán coast. These estuaries are the result of flooding of drainage systems that were incised during lowstands of sea level during the Quaternary Period when there were multiple cycles of sea-level rising and falling in response to the advance and retreat of glacial ice sheets. Estuaries may have a single river or multiple rivers emptying into them. These coastal water bodies are generally brackish and shallow, typically less than 5 m (16.4 ft) deep. Sediments in these coastal bays are dominated by mud.

3.2.5.6   Lagoons

The term lagoon is used to separate those coastal bays that have essentially no freshwater input or tidal flux from those that do (estuaries). Along the Gulf Coast, there are a few lagoons, most prominent of which are Baffin Bay and Laguna Madre of Texas. Smaller examples are also present on the coast of Mexico (Laguna Madre, Alvarado, Celestún) (Carranza-Edwards 2011).

Baffin Bay is a drowned fluvial system that was active in an earlier time when the climate in this area was much wetter. Laguna Madre is a long, shore-parallel coastal bay that reaches salinities near 100 parts per thousand (ppt) in some isolated areas. Both are quite shallow.

3.3   GENERAL CHARACTERISTICS OF SEDIMENTS

In this section, each major category of Gulf of Mexico sediment will be discussed in general, and then the sediments of all environments within the Gulf will be addressed. Because of the extensive geography being considered, some generalities will need to be presented. Every square meter of the Gulf floor, at all depths, contains some sediment (Figure 3.14).

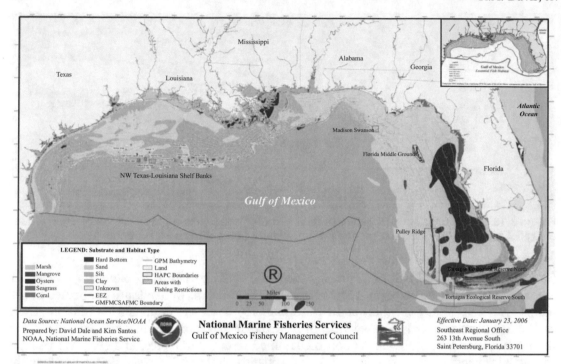

Figure 3.14. Map of the bottom surface sediment throughout the northern Gulf of Mexico (from NOAA).

Table 3.1. Percentage Range of the Constituents of Detrital Sand and Silt in Cores Taken from the Gulf of Mexico (from Davies and Moore 1970).

Category	Percent Range	Category	Percent Range
Quartz	1–67	Matrix (clay minerals)	1–72
Feldspar	0–7	Carbonate grains	2–75
Mica	0–4	Glass	0–83
Rock fragments	0–22	Accessories	0–6
Iron oxide grains	0–18		

Some of the dominant constituents (Table 3.1) can be associated with geographic areas. For example, the carbonate-rich sediments are typically associated with the Florida and Yucatán platforms. The glass fraction is associated with the area of Mexico near Veracruz where volcanic source rocks are abundant, and the mud (matrix) is most common in the deep basin, estuaries, and deltas.

3.3.1 Terrigenous Sediments

The term *terrigenous* comes from the Greek roots of *terra*, meaning land, and *genesis*, meaning origin. These sediments originate on land. They are eroded, transported by river systems to the coast, and become most of the Gulf of Mexico floor including all environments from the coast to the deep regions. Terrigenous sediments almost exclusively comprise silicate minerals. These are minerals that have their core in the elements silica and oxygen, with the relative abundance of each depending on the family within the silicate mineral spectrum. The most basic of these is quartz (SiO_2), which is one of the two most abundant terrigenous minerals in the Gulf. Quartz is very resistant to chemical erosion and is quite durable physically. As a result, quartz is able to withstand the rigors of erosion and transportation over long periods of time and distances of travel.

The other very abundant terrigenous species are clay minerals. Clay minerals are termed *layered silicates* because their crystallography causes them to split into thin sheets. Mica is an excellent example of a layered silicate. Most of the clay minerals are the weathering products of other minerals such as feldspars. There are multiple clay minerals depending on the number of layers in their crystal structure and the types of elements that are combined with the silicate structure. They include iron (Fe), aluminum (Al), magnesium (Mg), potassium (K), calcium (Ca), and sodium (Na).

It is possible to make some generalizations about the distribution of clay mineral species in the Gulf Basin. Four clay mineral species can be found in the basin: smectite, illite, kaolinite, and chlorite. Because of the range of clay mineral species over multiple physiographic provinces, it is best to consider their general distribution here. Smectite, which is the most common of the four, is relatively low in concentration on the shelf of Alabama and Florida. Illite is the next most common species in the Gulf; however, illite abundance shows a decrease from the shelf into the deep basin. Kaolinite decreases from east to west and is generally low on the central and western parts of the Gulf. Chlorite is the least abundant throughout the basin, less than 20 % in all environments (Wade et al. 2008).

The other two most common elements of terrigenous sediments are feldspar (a potassium silicate) and rock fragments (small pieces of rock of many compositions that comprise multiple mineral grains). Other resilient terrigenous mineral grains occur in very small percentages in sand. These are commonly called accessory minerals or heavy minerals (Figure 3.15), and most

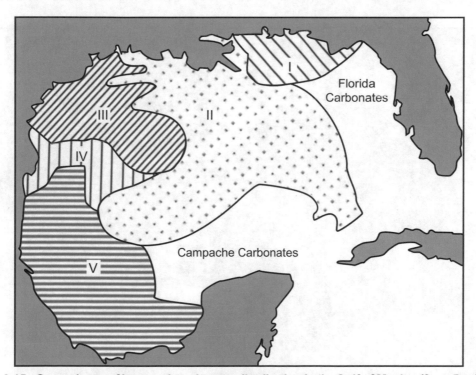

Figure 3.15. General map of heavy mineral group distribution in the Gulf of Mexico (from Davies and Moore 1970: reprinted with permission from The Journal of Sedimentary Research). Province I is from the Appalachians; kyanite and staurolite dominate. Province II is from the Mississippi River; augite, hornblende, and epidote dominate. Province III is from Central Texas with hornblende and epidote dominating. Province IV is Rio Grande; epidote, augite, and hornblende are dominant, and Province V is in Mexico; little is known about the heavies in Province V.

have a density (specific gravity) higher than quartz and feldspar. They include garnet, magnetite, zircon, limonite, rutile, and a few others.

3.3.2 Biogenic Sediments

The Gulf of Mexico is replete with organisms, many of which have skeletal components that contribute to the sediment. There are three categories of skeletal components: calcium carbonate, phosphatic skeleton, and siliceous materials. By far the most common is the calcium carbonate exoskeleton of invertebrates. Phosphatic skeleton is from fish, and siliceous material is small single-celled organisms and sponge spicules.

3.3.2.1 Calcium Carbonate

Many invertebrates have various types of skeletons of calcite, high-magnesium calcite, and aragonite. These compounds are all various types of calcium carbonate with some variation in crystallography and composition. They range from single-celled organisms to large invertebrates, including coral colonies and calcified green and red algae. In some of these organisms, such as gastropods (snails), the entire intact skeleton is included in the sediment. In others, such as echinoderms (starfish, sea urchins), the skeleton disarticulates and may become dozens of individual pieces. Regardless of size, the skeletal material can become a significant part of the sediment. In some places, such as the Florida Keys or the Yucatán Peninsula, the entire composition of the sediment may be skeletal carbonate (Figure 3.16). These carbonate exoskeletons are typically broken by waves, currents, and even by other organisms. Their abundance in

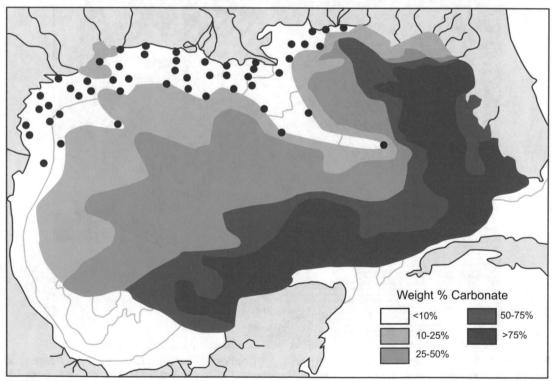

Weight % Carbonate

<10%	50-75%
10-25%	>75%
25-50%	

Figure 3.16. Map of the weight percent carbonate throughout the Gulf of Mexico (from Caso et al. 2004). *Black dots* are sample locations.

sediment ranges from 0 to 100 %. The particle size and the shape also range widely. As a consequence, the rate of transport of skeletal particles is hard to measure.

Calcium carbonate sediment is mostly found in shallow water, but there is also deepwater carbonate sediment (Figure 3.16). Calcium carbonate sediment comprises primarily planktonic foraminifera (single-celled animals) and submicroscopic algae called coccolithophores. These sediments, often called calcareous ooze sediments, are common on the abyssal plain of the Gulf. Such microscopic and submicroscopic skeletal particles can form limestones in the ancient record and become major petroleum producers.

3.3.2.2 Phosphate Skeletons

Fish skeletons tend to be phosphatic except for the otoliths (ear bones), which are calcium carbonate. Because fish skeletons are generally rather fragile and predators commonly consume the fish body, skeletal fragments of phosphatic composition are not common in sediments. Otoliths do tend to be preserved in sediments, but they are scarce in the overall volume of marine sediments.

3.3.2.3 Siliceous Skeletal Material

Three major categories of organisms have siliceous skeletal material: sponges, radiolarians, and diatoms. Sponges are soft benthic animals, but several of them have tiny siliceous spicules that help to support their soft structure. Radiolarians are planktonic, microscopic animals that are also siliceous. The other category is diatoms, which are photosynthetic, microscopic organisms. All of these siliceous organisms are quite small and are typically minor constituents of marine sediments in the Gulf of Mexico except for radiolarians.

3.3.3 Chemical Sediments

The direct precipitation of minerals from seawater is present in the Gulf but is not common or widespread. Evaporite minerals are limited to places where salinities reach more than 200 ppt. This would include local places in Laguna Madre and some sites on the Mexican coast. Gypsum and halite are the only evaporate minerals that are even somewhat common, and they are local and subject to dissolution.

Calcium carbonate is the other type of chemical sediment that is directly precipitated from seawater, in some cases with the aid of photosynthesis. Calcium carbonate can be very fine grained and is often referred to as lime mud. It is only common in Florida Bay. Ooids are sand-sized, spherical grains of calcium carbonate that are precipitated in thin layers over a nucleus. They are commonly limited to places where currents, typically tidally generated, are present. Ooids occur in tidal passes in the Florida Keys and off the east coast of the Yucatán Peninsula.

3.3.4 Sediment Grain Size

Sediment particles range widely in grain size. Because of this, the classification of grain size of sediment particles is based on $-\log_2$. This makes it possible to use a small number of categories to cover the entire range of sizes. Typically, sediments are categorized by both particle composition and size (e.g., quartz sand).

3.3.4.1 Gravel

Sediment particles larger than 2 millimeters (mm) (0.08 inches [in.]) are called gravel. They can range up to very large particles including boulders (greater than 25.6 centimeters [cm] [10 in.] in diameter). In fact, not much gravel is carried into the Gulf of Mexico because by the time eroded material makes its way down a long river, the size is reduced considerably. Some beaches have gravel composed of shells, and in some places, such as on the northwest coast of Cuba and parts of Mexico, gravel particles are eroded from rocks close to the beach and are still large. Gravel may also be produced as storms erode reefs. Gravel-sized particles in deep water are essentially all shell material.

3.3.4.2 Sand

Much of the terrigenous sediment present on the continental shelf of the Gulf is sand. Although commonly misinterpreted, sand is only a size term; it has nothing to do with the composition of the sediment. All sediment particles between 2.0 mm and 0.0625 (1/16) mm (0.08 and 0.0025 in.) regardless of origin or composition are called sand. The confusion between the two designations is that the sediment on most beaches, in many streams, and in sand boxes is within this grain size range and is mostly quartz. In many natural environments, sand is mixed with other particles, some larger and some smaller.

3.3.4.3 Silt

Silt is the grain size that is between very fine sand at 0.0625 (1/16) mm (0.0025 in.) and clay (4 μm [0.00016 in.]). Particles of this grain size are a minor component of most Gulf environments except for river deltas. Silt is mostly quartz with minor percentages of other nonlayered silicates.

3.3.4.4 Clay

Clay is another confusing term used in conjunction with sediments; it can mean clay minerals, as described above, or it can mean a grain size. Clay size actually means any sediment particle with a diameter smaller than 4 μm (0.00016 in.). Most of the clay-sized particles are also clay minerals, but some are not. These very small grains are easily transported by rivers and currents in the Gulf. As a consequence, they are very common throughout most of the various environments except where waves and currents are strong, such as in tidal inlets and along the beach/surf zone. These sediment particles are most abundant in estuaries, deltas, and the deep basin.

3.3.4.5 Mud

Although a commonly used term in colloquial English, *mud* is really an appropriate term in scientific literature. Mud—the mixture of silt and clay—is widely distributed in Gulf sediments. Because both silt and clay involve very small sediment particles that are commonly not separated in analysis, this combination term, *mud*, is used. This term will be used in the following discussions.

3.4 GENERAL SEDIMENT DISTRIBUTION

Because of the scale of the geography of the Gulf of Mexico, there have been few studies of the basin-wide distribution of sediment types. A recent effort in this direction by Balsam and

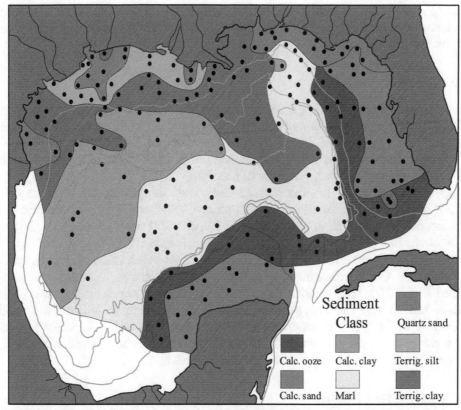

Figure 3.17. Map showing Gulf of Mexico sediment distribution along with sample sites (from Balsam and Beeson 2003: reprinted from Deep Sea Research Part I: Oceanographic Research Papers, Vol 50, Seafloor sediment distribution in the Gulf of Mexico, Figure 4, Copyright 2003, with permission from Elsevier). Contrast this map with that in Figure 3.10.

Beeson (2003) has synthesized sediment samples from the top of 186 cores that cover the entire Gulf (Figure 3.17).

From these samples, the authors have been able to produce relatively simple maps of various sediment characteristics. One of the least complicated maps to interpret is a map showing carbonate content (Figure 3.16). This map shows the influence of the Florida carbonate platform, the Mississippi Delta area, and the expected pattern of decreasing carbonate moving from the deep basin up onto the continental shelf.

3.4.1 Abyssal Plain

Sediments on the abyssal plain tend to be rather homogenous. They are a combination of calcareous ooze (formed by an accumulation of planktonic foraminifera) and thick turbidite sequences. Much of this turbidite material was transported through the Mississippi Canyon and across the Mississippi Fan (Bryant et al. 1991). There is also some clay mineral sediment, most of which had its origin at the mouth of the Mississippi River. Thin turbidite layers are interbedded with the calcareous ooze on the top of these salt dome knolls. The heavy mineral suite is typical of that from the Mississippi River with hornblende and epidote being dominant (Davies and Moore 1970; Davies 1972).

The deep-sea mud that is dominated by clay minerals shows that the most abundant species of clay mineral is smectite. Illite is the next most abundant (Sionneau et al. 2008). Chlorite and kaolinite are the minor clay mineral species.

3.4.2 Mississippi Fan

Most of the sediments on the Mississippi Fan (Figure 3.4) have their origin in the Mississippi River. Because of the slope of its surface, mass wasting is a major process for the delivery of these sediments (Coleman and Roberts 1991). Debris flows and turbidity currents are the primary methods for sediment delivery (Twichell 2011). These sequences contain wood fragments and shells of shallow-water organisms testifying to their shallow-water origin.

The upper sediments of the fan include fine sand, silt, and clay (mud). Layers of this material are covered by foraminiferal muds that are 20–50 cm (8–20 in.) thick with rates of accumulation being calculated at about 30 cm (1 ft) per 1,000 years (Huang and Goodell 1970).

Sediments of this huge accumulation are varied; some are turbidite units and some are thin layers. The thin layers of only a millimeter (0.04 in.) or so in thickness may be annual, almost like rings on a tree. On the other hand, a thick single layer might represent as long as a century (Bryant and Liu 2000). The turbidite units are the thickest of the sediment units, commonly greater than 5 cm (2 in.). This type of individual deposit can be up to 100 cm (40 in.) thick. The graded bedding in the turbidites contains medium to fine sand but is dominated by mud. The debris flow deposits may also include clay clasts. They typically show an erosional base, graded bedding, and the C, D, and E units of a Bouma sequence.

3.4.3 Continental Slope

The continental slope on the west edge of the Florida Platform and the Yucatán Platform margin is steep and is currently accumulating little sediment. The fans and sandy shelf-edge deltas accumulated during sea-level lows during the pre-Holocene era. The rough and irregular topography on the slope (Figure 3.6) causes much ponding of the sediment. Sediments from the Rio Grande source contain the highest percentage of quartz, and sediments from the Veracruz area contain the lowest percentage of quartz. The Mississippi River, which produces the most sediment (Figure 3.5), has an intermediate percentage of quartz in comparison with the other two source areas (Davies and Moore 1970). On both the Texas and Mexico slope in the western Gulf, the sediments are dominated by bluish to brownish fossiliferous mud (Morelock 1969). More specifically, the northwestern Gulf sediments are highly bioturbated hemipelagic muds, with some foraminifera interbedded with laminated silt and muds that are barren (Bryant et al. 2000).

At the shelf-slope break (Figure 3.18), there is not only a break in bathymetry but also a change in the attitude of the sediments. At this break, sediments dip seaward on the slope due to folding and faulting associated with salt diapirs (Morelock 1969). The upper Texas slope is covered with thick mud that is an extension of the sediment on the outer shelf. This mud is the reworked product of the fluvial-deltaic accumulations during lowstands of sea level (Pequegnat 1976). Freshwater shells have been found in these sediments at and near the shelf-slope break indicating that paleoshorelines were in the vicinity (Parker 1960). At the present time, essentially no sediment is being delivered to this environment.

Overall, the thickness of sediment on the slope is quite thin except for local slope fans (Figure 3.19). In the northwestern area of the Gulf, only about 70 cm (28 in.) of mud is present. This represents an average rate of accumulation of only 4.6 cm (2 in.) per 1,000 years (Bryant and Liu 2000); this rate of accumulation conflicts with a rate of about 20 cm (8 in.) per 1,000 years determined by H.H. Roberts of Louisiana State University. The slope sediment fans are

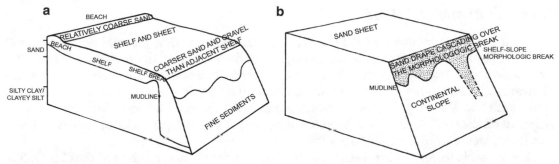

Figure 3.18. General diagram of the outer continental margin showing (a) the main sediment provinces and (b) the shelf-slope break (from Blake and Doyle 1983: reprinted with permission from the Society for Sedimentary Geology).

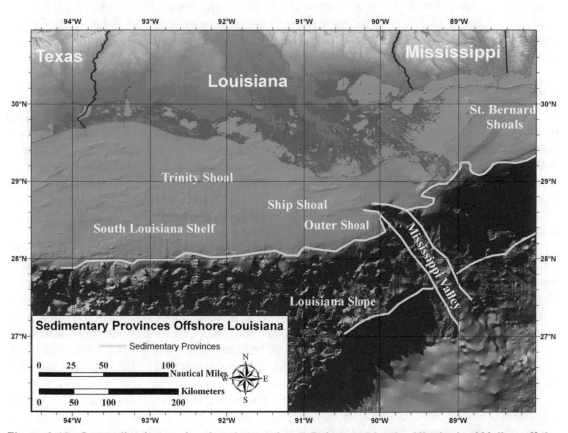

Figure 3.19. Generalized map showing the continental slope cut by the Mississippi Valley off the Louisiana coast with adjacent continental shelf and deep basin area (from Williams et al. 2011).

associated with Quaternary lowstands of sea level. Examples are on the Texas coast where such fans are tied to ancestral courses of the Brazos River and Colorado River (Abdullah et al. 2004). These fans begin at about the 200-m (656-ft) isobaths, essentially at the shelf-slope break, and descend down the slope. Sediments in these fans can be unexpectedly coarse and include some gravel.

3.4.4 Continental Shelf

Shelf surface sediments of the Gulf of Mexico tend to reflect a combination of runoff from the land and the nature of the geological underpinning of the particular region. The Florida Peninsula and the Yucatán Peninsula have similarities because both rest on a carbonate platform. The remainder of the shelf sediments in the Gulf is primarily the result of fluvial input, with the Mississippi River discharge dominating the northwestern Gulf. Much of the sediment on the shelf is directly or indirectly the result of the multiple cycles of sea-level change that took place during the Quaternary Period.

It is possible to designate six sediment provinces of the continental shelf of the U.S. Gulf of Mexico using detailed analysis of the silt fraction of bottom samples. The six sediment provinces are: (1) Apalachicola, (2) Mobile, (3) Mississippi, (4) Brazos–Colorado, (5) Guadalupe, and (6) Rio Grande (Mazullo and Peterson 1989). These provinces are based on the past and present locations of rivers that contributed these sediment grains primarily during low stages of sea level during the Quaternary and they have been reworked during subsequent sea-level rise. A study of 350 grab samples focused on grain roundness and shape (Mazullo and Peterson 1989). As expected, the sediment grains from the Mississippi River dominate the entire northern Gulf shelf but are most abundant in the west of the delta.

3.4.4.1 West Florida Peninsular Gulf

Sediments on the continental shelf off the Florida Peninsula are scarce beyond a depth of about 6 m (20 ft). Out to this depth they are shelly, quartz sand that has been reworked-from Quaternary cycles of sea-level change. This is a zone of transition between the quartz-dominated sediment on land and the carbonate-dominated sediment of the mid- and outer shelf (Brooks et al. 2003a). The most common minor constituent is phosphorite that is reworked from the Miocene deposits of the Florida Platform. The carbonate sediment is being produced biogenically within the shelf itself. The highest content of organic carbon is 5–6 % in the muddy sand and mud facies of the inner shelf (Brooks et al. 2003b).

Farther out on the shelf, it is possible to delineate bands of surface sediment facies that parallel the bathymetric contours (Reading 1978; Hine et al. 2003a). The inner shelf is dominated by quartz sand as described above, and the middle shelf is carbonate skeletal material (Figure 3.20). There is a belt of calcareous coralline (red) algae on the outer shelf. Just beyond that is a narrow belt of ooids (Figure 3.21), which must be relict deposits that formed during the recent lowstand of sea level. At the present time, this shelf is sediment starved because the estuaries remain void of sediment and virtually no sediment is being delivered to the shelf itself. The total modern sand sheet is about 8 m (26 ft) thick and composed of 90 % quartz and 10 % carbonate.

There are grain size trends that relate to the composition of the shelf sediment. The content of sand is 90 % or more most of the way across the shelf. Near the outer region, sand content decreases rather rapidly. The percent carbonate in the sediments gradually increases from the transition from quartz in the shoreface (where it is only 25 %) to near the edge (where it increases to nearly 100 %) (Doyle and Sparks 1980). As for the clay minerals, smectite content increases offshore and kaolinite decreases offshore. The heavy mineral suite associated with the quartz-rich area is characterized by zircon, tourmaline, garnet, and staurolite (Fairbanks 1962).

3.4.4.2 Florida Panhandle

The Florida panhandle shelf is quite different than that on most of the Florida Platform because it includes a relatively thick sequence of sediment that is the result of the dominance by

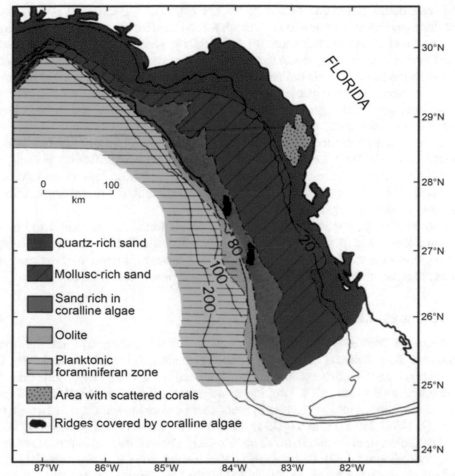

Figure 3.20. General map of the surface sediment facies on the west Florida shelf (from Reading 1978).

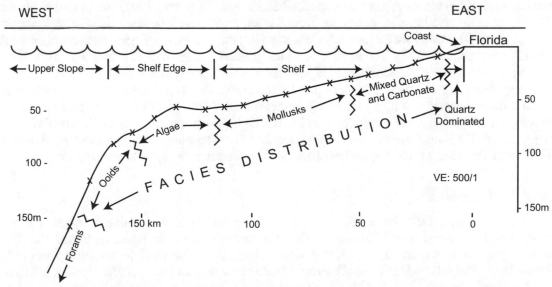

Figure 3.21. Profile diagram across the west Florida shelf showing sediment facies (Hine et al. 2003a, b). Reprinted from Marine Geology, Vol 200, Hine AC et al., The west-central Florida inner shelf and coastal system: A geologic conceptual overview and introduction to the special issue, Figure 5, Copyright 2003, with permission from Elsevier.

rivers and river deltas (Hine and Locker 2011). The origin of the sediment is the Apalachicola River and delta and other streams to the west that feed the Florida panhandle and the Alabama shelf (McKeown et al. 2004). The ancestral Apalachicola Delta is still visible in the bathymetry of this shelf and is a major contributor to the sediment on the present shelf as the Holocene sea-level rise has moved over it. This sediment has been reworked into numerous shoals and ridges as sea level advanced after the glacial maximum (Donoghue 1993). This area has been relatively sediment starved since the beginning of the Holocene. The sediment is dominated by terrigenous sand and contains minor amounts of shell.

The surface sediments here comprise fine and medium sand and are moderate to well sorted (McBride et al. 2004). Combined with the adjacent Alabama/Mississippi shelf, there is a trend in sediment texture from coarser- to finer-grained and from less sorted to more sorted. The outer edge of the shelf is dominated by carbonate sediment of a reefal origin with mixtures of terrigenous sand–silt–clay of inter-reef origin.

There are shelf-edge carbonate hardgrounds and bioherms that have up to 15 m (49 ft) of relief and are located at depths of 90–120 m (295–395 ft). These features probably reflect shoreline regions that existed during the last major sea-level lowstand (Bart and Anderson 2004). Overall carbonate content of this shelf area is typically less than 25 %.

3.4.4.3   Alabama–Mississippi

The Alabama–Mississippi continental shelf is a continuation of that off the Florida panhandle, and the sediments reflect that trend. The sediment is quartz-dominated, fine sand (Kopaska-Merkel and Rindsberg 2005). Sand-size sediment comprises more than 90 % of surface sediments (Bowles 1997). Looking more specifically at the shelf environments, the outer areas have a lime mud surface with relict reefs/carbonate buildups at two depth zones: 654–680 m (2, 145–2,230 ft) and 97–110 m (318–361 ft) (Roberts and Aharon 1994). Ludwick (1964), one of the original detailed studies on this shelf, showed distinct bands of sediments that parallel the shore (Figure 3.22). Ludwick also noted the presence of reefal materials at the outer portion of the shelf. Further work in 2001–2002 by the U.S. Geological Survey (USGS) revealed extensive hardgrounds in this area and some sandstones.

Shoreward of this carbonate region is a transition of terrigenous sand and mud. Further in on the shelf, there is topography that suggests relict barriers and shorelines. A detailed study by the Geological Survey of Alabama (Kopaska-Merkel and Rindsberg 2005) found that the inner shelf comprised five lithofacies with terrigenous sand, mud, and biogenic debris being the main constituents: (1) graded shelly sand, (2) clean sand, (3) dirty sand, (4) biogenic sediment, and (5) muddy sand (Figure 3.23). The sand is medium grained (mean of 0.43 mm [0.017 in.]). Shell content of the inner shelf sediment is higher than on the present beaches of Alabama.

A systematic study of the total organic carbon across the shelf found that the values range from a trace amount to a high of 2.9 % (Kennicutt et al. 1995). Most of the samples contained less than 1 % total organic carbon by weight. The high values were at or near the shelf edge in the head of De Soto Canyon. This study included five samplings over 26 months. A major finding of this study is the variability of the sediment content in as little as 6 months.

3.4.4.4   Louisiana

The Mississippi Delta has a major impact on the sediments of the continental shelf adjacent to the Louisiana coast. The active lobe of the delta extends across the entire shelf with the river discharging directly onto the continental slope (Figure 3.5). The shelf here ranges from only about 15 km (9.3 mi) wide off the Mississippi Delta to more than 150 km (93.2 ft) wide adjacent

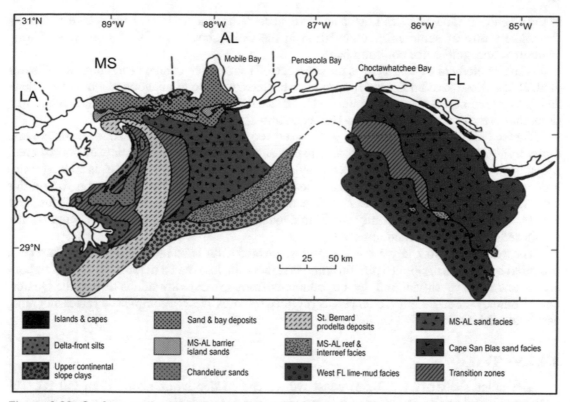

Figure 3.22. Surface sediment facies on the northeastern Gulf continental shelf (modified from Ludwick 1964).

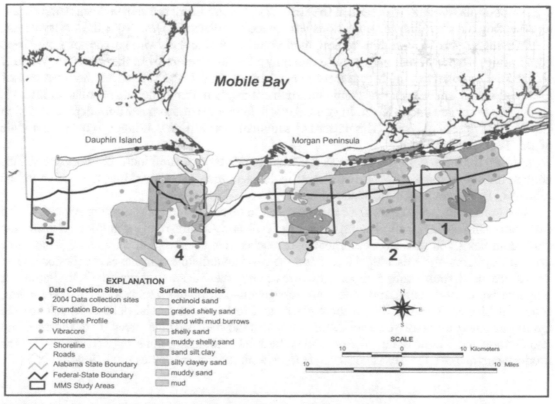

Figure 3.23. Map of shelf sediments produced by the Alabama Geological Survey in the search for beach nourishment sand (from Kopaska-Merkel and Rindsberg 2005).

to the Chenier Plain in western Louisiana. As expected, the modern lobe of the delta experiences the highest rate of sediment accumulation in the Gulf, about 1 m (3.3 ft) per year. Slump structures and gullies are common here.

A major element of this shelf is the presence of four large, elongate sand bodies (Williams et al. 2011). These sand bodies are the products of the reworking of abandoned lobes of the delta as it prograded across the continental shelf. The sand shoals represent old shoreline accumulations that were reworked during the Holocene transgression.

The sediment in these shoals is well-sorted terrigenous sand. These four shoals contain many millions of cubic meters of beach quality sand that has great potential for nourishment projects on the present, eroding barrier islands. Ship Shoal, the largest of these, is 50 km (31 mi) long and 7–12 km (4–7 mi) wide with a relief of up to 7 m (23 ft). The mean grain size on this shelf ranges from medium to fine sand. The nearshore area in the western portion of this shelf is fine, as is the sand from Trinity and Ship Shoal (Figure 3.12). The bottom sediments farther offshore in the west are medium sand.

The wide shelf to the west is essentially covered with a blanket of mud that has been provided by the Mississippi River. This mud is rather thin, less than 8 m (26 ft) thick throughout and is underlain by fluvial and deltaic sediments from sea-level lowstands during the Quaternary. Some of the sediment sequences, as revealed by cores, show a complex of facies including carbonate debris.

3.4.4.5 Texas

The inner shelf off the Texas coast will extend to depths of about 15 m (50 ft). The sediments here have various origins but two are major contributors—the Mississippi River and reworking of older sediments as the sea level rose over the past 8,000 years or so. Shells and shell debris are another significant component. The influence of the river diminishes from northeast to southwest along the Gulf. In the most eastern portion of the Texas shelf, mud is dominant or about equal with sand in the surface sediment. Local areas have linear sandy areas representing old shoreline accumulations left behind as relict sediments when the sea level rose.

Moving westward along this region, mud is still very abundant with patches of sandy mud dominating. Holocene sediment is quite thin only a few kilometers from the shoreline (White et al. 1985). The relatively high concentrations of muds tend to be related to the locations of lowstand deltas where mud was dominant, such as the paleo-Trinity delta. A similar situation is associated with the Brazos River. In general, sand dominates out to maximum depths of 5–8 m (16–26 ft) (White et al. 1988). This pattern of sediment distribution continues to near the middle of the Texas inner shelf.

Across the inner zone of the Texas shelf, the percentage of sand increases noticeably. This zone in the southern part of the Texas shelf includes numerous sandy ridges that are shoreline remnants from previous high stands of sea level.

Geographically, the Texas shelf is considered to be subdivided into three provinces: (1) the Colorado–Brazos delta complex, (2) the south Texas intra-deltaic ramp, and (3) the Rio Grande delta complex (Holmes 2011). Sediment that reaches this shelf region may come from three drainage systems–the Mississippi River and the two fluviodeltaic complexes mentioned above. There are numerous shore-parallel structures along the Texas continental shelf. Some are biogenic banks and reefs, and others are terrigenous sediments (Holmes 2011). The modern sediment blanket is rather thin in most places and rests on the fluvial-deltaic deposits of the Quaternary lowstands of sea level. Mud dominates the shelf surface except over the ancestral Rio Grande and Sabine deltas, where sand is the most abundant grain size (Berryhill 1975). The modern surface off the Texas coast tends to be a mud blanket (Eckles et al. 2004).

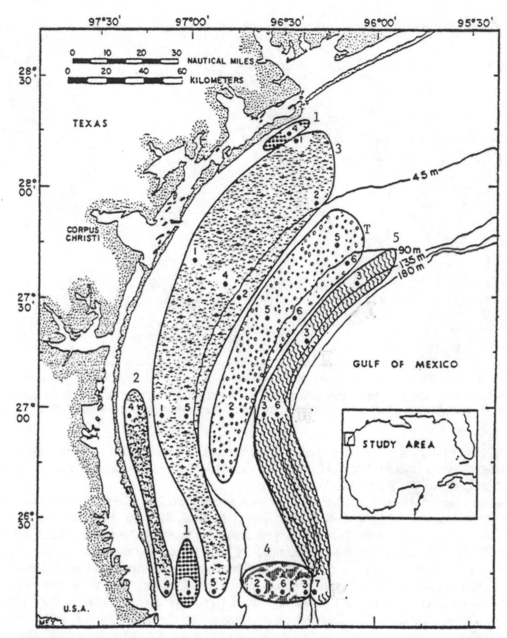

Figure 3.24. Bands of sediment on the continental shelf of south Texas show that most sediment textures and compositions on this shelf are coast-parallel (Behrens et al. 1981). From *Environmental Studies of a Marine Ecosystem: South Texas Outer Continental Shelf* edited by R. Warren Flint and Nancy N. Rabalais, Copyright 1981. Courtesy of the University of Texas Press. Details of sediment characteristics are shown in Table 3.9 of Behrens et al. (1981).

The inner shelf sediments are generally organized with parallel sediment types (Figure 3.24). The surf and nearshore zone tends to be sand dominated due to the high energy in this area. Moving farther offshore, this pattern is lost and more local variations are present. For example, the easternmost part of the Texas inner shelf is quite muddy with less than 10–12 % sand (White et al. 1987). Moving toward the west, sand dominates the shelf out to a depth of about 15 m

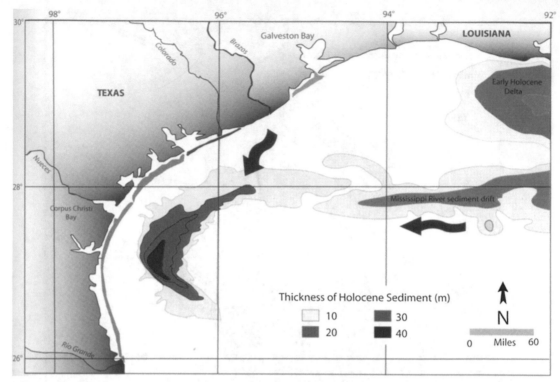

Figure 3.25. Map showing the distribution and thickness of Holocene sediment on the northwest Gulf of Mexico. The greatest thickness is on the south Texas shelf (from Holmes 2011).

(50 ft), then muddy sand and sandy mud dominate out to the mid-shelf. General mean grain size ranges from about 3.0Φ to 7.0Φ (Φ is $-\log_2$ in mm). There is a thick mud layer offshore in the central part of the Texas shelf (Figure 3.25).

Muddy sand and sandy mud with mean grain sizes mostly around 5.0Φ are found west of the ship channel into Houston (White et al. 1985). Down the coast, the inner shelf is quite muddy and is $6.0–7.0\Phi$. The sand content increases approaching and crossing the mouth of the Colorado River (White et al. 1988). Moving toward the Port Lavaca area, the sand percentage increases with mean grain size of very fine sand in the surf zone grading out to coarse silt. Continuing to the south, the pattern of grain size becomes relatively organized; the values are essentially parallel to the coast (Figure 3.24), from 3.5 to 6.5Φ (from White et al. 1983). In the mid-south coast of Texas, near Kingsville, the inner shelf sediments are dominated by sand with increasing mud offshore. All major categories of sediment texture are present on the inner shelf. The gravel is shells and shell debris. The sand fraction is 87% quartz, 6% feldspar, and 4% rock fragments; accessory minerals make up the remaining 3%. The heavy mineral content is about half black opaques (magnetite, etc.) along with tourmaline, hornblende, zircon and pyroxenes, and rutile. These heavies are similar to those described along the entire coast by Bullard (1942). The mud content increases when approaching the Rio Grande delta area (White et al. 1986).

The region in the central portion of the Texas shelf that shows what seems to be unusual bathymetry (Figure 3.10) also has an unusual sediment accumulation. This area of about 300 km^3 (72 mi^3) represents the second largest sediment depocenter on the Gulf of Mexico shelf (Figure 3.25) next to the Mississippi Delta (Holmes 2011). The thickness of this mud blanket is tens of meters, more than half of which was deposited during the past 3,000 years. It is interpreted that the origin of this late Holocene sediment is the production of mud from the Brazos, Colorado, and Mississippi Rivers.

The outer shelf sediments have a high abundance of calcium carbonate. This comes from the relict reefs and banks that developed during Quaternary lowstands of sea level. They have been partially reworked by the post-glacial rise in sea level and the debris incorporated into the outer shelf sediments (Rezak et al. 1985).

3.4.4.6 Mexico and Cuba

Sediments on the continental shelf of Mexico either reflect the nearby sources of terrigenous material or, as in the case of the Yucatán Peninsula, they are autochthonous carbonate sediments. Terrigenous sand and mud dominate the northern portion of the shelf.

Toward the south, volcanic rocks provide the source for shelf sediments and the composition reflects this. In the central Mexico coast near Veracruz, the shelf narrows in the area of volcanoes, and volcanic glass is a prominent component of the shelf sediment. Shelf sediment on the coast near Isla del Carmen falls below 50 % carbonate (Carranza-Edwards 2011). Moving into Campeche Bay, the sediments transition from terrigenous to carbonate. Sediments and bathymetry across the Campeche Bank of the Yucatán Peninsula are comparable to that of the west Florida shelf (Figure 3.26).

3.4.5 Mississippi Delta

The sediments of the Mississippi Delta have been well documented in the literature. There are essentially three primary geomorphic/physiographic provinces: the delta plain, the delta front, and the prodelta from land to the Gulf. The delta plain is the upper surface, much of which is supratidal or intertidal (Figure 3.27). This province is dominated by mud, with sand bodies representing modern and relict point bars on the numerous channels. The tremendous influence by human activities, primarily the petroleum industry, has had a major impact on this province. The rapid rise of relative sea level is causing the destruction of much of the wetlands portion of it. Sediments are subjected to widespread pollution from both river discharge and human activities. The delta front portion of this complex is dominated by sorted sand and occupies the outer edge of the delta where wave action dominates. These sands are worked by waves and currents into shore-parallel sediment bodies (Coleman and Roberts 1991).

The vast majority of the sediment volume in this delta is mud and comprises the prodelta province (Figure 3.27). These sediments accumulate rapidly and are generally saturated with water causing major instability problems. Failure and gravity slumping is widespread. Some diapirs—not only of salt but also mud—are present.

3.4.6 Beach Sediments

The nature and composition of beach sediment show a fair amount of commonality throughout the Gulf of Mexico. There are two main categories of beach sediment composition—terrigenous and carbonate—but some places show subequal mixtures. Sediment texture is typically well sorted and well rounded, except where the composition is bimodal with shells being a significant part of the composition. Carbonate-dominated sediments are in the Florida Keys, around the Yucatán Peninsula, and east of Havana on the Cuban coast.

In a few places, minor constituents show some concentrations due to the underlying geology. One is on the west-central coast of the Florida peninsula, where phosphorite is anomalously high due to the abundance of this mineral in the underlying Miocene strata. The thin dark layers on storm beach surfaces reflect the presence of this material.

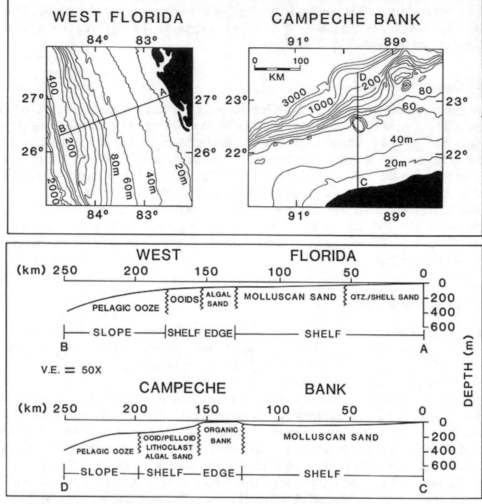

Figure 3.26. Similarities between the shelf and shelf-slope break on the west Florida Shelf and the Yucatán Shelf in Mexico (from Hine and Mullins 1983: reprinted with permission from the Society for Sedimentary Geology).

The concentration of shells in some local places provides what is an anomalous carbonate beach in an otherwise terrigenous-dominated beach environment. Examples include the southern portion of Sanibel Island, Florida and near the middle of Padre Island, Texas, not far north of the Mansfield Pass jetties.

Another anomaly in beach composition is the presence in some locations of what are commonly called *tar balls* which are small, pebble size clumps of oil-bound sand (Figure 3.28). The oil is from seeps on the floor of the Gulf. Tar balls are most common off the Texas coast and the coast of Mexico west of the Yucatán Peninsula.

3.4.7 Estuaries and Lagoons

Numerous estuaries line the Gulf Coast. They are generally somewhat similar in their origin in that they are drowned river systems. Most are muddy, shallow, and brackish. Tidal flux varies widely, but the tidal range is microtidal throughout the Gulf. The following discussion will address the nature of the sediment in each of the major estuaries.

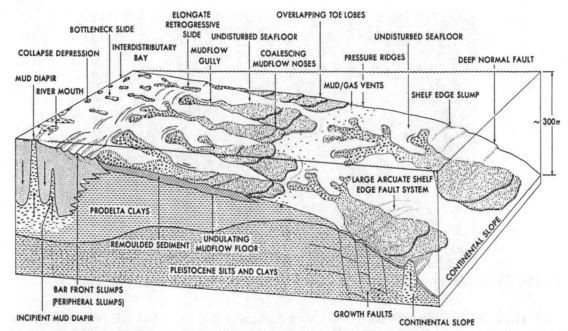

Figure 3.27. Block diagram of the various sedimentary facies on the Mississippi Delta (Coleman 1988). Republished with permission of the Geological Society of America from *Dynamic Changes and Processes in the Mississippi River Delta*, J.M. Coleman, Vol 100, pp 999–1015, 1988; permission conveyed through Copyright Clearance Center, Inc.

Figure 3.28. Small tar ball on the beach. This size of a ball will cover most of a person's heel when stepped on (photo courtesy of NOAA). These tar balls are from natural seeps and are common along the northern Gulf Coast especially in Texas.

3.4.7.1 Florida Bay

Florida Bay lies between the Florida Keys and the south Florida mainland. It is a triangle-shaped, shallow bay with scattered small mangrove islands. It is open to the Gulf on its west side (Figure 3.29) and receives its freshwater supply from the sheet flow that moves across the Everglades. This bay covers 1,393 km^2 (538 mi^2) and has an average depth below 2 m (6.6 ft). Sediments in this shallow bay are calcium carbonate and a combination of lime mud and

Figure 3.29. Satellite image of the Florida Keys and Florida Bay (courtesy of NASA).

skeletal material. These soft sediments are only a few meters thick and rest on the limestone surface of the Miami Oolite and the Key Largo formations (Enos and Perkins 1977) that form the basis of the Florida Keys.

3.4.7.2   Charlotte Harbor

Charlotte Harbor is one of the two large estuaries on the Gulf Coast of the Florida Peninsula. It has an area of 700 km² (270 mi²). Rapid growth from the 1950s to the present has increased the population to more than one million. It is served by the Myakka, Peace, and Caloosahatchee Rivers (Figure 3.30), which are presently carrying freshwater but little sediment. The sediment in this estuary has been studied in detail by Evans et al. (1989) and Brooks (2011). Modern (Holocene) sediments may be up to 3 m (10 ft) thick and are mostly fine quartz sand. Sandy shell is the dominant sediment in the tidal channels and passes. Phosphate minerals are up to 9 % by weight (Folger 1972) and are delivered by the Pease River from the central Florida phosphate-mining district. Mud is a minor constituent but is widespread. Mud is a combination of clay minerals, clay size quartz and calcite, and particulate organic matter (Huang and Goodell 1967).

3.4.7.3   Tampa Bay

Tampa Bay is an estuary surrounded by intensive and extensive development with a total of about three million residents. It is also a major tourist destination. The bay is supplied with runoff by the Manatee, Alafaya, and Hillsborough Rivers (Figure 3.31), but little sediment is being discharged by any of them (Brooks 2011). The USGS (2006) study of the bay includes detailed surface sediment analyses. Maps produced from this study show variation in the bay, but overall, the sediment is rather similar throughout. Grain size in most of the bay is fine and medium sand with muddy sediments concentrated in the northern portion of the two arms of the estuary. Mud is less than 10 % throughout the bay except for the upper part of the eastern bay where it reaches more than 50 % at some locations. The mud is a combination of clay minerals, fine quartz, and particulate organic matter. Carbonate composition provided by biogenic shells and debris is low except in the lower bay where it may exceed 50 % (USGS 2006).

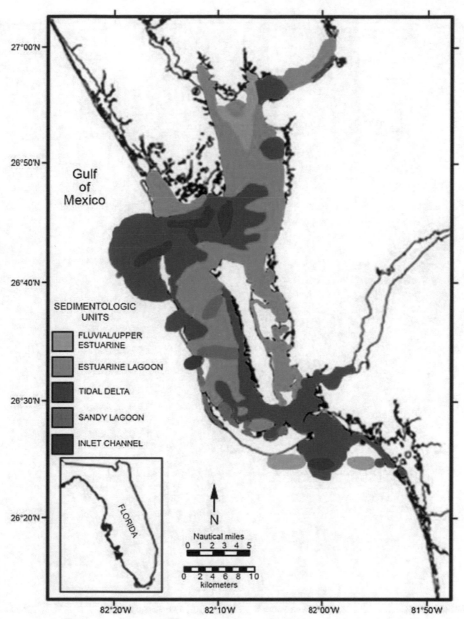

Figure 3.30. Map of sediment facies for Charlotte Harbor area. These facies can be related to sediment grain size (Evans et al. 1989). Reprinted from Marine Geology, Vol 88, Evans et al., Quaternary stratigraphy of the Charlotte Harbor estuarine lagoon system, southwest Florida: Implications of the carbonate-siliciclastic transition, Figure 7, Copyright 1989, with permission from Elsevier.

3.4.7.4   Apalachicola Bay

The sediment of Apalachicola Bay is dominated by mud with several areas of large and productive oyster beds (Figure 3.32). Much of the mud is actually deposited as oyster fecal pellets. The oyster reefs range in length from about 1 to 1.7 km (0.62–1.06 mi) and are oriented northwest–southeast. Quartz sand is a minor but widespread constituent.

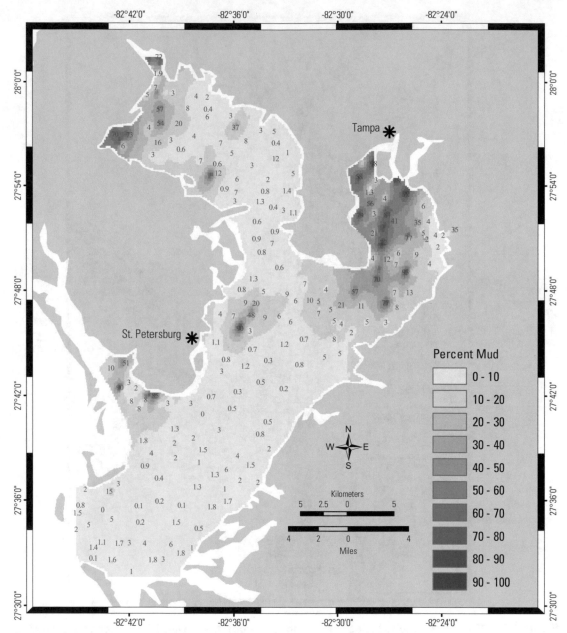

Figure 3.31. Image of Tampa Bay showing the percent of mud. The reciprocal can be considered as the sand percent because the gravel shell component is small (courtesy of the USGS).

3.4.7.5   Pensacola, East, and Escambia Bays

The Pensacola, East, and Escambia Bays have an unusual shape and are dominated by muddy sand (Figure 3.33). Like most coastal plain estuaries, sediments in this system are derived primarily from the rivers that empty into the estuaries. Both mud and sand are spotty in their distribution. Some sand is blown and washed over the adjacent barriers, but washover is infrequent because dunes along this part of the northern Gulf Coast are large. Washover occurs only where dunes are cut and washover channels develop.

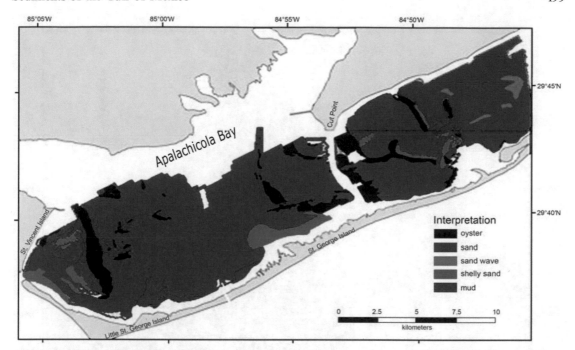

Figure 3.32. Map of Apalachicola Bay, Florida showing oyster reefs and sediment types (Twichell et al. 2010). Reprinted from Estuarine, Coastal and Shelf Science, Vol 88, Twichell et al., Geologic controls on the recent evolution of oyster reefs in Apalachicola Bay and St. George Sound, Florida, Figure 3, Copyright 2010, with permission from Elsevier.

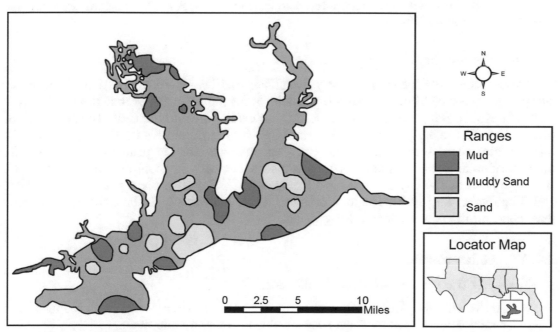

Figure 3.33. Map of Pensacola, East and Escambia Bays, Florida showing the grain size of the sediments (from Macauley et al. 2005).

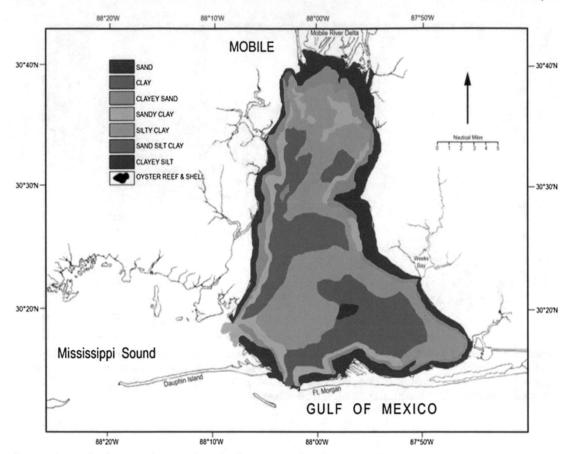

Figure 3.34. Surface sediment distribution map of Mobile Bay, Alabama, and the adjacent area (modified from Ryan 1969).

3.4.7.6   Mobile Bay

Large amounts of sediment—nearly five million metric tons—are carried into Mobile Bay every year. About 33 % of it remains in the delta at the mouth of the Mobile River, 50 % of it settles in Mobile Bay, and the remaining 15 % makes its way into the Gulf. The sediments of Mobile Bay (Figure 3.34) are rather uniform. Mud virtually dominates the estuary except for a thin margin where some sand is present. Clay minerals are the primary grain type with montmorillonite being dominant. There is also a decrease in montmorillonite abundance from the head of the bay down toward the open Gulf. The other clays present are kaolinite and illite (Isphording 1985). The total organic carbon content is high throughout; some areas have concentrations well above 2 % by weight.

3.4.7.7   Galveston Bay

Texas has five primary sources of sediments:

- Active streams: Most bays have active streams that carry terrigenous sediment in a combination of bed load and suspended load emptying into them.

- Erosion of shorelines: Some sediments are derived from the erosion of the shorelines of the bays, and most of these sediments come from bays that have small bluffs of Quaternary sediment, such as Lavaca Bay and Copano Bay.

- Tidal inlets: Tidal inlets enter and influence some of the bays; these inlets may transport marine sediment into the bay, generally accumulating in the form of a flood-tidal delta.

- Eolian and washover processes: Eolian and washover processes carry sand across the barrier islands and into the Gulfward margins of the bays.

- Biogenic shell material: The only nonterrigenous sediment that is common in the bays; biogenic shell material is found as both whole shells and as sand and gravel-sized debris. The bulk of this shell sediment is from oysters.

There is a general pattern to the sediments in the Gulf estuaries. They are all relatively low energy environments with low to modest energy caused by tidal flux. Waves are small with short periods. As a consequence, the standard pattern is somewhat target shaped with high sand content along the margins and mud in the center. Mud dominates and commonly covers about two-thirds to three-quarters of the area of the bay. Many of the estuaries have oyster reefs that cause local variations in the coarse fraction of the sediments. Shell debris in both gravel- and sand-sized particles is common in association with these reefs. The oysters are major factors in the sedimentation of the estuaries because of their huge capacity for filtering suspended sediment out of the water column and producing coarse silt and fine sand-sized pellets of mud.

The sediment distribution maps of the Texas estuaries are all taken from the sequence of publications of the University of Texas Bureau of Economic Geology by William A. White et al. (1983, 1985, 1986, 1987, 1988, 1989a, b). Sediment abundance is shown by the same symbols throughout (Figure 3.35).

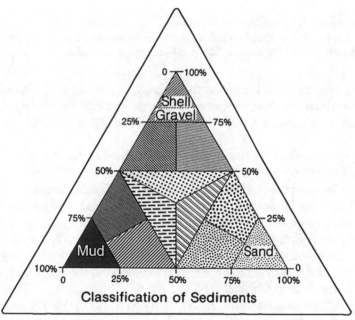

Figure 3.35. Triangular classification of surface sediments in the Texas estuaries with gravel, sand, and mud as the major categories. Figures 3.35, 3.36, 3.37, 3.38, 3.39, and 3.40 are all based on this classification (from White et al. 1983).

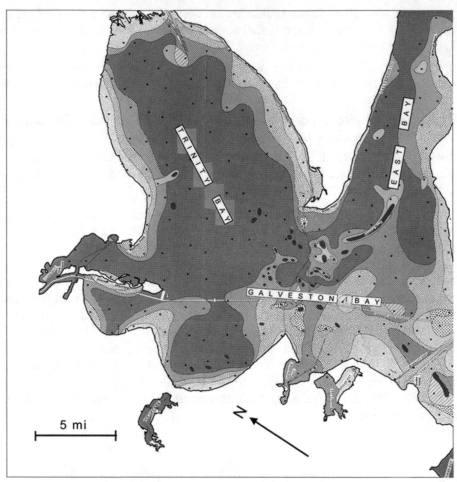

Figure 3.36. Surface sediment distribution map of the Galveston Bay complex. *Black dots* represent sample locations (from White et al. 1985). Black is oyster reefs.

Galveston Bay (Figure 3.36) is shallow throughout, averaging only about 2 m (6.6 ft) deep, but it has multiple deep shipping channels. Sediment sources are mainly from the barriers, especially the Bolivar Peninsula and the Gulf. Little sediment is provided by the Trinity River (Phillips 2005).

The topography of the bay is like a shallow bowl with relief caused by oyster reefs. This bay produces 80 % of the oyster meat in Texas. The oyster reefs also contribute significantly to bay sediments; both shells and pellets of mud are produced by these extremely active filter feeders.

Most of the sediment in Galveston Bay is mud, sandy mud, and muddy sand. The central part of the bay is mud (Figure 3.36). Sand is concentrated in the bay margin and associated with the flood-tidal delta near Bolivar roads and the Trinity River delta. Oyster reefs are widespread, and shell gravels are associated with them. Some of the sand in the bay is also derived from oyster shells. Human influence on sediment distribution is in the form of spoil mounds from dredging of the Houston Ship Channel (USEPA 1980).

The percent sand shows a general trend from high (60–100 %) at the shoreline, decreasing to about 20 % near the bay center (White et al. 1985). In general, the sand abundance, and therefore the grain size, decreases in a similar trend. Overall sand abundance is related to energy levels of both tidal currents and waves.

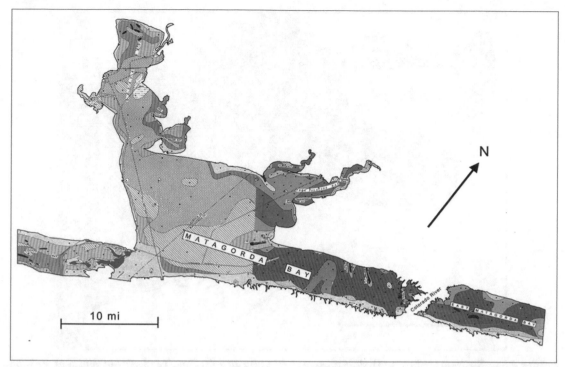

Figure 3.37. Surface sediment distribution map of Matagorda Bay area (from White et al. 1988, 1989a). Black is oyster reefs.

3.4.7.8   Matagorda Bay

Matagorda Bay is a rather narrow bay that separates the mainland from Matagorda Island. The Colorado River Delta has prograded across the bay over the past century, and the river now empties directly into the bay. This bay has a relatively complicated geography and includes both a typical estuarine morphology and a backbarrier portion that is bisected by the Colorado River delta (Figure 3.37). The Colorado River delta has prograded across the bay with the major channel emptying directly into the Gulf (Kanes 1970). This progradation has been extremely rapid and took place over a few decades during the twentieth century.

Unlike most of the Texas estuaries, Matagorda Bay does not have significant oyster reef development; oyster reefs are restricted to a few just west of the river delta that are oriented shore-normal and a few in East Matagorda Bay that are shore-parallel. The surface sediment pattern is similar to that of most bays of this coast. Mud dominates the area with increasing sand toward the shorelines. The relative abundance of sand reflects the location relative to wind direction and waves. The more protected areas show mud closer to the shoreline. The landward side of the Matagorda barrier has extensive sand due to washover and blowover.

There is little tidal flux in this system, and as a result, no sandy sediment accumulations are related to tidal flux except in the southwest corner of the bay. Both the constructed ship channel and Pass Cavallo are located here. The ship channel has small spoil banks, which are relatively high in sand, and the tidal inlet has a large flood-tidal delta that is sand dominated (Figure 3.37).

3.4.7.9   San Antonio Bay

San Antonio Bay is located at the mouth of the combined San Antonio and Guadalupe Rivers where a large bayhead delta—the Guadalupe Delta—has formed (Figure 3.38). The

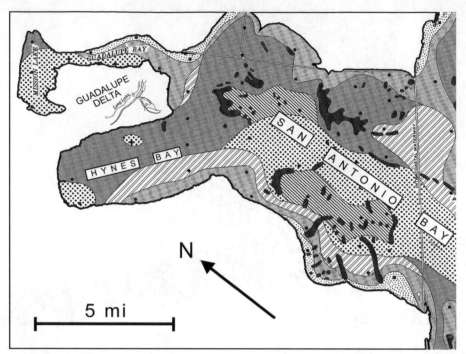

Figure 3.38. Surface sediment distribution map of San Antonio Bay, Texas (from White et al. 1988, 1989a, b). Black is oyster reefs.

Guadalupe River delta dominates the northwest part of San Antonio Bay and has shown significant progradation over the past century, but has somewhat stabilized during the past few decades. The delta sediments are more than 60 % plant remains (Donaldson et al. 1970); the bay sediments, as a whole, are only about 4 % plant remains.

This bay is the second most productive oyster region on the Texas coast. The reefs are scattered throughout the bay. This bay has no locations where tidal flux is a factor in sediment texture or accumulation. The bay is less than 2 m (6.6 ft) deep on average and is brackish. There are some low bluffs along the shoreline that produce sand when eroded.

An important aspect of San Antonio Bay is related to how the widespread oyster reefs influence sediment texture. The majority of the bay is regularly dredged by oystermen with their trawlers. This disturbance of bottom sediment tends to produce slightly coarser grain size because the dredging suspends fine sediment allowing it to travel to other parts of the bay (White et al. 1989a).

The entire bay is dominated by mud composed primarily of clay minerals. Morton (1972) studied 80 samples from throughout the bay for their clay mineralogy and found that smectite, illite, and kaolinite were the dominant species with smectite being the overwhelming majority.

3.4.7.10 Aransas and Copano Bays

Copano and Aransas Bays are similar to San Antonio Bay in their sediment composition and distribution. These bays are also important to the oyster industry in that reefs are extensive in both. They have the typical pattern of sediment texture distribution with fines in the middle and becoming coarser near the shoreline. Copano Bay has some low bluffs that are exposed to the prevailing wind and therefore are subject to some erosion. Extensive oyster dredging also influences the sediment texture here. Another factor is the petroleum industry's considerable

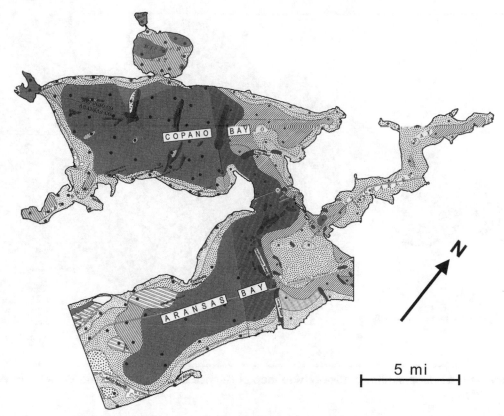

Figure 3.39. Satellite image of Aransas (*lower*) and Copano (*upper*) Bays (from White et al. 1983, 1989a). Black is oyster reefs.

drilling and related activity that has taken place in both bays. Drilling activity, construction, maintenance, post drilling activity, and the presence of the permanent structures in the bay all have an influence on the bays' surface sediments.

St. Charles Bay, which is a branch of Aransas Bay, is sand dominated as compared to other bays along this coast (Figure 3.39). Sand also dominates the northeast part of Aransas Bay adjacent to St. Charles Bay. The corresponding end of Copano Bay also shows a high percentage of sand. Both of these locations are on the downwind portion of the respective bays relative to the strong prevailing wind along this coast.

Both of these bays are major oyster areas. These animals are tremendous filter feeders and pass many liters of turbid water through their system every hour. As a consequence, the fine silt and clay-sized particles are aggregated into fecal pellets that are typically sand sized. Oyster shells are a common constituent and contribute most of the gravel fraction to the sediment. The sediment is mostly mud and is patchy because of the abundance of oyster reefs. There is some sand around the bay margins.

3.4.7.11 Corpus Christi Bay and Nueces Bay

The Nueces River is the primary source of freshwater and sediment to both the Corpus Christi Bay and Nueces Bay. The present sediment contribution of this river is minimal because of impoundments along its course. In the past, however, considerable terrigenous sediment was carried to these two bays by this river. Although the pattern of surface sediment on Corpus

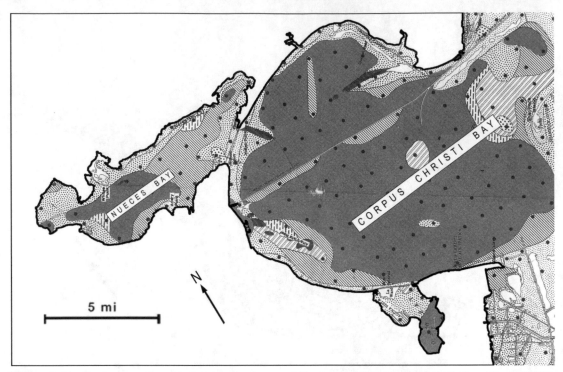

Figure 3.40. Surface sediment distribution map of Nueces and Corpus Christi Bays (from White et al. 1983).

Christi Bay is typical of this coast with a huge portion being mud dominated and a trend to more sand to the shoreline, Nueces Bay is different (Figure 3.40).

Mud comprises a fairly small portion of the surface sediment in Nueces Bay. Much of the bay is covered by sandy mud. The delta area is also high sand. Some of this can be attributed to the extensive dredging for oyster shell that has taken place in this bay (White et al. 1983). Oyster shell and oyster debris are fairly common in Nueces Bay but not in Corpus Christi Bay.

The constricted area between the two bays is also the site of the causeway across them. The combined effects of the constriction and the causeway structure have caused a relative coarsening of surface sediment. This has resulted from tidal flux through the constriction/ structure and its location on the downwind end of the bay where the fetch is maximum (Figure 3.40).

Corpus Christi Bay is much larger than Nueces Bay and is dominated by mud in its surface sediment, mostly with a mean grain size of greater than 7.0Φ (White et al. 1983). Sand increases to the shorelines. The only variations in this pattern are associated with a large oyster reef in the northeast part of the bay, the coarsening associated with dredge spoil along the ship channel, and the area near the population center associated with Corpus Christi. An oyster reef is also present in the northeast area.

3.4.7.12 Baffin Bay

Baffin Bay is unique among the coastal bays of the United States in that it receives little freshwater and virtually no terrigenous sediment at the present time. Its geography is an obvious drowned fluvial system, which is much more evident for Baffin Bay than for other

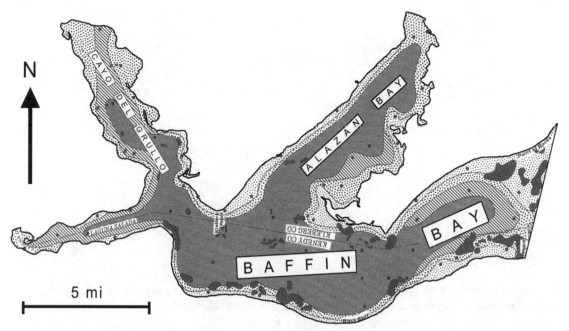

Figure 3.41. Surface sediment distribution map of Baffin Bay and part of adjacent Laguna Madre, Texas. It is apparent from the geography of the bay that it is a drowned fluvial system (from White et al. 1989b). Black is serpulid worm reefs.

bays along this coast. Because of its location in a semi-arid portion of the western Gulf, Baffin Bay is hypersaline.

Other reefs in this bay, called worm reefs, are constructed by the polychaete worm, *Sabellaria alveolata* (Andrews 1964). These worms do not secrete the tubes that they live in; they construct the tubes out of sand and shell grains. The worms are abundant in Baffin Bay (Figure 3.41). Their distribution is primarily along the shoreline and near the entrance.

The nonreefal portions of Baffin Bay are dominated by mud with sand percentage increasing toward the shore. Because the streams that feed Baffin Bay are intermittent, almost no terrigenous sediment is presently being contributed to the bay. The combination of environmental conditions does permit precipitation of carbonate sediment, including dolomite (Behrens and Land 1972). Ooid sand is also produced and is concentrated along the northern margins of the bay (White et al. 1989b). Exposed Pleistocene beach rock also contributes gravel-sized particles along the margin where this material is exposed.

The grain size of Baffin Bay sediments is primarily a relict condition from earlier times when the climate was wetter and the streams that fed the bay were regularly discharging water and sediment. The bay floor is mud, dominated by silt-sized grains. This sediment grades shoreward through sandy mud and then muddy sand.

The margins of the bay have scattered beach rock reflecting the nature of the climate and very low wave energy. Other indicators of such a climate and absence of terrigenous input is the presence of ooids scattered along relatively high-energy parts of the bay (Rusnak 1960; Behrens 1964).

Baffin Bay is very shallow with essentially no fluvial or marine input of runoff or sediment. Sand dominates along the bay margins, and mud is less than 20 % (White et al. 1989b). This bay originated as a fluvial system, as is shown by its outline. It is hypersaline, and as a consequence, sabellaria reefs have replaced the oyster reefs of other Texas estuaries. In addition, there are local populations of carbonate sediment and evaporate minerals.

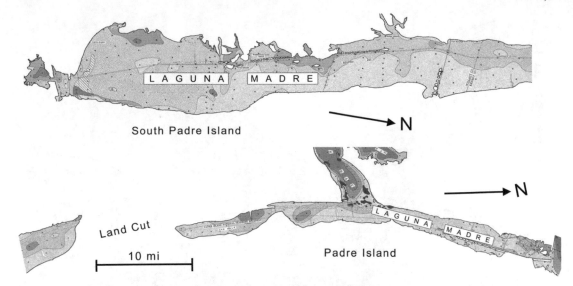

Figure 3.42. Surface sediment distribution map of Laguna Madre, Texas. Black is serpulid worm reefs. *Top map* is north and *bottom* is south (from White et al. 1986).

3.4.7.13 Laguna Madre

Most of the surface sediments in this lagoon are sand, which reflects the frequent and widespread washover/blowover from adjacent Padre Island (Figure 3.42). The land cut area is essentially pure sand and is the product of these processes. Grain size in Laguna Madre ranges from medium to fine sand with low amounts of mud. Laguna Madre also has muddy central areas with sand-dominant margins (White et al. 1986, 1989b). In places carbonate sediments, primarily ooids, are formed (Land et al. 1979). Locally there may be evaporate minerals that result from the high salinity. These are most common in the form of gypsum sand crystals that are present on the surface and at depths of greater than 4 m (13 ft) (McBride et al. 1992). This area has been arid and semi-arid for millennia resulting in little or no vegetation on either the barrier island or the mainland. The result has been considerable sand carried to the lagoon via eolian sediment transport.

3.4.7.14 Lagoons of Mexico and Cuba

The bays of the coast of Mexico are numerous and varied. They are essentially all lagoons in that salinity tends to be either high or shizohaline. The amount of runoff ranges widely; thus, a wide range in salinity and the sediment may be either terrigenous, carbonate, or a mixture of both. Coastal lagoons in Cuba are essentially absent.

Laguna Madre in Mexico is somewhat similar to that of the United States. It is shallow, hypersaline, and has some carbonate and evaporite precipitation in a terrigenous-dominated system. The fine, well-sorted sand is largely from washover and blowover of the barrier island on the Gulfside of the lagoon.

The Alvarado lagoon system in the Veracruz area is a good example of a Mexican Gulf Coast lagoon environment. It is a complex of four water bodies fed by a like number of lagoons providing terrigenous sediments with a small number of carbonates produced mostly by washover from storms. These lagoons are quite polluted. Like many of the coastal lagoons of Mexico, a substantial amount of pollution is the result of extensive mosquito spraying. The sediments have a fairly bimodal grain size with most of the open water locations being in the mud range with mean grain sizes in the 4.0–7.0Φ range and the coarse sediments in the −1.0–2.0Φ range. The coarse sediments are in places where currents are fast due to constrictions at inlets or channels between adjacent inlets (Rosales-Hoz et al. 1985).

The Celestun lagoon is an elongate water body separated from Campeche Bay by a long carbonate barrier. The sediment in the lagoon is nearly equally distributed between sand, silt, and clay-sized particles (Gonneea et al. 2004; Pech et al. 2007). This means that mud dominates.

Nichupte Lagoon is essentially at the boundary of the Gulf of Mexico and the Caribbean Sea. It is in the vicinity of Cancun on the Yucatán Peninsula. This coastal water body is probably the most polluted because of the high level of development in the vicinity. The sediments are all carbonate and are dominated by sand (65–75 %); mud is the remainder. Organic carbon is relatively high with values that range from 1.6 to 5.6 % with a mean of about 3.5 % by weight (Valdez-Lozano et al. 2006).

The coast of Cuba that is in the Gulf of Mexico is limited to the northwest portion of the country, essentially from the western tip of the country to the eastern tip of Varadero Beach, west of the small city of Matanzas. This coast includes five coastal bays about which there is virtually no information on their sediments. The following comments are based on the geomorphology of the area and their demographics. The five bays from west to east are Bahia Honda, Bahia de Cabanas, Bahia del Mariel, Havana, and Bahia de Matanzas.

Bahia Honda has no significant fluvial discharge and therefore not much sediment is being delivered. It has a bedrock shoreline and a modest amount of relief on its shoreline. Because the community on its border is industrial and there was a U.S. military base, it is assumed that the sediments are somewhat polluted. Bahia de Cabanas is quite similar to Bahia Honda in its general setting and geomorphology but is apparently more pristine in character. It has been designated as an excellent site for mariculture development (Texas A&M University—Corpus Christi, Harte Research Institute). Bahia del Mariel has an industrial port, and although it is smaller than the two previously addressed bays, it is similar in other respects and probably has somewhat polluted sediments. The most developed coastal bay is the harbor in Havana, which is very polluted. Bahia de Matanzas is a funnel-shaped bay that is served by three rivers. The combination of the rivers and the potential of tidal sediment delivery in a funnel-shaped estuary have probably led to a fair amount of sediment delivery.

3.5 SUMMARY

The Gulf of Mexico is essentially a small ocean basin. It contains all of the physiographic and geologic elements of a true ocean basin. The continental margin mimics that of an ocean basin with a continental shelf, continental slope/rise, and an abyssal plain. The slope/rise province is dissected by submarine canyons. Large deep-sea fans are also present. The deep basin is an abyssal plain environment with little relief.

The coastal zone of the Gulf includes a range of environments including estuaries, lagoons, fluvial deltas, barrier islands, and tidal inlets, which reflect the tectonic stability of the basin and the generally extensive coastal plain.

Nearly all sediments are either terrigenous or carbonate. Locally there may be evaporate accumulations. The terrigenous sediments are dominantly derived from fluvial discharge and then dispersed via various current systems along the coast or into deep water. A few locations in Mexico and Cuba have some terrigenous sediment, which eroded from bedrock exposures along the coast. The Florida coast and its carbonate sediments may be of biogenic origin or from direct precipitation.

Sediment textures range widely depending in part on the environment in which they accumulate. Terrigenous sand is rather limited to coastal, high-energy environments such as beaches and tidal inlets. Mud is typical of deltas, the outer shelf, and the deep sea. The coarsest sediments tend to be carbonate skeletal dominated beaches in Florida and Mexico.

REFERENCES

Abdullah KC, Anderson JB, Snow JN, Holdford JL (2004) The late Quaternary Brazos and Colorado deltas, offshore Texas—their evolution and the factors that controlled their deposition. In: Anderson JB, Fillon R (eds) Late Quaternary stratigraphic. Evolution of the Northern Gulf of Mexico margin. SEPM Spec Publ 79:237–270

Anderson JB, Fillon RH (eds) (2004) Late Quaternary stratigraphic evolution of the Northern Gulf of Mexico margin. SEPM Spec Publ 79, 311 p

Andrews PB (1964) Serpulid reefs, Baffin Bay, Southeast Texas. Depositional environments, south-central Texas coast. Field Trip Guidebook. Gulf Coast Association of Geological Societies, Corpus Christi, TX, USA, pp 102–120

Balsam WL, Beeson JP (2003) Sea-floor sediment distribution in the Gulf of Mexico. Deep-Sea Res 50:1421–1444

Bart PJ, Anderson JB (2004) Late Quaternary stratigraphic evolution of the Alabama and west Florida outer continental shelf. In: Anderson JB, Fillon RH (eds) Late Quaternary stratigraphic evolution of the Northern Gulf of Mexico margin. SEPM Spec Publ 79:45–53

Behrens EW (1964) Oolite formation in Baffin Bay and Laguna Madre, Texas. In: Depositional environments south-central Texas coast. Field trip guidebook. Gulf Coast Association of Geological Societies, Corpus Christi, TX, USA, pp 82–100

Behrens EW, Land LS (1972) Subtidal Holocene dolomite, Baffin Bay, Texas. J Sediment Petrol 42:155–161

Behrens EW, Bernard BB, Brooks JM, Parker PL, Scalan S, Winters JK (1981) Marine benthic environment. In: Flint RW, Rabalais NN (eds) Environmental studies of a marine ecosystem. University of Texas Press, Austin, TX, USA, pp 68–75

Berryhill HL (ed) (1975) Environmental studies, south Texas outer continental shelf, 1975. U.S. Bureau of Land Management, New Orleans, LA, USA, 303 p

Berryhill HL (1987) Late Quaternary facies and structures, northern Gulf of Mexico. Studies in Geology No. 23. American Association of Petroleum Geologists, Tulsa, OK, USA

Blake NJ, Doyle LJ (1983) Infaunal-sediment relationships at the shelf-slope break. In: Stanley DJ, Moore GT (eds) The shelf-slope break: Critical interface on continental margins. SEPM Spec Publ 33:381–390

Bouma AH, Stelting CE, Colman JM (1985) Mississippi Fan, Gulf of Mexico. In: Bouma AH, Normark WR, Barnes NE (eds) Submarine fans and related turbidite systems. Springer-Verlag, New York, NY, USA, pp 1434–1450

Bowles FA (1997) Sediment characteristics of toroidal volume sonar search (TVSS) test sites off Panama City, Florida. NRL/MR/7432-97-8058. U.S. Naval Research Laboratory, Stennis Space Center, MS, USA

Brooks GR (2011) Florida Gulf Coast estuaries: Tampa Bay and Charlotte Harbor. In: Buster NA, Holmes CW (eds) Gulf of Mexico: Origin, biota and waters, Geology, vol 3. Texas A&M University Press, College Station, TX, USA, pp 73–87

Brooks GR, Doyle LJ, Davis RA, DeWitt NT, Suthard BC (2003a) Patterns and controls of surface sediment distribution, west-central Florida inner shelf. Mar Geol 200:307–324

Brooks GR, Doyle LJ, Suthard BC, Locker SD, Hine AC (2003b) Facies architecture of the mixed siliciclastic/carbonate inner continental shelf of west-central Florida: Implications for Holocene barrier development. Mar Geol 200:325–349

Bryant WR, Liu JY (2000) Deepwater Gulf of Mexico environmental and socioeconomic data search and literature synthesis, vol I, Narrative report. OCS study MMS 2000-048. U.S. Department of the Interior, Minerals Management Service, Washington, DC, USA, pp 37–42

Bryant WR, Lugo J, Cordova C, Salvador A (1991) Physiography and bathymetry. In: Salvador A (ed) The Gulf of Mexico Basin, Boulder, Colorado. Geological Society of America, Geology of North America, Boulder, CO, USA, pp 13–30

Bryant WR, Bean D, Liu JY, Dunlap W, Dunlap W, Silva A (2000) Geotechnical stratigraphy of sediments of the northwest Gulf of Mexico. In: Proceedings, Offshore Technology Conference, May 1–4, Houston, TX, USA

Bullard FM (1942) Source of beach and river sands on Gulf Coast of Texas. Geol Soc Am Bull 53:1021–1043

Carranza-Edwards E (2011) Mexican littoral of the Gulf of Mexico. In: Buster NA, Holmes CW (eds) Gulf of Mexico: Origin, waters and biota, vol 3, Geology. Texas A&M University Press, College Station, TX, USA, pp 293–296

Caso M, Pisanty I, Ezcurra E (eds) (2004) Environmental analysis of the Gulf of Mexico. Secretaria de Medio Ambiente y Recursos Naturales Instituto National de Ecologia (Mexico), Instituto de Ecologia, A. C. Mexico. English edition: (trans: Withers K, Nipper M). Harte Research Institute, Corpus Christi, TX, USA

Coleman JM (1988) Dynamic changes and processes in the Mississippi River delta. Geol Soc Am Bull 100:999–1015

Coleman JM, Roberts HH (1991) Late Quaternary sedimentation. In: Salvador A (cd) The Gulf of Mexico Basin, Boulder, Colorado, vol J. Geological Society of America, The Geology of North America, Boulder, CO, USA, pp 325–352

Davies DK (1972) Deep sea sediments and their sedimentation, Gulf of Mexico. Am Assoc Petrol Geol Bull 56:2212–2239

Davies DK, Moore WR (1970) Dispersal of Mississippi sediment in the Gulf of Mexico. J Sediment Petrol 40:339–353

Davis RA (1988) Morphodynamics of the west-central Florida barrier system: The delicate balance between wave- and tide-domination. In: Van der Linden WJ, Cloetingh SA, Vandenberghe J/PWJ, Van De Graaff WJ (eds) Coastal lowlands, geology and geotechnology. Kluwer Academic Publishers, Dordrecht, The Netherlands, pp 225–235

Davis RA, FitzGerald DM (2003) Beaches and coasts. Blackwell Scientific, Boston, MA, USA

Donaldson AC, Martin RH, Kanes WH (1970) Holocene Guadalupe delta in Texas Gulf Coast. In: Morgan JP (ed) Deltaic sedimentation; modern and ancient. SEPM Spec Publ 15:107–137

Donoghue JE (1993) Late Wisconsin and Holocene depositional history, northeastern Gulf of Mexico. Mar Geol 112:185–205

Doyle LJ, Sparks TH (1980) Sediments of the Mississippi, Alabama, and Florida (MAFLA) continental shelf. J Sediment Petrol 50:905–916

Eckles BJ, Fassell ML, Anderson JB (2004) Late Quaternary evolution of the wave-storm-dominated Central Texas Shelf. In: Anderson JB, Fillon RH (eds) Late Quaternary stratigraphy evolution of the Northern Gulf of Mexico margin. SEPM Spec Publ 79:271–288

Enos P, Perkins RD (1977) Quaternary sedimentation in south Florida. Geol Soc Am Mem 147:198 p

Evans MW, Hine AC, Belknap DF (1989) Quaternary stratigraphy of the Charlotte Harbor estuarine lagoon system, southwest Florida: Implications of the carbonate-siliciclastic transition. Mar Geol 88:319–348

Fairbanks NC (1962) Heavy minerals from the Eastern Gulf of Mexico. Deep Sea Res 9:307–338

Folger D (1972) Estuarine sediments of the United States. Professional paper 742. USGS (U.S. Geological Survey), Washington, DC, USA, 94 p

Gonneea ME, Paytan A, Herrera-Silveira JA (2004) Tracing organic matter sources and carbon burial in mangrove sediments over the past 160 years. Coast Estuar Sci 61:211–227

Hine AC, Locker SD (2011) Florida continental shelf—great contrasts and significant transitions. In: Buster NA, Holmes CW (eds) Gulf of Mexico: Origin, waters and biota, vol 3, Geology. Texas A&M University Press, College Station, TX, USA, pp 101–127

Hine AC, Mullins HT (1983) Modern carbonate shelf-slope breaks. In: Stanley DF, Moore GT (eds) The shelfbreak: Critical interface on continental margins. SEPM Spec Publ 33:169–188

Hine AC, Locker SD, Brooks GR (2003a) Regional stratigraphic framework linking continental shelf and coastal sedimentary deposits of west-central Florida; Implications for Holocene barrier development. Mar Geol 200:351–378

Hine AC, Brooks GR, Davis RA, Duncan DS, Locker SD, Twichell DC, Gelfenbaum G (2003b) The west-central Florida inner shelf and coastal system: A geologic conceptual overview and introduction to the special issue. Mar Geol 200:1–17

Holmes CW (2011) Development of the northwestern Gulf of Mexico continental shelf and coastal zone as a result of the Late Pleistocene-Holocene sea-level rise. In: Buster NA, Holmes CW (eds) Gulf of Mexico: Origin, waters, and biota, vol 3, Geology. Texas A&M University Press, College Station, TX, USA, pp 195–208

Huang TC, Goodell HG (1967) Sediments of Charlotte Harbor, southwestern Florida. J Sediment Petrol 37:449–474

Huang TC, Goodell HG (1970) Sediments and sedimentary processes of eastern Mississippi core, Gulf of Mexico. Am Assoc Petrol Geol Bull 54:2070–2100

Isphording WC (1985) Chemistry and partitioning of heavy metals in Mobile Bay, Alabama. Project R/ER-14. Mississippi-Alabama Sea Grant Consortium, Ocean Springs, MS, USA, 21 p

Kanes WH (1970) Facies and development of the Colorado River delta in Texas. In: Morgan JP (ed) Deltaic sedimentation. SEPM Spec Publ 15:78–107

Kennicutt MC, Schroeder WW, Brooks JM (1995) Temporal and spatial variations in sediment characteristic on the Mississippi-Alabama continental shelf. Cont Shelf Res 15:1–18

Kopaska-Merkel DC, Rindsberg AK (2005) Sand-quality characteristics of Alabama beach sediment, environmental conditions, and comparisons to offshore sand resources. Open File Report 0508. Geological Survey of Alabama, Tuscaloosa, AL, USA, 75 p

Land LS, Behrens EW, Frishman SA (1979) The ooids of Baffin Bay, Texas. J Sediment Res 49:1269–1277

Locker SD, Hine AC, Brooks GR (2003) Regional stratigraphic framework and coastal sedimentary deposits of west-central Florida. Mar Geol 200:351–378

Logan BW, Harding JL, Ahr WM, Williams JD, Snead RG (1969) Carbonate sediments and reefs, Yucatan Shelf, Mexico. Am Assoc Petrol Geologists Mem 11:1–128

Ludwick JC (1964) Sediments in the northeastern Gulf of Mexico. In: Miller RL (ed) Papers in marine geology. Macmillan Publishing, New York, NY, USA, pp 204–238

Macauley J, Smith LM, Bourgeois P, Ruth B (2005) The ecological condition of the Pensacola Bay System, Northwest Florida. EPA/620/R-05/002. U.S. Environmental Protection Agency, Office of Research and Development, Gulf Breeze, FL, USA, 38 p

Martin RG, Bouma AH (1978) Physiography of the Gulf of Mexico. In: Bouma AH, Moore GT, Coleman JM (eds) Framework, facies, and oil-trapping characteristics of the upper continental margin. Am Assoc Petrol Geol Stud Geol 7:3–19

Mazullo J, Peterson M (1989) Sources and dispersal of late Quaternary silt on the northern Gulf of Mexico continental shelf. Mar Geol 86:15–26

McBride EF, Honda H, Avdel-Wahab AA, Dworkin S, McGilvery TA (1992) Fabric and origin of gypsum sand crystals, Laguna Madre, Texas. Trans Gulf Coast Assoc Geol Soc 42:543–551

McBride RA, Moslow TE, Roberts HH, Diecchio RJ (2004) Late Quaternary geology of the northeastern Gulf of Mexico shelf: Sedimentology, depositional history and ancient analogs of a modern sand shelf sheet of the transgressive systems tract. In: Anderson JB, Fillon RH (eds) Late Quaternary stratigraphic evolution of the Northern Gulf of Mexico margin. SEPM Spec Publ 79:53–83

McKeown HA, Bart PJ, Anderson JB (2004) High-resolution stratigraphy of a sandy, ramp-type margin—Apalachicola, Florida. In: Anderson JB, Fillon RH (eds) Late Quaternary stratigraphic evolution of the Northern Gulf of Mexico margin. SEPM Spec Publ 79:25–42

Morelock J (1969) Shear strength and stability of continental slope deposits, Western Gulf of Mexico. J Geophys Res 74:465–482

Moretzsohn F, Sánchez Chávez JA, Tunnell JW Jr (eds) (2015) GulfBase: Resource database for Gulf of Mexico Research. World Wide Web electronic publication. http://www.gulfbase.org/facts.php. Accessed 9 Aug 2015

Morton RA (1972) Clay mineralogy of Holocene and Pleistocene sediments, Guadalupe Delta of Texas. J Sediment Petrol 42:85–88

Morton RA, Gibeaut JC (1995) Physical and environmental assessment of sand resources, Sabine and Heald Banks: Second phase 1994–1995. The University of Texas at Austin, Bureau of Economic Geology. Final Report. U.S. Department of the Interior, Office of International Activities and Marine Minerals, Washington, DC, USA, 246 p

Parker RH (1960) Ecology and distributional patterns of marine macro-invertebrates, northern Gulf of Mexico, 1951–1958. In: Shepard FP, Phleger FB, van Andel TH (eds) Recent sediments of the Northwest Gulf of Mexico. American Association of Petroleum Geologists, Tulsa, OK, USA, pp 302–381

Pech D, Ardisson PL, Hernandez-Guevara NA (2007) Benthic community response to habitat variation: A case study from a natural protected area, the Celestun coastal lagoon. Cont Shelf Res 27:2523–2533

Pequegnat WE (1976) Ecological aspects of the upper continental slope of the Gulf of Mexico. U.S. Department of Interior, Bureau of Land Management, Washington, DC, USA, pp 122–131

Perlmutter MA (1985) Deep water clastic reservoirs in the Gulf of Mexico: A depositional model. Geol Mar Lett 5:105–112

Phillips JD (2005) A sediment budget for Galveston Bay, final report. Texas Water Development Board, Austin, TX, USA, 16 p

Reading HG (ed) (1978) Sedimentary environments and facies. Elsevier, New York, NY, USA, 557 p

Rezak R, Bright TJ, McGrail DW (1985) Reefs and banks of the Northwestern Gulf of Mexico; Their geological, biological and physical dynamics. Wiley, New York, NY, USA

Roberts HH, Aharon P (1994) Hydrocarbon-derived carbonate buildups of the northern Gulf of Mexico continental slope: A review of submersible investigations. Geo-Mar Lett 14:135–148

Rodriguez AB, Anderson JB, Siringan FP, Taviani M (1999) Sedimentary facies and genesis of Holocene sand banks on the East Texas inner continental shelf. In: Snedden J, Bergman K (eds) Isolated shallow marine sand bodies: Sequence stratigraphic analysis and sedimento-logic interpretation. SEPM Spec Publ 64:165–178

Rogers B, Kulp M (2009) Chapter G. The St. Bernard Shoals—an outer continental shelf sedimentary deposit suitable for sandy barrier island renourishment. In: Lavoie D (ed) Sand resources, regional geology, and coastal processes of the Chandeleur Islands coastal system—an evaluation of the Breton National Wildlife Refuge. U.S. Geological Survey Scientific Investigations Report 2009–5252. U.S. Geological Survey, Reston, VA, USA, pp 125–142

Rosales-Hoz L, Carranza-Edwards A, Alvarez-Rivera U (1985) Sedimentological and chemical studies in sediments from Alvarado lagoon system, Veracruz, Mexico. Anales del Centro de Ciencia del Mar Limonologia, Universidad Nacional Autónoma de México, Mexico City, Mexico

Rusnak GA (1960) Some observations on recent oolites. J Sediment Petrol 30:471–480

Ryan JJ (1969) A sedimentologic study of Mobile Bay, Alabama. Contribution 30. Florida State University, Department of Geology, Sedimentology Research Laboratory, Tallahassee, FL, USA, 110 p

Salvador A (1991) Origin and development of the Gulf of Mexico basin. In: Salvador A (ed) The Gulf of Mexico Basin. The Geology of North America. J. Geological Society of America, Boulder, CO, USA, pp 389–444

Sionneau T, Bout-Roumazeilles V, Biscaye PE, Van Vliet-Lanoe B, Bory A (2008) Clay mineral distributions in and around the Mississippi River watershed and Northern Gulf of Mexico: Sources and transport patterns. Quaternary Sci Rev 27:1740–1751

Turner RE (1999) Inputs and outputs of the Gulf of Mexico. In: Kumpf H, Steindinger K, Sherman K (eds) The Gulf of Mexico Large Marine Ecosystem: Assessment, sustainability and management. Blackwell Science, New York, NY, USA, 704 p

Twichell DC (2011) A review of recent depositional processes on the Mississippi Fan, eastern Gulf of Mexico. In: Buster NA, Holmes CW (eds) Gulf of Mexico: Origin, waters, and biota, vol 3, Geology. Texas A&M University Press, College Station, TX, USA, pp 141–154

Twichell DC, Edmiston L, Andrews B, Stevenson W, Donoghue J, Poore R, Osterman L (2010) Geologic controls on the recent evolution of oyster reefs in Apalachicola Bay and St. George Sound, Florida. Estuar Coast Shelf Sci 88:385–394

Uchupi EM (1975) Physiography of the Gulf of Mexico and Caribbean Sea. In: Nairn AEM, Stehli FG (eds) The ocean basin and margins, vol 3, The Gulf of Mexico and Caribbean. Plenum Press, New York, NY, USA, 706 p

USEPA (U.S. Environmental Protection Agency) (1980) A water quality success story: Lower Houston ship channel and Galveston Bay, Texas. Office of Water Planning and Standards, Washington, DC, USA

USGS (U.S. Geological Survey) (2006) usSEABED: Gulf of Mexico and Caribbean (Puerto Rico and U.S. Virgin Islands) offshore surficial sediment data release. USGS Data Release 146, Reston, VA, USA

Valdez-Lozano L, Chumacero M, Real E (2006) Sediment oxygen consumption in a developed coastal lagoon of the Mexican Caribbean. Indian J Mar Sci 35:227–234

Wade TL, Soliman Y, Sweet ST, Wolff GA, Presley BJ (2008) Trace elements and polycyclic aromatic hydrocarbons (PAHs) concentrations in deep Gulf of Mexico sediments. Deep Sea Res 55:2585–2593

White WA, Calnan TR, Morton RA, Kimble RS, Littleton TG, McGowen JH, Nance HS (1983) Submerged lands of Texas: Corpus Christi Area: Sediments, geochemistry, benthic macro-invertebrates, and associated wetlands. Bureau of Economic Geology, Austin, TX, USA

White WA, Calnan TR, Morton RA, Kimble RS, Littleton TG, McGowen JH, Nance HS (1985) Submerged lands of Texas: Galveston-Houston area: Sediments, geochemistry, benthic macroinvertebrates, and associated wetlands. Bureau of Economic Geology, Austin, TX, USA

White WA, Calnan TR, Morton RA, Kimble RS, Littleton TG, McGowen JH, Nance HS (1986) Submerged lands of Texas: Brownsville-Harlingen area: Sediments, geochemistry, benthic macroinvertebrates, and associated wetlands. Bureau of Economic Geology, Austin, TX, USA

White WA, Calnan TR, Morton RA, Kimble RS, Littleton TG, McGowen JH, Nance HS (1987) Submerged lands of Texas: Beaumont-Port Arthur Area: Sediments, geochemistry, benthic macroinvertebrates, and associated wetlands. Bureau of Economic Geology, Austin, TX, USA

White WA, Calnan TR, Morton RA, Kimble RS, Littleton TG, McGowen JH, Nance HS (1988) Submerged lands of Texas: Bay City-Freeport area: Sediments, geochemistry, benthic macroinvertebrates, and associated wetlands. Bureau of Economic Geology, Austin, TX, USA

White WA, Calnan TR, Morton RA, Kimble RS, Littleton TG, McGowen JH, Nance HS (1989a) Submerged lands of Texas: Port Lavaca Area: Sediments, geochemistry, benthic macro-invertebrates, and associated wetlands. Bureau of Economic Geology, Austin, TX, USA

White WA, Calnan TR, Morton RA, Kimble RS, Littleton TG, McGowen JH, Nance HS (1989b) Submerged lands of Texas: Kingsville area: Sediments, geochemistry, benthic macroinver-tebrates, and associated wetlands. Bureau of Economic Geology, Austin, TX, USA

Williams SJ, Kulp M, Penland S, Kindinger JL, Flocks JG (2011) Mississippi River delta plain, Louisiana coast, and inner shelf Holocene geologic framework, processes, and resources. In: Buster NA, Holmes CW (eds) Gulf of Mexico: Origin, waters, and biota, vol 3, Geology. Texas A&M University Press, College Station, TX, USA, pp 175–193

CHAPTER 4

SEDIMENT CONTAMINANTS OF THE GULF OF MEXICO

Mahlon C. Kennicutt II[1]

[1]Texas A&M University, College Station, TX 77843, USA
mckennicutt@gmail.com

4.1 INTRODUCTION

Sediments are an essential constituent of aquatic environments that are vital to the health of organisms and ecosystems. Many organisms live in, ingest or otherwise come into contact with sediments repeatedly during their life cycles. Anthropogenic—human-derived—chemicals introduced into sediments have the potential to harm the health of organisms and ecosystems and are collectively referred to as *contaminants*. Sediment contaminants of the greatest environmental concern are those with concentrations significantly enriched above natural levels due to human influences. In most instances, organisms are adapted to natural levels of chemicals in the environment, and in some cases, these chemicals are essential elements for survival and growth. In contrast, additions of chemicals that are by-products of human activities can degrade sediment quality in the short and long term. Potential adverse effects on biota caused by exposure to contaminants include, but are not limited to, death, disruption of physiology and reproduction, impairment of ecosystem functioning and structure, and ultimately, may impact human health. This chapter summarizes the origins, geographic distributions, and temporal trends in sediment contaminants in the northern Gulf of Mexico from the mid-1980s to early 2010. Contaminant dynamics in coastal sediments are compared to continental shelf/slope and abyssal sediments. *Coastal* refers to all land areas in close proximity to the ocean including estuaries, bays, sounds, wetlands, coral reefs, intertidal zones, sea grass beds, and nearshore oceanic areas. Offshore regions are those areas more distant from the shore including sediments on the continental shelf/slope and abyss. The transition from the nearshore to the offshore is a continuum, and these areas are often oceanographically coupled. However, differences in environmental conditions, the energetics of the settings, and the locations of sources of contaminants make the dynamics of contaminated sediments in these two regions distinct from one another. Most studies of contaminated sediments focus on nearshore, coastal environments in close proximity to human populations. There are limited studies of offshore regions remote from most human activities.

4.1.1 Classes of Contaminants

Chemical structure and reactivity are fundamental determinants of the fate and effects of contaminants in aquatic environments. Contaminants can be single chemical species or mixtures of compounds and are classified as organic (carbon containing) or inorganic (non-carbon containing) substances (USEPA 1989). Organic contaminants are classified based on vapor pressures and water solubility as volatile or semivolatile organic compounds (VOCs and SVOCs, respectively). VOCs are organic chemicals with high vapor pressures at standard

C.H. Ward (ed.), *Habitats and Biota of the Gulf of Mexico: Before the Deepwater Horizon Oil Spill*, DOI 10.1007/978-1-4939-3447-8_4

atmospheric pressure and temperature and are typically not acutely toxic but can cause chronic biological effects. VOCs include a diverse assortment of naturally occurring and human-made chemicals that preferentially occur in the atmosphere and are dissolved in water. They rarely occur in or are studied in coastal and marine sediments; therefore, VOCs are not considered further in this summary. SVOCs have low vapor pressures and varying water solubility and are a diverse assortment of naturally occurring and human-made chemicals that occur in the atmosphere, adsorb onto particulate matter in water, and often accumulate in sediments. Due to low water solubility (hydrophobicity), many SVOCs preferentially partition into lipid-rich biological tissues. Environmental exposure to SVOCs can lead to bioaccumulation in organisms and biomagnification via food webs. SVOCs have a wide range of toxicities and can have diverse biological and ecosystem effects. Inorganic contaminant chemicals are defined as non-carbon substances of a mineral origin and include metals and nutrients (USEPA 1989). Metals are metallic elements of high atomic weight that can cause acute and chronic toxicity in organisms. Contaminant nutrients (such as nitrate and phosphate) primarily occur dissolved in water. Nutrient contamination in the Gulf of Mexico is treated elsewhere in reference to water quality (see Chapter 2).

Contaminant SVOCs and metals often persist in the environment, accumulate in sediments over time and increase the potential for, and possibly the levels of, organismal and ecosystem exposure. Beyond the contaminants mentioned above, other human-derived chemicals have the potential to cause environmental degradation, including pharmaceuticals, household chemicals, and personal hygiene products; fire retardants (brominated compounds); and endocrine-disrupting or mimicking compounds. However, most monitoring programs rarely measure these chemicals in sediments, so their importance as contaminants remains largely unknown. Therefore, these contaminants are not included in this chapter.

4.1.2 Scope of the Summary

Reports of sediment contaminants in the periodic literature, monitoring programs, and assessments of national coastal conditions issued by federal, state, and local agencies and programs from the mid-1980s to early 2010 are summarized. Data collections are used to qualitatively describe the regional status and trends in sediment contaminants. Published articles illustrate conclusions drawn from regional monitoring and assessment programs. National assessments are produced by government and academic experts based on data and information from hundreds of documents. Region-wide monitoring and assessments were first initiated in the mid-1980s. The most recent national coastal assessment was completed in 2012. Thousands of sediments from the northern Gulf of Mexico have been analyzed for contaminant concentrations for more than 30 years. Assessment summaries often lag behind the date of data collection by several years due to the process involved. The interpretations from these syntheses are reported in this chapter but the underlying primary data are not reanalyzed.

This summary of sediment contaminant concentrations does not directly address sediment quality, toxicity, and/or biological effects (see Section 4.2.2). National coastal assessments employ additional chemical, toxicological, and biological measurements to assess overall sediment quality and ecological status. Inclusion of these additional variables is beyond the scope of this review, and the reader is referred to the integrated assessment of environmental quality and biological effects contained in the national coastal assessments. The presence or concentration of contaminants in sediments is usually insufficient to infer sediment quality or to predict adverse in situ biological effects because many factors affect the interaction of contaminants, organisms, and ecosystems. However, within national coastal assessments, sediment contaminant concentrations that have been empirically shown to elicit biological effects in the published literature are used to qualitatively describe occurrences and distributions. While these comparisons are useful to highlight higher versus lower areas of sediment contamination and to suggest the origins of contaminants, sediment quality or prediction of adverse biological effects or toxicity should not be inferred.

4.2 THE ORIGINS AND BEHAVIOR OF CONTAMINANTS IN THE ENVIRONMENT

Chemical contaminants commonly occur in sediments of rivers, lakes, and adjacent oceans (USEPA 1989). Some contaminants were released into the environment years ago and persist, while others continue to be released. Contaminants, in particular the SVOCs and metals that are the focus of this summary, are found in industrial and municipal discharges and emissions, urban and agricultural runoff, accidental spills, and wet and dry atmospheric deposition. Of the 11 environmental concerns identified in United States (U.S.) estuaries, 8 potentially involve contaminants and collectively affect all of the 28 estuaries considered (Figure 4.1).

Releases of contaminants to the environment can be intentional (e.g., permitted discharges) and/or accidental (e.g., spills). Contaminants enter marine environments through the air, dissolved in or absorbed on particles in water, or as solid or liquid discharges. Hydrophobic compounds released into air and water preferentially adsorb onto particulate matter, and often, some portion is eventually deposited in sediments. Contaminants discharged as solids and liquids can result in rapid incorporation into sediments. Concentrations and geographic distributions of sediment contaminants are heterogeneous due to spatial and temporal variations in inputs, sediment deposition and accumulation rates, variable susceptibility to contamination and rates of removal, variations in chemical form and physicochemical properties, and differences in water inflow rates and receiving water residence times. Considerations of contaminant SVOCs in sediments are restricted to those chemicals most commonly studied including hydrocarbons, pesticides, and polychlorinated biphenyls (PCBs) (see Appendix A for details of the origins, toxicity, and environmental fate of SVOCs). Hydrocarbons are generally measured as polycyclic aromatic hydrocarbons (PAHs) that are the portion of petroleum that accounts for most of its toxicity. The contaminant metals considered are those of greatest environmental concern including lead (Pb), mercury (Hg), arsenic (As), cadmium (Cd), silver (Ag), nickel (Ni), tin (Sn), chromium (Cr), zinc (Zn), barium (Ba), vanadium (V), and copper (Cu) (see Appendix B for details of the origins, toxicity, and environmental fate of metals).

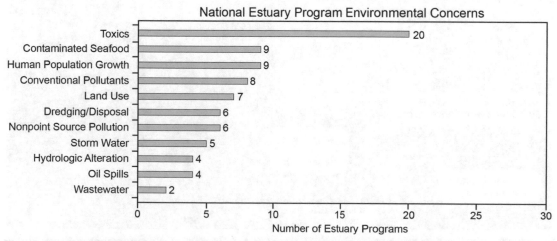

Figure 4.1. Environmental concerns for U.S. National Estuary Programs related to contaminants in sediments: *n* = number of affected estuaries out of a total of 28 (modified from USEPA 2006).

Some SVOCs (e.g., hydrocarbons) and essentially all metals naturally occur in sediments. Many SVOCs are human-made and do not naturally occur in the environment (i.e., pesticides and PCBs) though there are multiple sources attributable to humans. Natural occurrences (e.g., metals occur in crustal rocks and minerals) and processes (e.g., oil and gas seeps) that release chemicals to the environment must be considered when ascribing the origins and distributions of sediment contaminants. Sediment contaminants of most interest are those elevated above natural abundances.

The mode and composition of contaminant releases often determine their behavior, availability, and fate in the environment. Contaminants originate from point or non-point sources (USEPA 1989). Point sources are single, identifiable release locations that are limited in spatial extent, such as a discharge pipe or smoke stack. Point sources include discharges by municipal sewage treatment plants, overflows from combined sanitary and storm sewers, stormwater discharges from municipal and industrial facilities, and discharges from industrial and military complexes (USEPA 1989). Non-point sources are diffuse including river outflows, land runoff, precipitation, atmospheric deposition, drainage, and/or hydrologic modifications (USEPA 1989). Once released, contaminants interact with the environment based on their physicochemical properties, chemical form, biological reactivity, and the ambient conditions of the receiving media (Figures 4.2 and 4.3).

Contaminants released to the atmosphere can be bound to particles and transported long distances from the site of release. Environmental processes creating derivative or breakdown by-products can alter the chemical structure of contaminants. These alterations can cause

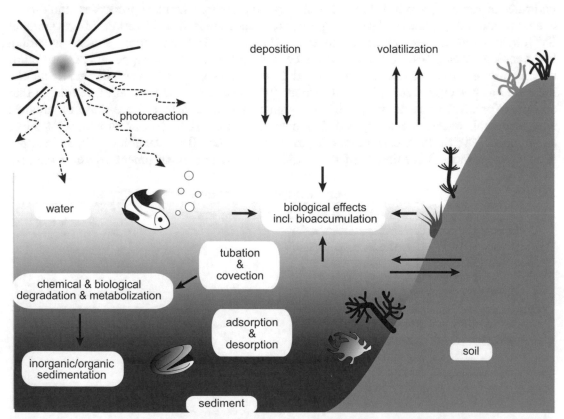

Figure 4.2. Distribution and fate of chemical contaminants in an aquatic environment (from Rydén et al. 2003 citing Römbke and Moltmann 1995). Republished with permission from the Baltic University Programme, Uppsala University and from the Taylor and Francis Group LLC Books whose permission was conveyed through the Copyright Clearance Center, Inc.

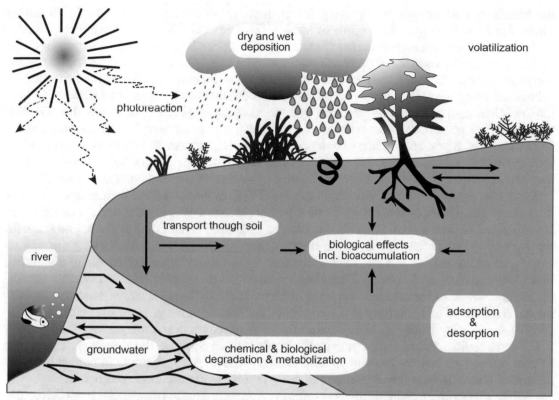

Figure 4.3. Distribution and fate of chemical contaminants in a terrestrial environment (from Rydén et al. 2003 citing Römbke and Moltmann 1995). Republished with permission from the Baltic University Programme, Uppsala University and from the Taylor and Francis Group LLC Books whose permission was conveyed through the Copyright Clearance Center, Inc.

redistribution in water, sediment, and/or biota due to changes in physicochemical properties (e.g., hydroxylation of PAHs increases water solubility). By-products can be more or less toxic than the parent substances. Contaminants that enter water may adsorb onto suspended particles, settle to the bottom, or be taken up by organisms. Resuspension of sediments can reintroduce contaminants to the overlying water column making sediments both a potential source and sink. Contaminants can accumulate in sediments over long periods of time due to periodic or chronic releases. Some processes, including microbiological metabolism, enzymatic detoxification, dissolution, and/or chemical breakdown, remove contaminants from sediments. In situ biological and ecosystem responses are the cumulative outcome of all stressors experienced, and sediments at many locations often contain a complex mixture of contaminants from multiple sources. The interactions of multiple stressors (chemical exposures and others) are poorly understood in most instances.

4.2.1 The Mississippi River

A defining characteristic of the northern Gulf of Mexico is the presence and influence of the Mississippi River, especially in regard to the origins and deposition of sediments. The Mississippi River is a major source of particulates and sediments to the northern Gulf of Mexico. As such, the Mississippi River is a major contributor of sediments in coastal areas near Mississippi River outflows, and river-derived sediments are widely distributed in the offshore continental shelf/slope/abyssal regions of the north-central Gulf of Mexico. In the mid-2000s the Mississippi River was transporting approximately 136 million metric tons (~150 million tons) of sediment per year to the Gulf of Mexico (Thorne et al. 2008). The quantity and quality of transported sediment has been affected by changes in land use and river management throughout the nineteenth and twentieth centuries. The supply of sediment from tributaries has markedly decreased due to the construction of dams and various diversions of the river and its tributaries over the years. Thorne et al. (2008) concludes that total suspended sediment loads on the lower Mississippi River have declined by approximately 80 percent (%) during the 1851–1982 time period. A comprehensive review of the distribution, sources, and fate of contaminants in the Mississippi River and its massive drainage basin (41 % of the 48 contiguous states of the United States) is beyond the scope of this summary. However general inferences about the importance of the Mississippi River as a source of sediment contaminants are considered.

A summary by the U.S. Geological Survey (USGS) in 1995 assessed contaminant levels in the Mississippi River for the 1987–1992 time period (Meade 1995). Contaminant concentrations in suspended sediments and bed sediments in the Mississippi River rapidly decreased from the northern to the southern regions of the river's drainage basin due to dilution with uncontaminated particulate matter and suspended sediment, evaporative losses, losses due to dissolution in water, chemical and microbial breakdown, and the geographic distribution of chemical discharges. While sediment loads are large, dilution and loss during riverine transport diminish the concentrations of contaminants in suspended sediment discharged into the Gulf of Mexico (Trefry and Presley 1976a, b). Metals naturally occur in sediment so most metal concentrations in offshore sediments are similar to crustal abundances. In contrast, the highest concentrations of contaminant metals are mostly found in coastal areas in close proximity to human activities that release metals. The few exceptions in the offshore region are discussed below.

While difficult to estimate, mass loadings of contaminants from Mississippi River discharges could be quite large compared with other sources (such as those in coastal areas) as the volume of sediment delivered to the Gulf of Mexico is large. However, sediment contaminant SVOC concentrations have been confirmed to be low in river-discharged sediments most likely due to dilution with uncontaminated sediments and predischarge losses. Metals naturally occur in sediments, so most river-derived sediment metal concentrations are similar to crustal abundances with little evidence of enhanced concentrations due to contamination, with a few notable exceptions.

4.2.2 Biological Effects Levels: Usage and Limitations

Chemical contaminant concentrations alone are usually insufficient to predict in situ biological responses or detrimental effects. Guidelines based on summaries of literature reports of sediment toxicity data have been developed to qualitatively assess whether sediment contaminant concentrations might be expected to cause biological effects (Long and Morgan 1990; Long et al. 1995). These guidelines are called effects range low (ERL) and effects range median (ERM) (Figure 4.4). The ERL criterion is the concentration of a chemical in sediments that resulted in biological effects approximately 10 % of the time based on the literature. The ERM criterion is the concentration of a chemical in sediments that resulted in biological effects approximately 50 % of the time based on the literature. Long et al. (1995) concluded that these sediment quality guidelines provide reasonably accurate estimates of chemical concentrations that are either nontoxic or toxic in laboratory bioassays. However, the reliability of predicting in situ biological response and sediment toxicity from ERL/ERM guidelines has been questioned (O'Connor et al. 1998). The ERL value is considered to be a concentration "at the low end of a continuum roughly relating bulk chemistry with toxicity" (Field et al. 2002; O'Connor 2004). Concentrations of more than one chemical above the ERL does not increase the probability of toxicity, and categorizing sediments on the basis of chemical concentrations with one or more

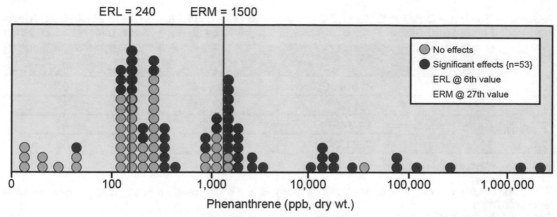

Figure 4.4. Derivation of ERL (effects range low) and ERM (effects range median) values for phenanthrene (a PAH); data showing no adverse effects are *green symbols* and those in which toxicity or some other measure of adverse effects were observed are *purple symbols* (from NOAA 1999).

ERL exceedances can lead to misperceptions of the probability that the sediments are in fact toxic. ERL/ERM guidelines are most useful when supporting data such as in situ biological analyses, toxicological assays, and other variables have been measured that confirm suspected cause-and-effect relationships.

Despite the limitations above, by convention, the *National Coastal Condition Reports* (NCCRs) assess coastal sediment quality based on the number of ERL/ERM exceedances (USEPA 2001, 2004, 2008, 2012). In this summary, ERL/ERM values are used only to draw attention to sites where contaminant concentrations exceed levels that may be of biological significance; however, cause-and-effect or toxicity is not inferred. The results of the NCCRs are summarized including exceedances of ERL and ERM values to qualitatively describe the distribution of higher and lower levels of contaminants in sediments and to assist in discerning the sources of the contamination but not for predicting sediment quality or toxicity. In those cases where NCCR conclusions about sediment quality/toxicity are inferred, additional variables have been taken into account, and attribution of degraded benthic conditions solely to elevated chemical concentrations is often not possible.

Furthermore, ERL values used to classify sediment metal concentrations (Cr, Cu, Ni, and As) are close to or less than natural background crustal values for Mississippi River suspended matter and Gulf of Mexico sediments (Table 4.1). In some cases, according to a peer reviewer of this chapter, Long et al. (1995) used concentrations of Cr and Ni that were determined with U.S. Environmental Protection Agency (USEPA) techniques that used an acid leach (without hydrofluoric acid) rather than total dissolution of the sediment to calculate ERL/ERM values. ERL/ERM values also have been revised over the years as additional data have become available, so reports may use differing values to assess exceedances. These limitations should be taken into account when considering the summaries in this chapter. Mapping and categorizing the number of sites that exceed ERL/ERM values are used to assess the location and origins of contaminants but not to infer or predict in situ biological effects or sediment toxicity.

4.3 COASTAL SEDIMENTS

Human population centers and industrial activities are concentrated in and near coastal areas. Coastal areas are often the sites of agricultural activities as well. Much of the water discharged by rivers and runoff from land surfaces enters or flows directly into aquatic

Table 4.1. Summary Data for ERLs and Concentrations of Cr, Cu, Ni, and As in Mississippi River-suspended Matter and Gulf of Mexico Sediments (prepared by an anonymous peer reviewer, 2012)

	Effects Range Low	Mississippi River Particles	Gulf of Mexico Sediment[a]
Cr (μg/g)	81 (none[b])	72[c], 74[d]	60–90
Cu (μg/g)	34 (70[b])	33[c], 21[d]	20–30
Ni (μg/g)	20.9 (30[b])	41[c], 37[d]	25–40
As (μg/g)	8.2 (33[b])	8[d]	10–15

Note: μg/g—microgram(s) per gram
[a]O'Connor (2004)
[b]Trefry and Presley (1976a), Presley et al. (1980), Trefry et al. (1985), Kennicutt et al. (1996); Continental Shelf Associates (2006)
[c]Trefry and Presley (1976b)
[d]Horowitz et al. (2001)

environments along coasts. In many cases high-energy discharges empty into much lower-energy settings instigating sedimentation and deposition. As such, coastal areas are major locations of contaminant accumulation in marine sediments.

Monitoring programs and national assessments over a 30-year period have concluded that contaminants are widespread in coastal, northern Gulf of Mexico sediments. Coastal condition, benthic condition, and sediment quality across the region have been judged good to poor during this time due to several factors including, but not limited to, the presence of sediment contaminants. The levels of contaminants in fish tissues have been rated as good to poor and contaminants have been widely detected in bivalve tissues, which demonstrates their bioavailability. Uptake alone does not infer adverse biological effects, and organisms can accumulate chemicals from sources other than contaminated sediments (Kimbrough et al. 2008). Benthic condition ratings were based on measures of infauna biodiversity, increased abundances of contaminant-tolerant species, and decreased abundances of contaminant-sensitive species. Benthic condition is a cumulative measure of all stressors, one of which can be the presence of elevated concentrations of contaminants suspected of causing adverse biological effects. Overall sediment quality is often judged based on measures of sediment toxicity, contaminants, and organic carbon content. The role and importance of sediment contaminants in degraded coastal benthic and sediment quality is often difficult to discern due to multiple stressors affecting a location. The following summary describes the origins and geographic distributions of contaminants in coastal northern Gulf of Mexico sediments from the mid-1980s to early 2010. Instances in which contaminants are important factors in degraded coastal, benthic, and sediment qualities are highlighted.

Coordinated efforts to monitor the condition of U.S. coastal regions were initiated in the 1980s (NOAA 1987, 1991; USEPA 2001, 2004, 2008, 2012). The NOAA National Status & Trends (NS&T) Program and the USEPA Environmental Monitoring and Assessment Program (EMAP) are two such efforts (NOAA 1987, 1991; USEPA 1999). The NCCRs "… describe and summarize the ecological and environmental conditions in U.S. coastal waters…." (USEPA 2001, 2004, 2008, 2012). The concentrations of contaminants in various matrices were seen as key indicators of condition. The USEPA Office of Wetlands, Oceans and Watersheds' Coastal Programs initiated these reports to provide a "comprehensive picture of the health of the nation's coastal waters" (USEPA 2001). The NCCRs are based on data collected from a variety of sources coordinated by USEPA and National Oceanic and Atmospheric Administration (NOAA) with input from the USGS, U.S. Fish and Wildlife Service (USFWS), and coastal states.

One aspect of these national assessments is a region-by-region consideration of sediment quality, sediment contamination, benthic quality, fish tissue contamination, and other measures of environmental quality. The northern Gulf of Mexico is one of the regions assessed. Interpretations of data by these programs are often summarized as national or regional averages, reported on a relative basis within studies, synthesized as ratings (good, fair, and poor) of composite indicators, compared to sediment concentrations suspected of causing biological effects (ERL/ERM values), and mapped based on numbers of exceedances of contaminant concentrations of environmental concern (see Section 4.2.2). To describe the status and trends in sediment contaminants in the northern Gulf of Mexico, these reports are summarized, but the underlying, primary data are not reanalyzed. Descriptions of subsets of data are presented as examples to clarify the underlying causes of trends in environmental quality.

4.3.1 NOAA National Status and Trends Program

The NOAA NS&T Program analyzes surface sediments from coastal and estuarine sites throughout the United States in support of data on contaminant concentrations in biological tissues. Sediments are intentionally collected distant from major points of contamination to quantify the combined influences of many point and non-point sources of chemicals in coastal areas (NOAA 1987, 1988, 1991; Wade et al. 1989; Sericano et al. 1990). Surficial sediments (the top 3 centimeters [cm] or 1.2 in.) are collected as part of the Mussel Watch Program and the Benthic Surveillance Project. The data from sediments collected in the 1980s are first considered; later data are included in NCCRs and other reports.

The NS&T Program data provide one of the earliest comprehensive overviews of contaminant concentrations and distributions in sediments along the coastal northern Gulf of Mexico. Concentrations of PAHs, pesticides, PCBs, and metals were measured in sediments using standard and calibrated methods. NS&T Program reports interpret sediment chemical concentrations on a relative basis and exclude concentrations ten times greater than the next highest concentration as outliers. Chemical concentrations in sediments were observed to have a lognormal distribution. Concentrations greater than the mean plus one standard deviation of all locations in the United States were termed *high* levels. High concentrations identify sediments affected by human activity but do not imply biological significance (NOAA 1991). Based on this definition, high concentrations of sediment contaminants were observed in bays sampled in the northern Gulf of Mexico from 1986 to 1989 (NOAA 1991) as follows:

- Florida Gulf Coast (17 locations sampled):
  - Tampa Bay—PAHs, PCBs, DDT (dichlorodiphenyltrichloroethane), Cd, and Pb
  - Apalachicola Bay—As
  - Panama City—PAHs, PCBs, DDT, and As
  - St. Andrew Bay—PAHs, PCBs, DDT, Hg, and Pb
  - Choctawhatchee Bay—PAHs, PCBs, DDT, Ag, and As
  - Pensacola Bay—As
- Alabama: (only Mobil Bay was sampled)—none
- Mississippi: Biloxi Bay—PAHs
- Louisiana: (13 locations sampled)—none
- Texas: (12 locations sampled)
  - Galveston Bay (Offatts Bayou)—DDT, Sn, and Zn

High concentrations were determined based on 213 sites sampled nationally. It was concluded that adding more sites would not meaningfully change calculated mean or high concentrations. High concentrations were associated with population centers, and sediment contaminant concentrations were generally below those expected to be of biological consequence (NOAA 1991). As indicated, sites were purposely chosen to be representative of the area; highly contaminated sites were purposely avoided.

Based on 301 samples collected in 1986 and 1987 by the NOAA NS&T Program, it was concluded that pesticides and PCBs were pervasive at low concentrations in sediments along the northern Gulf of Mexico coast (Sericano et al. 1990). DDT was detected in more than 88 % of the samples with the highest concentrations found in sediments in bays in Florida, Alabama, and Texas. The highest non-DDT pesticide concentrations in sediments were found in Choctawhatchee, Naples, Tampa, St. Andrew, and Rookery bays, Florida; Biloxi Bay and Breton Sound, Mississippi; Terrebonne and Barataria bays, Louisiana; and Galveston and Matagorda bays, Texas. Pesticide concentrations were similar to those previously reported for coastal sediments from the northern Gulf of Mexico. PCBs were commonly detected in sediments in the northern Gulf of Mexico bays with high concentrations in Texas and Florida bays. PCB concentrations in sediments were spatially heterogeneous within bays. While pesticides and PCBs were ubiquitous in sediments, concentrations were less than ERL values (note that highly contaminated sites were avoided). Tissues from nearby biological organisms exceeded sediment concentrations by several-fold, indicating bioaccumulation.

Other subsets of NS&T Program data have been analyzed to highlight the occurrences and distributions of specific chemicals in coastal environments. For example, Apeti et al. (2009) report that high sediment concentrations of Cd (high was defined as the highest 15 % of concentrations measured at 200 U.S. coastal sites) were located in Tampa Bay (Hillsborough Bay), Florida; the Mississippi River (Tiger Pass and Pass A Loutre), Louisiana; Breton Sound (Sable Island), Louisiana; Galveston Bay (Offats Bayou), Texas; Nueces Bay, Texas, and at a marina near Corpus Christi, Texas. Nevertheless, all sediment Cd concentrations in the northern Gulf of Mexico were below ERL values. Cd concentrations in bivalve tissues were poorly correlated with nearby sediment concentrations (adjusted for grain size) but significantly correlated with proximity to population centers. Diagenetic remobilization of Cd reduced concentrations in surficial sediments and may be one reason for the poor correlation between tissue and sediment concentrations.

4.3.2 USEPA Environmental Monitoring and Assessment Program

The USEPA Environmental Monitoring and Assessment Program (EMAP) measured PAHs, PCBs, pesticides, organotins, and metals in the northern Gulf of Mexico sediments from 1991 to 1995 (USEPA 1999; Maruya et al. 1997; Summers et al. 1994, 1995, 1996). Several bays identified as containing chemical contaminants in sediments corresponded with watersheds identified by the USEPA National Sediment Inventory as "areas of probable concern." At the time, several USEPA Superfund sites were located near these estuaries including Galveston Bay, Tampa Bay, and the Florida panhandle (Figure 4.5a).

Exceedances of ERL and ERM values were used to assess the potential for sediment contaminant concentrations to have biological effects. According to EMAP, ERL guidelines were exceeded by pesticide and metals concentrations at numerous locations while PAH and PCB concentrations exceeded ERL values at only a few locations (less than 1 % of area) across the northern coastal Gulf of Mexico. There was a fairly even geographical distribution of sites across the northern Gulf of Mexico from the Florida Gulf Coast to Corpus Christi Bay, Texas where contaminants exceeded ERL or ERM values (Figure 4.5b). Based on the percent area of

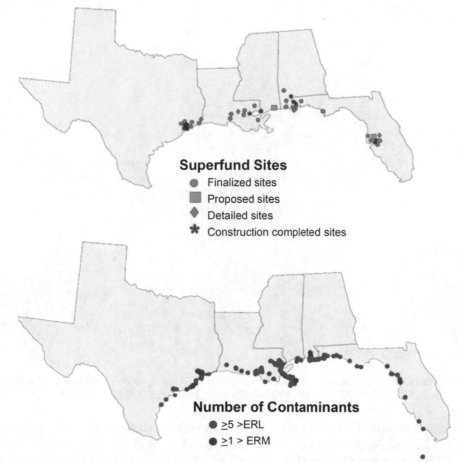

Superfund Sites
- ● Finalized sites
- ■ Proposed sites
- ◆ Detailed sites
- ✴ Construction completed sites

Number of Contaminants
- ● ≥5 >ERL
- ● ≥1 > ERM

Figure 4.5. (a) Location of coastal Superfund sites in the Gulf of Mexico as of 1999 and (b) sites where SVOCs or metals exceeded ERL or ERM values, 1991–1995 (modified from USEPA 1999).

coastal area judged to be contaminated, the majority of estuarine systems in the Gulf of Mexico were assessed as having fair to good sediment quality. However, several estuaries were identified as having predominantly contaminated sediments, for example Galveston Bay. At the time of the assessment, 50 % of chemical production and 30 % of the petroleum industry in the United States were located in and around Galveston Bay, Texas. Estuarine sediments were judged to be heavily impacted by urban and industrial activities. Galveston Bay has a long history of environmental issues due to expanding human demands and physical alterations of the bay and its watershed over many years. Sediment contaminant distributions indicated that locations in East Bay Bayou, Trinity Bay, marinas, and small lakes had as many as seven chemicals that exceeded ERL values (Figure 4.6). In East Bay Bayou the concentrations of several individual PAHs, including fluorene and phenanthrene, exceeded ERL values. Copper and chlordane concentrations exceeded ERL values at marinas and in Offats Bayou. Offats Bayou sediments also contained elevated concentrations of Pb, Zn, and DDT. As noted above, ERL values used to classify sediment Cu concentrations are close to or less than natural values.

A closer examination of EMAP sediment metal concentrations at 497 sites from 1991 to 1993 in estuaries of the northern Gulf of Mexico was conducted by Summers et al. (1996). Data were normalized to concentrations of Al to identify metals attributable to humans. Cr, Cu, Pb, Ni, and Zn concentrations were highly correlated with Al suggesting a predominantly natural

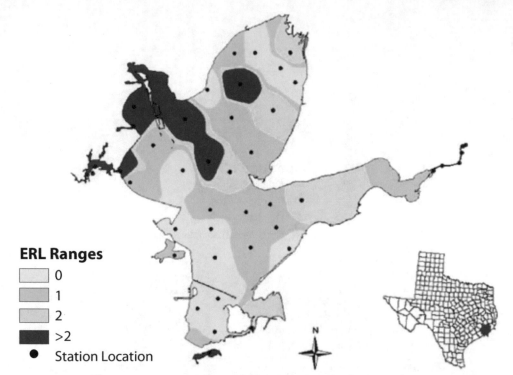

Figure 4.6. Distribution of locations in Galveston Bay, Texas, with chemical concentrations in sediments that exceeded ERL values, 1991–1995 (modified from USEPA 1999).

origin for these metals. This also supports the previous conclusions that ERL values used to classify Cr, Cu, and Ni sediment concentrations are close to or less than natural values and exceedances of these metals should be viewed with caution. As and Ag concentrations were moderately correlated with Al, suggesting a mixed natural and human origin for these metals. Hg and Cd concentrations were weakly correlated with Al suggesting a predominantly human origin for these metals. Of the sites with at least one metal elevated above natural levels, 39 % occurred near population centers, industrial discharge sites, or military bases (Figure 4.7). The remaining sites with at least one metal elevated above natural levels were located in the lower Mississippi River area (7 %) and near agricultural watersheds (54 %) suggesting that non-point sources were important.

4.3.3   USEPA National Coastal Condition Report I

The conditions of coastal sediments in the northern Gulf of Mexico were judged, based on a sediment contamination indicator, to be poor in the USEPA NCCR I report. The assessment is based on data collected from 1990 to 1997 from 500 locations (USEPA 2001). Sediment chemical concentrations exceeded ERL values at many locations; ERM values were exceeded at two locations: one in northern Galveston Bay and one in the Brazos River, Texas (Figure 4.8a). PAH and PCB sediment concentrations exceeded ERL values for less than 1 % of the locations. Pesticides exceeded ERL values for approximately 43 % of the locations. Metals exceeded ERL values for 37 % of the locations. Most of the pesticide ERL exceedances were for dieldrin and endrin (ERL values for these pesticides were near method detection limits). ERL values were exceeded for 12 and 4 % of locations due to DDT and chlordane concentrations, respectively. Enrichments in chemicals above natural levels were attributed to humans. Al concentrations are

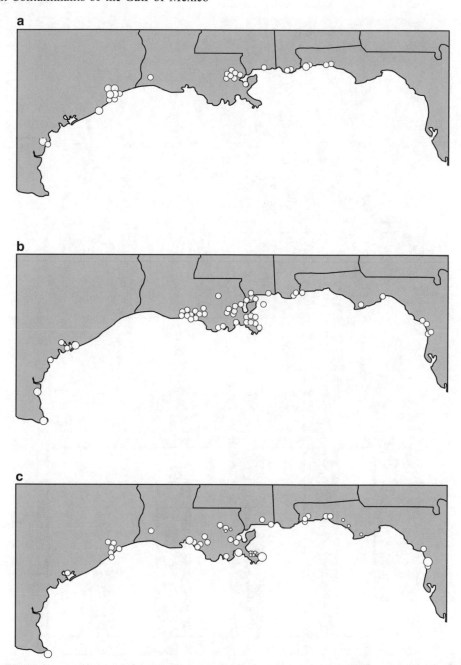

Figure 4.7. Distribution of elevated concentrations of metals in sediments in Gulf of Mexico estuarine sediments for (a) estuaries associated with discharges from human population centers, military installations, or industry and (b) estuaries associated primarily with discharges from agriculture watersheds. Circle sizes in (c) are proportional to the number of metals elevated at a site (the *largest circle* indicates eight elevated metals and *smallest circle* indicates one elevated metal) (modified from Summers et al. 1996).

used to determine the natural levels of metals in sediments because Al has few sources attributable to humans. Background Al/metal ratios were determined by analyzing uncontaminated sediments (Windom et al. 1989; Summers et al. 1996). Enrichments above natural levels ranged from 34 % for metals to 99 % for PAHs and PCBs (Figure 4.8b). PAH and PCB

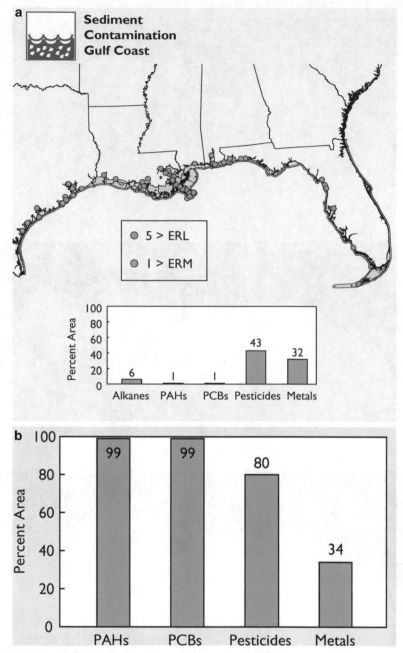

Figure 4.8. **(a) Distribution and percentages of locations that exceeded five or more ERL or one ERM value from NCCR I, 1990–1997 and (b) percentage of locations enriched in chemicals above natural levels (modified from USEPA 2001).**

enrichments are high as natural levels are zero (for PCBs) or near zero (for PAHs). Few details are provided about which metals exceeded ERL values. As noted above, ERL values used to classify sediment concentrations for Cr, Cu, Ni and As are close to natural values so excee-dances based on these criteria should be considered with caution. However, in this instance normalization of metal concentrations to Al does suggest that exceedances were most likely attributable to humans.

4.3.4 USEPA National Coastal Condition Report II

The 2004 NCCR II assessment was based on data collected from 191 locations in the year 2000 (USEPA 2004). The NCCR II sediment quality index included measures of sediment toxicity, sediment contamination, and total organic carbon content (TOC) (Figure 4.9a). Sediment quality was assessed as poor at a location if one of the component indicators was

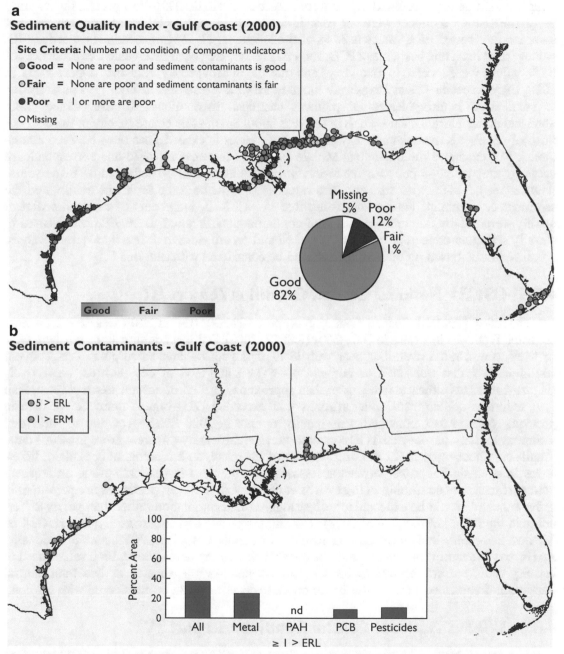

Figure 4.9. (a) Sediment quality index data for northern Gulf of Mexico sediments from NCCR II, 2000 and (b) locations where at least one sediment chemical concentration exceeded the ERM or more than five exceeded the ERL values. The *bar chart* shows the percent of locations where at least one or more sediment chemical concentration exceeded ERL values (modified from USEPA 2004).

categorized as poor, fair if the sediment contaminants indicator was rated fair, and good if all three component indicators were at levels unlikely to cause adverse biological effects. The conditions of coastal sediments in the northern Gulf of Mexico were judged to be fair, with 12 % of the area exceeding thresholds for sediment toxicity, sediment contaminants, and/or sediment TOC (Figure 4.9a). The sediment contaminants index was rated as follows: good (green) if no ERM values were exceeded and fewer than five ERL values were exceeded, fair (yellow) if five or more ERL values were exceeded and no ERM values were exceeded, and poor (red) if one or more ERM values were exceeded. As in the 1991–1997 period, the majority of ERL/ERM exceedances were due to sediment pesticide and metal concentrations. At least one metal exceeded ERL values in 28 % of the locations, 12–14 % of locations exceeded ERL values for at least one pesticide or PCB, and PAHs rarely exceeded ERL values. Exceedances of ERM values were located in Texas bays and one site in Mobile Bay, Alabama (Figure 4.9b). In 2000, ERM exceedances in Texas were much more widespread than in 1991–1997 possibly due to small-scale heterogeneities in sediment chemical distributions. No exceedances were observed along Florida's Gulf Coast in the year 2000. Small-scale heterogeneity in the distributions of chemicals in sediments may explain differences in exceedances rates between assessments. In Texas and the one site in Mobile Bay there is nearly a one-to-one correspondence between sites rated as poor for the sediment quality index and exceedances of ERM values (Figure 4.9b). This is also true for sites rated as fair for both the sediment quality and the sediment contaminant indices suggesting that in NCCR II, judgments of reduced sediment quality were mostly due to the presence of contaminants. As noted above, ERL values used to classify sediment concentrations for Cr, Cu, Ni and As are close to or less than natural values, so exceedances based on these criteria should be considered with caution.

4.3.5 USEPA National Coastal Condition Report III

The third National Coastal Condition Report—NCCR III—in 2008 was based on data collected from 487 locations in 2001–2002 (USEPA 2008). Sediment quality in the northern Gulf of Mexico was rated overall as poor with 18 % of the coastal area rated poor. The sediment contaminant index was rated as fair and poor for 1 and 2 % of coastal area, respectively (Figure 4.10). This indicates that greater than approximately 97 % of coastal area had fewer than five sediment contaminant concentrations that exceeded ERL values, many fewer than in previous years. Most poor sediment quality ratings in 2000–2002 were due to measured sediment toxicity or elevated TOC concentrations, significantly different from pre-2001 data. Small-scale heterogeneity in the distributions of chemicals in sediments may explain differences in exceedances rates between assessments. Reductions in pesticide usage on adjacent land surfaces may contribute to these differences as well; however, pesticides are persistent in sediments and would be expected to reflect long-term accumulation rather than yearly differences in inputs (Kimbrough et al. 2008). The authors conclude that in coastal northern Gulf of Mexico, sediments had elevated concentrations of metals, pesticides, PCBs, and, occasionally, PAHs, but concentrations were mostly below ERL values. As noted above, ERL values used to classify sediment concentrations for Cr, Cu, Ni, and As are close to or less than natural background values so exceedances based on these criteria should be considered with caution.

4.3.6 USEPA National Coastal Condition Report IV

The fourth National Coastal Condition Report—NCCR IV—was issued in 2012 based on data collected from 879 locations from 2003 to 2006 (USEPA 2012). Alabama, Mississippi, and Louisiana did not collect data in 2005 because of hurricanes Katrina and Rita. Sediment quality

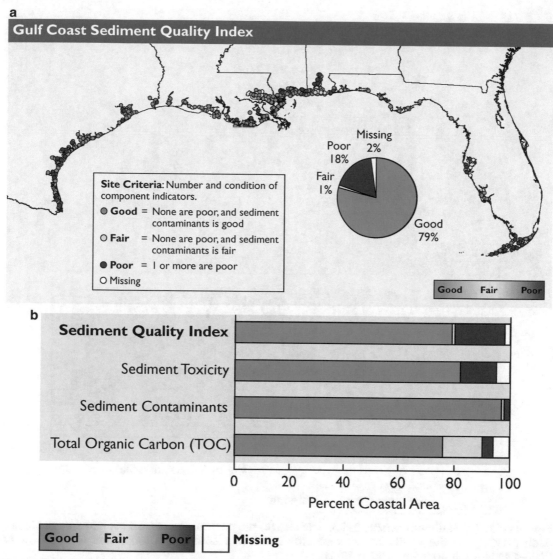

Figure 4.10. (a) Distribution of the sediment quality index ratings from NCCR III, 2001–2002 and (b) the percentage of coastal area achieving each ranking for the sediment quality index and component indicators (modified from USEPA 2008).

in the northern Gulf of Mexico was rated overall as poor with 19 % of the coastal area rated poor for at least one of the component indicators (Figure 4.11). The poor rating for the sediment quality index was mostly due to measured sediment toxicity, consistent with 2001–2002 data. Three locations in Florida Bay had high sediment concentrations of Ag that may have been the cause of the poor ratings for sediment toxicity. In all other instances, toxicity and sediment contamination were not well correlated. The authors suggested that the lack of a correlation of sediment toxicity and contamination may be due to toxicity caused by hydrogen sulfide or high salinity, grain size, contaminants not being bioavailable or not at lethal levels, or the presence of contaminants not measured.

The sediment contaminants indicator overall was rated good with 2 % and approximately 3 % of coastal area rated fair and poor, respectively, indicating approximately 95 % of the

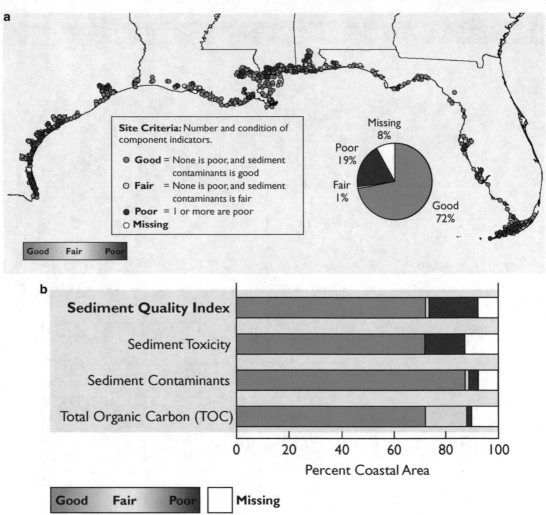

Figure 4.11. (a) Sediment quality index data and (b) the percentage of Gulf coastal area achieving each ranking for the sediment contaminants index and component indicators from NCCR IV, 2003–2006 (modified from USEPA 2012).

coastal area had fewer than five sediment chemical concentrations that exceeded ERL values. While the percentage of areas impacted by contaminants was similar in 2001–2002 and 2003–2006, the location and the cause of the impact were dissimilar except possibly in Mobile Bay. As noted above, sites rated poor were located in Florida Bay where sediment concentrations of Ag exceeded ERM values. Coastal areas rated fair for the sediment contaminants indicator were mostly located in Mobile Bay, Alabama. As in 2001–2002, the authors noted that northern Gulf of Mexico sediments had elevated concentrations of metals, pesticides, PCBs, and occasionally PAHs but concentrations were mostly below ERL values. Small-scale heterogeneity in the distributions of chemicals in sediments may explain differences in exceedances rates between assessments. As noted above, ERL values used to classify sediment concentrations for Cr, Cu, Ni, and As are near or less than natural values so exceedances based on these criteria should be considered with caution.

4.3.7 Gulf of Mexico Bays

Sediment contaminant concentrations in individual bays and estuarine complexes in the northern Gulf of Mexico provide a finer-scale view illustrating the importance of small-scale variations in chemical sources and the impact of local environmental conditions on geographic distributions. The following assessment is based on 2001–2002 data and includes some of the data used in NCCR assessments. The National Estuary Program (NEP) was established under Section 320 of the 1987 Clean Water Act Amendments as a USEPA effort to protect and restore the water quality and ecological integrity of major U.S. estuaries. There are 28 designated estuaries of national significance, and seven are located in the northern Gulf of Mexico (Figure 4.12). In general, the sediment contaminants indicator is rated as more degraded at this finer scale than at the regional scale, illustrating small-scale heterogeneities in chemical sources and sediment contaminant concentration distributions in coastal environments. This difference in ratings also illustrates that the selection of sampling sites can affect the data collected. Sampling locations in close proximity to shorelines and in shallower embayments highlight the steepness of spatial gradients in chemical concentrations as land-based sources are approached (see Regional Environmental Monitoring and Assessment Program [REMAP] data below as well).

The sediment quality index for the collective NEP estuaries of the Gulf of Mexico region was rated as fair to poor with 18 % of the estuarine areas rated as either fair or poor for the sediment quality indicator (Figures 4.13 and 4.14). As before, the sediment quality index was a composite indicator based on sediment toxicity, contaminants, and TOC content. Sediment contaminant index ratings were also defined the same: good (green) if no ERM values were exceeded and fewer than five ERL values were exceeded, fair (yellow) if five or more ERL values were exceeded, and poor (red) if one or more ERM values were exceeded. Northern Gulf of Mexico NEPs were rated fair for sediment contaminant concentrations with 11 % of the region's estuarine area rated poor (at least one sediment chemical concentration exceeded ERM values) (Figure 4.13). Most sediment quality ratings of poor were due to poor ratings for the sediment contamination indicator, although, on occasion, toxicity and TOC contributed to a poor rating. The largest numbers of locations with fair and poor ratings for the sediment contaminants indicator were located in Texas, including Galveston Bay and Corpus Christi Bay (Figure 4.14).

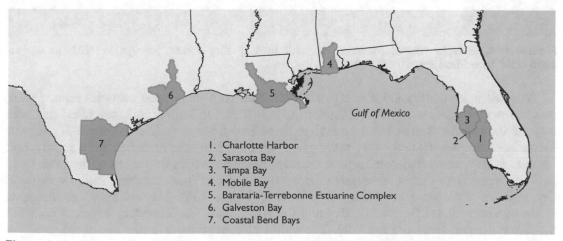

1. Charlotte Harbor
2. Sarasota Bay
3. Tampa Bay
4. Mobile Bay
5. Barataria-Terrebonne Estuarine Complex
6. Galveston Bay
7. Coastal Bend Bays

Figure 4.12. National Estuary Program estuaries (modified from USEPA 2006).

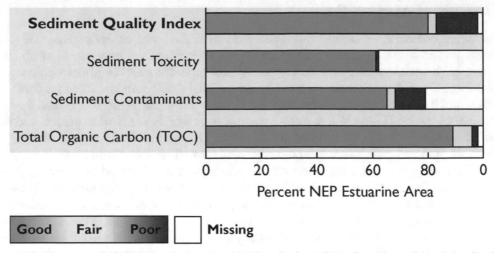

Figure 4.13. Percent of NEP estuary areas achieving each ranking for the sediment quality index and component indicators, 2000–2002 (modified from USEPA 2006).

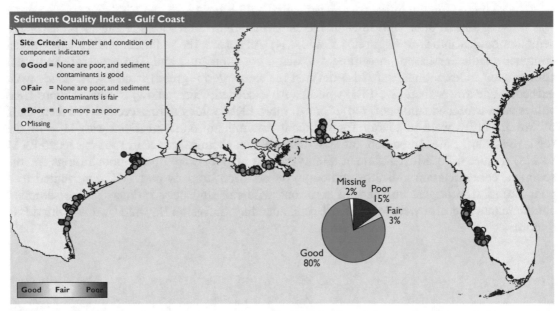

Figure 4.14. Geographic distribution of sediment quality index ratings in NEP estuaries, 2000–2002 (modified from USEPA 2006).

The sediment quality index ratings for individual Gulf of Mexico estuaries ranged from good to poor (Figure 4.15). The sediment quality index for Mobile Bay was rated fair with 9 % of the estuarine area rated poor (due to sediment toxicity). The Barataria-Terrebonne Estuarine Complex (BTEC) was rated good with 8 % of the estuarine area rated poor. Galveston Bay was rated fair to poor with approximately 5 % of the estuarine area rated poor. Coastal Bend Bays were rated poor with 38 % of the estuarine area rated poor. Contaminant data were not collected in Florida (Figure 4.15). The sediment contaminants index was rated good for Mobile Bay and the BTEC, fair for Galveston Bay, and poor for the Coastal Bend Bays (Figure 4.15). The sediment contaminants index was rated poor for 10 % of the Galveston Bay estuarine area.

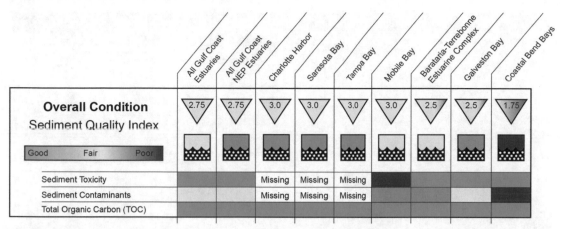

Figure 4.15. Comparison of indices for northern Gulf of Mexico estuaries, 2001–2002 (modified from USEPA 2006).

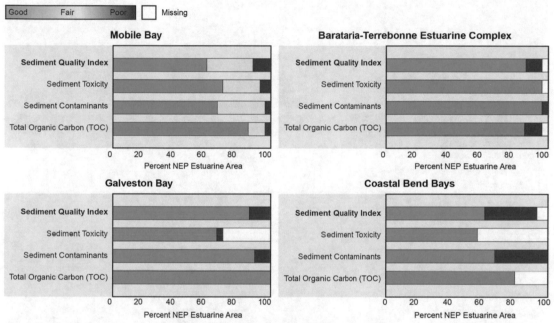

Figure 4.16. Percent of estuarine areas achieving each rating for the sediment quality index and components indicators, 2001–2002 (modified from USEPA 2006).

A closer review of the indicator rankings provides insight into the trends observed in individual estuaries. Correlation of the sediment quality index with the sediment contaminants index can be confounded by the inclusion of sediment toxicity and TOC indices (Figure 4.16). However, for Galveston and the Coastal Bend bays, most of the poor ratings were attributable to sediment chemical concentrations that exceeded ERM values for at least one chemical. The sediment contaminant index for Mobile Bay was rated good with 2 % of the estuarine area rated poor. Sediment quality in Mobile Bay was impacted at locations in the central portion of the bay primarily due to toxicity with occasional contributions from contaminants and/or TOC (Figure 4.17). Contaminants are known to accumulate in Mobile Bay sediments over time. Of the 23 USEPA Total Maximum Daily Load 303(d)-listed streams

238   M.C. Kennicutt II

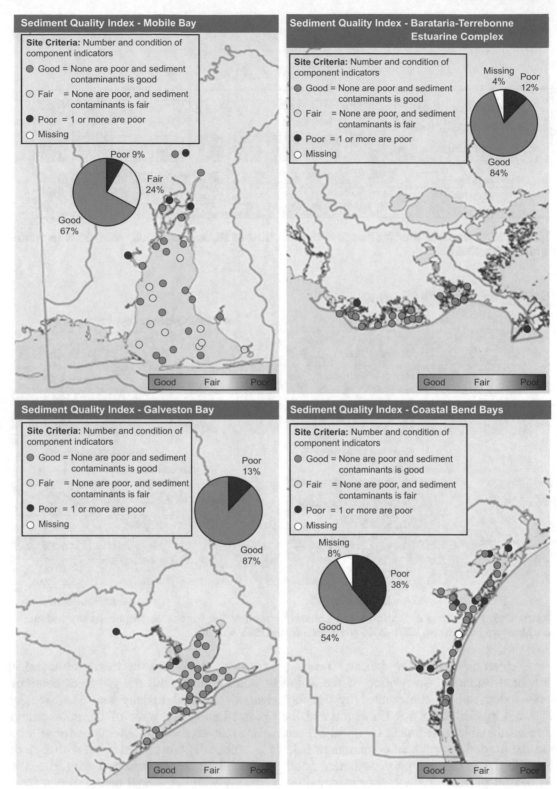

Figure 4.17. Sediment quality index ratings for Mobile Bay, the Barataria-Terrebonne Estuarine Complex, Galveston Bay and Coastal Bend Bays, 2000–2001 (modified from USEPA 2006).

located in the Mobile Bay NEP, 8 were impaired in part due to Hg contamination. Sediment contaminants influenced sediment quality at locations near the shore in the northern portions of Mobile Bay. The BTEC was rated good for the sediment contaminants with 4 % of the area rated poor (Figure 4.17). Two locations were rated poor mostly because of localized, elevated TOC concentrations. Galveston Bay was rated fair for sediment contaminants indicator despite 10 % of the estuarine area being rated poor. SVOCs and metals commonly have elevated concentrations in Galveston Bay runoff, freshwater inflow, and waste discharges and lower, relatively uniform, concentrations in sediments in the central part of the bay. The upper Houston Ship Channel generally had the highest concentration of chemicals. Coastal Bend Bays were rated poor for sediment chemical concentrations with 38 % of the estuarine area rated poor. Concentrations of As, Cd, Hg, and Zn were often elevated in Corpus Christi Bay sediments. The highest levels of pesticides were found in Baffin and Copano bays, Texas. Elevated levels of PAHs, metals, pesticides, and PCBs have been documented in sediments near stormwater outfall sites. A detailed study of Corpus Christi Bay documented elevated concentrations of PAHs, DDT, chlordane, PCBs, As, Pb, Al, Cu, Ni, Zn, Cd, and Cr in sediments near stormwater outfalls (Carr et al. 2000). Park et al. (2002) concluded that the atmosphere was a pathway for persistent, anthropogenic PAHs, PCBs, and pesticides to enter coastal environments. Gas exchange has been shown to be an important transport process for SVOCs between the atmosphere and surface waters. In Corpus Christi Bay urban and industrialized areas, atmospheric inputs of PAHs and PCBs were identified as a continuing source of contaminants (Park et al. 2002).

The Regional Environmental Monitoring and Assessment Program (REMAP) was initiated to test the applicability of the EMAP approach to describe ecological conditions at regional and local scales. In the northern Gulf of Mexico, Galveston Bay was selected as a site for a REMAP project and included the measurement of SVOCs and metals in sediments at 29 random locations (USEPA 1998). The study characterized the condition of Galveston Bay as a whole and four small bays in Galveston Bay and also studied the impact of marinas. Comparisons of EMAP and REMAP results were used to highlight those areas with contaminated sediments. Enrichment was determined using regression equations for each metal against Al concentrations in the sediments (Summers et al. 1996). As noted above, ERL values used to classify sediment concentrations for Cr, Cu, Ni, and As are near or less than natural values so exceedances based on these criteria should be considered with caution. In this instance, normalization to Al provides some assurance that elevated metal levels are due to human activities. In Galveston Bay, As, Cu, Pb, Ni, and Zn concentrations exceeded ERL values but not ERM values at one or more sites. Sites contaminated with the most metals were Offats Bayou, Clear Lake, Moses Lake/Dollar Bay, and two marina sites. The Galveston Bay area had high Cr and Ni sediment concentrations across a large area. The percent of area that exceeded ERL values for As were lower and for Zn were similar to EMAP results for the region. Cu exceedances of the ERL value were found at marina sites and in Offats Bayou but not in the randomly sampled area. Organotin concentrations in sediments exceeded the screening level of 1.0 parts per billion (ppb) in 52 % of the area compared to 31 % for EMAP results. Sites with high dieldrin and endrin concentrations in sediments were located in upper Galveston Bay, Clear Lake, and upper Trinity Bay. Dieldrin and endrin ERL values were exceeded at 17 and 5 % of the area, respectively, in Galveston Bay and 33 and 0 %, respectively for small bay and marina sites, which was lower than EMAP results for the region. Concentrations of other pesticides, including DDT, did not exceed ERL values in either study. Individual PAH concentrations exceeded ERL values in Trinity Bay near several active oil wells. PAH concentrations for sediments at three sites in Galveston were considerably higher than at the other sites. C3-fluorene concentrations exceeded ERL values in 3 % of

Galveston Bay sediments, similar to EMAP results for the region. PCB concentrations in Galveston Bay sediments did not exceed ERL values, compared to an EMAP rate of 1 % exceedances.

4.3.8 Temporal Variations

There is limited data that is useful for distinguishing temporal variations in concentrations of contaminants in coastal, northern Gulf of Mexico sediments. While data have been collected over many years, variations in analytical methods, reporting methods, sampling locations, and small-scale spatial heterogeneity confound detection of trends with time. Based on data collections in the 1980s and the early 1990s it was concluded that distributions and sediment concentrations of SVOCs were similar to previously reported concentrations (Summers et al. 1992, 1994, 1995; Wade et al. 1988). Based on NCCRs, a comparison of yearly sediment contaminant indicator ratings showed no significant temporal trend in the percent coastal area rated poor for 1991–2002 (Figure 4.18b). There was also no significant difference in the percent of area rated poor when the data were averaged for the years 1991–1994 and compared to the averages for the years 2000–2002. The percent of area rated good for sediment contaminant concentrations increased significantly ($R^2 = 0.77$; $p < 0.05$) from 1991 to 2002 (Figure 4.18a). Although the percent area rated as poor remained similar, the sediment contaminants indicator improved, as indicated by a significant decrease ($z = 3.96$; $p < 0.05$) in the combined total percent area rated poor and fair from 16.4 % in 1991–1994 to 5.9 % in 2000–2002. This trend is consistent with reductions in pesticide and ERM concentration exceedances. The incidence of PAH and PCB exceedances is similar during this timeframe or showed no consistent temporal trend. In NCCR IV, these comparisons were again made including additional data for 2003–2006 (USEPA 2012). Data for 2005–2006 were collected using a 2-year survey design, and the data were treated as a single year in trend analyses. The percent of area rated poor for the sediment contaminants indicator decreased from 13 % in 2000 to zero in 2004–2006 (Figure 4.18c). Small-scale heterogeneity in the distributions of contaminants in sediments may explain differences in exceedances rates between assessments.

ERL/ERM value exceedances from the NCCRs in the 1990s and 2000s are summarized in Table 4.2. The limitations of ERL values as predictors of sediment quality and/or toxicity should be considered in interpreting the significance of temporal trends. Also note that ERL/ERM exceedances from year to year do not always occur at the same location and the contaminants causing the exceedances can vary. In addition, the percent area impacted was calculated based on a varying number of exceedances (1–5) with fewer than 5 ERL exceedances resulting in a site being rated as good (i.e., 1–4 sediment contaminant concentrations may exceed ERL values at locations but they are not included in the percent of area impacted). A conservative approach was taken in estimating the area impacted using the highest possible percent as a less-than figure for the percent area impacted (i.e., the percent area for exceedances might be significantly lower). Based on these conservative estimates and limitations, pesticide and metal ERL/ERM exceedances decreased from 1990 to 2006 with the largest reductions from 2001 to 2006. PCB and PAH exceedances appear to be similar and low throughout the 1990–2006 time period. The percent of area rated good for the sediment contaminant indicator (fewer than 5 ERL contaminant exceedances) increased from 1991 to 2002, and the percent of area rated poor for the sediment contaminants indicator (no ERM exceedances for all contaminants) decreased from 13 % in 2000 to none in 2004–2006. Overall, these data suggest that the number of sites where pesticide and metal sediment contaminant concentrations exceed levels suspected of causing biological effects (and thus average concentrations) have decreased with time. PCB and PAH sediment concentrations rarely exceeded levels

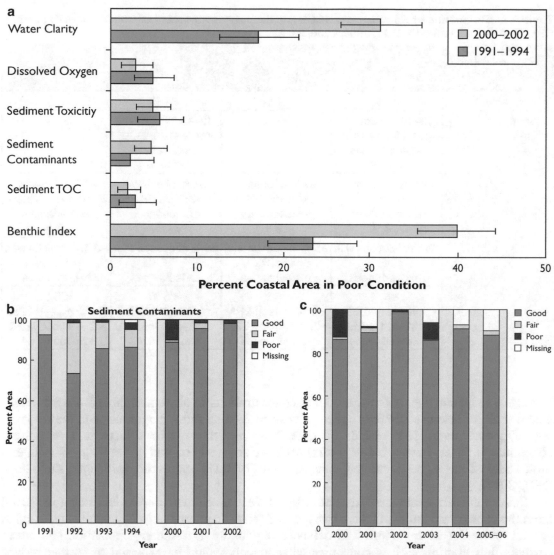

Figure 4.18. (a) Comparison of the average percent area rated poor for ecological indicators, 1991–1994 and 2000–2002; (b) the percent area in good, fair, poor, or missing categories for the sediment contaminants indicator, 1991–2002; and (c) the percent area rated good, fair, poor, or missing categories for the sediment contaminant index, 2000–2006 (modified from USEPA 2008, 2012).

suspected of causing biological effects from 1990 to 2006. Exceptions are areas known to be heavily and chronically contaminated (i.e., parts of Galveston Bay and Houston Ship Channel) and areas subject to major accidental spills or other high-level releases of contaminants.

4.4 CONTINENTAL SHELF/SLOPE AND ABYSSAL SEDIMENTS

In general, the concentrations of human-derived sediment contaminants are expected to decrease with distance off shore. Human activities that have the potential to contaminate sediments mostly occur in coastal areas and/or on adjacent land surfaces. In contrast, naturally

Table 4.2. Summaries of Results from NCCR I, II, III, and IV for Percent of Coastal Area Exceeding ERL/ERM Values (USEPA 2001, 2004, 2008, 2012)

NCCR Report (years of data collection)	Pesticides	Metals	PCBs	PAHs
I (1990–1997)	43 %	37 %	<1 %	<1 %
II (2000)	12–14 % with one pesticide or PCB exceedances <14 %	28 %	12–14 % with one pesticide or PCB exceedances ≪14 %	Rare <1 %
III (2001–2002)	97 % of coastal with <5 ERL exceedances <3 %	97 % of coastal with <5 ERL exceedances <3 %	97 % of coastal with <5 ERL exceedances ≪3 %	97 % of coastal with <5 ERL exceedances <3 %
IV (2003–2006)	95 % of coastal with <5 ERL exceedances	95 % of coastal with <5 ERL exceedances	95 % of coastal with <5 ERL exceedances	95 % of coastal with <5 ERL exceedances
[Note: 1 % of ERM exceedances were for Ag in a Florida Bay]	· <5 % <1 % exceed ERM <2 % with <5 ERL exceedances	<5 % <1 % exceed ERM <2 % with <5 ERL exceedances	≪5 % <1 % exceed ERM <2 % with <5 ERL exceedances	≪5 % <1 % exceed ERM <2 % with <5 ERL exceedances

occurring sediment metals occur at higher concentrations in abyssal sediments than in continental shelf and slope sediments due to slow accumulation rates, diagenesis, and the scavenging of metals over long periods of time; however, this offshore increase is a natural phenomenon. PAHs have a major natural offshore source in oil and gas seepage as well, and thus, some of the highest sediment concentrations of PAH are on the continental shelf/slope (NRC 2003).

Coastal environments are often sites of particle and sediment deposition. Being restricted from the open ocean limits offshore transport of coastal contaminants. Contaminants found in coastal areas can be transported to offshore regions by atmospheric circulation and ocean currents, but dilution with uncontaminated sediments would be expected to further reduce sediment contaminant concentrations. Seaward transport is most important in close proximity to river systems that outflow directly into the ocean. For example, the Mississippi River transports material significant distances offshore during periods of high outflow. Hydrocarbons and limited amounts of pesticides and PCBs can be directly released to continental shelf/slope and/or abyssal waters because of use on and discharges from offshore oil and gas platforms and from emissions and discharges from ships and accidental spills (NRC 2003). Surveys of contaminant concentrations in sediments on the continental shelf/slope and in the abyss of the northern Gulf of Mexico are limited. In cases where studies have been conducted, they are mostly directed at specific activities, such as oil and gas platform discharges, or unique environments, such as natural oil and gas seeps. The majority of studies measure only hydrocarbons and metals due to the general absence of pesticides and PCBs in the offshore region.

4.4.1 Natural Oil and Gas Seepage

Seepage of oil and gas is a natural phenomenon that occurs when deeply generated oil and gas migrates to the earth's surface (Wilson et al. 1974; NRC 2003). Deeply buried petroleum reservoirs and source rocks generate oil and gas that can migrate upward into marine sediments if pathways such as faults exist. Sediments overlying prolific oil and gas provinces are well known sites of natural oil and gas seepage worldwide (Wilson et al. 1974; Brooks et al. 1986). Over geologic time (millions of years) much larger amounts of petroleum have been lost to seepage than is trapped in reservoirs (Wilson et al. 1974). Offshore seeps are widespread in the northern Gulf of Mexico, accounting for approximately 95 % of the total oil input to the offshore region (Anderson et al. 1983; Kennicutt et al. 1983; Brooks et al. 1987; Wade et al. 1989; Sassen et al. 2003; NRC 2003). Most petroleum seepage in the Gulf of Mexico is located in the north-western offshore region (Figure 4.19). The full extent of oil and gas seepage in the Gulf of Mexico is difficult to quantify due to challenges in detection (e.g., subsea releases), differences in quantification methods (e.g., satellite observations and direct sampling by corer), dispersion by ocean currents, the paucity of geographic coverage, and variable and uncertain seep volumes and seepage rates (Wilson et al. 1974; De Beukler 2003). Within these limitations, natural seepage of oil has been estimated to exceed 127,000 metric tons (140,000 tons) annually in the northern Gulf of Mexico (NRC 2003).

The immediate effects of oil and gas seepage are mainly in sediments in close proximity to seeps (i.e., within a few hundred meters of the seep's surface expression [Wade et al. 1989]). The low-energy environment of offshore generally limits redistribution of sediments from

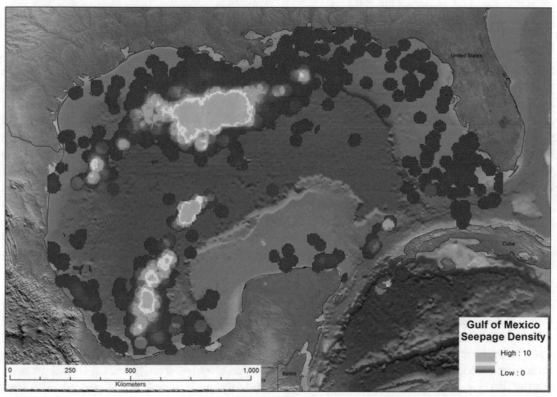

Figure 4.19. Oil and gas seepage in the Gulf of Mexico (determined from analysis of synthetic aperture radar, graphic provided by CGG's NPA Satellite Mapping, used with permission).

seeps; however, mass wasting and turbidity currents have the potential to transport sediments long distances. Most oil constituents have low water solubility and are less dense than water, so seeping oil mostly adsorbs onto the sediment it is seeping through or escapes into the water column rising to the sea surface to form slicks. Some seeps, becoming entrained in sediments on the sea floor, may not reach the water's surface. Those constituents of oil that are soluble in water (i.e., low molecular weight PAHs) can dissolve in the overlying water column (Wade et al. 1989). Adsorption onto sediments leads to heterogeneous and discontinuous distributions of oil in sea-bottom sediments. Oil released into marine environments by sea-bottom seepage undergoes similar physical and chemical processes (except evaporation and photooxidation due to the submerged location) as subaerial releases. Natural processes degrade and metabolize the oil, but oil can be replenished as long as the seep remains active. The persistence of natural oil seeps is estimated to be years and possibly centuries or longer as oil and gas have been generated in the deep subsurface over geologic time (Wilson et al. 1974). Sediments contaminated with petroleum by natural seeps can contain highly variable concentrations of oil due to the point-source characteristics of seepage. Petroleum concentrations in seep sediments can vary from trace amounts at the fringes of a seep to several percent by weight at an active seep or even a separate liquid phase. Oil and gas seep rates vary with time and seeps can be dormant for periods of time (no seepage). Microbial degradation of oil produces authigenic calcium carbonate minerals that can temporarily cap seeps. While seepage is common across the deep water of north-central Gulf of Mexico, the percentage of continental shelf/abyss sea-bottom area containing seep oil is estimated to be limited. A detailed review of oil and gas seepage in the Gulf of Mexico and additional references are provided in Chapter 5 on Oil and Gas Seeps in the Gulf of Mexico.

4.4.2 Other Contaminants Attributable to Humans

Contaminants in northern Gulf of Mexico continental shelf/abyss sediments can have origins in human activities such as discharges from offshore oil and gas platforms, emissions and releases from ships, and accidental spills that occur offshore. Materials can be transported by the atmosphere, oceanic currents, and rivers to offshore regions. The northern Gulf of Mexico is, and has been, one of the most prolific oil and gas provinces in the world for many years (Figures 4.20 and 4.21) (Energy Information 2009). In 2006, there were nearly 4,000 oil and gas platforms in the northern Gulf of Mexico, mostly offshore of Louisiana and Texas (Figure 4.20). In recent years, oil and gas exploration and production in the Gulf of Mexico has expanded outward onto the continental slope.

A survey of background PAH and metal concentrations in continental slope/abyss sediments (greater than 300 meter [m] (984 feet [ft]) water depth) in the northern Gulf of Mexico was conducted by Rowe and Kennicutt (2009). Sediment sampling sites were purposely located many kilometers from petroleum development activities. Sediment PAH concentrations were low (less than 1.0 parts per million [ppm]) at all 50 locations sampled (Figures 4.22 and 4.23). These sites are remote from natural oil and gas seepage and oil and gas exploration activities, and the PAH concentrations detected approached method detection limits. The composition of PAHs was indicative of petroleum and combustion (pyrogenic) sources. Combustion-derived PAH can originate from discharges from ships (e.g., stack emissions and bilge pumping) and oil and gas platforms, atmospheric deposition, and/or riverine transport. Ship traffic and platform operations were judged to be the most likely sources, as atmospheric deposition would produce similar PAH concentrations over large regions and this pattern was not observed. The geographic distribution of PAHs suggested an input of PAHs from the sediment plume of the Mississippi River as low PAH (less than 1.0 ppm) co-occurred with low Ba concentrations (less

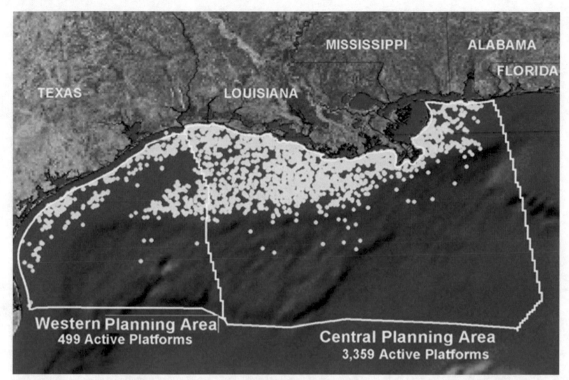

Figure 4.20. Map of the 3,858 oil and gas platforms in the Gulf of Mexico in 2006. The size of the dots used to note platform locations is highly exaggerated and the density of platforms is low. From NOAA (2012).

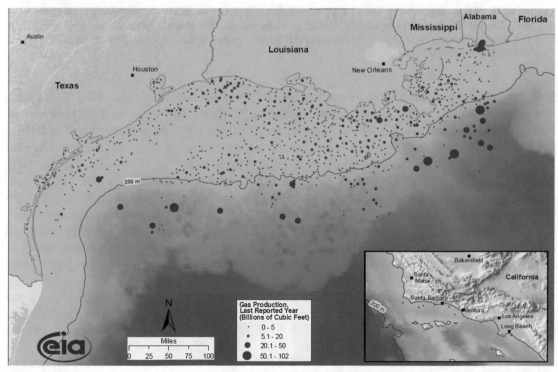

Figure 4.21. Map of offshore gas production in the Gulf of Mexico (from Energy Information 2009).

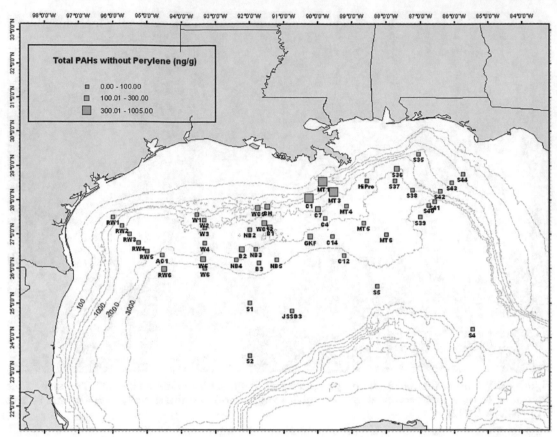

Figure 4.22. Concentration of total polycyclic aromatic hydrocarbons (PAHs: without perylene) in northern Gulf of Mexico continental slope/abyss sediments. Perylene is a naturally occurring PAH and is not suggestive of fossil fuel-derived PAH (ng/g = ppb; 1,000 ng/g = 1 ppm) (modified from Rowe and Kennicutt 2009).

than 0.5 part per thousand [ppt]—a marker for drill mud discharges) (Figure 4.24). A few sites had elevated Ba and PAH concentrations indicating platform discharges as a possible source. Median PAH concentrations were one-quarter that reported for coastal sediments. The highest total PAH sediment concentrations were four or more times lower than ERL values. Elevated PAH concentrations and compositions at three locations near the Mississippi Canyon were believed to be sourced in Mississippi River outflows. Average concentrations of Ag, Cd, Cu, Hg, Pb, and Zn were similar to average crustal abundances and sediments from the northern Gulf of Mexico, which are thought to be uncontaminated. Metal concentrations and ratios were similar to those for Mississippi River Delta sediment as well (Table 4.1). The few elevated metal concentrations were attributed to natural diagenetic remobilization processes. Enrichments of Ba compared to average crustal material and clay-rich sediments were traced to the presence of drilling muds (Figure 4.24).

These conclusions regarding the distribution and concentrations of naturally occurring metals in offshore northern Gulf of Mexico sediments have been confirmed by many other studies (Table 4.3) (Tieh and Pyle 1972; Tieh et al. 1973; Trefry and Presley 1976a, b; Presley et al. 1980; Trefry et al. 1985; Kennicutt et al. 1996). Presley et al. (1980) reports that sediments from a 1,500 square kilometer (km²) (579 square mile [mi²]) area of the Mississippi River Delta had Pb and Cd concentrations 10–100 % higher than background levels. Vertical

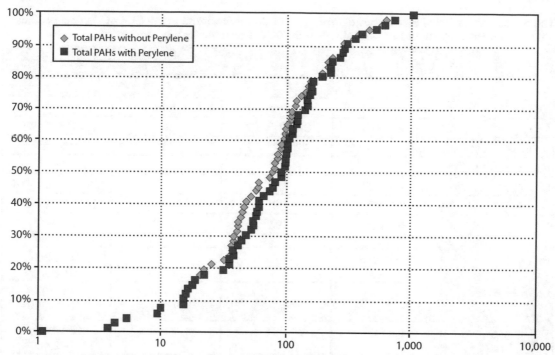

Figure 4.23. Frequency distribution of total PAH and total PAH without perylene concentrations (ppb) versus cumulative percentage in the northern Gulf of Mexico continental slope/abyss sediments. Perylene is a naturally occurring PAH and is not suggestive of fossil fuel-derived PAH (modified from Rowe and Kennicutt 2009).

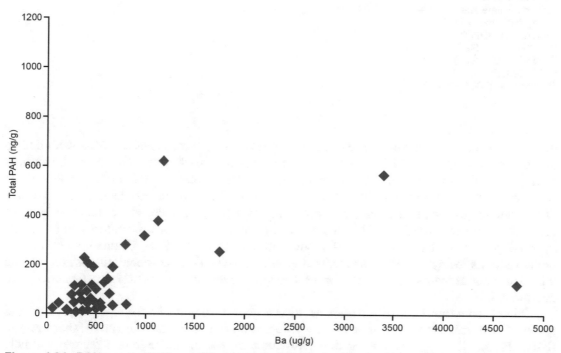

Figure 4.24. PAH concentrations without perylene (ng/g [ppb]) versus barium concentrations (µg/g [ppm]) in the northern Gulf of Mexico continental slope/abyss sediments. Perylene is a naturally occurring PAH and is not suggestive of fossil fuel-derived PAH (modified from Rowe and Kennicutt 2009).

Table 4.3. Examples of Background Contaminant Concentrations in the Gulf of Mexico Sediment, Mississippi River Particulates, and Mississippi Delta Particulates and Cores

	Gulf of Mexico Sediment	Mississippi River Particulates	Mississippi Delta Particulates	Mississippi Delta Sediment Cores
Metal				
Cr (µg/g)	60–90[a], 22–100[b]	72[c], 74[d], 80[e]	84[e]	
Cu (µg/g)	20–30[a]	33[c], 21[d], 45[e]	56[e]	23–27[e], 21–25[f]
Ni (µg/g)	25–40[a], 27–42[b]	41[c], 37[d], 55[e]	56[e]	32–42[e]
As (µg/g)	10–15[a]	8[d]		
Cd (µg/g)	0.2–0.3[b]	1.3[e]	1.5[e]	0.3–0.8[e], 0.2–0.4[f]
Pb (µg/g)	17–32[b]	46[e]	49[e]	24–41[e], 23–37[f]
Zn (µg/g)	50–70[b]	193[e]	244[e]	110–140[e], 125–155[f]
Ba (µg/g)	805–1,478[g]			400–1,300[f], 50–500[h]
V (µg/g)				14–50[h]
SVOC				
PAH (ng/g)	43–748[g]			200–800[f], 180–2,400[h]
PCB (ng/g)				0–22[f]
DDT (ng/g)				0–1.6[f], <10[h]

Note: ng/g—nanogram(s) per gram [ppb]; µg/g—microgram(s) per gram [ppm]
[a]Trefry and Presley (1976a), Presley et al. (1980), Trefry et al. (1985), Kennicutt et al. (1996), Continental Shelf Associates (2006)
[b]Uncontaminated far-field sites only CSA (2006)
[c]Trefry and Presley (1976b)
[d]Horowitz et al. (2001)
[e]Presley et al. (1980)
[f]Santschi et al. (2001)
[g]CSA (2006)
[h]Turner et al. (2003)

distributions of contaminant concentrations suggested that these elevated levels were due to human-derived inputs and had occurred over the previous 30–40 years. However, the authors found no indication of metal contamination in other areas of the delta or along the continental shelf of the northwest Gulf of Mexico. The authors conclude that there was little evidence of elevated metal concentrations (except Cd and Pb) attributable to humans along the northwestern Gulf of Mexico continental shelf. In contrast, introductions of metals in such places as Corpus Christi Harbor and the Houston Ship Channel were readily recognized as being elevated above background levels. For additional discussion of the input of contaminants related to the Mississippi River and a historical perspective, see Section 4.4.4.

Other than natural petroleum seepage, discharges associated with the extraction of oil and gas have the greatest potential to contaminate sea-bottom sediments in the offshore region (NRC 2003). The exploration for and extraction of petroleum in the offshore routinely discharges produced waters and drill muds and cuttings. In addition, runoff waters from structures, emissions from platforms, and accidental spills can release contaminants to the

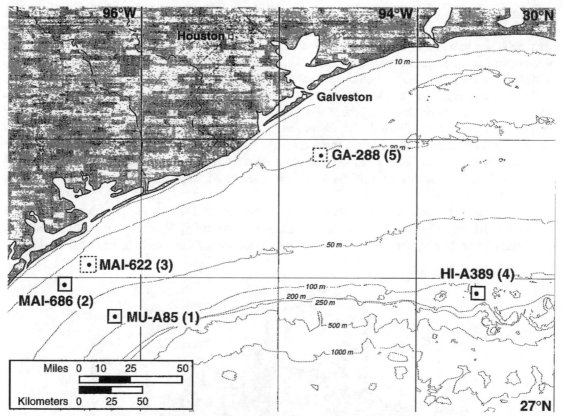

Figure 4.25. Location of the three GOOMEX study sites (sites 1, 2, and 4) (modified from Kennicutt 1995).

offshore environment. The volumes and frequencies of these releases are highly variable and generally low compared with coastal sources. Greater than 90 % of the petroleum released during extraction activities is due to produced water discharges which release low but continuous amounts of dissolved components and dispersed crude oil to the marine environment. Since produced waters mostly contain dissolved contaminants little is deposited in sea-bottom sediments in the deeper water regions of the offshore unless the discharge point is directly onto the sea bottom. In contrast, discharges of drill muds and cuttings during drilling operations are denser than water and often end up deposited on the sea floor close to the platform.

The Minerals Management Service (MMS) program—Gulf of Mexico Offshore Operations Monitoring Experiment (GOOMEX), Phase I: Sublethal Response to Contaminant Exposure measured contaminants in sediments at three offshore platforms (Kennicutt 1995). The GOOMEX study sites were located in the northwestern Gulf of Mexico in water depths from 29 to 125 m (95–410 ft) (Figure 4.25).

During offshore drilling activities, drill muds and cuttings can be discharged in large quantities near the sea surface or shunted to near-bottom waters (Kennicutt 1995). These discharges include a variable mix of drill muds (mainly barium sulfate but also chemical additives including diesel fuel). Cuttings from the drilled sections discharged at the same time can also have variable mineralogy. Drill cuttings are occasionally oil wet with petroleum. The three sites studied in GOOMEX, purposely located outside of the influence of the Mississippi River plume, were active oil and/or gas development and production platforms for more than 10 years. There are limitations in extrapolating the GOOMEX results to other platforms because each platform has varying drilling and discharge histories, contaminant concentrations in the discharge material, and amounts of material discharged. In addition, platforms are in widely varying water depths, the location of discharges can be different, and oceanographic settings are site dependent (i.e., current patterns can vary greatly). Two of the GOOMEX sites (HI-A389 and MU-A85) shunted platform discharges to the sea bottom. Most platform discharges are released at the sea surface so these two sites can be considered worst-case scenarios for contaminating the seabed. Discharges at the sea surface, especially in deeper water, are likely to be diluted and dispersed prior to sedimentation on the sea floor. The GOOMEX study concluded that the most common contaminants discharged at platforms were PAHs and metals, and those sediments in close proximity to the platform contained elevated concentrations. Pesticides and PCBs were not measured because little or no source was suspected at the site.

PAH concentrations were highest near the platforms and decreased rapidly with distance from the platform (Figures 4.26 and 4.27) (Kennicutt et al. 1996). Spatial patterns of PAHs and metals exhibited strong directional orientations reflective of the local current regime. With a

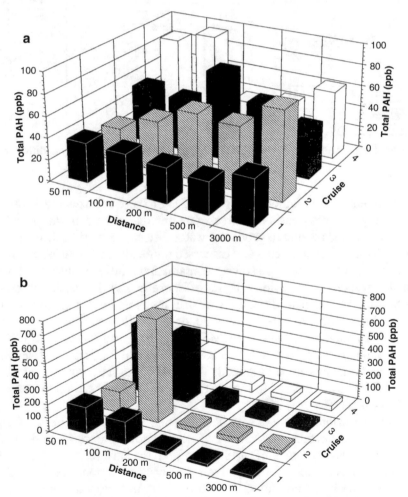

Figure 4.26. Variability in mean PAH concentrations (ppb) in sediments with distance from the platform by cruise at (a) MAI-686, (b) MU-A85, (c) HI-A389, and (d) ERL/ERM exceedances for metals for all cruises (modified from Kennicutt 1995; Kennicutt et al. 1996).

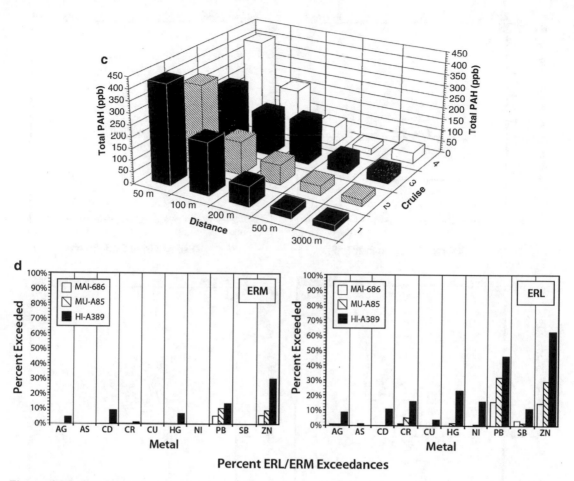

Figure 4.26. (continued)

few exceptions, most PAHs were biodegraded. Between cruises, variations at the sites were small, suggesting that contaminants in sediments were stable over a period of years in these low-energy environments. Compared to coastal sediments, PAH sediment concentrations in the vicinity of these offshore platforms were low and far below ERL values. No significant bioaccumulation of petroleum was observed in megafaunal invertebrates or fish near the platforms. Some sediment metal concentrations exceeded ERL and ERM values (Figure 4.26). The aerial extent of chemicals in sediments was contaminant dependent (i.e., PAHs to 200 m [656 ft]; Ba to greater than 500 m [1,640 ft]) (Figure 4.27).

Sediments at two of the three study sites (HI-A389 and MU-A85) exhibited gradients in Ba, Ag, Cd, Hg, Pb, and Zn concentrations with distance from the platform. Most decreases in metal concentrations correlated with decreasing Ba concentrations, a marker for drill muds, suggesting that Cu and Hg were constituents of the barite ore used in the drill muds. Cd, Pb, and Zn had no known non-drilling discharge sources. At HI-A389, Cd, Pb, and Zn sediment concentrations close to the platform were at levels that exceeded ERL and ERM values. Cr and iron (Fe) concentration distributions suggested there was a platform-related source for these metals other than drill mud and cuttings, possibly from platform drainage. Chemical concentrations were highest at sediment depths of 10–20 cm (4–8 in). At all three sites, Pb

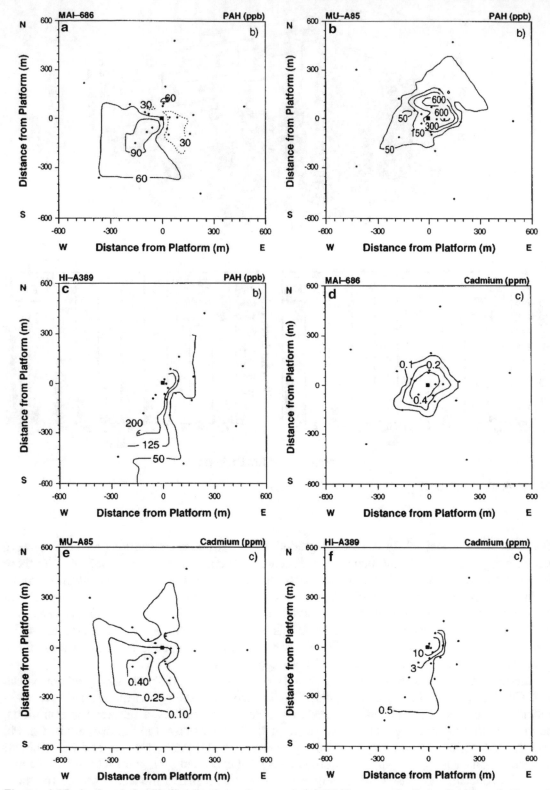

Figure 4.27. (a, b, c) Areal distribution of mean total PAH concentrations (ppb) and (d, e, f) cadmium concentrations (ppm) in sediments as a composite of four cruises at three platforms (MAI-686, MU-A85, HI-A389) in the Gulf of Mexico (modified from Kennicutt 1995; Kennicutt et al. 1996).

concentrations in sediments near the platform increased with time suggesting an ongoing release of Pb from the platform. At platforms in water depths deeper than 80 m (262 ft) contaminant concentrations were similar over a period of years. As noted above, ERL values used for Cr, Cu, Ni, and As are close to natural abundances, so exceedances based on these criteria should be considered with caution.

The GOOMEX study concluded that drill mud and cutting discharges by offshore platforms can lead to elevated contaminant concentrations in sediments within a few hundred meters of a platform. Sediments with elevated contaminant concentrations occur as thin veneers with the thickness of the veneer dependent on discharge volumes and rates and oceanographic conditions and may persist for years in low-energy environments. In higher-energy environments and instances of near-surface release, discharges are dispersed in the water before deposition in sediments and diluted with uncontaminated sediments. Metals and PAHs may be sequestered in mineral matrices (cuttings), limiting their bioavailability. Despite the large number of platforms in the northern offshore Gulf of Mexico, the density is low and the total area affected by the discharge of chemicals is expected to be a small percentage of the offshore area (e.g., within an order of magnitude, a rough estimate is that the total contaminated sediment surface area is 0.0006 % of continental shelf/slope surface area in the northern Gulf of Mexico—assuming approximately 200 m × 200 m (656 ft × 656 ft) surface zone of contaminated sediment is associated with a platform; approximately 4,000 platforms; a Gulf of Mexico surface area of approximately 1.55 million km^2 (600,000 mi^2) with approximately 50 % in the northern region and approximately 33 % of the area underlying the continental shelf/ slope area where platforms are located).

Between 2000 and 2002 the benthic impacts of drilling at four sites on the Gulf of Mexico continental slope were studied (Figure 4.28) (CSA 2006). The study was designed to document (1) drilling mud and cuttings accumulations, (2) physical modification/disturbance of the seabed due to anchors and their mooring systems, (3) debris accumulations, (4) physical/ chemical modification of sediments, and (5) effects on benthic organisms. All of the sites were in water depths greater than 1,000 m (3,280 ft) and included exploration and post-development sites. Sediments were collected within a 500 m (1,640 ft) radius of the platform and far from the platform to establish background concentrations. With two exceptions, sediment PAH concentrations ranged from 0.04 to 0.748 ppm dry weight. One station had a PAH concentration of 3.5 ppm and another station had a PAH concentration of 23.8 ppm in the top 2 cm (0.75 in.). The source of the PAHs was suggested to be from drilling or production activities. Concentrations of As, Cd, Cr, Cu, Pb, Hg, and Zn were elevated in sediments near the platforms when compared with those far from the platforms. In general, elevated concentrations of metals were associated with high Ba concentrations, but even elevated concentrations were within the expected range of background concentrations for uncontaminated marine sediments.

Transportation and consumption of petroleum is widespread in the Gulf of Mexico but these activities occur mainly in coastal areas. Large petrochemical and refining complexes are located along the Texas coast, making the northern Gulf of Mexico a major destination for seaborne and pipeline transportation of petroleum (Figure 4.28) (NRC 2003). The majority of petroleum consumption occurs on land, and little of the petroleum released by these activities reaches the offshore. A potential source of contamination in the offshore region is use of fuel and emissions by ship and boat traffic including commercial ocean transportation, fisheries vessels, and recreational fishing and tourism. The transportation of petroleum by tankers can result in releases of varying sizes from major spills to small, regular operational releases. These transportation releases occur wherever vessels travel or pipelines are located. Ship traffic is much less densely concentrated in the offshore than in coastal regions, and offshore inputs

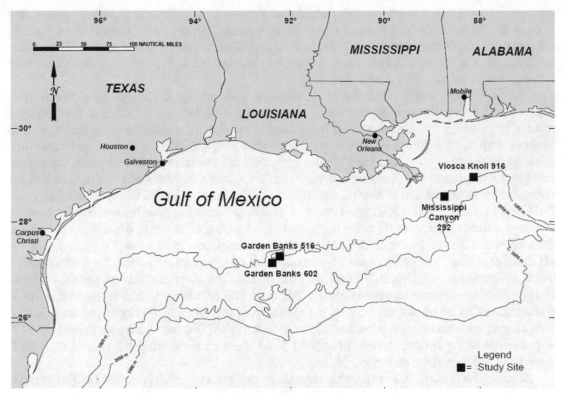

Figure 4.28. Sites for the 2000–2002 study of benthic impacts at four oil and gas platforms in the Gulf of Mexico (from Continental Shelf Associates 2006).

from transportation are low. Most petroleum releases from usage and transportation activities are less dense than water, occur at or near the air–sea interface, and are susceptible to removal by natural processes that limit deposition of petroleum in offshore sediments. In the offshore regions of the northern Gulf of Mexico, usage and transportation of petroleum are minor sources of petroleum contamination to sediments, especially compared to natural oil and gas seepage.

The offshore regions of the Gulf of Mexico have experienced oil spills over the years (Figure 4.29). For example, it is estimated that the IXTOC-I blowout in the southern Gulf of Mexico released 431,820 metric tons (476,000 ton) of petroleum over a period of approximately nine months in 1979 (NRC 2003). Oil spills can be sudden, one-time releases or can continue over time. As stated previously, in general, oil is less dense than water and oil spills result in sea surface slicks. However, over time oil weathers (e.g., loss of volatiles, microbial degradation) and incorporates enough denser sediment and particulate matter to sink to the sea bottom. Oil can also be attached to particulates that are ingested by zooplankton and excreted as fecal pellets that rapidly sink to the seafloor. Oceanic currents can keep the oil in suspension and prevent its accumulation on the bottom. In the few instances when the oil is heavier than water, the oil can sink directly to the bottom, especially in low-energy settings. Oil is sometimes released by blowouts during drilling of exploratory wells, pipeline leaks, and shipwrecks. Subsurface releases differ from surface releases in that the oil moves substantial distances beneath the surface before it rises to the surface. An NRC report (2003) concluded that the majority of the oil in most deep water releases rises to the surface having little effect on sea floor sediments. However, each oil spill can have highly differing scenarios and characteristics

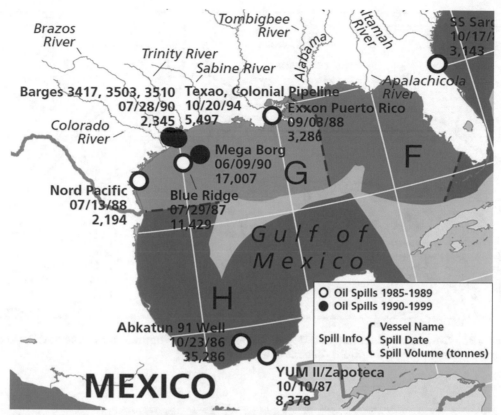

Figure 4.29. Distribution of selected oil spills in the Gulf of Mexico (volumes in tonnes/metric ton (nes) of oil spilled; 1 metric ton = 1.102 U.S. short ton) (modified from NRC 2003).

so these generalities should be applied with caution. Benthic tar mats have been observed in the northern Gulf of Mexico (Figure 4.30) (Alcazar et al. 1989). Tars were altered to varying degrees by microbial degradation and dissolution. Most of the benthic tars analyzed were significantly different chemically from oils produced in the northern Gulf of Mexico. These benthic tars appeared to be derived from oils produced in other areas of the world and transported into the Gulf of Mexico by humans or ocean currents.

4.4.3 Mass Loading of Petroleum Hydrocarbons

Average annual loadings of petroleum to the northern Gulf of Mexico have been summarized (NRC 2003). Although these estimates are based on calculations of inputs and not measurements of hydrocarbons in sediment, insight into the complexities of the origins and distributions of hydrocarbons in the northern Gulf of Mexico is provided. Natural oil and gas seeps are by far the predominant sources of loadings of petroleum to the continental shelf/slope region, and there is a near absence of seeps in the coastal regions of the northern Gulf of Mexico (Figure 4.31). Hydrocarbon seeps are concentrated in the north-central region of the Gulf of Mexico at the distal end of the continental shelf and along the continental slope. The high estimated annual petroleum seepage loadings for the offshore northeastern Gulf of Mexico are due to the inclusion of a few oil and gas seeps in the north-central Gulf of Mexico (NRC 2003). Few seeps are known in the northeastern Gulf of Mexico in the coastal or offshore regions, and these regional patterns mirror the distribution of known oil and gas reserves in the

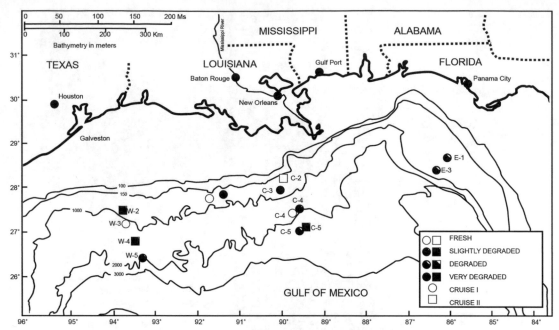

Figure 4.30. Geographic location and degree of degradation of benthic tars collected in trawls in the Gulf of Mexico (modified from Alcazar et al. 1989).

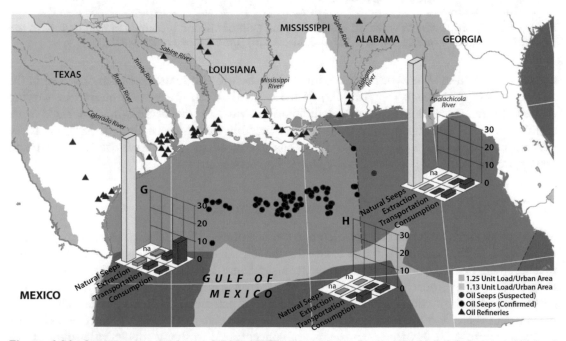

Figure 4.31. Average annual input (1990–1999) of petroleum hydrocarbons (kilotons; 1,000 ton) sources for the offshore (*left histogram*) and coastal (*right histogram*) Gulf of Mexico [*Key*: *yellow* = natural seeps, *green* = extraction of petroleum, *purple* = transportation, and *red* = usage; unit load/urban areas—hydrocarbons from urban areas in rivers entering Gulf of Mexico] (modified from NRC 2003).

northern Gulf of Mexico. While much less than seeps, the next highest estimated petroleum loadings in the offshore region are from extraction of petroleum and transportation activities. These loadings primarily are due to offshore oil and gas platform discharges and accidental spills included in transportation estimates. In the offshore region, estimated loadings due to petroleum usage are low. Comparing estimated petroleum loadings for extraction of petroleum and transportation in the northwestern and northeastern Gulf of Mexico, higher loadings are estimated for the northwest region where most offshore oil and gas platforms, refineries, and chemical plants are located. The highest estimated loading in the coastal northwestern Gulf of Mexico is for usage of fossil fuels, with transportation sources second highest. In the northeastern Gulf of Mexico, petroleum usage is the largest estimated loading of petroleum in the region both in the coastal and offshore regions (excluding seepage) with higher estimated loadings for usage sources in coastal areas. If all other estimated petroleum loadings across the northern Gulf of Mexico are summed, they are still far lower than estimated offshore seepage loadings. The dominant estimated anthropogenic petroleum loadings to coastal areas are from non-point petroleum usage sources in runoff from adjacent land surfaces. As mentioned, most petroleum refining and chemical operations are located in the northwestern Gulf of Mexico, accounting for the larger estimated loading inputs for petroleum usage in the northwestern as compared to the northeastern coastal regions. In addition there is significantly more onshore and coastal oil and gas production in the northwestern region of the Gulf of Mexico compared to the northeastern region. The Houston area is a major transportation hub for refineries and chemical plants. Therefore, most ship traffic in the Gulf of Mexico is destined for ports in the northwestern region, accounting for the higher transportation-related petroleum loadings in the northwestern as compared to the northeastern region. It is not possible to definitively correlate petroleum loadings and PAH concentrations in sediment, but the relative potential of sources of contamination are highlighted. It is noteworthy that PAH concentrations in coastal areas rarely exceed ERL values, and concentrations of PAH in the offshore region are even lower (with the exception of sites of natural oil and gas seepage). Natural removal processes are efficient and may account for the general lack of accumulation of PAHs in sediments to concentrations above ERL values as compared to persistent contaminants such as pesticides and metals. However, catastrophic spills/releases of large volumes of petroleum can and have resulted in significant environmental impact in coastal and offshore regions.

4.4.4 Temporal Variations

Studies of temporal variations in sediment contaminants on the continental shelf/slope and abyss of the Gulf of Mexico are limited. The analysis of dated cores provides a perspective on the origins of contaminants over time (Presley et al. 1980; Trefry et al. 1985; Santschi et al. 2001; Turner et al. 2003; Overton et al. 1986). Studies of suspended particulates, filtered water, and sediment collected in and near the Mississippi River indicated that more than 90 % of the metal load of the river was associated with particulate matter that was relatively constant in chemical composition (Presley et al. 1980). Mississippi River-suspended material was similar to average crustal rocks for Fe, Al, V, Cr, Cu, cobalt (Co), manganese (Mn), and Ni concentrations. In contrast, Zn, Cd, and Pb were enriched most likely due to the activities of humans. Profiles of contaminant concentrations in sediment cores from the Mississippi River Delta documented concentrations of metals and organic contaminants typical of uncontaminated Gulf of Mexico sediments (Santschi et al. 2001) (Table 4.2). Contaminant inputs, when present, were highest in the 1950s–1970s, decreasing to mostly background levels in more recent times.

In another study, PAHs indicative of pyrogenic sources, hopanes indicative of petrogenic hydrocarbons, and pesticides were studied in a series of sediment cores (Turner et al. 2003;

Overton et al. 1986). The results suggest low but chronic deposition of contaminants from oil and gas exploration activities, natural seeps, and agricultural pesticide usage (Turner et al. 2003). PCBs were not detected. Pesticide and PAH concentrations first increased above background concentrations after World War II. Hopanes, interpreted as indicators of petroleum, were present throughout the cores, and concentrations gradually increased after the 1950s as oil and gas activities intensified. Pesticide concentrations increased with the first use and peak application of pesticides. The authors suggest that decreases in annual usage, the phase-out of certain pesticides in the late 1970s and 1980s, flood events, microbial degradation, gravity-driven sediment transport, and post-depositional mixing from storm surges might account for variations in pesticide concentrations over time (Turner et al. 2003). Pyrogenic and petrogenic PAH contamination in sediments gradually increased from the time of first oil exploration activities in the 1950s. Pyrogenic (combustion sourced) PAH had an inverse relationship with petroleum-derived hydrocarbons. The authors suggest that heavier suspended particles settle out in the shelf area closest to the mouth of the river; therefore, concentrations of contaminants would be expected to be higher near the river's sediment plume and decrease as the sediment plume moves west and offshore (Turner et al. 2003). The highest concentrations of pyrogenic PAH occurred in areas of highest sediment deposition off the mouth of the Mississippi River. Pyrogenic PAH concentrations generally decreased in a westerly direction away from the river mouth. Anomalously high PAH concentrations were possibly associated with marsh-burning practices and estuarine runoff (Turner et al. 2003). The authors suggest that higher hopane concentrations had their origins in natural hydrocarbon seeps and/or oil and gas exploration activities. Sediment Ba distributions coincided with the use of barite as a drill mud in offshore oil and gas exploration. In contrast, sediment concentrations of V, a strengthening component of metal alloys, correlated with national consumption rates for steel. Cu, Cd, and Zn concentrations in sediments fluctuated coincidentally with V but not Ba. The method used in this study leached sediments with acid and peroxide so the sediments were not completely dissolved. V and Al concentrations are only a fraction of the total, and results should be interpreted with caution. The authors conclude that the dominant origin of contaminant metals in offshore sediments was riverine sources and not sources on the continental shelf. The authors further conclude that the Mississippi River was a regional source of pesticides and pyrogenic PAHs to offshore sediments and that atmospheric inputs were minimal. These results are consistent with other historical reconstructions of contaminant loadings (Barber and Writer 1998; Boehm and Farrington 1984; Carr et al. 1996; Overton et al. 1986).

4.5 SUMMARY

Sediments are vital to the health of aquatic environments. However, the presence of elevated concentrations of contaminants can adversely degrade sediment quality, which may affect organisms and ecosystems and possibly human health. The chemicals most widely found in sediments of the northern Gulf of Mexico that have the highest likelihood of causing detrimental biological effects are PAHs, pesticides, PCBs, Pb, Hg, As, Cd, Ag, Ni, Sn, Cr, Zn, and Cu. The potential for harmful effects by these chemicals is due to their toxicological and physicochemical properties, their widespread use and release by humans, bioavailability, accumulation in sediments and lipid-rich biological tissues, and persistence in the environment. Contaminants have been released to the Gulf of Mexico for many years and continue to be released by a wide range of human activities most highly concentrated in coastal areas. Accidental or intentional release of contaminants can be traced to population centers and urban-associated discharges; agricultural practices, industrial, military and transportation activities; and the exploration and production of oil and gas.

PAHs and some metals have natural as well as human-related sources. A portion of these chemicals ultimately end up in coastal, and to a much lesser extent, offshore sediments. Inputs of these chemicals to the environment are spatially and temporally variable in composition and concentrations. Sediments are integrators of input, breakdown, and removal processes. The mixture of contaminants and concentrations found in sediments at a location is often unique and variable over small spatial scales. In the 1980s, the NOAA NS&T Program (NOAA 1991) observed that the highest concentrations of contaminants in sediments were located close to population centers. In the 1990s, EMAP and REMAP (USEPA 1998, 1999) concluded that although measurable concentrations of contaminants were present in almost all estuaries of the northern Gulf of Mexico, less than 25 % of the estuarine area had contaminant concentrations that exceeded concentrations suspected of causing biological effects. NCCR I (USEPA 2001) concludes that the overall coastal condition was fair to poor, with 51 % of the estuaries of the northern Gulf of Mexico in good ecological condition and showing few signs of degradation due to contamination. Sediment quality at the remaining locations was judged poor, and contaminant concentrations exceeded levels suspected of causing biological effects at many locations. Most exceedances were for pesticides and metals while PCB and PAH exceedances occurred at less than 1 % of the locations. Enrichments in sediment contaminant concentrations were attributed to humans. In the 2000s, NCCR II (USEPA 2004) concluded that the overall condition of the northern Gulf of Mexico coast was fair. ERM exceedances occurred mainly in Texas and Mobile Bay, and no exceedances were observed along the Florida Gulf Coast. Pesticides and metals exceeded concentrations suspected of causing biological effects at some locations but few PCB and PAH exceedances were observed. In NCCR III (USEPA 2008), the sediment contaminant index was rated as fair and poor for 1 and 2 % of coastal areas, respectively, indicating that approximately 97 % of the coastal area had fewer than five chemicals that exceeded sediment concentrations suspected of causing biological effects. Elevated concentrations of pesticides and metals and occasionally PCBs and PAHs were observed in sediments but few exceeded concentrations suspected of causing biological effects. In NCCR IV (USEPA 2012), the sediment contaminants indicator was rated good with 2 % and approximately 3 % of coastal areas rated as fair and poor, respectively, indicating that approximately 95 % of the coastal areas had fewer than five chemicals that exceeded concentrations suspected of causing biological effects. Elevated concentrations of metals and pesticides, and occasionally PCBs and PAHs, in sediments were observed but few concentrations exceeded biological effect values. Finer-scale monitoring in bays documented steep gradients in contaminant concentrations close to shore near population centers and industrial complexes. The highest concentrations of contaminants in coastal sediments were generally restricted to hot spots of limited spatial extent associated with unique contaminant sources, but a few bays contained extensive areas of contaminated sediments.

Concentrations of contaminants rapidly decrease with distance offshore. Petroleum found in continental shelf/slope sediments is almost exclusively due to natural oil and gas seepage. Few releases of petroleum in the offshore region attributable to humans reach the underlying sediments. The one exception is the discharge of petroleum and metal-contaminated drill muds and cuttings from platforms. Deposits of contaminated sediments from these discharges are restricted to within a few hundred meters of the discharge point. They usually occur as thin veneers less than a few meters thick and become diluted with uncontaminated sediments with time due to the action of currents. Given the immense area of sea bottom in the offshore region, these localized, contaminated sediment deposits are expected to have limited impact. Contaminant concentrations in the offshore region are low, and PCBs and pesticides are generally absent. Contaminated sediments close to platforms measured over a period of years were similar, with a few exceptions such as increase in Pb concentrations and rates of microbial degradation of petroleum, most likely due to the low-energy setting and slower rates of removal processes. It is expected that offshore areas will remain relatively uncontaminated by chemicals attributable to humans for the foreseeable future.

In conclusion, sediment contaminants have threatened and will continue to threaten the quality of the environment in the coastal regions of the northern Gulf of Mexico, but much less so in the offshore region. Elevated concentrations of pesticides and metals in coastal areas are of most concern, but the mixture of chemicals and concentrations can be highly variable in time and space. In coastal areas, pesticides and metals account for most exceedances of concentrations suspected of causing biological effects, and these exceedances appear to be decreasing with time. Assessments suggest a decrease in contamination of coastal sediments in the northern Gulf of Mexico, but there is a high degree of spatial and temporal variability from location to location. The usage of some chemicals has been banned and/or decreased over time (e.g., certain pesticides), and sediment concentrations of these chemicals are expected to continue to decline. Continued reductions in emissions and discharges and remediation of contaminated sites can be expected to accelerate improvements in sediment contaminant levels thus reducing the role of sediment contaminants in degrading environmental quality in the northern Gulf of Mexico.

REFERENCES

Alcazar A, Kennicutt M, Brooks J (1989) Benthic tars in the Gulf of Mexico: Chemistry and sources. Org Geochem 14:433–439

Anderson RK, Scalan RS, Parker PL, Behrens EW (1983) Seep oil and gas in Gulf of Mexico slope sediment. Science 222:619–662

Apeti DA, Lauenstein GG, Riedel GF (2009) Cadmium distributions in coastal sediments and mollusks of the US. Mar Pollut Bull 58:1016–1024

Barber LB, Writer JH (1998) Impact of the 1993 flood on the distribution of organic contaminants in bed sediments of the Upper Mississippi River. Environ Sci Technol 32:2077–2083

Boehm PD, Farrington JW (1984) Aspects of the polycyclic aromatic hydrocarbon geochemistry of recent sediments in the Georges Bank region. Environ Sci Technol 18:840–845

Brooks JM, Kennicutt MC II, Carey BD Jr (1986) Offshore surface geochemical exploration. Oil Gas J 84:66–72

Brooks JM, Kennicutt MC II, Bidigare RR, Wade TL, Powell E, Denoux GJ, Fay RR, Childress JJ, Fisher CR, Rosman I, Boland G (1987) Chemosynthetic ecosystems, hydrates, and oil seepage on the Gulf of Mexico slope: An update. EOS 68:498–499

Carr RS, Long ER, Windom HL, Chapman DC, Thursby G, Sloane GM, Wolfe DA (1996) Sediment quality assessment studies of Tampa Bay, Florida. Environ Toxicol Chem 15:1218–1231

Carr RS, Montagna PA, Biedenbach JM, Kalke R, Kennicutt MC, Hooten R, Cripe G (2000) Impact of storm-water outfalls on sediment quality in Corpus Christi Bay, Texas, USA. Environ Toxicol Chem 19:561–574

CSA (Continental Shelf Associates) (2006) Effects of oil and gas exploration and development at selected continental slope sites in the Gulf of Mexico. Executive Summary, vol I. OCS Study MMS 2006-044, -45, and -46. U.S. Department of the Interior, Minerals Management Service, Gulf of Mexico OCS Region, New Orleans, LA, USA

De Beukler SM (2003) Remote sensing analysis of natural oil and gas seeps on the continental slope of the northern Gulf of Mexico. Master's Thesis, Texas A&M University, College Station, TX, USA

Energy Information Administration (2009) Annual energy outlook 2009 with projections to 2030. DOE/EIA-0383(2009), Washington, DC, USA. http://www.eia.gov/oiaf/aeo/pdf/0383 (2009).pdf

Field L, Macdonald D, Norton S, Ingersoll C, Severn C, Smorong D, Lindskoog R (2002) Predicting amphipod toxicity from sediment chemistry using logistic regression models. Environ Toxicol Chem 21:1993–2005

Horowitz AJ, Elrick KA, Smith JJ (2001) Estimating suspended sediment and trace element fluxes in large river basins: Methodological considerations as applied to the NASQAN program. Hydrol Proc 15:1107–1132

Kennicutt MC II (ed) (1995) Gulf of Mexico offshore operations monitoring experiment, phase I: Sublethal responses to contaminant exposure. Final report. OCS Study MMS 95-0045. U.S. Department of the Interior, Gulf of Mexico OCS Region, New Orleans, LA, USA. 748 p

Kennicutt C II, Keeney-Kennicutt WL, Presley BJ, Fenner F (1983) The use pyrolysis and barium distributions to assess the areal extent of drilling fluids in surficial marine sediments. Environ Geol 4:239–249

Kennicutt MC II, Boothe PN, Wade TL, Sweet R, Rezak FJ, Kelly JM, Brooks PBJ, Wiesenburg DA (1996) Geochemical patterns in sediments near offshore production platforms. Can J Fish Aquat Sci 53:2554–2566

Kimbrough KL, Johnson WE, Lauenstein GG, Christensen JD, Apeti DA (2008) An assessment of two decades of contaminant monitoring in the Nation's Coastal Zone. NOAA Technical Memorandum NOS NCCOS 74. National Oceanic and Atmospheric Administration, Silver Spring, MD, USA. 105 p

Long ER, Morgan LG (1990) The potential for biological effects of sediment-sorbed contaminants tested in the national status and trends program. NOAA Technical Memorandum NOS OMA 52. NOAA, Seattle, WA, USA

Long ER, MacDonald DD, Smith SL, Calder FD (1995) Incidence of adverse biological effects within ranges of chemical concentrations in marine and estuarine sediments. Environ Manag 19:81–97

Maruya K, Loganathan BG, Kannan K, McCumber-Kahn S, Lee RF (1997) Organic and organometallic compounds in estuarine sediments from the Gulf of Mexico (1993-1994). Estuaries 20:700–709

Meade RH (ed) (1995) Contaminants in the Mississippi River. Circular 1133. U.S. Geological Survey, Reston, VA, USA

NOAA (National Oceanic and Atmospheric Administration) (1987) National status and trends program—a preliminary assessment of findings of the Benthic Surveillance Project-1984. NOAA Office of Oceanography and Marine Assessment, Rockville, MD, USA. 81 p

NOAA (1988) National status and trends program—a summary of selected data on chemical contaminants in sediments collected during 1984, 1985, 1986, and 1987. NOAA Technical Memorandum NOS OMA 59, Rockville, MD, USA. 23 p plus appendices

NOAA (1991) National status and trends program—second summary of data on chemical contaminants in sediments from the national status and trends program. NOAA Technical Memorandum NOS OMA 38, Rockville, MD, USA. 29 p plus appendices

NOAA (1999) Sediment quality guidelines developed for the national status and trends program. http://ccma.nos.noaa.gov/publications/sqg.pdf

NOAA (2012) Office of Ocean Exploration and Research. http://oceanexplorer.noaa.gov/explorations/06mexico/background/oil/media/platform_600.html. Accessed 21 June 2015

NRC (National Research Council) (2003) Oil in the sea III: Inputs, fates, and effects. National Academy Press, Washington, DC, USA. 278 p

O'Connor T (2004) The sediment quality guideline, ERL, is not a chemical concentration at the threshold of sediment toxicity. Mar Pollut Bull 49:383–385

O'Connor T, Daskalakis K, Hyland J, Paul J, Summers K (1998) Comparisons of measured sediment toxicity with predictions based on chemical guidelines. Environ Toxicol Chem 17:468–471

Overton EB, Schultz MH, St. Pé KM, Byrne C (1986) Distribution of trace organics, heavy metals, and conventional pollutants in Lake Pontchartrain, Louisiana. In: Sohn ML (ed) Organic marine geochemistry. American Chemical Society, Washington, DC, USA, pp 247–270

Park J, Wade T, Sweet S (2002) Atmospheric deposition of PAHs, PCBs, and organochlorine pesticides to Corpus Christi Bay, Texas. Atmos Environ 36:1707–1720

Presley BJ, Trefry JH, Shokes RF (1980) Heavy metal inputs to Mississippi Delta sediments. Water Air Soil Pollut 13:481–494

Römbke J, Moltmann JF (1995) Applied Ecotoxicology. CRC Press LLC, Boca Raton, FL, USA. 304 p

Rowe GT, Kennicutt MC II (eds) (2009) Northern Gulf of Mexico continental slope habitats and benthic ecology study. Final Report. OCS Study MMS 2009-039. U.S. Department of the Interior, Minerals Management Service, Gulf of Mexico OCS Region, New Orleans, LA, USA. 456 p

Rydén L, Migula P, Andersson M (2003) Environmental science: Understanding, protecting and managing the environment in the Baltic Sea Region. The Baltic University Programme, Uppsala University. In Migula P, Kihlström JE, Rydén L, How Pollutants Affect Life: Toxicology and Human Health, Chapter 14, pp 418–443.

Santschi PH, Presley BJ, Wade TL, Garcia-Romero B, Baskaran M (2001) Historical contamination of PAHs, PCBs, DDTs, and heavy metals in Mississippi River Delta, Galveston Bay and Tampa Bay sediment cores. Mar Environ Res 52:51–79

Sassen R, Milkov AV, Roberts HH, Sweet ST, DeFreitas DA (2003) Geochemical evidence of rapid hydrocarbon venting from a seafloor-piercing mud diapir, Gulf of Mexico continental shelf. Mar Geol 198:319–329

Sericano JL, Atlas EL, Wade TL, Brooks JM (1990) NOAA's status and trends mussel watch program: Chlorinated pesticides and PCBs in oysters (Crassostrea virginica) and sediments from the Gulf of Mexico, 1986-1987. Mar Environ Res 29:162–203

Summers JK, Macauley JM, Heitmuller PT, Engle VD, Adams AM, Brooks GT (1992) Annual statistical summary: EMAP-Estuaries Louisianian Province-1991. EPA/600/ R-93/001. Environmental Protection Agency, Washington, DC, USA

Summers JK, Macauley JM, Heitmuller PT, Engle VD, Brooks GT, Babikow M, Adams AM (1994) Annual statistical summary: EMAP-Estuaries Louisianian Province-1992. EPA/620/ R-94/002. USEPA Office of Research and Development, Gulf Breeze, FL, USA

Summers JK, Paul JF, Robertson A (1995) Monitoring organic compounds in Gulf of Mexico sediments the ecological condition of estuaries in the United States. Toxicol Environ Chem 49:93–108

Summers JK, Wade TL, Engle VD (1996) Normalization of metal concentrations in estuarine sediments from the Gulf of Mexico. Estuaries 19:581–594

Thorne C, Harmar O, Watson C, Clifford N, Biedenharn D, Measures R (2008) Current and historical sediment loads in the lower Mississippi River. Final Report A074274. Contract 1106-EN-01. U.S. Army European Research Office of the U.S. Army, London, UK

Tieh TT, Pyle TE (1972) Distribution of elements in Gulf of Mexico sediments. In: Rezak R, Henry VJ (eds) Contributions on the geological and geophysical oceanography of the Gulf of Mexico. Gulf Publishing, Houston, TX, USA, pp 129–152

Tieh TT, Pyle TE, Egler DH, Nelson RA (1973) Chemical variations in sedimentary facies of an inner continental shelf environment, northern Gulf of Mexico. Sediment Geol 9:110–115

Trefry JH, Presley BJ (1976a) Heavy metals in sediments from San Antonio Bay and the northwest Gulf of Mexico. Environ Geol 1:283–294

Trefry JH, Presley BJ (1976b) Heavy metal transport from the Mississippi River to the Gulf of Mexico. In: Windom HL, Duec RA (eds) Marine pollutant transfer. Lexington Books, Lexington, MA, USA, pp 39–76

Trefry JH, Metz S, Trocine R, Nelsen T (1985) A decline in lead transport by the Mississippi River. Science 230:439–441

Turner RE, Overton EB, Rabalais NN, Sen Gupta BK (eds) (2003) Historical reconstruction of the contaminant loading and biological responses in the Central Gulf of Mexico shelf sediments. U.S. Department of the Interior, Minerals Management Service, Gulf of Mexico OCS Region, New Orleans, LA, USA. 140 p

USEPA (U.S. Environmental Protection Agency) (1989) Glossary of terms and acronym list. 19K-1002. Office of Communications and Public Affairs (A-107). Washington, DC, USA. 29 p

USEPA (1998) Galveston Bay 1993. EPA/906/R-98/002. Regional Environmental Monitoring and Assessment Program, Region IV Ecosystems Protection Branch, Washington, DC, USA

USEPA (1999) Ecological condition of estuaries in the Gulf of Mexico. EPA 620-R-98-004. U.S. Office of Research and Development, Gulf Ecology Division, Gulf Breeze, FL, USA

USEPA (2001) National coastal condition report. EPA/620-R-01-005. Office of Research and Development and Office of Water, Washington, DC, USA

USEPA (2004) National coastal condition report II. EPA/620-R 03-002. Office of Research and Development and Office of Water, Washington, DC, USA

USEPA (2006) National estuary program coastal condition report. EPA-842/B-06/001. Office of Water/Office of Research and Development, Washington, DC, USA

USEPA (2008) National coastal condition report III. EPA/842-R-08-002. Office of Research and Development and Office of Water, Washington, DC, USA

USEPA (2012) National coastal condition report IV. EPA/842-R-10-003. Office of Research and Development and Office of Water, Washington, DC, USA

Wade TL, Atlas EL, Brooks JM, Kennicutt MC II, Fox RG, Sericano J, Garcia-Romero B, Defreitas D (1988) NOAA Gulf of Mexico status and trends program: Trace organic contaminant distribution in sediments and oysters. Estuaries 11:171–179

Wade TL, Kennicutt MC, Brooks JM (1989) Gulf of Mexico hydrocarbon seep communities: part III. Aromatic hydrocarbon concentrations in organisms, sediments and water. Mar Environ Res 27:19–30

Wilson RD, Monaghan PH, Osanik A, Price LC, Rogers MA (1974) Natural marine oil seepage. Science 184:857–865

Windom HL, Schropp SJ, Calder FD, Ryan JD, Smith RG, Burney LC, Lewis FG, Rawlinson CH (1989) Natural trace metal concentrations in estuarine and coastal marine sediments of the southeastern United States. Environ Sci Technol 3:314–327

APPENDIX A: CHARACTERISTICS OF COMMON SVOC CONTAMINANTS

Table A.1. Characteristics of Common SVOC Contaminants ([a]ERL/ERM values from Long and Morgan 1990 (top) and NOAA 1999 (bottom); origins, toxicity, and fate are modified from Kimbrough et al. 2008)

Sources	Toxicity	Fate	Biological Effect Values (ppb)	
			ERL	ERM
Chlordanes (a group of organic pesticides called cyclodienes. It is a technical mixture whose principal components are alpha-chlordane, gamma-chlordane, heptachlor, and nonachlor)				
Chlordane, an insecticide, is a complex mixture of at least 50 compounds. It was used in the United States during 1948–1983 for agricultural and urban settings to control insect pests. It was also the predominant insecticide for the control of subterranean termites. Agricultural uses were banned in 1983, and all uses were banned by 1988	Exposure to chlordane can occur through eating crops from contaminated soil, fish, and shellfish from contaminated waters, or breathing contaminated air. Chlordane can enter the body by being absorbed through the skin, inhaled, or ingested. At high levels, chlordane can affect the nervous system, digestive system, brain, and liver, and is also carcinogenic. Chlordane is highly toxic to invertebrates and fish	Removal from both soil and water sources is primarily by volatilization and particle-bound runoff. In air, chlordane degrades as a result of photolysis and oxidation. Chlordane exists in the atmosphere primarily in the vapor-phase, but the particle-bound fraction is important for long-range transport. Chlordane binds to dissolved organic matter, further facilitating its transport in natural waters	0.5 [NA]	6 [NA]
Total DDT				
DDT is used against agricultural pests and mosquito control. Its use in the United States was banned in 1972, but it is still used in some countries today	Due to its environmental persistence and hydrophobic nature, DDT bioaccumulates in organisms. Many aquatic and terrestrial organisms are highly sensitive to DDT. As a result of DDT's toxic effects on wildlife, in particular birds, its usage was banned in the United States	DDT transforms to DDD and DDE, the latter being the predominant form found in the environment. Evaporation of DDT from soil followed by long distance transport results in its widespread global distribution. DDT and its transformation products are persistent and accumulate in the	3 [1.58]	350 [46.1]

(continued)

Table A.1. (continued)

Sources	Toxicity	Fate	Biological Effect Values (ppb)	
			ERL	ERM
		environment because they resist biodegradation. DDT that enters surface waters is subject to volatilization, adsorption to suspended particulates and sediment, and bioaccumulation. About half of the atmospheric DDT is adsorbed to particulates		
Dieldrin				
Dieldrin is defined as the sum of two compounds, dieldrin and aldrin. Dieldrin and a related compound (aldrin) were widely used as insecticides in the 1960s for the control of termites around buildings and general crop protection from insects. In 1970, all uses of aldrin and dieldrin were cancelled based on concern that they could cause severe aquatic environmental change and their potential as carcinogens. The cancellation was lifted in 1972 to allow limited use of aldrin and dieldrin, primarily for termite control. All uses of aldrin and dieldrin were again cancelled in 1989	Exposure to aldrin and dieldrin occurs through ingestion of contaminated water and food products, including fish and shellfish, and through inhalation of indoor air in buildings treated with these insecticides. Aldrin is rapidly metabolized to dieldrin in the human body. Acute and long-term human exposures are associated with central nervous system intoxication. Aldrin and dieldrin are carcinogenic to animals and classified as likely human carcinogens	Aldrin is readily converted to dieldrin, while dieldrin is resistant to transformation. Dieldrin bioaccumulates and is magnified through aquatic food chains and has been detected in tissue of freshwater and saltwater fish and marine mammals. Aldrin and dieldrin applied to soil are tightly bound, but may be transported to streams and rivers by soil erosion. Volatilization is the primary loss mechanism from soil. Dieldrin undergoes minor degradation to photodieldrin in marine environments	0.02 [NA]	8 [NA]

(continued)

Table A.1. (continued)

Sources	Toxicity	Fate	Biological Effect Values (ppb)	
			ERL	ERM
Polycyclic aromatic hydrocarbons (PAHs): NOTE: ERL/ERM values are available for individual PAH compounds)				
PAHs are found in creosote, soot, petroleum, coal, and tar. PAHs can also have natural sources (e.g., forest fires, volcanoes) in addition to anthropogenic sources (auto emissions, home heating, coal-fired power plants). PAHs are formed by the fusing of benzene rings during the incomplete combustion of organic materials. They are also found in oil and coal. The main sources of PAHs to the environment are forest fires, coal-fired power plants, and automobile exhaust and local releases of oil	Made up of a suite of hundreds of compounds, PAHs exhibit a wide range of toxicities. Human exposure to PAHs can come as a result of being exposed to smoke from forest fires, automobile exhaust, home heating using wood, grilling and cigarettes. Toxic responses to PAHs in aquatic organisms include reproduction inhibition, mutations, liver abnormalities and mortality. Exposure to aquatic organisms can come as a result of oil spills, boat exhaust and urban runoff	The fate and transport of PAHs is variable and dependent on the physical properties of each individual compound. Most PAHs strongly associate with particles; larger PAH compounds (high molecular weight) associate to a higher degree with particles relative to smaller PAH compounds (low molecular weight). Smaller compounds predominate in petroleum products whereas larger compounds are associated with combustion	4,000 [4,022]	35,000 [44,792]
Polychlorinated biphenyls (PCBs) (there are 209 possible PCB compounds, called *congeners* that were marketed as mixtures known as Aroclor)				
PCBs are synthetic organic chemicals composed of biphenyl substituted with varying numbers of chlorine atoms. They were manufactured between 1929 and 1977. PCB use was regulated in 1971; new uses were banned in 1976. PCBs were used in electrical transformers, capacitors,	The main human exposure route for PCBs is through eating contaminated seafood and meats. PCBs are associated with skin ailments, neurological and immunological responses and at high doses can decrease motor skills and cause liver damage and memory loss. Exposure of aquatic	PCBs are persistent in the environment and associate with particles in aquatic systems as a result of their strong hydrophobic nature. They are long lived in the environment; improper disposal and leakage is responsible for environmental introduction	50 [22.7]	400 [180]

(continued)

Table A.1. (continued)

| Sources | Toxicity | Fate | Biological Effect Values (ppb) | |
			ERL	ERM
lubricants, and hydraulic fluids. Other uses included paints, adhesives, plasticizers, and flame retardants. Manufacturing of PCBs for use as flame retardants and lubricants stopped in 1977. Currently, PCBs are predominately used in electrical applications and can still be found in transformers and electrical equipment	life to PCBs results in birth defects, lowered fecundity, cancer, and death. PCBs are hazardous because they are toxic, degrade slowly, and bioaccumulate			

[a]ERL—concentration of a chemical in sediments that resulted in biological effects approximately 10 % of the time based on literature
ERM—Concentration that resulted in biological effects approximately 50 % of the time based on literature

APPENDIX B: CHARACTERISTICS OF COMMON METAL CONTAMINANTS

Table B.1. Characteristics of Common Metal Contaminants ([a]ERL/ERM values from Long and Morgan 1990 (top) and NOAA 1999 (bottom); origins, toxicity, and fate are modified from Kimbrough et al. 2008)

| Origins | Toxicity | Fate | Biological Effect Values (ppm) | |
			ERL	ERM
Arsenic (As)				
Arsenic has natural and industrial sources. Products that contain arsenic include preserved wood, semiconductors, pesticides, defoliants, pigments, antifouling paints, and veterinary medicines. In the recent past, as	Arsenic is toxic at high concentrations to fish, birds, and plants. In animals and humans prolonged chronic exposure is linked to cancer. Inorganic arsenic, the most toxic form, represents approximately 10 % of total arsenic in bivalves. Less	Human activities have changed the natural biogeochemical cycle of arsenic leading to contamination of land, water, and air. Arsenic in coastal and estuarine water occurs primarily from river runoff and atmospheric deposition. The	33 [8.2]	85 [70]

(continued)

Table B.1. (continued)

Origins	Toxicity	Fate	Biological Effect Values (ppm)	
			ERL	ERM
much as 90 % of arsenic was used for wood preservation. Atmospheric sources of arsenic include smelting, fossil fuel combustion, power generation, and pesticide application	harmful organic forms, such as arsenobetaine, predominate in seafood	major source of elevated levels of arsenic in the nation is natural crustal rock		
Cadmium (Cd)				
Cadmium occurs naturally in the earth's crust as complex oxides and sulfides in ores. Products that contain cadmium include batteries, color pigment, plastics, and phosphate fertilizers. Industrial sources and uses include zinc, lead, and copper production, electroplating and galvanizing, smelting, mining, fossil fuel burning, waste slag, and sewage sludge. Anthropogenic emissions originate from a large number of diffuse sources	Cadmium is toxic to fish, salmonoid species, and juveniles are especially sensitive, and chronic exposure can result in reduction of growth. Respiration and food represent the two major exposure pathways for humans to cadmium	Cadmium has both natural and non-point anthropogenic sources. Natural sources include river runoff from cadmium-rich soils, leaching from bedrock, and upwelling from marine sediment deposits. Cadmium is transported by atmospheric processes as a result of fossil fuel burning, erosion, and biological activities. Land-based runoff and ocean upwelling are the main conveyors of cadmium into coastal environments. Elevated cadmium levels are primarily located in freshwater-dominated estuaries consistent with river transport of cadmium to coastal environments	5 [1.2]	9 [9.6]

(continued)

Table B.1. (continued)

			Biological Effect Values (ppm)	
Origins	Toxicity	Fate	ERL	ERM
Copper (Cu)				
Copper is a naturally occurring ubiquitous element in the environment. Trace amounts of copper are an essential nutrient for plants and animals. Anthropogenic sources include mining, manufacturing, agriculture, sewage sludge, antifouling paint, fungicides, wood preservatives, and vehicle brake pads. The United States ranks third in the world for utilization and second in production. The USEPA phase-out of chromated copper arsenate (CCA) wood preservatives and the 1980s restrictions on tributyltin marine antifouling paints have stimulated a transition to copper-based wood preservatives and marine antifouling paint	Copper can be toxic to aquatic organisms; juvenile fishes and invertebrates are much more sensitive to copper than adults. Although copper is not highly toxic to humans, chronic effects of copper occur as a result of prolonged exposure to large doses and can cause damage to the digestive tract and eye irritation. Atmospheric transport and deposition of particulate copper into surface waters may also be a significant source of copper to coastal waters	The most common form of copper in water is Cu (II); it is mostly found bound to organic matter. Transport of copper to coastal and estuarine water occurs as a result of runoff and river transport	70 [34]	290 [270]
Lead (Pb)				
Lead is a ubiquitous metal that occurs naturally in the earth's crust. Environmental levels of lead increased worldwide over the past century	Lead has no biological use and is toxic to many organisms, including humans. Exposure of fish to elevated concentrations of lead results in	Loadings of lead into coastal waters are primarily linked with wastewater discharge, river runoff, atmospheric deposition, and natural weathering of rock. Lead can	35 [46.7]	210 [218]

(continued)

Table B.1. (continued)

Origins	Toxicity	Fate	Biological Effect Values (ppm)	
			ERL	ERM
because of leaded gasoline use. Significant reductions in source and load resulted from regulation of lead in gasoline and lead-based paints. High levels found in the environment are usually linked to anthropogenic activities such as manufacturing processes, paint and pigment, solder, ammunition, plumbing, incineration, and fossil fuel burning. In the communications industry, lead is still used extensively as protective sheathing for underground and underwater cables, including transoceanic cable systems	neurological deformities and black fins in fish. Lead primarily affects the nervous system, which results in decreased mental performance and mental retardation in humans. Exposure to lead may also cause brain and kidney damage, and cancer	be found in air, soil, and surface water		
Mercury (Hg)				
Mercury is a highly toxic, nonessential trace metal that occurs naturally. Elevated levels occur as a result of human activity. In the United States, coal-fired electric turbines, municipal and medical waste incinerators, mining, landfills, and sewage sludge are the primary emitters of mercury into the air	Mercury is a human neurotoxin that also affects the kidneys and developing fetuses. The most common human exposure route for mercury is the consumption of contaminated food. Children, pregnant women or women likely to become pregnant are advised to avoid consumption of swordfish, shark, king mackerel, and	In the environment, mercury may change forms (between elemental, inorganic, and organic). Natural sinks, such as sediment and soil, represent the largest source of mercury to the environment. Estimates suggest that wet and dry deposition accounts for 50–90 % of the	0.15 [0.15]	1.3 [0.71]

(continued)

Table B.1. (continued)

Origins	Toxicity	Fate	Biological Effect Values (ppm)	
			ERL	ERM
	tilefish and should limit consumption to fish and shellfish recommended by FDA and USEPA	mercury load to many estuaries, making atmospheric transport a significant source of mercury worldwide. Long-range atmospheric transport is responsible for the presence of mercury at or above background levels in surface waters in remote areas		
Nickel (Ni)				
Nickel is a naturally occurring, biologically essential trace element that is widely distributed in the environment. It exists in its alloy form and as a soluble element. Nickel is found in stainless steel, nickel–cadmium batteries, pigments, computers, wire, and coinage; and is used for electroplating	Food is the major source of human exposure to nickel. Exposure to large doses of nickel can cause serious health effects, such as bronchitis, while long-term exposure can result in cancer. There is no evidence that nickel biomagnifies in the food chain	Nickel derived from weathering rocks and soil is transported to streams and rivers by runoff. It accumulates in sediment and becomes inert when it is incorporated into minerals. River and stream input of nickel are the largest sources for oceans and coastal waters. Atmospheric sources are usually not significant	30 [20.9]	50 [51.6]
Tin (Sn)				
Tin sources in coastal water and soil include manufacturing and processing facilities. It also occurs in trace amounts in natural waters. Concentrations in unpolluted waters	Humans are exposed to elevated levels of tin by eating from tin-lined cans and by consuming contaminated seafood. Exposure to elevated levels of tin compounds by humans leads to	Tin enters coastal waters bound to particulates, and from riverine sources derived from soil and sediment erosion. Bioconcentration factors for inorganic tin were reported to be 1900 and 3000	NA	NA

(continued)

Table B.1. (continued)

Origins	Toxicity	Fate	Biological Effect Values (ppm)	
			ERL	ERM
and the atmosphere are often near analytical detection limits. Tin has not been mined in the United States since 1993	liver damage, kidney damage, and cancer	for marine algae and fish. Inorganic tin can be transformed into organometallic forms by microbial methylation and is correlated with increasing organic content in sediment. Tin is regarded as being relatively immobile in the environment and is rarely detected in the atmosphere. It is mainly found in the atmosphere near industrial sources as particulates from combustion of fossil fuels and solid waste		
Zinc (Zn)				
As the fourth most widely used metal, zinc's anthropogenic sources far exceed its natural ones. The major industrial sources include electroplating, smelting, and drainage from mining operations. The greatest use of zinc is as an anticorrosive coating for iron and steel products (sheet and strip steel, tube and pipe, and wire and wire rope). Canada is one of the largest producers and exporters of zinc. The United States is the largest customer for	Zinc is an essential nutrient. Human exposure to high doses of zinc may cause anemia or damage to the pancreas and kidneys. However, zinc does not bioaccumulate in humans; therefore, toxic effects are uncommon and associated with excessively high doses. Fish exposed to low zinc concentrations can sequester it in some cases	Dissolved zinc occurs as the free hydrated ion and as dissolved complexes. Changes in water conditions (pH, redox potential, chemical speciation) can result in dissolution from or sorption to particles. In air, zinc is primarily found in the oxidized form bound to particles. Zinc precipitates as zinc sulfide in anaerobic or reducing environments, such as wetlands, and thus is less mobile, while remaining as the free ion at lower pHs. As a result of natural and	120 [150]	270 [410]

(continued)

Table B.1. (continued)

Origins	Toxicity	Fate	Biological Effect Values (ppm)	
			ERL	ERM
Canadian refined zinc, and the automobile industry is the largest user of galvanized steel		anthropogenic activities, zinc is found in all environmental compartments (air, water, soil, and biota)		
Butyltins				
Tributyltin is used as an antifouling agent in marine paints applied to boat hulls. Slow release from the paint into the aquatic system retards organism attachment and increases ambient environmental levels. The United States partially banned the use of tributyltin in 1988 for use on boats less than 25 m in length, drastically limiting use on many recreational vessels	Tributyltin is an extremely toxic biocide that is regulated as a result of its toxic effects (reproduction and endocrine disruption) on nontarget aquatic species. Organotin compounds are readily bioaccumulated by aquatic organisms from water but there is no evidence for biomagnification up the food chain. Sex changes have been shown to occur in gastropods exposed to elevated levels of tributyltin	Tributyltin is sparingly soluble in water and associates readily with suspended particles in the water column. Butyltins are persistent in the aquatic environment and accumulate in sediment; therefore, they will continue to be a source of butyltin to the aquatic environment. Tributyltin transforms to dibutyltin and then to monobutyltin. Releases of organotins to the atmosphere are not significant due to their low vapor pressure and rapid photodegradation	NA	NA

^aERL—concentration of a chemical in sediments that resulted in biological effects approximately 10 % of the time based on literature
ERM—Concentration that resulted in biological effects approximately 50 % of the time based on literature

CHAPTER 5

OIL AND GAS SEEPS IN THE GULF OF MEXICO

Mahlon C. Kennicutt II[1]

[1]Texas A&M University, College Station, TX 77843, USA
mckennicutt@gmail.com

5.1 INTRODUCTION

A seep is a natural phenomenon where gaseous or liquid hydrocarbons, or both, leak from the ground (Figure 5.1). Seeps can occur on land and beneath the ocean above subsurface petroleum sources and accumulations. Seeps can have biogenic or thermogenic origins. Biogenic hydrocarbons are mostly methane and result from bacterial metabolism. Thermogenic hydrocarbons result from organic matter exposure to high temperatures in the deep subsurface. These two distinct origins of hydrocarbons impart unique chemical and isotopic compositions. Deep-seated buoyant thermogenic hydrocarbons can migrate along geological layers, across strata via faults and fractures in rocks and sediments, or they can be exposed as outcrops of oil-bearing rocks. Most seeps are generally under low pressures that produce slow rates of release. Oil and gas seeps are common globally and have been exploited by humans since Paleolithic times (Chisholm 1911; Etiope 2015). Seeps are highly variable in composition and include gases, crude oil, liquid bitumen, asphalt, and tar. Thermogenic seeps are often accompanied by water and brine (salt) containing inorganic solutes dissolved from source formations and the strata they migrate through. Seeps were targets for early exploration and exploitation of petroleum and ultimately led to the modern oil and gas industry. It has long been recognized that surface seeps indicate the existence of petroleum beneath them and are the basis for widely used fossil fuel exploration techniques known collectively as *surface prospecting*. Seeps can be ephemeral or may persist for many years. Worldwide, natural seeps release vast amounts of oil and gas to the environment every year and have for millions of years (NASA 2000; Hunt 1996). Often, while the collective volumes are large, seeps generally release petroleum slowly enough to allow surrounding organisms to avoid, adapt to, and, in some instances, even thrive in their presence (Coleman et al. 2003).

Understanding the location, type, and volume of petroleum seepage is important as indicators of deeper petroleum reservoirs, the presence of faults, and geohazards. Seeps release oil to the sea and greenhouse gases to the atmosphere (Etiope 2009, 2012, 2015; Coleman et al. 2003; Ciais et al. 2013). Conversely, the geographic distributions of oil and gas production and reserves, subsurface geology and sedimentary basins, salt structures, sea-surface slicks, seep-related water column and seafloor features, gas hydrate, and cold-seep communities can be used to infer the presence of seeps.

5.2 HISTORY

5.2.1 History of Oil and Gas Seeps Worldwide

Over millions of years, the preserved remains of dead plants and animals have been buried deep in the Earth by overlying sediments. This burial results in rising temperatures and

© The Author(s) 2017

C.H. Ward (ed.), *Habitats and Biota of the Gulf of Mexico: Before the Deepwater Horizon Oil Spill*,
DOI 10.1007/978-1-4939-3447-8_5

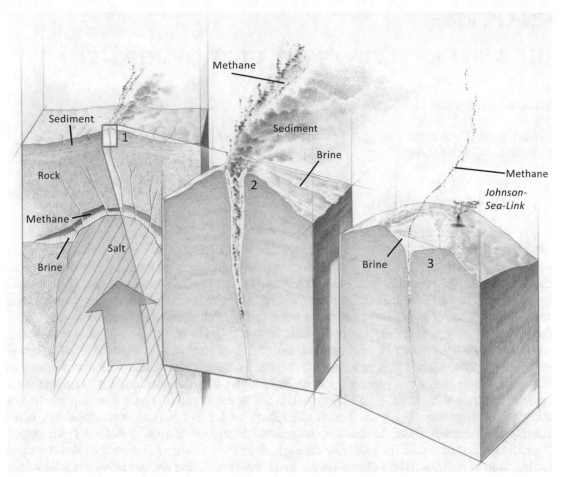

Figure 5.1. Schematic diagram of a typical marine seep location and associated features: (1) rising pillars of salt (diapirs) fracture the overlying strata creating migration pathways from deep-seated reservoirs to the near-surface; (2) the efflux of gases and fluids can disrupt and mix with overlying sediments creating seabed mounds and/or craters that are often associated with gas and/or liquid plumes in the overlying water column; and (3) seeping brines that are denser than sea water can accumulate in the depression forming a sea-bottom lake of high salinity water (MacDonald and Fisher 1996; Bruce Morser/National Geographic Creative, used with permission). *Johnson-Sea-Link* refers to a scientific research submersible (http://oceanexplorer.noaa.gov/technology/subs/sealink/sealink.html).

pressures due to Earth's internal residual heat from planetary accretion, radioactive decay, and increasing overburden. These conditions lead to the breakdown or cracking of complex biochemicals into lower-molecular-weight compounds including hydrocarbons. Once formed, liquids and gases are less dense than water or brine in the surrounding strata and tend to move upward under the force of buoyancy. If the petroleum fluids survive upward migration and avoid being trapped in the subsurface, they reach the surface forming petroleum seeps (Figure 5.2). Chisholm (1911) noted that "…bitumen, in its various forms, [is] one of the most widely-distributed of substances occurring in strata of every geological age from the lowest Archean rocks to those now in process of deposition…" Surface petroleum seeps have been part of the landscape throughout human history (Hunt 1996; Kvenvolden and Cooper 2003; Etiope 2015).

The first evidence of humans using petroleum from seeps dates to more than 40,000 years ago, associated with stone tools used by Neanderthals at sites in Syria (Hirst 2009; Etiope 2015).

Figure 5.2. Photographs of typical petroleum seeps on land: (a) natural oil (petroleum) seep near Korňa, Kysucké Beskydy, Western Carpathians, Slovakia. Flysch belt (photo from http://en.wikipedia.org/wiki/Petroleum_seep; attributed to Branork [own work: 2008], CC BY 3.0) and (b) tar volcano in the Carpinteria Asphalt mine (R. Arnold, USGS, https://commons.wikimedia.org/wiki/File:Tar_volcano_in_the_Carpinteria_Asphalt_mine.jpg).

The use of seeping petroleum as a sealant, adhesive, building mortar, incense, and decorative application on pots, buildings, or human skin has been documented worldwide (Krishnan and Rajagopal 2003). More than 5,000 years ago, ancient Sumerians, Assyrians, and Babylonians used asphalt from seeps along the Euphrates for waterproofing (PBS 2004). Ancient Egyptians used liquid oil for medicinal purposes and embalming (Harwell and Lewan 2002; Barakat et al. 2005; Rullkötter and Nissenbaum 1988). In North America, prehistoric Native Americans used tar as a glue to bind stone tools to wooden handles and as a waterproof caulking for baskets and canoes (Harris and Jefferson 1985). In 480 BC, Persian military forces used oil-soaked flaming arrows during the siege of Athens (PBS 2004). The first oil well is believed to have been drilled in 347 AD when the Chinese used bamboo poles to bore as deep as

244 meters (m) (800 feet [ft]) into the subsurface (Kuhn 2004). In the sixteenth century, oil imported from Venezuela was used to treat Holy Roman Emperor Charles V for gout. The word *petroleum*, Latin for rock oil, was first used by German mineralogist Georg Bauer in 1556 (PBS 2004). In the eighteenth century, Lewis Evans's "Map of the Middle British Colonies in America" noted the presence of petroleum seeps in Pennsylvania (PBS 2004). During the Revolutionary War, Native Americans taught George Washington's troops how to treat frostbite using seep oil, and Seneca Oil was advertised as a cure-all tonic. As early as 1815, some streets in Prague were lit with petroleum-fueled lamps (PBS 2004).

The modern history of petroleum exploitation is closely linked to petroleum seeps. Kerosene was produced from seepage oil in 1823. The process of refining kerosene from coal was developed in 1846 (PBS 2004; Kindersley 2007). This process was improved to refine kerosene from seeps in 1852. The first rock oil mine was dug in central Europe in 1853. In 1854, Benjamin Silliman was the first American to fractionate petroleum by distillation. These advances were rapidly adopted around the world (PBS 2004). The first commercial oil well was drilled in Poland in 1853 and the second in nearby Romania in 1857 at seep sites. This was followed by the opening of the world's first oil refineries (Stoicescu and Ionescu 2014). By the end of the nineteenth century, the Russian Empire led the world in petroleum production. In North America, the first oil well was dug in Oil Springs (named for a nearby seep) in Ontario, Canada, in 1858 (Kolbert 2007). The U.S. petroleum industry began in 1859 on Oil Creek (named for a nearby seep) near Titusville, Pennsylvania (PBS 2004). In the 1860s to the 1900s, sources of oil were discovered in association with petroleum seeps in Peru (1863), the Dutch East Indies (1885), and Persia (1908), as well as in the Americas in Venezuela, Mexico, and the Canadian province of Alberta. By 1910, some of these sites were being developed at an industrial level. In the late nineteenth century and early twentieth century, the demand for petroleum, created by improvements in the internal combustion engine and replacement of horse-drawn carriages, quickly outstripped the supply from seep-related sources. Surface seeps remained a primary indicator of deeper reservoirs of petroleum for many years until the advent of seismic technologies that could visualize the deep subsurface. The first commercial discoveries of oil using seismic methods were in 1924 in Mexico and Texas (Sheriff and Geldart 1995).

During World War I, oil was increasingly viewed as a strategic asset due to the use of oil-powered naval ships, new horseless army vehicles (such as trucks and tanks), and military airplanes (PBS 2004). Oil use during the war increased so rapidly that a severe shortage developed in 1917–1918. By the middle third of the twentieth century, transformative changes occurred in the oil industry. Beginning with Standard Oil's activities in Saudi Arabia, oil prospecting began a global expansion. The internationalization of oil exploration, production, and distribution played an important role in World War II. Superior access to oil aided the Allied effort. Scientific discoveries and inventions also created a vast market for petroleum products in plastics, synthetic chemicals, and other industries.

5.2.2 History of Oil and Gas Seeps in the Gulf of Mexico

Petroleum has seeped to the surface in the Gulf of Mexico region for many millions of years (Geyer 1980; Geyer and Giammona 1980—the source of the following summary; NASA 2000). The Karankawa Indians living on Padre Island in pre-Columbian times decorated pottery and waterproofed boats with seeping oil. In the sixteenth and seventeenth century, Spanish explorers caulked their ships with tar from the beaches of south Texas and Louisiana. Oviedo y Valdés referred to asphalt in the New World in 1533, and Sebastian Ocampo recorded the presence of liquid hydrocarbons in the Bay of Havana, Cuba, in 1508. In the late nineteenth and early twentieth century, large cakes of petroleum or asphalt were frequently found on the beaches between Sabine Pass and Matagorda, Texas. There have been numerous reports of sea-surface oil slicks in the Gulf of Mexico (Figure 5.3). In the early 1900s, there were regular reports of enormous patches of oil by ships navigating Gulf of Mexico waters, including oil bubbling to the surface. In 1933, there were 30 instances of oil seeps reported in the Gulf of Mexico. Prior to commercial offshore production of oil, a beach survey in 1955 reported tar with a presumed seep source on all beaches in the northern Gulf of Mexico from Mexico to central Florida.

In the Gulf of Mexico region, petroleum seeps on land are numerous and historically important. On January 10, 1901, a drill pipe spurting mud, gas, and oil blew out at Spindletop near Beaumont, Texas, (drilled on a seep) transforming Texas into a major petroleum producer (Figure 5.4). This was preceded by oil discoveries at seeps in East Texas in the late 1800s. Spindletop was the first salt-dome oil well and led to the first oil boom and numerous other onshore discoveries associated with petroleum seeps (Petty 2010). The companies established to develop Gulf of Mexico oil fields, including Gulf Oil, Sun Oil, Magnolia Petroleum, the Texas Company, and Humble Oil, are now major energy companies (Texas Almanac 2014). Between 1902 and 1912, wells were drilled in north-central Texas with discoveries in Brownwood, Petrolia, Wichita Falls, and west of Burkburnett. During the 1920s, numerous discoveries were made in east, west central, and the panhandle of Texas. During the following years, onshore oil discoveries were found across Texas and in other coastal states to the east, mostly confined to the northwestern region of the Gulf of Mexico. From the 1910s to 1930s, the use of piers, pilings, concrete platforms, barges, and artificial islands extended oil and gas exploration into coastal bays, lagoons, and offshore. By the 1940s, the first fixed platforms were being used, and the first oil discovery, drilled out of sight of land, was in 1947 off the Texas shore. In many instances, the origins of onshore oil and gas seeps were traced into the offshore region.

5.3 PREVALENCE

As the history above demonstrates, petroleum seeps have been reported worldwide and in the Gulf of Mexico for thousands of years. Vestiges of ancient seeps in the geological record at numerous locations worldwide demonstrate the common occurrence of seeps over geological time and that seeps persist for finite periods of time (Callender et al. 1990; Callender and Powell 1992; Campbell et al. 2002). Inventories likely underreport the prevalence of seeps for several reasons. Large areas of the Earth's surface—including remote and difficult-to-access areas such as Antarctica, the Arctic Ocean, the interiors of Africa and South America, and the deep sea—remain unexplored for seeps (Figure 5.5). Once released to the surface, seeping petroleum is subject to a range of processes that alter its composition, and some processes mask the presence of seeps. In addition, energy companies consider the location of seeps a competitive advantage, so many seeps go unreported in the open literature.

Based on surveys in 2009 and 2015, reported petroleum seeps were mostly located in the northern hemisphere as are a majority of the world's oil and gas reserves (Figures 5.5 and 5.6) (Etiope 2009, 2015). In these surveys, seeps were most often reported on land, which is likely due to the long history and relative ease of visual observations. Seeps in the ocean often require detection by satellite, airborne sensors, and/or direct sampling; although visual reports of surface oil slicks are numerous (Figure 5.3). Global inventories are few, mostly rely on visual detection, and often do not include seeps detected by a wide range of other indicators that signify their presence now or in the past. Areas where petroleum seeps occur may have multiple seeps. Estimating the volume of leakage is often difficult; thus, inventories mostly count the

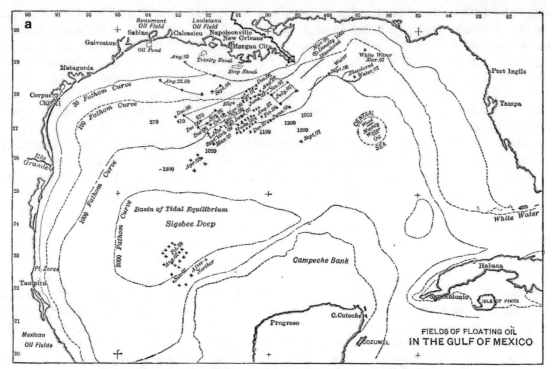

FIELDS OF FLOATING OIL IN THE GULF OF MEXICO.

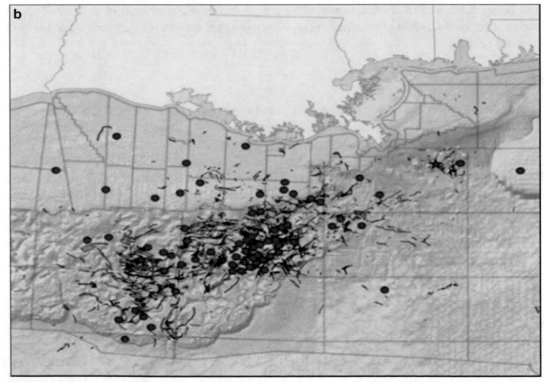

Figure 5.3. Maps of (a) historical reports of floating oil [*red dots* in (b)]; Soley (1910) and (b) oil slicks (*black lines*) in the northern Gulf of Mexico 1991–2009 (determined from analysis of synthetic aperture radar, graphic provided by CGG's NPA Satellite Mapping, used with permission).

Figure 5.4. Spindletop blows (photo from http://commons.wikimedia.org/wiki/File:Lucas_gusher. jpg).

number of seeps and do not attempt to estimate volumes. In contrast, marine seep volumes (as loadings) have received attention due to their important contribution to oil in the sea, and the volume of global gas seeps have been estimated as contributors to greenhouse gases (Coleman et al. 2003; Etiope 2015).

To assess the association of seepage and subsurface accumulations of petroleum, Schumacher (2012) compiled seepage survey results for more than 2,700 exploration wells and compared the results with subsequent drilling outcomes. Locations were in frontier and mature basins, onshore and offshore, and in a wide variety of geologic settings. Subsurface drilling targets were from 300 m (984 ft) to more than 4,900 m (16,076 ft), and there was a full spectrum of trap styles. The presence of seepage was inferred from soil gas, microbial, iodine, radiometric, and/or magnetic surface surveys. Eighty-two percent of wells associated with surface seepage anomalies were considered commercial discoveries, and 11 % of wells drilled without a documented surface seepage anomaly resulted in discoveries. The measure of association was economic viability determined by external factors, and not the presence or absence of petroleum in the subsurface. The sites chosen for analysis in this study were not random; they were based on conventional prospect evaluation methods. This study illustrates

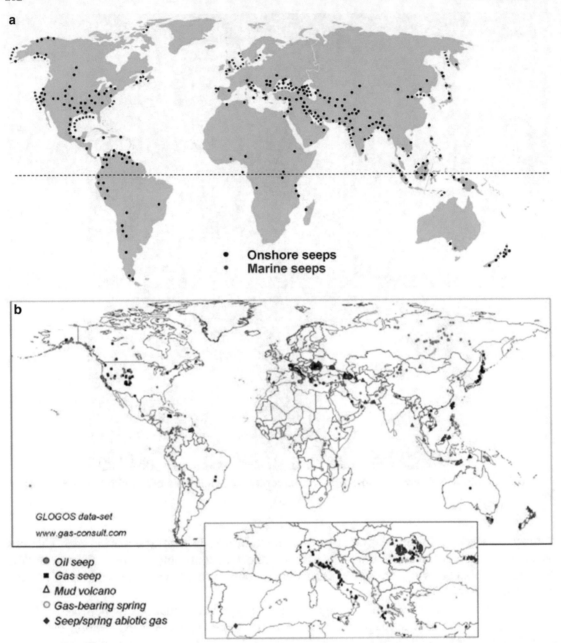

Figure 5.5. Global prevalence of petroleum seeps: (a) more than 1,150 seeps in 84 countries (Etiope 2009; examples in Russia, the Arctic Ocean, and Antarctica were not included in the survey) and (b) distribution of onshore seeps including about 2,100 seeps in 86 countries (from the global data set of onshore gas and oil seeps [GLOGOS]; Etiope, 2015; reprinted with permission of Springer).

that seeps are often only surveyed for in areas suspected of being oil and gas prospects, thus limiting geographic coverage.

Estimates of the volume of gases seeping on land are numerous (Etiope 2015). These estimates take into account spatial and temporal variability, susceptibility to rapid alteration once exposed at the surface, and release of gases and volatiles directly into the

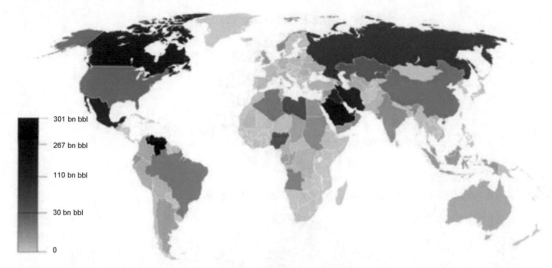

Figure 5.6. A map of world oil reserves (2013; graphic from http://en.wikipedia.org/wiki/List_of_countries_by_proven_oil_reserves).

overlying atmosphere. Marine oil seeps have been extensively studied and quantified as well (Figure 5.7) (Coleman et al. 2003). Compared to the long history of land-based observations of seepage, offshore seepage detection is a relatively recent development—one exception being historical reports of sea-surface slicks (Figure 5.3). Advances in techniques to detect seeps in the ocean and the expansion of oil and gas exploration into the offshore regions in the last 50–60 years greatly increased geographic coverage and the number of reported marine seeps. Annual oil seepage to the marine environment was estimated to be 600,000 tonnes (i.e., metric tons) (180 million gallons [gal]) globally and 160,000 tonnes (47 million gal) in North America from 1990 to 1999 (Kvenvolden and Cooper 2003; Coleman et al. 2003; Table 5.1). While variable, natural seeps are estimated to contribute about 45 % of the oil entering the marine environment worldwide and about 60 % in North American waters, with the remainder due to the extraction, transportation, and consumption of petroleum (Figure 5.7 and Table 5.1) (Coleman et al. 2003). In North American waters, the largest natural seeps are located in the Gulf of Mexico and offshore of southern California.

The immensity of the volume of oil released by seeps in the Gulf of Mexico is indicated by the larger relative contribution of petroleum seepage to oil in North American waters as compared to worldwide estimates (Coleman et al. 2003). The alteration of petroleum, once released to the environment, introduces considerable uncertainty in estimating the volume of petroleum seepage. Gaseous hydrocarbons are particularly susceptible to alteration after seepage and are rarely considered in global seep inventories since little is known about the rates and volumes of seepage, though gas seeps are known to be common (Kvenvolden and Cooper 2003). Most gas seepage is either dissolved in seawater or quickly metabolized by microbes, leaving scant evidence of its presence. Because methane is a greenhouse gas, the contribution of atmospheric methane from natural seepage (mostly biogenic in origin) has been estimated (Etiope 2015). While petroleum seeps have been reported extensively worldwide, global inventories remain incomplete and uncertainties in volume estimates are large.

Oil seeps account for approximately 95 % of the total oil input to northern Gulf of Mexico waters (Figure 5.8) (Coleman et al. 2003). As with global estimates, the full extent of oil and gas

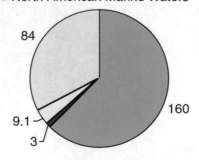

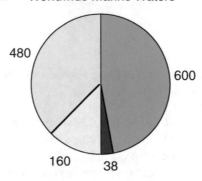

Figure 5.7. **Relative contribution of average, annual releases (1990–1999) of petroleum hydrocarbons (in kilotonnes) from natural seeps to the marine environment in (a) North American waters and (b) worldwide compared with other sources (republished with permission of Emerald Group Publishing Limited from Coleman et al. 2003; permission conveyed through Copyright Clearance Center, Inc.)**

Table 5.1. Average, Annual Releases (1990–1999) of Petroleum (oil) to the Marine Environment by Source (in thousands of tonnes, Coleman et al. 2003) (a tonne equals about 300 gal of oil, Kvenvolden and Cooper 2003)

Region	North America			Worldwide		
Source	Best Est.	Min.	Max.	Best Est.	Min.	Max.
Natural seeps	160	80	240	600	200	2,000
Extraction of petroleum	3.0	2.3	4.3	38	20	62
Transport-ation of petroleum	9.1	7.4	11	150	120	260
Consumption of petroleum	84	19	2,000	480	130	6,000
Total	260	110	2,300	1,300	470	8,300

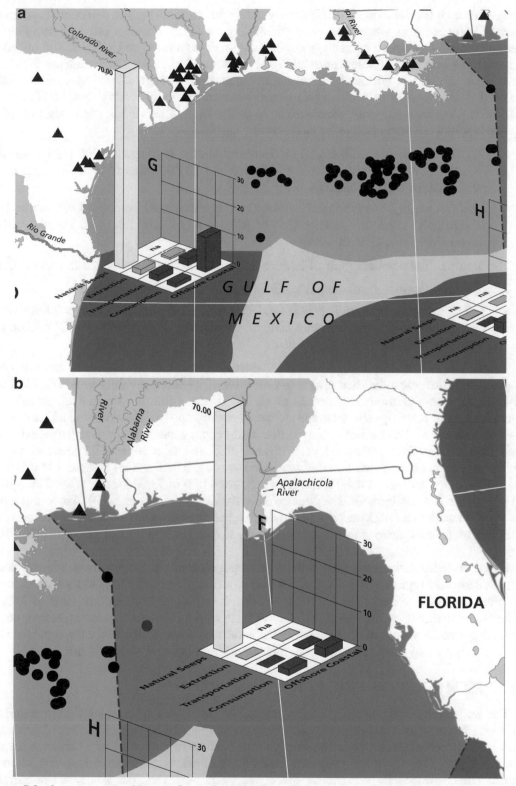

Figure 5.8. Average annual input of petroleum hydrocarbons in thousands of tonnes to the coastal Gulf of Mexico for 1990–1999 (*yellow* = natural seeps, *green* = extraction, *purple* = transportation, and *red* = consumption) (modified from Coleman et al. 2003).

seepage in the Gulf of Mexico is difficult to quantify due to challenges in detection (e.g., occurs subsea), differences in quantification methods (e.g., satellite observations and sampling by corer), dispersion by ocean currents, gaps in geographic coverage, and variable and uncertain seep volumes and rates (Coleman et al. 2003; Wilson et al. 1974; De Beukelaer 2003; De Beukelaer et al. 2003). Within these uncertainties, natural seepage of oil has been estimated to exceed 140,000 tonnes (42 million gal) annually in northern Gulf of Mexico waters (Coleman et al. 2003). Comparing overall petroleum input to the Gulf of Mexico in the 1990s, annual oil seepage inputs were estimated to be as follows:

- 140,000 tonnes (42 million gal) total natural annual loadings: 70,000 (21 million gal) tonnes in the northeastern Gulf of Mexico and 70,000 tonnes (21 million gal) in the northwestern Gulf of Mexico.

- 25,400 tonnes (7.62 million gal) total anthropogenic annual loadings: 4,400 tonnes (1.32 million gal) in the northeastern Gulf of Mexico and 21,000 tonnes (6.3 million gal) in the northwestern Gulf of Mexico.

The inputs to the northern Gulf of Mexico coastal waters were estimated to be as follows:

- Negligible total natural annual loadings (few known seeps)

- 17,740 tonnes (5.322 million gal) total anthropogenic annual loadings: 2,660 tonnes (798,000 gal) in the northeastern Gulf of Mexico and 15,080 tonnes (4.524 million gal) in the northwestern Gulf of Mexico.

A tonne equals about 300 gal of oil (Kvenvolden and Cooper 2003). Kvenvolden and Cooper (2003) provide a detailed review of estimates of oil seepage rates as of 1975, 1985, and 2000. In the latest estimates (Coleman et al. 2003), the authors note that the number of regions known to have significant seeps increased mainly due to detection by satellite remote-sensing techniques. The authors note further that seepage rates in the Gulf of Mexico are much higher than first estimated in 1975 and 1985 as the number of known seeps has significantly increased. Based on satellite remote sensing, MacDonald (1998) and MacDonald et al. (1993, 1996) estimated total seepage to be from 4,000 to 73,000 tonnes (1.2–21.9 million gal) per year in the northern Gulf of Mexico (Kvenvolden and Cooper 2003). Assuming a seep rate for the entire Gulf of Mexico is about double the northern Gulf of Mexico estimate, the total Gulf of Mexico seep rate is estimated to be about 140,000 tonnes per year (42 million gal).

Based on these estimates, most petroleum seepage occurs in the northwestern and north-central deepwater region of the Gulf of Mexico coincident with oil and gas production and is negligible in coastal waters. The high estimate for the offshore northeastern Gulf of Mexico region is due to one seep site reported in the far western part of the northeastern sector, but oil seeps are generally absent in the region. Similar estimates are less certain for the southern Gulf of Mexico, but many seeps are known in this region both onshore and offshore.

5.4 PETROLEUM GEOLOGY

The well-established principles of the geology of petroleum systems set the stage and the conditions for why and where petroleum seeps occur (Figures 5.9 and 5.10). Petroleum seeps result from direct migration from the source to the surface or as a result of a breach in the seal of a reservoir. The force of buoyancy and differentials in pressure drives migration of hydrocarbons toward the surface. It often has been observed that nearly all subsurface occurrences and accumulations of petroleum leak to some degree at some point in time. Oil and gas often migrate directly from subsurface sources without pooling, and some seeps are

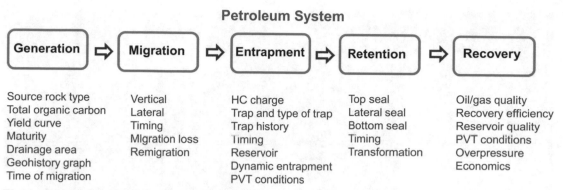

Figure 5.9. Elements of a petroleum system (M.H. Nederlof, reproduced with permission; http://www.mhnederlof.nl/petroleumsystem.html).

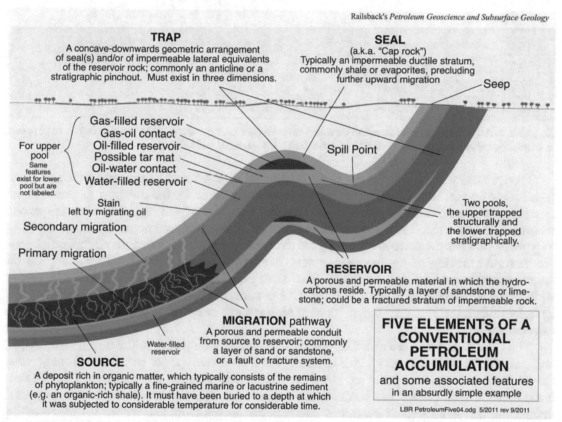

Figure 5.10. A petroleum system includes a mature source rock, migration pathway, reservoir rock, trap, and seal. The relative timing of formation of these elements and the processes of generation, migration, and accumulation are critical for hydrocarbon accumulation and preservation (from Railsback 2011; reprinted with permission).

outcrops of oil-bearing rocks (e.g., tar sands and exposed source rocks). In instances of seepage from underlying reservoirs, seals are breached due to overpressure and/or mechanical disruption (i.e., faults) by upward intrusions of less dense materials, such as salt diapirs and tectonic forces.

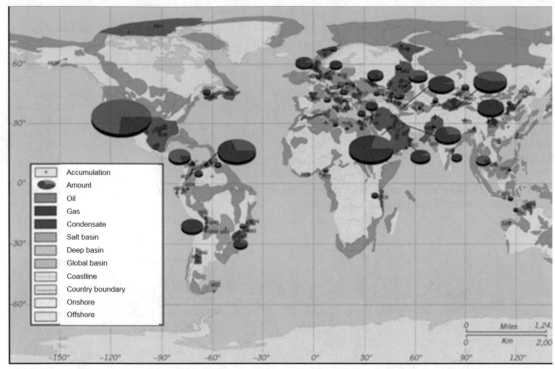

Figure 5.11. Worldwide distribution of deep oil and gas basins defined as occurring in reservoirs at burial depths deeper than ~4,572 m (~15,000 ft) (Cao et al. 2013).

Overpressure is caused by the rapid loading of fine-grained sediments, which prevents expulsion of water and equalization of the pressures created by the overburden. Intermittent resealing of breaches can occur and slowing or cessation of burial allows time for excess pressures to dissipate. Fluid expansion, which is a change in volume, is a second cause of overpressure. Overpressure is caused by the thermal expansion of water, clay dehydration, and the thermal cracking of source-rock organic matter to form oil and gas. Depending on the degree of overpressure and the mechanical strength of the encasing rocks, seepage can be widespread and diffuse. A slow seepage rate is commonly referred to as *microseepage*. In instances when the rocks fracture, focused high-volume seepage is commonly referred to as *macroseepage*.

The Gulf of Mexico is a prolific petroleum basin containing vast volumes of subsurface oil and gas (Figure 5.11) (Cao et al. 2013). The region is an archetype for petroleum seepage since the geologic history and setting are ideally suited for seepage. In fact, there may be no leakier basin of its size on the Earth's continental margins. The Gulf of Mexico has been a long-term depocenter that has received enormous sediment discharges from major river systems creating source and reservoir rocks (Figure 5.12). The Gulf of Mexico contains multiple, deeply buried source rocks that are thermally mature (Figure 5.13). Salt tectonics driven by the underlying Middle Jurassic Louann Salt created migration pathways from the source rock to reservoirs and to the surface (Figure 5.14).

5.4.1   Source Rocks and Petroleum Generation

A *source rock* is the rock from which hydrocarbons have been, or are capable of being, generated and is a necessary element of a viable petroleum system (Figure 5.15) (Hunt 1996).

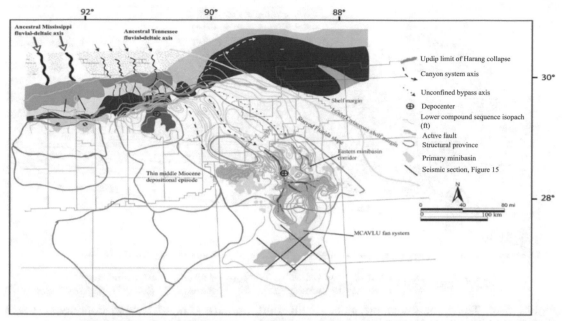

Figure 5.12. Map of Middle Miocene depocenters (Combells-Bigott and Galloway 2006; AAPG©2006, reprinted by permission of the AAPG whose permission is required for further use).

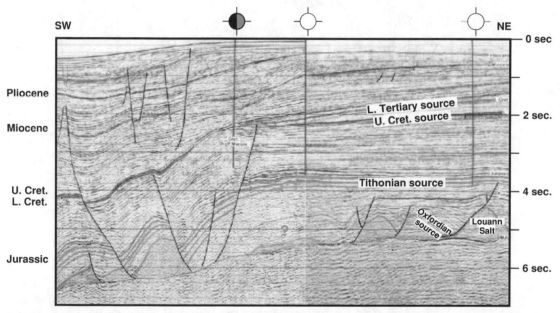

Figure 5.13. One of many wells that have penetrated multiple potential source rocks in the Gulf of Mexico (Hood et al. 2002; AAPG©2002, reprinted by permission of the AAPG whose permission is required for further use) [*L. Tertiary* Lower Tertiary, *U./L. Cret.* Upper/Lower Cretaceous, *sec.* seconds].

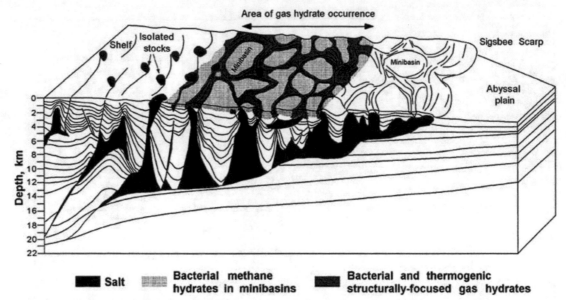

Figure 5.14. The subsurface structure of the northwestern Gulf of Mexico continental slope (Milkov and Sassen 2001; reprinted with permission from Elsevier).

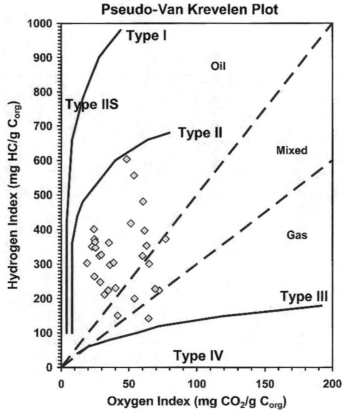

Figure 5.15. Source-rock types are classified by the content of hydrogen (H), carbon (C), and oxygen (O), which changes with maturation and oil and gas generation (photo from AAPG wiki, 2014; available at http://wiki.aapg.org/File:VanKrevelanDiagram.png).

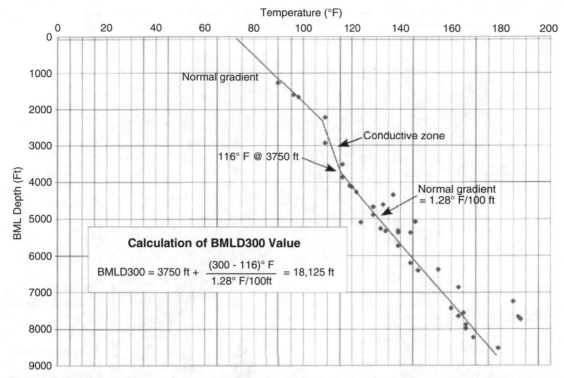

Figure 5.16. Subsurface temperature distributions are the end-result of thermal gradient, thermal conductivity, and heat flow. Subsurface temperature gradients are inversely proportional to the thermal conductivity of sedimentary rocks under conditions of constant heat flow. [BMLD300—below mud-line depth to the 300° isotherm; example from Eugene Island 282 in the Gulf Mexico, data from MMS Atlas of Northern Gulf of Mexico Oil Sands (2001)] Forrest et al. 2007; republished by permission of the Gulf Coast Association of Geological Societies, whose permission is required for further publication use, http://www.searchanddiscovery.com/documents/2007/07013forrest/images/forrest.pdf.

Source rocks are organic-rich sediments that have been deposited in a variety of environments including deepwater marine, lacustrine, and deltaic settings that preserve the remains of dead plants and animals. Anoxic or suboxic conditions at the time of deposition are often a requirement for preservation (Hunt 1996). Source rocks are classified by the type of kerogen (organic matter) they contain, which in turn determines the type of hydrocarbons generated as they thermally mature (Hunt 1996). Type I source rocks contain algal remains deposited under anoxic conditions in deep lakes and generate waxy oils. Type II source rocks contain marine planktonic and bacterial remains preserved under anoxic conditions in marine environments and produce both oil and gas. Type III source rocks contain terrestrial plant material that has been decomposed by bacteria and fungi under oxic or suboxic conditions and generate mostly gas and volatiles. Most coals and coaly shales are Type III source rocks. As a primary control on the type of petroleum generated at depth, the source-rock type determines the composition of gases and liquids available to migrate to the surface. Characterization of overlying seeps has been used prior to drilling to infer the presence, maturity, and type of source rocks more deeply buried in a basin.

When source rocks are buried by sediments, temperatures increase, and under suitable conditions the insoluble organic matter (kerogen) in the rock begins to thermally crack or breakdown (Figures 5.15 and 5.16) (Hunt 1996). This breakdown produces hydrocarbons from

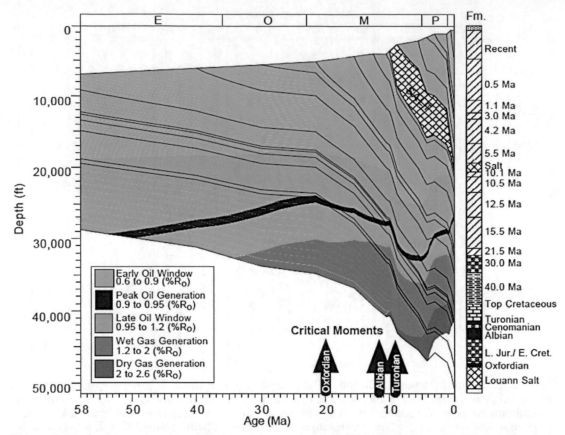

Figure 5.17. Geohistory and burial plot in the north-central Gulf of Mexico (McBride et al. 1999; used with permission of B.C. McBride) (*Ma* million years, *L. Jur.* Lower Jurassic, *E. Cret.* Early Cretaceous; upper scale: *E* Eocene, *O* Ordovician, *M* Miocene, *P* Paleocene; and %Ro is Vitrinite Reflectance).

the large and complex biomolecules of kerogen. High temperatures and deep burial can lead to the almost complete cracking of hydrocarbons to methane, producing dry gas (Figure 5.17) (McBride et al. 1999). Wet gas contains significant amounts of C_2 (ethane) to C_5 hydrocarbons (pentanes). Oil and gas generated from thermally mature source rocks are first expelled along with other pore fluids. This expulsion is due to the effects of internal source-rock overpressuring caused by hydrocarbon generation, as well as by compaction and is referred to as *primary migration*. Once released into porous and permeable carrier beds or into fault planes, oil and gas move upward toward the surface because of buoyancy and pressure; this upward movement is referred to as *secondary migration*.

Hood et al. (2002) described the Gulf of Mexico regional geologic framework based on two-dimensional (2D) and three-dimensional (3D) seismic data, identification and mapping of source intervals, and likely migration pathways to reservoirs. The compositions of more than 2,600 produced gas and oil samples and 3,000 seafloor seeps were used to describe source-rock characteristics, including organic matter type, depositional facies, level of maturation, and age. The major offshore hydrocarbon systems in the Gulf of Mexico were identified as the Lower Tertiary, Upper Cretaceous, and Upper Jurassic intervals (Table 5.1 and Figure 5.18). Eocene oil types correlated with source rocks and paleofacies distributions of Eocene deltaic systems. Eocene oils and gases occur on the Texas and Louisiana continental shelf and extend from

Table 5.2. Northern Gulf of Mexico Source Intervals (ages) and Source-Rock Correlations (Hood et al. 2002; AAPG©2002, reprinted by permission of the AAPG whose permission is required for further use)

Source Interval	Oil Types	Rock Oil Type
Lower tertiary (centered on Eocene)[a]	Tertiary marine	Tie with high maturity cores of south Louisiana multiple-maturity suites and south-central Louisiana offshore Texas (salt sheath)
	Tertiary intermediate	
	Tertiary terrestrial	
Upper cretaceous (centered on Turonian)[a]	Marine—low sulfur—no tertiary influence	Direct ties with mature source rocks: offshore-eastern Gulf of Mexico, onshore Tuscaloosa trend, and Louisiana and Mississippi Giddings trend, Texas
Lower cretaceous	Carbonate—elevated salinity—cretaceous	Direct ties with source rocks: South Florida Basin
Undifferentiated cretaceous		Calcareous—unidentified cretaceous—production from fractured lower cretaceous black shale—south Texas
Uppermost Jurassic (centered on Tithonian)[a]	Marine—high sulfur—Jurassic	Inferred tie to postmature, organic-rich calcareous shales of the eastern Gulf of Mexico and oils in lower cretaceous reservoirs on Florida shelf where the Turonian/Eocene section is immature
	Marine—moderately high sulfur—Jurassic	
	Marine—moderate sulfur—Jurassic	
Upper Jurassic (Oxfordian)	Carbonate—elevated salinity—Jurassic	Tie to postmature, organic-rich carbonates—Mobile Bay
Triassic (Eagle Mills)	Triassic—lacustrine	Tie to postmature, organic-rich cores—northeast Texas (paleontology and palynology confirm nonmarine source character)

[a]"Centered on" means that the source is largely within, and may not be restricted to, the designated interval.

onshore to the offshore Texas continental slope. Turonian oils were matched with offshore (east of the Mississippi River Delta) and onshore source rocks (e.g., Tuscaloosa and Giddings trends). Upper Cretaceous Eagle Ford source rocks currently are being developed as prolific shale-oil or gas-shale reservoirs. Based on seismic images, it is known that source rocks thin and ultimately pinch-out toward the basin. Oils and associated gases on the Gulf of Mexico upper slope are interpreted as originating from a Tithonian source. High maturity, organic-rich calcareous shales of the same age in the eastern Gulf of Mexico have been confirmed. Tithonian-sourced oils in Cretaceous reservoirs on the Florida Shelf and the Upper Cretaceous and Tertiary sections are immature. Oxfordian carbonate-sourced oils are common across the northwestern Gulf basin rim. Lower maturity hydrocarbons from this source seep to the surface in deep central Gulf of Mexico.

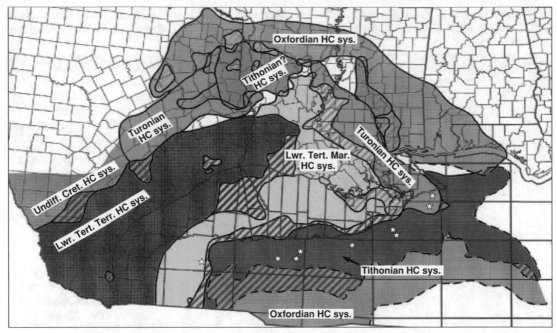

Figure 5.18. Hydrocarbon systems in the northern Gulf of Mexico basin. Each hydrocarbon system comprises a family of oils and gases with similar compositions interpreted as originating from a common source interval. Note that this map extends onshore (Hood et al. 2002; AAPG©2002, reprinted by permission of the AAPG whose permission is required for further use). [*HC sys.* hydrocarbon system, *Terr.* terrestrial, *Mar.* Marine, *Undiff. Cret.* Undifferentiated Cretaceous, *Lwr. Tert.* Lower Tertiary].

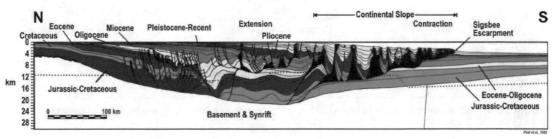

Figure 5.19. A north-south geologic cross section of the northern Gulf of Mexico basin illustrating the complex relationships between sediments and salt (*black*). The thick intraslope sedimentary basins, salt bodies, and numerous faults of the continental slope provide a geologic framework that favors leakage of subsurface fluids and gases to the modern seafloor. The cross section is an interpretation of two-dimensional seismic data calibrated using well data. From Fisher et al. 2007, citing Peel et al. 1995. Republished with permission of the American Association of Petroleum Geologists from Peel et al. 1995; permission conveyed through Copyright Clearance Center, Inc.

5.4.2 Migration Pathways

Migrating petroleum often crosses strata above deeper source intervals to travel to shallower more porous reservoir rocks or onward to the surface (Figure 5.19) (Peel et al. 1995). Salt movement, faulting or other tectonic activity, depending on geological setting, can create cross-stratum conduits. In the Gulf of Mexico, effective potential migration pathways intersect both the deep source intervals and younger reservoirs. Nearly 70 % of the world's proven hydrocarbon reserves are found in structures related to salt tectonics (Cao et al. 2013). Migration pathways may form during continued sedimentary loading, without external tectonic influences, due to gravitational instability (e.g., salt is less dense than surrounding rocks). However, active tectonics increases the likelihood of the development of salt structures (Figure 5.20). A salt body pushing through its overburden is known as *diapirism*. Many of the first oil discoveries were associated with salt domes. Salt diapirism is particularly important and prevalent in the Gulf of Mexico and is a major reason for the extensive petroleum seepage observed.

As elsewhere in the world, seepage and subsurface petroleum systems in the Gulf of Mexico are closely correlated. Petroleum seepage patterns and analysis in the offshore Gulf of Mexico have been used to extend mapping of hydrocarbon systems and maturity maps beyond subsurface core data (Hood et al. 2002). As described above, the basic requirements for petroleum to reach the surface are common in the deepwater region of the Gulf of Mexico. Multiple prolific source rocks are present that have been deeply buried by sediment deposition over geologic time. Burial results in maturation contributing to overpressuring that, combined with buoyancy, drives upward fluid migration. The same geological processes produce large sandstone bodies. These bodies serve as excellent high porosity reservoirs where some of the generated liquid and gaseous hydrocarbons are trapped. Salt tectonics involving the underlying Jurassic Louann Salt has created deep subsurface faults. These faults provide conduits for not only migration of petroleum into reservoir rocks but also breaching reservoir seals allowing seepage to the surface.

The Louann Salt is a widespread evaporite formation that formed in the Gulf of Mexico in the Middle Jurassic Epoch (Figure 5.21) (Hudec et al. 2013). The Louann Salt layer formed in a rift as the South American and North American Plates separated forming an embayment in the paleo-Pacific Ocean. The Louann Salt underlies much of the northern Gulf of Mexico from Texas to the Florida panhandle and extends beneath large areas of the Gulf of Mexico coastal plain of Mississippi, Louisiana, and Texas and southward into the deep sea. The geographic distribution of petroleum seeps in the Gulf of Mexico closely correlates with subsurface salt structures that create migration pathways. Salt structures are particularly prevalent in the northwestern Gulf of Mexico from the edge of the continental shelf along the continental slope and into the abyssal Sigsbee Escarpment. This is an area of intensive petroleum seepage. A series of additional salt basins and structures extend to the northeast and toward the basin in the southern Gulf of Mexico offshore of the Campeche Peninsula (Figure 5.21). Seeps have been reported in the southern and southwestern Gulf of Mexico, and major oil and gas accumulations have been found (Figures 5.22 and 5.23). Lava-like flows of solidified asphalt have been reported on the Campeche Knolls (a surface expression of deeper salt diapirs) in 3,000 m (9,843 ft) of water in this area (Figure 5.21) (MacDonald et al. 2004). Gas hydrate, thermogenic gases, biodegraded oil seeps (possibly from an Upper Jurassic source of moderate maturity), and cold-seep communities were reported at this site, confirming asphalt flows seen in a seafloor photograph taken in the 1970s (MacDonald et al. 2004).

5.5 BIOGEOCHEMISTRY

Petroleum seeps in the Gulf of Mexico are a highly variable mixture of chemical compounds reflective of the subsurface source materials and postseepage alteration processes. Seeps exhibit the full spectrum of alterations from pristine (e.g., unaltered) to severely biodegraded. Seeps can be 100 % methane, while in other instances, a complete suite of hydrocarbons typically found in oil is present. The chemical compositions of seeps have been determined by sampling and analysis of air, water, and sediments using seafloor

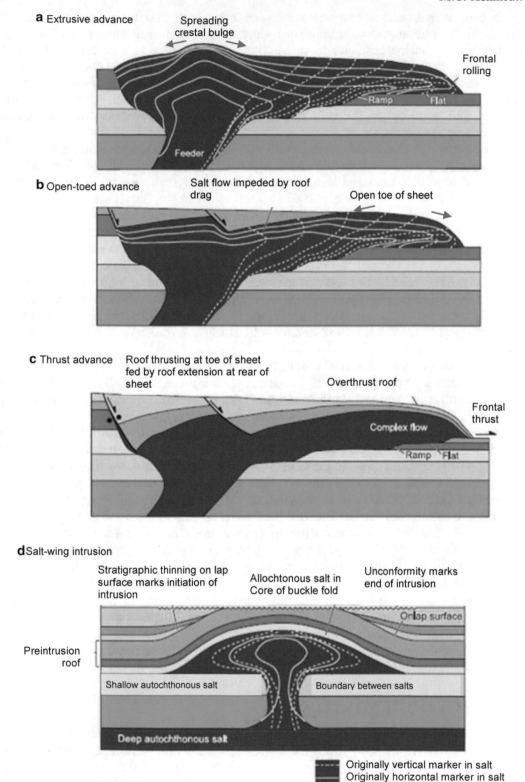

Figure 5.20. Four models of salt-sheet advance. *White lines* in the salt represent deformed markers from an originally rectangular grid. A line flanked with pairs of *black dots* indicates a salt weld (Hudec and Jackson 2006; AAPG©2006, reprinted by permission of the AAPG whose permission is required for further use).

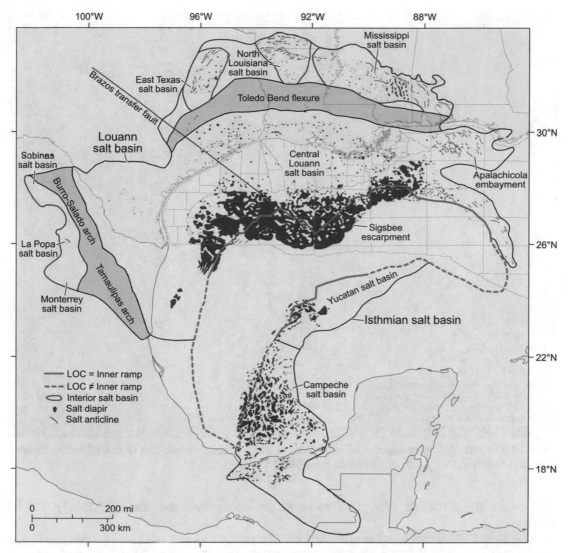

Figure 5.21. Salt basins of the Gulf of Mexico (salt structures are *red*); Hudec et al. 2013; AAPG©2013, reprinted by permission of the AAPG whose permission is required for further use.

coring devices and remotely operated vehicles and manned submersibles. Sea-surface slicks are collected with adsorbents and screens. Each petroleum seep has its own chemical signature.

5.5.1 Chemistry

Collectively, Gulf of Mexico seeps contain gaseous compounds with 1–5 carbon atoms, volatile compounds with 6–12 carbon atoms, and higher-molecular-weight hydrocarbons with 13 to more than 60 carbon atoms. Seeps can contain alkanes, branched alkanes, cycloalkanes, and aromatic (unsaturated) hydrocarbons. As with petroleum, heteroatomic compounds (containing oxygen, nitrogen and sulfur), resins, asphaltenes, metals and sulfur can be present in oil seeps as well. Complex biochemical-derived compounds that can be linked to known biological

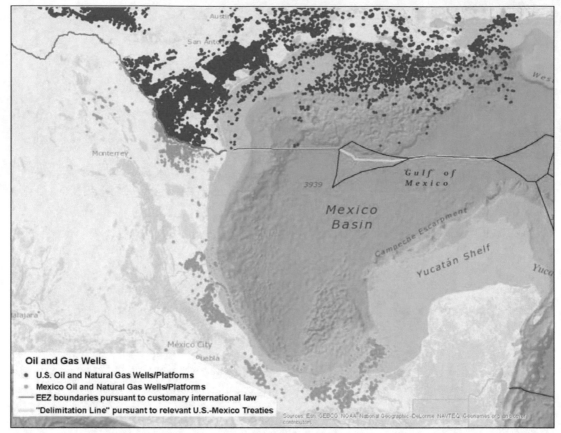

Figure 5.22. The U.S. and Mexico oil and natural gas activity around the Gulf of Mexico (Seelke et al. 2015) https://www.fas.org/sgp/crs/row/R43313.pdf.

precursors, the so-called *biological markers* or *biomarkers*, are also commonly present in seeps.

Gas seeps are widespread in the Gulf of Mexico but most have recent, microbiological origins and are often 100 % methane. Seep gases can be unbound (free), bound to mineral or organic surfaces, or entrapped in mineral inclusions (Abrams 2005; Abrams and Dahdah 2011). Gas seeps of microbial methane can be differentiated from deep-sourced thermogenic hydrocarbon gases based on compositional and stable and radiocarbon analyses (Figures 5.24 and 5.25). Microbes produce almost exclusively methane although some have suggested trace amounts of higher-molecular-weight gases may have a microbial origin (Sassen and Curiale 2006). Being of recent origin, biological methane can contain radiocarbon unless fossil organic matter is being metabolized (Figure 5.25). Thermogenic gases, other than highly mature thermogenic methane (dry gas), are often associated with appreciable amounts of ethane to butane gases.

Various compositional ratios of C_1–C_4 gases have been used to infer origins and maturity. Being derived from fossil carbon, thermogenic gases contain no radiocarbon (Figure 5.25). Thermogenic methane is enriched in ^{13}C relative to microbial-derived methane, with most stable carbon isotopic values ranging from −50 to −35 ‰ (parts per thousand—denoted as ‰— enrichments or depletions relative to a standard of known composition). Microbial gases stable carbon isotopic values vary from −120 to −60 ‰ (Whiticar 1999). Methane hydrogen stable

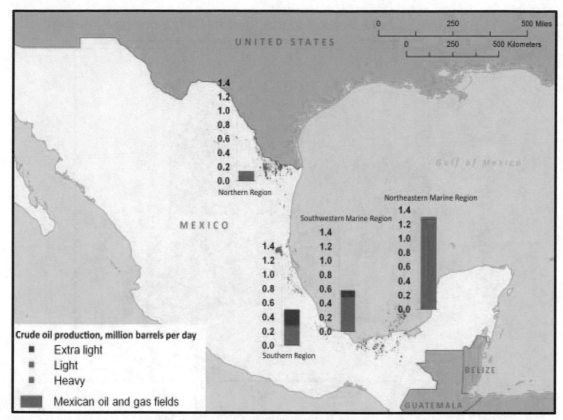

Figure 5.23. Mexico oil production in 2012 (millions of barrels per day; U.S. Energy Information Administration http://www.eia.gov/todayinenergy/detail.cfm?id=11251).

isotopic compositions (^{1}H, ^{2}H [deuterium]) provide additional information about the origins of gases. The hydrogen isotopic composition of methane derived from bacterial carbonate reduction ranges from −250 to −150 ‰, whereas values for methane derived from bacterial methyl-type fermentation range from −375 to −275 ‰. Thermogenic methane deuterium values range from −300 to −100 ‰ (Schoell 1980). Seep gases can be mixtures of multiple sources, and stable isotopic compositions can be altered by microbial oxidation confounding determination of original compositions.

It has been observed that migrated gasoline-range hydrocarbon compositions can vary from those found in reservoir oils (Abrams et al. 2009). The origins of gasoline-range (volatile) hydrocarbons (C_5 to C_{12}) in near-surface sediments are difficult to determine due to limited knowledge of inputs from recent organic matter. Seep gasoline-range hydrocarbons are often highly altered by microbes as a readily available source of labile reduced carbon.

The chemical and isotopic analyses of Gulf of Mexico oil seeps have been used to infer origins based on individual hydrocarbon concentrations and ratios; sums of homologue concentrations and ratios (e.g., alkanes and polycyclic aromatic hydrocarbons); stable carbon, hydrogen, and sulfur isotopic ratios; sulfur and metal content (e.g., Ni/V ratios); and biomarker compositions. Seep biomarker compounds provide information that can be used to correlate surface seep to subsurface oils and/or source rocks and indicate source-rock maturity and geologic age. Biomarkers commonly analyzed by gas chromatography/mass spectrometry include, but are not limited to, hopanes, steranes, tricyclic/tetracyclic terpanes, diasteranes, monoaromatic steroids, and triaromatic steroids. Low-intensity seeps can be overprinted by

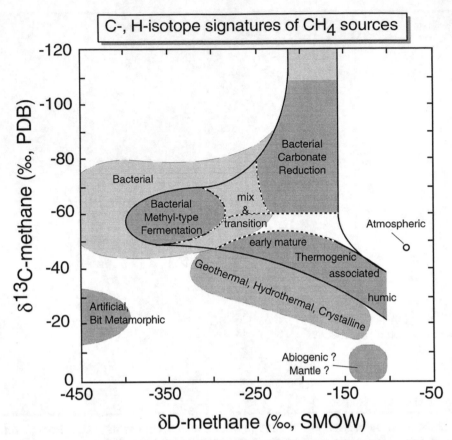

Figure 5.24. Methane stable hydrogen and carbon isotopic compositions vary with source, maturation, and alteration (Whiticar 1999; reprinted with permission from Elsevier).

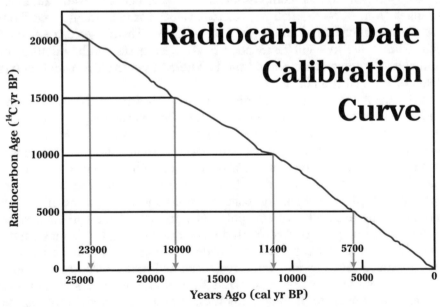

Figure 5.25. Comparison of radiocarbon dates to calendar dates. Actual ages are underestimated because the ratio of ^{14}C to ^{12}C changed over time (Reimer et al. 2004). (*cal yr BP* calibration year Before Present].

recent organic matter, which can obscure origins (Cole et al. 2001). Not all high-molecular-weight thermogenic hydrocarbons in recent sediments are due to oil seepage. Eroded material from surface exposures of thermally mature source rock can be redeposited in recent sediments.

5.5.2 Weathering

Once exposed to the near-surface environment, a range of physical, chemical, and biological processes can alter the physical and chemical properties of a seep from those of the subsurface source (Figure 5.26) (Coleman et al. 2003). Collectively, these processes are referred to as *weathering* and include evaporation, emulsification, dispersion, dissolution, and oxidation. These processes can occur in seafloor sediments, in the water column, at the sea surface, on land, and in the atmosphere. Some processes are mediated by microbiota. In the marine environment, seeping petroleum is also subject to various oceanographic processes including advection and spreading, dispersion and entrainment, sinking and sedimentation, partitioning, biological uptake and utilization, and stranding (Figure 5.27). All of these processes confound estimates of the original volume of seepage and can mask or displace seepage from its site of origin. Depending on the degree and type of alteration, some seep components and properties are preserved. These preserved properties can be used to infer the origin of the seep. Additionally, the progression of some changes in composition is predictable.

Evaporation is an important weathering process if seeping petroleum reaches the air/water or air/land interface. In particular, low-molecular-weight hydrocarbons (C_1 to C_{12}) are subject to evaporative loss (Coleman et al. 2003). Gases can reach the atmosphere with little or no alteration, depending on the physical setting, and compounds with higher molecular weights may be little altered by evaporative losses. Petroleum seeps can be a mixture of hundreds of compounds that vary from location to location and over time, and evaporative losses can be quite complex, variable, and often difficult to predict.

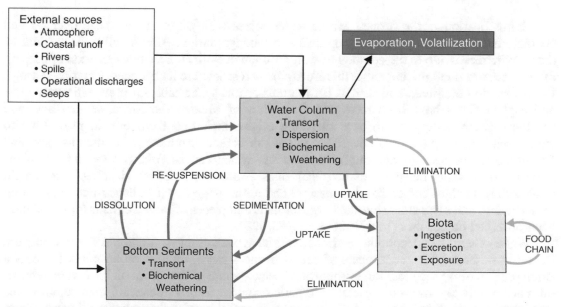

Figure 5.26. Conceptual model for the fate of petroleum in the marine environment (Coleman et al. 2003; republished with permission of Emerald Group Publishing Limited, permission conveyed through Copyright Clearance Center, Inc.)

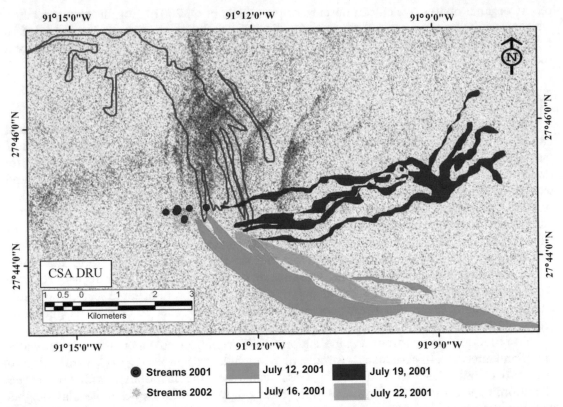

Figure 5.27. Sea-surface oil slicks in the north-central Gulf of Mexico associated with water column streams of gas bubbles detected in acoustic profiles emanating from the seafloor (*red dots*—2001 and *yellow stars*—2002). The *dark gray areas* are oil slicks detected on satellite images on 20 June 2002 (De Beukelaer et al. 2003; reprinted with permission of Springer).

   Emulsification is the process where water mixes with oil changing the properties and characteristics of seepage and susceptibility to biodegradation. Additionally, the volume of the seep increases due to the addition of water (Coleman et al. 2003). Emulsification of seeping oil requires turbulent mixing and is therefore mostly restricted to higher energy marine settings. Emulsions do not spread and tend to form lumps or mats. Tar balls, tar mats and pavements, and asphalt flows have been recovered from Gulf of Mexico shorelines, sea surface, and seafloor. These materials can have differing origins including formation in place (due to emulsification), seepage of oil degraded in the subsurface, eruptions of molten asphalt, and formation at the sea surface due to weathering, which can be followed by sinking to the seafloor once their densities exceed that of seawater (Alcazar et al. 1989; MacDonald et al. 2004). In the Gulf of Mexico, natural and anthropogenic tar balls commonly wash up on shorelines and after storm events. Large tar mats and pieces of tar pavements (or reefs) have been observed on beaches (Van Vleet et al. 1983, 1984).

   While generally hydrophobic, hydrocarbons have measurable solubility in water. Gases are the most water-soluble constituents of seeps. In most cases, dissolution accounts for only a small portion of oil seep loss but is important because some of the more soluble components of oil, particularly low-molecular-weight aromatic compounds (e.g., benzene, toluene, alkylated benzenes, and naphthalenes), are more toxic to aquatic species than aliphatic hydrocarbons (Coleman et al. 2003). Dissolution can be extensive in marine settings due to long-term exposure to seawater.

Two oxidative processes, photooxidation and biological oxidation, can alter seeps. Photooxidation includes a wide variety of light-catalyzed reactions. Photooxidation binds oxygen to carbon substrates transforming hydrocarbons into functionalized compounds such as alcohols, ketones, and organic acids that are more water-soluble than the original aliphatic hydrocarbons. If oxygen, light, and time are unlimited, the end products of photooxidation are carbon dioxide and water. Photooxidation is usually unimportant from a mass-balance consideration for seeps but may play an important role in the removal of dissolved hydrocarbons in high-light environments (e.g., on land or in shallow water). Some oxidized by-products are more toxic than precursor compounds (Coleman et al. 2003). The chemistry and extent of photooxidation of hydrocarbons can be quite complex. Its course and importance is dependent on a number of compositional and environmental variables.

5.5.3 Biochemistry

Biogeochemical processes are fundamental to, and a critical connection between, commonly expressed phenomena at petroleum seep sites. The primary effects of seeps are the introduction of reduced labile carbon as oil and gas and biological utilization of the labile carbon as an energy source. Other seep effects are those related to the toxicity of some petroleum constituents, and yet other processes involve the by-products (i.e., carbon dioxide and sulfide) and metabolites of hydrocarbon oxidation. Many of these processes are complex, unfold in a stepwise fashion with subsequent processes dependent on the preceding process, have rate-dependent or concentration-threshold limitations and often, these processes are not fully understood. These biogeochemical manifestations of oil and gas seeps have been widely used to recognize the presence of seeps in the absence of direct measurements of hydrocarbons.

A wide range of biota have the capacity to oxidize hydrocarbons, including bacteria, fungi, heterotrophic phytoplankton, and some higher organisms. There are two types of biological oxidation: metabolic detoxification after ingestion and microbial utilization. These two types have markedly differing biochemistries and end products. Metabolic detoxification of hydrocarbons by higher organisms exposed to aromatic hydrocarbons converts them to water-soluble compounds (e.g., alcohols, ketones, phenols, epoxides, and organic acids) that are excreted by the organisms as a protective mechanism. This process is biochemically complex and involves specialized enzymes (e.g., mixed function oxygenases). Not all organisms have the capacity to detoxify hydrocarbons. From a mass-balance perspective, metabolic detoxification of hydrocarbons is unimportant in removing seep hydrocarbons from the environment.

In contrast to metabolic detoxification, microbial oxidation, which occurs commonly, is important in removing hydrocarbons from the environment. Many seeps are highly altered by these processes. Microbial oxidation utilizes hydrocarbons as a carbon source to produce energy from the breaking of carbon bonds and is often referred to as *biodegradation*. Biodegradation causes two important effects: the effect of the by-products/metabolites of hydrocarbon oxidation and change in the residual oil and/or gas. As with photooxidation, the ultimate end products of biodegradation of hydrocarbons can be carbon dioxide and water, but a range of intermediates, such as organic acids, are also formed. The chemical and stable isotopic compositions of residual hydrocarbons are often altered.

Oil and gas seep environments are generally methane and sulfide rich (Joye et al. 2010). When oxygen is available, aerobic microbial oxidation can take place, and in the absence of oxygen, anaerobic microbial oxidation can occur (Figure 5.28) (Valentine and Reeburgh 2000; Coleman et al. 2003). These environmental settings have two distinct types of biogeochemistry and involve different species or consortia of bacteria. Microbial activity at seeps involves two

$$2CH_4 + 2H_2O \rightarrow CH_3CHOOH + 4H_2 \quad \text{(Methane oxidizers)}$$

$$4H_2 + SO_4^{2-} + H^+ \rightarrow HS^- + 4H_2O \quad \text{(Sulfate reducers)}$$

$$CH_3COOH + SO_4^{2-} \rightarrow 2HCO_3^- + HS^- + H^+ \quad \text{(Sulfate reducers)}$$

$$2CH_4 + 2SO_4^{2-} \rightarrow 2HCO_3^- + 2HS^- + 2H_2O \quad (Net)$$

Figure 5.28. Proposed mechanism steps 1–4 for the anaerobic oxidation of methane by archaea and sulfate reducing bacteria (Valentine and Reeburgh 2000).

primary mechanisms: hydrocarbon oxidation and sulfate reduction (Joye et al. 2010). Microorganisms oxidize methane and other hydrocarbons increasing bicarbonate (HCO_3^-) ion concentrations, which increase porewater alkalinity and enhance the precipitation of calcium carbonates and other minerals. Carbonate precipitates form crystals, nodules, and cemented sediments (hard bottom) (Boetius et al. 2000; Joye et al. 2004, 2010). Carbonate precipitation can fill sedimentary veins and fissures restricting or reducing seepage. Hydrocarbon-derived carbonates are common at active sites and are often preserved at past seep sites (Ritger et al. 1987; Greinert et al. 2001; Campbell 2006). Microorganisms reduce sulfate and produce hydrogen sulfide (H_2S) that is used as an energy source for free-living sulfur-oxidizing bacteria and the symbionts of cold-seep community megafauna (Barry et al. 1997; Fisher 1990; Levin 2005). The formation of hydrogen sulfide can also lead to the formation of elemental sulfur (S^o) and sulfur minerals such as pyrite (FeS_2) if iron is present (Sassen 1987; Sassen et al. 1988). This complex biogeochemistry produces characteristic mineral assemblages in sediments at seep sites. Microbial sulfate reduction at seeps sites can be highly temporally and spatially variable, with sites of complete depletion only meters away from sites of little depletion (Formolo and Lyons 2013). Biogeochemical sulfur cycling is complex, nonsteady state, and closely coupled with the availability of reactive iron. Pyrite is an early by-product of seepage close to the sediment–water interface. Rates of biodegradation are influenced by oxygen and nutrient concentrations; temperature, salinity, and pressure; the physical properties and chemical composition of the petroleum oxidized; and the energy level of the seep environment (e.g., replenishment of oxygen and nutrients which can be limiting and dilution and transport of by-products).

The effects of biodegradation on oil and gas chemical and stable isotopic compositions are well known. The effect of biodegradation on methane stable carbon and hydrogen isotopic compositions are illustrated in Figure 5.24. Residual methane becomes progressively enriched in ^{13}C and ^2H as the extent of degradation increases. In more complex mixtures of hydrocarbons, such as oil, the susceptibility of hydrocarbons to microbial oxidation is dependent on structure and molecular weight (Kennicutt 1988). In general, microbes oxidize small molecules with 20 carbon atoms or less before larger ones. Within the same molecular weight range, straight-chain aliphatics (normal alkanes) are more susceptible to oxidation than branched and cycloalkanes, which are more susceptible than aromatic hydrocarbons (Coleman et al. 2003). Microbes metabolize propane and n-butane more readily than other hydrocarbon gases. Thermogenic gas seeps that contain biodegraded gas are commonly depleted in propane relative to methane, and n-butane relative to isobutane. High-molecular-weight compounds are often preserved, but even these compounds can be altered if biodegradation is severe. This progressive loss of hydrocarbon types can result in residual oil containing mostly an unresolved complex mixture (UCM) of compounds in gas chromatographic analyses (Figure 5.29) (Abrams 2005; Sassen 1980; Kennicutt 1988). The UCM is largely uncharacterized but is believed to be a highly complex mixture of hundreds of ill-defined hydrocarbons. Unaltered oil seep hydrocarbons are characterized by near equal amounts of long-chained n-alkanes (C_{12+}) and the presence of isoprenoid hydrocarbons (C_{13}–C_{20} including pristane [C_{19}] and phytane [C_{20}]) (Figure 5.29). The UCM increases in prominence with degree of biodegradation because other components are preferentially removed. Gas chromatographic signatures can be overprinted by recent organic matter, and oil can be biodegraded in the subsurface prior to migration.

Worldwide, white and pigmented filamentous bacterial mats of several undescribed species of Beggiatoa have been observed at seep sites. These mats have been sampled at several petroleum seep sites in the Gulf of Mexico (Figure 5.30) (Sassen et al. 1993). Mats typically occur at the interface between reducing sediments and the overlying oxygenated water column. These mats are localized at seafloor seepage features taking advantage of the close proximity of anoxic and oxic conditions, but there is little evidence that these bacteria directly utilize hydrocarbons. Elemental sulfur (S°) is often visible within cells of Beggiatoa, and mat material is characterized by high sulfur content. Mats are part of a complex bacterial consortium in most sediments that contain oil and gas seepage. *Beggiatoa* spp. can oxidize hydrogen sulfide (H_2S) during the reduction of sulfate produced by the aerobic oxidation of hydrocarbons by other bacteria (Figure 5.31). Oxygen is depleted, and carbon dioxide produced during hydrocarbon oxidation can be incorporated during the biosynthesis of organic matter. These mats appear to retard the loss of hydrocarbons to the water column by creating a physical barrier to flow from the sediments. These processes are partly responsible for the highly biodegraded state of many seep oils.

5.5.4 Geochemistry

The mid-1980s marked the beginnings of extensive studies of unusual geologic characteristics at petroleum seep sites on the continental slope of the Gulf Mexico. Coring, manned submersibles, and remotely operated vehicles have collected seafloor samples for the investigation of the geochemistry, petrography, and structure of authigenic carbonates from the shallow slope to beyond the Sigsbee Escarpment (Figure 5.32) (Roberts and Aharon 1994; Roberts et al. 2009, 2010; Feng et al. 2010). These studies determined the geochemical origins of anomalously high seafloor reflectivity as being primarily due to lithification of sediments by authigenic carbonates (Roberts et al. 2009). Microbe-generated carbon dioxide initiates a cascade of microbe-mediated chemical reactions, including the precipitation of authigenic minerals that produce unique morphologies, mineralogies, and critical habitat for cold-seep fauna at petroleum seep sites.

The stable carbon and oxygen isotopic compositions of carbonates at seep sites confirm their close association with hydrocarbon-sourced carbon dioxide (Roberts et al. 2009, 2010a, b; Feng et al. 2010). The origin of the carbon dioxide incorporated into carbonates is indicated by $\delta^{13}C$ values and $\delta^{18}O$ values, which reflect the temperature and fluid source of the carbonates. Potential sources of carbon at seep sites include biogenic and thermogenic gases, oil, seawater carbon dioxide, and methanogenesis (Roberts et al. 2010). Seep carbonate stable isotopic compositions are highly variable indicating various admixtures of these multiple sources (Figure 5.33) (Roberts et al. 2010). Seep site carbonates are generally depleted in ^{13}C and enriched in ^{18}O compared to deep-sea carbonate minerals formed from seawater carbon dioxide. ^{18}O is enriched in gas hydrate, and the anomalously positive $\delta^{18}O$ of some seep carbonates suggests an origin related to the decomposition of gas hydrate. Unusually low

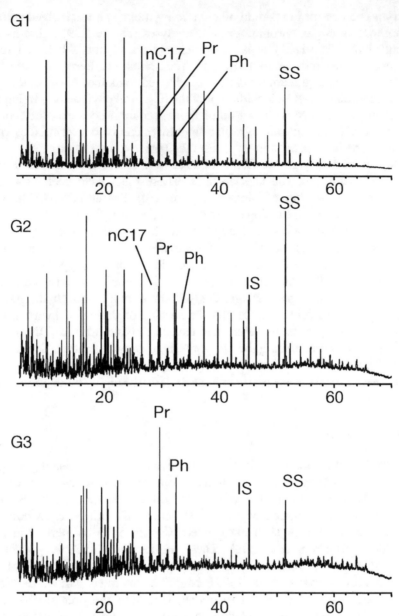

Figure 5.29. Oils of different biodegradation levels (G1, G2, and G3) showing systematic relative removal of *n*-alkanes (e.g., nC17) relative to the isoprenoid alkanes pristane (Pr) and phytane (Ph) (spiked standards are IS and SS) (reprinted by permission from Macmillan Publishers Ltd: Nature [Jones et al. 2008] ©2008). [Times on the chromatograms are displayed from 5 to 70 min from time of injection.]

$\delta^{18}O$ values (as low as 2.4 ‰ Standard Mean Ocean Water) are likely related to the expulsion of warm fluids at the seafloor during rapid flux events, though other processes also affect carbonate stable isotopic compositions (Bohrmann et al. 1998; Greinert et al. 2001; Sassen et al. 2004: Hesse 2003).

A study of the characteristics of about 100 seep-related carbonate rocks collected on the continental slope of the Gulf of Mexico revealed that the rocks were mostly high in Mg-calcite

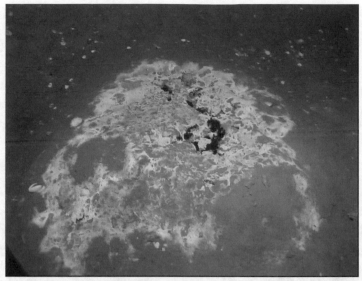

Figure 5.30. Seafloor gas-hydrate mound, seeping oil and brine and associated *Beggiatoa* microbial mat in the north-central Gulf of Mexico (Image courtesy of R. Weiland, BP America).

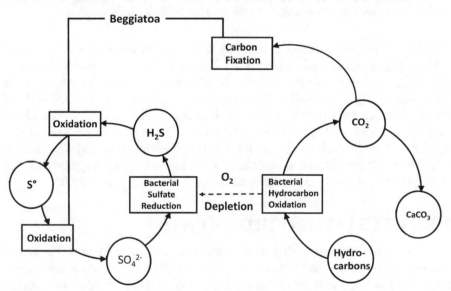

Figure 5.31. Summary of the proposed geochemical role of *Beggiatoa* in cold hydrocarbon seeps (reprinted from Sassen et al. 1993, with permission from Elsevier).

and aragonite and also contained significant amounts of dolomite (Roberts et al. 2009; Feng et al. 2010). These rocks have a range of morphologies including nodules, chimneys, slabs, blocks, mounds, and irregular-shaped aggregations (Figure 5.34) (Roberts et al. 2009).

The local chemical environment and the rate of hydrocarbon seepage control carbonate rock formation. Carbonate nodules of up to about 2 centimeters (cm) (0.8 inches [in.]) in diameter scattered throughout sediments are often incorporated into composite aggregates, occur in association with mussel and tubeworm communities, and form deep in sediments most likely in response to slow hydrocarbon flux rates (Figure 5.34). Chimneys as long 50 cm (19.7 in.)

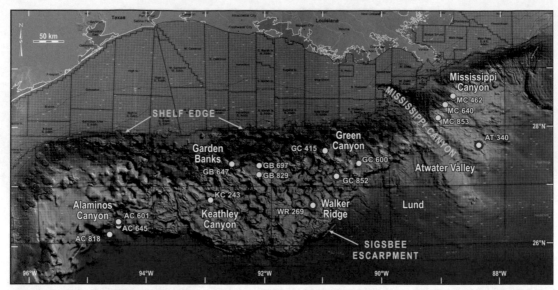

Figure 5.32. Sites on the northern Gulf of Mexico continental slope where manned submersible and remotely operated vehicle dives have been made and seep carbonates were collected. These collections have been extensively supplemented by coring at many other sites across the northern Gulf of Mexico (Roberts et al. 2009; republished by permission of the Gulf Coast Association of Geological Societies, whose permission is required for further publication use).

can occur as broken pipe-like shapes protruding from muddy sediments and may form due to focused vertical migration of hydrocarbon-rich fluids possibly associated with animal burrows. Slabs can be developed with rough surfaces, sometimes multiple layers and mostly composed of aragonite, suggesting precipitation from sulfate-rich porewaters. Carbonate blocks can be heterogeneous and up to several meters in diameter and length and contain void-lining aragonite-splay cements and brecciated structures of unknown origin (possibly related to abrupt expulsion of hydrocarbons or the decomposition of gas hydrate). Carbonate rocks can be rich in mussel and clam shells and contain relics of burrowing activity and occasionally have iron and manganese coatings (Callender et al. 1990, 1992; Feng et al. 2010).

5.6   TERRESTRIAL ENVIRONMENTS

Seeps on land can increase hydrocarbon concentrations in soils and the overlying atmosphere; enhance microbiologic activity; introduce minerals (such as uranium creating radiation anomalies); form calcite, pyrite, elemental sulfur, magnetic iron oxides, and sulfides; bleach red beds; alter clay minerals; affect soil electrochemical properties; and modify biogeochemical and geobotanical processes (Schumacher 2012). Liquid seepage is adsorbed onto soils, while gas seepage can move, mostly unaltered, directly into the atmosphere. Due to the relative lack of water at land sites of seepage compared to marine environments, emulsification is unimportant, but photooxidation can occur on direct exposure to sunlight. The geochemistry of mineral formation mediated by microbiota is similar to that described for marine settings and can encapsulate gases and liquids in mineral interstices. Due to changes in the chemistry of soils and the toxicity of some components of petroleum, land seeps can affect surrounding vegetation health and composition. Soil and air have been analyzed to detect seep-induced surface anomalies on land. Techniques also have been developed to detect these changes using airborne and satellite imagery, spectral reflectance, and other sensors. These surveys typically map

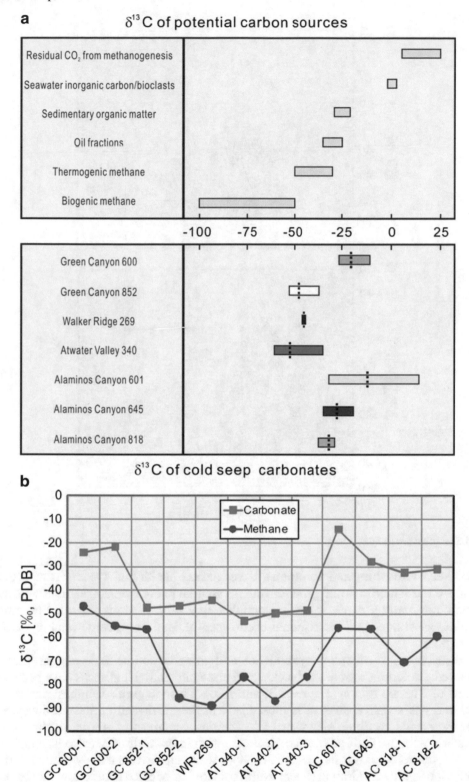

Figure 5.33. Comparisons of Gulf of Mexico $\delta^{13}C$ and $\delta^{18}O$ values of: (a) potential carbon sources and carbonates collected in >1,000-m water depth, (b) carbonates collected in <1,000-m depth, (c) >1,000-m water depth, and (d) seep carbonates and methane from sediment cores from the same sites (Roberts et al. 2010 [reproduced with permission of PERGAMON via Copyright Clearance Center, Inc.] and references therein).

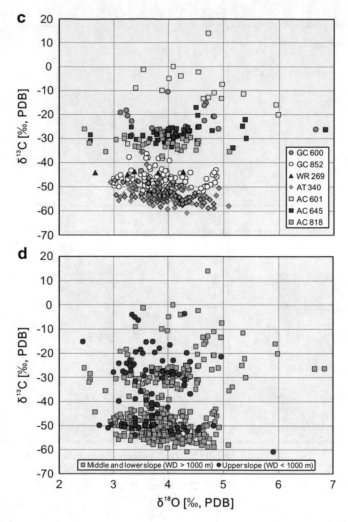

Figure 5.33. (continued)

suspected seep indicators and variations in vegetation health and types. There are various limitations to these methods and ground-truth is essential to confirm correlations with seepage. While there are many prospect-specific examples, few surveys are in the open literature that would allow assessment of the regional occurrence of land petroleum seeps in the Gulf of Mexico region.

The distribution of oil and gas production and potential source rocks suggest that numerous onshore petroleum seeps are to be expected and many individual seeps have been reported in the Gulf of Mexico region (Figures 5.35 and 5.36). These maps also suggest that geographically, oil and gas seepage on land is most likely in the north-central and northwestern Gulf coast. These trends continue southward into the onshore areas of northern Mexico and the southwestern offshore Gulf of Mexico. Sedimentation in the Gulf of Mexico basin is known to have been asymmetrical over its geological history with major rivers mainly located in the north (one exception is the Rio Grande River). Northern Gulf of Mexico source-rock horizons pinch-out toward the center of the basin. However, onshore from the southwestern Gulf of Mexico

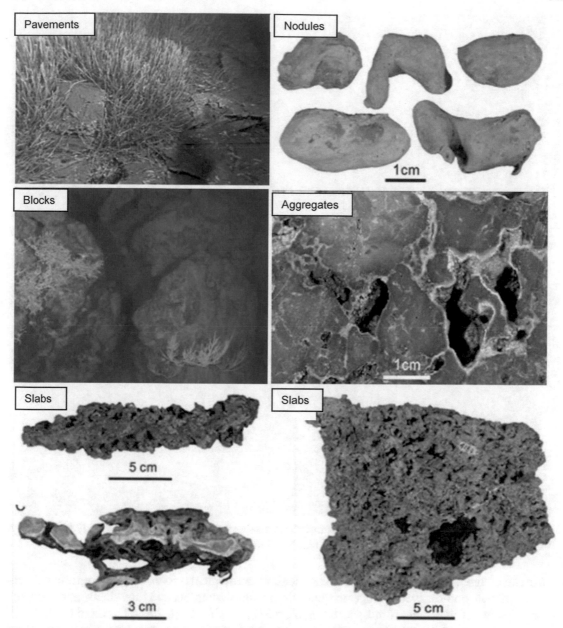

Figure 5.34. The range of morphologies of carbonate rocks collected at seep sites on the continental slope of the northern Gulf of Mexico (Feng et al. 2010; republished by permission of the Gulf Coast Association of Geological Societies, whose permission is required for further publication use).

and along the Campeche peninsula, major oil and gas discoveries have been made, salt basins occur, and petroleum seeps have been reported (Figures 5.21, 5.22, and 5.23).

5.7 MARINE ENVIRONMENTS

In marine environments, petroleum seeps interact with the surrounding environment creating, a range of associated phenomena (Kennicutt et al. 1987, 1988a). These phenomena can, but do not always, include bubble streams, acoustic plumes, hydrocarbon concentration

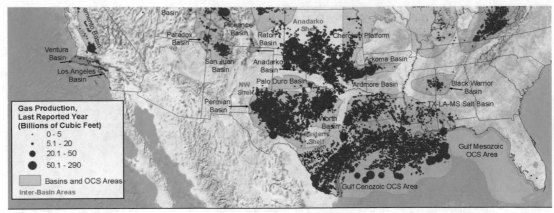

Figure 5.35. Map of Gulf coast onshore and offshore natural gas production in 2009 (Energy Information Administration, 2009 http://www.eia.gov/oil_gas/rpd/conventional_gas.jpg).

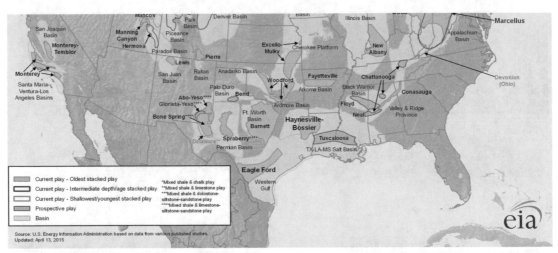

Figure 5.36. Map of potential onshore subsurface shale gas and oil sources (Energy Information Administration, 2015; http://www.eia.gov/oil_gas/rpd/shale_gas.jpg).

anomalies, encircling features (visual, chemical, mineralogical, biological), topographical features, hydrocarbon-derived authigenic minerals, bacterial mats, sea-surface slicks, and methane anomalies in the overlying atmosphere (Figure 5.37) (Foucher et al. 2009; Hovland et al. 2012). The processes manifested at seep sites are controlled by the type and rate of gas and fluid expulsion (Foucher et al. 2009). The presence of overlying seawater, and its dissolved inorganic constituents (i.e., sulfate), has distinct and important effects on the behavior and fate of petroleum seeps in marine environments (see Sections 5.5.3 and 5.5.4).

5.7.1   Sea-Surface Slicks and Water Column Plumes

In the 1960s and 1970s, several authors reported bubbles of gas rising to the surface in the Gulf of Mexico and the first water column concentrations were measured (Bernard et al. 1976; Brooks et al. 1974, 1979; Frank et al. 1970; Geyer 1980; Geyer and Giammona 1980; Swinnerton and Linnenbom 1967). At this time, others reported that bubbles rising from gas seeps could be detected by standard sonar equipment (Figure 5.38) (Pickwell 1967; McCartney and Bary 1965).

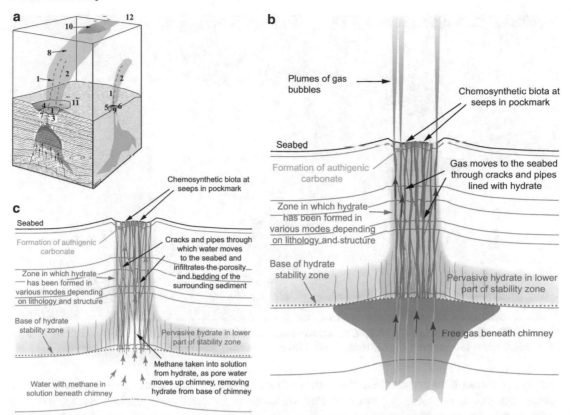

Figure 5.37. Possible processes at an active seep: (a) (1) bubble streams, (2) acoustic plumes, (3) elevated hydrocarbon concentrations, (4) encircling features (visual, chemical, mineralogical, biological), (5) topographic effects, (6) authigenic minerals, (7) bacterial mats, (8) upwelling seawater, (9) entrainment of seawater, (10) sea-surface slicks, (11) attraction of fish and other macrofauna, (12) methane anomalies in atmosphere (not all of these effects occur at all seeps, (reprinted from Hovland et al. 2012 with permission from Elsevier). (b) A pock mark and underlying chimney during active expulsion of free gas and progressive formation of gas hydrate and authigenic carbonates and (c) a pockmark when only methane saturated porewater is migrating through the system and gas hydrate is forming in the underlying sediments (not all of these effects occur at all seeps, Foucher et al. 2009).

Frank et al. (1970) suggested that surveying and mapping of concentrations of hydrocarbons in offshore, near-bottom waters could be used to detect oil and gas seeps. Gas seepage samples were collected at the sea surface by snorkel diving (Bernard et al. 1976). Eleven of the 14 samples collected were mostly methane of microbial origin, but three samples contained significant amounts of ethane and propane of thermogenic origin (Figure 5.39) confirming earlier reports by Brooks et al. (1974). The two gas seeps in this earlier study were 100 % methane, and they were determined to be microbiological in origin based on stable carbon isotopic analysis. These were some of the first studies to use molecular and stable isotopic compositions to differentiate microbial and thermogenic gas seepage in marine environments.

Hydrocarbon gas distributions in seawater have been surveyed by ships that deploy equipment, collectively called *sniffers*, to pump seawater to the surface for analysis. These techniques have found wide use in oil and gas exploration and in the Gulf of Mexico since the 1960s (Dunlap et al. 1960; Lamontagne et al. 1973, 1974; Bernard et al. 1976; Brooks and Sackett 1973; Sackett and Brooks 1973; Sackett 1977). Hydrocarbon sniffers consist of a gas extraction

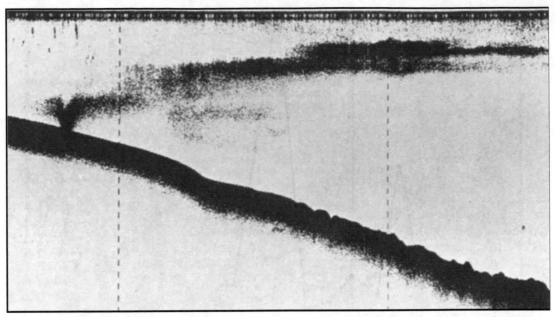

Figure 5.38. Echo sounder evidence of a naturally occurring petroleum seep in the Gulf of Mexico (reprinted from Geyer and Giammona 1980 with permission from Elsevier).

system, adsorbents to concentrate the hydrocarbons, and a gas chromatograph equipped with a flame ionization detector to separate and measure individual hydrocarbon gases. Modern sniffers employ real-time, hydrocarbon detection systems based on various concepts and are deployable on remotely operated and autonomous vehicles. Some sniffers can detect gaseous and liquid hydrocarbons and some have used low-flying airplanes to detect methane in the air overlying the ocean (and land); the use of sniffers in drones has been proposed. Most of the methane detected in the Gulf of Mexico water column is of recent microbiological origin.

When seeping gases and liquids survive transport through the water column, sea-surface slicks are formed above sea-bottom seeps, and on occasion, gas bubbles and oil droplets can be seen bursting at the surface (Figure 5.40) (Sassen et al. 2001a). Removal processes include dissolution in seawater and microbial oxidation. Liquids can become adsorbed on and commingled with particles in the water column (organic and inorganic) and may return to the seafloor (sometimes distant from its origins depending on oceanic currents) once particle density exceeds that of seawater. Gas bubbles rise more rapidly in a water column than do oil droplets, potentially leading to a fractionation of the seeping petroleum (Figure 5.41). Seepage water column plumes are readily detected by sea-bottom acoustic profilers, especially when gases are at concentrations high enough to form bubbles (Figure 5.38). Gaseous bubbles often entrain liquid petroleum creating sea-surface slicks. Most oil slicks form when oil droplets reach the surface and sometimes form pancakes and coalesce. Plumes have been sampled in situ with standard oceanographic water samplers.

Observations of sea-surface slicks were significantly expanded with the advent of remote-sensing (satellite) techniques (Figures 5.42 and 5.43) (MacDonald et al. 1993, 1996; MacDonald (1998); De Beukelaer 2003, De Beukelaer et al. 2003; NASA 2000). Various techniques, such as Landsat Thematic Mapper and Synthetic Aperture Radar, have been used to image hydrocarbon slicks on the sea surface. Hydrocarbons on the sea surface dampen ripples and reduce the reflectivity of water, so that slicks appear as dark patterns on the sea surface in satellite images. Remote-sensing techniques are efficient and repeatable and can cover large areas of the sea

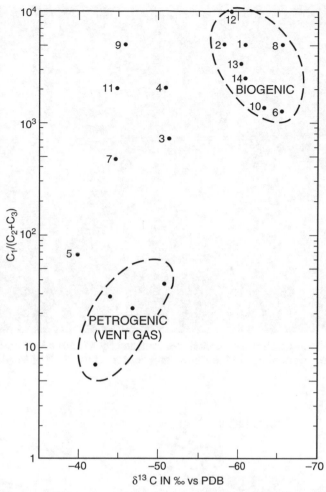

Figure 5.39. Molecular and stable carbon isotopic compositions of microbial and thermogenic seep hydrocarbon gases in the Gulf of Mexico water column (reprinted from Bernard et al. 1976 with permission from Elsevier). (C_1—methane, C_2—ethane, C_3—propane, PDB—Pee Dee Belemnite).

surface. The use of several independent detection techniques and multiple images over time increases confidence in differentiating surface petroleum slicks from other phenomena that create ephemeral organic oil films that form slicks, such as phytoplankton blooms.

5.7.2 Seafloor Sediments

The first reports of retrieval of oil-stained seafloor sediments in the Gulf of Mexico began appearing in the literature in the 1980s. Anderson et al. (1983) noted high concentrations of biodegraded oil, carbonate deposits, and organic sulfur in north-central Gulf of Mexico continental slope sediments recovered by coring. Chemical and stable carbon isotopic compositions indicated that the observed high concentrations of methane to pentane must have been produced thermally at depth beneath the seafloor and had reached the surface through faults and fractures associated with salt diapirs. The authors also noted anomalous seismic reflections that suggested the presence of gas hydrate. Since this first report, the seafloor of the Gulf of

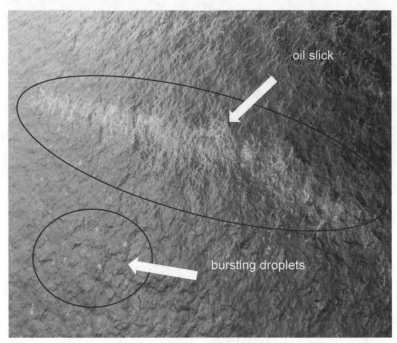

Figure 5.40. Aerial photograph of an oil slick surrounded by a field of individual droplets bursting at the surface to form "pancakes" in the deepwater region of the Gulf of Mexico (no attribution of photo).

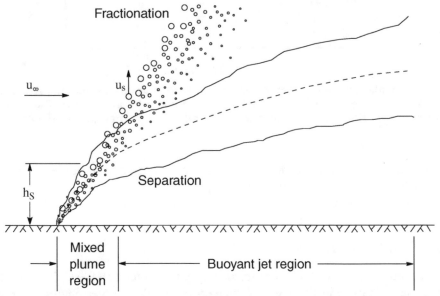

Figure 5.41. A bubble plume in a cross-flowing current. Gas bubbles separate from oil droplets as both rise through the water column (reprinted from Socolofsky and Adams 2002, ©International Association for Hydro-Environment Engineering and Research with permission of Taylor & Francis Ltd, www.tandfonline.com, on behalf of International Association for Hydro-Environment Engineering and Research). [h_s—separation height, U_∞—cross flow velocity, U_s—slip velocity].

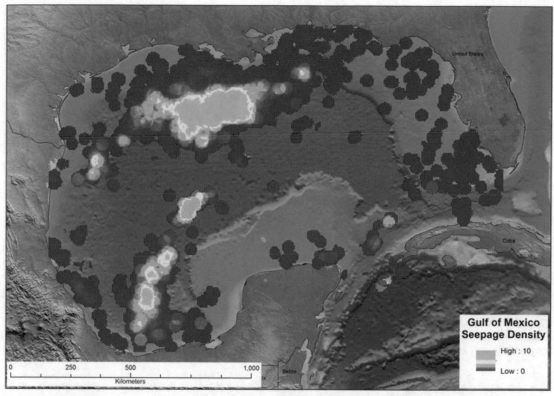

Figure 5.42. Oil and gas seepage in the Gulf of Mexico (determined from analysis of synthetic aperture radar, graphic provided by CGG's NPA Satellite Mapping, used with permission).

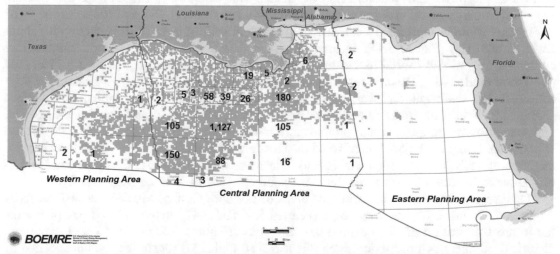

Figure 5.43. Number of persistent sea-surface slicks in lease block areas from 1991 to 2009. *Green squares* were active leases in 2011 (base map BOMERE http://img.docstoccdn.com/thumb/orig/ 79768556.png); sea slick data (determined from analysis of synthetic aperture radar, graphic provided by CGG's NPA Satellite Mapping, used with permission).

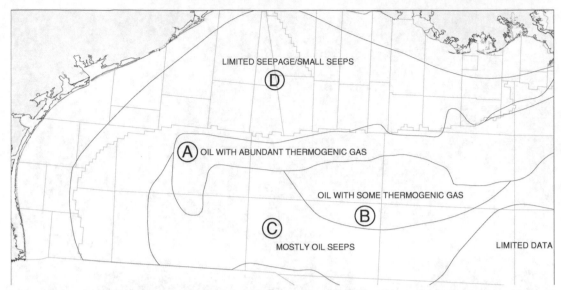

Figure 5.44. Gulf of Mexico seep distributions map based on 5,200 sea-bottom drop cores and sea-surface slicks identified by remote sensing: (a) abundant macroseeps of oil and thermogenic gas, (b) abundant macroseeps of oil with less thermogenic gas, (c) abundant macroseeps of limited thermogenic gas, and (d) limited microseepage (Hood et al. 2002; AAPG©2002, reprinted by permission of the AAPG whose permission is required for further use).

Mexico has been extensively sampled by various coring devices from ships, remotely operated vehicles, and manned submersibles.

Hood et al. (2002) used data from 3,000 sea-bottom cores to extend hydrocarbon-system maps and predictions of hydrocarbon type and properties beyond data based on subsurface cores. Regions of the northwestern Gulf of Mexico were categorized based on the distribution, type, and intensity of oil and gas seeps (Figure 5.44). In areas (A), (B), and (C) nearly 75 % of sea-bottom samples contained moderate or substantial quantities of oil compared with about 12 % in area (D) (Figure 5.44). In area (A) more than 25 % of the seafloor samples that contained oil had substantial associated thermogenic gas compared to less than 5 % in area (C) (Hood et al. 2002). Seismic records confirmed the associations (migration paths) between source rocks, shallower reservoir rocks, subsurface salt tectonics and faults, near-surface sedimentary wipe-out zones (gas-charged sediments or chimneys), pathways to the seabed, and seabed morphologies such as mounds and mud volcanoes (Figures 5.45 and 5.46) (Hood et al. 2002). Fisher et al. (2007) confirmed the locations of macroseepage in the Gulf of Mexico by mapping seabed cores with unambiguous indications of oil and gas in the upper 5 m (16.4 ft) of sediment, including a few locations in the southern Gulf of Mexico (Figure 5.47).

On a regional basis, a comparison of maps of northern Gulf of Mexico sea-surface slicks (Figures 5.42 and 5.43), seafloor seeps (Figures 5.44 and 5.47), active oil and gas platforms (an indirect indication of deep oil and gas reservoirs) (Figures 5.48 and 5.49), and maximum historical oil and gas production rates (Figures 5.50 and 5.51) illustrates the coincidence of petroleum seeps and deeply buried oil and gas source rocks and accumulations. Of particular note is the geographical coincidence of abundant oil and gas seepage areas (Figures 5.44 and 5.47) and the locations of deepwater oil and gas wells.

Petroleum seepage into seafloor sediments is manifested in a variety of characteristic mineral assemblages and morphologies (Figure 5.37) (Hovland et al. 2012; Boetius and Wenzhofer 2013 and others). Mineralogical changes are closely coupled with microbiological activity

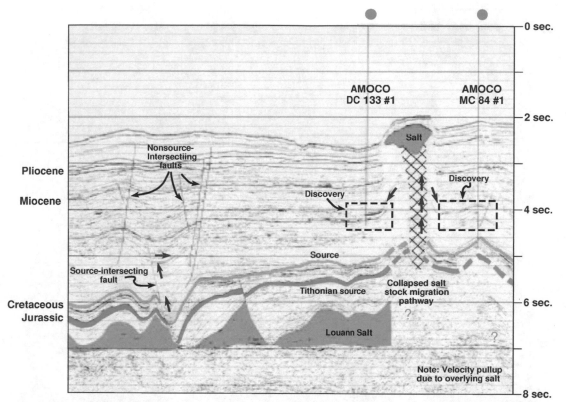

Figure 5.45. Example of a seismic line showing the source intervals and several potential hydro-carbon migration pathways (Hood et al. 2002; AAPG©2002, reprinted by permission of the AAPG whose permission is required for further use).

that produces excess carbon dioxide and bicarbonate ion favoring the formation of calcium carbonate and other authigenic minerals (see Sections 5.3.3 and 5.3.4). Mud volcanoes are large seabed structures (diameters of 1–10 kilometers [km]; 0.6–6.2 miles [mi]) formed by gas, pore fluid, and mud eruptions with a centrally pointed, flat, or crater-like top (Figure 5.52) (Prior et al. 1989; Milkov et al. 2003). Smaller depressions can form in the seafloor due to gas eruptions (10–1,000 m; 32.8–3,280.8 ft) called *pockmarks* (Figures 5.53 and 5.54) (Foucher et al. 2009). Structures below the seafloor can extend kilometers acting as migration pathways for seepage and are called *gas chimneys* (Foucher et al. 2009). Gas hydrate can form mounds as it expands and accumulates (Fisher et al. 2007; Boetius and Suess 2004). High-energy releases of petroleum can result in the formation of emulsions that mix water and sediments with seeping fluids. This often results in seepage being retained in sea-bottom sediments due to the increased density of the mixture; therefore, not all seafloor seeps result in sea-surface slicks.

High seafloor reflectivity or amplitude responses and acoustic wipe-out zones are caused by the influx of gases and liquids to the seabed, seafloor lithification, physical disruption of internal sediment layering, and gas-hydrate formation and decomposition (Roberts and Aharon 1994, Roberts et al. 1990, 1992, 2007, 2010; Gay et al. 2011). Since 1998, the Bureau of Ocean Energy Management (BOEM) has mapped over 31,000 seafloor acoustic amplitude anomalies in the deepwater northern Gulf of Mexico using 3D time-migrated seismic surveys (Figure 5.55) (Shedd et al. 2012).

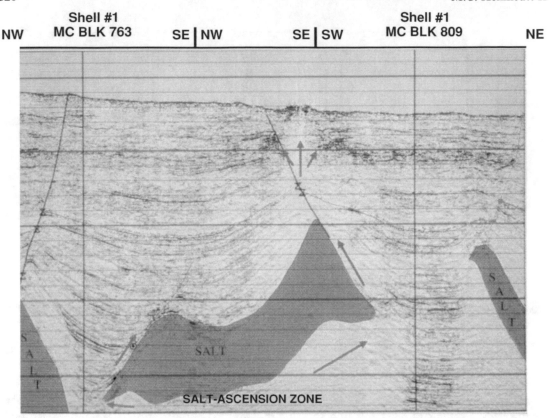

Figure 5.46. A seismic line showing a cross-strata migration pathway to a discovery in the north-central Gulf of Mexico. Hydrocarbon migration occurs up the collapsed salt stock along the salt–sediment interface into smaller faults that reach the seafloor (Hood et al. 2002; AAPG©2002, reprinted by permission of the AAPG whose permission is required for further use).

5.7.3 Gas Hydrate

Gas hydrates are the largest accumulations of natural gas on Earth and are known to form in outer continental margin sediments and permafrost (Figures 5.56, 5.57 and 5.58) (Collett et al. 2009; Pinero et al. 2013). In the 1930s, gas hydrates were suspected of causing blockages in pipelines but remained unknown in nature until the 1980s when deep-sea drilling recovered cores containing intact gas hydrate from the outer continental shelf/slope, hundreds of meters below the seafloor (Collett et al. 2009). Pinero et al. (2013) noted that gas hydrates "... have been recovered in more than 40 regions worldwide and their presence has been deduced from geophysical, geochemical, and geological evidences at more than 100 continental margin sites." In recent years, near-surface gas hydrate has been recognized as a novel source of gas seepage at some locations, including the Gulf of Mexico.

Most low-molecular-weight gases, including methane, form hydrates when high interstitial porewater concentrations occur at suitable temperatures and pressures. Gas hydrates are crystalline water-based solids that resemble ice, in which small nonpolar molecules (typically gas compounds), or polar molecules with large hydrophobic moieties, are trapped inside "cages" of hydrogen-bonded water molecules (Figure 5.59) (Collett et al. 2009; Boswell et al. 2012). Without the support of the trapped molecules, the lattice structure collapses into ice crystals or liquid water. Gas-hydrate decomposition is a phase change, not a chemical

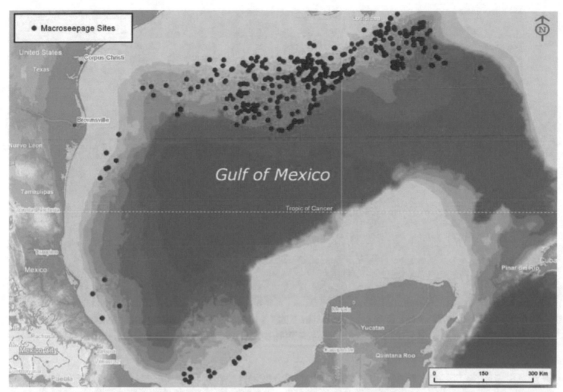

Figure 5.47. Gulf of Mexico locations (*red dots*) of piston cores that have significant levels of oil (greater than 300,000 total Scanning Fluorescence maximum intensity units) or methane (greater than 100,000 parts per million [ppm]) in the top 5 m of sediment (Fisher et al. 2007; Courtesy of TDI-Brooks International).

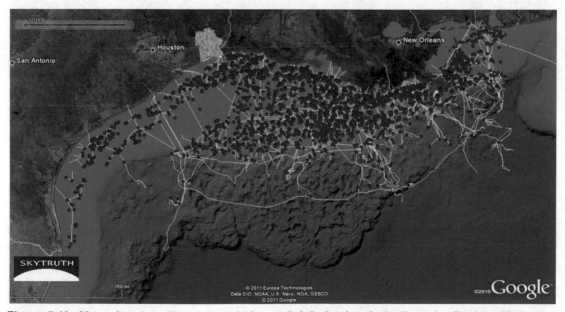

Figure 5.48. Map of active oil and gas platforms (*pink dots*) and pipelines (*yellow*) in 2011, map available at http://blog.skytruth.org/2011/04/gulf-of-mexico-deepwater-development.html. (For gas production map see Fig. 5.35).

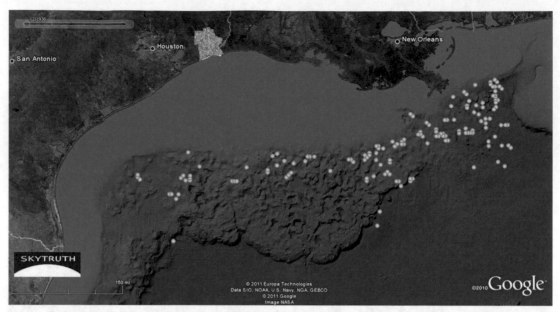

Figure 5.49. Deepwater oil and gas wells in 2011 (*yellow dots*) map available at http://blog.sky truth.org/2011/04/gulf-of-mexico-deepwater-development.html.

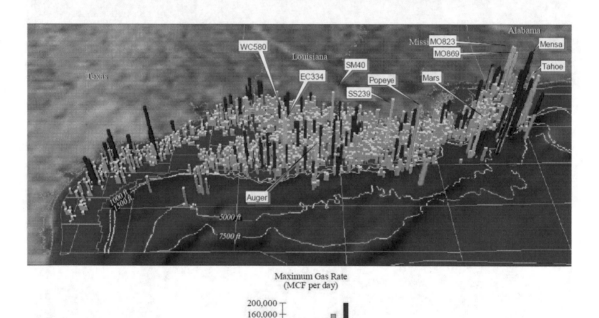

Figure 5.50. Maximum historical gas production rates for Gulf of Mexico wells (U.S. Department of the Interior, Minerals Management Service, Gulf of Mexico OCS Region, 2002; map available at http://www.geographic.org/deepwater_gulf_of_mexico/production_rates.html).

reaction. Recently, the concept of a gas-hydrate petroleum system has been proposed with similar requirements as conventional petroleum systems including (1) gas-hydrate pressure-temperature stability conditions, (2) a source of gas, (3) available water, (4) pathways for gas

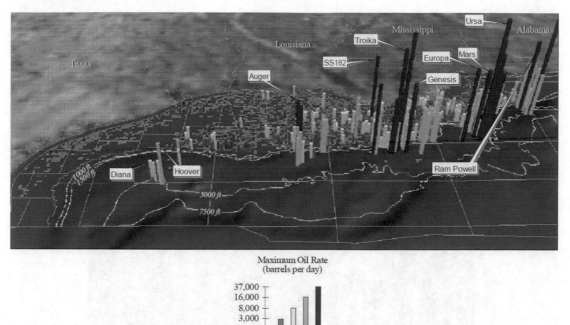

Figure 5.51. Maximum historical oil production rates for Gulf of Mexico wells (U.S. Department of the Interior, Minerals Management Service, Gulf of Mexico OCS Region, 2002; map available at http://www.geographic.org/deepwater_gulf_of_mexico/production_rates.html).

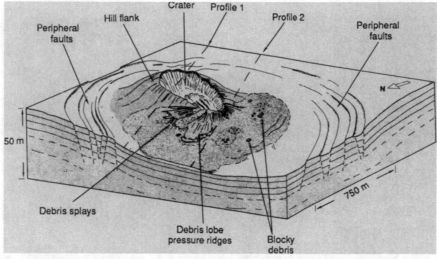

Figure 5.52. Block diagram of a seabed crater in the north-central Gulf of Mexico (Prior et al. 1989; reprinted with permission from the American Association for the Advancement of Science).

migration, (5) a suitable host sediment or reservoir, and (6) the requisite timing among system elements (Collett et al. 2009). A global inventory of methane in gas hydrates has recently been estimated based on theoretical considerations (Figure 5.57) (Pinero et al. 2013).

Until the 1980s, gas-hydrate deposits were believed to occur deep in the subsurface as inferred from seismic records based on bottom simulating reflectors, indicating a phase change

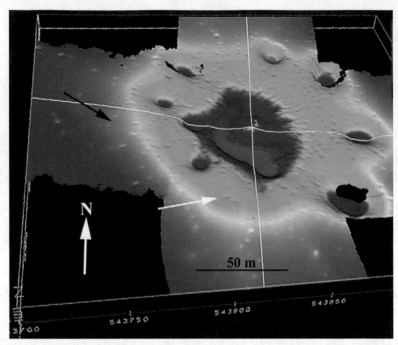

Figure 5.53. Perspective view of a shaded relief digital terrain model from one of a series of pockmarks on the Norwegian seafloor (Hovland et al. 2010; reprinted with permission from Elsevier).

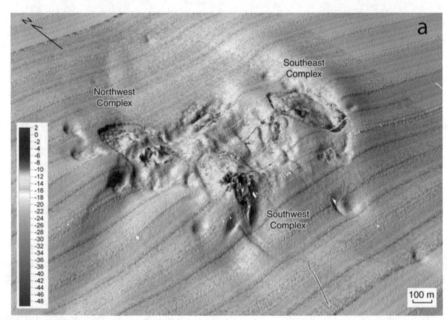

Figure 5.54. Sea-bottom morphological features at a mound in north-central deepwater region of the Gulf of Mexico (bathymetry overlain by acoustic backscatter data) with a scale showing elevation (Macelloni et al. 2010; republished by permission of the Gulf Coast Association of Geological Societies, whose permission is required for further publication use).

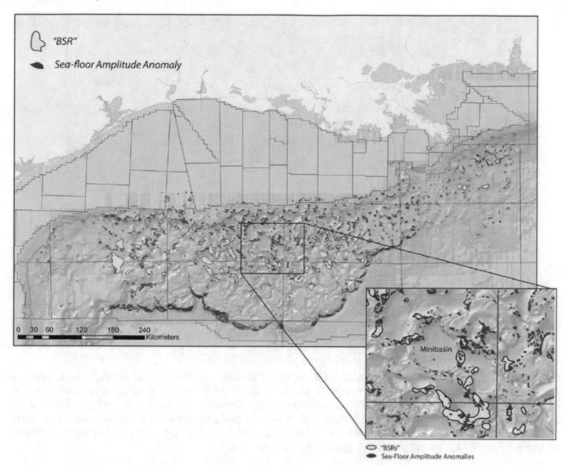

Figure 5.55. Map showing seafloor bathymetry (*gray*), seafloor seismic amplitude anomalies (*red*), and mapped bottom simulating reflectors (BSRs) (*yellow*) (Shedd et al. 2012, reprinted with permission from Elsevier; BOEM, http://www.boem.gov/Seismic-Water-Bottom-Anomalies-Map-Gallery/).

from gas to solid hydrate, which crosses subsurface strata (e.g., simulates the seafloor surface). Analyses also indicated that the gas in deep hydrates was solely methane of recent microbiological origin. This view changed when, for the first time, core samples of surface sediments in the Gulf of Mexico recovered thermogenic gas hydrates (Brooks et al. 1984, 1986, 1994) that contained substantial amounts of thermogenic methane and higher-molecular-weight hydrocarbon gases. Following the Gulf of Mexico discoveries, gas hydrates have been recovered from surface sediments cores in the Cascadia continental margin of North America, the Black Sea, the Caspian Sea, the Sea of Okhotsk, the Sea of Japan, and the North and South Atlantic Ocean (Collett et al. 2009).

Seeping hydrocarbon gases can crystallize as gas hydrate in layers, as nodules, and as exposed mounds and vein fillings in sediments (Figure 5.58) (Sassen et al. 2001a, b). Biogenic gas hydrate is white, while thermogenic gas hydrate can be stained with oil or encrusting bacteria giving the hydrate a yellow to orange color (Figure 5.58). Gas-hydrate-derived seepage is largely restricted to occurrences at shallow depths in sediments or outcroppings on the seafloor. In general, gas hydrates tend to accumulate in near-surface sediments, not decompose (Sassen et al. 2001c, 2004). It has been suggested that warmer bottom water temperatures in the

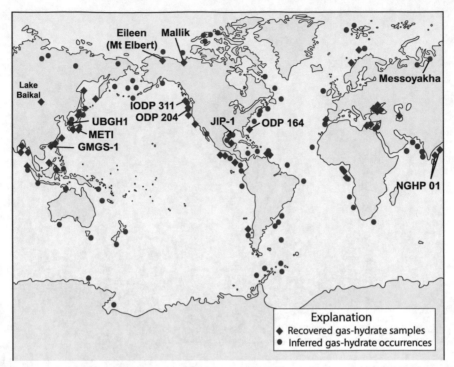

Figure 5.56. Location of sampled and inferred gas-hydrate occurrences in oceanic sediments of outer continental margins and permafrost regions. Most of the recovered gas-hydrate samples have been obtained during deep coring projects or shallow seabed coring operations. Most of the inferred gas-hydrate occurrences are sites at which bottom-simulating reflectors have been observed on available seismic profiles (Collett et al. 2009; AAPG©2009, reprinted by permission of the AAPG whose permission is required for further use).

ocean can initiate seepage from gas hydrate, which has generated interest in the stability and contribution of gas hydrate to atmospheric greenhouse gases (Sassen et al. 2004). Methane gas hydrate can occur as three different crystalline structures, and all have been observed in nature (Figure 5.59) (Brooks et al. 1984; Sassen and MacDonald 1994; Sassen et al. 2000).

Gas hydrates have been recovered at many sites on the continental slope and abyss in the Gulf of Mexico over the past three decades (Figure 5.60) (Boswell et al. 2012). The first documented physical retrieval of gas hydrate in the shallow subsurface of the Gulf of Mexico was in a Deep-Sea Drilling Project core from the Orca Basin in 1983 when small crystals were determined to be biogenic methane hydrates (DSDP96 in Figure 5.60) (Pflaum et al. 1986). As noted above, in 1984 the first retrieval of near-surface thermogenic gas hydrate in nature was reported in the deepwater region of the Gulf of Mexico (GC 185, Figure 5.60) (Brooks et al. 1984). Thermogenic gas hydrate was recovered from the upper few meters of bottom sediments associated with oil-stained cores in a water depth of 530 m (1,739 ft) close to the limit of gas-hydrate stability. Gas hydrate occurred sporadically associated with sediment seismic wipe-out zones within an area of at least several hundred square kilometers (100 km^2 = 38.6 mi^2). In 1994, the first recovery of structure H gas hydrate in nature was reported in the deepwater region of the Gulf of Mexico (Sassen and MacDonald 1994). In the following years, gas hydrates were shown to be associated with vents, carbonate hard grounds, and shallow fault systems at the margins of salt structures, and hydrate gases were correlated with deeper-reservoired gases (Brooks et al. 1986; Sassen et al. 1999a, b; Milkov and Sassen 2000, 2001; Milkov et al. 2000). In subsequent years, gas hydrate has been recovered many

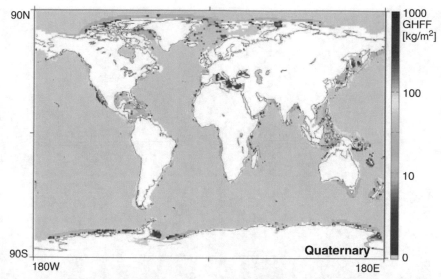

Figure 5.57. Global distribution and estimated quantities of methane gas hydrates based on theoretical steady-state considerations of control parameters (such as sediment organic carbon content, porosity, T/P, heat flow, fluid advection, and others). The "real" gas-hydrate distribution is expected to have a more patchy distribution, and these values are considered minimum estimates (Pinero et al. 2013; Creative Commons Attribution 3.0 License).

times by coring and submersible sampling of the seafloor, and region-wide in-place gas in the form of gas hydrate has been estimated (Figure 5.61) (Frye 2008).

5.7.4 Cold-Seep Communities

In the deep sea, a highly specialized ecology has developed that thrives and depends on petroleum seeps. The discovery of hydrothermal-vent communities in 1977 marked a major change in the understanding of life on Earth and how it might have evolved (Ballard 1977). Until this discovery, it was believed that the primary source of energy available to support life in the oceans was the sun through the process of photosynthesis. Deep-sea hydrothermal-vent communities are supported by alternative sources of energy from reduced chemicals escaping at the deep seafloor. The biological conversion of one or more carbon molecules (usually carbon dioxide or methane) and nutrients into organic matter using the oxidation of inorganic molecules (e.g., hydrogen gas, hydrogen sulfide) or methane as a source of energy is known as *chemosynthesis*. Shortly after these deep-sea discoveries, similar assemblages of organisms were recovered on the continental slope of the north-central Gulf of Mexico at a petroleum seep and at a brine seep at the base of the escarpment off the shore of western Florida (Kennicutt et al. 1985; Paull et al. 1984, 1985; Brooks et al. 1987a, b; Brooks et al. 1989). These unique biological assemblages have become known as cold-seep communities as contrasted to hydrothermal-vent communities.

Since the 1980s cold-seep communities have been discovered worldwide in locations with sufficient inorganic substrates to support life, including the Atlantic Ocean, the Pacific Ocean, the Mediterranean Sea, and recently, Antarctica. The common feature among cold-seep sites is the presence of hydrogen sulfide and methane and an interface with oxygenated water (Figure 5.62). Most sites, outside of the Gulf of Mexico, are associated with microbial methane seeps, gas seeps from gas hydrate, and brine seeps; thermogenic hydrocarbons are mostly absent. Cold-seep communities occur on the ocean's margins in areas of high primary

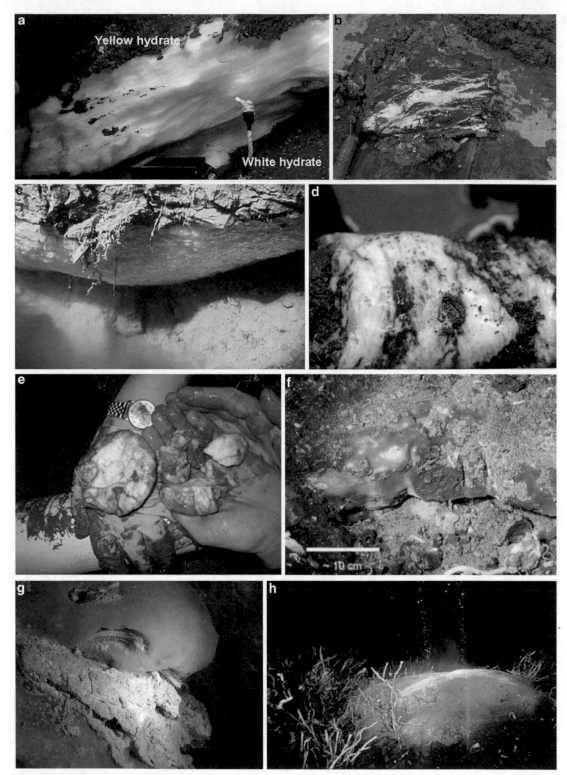

Figure 5.58. Examples of the various forms of gas hydrates in oceanic sediments: (a) Yellow and white hydrates layers (Barkley Canyon off the East Coast of the U.S. photo available at http://www. nurp.noaa.gov/Spotlight/GasHydrates.htm); (b) Gas hydrate embedded in the sediment of hydrate ridge, off Oregon, U.S. photo available at http://commons.wikimedia.org/wiki/File:Gashydrat_im_

Sediment.JPG; (c) Gas hydrate beneath a rock overhang (Blake Ridge, East Coast U.S. Image credit NOAA Deep East Exploration 2001) http://oceanexplorer.noaa.gov/okeanos/explorations/ex1304/background/coldseeps/welcome.html; (d) Gas hydrates in fractures (Photos: Tim Collett, USGS and 2006–2008 Canada-Japan Mallik Project, http://www.geoexpro.com/articles/2009/02/gas-hydrates-not-so-unconventional); (e) Gas hydrate from shallow sediments in the Gulf of Mexico (photograph by B. Winters, USGS). http://woodshole.er.usgs.gov/project-pages/hydrates/primer.html; (f) Gas-hydrate outcroppings in the Gulf of Mexico (R. Sassen, pers. comm.); (g) Gas hydrate and *Hesiocaeca methanicola* (Gulf of Mexico, Image courtesy of Deep East 2001, Ian MacDonald, NOAA/OER). http://oceanexplorer.noaa.gov/explorations/deepeast01/logs/sep23/media/icewormsmed.html); and (h) Tubeworms surrounding a hydrate mound in the Gulf of Mexico (MacDonald 2002).

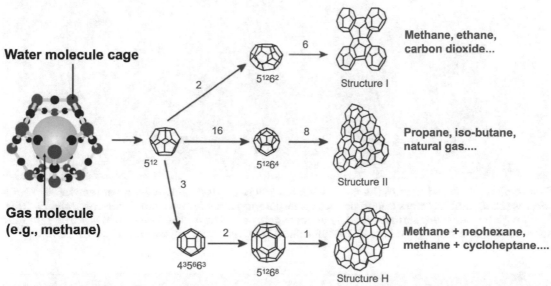

Figure 5.59. Gas-hydrate crystal structures. The three structure types that have been observed as gas hydrates are structures I, II, and H. Guest gas molecules in each hydrate structure are listed (Collett et al. 2009; AAPG©2009, reprinted by permission of the AAPG whose permission is required for further use). [The five types of water cages that make up the gas-hydrate structures are the pentagonal dodecahedron (512), the tetrakaidecahedron (51262), the hexakaidecahedron (51264), the irregular dodecahedron (435663), and the icosahedrons (51268). Representative guest gas molecules in each hydrate structure are listed].

productivity and tectonic activity where crustal deformation and compaction leads to expulsion of biogenic methane-rich fluids. The primary biota of cold-seep communities are a variety of bacteria and macroinvertebrates, although background deep-sea fauna are often observed (Figure 5.63). Bivalve species with symbionts of the genus *Bathymodiolus* including the families Solemyidae, Lucinidae, Vesicomyidae, Thyasiridae, and Mytilidae are commonly present (Callender et al. 1990, Callender and Powell 1992; Oliver et al. 2011). The other megafauna typical of cold-seep communities are tubeworms (Lamellibrachia and pogonophorans).

Since the initial discoveries, extensive surveys and studies have shown that cold-seep communities occur at most, if not all, deep-water macroseeps in the Gulf of Mexico. These sites are the most intensively studied and best understood cold-seep communities in the world (Fisher et al. 2007). By 2007, over 90 cold-seep communities had been discovered from the base of the Florida Escarpment in the east across the northern Gulf of Mexico to offshore southern

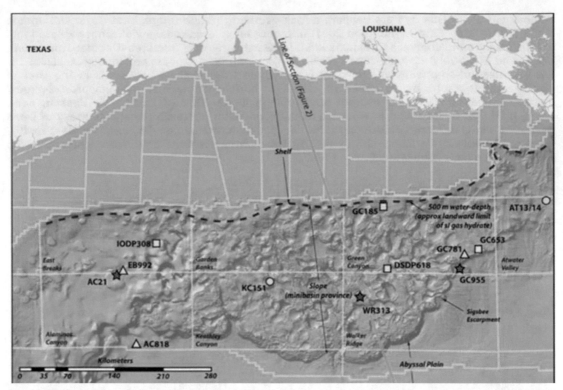

Figure 5.60. Selected sites in the Gulf of Mexico where gas hydrates have been recovered. *Yellow circles* (2005) and *red stars* (2009) denote drilling/coring sites for two joint industry projects. Other known gas-hydrate sites are marked by *yellow squares* and triangles (Boswell et al. 2012; reprinted with permission from Elsevier). [DSDP 96 drilled the Orca Basin shown as DSDP Site 618].

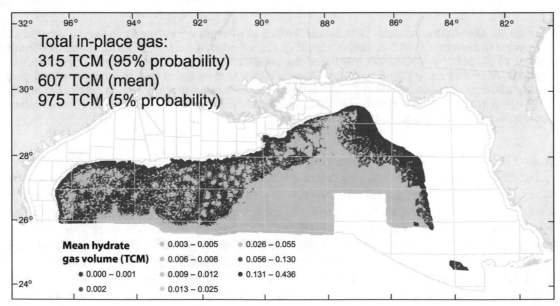

Figure 5.61. An assessment of mean in-place volume of gas (at STP) within hydrates (TCM = trillion cubic meters; 1 m^3 = 35.3 ft^3; STP = standard temperature and pressure, 1 atm and 20 °C [69°F]; Collett et al. 2009 modified from Frye 2008; AAPG$^{©}$2009, reprinted by permission of the AAPG whose permission is required for further use). The *shoreward boundary* represents the temperature and pressure limits to gas-hydrate stability.

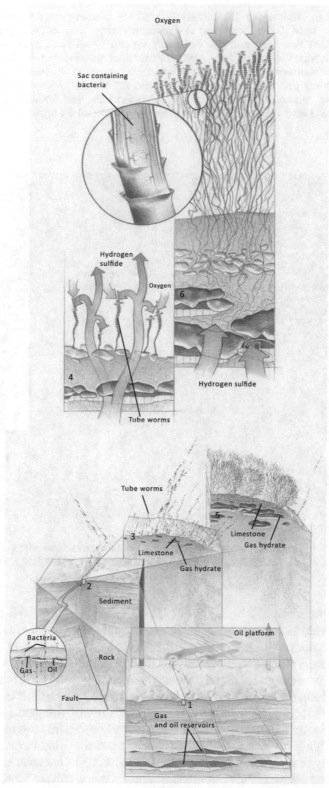

Figure 5.62. Schematic of the progression of the biology/ecology of a typical marine seep location: (1) gas and liquids seep via migration pathways from deep-seated reservoirs to near-surface sediments, (2) bacteria populations are enhanced in the sediments around the seep and may form

bacterial mats that trap oil and gas, (3) bacterial metabolism produces carbon dioxide and hydrogen sulfide which can lead to the formation of carbonate mineral substrates (limestone) that allow tubeworms to colonize the area and the cold temperature and high pressures can create gas hydrate, (4) chemosynthesis-based tubeworms and bivalves flourish in the hydrogen sulfide- and oxygen-rich environment, (5) as the tubeworm colony ages its foundations slowly solidify the sediments blocking the seepage, and (6) eventually the tubeworms acquire most of their nutritional chemicals through extensive root-like systems that extend into the underlying sediments (MacDonald and Fisher 1996; Bruce Morser/National Geographic Creative, used with permission).

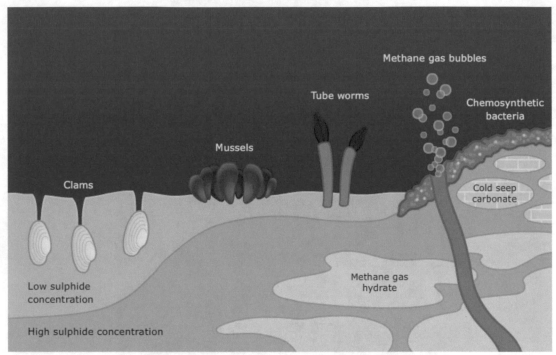

Figure 5.63. A typical biological assemblage associated with cold-seep sites. Image reprinted with permission of the University of Waikato. ©University of Waikato. All Rights Reserved. https://www. sciencelearn.org.nz/.

Texas in water depths from 290 to 3,300 m (951–10,827 ft) (Figure 5.64) (Fisher et al. 2007). Other studies extended the biogeographic range of cold-seep communities onto the abyssal plain and into the southern reaches of the Gulf of Mexico (MacDonald et al. 2004).

Surveys and process studies have established that the critical connection between geological and biogeochemical processes that make cold-seep communities viable is the presence of a wide range of microbes (Fisher et al. 2007). Cold-seep communities in the Gulf of Mexico are unique in that the methane that fuels these bacteria is predominantly thermogenic in origin, whereas at other worldwide sites, microbiologic methane is generally more important. Symbionts, microbial mats, and free-living bacteria are ubiquitous, serving as the primary producers of cold-seep food webs (MacAvoy et al. 2005; Fisher et al. 2007). Consortia of bacteria are capable of critical metabolic conversions such as oxidizing methane and reducing sulfate ions, thereby supporting macrofauna communities and producing critical hard substrate via carbonate precipitation (see Sections 5.5.3 and 5.5.4).

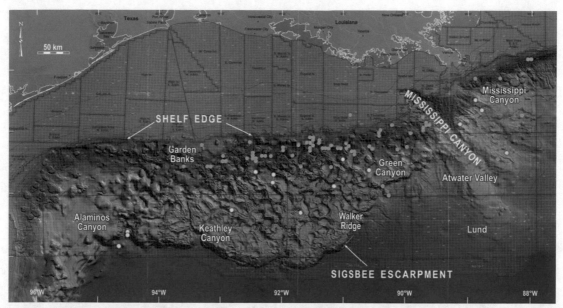

Figure 5.64. Multibeam bathymetric image of the northern Gulf of Mexico showing the location of seep sites where cold-seep communities have been confirmed by remotely operated vehicles or manned submersible dives (Fisher et al. 2007; Figure 1).

Carbonate formation stabilizes sediment and produces the essential substrate necessary for the attachment of various fauna. Sulfide is a required energy source for symbionts in tubeworms and some mussels and for free-living aerobic bacteria (see Sect. 5.5.4) (Joye et al. 2004). The copious and sustained supply of methane and sulfide at seep sites has led to the prevalence of mussels and tubeworms at Gulf of Mexico seeps (Figures 5.65, 5.66, and 5.67) (Kennicutt et al. 1988a, b; Kennicutt and Brooks 1990; MacDonald et al. 1990). Bathymodiolin mussels (*Bathymodiolus childressi*—that contain symbionts that utilize methane) are one of the dominant species, and two species of vestimentiferan tubeworms (*Lamellibrachia lumeysi* and *Seepiophila jonesi*—that contain symbionts that utilize sulfide) are abundant at Gulf of Mexico seep sites (Fisher et al. 2007). Tubeworms have no mouth, gut, or anus and rely on intracellular sulfide absorbing symbionts for the bulk of their nutrition (Nelson and Fisher 1995). Mussel symbionts passively take up methane from the surrounding seawater, whereas tubeworms have specialized blood hemoglobins that bind and actively transport oxygen (from the surrounding water) and sulfide (from sediment porewaters) (Fisher et al. 2007). Individual and aggregations of tubeworms can live for centuries, and the availability of hard substrate can restrict settlement and be growth limiting (Bergquist et al. 2000; Cordes et al. 2007a, b). Tubeworm aggregations begin to senesce and thin out as individuals die, possibly due to carbonate precipitation, resource depletion, and/or old age (Fisher et al. 2007). At some sites, hard and soft corals colonize carbonates; however, direct trophic ties between deep-sea corals and seep primary production have not been demonstrated. Other seep animals such as communities of symbiont containing vesicomyid clams (*Calytogena ponderosa* and *Vesicomya chordata*) are often present in low densities and were some of the first cold-seep species discovered in the Gulf of Mexico (Kennicutt et al. 1985; Brooks et al. 1987a, b). An unusual community of specialized polychaetes (ice-worms, *Hesiocaeca methanicola*) was found associated with exposed gas hydrate at several sites as well (Figure 5.58g) (Fisher et al. 2000).

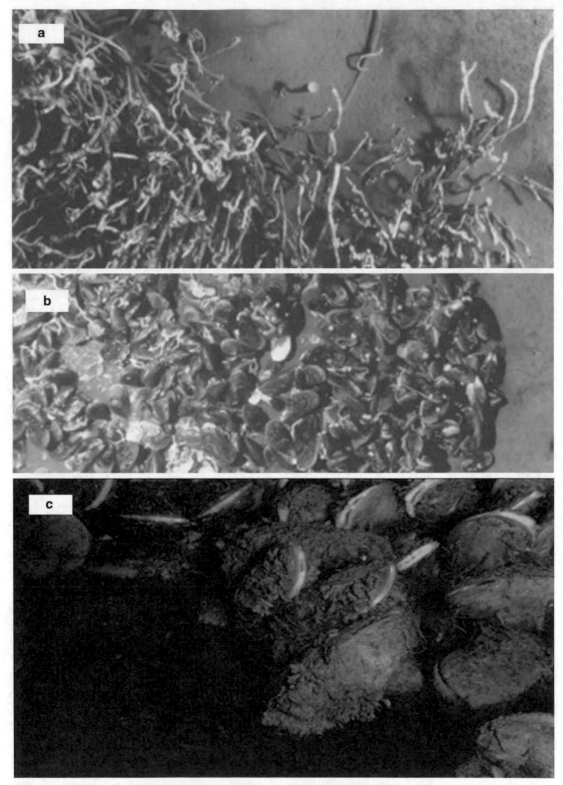

Figure 5.65. Typical Gulf of Mexico petroleum seep biological assemblages: (a) mussels, tube-worms, and background fauna closely associated with hard substrate derived from the oxidation of gas and oil (Kennicutt et al. 1988b; Figure 1). (b) Dense clusters of mussels associated with gas seeps (Kennicutt et al. 1988b; Figure 2). (c) Bathymodiolus mussels partly submerged in anoxic brine at the edge of a pockmark. Shells of dead mussel are submerged in the brine at the lower edge of the frame (MacDonald et al. 1990; reprinted with permission from The American Association for the Advancement of Science).

Figure 5.66. Aggregations of the tubeworms *Lamellibrachia luymesi* **and** *Seepiophila jonesi* **(photo courtesy of K. Luley; Cordes et al. 2009, reprinted with permission from Annual Reviews).**

Due to the limitations of most manned submersibles to a water depth of 1,000 m (3,281 ft) and difficulties in sampling the deep sea until recently, few seep sites were discovered in water depths greater than 1,000 m (3,281 ft) in the Gulf of Mexico (Brooks et al. 1990; MacDonald et al. 2003, 2004). Previous surveys and the brine-associated community known offshore Florida suggested that cold-seep communities might be present in water depths greater than 1,000 m (3,281 ft). Oil and gas seeps were known to extend to the abyssal plain, and there was no empirical evidence that water depth limited the occurrence of cold-seep communities. In 2006 and 2007, the presence of cold-seep communities was confirmed at 15 sites on the lower Louisiana slope in water depths greater than 1,000 m (3,281 ft), which significantly expanded the geographic range of sites in the Gulf of Mexico (Fisher et al. 2007; Roberts et al. 2010). These sites contained dense communities of tubeworms and mussels, communities of deep-living soft and hard corals, the largest mussel bed known in the Gulf of Mexico, an actively venting mud volcano, asphalt flows, a brine lake, and a variety of new species, including two in the genera Lamellibrachia and Escarpia. The same species of mussel found at shallower sites, *Bathymodiolus childressi*, was observed in water depths as great as 2,200 m (7,018 ft). Follow-up studies showed that these deeper living populations were genetically isolated from shallower ones (Cordes et al. 2007b). At water depths greater than 1,000 m (3,281 ft), *Bathymodiolus brooksi*, a mussel with both methanotrophic and chemoautotrophic symbionts, was also observed (Fisher et al. 1993), and at sites deeper than 2,200 m (7,018 ft), a third mussel species, *Bathymodiolus heckeri*, with symbionts that utilize reduced sulfur and carbon (methane and perhaps methanol) substrates for energy was the dominant mussel (Roberts et al. 2007, Duper-ron et al. 2007). At these deeper sites, several other types of biological communities were present including vesicomyid clams in low densities, high-density communities of symbiont containing pogonophoran tubeworms and large aggregations of heart urchins residing in highly reduced sediments (Fisher et al. 2007; Roberts et al. 2010). It is now believed that if the requisite environments are present, cold-seep communities can exist throughout the deep sea regardless of water depth.

Figure 5.67. Cold seep organisms and outcropping gas hydrate: (a) *Bathymodiolus* mussels closely associated with a methane gas brine pool; (b) outcropping gas hydrate tinged yellow/orange by associated oil; and (c) *Lamellibrachia* tubeworm cluster (MacDonald and Fisher 1996; Jonathan Blair/National Geographic Creative, used with permission).

If there were no oil and gas seeps in the Gulf of Mexico, many of the phenomena described above would be absent and the mass loading of petroleum to the northern Gulf Mexico would be greatly reduced from current estimates. The distribution of methane seeps would be largely the same since the origin of this methane is predominantly microbial reworking of recent organic matter. Liquid hydrocarbons would be exclusively due to anthropogenic inputs and concentrated in the coastal regions rather than the deep sea in the absence of seeps. It could be reasonably expected that tar balls and mats would be substantially reduced on beaches and elsewhere but still be present due to human activities. Shallow gas-hydrate occurrences would likely be absent as sediment gaseous hydrocarbon concentrations would rarely reach supersaturation. It would be expected that hard sea-bottom substrate occurrences would be reduced on average, but relic, shallow water, and erosion-exposed hard bottom would still be present. From an ecological standpoint, the picture is more complex in regard to an absence of seeps.

The predominant megafauna at cold-seep communities require elevated sulfide concentrations associated with seeps to support endosymbiosis. It is known that these communities have ceased to exist when seepage is no longer present. Since many cold-seep species are *endemic* (i.e., found only at seeps), Gulf of Mexico biodiversity would be decreased. Studies have shown that cold-seep communities are largely oases of life in an otherwise relatively uniform deep-sea environment. MacAvoy et al. (2005) concluded that some heterotrophic fauna collected in close association with cold-seep communities most likely obtain the bulk of their nutrition from chemosynthetic production through a combination of grazing on free-living bacteria and directly consuming faunal biomass. However, other background deep-sea fauna have been shown to contain little evidence of the utilization of cold-seep primary production, so the broader ecological importance of cold-seep communities to the deep sea remains largely a mystery (Carney 2010). On the other hand, Boetius and Wenzhofer (2013) concluded that, on a global basis, seep sites on continental slopes sustain some of the richest ecosystems in the deep sea and that cold-seep communities utilize about two orders of magnitude more oxygen per unit area than non-seep communities. Other studies have shown that cold-seep ecosystems contribute substantially to the microbial diversity of the deep sea. Hydrocarbon seeps have been described as "…geologically driven hot spots of increased biological activity on the seabed…" (Foucher et al. 2009), and it has become increasingly recognized that biological hot spots are critical to sustaining biodiversity. The differences in the larger Gulf of Mexico ecosystem that might be expected if there were no seeps is difficult to predict given the present state of knowledge but the effects are expected to be limited, as most Gulf of Mexico biomass and diversity occurs in coastal regions beyond the influence of seeps. However, oil and gas seeps are an intrinsic feature of the region and are expected to persist as long as oil and gas remains deep within the basin and finds its way to the surface.

5.7.5 Exemplar Sites

The prevalence, persistence, number, and volumes of petroleum seeps in the Gulf of Mexico have established the conditions for the common occurrence of sites that display a spectrum of characteristics typical of marine oil and gas seeps. The number of confirmed cold-seep communities in the Gulf of Mexico exceeds the combined number of all other sites identified in the world's ocean, and it is likely hundreds of other sites are yet to be discovered (Fisher et al. 2007). Extensive studies of Gulf of Mexico oil and gas seep sites over the last three decades have clarified some of the complex interactions of physical, chemical, biological, and ecological processes. The geological and biological manifestations of petroleum seeps on the seafloor are related not only to the composition of released gases and fluids but also the rate and history of seepage (Fisher et al. 2007). Seeps release oil, gas, brines, and occasionally

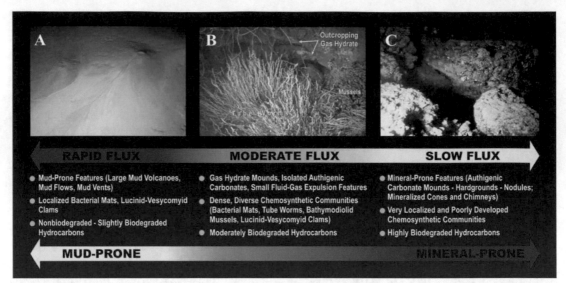

Figure 5.68. Relationships between rates of petroleum seepage features. (a) Rapid flux may produce mud volcanoes, mud vents, mudflows, and sparse cold-seep communities, (b) Moderate flux is commonly associated with surficial exposures of gas hydrate or gas hydrate in the shallow subsurface and well-developed cold-seep communities; and (c) Slow flux systems are likely to produce authigenic carbonates and other mineral assemblages; the supply of reduced chemicals is generally low and insufficient to support cold-seep communities (Fisher et al. 2007; Figure 4).

fluidized sediment. The rates of these releases vary from slow seepage to rapid venting (Figure 5.68). As fluxes of fluids and gases vary over time, the associated communities are distinct and evolve, and when seepage ceases, these communities do as well (Fisher et al. 1997, 2007).

One of the most detailed studies of petroleum seep sites in the Gulf of Mexico is that of Roberts et al. (2010). Seep-related seafloor features, mineralogical assemblages, and associated biological communities were characterized using 3D seismic survey data complemented by observations and the collection of shallow subsurface samples by manned submersible and remotely operated vehicles at 15 sites (Figure 5.69) (Roberts et al. 2010). These studies confirmed the close links between highly positive seafloor reflectivity, hard bottoms, hydrocarbon seeps or vents, authigenic minerals, gas hydrate, anoxic surface sediments, brine pools and flows, and cold-seep communities. Four exemplar sites from this study are presented to illustrate and describe the characteristics of the seafloor at petroleum seeps on the Gulf of Mexico continental slope (Figure 5.69 is a location map for the following summaries).

The Alaminos Canyon site, located in lease block 601 (AC601), exhibited surface amplitude anomalies in 3D seismic data that identified a brine lake and fluid-gas expulsion features associated with faults (Figure 5.70) (Roberts et al. 2010). The brine lake was circular (about 180 m [591 ft] in diameter) and averaged about 4 m (13 ft) deep (Figure 5.71). Terraced areas surrounding the brine lake contained small outcrops of authigenic carbonate suggesting lake levels were higher in the past. A clear interface was present between the brine lake surface and the surrounding seawater. The salinity of the lake was about twice that of seawater. White flocs floating within the brine were determined to be barite. Lake brine sulfate levels were about half that of seawater, but porewaters in lake-bottom cores contained no sulfate, and chloride-to-sodium ratios suggested halite was the source of the brine (Roberts et al. 2007a, b). The water column directly above the lake was supersaturated with methane. The brine itself had no animal

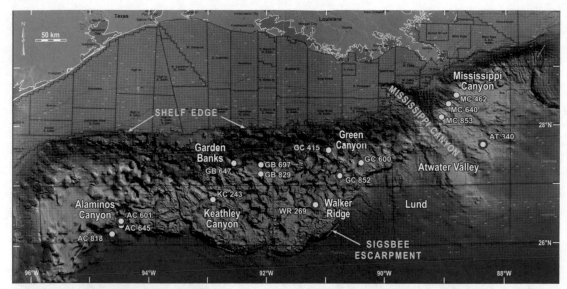

Figure 5.69. Shaded multibeam bathymetry relief map of the northern Gulf of Mexico continental slope with the locations of sites discussed below in *red circles* (BOEMRE oil and gas lease areas are in white lettering; Roberts et al. 2010 [reproduced with permission of Pergamon via Copyright Clearance Center, Inc.]).

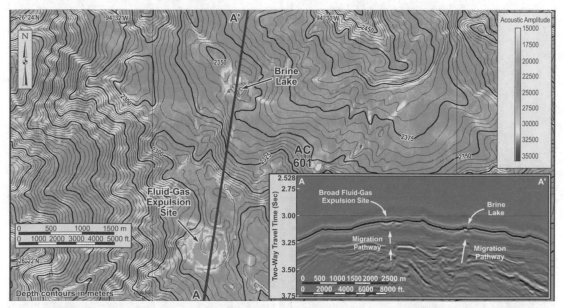

Figure 5.70. 3D seismic surface amplitude and bathymetry map illustrating a large fluid-gas expulsion area and a brine lake. The seismic profile (*inset*) shows that these features are above a breached subsurface anticlinal structure and faults that are migration pathways from the deep subsurface (Roberts et al. 2010; reproduced with permission of Pergamon via Copyright Clearance Center, Inc.).

life, but there were scattered clumps of mussels, a few tubeworms, and numerous urchin trails around the lake about 5–15 m (16.4–49.2 ft) from the lake shoreline (Figure 5.71). On higher ground further away from the lake, larger vestimentiferan tubeworm communities and

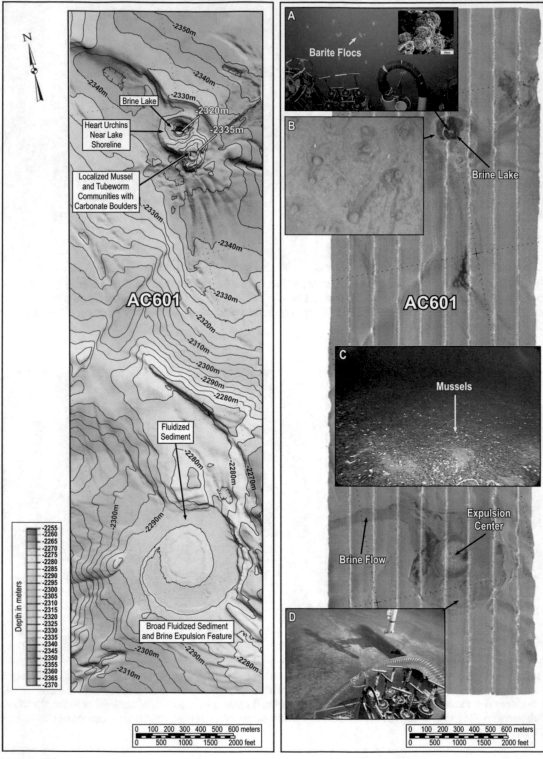

Figure 5.71. High-resolution autonomous underwater vehicle (AUV)-acquired multibeam bathymetry (*left*) and associated backscatter data (*right*) of the brine lake and large fluid-gas expulsion site in AC601. *Dark areas* on the backscatter image are highly reflective surfaces. (a) White flocs of barite are floating on the brine of the lake. (b) Heart urchins were observed above the brine shoreline on the lake's margin. (c) The high reflectivity (*dark*) on the backscatter image of the large, circular expulsion center is beds of living mussels. (d) Samples being collected of soft orange-stained mud in the southern part of the circular expulsion feature (Roberts et al. 2010; reproduced with permission of Pergamon via Copyright Clearance Center, Inc.).

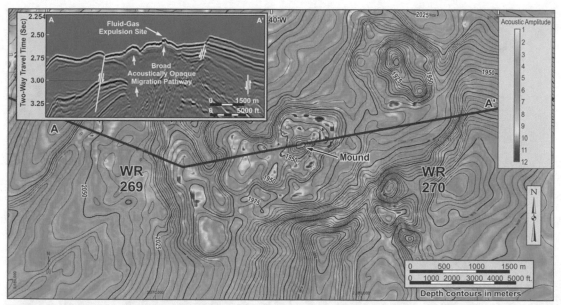

Figure 5.72. 3D seismic surface amplitude and bathymetry map of a series of mounds exhibiting high-positive surface amplitudes. The seismic profile (*inset*) shows a broad, acoustically transparent migration pathway beneath the mounds and above a salt diapir (not shown) in the shallow subsurface (Roberts et al. 2010; reproduced with permission of Pergamon via Copyright Clearance Center, Inc.).

extensive exposures of authigenic carbonate were observed. High-amplitude areas, representing zones of lithified seafloor, surround the lake (Figure 5.70). The reflective and high-positive amplitude area about 3 km (1.9 mi) south of the brine pool was a well-defined vertical migration pathway from the crest of an underlying breached anticline. Surficial brine flows, large areas of dark reduced mud, scattered clumps of living mussels, and light-gray fluidized mud suggested recent extrusions of sediment. Additional observations identified broad areas of red-stained sediment, a large mudflow along the southern rim of the feature, scattered clamshells, and a large bed of living mussels. The mussel bed was along the north-northwest rim of a circular expulsion feature. Carbonate outcrops were not obvious in the vicinity of the expulsion feature. High-positive 3D seismic surface amplitudes within the western and southern parts of the feature and apparent flow paths out of the feature suggested that mussel and clam beds may have developed on the surface of mudflows and were subsequently buried.

The Walker Ridge site, located in lease block 269–270 (WR269-270), exhibited a series of mounded features (Figure 5.72) (Roberts et al. 2010). The mounds are on the margin of an uplifted and compressed mini basin filled with Plio-Pleistocene turbidites, fans, and hemipelagic sediments. A north-south trending salt diapir underlies the uplifted eastern side of the basin, and migration pathways are linked from the seabed to the subsurface. Three mounded areas displayed a high surface amplitude response in 3D seismic data, and indications were that these areas are composites of smaller mounded features. The highest seafloor amplitudes were at the tops of the mounds (Figure 5.72). The easternmost mound in this grouping had the highest relief in the mound group. At the highest relief zones of the feature gas that was 99 % methane was observed venting. A broad and acoustically opaque area and the presence of bubble-phase gas and communities of pogonophorans, holothurians, and crustaceans (primarily crabs) were observed (Figure 5.73, inset). Large patches of dark, reducing sediments were observed on the mound's eastern flank (Figure 5.73). Scattered mussel shells and shell fragments were observed

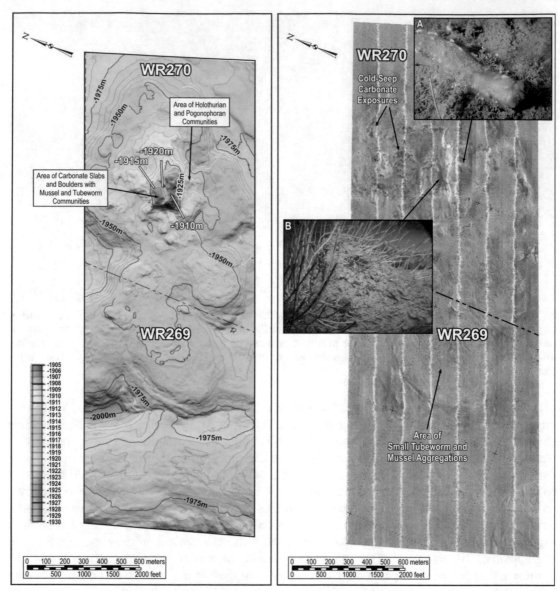

Figure 5.73. Multibeam bathymetry (*left*) and backscatter images (*right*) of the highest relief mound at the study site included (a) pogonophorans and holothurians communities on the east flank of the mound and (b) carbonates, vestimentiferan tubeworms, and mussels rimmed a slight depression at the crest of the mound (Roberts et al. 2010; reproduced with permission of Pergamon via Copyright Clearance Center, Inc.).

on the steep slope leading to the crest of the mound. There was a shallow depression on the crest of the mound (2–5 m [6.7–16.4 ft] deep) with a well-defined rim. Beds of large living mussels, vestimentiferan tubeworm colonies, slabs of carbonate and small gas seeps were also found. The carbonates and tubeworm colonies were mainly confined to the rim of the shallow depression, whereas mussels were at the rim and toward the middle of the depression.

The Green Canyon site, located in lease block 852 (GC852), exhibited a ridge crest and large cold-seep carbonate blocks and slabs (Figure 5.74). The ridge was supported by salt within a

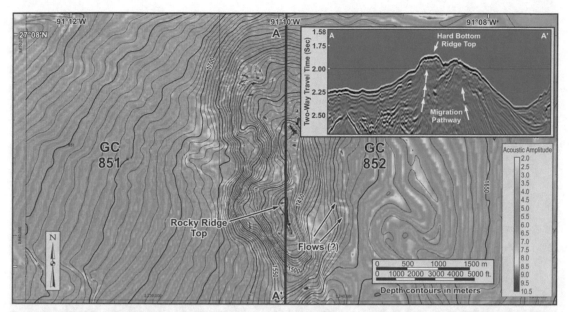

Figure 5.74. 3D seismic surface amplitude and bathymetry map illustrating high-positive amplitude anomalies on the top and on the flanks of a north-south-oriented submarine ridge. Highly
reflective areas of the ridge included widespread hard-bottom areas composed of cold-seep
carbonate slabs and large blocks. A seismic profile across the ridge (*inset*) defines migration
pathways for the transport of hydrocarbons (oil and gas) to the seafloor (Roberts et al. 2010;
reproduced with permission of Pergamon via Copyright Clearance Center, Inc.).

well-developed basin. The 3D seismic seafloor reflectivity and surface amplitude data suggested localized hard seabed conditions, flow deposits, and cold-seep communities on top of
the ridge. The cross-sectional shapes of the ridge and fluid-gas migration pathways to the
surface of the ridge crest are shown in Figure 5.74 (inset). The crest of the ridge varied in width
from about 100 to 300 m (328–984 ft). Sea-surface oil slicks were observed above the ridge.
Hard bottoms of authigenic carbonate slabs and mound-like structures of large cold-seep
carbonate blocks were observed on the southern ridge crest. Mussel beds were scattered
among the carbonate slabs and boulders and vestimentiferan tubeworms were widespread
along the southernmost ridge crest (Figure 5.75). Prolonged hydrocarbon seepage was indicated
by abundant authigenic carbonates at the southern ridge crest. Brine seepage was found
throughout the area and several small gas seeps were observed in mussel beds but no oil
seepage was observed. Gorgonians and scattered bamboo corals were widespread throughout
the southern ridge crest area. Huge carbonate blocks and a dense and diverse hard coral
community were observed at the shallowest point on the ridge (Figure 5.74). There was little
evidence of active seepage on the ridge.

The Atwater Valley site, located in lease block 340 (AT340) exhibited a cluster of high-
positive amplitude features (Figure 5.76) (Roberts et al. 2010). The low relief mound, east of the
Mississippi Canyon, transitions from a canyon to a submarine fan complex caused by an
underlying salt body in the shallow subsurface. The area included numerous small-scale surface
mounds, and bubble-phase gas was evident along the migration pathway. The 3D seismic
surface amplitude (reflectivity) maps showed high-positive surface amplitude features associated with mounds of up to 20 m (65.6 ft) relief (Figures 5.77 and 5.78). Abundant bathymodiolid mussel shells were cemented into authigenic carbonate. The diverse fauna at the site
included numerous bathymodiolid mussel beds, vestimentiferan tubeworm colonies, and sea
urchins and anemones were scattered among the carbonate blocks. Areas of brine seepage

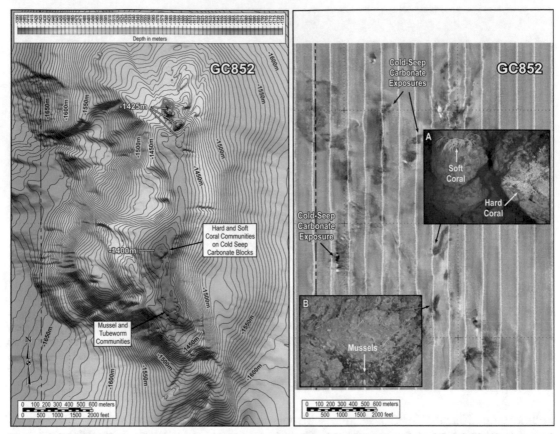

Figure 5.75. Multibeam bathymetry (*left*) and associated backscatter data (*right*) illustrates variable and small-scale relief along a ridge crest. Communities of mussels and tubeworms were found on the southern ridge crest and its upper flanks. At the highest point along the ridge, deepwater coral communities were found in association with a large area of cold-seep carbonate blocks. The most active seepage sites and cold-seep communities were on the southern ridge area. (a) Soft and hard coral communities on large cold-seep carbonate blocks (b) Beds of mussels between the cold-seep carbonate slabs and blocks of the southern ridge crest area (Roberts et al. 2010; reproduced with permission of Pergamon via Copyright Clearance Center, Inc.).

adjacent to the mound contained large communities of heart urchins in dark reducing sediments (Figures 5.77 and 5.78).

5.8   SUMMARY

Petroleum seepage is a prevalent, natural worldwide phenomenon that has occurred for millions of years and is especially widespread in the deepwater region of the Gulf of Mexico. As one of the most prolific oil and gas basins in the world, the Gulf of Mexico has abundant deep-seated supplies of oil and gas to migrate to the surface. The deepwater region of the Gulf of Mexico is an archetype for oil and gas seepage, and most of our knowledge of petroleum seeps is based on studies of the region. The essential geological conditions for seepage are met in many areas of the deepwater region of the Gulf of Mexico region including multiple deeply buried mature source rocks and migration pathways to the surface. The northern Gulf of Mexico basin has been a depocenter for massive amounts of sediments over geologic time, and salt tectonics are prevalent, setting boundaries on the geographic patterns of petroleum

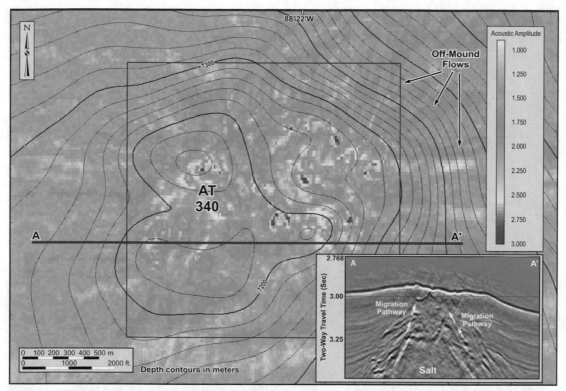

Figure 5.76. 3D seismic surface amplitude and bathymetry map showing a low-relief regional mound with surficial high-positive seafloor amplitude anomalies. Seafloor amplitude data showed apparent flows originating from the mound crest and extending to deeper water. An east-west seismic profile across the regional mound (*inset*) shows supporting by salt in the shallow subsurface and fluid-gas migration routes fat the salt margins of the mounded seafloor (Roberts et al. 2010; reproduced with permission of Pergamon via Copyright Clearance Center, Inc.).

seepage. Gulf of Mexico seeps are highly variable in composition and volume and include gases, volatiles, liquids, pitch, asphalt, tars, water, brines, and fluidized sediments. Seeps are dynamic over a range of temporal scales and can be ephemeral or persist for many years. In the Gulf of Mexico, seeps annually release vast amounts of oil and gas to the environment. In the Gulf of Mexico region, seeps occur on land; however, most petroleum seepage is in the northwest and north-central offshore regions. Collectively, petroleum seeps in the Gulf of Mexico are sources of highly variable mixtures of hydrocarbons, which are often altered by the weathering processes that occur after seepage. Seeps can be pristine to severely biodegraded. The prevalence, persistence, number, and volumes of petroleum seeps in the Gulf of Mexico display a spectrum of characteristics typical of petroleum seeps. Biogeochemical processes are the critical connections between commonly expressed phenomena at petroleum seep sites, including topographic features and authigenic minerals. The Gulf of Mexico continental slope and abyss are complex topographically with areas of high seafloor reflectivity and acoustic wipe-out zones caused by the active influx of gases and fluids, lithification, physical disruption of sediments, and gas-hydrate formation and decomposition. Gas seeps are widespread in the Gulf of Mexico and most have microbiological origins, but thermogenic gas seeps are also common. Gas hydrate occurs in near-surface sediments at water depths below about 500 m (1,640 ft), which defines their upper stability limit. Surveys and studies have shown that cold-seep chemosynthetic communities are common at macroseeps Gulf-wide, including on the

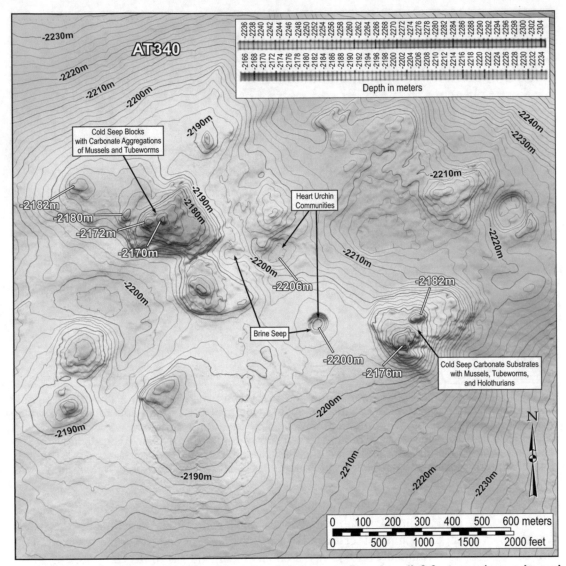

Figure 5.77. A multibeam bathymetry map showing small-scale relief features (mounds and depressions) composed of cold-seep carbonate slabs and blocks and brine seeps in the depressions (Roberts et al. 2010; reproduced with permission of Pergamon via Copyright Clearance Center, Inc.).

abyssal plain and in the southern Gulf of Mexico. Geological and biological manifestations at petroleum seeps on the seafloor are controlled by the composition of released gases and fluids and the rate and history of seepage. The rates of seepage of oil, gases, brines, and fluidized sediment vary from slow seepage to rapid venting. As these fluxes vary over time, cold-seep community assemblages evolve, and when seepage ceases, seep communities disappear. In the offshore Gulf of Mexico, the geographic distributions of source-rock horizons, salt basins, oil and gas production platforms, satellite and air-borne images of sea-surface oil slicks, regional oil and gas reserves, cold-seep communities, and gas hydrates illustrate the close association of petroleum seepage and these phenomena. Petroleum seepage in the Gulf of Mexico has occurred for millions of years and is widespread and active today.

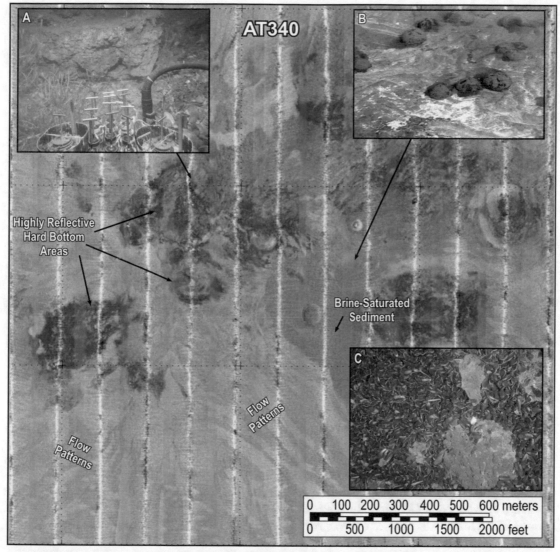

Figure 5.78. A backscatter image showing highly reflective hard-bottom and mounded areas. Active gas seepage associated with these mounds, numerous (a) living mussel beds, (b) urchin communities in areas of brine seepage, and (c) tubeworm communities among the carbonate blocks and slabs were observed. Flow lines in this backscatter image are brine flows associated with active fluid-gas expulsion (Roberts et al. 2010; reproduced with permission of Pergamon via Copyright Clearance Center, Inc.).

REFERENCES

Abrams MA (2005) Significance of hydrocarbon seepage relative to petroleum generation and entrapment. Mar Petrol Geol 22:457–477

Abrams MA, Dahdah N (2011) Surface sediment hydrocarbons as indicators of subsurface hydrocarbons: Field calibration of existing and new surface geochemistry methods in the Marco Polo area, Gulf of Mexico. AAPG Bull 95:1907–1935

Abrams MA, Dahdah N, Francu E (2009) Development of methods to collect and analyze gasoline plus range (C_5 to C_{12}) hydrocarbons from seabed sediments as indicators of subsurface hydrocarbon generation and entrapment. Appl Geochem 24:1951–1970

Alcazar A, Kennicutt M, Brooks J (1989) Benthic tars in the Gulf of Mexico: Chemistry and sources. Org Geochem 14:433–439

Anderson R, Scalan R, Parker P, Behrens E (1983) Seep oil and gas in Gulf of Mexico slope sediment. Science 222:619–621

Ballard RD (1977) Notes on a major oceanographic find. Oceanus 20:35–44

Barakat AO, Mostafa A, Qian Y, Kim M, Kennicutt MC II (2005) Organic geochemistry indicates Gebel El Zeit, Gulf of Suez, is a source of bitumen used in some Egyptian mummies. Geoarchaeology 20:211–228

Barry JP, Kochevar RE, Baxter CH (1997) The influence of pore-water chemistry and physiology in the distribution of vesicomyid clams at cold seeps in Monterey Bay: Implications for patterns of chemosynthetic community organization. Limnol Oceanogr 42:318–328

Bergquist DC, Williams FM, Fisher CR (2000) Longevity record for deep-sea invertebrate. Nature 403:499–500

Bernard BB, Brooks JM, Sackett WM (1976) Natural gas seepage in the Gulf of Mexico. Earth Planet Sci Lett 31:48–54

Boetius A, Suess E (2004) Hydrate ridge: A natural laboratory for the study of microbial life fuelled by methane from near-surface gas hydrates. Chem Geol 205:291–310

Boetius A, Wenzhofer F (2013) Seafloor oxygen consumption fuelled by methane from cold seeps. Nat Geosci 6:725–734

Boetius A, Ravenschlag K, Schubert C, Rickert D, Widdel F, Gieske A, Amann R, Jørgensen BB, Witte U, Pfannkuche O (2000) A microbial consortium apparently mediating the anaerobic oxidation of methane. Nature 407:623–626

Bohrmann G, Greinert J, Suess E, Torres M (1998) Authigenic carbonates from the Cascadia subduction zone and their relation to gas hydrate stability. Geology 26:647–650

Boswell R, Collett TS, Frye M, Shedd W, McConnell DR, Shelander D (2012) Subsurface gas hydrates in the northern Gulf of Mexico. Mar Petrol Geol 34:4–30

Brooks JM, Sackett WM (1973) Sources, sinks and concentrations of light hydrocarbons in the Gulf of Mexico. J Geophys Res 78:5248–5258

Brooks JM, Gormly JR, Sackett WM (1974) Molecular and isotopic composition of two seep gases from the Gulf of Mexico. Geophys Res Lett 1:312

Brooks JM, Bernard BB, Sackett WM, Schwarz JP (1979) Natural gas seepage on the South Texas Shelf. In: Offshore Technology Conference, Houston, TX, USA. OTC 3411:471–478

Brooks JM, Kennicutt MC II, Fay RR, McDonald TJ, Sassen R (1984) Thermogenic gas hydrates in the Gulf of Mexico. Science 225:409–411

Brooks JM, Cox HB, Bryant WR, Kennicutt MC II, Mann RG, McDonald TJ (1986) Association of gas hydrates and oil seepage in the Gulf of Mexico. Org Geochem 10:221–234

Brooks JM, Kennicutt MC II, Bidigare RR, Wade TL, Powell EN, Denoux GJ, Fay RR, Childress JJ, Fisher CR, Rossman IR, Boland GS (1987a) Hydrates, oil seepage, and chemosynthetic ecosystems on the Gulf of Mexico slope: An update. Eos Trans AGU 68:498–499

Brooks JM, Kennicutt MC II, Fisher CR, Macko SA, Cole K, Childress JJ, Bidigare RR, Vetter RD (1987b) Deep-sea hydrocarbon seep communities: Evidence for energy and nutritional carbon sources. Science 238:1138–1142

Brooks JM, Kennicutt MC II, MacDonald IR, Wilkinson DL, Guinasso NL, Bidigare RR (1989) Gulf of Mexico hydrocarbon seep communities: Part IV descriptions of known chemosynthetic communities. In: Offshore Technology Conference, Houston, TX, USA. OTC 5954:6

Brooks JM, Wiesenburg DA, Roberts H, Carney RS, MacDonald IR, Fisher CR, Guinasso NL, Sager WW, McDonald SJ, Burke RA, Aharon P, Bright TJ (1990) Salt, seeps and symbiosis in the Gulf of Mexico. EOS, Trans Am Geophys Union 71:1772–1773

Brooks JM, Anderson AL, Sassen R, Kennicutt MC II, Guinasso NL (1994) Hydrate occurrences in shallow subsurface cores from continental slope sediments. Ann NY Acad Sci 715:381–391

Callender WR, Powell EN (1992) Taphonomic signature of petroleum seep assemblages on the Louisiana upper continental slope: Recognition of autochthonous shell beds in the fossil record. Palaios 7:388–408

Callender WR, Staff GM, Powell EN, MacDonald IR (1990) Gulf of Mexico hydrocarbon seep communities V. Biofacies and shell orientation of autochthonous shell beds below storm wave base. Palaios 5:2–14

Campbell KA (2006) Hydrocarbon seep and hydrothermal vent paleoenvironments and paleontology: Past developments and future research directions. Palaeogeogr Palaeoclimatol Palaeoecol 232:362–407

Campbell KA, Farmer JD, Marais D (2002) Ancient hydrocarbon seeps from the convergent margin of California: Carbonates, fluids, and paleoenvironments. Geofluids 2:63–94

Cao B, Bai G, Wang Y (2013) More attention recommended for global deep reservoirs. Oil Gas J 111(9). http://www.ogj.com/articles/print/volume-111/issue-9/exploration-development/more-attention-recommended-for-global-deep-reservoirs.html, accessed December 15, 2015

Carney RS (2010) Stable isotope trophic patterns in echinoderm megafauna in close proximity to and remote from Gulf of Mexico lower slope hydrocarbon seeps. Deep-Sea Res II 57:1965–1971

Chisholm H (ed) (1911) Petroleum. Encyclopaedia Britannica, 11th edn. Cambridge University Press, Cambridge, UK. http://www.studylight.org/encyclopedias/bri/view.cgi?number=25685. Accessed 14 Sept 2014

Ciais P, Sabine C, Bala G, Bopp L, Brovkin V, Canadell J, Chhabra A, DeFries R, Galloway J, Heimann M, Jones C, Le Quéré C, Myneni RB, Piao S, Thornton P (2013) Carbon and other biogeochemical cycles. In: Stocker TF et al (eds) Climate change 2013: The physical science basis. Contribution of working group I to the fifth assessment report of IPCC. Cambridge University Press, Cambridge, UK

Cole GA, Yu A, Peel F, Taylor C, Requejo R, DeVay J, Brooks JM, Bernard BB, Zumberge, J, Brown S (2001) 21st Annual GCSSEPM Foundation Research Conference—Petroleum systems of basins: Global and Gulf of Mexico experience. Gulf Coast Section. www.gcssepm.org

Coleman J, Baker C, Cooper CK, Fingas M, Hunt G, Kvenvolden KA, Michel K, Michel J, McDowell J, Phinney P, Rabalais N, Roesner L, Spies RB (2003) Oil in the sea III: Inputs, fates, and effects. Committee on Oil in the Sea: Inputs and Effects, Ocean Studies Board and Marine Board, Divisions of Earth and Life Studies and Transportation Research Board, National Research Council. National Academies Press, Washington, DC, USA

Collett T, Johnson A, Knapp C, Boswell R (2009) Natural gas hydrates: A review. In: Collett T, Johnson A, Knapp C, Boswell R (eds) Natural gas hydrates: Energy resource potential and associated geologic hazards. AAPG Memoir 89:74 (Chapt. 1)

Combells-Bigott RI, Galloway WE (2006) Depositional and structural evolution of the middle Miocene depositional episode, east-central Gulf of Mexico. AAPG Bull 90:335–362

Cordes EE, Bergquist DC, Redding ML, Fisher CR (2007a) Patterns of growth in cold-seep vestimentiferans including *Seepiophila jonesi*: A second species of long-lived tubeworm. Mar Ecol 28:160–168

Cordes EE, Carney SL, Hourdez S, Carney R, Brooks JM, Fisher CR (2007b) Cold seeps of the deep Gulf of Mexico: Community structure and biogeographic comparisons to Atlantic and Caribbean seep communities. Deep-Sea Res Part I 54:637–653

Cordes EE, Bergquist DC, Fisher CR (2009) Macro-ecology of Gulf of Mexico cold seeps. Ann Rev Mar Sci 1:143–168

De Beukelaer SM (2003) Remote sensing analysis of natural oil and gas seeps on the continental slope of the northern Gulf of Mexico. Master's Thesis, Texas A&M University, College Station, TX, USA

De Beukelaer SM, MacDonald IR, Guinnasso NL, Murray JA (2003) Distinct side-scan sonar, RADARSAT SAR, and acoustic profiler signatures of gas and oil seeps on the Gulf of Mexico slope. Geo-Mar Lett 23:177–186

Dunlap HF, Bradley JS, Moore TF (1960) Marine seep detection—a new reconnaissance exploration method. Geophysics 25:275–282

Duperron S, Sibuet M, MacGregor BJ, Kuypers MM, Fisher CR, Dublier N (2007) Diversity, relative abundance and metabolic potential of bacterial endosymbionts in three Bathymodiolus mussel species from cold seeps in the Gulf of Mexico. Environ Microbiol 9:1423–1438

Etiope G (2009) A global dataset of onshore gas and oil seeps: A new tool for hydrocarbon exploration. Oil and Gas Business. http://www.earth-prints.org/handle/2122/6040

Etiope G (2012) Methane uncovered. Nat Geosci 5:373–374

Etiope G (2015) Natural gas seepage. The Earth's hydrocarbon degassing. Springer, Basel, p 199. doi:10.1007/978-3-319-14601-0

Feng D, Roberts HH, Di P, Chen D (2010) Characteristics of hydrocarbon seep-related rocks from the Deep Gulf of Mexico. Gulf Coast Assoc Geol Soc Trans 59:271–275

Fisher CR (1990) Chemoautotrophic and methanotrophic symbioses in marine invertebrates. Rev Aquat Sci 2:399–436

Fisher CR, Brooks JM, Vodenichar J, Zande J, Childress JJ, Burke RA Jr (1993) The co-occurence of methanotrophic and chemoautotrophic sulfur-oxidizing bacterial symbionts in a deep-sea mussel. Mar Ecol 14:277–289

Fisher CR, Urcuyo IA, Simpkins MA, Nix E (1997) Life in the slow lane: Growth and longevity of cold-seep vestimentiferans. Mar Ecol 18:83–94

Fisher CR, MacDonald IR, Sassen R, Young CM, Macko SA, Hourdez S, Carney RS, Joye S, McMullin E (2000) Methane ice worms: Hesiocaeca methanicola colonizing fossil fuel reserves. Naturwissenschaften 87:184–187

Fisher C, Roberts HH, Cordes EE, Bernard BB (2007) Cold seeps and associated communities in the Gulf of Mexico. Oceanography 20:118. http://dx.doi.org/10.5670/oceanog.2007.12

Formolo MJ, Lyons TW (2013) Sulfur biogeochemistry of cold seeps in the Green Canyon region of the Gulf of Mexico. Geochim Cosmoschim Acta 119:264–285

Forrest J, Marcucci E, Scott P (2007) Geothermal gradients and subsurface temperatures in the Northern Gulf of Mexico. Search and discovery article #30048. http://www.searchanddiscovery.com/documents/2007/07013forrest/images/forrest.pdf

Foucher JP, Westbrook GK, Boetius A, Ceranicole S, Dupre S, Mascie J, Mienert J, Pfannkuche O, Pierre C, Praeg D (2009) Structure and drivers of cold seep ecosystems. Oceanography 22:92–109

Frank DJ, Sackett WM, Hall R, Fredericks A (1970) Methane, ethane and propane concentrations in Gulf of Mexico. AAPG Bull 54:1933–1938

Frye M (2008) Preliminary evaluation of in-place gas hydrate resources: Gulf of Mexico Outer Continental Shelf. Minerals Management Service Report 2008-004. http://www.mms.gov/revaldiv/GasHydrateAssessment.htm. Accessed 14 Sept 2014

Gay A, Takano Y, Gilhooly WP III, Berndt C, Heeschen K, Suzuki N, Saegusa S, Nakagawa F, Tsunogai U, Jiang SY, Lopez M (2011) Geophysical and geochemical evidence of large scale fluid flow within shallow sediments in the eastern Gulf of Mexico, offshore Louisiana. Geofluids 11:34–47

Geology In (2015) The Petroleum System. http://www.geologyin.com/2014/08/petroleum-system.html. Accessed June 2015

Geyer RA (ed) (1980) Marine environmental pollution, 1. Hydrocarbons. Elsevier Oceanography Series. Elsevier Scientific, New York, NY, USA. 591 p

Geyer RA, Giammona CP (1980) Naturally occurring hydrocarbon seeps in the Gulf of Mexico and Caribbean Sea. Elsevier Oceanogr Ser 27:37–106

Greinert J, Bohrmann G, Suess E (2001) Gas hydrate-associated carbonates and methane-venting at hydrate ridge: Classification distribution and origin of authigenic lithologies. In: Paull CK, Dillon PW (eds) Natural gas hydrates: Occurrence, distribution, and dynamics. Geophys Monog Series 124:99–113

Harris J, Jefferson G (eds) (1985) Rancho La Brea: Treasures of the Tar Pits. Nat History Museum Los Angeles County Sci Ser 31:1–87

Harwell J, Lewan M (2002) Sources of mummy bitumen in ancient Egypt and Palestine. Archaeometry 44:285–293

Hesse R (2003) Pore water anomalies of submarine gas-hydrate zone as tool to assess hydrate abundance and distribution in subsurface: What have we learned in the past decade? Earth Sci Rev 61:149–179

Hirst K (2009) Bitumen—a smelly but useful material of interest http://archaeology.about.com/od/bcthroughbl/qt/bitumen.htm. Accessed 14 Sept 2014

Hood KC, Wenger LM, Gross OP, Harrison SC (2002) Hydrocarbon systems analysis of the northern Gulf of Mexico: Delineation of hydrocarbon migration pathways using seeps and seismic imaging, in Surface exploration case histories: Applications of geochemistry, magnetics, and remote sensing. In: Schumacher D, LeSchack LA (eds) AAPG Studies in Geology no. 48 and SEG Geophysical References Series no. 11, pp 25–40

Hovland M, Heggland R, De Vries MH, Tjelta TI (2010) Unit-pockmarks and their potential significance for predicting fluid flow. Mar Petrol Geol 27:1190–1199

Hovland M, Jensen S, Fichler C (2012) Methane and minor oil macro-seep systems—their complexity and environmental significance. Mar Geol 332–334:163–173

Hudec MR, Jackson MPA (2006) Advance of allochthonous salt sheets in passive margins and orogens. AAPG Bull 90:1535–1564

Hudec MR, Norton IO, Jackson MPA, Peel FJ (2013) Jurassic evolution of the Gulf of Mexico Salt Basin. AAPG Bull 97:1683–1710

Hunt J (1996) Petroleum geochemistry and geology. W.H. Freeman, New York, NY, USA. 743 p

Jones DM, Head IM, Gray ND, Adams JJ, Rowan AK, Aitken CM, Bennett Huang BH, Brown A, Bowler BFJ, Oldenburg T, Erdmann M, Larter SR (2008) Crude-oil biodegradation via methanogenesis in subsurface petroleum reservoirs. Nature 451:176–180

Joye SB, Boetius A, Orcutt BN, Montoya JP, Schulz HN, Erickson MJ (2004) The anaerobic oxidation of methane and sulfate reduction in sediments from Gulf of Mexico cold seeps. Chem Geol 205:219–238

Joye SB, Bowles MW, Samarkin VA, Hunter KS, Niemann H (2010) Biogeochemical signatures and microbial activity of different cold-seep habitats along the Gulf of Mexico deep slope. Deep Sea Res Pt II 57:1990–2001

Kennicutt MC II (1988) The effect of biodegradation on crude oil bulk and molecular composition. Oil Chem Pollut 4:89–112

Kennicutt MC II, Brooks JM (1990) Recognition of areas effected by petroleum seepage: Northern Gulf of Mexico continental slope. Geo-Mar Lett 10:221–224

Kennicutt MC II, Brooks JM, Bidigare RR, Fay RR, Wade TL, McDonald TJ (1985) Vent-type taxa in a hydrocarbon seep region on the Louisiana slope. Nature 317:351–353

Kennicutt MC II, Sericano JL, Wade TL, Alcazar F, Brooks JM (1987) High molecular weight hydrocarbons in Gulf of Mexico continental slope sediments. Deep Sea Res Part A 34:403–424

Kennicutt MC II, Brooks JM, Denoux GJ (1988a) Leakage of deep, reservoired petroleum to the near surface on the Gulf of Mexico Continental slope. Mar Chem 24:39–59

Kennicutt MC II, Brooks JM, Bidigare RR (1988b) Hydrocarbon seep communities: Four years of study. Oceanography 1:44–45

Kindersley D Ltd. (2007) Oil and natural gas. Presented by the Society of Petroleum Engineers. (ePub 2013) 978-1-4654-0441-1. http://www.energy4me.org/download/oil_gas_WEB.pdf. Accessed 14 Sept 2014

Kolbert E (2007) Unconventional crude. The New Yorker Magazine, p 46. http://www.new yorker.com/magazine/2007/11/12/unconventional-crude. Accessed 14 Sept 2014

Krishnan J, Rajagopal K (2003) Review of the uses and modelling of bitumen from ancient to modern times. Appl Mech Rev 56:149–214

Kuhn O (2004) Ancient Chinese drilling. CSEG Record 29(6):39–43

Kvenvolden KA, Cooper CK (2003) Natural seepage of crude oil into the marine environment. Geo-Mar Lett 23:140–146

Lamontagne RA, Swinnerton JW, Linnenbom J, Smith W (1973) Methane concentrations in various marine environments. J Geophys Res 78:5317–5324

Lamontagne RA, Swinnerton JW, Linnenbom J (1974) C_1–C4 hydrocarbons in the North and South Pacific. Tellus 26:71–77

Levin LA (2005) Ecology of cold seep sediments: Interactions of fauna with flow, chemistry and microbes. In: Gibson RN, Atkinson RJA, Gordon JDM (eds) Oceanography and marine biology: An annual review. Taylor & Francis, Boca Raton, FL, USA, vol 43, pp 1–46

MacAvoy SE, Fisher CR, Carney RS, Macko SA (2005) Nutritional associations among fauna at hydrocarbon seep communities in the Gulf of Mexico. Mar Ecol Prog Ser 292:51–60

MacDonald IR (1998) Natural oil spills. Sci Am 279:56–61

MacDonald IR (2002) Stability and change in Gulf of Mexico chemosynthetic communities, vol II, Technical report. Gulf of Mexico OCS Region OCS Study MMS 2002-036. U.S. Dept. of the Interior, Minerals Management Service, New Orleans, LA, USA. 456 p

MacDonald IR, Fisher C (1996) Life without light. National Geographic, October:86–97

MacDonald IR, Reilly JFI, Gullnasso NJ, Brooks JM, Carney RS, Bryant WR, Bright TJ (1990) Chemosynthetic mussels at a brine-filled pockmark in the Northern Gulf of Mexico. Science 248:1096

MacDonald IR, Guinasso NL, Ackleson SG, Amos JF, Duckworth R, Sassen R, Brooks JM (1993) Natural oil slicks in the Gulf of Mexico visible from space. J Geophys Res 98:16351–16364

MacDonald IR, Reilly J, Best SE, Venkataramaiah R, Sassen R, Guinasso NL, Amos J (1996) Remote sensing inventory of active oil seeps and chemosynthetic communities in the Northern Gulf of Mexico. In: Schumacher D, Abrams MA (eds) Hydrocarbon migration and its near-surface expression. AAPG Memoir 66, pp 27–37

MacDonald IR, Sager WW, Peccini MB (2003) Gas hydrate and chemosynthetic biota in mounded bathymetry at mid-slope hydrocarbon seeps: Northern Gulf of Mexico. Mar Geol 198:133–158

MacDonald IR, Bohrmann G, Escobar E, Abegg F, Blanchon P, Blinova V, Bruckmann W, Drews M, Eisenhauer A, Han X, Heeschen K, Meier F, Mortera C, Naehr T, Orcutt B, Bernard B, Brooks J, de Farago A (2004) Asphalt volcanism and chemosynthetic life in the Campeche Knolls, Gulf of Mexico. Science 304:999–1002

Macelloni L, Caruso S, Lapham L, Lutken CB, Brunner C, Lowrie A (2010) Spatial distribution of seafloor biogeological and geochemical processes as proxy to evaluate fluid-flux regime and time evolution of a complex carbonate/hydrates mound, Northern Gulf of Mexico. Gulf Coast Assoc Geol Soc Trans 60:461–480

McBride BC, Weimer P, Rowan MG (1999) The effect of allochthonous salt on the petroleum systems of Northern Green Canyon and Ewing Bank (offshore Louisiana), Northern Gulf of Mexico. AAPG Search and Discovery Article #10003. http://www.searchanddiscovery.com/documents/98004/index.htm, accessed December 21, 2016.

McCartney BS, Bary B (1965) Echo sounding on probable gas bubbles from the bottom of Saanich Inlet, British Columbia. Deep-Sea Res 12:285

Miles JA (1989) Illustrated glossary of petroleum geochemistry. Clarendon, Oxford, UK. 137 p

Milkov A, Sassen R (2000) Thickness of the gas hydrate stability zone, Gulf of Mexico continental slope. Mar Petrol Geol 17:981–991

Milkov AV, Sassen R (2001) Estimate of gas hydrate resource, northwestern Gulf of Mexico continental slope. Mar Geol 179:71–83

Milkov AV, Sassen R, Novikova I, Mikhailov E (2000) Gas hydrates at minimum stability water depth in the Gulf of Mexico: Significance to geohazard assessment. Trans Gulf Coast Assoc Geol Soc 50:217–224

Milkov AV, Sassen R, Apanasovich TV, Dadashev FG (2003) Global gas flux from mud volcanoes: A significant source of fossil methane in the atmosphere and the ocean. Geophys Res Lett 30:31–46

NASA/Goddard Space Flight Center--EOS Project Science Office (2000) Scientists find that tons of oil seep into the Gulf of Mexico each year. ScienceDaily, January 27. http://www.sciencedaily.com/releases/2000/01/000127082228.htm. Accessed 14 Sept 2014

Nelson DC, Fisher CR (1995) Chemoautotrophic and methanotrophic endosymbiotic bacteria at vents and seeps. In: Karl DM (ed) Microbiology of deep-sea hydrothermal vent habitats. CRC Press, Boca Raton, FL, USA, pp 125–167

Oliver G, Rodrigues C, Cunha MR (2011) Chemosymbiotic bivalves from the mud volcanoes of the Gulf of Cadiz, NE Atlantic, with descriptions of new species of Solemyidae, Lucinidae and Vesicomyidae. ZooKeys 113:1–38

Paull CK, Hecker B, Commeau R, Freeman-Lynde RP, Neumann C, Corso WP, Golubic S, Hook JE, Sikes E, Curray J (1984) Biological communities at the Florida Escarpment resemble hydrothermal vent taxa. Science 226:965–967

Paull C, Jull A, Toolin L, Linick T (1985) Stable isotope evidence for chemosynthesis in an abyssal seep community. Nature 317:709–711

PBS (Public Broadcasting System) (2004) Extreme oil. http://www.pbs.org/wnet/extremeoil/history/prehistory.html. Accessed 14 Sept 2014

Peel F, Travis CJ, Hossack JR (1995) Genetic structural provinces and salt tectonics of the Cenozoic offshore US Gulf of Mexico: A preliminary analysis. In: Jackson MPA, Roberts DG, Snelson S (eds) Salt tectonics: A global perspective. AAPG Memoir 65, pp 153–175

Petty O (2010) Oil exploration. Handbook of Texas online. Uploaded on June 15, 2010. Modified on December 16, 2010. Texas State Historical Association. http://www.tshaonline.org/handbook/online/articles/doo15. Accessed 14 Sept 2014

Pflaum R, Brooks J, Cox B, Kennicutt M, Sheu DD (1986) Molecular and isotopic analysis of core gases and gas hydrates. Deep Sea Drilling Project Leg 96. In: Reports of the DSDP, Washington, DC, USA. 96 p

Pickwell GV (1967) Gas bubble production by Siphonophores. Report NUWC TP 8. Naval Undersea Warfare Center, San Diego, CA, USA

Pinero E, Marquardt M, Hensen C, Haeckel M, Wallmann K (2013) Estimation of the global inventory of methane hydrates in marine sediments using transfer functions. Biogeosciences 10:959–975

Prior DB, Doyle EH, Kaluza MJ (1989) Evidence for sediment eruption on deep sea floor, Gulf of Mexico. Science 243:517–519

Railsback LB (2011) Petroleum Geoscience and Subsurface Geology. Prepared for GEOL 4320/6320 Petroleum Geology Course. http://www.gly.uga.edu/railsback/PGSG/PGSGmain.html, accessed December 14, 2015

Reimer PJ, Baillie MGL, Bard E, Bayliss A, Beck JW, Bertrand CJH, Blackwell PG, Buck CE, Burr GS, Cutler KB, Damon PE, Edwards RL, Fairbanks RG, Friedrich M, Guilderson TP, Hogg AG, Hughen KA, Kromer B, McCormac G, Manning S, Bronk Ramsey C, Reimer RW, Remmele S, Southon JR, Stuiver M, Talamo S, Taylor F, van der Plicht J, Weyhenmeyer CE (2004) IntCal04 terrestrial radiocarbon age calibration. Radiocarbon 46:1029–1058

Ritger S, Carson B, Suess E (1987) Methane-derived authigenic carbonates formed by subduction-induced pore-water expulsion along the Oregon/Washington margin. Geol Soc Am Bull 98:147–156

Roberts HH, Aharon P (1994) Hydrocarbon-derived carbonate buildups of the northern Gulf of Mexico continental slope: A review of submersible investigations. Geo-Mar Lett 14:135–148

Roberts HH, Aharon P, Carney R, Larkin J, Sassen R (1990) Sea floor responses to hydrocarbon seeps, Louisiana continental slope. Geo-Mar Lett 10:232–243

Roberts HH, Cook DJ, Sheeldo MK (1992) Hydrocarbon seeps of the Louisiana Continental Slope: Seismic amplitude signature and seafloor response. Gulf Coast Assoc Geol Soc 42:349–362

Roberts HH, Carney R, Kupchik M, Fisher C, Nelson K, Becker E, Goehring L, Lessard-Pilon S, Telesnicki G, Bernard BB (2007) ALVIN explores the deep northern Gulf of Mexico slope. EOS Trans Am Geophys Union 88:341–343

Roberts HH, Feng D, Shedd W, Chen D (2009) Pervasive authigenic carbonate deposition at hydrocarbon seeps of the northern Gulf of Mexico: Geomorphic, petrographic, and geochemical characteristics. Gulf Coast Assoc Geol Soc Trans 59:653–661

Roberts HH, Shedd W, Hunt J (2010) Dive site geology: DSV ALVIN (2006) and ROV JASON II (2007) dives to the middle-lower continental slope, northern Gulf of Mexico. Deep Sea Res Part II: Top Stud Oceanogr 57:1837–1858

Rullkötter J, Nissenbaum A (1988) Dead sea asphalt in Egyptian mummies: Molecular evidence. Naturwissenschaften 75:618–621

Sackett WM (1977) Use of hydrocarbon sniffing in offshore exploration. J Geochem Explor 7:243–254

Sackett WM, Brooks JM (1973) Sources and sinks of light hydrocarbons in the Gulf of Mexico. J Geophys Res 78:5248–5258

Sassen R (1980) Biodegradation of crude oil and mineral deposition in a shallow Gulf Coast salt dome. Org Geochem 2:153–166

Sassen R (1987) Organic geochemistry of salt dome cap rocks, Gulf Coast salt basin. In: Lerche I, O'Brien JJ (eds) Dynamical geology of salt and related structures. Academic Press, San Diego, CA, USA, pp 631–649

Sassen R, Curiale J (2006) Microbial methane and ethane from gas hydrate nodules of the Makassar Strait, Indonesia. Org Geochem 37:977–980

Sassen R, MacDonald IR (1994) Evidence of structure H hydrate, Gulf of Mexico continental slope. Org Geochem 22:1029–1032

Sassen R, Chinn E, McCabe C (1988) Recent hydrocarbon alteration, sulfate reduction and formation of elemental sulfur and metal sulfides in salt dome cap rock. Chem Geol 74:57–66

Sassen R, Roberts HH, Aharon P, Larkin J, Chinn EW, Carney R (1993) Chemosynthetic bacterial mats at cold hydrocarbon seeps, Gulf of Mexico continental slope. Org Geochem 20:77–89

Sassen R, Joye S, Sweet ST, DeFreitas DA, Milkov AV, MacDonald IR (1999a) Thermogenic gas hydrates and hydrocarbon gases in complex chemosynthetic communities, Gulf of Mexico continental slope. Org Geochem 30:485–497

Sassen R, Sweet ST, Milkov AV, DeFreitas DA, Salata GG, McDade EC (1999b) Geology and geochemistry of gas hydrates, central Gulf of Mexico continental slope. Trans Gulf Coast Assoc Geol Socs 49:462–468

Sassen R, Sweet ST, DeFreitas DA, Milkov AV (2000) Exclusion of 2-methylbutane (isopentane) during crystallization of structure II gas hydrate in sea-floor sediment, Gulf of Mexico. Org Geochem 31:1257–1262

Sassen R, Losh SL, Cathles LM III, Roberts HH, Whelan JK, Milkov AV, Sweet ST, DeFreitas DA (2001a) Massive vein-filling gas hydrate: Relation to ongoing gas migration from the deep subsurface in the Gulf of Mexico. Mar Petrol Geol 18:551–560

Sassen R, Sweet ST, DeFreitas DA, Morelos JA, Milkov AV (2001b) Gas hydrate and crude oil from the Mississippi Fan Foldbelt, downdip Gulf of Mexico Salt Basin: Significance to petroleum system. Org Geochem 32:999–1008

Sassen R, Sweet ST, Milkov AV, DeFreitas DA, Kennicutt MC II (2001c) Thermogenic vent gas and gas hydrate in the Gulf of Mexico slope: Is gas hydrate decomposition significant? Geology 29:107–110

Sassen R, Roberts HH, Carney R, Milkov AV, DeFreitas DA, Lanoil B, Zhang C (2004) Free hydrocarbon gas, gas hydrate, and authigenic minerals in chemosynthetic communities of the northern Gulf of Mexico continental slope: Relation to microbial processes. Chem Geol 205:195–217

Schoell M (1980) The hydrogen and carbon isotopic compositions of methane from natural gases of various origins. Geochim Cosmochim Acta 44:649–661

Schumacher D (2012) Pre-drill prediction of hydrocarbon charge: microseepage-based prediction of charge and post-survey drilling results. AAPG Datapages/Search and Discovery Article 90174. In: CSPG©2014 CSPG/CSEG/CWLS GeoConvention 2012, (Vision) May 14-18, 2012, Calgary, AB, Canada

Seelke CR, Villareal MA, Ratner M, Brown P (2015) Mexico's oil and gas sector: Background, reform efforts, and implications for the United States. Congressional Research Service 7-5700. ww.crs.gov, R43313. 21 p

Shedd W, Boswell R, Frye M, Godfriaux P, Kramer K (2012) Occurrence and nature of "bottom simulating reflectors" in the northern Gulf of Mexico. Mar Petrol Geol 34:31–40

Sheriff RE, Geldart LP (1995) Exploration seismology, 2nd edn. Cambridge University Press, New York, NY, USA, pp 3–6

Socolofsky SA, Adams EE (2002) Multi-phase plumes in uniform and stratified crossflow. J Hydraul Res 40:661–672

Soley JC (1910) Oil fields of the Gulf of Mexico. Sci Am Suppl 69:1933–1938

Stoicescu M, Ionescu E (2014) Romanian achievement in the petroleum industry. In: CBU international Conference on Innovation, Technology Transfer and Education February 3-5, 2014, Prague, Czech Republic

Swinnerton JW, Linnenbom VJ (1967) Gaseous hydrocarbons in sea water: Determinations. Science 156:1119–1120

Texas Almanac, Texas Historical Association http://www.texasalmanac.com/topics/business/oil-and-texas-cultural-history; http://www.tshaonline.org/handbook/online/articles/doo15. Accessed 14 Sept 2014

Valentine DL, Reeburgh WS (2000) New perspectives on anaerobic methane oxidation. Environ Microbiol 2:477–484

Van Vleet ES, Sackett WM, Weber FF Jr, Reinhardt SB (1983) Input of pelagic tar into the Northwest Atlantic from the Gulf loop current: Chemical characterization and its relationship to weathered IXTOC-I oil. Can J Fish Aquat Sci 40:12–22

Van Vleet ES, Sackett WM, Reinhardt SB, Mangini ME (1984) Distribution, sources and fates of floating oil residues in the Eastern Gulf of Mexico. Mar Pollut Bull 15:106–110

Whiticar MJ (1999) Carbon and hydrogen isotope systematics of bacterial formation and oxidation of methane. Chem Geol 161:291–314

Wilson RD, Monaghan PH, Osanik A, Price LC, Rogers MA (1974) Natural marine oil seepage. Science 184:857–865

APPENDIX A: GEOCHEMICAL DEFINITIONS

Table A.1. Geochemical Definitions (from Miles 1989)

Item	Definition
$\delta^{13}C$	• ^{13}C is an isotope of carbon with six protons and seven neutrons. Microbes preferentially reject ^{13}C with the result that microbial gas and carbonates are depleted in ^{13}C • The standard established for carbon-13 work was the Pee Dee Belemnite (PDB) and was based on a Cretaceous marine fossil, *Belemnitella americana*, which was from the Pee Dee Formation in South Carolina. This material had an anomalously high ^{13}C:^{12}C ratio (0.0112372), and was established as $\delta^{13}C$ value of zero. Use of this standard gives most natural materials a negative $\delta^{13}C$. The calculation is: $$\delta^{13}C_{sample} = \left(\frac{^{12}C/^{13}C_{sample}}{^{12}C/^{13}C_{PDB}} - 1 \right) \times 1000$$ • The standards are used for verifying the accuracy of mass spectroscopy; as isotope studies became more common, the demand for the standard exhausted the supply. Other standards, including one known as VPDB (for Vienna PDB) have replaced the original • Methane has a very light $\delta^{13}C$ signature: biogenic methane of about -60 ‰; thermogenic methane about -40 ‰ • More commonly, the ratio is affected by variations in primary productivity and organic burial. Organisms preferentially take up light ^{12}C, and have a $\delta^{13}C$ signature of about -25 ‰, depending on their metabolic pathway
^{14}C, carbon dating	• Carbon-14, ^{14}C, or radiocarbon, is a radioactive isotope of carbon with a nucleus containing six protons and eight neutrons. Its presence in organic materials is the basis of the radiocarbon dating method pioneered by Willard Libby and colleagues (1949) to date archaeological, geological, and hydrogeological samples • Three naturally occurring isotopes of carbon are on earth: 99 % of the carbon is ^{12}C, 1 % is ^{13}C, and ^{14}C in trace amounts, i.e., making up about 1 part per trillion (0.0000000001 %) of the carbon in the atmosphere. The half-life of ^{14}C is 5,730 ± 40 years. ^{14}C decays into ^{14}N through beta decay. The primary natural source of ^{14}C on Earth is cosmic ray action upon nitrogen in the atmosphere, and it is therefore a cosmogenic nuclide. However, open-air nuclear testing from 1955 to 1980 contributed to this pool • Radiocarbon dating is a radiometric dating method that uses ^{14}C to determine the age of carbonaceous materials up to about 60,000 years old. Willard Libby and his colleagues developed the technique in 1949 during his tenure as a professor at the University of Chicago. Libby estimated that the radioactivity of exchangeable ^{14}C would be about 14 disintegrations per minute (dpm) per gram of pure carbon, and this is still used as the activity of the modern radiocarbon standard • One of the frequent uses of the technique is to date organic remains from archaeological sites. Plants fix atmospheric carbon during photosynthesis, so the level of ^{14}C in plants and animals when they die approximately equals the level of ^{14}C in the atmosphere at that time. However, it decreases thereafter from radioactive decay, allowing the date of death or fixation to be estimated. The initial ^{14}C level for the calculation can either be estimated, or else directly compared with known year-by-year data from tree-ring data (dendrochronology) up to 10,000 years ago (using overlapping data from live and dead trees in a given area), or else from cave deposits (speleothems), back to about 45,000 years before the present. A calculation or (more accurately) a direct comparison of ^{14}C levels in a sample, with tree ring or cave-deposit ^{14}C levels of a known age, then gives the wood or animal sample age-since-formation • Oils and gases are always much older than 50,000 years and so are made of the so-called dead carbon

Item	Definition
Alkane C_{12} nC_{12}	• An alkane is a saturated hydrocarbon. Alkanes consist only of hydrogen and carbon atoms, all bonds are single bonds, and the carbon atoms are not joined in cyclic structures but instead form an open chain. They have the general chemical formula C_nH_{2n+2}. Alkanes belong to a homologous series of organic compounds in which the members differ by a molecular mass of 14.03u (mass of a methanediyl group, $-CH_2-$, one carbon atom of mass 12.01u, and 2 hydrogen atoms of mass $\approx$ 1.01u each). There are two main commercial sources: crude oil and natural gas • Each carbon atom has four bonds (either C–H or C–C bonds), and each hydrogen atom is joined to a carbon atom (H–C bonds). A series of linked carbon atoms is known as the carbon skeleton or carbon backbone. The number of carbon atoms is used to define the size of the alkane (e.g., C2-alkane, C18-alkane, and C28-alkane). Other terms include the addition of the term *normal* as in nC_2, nC_{18}, nC_{28}, etc.
Isoprenoids Pristane Phytane	• Isoprenoids are hydrocarbons that contain double bonds. Their general chemical formula is C_nH_{2n+2}. A common origin for pristane ($C_{19}H_{40}$) and phytane ($C_{20}H_{42}$) is the phytyl side chain of chlorophyll *a* in phototrophic organisms and bacteriochlorophyll *a* and *b* in purple sulfur bacteria. Reducing or anoxic conditions in the sediments promote the cleavage of the phytyl side chain to yield phytol, which undergoes reduction to dihydrophytol and then to phytane. Oxic conditions promote the competing conversion of phytol to pristane. A common precursor for both pristane and phytane is inferred by the similarity of their $\delta^{13}C$ values, which commonly differ by no more than 0.3 ‰ (Peters et al. 2007) • Pristane and phytane are resistant to biodegradation. The ratios between pristane/nC_{17} and phytane/nC_{18} are established in non-biodegraded samples. As biodegradation intensifies, nC_{17} and nC_{18} are preferentially depleted and the value of the ratio increases • In addition, the boiling points of pristane/nC_{17} and phytane/nC_{18} are very close so if an oil is subject to evaporation the ratios will stay constant. It is possible therefore to distinguish evaporation and biodegradation mechanisms in partially depleted oils
Biomarker	• Compounds, or characteristics of compounds, found in petroleum or rock extracts that indicate an unambiguous link with a natural product are known as biological markers, biomarkers for short. Diagenetic changes that occur in sediment may alter functional groups and bonds in the natural compound, but the carbon skeleton of the compound remains the same. The simplest compounds that are biomarkers are normal alkanes derived from plant waxes and fatty acids, isoalkanes, and isoprenoids. Chlorophyll decomposes to porphyrin and to pristane and phytane from the side chain

CHAPTER 6

COASTAL HABITATS OF THE GULF OF MEXICO

Irving A. Mendelssohn,[1] Mark R. Byrnes,[2] Ronald T. Kneib[3] and Barry A. Vittor[4]

[1]Louisiana State University, Baton Rouge, LA 70808, USA; [2]Applied Coastal Research and Engineering, Mashpee, MA 02649, USA; [3]RTK Consulting Services, Hillsboro NM 88042, USA; [4]Barry A. Vittor & Associates, Mobile, AL 36695, USA
imendel@lsu.edu

6.1 INTRODUCTION

The Gulf of Mexico (GoM) is the ninth largest body of water in the world (including ocean basins) with an outer shoreline extending approximately 6,077 kilometers (km) (3,776 miles [mi]) from the Florida Keys to the northwest coast of Cuba (Moretzsohn et al. 2012). The Gulf encompasses an area of approximately 1.5 million km^2 (0.58 million mi^2), and with an average depth of about 1,615 meters [m] (5,300 feet [ft]), it provides habitat for a myriad of marine, shoreline, and estuarine flora and fauna that occupy a diverse suite of coastal ecosystems (NOAA 2011). The Gulf is among the most biologically productive marine environments in the world, producing 78, 62, and 16 % of U.S. shrimp, oyster, and fishery landings, respectively (NOAA 2011). The productive value (market value) of the Gulf has been estimated at 124 billion U.S. dollars annually for Mexico and the United States from oil and gas, fisheries, ports and shipping, and tourism (Yoskowitz 2009). The Gulf provides a variety of important ecosystem services from regulating greenhouse gases to providing food to supporting recreational activities, all of which enhance the diverse social cultures within the GoM region (see Section 6.4.4).

The GoM shoreline includes a variety of coastal habitats, ranging from submerged seagrass beds to intertidal wetlands to supratidal sand dunes and maritime forests. Included in this habitat diversity are barrier islands, hypersaline lagoons, herbaceous marshes, forested wetlands of mangroves and cypress swamps, beaches, intertidal flats, oyster and coral reefs, subaquatic vegetation, and sponge beds (NOAA 2011). The following review emphasizes vegetated habitats of coastal strand beaches, as well as adjacent saline wetlands and subaqueous environments.

Coastal strand beaches, and adjacent marsh and subaqueous habitats of the GoM, extend as far north as approximately 30.5°N near the Florida Panhandle shoreline to as far south as 18°N along the Veracruz-Tabascan shoreline of Mexico. The westernmost extent is approximately 98°W along the Tamaulipas shoreline of Mexico, and the far eastern extent is along the Matanzas shoreline in Cuba at about 80.6°W (Figure 6.1). This geographic range spans geophysical boundaries and climatic zones (temperate, subtropical, and tropical), giving rise to physiographic and ecological classifications of shoreline habitats.

The primary goal of this paper is to provide a conceptual framework from which to understand the ecology of coastal habitats in the GoM including the physical and geological processes that control their formation. Given the importance of vegetated habitats along coastal beaches and marshes of the GoM, their documented societal value, and the present pressures for development, emphasis will be placed on vegetated shoreline habitats. These include intertidal wetlands, such as salt marshes and mangroves, intertidal to subtidal seagrasses and

C.H. Ward (ed.), *Habitats and Biota of the Gulf of Mexico: Before the Deepwater Horizon Oil Spill*,
DOI 10.1007/978-1-4939-3447-8_6

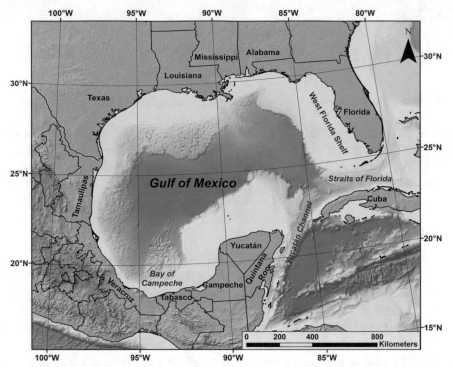

Figure 6.1. Geographic extent of the Gulf of Mexico (basemap from CEC, 2007; French and Schenk, 2005).

flats, and supratidal barrier strand habitats, including beaches, sand dunes, and maritime forests on barrier islands. The landward limit of this review is the boundary between salt and brackish marshes, although when relevant, processes, biota, and/or habitats farther landward will be discussed. The seaward limit to which physical and biotic processes are addressed is the extent of active littoral transport (approximate location of the 10-m [33-ft] depth contour). From a geological standpoint, Holocene processes and sedimentary deposits are emphasized.

Although much is known about vegetated marine habitats of the GoM, no single document has attempted to review the diverse geology and ecology of coastal shoreline habitats in this vast geographic region. This review provides a summary of the geological and ecological status of shoreline habitats in the GoM, emphasizing vegetated ecosystems. It provides a baseline and general understanding of the operative physical and geological processes influencing coastal habitat formation and evolution, as well as the ecological structure and function of habitats.

6.2   PHYSIOGRAPHIC FRAMEWORK

Gulf Coast margins are characterized by persistent geochemical and biological interactions where continental and marine waters mix, and there is a continual exchange of large amounts of sediment, organic matter, and energy with the open Gulf. Topographic features, coastal and nearshore circulation, tidal mixing, and freshwater inflow from rivers and groundwater all contribute to small-scale interactions that control coastal habitat distribution and response (see Section 6.4). Along the north and south margins of the Gulf, river systems deliver large quantities of organic matter, sediments, and nutrients, resulting in high rates of sediment deposition and primary productivity, along with episodic sediment resuspension and redistribution (Robbins et al. 2009). On the eastern and western Gulf margins, river input is relatively small, and Loop Current and upwelling processes predominate (Schmitz 2003; Hine and Locker 2011).

Various classification systems have been used to describe marine and terrestrial ecosystems within and adjacent to the GoM relative to watershed characteristics and oceanographic processes. Because the primary focus of this review is to describe the evolution of vegetated marine habitats of coastal strand beaches and adjacent wetlands (coastal habitat at the land–water interface), the marine ecoregion classification of Wilkinson et al. (2009) was used to illustrate natural environmental variability and the potential impact of human activities along the margins of the GoM. However, terrestrial ecoregions describe the inland character of subaerial coastal habitats at the marine boundary. As such, both systems are described below and referred to throughout the text when discussing shoreline change and coastal habitat distribution.

6.2.1 Marine Ecoregions

The GoM provinces of Robbins et al. (2009) largely overlap marine ecoregions established by Wilkinson et al. (2009), including (1) the South Florida/Bahamian Atlantic Marine Ecoregion, (2) the Northern GoM Marine Ecoregion, (3) the Southern GoM Marine Ecoregion, and (4) the Caribbean Sea Marine Ecoregion (Figure 6.2). Furthermore, Spalding et al. (2007) classified waters surrounding Cuba and a larger portion of the central Caribbean as the Greater Antilles Marine Ecoregion. The marine ecoregion classification was established to address ecosystem-based conservation and sustainable development strategies. Three levels were identified for each ecoregion, except for the Greater Antilles. Level I captures largest-scale ecosystem differences, such as large water masses and currents and regions of consistent sea surface

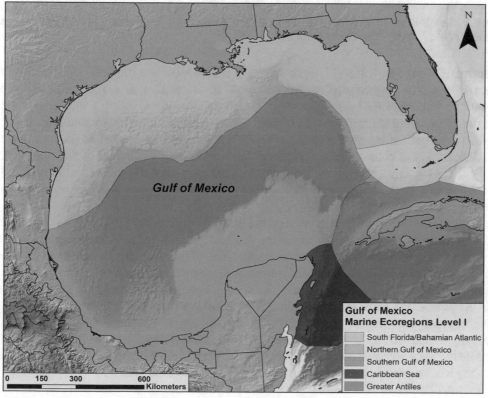

Figure 6.2. Level I marine ecoregions of the GoM (data from Spalding et al. 2007; Wilkinson et al. 2009; and basemap from CEC 2007; French and Schenk 2005).

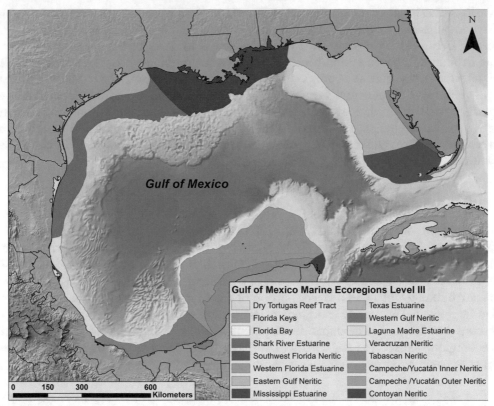

Figure 6.3. Level III marine ecoregions for the GoM (data from Wilkinson et al. 2009 and basemap from CEC 2007; French and Schenk 2005).

temperature (Figure 6.2). Level II reflects the break between neritic and oceanic areas and is delineated based on physiographic features (e.g., continental shelf and slope). This sub-level indicates the importance of depth for determining the location of benthic marine communities and primary physiographic features for controlling current flows and upwelling. Level III is limited to the continental shelf and based on differences within the neritic zone as determined by local water mass characteristics, regional landforms, and biological community type (Figure 6.3). Spalding et al. (2007), Wilkinson et al. (2009), Yáñez-Arancibia and Day (2004), and Yáñez-Arancibia et al. (2009) summarized these ecological regions and related environmental characteristics.

Although marine province and ecoregion characterizations for the GoM basin are generally consistent, marine ecoregion classification provides details more closely related to coastal habitat identification. The South Florida/Bahamian Atlantic region includes the southern tip of the Florida Peninsula, where groundwater discharge is important and sandy beaches, mangroves, seagrasses, and coral reefs dominate. This marine ecoregion extends from the Florida Keys north to southern Keewaydin Island (just south of Naples, Florida) and comprises four subregions that reflect physiographic and hydrologic complexities associated with this biologically unique area. Level III subregions include the Dry Tortugas Reef Tract, Florida Keys, Florida Bay, Shark River Estuarine, and Southwest Florida Neritic (Figure 6.3). Habitats of this region are often underlain by a calcium carbonate substrate, a driver of vegetation structure and function. Sea surface temperatures vary from 22.5 °C (72.5 °F) in the winter to 28 °C (82 °F) in the summer (Figure 6.4). Although the Cuban shoreline of the GoM is not included in this classification level, it is part of the Greater Antilles Marine Ecoregion (Level I) and is dominated by limestone substrate similar to that of southern Florida. Furthermore, physical processes and ecological characteristics along the northwestern Cuban shoreline are similar to those of the Florida Keys.

The Caribbean coast of Mexico is the northern portion of the Caribbean Sea Ecoregion, named the Contoyan Neritic sub-region (Wilkinson et al. 2009). The sub-region name reflects proximity to Isla Contoy, located just east of the Campeche/Yucatán Inner Neritic zone (Figure 6.3). The area generally has lower average sea surface temperatures (28 °C [82 °F] in summer and 22.5 °C [72.5 °F] in winter; Figure 6.4) and lower nutrient loading than the Southern GoM Marine Ecoregion. Coral reefs, carbonate beaches, mangrove forests, and seagrass meadows are common coastal habitats, and water flow through the Yucatán Channel has a primary influence on coastal and shelf ecosystems. Beaches are primary tourist attractions of economic importance to the region.

The Northern GoM Ecoregion is a warm-temperate area in the GoM basin that contains approximately 60 % of tidal marshes in the United States, freshwater inputs from 37 major rivers, and numerous nursery habitats for fish (Figure 6.2) (Wilkinson et al. 2009). Average sea-surface summer temperatures in this region range from 28 to 30 °C (82 to 86 °F), while winter temperatures range from 14 to 24 °C (57 to 75 °F) (Figure 6.4). This is generally a region of high nutrient loading and includes biotic communities such as mangroves, salt marshes, and seagrasses, coastal lagoons and estuaries, and low river basins. This ecoregion extends from southern Keewaydin Island on the west coast of Florida to just south of Barra del Tordo in the State of Tamaulipas, Mexico and comprises six subregions that reflect the influence of tropical currents from the Caribbean Sea through the Yucatán Channel, the Loop Current and associated warm-water eddies, freshwater contributions from major river systems and groundwater, and outflows through the Straits of Florida. Level III subregions include the Western Florida Estuarine, Eastern Gulf Neritic, Mississippi Estuarine, Texas Estuarine, Laguna Madre Estuarine, and the Western Gulf Neritic (Figure 6.3).

The Southern GoM Ecoregion encompasses tropical waters of Mexico that support a variety of coastal habitats, including coastal lagoons, estuaries, beaches and dunes, mangroves, seagrass beds, and coral reefs. Air temperatures vary little between winter and summer, averaging about 26 °C (79 °F), although sea surface temperatures vary between 24 and 28.5 °C (75 and 83 °F), respectively (Figure 6.4). This is also a region of generally high nutrient loading and some local upwelling. The continental margin in this region is very topographically diverse, including a relatively narrow continental shelf (6 to 16 km [3.8 to 10 mi] wide) in the southwestern portion of the ecoregion with beaches and estuaries composed primarily of reworked fluvial sediment, interspersed with coastal rocky outcrops (Moreno-Casasola 2007; Contreras-Espinosa and Castañeda-Lopez 2007). In contrast, the southeastern coast of Campeche and Yucatán is fronted by a wide and shallow carbonate continental shelf and carbonate sand beaches. Many of the same biotic communities present in the northern GoM are common in this ecoregion, although coastal salt marshes are almost completely replaced by mangroves, and coral reefs and seagrasses become important. The Southern Gulf ecoregion extends from Barra del Tordo, along all six Gulf-facing States in Mexico, to the northeastern end of the Yucatán Peninsula. Subregions include Veracruz Neritic, Tabasco Neritic, Campeche/Yucatán Inner Neritic, and Campeche/Yucatán Outer Neritic (Figure 6.3).

Marine ecoregions for Cuba were not classified beyond Level I (Greater Antilles; Spalding et al. 2007); however, coastal systems within the Central Caribbean Ecoregion described by Sullivan-Sealey and Bustamante (1999) (equivalent to the Greater Antilles Ecoregion of Spalding et al. 2007) were classified based on dominant community type. Coral reefs, seagrass beds, and mangrove-dominated habitat are common along the northwestern Cuba coast. Further discussion of this classification is presented below in Section 6.4.2.

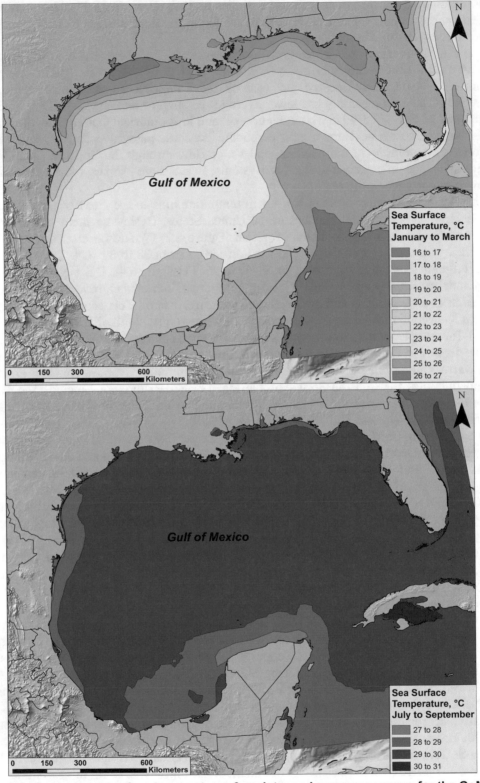

Figure 6.4. Average sea surface temperatures for winter and summer seasons for the GoM (data from Casey et al. 2010; basemap from French and Schenk 2005).

6.2.2 Terrestrial Ecoregions

Although oceanographic processes associated with specific marine ecoregions influence habitat development at the land–sea interface, geology, soils, and watershed characteristics associated with terrestrial ecoregions exert primary control on physiography of coastal habitats and nearshore water bottoms (Griffith et al. 2007). As such, coastal habitat descriptions within the context of GoM marine ecoregions may refer to terrestrial ecoregions when examining habitat distribution and change. Like marine ecoregions, their terrestrial counterparts portray areas within which relative homogeneity exists among physical and biological components of an ecosystem. Thirteen terrestrial ecoregions border the GoM from Florida to Cuba; four in the United States and ten in Mexico and Cuba (Figure 6.5).

The Southern Coast and Islands portion of the Southern Florida Coastal Plain Ecoregion extends from Keewaydin Island south to Key West and the Dry Tortugas (Griffith et al. 1997). The region includes the Ten Thousand Islands and Cape Sable, the islands of Florida Bay, and the Florida Keys (Figure 6.6). It is an area of mangrove swamps and coastal marshes, coral reefs, coastal strand vegetation on beach ridges, and limestone rock islands. The area has a nearly frost-free climate with mean annual temperature of 22 to 25 °C (72 to 77 °F) and mean annual precipitation of 1.34 m (4.4 ft) (Figure 6.7; Wiken et al. 2011). It is characterized by low-relief topography with wet soils. Relatively minor differences in elevation have significant impact on vegetation and diversity of habitat. Limestone underlies surficial sand and gravel and areas of peat and clay.

North of this area lies the Southwestern Florida Flatwoods portion of the Southern Coastal Plain Ecoregion (Level III), which includes barrier islands and Gulf coastal flatlands between

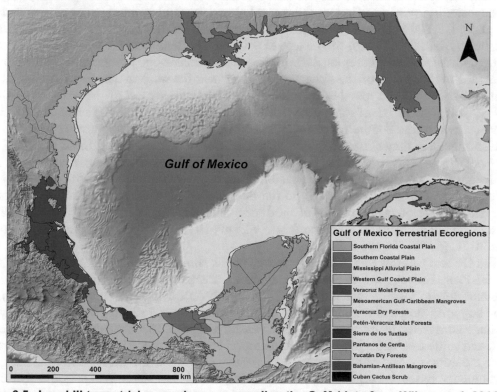

Figure 6.5. Level III terrestrial ecoregions surrounding the GoM (data from Wiken et al. 2011 and basemap from CEC 2007; French and Schenk 2005). Cuban ecoregions were developed by Olson et al. (2001).

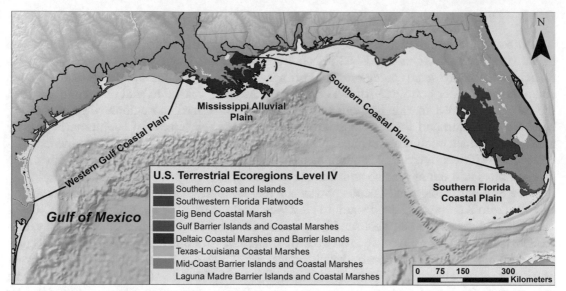

Figure 6.6. Level IV terrestrial ecoregions for the U.S. GoM coast (data from USEPA 2011 and basemap from Amante and Eakins 2009).

Anclote Key and Keewaydin Island (Figure 6.6) (Griffith et al. 1997). The terrain consists mostly of flat plains, and also includes sandy beaches, coastal lagoons, marshes, and swampy lowlands. The Pinellas Peninsula portion of the Southern Coastal Plain Ecoregion is underlain by deeply weathered sand hills of Miocene age in the north and Pleistocene-age sand, shell, and clay deposits in the south. Besides the coastal strand, natural vegetation consists of longleaf pine and pine flatwoods. The dominant characteristic of the region is the Clearwater/St. Petersburg urban area. North of Anclote Key is the Big Bend Coastal Marsh segment of the Southern Coastal Plain Ecoregion, where Miocene to Eocene-age limestone resides at or near the surface to the mouth of the Ochlockonee River near the western margin of Apalachee Bay (Figure 6.6). Coastal salt marshes and mangroves characterize most of the coast.

The Gulf Barrier Islands and Coastal Marshes Ecoregion (Level IV) represents the westernmost extent of the Southern Coastal Plain Ecoregion (Figure 6.6). This area contains salt and brackish marshes, dunes, beaches, and barrier islands that extend from Saint George Sound near Apalachicola Bay to western Mississippi Sound at the Pearl River. Quaternary quartz sand, shell fragments, silt, clay, muck, and peat are primary physical components of coastal deposits. Cordgrass and saltgrass are common in the intertidal zone, while coastal strand grasses and pine scrub vegetation occur on parts of the dunes, spits, and barrier islands (Griffith et al. 2001). Average annual precipitation is approximately 1.5 m (4.9 ft), and average annual temperature is about 20 °C (68 °F) (Figure 6.7).

The Deltaic Coastal Marshes and Barrier Islands component (Level IV) of the Mississippi Alluvial Plain Ecoregion (Level III) encompasses brackish and saline marshes of the south Louisiana deltaic plain between the Pearl River and Vermilion Bay (Daigle et al. 2006). The region supports vegetation tolerant of brackish or saline water including salt marsh cordgrass, marshhay cordgrass, black needlerush, and coastal saltgrass. Black mangrove occurs in a few areas, and some live oak is found along old natural levees. Barrier islands in this region are low relief, medium to fine sand deposits with beach grasses in elevated dune and backshore environments. Extensive organic deposits lie mainly at or below sea level in periodically flooded settings, and inorganic silts and clays are soft and generally have high water content. Wetlands and marshes act as a buffer to help moderate flooding and tidal inundation during storm events. Flood control levees and channelization of the Mississippi River have led to a reduction

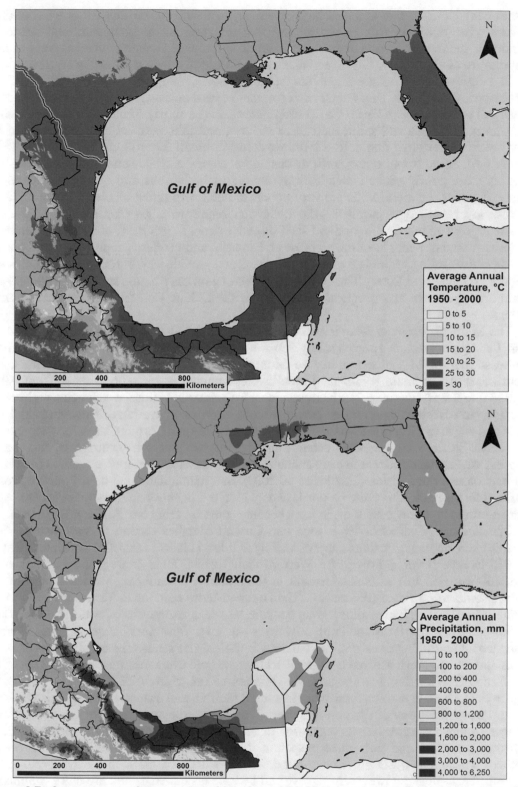

Figure 6.7. Average annual temperature (data from CEC 2011a) and precipitation (data from CEC 2011b) for terrestrial areas adjacent to the GoM. Basemap credits: © 2014 ESRI, DeLorme, HERE.

in sediment input to marshes and bays, resulting in delta erosion and accelerated relative sea-level rise (due primarily to subsidence) that threaten the environmental and economic stability of the region. This ecoregion has a humid subtropical climate with an average annual temperature of about 21 °C (70 °F) and mean annual rainfall of about 1.7 m (5.6 ft) (Figure 6.7).

In southwestern Louisiana and southeastern Texas, marginal deltaic deposits of the Mississippi-Atchafalaya River system form the Texas-Louisiana Coastal Marshes section (Level IV) of the Western Gulf Coastal Plain Ecoregion (Level III). The region is characterized by extensive brackish and saline marshes, few bays, and thin, perched barrier beaches at the GoM marsh-water edge that extend from western Vermilion Bay (LA) to High Island, Texas (Figure 6.6). Streams and rivers north of this region supply nutrients and sediments to coastal marshes from poorly consolidated Tertiary coastal plain deposits and Quaternary alluvium, terrace deposits, or loess. Soils are very poorly drained with muck or clay surface textures. Narrow, low relief ridges paralleling the modern shoreline are called *cheniers*, for the live oak trees that grow on these relic sand and shell shoreline deposits (McBride et al. 2007). Extensive cordgrass marshes occur in more saline areas. Estuaries and marshes support abundant marine life, supply wintering grounds for ducks and geese, and provide habitat for small mammals and alligators (Daigle et al. 2006). This coastal marsh ecoregion has a humid subtropical climate and average temperature and precipitation similar to the Deltaic Coastal Marshes ecoregion to the east.

The Mid-Coast Barrier Islands and Coastal Marshes portion of the Western Gulf Coastal Plain Ecoregion extends approximately 350 km (217 mi) from the Bolivar Peninsula on the southeast margin of Galveston Bay to Mustang Island, just south of Port Aransas, TX (Figure 6.6). The climate is sub-humid and average annual precipitation ranges from 0.9 to 1.2 m (3.0 to 3.9 ft) (Figure 6.7). The region includes primarily Holocene sediments with saline, brackish, and freshwater marshes, barrier islands with minor washover fans, and tidal flat sands and clays. In estuarine areas between Matagorda Bay and Corpus Christi Bay, some older Pleistocene barrier island deposits occur. Smooth cordgrass, marshhay cordgrass, and coastal saltgrass vegetation dominate in more saline zones. Barrier islands support extensive foredunes and back-island dune fields (Griffith et al. 2007). Salt marsh and wind-tidal flats are mostly confined to the backside of the barrier islands with fresh or brackish marshes associated with river-mouth delta areas. Black mangroves become common from San Antonio Bay south.

The Laguna Madre Barrier Islands and Coastal Marshes section of the Western Gulf Coastal Plain Ecoregion extends approximately 200 km (124 mi) from Corpus Christi Bay to the Rio Grande River; however, the Western Gulf Coastal Plain Ecoregion encompasses an extra 250 km (155 mi) of coastal habitat in the State of Tamaulipas, Mexico from the Rio Grande south to La Pesca (Figure 6.5). The Laguna Madre sections in Texas and Mexico are distinguished by their hypersaline lagoon systems, vast seagrass meadows, wide tidal mud flats, and long, narrow barrier islands with numerous washover fans. Surficial geology is primarily Holocene alluvium, beach ridges, and barrier island-tidal flat sands. The coastal zone of south Texas and northeastern Mexico has a semi-arid climate and average annual precipitation of 0.7 to 0.8 m (2.3 to 2.6 ft); average annual temperatures range from 22 to 25 °C (72 to 77 °F) (Figure 6.7). There is extreme variability in annual rainfall, and evapotranspiration is generally two to three times greater than precipitation. Tropical storms and hurricanes can bring large changes to this ecoregion. Grass vegetation of barrier island systems consists mostly of bitter panicum, sea oats, and gulf dune paspalum. Marshes generally are less extensive on the southern Texas and northern Mexico coast. A few stands of black mangrove occur along the south Texas coast; however, mangrove and herbaceous marsh habitat are more common in the Mexican part of this ecoregion along the fringes of backbarrier lagoons. Along the Tamaulipas coast, beaches are low profile and sand rich with narrow or no lagoons.

As no major rivers flow into the Texas Laguna Madre, the lagoon water can be hypersaline. Combined with the Laguna Madre of Tamaulipas, Mexico, it is the largest hypersaline system in the world (Tunnell 2002a). The shallow depth, clear water, and warm climate of this lagoon are conducive to seagrass production. Nearly 80 % of all seagrass beds in Texas are now found in the Laguna Madre (Tunnell 2002a).

The Veracruz Moist Forests Ecoregion along the eastern coast of Mexico extends from La Pesca to the Farallón Lagoon in Veracruz (Figure 6.5). This ecoregion encompasses lowlands of the eastern slopes of the Sierra Madre Oriental. It is composed of sedimentary rocks from the Cretaceous period, and the soils are shallow but rich in organic matter. The climate is tropical humid, with rain during 7 months of the year. Mean annual temperatures fluctuate between 20 and 24 °C (68 and 75 °F), and average annual precipitation ranges between 1.1 and 1.6 m (3.6 and 5.2 ft) (Figure 6.7) (WWF 2014a). Numerous fluvial systems drain geologic deposits that provide sediment and water to coastal saltwater lagoons and Gulf beaches (Contreras-Espinosa and Castañeda-Lopez 2007; Moreno-Casasola 2007). The ecoregion encompasses a variety of coastal physiography from sandy beaches and lagoons to rocky cliffs composed primarily of Mesozoic and Cenozoic sedimentary rocks. Between Laguna de Tamiahua to the rocky headland at Playa Munéco, clastic sediment beaches are supplied by upland sedimentary sandstone, shale, and limestone (Wiken et al. 2011). Extensive coastal sand dunes are common and sandy/cobble pocket beaches exist between rocky headlands. Mangroves are common in coastal lagoons and estuaries.

The Veracruz Dry Forests Ecoregion is located in central Veracruz, surrounded by tropical forest ecoregions (Figure 6.5). The region is located in the coastal plain of central Veracruz, north of the Santa Martha and San Andrés volcanoes. The climate is tropical dry due to the influence of the Chiconquiaco Sierra Mountains. The soils are calcareous and derived from sedimentary rocks, and the area is relatively humid (<1 m/year [3.3 ft/year] rainfall). These characteristics allow the development of a dry forest along the coast, near Veracruz City. The forests constitute the preferred habitat for many birds, including migratory species that use coastal environments of the region as a stopover during their migratory route (WWF 2014b).

Adjacent and south of the Veracruz dry forests is the Petén-Veracruz Moist Forests Ecoregion. This moist forest ecoregion consists of a mixture of wetlands, riparian habitats, and moist forests that extend from southern Veracruz and into the State of Tabasco (Figure 6.5). Soils of this ecoregion are some of the most productive in the country, resulting in high species richness and high desirability for local agriculture. As such, much of the natural habitat has been cultivated for agriculture, and it is estimated that only a small percentage of the original habitat remains (Hogan 2013a). Beach and estuarine deposits in the Petén-Veracruz Moist Forests Ecoregion are influenced by fluvial systems that primarily drain Cenozoic sedimentary sandstones. Quaternary alluvial, marsh, and lacustrine deposits are common near the coast. The Papaloapan watershed is a dominant physiographic feature in this ecoregion (Wiken et al. 2011).

The Sierra de los Tuxtlas small coastal ecoregion is bounded on landward sides by the Petén-Veracruz Moist Forests Ecoregion. Formed from volcanic activity, coastal deposits are primarily rocky cliffs and sandy pocket beaches between rocky headlands. Upland environments are thickly forested and the area is recognized as an important zone for migratory birds (Valero et al. 2014).

The Pantanos de Centla Ecoregion is located in the eastern part of Tabasco and the western portion of Campeche south and west of Laguna de Términos (Figure 6.5). The ecoregion is biologically rich and contains almost 12 % of aquatic and sub-aquatic vegetation in Mexico. Soils of this ecoregion are quite productive and species richness is high. Deltaic deposits and extensive marsh habitat are primary components of the Centla region of Tabasco from the

Grijalva-Usumacinta watershed. Lowlands fringing Laguna de Términos (Campeche) contain large expanses of mangroves (ParksWatch-Mexico 2003).

The Mesoamerican Gulf-Caribbean Mangroves Ecoregion resides at various locations along the Mexican GoM coast, primarily associated with saltwater lagoons and estuaries (Figure 6.5). Mangroves north and west of the Alvarado Lagoon (Veracruz) thrive in coastal areas exposed to riverine water and sediment input throughout the states of Tamaulipas and Veracruz. Mangroves grow on flat terrain and are influenced by the Tonala River near the border between Tabasco and Veracruz, the Papaloapan in northern Veracruz, and the Pánuco River near the border between Tamaulipas and Veracruz. Mangroves grow on clay soils that are deep and rich in organic matter. The climate is tropical sub-humid with summer rains; temperature oscillations are very slight, and the levels of humidity are relatively high with between 1.2 and 2.5 m (3.9 and 8.2 ft) of annual rainfall. Red, black, and white mangroves are the dominant species, and as with most mangrove areas, local herbaceous flora is not abundant because they are generally intolerant to frequent flooding (Hogan 2013b; WWF 2014c).

Mangrove habitat flourishes surrounding Laguna de Términos in the State of Tabasco, Mexico. The delta of the Usumacinta and Grijalva Rivers supports mangrove habitat in this region as well. Soils are deep and rich in organic matter, which make them among the most productive soils in Mexico. The climate is warm and humid with abundant rain in summer, and this mangrove ecoregion is one of the wettest, with about 1.6 m (5.2 ft) of rain annually. Usumacinta mangroves and the nearby floodlands are considered the most important wetlands of the country, referred to as the Pantanos de Centla (Figure 6.5). The Grijalva-Usumacinta fluvial system and deltaic plain supply the largest discharge of fresh water to the southern GoM. Intrusions of salt water during the dry season allow mangroves to form up to 30 km (18.6 mi) inland. Vegetation is established in soils with very high organic matter content. Red, white, and black mangroves are key species in the community (WWF 2014d).

Homogenous limestone layers from Tertiary and Quaternary periods characterize the western portion of the Yucatán Peninsula, where the Yucatán Dry Forest Ecoregion abuts the coast near the city of Campeche (Figure 6.5) (WWF and Hogan 2013). The area is relatively dry, with average annual rainfall of about 0.5 m (1.6 ft) and average annual temperatures between 24 and 26 °C (75 and 79 °F) (Figure 6.7) (Wiken et al. 2011). Mangroves dominate coastal vegetation and very little surface water drains to the coast; drainage is primarily subterranean. Beach sand is primarily limestone particles (Moreno-Casasola 2007). Petenes mangroves characterize the northwestern edge of the Yucatán Peninsula (WWF and Hogan 2014a). The area is continuously flooded, though rivers are absent from this portion of the Mesoamerican Gulf-Caribbean Mangroves Ecoregion. Instead, springs form in the bottom of the mangroves, providing fresh water to help regulate salinity and raise nutrient concentrations. The Celestún Lagoon is the most important hydrologic feature within Petenes mangroves portion of the ecoregion. Soils form on a karstic limestone platform and are shallow in some areas and deep in others. Different types of mangroves grow in this area depending on the levels of salinity and the amount of nutrients present. Coastal fringe mangrove habitat contains greater nutrients and is composed of taller trees (15 to 20 m [49 to 66 ft]) as compared with pygmy mangrove habitat inland of the fringe where shorter trees (less than 5 m) dominate. Both types of mangrove habitat contain primarily red and white tree species; black mangroves are scarce because they are relatively intolerant of persistent floods.

The eastern Yucatán Peninsula has similar physiographic and ecologic characteristics. It has a mean annual temperature of 26 °C (79 °F), and there are warm, sub-humid climates with intermediate rains (Wiken et al. 2011). Mangroves dominate coastal vegetation, and white limestone sand beaches are present. Drainage is completely subterranean, and carbonate rocks are of Upper Tertiary origin.

Two terrestrial ecoregions dominate coastal habitats in western Cuba adjacent to the GoM. The Cuban Cactus Scrub Ecoregion is always associated with dry coastal climates and is located in patches along west coast shorelines (Figure 6.5) (WWF 2014e). The ecoregion has a desert-like appearance with average annual precipitation of 0.8 m (2.6 ft) or less and average temperatures of 26 °C (79 °F) (Figure 6.7). The principal soil type is derived from coralline limestone rock and has a karstic structure. Beaches are generally narrow and are composed of coralline sand and pebble fragments. Although the Bahamian-Antillean Mangroves Ecoregion is primarily associated with the Bahamas islands, coastal habitat on the submerged limestone bank along the northwestern Cuban shoreline is included in this ecoregion (Figure 6.5) (WWF and Hogan 2014b). Porous limestone substrate and relatively low precipitation means no major rivers supply nutrients and sediment to the coast. Coral reefs and carbonate islands are common seaward of the mainland coast, and mangroves thrive in these environments. Mainland beaches are composed of coralline sand and carbonate shell deposits, and seagrass beds in association with mangroves are common.

6.3 PHYSICAL SETTING

River-dominated shelves and energetic tropical cyclonic events that control the development of coastal habitats characterize the GoM ocean basin. Gulf waters are bordered by the United States to the north (Florida, Alabama, Mississippi, Louisiana, Texas), six Mexican states to the south and west (Tamaulipas, Veracruz, Tabasco, Campeche, Yucatán, Quintana Roo), and the island of Cuba to the southeast (Figure 6.1). The Gulf basin extends approximately 1,600 km (994 mi) from east to west and about 900 km (560 mi) from north to south. The Gulf-facing shoreline from Cape Sable, Florida to the tip of the Yucatán peninsula extends approximately 5,700 km (3,542 mi), with another 380 km (236 mi) of shore on the northwest margin of Cuba. When bays and other inland waters are included, total shoreline length increases to at least 27,000 km (16,777 mi) in the United States alone (Moretzsohn et al. 2012). Based on bathymetric contours for the Gulf basin, shallow and intertidal regions (<20 m [66 ft] deep) represent about 11 % of the Gulf basin, whereas shelf, slope, and abyssal regions comprise approximately 25, 38, and 26 %, respectively (Figure 6.8). Average water depth for the basin is on the order of 1,615 m (5,300 ft), and maximum depth is about 4,400 m (14,435 ft) (Sigsbee Deep).

Water and sediment presently are supplied to the Gulf by more than 150 rivers, including 20 major river systems (Robbins et al. 2009). Freshwater inflow to the Gulf is approximately 10^{12} cubic meters per year (m^3/year) (35 × 10^{12} ft^3/year), with about 63 % of the total arriving via the Mississippi-Atchafalaya watershed (Moretzsohn et al. 2012). Other U.S. rivers contribute another 14 %, and the remaining 23 % is supplied from Mexico and Cuba. South Texas receives the least rainfall among Gulf coastal areas. Groundwater contributions are significant in many areas, especially the eastern and southern margins of the Gulf.

Thirty-nine major estuarine systems exist along the Gulf coastline, of which 82 % are located within the Northern Gulf Marine Ecoregion and 18 % along the Southern Gulf coast (Wilkinson et al. 2009; Moretzsohn et al. 2012). Marine-dominated bays occur in the eastern Gulf, whereas river-dominated estuaries characterize the northern Gulf and coastal lagoons are common in the Southern Gulf (Moretzsohn et al. 2012). More than 14,500 km^2 (5,600 mi^2) of estuarine wetlands reside along Gulf coastlines. Approximately one-third consist of forested mangrove wetlands, with the remainder being herbaceous marsh (Wilkinson et al. 2009). Tidal influence on estuaries is relatively uniform (in contrast to freshwater influence), with tide ranges generally less than 1 m (3.3 ft) (Stumpf and Haines 1998).

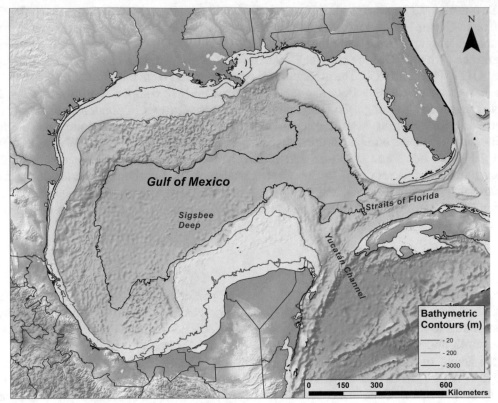

Figure 6.8. Map illustrating primary depth contours defining the GoM basin (contour data from Becker et al. 2009; basemap from Amante and Eakins 2009).

The following sub-sections provide a summary of geologic controls regarding formation of the GoM Basin, terrestrial and watershed controls on coastal habitat formation, primary oceanographic processes influencing basin-wide circulation patterns and coastal habitat evolution, and historical shoreline change patterns relative to longshore sediment transport magnitude and direction. These geological and physical processes are the primary factors influencing the spatial distribution of coastal habitats and their ecology (Sections 6.4.3 and 6.5) within an ecoregion context (Section 6.2).

6.3.1 Formation of the Gulf of Mexico Basin

The GoM has been described as a relatively small oceanic basin that evolved in response to separation of the North and South American plates by crustal extension and seafloor spreading during the Mesozoic breakup of Pangea (Galloway 2011). As such, topographic relief and bathymetry reflect the overall geologic structure of the basin. Furthermore, physiography of the Gulf basin has been influenced by sea-level changes in response to alternating glacial and interglacial periods on the North American continent. Sea-level changes driven by episodic influxes of meltwater generally controlled drainage systems of the region, the morphology of coastal plain alluvial systems, and sediment volumes supplied to the basin (Bryant et al. 1991).

The general geographic limits of the GoM basin correspond with structural features (Figure 6.9). The Florida and Yucatán carbonate platforms mark the eastern and southern flanks of the basin. The western flank of the basin corresponds to the location of the Chiapas massif and the Sierra Madre Oriental of Mexico, whereas the northern border flanks the

Ouachita orogenic belt, the Ouachita Mountains, the central Mississippi deformed belt, and the southern reaches of the Appalachian Mountains (Salvador 1991a). Along the north and northwest margins of the basin, the coastal plain and continental shelf of the GoM are widest and have a relatively gentle slope toward the center of the Gulf, similar to the slope of the basement in the region. In eastern Mexico, the coastal plain and shelf are quite narrow and steep, just like the basement surface (Bryant et al. 1991). Adjacent to the east and southeast margins of the basin, some of the deepest parts of the GoM rise rapidly at the submarine escarpments fronting the Florida and Yucatán platforms, under which basement rocks are flat and featureless.

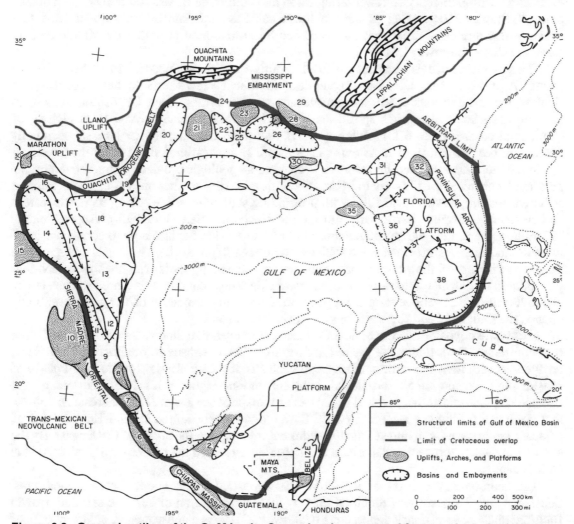

Figure 6.9. General outline of the GoM basin. Second-order structural features include: (1) Macuspana basin; (2) Villahermosa uplift; (3) Comalcalco basin; (4) Isthmus Saline basin; (5) Veracruz basin; (6) Cordoba platform; (7) Santa Ana massif; (8) Tuxpan platform; (9) Tampico-Misantla basin; (10) Valles-San Luis Potosi platform; (11) Magiscatzin basin; (12) Tamaulipas arch; (13) Burgos basin; (14) Sabinas basin; (15) Coahuila platform; (16) El Burro uplift; (17) Peyotes-Picachos arches; (18) Rio Grande embayment; (19) San Marcos arch; (20) East Texas basin; (21) Sabine uplift; (22) North Louisiana salt basin; (23) Monroe uplift; (24) Desha basin; (25) La Salle arch; (26) Mississippi salt basin; (27) Jackson dome; (28) Central Mississippi deformed belt; (29) Black Warrior basin; (30) Wiggins uplift; (31) Apalachicola embayment; (32) Ocala uplift; (33) Southeast Georgia embayment; (34) Middle Ground arch; (35) Southern platform; (36) Tampa embayment; (37) Sarasota arch; (38) South Florida basin (republished with permission of the Geological Society of America from Salvador 1991a; permission conveyed through Copyright Clearance Center, Inc.).

The Late Triassic breakup of Pangea preceded the formation of the GoM Basin, which began about 230 million years ago with the collapse of the Appalachian Mountains (Bird et al. 2011). As a result of rifting within the North American Plate during the Middle to Late Jurassic, it began to crack and drift away from the African and South American plates (Salvador 1991b). Although differing evolutionary models for the basin exist, most researchers believe that counterclockwise rotation of the Yucatán Peninsula block away from the North American Plate, involving a single ocean-continent transform boundary, led to the formation of the basin (Bird et al. 2011). Opening of the Gulf required approximately 500 km (310 mi) of extension accompanied by southward migration and counterclockwise rotation of the Yucatán block (Galloway 2011). Most of the structural basin is underlain by transitional crust that consists of continental crust that was stretched and attenuated primarily by Middle to Late Jurassic rifting (Galloway 2011).

The separation of what became North and South America produced a narrow belt of ocean about 170 million years ago. Initial conditions in what is now the GoM basin consisted of shallow, hypersaline seas in which extensive salt deposition took place. Deposition of Louann salt and associated evaporites spread across the hypersaline basin formed by stretching of continental crust (Figure 6.10) (Galloway 2011). Salt deposition during the Jurassic eventually resulted in the formation of numerous salt domes that are scattered throughout the GoM.

Since the Late Jurassic, the basin has been a stable geologic province characterized by the persistent subsidence of its central part, likely due at first to thermal cooling and later to sediment loading as the basin filled with prograding sediment wedges along its northwestern and northern margins, particularly during the Cenozoic (Salvador 1991b). Approximately 155 million years ago, the Yucatán Peninsula and the Florida Peninsula were connected landmasses and the ancestral GoM was a shallow marine sea (Figure 6.11). The coast in Mexico and Texas was inland of the present coast and was dominated by reefs with shallow basins that precipitated evaporite minerals on their landward side. These conditions required sea level to be about 100 m (328 ft) above its present position. Persistent subsidence of the basin eventually opened the Gulf between the Yucatán and Florida peninsulas.

Carbonate deposition in the Middle Cretaceous (about 100 million years ago) included large reef complexes throughout the basin. Landward of these deposits in the northeastern GoM, terrigenous sediment from the southern Appalachians provided clastics for the initial phase of coastal plain development and fluvial delta formation (Figure 6.12). Near the end of the Cretaceous, tectonic activity caused ocean basins to experience a significant increase in volume that produced falling sea level in the Gulf. Lowered sea level resulted in significant erosion of adjacent landmasses, causing substantial sediment transport to the northern GoM coast. By the end of the Early Cretaceous, deposition and subsidence created the modern morphology of the Gulf Basin.

The sedimentary section of the GoM was deposited under stable tectonic conditions. Subsidence of the basin was modified only by local deformation of Jurassic salt and growth faulting adjacent to primary depocenters (Galloway 2011). As a consequence, environments of deposition and lithologic composition of the sedimentary sequence persisted from Late Jurassic to present. Overall, three distinct provinces were formed in the sedimentary sequence of the GoM basin: (1) carbonate and evaporite deposits associated with Florida and Yucatán platforms; (2) carbonates and fine-grained terrigenous sediment along the Tamaulipas, Veracruz, Tabasco, and Campeche coasts of Mexico; and (3) coarse-grained terrigenous sediment in the northern GoM, indicating the importance of fluvial input from the continental interior to the area between eastern Mexico and northern Florida (Salvador 1991b; Galloway 2011).

Although the basin was stable, uplift of the Appalachians during the Miocene produced extensive fluvial sediment that was transported to the northern Gulf coast. Large deltaic

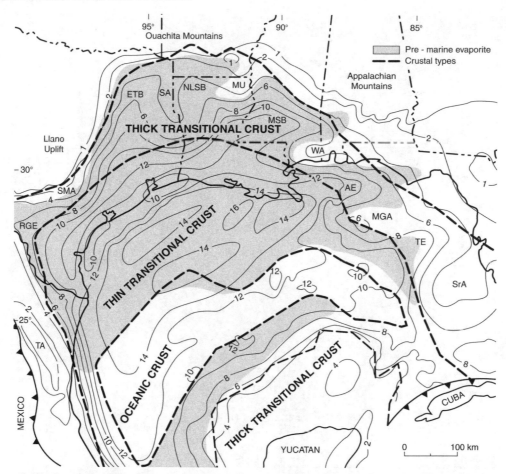

Figure 6.10. Crustal types, depth to basement, and original distribution of Jurassic Louann salt beneath the GoM basin. Principal basement structures include: *SrA* Sarasota Arch, *TE* Tampa Embayment, *MGA* Middle Ground Arch, *AE* Apalachicola Embayment, *WA* Wiggins Arch, *MSB* Mississippi Salt Basin, *MU* Monroe Uplift, *NLSB* North Louisiana Salt Basin, *SA* Sabine Arch, *ETB* East Texas Basin, *SMA* San Marcos Arch, *RGE* Rio Grande Embayment, *TA* Tamaulipas Arch (from Galloway 2011; republished with permission of the Texas A&M University Press).

systems were developed along the northern coast of the basin. Sea level was tens of meters above the present position. During the Pliocene, terrigenous sediments from the mainland dominated the northern Gulf. The Yucatán platform remained controlled by carbonate sedimentation because of a lack of siliciclastic sediment (Figure 6.13). The shoreline had a configuration similar to present time but at a more landward position (Salvador 1991b).

6.3.2 Terrestrial Controls on Coastal Evolution

Two primary factors control the development of terrestrial habitat at the interface between marine and freshwater environments: geology of coastal deposits and watershed contributions. Coastal geology refers to existing deposits that are subject to erosion and transport by modern marine processes and watershed input refers to terrestrial supply of sediment, nutrients, and fresh water to estuarine and fluvial deltaic deposits, and neritic carbonate environments, at the marine land–water interface. Coastal habitats of the GoM reflect the variety of geologic controls and watershed processes operating along the modern Gulf shoreline. Figure 6.14

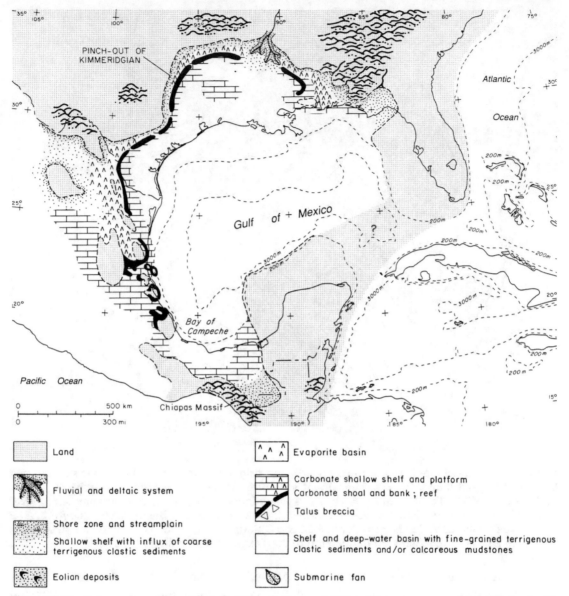

Figure 6.11. Paleogeography of the GoM basin around 155 million years ago (republished with permission of the Geological Society of America from Salvador 1991b; permission conveyed through Copyright Clearance Center, Inc.).

depicts the age and type of geologic deposits coincident with the land–water interface around the GoM. Most low-lying shorelines are composed of Quaternary sedimentary deposits of carbonate and terrigenous origin. However, Eocene, Oligocene, and Miocene deposits are common along the Big Bend coastline of northwestern Florida and along the Tamaulipas and Veracruz coast of Mexico. Furthermore, Quaternary volcanic rocks intersect the land–water interface in two locations along the Veracruz coast (Palma Sola and Tuxtla). Finally, Cretaceous carbonate deposits are encountered along the northwest margin of Cuba (Figure 6.14). Most prominent Quaternary deposits are those associated with riverine deltas in the northern GoM marine ecoregion (e.g., Rio Grande, Mississippi, and Apalachicola) and carbonate deposits along the southwest Florida coast, the Yucatán Peninsula, and the northwest coast of Cuba.

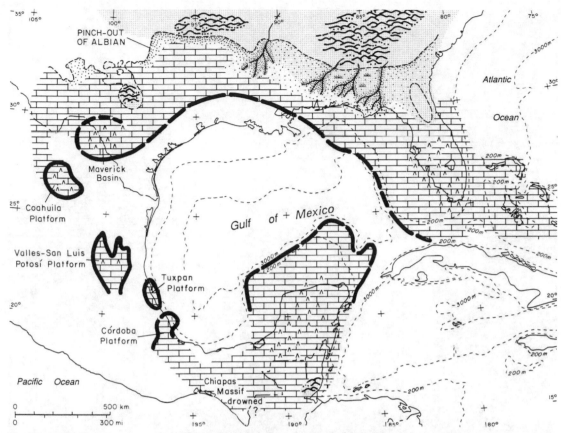

Figure 6.12. Paleogeography of the GoM basin around 100 million years ago (republished with permission of the Geological Society of America from Salvador 1991b; permission conveyed through Copyright Clearance Center, Inc.) For explanation of patterns, see Figure 6.11.

Although geologic deposits with direct exposure at the marine land–water interface have significant impact on coastal habitat formation, freshwater input from riverine watersheds and coastal groundwater sources provide vital nutrients and sediment to estuaries and outer coast shoreline habitat. Riverine contributions to coastal habitat in Mexico are important in the States of Tamaulipas, Veracruz, and Tabasco, but annual freshwater and sediment input to the Gulf from the United States vastly exceeds input from Mexico. Figure 6.15 illustrates primary watersheds adjacent to the GoM, showing the spatial extent of each watershed. The Mississippi-Atchafalaya watershed drains nearly two times as much area as all other watersheds combined. Furthermore, average discharge from this watershed contributes about 63 % of freshwater input to the Gulf annually. Table 6.1 provides a summary of freshwater discharge to the Gulf by watershed. Groundwater contributions to coastal habitat evolution are relatively minor but important for carbonate environments of the Yucatán Peninsula and the southwest Florida coast. In both areas, precipitation associated with tropical cyclones and other rain events provides the primary source of fresh water to recharge carbonate aquifers (Beddows et al. 2007).

The interaction among fluvial water/sediment supply, coastal geology, and marine physical processes produces the variety of coastal depositional environments bordering the GoM. Although marine ecoregions provide a reasonable framework for describing primary coastal habitats at the land–water interface, terrestrial ecoregions emphasize land-based characteristics above sea level (see Section 6.2.2). Section 6.4.2 presents depositional characteristics of

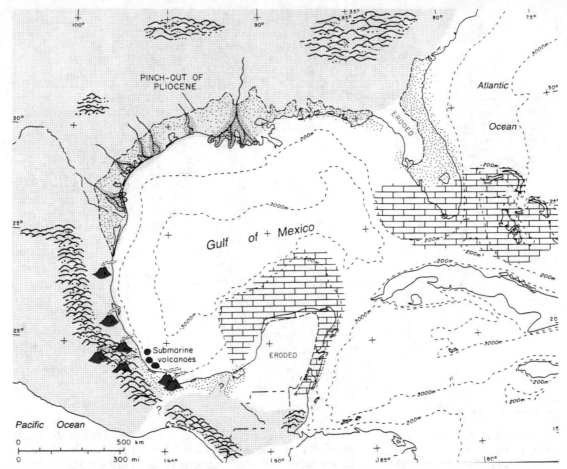

Figure 6.13. Paleogeography of the GoM basin around 5 million years ago (republished with permission of the Geological Society of America from Salvador 1991b; permission conveyed through Copyright Clearance Center, Inc.). For explanation of patterns, see Figure 6.11.

vegetated marine habitats and adjacent subaqueous environments along the Gulf shoreline that provide more detail regarding habitat type and distribution than discussed previously under terrestrial ecoregions. First, the distribution of dominant coastal depositional systems will be presented within the context of coastal processes controlling sediment transport and deposition.

6.3.3 Oceanographic Processes

The formation and evolution of coastal habitats within the Gulf are a direct response to water, sediment, and nutrient input to the basin relative to physical oceanographic processes that control erosion and deposition at the land–water interface in response to long- and short-term fluctuations in water level. Far-field forces such as basin-scale circulation, tide dynamics, and eustatic sea-level rise exert significant control on long-term habitat evolution, whereas intense periodic events such as storms and floods present short-term perturbations to the coast that can create habitat in a given locale as fast as it is destroyed in another. As such, coastal habitats are always changing in response to physical disturbances. The following section summarizes dominant physical processes in the GoM.

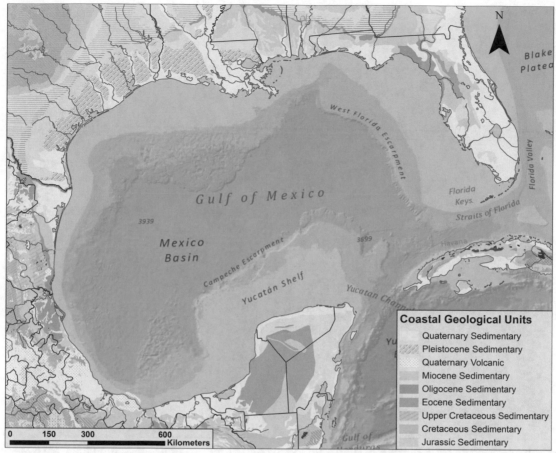

Figure 6.14. Terrestrial geologic deposits bordering the GoM (geology data from Garrity and Soller 2009). Basemap credits: ESRI, GEBCO, NOAA, CHS, CSUMB, National Geographic, DeLorme, and NAVTEQ.

6.3.3.1 Meteorological Conditions

The GoM is influenced by a maritime subtropical climate controlled primarily by clockwise circulation around a high barometric pressure area known as the Bermuda High. This pressure system dominates circulation throughout the year, weakening in the winter and strengthening in the summer. The Gulf is located southwest of this center of circulation, resulting in a predominantly southeasterly flow throughout the GoM. Two types of cyclonic storms may be superimposed on this circulation pattern depending on time of year. During winter months (December through March) when strong north winds bring drier air into the region, cold fronts associated with cold continental air masses primarily influence northern Gulf coastal areas, but also reach the southern GoM. Tropical cyclones develop and/or migrate into the GoM during warmer months (June through October). These storms may affect any area of the Gulf and substantially alter local wind circulation. Severe weather events such as thunderstorms, lightning, floods, and tornadoes are common in the Gulf as well. While tornadoes and floods are primarily inland weather hazards, the coastal zone is most vulnerable to hurricanes and their accompanying impacts such as storm surge.

For coastal areas along the GoM, prevailing wind directions are generally from the southeast and south, except for the coastal areas in the northeastern Gulf, where the prevailing winds are from the north (BOEM 2011). Average wind speeds from shoreline and buoy stations

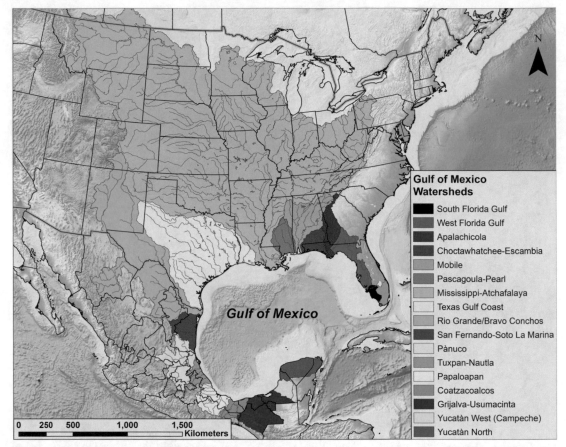

Figure 6.15. **Primary watersheds supplying freshwater, nutrients, and sediment to the GoM (watershed data from CEC 2010; basemap from Amante and Eakins 2009).**

are relatively uniform, ranging from 5.2 to 6.4 m/s (17.1 to 21.0 ft/s). In general, wind speeds are highest in winter months and lowest in summer months. In coastal areas, sea breezes may become the primary circulation feature during summer months. The humid subtropical climate of the GoM exhibits abundant and fairly well distributed precipitation throughout the year. Precipitation in coastal cities along the Gulf tends to peak in summer months. As such, relative humidity in coastal areas is high. Lower humidity occurs during late fall and winter when cold, continental air masses regularly bring dry air into the northern Gulf. Maximum humidity occurs during spring and summer when prevailing southerly winds introduce warm, moist air. Typically, highest relative humidity occurs during the coolest part of the day (around sunrise), while lowest relative humidity occurs during the warmest part of the afternoon. Climate in the southwestern GoM is relatively dry. Overall, the subtropical maritime climate is a dominant feature driving weather patterns in this region. As such, the GoM climate shows very little diurnal or seasonal variation.

6.3.3.2   Tides

Astronomical tide range throughout the GoM is relatively small (generally less than 1 m [3.3 ft]), but what it lacks in magnitude is compensated for by variety of tide types. While semidiurnal tides (two highs and two lows per day) are dominant along most coasts, GoM water levels are controlled by diurnal tides (one high and one low per day) due to the near resonance of Gulf water with diurnal tidal forcing (Kantha 2005). Diurnal tide in the GoM is driven by

Table 6.1. Drainage Characteristics for Primary Fluvial Basins Draining to the GoM

Watershed	Drainage Area (km²)	Average Daily Discharge (m³/s)	% Total Discharge	Primary River(s) (water years)	Source
South Florida Gulf	30,960	46	0.1	Caloosahatchee (1966–2011)	USGS (2012a)
West Florida Gulf	67,370	401	1.2	Ochlockonee (1926–2011), Peace (1932–2011), Suwanee (1931–2011), & Withlacoochee (1928–2011)	USGS (2012b, c, d, e)
Apalachicola	52,200	683	2.0	Apalachicola (1978–2011)	USGS (2012f)
Choctawhatchee-Escambia	37,230	389	1.1	Choctawhatchee (1931–2011) & Escambia (1988–2011)	USGS (2012g, h)
Mobile	114,450	1,709	4.9	Tombigbee (1961–2011) & Alabama (1976–2011)	USGS (2012i, j)
Pascagoula-Pearl	51,520	518	1.5	Pascagoula (1994–2011) & Pearl (1939–2011)	USGS (2012k, l)
Mississippi-Atchafalaya	3,282,169	21,940	63.3	Mississippi & Atchafalaya (1980–1996)	Battaglin et al. (2010)
Texas Gulf Coast	484,678	1,081	3.1	Calcasieu (1923–2011), Sabine (1961–2011), Neches (1951–2011), Trinity (1924–2011), Brazos (1967–2011), Colorado (1948–2011), Guadalupe (1935–2011), San Antonio (1924–2011), & Nueces (2000–2011)	USGS (2012m, n, o, p, q, r, s, t, u)
Rio Grande/Bravo-Conchos[a][b]	558,360	177	0.5	Rio Grande	NWCM (2010)
San Fernando-Soto La Marina[a]	54,720	115	0.3	San Fernando & Soto La Marina	NWCM (2010)
Pànuco[a]	97,820	645	1.9	Pànuco	NWCM (2010)
Tuxpan-Nautla[a]	26,190	384	1.1	Tuxpan, Cazones, Tecolutla, & Nautla	NWCM (2010)
Papaloapan[a]	57,480	1,565	4.5	La Antigua, Jamapa, & Papaloapan	NWCM (2010)
Coatzacoalos[a]	29,770	1,252	3.6	Coatzacoalcos & Tonala	NWCM (2010)
Grijalva-Usumacinta[a][b][c]	103,300	3,727	10.8	Grijalva-Usumacinta & Candelaria	NWCM (2010)
Yucatán West (Campeche)	21,620	N/A	0	Groundwater only	NWCM (2010)
Yucatán North	56,270	N/A	0	Groundwater only	NWCM (2010)

(continued)

Table 6.1. (continued)

Watershed	Drainage Area (km²)	Average Daily Discharge (m³/s)	% Total Discharge	Primary River(s) (water years)	Source
Total	*5,126,107*	*34,633*	*100.0*		
United States	4,120,577	26,768	77.3		
Mexico	1,005,530	7,865	22.7		

[a]The data on average daily discharge represent the mean value of their historical registry
[b]The mean daily discharge includes imports from other countries
[c]The watershed area refers only to the Mexican portion

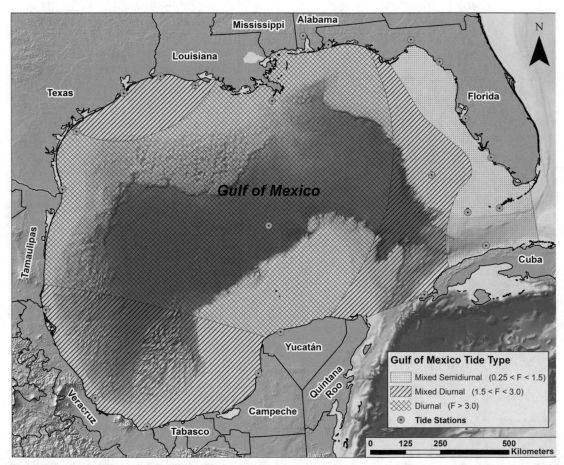

Figure 6.16. Spatial distribution of tide type based on water level form number for the GoM (modified from Kjerfve and Sneed 1984; basemap from French and Schenk 2005). Tide stations from which harmonic constituents were used to create the map are illustrated as *green dots*.

in-phase co-oscillations of the Atlantic Ocean and the Caribbean Sea through the Straits of Florida and the Yucatán Channel and exhibits a natural period of oscillation from 21 to 28.5 h (Reid and Whitaker 1981; Seim et al. 1987). Whereas direct tidal forcing explains about 13 % of the diurnal water level variance, more than half of the semidiurnal water level variance is in response to direct tidal forcing (Kjerfve and Sneed 1984). Because the semidiurnal (M_2) tide is dominant in the North Atlantic, it influences tides in the Gulf via flows through the Straits of Florida and indirectly through the Yucatán Channel. Even though semidiurnal tides tend to get amplified across wide continental shelves, only tides in the eastern Gulf from Apalachicola Bay south along the West Florida Shelf are measurably influenced by the semidiurnal signal (Kantha 2005).

Although astronomical tides often are considered unimportant for the GoM, many studies have measured and analyzed tide and current data for the Gulf (e.g., Marmer 1954; Seim et al. 1987; DiMarco and Reid 1998; He and Weisberg 2002). Dominant constituents were found to be the luni-solar diurnal (K_1), principal lunar diurnal (O_1), and the principal lunar semidiurnal (M_2). Along with the principal solar semidiurnal (S_2) tidal component, He and Weisberg (2002) found these tidal constituents accounted for 90 % of the tidal variance along the West Florida Shelf. The distribution of tide type within the Gulf was determined by Kjerfve and Sneed (1984) and Seim et al. (1987) using the water level form number (F) of Defant (1960). A common way of defining form number or amplitude ratio is

$$F = (K_1 + O_1)/(M_2 + S_2)$$

when $F < 0.25$, tide is classified as semidiurnal. Within the range $0.25 < F < 1.5$, tide is mixed but primarily semidiurnal. For the range $1.5 < F < 3.0$, tide is mixed but primarily diurnal, and when F exceeds 3.0, tide is classified as diurnal. Figure 6.16 illustrates the distribution of tide type within the GoM, indicating a dominant diurnal signal.

6.3.3.3 Circulation

The GoM has been characterized as a two-layered circulation system with a surface layer up to 1,000 m (3,300 ft) deep and a bottom layer reaching the ocean floor at depths of approximately 4,000 m (13,120 ft) (Lugo-Fernandez and Green 2011). Circulation patterns in the Gulf are the result of complex interactions among bathymetry and forcing mechanisms such as wind, atmospheric conditions, water density (variations in temperature and salinity), and the Loop Current (c.g., Ocy et al. 2005; Sturges and Kenyos 2008). Even though the Loop Current and associated eddies are dominant circulation features in the GoM, Cochrane and Kelly (1986) identified a cyclonic (rotating counter-clockwise) gyre present over the Texas-Louisiana continental shelf in response to prevailing wind stress. On the inner shelf, currents flow west-southwest, and a corresponding countercurrent along the shelf break completes the gyre system (Figure 6.17) (Nowlin et al. 1998; Zavala-Hidalgo et al. 2003).

Although circulation on the Mississippi-Alabama-Florida (MAFLA) shelf is variable due to interactions among the Loop Current and associated intrusions, tides, winds, and freshwater inflow, Kelly (1991) documented a dominant westward wind-driven flow on the inner shelf and an eastward return flow over the middle and outer shelf, creating a pattern of complex cyclonic and anticyclonic eddy pairs with strong inter-annual variability (Brooks and Giammona 1991; Jochens et al. 2002). Flow structure on the west Florida continental shelf consists of outer shelf, middle shelf, and coastal boundary layer regimes. The Loop Current and associated eddies more directly affect circulation on the outer shelf, whereas in water depths less than 30 m (98 ft), wind-driven flow is predominantly alongshore with a weak, southward-directed mean surface flow. In the coastal boundary layer, longshore currents driven primarily by winds and tides dominate cross-shelf flows.

The Loop Current is a horseshoe-shaped circulation pattern that enters the Gulf through the Yucatán Channel and exits through the Florida Straits (Figure 6.17) (BOEM 2011). The extent of intrusions of the Loop Current into the Gulf varies and may be related to current location on the

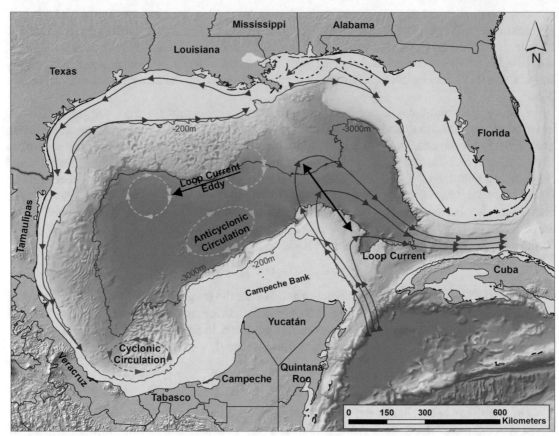

Figure 6.17. Generalized circulation patterns for the GoM (modified from BOEM 2011 and MMS 2007; contour data from Becker et al. 2009; basemap from French and Schenk 2005).

Campeche Bank at the time it separates from the bank. The Loop Current encompasses approximately 10 % of the GoM (Lugo-Fernandez and Green 2011), has surface current speeds up to 1.8 m/s (5.9 ft/s) (Oey et al. 2005), and exists to depths of 800 m (2,625 ft) (Nowlin et al. 2000; Lugo-Fernandez 2007). Water entering the Gulf through the Yucatán Channel typically is warmer and saltier than GoM waters, which generates energetic conditions that drive circulation patterns in the Gulf (Lugo-Fernandez 2007; Jochens and DiMarco 2008; Lugo-Fernandez and Green 2011). Location of the Loop Current varies, as it periodically extends to the northwest and onto the continental slope near the Mississippi River Delta (Oey et al. 2005). As the Loop Current spreads north to approximately 27°N, instability causes formation of anticyclonic warm-core eddies (closed, clockwise-rotating rings of water) shed from the Loop Current (Vukovich 2007). Even though the physical mechanisms that trigger eddy formation are not fully understood (Chang and Oey 2010; Sturges et al. 2010), the period between eddy separations ranges from 0.5 to 18.5 months (e.g., Vukovich 2007). Loop Current eddies typically have a diameter of 300 to 400 km (186 to 249 mi), surface current speeds between 1.5 and 2 m/s (4.9 and 6.6 ft/s), and west-southwest migration speeds ranging from 2 to 5 km/day (1.2 to 3.1 mi/day) (Brooks 1984; Oey et al. 2005).

Cold-core cyclonic (counter-clockwise rotating) eddies have been observed in the Gulf as well. These cyclones surround a central core of seawater that is cooler and fresher than adjacent waters. Cyclonic circulation is associated with upwelling, which brings cooler, deeper water toward the surface. A cyclone can form north of a Loop Current eddy encountering northern

GoM bathymetry due to off-shelf advection (Frolov et al. 2004). Schmitz (2005) has also associated cyclones with the Loop Current. Small cyclonic eddies around 50 to 100 km (31 to 62 mi) wide have been observed over the continental slope off Louisiana (Hamilton 1992). These eddies can persist for 6 months or longer and are relatively stationary.

In addition to currents associated with the Loop Current and meso-scale eddies, two other significant circulation features have been reported in the GoM (MMS 2007). The first is a permanent anticyclonic feature oriented approximately east-northeast and aligned with 24°N in the western half of the Gulf (Monreal-Gomez et al. 2004). The generating mechanism for this anticyclonic circulation and associated western boundary current along the coast of Mexico is a point of debate (Sturges and Blaha 1975; Elliott 1979, 1982; Blaha and Sturges 1981; Sturges 1993); however, the feature is suspected of being wind driven (Oey 1995). The second circulation feature is a cyclonic gyre centered in the Bay of Campeche, also thought to be wind driven (Figure 6.17) (Vazquez de la Cerda 1993; Nowlin et al. 2000; Monreal-Gomez et al. 2004).

6.3.3.4 Wind Waves

Wave climate is one of the primary factors controlling sediment transport, deposition, and erosion in coastal habitats, and is defined as the average wave condition over a period of years based on wave height, period, direction, and energy. In coastal and nearshore environments, wind speed and direction, and nearshore bathymetry, are the primary forcing mechanisms of wave climate. Changing geomorphic characteristics of coastal habitats are dependent upon short-term fluctuations in wave climate, long-term cycles of wind and wave activity (including the effects of frontal passages and hurricanes), and the availability of sediment and fresh water to deltaic, estuarine, and marine coastal settings. Wind directions and intensities vary seasonally with southerly winds prevailing most of the year. During winter months, wind-circulation patterns and low barometric pressures preceding the passage of cold fronts can cause strong onshore winds and increased wave heights that typically erode beaches. After frontal system passage, wind direction shifts and northerly winds can generate waves that erode north-facing shorelines at many locations.

Various moored buoys and coastal wave gauges are situated throughout the GoM (Figure 6.18). Average deep-water wave heights range from 0.5 m (1.6 ft) in summer months to 1.5 m (4.9 ft) in winter months (NDBC 2012). However, most fair-weather average significant wave heights in Gulf coastal environments are less than 0.6 m (2.0 ft) high (Li 2012; BOEM 2011). Average fair-weather wave periods are on the order of 3.5 to 4 s. Although fair-weather waves contribute to coastal habitat evolution throughout the Gulf, greatest sediment redistribution along the coast occurs during tropical cyclones and winter cold fronts for this storm-dominated region.

6.3.3.5 Tropical Cyclones

A tropical cyclone is a warm-core, low-pressure system (organized system of clouds and thunderstorms) without an associated frontal weather zone. These systems develop over tropical and subtropical waters and have a closed low-level circulation (includes tropical depressions, tropical storms, and hurricanes) (NHC 2012). Tropical cyclones affecting the Gulf originate over portions of the Atlantic basin, including the Atlantic Ocean, the Caribbean Sea, and the GoM. They occur as early as May and as late as December, but most frequently from mid-August to late October (Figure 6.19) (NHC 2012). On average, about 11 tropical cyclones occur in the Atlantic Basin annually, many of which remain over the ocean and never impact U.S. coastlines. Approximately six of these storms become hurricanes each year (Blake

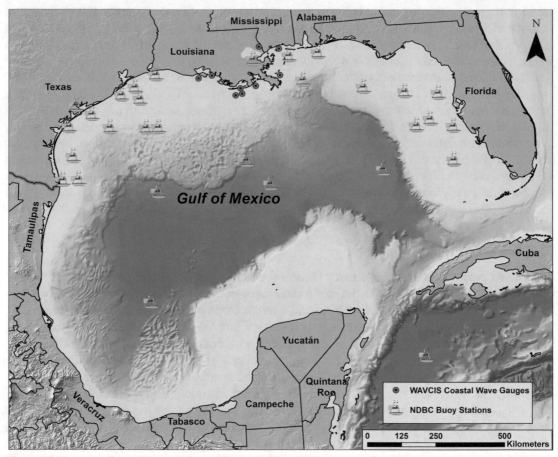

Figure 6.18. Location of National Data Buoy Center (NDBC) wave buoys and WAVCIS coastal wave gauges in the GoM (data from NDBC 2012 and WAVCIS 2012; basemap from French and Schenk 2005).

et al. 2007). Historical data indicate that hurricane tracks are relatively predictable based on a storm point of origin. Figure 6.20 illustrates the likelihood of hurricane occurrence for August, September, and October of any given year relative to storm origin and tracking. Data illustrate that hurricanes formed in the southern Caribbean in September have the greatest chance of impacting coastal habitat within the GoM, followed by August storms formed in the eastern Atlantic (Figure 6.20).

Gulf coastal areas generally experience hurricane return periods ranging from 7 to 20 years for hurricanes passing within 100 km (62 mi) of a given location (Keim et al. 2007; NHC 2012). Hurricanes and tropical storms can increase surface current speeds to between 1 and 2 m/s (3.3 and 6.6 ft/s) in nearshore and continental shelf regions (Nowlin et al. 1998; Teague et al. 2007). Recorded offshore wave heights during major hurricanes have exceeded 30 m (98 ft) (MMS 2005), attesting to the impact these storms can impose on coastal habitat. Furthermore, hurricane storm surges have been reported to range between 2 and 8 m (6.6 and 26.2 ft) for hurricanes throughout the Gulf, inundating large expanses of coastal marine and freshwater habitat (Fritz et al. 2007; Sullivan 2009).

Numerous studies have documented the destructive nature of hurricanes on coastal and nearshore habitat (e.g., Meyer-Arendt 1993; Cahoon 2006; Morton and Barras 2011). However, storm events may rejuvenate coastal marshes by delivering sediment that raises soil elevations and stimulates organic matter production (e.g., Turner et al. 2006; McKee and Cherry 2009).

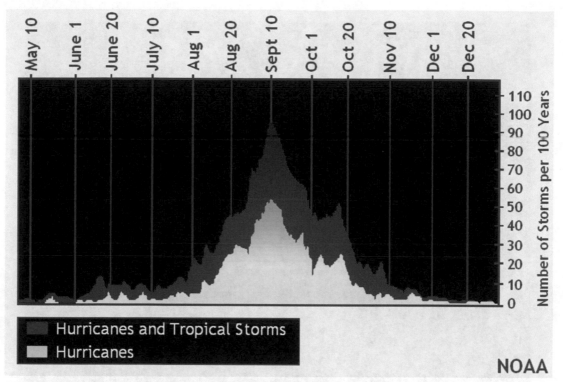

Figure 6.19. Historical distribution of tropical cyclones in the Atlantic Basin, with peak occurrence between 20 August and 1 October (from NHC 2012).

Barrier strand deposits generally absorb the brunt of destructive storm forces as these sand deposits provide the first line of defense to storm energy. Consequently, beach erosion and overtopping during storm surge may result in significant geomorphic change in barrier strand environments and adjacent salt marshes, but erosion and resuspension of coastal and estuarine sediment during storms often leads to nourishment of interior marshes via fine-grained sediment deposition.

6.3.3.6   Relative Sea-Level Rise

Long-term changes in coastal habitat type and extent are controlled by rate at which sediment is supplied to the coastal zone relative to sea level. When sea-level rise exceeds sediment deposition and organic matter accumulation required to maintain wetlands at or above water level, land loss predominates. As sea level has risen throughout the Gulf over the past 15,000 years, previously exposed upland environments on the modern continental shelf surface were inundated and reworked by waves and currents, not unlike the slow but steady submergence of coastal uplands that continues today (Balsillie and Donoghue 2011; Davis 2011a). Rates of coastal inundation and subaerial deposition fluctuate in space and time, but the fate of coastal habitats is dependent on long-term sea-level trends. Douglas (2005), Balsillie and Donoghue (2011), and Davis (2011a), as well as many others, provide detailed discussions on geologic and historical variations in sea-level change throughout the Gulf relative to coastal habitat evolution. For the following discussion, 21 tide gauge time series are used to document variations in relative sea-level rise around the GoM as a function of geographic setting (Figure 6.21).

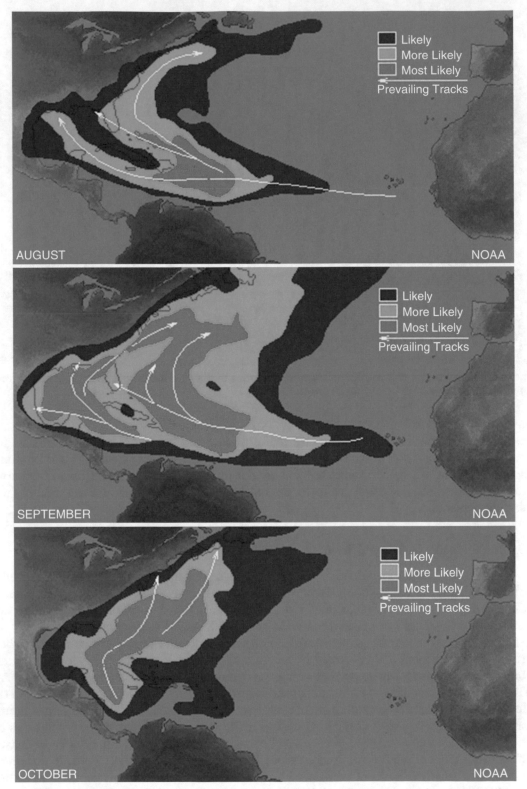

Figure 6.20. Climatological areas of origin and typical hurricane tracks for August through October (from NHC 2012).

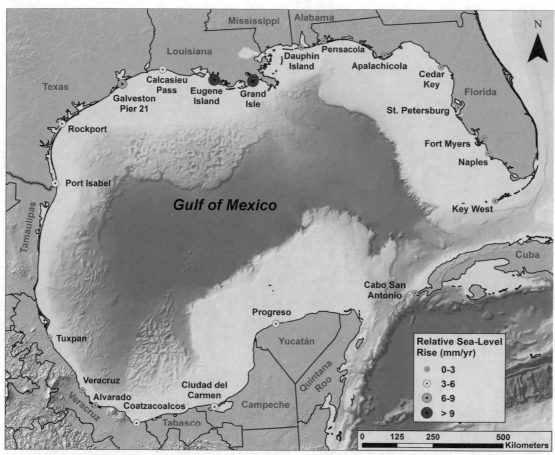

Figure 6.21. Distribution of tide gauge stations around the GoM illustrating sea-level rise trends (data from the Permanent Service for Mean Sea Level [PSMSL] database (see Woodworth and Player 2003) and U.S. Army Corps of Engineers, New Orleans District (USACE 2014); basemap from French and Schenk 2005).

Due to a variety of geologic controls in Gulf coastal environments, sea-level changes vary significantly. Carbonate geology of the Florida Gulf Peninsula provides a stable platform upon which sea level rises at a rate similar to eustatic (global) change due to a lack of sediment runoff from the continent and distance from areas of tectonic activity in the Earth's crust (Davis 2011a). Recent sea-level changes recorded in tide gauge time series data are relatively small but sea level is rising at a rate of about 1.6 to 2.5 mm/year (0.06 to 0.1 in/year) (Figure 6.22), very similar to the present rate of global sea-level rise (about 2 mm/year [0.08 in/year]) (Douglas 2005). As such, the Florida Gulf Peninsula provides baseline conditions upon which sea-level changes can be compared with other coastal locations in the Gulf.

Although coastal habitats along the Florida Panhandle, Alabama, and Mississippi are primarily wave-dominated barrier beaches and backbarrier estuarine marshes that are supplied by significant riverflows into estuaries and the Gulf (Isphording et al. 1989; Isphording 1994), tide gauge data for the northeast Gulf coast illustrate sea-level change trends consistent with eustatic sea-level rise (Figure 6.23). In fact, tide gauge data for Apalachicola illustrate a lower rise rate (1.5 mm/year [0.06 in/year]) than any recorded changes along the west coast of Florida, even though the gauge is located in close proximity to the Apalachicola River Delta. One might expect sediment compaction in this area to contribute significantly to the present rate of sea-level rise; however, deltaic sediment deposits are relatively thin (Twichell et al. 2007) and

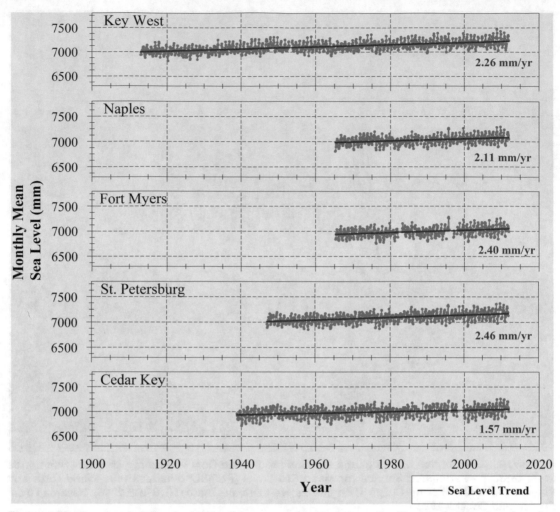

Figure 6.22. Sea-level change rates for tide gauges located along the Florida Gulf Peninsula using time series of monthly water levels from the PSMSL database.

the tide gauge is situated near a stable Pleistocene interfluve adjacent to the Apalachicola River. As one moves west toward Pensacola Bay and the entrance to Mobile Bay (Dauphin Island), relative sea-level rise increases to about 2.9 mm/year (0.11 in/year), reflecting gauge proximity to thicker sequences of Holocene sediment infilling drowned river valleys (Hummell and Parker 1995).

Relative sea-level rise on the Mississippi River Deltaic Plain is the highest of any location in the GoM primarily due to compactional subsidence of thick Holocene sediment and peat deposits that filled the Mississippi River valley during the most recent rise in sea level (Törnqvist et al. 2008). Subsidence, in addition to eustatic sea-level rise and reduced sediment supply associated with levee fortification of the river since the 1920s, has resulted in dramatic land loss in coastal Louisiana since the 1930s (Blum and Roberts 2009). Although only two NOAA tide gauge records have been used to characterize relative sea-level rise on the delta plain since the 1940s (Figure 6.24), various U.S. Army Corps of Engineers water-level gauges on the delta plain support the trend documented at these sites (e.g., Penland and Ramsey 1990). Relative sea-level rise at the Grand Isle gauge is approximately 9 mm/year (0.35 in/year), about 4.5 times greater than eustatic sea-level rise. The Eugene Island gauge recorded an even higher

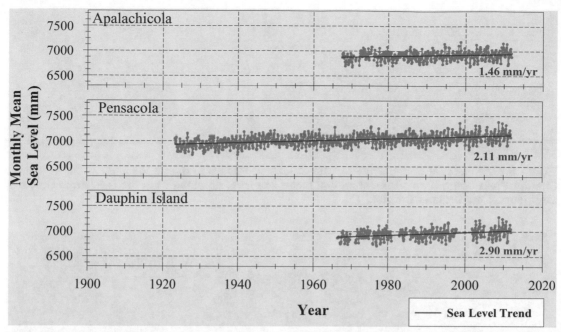

Figure 6.23. Sea-level change rates for tide gauges located along the northeastern GoM coast using time series of monthly water levels from the PSMSL database.

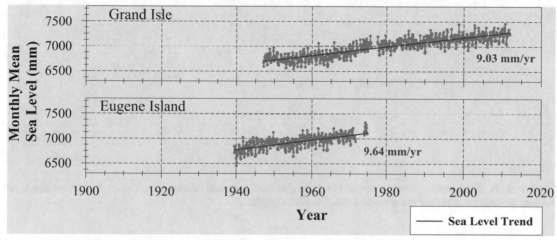

Figure 6.24. Sea-level change rates for tide gauges located along the Louisiana Deltaic Plain coast using time series of monthly water levels from the PSMSL database.

rate of relative sea-level rise (9.6 mm/year [0.38 in/year]), but record length is about half that of Grand Isle. Even though subsidence has been active since sedimentation at the river mouth was initiated, prior to dam construction within the watershed and levee construction for flood control, sediment loads were sufficient to create thousands of square kilometers of vegetated wetlands and barrier beaches. As such, a prograding delta complex and marginal deltaic wetlands flourished. Although Holocene deltas experienced landloss due to river abandonment in the past, only after civil works projects constricted sediment yield to within the confines of the dam/levee systems did delta-scale wetland losses become a chronic problem.

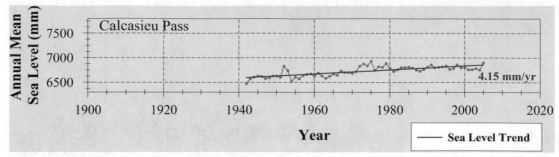

Figure 6.25. Sea-level change rate for the Calcasieu Pass tide gauge located on the Louisiana Chenier Plain using time series of annual mean sea level obtained from the U.S. Army Corps of Engineers, New Orleans District.

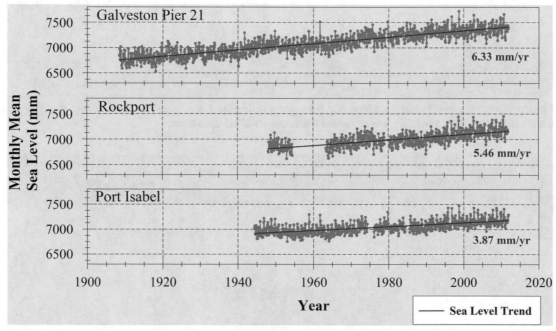

Figure 6.26. Sea-level change rates for tide gauges located along the Texas coast using time series of monthly water levels from the PSMSL database.

As recorded at the Calcasieu tide gauge, relative sea-level rise remains high along the LA/TX Chenier Plain (Figure 6.25), but less than half that recorded along the delta plain and about 2 mm/year (0.08 in/year) less than the rate recorded for Galveston. Although relative sea-level rise is high at Galveston, due in part to groundwater withdrawal in the Houston area (Gabrysch 1984), as one proceeds southwest along the Texas coast toward Rockport and Port Isabel, a reduction in relative sea-level rise is documented (Figure 6.26). Between Galveston Island and Port Isabel, relative sea-level rise decreased from 6.3 to 3.9 mm/year (0.25 to 0.15 in/year), both greater than eustatic sea-level rise and change trends in the eastern GoM. The Texas coastal plain includes a number of river systems that have contributed sediment to the coast. As such, compaction of fluvial sediment deposits may be contributing to higher relative sea-level rise in coastal Texas.

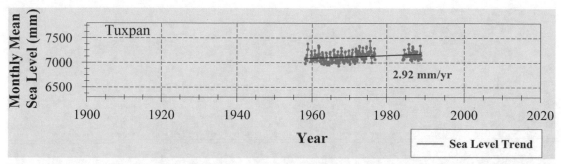

Figure 6.27. Sea-level change rate for the tide gauge located along the Veracruz coast at Tuxpan using time series of monthly water levels from the PSMSL database.

Figure 6.28. Aerial view of Tuxpan Beach with elevated upland areas producing a more stable and forested coastal setting. Image credit: ArcGIS World Imagery.

Approximately 580 km (360 mi) south of the Rio Grande in Tuxpan, Veracruz (Mexico), short-term tide gauge records indicate a sea-level rise rate of about 2.9 mm/year (0.11 in/year) (Figure 6.27), similar to that recorded at Dauphin Island, AL. The rise rate is about 1 mm/year (0.04 in/year) less than that recorded in south Texas on the northern margin of the Rio Grande delta where upland runoff had a significant impact on coastal sedimentation. Even though coastal deposits north of Tuxpan to the Rio Grande primarily are composed of terrigenous clastic sediments from upland sources that commonly form barrier islands and lagoons, beaches narrow with distance south of the Rio Grande resulting in mainland beach morphology and a more stable coast toward Tuxpan (Figure 6.28) (Carranza-Edwards et al. 2007). Sea-level

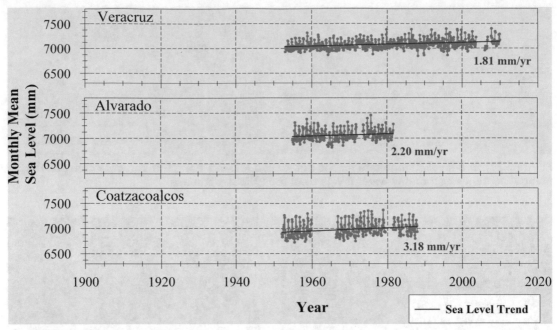

Figure 6.29. Sea-level change rates for tide gauges located along the Veracruz coast near Tuxtlas using time series of monthly water levels from the PSMSL database.

rise may reflect this southward change in coastal geomorphology adjacent to the east Mexico shelf.

South of Tuxpan for about 480 km (298 mi) to Coatzacoalcos is the most geologically diverse coastal region of the southern GoM. The area includes low-lying sandy beaches backed by lagoons and wetlands, bluffed mainland beaches, and rocky volcanic coasts with sandy pocket beaches between rock headlands. Relative sea-level rise for this coastal segment (Veracruz, Alvarado, and Coatzacoalcos) is between 1.8 and 3.2 mm/year (0.07 and 0.13 in/year) (Figure 6.29). The rocky coasts of Veracruz and Alvarado provide a stable platform upon which to record eustatic sea-level changes (1.8 and 2.2 mm/year [0.07 and 0.09 in/year]), but the coast south of the volcanic Los Tuxtlas area is more influenced by fluvial sedimentation from the Coatzacoalcos River and tributaries. Fluvial deposition and Holocene sediment compaction may have contributed to increased relative sea-level rise rates at the Coatzacoalcos gauge.

The southern Gulf coast between Coatzacoalcos and Ciudad del Carmen encompasses the entire Tabascan coast, as well as the eastern section of Veracruz and western Campeche. Coastal geomorphology is controlled by fluvial sedimentation from the Coatzacoalcos and Grijalva-Usumacinta River systems. Deltaic environments associated with the Grijalva-Usumacinta and San Pedros Rivers contain some of the most extensive marshes in Gulf coastal Mexico known as the Centla Marshes (Moreno-Casasola 2007). Deltaic settings provide for greatest magnitudes of relative sea-level rise due to compactional subsidence. However, the closest tide gauge to these active deltaic environments is at Ciudad del Carmen (Campeche), just east of the Grijalva-Usumacinta delta and marginal deltaic beach ridge plain adjacent to Isla del Carmen. Relative sea-level rise at this location (Figure 6.30) is slightly greater than that recorded at Coatzacoalcos (Figure 6.29), and both rates exceed present eustatic sea-level rise by at least 1.2 mm/year (0.05 in/year).

Farther east along the Yucatán Peninsula, one tide gauge is available to describe the relative sea-level history of this predominantly carbonate environment. River runoff from this area

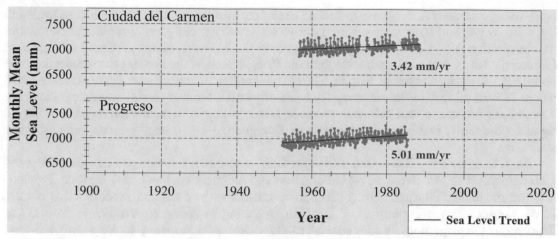

Figure 6.30. Sea-level change rates for tide gauges located along the Campeche and Yucatán coast using time series of monthly water levels from the PSMSL database.

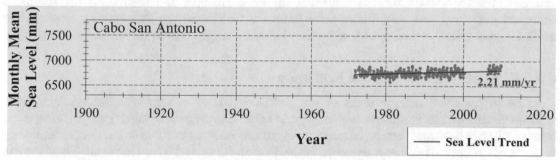

Figure 6.31. Sea-level change rate for the tide gauge located along the northwestern coast of Cuba using time series of monthly water levels from the PSMSL database.

does not exist; instead, all exchange of fresh water between upland and the Gulf is accomplished via groundwater (Isphording 1975). As such, one might expect this area to be a relatively stable platform upon which to monitor sea-level rise. However, a 38-year record of water level changes at Progreso along the northern Yucatán coast indicated a 5 mm/year (0.20 in/year) rise in sea level (Figure 6.30), the highest rate of sea-level rise along the Gulf coast of Mexico.

The final gauge used to document variations in relative sea-level rise within the Gulf is located at Cabo San Antonio, along the northwestern coast of Cuba (Figure 6.31). Similar to the Yucatán Peninsula and southwestern Florida coast, the geologic setting is primarily carbonate, and clastic sediment is composed of shell, coral, and other limestone fragments. A 38-year time series of water level measurements indicates a relative sea-level rise rate of about 2.2 mm/year (0.09 in/year), very similar to that recorded for eustatic sea-level rise. This rate is almost equivalent to that recorded at Key West (2.3 mm/year), about 400 km (250 mi) northeast across the Florida Straits. The consistency in sea-level rise trends between these sites leads to questions regarding measurements at Progreso, an area of similar geologic setting.

6.3.4 Shoreline Change and Longshore Sediment Transport

Although three distinct sedimentary provinces characterize the modern GoM basin (Section 6.3.1), a variety of coastal depositional systems have evolved along the 6,077 km (3,776 mi) land–water interface in response to upland drainage; groundwater supply; sediment availability; wind, wave, and current processes; relative sea-level rise; and physiographic characteristics of margin deposits. Carbonate deposits dominate the Mexican States of Campeche (east of Laguna de Términos), Yucatán, and Quintana Roo, as well as the northwestern coast of Cuba and the southwestern coast of Florida. Terrigenous sediment is dominant in the northern GoM where 77 % of all fluvial flow entering the basin originates. Smaller fluvial watersheds along the Tamaulipas, Veracruz, and Tabascan coasts of Mexico contribute the remaining 23 % of fluvial input to the Gulf, resulting in a mixture of fine-grained terrigenous clastics and carbonate sediment.

Shorelines fronting coastal habitats in the GoM evolve as a function of geologic setting and climatological factors affecting the balance between sediment erosion and deposition. Previous sub-sections under Physical Setting (Section 6.3) describe the dominant processes that control land changes along the margins of the Gulf, resulting in sediment erosion, transport, and deposition. On a geologic scale, coastal habitats evolve in response to long-term sea-level changes relative to sediment supply and land movements. Although historical changes in coastal habitats (century time scale) are influenced by these same processes, storm and wave energy controls sediment transport magnitude and direction, resulting in shoreline and habitat change. This section documents historical shoreline changes and associated net sediment transport pathways and magnitudes throughout the GoM over the past century or so. When available, a qualitative description of interior habitat changes is provided in Section 6.4.2.

6.3.4.1   South Florida Marine Ecoregion

One of the most diverse areas of the GoM coast is associated with habitats along the southwestern Florida peninsula where groundwater discharge has significant influence on habitat distribution and sandy beaches, mangroves, seagrasses, and coral reefs dominate. Specific shorelines of interest encompass the Florida Keys and Ten Thousand Islands areas of southwest Florida (Figure 6.32). The Florida Keys is an arcuate complex of Pleistocene coral reef islands and ooid shoals that accumulated approximately 120,000 years ago when sea level was 2 to 3 m above its present position (Hine and Locker 2011). These islands are bedrock based and are separated by tidal passes. Individual keys (islands) are stable but very low in elevation, making them vulnerable to storm surge during tropical storms and hurricanes. Landward of the keys is Florida Bay, a very shallow bay with a soft, carbonate mud bottom (Davis 2011b). The mud is quite thin (<1 m [3.3 ft] thick) and is deposited on Pleistocene limestone of the Key Largo Limestone and the Miami Oolite formations (Hine and Locker 2011). Mud deposits generally are quite cohesive, resulting in only minor sediment resuspension due to tidal currents; however, resuspension does occur during non-tidal wind events (Enos and Perkins 1979).

There are approximately 58 km (36 mi) of beaches in the Florida Keys, extending from the head of Florida Bay southwest to the Dry Tortugas (Clark 1990). Florida Keys beach sand is derived from erosion of limestone, precipitation of aragonite particles from seawater, and fragments of corals, shells, and calcareous algae (Clark 1990). Although shoreline change estimates are not well documented, historical analyses of beach erosion have been completed at a few locations along the Florida Keys (Clark 1990; FDEP 2012a). In addition, aerial photography documents numerous erosion control structures that were constructed to protect against beach erosion in this area. Beach erosion along the Keys primarily is associated with tropical cyclones and geomorphic changes associated with natural variations in littoral sediment transport. However, most of the 16.4 km (10.2 mi) of critically eroding beaches (Figure 6.33) can be associated with coastal protection structures (e.g., seawalls, revetments, groins) located at

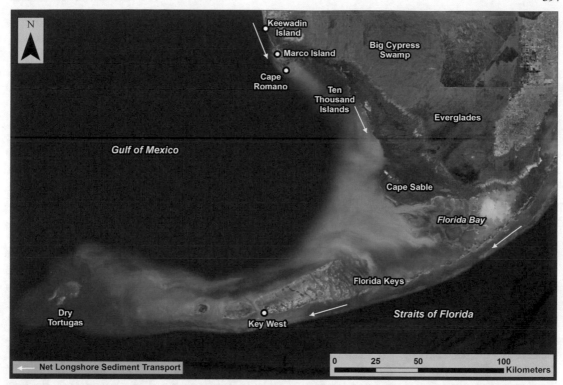

Figure 6.32. Image illustrating the Florida Keys and Ten Thousand Islands within southwestern Florida. Net longshore sediment transport direction is indicated with *arrows* (data from Clark 1990; Dean and O'Brien 1987). Image credit: Microsoft Bing Maps Aerial.

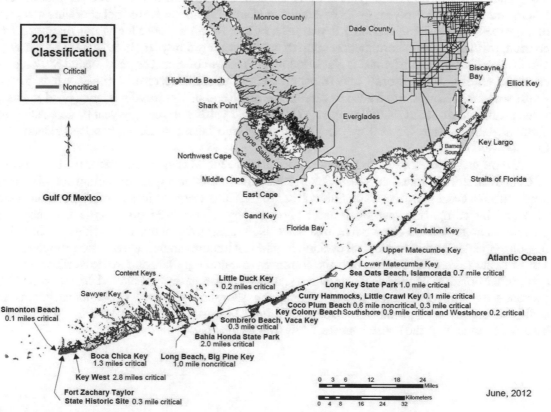

Figure 6.33. Critically eroding beaches along the Florida Keys (from FDEP 2012a).

the ends of many small pocket beaches (FDEP 2012a). The Florida Department of Environmental Protection (FDEP), Bureau of Beaches and Coastal Systems, defined a critically eroding beach as a segment of shoreline where natural processes or human activity have caused or contributed to erosion and recession of beach or dune systems to such a degree that upland development, recreational interests, wildlife habitat, or important cultural resources are threatened or lost. For beaches fronting the Straits of Florida, net littoral sand transport is to the southwest.

North of Florida Bay to Marco Island are the predominantly vegetated shorelines of Cape Sable and the Ten Thousand Islands, an area containing numerous mangrove-covered islands and marsh habitat (Figure 6.32). Tidal channels separate the series of small islands, and oyster reefs are common in brackish waters that result from freshwater runoff from Big Cypress swamp and the Florida Everglades (Davis 2011b). Marsh habitats are the result of gradual deposition of sediment over the inner shelf during the late Holocene following early Holocene transgression (Parkinson 1989). South of Marco Island and Cape Romano, there is a noticeable transition from dominantly terrigenous sand to biogenic sediment. Beaches generally are absent with only a few local accumulations of shell and skeletal debris (Davis 2011b). The coast is quite stable due to an abundance of mangrove vegetation. Although hurricanes are common in this area, their impact has had little influence on coastal geomorphology (Davis 1995). Furthermore, because of its remote location, there is relatively little human impact on the coastal system.

According to Clark (1990), GoM beaches in southwestern Florida (north of Florida Bay) include about 42 km (26 mi) of sandy shoreline. Average beach width is on the order of 8 to 15 m (26 to 49 ft) and sediment composition is predominantly carbonate. Figure 6.34 illustrates historical shoreline changes south of Gordon Pass (north end of Keewaydin Island) to the Marco Island area between the 1970s and 2000s. Although critically eroding beaches have been identified along both islands, beach nourishment in historically eroding areas has been an effective management technique for mitigating chronic erosion, resulting in a net sand surplus along much of Marco Island (Figure 6.34). Shoreline change since the 1970s for Keewaydin Island was about −0.4 m/year (−1.3 ft/year), and Marco Island illustrated net shoreline advance of approximately 5.7 m/year (18.7 ft/year). The Cape Romano shoreline is not managed for erosion, resulting in net shoreline recession of approximately 5 m/year (16 ft/year) between 1978 and 2010. This segment of coast is classified by the Bureau of Beaches and Coastal Systems as critically eroding. Net littoral sand transport along the southwestern Florida coast is to the south-southeast. Longshore transport rates at the north end of Keewaydin Island (Gordon Pass) were estimated at about 54,000 m^3/year (71,000 cubic yards per year [cy/year]), decreasing to about 42,000 m^3/year (55,000 cy/year) south of Marco Island at Caxambas Pass (Dean and O'Brien 1987).

Although limited studies document historical shoreline/wetland changes for the coast south of Cape Romano, Wanless and Vlaswinkel (2005) illustrated the impact of human activities and hurricane processes on the Cape Sable area. Significant changes in shoreline position were recorded by comparing historical aerial photography. Figure 6.35 documents net shoreline position change for the Cape Sable area since 1928 illustrating natural variations in shoreline response primarily due to tropical cyclone impacts. Although shoreline recession ranges from 1 to 4 m/year (3.3 to 13.1 ft/year) near the entrances to Lake Ingraham and in the Northwest Cape area, other portions of the coast exhibit net stability in this relatively sheltered coastal area. The presence of truncated ridge deposits along the shoreline suggests geologic variations in sediment supply and possibly transport direction; however, net transport direction during historical times is to the south-southeast toward Florida Bay.

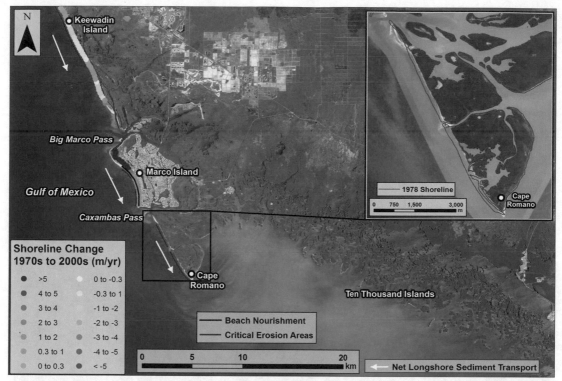

Figure 6.34. Shoreline change from Keewaydin Island to Cape Romano. Most critically eroding shorelines occur in the Cape Romano area at rates of about 5 m/year. *Sources*: **Shoreline change data, Absalonsen and Dean (2010); Cape Romano shoreline position (1978), NOAA (2013a); beach nourishment locations, Miller et al. (2004), FDEP (2008); critical erosion areas, FDEP (2012a). Image credit: Microsoft Bing Maps Aerial.**

6.3.4.2 Northern Gulf of Mexico Marine Ecoregion

The Northern GoM Marine Ecoregion extends from Keewaydin Island on the west coast of Florida to just south of Barra del Tordo in the State of Tamaulipas, Mexico, and includes barrier beaches and coastal marshes of Florida, Alabama, Mississippi, Louisiana, and Texas (Figure 6.2). This area encompasses a variety of coastal geological deposits formed by the interaction between fluvial drainage systems and coastal processes in the GoM. Most coastal depositional systems are composed of terrigenous clastic sediment; however, karstic shoreline deposits are dominant in the Big Bend area of Florida (Hine 2009). Shoreline changes throughout this region are a function of sediment supply, changes in relative sea level, and the level of energy associated with dynamic coastal processes (winds, waves, and currents under normal and storm conditions). Eight geographic areas are used to illustrate patterns of shoreline change within the Northern Gulf Ecoregion: (1) Central West Florida Barrier Islands, (2) Big Bend Coast, (3) Northeastern Gulf Barrier Islands, (4) Mississippi River Delta Plain Coast, (5) Chenier Plain Coast, (6) Texas Mid-Coast Barrier Islands, (7) Laguna Madre Barrier Islands, and (8) Laguna Morales Barrier Beaches.

6.3.4.2.1 Central West Florida Barrier Islands

The barrier-inlet system along the central west Florida coast consists of approximately 27 barrier islands and inlets extending from Gordon Pass (just north of Keewaydin Island) to Anclote Key, northwest of Tampa. The islands range from a few kilometers to tens of kilometers long and all were formed in the past 3,000 years. According to Davis (2011b), no

Figure 6.35. Shoreline change for the Northwest, Middle, and East Cape portions of Cape Sable. Most eroding shorelines occur adjacent to the entrances to Lake Ingraham and along the south-ernmost portion of the Northwest Cape (shoreline position data from NOAA 2013b). Image credits: Microsoft Bing Maps Aerial (main); ArcGIS World Imagery (inset); John Strohsahl (2008) (photo inset), used with permission.

significant terrigenous sediment is transported to the coast in this area; barrier island formation results from reworking of pre-Holocene deposits over the past 3,000 years. Large quantities of sediment from reworking of inner shoreface deposits have been transported landward during historical time (Hine et al. 1987; Hine and Locker 2011). The prism of sediment that includes the barrier island system begins at a water depth of about 6 m (20 ft) and extends landward with maximum thickness at the dunes where it reaches an elevation of only 4 to 5 m (13 to 16.4 ft) in most places (Davis et al. 2003). According to data in Table 6.1, discharge from watersheds in this area is relatively minor, indicating that modern drainage systems do not deliver significant amounts of sediment to the coast.

The balance between tide and wave energy controls morphodynamics of the central West Florida barrier islands (Davis 2011b). The net direction of littoral sand transport along the coast is to the south; however, transport reversals do exist in several locations due to changes in shoreline orientation (Davis 1999). Additionally, bedrock outcrops on the inner shoreface cause wave refraction that contributes to reversals in transport. According to Dean and O'Brien (1987), longshore transport rates vary between 35,000 and 85,000 m³/year (46,000 and 111,000 cy/year).

Most of the central West Florida barrier islands have been developed for residential and commercial activities. Coastal protection structures are prevalent on the islands, often resulting in buildings being situated too closely to the shoreline (Davis 2011b). As such, beach erosion

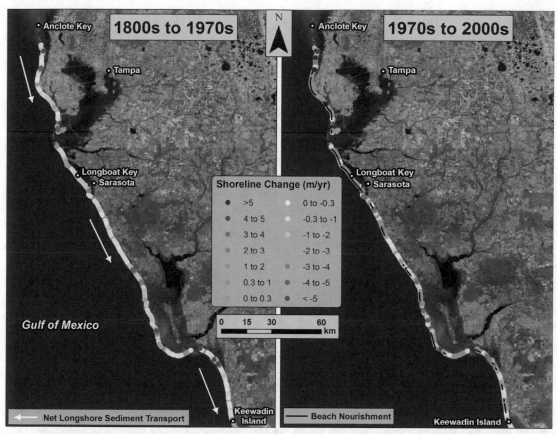

Figure 6.36. Shoreline change for the central West Florida barrier island coast. Long-term and recent shoreline changes illustrate the impact of beach nourishment throughout this coastal region. Shoreline change data from Absalonsen and Dean (2010); beach nourishment data from Miller et al. (2004) and FDEP (2008). Image credit: Microsoft Bing Maps Aerial.

near these structures has been alleviated by beach nourishment, which has been an integral part of beach management activities since the 1970s (Figure 6.36). The highest rates of erosion in this area typically are located near tidal inlets. Overall, average rates of shoreline change were approximately zero between the mid-1800s and the 1970s, even though net change along the islands ranged from 9 m (30 ft) of erosion to 9 m (30 ft) of deposition. Between the 1970s and 2000s, beach nourishment was an integral component of beach management along the islands, and net deposition prevailed at an average rate of about 0.9 m/year (3.0 ft/year) (Figure 6.36) (data from Absalonsen and Dean 2011). Although beach erosion hot spots are common along the islands and beach nourishment has been successful at mitigating erosion, Davis (2011b) indicates that natural accretion has occurred in several places along the islands. Furthermore, tropical cyclone impacts along the central West Florida barrier beaches have been reduced by the presence of a shallow and gently sloping shoreface which limits large waves from reaching subaerial beaches (Davis 2011b). Land loss in the bays and lagoons is minor because these water bodies generally are small or are already protected by erosion control structures such as bulkheads (Doyle et al. 1984).

6.3.4.2.2   Big Bend Coastal Marshes

The Big Bend region of Florida is typified by a shallow sloping submarine surface, general lack of wave activity, and lack of sediment supply. These three characteristics have created an

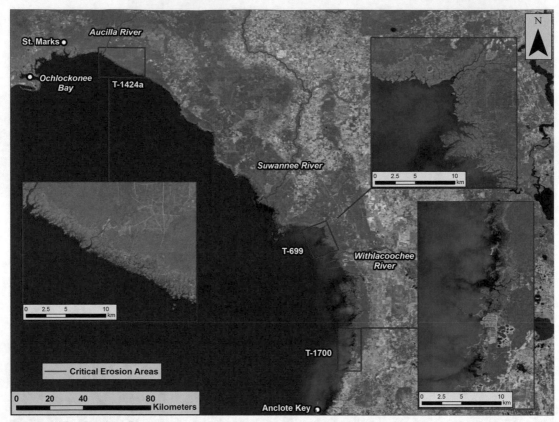

Figure 6.37. **Big Bend coastal marshes along the northwestern Florida peninsula illustrating few critical erosion areas in a low energy marsh environment.** *Inset* **areas illustrate regions for coastal change assessment in Figures 6.39 through 6.41. Critical erosion areas defined by FDEP (2012a). Image credit: Microsoft Bing Maps Aerial.**

extensive salt marsh system that rims the Big Bend coast north of Anclote Key to Ochlockonee Bay (Figure 6.37). This swath of coastal wetlands is a mixture of marsh, mangrove, and hammock vegetation, influenced by porous limestone bedrock (FDEP 2012b).

The geology of the Big Bend region is characterized by karstified Eocene and Oligocene limestone deposits over which thin muddy marsh dominated by *Juncus* sp. flourishes (Figure 6.38). According to FDEP (2012b), fluctuations in sea level during glaciation caused infilling of karstic features with Holocene and Pleistocene quartz sands and sandy clays. Holocene intertidal calcitic mud commonly overlies Pleistocene sand, and organic material derived from decaying marsh grasses intermixed with sand form the surface layer in coastal marshes. Although the Big Bend coastal area is considered sediment starved, Holocene sediment deposition continues along rivers such as the Aucilla, Suwannee, and Withlacoochee (FDEP 2012b). Big Bend karstic features generate a tight connection between the Floridian aquifer system and surface waters of the region. Because of the low topographic gradient on the limestone surface, the Big Bend area has low wave energy at the coast, similar to that of an incipient epicontinental sea (Hine 2009).

Earlier observations of coastal change in the Big Bend area by Tanner (1975a) indicated that marshes in the vicinity of Ochlockonee Bay have been stable or receding at slow rates since 1950 (on the order of 0.2 m/year [0.7 ft/year]). Tanner (1975a) also noted that average wave breaker heights in the "zero energy" coast (St. Marks to Anclote Key) were less than about 4 cm (1.6 in), that there were no integrated littoral drift cells, and that marshes along the GoM shoreline were

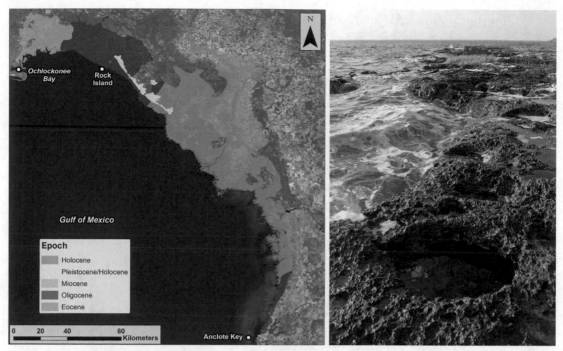

Figure 6.38. Distribution of Eocene-age and Oligocene-age limestone in the Big Bend area (*left*; geologic data from Scott et al. 2001). Image credit: Microsoft Bing Maps Aerial. Photograph of exposed karst surface near Rock Island (*right*; photo by Doug Alderson, used with permission).

well developed, suggesting that wave attenuation over a wide nearshore shelf decreases sediment transport energy to near zero. This implies that shoreline recession in coastal marshes is driven by submergence associated with relative rising sea level rather than erosion due to variations in wave energy.

In a more recent analysis, Raabe et al. (2004) documented coastal change in the Big Bend area using historical maps and aerial imagery. Inset locations shown in Figure 6.37 are used to illustrate changes in Big Bend historical record. Figure 6.39 documents shoreline and habitat change for a portion of the southern Big Bend for the period 1896 to 1995. Although conversion from marsh to water (blue) is present throughout the area, greatest loss of tidal marsh is present north of the Weeki Wachee River. Raabe et al. (2004) conducted field surveys of this area and found large mudflat areas with salt marsh rhizome remnants on the surface. Hernando Beach provides an example of coastal wetland loss due to development, and coastal forest retreat and oyster bar submergence illustrates the influence of slowly rising sea level during the period of record.

Figure 6.40 illustrates a comparison of 1858 and 1995 shorelines for the marshes between Withlacoochee Bay and Waccasassa Bay. Rapid expansion of tidal marsh inland 1 km (0.6 mi) or more over a gently sloping exposed limestone platform replaced coastal forest habitat as slowly rising marine waters submerged inland habitat (Raabe et al. 2004). Minor amounts of shoreline erosion were documented along outer margins of the marine marsh; however, marine submergence under rising sea level appears to be the dominant factor influencing coastal change in this area. According to Raabe et al. (2004), a number of natural and anthropogenic factors may have contributed to the inland expansion of coastal marsh, including soil damage during tree harvest, dissolution of limestone, change in freshwater flow from the Waccasassa River, and concentrated storm surge in the Waccasassa embayment that would focus marine

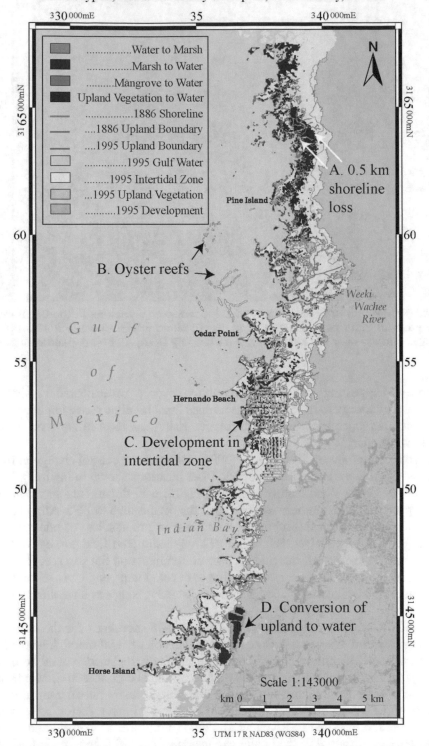

Intertidal Zone Changes from 1886 to 1995 for Topographic Survey T-1700
From Bayport, Hernando County to Aripeka, Pasco County, Florida

Figure 6.39. Coastal change for T-sheet 1700 (see Figure 6.37) between Horse Island and the Pine Island area documenting submergence of the intertidal zone between 1896 and 1995 (from Raabe et al. 2004).

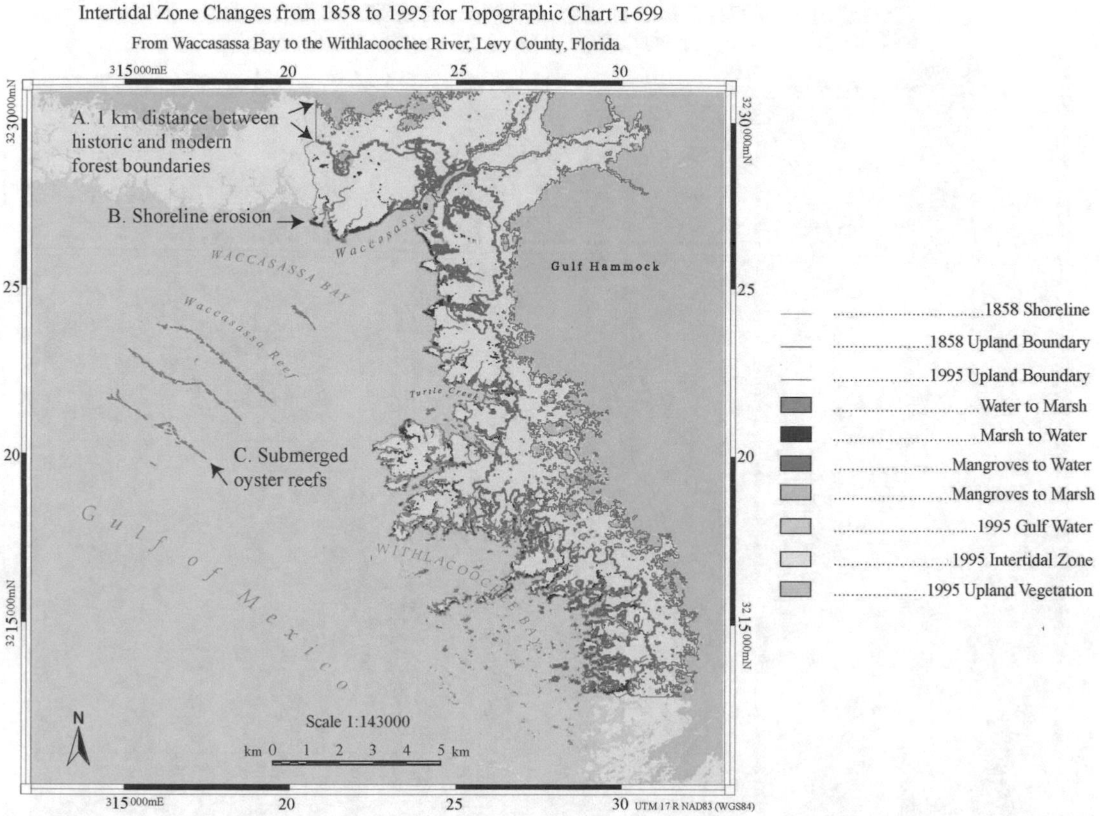

Figure 6.40. Coastal change for T-sheet 699 (see Figure 6.38), Withlacoochee Bay to Waccasassa Bay, documenting inland expansion of marsh from marine submergence, 1858 to 1995 (from Raabe et al. 2004).

energy and flooding inland. All of these factors may exacerbate the impact of rising sea level in the area.

Along the northwest portion of the Big Bend coast, between the Fenholloway River and the Aucilla River, is an area illustrating changes most common to the Big Bend marshes and coastal forests. Figure 6.41 shows relatively small losses along the marine marsh boundaries but rather significant inland recession of the coastal forest boundary as tidal marshes expand inland. According to Raabe et al. (2004), increased tidal flooding has resulted in loss of hammocks in tidal marsh and widespread inland recession of the upland forest boundary. Although marsh shoreline recession is most common along the coast, small areas of shoreline advance are present, primarily the result of high marsh bank slumping and recolonization by low marsh species (Raabe et al. 2004).

Overall, Big Bend shoreline change documents relatively minor movement in both directions with significant growth of intertidal marsh over adjacent uplands in response to sea-level rise over an approximate 100-year period. As documented by Raabe et al. (2004), dieback of coastal forests is common in the low-gradient Big Bend area as marine water submerges the limestone surface under rising seas.

6.3.4.2.3 Northeastern Gulf Barrier Islands and Beaches

The barrier island-inlet system of the northeastern GoM extends from the western margin of Ochlockonee Bay, FL (eastern margin of the Apalachicola River Delta) west to Cat Island,

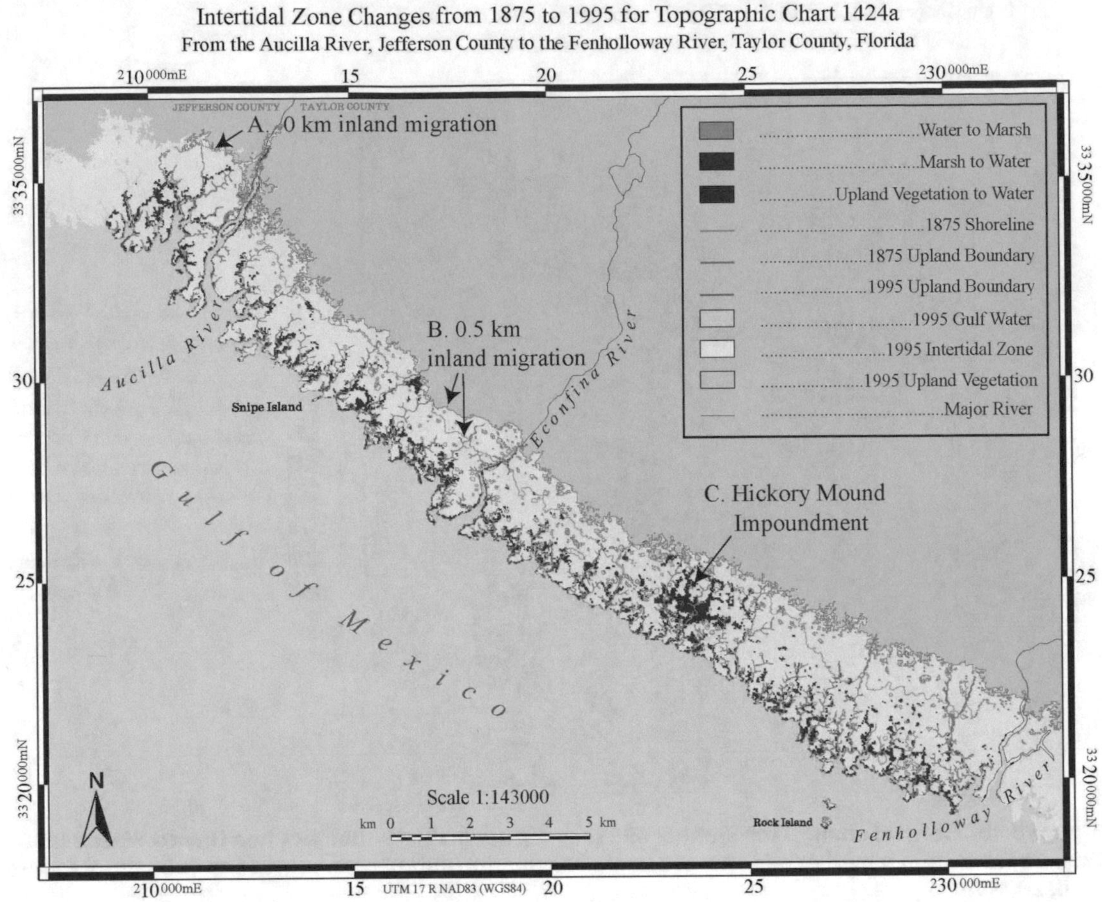

Intertidal Zone Changes from 1875 to 1995 for Topographic Chart 1424a
From the Aucilla River, Jefferson County to the Fenholloway River, Taylor County, Florida

Figure 6.41. Coastal change for T-sheet 1424a (see Figure 6.38), Fenholloway River to Aucilla River, documenting small to moderate changes along the marine and coastal forest boundaries, 1875 to 1995 (from Raabe et al. 2004).

MS (Figure 6.42). Geomorphic features include barrier islands, sand spits, mainland beaches, and inlet systems of various sizes. Shorelines of the Apalachicola River Delta vary in orientation, resulting in an array of sand transport directions and magnitudes relative to dominant wave approach. Broad and gently sloping inner continental shelf deposits seaward of the delta result in relatively low littoral transport rates versus those present along the east-west barrier strandplain west of the delta (Davis 2011b). Overall, the dominant direction of longshore sand transport is from east to west, and transport magnitudes vary based on shoreline orientation.

Historical shoreline change along most of the northeastern GoM beaches has been net erosional since the mid-1800s, primarily the result of tropical cyclone impacts. Storm-driven wave and current processes are the primary erosional forces responsible for instantaneous geomorphic changes, whereas more frequent climatological occurrences that produce normal wave and current processes rework storm-induced beach changes, resulting in long-term coastal evolution. Overall, shoreline recession is dominant throughout this portion of the GoM; however, beach nourishment since the 1970s has mitigated erosion hot spots, augmenting the littoral transport system and reducing erosion. Although sea-level rise for this section of coast is slightly greater than the eustatic rate (see Section 6.3.3.6), it has not caused significant shoreline recession during the period of record (Davis 2011a; Byrnes et al. 2012).

Figure 6.42. Location of shoreline reaches for the Northeastern Gulf Barrier Islands and Beaches region extending from Ochlockonee Bay, FL to Cat Island, MS. Image credit: Microsoft Bing Maps Aerial.

Based on geomorphic characteristics, shoreline change and longshore transport are summarized for three distinct areas of the northeastern GoM (Figure 6.42). The coast between Ochlockonee Bay and St. Joseph Peninsula is characteristic of deltaic and marginal deltaic environments of the Apalachicola River delta (Figure 6.43). Shoreline orientation varies significantly, and patterns of sand transport and beach change reflect shoreline orientation relative to incident waves. Although reversals in net littoral sand transport are common for this section of the coast, net longshore sand transport is from east to west.

According to Dean and O'Brien (1987), net longshore transport along Dog Island and St. George Island (south of Apalachicola) is to the west at a rate of about 130,000 m^3/year (170,000 cy/year), even though transport at the eastern end of Dog Island is to the east. As shoreline orientation shifts to more southerly in the St. Vincent Island area (east of St. Joseph Peninsula), west-directed transport decreases to about 90,000 m^3/year (118,000 cy/year). North of Cape San Blas (southern point of St. Joseph Peninsula), the shoreline faces a more westerly direction and net longshore transport is to the north-northwest at approximately 130,000 m^3/year (170,000 cy/year) (Dean and O'Brien 1987). Historical shoreline change rates for the Apalachicola delta coast vary from 8.2 to −8.2 m/year (26.9 to −26.9 ft/year) between the mid-1800s and 1970s/1980s (Figure 6.43). However, net shoreline recession was dominant at a rate of about 0.2 m/year (0.7 ft/year). Although beach nourishment was completed along the southern extent of St. Joseph Peninsula in 2009, net shoreline recession rates for the period 1970s/1980s to 2009 increased to an average of 0.6 m/year (2.0 ft/year), perhaps due to increased storm impacts since the 1970s.

The next segment of coast is concave and extends from Port St. Joe (near the northern end of St. Joseph Peninsula) to Mobile Point on the eastern side of Mobile Pass (Figure 6.42). Although net longshore sand transport is to the east at about 100,000 m^3/year (131,000 cy/year) along a short length of beach at the eastern end of this 300-km (186-mi) segment of coast (near Mexico Beach Inlet), net transport for the western 270 km (168 mi) of beach is to the west at rates between 115,000 and 400,000 m^3/year (150,000 and 523,000 cy/year) (Dean and O'Brien 1987; Byrnes et al. 2010). Seven inlets interrupt sand transport between Port St. Joe and Mobile Pass, and all but three are maintained by the U.S. Army Corps of Engineers, Mobile District. Mexico Beach Inlet in Florida, a natural entrance that exchanges water and sediment between the GoM and Saint Andrew Sound, is maintained by the City of Mexico Beach, and Little Lagoon Pass is maintained by the State of Alabama. Historical shoreline change rates for the 1800s to 1970s/1980s illustrate hot spots of erosion and accretion east of St. Andrew Bay Entrance that range from −8.4 to 7.2 m/year (−27.6 to 23.6 ft/year) (Figure 6.44); however, most beaches document shoreline changes between −1 and 1 m/year (−3.3 and 3.3 ft/year).

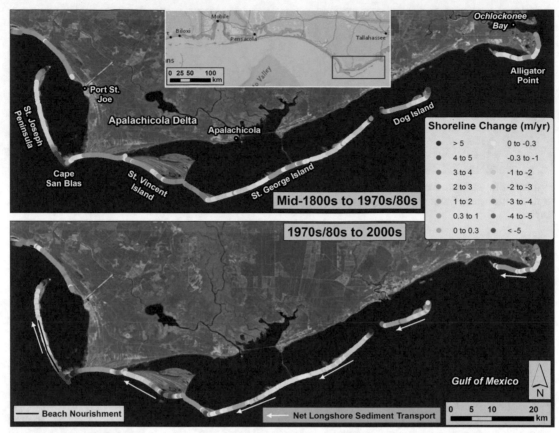

Figure 6.43. Historical shoreline change for sandy beaches for the Apalachicola River delta region. Variations in net shoreline change between the mid-1800s and 1970s/1980s are illustrated in the *top panel*, whereas net shoreline changes between the 1970s/1980s and 2000s are shown in the *bottom panel* relative to beach nourishment (*black line* segments) and the direction of net littoral sand transport (*white arrows*). Shoreline change data from Absalonsen and Dean (2010) and Miller et al. (2004). Beach nourishment data from Miller et al. (2004) and FDEP (2008). Image credits: Microsoft Bing Maps Aerial (main); ArcGIS National Geographic World Map (overview).

Overall, net shoreline recession of −0.1 m/year (−0.3 ft/year) was recorded for this 300-km (186-mi) coastal segment. Between the 1970s/1980s and 2000s, sand nourishment was imposed along a number of beaches (FDEP 2008), contributing to a shift in net shoreline change to 0.1 m/year (0.3 ft/year) (Absalonsen and Dean 2011) (Figure 6.44).

The westernmost 100 km (62 mi) of the northeastern Gulf barrier islands and beaches encompasses the barrier islands fronting Mississippi Sound (Figure 6.42). The barrier islands extend from Dauphin Island (AL) to Cat Island (MS) and provide the first line of protection to mainland Mississippi and Alabama from storm waves and surge. The islands are composed of beach sand derived from updrift beaches east of Mobile Pass and from ebb-tidal shoals at the entrance. Four tidal passes between the islands promote exchange of sediment and water between marine waters of the GoM and brackish waters of Mississippi Sound (Figure 6.45). Tidal passes also interrupt the flow of littoral sand to the west from Mobile Pass ebb-tidal shoals and Dauphin Island. Mobile Pass, Horn Island Pass, and Ship Island Pass are federally maintained navigation channels since the early 1900s (Byrnes et al. 2010, 2012).

Byrnes et al. (2010) and Byrnes et al. (2013) document long-term beach changes for the Mississippi Sound barrier islands, emphasizing the dominance of east-to-west longshore transport processes on erosion and deposition along the coast. Net shoreline recession of about

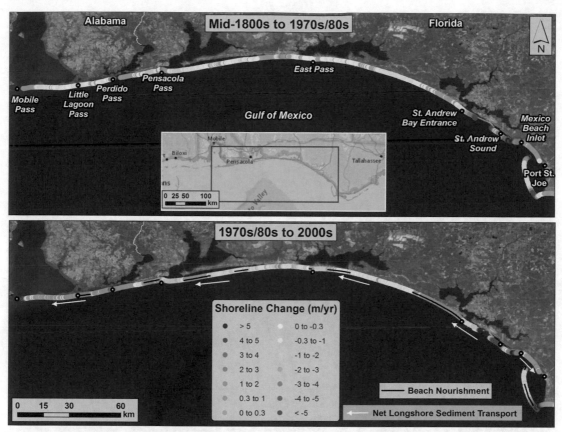

Figure 6.44. Historical shoreline change for sandy beaches from Port St. Joe to Mobile Pass. Variations in net shoreline change between the mid-1800s and 1970s/1980 are illustrated in the *top panel*, whereas net shoreline changes between the 1970s/1980s and 2000s are shown in the *bottom panel* relative to beach nourishment (*black line* segments) and the direction of net littoral sand transport (*white arrows*). Shoreline change data from Absalonsen and Dean (2010) and Byrnes et al. (2010). Beach nourishment data from Miller et al. (2004) and FDEP (2008). Image credits: Microsoft Bing Maps Aerial (main); ArcGIS National Geographic World Map (overview).

1.5 m/year (4.9 ft/year) was documented for Gulf facing beaches for the period 1847 to 1981/1986 (Figure 6.45); storm processes and inlet dynamics in the dominant east-west littoral transport environment control shoreline position change. Shoreline recession since 1981 increased to 2.4 m/year (7.9 ft/year), perhaps due to an increase in tropical cyclone impacts during this 30-year period. Cross-shore island changes are particularly important along central Dauphin Island and along East Ship Island where long-term rates of change have been documented at up to −3 m/year (−10 ft/year) and −6 m/year (−20 ft/year), respectively (Byrnes et al. 2012). However, lateral island migration (from east to west) controls long-term island morphologic changes at rates between 10 and 50 m/year (33 and 164 ft/year), emphasizing the dominance of net longshore transport processes (Figure 6.46) (Byrnes et al. 2013). The systematic pattern of updrift erosion and downdrift deposition illustrates sand movement from east to west and promotes westward migration, and has reduced island areas by about one-third since the 1850s (Byrnes et al. 2012).

As illustrated in Figure 6.46, littoral sand transport along the Mississippi Sound barrier islands is predominantly from east to west in response to prevailing winds and waves under normal and storm conditions from the southeast. Reversals in longshore transport occur at the eastern ends of the islands, but their impact on net sediment transport is localized and minor

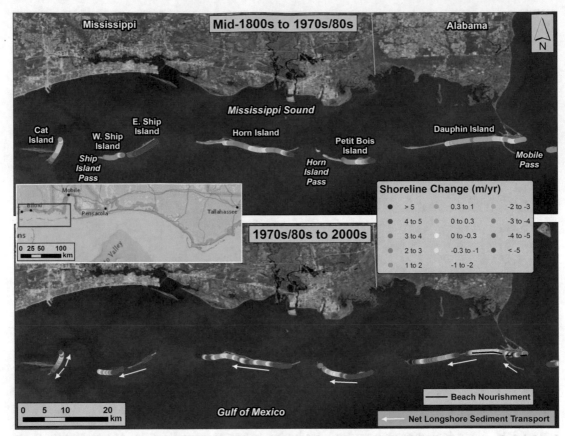

Figure 6.45. Historical shoreline change for sandy beaches from Mobile Pass to Cat Island. Variations in net shoreline change between the mid-1800s and 1970s/1980 are illustrated in the *top panel*, whereas net shoreline changes between the 1970s/1980s and 2000s are shown in the *bottom panel* relative to beach nourishment (*black line* segments) and the direction of net littoral sand transport (*white arrows*). Shoreline change data from Byrnes et al. (2010) and Byrnes et al. (2013). Beach nourishment data from Miller et al. (2004). Image credits: Microsoft Bing Maps Aerial (main); ArcGIS National Geographic World Map (overview).

relative to dominant transport processes from the southeast. Net longshore transport magnitude was estimated using historical survey datasets encompassing an approximate 90-year period to quantify sand flux along the barrier-inlet system (littoral sediment budget). According to Byrnes et al. (2013), longshore sand transport magnitudes range from about 230,000 m^3/year (300,000 cy/year) along the western end of Dauphin Island to approximately 320,000 m^3/year (420,000 cy/year) along Horn Island to 110,000 m^3/year (145,000 cy/year) near Ship Island (Figure 6.47).

6.3.4.2.4 Mississippi River Deltaic Plain

The Mississippi River deltaic plain extends from the Chandeleur Islands to Southwest Pass (west margin of Marsh Island) (Figure 6.48). Mississippi River delta growth over the past 7,000 years has produced millions of acres of wetlands that form and degrade as the river switches course every 1,000 to 2,000 years. Channel gradients become so low that hydraulic flow inefficiencies result in river channel realignment to a more efficient route to the Gulf (Roberts 1997). As delta lobes are abandoned (that is, fluvial processes no longer contribute significantly to sedimentation and land building), erosive wave and current forces begin to rework the outer margins of the delta. Erosion and sediment reworking are exacerbated by

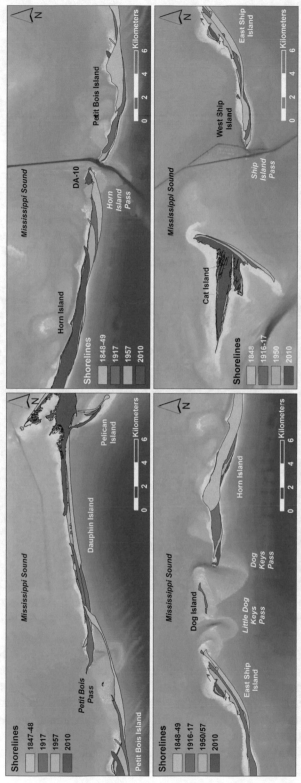

Figure 6.46. Shoreline change for the Mississippi Sound barrier islands illustrating alongshore erosion and deposition trends that indicate net east to west littoral transport (from Byrnes et al. 2013; used with permission of the Journal of Coastal Research).

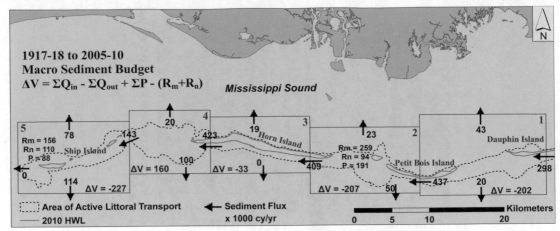

Figure 6.47. Macro-scale sediment budget for the Mississippi Sound barrier island chain, 1917–1918 to 2005–2010. *Arrows* illustrate the direction of sediment movement throughout the system and *black numbers* reflect the magnitude of net sediment transport (from Byrnes et al. 2013; used with permission of the Journal of Coastal Research).

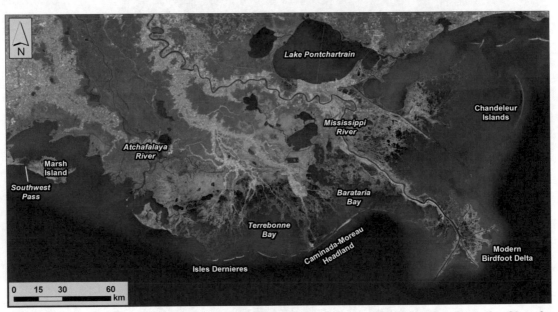

Figure 6.48. Location diagram for the Mississippi River deltaic plain extending from the Chandeleur Islands west to Southwest Pass. Image credit: Microsoft Bing Maps Aerial.

compactional subsidence, as the primarily depositional system evolves (Williams et al. 2011). Eventually, headland beaches and barrier islands are formed as transgression proceeds on the sediment-starved abandoned delta lobe (Kulp et al. 2005). Headland beaches and barrier islands formed along the outer margin of the Mississippi River delta plain reflect various stages of delta lobe evolution, and because the natural source of river sediment has been reduced from interior watersheds via dams and isolated from the modern deltaic plain via levees, deltaic habitats are rapidly deteriorating. Coastal habitats are particularly vulnerable to change where direct exposure to storm waves and currents results in rapid shoreline changes and significant sediment transport rates.

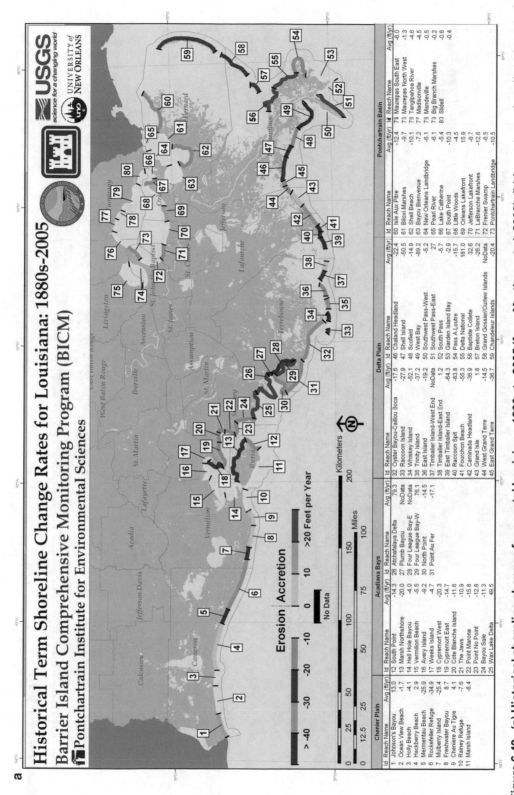

Figure 6.49. (a) Historical shoreline changes for coastal Louisiana, 1800s to 2005 (from Martinez et al. 2009). (b) Long-term shoreline changes for coastal Louisiana, 1930s to 2005 (from Martinez et al. 2009).

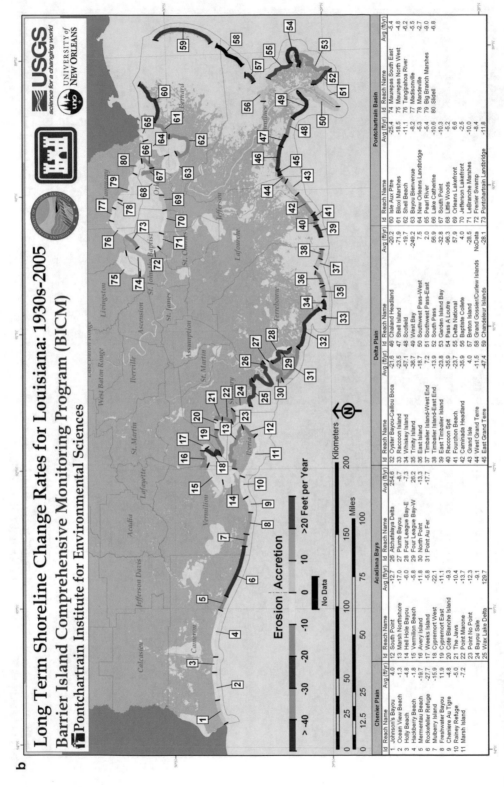

Figure 6.49. (continued)

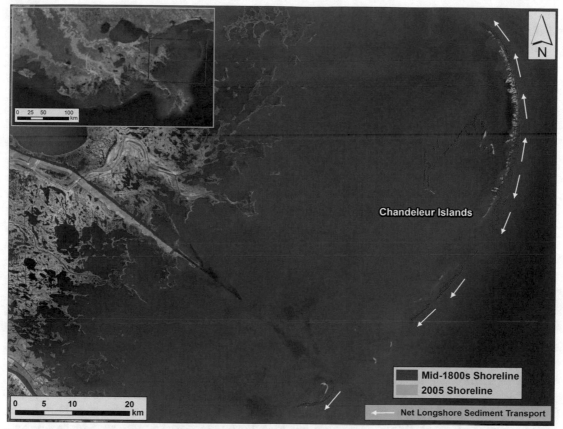

Figure 6.50. Deterioration and rapid shoreline recession along the Chandeleur barrier island system, mid-1800s to 2005 (data from Martinez et al. 2009). *White arrows* show the direction of net littoral sand transport. Image credit: Microsoft Bing Maps Aerial.

In response to high subsidence rates, diminished sediment supply to coast habitats, and continued exposure to storm waves and currents, the Mississippi-Atchafalaya River deltaic plain experiences the highest rates of laterally continuous shoreline retreat and land loss in the GoM (Penland et al. 1990; Miner et al. 2009). While land loss associated with shoreline change along the Gulf shore and around the margins of large coastal bays is extreme, loss of the interior wetlands is even more extensive due to submergence and deterioration of the Mississippi River delta plain. Wetland erosion along the Louisiana deltaic shoreline (excluding accretion along the modern delta fringe) averaged about 4.8 m/year (15.7 ft/year) between 1855 and 2005; however, the rate of erosion increased to approximately 14.1 m/year (46.3 ft/year) between 1996 and 2005 (Martinez et al. 2009). Highest rates of Gulf shoreline recession along the Mississippi River deltaic plain coincide with subsiding marshes and migrating barrier islands such as the Chandeleur Islands, Caminada-Moreau headland, and the Isles Dernieres (Figure 6.49).

The Chandeleur Islands barrier system represents the final stage of delta lobe deterioration where transgressive sand deposits reside along the outer margin of a submerged delta lobe under rapid shoreline recession and frequent overwash (Figure 6.50). Historical shoreline recession (1855 to 2005) for this segment of coast was 6.4 m/year (21.0 ft/year) (Figure 6.49a); between the 1930s and 2005, the rate increased to −8.6 m/year (−28.2 ft/year) (Reaches 57 to 59 on Figure 6.49b). Most sand transport within this low-profile barrier island system is directed landward during storm events (washover); however, longshore transport is characterized as bi-directional (north and south of the central portion of the island chain), and net rates estimated using wave modeling varied between 60,000 and 130,000 m³/year (78,000 and 170,000 cy/year) (Ellis and Stone 2006; Georgiou and Schindler 2009).

The Plaquemines barrier system protects Barataria Bay from Gulf waves and currents and extends from Sandy Point (east) to West Grand Terre Island at Barataria Pass (Figure 6.51). Longshore transport is eastward from Barataria Pass and westward from Sandy Point, converging near the eastern end of East Grand Terre Island (Figure 6.51; USACE 2012). Annualized maintenance dredging from the bar channel at Barataria Pass (1996 to 2007) was approximately 140,000 m³/year (183,000 cy/year) (USACE 2010). Of this quantity, about 90,000 m³/year (118,000 cy/year) was sand; however, this quantity is an estimate of gross transport to the pass from east and west. Georgiou et al. (2005) estimated that approximately 10,000 m³/year (13,000 cy/year) of sand was transported westward along the Plaquemines shoreline based on survey data, and USACE (2012) estimated sand transport along Shell Island at approximately 33,000 m³/year (43,000 cy/year) westward. Historical shoreline change rates average about −7.0 m/year (−23.0 ft/year) (1884 to 2005); however, shoreline recession rates increased to approximately 8.1 m/year (26.6 ft/year) between the 1930s and 2005 (Reaches 44 to 48, Figure 6.49b; Martinez et al. 2009).

The Bayou Lafourche barrier system extends approximately 60 km (37 mi) from Barataria Pass (eastern end of Grand Isle) to Cat Island Pass at the western end of Timbalier Island (Figure 6.51). The Caminada-Moreau Headland is included in this coastal segment and contains some of the highest rates of shoreline recession in south Louisiana (11.2 m/year [36.7 ft/year]; Reach 42, Figure 6.49a). Timbalier Island has experienced rapid lateral migration to the west, reflecting the dominant direction of longshore transport west of the Caminada-Moreau Headland (McBride et al. 1992). Based on shoreline change analyses and nearshore sedimentation trends, Georgiou et al. (2005) estimated net longshore transport for this area to be approximately 146,000 m³/year (191,000 cy/year) eastward. According to Rosati and Lawton (2011), net westward transport of maintenance dredging material from Cat Island Pass (Houma Navigation Canal) was about 100,000 m³/year (130,000 cy/year) toward the Isles Dernieres. However, Georgiou et al. (2005) estimates that a maximum of 50,000 m³/year (65,000 cy/year) of sand moves westward along the Timbalier Islands. Based on data from Martinez et al. (2007), historical shoreline change for the Bayou Lafourche barrier shoreline was about −8.8 m/year (−28.9 ft/year) (1884–2005). Shoreline recession rates decreased to about 5.8 m/year (19.0 ft/year) between the 1930s and 2005 (Figure 6.49b).

The westernmost barrier island system along the south Louisiana coast is the Isles Dernieres. In the mid-1800s, the Isles Dernieres (then known as Last Island) was home to the first coastal resort in Louisiana (Davis 2010). At that time, the island was continuous, about 50-km (31-mi) long, and approximately 1 km (0.6 mi) wide. The hurricane of 1856 destroyed the resort community and the island has continued to deteriorate since that time. Although the east-to-west longshore sediment transport pathway is well defined for the Isles Dernieres (Figure 6.51), littoral transport rates estimated using wave modeling routines vary from about 33,000 m³/year (43,000 cy/year) (Georgiou et al. 2005) to 60,000 m³/year (78,000 cy/year) (Stone and Zhang 2001). Based on the sediment budget for Cat Island Pass (Rosati and Lawton 2011) developed using survey data, net transport quantities of Stone and Zhang (2001) and Georgiou et al. (2005) likely underestimate annualized transport rates. Historical shoreline change rates (−11.3 m/year [−37.1 ft/year]; 1887–2005) are of similar order to those recorded for the Caminada-Moreau headland. For the 1930s to 2005 period, recession rates increased slightly to 12.0 m/year (39.4 ft/year) (Reaches 33 to 36; Figure 6.49b).

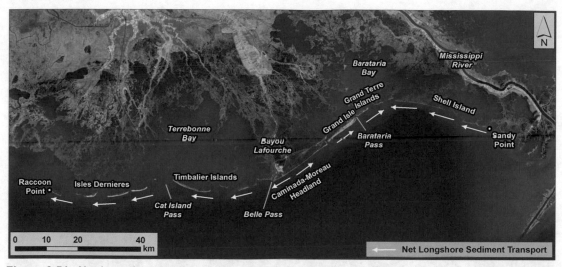

Figure 6.51. Net longshore sediment transport pathways for the barrier island shoreline between Sandy Point and Raccoon Point fronting the Mississippi River deltaic plain. Image credit: Microsoft Bing Maps Aerial.

6.3.4.2.5 Mississippi River Chenier Plain

The Chenier Plain coast of southwestern Louisiana and southeastern Texas is a unique marginal-deltaic depositional environment indirectly influenced by high levels of riverine input from the Mississippi-Atchafalaya River system. The area extends from Southwest Pass (LA) to Rollover Pass (TX) (Figure 6.52). The Chenier Plain coast is approximately 200 km (124 mi) long and extends up to 30 km (19 mi) inland from the GoM. Chenier Plain deposits are composed primarily of mud, interspersed with thin sand- and shell-rich ridges. Coastal deposits were formed from sediments supplied by longshore transport of primarily fine-grained Mississippi-Atchafalaya River sediment (Hoyt 1969) when the river mouth was oriented to the west. When the river mouth was located eastward and sediment supply to the Chenier Plain was limited relative to erosive wave energy, previously deposited mud-rich sediment was reworked by coastal processes, concentrating coarse-grained sediments and forming shore-parallel ridges (Penland and Suter 1989). Subsequent shifts in sediment supply created the alternating ridge and swale topography so common to the Chenier Plain (McBride et al. 2007).

Although no direct measurements of littoral sediment transport have been made along mixed sediment coastal and nearshore deposits of the Chenier Plain, Holocene geomorphic records illustrate an east to west longshore transport direction (McBride et al. 2007). Only three primary waterways interrupt longshore transport along the Chenier Plain coast, two of which have significant inland bays (Calcasieu and Sabine). All three waterways are structured with jetties that illustrate net longshore sediment transport direction (sand accumulation at the eastern jetties). Sediment transport magnitude is more difficult to estimate; however, net transport quantities estimated by Georgiou et al. (2008) between Calcasieu Pass and Sabine Pass using numerical modeling were reported as a maximum of about 40,000 m^3/year (52,000 cy/year). Shepsis et al. (2010) used survey data and numerical modeling to estimate a net west-directed longshore transport rate of approximately 70,000 m^3/year (92,000 cy/year) for the same coastal segment. Furthermore, Taylor Engineering (2010) documented a series of longshore sand transport rates for the Rollover Pass area that ranged between 44,000 and 73,000 m^3/year (96,000 cy/year) to the southwest.

Shoreline change along the Louisiana Chenier Plain coast is dominated by erosion between Southwest Pass and the Mermentau River Outlet at a rate of about 5.3 m/year (17.4 ft/year)

Figure 6.52. Historical shoreline change trends for the Mississippi-Atchafalaya River Chenier Plain coast. Shoreline change data from Martinez et al. (2009) and Paine et al. (2011). *White arrows* show the direction of net littoral sand transport. Image credit: Microsoft Bing Maps Aerial.

(Figure 6.52) (Byrnes et al. 1995; Martinez et al. 2009). However, a 23-km (14-mi) segment of coast east and west of Freshwater Bayou illustrates net shoreline advance between 1884 and 2005 (2.9 m/year [9.5 ft/year]; Figure 6.52), perhaps reflecting sediment supplied to this area by the Atchafalaya River (Huh et al. 1991). West of this deposition zone to a position 7.5 km (4.7 mi) west of the Mermentau River Outlet is a 68-km (42 mi) shoreline segment that illustrates greatest historical recession rates along the Chenier Plain (8.7 m/year [28.5 ft/year]). Further west of this point to Sabine Pass, net deposition and shoreline advance (1.6 m/year [5.2 ft/year]) becomes dominant (Byrnes et al. 1995). This alternating trend of shoreline recession and advance shifts to net recession west of Sabine Pass to Rollover Pass, where beach erosion dominates shoreline dynamics (Figure 6.52), particularly when tropical cyclones impact the area (Byrnes and McBride 2009). Thin sand and shell beaches, perched on inland herbaceous marsh deposits, exist along the entire coast, and net shoreline recession rates average about 2.6 m/year (8.5 ft/year). Overall, temporal and spatial trends in shoreline response illustrate increasing shoreline recession with time (Byrnes et al. 1995). Besides being a function of incident wave energy, shoreline change data indicate that factors such as shoreline orientation to dominant wave processes, sediment supply, and engineering structures have a profound influence on coastal response.

6.3.4.2.6 Texas Mid-Coast Barrier Islands

Barrier beaches along the central Texas coast extend approximately 300 km (186 mi) southwest between Rollover Pass and Packery Channel (North Padre Island) (Figure 6.53). The area between Rollover Pass and San Luis Pass encompasses Bolivar Peninsula and Galveston Island, a zone of sandy beaches and dune systems with ridge and swale topography (Bernard et al. 1970). In historical times, navigation structures at Bolivar Roads (Houston Ship Channel Entrance) have influenced sediment transport pathways along the southeast Texas coast. In addition, the Galveston seawall and groin system on the eastern part of Galveston Island, while protecting the island, has limited sediment to downdrift beaches, resulting in a net

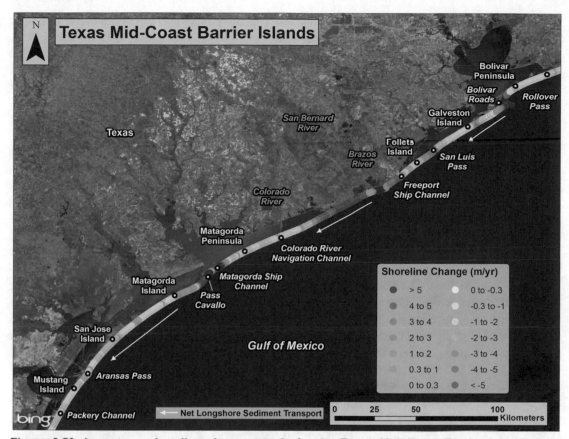

Figure 6.53. Long-term shoreline change trends for the Texas Mid-Coast Barrier Islands (mid 1800s to 2007 for the area between Rollover Pass and San Luis Pass; 1930s to 2007 for the area southwest of San Luis Pass to Packery Channel). Shoreline change data from Paine et al. (2011). *White arrows* show the direction of net littoral sand transport. Image credit: Microsoft Bing Maps Aerial.

deficit to the sediment budget along south Galveston Island. These structures serve to compartmentalize the coast by blocking southwest-directed longshore sand transport to downdrift beaches. As a result of these structures and natural processes, approximately 88 % of the coast in this area illustrates long-term shoreline recession (Figure 6.53) (Paine et al. 2011). Net shoreline recession for the period 1882 to 2007 was about 0.2 m/year (0.7 ft/year). Although shoreline recession is dominant, small areas of net deposition occur at shoreline segments adjacent to the north and south jetties at Bolivar Roads, and the southwestern end of Galveston Island (Paine et al. 2011). Longshore sand transport measurements obtained by Rogers and Ravens (2008) for the surf zone on Galveston Island ranged from 86,000 m^3/year (112,000 cy/year) to 231,000 m^3/year (302,000 cy/year).

The coast southwest of San Luis Pass to Pass Cavallo encompasses the headland of the Brazos and Colorado River deltas and associated barrier peninsulas called Follets Island and Matagorda Peninsula. Sediments eroded by waves reworking muddy and sandy deltaic headland deposits supplied sandy sediment to beaches adjacent to the headland deltas. Three navigation channels have been controlled with jetties along this section of coast, resulting in disruption of natural littoral transport to downdrift beaches. These include the Freeport Ship Channel jetties just north of the Brazos River entrance, the relatively short jetties that extend seaward from the Colorado River Navigation Channel entrance, and the Matagorda Ship

Channel jetties. These structures and channels have effectively compartmentalized sediment transport patterns along this section of coast (Paine et al. 2011). According to Paine et al. (2011), approximately 85 % of this coastal segment recorded shoreline recession. South of the San Bernard River to Pass Cavallo, average long-term recession rates averaged about 1.2 m/year (3.9 ft/year), whereas north of this point to San Luis Pass, shoreline recession averaged about 0.2 m/year (0.7 ft/year) (Figure 6.53). Areas of significant long-term shoreline recession include Follets Island, the Brazos headland, and a segment of Matagorda Peninsula southwest of the Matagorda Ship Channel. Beaches illustrating net shoreline advance are focused along short segments of the Matagorda Peninsula, including 3 km (1.9 mi) of beach northeast of the Colorado River mouth, a 5.5 km (3.4 mi) segment adjacent to the north jetty at the Matagorda Ship Channel, and a 2 km (1.2 mi) long segment at the southwestern tip of Matagorda Peninsula (Figure 6.53) (Paine et al. 2011). Net longshore sand transport between San Luis Pass and the Brazos River is consistent with transport direction and rates for Galveston Island. South of the Brazos headland along the Matagorda Peninsula, Heilman and Edge (1996) and Thomas and Dunkin (2012) estimated net longshore transport at between 38,000 and 250,000 m³/year (50,000 and 327,000 cy/year) to the southwest.

Southwest of Pass Cavallo to Packery Channel, long-term shoreline recession is prevalent along most beaches (0.8 m/year [2.6 ft/year]; Figure 6.53). Coastal engineering structures that impact sand transport for this shoreline segment include jetties at the Matagorda Ship Channel entrance that restrict sand transport to Matagorda Island, jetties at Aransas Pass that interrupt sand transport between San Jose and Mustang Islands, and the small Packery Channel jetties (Paine et al. 2011). Paine et al. (2011) documented net shoreline recession along about 80 % of this shoreline segment. However, approximately half the Gulf shoreline of Matagorda Island has advanced at relatively low rates since 1937. Highest rates of net shoreline recession (averaging 9.7 m/year [31.8 ft/year]) were recorded along a 6 km (3.7 mi) segment of Matagorda Island southwest of Pass Cavallo (Figure 6.53) (Paine et al. 2011). Net recession rates greater than 1 m/year (3.3 ft/year) were measured along most of San Jose Island, the central portion of Mustang Island, and the southern end of Mustang Island. Net shoreline recession rates elsewhere were less than 1 m/year (3.3 ft/year).

Although limited information is available regarding longshore sand transport rates, the predominant transport direction appears southwestward north of Packery Channel and variable south of this point. As such, net transport rates decrease to the southwest as the difference between northeast- and southwest-directed transport becomes minimized. Based on wave simulations, Kraus and Heilman (1997) determined the net longshore sand transport rate for Mustang and north Padre Islands to be about 34,000 to 53,000 m³/year (39,000 to 69,000 cy/year) to the southwest. However, deposition at the Aransas Pass jetties between 1866 and 1937 suggests net northward transport (Figure 6.54). Conversely, Morton and Pieper (1977) document deposition at the southern end of San Jose Island, southward channel migration at Aransas Pass, and shoreline recession along the north end of Mustang Island prior to jetty construction as evidence of net southwest longshore transport. Williams et al. (2007) documented deposition adjacent to the Packery Channel jetties as nearly symmetrical with slightly greater deposition south of the jetty (Figure 6.54). Based on these and other observations, the coast southwest of Aransas Pass to Padre Island National Seashore appears to be a nodal area for changes in the dominant direction of littoral sand transport (McGowen et al. 1977).

6.3.4.2.7 Laguna Madre Barrier Islands

The Laguna Madre of Texas and Tamaulipas is separated by the Rio Grande Delta at the United States–Mexico border and bounded by barrier islands and peninsulas along the GoM coast and mainland deposits along its western margin. The Laguna Madre extends

Figure 6.54. Patterns of deposition adjacent to the jetties at Aransas Pass and Packery Channel documenting variable transport directions. Shorelines from Miller et al. (2004). Image credit: ArcGIS World Imagery.

approximately 445 km (277 mi) from Corpus Christi Bay to La Pesca at the mouth of the Rio Soto la Marina (Figure 6.55). The Texas and Tamaulipas lagunas each encompass approximately 185 km (115 mi) of coast, and the Rio Grande Delta occupies about 75 km (47 mi) between the lagunas (Tunnell 2002b). The delta lobe protrudes about 35 km (22 mi) into the Gulf relative to shoreline orientation adjacent to the delta. Padre Island extends the entire length of the Texas Laguna Madre, except for an inlet cut through southern Padre Island in 1962 called Mansfield Channel (Figure 6.55). The southern terminus of the Texas Laguna Madre is marked by Brazos-Santiago Pass, which connects Port Isabel to the GoM. Brazos Island State Park (Boca Chica beach) is located along the southern 12 km (7.5 mi) of Texas coast that terminates at the Rio Grande River mouth.

Along the Tamaulipas coast, a deltaic headland/peninsular beach called Barra el Conchillal protects the northern portion of the Mexican Laguna Madre from Gulf waves and currents. This relatively low-profile beach averages approximately 2 km (1.2 mi) wide where it fronts northern Laguna Madre and extends approximately 115 km (71 mi) from the Rio Grande to Boca de Sandoval. Three washover barrier islands, with widths of 500 m (1,640 ft) or less, protect Laguna Madre south of Boca de Sandoval to the mouth of Rio Soto la Marina at La Pesca. Between Boca de Sandoval and Boca de Catán, Barra los Americanos and Barra Jesus Maria encompass about 56 km (35 mi) of coast marked by ephemeral inlets and washover features formed during storm events (Figure 6.55; Tunnell 2002b). The southernmost 78 km (48 mi) of barrier shoreline fronting Laguna Madre (Barra Soto la Marina) extends to the jetties

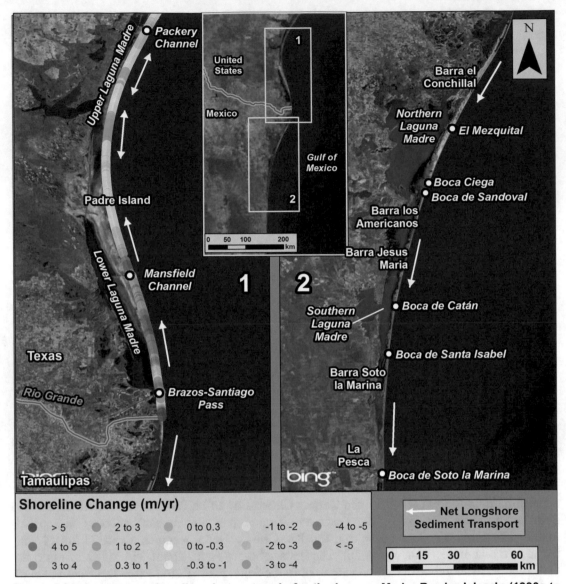

Figure 6.55. Long-term shoreline change trends for the Laguna Madre Barrier Islands (1930s to 2007). Shoreline change data from Paine et al. (2011). *White arrows* show the direction of net littoral sand transport. Image credit: Microsoft Bing Maps Aerial.

at the mouth of Rio Soto la Marina. Although all beaches along the Tamaulipas coast are prone to washover during storms, beach widths tend to decrease from the Rio Grande south and beach face slopes increase (Carranza-Edwards et al. 2007). Fine-grained terrigenous sands are dominant, with grain size increasing as beach slopes become steeper.

Longshore sand transport for Padre Island beaches varies depending on shoreline orientation. Although literature indicates that net littoral transport along northern Padre Island is to the south, sedimentation at the Packery Channel jetties indicates a nearly symmetrical deposition pattern, suggesting variable transport direction (Figure 6.54). Because transport direction varies depending on season and year along this section of coast, net transport rates are relatively low (Williams et al. 2007). Based on 8 years of wave data, Kraus and Heilman (1997) calculated net southward transport near Packery Channel at an average rate of 34,000 m^3/year (44,500 cy/

Figure 6.56. Shoreline offset at Mansfield Channel jetties illustrating net longshore transport to the north along South Padre Island, Texas. Image credit: ArcGIS World Imagery.

year). When shoreline orientation shifts from southwest to southeast along central Padre Island, net northward transport is well illustrated at jettied entrances (Figure 6.56). Heilman and Kraus (1996) calculated average net longshore transport rates along South Padre Island to be about 115,000 m^3/year (150,000 cy/year) to the north. South of the Rio Grande, along the deltaic headland beach of Barra El Conchillal, net transport direction shifts southward based on shoreline orientation and dominant wave climate. This pattern of transport continues south to La Pesca. Although no information is available on net transport rates for beaches fronting the Tamaulipas Laguna Madre, deposition patterns at jettied entrances document the net direction of longshore transport (Figure 6.57).

Between Packery Channel and Mansfield Channel, a longshore sand transport convergence zone shifts north and south depending on annual variation in wave energy relative to shoreline orientation. As such, a 38-km (24-mi) section of beach along north central Padre Island illustrates net accretion (~0.1 m/year [0.3 ft/year]) since 1937 (Figure 6.55). Conversely, the 20-km (12-mi) shoreline segment to the north toward Packery Channel and the 53-km (33-mi) segment south toward Mansfield Channel recorded net shoreline recession of about 1.0 m/year (3.3 ft/year) and 0.9 m/year (3.0 ft/year), respectively. The shoreline recession rate increased substantially for the 7-km (4.3-mi) segment north of Mansfield channel to about 4.1 m/year (13.5 ft/year), perhaps due to interruption of north-directed longshore sediment transport by the jetties at Mansfield Channel entrance. South of the channel, sand deposition within 1.5 km (0.9 mi) of the south jetty resulted in beach accretion and shoreline advance of about 1.9 m/year (6.2 ft/year). However, south of this deposition zone for approximately 50 km (31 mi), shoreline recession was prevalent at an average rate of about 3.1 m/year (10.2 ft/year). Only the southern 5 km (3.1 mi) of beach fronting South Padre Island was net depositional (1.6 m/year [5.2 ft/year]), likely the result of beach nourishment. South of Brazos Santiago Pass, the coast was net depositional during the Holocene as fluvial sediment from the Rio Grande supplied sand to form barrier islands (Paine et al. 2011). Since 1937, the northern 4.5 km (2.8 mi) of beach recorded net deposition from north-directed longshore sand transport, resulting in average shoreline advance of 0.8 m/year (2.6 ft/year). Conversely, the southern 7.5 km (4.8 mi) of beach to the Rio Grande documented shoreline recession of about 2.9 m/year (9.5 ft/year).

Figure 6.57. Shoreline offset at the El Mezquital and Boca de Soto la Marina entrances along the Tamaulipas Laguna Madre coast illustrating net longshore transport to the south. Image credit: Microsoft Bing Maps Aerial.

Quantitative shoreline change data are not available for the Tamaulipas Laguna Madre beaches, however, Moreno-Casasola (2007) stated that the barrier island coast south of the Rio Grande is presently eroding or migrating landward due to storm impacts, rising sea level, and limited new sand supply to the coast. Beaches along this coastal segment are low profile and highly susceptible to storm overwash. Relatively low net recession rates have been observed along most of this coastal segment (Carranza-Edwards 2011).

6.3.4.2.8 Laguna Morales Barrier Beaches to Barra del Tordo

This 85-km (53-mi) segment of coast extends from Boca de Soto la Marina at La Pesca to Barra del Tordo near the mouth of the Rio Carrizales (Figure 6.58). Narrow lagoons and waterways back beaches along this section of coast from Laguna Morales in the north to the estuary at Barra del Tordo. Beaches are relatively narrow and similar to those in the southern portion of the Laguna Madre region (Carranza-Edwards et al. 2007). Net longshore sand transport is to the south and onshore; however, deposition patterns at the mouth of Rio Carrizales, where a single jetty currently exists along the south side of the entrance, indicates that north and south transport is fairly balanced. Figure 6.58 illustrates sand spit development at the mouth of Rio Carrizales prior to jetty placement along the southern shoreline.

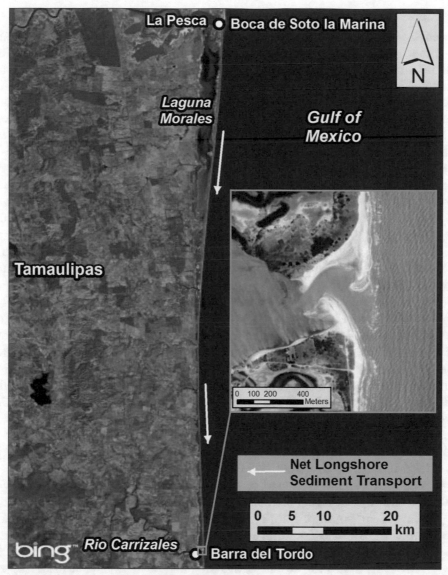

Figure 6.58. North-south shoreline between Boca de Soto la Marina and Barra del Tordo illustrating bi-directional transport at Rio Carrizales in an overall net south-directed longshore transport system. *White arrows* **show the direction of net littoral sand transport. Image credit: Microsoft Bing Maps Aerial.**

6.3.4.3   Southern Gulf of Mexico Marine Ecoregion

The Southern GoM Marine Ecoregion extends from Barra del Tordo at the mouth of Rio Carrizales along the southern GoM shoreline through Veracruz, Tabasco, and Campeche to the northeastern tip of the Yucatán Peninsula (Figure 6.2), a shoreline distance of about 1,700 km (1,056 mi). These shorelines encompass a variety of coastal geological deposits primarily formed by the interaction between fluvial drainage systems and coastal processes in the GoM. Most coastal depositional systems are composed of terrigenous clastic sediment; however, limestone shoreline deposits are dominant east of Isla del Carmen along the Yucatán Peninsula. Furthermore, volcanic headlands exist along the Veracruz coast adjacent to barrier beaches and deltaic deposits. Three geographic areas are used to illustrate patterns of shoreline change within the Southern Gulf Ecoregion: (1) Veracruz Neritic Barrier Shoreline, (2) Tabascan Neritic Rocky and Deltaic Shoreline, and (3) Campeche/Yucatán Carbonate Beach.

6.3.4.3.1 Veracruz Neritic Barrier Shoreline

Between Barra del Tordo and Tuxpan, the coast is composed of terrigenous clastic beaches, primarily sourced by Rio Panuco, that commonly form as barrier islands. The largest barrier island along this section of coast is Cabo Rojo, an island with extensive ridges and active dune fields (Figure 6.59). According to Stapor (1971), Rio Panuco is the primary source of sediment via southerly longshore transport leading to the development of Cabo Rojo. Sand beaches are generally wide and accretionary, and dune elevations are several meters high along most of the island. Between Barra del Tordo and Tampico, barrier islands are low profile, and beach widths are relatively narrow (<40 m [131 ft]) (Carranza-Edwards et al. 2007). Beaches are composed of terrigenous sand but shell fragments are frequently present. Three structured entrances that indicate net transport to the south are present along this section of coast. Croonen et al. (2006) analyzed the rate at which sand accumulated along the north jetty at the Port of Altamira and estimated south-directed transport at 300,000 m^3/year (392,000 cy/year). The jetty is a significant littoral barrier for sand transport to down-drift beaches, thereby creating a narrow, erosive barrier island protecting the lagoon south of the Port. Shoreline recession rates in this area were reported at 5 to 10 m/year (16.4 to 32.8 ft/year) (Croonen et al. 2006).

Between the Tampico Harbor jetties and Tuxpan, the most prominent coastal feature is Cabo Rojo, an extensive late-Quaternary barrier island extending approximately 100 km (62 mi) along the Gulf margin of Laguna Tamiahua (Figure 6.59) (Stapor 1971). Beaches are low profile and wide between Cabo Rojo and Tuxpan with extensive dune ridges behind the beaches. Net sand transport along the coast is to the south, as indicated by excess deposition along the north jetties at the Laguna Tamiahua and Tuxpan (Rio Pantepec) entrances and the prograding beach ridge plain along the southern leg of the cape. Although a net depositional feature, Cabo Rojo has experienced net erosion over the past few decades at rates of approximately 1 m/year (3.3 ft/year) (Peresbarbosa-Rojas 2005). Although net longshore transport quantities have not been estimated for this coastal segment, deposition patterns at jettied entrances suggest that transport rates are less than that identified for the beaches north of Tampico.

Except for a 7 km (4.3 mi) section of coast north of the mouth of Rio Cazones (Veracruz), where volcanic outcrops intersect the coast, beaches extending from Tuxpan to Playa Punta Delgada (50 km [31 mi] south of Nautla) are characterized as low, sandy mainland deposits that are relatively narrow. Rio Tecolutla and Rio Nautla supply relatively large volcanoclastic sediment loads directly to beaches along this section of coast (Figure 6.59) (Okazaki et al. 2001). However, dunes are absent in this area and beaches appear primarily erosional. Shorelines between Playa Punta Delgada and Playa Salinas are composed of bluffs and rocky points of volcanic origin (referred to as the Trans-Mexican Volcanic Belt), interspersed with small lagoons and narrow flood plains (Moreno-Casasola 2007). Sandy beaches are observed throughout this section of coast, and active dune fields are prominent north of Veracruz to Laguna de Farallón (Carranza-Edwards 2011). Although less common, rocky headlands persist as far north as Playa Punta Delgada, interrupting littoral sand transport along beaches. Most beaches within this ecoregion are undergoing erosion, as illustrated by active erosion or scarping of the primary dune ridge along the coast (Tanner 1975b). Sediment transport is primarily to the south but is variable in response to localized fluvial inputs, lithologic boundaries, and sedimentation accumulation landforms (Psuty et al. 2008, 2009).

6.3.4.3.2 Tabascan Neritic Rocky and Deltaic Shoreline

This 570-km (354-mi) shoreline segment has the greatest variety of shoreline types and extends from southeastern Veracruz through Tabasco to southwestern Campeche (Figure 6.60). Coastal areas in Veracruz, particularly the barrier beaches in the Alvarado region, are low lying

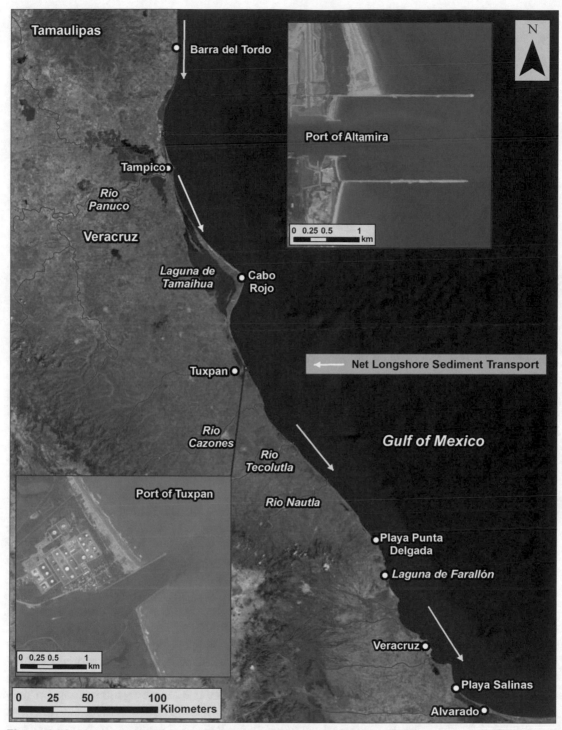

Figure 6.59. Veracruz Neritic shoreline between Barra del Tordo and Playa Salinas illustrating a net south-directed longshore transport system. *White arrows* show the direction of net littoral sand transport. Image credit: Microsoft Bing Maps Aerial.

Figure 6.60. Tabascan Neritic shoreline between Playa Salinas and Isla Aguada illustrating varia-bility in the net longshore transport system. *White arrows* **show the direction of net littoral sand transport. Image credit: Microsoft Bing Maps Aerial.**

and vulnerable to storm surge and rising sea level (Moreno-Casasola 2007). Southeast of this area, between Punta Puntilla and Laguna Ostión, the coast is a mixture of low-lying sandy beaches and rocky headlands within Los Tuxtlas Biosphere Reserve (associated with the San Andres Tuxtla volcanic massif). East of Laguna Ostión to Coatzacoalcos, beaches are low profile and extensive dune fields are present. Between Rio Coatzacoalcos and Rio Tonalá, numerous small rivers supply clastic sediment to the coast; however, most beaches are erosional (Carranza-Edwards 2011). Sediment within the Tabascan coastal zone is terrigenous, primarily sourced from the Tonalá, Grijalva, Usumacinta, and San Pedro y San Pablo Rivers (Thom 1967). Sand grain size varies from fine to very fine, and heavy mineral concentrations are common (Carranza-Edwards et al. 2007). Beach ridges are associated with deltaic deposition during an accretionary phase of development when sediment loads were high. However, historical changes in coastal evolution have been dominated by beach erosion (Tanner and Stapor 1971). Deltaic shorelines extend east into Campeche, terminating at the channel between Zacatal and Isla del Carmen at Laguna de Términos (Figure 6.60). Isla del Carmen, a barrier island fronting Laguna de Términos, is located in the transition area between limestone of the Yucatán Peninsula and alluvial terrain of deltaic deposits to the west (Contreras-Espinosa and Casta-ñeda-Lopez 2007).

Based on aerial imagery and Stapor (1971), net longshore sediment transport rates vary in this east-west oriented coastal segment depending on local shoreline orientation and sediment supply from the river systems. Near Alvarado and the Papaloapan River system, net transport is to the east. This trend continues along the Tuxtlas shoreline, only to be interrupted by rapid changes in shoreline orientation at headland outcrops. Pocket beaches often are shielded from wave approach depending on headland size and orientation, meaning longshore transport may vary significantly relative to open-coast sandy beaches. South of Laguna Ostión, net transport is from west to east until the jetties at Coatzacoalcos. East of the jetties, transport appears balanced with slightly greater transport from east to west. However, at the entrance to Laguna del Carmen at Sánchez Magallanes (~60 km [37 mi] east of Coatzacoalcos), the offset in sand deposition at the east and west jetties illustrates dominant littoral transport from east to west (Figure 6.61). This pattern of transport continues to the mouth of Laguna de Términos. Net transport rate estimates do not exist for this area.

Figure 6.61. Shoreline offset at the Laguna del Carmen jetties at Sánchez Magallanes illustrating net west-directed longshore sand transport (A; Image credit: Microsoft Bing Maps Aerial). Down-drift beach erosion west of the jetties has resulted in significant property damage (B; photo from Hernández-Santana et al. 2008, used with permission).

Although quantitative shoreline change information is not available for the coast between Playa Salinas and Rio Tonalá, Carranza-Edwards et al. (2007) indicated that coastal processes for most sandy beaches in this area are net erosional. East of Rio Tonalá, Tanner and Stapor (1971) recorded erosion along the seaward edge of the beach-ridge plain where younger beach ridges are truncated or scarped rather than tapered. Furthermore, trunks of dead trees were found in the surf zone as a result of beach erosion and shoreline recession. Ortiz-Pérez (1992) and Ortiz-Pérez and Benítez (1996) used historical maps to compare shoreline positions for the periods 1943 to 1958 and 1972 to 1984 to illustrate that shoreline recession is widespread for the deltaic shorelines of Tabasco and Campeche. At the mouth of Rio San Pedro y San Pablo, they found net shoreline recession was dominant at about 8 m/year. Hernández-Santana et al. (2008) supplemented these data with a 1995 shoreline and documented change between Rio Tonala and the Rio San Pedro y San Pablo entrance from 1943 to 1995. Estimates of shoreline change between 1984 and 1995 for the mouth of the Rio San Pedro y San Pablo were consistent at about −8 to −9 m/year (−26 to −30 ft/year) (Figure 6.62). Comparison of shoreline position for 1972, 1984, and 1995 at other coastal locations illustrated shoreline recession for most of the Tabascan/Campeche deltaic coast.

Ortiz-Pérez et al. (2010) updated previous shoreline change studies to include a 2008 aerial imagery shoreline. Figure 6.63 documents net changes quantified for nine segments of coast east of Rio Tonalá for the period 1995 to 2008. The western portion of segment 1 shows shoreline advance (0.97 m/year [3.2 ft/year]) near Rio Tonalá and the eastern side indicates net erosion (0.5 m/year [1.6 ft/year]). Sánchez Magallanes is located on the western margin of the jettied entrance to Laguna del Carmen (Figure 6.64), where west-directed longshore sand transport is blocked by the east jetty (Figure 6.61). This interruption in littoral transport produces significant net erosion immediately downdrift of the entrance (3 to 5 m/year

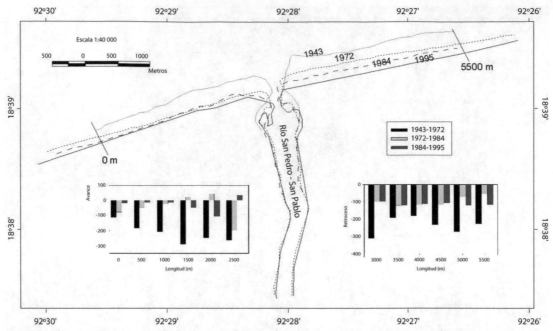

Figure 6.62. Shoreline position change adjacent to Rio San Pedro y San Pablo for the periods 1943–1972, 1972–1984, and 1984–1995 (from Hernández-Santana et al. 2008), used with permission.

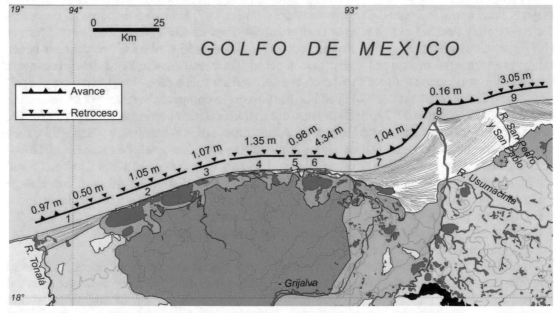

Figure 6.63. Net annual rates of shoreline advance (avance) and recession (retroceso) from Rio Tonalá to the Rio San Pedro y San Pablo delta plain, 1995 to 2008 (from Ortiz-Pérez et al. 2010, used with permission). *Arrows* pointing to the Gulf indicate shoreline advance; *arrows* pointing toward land imply recession.

Figure 6.64. Shoreline change locations for the Tabasco/Campeche coast. Image credit: Microsoft Bing Maps Aerial.

[9.8 to 16.4 ft/year] from 1972 to 2005) (Hernández-Santana et al. 2008), resulting in net erosion (0.5 m/year [1.6 ft/year]) for about 19 km (11.8 mi) west of the jetties. Although net deposition does occur adjacent to the east jetty, the next 28.5 km (17.7 mi) of coast (segment 2) east of the entrance is net erosional at approximately 1.05 m/year (3.4 ft/year) (Figure 6.63). From Boca Panteones east to Barra Tupilco (~17 km [10.5 mi]; segment 3) (Figure 6.64), shoreline recession is dominant at about 1.07 m/year (3.5 ft/year). The magnitude of erosion increases slightly along the 24.8 km (15.4 mi) shoreline cast of Barra Tupilco (segment 4) to approximately 1.35 m/year (4.4 ft/year) but increases to 4.34 m/year (14.2 ft/year) over the next 5 km (3.1 mi) near Puerto Dos Bocas (Ortiz-Pérez et al. 2010). Between Rio Gonzaléz and the eastern flank of the Rio Grijalva delta (~73 km [45 mi]; segments 7 and 8), the shoreline experiences net advance of between 0.16 and 1.04 m/year [0.5 and 3.4 ft/year]. However, the coast adjacent to Rio San Pedro y San Pablo and east about 20 km (12.4 mi) eroded at about 3.05 m/year (10.0 ft/year) (segment 9) between 1995 and 2008 (Ortiz-Pérez et al. 2010) and has been consistently eroding since at least 1943 (Figure 6.62) (Hernández-Santana et al. 2008).

Torres-Rodríguez et al. (2010) and Bolongaro Crevenna Recaséns (2012) evaluated erosion trends along the Campeche coast at selected locations east of the Rio San Pedro y San Pablo between 1974 and 2002/2008. Seven locations were used to document erosion trends, including shorelines adjacent to the Rio San Pedro y San Pablo mouth that overlap with shoreline change information compiled by Ortiz-Pérez et al. (2010) (Figure 6.64). Bolongaro Crevenna Recaséns (2012) documents a change rate of −4.8 m/year (−15.7 ft/year) between 1974 and 2006 and Ortiz-Pérez et al. (2010) calculated a rate of about −3.1 m/year (−10.2 ft/year) for the period 1995 to 2008. Although rates differ, variations in time interval and/or beach extent perhaps had the greatest influence on change rates. East of Rio San Pedro y San Pablo, the Nitrogenoducto area illustrated shoreline recession of about 0.7 m/year (2.3 ft/year) (1974 to 2004) whereas the Atasta shoreline area recorded −14.3 m/year (−46.9 ft/year) (1974 to 2008) (Bolongaro Crevenna Recaséns 2012). The very eastern portion of the delta plain near Playa la Disciplina and the channel to Laguna de Términos recorded a change rate of −17.1 m/year (−56.1 ft/year) between 1974 and 2008 (Torres-Rodríguez et al. 2010). The large change rates at Atasta and Playa la Disciplina reflect the influence of hurricanes impacting this area in 2005 and 2007. Additionally, three shoreline areas were evaluated for Isla del Carmen at Playa Norte, Club de

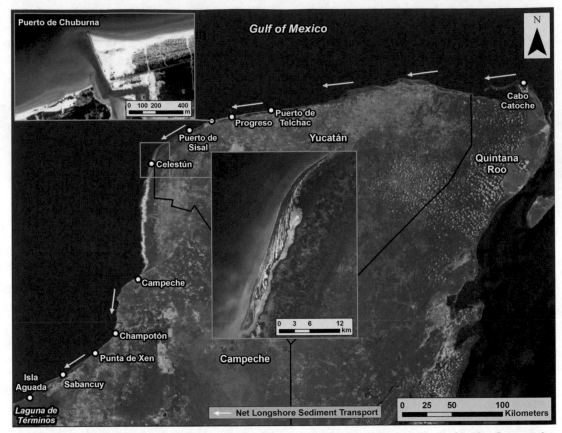

Figure 6.65. Index map illustrating net longshore sediment transport pathway for the Campeche-Yucatán coast (Image credit: Microsoft Bing Maps Aerial). Two *inset* images indicate the net direction of transport via sand spit growth and differential shoreline change at structured entrances. (Image credit: ArcGIS World Imagery).

Playa, and Cases (Figure 6.64). Large change variations existed but the Gulf facing shoreline was net erosional at all locations and the eastern two locations illustrated greatest change (−5.2 m/year [−17.0 ft/year] at Club de Playa; −3.6 m/year [−11.8 ft/year] at Cases). Playa Norte is at the eastern end of Isla del Carmen, the downdrift end of the longshore transport system, and had the smallest net erosion rate (0.3 m/year [1.0 ft/year]) (Bolongaro Crevenna Recaséns 2012).

6.3.4.3.3   Campeche/Yucatán Carbonate Beach

The Campeche-Yucatán carbonate beaches extend approximately 700 km (435 mi) from Isla Aguada at the eastern margin of Laguna de Términos to the northeastern end of Yucatán Peninsula near Cabo Catoche (Figure 6.65). The coast in this area is primarily a low-relief limestone platform through which rainfall filters and supplies coastal habitats with fresh water. Between Isla Aguada and Champotón, calcareous sand beaches are narrow and low relief. North of Champotón to Campeche, the coast is primarily limestone rock. Concrete bulkheads and other coastal structures protect the city of Campeche from flooding and erosion, and narrow calcareous sand and rock beaches are common south of the city. North of Campeche to Celestún, the shoreline is protected from energetic Gulf waves and the dominant shoreline type is mangrove. The northern Yucatán coast includes a beach-ridge plain overlying the limestone platform of the Yucatán Peninsula. Calcareous sand beaches along the Yucatán coast protect shallow and narrow lagoons from GoM waves and currents (Meyer-Arendt 1993). In many locations along northern Yucatán, beaches are quite narrow and low-relief dunes are common. River runoff is not present in this area, so beach sand is composed of carbonate particles derived from limestone deposits, coral reefs, and shells. As such, organic content in coastal waters is low and water clarity is excellent.

Carbonate sand beach ridges along the northern Yucatán coast reflect a period of sand abundance and accretion during the Holocene, but the present lack of sand in the littoral transport system has resulted in net erosion in recent years (Meyer-Arendt 1993). The dominant east-to-west longshore sediment transport system has produced several westward-curving sand spits (e.g., Celestún) and shoreline offsets at shore-perpendicular structures (e.g., jetties, groins) (Figure 6.65). Along the north-south shoreline between Celestún and Isla Aguada, net sand transport direction is to the south-southwest. The only section of coast where longshore transport is not a significant coastal process is along the low-energy coast between Celestún and Campeche where mangroves are dominant. Estimates of longshore transport magnitude are not available for the area between Isla Aguada and Celestún; however, shoreline change rates for the sandy beaches between Isla Aguada and Champotón are consistent with change rates along the northwestern Yucatán coast where transport rates vary from approximately 48,000 to 60,000 m^3/year (63,000 to 78,000 cy/year) (Appendini et al. 2012).

Between Celestún and Cabo Catoche, numerous coastal communities and industrial ports are present among the carbonate beaches and shallow coastal lagoons. Navigation structures associated with port development have resulted in large differences in shoreline position on either side of entrances (e.g., Puerto de Sisal, Puerto de Chuburná, Puerto de Telchac) indicating the dominant direction of littoral transport. The net direction of longshore sand transport in this area is illustrated well based on sand accumulation at shore-perpendicular structures and the natural growth of sand spits; however, the magnitude of net longshore transport requires knowledge of wave and current processes or a time series of shoreline and hydrographic surveys for documenting long-term sediment erosion and accretion patterns. Long-term regional survey datasets are not available for the northern Yucatán coast, so Appendini et al. (2012) used 12 years of wave hindcast data to estimate potential longshore sediment transport rates. The reliability of transport estimates was verified by comparing calculated rates with infilling rates at a shore-perpendicular structure that acts as a total littoral barrier to longshore transport. Based on transport simulations, Appendini et al. (2012) determined a range in transport from approximately 20,000 to 80,000 m^3/year (26,000 to 105,000 cy/year). Figure 6.66 illustrates variability in potential longshore sand transport rates for the northern coast of the Yucatán Peninsula, suggesting that approximately 60,000 m^3/year (78,000 cy/year) is being transported from the northwestern coast toward Celestún, creating an extensive sand spit deposit (Figure 6.65).

Torres-Rodríguez et al. (2010) document shoreline changes along the Campeche coast between Isla Aguada and Champotón for the period 1974 to 2002/2008. Greatest rates of change were recorded for a 10 km (6.2 mi) beach segment at Sabancuy (−6.8 m/year [−22.3 ft/year]) where jetties protecting navigation between Estero Sabancuy and the Gulf caused significant erosion downdrift of the entrance (Torres-Rodriguez et al., 2010). About 35 km (21.7 mi) north of this area, Torres-Rodríguez et al. (2010) document shoreline recession of about 4.4 m/year (14.4 ft/year) near Punta de Xen. Near Champotón, shoreline recession decreased to about 2.4 m/year (7.9 ft/year); however, much of the coast is rocky, implying a more stable shoreline type. The most stable carbonate beaches along the Campeche coast were identified near Isla Aguada where net shoreline recession rates of 0.2 m/year (0.7 ft/year) were calculated near the southwestern end of the littoral drift zone (Torres-Rodríguez et al. 2010). Sand accumulation from longshore transport perhaps resulted in lower net shoreline recession relative to updrift beaches.

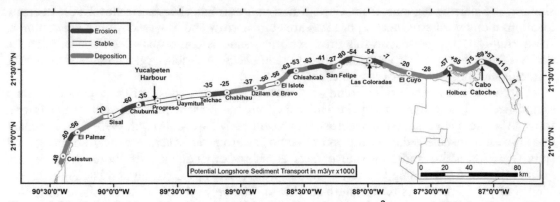

Figure 6.66. Potential longshore sediment transport estimates (m³/year × 1,000) for the northern coast of the Yucatán Peninsula (modified from Appendini et al. 2012). Areas of erosion and deposition are identified based on gradients in longshore transport rates. Negative and positive values represent westward and eastward transport, respectively.

Along the northern Yucatán coast, long-term accretion has been the primary process associated with barrier beach formation during Holocene time. However, during historical times, beach erosion has been the principal geomorphic response to coastal processes shaping the coast. Despite past accretion trends, dune scarping is common at several locations along the northern Yucatán coast, especially east of Progreso near Puerto de Chuburná (Meyer-Arendt 1993). Although shoreline recession rates of about 1.8 m/year (5.9 ft/year) have been documented by Gutierrez-Espadas (1983) for a 110-year period for this area, short-term rates averaged about 0.3 to 0.6 m/year (1.0 to 2.0 ft/year) for the period 1948 to 1978 (Meyer-Arendt 1993). Greatest shoreline changes along the northern Yucatán coast occur in response to jetty construction at harbor entrances and in association with sand spit growth (e.g., near Celestún). Natural coastal erosion generally is attributed to the passage of nortes (winter cold fronts) and hurricanes; normal waves and currents contain relatively low energy not capable of producing significant sand transport or shoreline changes.

6.3.4.4   Caribbean Sea Marine Ecoregion: Cabo Catoche to Cancún

The northeast outer coast of Quintana Roo, from Cabo Catoche to Punta Cancún, is composed of Holocene carbonate sediment derived from marine and coral reef limestone of Upper Pleistocene and Holocene age (Ward 2003). Coastal ecosystems include coral reefs, beaches and dunes, and coastal lagoons (Figure 6.67). Beaches are composed of fine, well-sorted sand, primarily derived from ooliths with skeletal mollusk detritus and coral fragments, and sand sources are from reef degradation and onshore sand transport (Morán et al. 2007; González-Leija et al. 2013). Prominent features along the coast include barrier islands and sand spits connected to the mainland that create coastal lagoons (e.g., Isla Blanca), and offshore islands (Isla Contoy and Isla Mujeres) that formed partly as remnants of Upper Pleistocene dune ridges (Ward 2003). This part of the Quintana Roo coast is particularly vulnerable to tropical cyclones and nortes. As such, the coast is typically erosional (Molina et al. 2001). When storms impact the area, limestone rock ledges are often exposed until coastal sand transport processes cover the rock ledges during post-storm depositional periods. Most beaches in this region are narrow (40 to 400 m [130 to 1,300 ft] wide) and have low elevations (Molina et al. 2001).

According to Carrillo et al. (2015), the general pattern of surface water currents along the Quintana Roo coast is from south to north, as is the Yucatán Current (Figure 6.68). Although reversals in longshore transport south of Cancún occur in the nearshore reef between rock

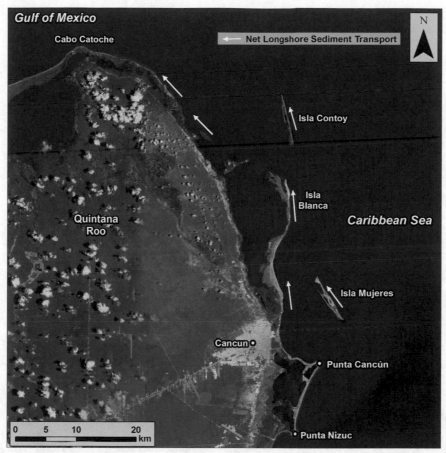

Figure 6.67. Index map illustrating net longshore sediment transport pathways for the northeast Quintana Roo coast. Image credit: Microsoft Bing Maps Aerial.

headlands, the Yucatán Current, coupled with wave refraction, produces a net northward current for all littoral areas along the northeast coast of Quintana Roo (Krutak and Gío-Argáez 1994). Longshore sand deposition resulted in the formation of numerous Holocene beach ridges along the northern coast of Isla Blanca, illustrating the dominant direction of littoral transport. Overall, this information is consistent with the potential longshore sediment transport modeling estimates of Appendini et al. (2012); however, predicted transport for approximately 15 km (9.3 mi) east of Cabo Catoche is to the southeast, not to the north (see Figure 6.66). Review of aerial imagery for this section of coast indicates that a reversal in longshore transport is evident based on sand deposition patterns at entrances between the islands. This localized departure from overall transport trends does not diminish the fact that both studies recognize a dominant south to north, then east to west, longshore transport pathway for coastal flows around Cabo Catoche. Although transport direction in this area is well documented, the magnitude of longshore sand transport is lacking.

The beach south of Punta Cancún was evaluated for shoreline change by Dibajnia et al. (2004) to document erosion trends relative to proposed beach replenishment in the area. They identified variations in shoreline response, illustrating net shoreline recession of about 1.5 m/year (4.9 ft/year) for beach extending approximately 3 km (1.9 mi) south of Punta Cancun, 0.5 m/year (1.6 ft/year) for the 6-km (3.7-mi) long central beach segment, and about 2.6 m/year (8.5 ft/year) for the 2 km (1.2 mi) segment south to Punta Nizuc for the period 1989 to 2000.

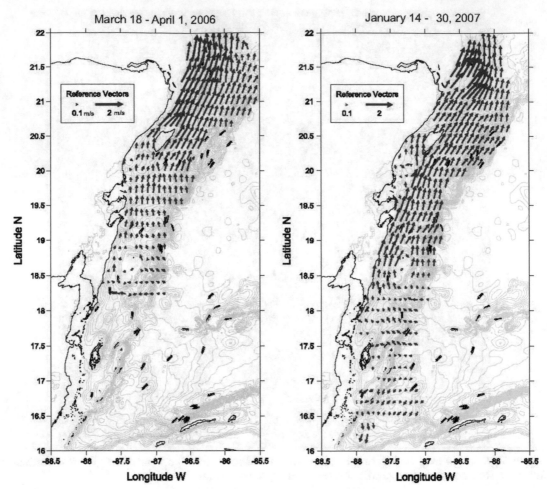

Figure 6.68. Pattern of surface water currents along the Quintana Roo coast (reprinted from Carrillo et al., 2015, with permission from Elsevier).

Based on change measurements, Dibajnia et al. (2004) estimated beach losses at approximately 33,000 to 76,000 m³/year (43,000 to 99,000 cy/year). If one assumes beach changes primarily are associated with longshore transport processes, estimated quantities can be used to approximate littoral transport rates. Because equivalent shoreline studies are not available north of this area, and exposure to waves and currents is similar for both regions, the estimates of Dibajnia et al. (2004) may provide a proxy for beaches along the northeastern Quintana Roo coast.

6.3.4.5   Greater Antilles Marine Ecoregion: Northwestern Cuba

The northwestern coast of Cuba, between Cabo San Antonio and Havana, has a coastline length of approximately 350 km (217 mi) and is highly diverse in terms of geology, soils, and plant communities (González-Sansón and Aguilar-Betancourt 2007). West of Havana to Bahía Honda, the coast is characterized by low cliffs and sandy beaches with inlets and bays (Figure 6.69) (Rodríguez 2010). Moving west of Bahía Honda, a chain of coral reefs and cays becomes the Archipelago de los Colorados, sheltering the inland coast and fringing mangrove habitat (González-Sansón and Aguilar-Betancourt 2007). Mangrove habitat flourishes when protected by sandy barrier beaches and spits or fringing coral reefs. Although narrow, fringing beaches are present between Havana and Mariel, sandy carbonate beach

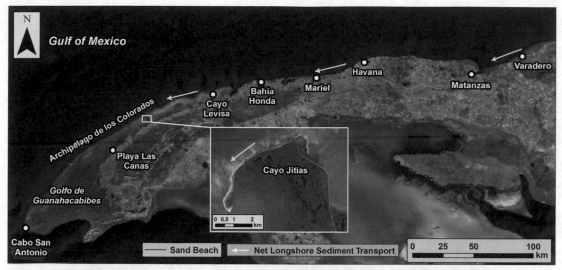

Figure 6.69. Index map illustrating the location of sand beaches and net longshore sediment transport pathway for northwest Cuba. Image credit: Microsoft Bing Maps Aerial.

environments become more common west of Mariel where source material for carbonate sands (degrading reefs) becomes more abundant. Along the southwestern end of Golfo de Guanahacabibes is a limestone peninsula with rocky beaches with narrow carbonate sand deposits.

The longshore sand transport system in this area is very complicated by the presence of coral reefs and limestone rock shores that dissipate and reflect wave energy depending on distance from shore and orientation relative to dominant wave approach. However, where sand beaches are present, the dominant direction of transport is from east to west due to predominant winds and waves from the east-northeast (Figure 6.69) (UNEP/GPA 2003; González-Sansón and Aguilar-Betancourt 2007). Sand spits at inlets and along the western ends of cays support this direction of net transport. Although longshore sand transport magnitudes are not available for the northwest coast of Cuba, predicted annualized sand transport rates for beaches at Varadero (east of Havana) are estimated at 89,000 to 134,000 m³/year (116,000 to 175,000 cy/year) (Kaput et al. 2007). Beaches along the northwest coast are more protected from predominant waves than those at Varadero, so net littoral transport rates are perhaps lower than those simulated by Kaput et al. (2007).

Shoreline change rates for northwestern Cuba beaches are lacking as well. Again, if measurements made for Varadero beaches over the past 30 years are indicative, the net rate of shoreline recession would be approximately 1.2 m/year (3.9 ft/year) (Kaput et al. 2007). Dead trees and stumps exposed on the beaches along the northwestern coast indicate chronic beach erosion (Figure 6.70); however, change rates are difficult to estimate. Using similar logic as stated for estimating net littoral transport rates, long-term shoreline recession rates would perhaps be on the order of 1 m/year (3.3 ft/year).

6.4 BIOGEOGRAPHIC SETTING

The coastal strand and its associated vegetated marine habitats consist of several characteristic habitats that are qualitatively similar throughout the world. These habitats include supratidal barrier islands and beaches; intertidal saline wetlands, including salt marshes and mangroves; subtidal seagrasses; and intertidal flats and subtidal soft bottoms (Christensen 2000; Mendelssohn and McKee 2000; Hester et al. 2005; and references therein). These habitats provide a suite of societal benefits as described in Section 6.4.4 of this chapter.

Figure 6.70. Erosion along the northwest coast of Cuba; (a) Playa Las Canas (UNEP/GPA 2003) and (b) Cayo Levisa (photo credit: van Lieshout (2007), used with permission).

6.4.1 Introduction to Vegetated Marine Habitats

6.4.1.1 The Barrier Strand

The barrier strand, composed of shore-parallel accumulations of sand and shell in the form of barrier islands, beaches, and related shoreline types, is best considered a habitat-complex. Several unique habitats, such as beach, dune, swale, maritime shrub and forest, salt pan, back barrier marsh, and submerged seagrass occur as part of the barrier strand complex

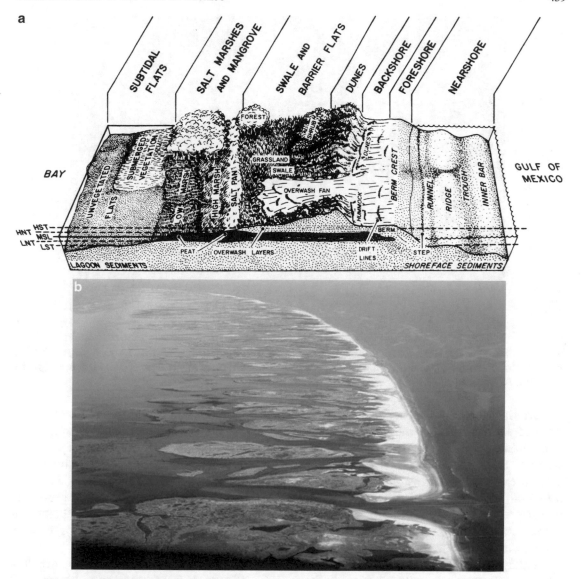

Figure 6.71. (a) Barrier strand habitats in the northern GoM (from Mendelssohn et al. 1983; used with permission from ASCE). (b) Oblique aerial photo of a segment of the Chandeleur Islands (photo credit: I. A. Mendelssohn).

(Figure 6.71a, b). Although the physiography of the barrier strand may differ in specific geographical locations in the GoM, the habitats therein and the primary factors controlling their biotic communities are quite similar. This introduction draws examples from the Deltaic Coastal Marshes and Barrier Islands Terrestrial Ecoregion (Figure 6.6).

The beach habitat is a strip of generally sandy substrate that extends from the low tide line to the top of the foredune, or in the absence of a foredune, to the farthest inland reach of storm waves (Barbour 1992) (Figure 6.72). This habitat is characterized by shifting sands, intense salt-spray, periodic saltwater inundation, and sand-washover. Only those plant species highly adapted to these stressors (e.g., *Cakile edentula* (sea rocket)) can survive on the beach.

Figure 6.72. Beach habitat includes the foreshore and backshore of the barrier strand and is subject to periodic wave runup, shifting sands, and saltwater from salt spray and surf (photo credit: I. A. Mendelssohn).

Landward of the beach, sand dunes, which can vary greatly in height, form as accumulations of aeolian transported sand and fine shell (Figure 6.73). Some dunes remain unvegetated and mobile, while those that are more stable become vegetated, which further promotes stability. Dune vegetation is usually distinct from beach vegetation. Because sand dune habitat seldom experiences saltwater inundation, the substrate, although infertile, has little salt accumulation, and thus, plant salt tolerance is not necessary. However, salt spray, the salt-laden aerosol generated from onshore winds blowing across breaking waves, is a common environmental stressor on primary dunes, and vegetation, like *Uniola paniculata* (sea oats), must be adapted to this stressor to survive in the sand dune habitat.

Landward of the primary dune, and between secondary and tertiary dunes, are low elevation depressions called swales or dune slacks (Figure 6.74). Swales have greater soil moisture than beach or dune habitats, and the types of vegetation occurring in swales are more flood-tolerant than beach and dune vegetation. Because of generally greater plant growth in the swale habitat and the lesser probability of plant-derived litter being removed by tides, soils in the swale are relatively high in organic matter (compared to the dune and beach), and therefore, have a greater water holding capacity and are more fertile for plant growth (Dougherty et al. 1990). Many of the mostly herbaceous plants that dominate the swale occur only, or primarily, in this habitat. On wider and more stable barrier islands, protected portions of the swale are usually dominated by shrubs and trees, e.g., *Myrica cerifera* (wax myrtle) and *Quercus virginiana* (live oak), respectively, and have been termed maritime forests (Christensen 2000).

On larger, more stable barrier islands, dune and swale topography often repeats multiple times, but when moving landward, elevation decreases and seawater inundation from backbarrier lagoons and bays occurs. This portion of a barrier island system is dominated by backbarrier salt marshes and in the more tropical climates, mangroves (Figure 6.75a, b). Salt

Figure 6.73. Dune habitat is characterized by accumulations of sand, either mobile or stabilized, depending on the extent of vegetation cover. As such, vegetation must be adapted to sand burial and salt spray, as well as moisture deficiency, to survive (photo credit: I. A. Mendelssohn).

Figure 6.74. Swale habitat is an interdunal topographical depression that occurs landward of the primary dune. Because the environment is more benign here, species diversity is generally high (photo credit: I. A. Mendelssohn).

Figure 6.75. Backbarrier marsh (a) occurs on the landward side of a barrier island/beach and is composed of both regularly flooded low marsh, dominated by *Spartina alterniflora* or *Avicennia germinans*, and (b) infrequently flooded high marsh, dominated by *S. patens* and *Distichlis spicata*, among other species (photo credit: I. A. Mendelssohn).

pan habitat (Figure 6.76) generally occurs between the swale and the backbarrier wetlands. This is an area where infrequent tidal incursions result in salt accumulation in the soil and thus high soil salinities. Where salinities are exceptionally high (more than twice sea-strength), salt pans can be devoid of vegetation. However, more often than not, sparse populations of the most salt-tolerant halophytes dominate salt pans. At somewhat lower elevations, tidal incursions occur more frequently, but still not on a daily basis. This is the high marsh, which consists of

Figure 6.76. Salt pan habitat has hypersaline soils in which few plant species can survive, and those that do are stunted and of low productivity (photo credit: I. A. Mendelssohn).

salt-tolerant plants that can only withstand intermittent flooding, usually only on spring or wind tides. Further bayward is the low marsh, where tidal inundation occurs daily. Salt marshes and mangroves that occur in regularly flooded portions of backbarrier environments reach their greatest development here. Intertidal flats are only exposed at very low water, and therefore are generally unvegetated by macrophytes.

Within shallow waters landward of the barrier strand, seagrass beds may occur where turbidity conditions permit. Their presence is determined primarily by water clarity and low-nutrient conditions. Also associated with the barrier strand are intertidal flats (Figure 6.77), which occur throughout the GoM, and are herein considered an independent coastal habitat (see Section 6.5.6). Often they are associated with barrier islands, but they also occur along shorelines in bays and lagoons.

6.4.1.2   Marine Intertidal Wetlands

Salt marshes, mangroves, and reed beds generally are low-energy coastal shoreline intertidal wetlands. Salt marshes are dominated by halophytic forbs, graminoids, and shrubs that periodically flood with seawater as a result of lunar (tidal) and meteorological (primarily wind) water level changes. Like other wetlands, salt marshes are characterized by a pronounced hydrology, soil development under flooded conditions (hydric soils), and the dominance of vegetation (hydrophytes) adapted to saturated soil conditions (Lyon 1993). Salt marshes (Figure 6.78a) usually dominate in temperate climates, but to a lesser degree are also found in subtropical and tropical environments (Costa and Davy 1992). Mangrove habitats (Figure 6.78b), which primarily occur in tropical and subtropical climates, share many of the same characteristics, but are dominated by woody plant species. The word, mangrove, is an ecological term used to describe salt- and flood-tolerant trees and shrubs that inhabit the intertidal zone (Mendelssohn and McKee 2000). In addition to the typical saline wetlands that occur along

Figure 6.77. Unvegetated tidal flats, adjacent to vegetated salt marshes, are exposed at low tides and provide habitat for wading birds and benthic fauna (photo credit: I. A. Mendelssohn).

the GoM coastline, reed beds, dominated by *Phragmites australis*, are a unique habitat of the northern GoM. The largest expanse of coastal reed beds in North America occurs along the coastal shorelines of the Mississippi River Birdfoot Delta (Figure 6.78c). Because it is a shoreline coastal habitat and occupies the position of saline wetlands elsewhere in the Gulf, it is included in this review of coastal habitats. The *Phragmites* reed habitat at the terminus of the Mississippi River is structured by the Mississippi River and the high subsidence rates that occur there. Salinities are fresh to intermediate and both native and European strains of *Phragmites australis* occur (Lambertini et al. 2012).

6.4.1.3   Seagrass Beds

Seagrass beds or meadows are primarily composed of clonal marine flowering plants that occur in shallow, generally soft-sediment habitats along the shores of bays and estuaries in temperate and tropical environments (Williams and Heck 2001) (Figure 6.79). Seagrasses comprise a very important vegetative habitat in the GoM. These flowering angiosperms are entirely restricted to underwater habitats where water clarity, salinity, and substrate are suitable. They often are referred to as "submerged aquatic vegetation" or SAV. Five genera occur in the Gulf, including *Thalassia*, *Halodule*, *Syringodium*, *Halophila*, and *Ruppia*. *Ruppia maritima* is generally associated with low-salinity brackish waters in bays and estuaries and is not addressed in this chapter. Estimates of the areal extent of seagrass beds in the GoM range from approximately 17,000 km^2 (4,250,000 acres) to 19,000 km^2 (4,695,000 acres) (Table 6.2) (Onuf et al. 2003; Handley et al. 2007). They are unevenly distributed with sizable areas occurring along Cuba's northwestern coast, the southern tip and Big Bend areas of Florida, the southern Texas coast, and Mexico's Yucatán Peninsula. Lesser amounts of seagrasses are found along the northern GoM from the Florida Panhandle to north Texas. Areas of seagrass also occur in the Mexican states of Tamaulipas, Tabasco, and Veracruz.

Figure 6.78. Oblique aerial photographs of (a) salt marsh dominated by *Spartina alterniflora*, (b) mangrove islands dominated by *Avicennia germinans*, and (c) *Phragmites australis*-dominated reed beds, all located in coastal Louisiana (photo credit: I. A. Mendelssohn).

Figure 6.79. Mixed meadow of seagrasses from the Big Bend area of Florida (photo credit: Barry A. Vittor & Associates).

Table 6.2. Areal Estimates of Seagrass Extent for the GoM

Location	Area, Ha (Acres)	Source(s)
NW Cuba	205,500 (507,790)	a
Tamaulipas, MX	35,700 (88,215)	a
Tabasco, MX	810 (2002)	a
Yucatán Peninsula, MX	591,100 (1,460,640)	a
Texas, excl. Laguna Madre	16,763 (41,422)	b, c
Texas, Laguna Madre	70,817 (174,987)	d
Louisiana	12,525 (30,949)	e
Mississippi	1,280 (3,164)	f
Alabama	682 (1,685)	g
Florida, Panhandle	15,864 (39,200)	h
Florida, Springs Coast	250,100 (618,000)	h
Florida, South	526,100 (1,299,993)	h
Totals	1,727,241 (4,268,048)	

(a) Onuf et al. (2003); (b) Pulich (2001); (c) Pulich et al. (1997); (d) Handley et al. (2007); (e) NOAA (2004); (f) Barry A. Vittor and Associates, Inc. (2010); (g) Barry A. Vittor and Associates, Inc. (2009); (h) Yarbro and Carlson (2011)

6.4.1.4 Intertidal Flats and Subtidal Soft Bottoms

Although GoM intertidal flats and subtidal soft-bottom habitats lack rooted vascular vegetation, they represent a significant interface between vegetated coastal habitats and nearshore waters. These sedimentary habitats adjoin or surround seagrass meadows, salt marshes, and mangroves, and many of their motile fauna move freely between vegetated and non-vegetated habitats. Non-vascular plants (marine macroalgae or "seaweeds") do occur in intertidal and subtidal areas, but are a minor component of those habitats. Epibenthic and benthic fauna obtain some nutrients from seaweeds but primarily convert organic production by vegetated habitats to forms available to epifauna and nekton. This trophic linkage is critical to fishery resources in the GoM.

Intertidal flats are less prominent in the GoM than along the Atlantic and Pacific coasts because amplitudes of lunar tides are much lower in the Gulf, and exposure of these habitats results mainly from wind-driven tidal action, especially during winter months. Subtidal soft-bottom habitats encompass those substrates that are deeper than the beach swash zone, and for this chapter, extend seaward to a depth of less than 10 m.

Gulf seaweeds are associated primarily with hard substrates, but genera such as *Avrainvillea*, *Caulerpa*, *Halimeda*, *Penicillus*, and *Udotea* include species that are found mainly on sand or mud surfaces. *Ceramium* may occur in seagrass beds as well as on hard bottom. Other taxa, including *Ulva*, can occur on intertidal flats and subtidal soft-bottoms where there are hard surfaces on which to attach. Fredericq et al. (2009) listed 673 seaweed species in the GoM; however, only 50 of these species occur predominantly on sand or mud bottoms. Only three of these soft-bottom species have been reported from the Mississippi Estuarine or Texas Estuarine Ecoregions; most are found in the Eastern Gulf Neritic Ecoregion. Because seaweeds generally exhibit patchy distributions, no estimates of surface area coverage are available for the GoM.

6.4.2 Depositional Characteristics of Vegetated Marine Habitats

The ecological structure and function of coastal flora and fauna in the GoM varies in response to spatial changes in depositional environments and climatic regime (temperate to subtropical to tropical). As a result, the distribution of vegetated marine habitats and their depositional environments within the GoM can best be summarized from an ecoregion perspective, given that marine and terrestrial ecoregions are in large part climatically driven (Wilkinson et al. 2009; Yáñez-Arancibia and Day 2004). At the broadest geographical scale, coastal habitats of the GoM occur in five primary marine ecoregions: (1) South Florida/Bahamian Atlantic, (2) Northern GoM, (3) Southern GoM, (4) Caribbean Sea, and (5) Greater Antilles (Figure 6.2). Because coastal habitats represent transitional environments between marine and terrestrial ecosystems (see Section 6.2), ecoregions based on terrestrial characteristics will be used as necessary when describing coastal habitats and their distribution.

6.4.2.1 South Florida/Bahamian Atlantic Marine Ecoregion

The most diverse area of the GoM coast is that of the southwestern Florida peninsula, where coastal habitats of the South Florida/Bahamian Atlantic Marine Ecoregion occur. The variety of habitats in this area is immense, where groundwater discharge is important and sandy beaches, mangroves, seagrasses, and coral reefs dominate. This marine ecoregion extends from the Florida Keys north to southern Keewaydin Island (just south of Naples, Florida) and comprises the Southern Coast and Islands Terrestrial Ecoregion. This part of the southwest Florida coast has many physiographic and hydrologic complexities associated with this biologically unique area. The entire Mesozoic and most of the Cenozoic geological sequence associated with the Florida peninsula is composed of carbonate rock (Hine and Locker 2011). As such, habitats of this region are often underlain by a calcium carbonate substrate, a driver of vegetation structure and function. As a consequence of this carbonate underpinning, the southwest Florida area is quite stable with little sediment compaction or subsidence.

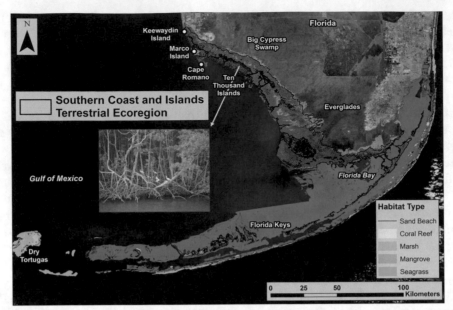

Figure 6.80. Coastal habitats for and adjacent to the Southern Coast and Islands Terrestrial Ecoregion (data from Beck et al. 2000; FFWCC-FWRI 2003; Giri et al. 2011a; IMaRS/USF et al. 2010). Image credit: Microsoft Bing Maps Aerial. Inset photo: Chauta 2012, used with permission.

The Southern Coast and Islands Terrestrial Ecosystem portion of the South Florida/Bahamian Atlantic Marine Ecoregion encompasses the Florida Keys and Ten Thousand Islands areas of southwest Florida (Figure 6.6). This highly diverse marine vegetated ecosystem consists of mangroves, seagrass beds, coral reefs, and marshes (Griffith et al. 2002) (Figure 6.80). Seagrass habitat has been cited as the largest in the northern hemisphere and is dominated by species such as *Thalassia testudinum* (turtlegrass), *Halodule wrightii* (shoalweed), and *Syringodium filiforme* (manatee grass) (Yarbro and Carlson 2011). Mangroves that dominate intertidal wetlands in the region consist of four primary tree species: *Rhizophora mangle* (red mangrove), *Avicennia germinans* (black mangrove), *Laguncularia racemosa* (white mangrove), and *Conocarpus erectus* (buttonwood). The southwestern Florida coast is characterized by a subtropical climate, modulated by the Gulf Stream, cold fronts, and hurricanes.

The Ten Thousand Islands area north of Florida Bay to Marco Island is characterized by numerous mangrove-covered islands (Figures 6.80 and 6.81). Beaches generally are absent along the southwestern coast with only a few local accumulations of shell and skeletal debris (Davis 2011b). The coast is quite stable due to an abundance of mangrove vegetation. Although hurricanes are common in this area, their impact has had little influence on coastal geomorphology (Davis 1995).

Ecosystem changes for the Southern Coast and Islands ecoregion have been documented using core data and information on historical hydrologic changes in the Everglades. Willard et al. (2001) and Wingard et al. (2007) documented long-term increases in salinity in Florida Bay and the Ten Thousand Islands area due to a combination of sea-level rise and hydrologic changes in the Everglades. These hydrologic changes produced a shift in wetland habitat from brackish/fresh-water marshes to dwarf mangrove stands. Although historical shoreline/wetland changes are primarily related to storm events and human activities (Section 6.3.4.1), Davis (2011a) suggests minimal long-term changes may be expected due to the stability of carbonate substrate in this relatively low-energy coastal region.

Figure 6.81. Image from Google Earth illustrating the Ten Thousand Islands area of southwest Florida. Map data: Google, U.S. Geological Survey.

6.4.2.2   Northern Gulf of Mexico Marine Ecoregion

The Northern GoM Marine Ecoregion is the most geographically expansive of the GoM ecoregions and extends from southern Keewaydin Island on the west coast of Florida to just south of Barra del Tordo in the State of Tamaulipas, Mexico, and includes coastal areas of Alabama, Mississippi, Louisiana, and Texas (Figure 6.2). Climate within this region is temperate to subtropical, with relatively distinct seasonal patterns in temperature resulting from temperate cold fronts during the winter and warm tropical currents in the summer. The region generally has high nutrient loading and includes biotic communities such as mangroves, salt marshes, and seagrasses; coastal lagoons and estuaries; and low river basins. It contains approximately 60 % of tidal marshes in the United States, freshwater inputs from 37 major rivers, and numerous nursery habitats for fish (Wilkinson et al. 2009).

6.4.2.2.1   Southwestern Florida Flatwoods Terrestrial Ecoregion

The barrier-inlet system along the central west Florida coast consists of approximately 27 barrier islands and inlets extending from Keewaydin Island (just south of Naples) to Anclote Key, just northwest of Tampa (Figure 6.82). Coastal habitats in this subtropical area include seagrasses, mangroves, and barrier islands and beaches. This ecoregion is underlain by a carbonate limestone on which sand and silts support large seagrass beds, dominated by *Thalassia*, that are key nursery, spawning, and feeding habitats for a variety of fish species (Zieman and Zieman 1989). Groundwater discharge is a notable source of freshwater and nutrients in the area. Mangroves are important intertidal wetland plants, but lose dominance at higher latitudes because of their relatively low cold tolerance (Mendelssohn and McKee 2000).

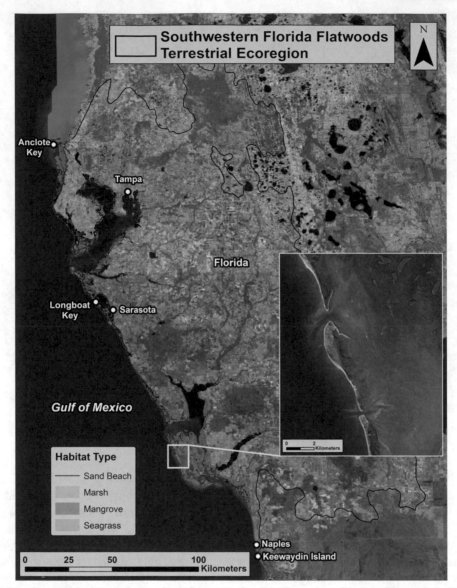

Figure 6.82. Coastal habitats for and adjacent to the Southwestern Florida Flatwoods Terrestrial Ecoregion (data from Beck et al. 2000; FFWCC-FWRI 2003; Giri et al. 2011a). Image credit: Microsoft Bing Maps Aerial.

Extensive barrier islands and the replacement of mangroves with herbaceous salt marshes characterize the northern reaches of this ecoregion.

Davis (2011b) refers to this area as the most morphologically complex barrier system in the world. The barrier islands range from 1 kilometer to tens of kilometers long, and inlets include a wide variety of sizes and morphologies under natural and engineered conditions. This coastal segment is classified as microtidal (range <2 m [6.6 ft]) with a mean annual wave height of less than 0.5 m (1.6 ft). Furthermore, this part of the central West Florida coast generally has avoided significant hurricane landfall compared with the northern Gulf.

Tidal inlets associated with the central West Florida barrier island system show a wide variety of scales and morphologies. Tide-dominated inlets tend to be stable and have existed throughout the historical record. Mixed-energy inlets respond to a general balance in tide and

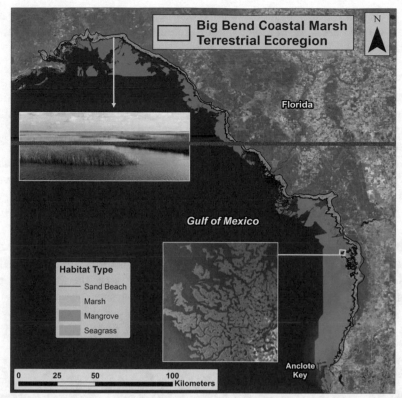

Figure 6.83. Coastal habitats for and adjacent to the Big Bend Coastal Marsh Terrestrial Ecoregion (data from Beck et al. 2000; FFWCC-FWRI 2003; Giri et al. 2011a). Image credits: Microsoft Bing Maps Aerial (main); ArcGIS World Imagery (inset); FL Department of Environmental Protection, http://www.dep.state.fl.us/coastal/images2/spartina_marsh.jpg (inset photo).

wave energy, whereas wave-dominated inlets typically are unstable and tend to close due to the dominance of wave transport energy relative to the flushing capabilities of tidal flow. Overall, tidal prism and the volume of water that flows into and out of backbarrier estuaries/bays during each tidal cycle control the scale and stability of inlets. Flood tidal shoals are the largest sand bodies extending into estuaries in this region, often formed during hurricanes through breaching of barrier islands (Davis 2011b).

6.4.2.2.2 Big Bend Coastal Marsh Terrestrial Ecoregion

North of Anclote Key, barrier islands cease to exist as sediment supply to the coast is negligible and coastal habitats are characterized by open-water marsh (primarily *Juncus*) in a tide-dominated environment (Figure 6.6). Coastal marshes experience spring tides up to 1.3 m (4.3 ft) and average wave heights of <0.3 m (1 ft). Because the coast is sediment starved, extensive limestone outcrops exist in subtidal and supratidal environments. As such, the Big Bend region of Florida has extensive seagrass beds, some extending into relatively deep water >12 m (39 ft) (Figure 6.83). Open-coast marshes that characterize the area can extend several kilometers inland, covering a karstic limestone surface along the coast (Hine et al. 1988). According to Davis (2011a), the Big Bend coastal area is similar to an open-water estuary with large freshwater discharge from springs that form the headwaters of rivers that empty into the Gulf. Linear oyster reefs that are fixed on the Tertiary limestone that crops out at the surface dominate the shallow inner shelf in this region. The presence of open-coast marshes indicates the degree to which wave and current processes rework coastal deposits. The broad, shallow shelf provides significant protection to coastal environments, and storms have had only minor impacts on the area, primarily by adding sediment to the marsh surface (Goodbred et al. 1998; Davis 2011a).

6.4.2.2.3   Gulf Barrier Islands and Coastal Marshes Terrestrial Ecoregion

Barrier islands and marshes of the northeastern GoM extend from the Apalachicola River Delta west to the Pearl River (Mississippi) (Figure 6.6). Coastal depositional systems include barrier islands, sand spits, mainland beaches, and backbarrier marshes. Inlets of various sizes separate coastal strand environments along this 550 km (342 mi) stretch of coast (Figure 6.84). A variety of shoreline orientations and ranges in shoreface slopes produce a complex pattern of longshore sediment transport associated with wave refraction patterns. However, a general east-west net transport direction is dominant along the coast (Section 6.3.4.2.3; Byrnes et al. 2010; Byrnes et al. 2012; Morang et al. 2012). Coastal sediments are terrigenous and derived from mainland and shoreface erosion under rising sea level.

Coastal habitats of this ecoregion are characterized by a series of barrier islands and beaches that are separated from narrow mainland salt marshes by elongate sounds (Figure 6.84). Barrier islands and beaches are well developed with relatively large dune fields on which *Uniola paniculata* (sea oats) often dominates. Mainland salt marshes are generally infrequently flooded and *Juncus roemerianus* (black needlerush) is the primary salt marsh plant species, as is the case for the salt marshes of the Florida panhandle. Only one species of mangrove is

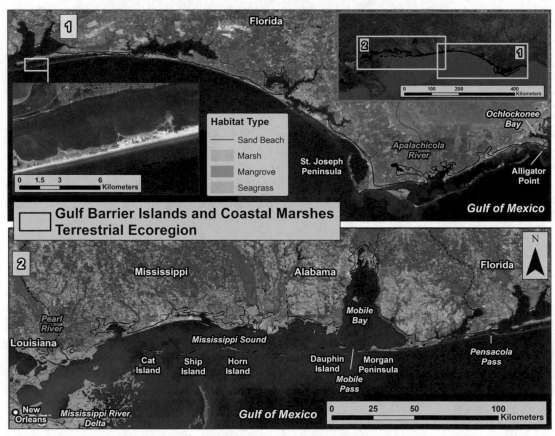

Figure 6.84. Coastal habitats for and adjacent to Gulf Barrier Islands and Coastal Marshes Terrestrial Ecoregion (data from Beck et al. 2000; FFWCC-FWRI 2003; Giri et al. 2011a; NOAA et al. 2004; NOAA et al. 2007; NOAA and DHS 2009). Image credits: Microsoft Bing Maps Aerial (main, inset); ArcGIS World Imagery (overview).

present, *Avicennia germinans*, which is the most cold tolerant of the four new world mangrove species commonly found in south Florida (Sherrod and McMillan 1985). The northern limit of the black mangrove in the GoM occurs on Horn Island, Mississippi. Seagrass beds are also limited in this ecoregion because of a lack of clarity in coastal waters.

Along the eastern boundary of this region, barrier islands fringe the Apalachicola Delta, a large promontory that abruptly changes shoreline orientation west of the Big Bend. As such, the delta marks the western limit of the low wave-energy coast of the Florida Gulf Peninsula. Between Alligator Point, just west of Ochlockonee Bay on the eastern margin of the Apalachicola Delta, and Pensacola Pass (about 330 km [205 mi]), white sandy barrier island and mainland beaches characterize what is known as the Florida Panhandle coast (Davis 2011b). Inland bays and lagoons provide estuarine habitat for herbaceous marshes and seagrass meadows. The inner shelf adjacent to the Apalachicola Delta coast is broad and gently sloping; however, the shoreface west of this area is steeper and wider. Consequently, wave energy at this coast generally is higher. Beach erosion along southeast facing shorelines often is coupled with deposition along the southwest margin of barrier beaches (Donoghue et al. 1990). Littoral sediment transport along the coast and deposition and erosion patterns in bays are controlled by storm processes associated with tropical cyclone and winter cold front passage (Stone et al. 2004).

The western extension of the Florida Panhandle coast encompasses the Morgan Peninsula coast between Pensacola Pass and Mobile Bay entrance. Morgan Peninsula, the most prominent geologic feature along this 75 km (47 mi) coastal segment, forms the southeastern terminus of Mobile Bay and consists of an extensive beach backed by parallel dunes and numerous sub-parallel beach ridges, formed as a result of west-directed net longshore sediment transport processes (Bearden and Hummell 1990; Stone et al. 1992). The eastern Alabama coast is similar to Florida Panhandle coast where sandy barrier beaches are close to but separated from the mainland by lagoons.

Seafloor topography and Holocene sediment distribution on the Alabama shelf reflect a combination of processes, including regression during the late-Pleistocene and reworking of the exposed shelf surface by ancient fluvial systems, and reworking of the exposed shelf surface by coastal processes during the subsequent Holocene rise in sea level (Parker et al. 1997). Redistribution of sediment by waves and currents during transgression partially or totally destroyed geomorphic features associated with Pleistocene fluvial environments. Concurrently, these same processes formed modern shelf deposits as subaerial coastal features became submerged and reworked during relative rising sea level. As such, much of the shelf offshore Alabama and the Florida Panhandle is sand (Byrnes et al. 2010).

Along the western quarter of the Gulf Barrier Islands and Coastal Marshes Terrestrial Ecoregion (Figure 6.84), adjacent to the eastern margin of the Mississippi River delta (i.e., the St. Bernard delta complex), resides Mississippi Sound and barrier island coastal habitat. The barrier islands extend approximately 100 km (62 mi) from Dauphin Island (AL) to Cat Island (MS) and provide the first line of protection to mainland Mississippi and Alabama from storm waves and surge. The islands are composed of beach sand that is derived from updrift beaches east of Mobile Pass and from ebb-tidal shoals at the entrance. Four tidal passes between the islands promote exchange of sediment and water between marine waters of the GoM and brackish waters of Mississippi Sound. Tidal passes also interrupt the flow of littoral sand to the west from Mobile Pass ebb-tidal shoals and Dauphin Island.

According to Otvos and Carter (2008) and Otvos and Giardino (2004), the Mississippi Sound barrier islands formed during a deceleration in sea-level rise approximately 5,700 to 5,000 years ago. At that time, the core of Dauphin Island at its eastern end was the only subaerial feature in the location of the modern barrier island system through which predominant west-directed littoral sand transport from the Florida panhandle via Mobile Pass ebb-tidal

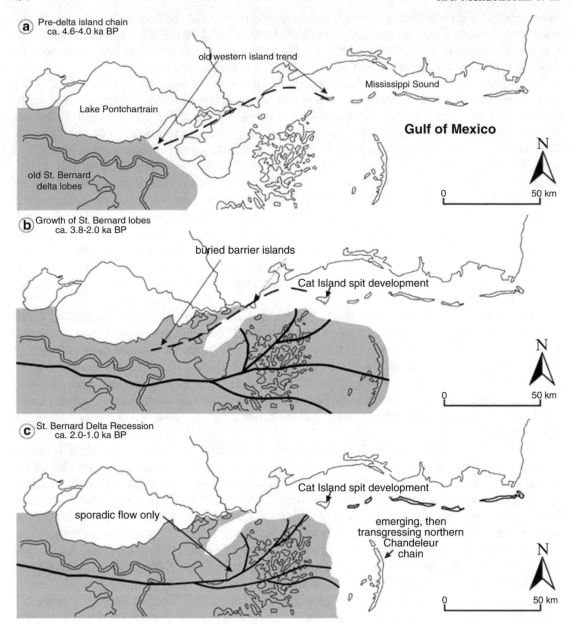

Figure 6.85. Barrier Island and St. Bernard delta lobe development as envisioned by Otvos and Giardino (2004) (reprinted with permission from Elsevier).

shoals could transit and deposit as elongate sand spits and barrier islands. The laterally prograding barrier island system originally extended west to the Mississippi mainland shoreline near the Pearl River, marking the seaward limit of subaerial deposition and the formation of Mississippi Sound.

Beginning approximately 3,500 years ago, the Mississippi River flowed east of New Orleans toward Mississippi Sound, creating the St. Bernard delta complex (Figure 6.85) (Otvos and Giardino 2004). Delta deposition extended over the western end of the Mississippi barrier island system, west of Cat Island. By about 2,400 years ago, fluvial sediment from the expanding St. Bernard delta created shoals as far west as Ship Island (Otvos 1979), changing wave propagation patterns and diminishing the supply of west-directed littoral sand to Cat Island. With changing wave patterns and reduced sand supply from the east, the eastern end of Cat Island began to erode, resulting in beach sand transport perpendicular to original island orientation (Rucker and Snowden 1989; Otvos and Giardino 2004). Persistent sand transport from the east has been successful at maintaining island configuration relative to rising sea level for much of the barrier system; however, reduced sand transport toward Ship Island has resulted in increased island erosion and segmentation from tropical cyclones (Rucker and Snowden 1989).

Mississippi Sound is considered a microtidal estuary because its diurnal tide range is only about 0.5 m (1.6 ft). The Sound is relatively shallow and elongate (east-west) with an approximate surface area of 2,000 km^2 (772 mi^2) (Kjerfve 1986) and a tidal prism of about 1.1×10^9 m^3 (1.4×10^9 cy). Although tidal currents account for at least 50 % of flow variance, the Sound responds rapidly to meteorological forcing, as evidenced by subtidal sea-level variations of up to 1 m (3.3 ft) and persistent net currents in the tidal passes (Kjerfve 1986). The relatively shallow and large area of the Sound creates strong currents in tidal passes between the barrier islands, ranging from 0.5 to 1.0 m/s (1.6 to 3.3 ft/s) and 1.8 to 3.5 m/s (5.9 to 11.5 ft/s) on flood and ebb tides, respectively. Overall, circulation within Mississippi Sound is weak and variable, and the estuary is vertically well mixed.

Barrier islands protecting Mississippi Sound experience a low-energy wave climate. Littoral sand transport along the islands is predominantly from east to west in response to prevailing winds and waves from the southeast. Reversals in longshore transport occur at the eastern ends of the islands but their impact on net sediment transport is localized (Byrnes et al. 2012). Although beach erosion and washover deposition are processes that have influenced island changes, the dominant mechanism by which sand is redistributed along the barrier islands and in the passes is by longshore currents generated by wave approach from the southeast (primarily storms).

6.4.2.2.4   Deltaic Coastal Marshes and Barrier Islands Terrestrial Ecoregion

The Mississippi River Delta Plain consists of large expanses of coastal wetlands within a geomorphologic framework of lakes, estuaries, and natural levee systems associated with active and abandoned distributaries (Figure 6.86). Locally, barrier island systems form the seaward edge of the delta plain, constituting an important component of the delta-plain ecosystem due to the habitat they provide, their storm-surge buffering capabilities, and their role in regulating marine to estuarine gradients (Kulp et al. 2005). Modern depositional models describe the Holocene history of the Mississippi River Delta Plain as a dynamic, multistage process that reflects the collective influence of changes in patterns of local relative sea-level rise and fluvial-sediment dispersal (Penland et al. 1988; Boyd et al. 1989). Sedimentary deposits of the Holocene delta plain consist of fine-grained sediment deposited within a variety of fluvial, deltaic, and coastal depositional environments. These sedimentary deposits formed in response to deltaic progradation and abandonment, resulting in an assemblage of overlapping regressive and transgressive units that consist of unconsolidated fluvial sediment (Kulp et al. 2005).

The present Mississippi River delta consists of two active delta complexes (Balize and Atchafalaya) and several inactive delta complexes (Figure 6.87). A delta complex encompasses the sedimentary deposits from a sequence of smaller delta lobes that are linked to a common distributary (Kulp et al. 2005). According to Roberts (1997), deposition within a delta complex generally occurs for approximately 1,000 to 2,000 years. During delta expansion, wetlands fringing the delta front and distributary network grow laterally, creating wetland habitat dominated by fluvial distributaries and bays adjacent to active distributary networks. Aerial

Figure 6.86. Coastal habitats for the Deltaic Coastal Marshes and Barrier Islands Terrestrial Ecoregion (data from Beck et al. 2000; Giri et al. 2011a; NOAA et al. 2004; NOAA and DHS 2009). Image credit: Microsoft Bing Maps Aerial.

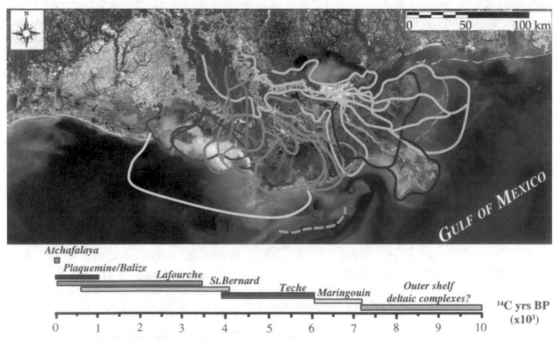

Figure 6.87. Distribution and chronology of Holocene Mississippi River delta complexes (from Kulp et al. 2005; used with permission of the Journal of Coastal Research).

expansion of a delta complex produces elongated distributary networks, which lead to a reduction in hydraulic gradient and eventual abandonment of the delta for a shorter, more hydraulically efficient route. Distributary switching and delta abandonment are natural processes by which marine inundation and delta erosion commence as a result of decreased

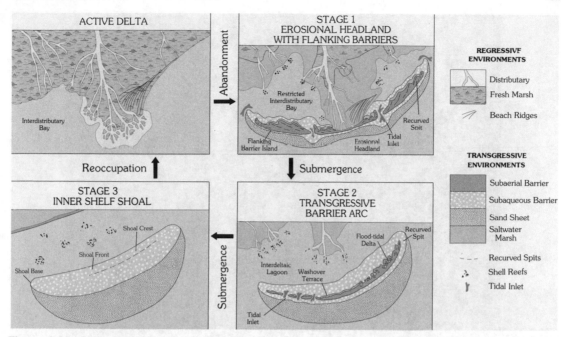

Figure 6.88. Conceptual model of delta lobe evolution. Distributary abandonment results in erosion and reworking of the delta lobe, ultimately forming an inner-shelf, sand-rich shoal (from Penland et al. 1992).

sediment supply and substrate compaction (Figure 6.88) (Roberts 1997; Williams et al. 2011). At abandoned deltaic headlands, relative sea-level rise results in erosional headland retreat as marine processes rework the shoreline. Sediment is dispersed laterally by waves and contributes to construction and nourishment of flanking beaches, beach ridges, and marginal deltaic deposits.

As a result of high subsidence rates and diminished sediment supply to the coast from a controlled river system, the Mississippi-Atchafalaya River Deltaic and Chenier Plains experience the highest rates of laterally continuous shoreline retreat and land loss in the GoM. While land loss associated with shoreline change along the Gulf shore and around the margins of large coastal bays is extreme, loss of the interior wetlands is even more extensive due to submergence and destruction of the Mississippi River Delta Plain (Penland et al. 1990). The result has been substantial land loss on the delta plain since the 1930s (Figure 6.89).

6.4.2.2.5 Texas-Louisiana Coastal Marshes Terrestrial Ecoregion

The Texas-Louisiana Coastal Marshes Terrestrial Ecoregion encompasses marginal deltaic depositional environments indirectly influenced by high levels of riverine input from the Mississippi-Atchafalaya River system. The region includes coastal habitats of southeastern Texas and southwestern Louisiana, an area known as the Chenier Plain (Figure 6.6). Coastal waters in this ecoregion generally are variable in salinity, and water clarity is low because of sediment load. Bottom sediments tend to be fine clays and muds, and conditions are ideal for growth of marshes and oyster reefs (Beck et al. 2000) (Figure 6.90).

The Chenier Plain extends approximately 200 km (124 mi) from Southwest Pass at Vermilion Bay to eastern Texas (Figure 6.91). This Late-Holocene, marginal-deltaic environment is up to 30 km (19 mi) wide and is composed primarily of mud deposits that are capped by marsh and interspersed with thin sand- and shell-rich ridges known as cheniers. In the Chenier Plain, oak trees line these ridges, which are better drained and topographically higher than the surrounding marsh.

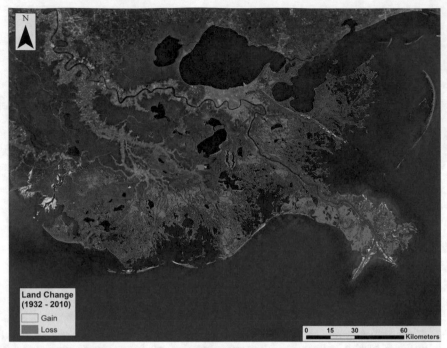

Figure 6.89. Wetland change on the Louisiana deltaic plain, 1932 to 2010 (data from Couvillion et al. 2011). Image credit: ArcGIS World Imagery.

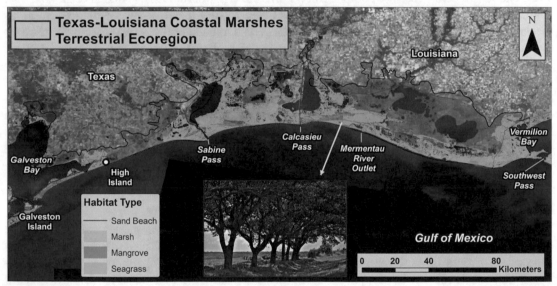

Figure 6.90. Coastal habitats for the Texas-Louisiana Coastal Marshes Terrestrial Ecoregion (data from Beck et al. 2000; BEG 1995; Giri et al. 2011a; NOAA et al. 2004). Image credit: Microsoft Bing Maps Aerial. Inset photo: White 2011, used with permission.

The Chenier Plain evolved during the Holocene as a series of progradational mudflats that were intermittently reworked into sandy or shelly ridges to form the modern Chenier Plain physiography (Gould and McFarlan 1959; Byrne et al. 1959; McBride et al. 2007). Numerous cycles of deposition and erosion created alternating ridges separated by marshlands. Sediment of the Chenier Plain has been primarily supplied by longshore transport of fine-grained

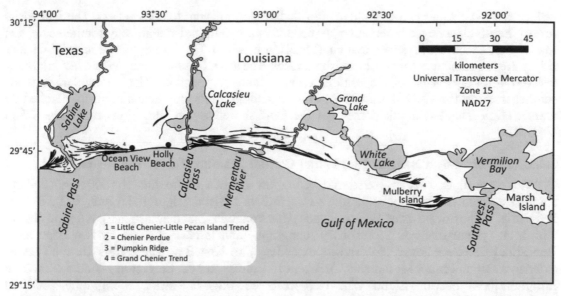

Figure 6.91. General geomorphic characteristics of the Mississippi River Chenier Plain (modified from McBride et al. 2007).

Mississippi River sediments (Hoyt 1969). These sediments, transported by westward-flowing nearshore currents, were eventually deposited along the Chenier shoreline as mudflats that built seaward. When deposition ceased or declined because of a shift in Mississippi River delta depocenters in the east, the previously deposited mud-rich sediment was reworked by coastal processes, concentrating coarse-grained sediments and forming shore-parallel ridges (Penland and Suter 1989). Renewed mudflat progradation, stemming from the introduction of new sediment by Mississippi River distributaries, resulted in isolation of these ridges by accretion of new material on the existing shoreline. Thus, repeated seaward growth and retreat along the Chenier Plain is a consequence of deltaic deposition farther east and the periodic cessation of sediment supply to the Chenier Plain as deltaic depocenters become abandoned and Chenier coast marine processes dominate. Currently, the Atchafalaya River is supplying the Chenier Plain with fine sediments by westward-directed longshore transport (Kineke et al. 2006). Distinct ridges, most of which represent relict shoreline positions, are interspersed in the mud-dominated coastal depositional system. Ridges typically are oriented shore-parallel to sub-parallel, are approximately 10 to 90 km (6.2 to 56.0 mi) long, 1 to 5 m (3.3 to 16.4 ft) thick, and 1 km (0.6 mi) wide (McBride et al. 2007).

Marginal deltaic coastal habitats evolved in a low-energy, microtidal, storm-dominated environment that experiences episodic sediment supply. Mean spring tide is mainly diurnal, ranging from 0.6 to 0.8 m (2.0 to 2.6 ft). Dominant nearshore currents are to the west and are controlled by winds and waves that are predominantly from the southeast (McBride et al. 2007). According to tide gauge data, the average rate of relative sea level rise for the Chenier Plain is 4.15 mm/year (0.16 in/year) (Figure 6.25), most of which can be attributed to compactional subsidence of Holocene sediment.

The upper Texas coast extends about 141 km (88 mi) from Sabine Pass to San Luis Pass. From a geologic perspective, Galveston Island is included with barrier island deposits south of Galveston Bay. Beach and marsh deposits east of Galveston Bay are more closely aligned with Chenier Plain deposits of southwestern Louisiana. Like southwestern Louisiana, the eastern portion of the upper Texas coast is characterized by a modern strandplain-chenier system with well-preserved chenier ridges with marsh-filled swales adjacent to Sabine Pass. These deposits reflect late-Holocene sedimentation associated with marginal deltaic environments of the Mississippi/Atchafalaya River system (McBride et al. 2007). Swales between relic chenier ridges are the sites of extensive brackish marshes. The strandplain-chenier system has gradually evolved through cycles of deposition, erosion, and compaction. The strandplain extends southeast along the Gulf shore toward High Island as thin sandy beach deposits perched on marsh. High Island is a salt dome near the Gulf shoreline with elevations exceeding 7.5 m (24.6 ft).

6.4.2.2.6   Mid-Coast Barrier Islands and Coastal Marshes Terrestrial Ecoregion

This central east Texas terrestrial ecoregion extends approximately 300 km (186 mi) southwest from the Bolivar Peninsula to north Padre Island (Figures 6.6 and 6.92). Bolivar Peninsula, to the northeast of Bolivar Roads (Houston Ship Channel Entrance), is a sandy beach and dune system that has accretionary topography and is characterized by two large relict flood-tidal shoal/washover fan deposits extending into East Bay. These fans are the sites of extensive salt and brackish marshes. Adjacent to Bolivar Roads, Galveston Island is a modern progradational barrier island with well-preserved ridge-and-swale topography (Bernard et al. 1970). Relict beach ridges and intervening swales have an orientation roughly parallel to the present island shoreline marked by the Gulf beach. Bayward of the ridge and swale features on Galveston Island are numerous truncated channels, the remnants of past tidal inlets and

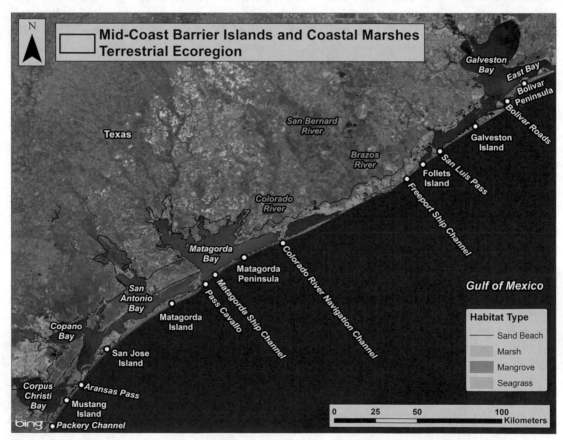

Figure 6.92. Coastal habitats for the Mid-Coast Barrier Islands and Coastal Marshes Terrestrial Ecoregion (data from Beck et al. 2000; BEG 1995; Giri et al. 2011a). Image credit: Microsoft Bing Maps Aerial.

storm washover channels along with extensive marshes. Galveston Island is relatively wide along its northeastern half and tapers and narrows toward San Luis Pass to the southwest (White et al. 2004b). Landward of Galveston Island is Galveston Bay. Although impacted by human activities including the Houston Ship Channel and extensive industrial and petrochemical activities, Galveston Bay has extensive intertidal wetlands dominated by *Spartina alterniflora*. Seagrasses are of lesser importance in this bay. South of Galveston Bay, the barrier strand continues with a series of backbarrier lagoons and, in some cases, adjacent bays. Coastal habitats including salt marshes, mangroves, seagrasses, tidal flats, and barrier beaches and associated dunes and swales are present (Figure 6.92). Although riverine freshwater input has been altered in many of these areas, hypersaline conditions do not normally occur because of sufficient rainfall. However, this situation progressively changes approaching Laguna Madre.

The segment of coast between San Luis Pass and Pass Cavallo encompasses the headland of the Brazos and Colorado River deltas with flanking barrier peninsulas called Follets Island and Matagorda Peninsula (about 143 km [89 mi] long). Primary natural geomorphic features along the shoreline include the Brazos and Colorado deltaic headlands, consisting of muddy and sandy sediments deposited by the Brazos and Colorado Rivers and overlain by a discontinuous, thin veneer of sandy beach deposits; a narrow, sandy peninsula extending northeastward from the Brazos headland toward San Luis Pass; and a narrow, sandy peninsula extending southwestward from the Colorado headland toward Pass Cavallo (Paine et al. 2011).

Sediments eroded by waves reworking deltaic headland deposits supplied sandy sediment to the flanking barrier peninsulas. Furthermore, the Brazos and Colorado Rivers supply sediment to the coast from their drainage basins. The drainage basin of the Brazos River encompasses approximately 116,000 km^2 (44,800 mi^2) of Cretaceous, Miocene, and Pleistocene sedimentary deposits, but the river capacity for carrying sediment to the coast during major floods has been reduced by completion of several dams and reservoirs between 1941 and 1969 (Paine et al. 2011). The drainage basin of the Colorado River is slightly smaller (103,000 km^2 [40,000 mi^2]), and nine dams completed in the upper and central basin between 1937 and 1990 have reduced its sediment-carrying capacity.

Further south, between Pass Cavallo and Packery Channel, much of the coast illustrates net shoreline recession. This section of shore includes Matagorda Island, San Jose Island, and Mustang Island. These sand-rich islands are characterized by broad sandy beaches and dune systems that reflect the position of the islands within a longshore current convergence zone between the Brazos/Colorado and Rio Grande deltaic headlands (White et al. 2002). Although tidal inlets separate these islands, no rivers supply water/sediment directly to the Gulf. Instead, rivers provide freshwater and sediment to the headwaters of Corpus Christi Bay, Copano Bay, and San Antonio Bay.

6.4.2.2.7 Laguna Madre Barrier Islands and Coastal Marshes Terrestrial Ecoregion

This ecoregion encompasses parts of Texas and Mexico included in the Western Gulf Coastal Plain Terrestrial Ecoregion and the Texas Estuarine, Laguna Madre Estuarine, and Western Gulf Neritic Marine Ecoregions (see Figures 6.6 and 6.3, respectively). The southern Texas coast comprises about 183 km (114 mi) of beach where the principal natural geomorphic feature is Padre Island, a long Holocene barrier island system with a well-developed dune system (Figure 6.93) that extends from Packery Channel near Corpus Christi Bay south to a narrow peninsula at Brazos Santiago Pass (White et al. 2007) (Figure 6.94). Padre Island developed initially as a spit extending from the relict Rio Grande Holocene deltaic system that has been eroding for hundreds of years. The Rio Grande enters the GoM along the border with Mexico and has created a large fluvial-deltaic headland that forms the southern boundary of a regional longshore current cell bound on the north by the Brazos-Colorado headland.

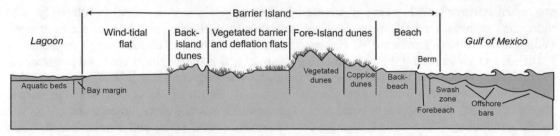

Figure 6.93. Generalized barrier island profile for Padre Island illustrating prominent features (modified after Paine et al. 2011).

The Rio Grande has a large drainage basin (558,400 km^2 [215,600 mi^2]) that extends into Mexico, New Mexico, and Colorado, but dams constructed in the middle and lower parts of the basin, combined with extensive irrigation use of Rio Grande water on the coastal plain, have reduced sediment delivered to the coast (Paine et al. 2011). Most of Padre Island is undeveloped, except for the town of South Padre Island. Engineering structures for this stretch of coast include the jetties and channels at Brazos Santiago Pass and the shallower Mansfield Channel.

The Laguna Madre of Texas and Tamaulipas (Mexico) is the only set of coastal, hypersaline lagoons on the North American continent. Extending along approximately 485 km (301 mi) of shoreline in south Texas and northeastern Mexico, the lagoons are separated by 85 km (53 mi) of Rio Grande Delta. The Laguna Madre system lacks significant precipitation, riverine input, and tidal flux, and in combination with high evapotranspiration rates and shallow depths, results in a classic hypersaline lagoon. The Texas lagoon is about 190 km (118 mi) long and the Mexico lagoon is about 210 km (130 mi) long, and each contains extensive tidal flats (Figure 6.94). Adjacent coastal habitats reflect this arid and hypersaline environment. Because the climate is harsh north and south of the Rio Grande, many bayshores are fringed by sparse vegetation and open sand flats, and barrier islands are characterized by sparsely vegetated dune fields. Extreme salinities have been moderated in recent decades due to channel dredging and the cutting of passes in the Texas Laguna Madre (Beck et al. 2000). The lagoons are protected on the east by barrier islands and peninsulas, and on the mainland side by large cattle ranches, farmlands, and the brush country. Laguna Madre also has the most extensive wind-tidal flats and clay dunes in North America (Beck et al. 2000).

The coast from Brazos Island State Park in Texas to Barra del Ostión in Mexico is dominated by deltaic sediment from the Rio Grande. This area also is referred to as the Mexican Laguna Madre region, where riverine sediment is dominant along the mainland coast of the lagoon and reworking of deltaic deposits by coastal waves and currents along the GoM provides vast quantities of sand to barrier beaches along the Tamaulipas coast (Moreno-Casasola 2007). Furthermore, the Mexican Laguna Madre in Tamaulipas consists of extensive barren tidal flats from which salt is commercially collected (Tunnell 2002a). Moving south from the Rio Grande, beach widths generally decrease and beach slopes increase. Terrigenous particle size is smaller on gentle slopes and larger on steep slopes. The predominant sediment size along Tamaulipas beaches is fine-grained sand, and sand distribution tends to be well sorted (Carranza-Edwards et al. 2007). Beaches in this region tend to be erosional (Figure 6.95).

The Laguna Madre is a region of high humidity but low precipitation, and consequently, emergent salt marshes fringing the Laguna Madre are dominated by succulent halophytes (salt loving plants) that have very high salt tolerances. Taxa such as *Salicornia* (glasswort), *Batis* (saltwort), *Distichlis* (saltgrass), *Borrichia* (sea oxe-eye), and *Limonium* (sea lavender), all common salt pan inhabitants, dominate the hypersaline wetlands of the Laguna Madre. Black

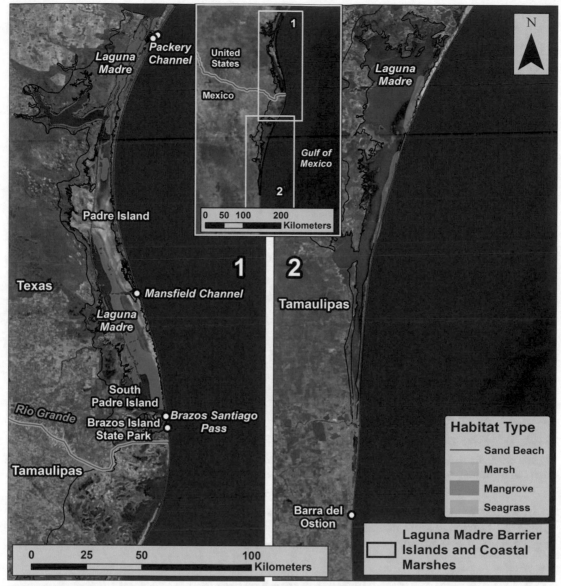

Figure 6.94. Coastal habitats for the Laguna Madre Barrier Islands and Coastal Marshes Terrestrial Ecoregion (data from Beck et al. 2000; BEG 2000; Giri et al. 2011a; Green and Short 2003). Image credit: Microsoft Bing Maps Aerial.

mangroves dwarfed by the hypersaline conditions also occur. In addition to hypersaline marshes, extensive fringing tidal flats, which are virtually unvegetated except for cyanobacteria algal mats, are common in the Laguna Madre. Interestingly, seagrass beds are much more abundant in the Laguna Madre than in other Texas bays due to clear and shallow waters of the former, resulting from the absence of riverine sediment input and the presence of a sandy lagoonal substrate. Barrier islands in this region are relatively simple compared to those on the Atlantic Coast (Judd 2002) and lack the multi-layer shrub-tree canopy structure of barrier islands in much of northern and eastern GoM. For example, virtually all plant species on southern Padre Island are herbaceous, although woody black mangroves occur sporadically. *Opuntia* spp. (prickly pear cactus) and *Prosopis glandulosa* (mesquite) also occur as individuals

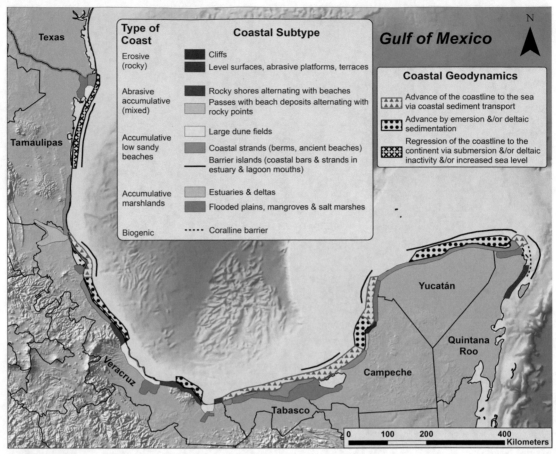

Figure 6.95. Distribution of coastal types and geological sediment trends for the southern GoM coast of Mexico (modified from Moreno-Casasola 2007; basemap from French and Schenk, 2005).

on these barrier strands. The live oak, *Quercus virginiana*, which is considered the climax habitat of barrier islands in the rest of the GoM, is absent except for a small stand on the Laguna Madre side of northern Padre Island (Judd 2002).

6.4.2.3 Southern Gulf of Mexico Marine Ecoregion

The Southern GoM Marine Ecoregion extends from approximately Barra del Tordo (about 40 km [25 mi] south of the terminus of the Laguna Madre Ecoregion) south and then east along the southern GoM shoreline to the northeastern tip of the Yucatán Peninsula (Figure 6.2), a shoreline distance of approximately 1,700 km (1,056 mi). This ecoregion includes the shorelines of Veracruz, Tabasco, Campeche, and Yucatán. Shorelines encompass a diverse suite of coastal habitats that include barrier beaches and islands, deltaic systems, coastal lagoons, estuaries, mangroves, seagrass beds, and coral reefs. Although climate in this area is primarily tropical, low-pressure cold fronts (locally called nortes) episodically traverse the region during autumn, winter, and spring, producing cooler conditions. High aquatic productivity in this region is thought due to wind-driven nutrient upwelling and freshwater input to the Gulf from the Usumacinta-Grijalva River, the second largest river system in the GoM (Table 6.1).

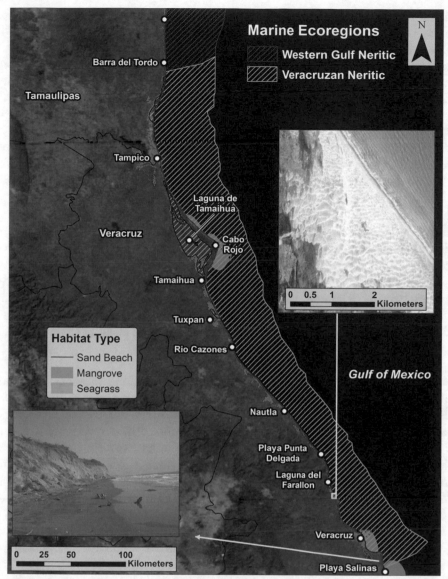

Figure 6.96. Coastal habitats for the Veracruz Neritic Marine Ecoregion (data from Giri et al. 2011a; Green and Short 2003). Image credits: Microsoft Bing Maps Aerial (main); ArcGIS World Imagery (inset).

6.4.2.3.1 Veracruz Neritic Marine Ecoregion

The northern boundary of the Veracruz Neritic Marine Ecoregion begins just south of Barra del Tordo, where the arid environment of Laguna Madre and the Rio Grande basin gives way to higher precipitation coastlines of the Veracruz barrier beaches (Britton and Morton 1989). Summer rainfall increases greatly, allowing a moderately diverse tropical flora to occur. Coastal topography of central Veracruz consists of fluvial and marine sediment draped around volcanic promontories (Psuty et al. 2008). Between Barra del Tordo and Tuxpan, the coast is composed of terrigenous clastic beaches that commonly form as barrier islands. The most extensive barrier island along this section of coast is Cabo Rojo, an island with extensive ridges and active dune fields, the highest of the western GoM (Figure 6.96). Rio Panuco is the primary source of sediment via southerly longshore transport leading to the development of Cabo Rojo

(Stapor 1971). The island protects Laguna de Tamaihua, where mangroves are common along the shoreline, and extends from Tampico to Tamaihua (about 120 km [75 mi]). Seagrass is present along the Gulf shoreline of Cabo Rojo and seaward of the beaches fronting Veracruz and Playa Salinas. Sand beaches are generally wide and accretionary, and dune elevations are several meters high along most of the island. North of Tampico to Barra del Tordo, low-profile barrier islands with relatively narrow beach widths protect shallow, narrow lagoons (Carranza-Edwards et al. 2007). Beaches are composed of terrigenous sand, and shell fragments are frequently present.

Beaches extending from Tuxpan to Playa Punta Delgada (50 km [31 mi] south of Nautla) are characterized as low, sandy mainland deposits that are relatively narrow, except for a 7-km (4.3-mi) section of coast north of the mouth of Rio Cazones (Veracruz), where volcanic outcrops intersect the coast. Dunes are absent in this area and beaches appear primarily erosional. Coastal habitat between Playa Punta Delgada and Playa Salinas is composed of bluffs and rocky points of volcanic origin, interspersed with small lagoons and narrow flood plains (Moreno-Casasola 2007). Sandy beaches are observed throughout this section of coast, and active dune fields are prominent north of Veracruz to Laguna de Farallón. Although less common, rocky headlands persist as far north as Playa Punta Delgada, interrupting littoral sand transport along beaches. The port of Veracruz occurs along this shoreline, but in a relatively low relief section. Most beaches within this ecoregion are undergoing erosion, as illustrated by active erosion or scarping of the primary dune ridge along the coast (Tanner 1975b).

6.4.2.3.2   Tabascan Neritic Marine Ecoregion

Tuxtlas Volcanic Coast. A prominent volcanic feature along the coastal portion of the Tabascan Neritic Marine Ecoregion in the State of Veracruz is an area known as Sierra de los Tuxtlas (Figure 6.97). The coastal area west of Tuxtlas is known as the Papaloapan region where an extensive sand barrier protects the Alvarado estuarine system (Figure 6.97) (Moreno-Casasola 2007). The 70 km (43 mi) stretch of coast between Playa Salinas and Punta Puntilla contains relatively wide sandy beaches with elevated dune fields that extend up to several kilometers inland. It is classified as a stable to accreting coast (Figure 6.95); however, Tanner (1975b) documented dune scarping by waves 2 to 3 km (1.2 to 1.9 mi) south of Alvarado Lagoon. The area between Punta Puntilla and Playa Linda (Los Tuxtlas region) is characterized by mixed abrasive-accumulative coastlines, alternating between projections of volcanic rocks and sandy beaches. Within this matrix of coastal geologic deposits are Laguna de Sontecomapan and a prominent sandy beach fronting the lagoon. Moving east along the coast from Laguna del Ostión, an abrupt change in shoreline orientation is encountered at the lagoon entrance to the Gulf, just west of Coatzacoalcos.

Tabascan Barrier Beaches and Marshes. The area east of Laguna del Ostión to Isla Aguada (Campeche) is within the Tabascan Neritic Marine Ecoregion where riverine input to the coast influences the sedimentological character of beaches. The non-calcareous deltaic shoreline extends along the southernmost arc of the GoM to a point just north of Laguna de Términos, where bedrock gradually changes to limestone of the Yucatán (Britton and Morton 1989). West of Laguna de Términos, coastal deposits are dominated by deltaic sedimentation from the Grijalva, Usumacinta, and San Pedro Rivers (Figure 6.95) (Thom 1967). As fluvial sediment accumulated at the Gulf shoreline, waves and currents redistributed sediment as ridges along the eastern Campeche and Tabascan coast. Modern sedimentation processes in the eastern portion of this area are dominated by fluvial input from the Grijalva and Usumacinta Rivers, the two longest rivers in Mexico, as they meander through mountainous uplands and lowlands of the Centla Marsh Biosphere Reserve. Beaches along the Tabascan lowlands are composed of light brown to gray, fine-grained clastic sediment of riverine origin, in contrast to

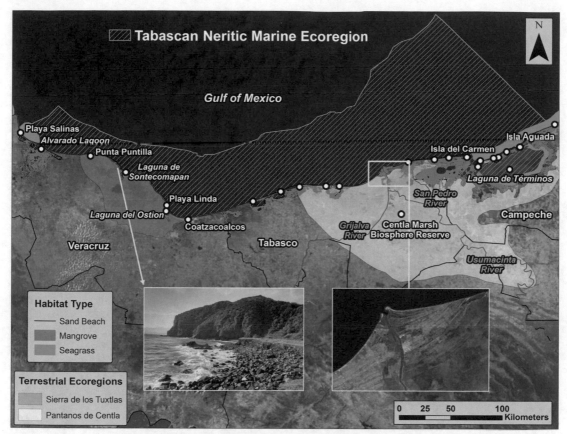

Figure 6.97. Coastal habitats for the Tabascan Neritic Marine Ecoregion (data from Giri et al., 2011a; Green and Short, 2003). Image credits: Microsoft Bing Maps Aerial (main); Ela 2016, distributed under a CC-BY 2.0 license (left inset); ArcGIS World Imagery (right inset).

the bright white calcareous sand of the Yucatán/Campeche area. Isla del Carmen, a barrier island fronting Laguna de Términos, is located in the transition area between limestone of the Yucatán Peninsula and alluvial terrain of deltaic deposits to the west (Figure 6.98) (Contreras-Espinosa and Castañeda-Lopez 2007). Beaches are wider and more elevated in the Isla Aguada transition area than beaches to the east, but carbonate sediment composition is very similar for both areas. Moreno-Casasola (2007) indicates that deposition in coastal beach and marsh habitat dominates Holocene sedimentation patterns along the coast (Figure 6.95).

Large wetland and barrier beach systems are associated with Tamiahua Lagoon, Alvarado Lagoon, Términos Lagoon, and lagoons adjacent to the west and north coasts of the Yucatán (Herrera-Silveira and Morales-Ojeda 2010). The most extensive mangrove stands in the GoM occur along the southern GoM shorelines (Dugan 1993; Thom 1967). Of all the coastal systems in the Southern GoM Ecoregion, Términos Lagoon (Laguna de Términos) has probably received the most scientific attention. Barrier islands and beaches, seagrass beds, mangroves, and even freshwater marshes are found in the Términos ecosystem (Figure 6.98), which occupies approximately 1,500 km² (580 mi²). These are some of the most productive natural habitats in the southern GoM.

Coastal processes along the Tabascan shore and beach-ridge plain are presently causing beach erosion along most of the coast. Tanner and Stapor (1971) recorded erosion along the seaward edge of the beach-ridge plain where younger beach ridges are truncated or scarped rather than tapered. Furthermore, trunks of dead trees were found awash in the surf zone as a

Figure 6.98. Habitats associated with Laguna de Términos in the Southern GoM Marine Ecoregion (data from Giri et al. 2011a; Rojas-Galaviz et al. 1992). Black arrows indicate water circulation pattern. Image credit: ArcGIS World Imagery.

result of beach erosion and shoreline recession. Finally, Tanner and Stapor (1971) found no evidence of beach ridges presently forming, implying that coastal erosion is a dominant process along the Tabascan shore. Although erosion along the beach-ridge plain does not appear extensive, beach ridges are eroding rather than growing.

6.4.2.3.3 Campeche/Yucatán Inner Neritic Marine Ecoregion

The Campeche-Yucatán carbonate beaches and mangroves are located adjacent to the Campeche-Yucatán Inner Neritic and Contoyan Neritic Marine Ecoregions (Figure 6.3). The coast extends approximately 700 km (435 mi) from Sabancuy, just north of Términos Lagoon, to the northeastern end of the Yucatán Peninsula near Holbox Lagoon (Figure 6.99). The Yucatán Peninsula is mainly a low-relief karst limestone platform. Few streams and no rivers drain the flat land or reach the sea, but rainfall filters through porous limestone and is stored underground (Britton and Morton 1989). Along the northern Yucatán coast, calcareous sand beach deposits provide low-relief coastal strands often fronting shallow and narrow lagoons (Meyer-Arendt 1993). Seagrass fronting Gulf beaches is dominant along the entire coast. Beaches can be quite narrow, and low-relief dunes are common. This area has limited mangrove habitat due to low precipitation and little terrestrial freshwater runoff. However, mangrove habitat can occur locally where lagoons persist, such as Rio Lagartos and Holbox Lagoons along the northeastern tip of the Yucatán peninsula and the coast between Campeche and Celestún (Britton and Morton 1989; Herrera-Silveira and Morales-Ojeda 2010).

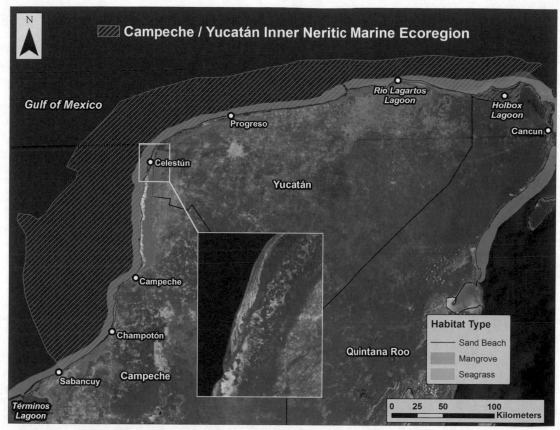

Figure 6.99. Coastal habitats for the Campeche/Yucatán Inner Neritic Marine Ecoregion (data from Giri et al., 2011a; Green and Short, 2003). Image credits: Microsoft Bing Maps Aerial (main); ArcGIS World Imagery (inset).

North of Sabancuy to Champotón, seagrass beds are common in nearshore areas and mangroves populate lagoonal areas landward of the beach. Calcareous sand beaches become wider in this area but relicf remains low (Figure 6.99) (Moreno-Casasola 2007). A few limestone cliffs are present along the coast between Champotón and Campeche, but most limestone shores in this ecoregion are low, narrow platforms that have elevations approximately 2 m (6.6 ft) above the surrounding sand veneer (Britton and Morton 1989). Concrete bulkheads and other coastal structures protect the city of Campeche from flooding and erosion, and narrow calcareous sand and rock beaches are common south of the city. North of the city of Campeche to Celestún, the inner coast is dominated by mangroves and low-relief calcareous lagoonal deposits landward of the shoreline, and the nearshore area has extensive seagrass beds.

As orientation of the coast shifts from north-south to east-west, a large calcareous sand peninsula at Celestún marks the terminal location to dominant westward longshore sand transport adjacent to the primarily low-energy mangrove coast to the south (Figure 6.99). This location is nearly coincident with the boundary between the Mexican States of Yucatán and Campeche, and is characterized by low precipitation (less than 50 cm/year [1.6 ft/year]) and shallow lagoons, which during drought, evaporate and form salt pans. The lagoons become hypersaline when precipitation allows. In spite of these conditions, much of the region north of Celestún to Progreso consists of relatively extensive, low stature mangroves (Britton and Morton 1989; Herrera-Silveira and Morales-Ojeda 2010).

Figure 6.100. Coastal habitats for the Contoyan Neritic Marine Ecoregion (data from Giri et al., 2011a; IMaRS/USF et al., 2010; Green and Short, 2003). Image credits: Microsoft Bing Maps Aerial (main); Maas 2006, distributed under a CC-BY 2.0 license (inset).

6.4.2.4   Caribbean Sea Marine Ecoregion

6.4.2.4.1   Contoyan Neritic Marine Ecoregion

The Contoyan Neritic region (extends from the northern part of the Yucatán Peninsula adjacent to Holbox Lagoon to Cancun; named after Isla Contoy) along the northeastern margin of the Yucatán Peninsula is characterized by coral reefs, seagrass meadows, and mangrove forests (Figure 6.100). Coralline beaches are narrow and dunes are low and not very extensive due to the presence of thick mangrove wetlands (Moreno-Casasola 2007). Lagoons in the region are shallow and often contain extensive seagrass beds and mangrove habitat. Low annual rainfall combined with severe dryness has eliminated rivers from the landscape. As such, freshwater necessary for productive mangrove ecosystems comes from springs (groundwater). Figure 6.95 illustrates that the barrier beach shoreline along the northern Yucatán Peninsula is net erosional, but beaches along the northeast margin of the Yucatán are net depositional, primarily due to longshore sedimentation processes (Moreno-Casasola 2007).

6.4.2.5   Greater Antilles Marine Ecoregion

Although Cuba was not specifically classified by Wilkinson et al. (2009), a quite comprehensive classification of marine ecoregions by Spalding et al. (2007) placed Cuba in their Greater Antilles Marine Ecoregion. The Cuban archipelago is typically Caribbean with regard to its marine ecosystems (González-Sansón and Aguilar-Betancourt 2007), composed primarily of small islands, mangroves, coral reefs, and seagrasses (Figure 6.101). Much of the underlying substrate for coastal habitats in this ecoregion is mixed calcium carbonate sands over which organic plant materials create mangrove swamps. The nearshore subtidal seafloor generally

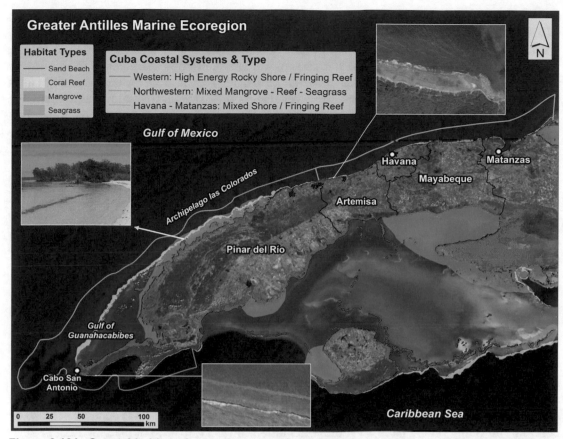

Figure 6.101. Coastal habitats for northwestern Cuba (data from Giri et al., 2011a; IMaRS/USF et al., 2010; Green and Short, 2003; Sullivan-Sealey and Bustamante, 1999). Image credits: Microsoft Bing Maps Aerial (main); ArcGIS World Imagery (bottom and upper right insets); Ji-Elle 2015, distributed under a CC-BY 2.0 license (left inset).

consists of unconsolidated sediment, either devoid of vegetation or forming large seagrass meadows dominated by *Thalassia testudinum* (turtle grass) or rocky bottom with extensive corals. Mangroves (*Rhizophora mangle*, *Avicennia germinans*, *Laguncularia racemosa*, and *Conocarpus erectus*) also are prevalent in protected, intertidal habitats along the northwestern Cuban shoreline (Figure 6.101) (Green and Short 2003; Sullivan-Sealey and Bustamante 1999). The Greater Antilles Marine Ecoregion has a wet-tropical climate characterized by a rainy season (May to October) and a dry season (January to March), interrupted by random, large-scale disturbances, primarily hurricanes and tropical storms. Similar to the Southern GoM Ecoregion, the northwestern Cuban coast is subject to nortes that punctuate the dry season. Predominant winds blow from the east and northeast.

Sullivan-Sealey and Bustamante (1999) describe four depositional systems encompassing the northwestern and southwestern Cuban coast. The Western High Energy Rocky Shore/ Fringing Reef Coastal System faces Yucatán Channel, where water flowing from the Caribbean Basin funnels to the eastern GoM and the Florida Straits, forming the Loop Current in the GoM (Figure 6.101). The coastline to the south is mostly rocky with long sandy beaches facing a narrow shelf that drops steeply to the southern entrance of the Yucatán Channel (Sullivan-Sealey and Bustamante 1999). Reefs fringe the entire edge of the shelf (Figure 6.101). Beaches along most of the western coast of Cuba are relatively stable due to the presence of offshore reefs to dissipate wave energy.

The Northwestern Mixed Mangrove-Reef-Seagrass Coastal System has a coastline length of about 375 km (233 mi), a mangrove-coastline length of about 355 km (221 mi), and is highly diverse in terms of geology, soils, and plant communities. Mountains of moderate height, sandy plains, lagoons, marshes, and flat and conical karst outcrops characterize the landscape (Borhidi 1996). This region includes an offshore barrier reef and an extensive shelf that is particularly wide in the Gulf of Guanahacabibes (Figure 6.101) (Sullivan-Sealey and Bustamante 1999). The shallow water Gulf contains numerous mangrove cays, seagrass beds, and patch reefs that extend to westernmost Cuba near Cabo San Antonio. Barrier reefs run along the outer border of the shelf, parallel to the Archipelago las Coloradas, which is composed of hundreds of mangrove cays.

The Havana-Matanzas Mixed Shore/Fringing Reef coastline is a coral reef dominated system that has a coastline length of 280 km (174 mi), of which about 30 km (19 mi) is populated with mangroves (Figure 6.101). This mixed-shore fringing reef system has an extensive rocky shore with terraces and cliffs with extended beaches (Sullivan-Sealey and Bustamante 1999). The coastal system is relatively narrow, and the continental shelf seaward of the coast is 1 to 3 km (0.6 to 1.9 mi) wide. The largest Cuban coastal population centers (Havana and Matanzas) are located within this coastal system.

6.4.3 Introduction to Aquatic Fauna of Vegetated Marine Habitats

Faunal components of vegetated marine habitats considered in this section, as well as adjacent intertidal flats and subtidal soft bottoms, are primarily macrobenthic epifauna (living on the sediment surface), infauna (living within the sediments), and nekton (natant or swimming organisms). The habits and distributions of these faunal components often overlap in coastal habitats. Some nekton are associated with the surface and mid-level depths of the water column, but many others have a distinct orientation toward the bottom, placing them in close proximity to the macrobenthic invertebrate assemblages. These demersal forms (e.g., flatfishes, gobies, natant decapod crustaceans) may also be categorized among epifaunal assemblages that dwell largely on the surfaces of sediments, submerged vegetation, or other structural elements in wetlands. This section does not include benthic meiofauna (organisms that pass through a 0.5 mm (0.02 in) mesh sieve usually used to collect macrofauna) nor does it include nektonic taxa (e.g., sea turtles, dolphins) that are the focus of other contributions to this collection of white papers.

Invertebrate assemblages of the GoM have been described in numerous reports and publications. Large-scale ecosystem surveys, such as the Bureau of Land Management (now Bureau of Ocean Energy Management [BOEM]) benchmark programs in the South Texas Outer Continental Shelf (STOCS) (Flint and Rabalais 1980), Mississippi-Alabama-Florida (MAFLA) (Dames and Moore 1979), and Southwest Florida Shelf (SOFLA) (Woodward-Clyde Consultants 1983), included some inshore sampling and characterized assemblages comprising a large array of decapod and stomatopod crustaceans, relatively small crustaceans such as cumaceans and amphipods, mollusks (especially gastropods), echinoderms, cnidarians, and some polychaetous annelids.

Defenbaugh (1976) grouped the epifauna of the northern Gulf into 12 assemblages. In zones immediately seaward of the coastal strand, decapods such as the portunid *Callinectes similis*, spider crab (*Libinia*), shame-faced crab (*Calappa*), purse crab (*Persephone*), and hermit crab (*Pagurus*) are common scavengers. Mud shrimp (e.g., *Callianassa*) form burrows in silty sand substrates while the stomatopod *Squilla* is more motile and carnivorous. Sea pansies (*Renilla*) are less common but noteworthy indicators of higher salinity waters. The gastropods *Nassarius*, *Littoridina*, and *Cantharus* and the bivalves *Mulinia* and *Nuculana* are common inhabitants of muddy sand and sand substrates throughout the GoM. Mollusks are generally most diverse in the southern Gulf, where sediments contain more carbonate and fewer large rivers discharge into the coastal area, but most dominant taxa in the southern Gulf also are found in other Gulf coastal habitats (Solis-Marin et al. 1993). Echinoderms such as the ophiuroid *Hemipholis* and the asteroids *Astropecten* and *Luidia* are associated with muddy sand and sand sediments throughout the Gulf. The echinoids *Diadema* and *Encope* are typical of subtidal waters in the Southern GoM Ecoregion (Solis-Marin et al. 1993). Figure 6.102 illustrates the distributions of three echinoid species in the GoM. The habitats of these species range into greater water depths than coastal wetlands but echinoids are common in clear, shallow waters off sandy beaches and in seagrass beds. Few echinoderms are found in littoral mud habitats, although some ophiuroids are detritivores and burrow in soft sediment.

Some epifaunal invertebrates, such as the penaeids *Farfantepenaeus aztecus*, *Farfantepenaeus duorarum*, and *Litopenaeus setiferus* also are nektonic and occur throughout the GoM, migrating offshore to spawn. Prevailing currents and behavioral adaptations allow their larvae to return to the estuaries that serve as primary nursery grounds. Blue crab (*Callinectes sapidus*), another key commercial epifaunal nektonic invertebrate species, exhibits similar migratory behavior. Coastal wetlands serve as the principal nursery areas for many commercially harvested decapod crustaceans, including penaeid shrimp (Figure 6.103).

Distributions of the juveniles of these species are closely linked with coastal wetlands. While some species, such as *F. aztecus* (Figure 6.103a) and *L. setiferus* (Figure 6.103c) are common in most coastal wetlands throughout the GoM, others such as *F. duorarum* (Figure 6.103b) seem dependent on specific wetland types (e.g., seagrass beds). Epifaunal invertebrate assemblages in vegetated habitats such as seagrass meadows generally exhibit higher densities and diversity than those on adjacent unvegetated soft bottoms; those metrics are often significantly correlated with aboveground plant biomass (Heck and Wetstone 1977).

Coastal benthic macroinfauna are among the best-known groups of marine invertebrates because they feed on detrital material produced in coastal wetlands and convert it to biomass production usable by secondary consumers of commercial value such as penaeid shrimp (e.g., Zimmerman et al. 2000). Infauna also are important indicators of habitat quality and the effects of environmental perturbation because they represent an integration of chronic and persistent natural and anthropogenic conditions (Rakocinski et al. 1998). Benthic surveys often are conducted to address specific potential or actual environmental impacts in the GoM, including effects of navigation dredging, oil spills (especially IXTOC in 1979) (Boehm and Fiest 1982), petroleum exploration and production, brine discharges from salt caverns, and effluent outfalls. Shallow-water benthic assemblages are sometimes categorized by substratum type (i.e., mud, sandy mud, muddy sand, or sand assemblages), but there are many species that occupy a wide range of sediment types. Mud habitats are depositional areas that support an infaunal assemblage adapted to elevated organics and periodic dissolved oxygen (DO) depletion; at the other extreme, sand habitats are characterized by species that require higher DO concentrations and greater flushing, with fewer burrowing taxa such as deposit-feeding polychaetes. The amphipods *Ampelisca abdita* and *A. cristata* occur mainly on silty-sand bottoms, but differences in the species' distributions within the GoM (Figure 6.104) suggest that habitat factors other than sediment type are also important. Uebelacker and Johnson (1984) described 593 polychaete species alone on the continental shelf of the United States regions of the Gulf; most of these were reported from coastal waters. They noted that some common polychaete species exhibited a faunal break east of Mobile Bay; some syllids were only found east of this area while some magelonids and ampharetids were only found west of the break. Other polychaetes exhibited disjunct distributions and were present in both the Eastern Gulf Neritic and Texas Estuarine subregions but not in the Mississippi Estuarine

[A] *Lytechinus variegatus.*

Photo credit: Hillewaert 2011;
© Hans Hillewaert/CC BY-SA 4.0

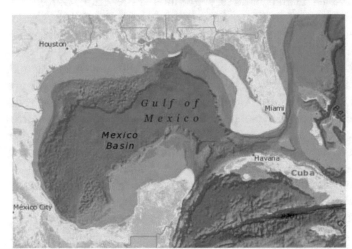

[B] *Diadema antillarum.*

Photo credit: Smith 2003;
© Daniel P.B. Smith/GNU Free
Documentation License.

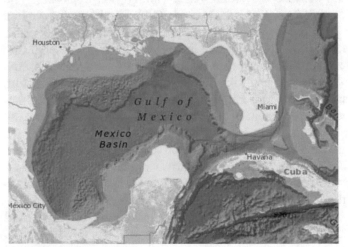

[C] *Encope michelini.*
Photo credit: 3D reconstruction
based on high-resolution X-ray
CT data, reprinted courtesy of
Dr. Louis Zachos and
DigiMorph.org.

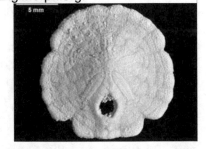

Figure 6.102. Distributions in the GoM of three echinoids found in seagrass, reefs [A, B] or sandy unvegetated sediments including beaches [C]. The GoM base map and species distributions were modified after data downloaded from http://www.eol.org on 21 March 2014.

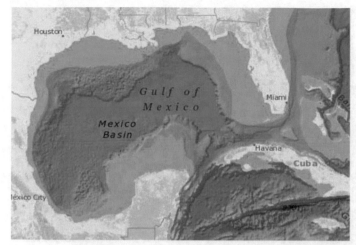

[A] *Farfantepenaeus aztecus.*

Photo credit: downloaded from http://txmarspecies.tamug.edu/invertdetails.cfm?scinameID=Farfant epenaeus aztecus in March 2014.

[B] *Farfantepenaeus duorarum.*

Photo credit: downloaded from http://txmarspecies.tamug.edu/invertdetails.cfm?scinameID=Farf antepenaeus duorarum in March 2014.

[C] *Litopenaeus setiferus.*

Photo credit: downloaded from http://txmarspecies.tamug.edu/invertdetails.cfm?scinameID=Litop enaeus setiferus in March 2014.

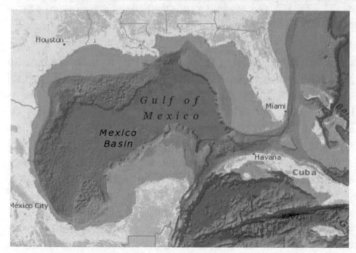

Figure 6.103. Distributions of important nursery areas for juveniles of three species of penaeid shrimp in the GoM. The GoM base map was modified from http://www.eol.org on March 21, 2014 and species distributions were modified from maps of penaeid nursery areas downloaded from http://www.ncddc.noaa.gov/website/DataAtlas_1985/atlas.html in March 2014. Photo images from the Identification Guide to Marine Organisms of Texas web site, http://txmarspecies.tamug.edu/index.cfm, used with permission.

[A] *Ampelisca abdita.*

Photo credit: California Academy of Sciences; downloaded from http://www.water.ca.gov/bdma/BioGuide/BenthicBioGuide.cfm#AA in March 2014.

[B] *Ampelisca cristata.*

Photo credit: EcologyWA, downloaded from https://www.flickr.com/photos/ecologywa/16951184998/in/photolist-rPVfJj-x4MPmt in February 2017.

Figure 6.104. Distributions of two benthic amphipods associated with intralittoral bay, estuarine and beach island habitats in the GoM. The GoM base map and distributions were modified after data downloaded from http://www.eol.org on 22 March 2014.

subregion. Tubicolous filter feeders and surface-dwelling carnivores are more abundant in sand habitats, and diversity overall is higher in the Southern GoM Ecoregion and in the South Florida/Bahamian Atlantic Ecoregion.

Seagrasses, salt marshes, and mangroves provide habitat for diverse assemblages of infaunal organisms, especially crustaceans, mollusks, and polychaetes. Diversity and abundance of infauna in seagrass meadows are greater than in surrounding non-vegetated areas (Lewis 1984), while salt marsh and mangrove infaunal assemblages generally exhibit lower diversity and abundance than adjacent mudflats, possibly due to the presence of thick roots, dense rhizome mats, and dense organic sediments (Sheridan 1997). However, lower levels of diversity and abundance in marsh and mangrove habitats also may be attributed to lesser degrees of inundation and oxygenation. Dominant species of infauna in these habitats are generally ubiquitous in the GoM, with very little difference among assemblages in the Southern GoM Ecoregion, Northern GoM Ecoregion, and South Florida/Bahamian Atlantic Ecoregion. Fiddler crabs (*Uca* spp.) can be found within any saline or brackish marsh as well as mangrove-dominated areas, but within such common genera, there can be distinctly different distributions among species (Figure 6.105).

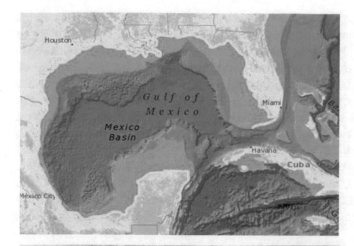

[A] *Uca rapax*. Photo credit: Richard Heard. Distribution modified from: www.usm.edu/gcrl/public/gulf.creatures/fiddler.crabs.php in March 2014.

[B] *Uca longisignalis*. Photo credit: Richard Heard. Distribution modified from: www.usm.edu/gcrl/public/gulf.creatures/fiddler.crabs.php in March 2014.

[C] *Uca speciosa*. Photo credit: Richard Heard. Distribution modified from: www.usm.edu/gcrl/public/gulf.creatures/fiddler.crabs.php in March 2014.

Figure 6.105. Distributions of three species of fiddler crabs (*Uca* spp.) commonly found in salt marsh and/or mangrove habitats of the GoM. The GoM base map and distributions were modified after data downloaded from http://www.eol.org on 22 March 2014. Photographs by Richard W. Heard, University of Southern Mississippi Gulf Coast Research Laboratory Campus, used with permission (Heard, 1982).

The term *nekton* describes an animal type that resides in water all or most of the time and is capable of self-directed propulsion through that medium even against currents. The ability to achieve deliberate and sustained horizontal movements in a dynamic fluid environment separates this group of aquatic organisms from plankton and places a lower limit on the size of nekton at about 2 cm in most estuarine/marine circumstances (Aleyev 1977). Although fishes usually comprise the highest species diversity among nekton, coastal habitats of the GoM are used by a variety of other groups classified as nekton, including some natant decapod crustaceans (penaeid and caridean shrimps, portunid crabs and lobsters), molluscs (squid, octopus, scallops), reptiles (turtles, alligators and crocodiles) and mammals (dolphins, whales and manatees). No nekton studies have targeted the full suite of nekton species (invertebrates, fishes, reptiles and mammals) that occur in coastal wetlands of the GoM. This discussion focuses primarily on the fishes and decapod crustaceans of vegetated marine habitats because these nekton groups are the most abundant and species-rich, but information on other groups is provided where appropriate.

There are more than 1,500 fish species, 150 natant decapod crustacean species, and less than 100 cephalopods represented among the GoM nekton, but as with the macrobenthos, relatively few species are endemic or even characteristic of the GoM (McEachran 2009; Felder et al. 2009; Judkins et al. 2009). The GoM nekton communities are derivatives of assemblages found in the Carolinian Atlantic and the Caribbean Sea. Fishes of the GoM include fewer than 5 % endemics, and many of these have sibling species in adjacent waters (McEachran 2009). However, a substantially higher proportion of such endemic species among fish families are typically associated with the shallow, intertidal vegetated coastal habitats of the GoM, particularly within the Eastern Gulf Neritic, Mississippi Estuarine and Texas Estuarine regions. For example, among the 28 species within the fish families Poeciliidae (live-bearers), Fundulidae (fundulids), and Cyprinodontidae (killifishes), which are commonly found in coastal wetland habitats of the GoM, 20, 46, and 60 %, respectively, are endemic. This is approximately an order of magnitude more than the average proportion of endemics among GoM fishes. One likely reason for the higher rate of endemism among these families is that species tend to be small, lack a planktonic life stage, and are not strong swimmers, so they do not travel extensively over their usually brief lifespans (1 to 3 years). All of these species are closely associated with coastal wetland habitats and never venture far from shore. Some are tolerant of a wide range of environmental conditions and are broadly distributed throughout the coastal wetlands of the GoM, while others may be so closely tied to specific habitats that their ranges are very limited (Figure 6.106). There is no single principal reason for constrained GoM distributions of small nekton species with relatively weak swimming abilities.

For example, the goldspotted killifish, *Floridichthys carpio*, is a very hardy species that inhabits only the quiet, shallow waters of mangroves, marshes, and coastal impoundments along the western coast of Florida and the Yucatán (Figure 6.107a) while the dwarf seahorse, *Hippocampus zosterae*, tolerates a narrow range of environmental conditions and is restricted largely to seagrass habitats (Figure 6.107b). This dependence on a single habitat type exposes seahorse populations to increased risk associated with habitat degradation (Musick et al. 2000; Hughes et al. 2009), in addition to negative pressures connected to their commercial exploitation in the GoM to meet demand in the aquarium trade and overseas medicinal markets (Baum and Vincent 2005).

Greater mobility of most other nekton, coupled with the location of coastal wetlands near the boundary of freshwater and marine environments, results in spatially and temporally dynamic nekton assemblages that may draw representatives from a range of marine, brackish, and freshwater groups within each ecoregion of the GoM. Consequently, most nekton assemblages found in these transitional habitats comprise a limited number of small, stress-tolerant

[A] *Fundulus grandis.*

Photo credit: W.M. Howell/ R. L. Jenkins (http://www.fishbase.org/summary/Fundulus-grandis.html); Downloaded March 12, 2013. Used with permission.

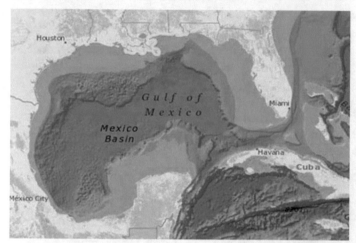

[B] *Fundulus jenkinsi.*

Photo credit: G.L. Grammer (http://www.fishbase.org/summary/Fundulus-jenkinsi.html). Downloaded March 12, 2013. Used with permission.

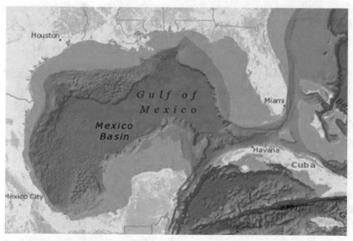

[C] *Fundulus grandissimus.*

Photo credit: T. Aarud (http://www.fishbase.org/summary/Fundulus-grandissimus.html). Downloaded March 12, 2013. Used with permission.

Figure 6.106. Distributions of three fundulids (killifishes) found in vegetated coastal wetlands (marshes, mangroves, and seagrass beds) of the GoM. *F. grandis* [A] also occurs along the Atlantic coast of Florida, but *F. jenkinsi* [B] and *F. grandissimus* [C] are GoM endemics. The GoM base maps and distributions were modified after data and references downloaded from http://www.eol.org on 10 March 2014.

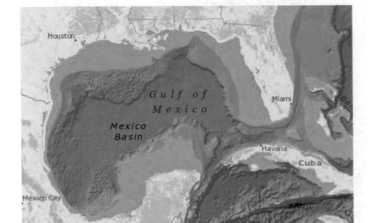

[A] *Floridichthys carpio.*

Photo credit: L. Holly Sweat, Smithsonian Marine Station at Fort Pierce, FL; downloaded from ...

http://www.sms.si.edu/irlspec/imag es/Florid_carpio.jpg in March 2014.

[B] *Hippocampus zosterae.*

Photo credit: Smithsonian Institution, Natl. Mus. Nat. Hist., Dept. of Vert. Zool., Div. of Fishes.

Figure 6.107. Examples of two habitat-restricted nekton species in the GoM. The GoM base map and species distributions were modified after data downloaded from http://www.eol.org on 21 March 2014.

species of year-round estuarine residents (Figure 6.106), as well as a complement of transient species (Figures 6.104 and 6.108) composed of seasonally abundant juvenile fishes and decapod crustaceans whose entire habitat within the GoM is more extensive, but encompasses coastal wetlands (Kneib 1997, 2000; Minello 1999; Heck et al. 2003).

Short-term (e.g., diel or tidal) and long-term (e.g., seasonal or ontogenetic) migrations also commonly occur between adjacent coastal wetlands (e.g., mangroves and coral reefs), with nekton providing a source of connectivity and the transfer of production among otherwise isolated environments comprising more sessile species (e.g., Kneib 1997, 2000; Deegan et al. 2000; Ellis and Bell 2008; Hammerschlag and Serafy 2009; Jones et al. 2010).

Major transfers of production from coastal wetlands occur when large numbers of species that use these habitats as nurseries (e.g., Figures 6.103 and 6.108) undertake offshore or coastal migrations as schooling species mature from juveniles to adults, or when coastal predators (e.g., Figure 6.109) forage on small resident nekton and benthic/epibenthic invertebrates in shallow coastal wetlands. Common predatory nekton associated with coastal wetlands in the GoM also exhibit a range of tolerances and preferences for certain environmental conditions.

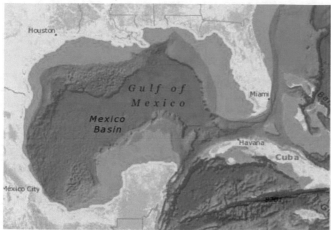

[A] *Anchoa mitchilli.*

Photo credit: W.M. Howell/ R.L. Jenkins; downloaded from http://www.fishbase.org/summary/5 45 in March 2014. Used with permission.

[B] *Mugil cephalus.*

Photo credit: J.E. Randall; downloaded from http://www.fishbase.org/summary/ Mugil-cephalus.html in March 2014. Used with permission.

[C] *Brevoortia patronus.*

Photo credit: Joel Boumje, free and unrestricted use of image provided by author through Wikimedia Commons.

Figure 6.108. Distributions of three species of abundant schooling nekton in the GoM. The GoM map and distribution shown in [A] were modified after data and references downloaded from http:// www.eol.org on 10 March 2014. Distributions for [B, C] were modified after distribution maps downloaded from http://www.ncddc.noaa.gov/website/DataAtlas1985/atlas.html in March 2014 and show only principal spawning/nursery areas for these species, which otherwise range throughout the GoM.

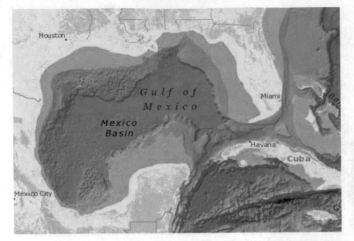

[A] *Cynoscion nebulosus.*

Photo credit: Donald Flescher,
NOAA Fisheries;
http://www.nefsc.noaa.gov/rcb/
photogallery/flescher.html,
accessed February 20, 2017.

[B] *Sciaenops ocellatus.*

Photo credit: Texas Parks & Wildlife;
downloaded from
http://www.tpwd.state.tx.us/huntwild/
wild/images/fish/reddrum2.jpg
downloaded March 2014.

[C] *Centropomus
undecimalis.*

Photo credit: Matthew Hoelscher;
downloaded from
http://commons.wikimedia.org/wiki
/File:Centropomus.jpg in March
2014. Distributed under a CC BY-SA
2.0 license.

Figure 6.109. Three common predatory nekton species in coastal wetland habitats of the GoM. The GoM base map was modified after data downloaded from http://www.eol.org on 10 March 2014. Distributions were modified after maps downloaded from http://www.ncddc.noaa.gov/website/ DataAtlas1985/atlas.html in March 2014.

For example, spotted seatrout (*Cynoscion nebulosus*) and red drum (*Sciaenops ocellatus*) may be widely distributed throughout the Gulf (Figure 6.109a, b) and elsewhere, while snook (*Centropomus undecimalis*) prefer clear waters associated with seagrass and mangrove habitats mostly in the southern GoM (Figure 6.109c).

In general, species richness of plants and animals in the GoM tends to be inversely related to water depth and is greatest along the south Florida coast, north through the Eastern Gulf Neritic (NOAA 2011). Approximately 40 % of fishes, 60 % of natant decapod crustaceans, and 12 % of cephalopods known from the GoM are from bay, nearshore, beach, coral reef, or estuarine habitats (McEachran 2009; Felder et al. 2009, Judkins et al. 2009) where they could be considered a source assemblage of nekton species for coastal wetland habitats. As with macrobenthos, most nekton families, and many species, are ubiquitous within the GoM. Fishes in the families Sciaenidae (drums), Ariidae (sea catfish), Gobiidae (gobies), Engraulidae (sardines), Clupidae (herrings), Mugilidae (mullets), and Sparidae (porgies) are among the most widely distributed according to trawl samples within Gulf estuaries (see McEachran 2009), but these groups are not always abundant in coastal wetland habitats, which are not usually sampled by trawling. Among the natant decapod crustaceans, species representing the families Penaeidae (penaeid shrimps), Palaemonidae (grass shrimps), and Portunidae (swimming crabs) are among the most widespread and abundant but there are major gaps in knowledge concerning these and other crustacean groups, particularly in the southern GoM (Felder et al. 2009). Only a few species of cephalopods, mostly within the family Loliginidae (inshore squids), are widespread in shallow estuarine waters associated with coastal wetlands (Judkins et al. 2009).

Table 6.3 is a summary of nekton families comprising the most abundant species closely associated with shallow coastal wetland habitats within most of the major nearshore ecoregions of the GoM. It suggests that fishes dominate coastal wetland assemblages in most ecoregions except in the Mississippi and Texas Estuarine regions, where natant decapod crustaceans may be far more abundant. Sedimentary environment characteristics and influence of freshwater riverine input and nutrients from extensive watersheds supplying these regions (Figure 6.15, Table 6.1) may favor the production of crustaceans, or the observation could be related to differences in the emphasis of research efforts within each region. For example, the 22 studies summarized in Minello (1999), which represent nekton samples primarily in the Mississippi and Texas Estuarine regions (Table 6.3) were all collected with small enclosure samplers usually deployed from the bow of small boats primarily sampling the edges of tidal marshes (e.g., Baltz et al. 1993; Minello et al. 1994). This presents a perspective on the nekton assemblages that may differ from that obtained using other methods applied in flooded intertidal habitats and smaller channels/ponds within interior marshes (e.g., Weaver and Holloway 1974; Herke and Rogers 1984; Felley 1987; Peterson and Turner 1994; Rozas and Minello 2010). Samples collected from the flooded interior portions of tidal marshes in the northern GoM are dominated by few very abundant species in the families Cyprinodontidae, Fundulidae, and Palaemonidae (Rozas 1993).

Nekton sample collection in the Campeche/Yucatán Inner Neritic Ecoregion used beach seines and trawls with a focus on fishes, and as such, did not report the abundance of any decapod crustaceans occurring within samples. However, penaeid shrimps (*Litopenaeus* spp., *Farfantepenaeus* spp.) and Mayan octopus (*Octopus maya*) support valued fisheries harvests presumably associated with the nursery function of lagoonal estuaries along the Mexican coast (Yáñez-Arancibia and Day 2004; Yáñez-Arancibia et al. 2009). Consequently, it seems reasonable to infer that natant decapod crustaceans are an important component of nekton in coastal wetlands of these regions as well. Shrimps in the family Hippolytidae are commonly associated with seagrass beds and while they appear to be abundant among the nekton of southern Florida, one might also expect this group to be well represented in the seagrass-dominated lagoonal

Table 6.3. Dominant Nekton Families in or Adjacent to Vegetated Marine Habitats (mangroves, marshes, seagrass) Within GoM Ecoregions

Nekton Family	Ecoregion – Level III				
	Florida Bay	Western Florida Neritic	Eastern Gulf Neritic	Mississippi/ Texas Estuarine	Campeche/ Yucatán Inner Neritic
Fishes					
Ariidae (sea catfishes)			■		
Atherinidae (silversides)	■	■	■	■	
Belonidae (needlefishes)			■		■
Carangidae (jacks)			■		
Cichlidae (cichlids)		■	■		
Clupeidae (herrings)			■	■	■
Cyprinodontidae (killifishes)	■	■		■	■
Engraulidae (anchovies)	■	■	■	■	■
Fundulidae (funduluids)	■	■		■	
Gerridae (mojarras)	■	■		■	■
Gobiidae (gobies)	■	■	■	■	
Mugilidae (mullets)	■	■	■	■	■
Poeciliidae (livebearers)		■		■	
Polynemidae (threadfins)			■		
Sciaenidae (drums)			■	■	■
Sparidae (porgies)	■				■
Syngnathidae (pipefishes)	■	■			
Tetradontidae (puffers)					■
Natant Decapod Crustaceans					
Hippolytidae	■				
Palaemonidae			■	■	
Penaeidae (penaeid shrimps)			■	■	■
Portunidae (swimming crabs)			■	■	
# of species required to achieve > 85% of total nekton abundance	41	28	26	29	33
References	2, 7, 11	5, 9	3, 8, 10	4	1, 6, 12

Only studies reporting numerical abundance were included. Dominant families were those that together accounted for >85 % of the nekton abundance within a study. Total number of species within families accounting for >85 % of individuals is documented in the table.
(1) Arceo-Carranza and Vega-Cendejas (2009); (2) Ley et al., (1999); (3) Livingston et al., (1976); (4) Minello (1999) (summary of 22 studies); (5) Mullin (1995); (6) Peralta-Meixueiro and Vega-Cendejas (2010); (7) Sheridan et al. (1997); (8) Stevenson (2007); (9) Krebs et al. (2007); (10) Subrahmanyam and Coultas (1980); (11) Thayer et al. (1987); (12) Vega-Cendejas and Hernández de Santillana (2004).

systems of the southern GoM ecoregion (e.g., Barba et al. 2005). The presence of schooling species in the fish families Atherinidae (silversides), Clupeidae (herrings), and Engraulidae (anchovies) listed among the dominants suggests that samples were collected in, or very near, open water and not within structurally complex habitats (e.g., mangrove prop roots or among the stiff, dense stems of emergent marsh vegetation) where fishes would be unable to maintain the group integrity of a school.

Relatively few species (averaging 26 to 41 per study) seem to compose the bulk of nekton assemblages from coastal wetland habitats in the GoM (Table 6.3). Most individuals are the juveniles of estuarine transient species (e.g., mullets, menhaden, drums, penaeid shrimps, portunid crabs), and all life stages of small estuarine resident species (e.g., gobies, killifishes, livebearers, grass shrimp) (Rozas 1993; Rozas and Reed 1993). At least one of these (i.e., *Fundulus jenkinsi*) is considered a species of special concern (Lopez et al. 2011) due to its apparent limited distribution (Figure 6.106b).

Although more species are associated with the southern neotropical portions of the GoM (Florida Bay and Campeche/Yucatán Inner Neritic) than the temperate northern regions, it should be noted that no attempt was made to standardize the collecting methods or focus of studies across regions (Table 6.4). Still, such a pattern would be expected and corresponds with the general spatial pattern of species richness around the GoM (NOAA 2011).

Many species of nekton are widespread within the GoM but only a few are both ubiquitous and abundant. *Anchoa mitchilli* (Figure 6.108a) is a clear standout among the fishes and is a dominant nekton component in all regions. Others (e.g., *Brevoortia patronus*) are distributed throughout the GoM (Figure 6.108c), but are among the most abundant nekton only within the Mississippi/Texas Estuarine Ecoregion (Table 6.4), suggesting a connection between riverine discharges and production of certain groups. The Mississippi/Texas Estuarine also includes a relatively high species richness of demersal gobiids (e.g., *Gobiosoma* spp., *Gobionellus* spp.) among the dominant nekton. Although gobies are common in estuarine habitats almost everywhere, their abundance in the Mississippi/Texas Estuarine region is noteworthy. A similarly high species richness of engraulids (*Anchoa* spp.) occurs in the Campeche/Yucatán Inner Neritic. Tropical waters of southern Florida and Yucatán include the greatest number of fish species that are not dominant elsewhere (Table 6.4). Greater species richness and lower abundance of fishes in these tropical regions may explain this observation (i.e., more species are required to account for at least 85 % of the individuals). However, these areas also contain extensive seagrass beds and/or coral reefs, which contribute substantially to diversity of fishes found in adjacent coastal wetlands, such as mangrove forests. Snappers (e.g., *Lutjanus* spp.) and mojarras (e.g., *Eucinostomus* spp.) tend to be among the dominants in mangroves. Pipe-fishes (e.g., *Syngnathus* spp.), sea horses (e.g., *Hippocampus* spp.) and porgies (e.g., *Lagodon rhomboides*) are dominant in areas where extensive seagrass habitats exist, such as in the Florida Bay (Table 6.4). The Mississippi/Texas Estuarine appears to include more dominant natant decapod crustaceans than are counted among the dominants in other regions. The lack of dominance among the fundulids (*Fundulus* spp.) collected from the Mississippi/Texas Estuarine is surprising, given that this region of the northern GoM contains most of the tidal marsh, which is usually the principal habitat of fundulids (Table 6.4, Figure 6.106). One possible explanation is that a majority of nekton samples from this region have been collected at the interface between vegetated tidal marshes and adjacent open waters (i.e., marsh edge), and fundulids may be more closely associated with the interior portions of shallow vegetated coastal habitats (e.g., Peterson and Turner 1994).

A pairwise comparison of the percentage of abundant fish species shared in common between ecoregions (Table 6.5) shows that the most abundant fishes of the neotropical southern GoM (Campeche/Yucatán Inner Neritic) are relatively distinct from those in all other regions, including southern Florida. The neotropical environment of Florida Bay and the Florida Keys share a substantial number (about 25 %) of abundant species with temperate wetlands of the Western Florida Estuarine and Eastern Gulf Neritic, but less similarity in the most abundant fishes is found on the coast of the Mississippi/Texas Estuarine compared with other regions. This is likely due to the higher diversity of habitat types and species found in the eastern GoM compared with more productive regions of the northern GoM, which tend to be dominated by tidal marshes and fewer nekton species at higher densities. Coastal currents (Figure 6.17) may contribute to the similarity in nekton assemblages along the west coast of the Florida peninsula,

Table 6.4. Species Comprising the Dominant Nekton Families Accounting for >85 % of Individuals in Field Studies Within or Immediately Adjacent to Vegetated Marine Habitats (e.g., marshes, mangroves, seagrass beds) in Each of the Listed Ecoregions.

	Ecoregion				
	Florida Bay	Western Florida Estuarine	Eastern Gulf Neritic	Mississippi/ Texas Estuarine	Campeche/ Yucatán Inner Neritic
Fishes					
Adinia xenica	■	■	■		
Anchoa cayorum	■				
Anchoa cubana					■
Anchoa hepsetus	■				■
Anchoa lamprotaenia					■
Anchoa lyolepis					
Anchoa mitchilli	■	■	■	■	■
Archosargus probatocephalus	■				■
Archosargus rhomboidalis					■
Atherinomorus stipes	■				
Bairdiella chrysoura		■			
Bathygobius soporator	■		■	■	
Belonesox belizanus	■				
Brevoortia patronus				■	
Calamus arctifrons	■				
Chilomycterus schoepfi					■
Ctenogobius smaragdus		■			
Cynoscion arenarius					
Cynoscion nebulosus		■	■		
Cyprinodon artifrons					■
Cyprinodon variegatus	■	■			
Diapterus auratus					■
Diapterus rhomboides					■
Dorosoma cepedianum				■	
Eucinostomus argenteus	■				■
Eucinostomus gula	■	■			■
Eucinostomus harengulus		■			
Eucinostomus melanopterus					
Eugerres plumieri					
Evorthodus lyricus				■	
Floridichthys carpio	■				
Floridichthys polyommus					■
Fundulus confluentus	■				

(continued)

Table 6.4. (continued)

Nekton Family	Ecoregion				
	Florida Bay	Western Florida Neritic	Eastern Gulf Neritic	Mississippi/ Texas Estuarine	Campeche/ Yucatán Inner Neritic
Fundulus grandis	■	■	■		
Fundulus grandissimus					■
Fundulus jenkinsi		■			
Fundulus majalis					■
Fundulus persimilis					■
Fundulus seminolis	■				
Fundulus similis	■		■		
Gambusia spp.	■				
Gambusia affinis			■		
Gambusia holbrooki		■			
Garmanella pulchra					■
Gerres cinereus	■				
Gobioides broussoneti				■	
Gobionellus boleosoma				■	
Gobionellus oceanicus				■	
Gobionellus shufeldti				■	
Gobiosoma bosc	■			■	
Gobiosoma robustum	■				
Harengula jaguana					■
Hippocampus erectus	■				
Hippocampus zosterae	■				
Heterandria formosa		■			
Hypoatherina herringtonensis					
Lagodon rhomboides	■				■
Leiostomus xanthurus		■	■		
Lucania parva	■				■
Lophogobius cyprinoides	■				
Lutjanus apodus	■				
Lutjanus griseus	■				
Lutjanus jocu	■				
Menidia martinica	■				
Menidia peninsulae			■		
Menidia spp.	■				
Menticirrhus americanus		■			
Microgobius gulosus	■	■		■	
Microgobius thalassinus			■	■	
Micropogonias undulatus			■		
Mugil cephalus			■		
Mugil curema					
Opisthonema oglinum					■
Poecilia latipinna	■	■			
Pogonias cromis					
Saratherodon melanotheron		■			
Sciaenops ocellatus		■	■		
Strongylura marina					■
Strongylura timucu					■
Strongylura notata					■

(continued)

Table 6.4. (continued)

	Ecoregion				
	Florida Bay	Western Florida Estuarine	Eastern Gulf Neritic	Mississippi/ Texas Estuarine	Campeche/ Yucatán Inner Neritic
Sphoeroides nephalus					[aqua]
Sphoeroides spengleri					[aqua]
Sphoeroides testudineus					[blue]
Syngnathus dunkeri	[aqua]				
Syngnathus floridae	[aqua]				
Syngnathus louisianae	[aqua]				
Syngnathus scovelli	[blue]				
Species abundant only in this Ecoregion	19	8	4	8	22
Natant Decapod Crustaceans					
Callinectes ornatus				[aqua]	
Callinectes sapidus		[blue]	[blue]	[blue]	
Callinectes similis					
Hippolyte zosericola	[aqua]				
Hippolyte curacaoensis	[blue]				
Farfantepenaeus aztecus			[aqua]	[blue]	
Farfantepenaeus duoarum					
Leander tenuicornis					
Litopenaeus setiferus			[blue]		
Macrobrachium ohione					
Palaemonetes intermedius					
Palaemonetes paludosus					
Palaemonetes pugio			[blue]		
Palaemonetes transverus					
Palaemonetes vulgaris			[aqua]		
Tachypenaeus constrictus					
Thor floridanus	[aqua]				
Tozeuma carolinense	[aqua]				
Species abundant only in this Ecoregion	4	0	0	7	0

Referenced studies are the same as in Table 6.3. *Aqua* shading indicates the species was among those in a family considered abundant (not necessarily that the species itself was abundant) and *blue* shading indicates a species that was the most abundant in a given family in at least one study within the indicated ecoregion.

while the Mississippi River may function as a physical barrier to east-west movement of certain nekton species associated with shallow coastal waters.

Quantitative information on nekton from vegetated marine habitats in the extreme southeastern portion of the GoM along the northwestern coast of Cuba is scarce, so data were not included in Tables 6.3 through 6.5. However, Ortiz and Lalana (2005) provide some useful qualitative insights from their general description of the marine biodiversity of the Cuban Archipelago. The families and species reported as noteworthy in seagrass beds, mangroves, and coastal lagoons include fishes in the families Lutjanidae (snappers), Serranidae (sea basses), Atennariidae (frogfishes), Ogocephalidae (batfishes), Synodontiae (lizardfishes), Pomadacidae (damselfishes), Gerridae (mojarras), Mugilidae (mullets), and Centropomidae (snooks). Other

Table 6.5. Matrix of Pairwise Comparisons Between Indicated Ecoregions Showing the Percentage of Species from Abundant Fish Families That are Shared in Common. Referenced studies are the same as in Table 6.3.

Ecoregion	Ecoregion				
	Florida Bay	Western Florida Estuarine	Eastern Gulf Neritic	Mississippi/ Texas Estuarine	Campeche/ Yucatán Inner Neritic
Florida Bay		25.5%	25.0%	13.0%	12.9%
Western Florida Estuarine			35.3%	7.7%	9.3%
Eastern Gulf Neritic				9.6%	8.3%
Mississippi/Texas Estuarine					6.7%

fishes associated with shallow subtidal flats included Scaridae (parrotfishes)—especially adjacent to coral reefs—and Dasyatidae, specifically the bluntnose stingray (*Dasyatis say*). Except for frogfishes and batfishes, which are rarely reported as abundant or important in other regions of the GoM, the nekton of the Cuban coast, at least at the family level, is similar to that of the southern Florida and Yucatán assemblages, with substantial contributions from coral reef and mangrove nekton assemblages (e.g., snappers, mojarras, damselfishes). Likewise, nektonic decapod crustaceans associated with shallow macroalgal beds, mangroves, and coastal lagoons included Portunidae (crabs in the genera *Portunus* and *Callinectes*) and shrimps in the family Penaeidae, with specific mention of *Farfantepenaeus notialis* and *Litopenaeus schmitti* (Ortiz and Lalana 2005). Most of these are either the same or sibling species that occur throughout the GoM (e.g., the northern white shrimp, *Litopenaeus setiferus* and the southern white shrimp, *L. schmitti* are sibling species as are the northern pink shrimp *Farfantepenaeus duorarum* and the southern pink shrimp *F. notialis*).

Some nekton species are restricted to narrow regional coastal reaches by their habitat requirements or physiological tolerances to variable environmental factors. For example, the American crocodile (*Crocodylus acutus*) occurs in the neotropical regions of the southern GoM, primarily from the Florida Keys, Florida Bay, Shark River Estuarine, and southwest through the Veracruzan Neritic, including a large population in Cuba. Crocodiles are limited to the southern GoM largely because of a low tolerance for cold even for short periods (Kushlan and Mazotti 1989). The related American alligator (*Alligator mississippiensis*), which can tolerate water temperatures below 8 °C (46 °F) for extended periods (Lance 2003) is distributed in the GoM throughout inshore coastal wetlands from south Florida north and west through the Texas Estuarine Ecoregion. Although alligators are more widely distributed within the GoM, crocodiles have a higher salinity tolerance and are more likely to be abundant in saline wetlands within their range, including mangrove habitats throughout the southern Gulf. Both species of crocodilians are top predators within the region, feeding on a diverse diet that includes other nekton (especially fishes) as well as terrestrial mammals, reptiles, and insects.

The diamondback terrapin (*Malaclemys terrapin*) is another nektonic reptile that is even more characteristic of tidal marshes and mangroves of the GoM than crocodilians, and is considered by some to be among the imperiled species of special regional interest (Beck et al. 2000). The distribution of terrapin subspecies within the GoM is particularly interesting because the subspecies appear to follow the distribution of Level III Ecoregions shown in Figure 6.3. Although there are seven recognized subspecies of diamondback terrapin, only four of these occur within the GoM (Ernst and Lovich 2009). *M. terrapin rhizophorarum* (mangrove diamondback terrapin) is restricted to mangrove habitats of the Florida Keys, Florida Bay, and the Shark River Estuarine Ecoregions (Ernst and Lovich 2009; Hart and McIvor 2008). *M. t. macrospilota* (ornate diamondback terrapin) occurs primarily within the marshes of the Western Florida Estuarine and Eastern Gulf Neritic. *M. t. pileata* (Mississippi diamondback terrapin) ranges within the tidal marshes of the Mississippi Estuarine. The fourth subspecies, *M. t. littoralis* (Texas diamondback terrapin) occupies the Texas Estuarine from western Louisiana to Corpus Christi, Texas. The conformity between the distributions of the subspecies of diamondback terrapins and Level III Ecoregions within the GoM is matched by few other nekton. Diamondback terrapins consume a variety of estuarine invertebrates including snails, crustaceans, and bivalves. Although strong swimmers, they tend to have limited home ranges, which may help to explain how the distinct subspecies persist.

Water depth, salinity, seasonal temperatures, dissolved oxygen, freshwater inputs, sediment type, availability of physical or biogenic structure (Day et al. 1989), as well as the size and spatial configuration of aquatic habitats within the coastal landscape (Boström et al. 2011), are among the multiple interacting factors controlling the composition and structure of nekton assemblages within coastal wetlands. Environmental variability on multiple spatial and temporal scales is a hallmark of estuarine systems, but the high mobility of nekton allows assemblages to persist by emigrating in response to unfavorable environmental conditions that might develop over the short-term or on limited spatial scales, and quickly immigrating to repopulate the same areas when conditions improve (Hackney et al. 1976; Day et al. 1989; Tyler et al. 2009).

Water depth usually affects the size of the species or life stages of nekton found in coastal wetlands. Shallow waters associated with most coastal wetlands generally are dominated by smaller (mostly <15 cm) individuals. Mean size and species richness of nekton assemblages tends to decrease from deeper to shallower waters, as does swimming ability, but densities often increase along the same depth gradient, with greater nekton densities occurring in shallow water (e.g., Peterson and Turner 1994; Eggleston et al. 2004; Ellis and Bell 2004). Within shallow vegetated habitats of the coast, the fish families Fundulidae (fundulids), Cyprinodontidae (killifishes), and Poecilidae (live-bearers) are abundantly represented (e.g., Rozas 1993; Peterson and Turner 1994). Water depth and physical structure (emergent and submergent plants and reefs) attract a subset of the Penaeidae (white, brown, and pink shrimp), Palaemonidae (grass shrimp), and Portunidae (swimming crabs such as the blue crab), at least near the edges of intertidal wetland habitats (e.g., Minello et al. 2008).

Aquatic accessibility to coastal wetlands is a key factor controlling the composition and abundance of nekton assemblages, particularly in intertidal habitats (Rozas 1995; Kneib 1997; Minello et al. 2012). Several factors may affect the accessibility of coastal wetlands to nekton including the frequency and duration of tidal or storm-driven inundation of intertidal habitats (e.g., marshes, mangroves, tidal flats) and the presence of structural landscape features (e.g., passes, creek channels, and ditches) that facilitate nekton movements (Saucier and Baltz 1993; Raynie and Shaw 1994) among otherwise isolated aquatic elements (e.g., lagoons, ponds, and impoundments) embedded within coastal landscapes (Knudsen et al. 1989; Herke 1995). Unlike most coasts, which experience semidiurnal tides (i.e., two high and two low tides daily), much of the GoM experiences diurnal tides (i.e., 1 high and 1 low tide daily) as illustrated in Figure 6.16. Mixed tides have the characteristic of exhibiting appreciably different amplitudes in successive high and low water events and may be either diurnal or semidiurnal. All tides within the GoM are considered microtidal in that tidal amplitude is considerably <2 m. Note that tides along the west coast of Florida, as well as most of the Cuban coast, are semidiurnal while all other portions of the GoM experience diurnal tides. Increased accessibility to intertidal habitats associated with twice daily high tides (semidiurnal) in the eastern GoM may explain at least some of the greater similarities in dominant nekton species shared by these regions (Table 6.5).

The dominance of small amplitude diurnal tides within the GoM may restrict the extent to which nekton have access to coastal wetlands and sometimes limit the effective use of these habitats to edges adjacent to open water (Baltz et al. 1993; Minello et al. 1994) or to habitats that remain submerged, such as subtidal seagrass beds and permanent or ephemeral ponds and impoundments. Even in the latter case, physical access routes in the form of passes between barrier islands into lagoons or embayments, or channels connecting natural ponds or artificial impoundments to open estuarine waters, are essential for immigration and emigration of most transient species of nekton that use these habitats as juvenile nurseries but spawn elsewhere (Day et al. 1989; Raynie and Shaw 1994; Herke 1995).

The association between productivity of inshore waters and nutrient dynamics of vegetated marine habitats has long been recognized (Odum 2000; Chesney et al. 2000; Beck et al. 2001), as have relationships between the area of vegetated coastal wetlands and fisheries production (e.g., Turner 1977, 1992), particularly in the northern GoM. However, in the neotropical southern GoM, the area of emergent vegetated wetlands appears to be less important in controlling fishery production than river discharge and freshwater inputs (Deegan et al. 1986; Yáñez-Arancibia and Day 2004), which are delivered to the coastal wetlands via relatively small watersheds compared to those in the northern GoM (Figure 6.15, Table 6.1). Secondary productivity in the GoM, as elsewhere, is driven by primary productivity and water quality, which control habitat quality and the production of higher trophic levels such as nekton (Yáñez-Arancibia and Day 2004). Although some coastal wetland nekton species have digestive tracts capable of assimilating energy from diets of algae and detritus (e.g., *Cyprinodon variegatus*, *Poecilia latipinna*, *Mugil cephalus*, *Brevoortia patronus*) (Odum and Heald 1972; Deegan et al. 1990), many supplement their diet by feeding on small invertebrates (Harrington and Harrington 1961, 1982). For the most part, nekton found in coastal wetlands are omnivorous and opportunistic, relying primarily on small surface-dwelling or epibenthic invertebrates as their primary food source (Stoner and Zimmerman 1988; Kneib 1997; Llansó et al. 1998). These benthic invertebrate food resources are capable of using algal and microbial assemblages associated with detritus as their primary energy source (see Figure 1 in Kneib 2003), and thus are likely to provide the most important links between coastal wetland primary production and nekton populations.

The role of different coastal wetland habitat types (e.g., seagrass, salt marsh, mangrove) in support of nekton secondary production remains a topic of some debate, but it does not appear that all types of wetland habitats contribute equally to estuarine nekton production. Beck et al. (2001) hypothesized that seagrass, marsh, and oyster reef habitats serve a nursery role in contributing to the production of nekton, but mangroves, tidal flats, and intertidal beaches do not provide a significant source of nekton production, though may serve a role as predator refugia for some species.

6.4.4 Ecosystem Services and Societal Benefits of Vegetated Marine Habitats

Natural ecosystems provide a suite of goods and services that have societal benefits (Costanza et al. 1997). These benefits are especially important relative to coastal ecosystems given that 41 % of the world population lives within 100 km (62 mi) of the coast (Martínez et al. 2007). Ecosystems of the GoM are no exception in providing goods and services that support human populations.

Table 6.6. Ecosystem Services of the GoM (from Yoskowitz et al. 2010; republished with permission of the Texas A&M University Press)

Ecosystem Services of the GoM by Service Level		
Level 1	Level 2	Level 3
Ecosystem Foundation or Support Services	Provisioning Services—Goods and Services Produced by, and Dependent on, Support Services	Outcomes and Benefits to Society
Nutrient Balance Hydrological Balance Biological Interactions Soil and Sediment Balance	Pollution Attenuation Air Supply Water Quantity Water Quality Food Raw Materials Medicinal Resources Gas Regulation Ornamental Resources Climate Regulation	Hazard Moderation Aesthetics and Existence Spiritual and Historic Science and Education Recreational Opportunities

Although there are many definitions for ecosystem services, the Gulf of Mexico Ecosystem Services Workshop (Yoskowitz et al. 2010) specifically defined GoM ecosystem services as "…the contributions from Gulf of Mexico marine and coastal ecosystems that support, sustain, and enrich human life." The central concept of this definition, and most others commonly used, is the emphasis on services that support human well-being and the identification of different classes of ecosystem services such as: (1) Ecosystem Foundation or Support Services, which are regulatory in nature and consist of processes that maintain the structure and function of ecosystems, (2) Provisioning Services, which are goods and services produced by or dependent on the support services, and (3) Outcomes and Benefits to Society, which include a suite of direct societal benefits (Table 6.6).

This organization has the advantage of being hierarchical in nature. Level I (Support Services) provides the foundation upon which all other ecosystem services depend. The higher the level, the more closely linked things are to direct human benefits. The Millennium Ecosystem Assessment (WHO 2005) uses a similar classification that groups ecosystem services into Supporting, Regulating, Provisioning, and Social and Cultural Services.

Nineteen ecosystem services provided by the GoM can be segregated by coastal habitat and prioritized as illustrated in Table 6.7. Specific ecosystem services provided by any particular coastal habitat vary with habitat. For example, ecosystem services performed by salt marshes are qualitatively and quantitatively different from those provided by barrier strand dunes or maritime forests. The importance of each service for a particular habitat is indicated. Although it can be argued whether or not the list is complete and/or the priorities correct, the table provides a summary from 30 coastal scientists and resource managers relative to their perceptions of ecosystem services provided by a suite of coastal habitats of which those presented in Table 6.6 are just a subset.

The goods and services provided to society by one particular coastal habitat, mangrove forests, have been studied and reviewed (Ewel et al. 1998). Although their relative importance varies among forest types and geographic locations, the primary goods and services include shoreline stabilization, buffering storms and hurricanes, sediment trapping, sinks for nutrients and carbon, nursery grounds for commercially important fisheries, wildlife habitat, and recreation opportunities. All mangrove forests contribute to soil formation and help stabilize

Table 6.7. Coastal Habitats and Their Ecosystem Services (modified from Yoskowitz et al. 2010)

Ecosystem Services	Dune/Beach	Salt Marsh	Mangrove	Seagrasses	Intertidal Flat	Subtidal Flat
1. Nutrient balance		6		6	5	2
2. Hydrological balance						
3. Biological interactions	6	1	1	?	2	1
4. Soil & sediment balance	3	11	3	4	1	3
5. Pollution attenuation			7			
6. Air supply					6	
7. Water quantity		10				
8. Water quality	7	9	6	3		
9. Food		4	8	1	4	5
10. Raw materials					8	4
11. Medicinal resources						
12. Gas regulation		8				
13. Ornamental resources						
14. Climate regulation		7				
15. Hazard moderation	1	2	2	8	9	
16. Aesthetics & existence	2	5	4	7	7	6
17. Spiritual & historic						
18. Science & education	5					
19. Recreational opportunities	4	3	5	5	3	7

coastlines; however, fringe forests dominated by *Rhizophora mangle* (e.g., in Florida) may be especially important in this regard. Sediment trapping is a related function most often attributed to riverine forests (e.g., the Shark River in the Everglades, Florida) (Ewel et al. 1998), but the scrub mangrove habitats found in the Mississippi River Delta along secondary waterways may also capture sediment (Perry and Mendelssohn 2009). Depending on geomorphology and hydrodynamics, mangroves may act as sinks or sources for nutrients and carbon. Basin forests are thought to be sinks for organic matter and nutrients (Twilley 1985; Twilley et al. 1986). Scrub or dwarf forests may also be sinks due to their restricted hydrology. Forest types with more open exchange (fringe, overwash island) may be sources of nutrients and carbon to adjacent estuaries. Mangrove forests are also thought to protect human communities against storm surge, with the trees contributing to wave attenuation (Bao 2011). Additionally, mangrove forests serve as nurseries and refuge for a variety of marine organisms of commercial or sport value, such as snapper (*Lutjanus* spp.), tarpon (*Megalops atlanticus*), barracuda (*Sphyraena barracuda*), jack (*Caranx* spp.), sheepshead (*Archosargus probatocephalus*), and red drum (*Sciaenops ocellatus*). In addition to serving as habitat for a variety of wildlife such as birds, reptiles, and mammals, mangrove forests also provide habitat for threatened or endangered species such as the West Indian manatee and American crocodile. Mangrove forests are important in terms of aesthetics and tourism; many people visit these areas to engage in fishing, boating, bird watching, and snorkeling.

Various scientists have identified the ecosystem services ascribed to coastal habitats differently. For example, Peterson et al. (2008) listed the following ecosystem services for tidal marshes, which include salt marshes: habitat and food web support, buffer against storm wave damage, shoreline stabilization, hydrological processing (flood water storage), water quality, biodiversity preservation, carbon storage, and socioeconomic services for humans. Many of these services are similar to those listed for salt marshes in the Yoskowitz et al. (2010) classification (Table 6.6). Costanza et al. (1997) estimated the economic value of tidal marshes and mangroves at $9,990/ha/year. Seagrass habitats were valued even higher at $22,832/ha/year. The coastal barrier strand, although not given a monetary evaluation, *per se*, provides a number of ecosystem services including protection of the mainland from storms and waves; buffering of wave energy to allow for formation of marshes and estuaries; creation of habitat for a variety of fish, shellfish, waterfowl and shorebirds, furbearing mammals, and endangered species such as sea turtles; recreation; vacation and retirement living; and economic benefits for tourism for coastal communities (Wells and Peterson 1982). In total, the ecosystem services provided by coastal habitats, including tidal marshes, mangroves, and the offshore coastal zone, were estimated at $63,563/ha/year (Costanza et al. 1997).

6.5 COASTAL HABITAT ECOLOGY

Coastal habitats that occur in the GoM represent a relatively finite list and are similar to those occurring worldwide. Factors such as climate, wave energy, water clarity, salinity, submergence, propagule availability, among others, determine the specific coastal habitat present in any particular geographic location and the flora and fauna comprising these habitats. In addition, factors such as disturbance type and frequency, biotic interactions such as herbivory, soil chemical condition, and others modulate many of the large-scale controls.

Coastal habitats are generally characterized by their dominant vegetation type. For example, mangrove trees define mangrove habitat, while seagrasses identify the seagrass habitat; halophytic graminoids and forbs distinguish a salt marsh. Barrier islands, in contrast, are primarily identified by their geomorphological characteristics (e.g., beach, dune, swale, etc.). Regardless, coastal habitats are important and conspicuous biogeomorphic features in the GoM. Intertidal wetlands are found throughout the GoM, but as mentioned briefly before, salt marshes dominate in the more temperate environments of the GoM, and mangroves dominate in more tropical settings. Mangroves, which are intertidal tropical and subtropical trees, are restricted to certain parts of the GoM by temperature. They dominate in the Southern GoM and Greater Antilles (Cuba) Marine Ecoregions, and become less prevalent and of lower stature in the more temperate regions of the Gulf (Mendelssohn and McKee 2000). The geographic limit of mature mangrove stands in the GoM is approximately 29.2°N latitude in coastal Louisiana in the Northern GoM region. Here, both plant communities co-occur (Patterson and Mendelssohn 1991); this ecotone also exists along the Florida and Texas shorelines. Mangrove plants also occur on the Chandeleur Islands (~29.8°N latitude) in Louisiana, and recent observations[1] have identified black mangroves on Horn Island, Mississippi (~30.2°N latitude), which, if persistent, is the farthest northern population in the GoM. Seagrass habitats in the GoM also have a somewhat restricted distribution due to low water clarity and/or low temperatures in much of the northern GoM. Seagrasses reach their dominance in clear waters of the subtropical and tropical southern Gulf, and their distribution is further limited in the Northern GoM Ecoregion by high turbidity associated with Mississippi River-influenced coastal waters (Hale et al. 2004). Barrier islands and beaches, as well as tidal flats, occur throughout the GoM wherever physical conditions allow. The following is a description of the major coastal strand habitats and their associated wetlands.

6.5.1 Barrier Strand Habitats

6.5.1.1 Dominant Forcing Functions

The barrier strand is a stressful environment where factors such as salt spray from saline waters of the GoM, soil moisture deficiencies, limited nutrient supply, and soil instability may negatively affect biota, especially barrier strand vegetation (Barbour et al. 1985; Packham and Willis 1997). Salt spray occurs when effervescence in the surf propels droplets into the air where they are concentrated and transported inland by the wind. The active agent in salt spray is the chloride ion, which enters the windward portions of plant parts through cracks and lesions in the epidermis. The degree of injury is related to the wind speed above the critical value of 7 m/s, where an abrupt increase in salt spray intensity occurs as turbulent air flow increases. In addition to affecting growth, salt spray is the primary environmental factor determining the distribution, architecture, and zonation of maritime plant species (Christensen 2000). Many plants that grow on foredunes (e.g., *Uniola paniculata* [sea oats]) are resistant to salt entry and can survive the intense salt-spray zones of the barrier strand. Plants that are less well adapted (e.g., *Andropogon* (=*Schizachyrium*) spp. [broomsedge]) are found in the lee of dunes or other vegetation. Salt spray is an important factor, along with sand burial, in preventing the establishment of some annual species (Van der Valk 1974; Miller et al. 2008).

Although dune species may be stressed by water deficits, especially on tall sand dunes, freshwater availability is greater than one might expect. Sand below the top few centimeters of a dune is often moist, even though the soil surface is dry. In fact, it has been suggested that the dry surface acts as a vapor trap, which impedes drying of deeper substrate. The water table, which may be several meters from the active root zone depending on the size of the dune, acts as an indirect source of water via vapor phase diffusion upward to the rooting zone. Because the capillary rise of water from a free water surface in very fine sand is not more than 40 cm, the water table in a dune of only a few meters can make no direct contribution to the moisture requirements of most dune plants. Rainfall and condensation provide important sources of water to dune vegetation. Regardless of the source of water, dune plants have evolved mechanisms to control their water requirements and acquire water. Many beach and dune species control water loss via a number of mechanisms including sunken stomates, strong stomatal control, and waxy leaf surfaces. Also, numerous beach and dune species are succulent and accumulate water in their leaf tissue. Still other plant species, especially dune grasses, have

---

[1] http://blog.al.com/live/2012/07/mangrove_trees_show_up_on_horn.html

a high capacity for the acquisition of water via deep roots that penetrate into moist soil. Because of these multiple adaptations to conserve and acquire water, water deficiency stress is not generally a major constraint to barrier strand species (Barbour et al. 1985).

A primary limitation to plant growth and expansion is the relatively nutrient-deficient sandy soils that compose the barrier strand. Major nutrient inputs to the dune system are salt spray, precipitation, and nitrogen fixation by both symbiotic and free-living bacteria. The mineralization of organic matter in the dunes is of limited importance because aeolian processes remove most lightweight organic matter; however, in protected swales and back-barrier marshes, soil organic matter may accumulate. Nitrogen is generally the primary plant-limiting nutrient, although phosphorus can be of secondary importance (Dahl et al. 1974; Dougherty et al. 1990; Laliberté et al. 2012). In fact, research on nutrient limitations of European dunes and swales indicates that phosphorus often co-limits primary productivity, especially in early stages of dune development (Lammerts et al. 1999).

Soil instability, and resulting sand burial, is another problem that dune vegetation encounters (Maun and Perumal 1999). Plants have a more difficult time becoming established in shifting windblown sand than in a stable substrate and can easily be buried with sand in large mobile dune fields. Dune plants, in particular, have adapted to this environment by developing the capacity to grow upward through considerable accumulations of sand. In fact, moderate sand burial has a stimulatory effect on the growth of dune grasses, but too much sand burial can cause plant mortality. In general, however, perennial grasses are more resistant to sand burial than annual forbs (20 cm limit for annuals and more than a meter for grasses) (Van der Valk 1974).

Although less investigated, herbivory is another factor that can limit the growth and expansion of dune vegetation (Hester et al. 1994). Grazing by rabbits, deer, nutria, and other mammals can dramatically reduce the structure of vegetation. However, this disturbance is often missed in the absence of adjacent areas where herbivores are excluded.

6.5.1.2 Plant Communities and Associated Vegetation

Because barrier strand vegetation throughout the GoM is subject to similar environmental stressors, as described above, plant form and habitat structure vary little. Even species composition can be quite similar, especially within the same latitudinal bands. Beach species are often prostrate herbaceous perennials capable of vegetative reproduction by stolons or rhizomes. Leaves are frequently small and lobed, with waxy surfaces and exhibiting succulence to various degrees. These are adaptations to plant water loss and/or low water availability, whether the cause is high transpiration, low water availability, soil salinity, or a combination (Barbour et al. 1985). Dune species are often grasses, like *Uniola paniculata* (sea oats) or *Panicum amarum* (bitter panicum), whose long roots can tap moisture deep in the soil, and whose rapid growth rates allow for tolerance to sand burial. Non-grass herbs, like *Hydocotyle* spp. (pennywort), found in the dune environment often have shallow roots to readily absorb frequent but short episodes of precipitation and strong stomatal control to reduce water loss. A mixture of graminoids and herbaceous dicots usually dominates swales. Because swales are generally protected from many of the stressors influencing beach and dune species, they do not show these same adaptations. However, swales often have higher water tables, and species such as *Spartina patens*, *Schoenoplectus olneyi*, and *Andropogon* (=*Schizachyrium*) *scoparius* (shore little bluestem) tolerate high soil moisture and even flooding. The swale habitat is the location where maritime forests and shrub thickets occur. Trees such as pines (*Pinus* spp.) and live oak (*Quercus virginiana*), and shrubs like *Myrica cerifera* (wax myrtle) and *Baccharis halimifolia* (groundsel bush), dominate swales located on more stable barrier islands and beaches. Backbarrier salt marshes, dominated by *Spartina alterniflora* (smooth cordgrass)

Table 6.8. Characteristic and Distinguishing Beach Flora in Each of the Four Regions of the Northern GoM as Identified by Barbour et al. (1987)

Texas	Louisiana	Mississippi, Alabama, and Florida Panhandle	South Florida
Croton punctatus Ipomoea stolonifera Panicum amarum	Spartina patens Cenchrus incertus Sporobolus virginicus	Uniola paniculata Schizachyrium maritimum Chrysoma pauciflosculosa Paronychia erecta	Iva imbricata Opuntia spp. Paspalum distichum Scaevola plumieri

and *S. patens* (wiregrass), and where climate allows, *Avicennia germinans*, are frequent occurrences, as are seagrass beds.

The beach and foredune vegetation on the backshores of barrier strands in the GoM can be divided into four geographic clusters: (1) a western region consisting of shorelines south of Galveston, Texas, (2) a northwest region of Louisiana beaches, (3) a northeast region consisting of Mississippi, Alabama, and the Florida panhandle, and (4) the south Florida beaches (Barbour et al. 1987). These groupings are separated by geographical and environmental discontinuities, such as differences in parent material of the sand and geological stability, as well as the influence of the Mississippi River. The beach survey of Barbour et al. (1987), which covered a shoreline distance of 2,500 km (1,550 mi), found that the northern GoM from the Texas-Mexico border to south Florida was dominated by a changing mixture of approximately a dozen plant species in nine genera. Only five of these dominants, and nine species overall, occurred in all four regions. *Uniola paniculata* was the dominant, except along the Louisiana coastline, where *Spartina patens* (wiregrass) replaced it. Other widespread beach species throughout the GoM were *Ipomoea stolonifera* (fiddle leaf morning glory), *Croton punctatus* (beach tea), *Sporobolus virginicus* (seashore dropseed), and *Heterotheca subaxillaris* (camphorweed), with the dune grass *Panicum amarum* (bitter panicum) prevalent, but decreasing in an eastward direction. Table 6.8 presents the distinguishing beach species for each region.

Considerable local variation occurs depending on the age and successional stage of the barrier strand. Figure 6.110 presents an elevation-vegetation transect across a young (12 years from formation) segment of Crooked Island West in northwest Florida (Johnson 1997). The embryo dunes along this profile are dominated by grasses, such as *Panicum amarum* var. *amarulum* (=*P. amarulum*, coastal panicgrass). More mature and stable shorelines formed as long as 53 years before the study on Crooked Island West show a transition from grasses to shrubs as dominants. The oldest and most stable dune ridges (some older than 100 years) are dominated by shrub species (Johnson 1997). One or two species dominate each community across the island: Foredunes—*Panicum amarum* var. *amarulum* and *Uniola paniculata* (with *Iva imbricata* (seacoast marsh elder) and *Schizachyrium maritimum* (gulf bluestem) as frequent associates); Swales—although diverse, species such as *Fimbristylis castanea* (marsh fimbry) and *Paspalum distichum* (knotgrass) are prevalent, as well as *Andropogon virginicus* (broomsedge) and *Dichanthelium aciculare* (needleleaf rosette grass); Maritime Forests—*Pinus clausa* (sand pine) and *P. elliottii* (slash pine) communities with *Ilex glabra* (inkberry) and *I. vomitoria* (yaupon), and many other small trees and shrubs in the understory.

Barrier strand communities associated with barrier islands and beaches of the Mississippi River Deltaic Complex in Louisiana are distinctly different from those to the east. Because of a limited sand supply, frequent winter cold fronts and episodic hurricanes, and rapid subsidence

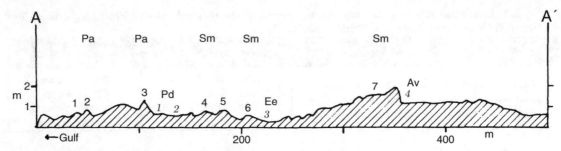

Figure 6.110. Vegetation—elevation profile along a dune-swale transect (A–A′) (from Johnson 1997; used with permission of the Journal of Coastal Research) on Crooked Island, located east of Panama City in northwest Florida. Letters indicate dominant plant species on each numbered ridge and swale (*Pa* = *Panicum amarum* var. *amarulum, Sm* = *Schizachyrium maritimum, Pd* = *Paspalum distichum, Ee* = *Eragrostis elliotii, Av* = *Andropogon virginicus*).

of the coastal deltaic landscape, barrier strand development is quite limited. Sand dunes are generally small in stature (Figure 6.111) and subject to frequent overwash. Consequently, beaches are predominantly erosional and relatively narrow. Some of these environmental and geologic features are, in part, responsible for the almost complete absence of *Uniola paniculata* (sea oats) along the Louisiana barriers, west of the Mississippi River (Hester and Mendelssohn 1991). Figures 6.111 and 6.112 present many of the common coastal strand species found in Louisiana.

Shorelines of southeastern Texas are very similar to those in southwestern Louisiana, but progressing southward, differences develop. Sand dunes and beaches become larger and more expansive and *Uniola paniculata* (sea oats) again gains dominance. Common plant species on Padre Island are provided in Table 6.9 (Smith 2002).

Beaches and barrier islands occur throughout the southern GoM (Figure 6.95). Coastal strand vegetation of this region has been described in a series of papers (Moreno-Casasola and Espejel 1986; Moreno-Casasola 1988, 1993, 2007; Silvia et al. 1991). As expected, the barrier strand flora of northern Tamaulipas is similar to that of southern Texas. Just south of the United States-Mexico border at Playa Washington, *Uniola paniculata* (sea oats) and *Ipomoea pes-caprae* (goat foot morning glory or bayhops) frequently occur along exposed parts of the dune and are sometimes replaced by *Croton punctatus* (beach tea) and *Scaevola plumieri* (gullfeed). Landward of this zone, *Croton* mixes with other species like *Clappia suaedifolia* (fleshy claydaisy), *Phyla cuneifolia* (wedgeleaf), *Sabatia arenicola* (sand rose gentian), and others. In southern Tamaulipas and northern Veracruz, dunes generally reach a height of 3–5 m (10 to 16 ft), with the exception of 30 m (98 ft) dunes in Cabo Rojo, and include the same species as previously mentioned, plus others like *Sesuvium portulacastrum* (shoreline sea purslane), *Coccoloba uvifera*, and *Canavalia maritima*. In general, tropical species like those present in south Florida are more prevalent. Figure 6.113 presents a vegetation profile at Bocatoma, Tamaulipas (Moreno-Casasola 1993). *Uniola* does not occur here, but rather *Sporobolus virginicus* (seashore dropseed) becomes the primary beach and dune grass. *Lippia* (=*Phyla*) *nodiflora* (frog fruit or fogfruit) is a typical swale species, and the mangrove associate, *Conocarpus erecta* (=*C. erectus*), (buttonwood or button mangrove), dominates the lagoonal shoreline.

One of the most interesting features of the Tamaulipas shoreline is Cabo Rojo, which has been described as a tombolo extending into the sea (Britton and Morton 1989). Because of the difference in shoreline orientation between the northern and southern sections of Cabo Rojo, the northern section of the barrier strand receives the full force of frequent winter nortes, while

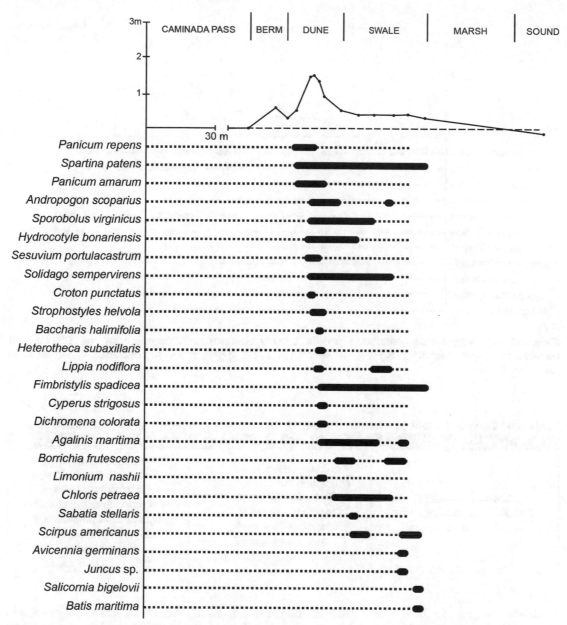

Figure 6.111. Vegetation—elevation profile across one section of the Caminada-Moreau Beach, west of Grand Isle, LA (modified from Mendelssohn et al. 1983). (*Andropogon scoparius = Schizachyrium scoparium*; *Scirpus americanus = Schoenoplectus olneyi*; *Lippia = Phyla*).

the southern section is somewhat protected. This difference both affects topography and species composition. Because of the extensive presence of aeolian sands on the northern leg, the dunes here can reach more than 30 m (98 ft). Strong winter winds and wave energy create a steep beach backed by sand dunes. Stable vegetated dunes occur behind the primary dune line, forming shrub thicket and forest habitats. *Coccoloba uvifera* (seagrape) is the most common leading species in the northern section. The southern section is composed of a series of old beach ridges that shield the strand from winds and accumulating sand. As a result, dunes are virtually absent, and *Ipomoea pes-caprae* (goatfoot morning glory) and *Croton punctatus*

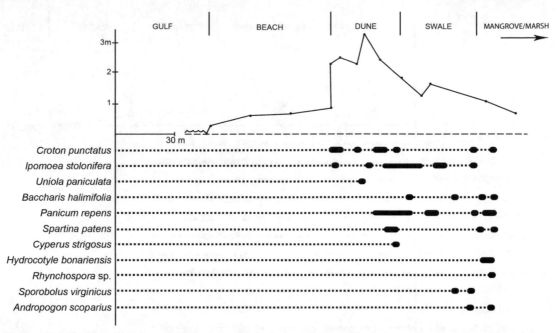

Figure 6.112. Vegetation—elevation profile across the northern segment of the Chandeleur Islands, Louisiana (modified after Mendelssohn et al. 1983) (*Andropogon scoparius* = *Schizachyrium scoparium*).

Table 6.9. Common Plant Species on the Barrier Strand of Padre Island (data from Smith 2002)

Backshore	Primary Dunes	Low Coastal Sand and Swales
Uniola paniculata Sesuvium portulacastrum Sporobolus virginicus	Uniola paniculata Paspalum monostachyum Paspalum setaceum Oenothera drummondii Ipomoea pes-capre Ipomoea imperati Chamaecrista fasciculata	Paspalum monostachyum Eragrostis secundiflora Fimbristylis castanea Heliotropium curassavicum Hydrocotyle bonariensis Erigeron procumbens Phyla nodiflora Stemodia tomentosa

(beach tea) dominate the leading vegetation on the beach. Plant diversity of strand vegetation in the northern section (25 species) is much greater than in the southern section (12 species) probably because the frequent washovers and disturbances in the northern leg create greater habitat heterogeneity and more microenvironments suited for more species (Poggie 1962).

The Veracruz shoreline is complex and gives rise to a variety of barrier strand physiognomies from narrow beaches in some areas in the northern part to the enormous dune systems along the central Veracruz shoreline (Britton and Morton 1989). Sands can vary from primarily light-colored quartz to dark, heavy mineral sand, derived from volcanic rocks. *Uniola paniculata* (sea oats), a dominant dune grass in the northern GoM, basically disappears south of the state of Tamaulipas. Plant zonation is generally distinct with definable plant communities extending from the beach landward to the large fixed dunes and semi-deciduous tropical forests (Figure 6.114a). Numerous microenvironments occur within such large coastal strands,

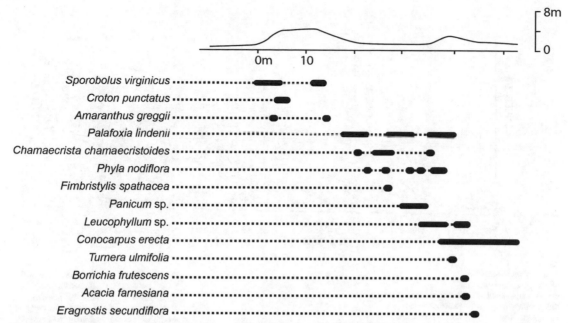

Figure 6.113. Vegetation profile at Bocatoma, Tamaulipas, Mexico (*Conocarpus erecta = C. erectus*) (modified after Moreno-Casasola 1993).

including wet swales and inter-dunal lagoons. Common species and their zonation are depicted in Figure 6.114b (Moreno-Casasola 1993, 2007).

Further to the east is the state of Tabasco, which has a relatively small coastline. Here, a complex of active and abandoned river channels and their associated deltas characterize the coastal plain. Quartz-sand beaches occur along the shoreline between the river mouths. Although dunes occur in scattered places, the area is characterized by a low-elevation beach-ridge system (Figure 6.115) (Moreno-Casasola 1993). In some areas (e.g., San Pablo), sand dune-ridges are backed by mangroves, which are further fringed by marsh shrubs, e.g., *Borrichia frutescens* (marsh elder) and *Hibiscus tiliaceus* (sea hibiscus). In the wet swales between beach ridges, a distinct community of low palms such as *Bactris* (bactris palm) and *Paurotis* (Everglades palm) alternate with solid stands of *Xylosma* sp. (logwood).

The State of Campeche, on the Yucatán peninsula, is characterized by its karst basement material and its almost continuous low-elevation barrier beach composed of shell and other calcareous materials (Moreno-Casasola 1993). The beach is often separated from the mainland by shallow, but wide lagoons and salt flats. The sand flats flood during the winter when nortes push seawater through the inlets. Where calcareous sands dominate, the coastal vegetation becomes more Caribbean-like with inclusions of *Coccoloba uvifera* (seagrape), *Scaevola plumieri* (gullfeed), *Suriana maritima* (bay cedar), and others (Figure 6.116).

Along the northern Yucatán shorelines, beach sand is primarily calcareous, and the beach is narrow with a parallel ridge (1 to 2 m [3.3 to 6.6 ft]). As described by Moreno-Casasola (1993), a vegetation gradient exists from beach to mainland. Pioneer vegetation consists of species such as *Chamaesyce buxifolia* (coastal beach sandmat), *Croton punctatus* (beach tea), *Scaevola plumieri* (gullfeed), *Sesuvium portulacastrum* (shoreline seapurslane), *Suaeda linearis* (annual seepweed), and *Tournefortia gnaphaloides* (sea rosemary). The pioneer zone ends at a shore-parallel thicket dominated by *Suriana maritima*. The landward swale consists of species such as

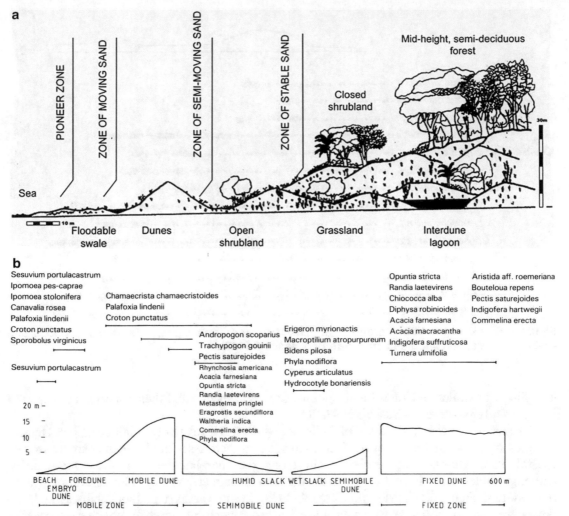

Figure 6.114. (a) Idealized vegetation profile of a mature dune system from beach to maritime forest (from Moreno-Casasola 2007). (b) Plant species composition and distribution of a dune system in the central part of Veracruz, Mexico (from Moreno-Casasola 1993; used with permission).

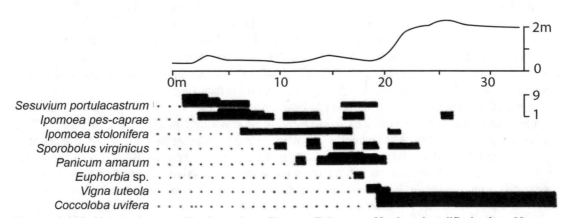

Figure 6.115. Vegetation profile from Las Flores, Tabasco, Mexico (modified after Moreno-Casasola 1993).

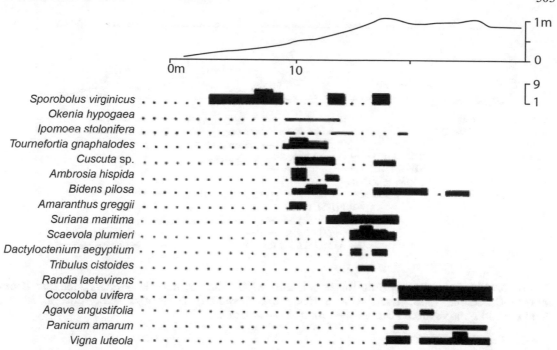

Figure 6.116. Vegetation profile from Champotón, Campeche, Mexico (modified after Moreno-Casasola 1993).

Hymenocallis littoralis (beach spiderlily), *Agave silvestris* (agave), *Scaevola plumieri*, and others. Along other parts of the northern Yucatán, *Coccoloba uvifera* (seagrape) is the dominant species, gradually increasing in height landward from the beach. Species of wild cotton (*Gossypium hirsutum* and *G. punctatum*) are interesting inclusions in this flora. *Gossypium punctatum*, closely related to commercial cotton, grows on the outer beach ridges and overlaps with the distribution of *Coccoloba uvifera* (seagrape) (Sauer 1967). Figure 6.117 presents a generalized vegetation profile for the northern Yucatán coastline.

Along the northwestern coast of Cuba, beach and dune habitats are especially well developed in the Guanahacabibes Peninsula and the shoreline between Havana and Varadero (Borhidi 1996) (Figures 6.101 and 6.118). This coast consists mainly of Pliocene limestone, which is seldom interrupted by muddy or sandy beaches. Flat karsts and cliffs are most common, with some rocky hills. The vegetation in this region consists of coastal thickets, dry evergreen forests and shrubwoods, fragments of semi-deciduous forests on the slopes, and small stands of mangroves. The dominant pioneer species of the strand line are *Ipomoea pes-capre* (goat foot morning glory) and *Canavalia maritima* (baybean). Landward of the pioneer species, but still on the beach, are combinations of species such as *Sporobolus virginicus* and *Baccharis halimifolia* (groundsel bush), *Borrichia arborescens* (tree seaside tansy), *Tournefortia gnaphaloides* (sea rosemary), *Spartina juncea* (=*S. patens*), and others. Many of these species also occur on the northern shoreline of the Yucatán. The primary dunes are often covered by the shrub seagrape, *Coccoloba uvifera*. Further landward the coastal gradient terminates with dry coastal evergreen shrubs (Figure 6.118) (Borhidi 1996).

A somewhat unique coastal habitat present along the shoreline is the coastal rock pavement community. Although widespread along the southern coast of eastern and central Cuba, it also occurs at Havana and Matanzas. The more open pioneer community is composed of succulent

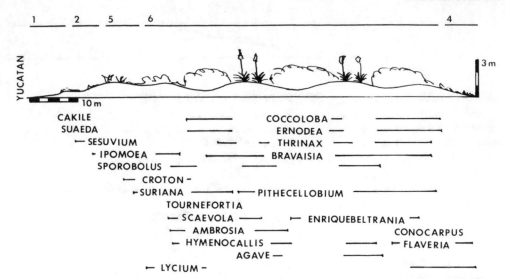

Figure 6.117. Vegetation profile for beach and wide ridge system in Yucatán, Mexico (from Moreno-Casasola 1993; used with permission). 1=beach; 2=embryo dune and foredune; 5=sheltered zone; 6=fixed dunes; 4=humid and wet slacks.

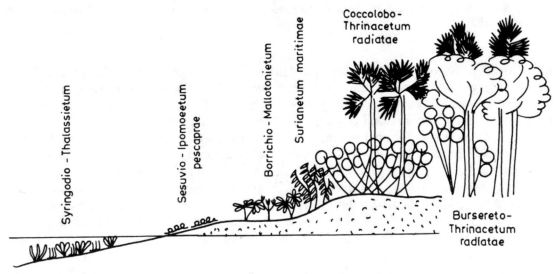

Figure 6.118. Vegetation of beach and dune habitats in the Casilda Peninsula, near Trinidad, Cuba (from Borhidi 1996, used with permission).

creepers such as *Lithophila muscoides* (talustuft), *Trianthema portulacastrum* (desert horse purslane), and *Sesuvium* spp. Landward is the coastal rocky shrub zone composed of *Rachicallis americana* (seaside rocket shrub), *Borrichia arborescens* (tree seaside tansy), *Conocarpus erecta* (mangrove button), *Opuntia dillenii* (erect pricklypear), and others. On cliffs exposed to salt spray and winds, *Rachicallis* sp. (seaside rocket shrub) and *Conocarpus* form a community. Sometimes *Coccoloba uvifera* (sea grape) will occur further landward on shallow sands, or thorn shrubs, dominated by species of *Mimosa* (mimosa), may dominate (Figure 6.119) (Borhidi 1996).

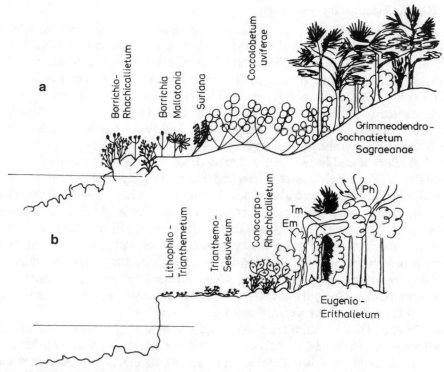

Figure 6.119. Vegetation of coastal rock pavement communities. (a) Aroya Blanco near Jibacoa, Cuba and (b) Punta Guanal near Matanzas, Cuba. *Em Eugenia maleolens, Tm Tabebuia myrtifolia, Ph Piscidia havanensis* **(from Borhidi 1996, used with permission).**

6.5.1.3   Fauna: Swash Zone and Shallow Tidal Pass Habitats

Relatively few studies in the past 20 to 30 years have focused on faunal assemblages associated with shallow (3 to 15 m [10 to 50 ft]) water swash zone habitats in the GoM. These sandy habitats are pervasive on mainland beaches from Texas to Florida and beaches found on the Louisiana, Mississippi, Alabama, and Florida barrier islands. Sandy beach and beach flat habitats in the Northern GoM are under continual pressure due to population growth along coastal areas, coastal resource utilization, recreational development, shoreline manipulation, tropical storms, and sea-level rise. Sandy sediments define coastal beaches; their geomorphology can be either narrow and steep (reflective) or wide and flat (dissipative) (Aagaard et al. 2013; Schlacher et al. 2008). Erosional beaches are typical in the Northern GoM (Buster and Holmes 2011). Their geological origin and the sorting effects of waves and currents influence the particle size of beach sand. Short-term geomorphic dynamics of sand beaches are typically linked to a source of sand and the energy to move it (Aagaard et al. 2013). Sand transport is greatest in the exposed surf zone and sand storage greatest in coastal dunes and nearshore sandbars. The intertidal and subtidal beach habitats and the shallow tidal pass habitats represent harsh habitats for organisms and often are characterized by steep gradients of environmental factors including wave action, currents, water depth, sediment composition, temperature, food availability, and regional/seasonal climatic factors (e.g., hurricanes) (Rakocinski et al. 1993; Schlacher et al. 2008).

6.5.1.3.1 Biotic Community Structure

A number of large-scale surveys and summaries have listed marine invertebrate species that occur in habitats throughout the GoM, including shallow, swash zone habitats (e.g., Rakocinski et al. 1991, 1993, 1998; Felder and Camp 2009). Barry A. Vittor and Associates, Inc. (2011) studied beach zone macroinfauna for the U.S. Army Corps of Engineers (USACE) Mississippi Coastal Improvements Program (MsCIP). The USACE Mississippi Sound and Adjacent Areas study (Shaw et al. 1982) broadly characterized benthic habitats based on sediment texture and macroinvertebrate assemblages, and feeding guilds present. Considerable variability in faunal assemblages occurred in similar sediment types. For example, sandy sediments of shallow sound habitats were characterized as having a macroinvertebrate assemblage dominated by the small bivalve, *Gemma gemma*, the polychaete, *Paraonis fulgens*, and the amphipod, *Lepidactylus triarticulatus*. This habitat had the lowest average taxa richness, the highest station mean density, and the lowest taxa diversity. The large variability in taxa richness and abundance seen between stations was due to the clumped distribution of *G. gemma* and *L. triarticulatus*. In contrast, a shallow tidal pass habitat with >95 % sand was characterized as having a macroinvertebrate assemblage dominated by surface and subsurface deposit feeders, including the polychaetes *Polygordius* spp., *Mediomastus* spp., and *Spiophanes bombyx*; the chordate *Branchiostoma* spp.; the crustacean *Acanthohaustorius* spp.; and suspension/filter feeders such as the bivalve *Crassinella lunulata*.

Rakocinski et al. (1991) studied the macroinvertebrate assemblages associated with barrier islands bordering the mainland of Mississippi, Alabama, and Florida. The Mississippi and Alabama barrier islands provide a wide range of environmental conditions for macroinvertebrate assemblages, the most influential being protected beaches on the north or "sound" sides of the islands versus exposed beaches located on the south or GoM sides of the islands. Early studies have also shown that macroinvertebrate assemblages on barrier island beaches have lower taxa richness and abundance than mainland beach habitats. A variety of environmental variables play a role in determining the macroinvertebrate assemblage in a given barrier island habitat, including wave action, sediment properties (primarily the percentage of sand), turbulence, salinity, dissolved oxygen (the occurrence of hypoxia), water depth, the frequency of tropical storms//hurricanes, and seasonal variability in these factors. Rakocinski et al. (1993) also studied benthic habitats seaward from the swash zone at Perdido Key (Florida) in an attempt to determine zonation patterns in macroinvertebrate assemblages. The authors sampled at 0, 25, 50, 75, 100, 150, 300, 500, and 800 m along a transect perpendicular to the beach. Crustaceans and polychaetes made up 75 % of the total number of individuals and species, with taxa richness and abundance increasing with depth (seaward). Total densities increased an order of magnitude from the shore to the deeper seaward stations and ranged from 2,000 individuals/m^2 to 20,000 individuals/m^2. The authors identified four unique zones along the depth gradient and land/sea interface: (1) the swash zone had a macroinvertebrate assemblage composed of motile, burrowing and/or tube-dwelling suspension feeders of medium body size; the dominant taxa were the polychaete, *Scolelepis squamata*, the decapod crustacean, *Emerita*, and the bivalve, *Donax*; (2) an inner subtidal zone which ranged from the shoreline to 100 m (330 ft) with depths <2 m (6.6 ft) and including nearshore troughs and sand bars; this habitat was dominated by small to large deposit and suspension feeding crustaceans and polychaetes; (3) a subtidal transition zone which ranged from 10 to 300 m (33 to 984 ft) offshore with depths of 2 to 4 m (6.6 to 13.1 ft); the macroinvertebrate assemblage was dominated by small and large bodied polychaetes; and (4) an outer subtidal zone which ranged from 300 to 800 m (984 to 2,625 ft) offshore, with depths between 4 and 6 m (13.1 and 19.7 ft); and a macroinvertebrate assemblage dominated by polychaetes, gammarid amphipods, gastropods, and the chordate *Branchiostoma*.

The macroinvertebrate assemblages at the shallow sand pass stations associated with the Mississippi barrier islands (Barry A. Vittor and Associates, Inc. 2011) were similar to the Shallow Sound Sand and Tidal Pass habitats characterized in the MSAW study (Shaw et al. 1982), and the Shallow Subtidal and Inner Subtidal (shoreline to 100 m [328 ft], depths <2 m [6.6 ft]) habitats recognized by Rakocinski et al. (1991, 1993). The macroinvertebrate assemblages characteristic of the Inner Subtidal habitat recognized by Rakocinski et al. (1993) were also similar to assemblages associated with the barrier islands in the MsCIP study with a dominance of polychaetes, haustorid amphipods, and bivalves; in addition, macroinvertebrate assemblages in the Shallow Subtidal habitats recognized by Rakocinski et al. (1991) were similarly dominated by polychaetes and amphipods.

Taxa richness and density data collected from sandy beach stations at distances of 3, 6, and 16 m (9.8, 19.7, and 52.5 ft) from shore in the MsCIP study had low taxa richness, extremely variable densities based on the patchy distribution of several habitat-specific macroinvertebrate taxa, and no discernible seasonal patterns. One factor that consistently separated macroinvertebrate assemblages on Petit Bois, Horn, and Ship Islands was whether or not the stations were located on the Mississippi Sound side of the islands or on the Gulf side. Stations located on the Mississippi Sound side of the islands had two to four times as many taxa and an order of magnitude higher densities than stations located on the GoM side of the islands. These data were similar to those found by Rakocinski et al. (1991) for Alabama, Mississippi, and Florida barrier islands with exposed GoM beaches and protected Sound beaches.

Epifaunal organisms associated with swash zone and shallow tidal pass habitats are typically opportunistic, large, active predatory and grazing organisms. The swash zone habitats in the GoM are dominated by various highly mobile decapod taxa (hermit crabs, *Pagurus*; blue crabs, *Callinectes*; ghost crabs, *Ocypode*; pinnixid crabs; portunid crabs, *Arenaeus*), shallow burrowing decapods (mole crabs, *Emerita*), echinoderms (sand dollars, *Mellita*), bivalves (*Donax*), and various gastropods (naticid moon snails; olives, *Olivella*).

Nekton assemblages associated with the waters surrounding the beaches along the barrier strand habitat are generally dominated by very few species, most of which are larval and juvenile life stages (Ruple 1984; Ross et al. 1987). Samples collected over several years along Horn Island, part of a barrier chain along the Mississippi-Alabama coast, included >75 species of fishes and natant decapod crustaceans, but >95 % of the individuals were represented by only four fish families (Clupeidae—herrings, Engrualidae—anchovies, Sciaenidae—drums, Carangidae—jacks) and one family (Portunidae) of natant decapod crustacean (Modde and Ross 1980; Ross et al. 1987). Only a few species within each family were abundant. *Harengula jaguana* (scaled sardine) dominated the clupeids. *Anchoa lyolepis* (dusky anchovy), *A. hepsetus* (striped anchovy), and *A. mitchilli* (bay anchovy) comprised almost all of the engraulids. The most abundant carangids were *Trachinotus carolinus* (Florida pompano) and *Caranx hippos* (crevalle jack), and the sciaenids were mostly *Menticirrhus littoralis* (gulf kingfish) and *Leiostomus xanturus* (spot). *Callinectes sapidus* (blue crab) was the most abundant portunid. A similar assemblage of surf zone fishes occurred along the beaches of Padre Island, Texas in the northwest GoM (Smith and Smith 2007), where in addition to the species listed above, mullet (*Mugil cephalus* and *M. curema*) were among seasonal dominants (*M. cephalus* in winter and *M. curema* in spring).

Although surf zone nekton have not been a focus of many studies throughout the GoM, samples collected from barrier strands along the Atlantic coasts of the United States (e.g., Layman 2000; Wilbur et al. 2003) and South America (e.g., Monteiro-Neto et al. 2003) are remarkably similar, even with respect to the dominant species. For example, *Trachinotus carolinus* is a prominent carangid and *Menticirrhus* spp. represent most of the sciaenids along barrier strand beaches at all of these locations. In some areas, mullet (Mugilidae) and silversides (Atherinidae) are also abundant (Layman 2000; Monteiro-Neto et al. 2003).

Shallow waters have been recognized as potentially important predator refugia for coastal marine and estuarine species, particularly in areas where SAV has been reduced or is absent (Ruiz et al. 1993). The shallow waters along barrier strand beaches of the GoM may serve a similar function, but a number of factors affect the occurrence and abundance of nekton along beach habitats, including seasonal reproductive patterns (Modde and Ross 1980; Gibson et al. 1993; Monteiro-Neto et al. 2003), diurnal or tidal foraging activity (Robertson and Lenanton 1984; Ross et al. 1987; Gibson et al. 1996), wind direction and intensity (Ruple 1984), and changes in beach configuration or composition of sediments resulting from storms or anthropogenic activities such as beach nourishment (Wilbur et al. 2003). Short-term episodic changes in physical attributes of nearshore waters, such as the onshore movement of hypoxic bottom waters, may drive onshore migrations of nekton populations as occurs at infrequent intervals in Mobile Bay, where the well-known summer phenomenon has been termed "Jubilee" (Loesch 1960; May 1973).

The physically dynamic nature of the barrier strand often results in the creation and extirpation of ponded aquatic habitats at different distances inland from the surf zone. These semi-permanent pools serve as habitat for a sometimes-ephemeral assemblage of nekton (Ross and Doherty 1994). Depending on distance from the shore, frequency of aquatic connections with the surf zone, and colonization dynamics, these assemblages are either dominated by nekton commonly found in back-barrier marsh habitats (e.g., Cyprinodontidae, Fundulidae, Poeciliidae), which exhibit a moderate level of stability, or a much less persistent assemblage of accidental colonists (Engraulidae, Sciaenidae, Carangidae, Clupeidae, Mugilidae) from the surf zone. Pools containing marsh colonists include reproductively active adults, thereby maintaining a persistent assemblage over time. However, surf zone colonists are represented only by juveniles that are unlikely to survive, and hence, form only ephemeral nekton assemblages.

6.5.2 Salt Marshes

6.5.2.1 Dominant Forcing Functions

Salt marshes generally occur along shorelines with sufficient protection from wave action, e.g., in protected shallow bays and estuaries, lagoons, and on the landward sides of barrier islands. Excessive wave action prevents establishment of seedlings, exposes the shallow root systems, and limits deposition of fine sediments that promote plant growth. Salt marshes are more extensive along low-relief coastlines where tidal intrusion reaches far inland and where there is abundant availability and accumulation of silts and clays, such as found in the north central GoM.

The hydrologic regime exerts a tremendous influence on the structure and function of wetlands, including salt marshes. Hydrology affects abiotic factors such as salinity, soil moisture, soil oxygen, and nutrient availability, as well as biotic factors such as dispersal of seeds. These factors, in turn, influence the distribution and relative abundance of plant species and ecosystem productivity. The tides constitute both a stress and a subsidy (Odum and Fanning 1973) for salt marsh development (Mendelssohn and Morris 2000). Tidal inundation leads to soil anaerobiosis and, depending on the flood tolerance of species, may inhibit survival, growth, and expansion. For salt marsh species, effects of low oxygen may limit vegetative spread via rhizomes (underground stems) and/or seed germination. Tides also import high concentrations of potentially toxic ions such as Na^+ and Cl^-. Tidal fluctuation, however, acts as a subsidy to salt marsh systems by importing nutrients, aerating the soil porewater, flushing accumulated salts and reduced compounds (e.g., hydrogen sulfide) that are phytotoxic, and enhancing the dispersion of seeds and/or vegetative fragments. The tidal subsidy effect is readily apparent along hydrodynamically active creekbanks, where marsh grasses, like *Spartina alterniflora*, are taller and more productive than in the interior of the marsh, where belowground tidal water movement is minimal (Mendelssohn and Seneca 1980; Howes et al. 1986).

Although salt marshes achieve best development on fine-grained sediments, they occur on a variety of substrates, including sands and volcanic lava. Terrigenous sediments are carried by rivers from inland areas to be deposited along the GoM or may originate from adjacent eroding shorelines. Fine silts and clays contain abundant exchangeable ions that fertilize and enhance productivity of the plants. Marshes may also develop on sandy substrata, particularly in stable, sheltered areas where the sand mixes with silt or organic matter (Chapman 1976). In the case of autochthonous deposits, the marsh vegetation itself contributes to sedimentation and soil development through production of organic matter, primarily below ground (Nyman et al. 1993; Turner et al. 2000). The organic matter content of soils may vary from <10 to >90 %, depending on the relative contribution of organic versus mineral deposits. High rates of root production combined with slow decomposition rates in the anaerobic soil environment may promote large accumulations of organic matter. Other biogenic deposits include carbonate skeletons of calcareous algae (e.g., *Halimeda* spp.), which are the major source of sand in the Caribbean, and shells of oysters and other invertebrates, which can be important constituents of salt marsh sediments.

Salt marsh soils are typically saline, but salinity varies depending on freshwater input, the ratio of rainfall to evapotranspiration, and hydrology (Thibodeau et al. 1998). In the low marsh, regular tidal inundation maintains salinities near that of seawater. At higher elevations, the interaction between frequency and duration of tidal flooding, on one hand, and evapotranspiration and freshwater runoff, on the other, results in substantial variability in soil salinity. During periods of high rainfall or in regions receiving freshwater runoff, salinities may be low between tidal flooding events. Salt marshes immediately adjacent to the Mississippi River, for example, may experience wide fluctuations in porewater salinity with average salinities less than 15 ‰ (ppt) (Mendelssohn and Kuhn 2003). Areas with high evapotranspiration rates and irregular tidal flushing develop hypersaline conditions with porewater salinities sometimes exceeding 70 ‰. Salt marsh plants are able to survive and grow at elevated salinities due to a number of unique adaptations. Localized freshwater discharges in seasonally dry regions may also prevent hypersaline conditions and promote vegetative development. However, along some arid tropical and subtropical coasts, for example the Laguna Madre of southern Texas and northern Mexico, extended periods of hypersaline conditions may stunt or even prevent the survival of perennial vegetation.

Inundation of salt marsh soils with water leads to anaerobic conditions due to a 10,000-times slower diffusion rate of oxygen in aqueous solution compared to air (Gambrell and Patrick 1978). Once oxygen is depleted by soil and plant root respiration, it is not quickly replaced and anaerobic conditions prevail. In the absence of oxygen, soil microorganisms utilize alternate oxidants (NO^{3-}, Mn^{+4}, Fe^{+3}, SO_4^{2-}) as electron acceptors. This process results in an increased soil oxygen demand, variation in availability and form of plant nutrients, and a build-up of toxic, reduced compounds in the soil. Soil Eh is a measure of the intensity of soil reduction, and low (≤ -100 millivolts [mV]) values are characteristic of strongly reducing conditions. Values ranging from $+300$ to -250 mV are typical of flooded soils and vary depending on soil texture, concentrations of redox elements, and flooding regime. The oxidation–reduction status of marsh soils is influenced by the presence of plant roots (Mendelssohn and Postek 1982; McKee et al. 1988). Leakage of oxygen from the plant roots into the surrounding soil creates an oxidized rhizosphere in which redox potentials can be higher than in the bulk soil. Thus, the growth of salt marsh vegetation is influenced by the anoxic condition of the soil substrate, but the plants themselves also modify the oxidation–reduction status.

The nutritional status of salt marshes is greatly influenced by tidal and riverine processes. Tides distribute mineral sediment and affect the redox status of the substrate, which in turn controls nutrient transformations, form, and/or availability. Rivers deliver nutrient-rich sediments to coastal salt marshes, resulting in some of the highest productivities (Sasser et al. 1995). The primary productivity of the vast majority of salt marshes is nitrogen limited (Mendelssohn and Morris 2000). Availability of phosphorus in anaerobic sediments typically exceeds that of ammonium, the dominant nitrogen form (Mendelssohn 1979), and is therefore of lesser importance. Numerous fertilization experiments in salt marshes have consistently demonstrated that nitrogen is the primary growth-limiting nutrient (see Mendelssohn et al. 1982 and references therein), although phosphorus can limit plant growth in sandy environments where phosphorus availability is low (Broome et al. 1975).

Another important controller of plant production, in addition to nutrients, is phytotoxin accumulation, which can occur in anaerobic sediments. In the marine environment a major phytotoxin produced under anaerobic conditions is hydrogen sulfide, which results from the bacterial reduction of sulfate to sulfide. Sulfate is the second most abundant anion in seawater and begins to be reduced under anaerobic conditions after NO^{3-}, Mn^{+4}, and Fe^{+3} have been reduced. The reduction of sulfate is carried out by true anaerobes, e.g., *Desulfovibrio*, and is thus dependent on anoxic conditions. Considerable research has demonstrated that sulfide is a primary driver of salt marsh primary productivity by impairing nitrogen uptake and assimilation (Mendelssohn and Morris 2000 and references therein).

On the broadest scale, climate, in particular temperature and rainfall, are primary controllers of species distribution and productivity. In the GoM, salt marshes are restricted in both growth and distribution in arid regions that generate high soil salinities. High temperature, per se, is not a direct constraint on the distribution of salt marsh vegetation but, as discussed previously, allows for development of mangroves, which outcompete salt marsh plants and thereby prevent salt marsh dominance (Mendelssohn and McKee 2000).

6.5.2.2 Vegetation

6.5.2.2.1 Structure and Zonation

Salt marsh communities are relatively species-poor and, in fact, along some shorelines of the northern GoM consist of monospecific stands of *Spartina alterniflora* (smooth cordgrass). Species richness generally decreases with increasing salinity (Mendelssohn and McKee 2000). For example, while as many as 93 species have been documented in Louisiana's coastal freshwater marshes, species richness in nearby salt marshes does not exceed 17 and, as previously mentioned, most individual salt marshes contain far fewer species (Chabreck 1972).

Most salt marshes are composed of plant communities dominated by graminoids such as grasses, sedges, and rushes; non-graminoid herbaceous communities dominated by forbs and succulents; and dwarf-shrub communities, especially common along arid and semi-arid coasts (Adam 1990). Unlike forests, which contain a number of strata, the vertical structure of salt marshes is relatively simple. Minor strata development is generated by different plant growth forms and the presence of benthic and epiphytic algae, where light penetration through the canopy allows.

Two physiographic zones, differing in hydrology and resulting soil and vegetation, occur in salt marshes. The low marsh, or regularly flooded marsh, is inundated by each tidal event, once or twice a day depending on whether the tides are diurnal or semidiurnal, respectively. The high marsh, sometimes referred to as the irregularly flooded marsh, is higher in elevation than the low marsh and thus is flooded less frequently, sometimes only during spring tides or extreme wind tides. Species richness tends to increase along an elevation gradient from the sea to the marsh/terrestrial ecotone. The low marsh has very low species richness, sometimes with only one species present, whereas the high marsh often exhibits a much greater number of species, especially where freshwater runoff from adjacent uplands occurs. The highest elevations of the salt marsh can develop into hypersaline areas called salt pans. The salt pan is inundated only by the highest spring tides, and then may not be inundated again until the next spring tide. As a result, salt often accumulates to lethal or near-lethal levels due to evapotranspiration in the absence of tidal dilution and leaching. Consequently, salt pans are often devoid of vegetation or are characterized by stunted halophytes and low species richness (Hoffman and Dawes 1997).

Zonation of species is a frequently observed characteristic of plant communities in habitats with strong physical and/or chemical gradients. In wetlands, spatial segregation of species often occurs in conjunction with elevation gradients that determine depth and duration of flooding and edaphic conditions influencing plant growth (Pielou and Routledge 1976; Vince and Snow 1984). Much work has centered on the role of abiotic factors as determinants of plant growth and distribution. However, the capacities of species to tolerate environmental conditions along elevation gradients greatly overlap, suggesting that factors other than environmental must play a role in generating zonation. In fact, biotic factors such as dispersal, competition, and herbivory may play a major role, along with abiotic constraints, in determining actual zonation (Pennings et al. 2005; Keddy 2010).

In salt marshes, species zonation is generally a ubiquitous feature, although species within a zone may vary from one geographical location in the GoM to another. However, where elevation gradients are shallow and/or occur over large distances, such as in the Mississippi River delta, zonation is visually less apparent, although quantifiable at larger spatial scales. Plant salt marsh zonation occurs along the elevation gradient from the seaward limit of the wetland to the terrestrial border. This elevation gradient is a complex gradient composed of multiple environmental factors that vary in time and space. The two most important abiotic factors controlling zonation along this gradient are inundation and salinity. Salt marsh species exhibit differential tolerances to these stressors. For example, *Spartina alterniflora*, a low marsh dominant, is more flood-tolerant than *S. patens*, a high marsh dominant, as documented in a Virginia salt marsh (Gleason and Zieman 1981). However, the species' tolerance limits to both inundation and salinity overlap considerably so that, for example, where inundation and salinity stresses are minimal many of these species could theoretically coexist. Thus, as briefly mentioned previously, abiotic factors alone cannot completely explain the observed zonation in salt marshes.

Competition also influences species zonation. Bertness and Ellison (1987) demonstrated in a New England salt marsh that zonation of *Spartina alterniflora* and *S. patens* is controlled by both environmental tolerances and competition. Competition between the species plays a more important role at the less stressful landward boundary of the marsh while abiotic factors control species pattern along the more stressful seaward end of the elevation gradient. For example, *Spartina patens* (wiregrass or saltmeadow cordgrass) does not occur at the most seaward limit of salt marshes because it cannot tolerate the inundation conditions. In contrast, *S. alterniflora* cannot exist at higher elevations because it is outcompeted by *S. patens*. As a result, competitive subordinates, in this example, *S. alterniflora*, are displaced to the more stressful zones of the gradient, while competitive dominants, in this case, *S. patens*, occupy the more benign areas. Similar conclusions were drawn from a number of studies throughout North America and elsewhere (Snow and Vince 1984; Bertness and Ellison 1987; Pennings et al. 2005).

Disturbance in the form of wrack deposition or herbivory can also influence zonation patterns in salt marshes. Bertness and Ellison (1987), for example, found that the pattern of species occurrence in a New England high marsh was generated by tidal deposition of large mats of dead plant material (wrack), causing differential plant mortality. *Spartina alterniflora* and *Distichlis spicata* (saltgrass) are more tolerant of wrack burial than other marsh plants and their relative abundance increases in disturbed areas. When the disturbance is more severe and of longer duration, all the underlying vegetation can be killed by the wrack and bare patches are generated. *Distichlis spicata*, *Salicornia europaea* (glasswort), and *Spartina alterniflora* rapidly colonize these patches and dominate compared to adjacent non-disturbed areas. However, over time, these disturbance communities are outcompeted and replaced by the surrounding communities of *Spartina patens* and *Juncus gerardii* (saltmeadow rush). This pattern mosaic can reoccur or even persist if wrack disturbance is frequent. Disturbances, such as wrack deposition, also promote greater plant species richness by opening gaps in the canopy and thereby facilitating species recruitment and establishment (Ellison 1987; Bertness 1992).

6.5.2.2.2   Salt Marsh Zonation and Distribution in the Gulf of Mexico

The GoM contains the largest area of salt marshes in North America, 55 % of the United States total (Mendelssohn and McKee 2000). Although the majority of these salt marshes occur in the northern GoM, salt marshes occur sporadically in the more southerly locations of the Gulf. The plant species composition and salt marsh area vary greatly due to a combination of factors including differential climate, tidal range, local relief, and wave energy.

Although salt marshes are limited within the South Florida Ecoregion, they do occur, often in association with mangroves, in areas of disturbance, or associated with salt pans (Figure 6.120). Where mangroves dominate the shoreline, salt marsh vegetation generally occurs along the seaward and landward intertidal fringes (Montague and Wiegert 1990). At the landward edge, where seawater inundation is infrequent, narrow bands or larger of *Juncus roemerianus* (black needlerush) and high marsh plant communities often occur. Farther landward, high marshes can become salt pans with little vegetation or dominated by *Cladium jamaicense* (sawgrass) in the presence of freshwater. In contrast, at the seaward edge of *Rhizophora mangle* (red mangrove) forests, a narrow fringe of *Spartina alterniflora* (smooth cordgrass) can occur (Figure 6.121) (Montague and Wiegert 1990). In the Ten Thousand Islands region of southwestern Florida (Figures 6.80 and 6.81), mangrove coverage has increased by approximately 35 % over 78 years, probably due to sea-level rise and possibly altered freshwater input (Krauss et al. 2011). Hence, the prevalence of coastal herbaceous marsh in the South Florida Ecoregion may be at risk.

Salt marshes of the eastern GoM (western Florida, Alabama, and Mississippi) are primarily irregularly flooded marshes dominated by *Juncus roemerianus*. Twenty-eight percent of U.S. *J. roemerianus* marshes occur in the eastern region of the GoM, an area containing only 8 % of U.S. marshland (Stout 1984). Other common salt marsh species in this region include *Spartina alterniflora*, *S. patens*, *S. cynosuroides* (big cordgrass), *Distichlis spicata*, *Salicornia* spp., *Schoenoplectus americanus* (=*Scirpus olneyi*) (chairmaker's bulrush or three-square), and *Schoenoplectus robustus* (=*Scirpus robustus*) (sturdy bulrush or leafy three-square) (Figure 6.122). *Spartina alterniflora* frequently occurs as a narrow fringe seaward of the *Juncus* zone, and *Distichlis spicata* and *S. patens* may occur at higher elevations landward of *Juncus* (Figure 6.122). About half of all salt marshes in Florida occur between Tampa Bay and the Alabama border (Montague and Odum 1997). This region, called the Big Bend area, where wave energy is low, shoreline relief is shallow, and tide range relatively high, has the greatest development of salt marshes in Florida. Similar to Alabama and Mississippi, the salt marshes here are irregularly flooded and dominated by *J. roemerianus*. In fact, about 60 % of northwest

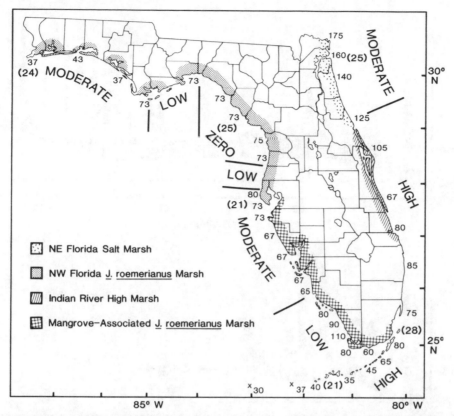

Figure 6.120. Salt marsh distribution in Florida and physical features of the coast; tidal range in cm (*small numbers*), relative wave energy (*block letters*), and relative sea-level rise in cm per century (numbers in *parentheses*) (from Montague and Wiegert 1990; Figure 14.2. *Occurrence of salt marshes in Florida and physical features of the coast*, from "Salt Marshes" by Clay L. Montague and Richard G. Wiegert in Ecosystems of Florida, Edited by Ronald L. Myers and John J. Ewel. Gainesville: University Press of Florida, 1990, pp. 483. Reprinted with permission of the University Press of Florida.).

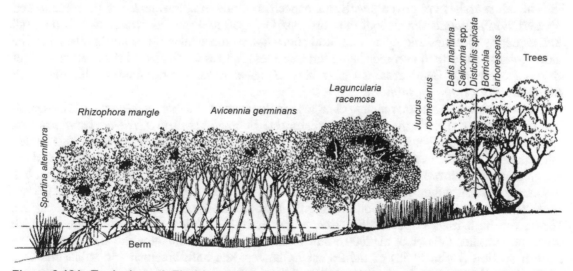

Figure 6.121. Typical south Florida mangrove-associated salt marsh. Notice *Spartina alterniflora* on the seaward fringe and *Juncus roemerianus* on the landward fringe (modified from Montague and Wiegert 1990).

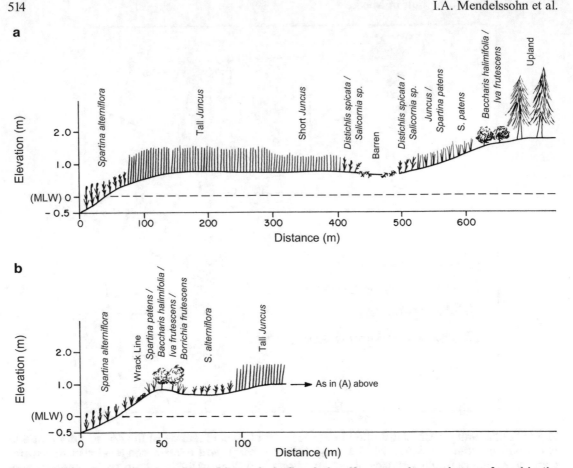

Figure 6.122. Generalized profiles of irregularly flooded gulf coast salt marshes as found in the northeastern GoM for (a) protected low energy shorelines and (b) open moderate energy shorelines (modified from Stout 1984).

Florida salt marshes are covered with monospecific stands of *J. roemerianus* (Montague and Wiegert 1990). *Juncus* throughout the northeast GoM often occurs as two growth forms: tall *Juncus* near shorelines and open water and short *Juncus* more inland. Further landward of the short *Juncus* is a suite of common high marsh species (Figures 6.122a, b). At the southern extent of the Florida Big Bend area at Cedar Key, *J. roemerianus* co-dominates with the black mangrove, *Avicennia germinans*.

Physiognomy of coastal marshes changes greatly west of the Pearl River at the Mississippi-Louisiana border. This is the Mississippi Estuarine Ecoregion, where *J. roemerianus* loses its dominance in the low-lying deltaic marshes of Louisiana (Figure 6.123). Here, regularly flooded salt marshes, the largest areal extent in the continental United States, are dominated by *S. alterniflora*, *Avicennia germinans*, and *Juncus roemerianus*. *Spartina patens* and *Distichlis spicata* are often subdominant species, depending on local topography and salinity (Visser et al. 1998). In the higher elevation Chenier Plain of southwestern Louisiana and southeastern Texas, the high marsh species, *Spartina patens* and *Distichlis spicata*, dominate brackish coastline marshes (Visser et al. 2000). Salt marsh vegetation dominates the shorelines of the eastern section (Delta Plain) of the Louisiana coast while both brackish and saline marshes occur along the western Louisiana coastline (Chenier Plain) (Figure 6.123). Westward flow of

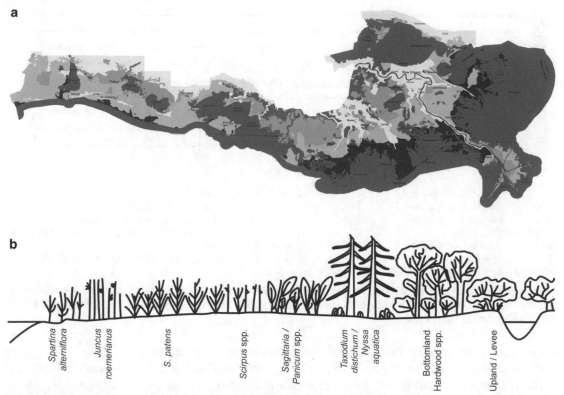

Figure 6.123. (a) Vegetation types of Louisiana coastal wetlands (*red* = saline marsh, *blue* = brackish marsh, *pink* = intermediate marsh, *green* = fresh marsh, and *yellow* = other) (modified from Sasser et al. 2008). (b) Idealized vegetation profile across the diverse marsh types of coastal Louisiana (modified from Mendelssohn and McKee 2000).

freshwater from the Atchafalaya River is a primary controlling factor, as is freshwater from the Mississippi River, in reducing coastal salinity enough to allow the occurrence of intermediate salinity marshes along the shorelines of the Birdfoot Delta.

Visser et al. (1998) classified saline coastal marshes of the Delta Plain into two primary types: polyhaline mangrove and polyhaline oystergrass (also commonly known as smooth cordgrass). Polyhaline mangrove is characterized by the presence of *Avicennia germinans*, but is equally dominated by *Spartina alterniflora* and *Batis maritima* (turtleweed), a common associate of the black mangrove. Polyhaline oystergrass is always dominated by *Spartina alterniflora*, and sometimes co-dominates with *J. roemerianus*. In contrast, the coastal marshes of the Chenier Plain are divided into two primary types: mesohaline wiregrass and mesohaline mixture, both of which have brackish salinities (Visser et al. 2000). The mesohaline wiregrass type is co-dominated by *Spartina patens* and *Schoenoplectus americanus* (=*Scirpus olneyi*). This marsh type is mostly found along the fringing marshes of Vermillion Bay, but is also present along the eastern shore of Calcasieu Lake and south of Sabine Lake (Visser et al. 2000). The mesohaline mixture type is co-dominated by *Spartina alterniflora* and *Distichlis spicata*, but *S. patens* also frequently occurs. Visser et al. (2000) found this vegetation type in the marshes fringing the GoM and the western shoreline of Calcasieu Lake.

Salt marshes along the central Texas shoreline, for example in Galveston and adjacent East and West Bays, are once again dominated by *S. alterniflora*. Galveston Bay is unique in Texas in having relatively extensive emergent marshes instead of submerged seagrass beds as the

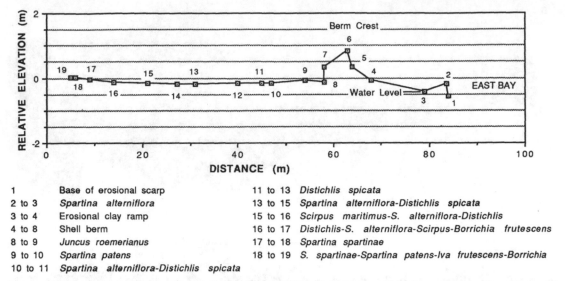

1	Base of erosional scarp	11 to 13	*Distichlis spicata*
2 to 3	*Spartina alterniflora*	13 to 15	*Spartina alterniflora-Distichlis spicata*
3 to 4	Erosional clay ramp	15 to 16	*Scirpus maritimus-S. alterniflora-Distichlis*
4 to 8	Shell berm	16 to 17	*Distichlis-S. alterniflora-Scirpus-Borrichia frutescens*
8 to 9	*Juncus roemerianus*	17 to 18	*Spartina spartinae*
9 to 10	*Spartina patens*	18 to 19	*S. spartinae-Spartina patens-Iva frutescens-Borrichia*
10 to 11	*Spartina alterniflora-Distichlis spicata*		

Figure 6.124. Salt marsh profile at Smith Point, East Bay, Texas showing relative elevations of plant communities (from White and Paine 1992; note, *Scirpus maritimus* is synonymous with *Schoenoplectus maritimus*).

major estuarine vegetation type. Salt marsh coverage is estimated at 120 km^2 (29,700 ac) (U.S. Department of Commerce 1989), although total wetland area is 1,594 km^2 (394,000 ac) (Moretzsohn et al. 2012). Of these, brackish marshes compose the greatest proportion at 65 to 70 %, with salt marshes composing approximately 25 to 30 % and fresh marshes 5 to 10 % (White and Paine 1992). For salt marshes, plant species such as *S. alterniflora*, *Batis maritima*, *Salicornia* spp., and *Juncus roemerianus* are most common in the lower elevation, more frequently flooded areas (Figure 6.124). In the high marsh, where flooding frequency is sporadic, species such as *Distichlis spicata*, *Salicornia bigelovii* (dwarf saltwort), *S. perennis* (=*Sarcocornia perennis*) (chickenclaws), *Monanthochloe littoralis* (shoregrass), and *Batis maritima* tend to dominate (Shew et al. 1981; U.S. Department of Commerce 1989; White and Paine 1992) (Figure 6.124). *Iva frutescens* (marsh elder) is locally abundant at higher elevations (White and Paine 1992). *Spartina patens* and *S. spartinae* (gulf cordgrass) also occur in Galveston Bay salt marshes, but are more prevalent in brackish marshes (U.S. Department of Commerce 1989) (Figure 6.124).

The intertidal *Spartina alterniflora*-dominated marshes of central Texas are replaced in southern Texas and northern Mexico by succulent-dominated, hypersaline marshes. Succulent species such as *Batis maritima*, *Borrichia frutescens* (bushy seaside tansy), *Suaeda maritima* (herbaceous seepweed), *Sesuvium portulacastrum* (shoreline seapurslane), and others are dominant (Mendelssohn and McKee 2000). In the Texas Laguna Madre (Laguna Atascosa National Wildlife Refuge), salt marsh species are clearly zoned along elevation gradients with *Batis maritima* at lower elevations and *Spartina spartinae* at higher elevations (Judd and Lonard 2002). *Monanthochloe littoralis*, found farthest from the shoreline, occurs at intermediate elevations (Judd and Lonard 2002). Based on importance values, four species dominate the salt marsh shoreline of the Laguna Atascosa National Wildlife Refuge in southernmost Texas, *Spartina spartinae*, *Borrichia frutescens*, *Monanthochloe littoralis*, and *Sporobolus virginicus* (seashore dropseed), although as many as 32 species can be found in these salt marshes (Judd and Lonard 2002). Associations consisting of *Suaeda nigra* (=*Suaeda moquinii*) (Mojave seablite) and *Salicornia ambigua* (=*Salicornia perennis*) occur in some of the highest salinity regions of the Laguna Madre in Mexico (Contreras-Espinosa and Castañeda-Lopez 2007). This association can be replaced by *Batis maritima*, and sometimes, by *Distichlis spicata* and *Monanthochloe littoralis*. In somewhat better-drained areas, *Spartina spartinae* and *Spartina densiflora* (denseflower cordgrass) are more frequent (Contreras-Espinosa and Castañeda-Lopez 2007). Costa and Davy (1992) also listed many of the same species as prevalent in salt marshes near Veracruz. Salt marsh area decreases greatly in the southern GoM, but salt marshes still occur in pockets adjacent to mangroves—especially at higher elevations (Olmsted 1993).

Olmsted (1993) estimated the extent of major wetlands of Mexico, including coastal lagoons; fresh, brackish, and salt marshes; mangrove swamps; freshwater lakes; and riverine forests, at 3,318,500 ha. At this time, an accurate estimate of salt marsh area for Mexico is not available, but salt marshes, per se, likely comprise only a small percentage of the total, which includes both freshwater and saline wetland types, like mangroves. The largest continuous wetland in the southern GoM is located in Tabasco and Campeche and is approximately 1,400,000 ha. Other large wetlands are located in Quintana Roo and the Yucatán (335,000 and 184,000 ha, respectively). These wetlands, together with those in Tabasco, Veracruz, Campeche, and Chiapas, make southeastern Mexico the most significant wetland region of Mexico (Olmsted 1993). Although salt marshes in tropical latitudes are often outcompeted and replaced by mangroves, West (1977) has cited three environmental situations where salt marsh species may exist, usually on the margins or within mangrove woodlands: (1) colonizing recently formed mudflats that fringe mangrove woodlands, (2) occupying saline soils on the inner edge or within the mangrove woodland, and (3) colonizing disturbed areas within a mangrove woodland.

In more tropical Mexico, brackish and saline marshes are commonly found in association with mangroves along the Gulf and Caribbean coasts, especially near coastal lagoons or near river deltas with low sediment load. Figure 6.125 presents vegetation zonation along a shoreline northwest of Laguna de Mecoacán in Tabasco, Mexico. Here, mangrove and herbaceous habitats occur adjacent to each other.

Infrequently inundated hypersaline salt flats, although small in areal extent, also occur along the coastlines of Mexico. Salt flats on the northern Gulf coast of Mexico contain three associations: (1) *Suaeda nigra* and *Salicornia ambigua*; (2) *Batis maritima*, *Borrichia frutescens*, *Clappia suaedifolia* (fleshy clapdaisy), and *Maytenus phyllanthoides* (Florida mayten); and (3) *D. spicata* and *Monanthochloe littoralis* (Olmsted 1993). On the Yucatán Peninsula, dominant plant species on salt flats include *Salicornia* spp., *B. maritima*, *Suaeda linearis* (annual seepweed), and *Sesuvium portulacastrum* (Johnston 1924).

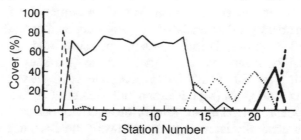

Figure 6.125. Cover of plant species along a mangrove-marsh transect near Laguna de Mecoacán, Tabasco, Mexico. (*Rhizophora mangle* = fine broken line, *Avicennia germinans* = fine line, *Batis maritima* = dotted line, *Spartina spartinae* = bold line, *Pithecellobium lanceolatum* = bold broken line (modified from Lopez-Portillo and Ezcurra 1989)).

6.5.2.2.3   Salt Marsh Primary Productivity

One of the most important and best-quantified functions of salt marshes is primary productivity, the rate of organic matter production per unit surface area per unit time. Factors controlling the primary productivity of coastal salt marshes dominated by *Spartina alterniflora*, the primary intertidal herbaceous salt marsh plant in the northern GoM, have been extensively reviewed (Mendelssohn et al. 1982; Smart 1982; Howes et al. 1986; Mendelssohn and Morris 2000). In general, the primary factors determining the growth of this species are salinity and soil waterlogging, both of which affect plant nitrogen utilization and allocation. Prolonged flooding results in soil anoxia, biochemically reduced soil conditions, and the accumulation of hydrogen sulfide in coastal salt marshes (DeLaune et al. 1983; Mendelssohn and McKee 1988). Soil anaerobiosis and phytotoxin accumulation inhibit the uptake of ammonium–nitrogen, the primary nutrient limiting plant growth in these systems (Morris 1984; DeLaune et al. 1984; Koch et al. 1990). Additionally, the roots may become deficient in oxygen and exhibit limited aerobic respiration and reduced energy for nutrient uptake (Mendelssohn et al. 1981; Koch et al. 1990). Hydrogen sulfide accumulation further inhibits root energy production and, hence, exacerbates plant nitrogen deficiencies (Koch et al. 1990). Elevated salinities can also negatively affect the growth of *S. alterniflora* by competitively inhibiting ammonium uptake (Morris 1984). In addition, nitrogen-containing cellular organic compounds (osmotica) that aid in maintaining plant water status are synthesized at elevated salinities; the allocation of nitrogen to osmotica production decreases the amount of nitrogen available for growth (Cavalieri and Huang 1979). Considerably less is known about the factors controlling the production of other salt marsh plant species. However, since hydrology and salinity are recognized as primary forcing functions in coastal salt marsh systems (Mitsch and Gosselink 1993), the primary productivity of other marsh species are likely controlled, at least qualitatively, by similar factors.

In addition to environmental controls, biotic factors may also influence salt marsh productivity. Bertness (1984, 1985) has shown that both fiddler crabs and mussels can enhance the production of *Spartina alterniflora*, the former by aerating the soil and the latter by increasing soil fertility through the production of feces and pseudofeces. Moderate grazing by snow geese in Canadian salt marshes also stimulates vascular plant primary production via the input of nitrogen from feces (Hik and Jefferies 1990); however, intense grazing can result in the denuding of the marsh (Abraham et al. 2005). There is also evidence for autogenic control of plant productivity. The accumulation of peat by the vegetation over time can inhibit plant growth possibly due to the increased hardness of the substrate and/or to the lower fertility of a peaty soil (Bertness 1988). These biotic controls on salt marsh plant production have generally been overlooked and require corroboration in other marsh types.

Coastal salt marshes are one of the most productive ecosystems in the world (Dring 1982). Although primary producers include the emergent vascular plants and benthic and epiphytic algae, most research has concentrated on quantifying the productivity of the emergent vascular vegetation. Rates of primary production can vary greatly depending on the methodology, plant species, environmental condition, latitude, grazing pressure, and temporal variability. For example, mean regional aboveground productivities of salt marshes in North America range from 76 g dry matter/m^2/year (30 g C/m^2/year) in Alaska to 1,976 g dry matter/m^2/year (812 g C/m^2/year) in the north central GoM (Mendelssohn and McKee 2000; Mendelssohn and Morris 2000). Although considerable overlap in primary productivity occurs for various geographic regions, productivities generally increase in a southward direction in North America. In the GoM, per se, salt marsh primary productivity varies longitudinally, with generally lower productivities in the western GoM (673 to 1,283 g dry matter/m^2/year) than in the central GoM (1,578 to 2,374 g dry matter/m^2/year) (95 % confidence intervals as modified from

Table 6.10. Range of Aboveground Primary Productivity Values for Salt Marshes Dominated by *Spartina alterniflora* **and** *Juncus roemericanus* **in the Northern GoM (data from Eleuterius 1972; Kirby and Gosselink 1976; Turner and Gosselink 1975; Turner 1976; Kruczynski et al. 1978; de la Cruz 1974; Hackney and Hackney 1978; Hackney et al. 1978; Hopkinson et al. 1978; Stout 1984; Webb 1983)**

Location	Aboveground Net Primary Productivity (g/m²/year)	
	Spartina alterniflora	*Juncus roemerianus*
Florida	130–1,281	245–949
Alabama	175–2,029	580–3,078
Mississippi	1,084–1,964	372–2,000
Louisiana	754–2,658	991–3,416
Texas	438–1,846	–

Mendelssohn and McKee 2000). Belowground rates of production are highly variable, possibly due to smaller sample sizes than for aboveground estimates as well as greater inherent variation in the data (Mendelssohn and McKee 2000). Regardless, belowground productivities are as high or even higher than aboveground (Mendelssohn and McKee 2000) (Table 6.10). For example, in one Louisiana salt marsh, belowground production was 11,676 g dry matter/m²/year compared to 1,821 g dry matter/m²/year for aboveground production (Darby and Turner 2008), although this belowground estimate was exceptionally high.

6.5.2.3 Fauna

Salt marshes may exhibit high infaunal and epifaunal invertebrate species diversity, compared to other habitats such as freshwater marshes (Odum 1988), but faunal species diversity in salt marsh systems varies across multiple temporal and spatial scales. For example, infaunal and epifaunal forms may show strong patterns of intertidal zonation, and epifaunal species richness may increase during warmer months due to seasonal abundance peaks of nektonic epifauna such as penaeid shrimps and blue crabs.

The effects of ecological stressors vary across the salt marsh landscape (Fleeger et al. 2008). Duration of tidal inundation (i.e., relative intertidal elevation) is a key factor determining infaunal and epifaunal community changes across the marsh landscape, and also controls intertidal access for nekton. Consequently, distributions of many salt marsh infauna and epifauna vary along an elevation continuum of habitat change on which are superimposed effects of cyclic patterns of environmental change, reproductive events, and predator–prey interactions at multiple temporal scales (e.g., diel, tidal, seasonal).

Most salt marshes in the GoM are located within the northern region and so, not surprisingly, much of the research on the fauna in this habitat has been conducted between Florida's Big Bend area of the Eastern Gulf Neritic Ecoregion to Corpus Christi in the Texas Estuarine Ecoregion. Across all GoM ecoregions, diversity tends to be greater in lower salt marsh elevations compared to the less frequently flooded areas of the high marsh. Subrahmanyam et al. (1976) found that low marsh zones had significantly more invertebrate species than did the upper marsh in *Juncus*-dominated marshes of the Eastern Gulf Neritic Ecoregion. Similarly, Humphrey (1979) found that the low *Juncus* marsh of the Mississippi Estuarine Ecoregion contained the greatest diversity and densities. Comparison of the two studies, however, reveals that low marsh diversity (H') was higher in the Subrahmanyam et al. (1976) study ($H' = 2.49$) compared to that found in Mississippi (0.77) (Humphrey 1979). Stout (1984) suggested that the

Mississippi marsh may represent the lower end of salinity tolerances for many estuarine and marsh organisms, but is still too salty for most freshwater or terrestrial species. Macroinvertebrate Shannon-Wiener H' diversity in an Alabama study (Ivester 1978) found that diversity in *Juncus* (0.69) and *Distichlis* (0.66) marshes was similar to Humphrey's (1979) findings in Mississippi. Alabama *Spartina* marshes had low macroinvertebrate diversity (0.37), with a decline in diversity in both late winter-early spring and in early fall (Ivester 1978). These diversity values for Alabama likely are underestimated because oligochaetes and insect larvae were not identified to species.

Subrahmanyam et al. (1976) reported a low marsh/upper marsh community dominated by the marsh periwinkle *Littoraria irrorata*, the isopod *Cyathura polita*, and tanaid crustaceans, with several abundant polychaetes (*Scoloplos fragilis*, *Nereis succinea*, and *Laeonereis culveri*). Mollusk populations increased toward a landward salt flat. The high marsh community had abundant fiddler crabs (*Uca* spp.) and the gastropods *Melampus bidentatus* and *Cerithisdea scalariformis*. At Bay St. Louis, Mississippi, dominant species by both density and biomass in the low needlerush marsh were the bivalves, *Polymesoda caroliniana* and *Geukensia demissa*, and the snail, *Neritina reclivata* (Humphrey 1979). The high marsh zone was dominated by the snail, *M. bidentatus*, fiddler crabs, and *P. caroliniana*.

Whaley and Minello (2002) examined the fine-scale (1 to 10 m) distributions of infauna in relation to the edge of a salt marsh in the Texas Estuarine Ecoregion. Surface-dwelling annelid worms and peracarid crustaceans were most abundant in low elevation sediments near the marsh edge for most sampling periods. Distributions of common surface-dwelling species were often unrelated to elevation but almost always negatively related to distance from the marsh edge. Abundances of near-surface direct deposit feeders and omnivores were related to both distance from the edge and elevation. In contrast to surface dwellers, densities of abundant subsurface deposit feeders (mainly oligochaetes) were frequently greatest in sediments located away from the marsh edge (Whaley and Minello 2002).

The relative value of salt marsh habitats for juvenile fishery species appears to be related to two environmental characteristics: the amount of marsh/water interface and the elevation of the marsh surface (Minello et al. 1994; Whaley and Minello 2002). Thus, there is decreasing use of the vegetated marsh by nekton with increasing distance from the marsh edge. Partyka and Peterson (2008) suggested that the faunal response to the presence of salt marsh habitat is more dependent on characteristics of the broader landscape that provide access to the shallow intertidal marsh surface and intertidal and subtidal creeks than on characteristics of the vegetated marsh.

Heard (1979) compiled a guide to 88 species of marine and estuarine invertebrates reported in marshes of the northeastern GoM, specifically along the Mississippi-Alabama coast and immediately adjacent areas of Florida and Louisiana. He identified three major groups of salt marsh infauna and benthic epifauna: polychaete worms, bivalve and gastropod mollusks, and crustaceans. Stout (1984) noted that the Heard (1979) listing would be greatly expanded with the addition of oligochaetes and insects, which are abundant in GoM salt marshes. The insect fauna of northeastern GoM marshes comprise aquatic species that also occur in freshwater systems, and include fly (dipteran) larvae (especially those of the Culicidae, Chironomidae, and Ceratopogonidae), heteropterans, coleopterans, and certain trichopteran larvae (Stout 1984).

Many salt marsh invertebrates occur across all ecoregions of the GoM coast, including taxa such as fiddler crabs (*Uca* spp.) (Figure 6.105) and nektonic taxa such as penaeid shrimps (Figure 6.103), and blue crabs (*Callinectes* spp.). In addition to widely occurring taxa, studies of salt marsh invertebrates from different regions of the northern GoM have found differences in community composition for some of the most abundant species.

Table 6.11. Abundant Invertebrates in Marsh Systems of the Northern GoM.

Region	Florida	Alabama	Mississippi	Texas
Common infauna and epifauna	*Littoraria irrorata* (G)	*Littoraria irrorata* (G)	*Polymesoda caroliniana* (B)	*Streblospio benedicti* (P)
	Scoloplos fragilis (P)	*Guekensia demissa* (G)	*Guekensia demissa* (B)	*Capitella capitata* (P)
	Nereis succinea (P)	*Polymesoda caroliniana* (B)	*Neritina reclivata* (G)	*Hargeria rapax* (T)
	Cyathura polita (I)	*Neritina reclivata* (G)	*Melampus bidentatus* (G)	*Corophium* spp. (A)
Data source	Subrahmanyam et al. (1976)	Ivester (1978)	Humphrey (1979)	Whaley and Minello (2002)

Key: *A* amphipod, *B* bivalve, *G* gastropod, *I* isopod, *P* polychaete, *T* tanaid

Table 6.11 lists abundant invertebrates in marsh systems of the northern GoM. Subrahmanyam et al. (1976) sampled infauna and epifauna in two Florida black needlerush (*Juncus roemerianus*) marshes, at Wakulla and St. Marks. Four major groups comprised the invertebrate community, including crustaceans (44 %), mollusks (31 %), annelids (24 %), and insect larvae (1 %). Numerically dominant taxa included the marsh periwinkle (*Littoraria irrorata*), the isopod *Cyathura polita*, and the polychaete *Scoloplos fragilis*.

Pure stands of black needlerush, intertidal smooth cordgrass (*Spartina alterniflora*), and salt grass (*Distichlis spicata*) were sampled by Ivester (1978) for invertebrate community comparisons along a salt flat in Alabama. A total of 19 taxa were identified, along with unidentified oligochaetes and insects. Six species represented over 90 % of total numbers of the community in each zone. Marsh periwinkle (*L. irrorata*) and the mussel, *Guekensia demissa*, were important only in smooth cordgrass. Marsh periwinkle was replaced by the gastropod *Melampus bidentatus* and increased numbers of the gastropod *Neritina reclivata* in black needlerush and salt grass stands. The bivalve *Polymesoda caroliniana* was prevalent in needlerush, as seen in both north Florida (Subrahmanyam et al. 1976) and Mississippi (Humphrey 1979). Oligochaetes dominated each of the three plant community types, ranging in relative abundance from 80 % in smooth cordgrass to 53 % in salt grass.

Epifauna utilize salt marsh tidal creeks in addition to the vegetated marsh surface, often entering vegetated areas after inundation by tides. Nektonic epifauna typically include grass shrimp (*Palaemontes* spp.), white shrimp (*Litopenaeus setiferus*), brown shrimp (*Farfantepenaeus aztecus*), and blue crab (*Callinectes sapidus*). These important decapod fishery species are seasonally abundant in salt marshes of the northern GoM coast (Zimmerman and Minello 1984; Thomas et al. 1990; Peterson and Turner 1994). Remarkably similar nekton assemblages occur in the tidal channels and ponds of *Spartina*-dominated marshes of Louisiana and Texas and the *Juncus*-dominated marshes in the northeastern GoM (Subrahmanyam and Drake 1975; Subrahmanyam and Coultas 1980).

Most nekton studies conducted in marshes of the GoM have emphasized samples collected from the interface (edge, with 5 to 10 m) between vegetated marsh and adjacent open water habitats and usually used drop samplers deployed from the bow of a boat (Baltz et al. 1993; Minello et al. 1994; Minello 1999; Minello and Rozas 2002; Zeug et al. 2007). Nekton samples collected at or near the marsh edge included species that use the flooded interior marsh surface as well as those that tend to be associated with nearshore shallow water. Thus, samples collected only from the marsh edge may not distinguish among assemblages of nekton that actually make extensive use of the majority of the interior marsh (e.g., resident species) and those visiting the periphery of that habitat (e.g., transients or schooling species).

The perception and emphasis on the dependence of nekton on marsh edge in the GoM has been incorporated into models of penaeid shrimp production that have linked changes in edge and wetland loss with growth rates (Haas et al. 2004), habitat fragmentation (Roth et al. 2008), and future shrimp harvests (Browder et al. 1989). Other research has recognized that the extent to which the marsh surface is used by nekton in the GoM also depends on the frequency and duration of tidal inundation (Rozas 1995), and that even deteriorating tidal marshes undergoing submergence in both Louisiana and Texas continue to be used extensively by nekton during the process by which the vegetated wetland is replaced by open water (Rozas and Reed 1993).

A few studies in the GoM have attempted to address edge-bias in marsh nekton collections by using different sampling methods. For example, bottomless lift nets were designed to provide quantitative samples of nekton from any location on the flooded marsh surface (Rozas 1992) and when used to sample nekton from intertidal marsh habitats at different relative tidal elevations (high, medium, and low) result in a very different view of nekton assemblages (Rozas and Reed 1993) than collections from the marsh edge (Minello 1999). Samples collected at the marsh edge often are dominated by decapod crustaceans and include relatively low densities of many fish species, but sometimes include high densities of schooling fishes such as clupeids (Table 6.12, Figure 6.108). Samples collected on the marsh surface contained approximately equal densities of natant decapod crustaceans and fishes, but with the fishes being dominated by only a few species of resident fundulids and cyprinodontids (Figures 6.106 and 6.107a), particularly in the high marsh habitats that may be infrequently flooded. Low marsh collections with lift nets comprised nekton assemblages that were most similar to the edge marsh samples collected by drop traps (Table 6.12), suggesting that the differences represented real patterns of zonation and not simply differences in the effectiveness of the sampling gear types.

Peterson and Turner (1994) used flume nets of different lengths and seine samples collected at high and low tide to address this issue as well. Their findings suggested a zonation pattern consisting of four groups (Figure 6.126). Group A consisted of resident species that quickly use the interior marsh whenever it is inundated, gaining access and moving through the habitat by way of small channels and low-lying microtopographic characteristics of the marsh surface. Fishes dominate this group and comprise representatives of the Fundulidae, Cyprinodontidae, and Poeciliidae. Group B had interior marsh users, which may require slightly deeper inundation of the marsh surface and tend to return to more permanent water at the creek edge at low tide. This group included the larger fundulid, *Fundulus grandis*, caridean shrimp in the genus *Palaemonetes*, and smaller individuals of the portunid crab, *Callinectes sapidus*. Some of the species in this group, particularly the fundulids, exhibited foraging habits (Rozas and LaSalle 1990; Lopez et al. 2010a, b) and life history characteristics such as reproductive activity synchronized with tidal cycles (Greeley and MacGregor 1983) and delayed hatching of eggs (Harrington 1959) that suggest specific adaptations to the intertidal marsh environment. Group C contained species that are commonly associated with the flooded edges of tidal marshes but rarely venture more than a few meters into the vegetation. These could include some of the Gobiidae (e.g., *Gobiosoma bosc* and *Gobionellus boleosoma*) as well as schooling species of Atherinidae (e.g., *Menidia beryllina*) or Mugilidae (e.g., *Mugil cephalus*), and the penaeid shrimps (e.g., *Farfantepenaeus aztecus* and *Litopenaeus setiferus*). Piscivorous predators such as spotted seatrout (*Cynoscion nebulosus*) would likely also find marsh edge habitats to be advantageous in ambushing prey. Group D comprised a group of shallow subtidal estuarine species that rarely made direct use of the vegetated tidal marsh, but instead were commonly found in the open water adjacent to vegetated wetlands and beaches, including a group of

Table 6.12. Densities (per m^2) of the Most Abundant Nekton in Collections from Drop Traps at Marsh Edges (*Spartina* and mixed vegetation) Reported by Minello (1999) and From Lift Nets in Marsh Vegetation at Different Relative Tidal Elevations (high, medium, low) Reported in Rozas and Reed (1993)

	Edge Marsh			Marsh Elevations		
	Spartina	Mixed		High	Medium	Low
Fishes						
Adinia xenica	0.1	0.1		4.2	2.2	0.8
Brevoortia patronus	0.8	5.0		0.0	0.0	0.0
Cynoscion nebulosus	0.2	< 0.1		< 0.1	< 0.1	< 0.1
Cyprinodon variegatus	0.2	1.1		6.0	2.8	0.9
Evorthodus lyricus	0.0	0.0		0.2	< 0.1	0.2
Fundulus grandis	0.4	1.0		10.7	2.3	1.3
Gobionellus boleosoma	0.9	0.1		< 0.1	0.0	0.0
Gobiosoma bosc	2.7	4.0		< 0.1	0.1	0.2
Lagodon rhomboides	1.3	0.1		< 0.1	0.0	0.0
Lucania parva	0.5	1.0		0	0	< 0.1
Menidia beryllina	0.6	0.8		0.3	1.2	0.4
Mugil cephalus	0.2	0.2		1.1	2.4	2.5
Natant Decapod Crustaceans						
Callinectes sapidus	6.2	2.7		2.0	1.0	1.7
Hippolyte zostericola	0.7	0.0		0.0	0.0	0.0
Farfantepenaeus aztecus	7.5	2.6		0.3	0.5	1.4
Farfantepenaeus duorarum	1.0	0.5		0.0	0.0	0.0
Litopenaeus setiferus	5.5	1.5		0.1	0.4	1.7
Palaemonetes pugio	58.8	25.7		10.5	5.4	14.6

Red shaded cells show the top six species in each marsh type

flatfishes (e.g., *Symphurus plagiusa* and *Achirus lineatus*) and common schooling species found in GoM estuaries such as the ubiquitous engraulid, *Anchoa mitchilli*, and the clupeid, *Brevoortia patronus*.

Natural (e.g., ponds and tidal channels) and anthropogenic (e.g., impoundments and oil/gas pipeline canals) aquatic features embedded within the matrix of the intertidal vegetated marsh landscape function as important habitat for marsh nekton (Peterson and Turner 1994; Rozas and Reed 1994; Akin et al. 2003; Rozas and Minello 2010). Nekton assemblages found in these tidal marsh sub-habitats are affected by size and depth of the habitat (Akin et al. 2003; Rozas and Minello 2010), temperature and salinity (Herke et al. 1987; Akin et al. 2003), and perhaps most importantly, accessibility to open water (McIvor and Rozas 1996). Blocked or restricted channels that impair movement of nekton among adjacent habitats (e.g., bays, lagoons, and the coastal neritic zone) can be expected to have a more limited nekton assemblage than those with open and free connections to other aquatic habitats (Neill and Turner 1987; Herke 1995). There

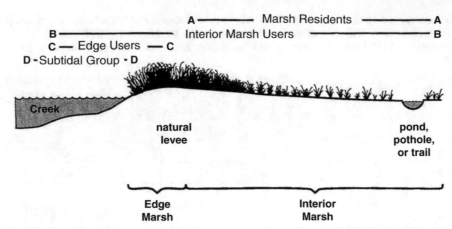

Figure 6.126. Zonation patterns of different nekton groups in tidal marshes of Louisiana (after Peterson and Turner 1994; republished with permission of Springer Science and Bus Media BV, provided by Copyright Clearance Center, Inc.). Species that use the interior marsh extensively tend to be "true residents" that are primarily fishes in the families Fundulidae, Cyprinodontidae, and Poeciliidae, while the marsh edge is visited by a diverse group of fishes and abundant natant decapod crustaceans, many of which commonly occur in other estuarine habitats within the GoM.

appear to be no substantial differences between nekton assemblages using anthropogenic aquatic habitats such as open pipeline canals and natural marsh channels (Rozas 1992; Granados-Dieseldorff and Baltz 2008). The species composition and spatial distribution of nekton using pipeline canals and marsh creek channels is similar to that shown in Figure 6.126, with the same suite of fish and natant decapod crustacean taxa dominating on the adjacent interior and edge marsh habitats and a similar subtidal group found in the deeper portions of the channels. Marsh ponds also show a similar nekton assemblage structure, with small resident fish species and natant decapod crustaceans associated with pond vegetation and schooling species such as *Anchoa mitchilli* and *Brevoortia patronus* (Figure 6.108) sometimes abundant in the adjacent unvegetated open waters of larger ponds (Rozas and Minello 2010), lagoons (Rozas et al. 2012), or embayments (Akin et al. 2003).

The presence of SAV, such as *Ruppia maritima* (Akin et al. 2003; Rozas and Minello 2010) in brackish canals and ponds, or *Thalassia testudinum* and *Halodule wrightii* in more saline bays and lagoons (Rozas et al. 2012), creates a hybrid emergent/SAV habitat that can affect the species composition of the nekton assemblage even in an adjacent tidal marsh. For example, pink shrimp (*Farfantepenaeus duorarum*) and hippolytid shrimp (e.g., *Hippolyte zostericola*, *Tozeuma* spp.), pinfish (*Lagodon rhomboides*), pipefish (*Syngnathus* spp.), and gobiid species tend to be more abundant in habitats associated with seagrasses (Rozas et al. 2012), and so could occur in samples from the adjacent marsh.

Salt marsh is one of the most biologically productive ecosystems in the world (Teal 1962). The high primary productivity that occurs in the marsh provides the base of the food chain supporting invertebrate detritivores or omnivores that provide an important link to higher-level consumers in the food web (Stout 1984). Infauna densities and biomass are positively associated with percent organic matter in salt marshes (Minello and Zimmerman 1992; Levin et al. 1996, 1998). Many small crustaceans and annelids consume organic detritus, and these fauna represent a trophic pathway to higher predators, most of which do not have the ability to derive much nutrition directly from detritus (Kneib 2003). Nekton associated with salt marshes can have a prominent role in the export of energy and materials from these productive coastal

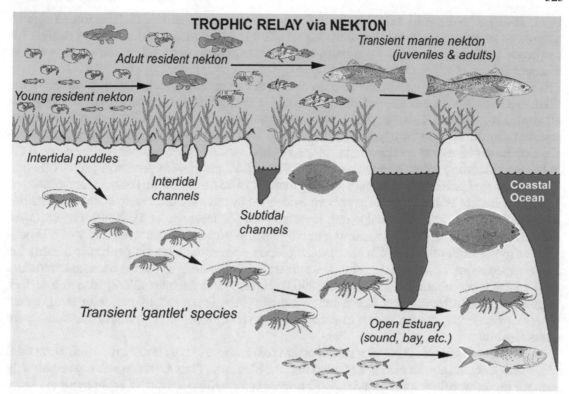

Figure 6.127. Transferring coastal wetland production from shallow to deeper waters via the trophic relay as depicted for salt marshes in the southeastern United States (from Kneib 2000; reprinted with permission of Springer). Key resident species include palaemonid shrimps as well as fundulid, cyprinodontid and poecilid fishes. Gauntlet species tend to be subject to predation both when immigrating to estuaries as early life stages from coastal spawning grounds and when emigrating as juveniles and adults. Typical gauntlet species for the region would be penaeid shrimps and gulf menhaden as depicted in the figure. Important transient predators include fishes in the families Sciaenidae, Ariidae, and Pleuronectiform flatfishes.

wetlands to open estuaries and the coastal ocean through multiple mechanisms. Some species, such as Gulf menhaden (Deegan 1993; Deegan et al. 2000) or natant decapod crustaceans (Zimmerman et al. 2000), which use coastal estuaries and wetlands as nurseries during their juvenile life stages, often show ontogenetic patterns of emigration to nearshore or offshore coastal water effectively transferring substantial biomass and energy between habitats. Permanent estuarine resident species (e.g., fundulid fishes and palaemonid shrimps) also contribute to these production transfers through spatially overlapping distributions along a stream-order gradient (Granados-Dieseldorff and Baltz 2008) that involve size-specific predator–prey interactions in which smaller nekton from shallow water are consumed by larger nekton (Figure 6.109) that move with the tide or seasons into adjacent deeper habitats (Figure 6.127) in a process often referred to as the trophic relay (Deegan et al. 2000; Kneib 1997, 2000).

Predator–prey interactions driving production transfers in salt-marsh ecosystems are mediated primarily by tidal dynamics, local physiography (i.e., landscape structure), and vegetation density (Moody 2009). Many salt marshes are drained by a network of tidal creeks, and the transfer of marsh production to aquatic estuarine predators appears to be facilitated by tidal flushing that provides nekton access to small detritivores, herbivores, and omnivores residing in the marsh (Zimmerman and Minello 1984; Kneib 1987; Rozas 1995; Zimmerman et al. 2000). Collectively, tidal creeks and the adjacent intertidal marsh function as tightly-connected nursery areas providing both high density food resources and structural refugia for the juvenile life stages of many commercially harvested fishes, decapod crustaceans, as well as a larger subset of other GoM taxa.

Various aspects of the predator–prey interactions controlling production transfers from tidal marshes have been studied for decades in the GoM and elsewhere (Kneib 1997). For example, Rozas and LaSalle (1990) examined the foraging habits of the abundant marsh resident fundulid, *Fundulus grandis*, and found that the guts of specimens leaving a Mississippi intertidal marsh on the ebbing tide were substantially fuller than when they entered the habitat on flooding tides. In addition to short-term tidal patterns in predator–prey dynamics, strong seasonal patterns of foraging activity thought to be driven by the seasonal abundance of juvenile transient nekton species have been evidenced by intra-annual variability in the densities of infaunal and small epifaunal marsh invertebrates (Zimmerman et al. 2000). For example, Subrahmanyam et al. (1976) measured a maximum late winter invertebrate density of 578 individuals/m^2 in tidal marshes of northwestern Florida corresponding with peak recruitment for certain species, and minimal densities in summer (375 individuals/m^2) when potential predators tend to be most abundant (Akin et al. 2003). Whaley and Minello (2002) also found that populations of infaunal prey fluctuated seasonally in a Texas salt marsh, with the greatest densities occurring during winter and early spring when epibenthic predator densities were generally low.

Predatory decapods likely play an important role in these energy transformations from tidal marshes. For example, Kneib (1986) found that the blue crab, *Callinectes sapidus*, was probably a major predator of adult mummichogs (*Fundulus heteroclitus*) in North Carolina salt marshes. Given the higher densities of blue crab in marshes of the GoM (Zimmerman et al. 2000) when compared to the U.S. Atlantic coast, one can reasonably infer a similar predator–prey relationship exists between blue crabs and small resident nekton in the GoM. The inference is supported by the findings of West and Williams (1986) who observed that blue crabs preferentially selected marsh periwinkles (*Littoraria irrorata*) and gulf killifish (*Fundulus similis*) over infaunal prey (the ribbed mussel *Geukensia demissa*) in an Alabama salt marsh. In a study of habitat use by decapod crustaceans among transplanted and natural smooth cordgrass marshes in Galveston Bay, Minello and Zimmerman (1992) found that grass shrimp (*Palaemonetes pugio*) and juvenile brown shrimp (*Farfantepenaeus aztecus*) were positively correlated with densities of macroinvertebrate prey in sediment cores. Whaley and Minello (2002) suggested that there was a strong trophic link between infauna and nekton near the marsh edge, and that this relationship contributed to the high fishery productivity derived from GoM marshes.

6.5.3 Mangroves

Mangroves generally displace intertidal coastal salt marshes in the Southern GoM Ecoregion, with exceptions noted above. With a longer growing season and warmer conditions, the tree stature of mangroves allows them to outcompete shorter salt marsh vegetation for light. In the absence of frost, mangroves eventually become dominant.

The term *mangrove* refers to an ecological group of salt- and flood-tolerant trees and shrubs that inhabit the intertidal zone (Tomlinson 1994). Synonymous terms that refer to the entire assemblage include mangrove community, mangrove ecosystem, mangrove swamp, mangrove forest, and mangal. Mangrove species may not be closely related taxonomically, with members of a community often from different plant families. Mangroves, however, have various morphological and physiological adaptations in common, which allow avoidance or tolerance of the anoxic and saline soils typical of the mangrove habitat. According to

Tomlinson (1994), true mangrove species are further distinguished by their complete fidelity to the intertidal habitat and by their taxonomic isolation (at least at the generic level but often at the subfamily or family level). Mangrove "associates," which may be herbaceous, epiphytic, or arboreal species, are found within the mangrove habitat but may also occur in more upland habitats and play a minor role in mangrove forest structure.

There are 65 species of mangroves worldwide (excluding hybrids), but only five species occur in North America (excluding Central America) (Spalding et al. 2010). Of these five species, only four occur in the GoM: *Rhizophora mangle* (red mangrove), *Avicennia germinans* (black mangrove), *Laguncularia racemosa* (white mangrove), and *Conocarpus erectus* (buttonwood). These species have a wide distribution, occurring in Florida, the Caribbean, and Mexico. Other plant species found in association with mangroves in the GoM include *Batis maritima*, *Sesuvium portulacastrum* (shoreline seapurslane), *Salicornia* spp. (glasswort), *Sporobolus virginicus* (seashore dropseed), *Monanthochloe littoralis* (shore grass), *Paspalum vaginatum* (seashore paspalum), *Distichlis spicata* (saltgrass), *Spartina alterniflora* (smooth cordgrass), and *S. patens* (wiregrass or saltmeadow cordgrass).

6.5.3.1 Dominant Forcing Functions

Similar to salt marshes, the hydrologic regime exerts a tremendous influence on the structure and function of mangroves. Hydrology affects abiotic factors such as porewater salinity, pH, oxygen, and phytotoxin accumulation, and soil factors, such as organic matter, texture, and nutrient availability, as well as biotic factors such as dispersal of seeds (Marchand et al. 2008). Disturbances from insect herbivores and woodborers are additional disturbances seen by mangroves (Feller and McKee 1999; Feller et al. 2007). Climate-related factors such as hurricanes, drought, and sea-level rise as well as human-pressures associated with the expanding human footprint also are important forcing functions. Given that the majority of mangrove-associated forcing functions are identical to those occurring in salt marshes and described in detail in Section 6.5.2.1, we herein only provide a brief summary. However, see Krauss et al. (2008) for specific examples of dominant forcing functions in mangroves.

6.5.3.2 Vegetation

6.5.3.2.1 Structure and Zonation

Mangrove forests are often classified into six basic types: overwash island, fringe, basin, riverine, dwarf, and hammock (Lugo and Snedaker 1974) (Figure 6.128). All of these types can be found in Florida, and each is characterized by certain tidal characteristics, hydroperiod, and forest structure. Riverine forests, which occur along tidal rivers and creeks with high input of freshwater, sediment, and nutrients, exhibit the highest productivity of all six types (see below). The high productivity and dynamic hydrology lead to high rates of organic matter export. Forest stands found along portions of the Shark River in Florida fall into the riverine category. The fringe forest type develops along the seaward edge of protected shorelines, has an open exchange with the sea and is well flushed by the tides. Fringe forests experience sea-strength salinity and receive fewer nutrients than riverine forests. Consequently, their productivity is somewhat lower. Overwash islands, which are sometimes considered to be a special case of fringe forest, experience higher tidal velocities that "overwash" the island and flush out accumulated litter. The overwash island type is found throughout the Ten Thousand Island region of Florida. Basin mangrove forests develop in topographic depressions, typically inland of fringing or riverine forests. Water movement is less, with tidal inundation occurring seasonally or with spring or storm tides. Once inundated by tides or freshwater, basin forests

		RIVERINE	FRINGE	BASIN	SCRUB/DWARF
FOREST PHYSIOGNOMY & PRODUCTIVITY	Profile				
	Canopy Height (m)	12.64 ± 1.43	7.65 ± 0.94	12.14 ± 1.29	0.83 ± 0.09
	Litter Production (MT/ha/yr)	12 ± 10	9 ± 0.7	6.6 ± 0.7	1.9 ± 0.6
HYDROLOGY	Water Source	Ocean Tides & Stream Flow	Ocean Tides	Ocean Tides & Saline Groundwater	Ocean Tides & Saline Groundwater
	Hydroperiod — Duration	Hours-Days	Hours	Days-Months	Perennial
	Hydroperiod — Frequency	Daily Or Seasonal	Daily	Seasonal	Continuous
	Hydroperiod — Depth	Shallow-Deep	Shallow	Shallow	Shallow
SOIL CHEMISTRY	Salinity (psu)	0/26	33/38	25/60	33/46
	Redox Potential (mV)	-48/+116	-96/+103	+87/+279	-244/-105
	Sulfide (mM)	0.0/0.2	0.1/0.3	0.1/0.2	0.9/2.2

Figure 6.128. A summary of four primary mangrove forest types originally described by Lugo and Snedaker (1974) (modified after Mendelssohn and McKee 2000).

may remain flooded for extended periods and leaf litter often accumulates on the forest floor. Basin mangroves may exhibit high productivity but low export of organic matter. The basin mangrove forest, often dominated by *A. germinans*, can be found along the southwest coast of Florida and wherever mangrove development is expansive, such as much of the southern GoM shoreline. Small trees characterize dwarf or scrub mangrove forests, often less than 1.5 m (4.9 ft) tall, with low density and extended hydroperiod. Hydrologic energy is low, and the near-continuous flooding leads to slow growth and low productivity. Extensive stands of scrub mangroves occur throughout the GoM. At higher latitudes in the northern GoM, forest stands are often scrub-like, with trees never exceeding heights of 3 to 4 m (9.8 to 13.1 ft). These stands, however, often predominate along tidal creeks and shorelines where they function more like fringe forests with a more open exchange of water and nutrients. Their short stature is caused by cold temperatures and periodic freezes, which limit plant growth and cause pruning of distal branches.

Mangrove zonation occurs where more than one species inhabit a shoreline and where a strong environmental gradient exists. In the northern Gulf, mangrove stands display typical zonation patterns found throughout the Neotropics (McKee 2012). Along shorelines and creekbanks *R. mangle* often predominates, but landward zones may contain monospecific stands of *A. germinans* or mixed stands of *R. mangle*, *A. germinans*, and/or *L. racemosa*. However, specific zonation patterns vary with local conditions and species composition. Spatial patterns of species dominance vary depending on propagule dispersal and seedling survival (Rabinowitz 1978; McKee 1995a, b; Sousa et al. 2007), physiochemical conditions (flooding, salinity, nutrients) (Krauss et al. 2008), competition, and disturbance history (Lopez-Portillo and Ezcurra 1989). During dispersal or stranding stages, mangrove propagule viability may be compromised by damage inflicted by herbivorous crabs or snails, and survival rates may influence species dominance patterns (Smith et al. 1989; Patterson et al. 1997; Cannicci et al. 2008).

Within the Deltaic Marshes of the Mississippi Estuarine Ecoregion, at the northern limits of mangrove distribution and where *Spartina alterniflora* salt marshes dominate, zonation of mangrove stands is simple, but striking. The pattern most frequently encountered is created by monospecific bands (10 to 20 m [33 to 66 ft] wide) of *A. germinans* along creekbanks, abutting large expanses of salt marsh dominated by *S. alterniflora* in the marsh interior. There also may be occasional patches of *A. germinans* occurring in the marsh interior and, if not killed by freezes, these may ultimately coalesce to form a larger, monospecific stand. Some research has examined factors influencing this spatial pattern in salt marsh–mangrove zonation (Patterson et al. 1993, 1997). In coastal Louisiana, snails attack the propagules of *A. germinans*, and the damage contributes to lower survival and establishment in the marsh interior (Patterson et al. 1997). Propagules also tend to strand at higher elevations on creekbanks in Louisiana marshes and sustain less damage by predators compared to those stranding at lower elevations. Those seedlings that do become established in salt marshes may be suppressed by competition from grasses such as *S. alterniflora* (McKee and Rooth 2008). Only when *S. alterniflora* is disturbed, reducing competition or creating bare patches of ground, can *A. germinans* recruit to the sapling stage (McKee et al. 2004). However, herbaceous vegetation may act as nursery species, promoting mangrove establishment and survival in highly stressful environments (McKee et al. 2007b).

6.5.3.2.2 Distribution

Total mangrove area in the United States is estimated to be 3,030 km^2 (749,000 ac) (Spalding et al. 2010). The largest expanse occurs along the southwestern coast of Florida around Florida Bay and Ten Thousand Islands where mangrove extent reaches 10 to 20 km (6.2 to 12.4 mi) inland in the Shark River Estuary region of the Everglades (Figure 6.129). Extensive stands also occur farther north in the Rookery Bay Estuary near Naples and around Tampa Bay and Charlotte Harbor. Above Tampa Bay, mangrove stands in Florida diminish and are gradually replaced

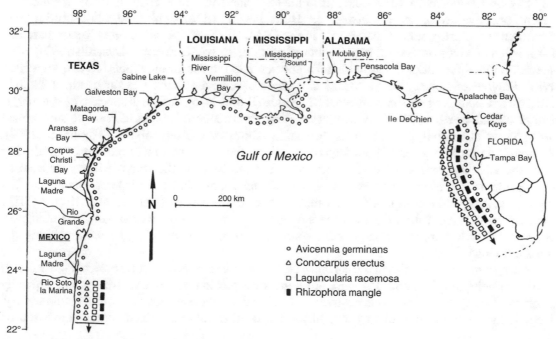

Figure 6.129. Distribution of mangrove species in the northern GoM (from Sherrod and McMillan 1985; used with permission).

with salt marsh vegetation. With increasing frequency of winter frosts, mangroves cannot persist, especially those species such as *R. mangle*, which are sensitive to colder temperatures. The only mangrove species found along the northernmost region of the GoM is *A. germinans*, which is the most cold tolerant of these four species (Markley et al. 1982). Scattered individuals of *R. mangle* and *L. racemosa* are periodically observed at mid-latitudes in the northern GoM, e.g., Cedar Keys (Markley et al. 1982; McMillan and Sherrod 1986; Stevens et al. 2006), but they often never exceed the juvenile stage of development (Zomlefer et al. 2006). No substantial mangrove stands have been reported between the Cedar Keys and the Mississippi River Delta. Mangrove populations do, however, exist along the Louisiana coast (Patterson et al. 1993; McKee et al. 2004; Perry and Mendelssohn 2009; Giri et al. 2011b). Mangrove area in Louisiana has varied from 2,180 ha (5,400 ac) in 1983 to 57 ha (141 ac) in 1986 with current stands (2010) estimated at 434 ha (1,072 ac) (Giri et al. 2011b). Scattered populations of *A. germinans* occur in Texas, e.g., around South Padre Island, but current estimates of area are not available (Sherrod and McMillan 1981; McMillan and Sherrod 1986; Everitt et al. 1996).

Mangroves are generally limited to tropical and subtropical climates between 32°N and 28°S latitudes (Lugo and Patterson-Zucca 1977; Tomlinson 1994; Duke et al. 1998). Their distributional limits usually correlate with the 20 °C (68 °F) winter isotherm of seawater (Duke et al. 1998). Sensitivity to freezing temperatures is widely regarded as the primary constraint on distributional limits (Sherrod and McMillan 1985; McMillan and Sherrod 1986; Sherrod et al. 1986; Kao et al. 2004; Stevens et al. 2006; Stuart et al. 2007). Latitudinal limits in the Northern Hemisphere vary spatially and temporally because of local variation in air and water temperatures (Stevens et al. 2006). Historically, the northernmost limit of mangroves along the Florida Gulf coast has been the Cedar Keys in the Big Bend region (29°08′N). Closed-canopy, monospecific stands of *A. germinans* occur at Cedar Keys, presumably because freeze intensity is not as great due to the insulating effect of surrounding water (Lugo and Patterson-Zucca 1977; Stevens et al. 2006). Stands of *A. germinans* have been reported as far north as 28°18′N on the northern GoM coast in Louisiana (Sherrod and McMillan 1985). However, a recent observation of *A. germinans* on Horn Island, Mississippi (~30.2°N latitude) may document the northernmost mangroves in the GoM.[2] In Texas, *A. germinans* has occurred as far north as Galveston Island, but may not persist due to winter freezes (McMillan 1971). This species is more abundant at Laguna Madre and south to the Rio Grande (Sherrod and McMillan 1981). Populations of *L. racemosa* and *R. mangle* have been reported as far north as 29°10′N latitude at Cedar Key on the west coast of Florida (Rehm 1976). A fourth species, *C. erectus*, is found as far north as 28°50′N in Florida. The persistence of *A. germinans* at subtropical latitudes is attributed in part to its coppicing (stump sprouting) ability (Tomlinson 1994). Although periodic freezes kill portions of the shoot, the trees are able to recover by producing new shoots from reserve meristems located near the base of the trunk. More recent work comparing temperature tolerance of *A. germinans* found that the dispersal stage had the highest survivorship compared to stranded or seedling stages (Pickens and Hester 2011). Consequently, cold tolerance of dispersing propagules, in combination with the ability to coppice following freeze damage, both contribute to persistence of *A. germinans* along the northern GoM.

Mangrove communities in the northern GoM are characterized by few species (Figure 6.129) and relatively simple stand structure. In southwest Florida, mangrove forests are typically composed of two or three species with few, if any, mangrove associates in the understory. Depending on stand age, disturbance history, and other factors, mixed stands composed of

---

[2] http://blog.al.com/live/2012/07/mangrove_trees_show_up_on_horn.html

R. mangle, *A. germinans*, and *L. racemosa* may be found in the Everglades and Ten Thousand Islands. In some areas, there may be stands of tall *R. mangle* along tidal creeks or on small islands. In interior areas, dwarf stands of red mangrove may be extensive. At higher latitudes, monospecific stands of *A. germinans* occur (e.g., Louisiana), but pale in aerial extent compared to co-occurring salt marshes.

Extensive stands of mangrove forests occur on both coasts of Mexico and together total more than 488,000 ha. South of the Rio Grande, in the southernmost part of Laguna Madre de Tamaulipas near La Pesca, the climate is mostly tropical and winter freezes are rare. Here, all three mangrove species occur. *Rhizophora*, as is typical for new world mangroves, lines the seaward shoreface. The *Rhizophora* zone here is not well developed, but it is characteristic (Britton and Morton 1989). Because of the steep sloping foreshore in this area, both *Avicennia* and *Laguncularia* mix in a narrow band landward of *Rhizophora*. Hence, at this location the typical zonation of *Rhizophora*, *Avicennia*, and *Laguncularia* is not generally present. Seaward of the *Rhizophora* fringe, extensive beds of the seagrass, *Syringodium filiforme* (manatee grass), dominate the subtidal (Britton and Morton 1989).

The five Mexican states along the southern GoM have a total of 194,043 ha of mangroves (Loza 1994). This is 40 % of the total mangrove area in Mexico. The most well-developed mangrove forests occur in Campeche (80,369 ha), much of which occur in the Laguna de Términos, which has a total area of 130,000 ha (Flores-Verdugo et al. 1992). This area is subject to a dry season from February to May, a tropical rainy season from June to October, and a season of frequent cold front passages (nortes) from October to February. Freshwater input into the lagoon is from four rivers and annual precipitation ranges from 110 to 200 cm (Rojas-Galaviz et al. 1992). The lagoon is an area of high habitat diversity as a result of a relatively heterogeneous environment underpinned by patterns in wind, freshwater input, and water circulation (Figure 6.98).

The mangrove habitat of the northern Yucatán, which in many ways is similar to hypersaline lagoons of Texas and northern Mexico, is, in contrast, quite different from mangrove habitat in the Laguna de Términos, where freshwater input is plentiful. The lagoon system of the northern Yucatán has been described as one of the most biologically depauperate tropical marine environments on Gulf shorelines (Britton and Morton 1989). The combination of an arid environment (low precipitation and high evapotranspiration) and infrequent tidal inundation, due to dune ridges and sills that retard water exchange, results in what amounts to evaporation ponds that concentrate salt far above sea strength. It is within this lagoonal system that small circular mangrove stands are scattered across lagoonal bottoms. These are topographical highs that are apparent even when the lagoons are dry, but in the presence of water, resemble small islands. Two variations on this theme occur. Along the margins of the lagoon, *Avicennia* dominates the centers of the hummocks where elevation is highest due to accumulated sand and organic matter. Fringing the *Avicennia*, but at somewhat lower elevations, are tufts of *Monanthochloe littoralis* (shoregrass), *Sesuvium portulacastrum* (shoreline seapurslane), *Salicornia virginica* (Virginia glasswort), and *Batis maritima* (turtleweed)—common hypersaline salt marsh herbs that often occur in salt pans. *Avicennia*'s high salt tolerance allows it to survive under these conditions. On hummocks located closer to the center of lagoons, large masses of sand and organic matter have accumulated. Here *Avicennia* fringes the outer periphery of the hummocks, but *Rhizophora* dominates the center. The higher elevations in the center of these hummocks may allow *Rhizophora* to avoid the highest soil salinities, thus being able to survive; *Rhizophora* has a lower salt tolerance than *Avicennia* (Tomlinson 1994). Herbaceous plant species similar to those found around the *Avicennia* hummocks occur here as well. A unique faunal feature of the hypersaline lagoons in the Yucatán is the greater flamingo. These beautiful birds feed upon small brine shrimp and other invertebrates adapted to hypersaline conditions.

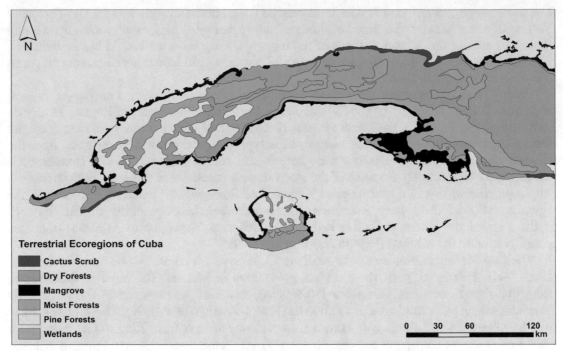

Figure 6.130. Terrestrial ecoregions of Cuba (data from Olson et al. 2001).

In the Greater Antilles Ecoregion, the largest tracts of mangroves occur in Cuba (532,400 ha) (Carrera and Santander 1994), although the northwestern portion of Cuba that borders the GoM includes only a small fraction of the total (Figure 6.130). Many of the mangroves inhabit the numerous small islands or cays in the Golfo de Guanahacabibes eastward through the Archipelago de los Colorados. The Guanahacabibes Peninsula at the far western extent of Cuba is characterized by flat karsts composed of Quaternary coral limestone. Broad mangrove habitat occurs on the peaty silt deposits on the northern shore, although the largest mangrove area in western Cuba occurs along the southern shoreline of Cuba on the Zapata Peninsula. Further to the east along Golfo de Guanahacabibes is the seashore area between Bahia Honda and Varadero, the latter being the approximate eastern boundary of the operationally defined Cuban GoM shoreline. Northwestern Cuba has a seasonal tropical climate characterized by a rainy season (May to October) and a dry season (November to April). Between January and March, frequent cold fronts (nortes) occur, similar to the Yucatán climate. The four primary mangrove trees are the same as along the Mexican GoM shoreline: *Rhizophora mangle*, *Avicennia germinans*, *Lagucularia racemosa*, and *Conocarpus erectus*. Herbaceous species are also similar, such as *Batis maritima*. However, in Cuba mangrove associates such as *Dalbergia ecastaphyllum* (coinvine), *Acrostichum aureum* (golden leather-fern) and others (Carrera and Santander 1994) are also found. Mangroves occur along shallow muddy beaches protected by coral reefs or in embayments. As in the rest of the Caribbean, *Rhizophora mangle* is the most seaward species, forming a zone between low tide and mean tide. In contrast, *Avicennia germinans* is found between mean tide and high tide. *Laguncularia racemos* is often intermingled with *Avicennia*. The *Rhizophora* community is usually free of other plant species, while the *Avicennia* community often includes species such as *Agrostichum* spp., *Batis maritima*, *Lycium carolinianum* (Carolina desert-thorn), *Cynanchum salinarum*

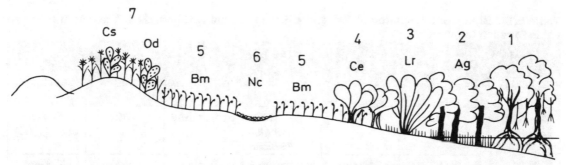

Figure 6.131. Zonation along a mangrove shoreline at La Isabela, Las Villas province, Cuba.
1 = *Rhizophora mangle*, 2 = *Avicennia germinans* (Ag), 3 = *Laguncularia recemosa* (Lr), 4 = *Conocarpus erectus* (Ce), 5 = *Batis maritima* (Bm), 6 = *Notoc commune* (Nc), 7 = *Chloris sagraeana* (Cs) sand dune vegetation; *Od* = *Opuntia dillenii* (from Borhidi 1996; used with permission).

(swallow-wort), and others. Finally, at the highest intertidal elevation is *Conocarpus erectus*, which is the most salt tolerant of the species, and can be found in monoculture or mixed with *Avicennia* and *Conocarpus* (Borhidi 1996). This shoreline zonation is presented in Figure 6.131. Landward of the *Conocarpus* zone are often expansive salt pans, only reached by the highest tides. The mangrove associates, *Batis maritima*, *Suaeda linearis* (annual seepweed), and *Salicornia* (=*Sarcocornia*) *perennis* (chickenclaws) are found here (Borhidi 1996).

6.5.3.2.3 Primary Productivity

Mangrove forests are highly productive ecosystems with net primary production rates reaching 13 metric tons/ha/year in some neotropical forests (McKee 2012). Productivity of mangroves is highest in lower latitudes (0 to 20°N), however, and decreases toward the subtropics. Most of the data on mangrove biomass production is based on annual litterfall rates, determined by monthly collections of leaf, wood, and reproductive materials that have fallen into litter traps. This approach does not include wood produced in tree trunks, above- and below-ground root production, or net production by other autotrophs (e.g., epiphytic and benthic algae). Thus, although litterfall rates provide a relative indication of primary production, they are underestimates of the net primary production by the ecosystem. However, most estimates of mangrove productivity are based on this method and will be emphasized here.

In general, litterfall rates vary among forest types: Dwarf (120 g/m^2/year), basin (730 g/m^2/year), fringe (906 g/m^2/year), and riverine (1,170 g/m^2/year) (Twilley and Day 1999). In the northern GoM, values for annual litterfall ranged from 50 to 1,724 g/m^2/year with an overall mean of 736 g/m^2/year (Table 6.13). Highest values are reported for overwash islands (1,132 g/m^2/year) and a restored forest (1,099 g/m^2/year), intermediate values for basin and fringe forests (295 to 906 g/m^2/year), and lowest values for scrub and tidally restricted forests (101 to 250 g/m^2/year). These values fall within the range reported for mangrove forests at more tropical latitudes and for the same forest types. Mangrove productivity along the Mexican GoM has best been documented in the Laguna de Términos. Species of mangrove here include *R. mangle*, *A. germinans*, *L. racemosa*, and *C. erectus*. The structure and function of the mangrove habitat have been described in a suite of publications (e.g., Rojas-Galaviz et al. 1992). Net primary productivity was much higher at a riverine site (2,458 g/m^2/year) compared to a fringing mangrove location (1,606 g/m^2/year) (Table 6.14), apparently due to greater nutrient and freshwater input at the riverine location. Also, higher seasonal productivity values occur during the rainy season and lowest during the period of frequent nortes.

Table 6.13. Biomass Production of Mangrove Forests in the Northern GoM (maximum reported value for each location)

Location	Latitude	Longitude	Dominant Species	Forest Type	Biomass Production (g/m²/year)	Source
Litterfall						
Rookery Bay & Estero Bay, FL	26°02′N	81°45′W	A. germinans R. mangle L. racemosa	Basin mixed	810	Twilley et al. (1986)
	25°02′N	81°34′W	A. germinans	Basin monosp.	444	Twilley et al. (1986)
Cockroach Bay Tampa Bay, FL	27°41′N	82°31′W	R. mangle	Overwash island	1,132	Dawes et al. (1999)
Rookery Bay, FL	26°3′N	81°42′W	A. germinans R. mangle	Basin mixed	1,724	McKee and Faulkner (2000)
			L. racemosa	Restored	1,108	McKee and Faulkner (2000)
Windstar, FL	26°7′N	81°47′W	A. germinans R. mangle	Basin mixed	1,065	McKee and Faulkner (2000)
			R. mangle L. racemosa A. germinans	Restored	1,170	McKee and Faulkner (2000)
Rookery Bay, FL	26°3′N	81°42′W	A. germinans R. mangle L. racemosa	Basin mixed	1,278	Raulerson (2004)
			R. mangle	Fringe	1,241	Raulerson (2004)
			L. racemosa	Restored	1,205	Raulerson (2004)
Windstar, FL	26°7′N	81°47′W	A. germinans R. mangle L. racemosa	Basin mixed	986	Raulerson (2004)
			R. mangle	Fringe	1,132	Raulerson (2004)
			R. mangle L. racemosa A. germinans	Restored	913	Raulerson (2004)
Rookery Bay, FL	26°3′N	81°42′W	A. germinans R. mangle L. racemosa	Basin mixed	264	Giraldo (2005)
			A. germinans	Basin monosp.	145	Giraldo (2005)
			R. mangle	Fringe	247	Giraldo (2005)
			R. mangle	Scrub	101	Giraldo (2005)

(continued)

Table 6.13. (continued)

Location	Latitude	Longitude	Dominant Species	Forest Type	Biomass Production (g/m²/year)	Source
Windstar, FL	26°7′N	81°47′W	A. germinans R. mangle L. racemosa	Basin mixed	220	Giraldo (2005)
			R. mangle	Fringe	192	Giraldo (2005)
Captiva Island, FL	26°42′N	82°14′W	A. germinans R. mangle L. racemosa	Unrestricted tide	151	Harris et al. (2010)
				Restricted tide	50	Harris et al. (2010)
Sanibel Island, FL	26°34′N	82°12′W	A. germinans R. mangle L. racemosa	Unrestricted tide	900	Harris et al. (2010)
Sanibel Island, FL	26°34′N	82°12′W	A. germinans R. mangle L. racemosa	Restricted tide	450	Harris et al. (2010)
Root production						
Rookery Bay, FL	26°3′N	81°42′W	A. germinans R. mangle	Basin mixed	610	McKee and Faulkner (2000)
			L. racemosa	Restored	797	McKee and Faulkner (2000)
Windstar, FL	26°7′N	81°47′W	A. germinans R. mangle	Basin mixed	453	McKee and Faulkner (2000)
			R. mangle L. racemosa A. germinans	Restored	412	McKee and Faulkner (2000)
Rookery Bay, FL	26°3′N	81°42′W	A. germinans R. mangle	Basin mixed	182	Giraldo (2005)
			A. germinans	Basin monosp.	198	Giraldo (2005)
			R. mangle	Fringe	200	Giraldo (2005)
			R. mangle	Scrub	211	Giraldo (2005)
Windstar, FL	26°7′N	81°47′W	A. germinans R. mangle L. racemosa	Basin mixed	144	Giraldo (2005)
			R. mangle	Fringe	144	Giraldo (2005)

(continued)

Table 6.13. (continued)

Location	Latitude	Longitude	Dominant Species	Forest Type	Biomass Production (g/m²/year)	Source
Bayou Lafourche, LA	29°10′N	90°14′W	*A. germinans*	Scrub	346	Perry and Mendelssohn (2009)
Shark River, Everglades, FL	25°22′N	81°01′W	*R. mangle* *L. racemosa* *A. germinans*	Riverine	643	Castaneda-Moya et al. (2011)
Taylor Slough, Everglades, FL	25°44′N	80°51′W	*R. mangle*	Scrub	407	Castaneda-Moya et al. (2011)

Table 6.14. Structure and Productivity of Mangrove Stands at a Fringing Mangrove Site (Estero Pargo) and a Riverine Mangrove Site (Boca Chica) in Términos Lagoon (from Rojas-Galaviz et al. 1992)

	Estero Pargo	Boca Chica
Structural characteristics		
Mean canopy height (m)	6	20
Stem density	7,510	3,360
Basal area (m²/ha)	23	34
Complexity index	69	32
Net primary productivity (g/m²/year)		
Woody growth	772	1,206
Leaves	594	881
Fruits	192	253
Branches	48	118
Total	834	1,252
Net annual primary productivity	1,606	2,458

Very few estimates of belowground production have been reported due to the difficulties involved in measuring root production and turnover. A few estimates of root accumulation rates have been made through the use of ingrowth bags. The bags, constructed of flexible mesh material, are filled with root-free sediment or another standardized substrate, and inserted vertically into the ground. After a time interval, the bags are retrieved and the ingrown root mass is measured. This approach reflects the net effect of root production, turnover, and decay during the time interval. Thus, the values are an underestimate of root production, but allow a relative comparison of root matter accumulation in the soil. Values reported for Florida mangroves are slightly less than rates of litterfall, ranging from 144 to 797 g/m²/year with an overall mean of 365 g/m²/year (Table 6.15). Highest values were reported for a riverine soil (643 g/m²/year) and a restored forest (605 g/m²/year), with intermediate values for basin mixed

Table 6.15. Abundant Macro-Invertebrate Groups Associated with Mangrove Systems in the GoM

Infauna	Encrusting Epifauna	Motile Epifauna
Oligochaeta	Oysters (M)	Gastropods (M)
Capitellidae (P)	Barnacles (C)	Hermit crabs (C)
Nereididae (P)	Sponges	Brachyuran crabs (C)
Spionidae (P)	Tunicates (Ch)	Penaeid shrimps (C)
Amphipoda (C)		
Tanaidacea (C)		
Alongi and Christoferssen (1992), Sheridan (1997), Dittmann (2001), Ellison and Farnsworth (2001), Lee (2008), Metcalfe and Glasby (2008)	Nagelkerken et al. (2008)	Henriques (1980), Alongi and Christoferssen (1992), Caudill (2005), Nagelkerken et al. (2008)

Key: C = crustacean, Ch = chordate, M = mollusk, P = polychaete

forests (347 g/m^2/year), and lowest values for fringe, basin monospecific, and scrub/dwarf forests (172 to 321 g/m^2/year).

6.5.3.3 Fauna

Information on the infauna and epifauna of GoM mangrove systems is relatively sparse compared with mangal systems in the tropical Indo-Pacific (see mangrove reviews by Nagelkerken et al. 2000; Faunce and Serafy 2006). This is due, in part, to the restricted regional distribution of this wetland type in the GoM and the fact that mangroves are almost always contiguous to other vegetated wetlands (e.g., seagrasses and tidal marshes) where benthic assemblages are somewhat less complicated to sample (i.e., woody structures such as prop roots and pneumatophores associated with mangroves make sampling the benthos more difficult; see Lee 2008). The numerically dominant benthic invertebrates in mangrove systems of the GoM are small crustaceans (e.g., amphipods and tanaidaceans) and polychaetes, which may occur at high densities (e.g., >52,000 individuals/m^2) but low biomass (e.g., ≤8.2 g wet/m^2) and diversity when compared to seagrass (e.g., ≤24,000 individuals/m^2 and ≤87.4 g wet/m^2) or even shallow unvegetated habitats (Sheridan 1997; Escobar-Briones and Winfield 2003). Most studies of mangrove invertebrate assemblages in the GoM have been conducted in southern Florida (e.g., Sheridan 1992, 1997; Vose and Bell 1994) and Mexico (e.g., Vega-Cendejas and Arreguín-Sánchez 2001; Lango-Reynoso et al. 2013). Much of this work has focused on the value of mangrove habitat as foraging areas and nurseries for marine and estuarine nekton (Faunce and Serafy 2006). Unlike other coastal wetland habitats in the GoM, the role of mangroves as important nurseries for nektonic fishes and decapod crustaceans appears to be equivocal (Barbier and Strand 1998; Sheridan and Hays 2003).

Detritus has long been considered the base of mangrove food webs (Odum and Heald 1972) and that notion has persisted in more recent studies of trophic structure and dynamics in GoM mangrove systems, especially in Mexico (Rosado-Solórzano and Guzmán del Próo 1998; Vega-Cendejas and Arreguín-Sánchez 2001; Rivera-Arriaga et al. 2003). Mangrove leaf detritus is more refractory than seagrass or marsh plant detritus and alternative perspectives on the base of GoM mangrove food webs have also emerged, placing emphasis on the importance of benthic algae and phytoplankton (Dittel et al. 1997) or detritus from other sources of riparian vegetation (Mendoza-Carranza et al. 2010). Regardless of the source of detritus or primary production, there appears to be general agreement that most energy transformations to higher trophic levels such as fishes and nektonic decapods in mangrove systems occur through small benthic and epibenthic invertebrate prey resources (mostly micro- and macro-crustaceans, polychaetes, and gastropods) that can use detritus as a source of nutrition (Vose and Bell 1994; Sheridan 1997; Llansó et al. 1998; Vega-Cendejas and Arreguín-Sánchez 2001; Sheridan and Hays 2003). The feeding activities of intertidal and semiterrestrial sesarmid crabs (e.g., *Aratus pisonii*) and gastropods (e.g., *Melampus coffeus*) found in GoM mangroves often function to process larger fractions of living and dead mangrove leaves into smaller particles that can be consumed by microcrustaceans and annelids (Beever et al. 1979; Erickson et al. 2003; Proffitt and Devlin 2005). Mangroves, like tidal marshes, also provide habitat for a guild of omnivorous xanthid crabs (e.g., *Eurytium limosum*) that can be involved in complex trophic interactions, which may ultimately terminate in the transfer of mangrove production to foraging nekton (Sheridan 1992; Vose and Bell 1994; Llansó et al. 1998).

In many areas of the world where mangroves are a dominant wetland type, macroinvertebrate community structure is influenced by the degree of tidal inundation, availability of organic matter, and sediment characteristics (Lee 2008). There is at least some evidence that the same factors affect assemblages in the GoM (e.g., Sheridan 1997; Vose and Bell 1994). Infaunal and epifaunal assemblages in high and low intertidal mangroves are often distinctly different, due in part to horizontal variation (landward–seaward) in environmental conditions (Nagelkerken et al. 2008). Metcalfe and Glasby (2008) found the highest diversity and abundance of polychaetes in the soft, unconsolidated substrates of seaward assemblages, with these measures decreasing progressively in the landward assemblages. Ellison and Farnsworth (1992) documented epifaunal assemblages on mangrove prop roots at six sites in Belize that ranged from nearshore to offshore sites. Species richness of all encrusting epifauna increased with distance offshore and duration of tidal inundation, with only two sponge species present in the most nearshore site, while there were nine at the most offshore site.

Structurally complex wetland habitats are usually associated with an enriched local faunal diversity (Heck and Wetstone 1977; Summerson and Peterson 1984; Thompson et al. 1996; Cocito 2006; Eriksson et al. 2006; Van Hoey et al. 2008). Biodiversity of some groups of molluscs and crabs may be enhanced by the presence of mangrove (Henriques 1980; Ellison and Farnsworth 2001) but, in general, infaunal diversity is relatively low in mangrove systems. There have been numerous studies of infaunal diversity in mangroves compared to adjacent vegetated and unvegetated habitats in Australia and New Zealand, and these have confirmed the relatively low diversity of infauna assemblages in mangroves (Dittmann 2001; Ellis et al. 2004; Alfaro 2006). Mangrove infaunal assemblages of the GoM show similar low diversity (Sheridan 1997). Lower diversity in some mangrove sediments may have resulted in part from organically enriched silt, which was unsuitable for certain taxa, such as many larger suspension feeders (Ellis et al. 2004).

Table 6.15 lists the most abundant invertebrate groups found in mangrove systems. Mangrove infauna and epifauna are relatively poorly known compared to other components of the mangrove ecosystem, such as floristics and trophic ecology (Lee 2008). Some of the abundant groups of mangrove infauna, such as polychaetes, have been much less frequently studied than other groups, such as brachyuran crabs. For the poorly studied groups, little is known of their overall diversity, abundance, and functional role in mangroves (Nagelkerken et al. 2008).

Microhabitats occupied by mangrove invertebrates include the hard substrata offered by the trunk, aerial roots, and foliage of mangrove trees, and the surrounding soft, unconsolidated sediments (Ellison and Farnsworth 1992). In addition, motile epifauna are attracted to inundated areas of mangroves for foraging and refuge from predators. The peaty mucks in which mangroves grow have a high silt content and are inhospitable to most larger suspension-feeding invertebrates (Ellison and Farnsworth 2001). Consequently, mangrove infauna tends to be small, surface-dwelling annelids (polychaetes and oligochaetes) and crustaceans (Lee 2008). Alongi and Christoferssen (1992) found that small, surface-dwelling polychaetes and amphipods were dominant mangrove taxa, followed by tanaid crustaceans. Similarly, annelids, and tanaids were the dominant infaunal taxa in mangroves of Rookery Bay, Florida (Sheridan 1997).

A variety of encrusting and sessile benthic invertebrates, including oysters, barnacles, sponges, and tunicates, occupy mangrove prop root and pneumatophore surfaces in intertidal and subtidal areas. Invertebrates such as the isopod *Sphaeroma terebrans* burrow into the roots of mangroves in south Florida (Brooks and Bell 2002). Of the motile epifauna, gastropods and semiterrestrial crabs (e.g., ocypodids, grapsids, and some xanthids) tend to be most abundant in mangrove ecosystems (Nagelkerken et al. 2008; Beever et al. 1979; Erickson et al. 2003; Proffitt and Devlin 2005). Gastropods occupy a wide range of ecological niches in mangroves and include herbivorous grazers, deposit feeders, and predators (Nagelkerken et al. 2008). Various crabs, including hermit crabs and the tree crab, *Aratus pisonii*, are also conspicuous community constituents. Erickson et al. (2003) reported that *A. pisonii* feeds on the leaves of mangroves in Florida Bay, while Alongi and Christoferssen (1992) found that juvenile crabs and penaeid shrimps were common epifauna occupying mangrove sediments.

Mangroves in most areas, including the GoM, rarely occur in the absence of other adjacent and structurally complex shallow estuarine (e.g., seagrass and marsh) or marine (e.g., coral reef) habitats with which the mangrove habitat shares elements of common highly mobile nekton assemblages (Sheaves 2005). Consequently, mangroves do not appear to be associated with a "characteristic" faunal assemblage. Where present along Gulf shorelines, mangroves attract populations of locally occurring epifauna. In a Louisiana black mangrove stand, Caudill (2005) found that the most abundant epifaunal species in lift nets were grass shrimp (*Palaemonetes* spp.), which contributed 53.5 % of all collected individuals, followed by white shrimp (*Litopenaeus setiferus*) and blue crabs (*Callinectes sapidus*). These species are consistently among the numerically dominant nekton occurring in Gulf coast marine wetlands. Structural complexity may increase apparent biomass production in mangroves by attracting motile fauna (Tolley and Volety 2005; Eriksson et al. 2006). Spatially complex habitats such as submerged mangrove trunks and roots mitigate predator–prey interactions by providing places for prey to hide or escape from predators (Figure 6.132). Juvenile brown shrimp and blue crabs, for example, primarily seek and occupy vegetated habitats (Zimmerman and Minello 1984; Peterson and Turner 1994; Howe et al. 1999; Rakocinski and Drury 2005; Moksnes and Heck 2006).

The composition of nekton assemblages found in mangroves appears to be strongly influenced by species that occur in adjacent habitats. In upstream mangrove forests adjacent to uplands, the nekton assemblages tend to be dominated by small estuarine resident species similar to those commonly found in interior tidal marsh and marsh edge habitats, but in downstream fringing mangrove habitats adjacent to seagrasses or coral reefs, juveniles of species in the families Lutjanidae (snappers) and Haemulidae (grunts), which occur as adults on reefs or in seagrass habitats, are added to the nekton assemblage (Ley et al. 1999; Serafy et al. 2003). Such observations have led to a considerable focus on habitat connectivity and the potential role of mangroves as a sub-habitat within a more complex habitat matrix upon which some species are dependent at different stages in their life histories (Nagelkerken et al. 2001, 2002; Mumby 2006; Jones et al. 2010).

The basic nekton assemblage structure of mangroves in much of the GoM appears to be very similar to that of tidal marshes, with the shallow waters inundating the structurally complex elements of prop roots and pneumatophores offering attractive habitat (Figure 6.132),

Figure 6.132. Illustration of nekton attraction to submerged structural features of mangrove habitat.   (http://www.naturefoundationsxm.org/education/mangroves/red_mangrove_illustration. gif; used with permission). Mangrove aerial roots may provide food, in the form of epiphytes and epifauna, habitat, and protection from predation for many nektonic species.

mostly for small estuarine resident species (Table 6.16). Although many of the dominant species may vary across regions, the most abundant fish families in mangroves are Cyprinodontidae (killifishes), Fundulidae (fundulids), Poeciliidae (live-bearers), and Gobiidae (gobies), as well as the juveniles of a number of transient species, often representing the fish families Gerridae (mojarras), Mugilidae (mullets), Sciaenidae (drums), and Sparidae (porgies). At the edges of the embayments and channels immediately adjacent to the mangrove vegetation, schooling species in the fish families Engraulidae (anchovies), Atherinidae (silversides), and Clupeidae (herrings) are often common (Figure 6.108), as are the much less abundant larger predatory species they attract, such as Sphyraenidae (barracudas), Eleopidae (tarpons), and Centropomidae (snooks) (Figure 6.109c).

The absence of natant decapod crustaceans from some studies is noteworthy (Table 6.16) and has two likely explanations. First, the methods used to collect nekton from among the sturdy prop roots and pneumatophores of mangroves often rely on the application of rotenone which, especially in its most commonly used formulation, is much more effective on fishes than crustaceans (Robertson and Smith-Vaniz 2008), so it is not surprising that fishes predominate in the collections. Second, natant decapod crustaceans may not be as abundant in mangroves as in other coastal wetland habitats. Some researchers have noted that penaeid shrimp were rarely collected or observed in mangroves, but were more commonly associated with adjacent seagrass beds (Thayer et al. 1987), even though there is some evidence that juvenile penaeid shrimp appear to satisfy at least a portion of their nutritional requirements from mangrove-associated bacteria and benthic macrofauna (Nagelkerken et al. 2008).

A variety of unconventional methods for sampling nekton have been applied in highly structured wetlands such as mangroves, and sometimes the findings provide complementary representations of the fauna because no two methods are equally effective in capturing nekton

Table 6.16. Nekton Species Comprising at Least 90 % of the Total Individuals in Mangrove Samples Collected with Block Nets or Enclosure Nets from Prop Roots in Florida (FL Bay 1: Thayer et al. 1987; FL Bay 2: Ley et al. 1999; Placido Bayou, FL: Mullin 1995; Tampa Bay, FL: Krebs et al. 2007) or Lift Nets Among the Pneumatophores of Black Mangroves in Louisiana (Caudill 2005)

	Enclosure / Block Nets					Lift Nets
	FL Bay 1	FL Bay 2	Placido Bayou, FL	Tampa Bay, FL		Caminada Bay, LA
Fishes						
Atherinidae						
Atherinomorus stipes	X	X				
Membras martinica	X					
Menidia penninsulae	X					
Menidia beryllina						X
Menidia spp.		X	X	X		
Cichlidae						
Sarotherodon melanotheron		X	X			
Unidentified cichlid				X		
Clupeidae						
Brevoortia gunteri			X			
Brevoortia patronus				X		
Harengula humeralis	X					
Harengula jaguana	X	X				
Jenkensia lamprotaenia	X					
Opisthonema oglinum		X				
Cyprinodontidae						
Floridichthys carpio	X	X	X	X		
Cyprinodon variegatus	X	X	X	X		X
Engraulidae						
Anchoa mitchilli	X	X	X	X		
Anchoa hepsetus	X					
Anchoa cayorum		X				
Fundulidae						
Lucania parva	X	X	X			
Fundulus grandis	X	X	X	X		X
Fundulus similis	X	X		X		
Fundulus confluentus	X	X		X		
Fundulus jenkinsi			X			
Adinia xenica		X		X		
Gerridae						
Diapterus plumeri		X	X	X		
Eucinostomus gula	X	X				
Eucinostomus argentus	X	X	X			
Eucinostomus harengulus		X		X		

(continued)

Table 6.16. (continued)

	Enclosure / Block Nets					Lift Nets
	FL Bay 1	FL Bay 2	Placido Bayou, FL	Tampa Bay, FL		Caminada Bay, LA
Eucinostomus spp.				X		
Gerres cinereus		X				
Gobiidae	███	███	███	███		███
Bathygobius soporator	X					
Ctenogobius smaragdus	X			X		
Evorthodus lyricus						X
Gobiosoma bosc	X	X				
Gobiosoma robustum	X	X	X			
Gobiosoma spp.				X		
Gobionellus boleosoma						X
Lophogobius cyprinoides	X	X				
Microgobius gulosus	X	X	X	X		
Mugilidae	███	███	███	███		███
Mugil cephalus	X	X		X		X
Mugil curema	X	X				
Mugil gyrans				X		
Mugil liza		X				
Poeciliidae	███	███	███	███		
Belonesox belizanus		X				
Gambusia holbrooki				X		
Gambusia sp.	X	X				
Poecilia latipinna	X	X	X	X		
Sciaenidae	███		███	███		
Cynoscion nebulosus	X			X		
Leiostomus xanthurus			X	X		
Pogonias cromis	X			X		
Sciaenops ocellatus	X			X		
Sparidae	███		███	███		███
Archosargus probatocephalus	X			X		
Lagodon rhomboides	X		X	X		X
Natant Decapod Crustaceans						
Portunidae				███		███
Callinectes sapidus				X		X
Penaeidae				███		███
Farfantepenaeus aztecus						X
Farfantepenaeus duorarum				X		
Litopenaeus setiferus						X
Palaemonidae						███
Palaemonetes spp.						X

Red shaded cells indicate the presence of dominant families at each location

of all species, sizes, or at all locations (Loftus and Rehage 2005). For example, rotenone in aqueous solution, as it is commonly used in marine sampling programs in coral reef and mangrove habitats, is strongly selective in affecting fishes, but has little or no effect on most natant decapod crustaceans or other types of nekton (Robertson and Smith-Vaniz 2008). Consequently, nekton-sampling programs relying on rotenone are biased toward fishes, even if they may not be the most abundant nektonic organisms in the assemblage.

Table 6.17 compares the nekton assemblage from a mangrove habitat in Florida Bay using a combination of nets and rotenone poisoning with results from visual censusing and video recording. Only half of the top ten most abundant fish families appear in all three lists, and these were represented by species considered as resident nekton (families Gerridae, Cyprinodontidae, Fundulidae, Poecilidae, and Gobidae). However, there are also some striking differences in the importance of dominant groups. Engraulids (anchovies) were the top-ranked family captured in nets but accounted for only 1.4 % of fishes in the video recording and do not even rank in the top ten families in the visual census. Also, in the visual census, the atherinids (silversides) comprised nearly 84 % of the fishes, but were not detected in the video recording data. Species diversity and evenness were greater in the net sample than in either the visual census or the video recording collections. For example, only two families represented by six species accounted for nearly 88 % of all nekton in the visual census, but five families and 15 species were required to account for a similar percentage of nekton in the net samples. The video recording samples were intermediate in this regard, requiring seven families and ≥ 11 species to account for 88 % of the individuals. The video data also included a high percentage (9.6 %) of juvenile individuals that could not be identified to either species or family. The three methods clearly paint a very different picture of the nekton assemblage in this single mangrove system. The enclosure nets with rotenone likely represented the species found among the mangrove prop roots while the visual census may have better represented the nekton assemblage immediately adjacent to the mangrove forest *per se*.

There is considerable debate over the relative contribution of mangrove habitat in supporting estuarine nekton assemblages. While some studies have suggested a positive relationship between the areal extent of fringing mangrove forests and regional fisheries production (Aburto-Oropeza et al. 2008), others have questioned the importance of mangroves as nursery habitat for nekton (Sheridan and Hays 2003; Faunce and Serafy 2006). Unlike other coastal wetland habitats in which production by the dominant plant species tends to support a trophic structure on which nekton assemblages derive considerable nutrition, mangroves generally do not directly contribute much trophic support for nekton (Stoner and Zimmerman 1988; Sheridan and Hays 2003). Alternative sources of trophic support from nearby habitats such as seagrasses may contribute more to the diets of nekton found in mangroves (Nagelkerken and van der Velde 2004). Nonetheless, mangrove habitats seem to provide at least some nursery functions for both recreationally and commercially important fishery organisms and their food resources (Odum et al. 1982).

6.5.4 *Phragmites* Reed Beds

Phragmites australis (common reed) is a warm-season, rhizomatous, stoloniferous perennial grass that grows in tidal and non-tidal habitats throughout the world (Chambers et al. 1999; USDA 2002). In the northern GoM, *Phragmites australis* (hereafter *Phragmites*) inhabits stream banks and interior marsh locations from Texas to Florida, but occurs in greatest abundance along the Balize, or Birdfoot, Delta and Chenier Plain regions of Louisiana (Tiner 2003; Rosso et al. 2008; Stanton 2005). *Phragmites* has been part of the marsh communities of North America (including the Gulf Coast) for millennia (Lamotte 1952), but has drastically

Table 6.17. Top Ten Fish Families in Mangrove Prop Roots in Florida Bay Based on Sampling With: (1) Enclosure Nets + Rotenone, (2) Visual Censusing (Ley et al. 1999), and (3) Video Recording (Ellis and Bell 2008)

Rank Order	Fish Family	Abundance	Cumulative %	Number of Species
Enclosure nets + rotenone				
1	Engraulidae – anchovies	18,598	21.3	2
2	Atherinidae – silversides	17,683	41.6	3
3	Cyprinodontidae – killifishes	15,277	59.1	2
4	Poeciliidae – livebearers	13,410	74.4	3
5	Fundulidae – fundulids	11,753	87.9	5
6	Gobiidae – gobies	3,915	92.4	4
7	Gerridae – mojarras	3,321	96.2	4
8	Cichlidae – cichlids	1,066	97.4	2
9	Belonidae – needlefishes	1,040	98.6	2
10	Batrachoididae – toadfishes	572	99.3	2
	TOTAL above and others	**87,257**	**100.0**	**48**
Visual Census				
1	Atherinidae – silversides	407,772	83.9	3
2	Lutjanidae – snappers	19,186	87.8	3
3	Peociliidae – livebearers	15,377	91.0	3
4	Fundulidae – fundulids	12,149	93.5	3
5	Gerridae – mojarras	11,122	95.9	4
6	Cyprinodontidae – killifishes	9,004	97.7	2
7	Haemulidae – grunts	2,901	98.3	2
8	Belonidae – needlefishes	1,974	98.7	1
9	Gobiidae – gobies	1,743	99.9	3
10	Muglidae – mullets	1,377	99.3	2
	TOTAL above and others	**485,846**	**100.0**	**42**
Video Record				
1	Gerridae – mojarras	4,121	46.2	1?
2	Sparidae - porgies	1,715	65.4	2
3	Haemulidae – grunts	1,354	80.6	2?
4	Lutjanidae – snappers	294	83.9	2
5	Peociliidae – livebearers	198	86.1	2
6	Gobiidae – gobies	151	87.8	1?
7	Engraulidae – anchovies	122	89.2	1?
8	Fundulidae – fundulids	61	89.9	1
9	Cyprinodontidae – killifishes	39	90.3	1
10	Ephippidae - spadefishes	4	90.3	1
	TOTAL above and others	**8,919**	**100.0**	**16**

increased its distribution over the past century (Chambers et al. 1999). This range expansion has occurred at the expense of other species, as *Phragmites* often forms large, dense, near-monotypic stands that outcompete other species for resources (Phillips 1987; Minchinton and Bertness 2003). Increased *Phragmites* density also alters ecosystem services and reduces available habitat for many species of wading birds, fish, and mammals that utilize wetlands (Hauber et al. 1991; Chambers et al. 1999). However, *Phragmites* marshes do provide habitat for a variety of species, and may play an important role in stabilizing some of the most erodible wetland habitats in the northern GoM (Rooth and Stevenson 2000).

6.5.4.1 Dominant Forcing Functions

The primary drivers that control the ecological structure and function of *Phragmites* beds in the central GoM (i.e., Mississippi River Delta) are similar to those discussed for salt marshes and mangroves, with a few notable exceptions. Salinity in *Phragmites* marshes at the mouth of the Mississippi River is low to absent due to freshwater input from the river. Marshes here are classified as the intermediate type, between fresh and brackish, or fresh. However, salinity is an important environmental factor outside of the Birdfoot Delta, such as in the Chenier Plain region, where *Phragmites* stands are normally exposed to brackish water conditions (Stanton 2005). Also, reed beds at the mouth of the Mississippi River are continuously flooded with a few meters of water during the spring flood period. Water may not drain from the beds during this period. However, *Phragmites* has an efficient internal aeration system that provides oxygen to underground plant organs (Brix et al. 1996), allowing this species to not only tolerate these conditions, but also thrive. The input of nutrient-rich sediments from the Mississippi River may assist in this regard. The high land subsidence rate is another driver impacting the attached reed beds in the region. In the Birdfoot Delta, relative sea-level rise (the combination of subsidence and global sea-level change) is more than a centimeter a year (Penland and Ramsey 1990). This process has been a primary driver of wetland loss in the region, as has the natural abandonment of the sub-deltas comprising the Birdfoot Delta and the infilling of their distributary channels (Wells and Coleman 1987), and is an added force for ecological change.

6.5.4.2 Vegetation

6.5.4.2.1 Origin and Structure

Phragmites australis occupies a range of habitats in the northern GoM and can proliferate in a wide variety of water depths (Hauber et al. 1991; White et al. 2004a). Historically, *Phragmites* was only a minor component of tidal marsh vegetation assemblages and was confined to higher elevations at the upland boarders of marshes and areas along creek banks (Niering et al. 1977; Warren et al. 2001). However, recent expansion of *Phragmites* into more diverse habitats has been occurring along the U.S. Atlantic and Gulf coasts (Chambers et al. 1999; Peterson and Partyka 2006; Hauber et al. 2011). Reasons for the *Phragmites* expansion vary by location but can include tidal restrictions, habitat modification, disturbance, and invasion by a more competitive European population (Chambers et al. 1999; Burdick et al. 2001; Howard et al. 2008; Hauber et al. 2011). For example, in coastal Mississippi Peterson and Partyka (2006) found that *Phragmites* was widespread along creek banks in up-estuary/low salinity environments where there was little anthropogenic disturbance, but *Phragmites* also occurred in high elevation/high salinity areas that were heavily modified by man.

The situation in Louisiana is somewhat unique, particularly in the Mississippi River's Birdfoot Delta, where distinct genotypes of *Phragmites* can occupy different areas within the same marsh (White et al. 2004a; Hauber et al. 2011; Lambertini et al. 2012) (Figure 6.133).

Figure 6.133. *Phragmites australis* **forms circular patches in the Mississippi River Birdfoot Delta. Molecular genetic analyses have demonstrated that these patches, which have identifiable aerial signatures (color and height), are genetically distinct populations (Lambertini et al. 2012) (photo credit: I. A. Mendelssohn).**

Phragmites is the dominant vegetation type in the outer two-thirds of the Birdfoot Delta, yet within these *Phragmites* areas are patches of the reed that are both phenotypically and genetically distinct (Hauber et al. 1991, 2011; White et al. 2004a) (Figure 6.133). However, the occurrence of *Phragmites* on the outer portions of the delta is a mixed blessing. On one hand, *Phragmites* populations stabilize an easily erodible landscape (Rooth and Stevenson 2000) and protect the more diverse and fragile interior marshes of the delta (Hauber et al. 2011). On the other hand, *Phragmites* populations provide little in the way of habitat and food for migrating waterfowl that overwinter in the Mississippi River delta (Hauber et al. 1991).

The *Phragmites* marshes of the Mississippi River's Birdfoot Delta are unique in that they contain the most phenotypically and genetically diverse *Phragmites* populations in the world (Hauber et al. 2011). The most common phenotype, known as the Delta phenotype, tends to occur in the outer portions of the delta and is considered the oldest *Phragmites* lineage in the Birdfoot Delta (Hauber et al. 1991; Fournier et al. 1995; White et al. 2004a). More recently the interior marshes of the delta have been colonized by at least two other lineages, the Gulf Coast subspecies and the introduced haplotype M, which has Eurasian origins (Hauber et al. 2011). Another *Phragmites* haplotype, AD, was also recently discovered by Hauber et al. (2011), but its distribution is presently unknown.

6.5.4.2.2 Ecosystem Function

The processes responsible for controlling ecosystem function in *Phragmites* marshes along the Gulf Coast have gone largely unstudied. In fact, much of the data regarding primary production, decomposition, nutrient cycling, and elevation change comes from a single source—Stanton (2005). *Phragmites* is a tenacious ecosystem engineer that can colonize a wide range of water depths (White et al. 2004a), and build soil upwards through the accumulation of organic and inorganic materials to increase soil elevation and potentially reduce flooding stress (Rooth and Stevenson 2000; Stanton 2005). For example, Stanton (2005) found that elevation in the center of a 40-year-old *Phragmites* colony in Louisiana was 10 cm higher than the surrounding marsh, which corresponded to an increase in peat thickness of 10 cm relative to the surrounding marsh. Interestingly, the elevation increase did not lead to a change in the interstitial water chemistry of the *Phragmites* colony, although this was likely due to the hydrology of the marsh being manipulated by control structures (Stanton 2005).

Organic matter decomposition in wetlands depends on flooding frequency, flooding duration, soil temperature, soil redox potential, and organic matter quality (Brinson 1977; Neckles and Neil 1994; Windham 2001). *Phragmites* litter is of particularly poor quality (high carbon to nitrogen ratio) making decomposition rates in *Phragmites* marshes slow (Windham 2001) and the accumulation of organic matter rapid (Stanton 2005).

Above- and below-ground biomass production rates in *Phragmites* marshes can be very high due to the large stature of the plant and its ability to outcompete other species for resources (Burdick and Konisky 2003; Stanton 2005; Howard and Rafferty 2006). Above-ground biomass production in *Phragmites* marshes varies depending on stem densities and heights, but is proportional to stem diameter (Stanton 2005), and stem heights can increase in response to flooding (Howard and Rafferty 2006). As *Phragmites* colonies age, stem densities tend to increase, but stem height and diameters tend to decrease (Stanton 2005). Aboveground productivity determined in Louisiana ranged from 990 to 2,318 g dry mass/m^2/year (Hopkinson et al. 1978). Belowground, *Phragmites* produces roots and rhizomes that often extend greater than 50 cm below the soil surface (Windham 2001). This allows *Phragmites* to utilize resources unavailable to many other common marsh species, put more energy toward aboveground growth, capture more light for photosynthesis, and ultimately outcompete and displace other marsh species (Stanton 2005).

6.5.4.3 Fauna

The fauna of *Phragmites* marshes has been the focus of few studies, particularly in the GoM. Much of the available data on benthic, epibenthic and nektonic fauna associated with *Phragmites* has been collected in the northeastern United States, but it is not clear how widely the results of these studies may apply to *Phragmites* environments in other geographic regions (Meyerson et al. 2000). Nevertheless, there are likely to be similarities in faunal assemblage structure in GoM *Phragmites* marshes when compared with GoM smooth cordgrass and needlerush marshes.

Phragmites australis occurs as various haplotypes with different growth forms and habits (Howard et al. 2008) that vary with respect to potential effects on faunal assemblages. Generally, the most important impacts of *Phragmites* in coastal wetlands occur in association with a very robust Eurasian haplotype that has an aggressive growth habit (Burdick and Konisky 2003; Philipp and Field 2005; Howard et al. 2008). The invasive Eurasian variety of the common reed is very productive in terms of both above and below ground biomass generated annually (Windham 2001), which alters soil properties, increases elevation, and reduces microtopographic relief of intertidal wetlands (Windham and Lathrop 1999).

Posey et al. (2003) found that while vegetation type (i.e., smooth cordgrass or *Phragmites*) had a detectable effect on benthic invertebrate assemblage structure, microhabitat characteristics, such as local topography, had a stronger relation to faunal abundance patterns. Infaunal and epifaunal assemblages in high and low intertidal salt marsh (Stout 1984) and mangroves (Nagelkerken et al. 2008) are often distinctly different depending on the degree and duration of tidal flooding. Angradi et al. (2001) found that invertebrate abundance and assemblage composition did vary with distance from the marsh edge of *Phragmites* marsh in southern New Jersey. It is assumed that benthic community structure in Gulf *Phragmites* marshes is influenced to some degree by tidal inundation.

Angradi et al. (2001) found that invertebrate taxon richness was significantly higher in *Spartina alterniflora* marsh compared with *Phragmites* marsh. Moreover, dominance by the most abundant taxa was greater in *Phragmites* marsh (>85 %) at most sampling locations, also indicating lower benthic diversity in *Phragmites* (Angradi et al. 2001). Yuhas et al. (2005), however, found no clear pattern of difference in taxa abundance and richness comparing *Phragmites* and *S. alterniflora* marshes in New Jersey, though they only sampled creek bank and the marsh edge and not the interior vegetated marsh. Taxonomic diversity of invertebrate assemblages in GoM *Phragmites* marshes is likely to vary by location and time of year.

Table 6.18 lists the most abundant invertebrates found in *Phragmites* systems. Infaunal assemblages in *Phragmites* marsh are broadly similar to those of salt marsh. For example, Angradi et al. (2001) found that the most abundant infauna in *Phragmites* included oligochaetes and the polychaete *Manayunkia aestuarina*. Fell et al. (1998) sampled sites along the Connecticut River and found that certain high-marsh invertebrates (snails, amphipods and isopods) were common to abundant in salt marshes with and without *Phragmites*. Posey et al. (2003) found that *Phragmites* marshes in the Chesapeake Bay were numerically dominated by the polychaetes *Capitella capitata*, *Hobsonia florida* and *Laeonereis culveri*, oligochaetes, and chironomid fly larvae. They found that a typical mesohaline assemblage numerically dominated the infaunal community. Yuhas et al. (2005) collected benthic samples in *Phragmites* marshes in New Jersey, and found that oligochaetes were the most abundant infauna, comprising 24.4 % of all collected individuals.

Where present, intertidal *Phragmites* habitats along GoM shorelines can be expected to provide habitat for populations of locally occurring epifauna and nekton as is the case in the northeastern United States, but there is considerable debate over the relative habitat value of *Phragmites* compared to other marsh types. Able and Hagan (2000) examined decapod

Table 6.18. Abundant Invertebrates in *Phragmites* Systems

Infauna	Epifauna
Oligochaeta	Blue crab (C)
Capitella capitata (P)	Fiddler crab (C)
Hobsonia florida (P)	Mud crab (C)
Laeonereis culveri (P)	
Amphipoda (C)	
Isopoda (C)	
Chironomidae (Di)	
Fell et al. (1998), Angradi et al. (2001), Posey et al. (2003), Yuhas et al. (2005)	Able and Hagan (2000)

Key: *C* crustacean, *Di* dipteran, *P* polychaete

crustacean use of *Phragmites* and *S. alterniflora* marsh in the brackish water reaches of the Mullica River (NJ). Fiddler crabs (*Uca* spp.) and mud crabs (*Rhithropanopeus harrisii*) were more abundant in *Phragmites* (Able and Hagan 2000). Angradi et al. (2001) found that a *Spartina* marsh had greater production of benthic infauna than a *Phragmites* marsh, with higher overall abundance of benthic invertebrates. Posey et al. (2003) found only a small effect on faunal abundance patterns, with most species exhibiting slightly higher mean density in smooth cordgrass compared to adjacent *Phragmites* marshes. It has been suggested that fewer refugia from predators during high tide in *Phragmites* marsh may explain some of the differences in faunal abundance and community structure in comparison to *Spartina* habitat (Angradi et al. 2001).

The relative value to nekton of intertidal areas dominated by *Phragmites* compared to other marsh plant species remains unclear. Samples collected by trawling adjacent to subtidal creeks (Grothues and Able 2003) or using block net-type gear in intertidal creeks (Kimball et al. 2010) or on intertidal vegetated edges (Meyer et al. 2001) have not detected substantial differences in nekton assemblages in *Phragmites*-dominated sites compared to marsh habitats dominated by other plant species (e.g., *Spartina*). Also, the gut contents of at least one resident fundulid fish species collected from the inundated surface of *Phragmites* marshes contained similar invertebrate prey items as those foraging in marshes dominated by other vegetation types (Fell et al. 1998) and stable isotope analyses of several transient species of estuarine fishes have shown that *Phragmites* contributes to trophic support (Wainright et al. 2000; Litvin and Weinstein 2003; Weis and Weis 2003; Mendoza-Carranza et al. 2010; Weinstein et al. 2010).

Other studies have revealed a somewhat different picture. For example, Raichel et al. (2003) found that adults of the fundulid fish, *Fundulus heteroclitus*, were equally abundant in marshes dominated by *Phragmites* compared to those dominated by *Spartina*, but larvae and early juveniles of the fish were significantly less abundant in *Phragmites*. Furthermore, they also found that abundance of potential prey resources (e.g., copepods and other small crustaceans) for these early stages of resident nekton was significantly lower in *Phragmites* relative to *Spartina* marsh.

A key element that seems to drive the observed differences between *Phragmites* reed beds and other marsh types with respect to effects on benthic and nektonic fauna is that the robust rhizome and root growth of *Phragmites* affects elevation, microtopographic features and hydrologic characteristics of the intertidal marsh (Weinstein and Balletto 1999; Osgood et al. 2003; Buchsbaum et al. 2006). Many fundulid fishes spawn in intertidal marshes and often rear their young in shallow intertidal pools and puddles on the marsh surface. The robust growth of rhizomes and roots in *Phragmites* beds raises and flattens the marsh surface (Figure 6.134), reducing the availability of intertidal spawning and rearing sites for resident fishes (Able et al. 2003; Hunter et al. 2006) as well as the production of early life stages (Hagan et al. 2007).

Jivoff and Able (2003) used an otter trawl to sample tidal creeks adjacent to *Phragmites*- and *Spartina*-dominated marshes near high tide and found a tendency toward greater abundance of adult and fewer small recruit blue crabs (*Callinectes sapidus*) associated with *Phragmites* beds. They suggested that the observed differences in size-specific abundance were due to the effect of marsh surface vegetation type on high tide use of the marshes. Specifically, they proposed that smaller blue crabs made greater use of the less densely vegetated and more tidally inundated *Spartina* marsh surfaces compared to *Phragmites* sites. Of course, this assumes that greater abundance of blue crab life stages in tidal creeks at high tide occurs as a result of the inaccessibility of the marsh surface.

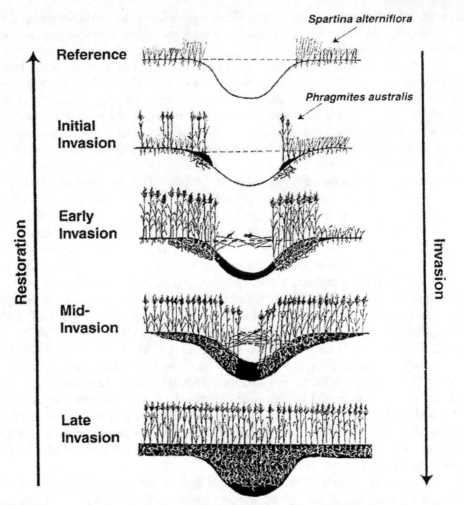

Figure 6.134. Changes in the micro-topographic features of the vegetated marsh surface with the growth and expansion of *Phragmites australis* as it replaces *Spartina alterniflora*, filling in small intertidal creeks and aquatic microhabitats and flattening the marsh surface (from Able et al. 2003; republished with permission of Springer Science and Bus Media BV, permission conveyed through Copyright Clearance Center, Inc.).

Questions regarding the quality of potential food resources derived from *Phragmites* production also have been raised. Stable isotope studies conducted primarily in New Jersey have indicated that an abundant resident fundulid fish (*Fundulus heteroclitus*) derives most of its nutrition for growth, reproduction and survival from a combination of primary producers including benthic microalgae and extant marsh grasses such as *Spartina* and *Phragmites*. However, in *Phragmites* reed beds there is reduced production of benthic microalgae due to shading effects of the robust aboveground growth of the reed relative to other marsh grasses (Currin et al. 2003). Consequently, the food webs associated with *Phragmites* are more dependent on detrital pathways based on reed decomposition.

Recent stable isotope studies have shown that there are potentially important differences in the allocation of lipids and free fatty acids when fish depend on food webs based on *Phragmites* production rather than benthic microalgae or other marsh plants. In particular, triacylglycerols which are essential for reproduction and survival during times when food resources may be limited (e.g., winter) are substantially lower in tissue samples of not only resident fundulids (Weinstein et al. 2009), but also certain marsh transient species such as *Morone americana* (white perch) (Weinstein et al. 2010). These findings have led some to the conclusion that *Phragmites* reed beds provide an inferior source of trophic support for nekton compared to food webs based on noninvasive primary producers.

Although the studies described in this section were conducted in marshes along the Atlantic coast, very similar species of benthic invertebrates and nekton, including a sibling suite of fundulid (*Fundulus grandis*, *F. pulvereus*, *F. jenkinsi*) species in the northern GoM (Figure 6.106) could reasonably be expected to respond to *Phragmites* reed beds in ways that are similar to responses observed in the common fauna along the U.S. Atlantic coast. *Fundulus jenkinsi* (salt marsh topminnow) may be of particular interest in this regard because it is a federally listed species of concern in the GoM and is sensitive to the types of physical marsh surface features (e.g., stem density, bank slope, and marsh elevation) (Lopez et al. 2010a, b, 2011) that are most affected by the common reed *Phragmites australis*. At a broader spatial scale, faunal assemblages of *Phragmites* marshes within the GoM are likely sensitive to their location within the coastal landscape, so that the composition and functioning of the faunal components are likely influenced by connectivity to adjacent habitats (Partyka and Peterson 2008). However, the overall contribution of *Phragmites* to secondary production in coastal areas has not been adequately investigated.

6.5.5 Seagrass Meadows

Seagrasses are unusual among the vascular flowering plants (i.e., angiosperms) in being entirely restricted to underwater marine habitats (den Hartog 1970). The more general term *submerged aquatic vegetation* or SAV is occasionally used interchangeably, although with the understanding that it can refer to freshwater species as well. Species occurring primarily in brackish or fresh water are not considered in this discussion.

Evolutionarily, the seagrasses are not true grasses (i.e., family *Poaceae*), but instead represent a diverse taxonomic group of four phylogenetically related plant families all belonging to the order Alismatales (Green and Short 2003; APG III 2009; Stevens 2001). There are approximately 60 known species of seagrasses worldwide with the majority of species being placed in three primary families: Hydrocharitaceae (3 genera with 17 species), Cymodoceaceae (5 genera and 16 species), and Zosteraceac (2 genera with 14 species). The small family Posidoniaceae is represented by a single genus (Posidonia) with only 2 species (Green and Short 2003). Some species of *Ruppia* (family Ruppiaceae) are treated as seagrasses by various authors, although they do not typically occur in the higher salinity waters considered in this summary.

6.5.5.1 Dominant Forcing Functions

Yáñez-Arancibia and Day (2004) divide the GoM into several different ecological regions and subregions based on the interactions of various physiographic, oceanographic, and biogeographic features including climate, geomorphology, freshwater input, and coastal drainage patterns (i.e., hydrologic units), physical chemistry, wildlife, estuarine vegetation, and human influences (Yáñez-Arancibia and Day 2004; Wilkinson et al. 2009). A description of seagrass communities of the South Florida/Bahamian Atlantic Marine Ecoregion, Northern GoM Marine Ecoregion (Eastern Gulf Neritic, Mississippi Estuarine, Texas Estuarine), and Southern GoM Marine Ecoregion, as well as their distribution, are provided below.

Broad patterns in species composition and the spatial positioning of seagrass beds are apparent across the ecoregions of the GoM. These patterns generally correlate with latitude, although many other complex factors likely contribute, including both large-scale and local differences in geomorphology, salinity, and hydrology.

6.5.5.2 Vegetation

6.5.5.2.1 Structure and Zonation

Seagrass species exhibit autecological (relationship of an individual species to its environment) differences in their natural history that affect their spatial distribution within beds. Where multiple seagrass species co-occur, a general pattern of zonation can be observed. For example, a survey of SAV distribution in East Bay (Bay County, Florida) conducted during 2011 documented three seagrass species. *Halodule wrightii* (shoalweed) dominates the SAV community closest to shoreline, and is most frequently found at depths of 1 to 3 ft. *Thalassia* becomes prevalent at depths of 3 to 6 ft. *Syringodium filiforme* (=*Cymodocea filiformis*) (manatee grass) is often interspersed with *Thalassia* at shallower depths, but becomes the dominant species at depths greater than 5 ft. No SAV was observed at depths greater than 8 ft. *Halophila engelmanni* (Engelmann's seagrass) is often found at great depths. In the Big Bend area of Florida, it can be found in monotypic stands away from the primary grass beds down to a depth of 20 m (66 ft) (Continental Shelf Associates, and Martel Laboratories 1985, cited in Zieman and Zieman 1989). *Halophila decipiens*, another species adapted to low-light conditions, covers approximately 20,000 km^2 (4,900,000 ac) of seagrass habitat off the west coast of Florida (Hammerstrom et al. 2006). Seagrass meadows typically contain a variety of rhizophytic and drift algae. Mattson (2000) summarized macroalgae associated with the Big Bend area of Florida, including *Caulerpa* spp. (rhizophytic forms) and *Hypnea* spp. (drift algae).

The landscape position of seagrass beds differs among the ecoregions. On the west coast of Florida in the South Florida/Bahamian Atlantic and Eastern Gulf Neritic ecoregions (Yáñez-Arancibia and Day 2004; Wilkinson et al. 2009) seagrasses are frequently found fronting the GoM, especially around the Florida Bay, Springs Coast, and Big Bend areas and also the eastern Florida Panhandle including Apalachee Bay (Yarbro and Carlson 2011; Onuf et al. 2003; Zieman and Zieman 1989; Iverson and Bittaker 1986). Moving westward along the coastline of the Panhandle, seagrasses gradually become more associated with sheltered embayments and areas behind protective barrier islands (e.g., Apalachicola Bay, St. Joseph Bay, St. Andrews Bay, Choctawhatchee Bay, and Santa Rosa Sound) (Yarbro and Carlson 2011). In the Mississippi Estuarine Ecoregion (Yáñez-Arancibia and Day 2004), which includes all of Alabama, Mississippi, and Louisiana, seagrasses are restricted to areas behind barrier islands (Onuf et al. 2003). Similarly, seagrasses are also found in the protective bays and coastal lagoons in the Texas Estuarine Ecoregion where no beds occur in the Gulf proper (Onuf et al. 2003). In the Southern GoM Ecoregion and along the northwestern coast of Cuba, seagrasses occur in several embayments as well as in the Gulf itself (Onuf et al. 2003).

There also are differences in species composition of seagrass beds among the subregions. *Thalassia*, *Halophila*, and *Syringodium* tend to be much more common at lower latitudes in Florida and Texas, and these species gradually become less prevalent as one moves northward into Alabama, Mississippi, Louisiana, and north Texas (i.e., the Mississippi Estuarine Ecoregion and portions of the Texas Estuarine Ecoregion). In this broad central GoM area, *Halodule wrightii* predominates and the three other species are largely absent. When present, they generally represent a minor component of the seagrass community, unlike the southern areas of Florida and Texas, and Mexico and Cuba, where they occur abundantly. This distributional pattern based on latitude probably reflects a number of various physiographic processes. Examples include regional differences in climate, salinity, and hydrology (e.g., variations in freshwater input especially from the Mississippi River), turbidity (which relates to hydrology), the geochemistry and texture of bottom sediments (muddy silts in the upper Gulf near the Mississippi River versus gravel, shell, and sands in the Eastern Gulf Neritic), shelf geomorphology, nutrient loads, and geologic histories (Wilkinson et al. 2009; Onuf et al. 2003). Zieman and Zieman (1989) and references therein report on species-specific differences in seagrasses that are related to species substrate preferences, depth and light regimes, and salinity tolerance; all of these factors are likely responsible for current day distributional patterns in the GoM.

6.5.5.2.2 Distribution

Seagrass beds in the South Florida/Bahamian Atlantic Ecoregion are characterized by *Thalassia testudinum*, *Syringodium filiforme*, and *Halodule wrightii*; *Halophila* species (e.g., *H. engelmanni* and *H. decipiens*) are abundant primarily in deeper or more turbid waters (Yarbro and Carlson 2011). This ecoregion includes Florida Bay and Florida Keys, and covers nearly 6,000 km^2 (1,480,000 ac) or over 55 % of the seagrasses in Florida's GoM coastal waters (Onuf et al. 2003). Distributions and abundances of seagrasses in Florida Bay and the Florida Keys have been studied extensively; Fourqurean et al. (2002) described the results of three monitoring programs focused on changing habitat conditions and die-off of some seagrasses in that area. Seagrass communities are similar in northwestern Cuba, and are dominated by *Thalassia*, *Syringodium*, *Halophila engelmanni*, and *Halophila decipiens*; perhaps 2,000 km^2 (494,000 ac) of grassbeds occur in this area (Onuf et al. 2003).

In the Eastern Gulf Neritic Ecoregion of the GoM, *Thalassia testudinum*, *Syringodium filiforme*, and *Halodule wrightii* are the most frequently encountered species of seagrass (Zieman and Zieman 1989; FNAI 2010; Yarbro and Carlson 2011). *Halophila engelmanni* is generally considered an uncommon and minor component of marine SAV beds in the northern reaches of the Florida Gulf Coast, where it is often found intermixed with other species (Zieman and Zieman 1989; FNAI 2010; Yarbro and Carlson 2011). Approximately 15,864 ha (39,200 acres) of SAV have been mapped in Florida waters, north of Crystal River to Escambia Bay (Yarbro and Carlson 2011). SAV is most abundant in the Big Bend area and in the St. Andrew Bay system.

Along the northern Gulf in the centrally located Mississippi Estuarine Ecoregion (Yáñez-Arancibia and Day 2004), *Halodule wrightii* predominates as the major species of seagrass (Barry A. Vittor and Associates, Inc. 2004, 2009; Onuf et al. 2003). The other species (e.g., *Thalassia*, *Syringodium*, and *Halophila*) reach the northern extent of their distribution and are generally rare in occurrence (Onuf et al. 2003). Minor areas of *Thalassia testudinum* are documented in Louisiana, Mississippi, and Alabama. In Alabama only one small extant population (0.02 ha) is currently known to exist (Barry A. Vittor and Associates, Inc. 2004, 2010; Heck and Bryon 2005). *Syringodium filiforme* and *Halophila engelmanni* have been recorded from Mississippi and Louisiana waters (Onuf et al. 2003); these two species have not been documented in Alabama and likely do not occur in the state (Kral et al. 2011). Approximately 14,487 ha (35,747 acres) of SAV occur in this subregion: the greatest coverage is found in the area of the Chandeleur Islands, along the eastern coast of Louisiana (NOAA 2004). Seagrass abundance in this ecoregion is highly variable, but all areas have experienced significant declines in SAV during the past 50 years, with only occasional periods of re-growth or expansion.

Within the Texas Estuarine Ecoregion, there is a gradual shift in species composition of seagrass beds moving southward down the GoM coastline. In the upper reaches of the ecoregion, *Halodule* is practically the only species present, continuing the pattern seen in the adjacent Mississippi Estuarine Ecoregion. Along the upper Texas coast *Thalassia testudinum* is found at only a single location near the west end of Galveston Bay. Moving southward, there is a transition of *Halodule*-dominated beds to *Thalassia*, which becomes increasingly prevalent in the central and southern Texas coast. At Aransas Pass, for example, *Thalassia* is the dominant species and comprises nearly 25 % of the bay bottom. *Thalassia* is even more dominant along the lower Texas coast. At the southern end of Laguna Madre, it comprises over 90 % of the seagrass beds near the Gulf outlet there (Onuf et al. 2003). *Halophila engelmanni*, along with *Ruppia maritima*, is found sporadically across the entire Texas coast. Seagrasses in Texas are spatially confined to protected areas located behind the state's coastal barrier islands and are not found seaward along its frontage with the GoM. While roughly 87,580 ha (216,410 acres) of SAV occur in Texas coastal waters, most (over 95 %) are found south of Matagorda and Galveston Bays (Handley et al. 2007); approximately 1,310 ha (3,237 acres) of SAV occur in those embayments. Estimates of seagrasses have varied widely: NOAA (2004) estimated that there are nearly 123,834 ha (306,000 acres) of seagrasses in Texas, or approximately 30 % more than described by Onuf et al. (2003) and 40 % more than reported by Handley et al. (2007).

According to Onuf et al. (2003), seagrass distributions within the Southern GoM Ecoregion were described mainly by studies that date to the 1950s. The same five genera reported in the rest of the GoM (*Thalassia, Syringodium, Halodule, Halophila, Ruppia*) also occur along the coast of Mexico. In the State of Tamaulipas, *Halodule wrightii* is the dominant species in the hypersaline Laguna Madre, and comprises approximately 18 % (35,700 ha) of the Lagoon's extent. Tabasco and Veracruz contain fewer seagrasses: Onuf et al. (2003) reported that only *H. wrightii* and *Ruppia maritima* occur in the coastal estuaries of Tabasco, probably due to heavy sediment loads and elevated turbidities caused by discharges from the Grijaval-Usumacinta River system. *Thalassia testudinum, Syringodium filiforme, Halodule wrightii, Halophila engelmanni*, and *Halophila decipiens* are found in association with a large coral reef system that fronts the state's main port. Shallower waters contain primarily *Halodule wrightii*, while *Halophila decipiens* has been found to a depth of over 10 m. The Yucatán Peninsula contains extensive seagrass beds populated by *Thalassia testudinum, Syringodium filiforme*, and *Halodule wrightii*; *Halophila* is least abundant. *Thalassia* is especially dominant in areas around coral reef lagoons.

6.5.5.3   Fauna

Seagrasses are important as both habitat for adult animals, as well as a nursery habitat for post-larval and juvenile individuals. The faunal assemblages found in these seagrass beds consist of groups of animals with many different life forms and ecological characteristics. The assemblages have been subdivided into several categories based on where the animal spends most of its time (Kikuchi and Peres 1977). Epifaunal species live on leaves, and include microfauna and meiofauna, sessile fauna, mobile creeping and walking epifauna (e.g., gastropods), and swimming epifauna (e.g., caridean shrimp), which may also be considered as nekton. Infaunal species include burrowers and tube–dwellers as well as those animals creeping or crawling at the sediment water interface (Orth et al. 1984). The seagrass epifauna and infauna are primarily composed of crustaceans, mollusks, and polychaetes. The final category of seagrass fauna is made up of mobile nektonic species living freely under and over the leaf canopy (e.g., fishes, marine mammals, and marine reptiles).

Seagrasses and macroalgae commonly form an extensive bottom cover in the brackish and saline shallow coastal lagoon bay systems that are prominent features throughout much of the GoM (Sheridan and Minello 2003; Contreras-Espinosa and Warner 2004; Rozas et al. 2012). Seagrass meadows rarely occur in physical isolation from other coastal wetland habitats so it is not surprising that their faunal assemblages include many species in common with other adjacent coastal wetlands such as mangroves, tidal marshes, intertidal and subtidal flats, or

Figure 6.135. Generalized illustration of the variety of nekton species and size classes associated with seagrass meadows. The depiction is from an Australian seagrass habitat but can be generalized to the GoM (McKenzie et al. 2006–2012; www.seagrasswatch.org/seagrass.html, reprinted with permission).

with coral reefs, particularly in the southern GoM (e.g., Yáñez-Arancibia et al. 1993; Ortiz and Lalana 2005).

Most seagrass habitat is largely subtidal and, unlike intertidal mangrove and marsh habitats, is always accessible to aquatic fauna. Consequently, seagrass assemblages include a diverse group of benthic, epibenthic, and nektonic species and size classes of organisms (Figure 6.135). Primary production in seagrass ecosystems can be quite high reaching up to 8 g C/m^2/day for seagrasses alone (Zieman and Wetzel 1980). Studies of seagrass systems have also indicated the important role that epibenthic algae play in the total primary productivity of seagrass systems. These epiphytes have been found to be very productive and important sources of high-quality food for benthic marine consumers (Fry 1984; Kitting et al. 1984; Moncreiff et al. 1992; Williams and Heck 2001). The primary production of epiphytic algae can represent a substantial percentage of the total primary production of a seagrass meadow, sometimes matching or exceeding that of the seagrasses (Morgan and Kitting 1984; Mazella and Alberte 1986; Thom 1990; Williams and Heck 2001). Epiphytic algae have been found to be important determinants of epifaunal abundance (Hall and Bell 1988, 1993; Edgar 1990; Williams and Heck 2001). The high primary productivity of seagrasses and their associated epiphytic algae form the basis of complex food webs involving mammals, reptiles, fishes, crustaceans, mollusks, polychaetes, echinoderms, sponges, bryozoans, cephalochordates, and phytoplankton, as well as the algae and seagrasses themselves (Figure 6.136). Among the invertebrates, crustaceans and mollusks are very important groups in seagrass food webs, with taxa that play key roles in several trophic levels. Polychaetes also make up a high percentage of most seagrass faunal assemblages. The dominant fishes and natant decapod crustacean components of seagrass systems are strongly affected by seasonal recruitment patterns (Livingston et al. 1976), salinity (Arceo-Carranza and Vega-Cendejas 2009), proximity to inlets and passes (Reese et al. 2008), diel and tidal activity patterns of individual species (Sogard et al. 1989;

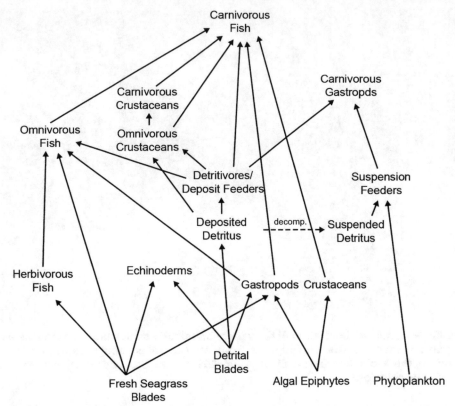

Figure 6.136. Generalized seagrass meadow food web (modified from Greenway 1995).

Hammerschlag and Serafy 2009) as well as a suite of short-term (Renaud 1986; Roth and Baltz 2009) and chronic water quality variables (Livingston 1984; O'Connor and Whitall 2007).

There is a paucity of available research related to the infaunal assemblages inhabiting the relatively isolated seagrass beds of Alabama, Louisiana, and northern Texas, which accounts for a lack of species-specific distributional information for the Mississippi and Texas Estuarine subregions. However, most seagrass research on benthic invertebrates in the northern Gulf is derived from areas of extensive seagrass meadows in Florida. In the northern GoM, these areas are represented by Gulf-fronting seagrass beds within the Eastern Gulf Neritic subregion (Figure 6.3), which includes a large portion of the Florida Gulf coast from its Springs Coast area northward through the Big Bend and westward along the northern panhandle to the vicinity of Escambia Bay near the Alabama–Florida stateline. This subregion is characterized by an extensive shelf system with nearshore substrates consisting primarily of sand, gravel, and shell with areas of limestone (Wilkinson et al. 2009). *Thalassia testudinum*, *Syringodium filiforme*, and *Halodule wrightii* are frequently encountered species in this subregion (Zieman and Zieman 1989; FNAI 2010; Yarbro and Carlson 2011).

Expansive seagrass beds that exist in the Eastern Gulf Neritic provide habitat for some organisms that are common to this subregion, but that may be rare in other areas of the northern Gulf. Examples are the bay scallop (*Argopecten irradians*), the green sea urchin (*Lytechinus variegatus*) (Figure 6.102a), and the sea star (*Echinaster serpentarius*) (Zieman and Zieman 1989). The limited distributions of these and other organisms in the northern Gulf is likely influenced by salinity, and the fact that a large portion of the northern Gulf is heavily impacted by freshwater input from the Mississippi River. Bay scallops for example, are usually associated with seagrass beds in salinities greater than 25 ppt (Nelson 1992).

A variety of crustacean taxa representing both infaunal and epifaunal groups are prevalent within seagrass meadows (Table 6.19). Copepods, ostracods, and amphipods comprise the majority of the infaunal crustacean taxa. Epifaunal crustaceans include several taxa of isopods,

Table 6.19. Common Invertebrates in GoM Seagrass Systems

Taxon	Marine ecoregion				
	South Florida	Eastern Gulf Neritic	Mississippi Estuarine	Texas Estuarine	Southern GoM
Annelda					
Arabella iricolor	X	X	X	X	X
Armandia agilis	X	X	X	X	X
Bhawania goodei	X	X	X	X	X
Capitella capitata	X	X	X	X	X
Cossura candida					X
Glycinde solitaria		X	X	X	X
Laeonereis culveri	X	X	X	X	X
Malacoceros vanderhorsti	X	X	X	X	X
Parandalia vivanneae					X
Pista cristata	X	X	X	X	X
Spio pettibonae	X	X	X	X	X
Streblospio benedcti	X	X	X	X	X
Mollusca-Bivalvia					
Americardia guppyi	X	X	X	X	X
Argopecten gibbus	X	X	X	X	X
Argopecten irrdians		X	X	X	
Macoma constricta	X	X	X	X	X
Pitar simpsoni	X	X			X
Semele bellastriata	X	X	X	X	X
Tagelus divisus	X	X	X	X	X
Tellina tampaensis					X
Mollusca-Gastropoda					
Bulla striata	X	X	X	X	X
Caecum plicatum	X	X	X	X	X
Calliostoma pulchrum	X	X	X	X	X
Crepidula planum	X				X
Haminoea glabra					X

(continued)

Table 6.19. (continued)

Taxon	Marine ecoregion				
	South Florida	Eastern Gulf Neritic	Mississippi Estuarine	Texas Estuarine	Southern GoM
Neritina virginea	X	X	X	X	X
Nitidella nitida	X				X
Petalifera ramose	X				X
Tectura antillarum	X				X
Mollusca-Cephalopoda					
Octopus maya					X
Pickfordiateuthis pulchella	X				X
Crustacea-Decapoda					
Alpheus heterochaelis	X	X	X	X	X
Armases cinereum	X			X	X
Callinectes sapidus	X	X	X	X	X
Cardiosoma guanhumi	X	X		X	X
Clibanarius vittatus	X	X	X	X	X
Dyspanopeus texana	X	X		X	X
Macrobrachium acanthurus	X	X	X	X	X
Tozeuma carolinense	X	X	X	X	X
Crustacea-Others					
Americamysis almyra	X	X	X	X	X
Ampelisca holmesi	X				
Apocorophium louisianum	X	X	X	X	X
Batea catharinensis	X	X	X	X	X
Cyathura polita		X	X	X	
Erichthonius brasiliensis	X	X	X	X	X
Leucothoe spinicarpa	X	X	X	X	X
Paracereis caudata	X				
Photis macromanus		X	X	X	
Echinodermata					
Echinaster serpentarius	X	X	X	X	X
Lytechinus variegatus	X	X	X	X	X

References: Hall and Bell (1993), Heck and Valentine (1995), Leber (1985), Lewis (1984), Livingston (1984), Mendoza-Carranza et al. (2010), Sheridan (1997), Stoner (1980a), Virnstein et al. (1983), Zieman and Zieman (1989)

copepods, ostracods, and amphipods (Hemminga and Duarte 2000). Hall and Bell (1993) found that harpacticoid copepods and nauplii were the most abundant meiofaunal taxa on seagrass blades at Egmont Key, FL. Based on the apparent ubiquitous nature of this taxonomic group, it is likely that harpacticoid copepods play an important role in seagrass ecosystems in the GoM. As previously noted, crustaceans play key roles in multiple trophic levels within seagrass systems, and in addition to the important epiphytic taxa, larger mobile crustaceans (particularly decapods such as shrimp and crabs) are also common within seagrass meadows. Crustaceans derive their nutrition from a wide range of food sources available in the seagrass beds. Some taxa (such as brachyuran crabs) are at least partially herbivorous, consuming live seagrass tissue (including both leaves as well as root and rhizome material) and epiphytic algae (Leber 1985; Livingston 1984; Woods and Shiel 1997), while others are largely predatory. The carnivorous pink shrimp (*Farfantepenaeus duorarum*), for example, is a dominant decapod predator in seagrass meadows (Figure 6.103b), preying on caridean shrimp, amphipods, bivalve mollusks, polychaetes, gastropods, crabs, and detritus (Livingston 1984). The majority of smaller crustaceans inhabiting seagrass beds rely on algae or detrital particles as a food source (Klumpp et al. 1989).

Crustaceans are an important food source for fishes foraging in seagrass beds, and may play a prominent role in the energy transfer from primary producers to higher trophic levels, via the food web that links microalgae and detritus through epifaunal crustaceans to smaller fish, and ultimately to larger fish predators (Edgar and Shaw 1995). Crustaceans are very important in another aspect as well; several decapod species that spend at least part of their life cycle in seagrass beds are highly valued as food sources for humans. These species are the basis for extensive fisheries in warm-temperate and tropical areas throughout the world. In the GoM, important commercial fisheries are in place for penaeid shrimp and blue crabs (*Callinectes sapidus*). Among the crustaceans associated with seagrasses, decapods have received the most scientific attention, due to their ecological significance and commercial value (Perry 1984; NMFS 1988; Orth and van Montfrans 1984; Olmi and Orth 1995; Heck et al. 2001). Seagrass beds have long been recognized as an important nursery habitat, providing structure where settlement and growth of decapod post-larvae and juveniles may occur (Lewis and Stoner 1981; Lewis 1984; Zieman and Zieman 1989; Fonseca et al. 1996; Orth et al. 1996; Bell et al. 2001; Heck et al. 2003; Dawes et al. 2004).

Mollusks (gastropods and bivalves) also make a significant contribution to the fauna of seagrass beds in the GoM (Table 6.19). Gastropod species employ a variety of feeding strategies. While most feed on microalgae and detritus particles present on the sediment and leaf surfaces (Hemminga and Duarte 2000), others are carnivorous, feeding on other members of the faunal community. Bologna and Heck (1999), for example, observed large predatory gastropods (e.g., whelks [*Busycon* spp.]), horse conchs (*Pleuroploca gigantea*), and tulip snails (*Fasciolaria* spp.) preying on bay scallops in seagrass beds in the northern GoM. Bivalves can be suspension feeders, deposit feeders, or both. True suspension feeders collect food by filtering particles from the water column. The availability of this food depends on waterflow and water-column mixing. Research suggests that suspension-feeding bivalves grow faster when near-bottom water velocities are higher (Hemminga and Duarte 2000). This waterflow is reduced inside seagrass canopies. Therefore, bivalves inhabiting the interior portions of seagrass beds might experience a reduced food supply. The reduction of food supply, and subsequent lower growth rate of suspension feeders, is potentially balanced by positive aspects of living in the seagrasses. Possible positive aspects include decreased chance of dislodgement of the animals during storms (Reusch and Chapman 1995), a reduction of predation intensity within the seagrass canopy (Irlandi and Peterson 1991), and larger degree of sediment stability within the seagrass beds (Irlandi 1994, 1996).

While many studies have illustrated the importance of seagrass beds to the survival rates of infaunal bivalves (Blundon and Kennedy 1982; Peterson 1982, 1986; Coen and Heck 1991), less is known about the effects that the presence of these bivalves have on the seagrasses themselves. Some more recent studies have suggested a mutualistic relationship between suspension feeding bivalves and seagrasses. Peterson and Heck (2001) tested two possible beneficial effects of the presence of tulip mussel aggregates (*Modiolus americanus*) in *Thalassia testudinum* beds. First, the suspension-feeding bivalves filter particulates from the overlying water column, and excrete nutrients in the form of ammonium and phosphorus, enriching the sediments in the seagrass bed. Additionally, mussel aggregates provide increased structural complexity that may provide a refuge from predation for epiphytic grazer species (e.g., small gastropods and amphipods). Higher densities of grazer species may lead to increased grazing activity on seagrass epiphytes, which consequently, could lead to an increase in light absorption of seagrass leaves. These two possible mechanisms (nutrient enrichment and increased light absorption), both due to the presence of suspension-feeding bivalves, can positively affect seagrass productivity.

As with commercially important crustaceans, which spend at least a portion of their life cycle associated with seagrass beds, some mollusk species are prized for their value as a source of seafood. An example is the bay scallop, *Argopecten irradians*. An important commercial and recreational fishery is established for bay scallops in the GoM (primarily in the Big Bend area of Florida). Scallops are intimately tied to seagrass systems, which they utilize as a primary settlement site as well as a refuge from predation (Gutsell 1930; Eckman 1987; Bologna and Heck 1999).

Polychaetes are an important part of the faunal community within all soft bottom habitats, including seagrass meadows (Table 6.19). Although there are no commercially important polychaete species and the majority of seagrass research focuses on decapods and mollusks, polychaetes represent a key part of seagrass food webs, and make up a relatively high percentage of total seagrass faunal abundance (Stoner 1980a; Lewis and Stoner 1983). Due to their generally high fecundity, polychaetes can exhibit seasonal pulses of abundance in temperate habitats (Orth and van Montfrans 1984). Dominant polychaete families in GoM seagrass beds include Nereidae, Capitellidae, Syllidae, Spionidae, Cirratulidae, Terebellidae, Sabellidae, and Maldanidae (Gloeckner and Luczkovich 2008). Although the majority of polychaetes are burrowing infaunal species, some are epiphytic, building tubes on seagrass leaves. Polychaetes represent a variety of feeding guilds, including predators, but most are suspension or deposit feeders.

Seagrass meadows differ from emergent vegetated wetlands in a number of important features that may be reflected in their nekton assemblages. For example, certain groups of natant decapod crustaceans (e.g., Hippolytidae) and fishes (e.g., Syngnathidae) are more common and abundant in this habitat than in other coastal wetlands (Minello 1999). Seagrasses lack the rigid physical structure of emergent intertidal vegetation, so the habitat does not constitute a substantial physical barrier to either the movement of nekton or collecting gear (trawls or seines) normally used to sample nekton in estuaries. As observed in other coastal wetland habitats, differences in the sampling methodology of seagrass-dominated systems (Pérez-Hernández and Torres-Orozco 2000) and differences in targeted groups (e.g., fishes, decapod crustaceans, or species of commercial or recreational importance)

(Gilmore 1987) can have a strong influence on the results and conclusions of individual nekton studies. The numerically abundant resident species of nekton (e.g., Gobiidae, Syngnathidae, and Hippolytidae) are small and often cryptic in habit and so may be underrepresented in seagrass nekton samples that use conventional gears (Gilmore 1987; Zieman and Zieman 1989).

The widespread use of small quantitative sampling devices such as throw traps and drop samplers has been popular largely in the northern and eastern GoM, while most nekton samples from seagrass-dominated habitats in the southern GoM are collected by trawling or seining. Also, there are regional differences in the types of species included among nekton samples from seagrass meadows in the United States, where natant decapod crustaceans are included, compared to collections in the southern GoM, which have focused on fishes or species of commercial importance. Effects of this regional difference in sampling approach and emphasis may explain some principal differences in the dominant nekton families and species reported from marine ecoregions around the GoM (Table 6.20).

The most obvious feature of these data is the apparent lack of natant decapod crustaceans and greater number of abundant fish families in seagrass-dominated lagoonal systems along the Mexican coast. Even though it has been suggested that fishes comprise most of the nekton assemblages of lagoon-estuarine habitats in the southern GoM (Yáñez-Arancibia et al. 1994), natant crustaceans are generally not reported in many studies within this region. The use of smaller quantitative sampling gear types in the northern GoM results in the capture of smaller resident nekton that numerically dominate these systems, particularly natant crustaceans representing the families Hippolytidae and Palaemonidae. It appears that palaemonids tend to dominate the natant decapod crustacean assemblage in temperate seagrass meadows near tidal marshes (Minello 1999; Rozas et al. 2012), but hippolytid shrimps are the numerical dominants in tropical and subtropical portions of the GoM where marshes are not as extensive (Sheridan et al. 1997; Sheridan and Minello 2003).

In contrast, trawls and seines are the gears of choice for sampling nekton in the southern GoM, where fishes and natant decapod crustaceans tend to be reported in separate studies rather than as elements of a nekton assemblage. Natant decapod crustaceans are important and abundant components of seagrass nekton assemblages in the southern GoM, but their importance must be inferred from separate studies that have targeted decapod crustaceans or considered the diets of predatory fishes. For example, penaeid shrimps are the most valuable artisanal fisheries in the coastal lagoons of the southern GoM but focused studies on this group are relatively rare (Pérez-Castañeda and Defeo 2001; May-Kú and Ordóñez-López 2006). In the seagrass- and macroalgal-dominated lagoons of the southern GoM, the predominant penaeids are the spotted pink shrimp (*Farfantepenaeus brasiliensis*), the southern pink shrimp (*Farfantepenaeus notialis*) and the northern pink shrimp (*Farfantepenaeus duorarum*) (Pérez-Castañeda and Defeo 2001; May-Kú and Ordóñez-López 2006). In contrast, the northern brown shrimp (*Farfantepenaeus aztecus*) and northern white shrimp (*Litopenaeus setiferus*) tend to be the most abundant penaeid species in areas of extensive salt marsh habitat (e.g., Louisiana and Texas) (Minello 1999), and the northern pink shrimp (*Farfantepenaeus duorarum*) is most often associated with seagrass meadows wherever they occur (Figure 6.103b), from the Florida Keys through the northern GoM (Bielsa et al. 1983; Rozas et al. 2012).

The sparse information available on the nekton of the northwestern Cuban coast does not mention many natant decapod crustaceans as being abundant in seagrass meadows, but

Table 6.20. Nekton Species Comprising at Least 90 % of the Total Individuals in Seagrass Samples from the Indicated Marine Ecoregions of the GoM

	Marine Ecoregion				
	Florida Bay	Eastern Gulf Neritic	Mississippi Estuarine	Texas Estuarine	Campeche/ Yucatán Inner Neritic
Fishes					
Achiridae					
Achirus lineatus					X
Ariidae					
Ariopsis felis	X				
Atherinidae					
Menidia colei					X
Menidia penninsulae	X				
Menidia beryllina			X		
Belonidae					
Strongylura notata					X
Clupeidae					
Harengula jaguana					X
Cyprinodontidae					
Floridichthys carpio	X				
Floridichthys polyommus					X
Cyrprinodon antifrons					X
Cyprinodon variegatus			X		
Engraulidae					
Anchoa mitchilli	X				X
Fundulidae					
Lucania parva	X	X	X		X
Fundulus perisimilis					X
Gerridae					
Eucinostomus gula	X				X
Eucinostomus argenteus		X			X
Gobiidae					
Gobiosoma bosc			X		
Gobiosoma robustum	X			X	
Ctenogobius boleosoma		X		X	
Microgobius gulosus	X				
Haemulidae					
Haemulon bonariense					X
Mugilidae					
Mugil trichodon					X
Poeciliidae					
Heterandria formosa			X		
Poecilia latipinna			X		
Poecilia velifera					X
Sciaenidae					
Bairdiella chrysoura	X				

(continued)

Table 6.20. (continued)

	Marine Ecoregion				
	Florida Bay	Eastern Gulf Neritic	Mississippi Estuarine	Texas Estuarine	Campeche/ Yucatán Inner Neritic
Sparidae					
Lagodon rhomboides	X	X		X	X
Syngnathidae					
Syngnathus scovelli	X		X		
Tetradontidae					
Sphoeroides testudineus					X
Natant Decapod Crustaceans					
Portunidae					
Callinectes sapidus		X	X	X	
Callinectes similis				X	
Penaeidae					
Penaeidae - Unidentified				X	
Farfantepenaeus aztecus			X		
Farfantepenaeus duorarum	X	X		X	
Litopenaeus setiferus			X	X	
Palaemonidae					
Palaemonetes spp.			X	X	
Palaemonetes pugio		X	X	X	
Palaemonetes intermedius		X		X	
Palemon floridanus		X			
Periclimenes americanus	X				
Periclimenes longicaudatus		X			
Hippolytidae					
Hippolyte zostericola		X		X	
Thor floridanus	X				
Tozeuma carolinense		X		X	
References Cited:	8,13,14	6,11	3,5,7	2,4,10,12	1,9,15

Red shaded cells indicate the presence of dominant families at each location. The most abundant species reported in each family are indicated by an "X". Studies in southern GoM ecoregions reported only fishes but natant decapod crustaceans are important fisheries species in the region.
References: (1) Arceo-Carranza and Vega-Cendejas (2009), (2) Burfeind and Stunz (2006), (3) Kanouse et al. (2006), (4) King and Sheridan (2006), (5) La Peyre and Gordon (2012), (6) Livingston (1984), (7) Mairaro (2007), (8) Matheson et al. (1999), (9) Peralta-Meixueiro and Vega-Cendejas (2011); (10) Reese et al. (2008); (11) Rozas et al. (2012), (12) Sheridan and Minello (2003), (13) Sheridan et al. (1997), (14) Thayer et al. (1987), (15) Vega-Cendejas and Hernández de Santillana (2004)

identifies areas of macroalgal beds and coastal lagoons surrounded by mangroves as a principal habitat for portunid crabs, including *Callinectes sapidus* and *C. ornatus*. The same coastal lagoons are identified as important habitat for penaeid shrimps *Farfantepenaeus notialis* and *Litopenaeus schmitti* (Ortiz and Lalana 2005).

The smaller and most abundant natant decapod crustaceans reported from seagrass systems in the United States are rarely reported in nekton studies from the Mexican and Cuban coasts of the GoM. However, caridean shrimps, particularly the Hippolytidae, are reported to be among the most abundant crustaceans within SAV in the major lagoonal systems of the southwestern GoM (Negreiros-Fransozo et al. 1996). Also, the diets of common fishes, such as the sparid *Lagodon rhomboides* and the batrachoidid *Opsanus phobetron*, from GoM lagoons along the Yucatán Peninsula include caridean shrimps, presumably Hippolytidae or Palaemonidae (Canto-Maza and Vega-Cendejas 2007, 2008). Consequently, it is reasonable to infer that natant decapod crustaceans may be of greater importance among the nekton assemblages of seagrass meadows in the southern GoM than suggested by the literature on fish assemblages from these systems.

Seagrass meadows have different trophic dynamics than other coastal wetland habitats in that seagrass primary production is consumed by some nekton as both live and dead (detritus) material (Heck and Valentine 2006), whereas the trophic role of emergent marsh plant production in the support of nekton populations is largely through a detrital pathway in marshes and the role of mangrove primary production may not contribute significantly to trophic support of nekton (Beck et al. 2001). Some nekton, such as the pinfish (*Lagodon rhomboides*) (Livingston 1982; Montgomery and Targett 1992) and a few larger nektonic herbivores, such as green sea turtles (*Chelonia mydas*) and manatees (*Trichechus manatus*) consume live seagrass and contribute to recycling of nutrients and maintenance of seagrass productivity (Thayer et al. 1984; Heck and Valentine 2006).

Seagrass meadows have long been recognized as important foraging sites for the juveniles of predatory fishes, especially members of the Sciaenidae (drums) and Lutjanidae (snappers) (Gilmore 1987; McMichael and Peters 1989; Rutherford et al. 1989; Rooker et al. 1999), but these species are commonly found in other coastal wetlands as well, and so are not characteristic of seagrass meadows *per se*. The smaller species of fishes, such as members of the Gobiidae and Fundulidae (especially *Lucania parva*) and natant decapods such as shrimps in the family Hippolytidae, are widely distributed and abundant in seagrass habitats within all regions of the GoM (Table 6.20). Many members of the Syngnathidae (pipefishes), such as the dwarf seahorse (*Hippocampus zosterae*) (Figure 6.107b) and fringed pipefish (*Anarchopterus criniger*) are even more dependent upon seagrass habitats within the GoM and have been identified by some as species of special concern (Beck et al. 2000). It has been suggested that the refuge value of seagrass structure is a key feature in maintaining these assemblages of smaller nekton (Heck et al. 2003).

6.5.5.4 Ecosystem Services and Function

Seagrasses represent a valuable natural resource to human culture and society. The benefits provided by their ecological services are innumerable. Some studies have estimated the economic contribution of seagrasses to be worth 20 billion dollars annually (Orth et al. 2006; Costanza et al. 1997; Yarbro and Carlson 2011). Seagrasses also constitute a significant ecological and functional guild, one that serves many diverse roles in marine environments. Seagrasses act as essential nursery habitats for many economically important fish and shellfish species and thus are vital to recreational and commercial fisheries (Hemminga and Duarte 2000; Beck et al. 2001; Heck et al. 2003; Yarbro and Carlson 2011). They provide crucial food resources for waterfowl, sea turtles, fishes, and other wildlife (Hemminga and Duarte 2000). Seagrasses offer habitat for endangered marine species (Orth et al. 2006). Structurally, the rhizomes and roots of seagrasses can stabilize sediments (Orth et al. 2006; Hemminga and Duarte 2000) and seagrass beds provide shoreline protection *via* wave attenuation

(Koch et al. 2009). Seagrasses also play a crucial role in natural biogeochemical processes, receiving, transforming, and exporting various compounds trophically through marine ecosystems (Orth et al. 2006). Their presence contributes significant amounts of organic carbon to marine food webs in the form of detrital material (Orth et al. 2006), and as such, are an important component of nutrient cycling in the environment. A large portion of this carbon may be transported to the food-limited deep sea where it becomes a vital contributor of organic material to these systems (Suchanek et al. 1985).

Seagrass meadows are productive ecosystems. Estimates of primary production in seagrass meadows have indicated an average net production of approximately 1,012 g dry weight/m^2/year, when production of both above-ground and below-ground components are considered (Duarte and Chiscano 1999; Hemminga and Duarte 2000). This estimate of primary productivity places seagrass meadows among the most productive ecosystems in the biosphere (Hemminga and Duarte 2000). When these estimates are scaled to the estimated global cover of seagrasses, the result is a contribution of about 1.13 % of the total marine primary production (Duarte and Cebrián 1996; Hemminga and Duarte 2000). Unlike phytoplankton, where most of the primary production is used up in the marine system, much of the seagrass production is either stored in the sediments or exported to neighboring ecosystems (Duarte and Cebrián 1996; Duarte and Agusti 1998; Hemminga and Duarte 2000). It is currently believed that approximately 16 % of seagrass production is stored in the sediments, representing a net sink of carbon in the ecosystem. The carbon stored in the sediments annually by seagrasses is estimated to be in the order of 0.08×10^{15} g C/year (about 12 % of the total carbon storage in marine ecosystems) (Duarte and Cebrián 1996; Duarte and Chiscano 1999; Hemminga and Duarte 2000). Therefore, seagrass meadows represent important parts of the marine carbon cycle and are responsible for a significant portion of the net CO_2 uptake by marine biota.

The high primary productivity of seagrass systems provides an abundant supply of organic matter that can be used as the basic energy source for food webs (Zieman and Wetzel 1980; Williams and Heck 2001). When considering the secondary production provided by these seagrass systems, it is important to understand the important contribution that epiphytic and benthic algae make to the seagrass system in terms of their production. Stable isotope studies conducted over the past 10 to 20 years have led to a paradigm shift in our view of seagrass trophic dynamics. At one time, seagrasses were thought to be the most important material for secondary production. It is now believed that benthic microalgae are the primary source of organic matter to higher trophic levels in seagrass food webs (Fry 1984; Kitting et al. 1984; Morgan and Kitting 1984; Mazella and Alberte 1986; Kenworthy et al. 1987; Dauby 1989, 1995; Thom 1990; Moncreiff et al. 1992; Loneragan et al. 1997; Yamamuro 1999; Lepoint et al. 2000; Williams and Heck 2001; Mateo et al. 2006). Valentine and Duffy (2006) suggested that seagrass food webs contain two key conduits for the transfer of primary production to higher order consumers. Seagrass grazing ecosystems are characterized by moderate to intense grazing on living seagrass tissue (leaves and rhizomes) by abundant large vertebrate, and some invertebrate, herbivores. This grazing results in low seagrass biomass and a direct conversion of seagrass production into vertebrate biomass. Seagrass detrital ecosystems are primarily devoid of large vertebrates, and herbivory is dominated by small invertebrate grazers that feed preferentially on epiphytic algae. This strategy indirectly enhances seagrasses, resulting in high seagrass biomass, much of which enters the detrital food chain (Valentine and Duffy 2006).

Historically, most seagrass ecosystems, especially in the tropics, were believed to be seagrass grazing ecosystems. More recently, due in large part to human impacts, reductions in abundances of large vertebrate herbivores such as green sea turtles, sirenians (manatees and dugongs), and waterfowl, have decreased the importance of these large herbivores as primary consumers in seagrass systems (Jackson et al. 2001; Valentine and Duffy 2006). Consequently, many present-day seagrass systems are of the detritus-based type, with little seagrass production being grazed directly (Robertson et al. 1982; Chin-Leo and Benner 1991; Ziegler and Benner 1999; Cebrián 1999, 2002; Mateo et al. 2006; Valentine and Duffy 2006).

Decomposition in situ appears to be the most probable fate for both seagrass leaf detritus and below-ground (rhizomes and roots) production (Mateo et al. 2006). The total amount of seagrass production (both above- and below-ground) that is decomposed is generally large. Research suggests that the amount of seagrass detritus that is transferred to decomposers and detritivores tends to be larger than for many other aquatic and terrestrial producers (Cebrián 1999, 2002; Mateo et al. 2006). It appears that most seagrass production is supported through internal nutrient recycling, and also that seagrass meadows maintain high levels of secondary production by microbial decomposers and invertebrate detritivores. Therefore, research suggests that the abundant faunal communities that are normally associated with seagrass beds are supported primarily through the detritus-based food chain (Mateo et al. 2006). Estimates of seagrass decomposition (and rates of decomposition) can be highly variable and are affected by several factors, including environmental physical conditions (water temperature, sediment oxygen content, water nutrient content, desiccation), the nutrient content of the detritus, and methodological approach used (Harrison 1989; Mateo et al. 2006).

Seagrasses play an important role in global carbon and nutrient cycling. Seagrass biomass, along with that of macroalgae within seagrass beds, has been identified as a substantial sink for carbon in the ocean (Smith 1981; Mateo et al. 2006). The majority of seagrass biomass ends up as detritus (Cebrián 1999, 2002; Mateo et al. 2006; Valentine and Duffy 2006). As a result, the amount of seagrass carbon available to be stored in the sediments can be large. In fact, research suggests that the carbon resulting from seagrasses represents approximately 12 % of the total carbon storage in the ocean, despite the fact that seagrass production represents only a small percentage (1 %) of the total oceanic production (Duarte and Cebrián 1996). In addition to burial of nutrients, seagrass beds require high nitrogen incorporation and likely play an important role in the cycling of nitrogen in shallow estuarine systems (Kenworthy et al. 1982; Bethoux and Copin-Montégut 1986; Hemminga et al. 1991; Lee and Dunton 1999).

Seagrass beds have been found to support higher faunal density and species diversity than unvegetated areas in the same environment (Orth 1977; Heck and Orth 1980; Stoner 1980a, b; Virnstein et al. 1983; Lewis 1984; Orth et al. 1984; Heck and Crowder 1991; Heck et al. 1997; Williams and Heck 2001). Furthermore, an increase in seagrass biomass results in an increase in habitat complexity or heterogeneity, which provides microhabitat space in the grass bed that is not found in the surrounding bare substratum (Stoner 1980a, b; Coen et al. 1981; Lewis and Stoner 1983; Lewis 1984). Consequently, aboveground plant biomass often is significantly correlated with invertebrate species number and abundance (Heck and Wetstone 1977; Stoner 1980a; Lewis 1984). This is especially true of epiphytic species. The high abundance of epiphytes present in seagrass meadows provides the primary pathway to higher trophic levels (Virnstein et al. 1983) via decapod crustaceans and other predators. Some infaunal species may exhibit an inverse relationship between abundance and macrophyte biomass. Thick roots and/or heavy rhizome mats may prevent certain types of infauna from inhabiting dense seagrass beds, which leads to the observed decrease in some infaunal species with increased seagrass biomass (Stoner 1980a; Brenchley 1982; Orth et al. 2006).

6.5.6 Intertidal Flats and Subtidal Soft Bottoms

6.5.6.1 Dominant Forcing Functions

The dominant forcing functions affecting faunal assemblages in unconsolidated soft sediments are components of the physical environment, including the prevailing hydrodynamic and sedimentary regimes. Biological interactions, such as competition and predation, occur to varying degrees within the constraints of the physical environment. Intertidal and subtidal flats occur along a gradient of inundation and physical exposure to wind and wave energy.

Unlike the exposed beaches of barrier islands, flats tend to have little or no slope and experience considerably less wave action, especially when facing a bay. As a shore becomes more protected from wave action, sediment particle size becomes finer and there is an accumulation of organic materials. Consequently, the sediments of flats grade from sandy to muddy along a decreasing gradient of wave and wind exposure. Water movement is minimal across mud flats, which can be a more stable substratum for benthic faunal assemblages. This stability is a favorable environment for organisms that construct permanent burrows. However, the presence of fine sediments combined with little or no slope means that pore water is retained, resulting in poor exchange between pore water and overlying water. These conditions favor the growth of dense microbial assemblages and often result in depletion of oxygen and even anaerobic conditions in the sediment below the first several centimeters. Low oxygen content in pore water may limit chemical and biological degradation processes, affecting the development and the productivity of the mud flat benthic community.

All marine and estuarine sediments are anoxic at some depth below the sediment–water interface. The boundary zone separating upper sediments dominated by aerobic processes from subsurface anaerobic sediments is defined as the redox potential discontinuity (RPD) (Fenchel 1969). Coarse sediments such as sand, gravel, or shell fragments allow more current flow into and through the substratum allowing for the RPD layer to penetrate deeper into these types of sediments. In muddy and silty habitats, subsurface hydrology is further limited due to occlusion of interstitial spaces, which allows oxygen to diffuse only a few millimeters into the sediment (Revsbech et al. 1980). Environments with more shallow RPDs tend to support deposit-feeding taxa that are able to maintain some form of hydrologic contact with the sediment–water interface by the manufacture of tubes or construction of burrows for irrigation. Burrowing and irrigation activity of infauna can distribute oxygen much deeper into the sediment (Rhoads et al. 1977). Other factors that affect the position and thickness of the RPD are the oxygen content of bottom water, sedimentation of organic matter, sediment grain size, and temperature (Vismann 1991; Diaz and Rosenberg 1995). Controlling for differences in sediment type, habitats with thinner RPDs tend to be associated with some type of environmental instability or stress, while habitats with deeper RPDs usually have flourishing epifaunal and infaunal assemblages.

Infauna that inhabit soft sediments in the GoM comprise assemblages that exhibit spatial and seasonal variability in their distributions (Boesch 1972; Dames and Moore 1979; Tenore 1985; Weston 1988; Byrnes et al. 1999). Shallow coastal waters are characterized by a variety of environments having great diurnal, seasonal, and annual fluctuations in their chemical, hydrographic, and physical properties. These factors contribute to the temporal variability of population occurrence and individual abundance of marine invertebrates (Flint and Holland 1980; Byrnes et al. 1999). Patterns of reproductive periodicity in marine systems apparently are related to ambient climatic conditions, primarily temperature, for most marine invertebrates (Sastry 1978). In tropical zones, seasonality is less pronounced.

Within seasons, benthic community structure in subtidal sediments is determined largely by disturbances and physical stresses, including riverine inputs, sedimentation, and currents (Oliver et al. 1980; Probert 1984; Hall 1994; Thrush et al. 1996). Changes in infaunal assemblage composition along broad depth gradients have been noted in numerous studies of shelf ecosystems, including in the GoM. Relatively shallow areas of the coastal strand and inner shelf comprise a turbulent zone (Day et al. 1971), where benthic fauna are adapted to unstable sediments.

Benthic boundary layer hydrodynamic flow is a significant factor regulating the composition of soft sediment invertebrate assemblages (Nowell and Jumars 1984; Hall 1994; Snelgrove and Butman 1994; Newell et al. 1998; Crimaldi et al. 2002; Hentschel and Herrick 2005). Hydrodynamic forcing has important effects on sediment regime (particle size, degree of sorting, organic content), sediment stability, and pore water oxygenation (Hall 1994), all of which affect habitat suitability for members of the various invertebrate guilds. Contrasting different sedimentary habitats in terms of how they determine infaunal community patterns can be complex because, in addition to sediment regime, other important parameters vary with hydrodynamic condition (Snelgrove and Butman 1994).

The influence of sedimentary regime on benthic community composition has been recognized since the pioneer studies of Peterson (1913), Thorson (1957), and Sanders (1958). Benthic faunal assemblages comprise taxa that are adapted to particular sedimentary habitats through behavioral, morphological, physiological, and reproductive adaptations. Fine-textured sediments are generally characteristic of depositional environments, where occluded interstitial space and accumulated organic material support surface and subsurface deposit feeders. Coarse sediments in high water current habitats, where finer particles are maintained in suspension in the water column, favor the occurrence of suspension-feeding taxa and facilitate feeding by carnivorous fauna that consume organisms occupying interstitial spaces (Fauchald and Jumars 1979).

6.5.6.2 Vegetation

Although intertidal flats and subtidal soft bottoms are generally characterized by the absence of rooted vegetation and might not be normally considered in an overview of vegetated coastal habitats, their close spatial association with vegetated coastal habitats and their inherent ecological importance make them worthy of discussion in the context of coastal habitats. Of course, these habitats are not completely devoid of photosynthesizing organisms. Primary producers, in the form of benthic microalgae (diatoms), cyanobacteria (blue-green algae) and macrophytic algal species (e.g., the green alga *Ulva* and *Enteromorpha*) are integral components of most coastal marine flats along the shorelines and within the many protected tidal lagoons of the GoM. The only primary producers on exposed beaches are benthic diatoms and swash-zone phytoplankton, which are often patchy in distribution and can exhibit vertical migration within sediments. Coastal sandflats generally have low productivity (McLachlan 1996). Allochthonous sources (originating from outside sources) of organic material, such as macroalgae (e.g., *Sargassum*) and estuarine plant detritus, provide episodic, localized enrichment when transported to intertidal and subtidal flats by currents and tides.

6.5.6.3 Fauna

The fauna of mud and sand flats are either opportunistic generalists that occupy a variety of habitats or specialists found only within a particular habitat type (Shaw et al. 1982). Infauna of the GoM occurs in distinct assemblages that are associated with certain sedimentary regimes and water depths (Dames and Moore 1979; Flint and Holland 1980; Baker et al. 1981; Shaw et al. 1982; Harper 1991). Although some invertebrate taxa occur across a range of sedimentary habitats, most species predominate in areas with particular sediment characteristics. The spatial distribution and size of habitats in a subtidal landscape play an important role in the functioning and structure of benthic communities (Thorson 1957; Andrew and Mapstone 1987; Morrisey et al. 1992; Rakocinski et al. 1998; Zajac et al. 1998, 2003; Pineda 2000; Thrush et al. 2000, 2005; Levinton and Kelaher 2004). Subtidal benthic assemblages on the shallow shelf also may be influenced by proximity to estuarine outflow. Generally, in coastal areas, inshore estuarine endemics and euryhaline opportunists grade into fully marine assemblages of the shelf (Boesch 1977). The nature of cross-shelf faunal change depends on local hydrographic and hydrologic environment, including the rate and volume of riverine input (e.g., silts and organic fines) to adjacent shelf areas.

Species diversity in tidal flat systems is generally lower than occurs in the subtidal environment, due in part to continually changing physical parameters, such as tidal fluctuation. Although diversity may be relatively low on tidal flats, these systems are highly productive in terms of invertebrate biomass.

The dominant groups of infauna found on mud flats (polychaetes, bivalves, and crustaceans) are similar to those on sand beaches, but the specific taxa are different in response to adaptations necessary for life in a habitat with fine sediments and anaerobic pore water conditions. Most organisms inhabiting the mud flat are either adapted to burrowing into and through the soft substrate or build and live in tubes in or on the substrate. Deposit-feeding organisms, such as the polychaete *Capitella*, burrow through the substrate, ingest sediment, and digest the organic matter with the help of bacteria; the polychaete, *Arenicola*, builds a u-shaped burrow with one arm of the burrow open to the surface and one filled with sediment to feed upon. Deposit-feeding bivalves are also common on mud flats (e.g., tellinid clams). Clams on mud flats are typically buried in the sediment, but have long siphons that extend to the surface for deposit feeding. Common polychaetes include bloodworms (*Glycera*) and clam worms (*Nereis*).

Dittmann (2000) found that benthic fauna in a tropical tidal flat showed a zoned distribution between the high and low tide marks. Defined groups were found, corresponding to a zonation of distinct assemblages at the high intertidal mudflat, the mid-intertidal *Callianassa* and sandflat sites, and the lower intertidal sandflat (Dittmann 2000). Bourget and Messier (1983) found that intertidal biomass was highest in the lower half of the intertidal zone compared to the upper half. Alternatively, Brown (1982) found that mean body size of the polychaete *Scoloplos fragilis* varied spatially across a tidal flat system, with body sizes significantly larger in the high-tide zone compared to the low-tide zone. The relative importance of physical versus biological controls on faunal distributions across tidal flat systems remains poorly understood.

Generally, shallow subtidal and inner shelf infaunal assemblages are dominated by polychaetes in terms of overall abundance (Day et al. 1971; Tenore 1985; Weston 1988; Barry A. Vittor and Associates, Inc. 1991). Other important groups of coastal infauna include amphipods and bivalves. Notable studies of benthic infauna in the Gulf include baseline investigations such as STOCS (Flint and Rabalais 1980), MAFLA (Dames and Moore 1979), Mississippi Sound and adjacent area study (Shaw et al. 1982), SOFLA (Woodward-Clyde Consultants 1983), and NOAA investigations of Florida Bay and Florida Keys (Barry A. Vittor and Associates, Inc. 1999). These studies showed that polychaetes typically account for half of all infaunal taxa, while mollusks and crustaceans each account for less than 25 % of the taxa.

Uebelacker and Johnson (1984) noted that some common polychaete species exhibited a faunal break east of Mobile Bay: some syllids only were found east of this area while some magelonids and ampharetids only were found west of the break. Other polychaetes exhibited disjunct distributions and were present in both the Eastern Gulf Neritic and Texas Estuarine Area subregions but not in the Mississippi Estuarine subregion.

Shaw et al. (1982) performed a large baseline survey of infauna that included the barrier strand areas of Alabama and Mississippi out to Gulf depths of 30 m (98 ft). Infauna of the clean sand habitat in tidal passes included the archiannelid *Polygordius*, cephalochordate *Branchiostoma caribaeum*, polychaetes *Mediomastus* spp. and *Spiophanes bombyx*, and the burrowing amphipod *Acanthohaustorius*. Offshore (shallow Gulf) assemblages varied with sedimentary habitat type. Mud habitats supported polychaetes such as *Magelona* cf. *phyllisae*, *Mediomastus* spp., *Diopatra cuprea*, and *Myriochele oculata*, and the cumacean *Oxyurostyllis smithi*. Offshore sand had assemblages dominated by *Polygordius*, *B. caribaeum*, and the polychaetes *Lumbrineris* spp., *Mediomastus* spp., and *Paraprionospio pinnata*.

Coastal Louisiana invertebrate assemblages include widespread taxa such as the polychaetes *Paraprionospio pinnata*, *Magelona* cf. *phyllisae*, and *Sigambra tentaculata* (Baker et al. 1981; Gaston and Edds 1994), that commonly occur inshore to mesohaline (18 to 5 ppt) environments (Shaw et al. 1982; Gaston et al. 1995). The Southwest Research Study (Baker et al. 1981) collected infauna from the central Louisiana shelf (inshore to 90 m [295 ft] depths) and found that sand habitats supported amphipods *Ampelisca verrilli* and *Photis macromanus*, and the polychaetes *Ceratonereis irritabilis*, *Prionospio cristata*, and *Glycera americana*.

Benthic community analysis of the Laguna Madre (Texas) conducted for the U.S. Army Corps of Engineers showed that Upper Laguna Madre assemblages were dominated mainly by polychaetes, while the Laguna Madre south of Baffin Bay was characterized primarily by several mollusk species (Barry A. Vittor and Associates, Inc. 1996). However, nearly all numerically important species in the Laguna Madre were typical of marine waters throughout the Gulf, including *Capitella capitata*, *Streblospio benedicti*, *Prionospio heterobranchia*, *Grandidierella bonnieroides*, and *Mulinia lateralis*. Similar infaunal assemblages were described in Laguna de Términos, which is located in a sedimentary transition zone on the Campeche coast: 173 species of mollusks and over 120 species of polychaetes have been identified, including *M. lateralis*, *Abra aequalis*, *Macoma constricta*, *C. capitata*, *S. benedicti*, and *Mediomastus californiensis* (Contreras-Espinosa and Castañeda-Lopez 2007).

Epifaunal benthic species assemblages in the coastal waters of the northern Gulf are fairly uniform across the region and are distinguished mainly by large differences in sediment texture/type and salinity (Defenbaugh 1976). Carbonate-dominated sand sediments, such as those found along the Florida Panhandle and the Eastern Gulf Neritic subregion, are populated by many species also found farther to the west, including the sand dollar (*Mellita quinquiesperforata*), the starfish (*Luidia clathrata*), rock shrimp (*Sicyonia brevirostris*), and the spider crab (*Libinia dubia*). However, a change in epifaunal assemblages occurs between shallow sand habitats of the Florida Panhandle and muddy sand sediments west of Mobile Bay. Species not commonly found west of the Bay include *Encope michellini* (sand dollar) (Figure 6.102c), *Arbacia punculata* (sea urchin), the decapod crabs *Podachela riisei*, *Ovalipes guadalupensis*, *Iliacantha intermedia*, *Calappa flamea*, *Stenorhychus seticornis*, and *Parthenope serrata*, the cnidarian *Calliactis tricolor*, and the poriferan *Cliona celata* (Barry A. Vittor and Associates, Inc. 1986).

Brittle stars (*Hemipholis elongatus* and *Ophiolepis elegans*) are very abundant in subtidal flats near tidal inlets, where they feed on detritus borne by tidal currents. Penaeid shrimps are also present on these sand sediments, but at far lower densities than found on muddy sediments (Swingle 1971). In addition to echinoderms, dominant epifaunal species include the cnidarian

Renilla mulleri (sea pansy), the gastropods *Sinum perspectivum* and *Cantharus cancellarius*, the bivalve *Chione clenchi*, the stomatopod *Squilla empusa*, and the decapods *Persephone crinata, Hepatus epheliticus, Callinectes similis*, and *Pagurus pollicaris*.

The nekton assemblages of sandy and muddy tidal flat habitats in the GoM often have been sampled to make comparisons with the nekton assemblages of adjacent vegetated (marsh, mangrove, seagrass) habitats. Consequently, the same sampling gear has been used in both vegetated and shallow unvegetated estuarine bottom habitats. If it can be assumed that effectiveness of the collecting methods are similar, unvegetated flats almost invariably yield lower densities of nekton, particularly small natant decapod crustaceans, than occur in more structurally complex coastal wetlands (Orth et al. 1984; Minello 1999; Beck et al. 2001; Heck et al. 2003). As in all of the other coastal wetlands discussed here, proximity to adjacent wetland types (e.g., seagrass, tidal marsh, mangrove) influences the species composition of the nekton assemblage in intertidal flats and on subtidal soft bottoms within the GoM. This can be illustrated by a comparison of dominant nekton families and species in quantitative collections from the U.S. coasts of the GoM (Table 6.21).

Decapod crustaceans dominate numerically the collections represented in the table in all northern and southeastern GoM regions represented, and fishes appear to be better represented on intertidal and subtidal flats in areas with adjacent marsh habitat. Where seagrasses are a dominant habitat adjacent to the flats, the hyppolytid shrimps are a major component of the natant decapods, but where marshes are adjacent to the flats, palaemonid and penaeid shrimps are more abundant. Except for the gobiid and engraulid fishes, which are common components of the nekton in most regions of the GoM, there is little commonality in the dominant fish families or species reported among individual studies. One possible reason for this is that most nekton using intertidal or shallow subtidal flats to forage or escape predators are constantly moving, so the assemblage at a particular location may change quickly.

The data summarized in Table 6.21 represents only a small portion of the nekton assemblage that occurs on the intertidal and subtidal flats because most of the quantitative samples represented in the table used collecting gear of relatively small sample unit size (e.g., 1 m^2), which is less effective in capturing larger nekton, highly mobile schooling species, or species that may be common and even abundant at the broader spatial scales sampled by trawls within the estuaries. Many of these species forage in shallow flats, where benthic invertebrate prey can be abundant, or are in the process of moving among more structurally complex coastal wetland habitats.

Among the nekton that are widely distributed on intertidal and subtidal flats are several common species in the families Sciaenidae (drums) and Ariidae (sea catfishes), which occur in most estuaries throughout the GoM (Nelson 1992; Gilmore 1987; Yáñez-Arancibia and Lara-Dominguez 1988). Because of their economic importance in commercial and recreational fisheries, the Sciaenidae, Penaeidae, and Portunidae are among the best studied of the nekton commonly found on estuarine flats (e.g., Gilmore 1987; McMichael and Peters 1989; Grammer et al. 2009; Rooker et al. 1999; Pérez-Castañeda and Defeo 2001; Luna et al. 2009). Other epibenthic species of nekton that are particularly well adapted to intertidal and subtidal flat habitats in GoM estuaries are the stingrays (e.g., *Dasyatis sabina, D. say*) and several species within the flatfish families Achiridae (e.g., *Achirus lineatus*), Cynoglossidae (e.g., *Symphurus plagusia*), and Paralichthyidae (e.g., *Citharichthys spilopterus*), which are widely distributed in both the northern and southern GoM (Contreras-Espinosa and Castañeda-Lopez 2007). These species are not considered of much economic importance, so their requirements and functional roles within this system have not been as clearly defined.

Table 6.21. Nekton Species Comprising at least 90 % of the Total Individuals in Samples From Unvegetated Shallow Subtidal and Intertidal Flats in the GoM (most prominent adjacent wetland type is shown in parentheses)

	Marine Ecoregion				
	Florida Bay (seagrass)	Eastern Gulf Neritic (marsh/seagrass)	Mississippi Estuarine (marsh)	Texas Estuarine (marsh/seagrass)	Texas Estuarine (seagrass)
Fishes					
Batrachoididae			■		
Opsanus beta			X		
Clupeidae			■		
Brevoortia patronus			X		
Cynoglossidae		■			
Symphurus plagiusa		X			
Engraulidae			■	■	■
Anchoa mitchilli			X	X	X
Fundulidae	■				
Lucania parva	X				
Gerridae		■			
Eucinostomus argenteus		X			
Gobiidae	■	■	■	■	
Gobiosoma bosc			X	X	
Gobiosoma robustum	X				
Ctenogobius boleosoma		X	X		
Gobiesocidae			■		
Gobiesox strumosus			X		
Mugilidae		■			
Mugil curema		X			
Sciaenidae		■		■	
Bairdiella chrysoura		X			
Leiostomus xanthurus				X	
Ophichthidae			■		
Myrophis punctatus			X		
Natant Decapod Crustaceans					
Hippolytidae	■				■
Thor floridanus	X				
Tozeuma carolinense					X
Palaemonidae		■	■	■	
Palaemonetes intermedius		X			
Palaemonetes pugio				X	
Palaemonetes spp.			X		
Penaeidae	■	■	■	■	■
Farfantepenaeus aztecus			X	X	X
Farfantepenaeus duorarum	X	X			
Portunidae		■	■	■	■
Callinectes sapidus		X	X	X	
Callinectes similis					X
Data References:	Sheridan et al., 1997	Rozas et al., 2012	Mairaro, 2007; Plunket, 2003	Minello, 1999	Sheridan and Minello, 2003

Red shaded cells indicate the presence of dominant families at each location. The most abundant species reported in each family are indicated by an "X"

Most marine food webs share fundamental structural and ordering characteristics with those of estuarine, fresh water, and terrestrial systems (Dunne et al. 2004). Benthic invertebrates have an important role in transferring energy from detrital production to higher trophic levels (Newell et al. 1998). Decapods are among the chief consumers of the benthos, and in general are opportunistic predators. Many decapods feed on the predominant invertebrates of coastal sediments, including polychaetes, bivalves, echinoderms, and smaller crustaceans (Stehlik 1993). Blue crabs (*Callinectes sapidus*) consume locally abundant infauna, epifauna, and fish (Tagatz 1968). Where abundant, blue crabs play a major role in energy transfer within estuaries (Baird and Ulanowicz 1989). Blue crabs and other decapods provide an important link between benthic secondary production and higher trophic levels.

The diet of many of the most common demersal fishes consists of benthic invertebrates (Grosslein 1976). Fishes such as flounders, skates (*Raja* spp.), and spot (*Leiostomus xanthurus*) are predominantly bottom feeders that consume infaunal and epibenthic crustaceans and polychaetes. Amphipods are known to be important in the diets of some demersal fishes, including Atlantic croaker (*Micropogonias undulatus*). The affinity of certain demersal fishes for particular sediment types often is related to the types of prey items supported by those sediments (Rogers 1977). Decapods are a primary component of the diets of demersal fishes (Bowman et al. 2000).

Epifauna associated with intertidal flats are predominantly mobile predatory species such as portunid crabs (e.g., *Callinectes sapidus*) that consume small bivalves, polychaetes, and crustaceans. Other mud and sandy-mud associated epifauna include brown shrimp (*Farfantepenaeus aztecus*), white shrimp (*Litopenaeus setiferus*), swimming crabs such as *Portunus gibbesii* and *Portunus spinimanus*, the mantis shrimp (*Squilla empusa*), the gastropods *Pollinices lunulata* and *Nassarius acutus*, and the bivalves *Nuculana concentrica* and *Macoma tageliformis*. Intertidal flats also provide migratory corridors for taxa such as penaeid shrimps and blue crab, which feed on polychaete worms and other tubicolous infauna as they move from estuarine nurseries into nearshore spawning grounds (Franks et al. 1972).

6.6 DISTURBANCES AND ECOLOGICAL IMPACTS

The ecological structure and function of coastal habitats discussed previously are the result of the interaction among environmental characteristics of habitats, the biology of the species occurring within habitats, and the numerous disturbances that periodically impact and often reset ecosystem processes. Coastal habitat disturbances vary widely, being both natural and anthropogenic, and acute and chronic. Here, we provide a brief overview of many primary disturbances structuring coastal habitats mentioned within this chapter. Our discussion is not meant to be exhaustive, but rather serves as a summary of diverse disturbances and their ecological impacts.

6.6.1 Natural Disturbances

Disturbance is a physical event that disrupts at least some aspects of the physical or biological structure of an ecosystem, and consequently, plays an important role in restructuring the ecosystem and altering its ecological functions. Disturbance can be natural, as in the case of tropical storms that cause coastal marsh and beach erosion, or human-induced, such as introduction of excessive nutrients to coastal waters and resulting depletion of dissolved oxygen. The relative influence of different types of natural disturbance varies with geographic location; northern populations of mangroves, for example, are subject to freeze damage while southern populations, especially outside of the United States, experience human impacts, such as harvesting, as well as the natural impacts from frequent tropical storms and hurricanes. This section will address small- and large-scale natural disturbances in coastal habitats.

6.6.1.1 Tropical Cyclones

Coastal habitats, such as beaches, salt marshes, and mangroves, can experience severe alteration from hurricanes and other tropical storms. Hurricane Andrew, which made landfall in Florida and Louisiana in late August 1992, removed sections of salt marsh and deposited them in shallow ponds and bays (Cahoon 2006). Hurricane Katrina similarly impacted brackish marshes east of the Mississippi River resulting in more than 80 ha (200 acres) of land loss (Morton and Barras 2011). Sediment and wrack deposition during hurricanes can bury marsh vegetation and result in plant mortality (Guntenspergen et al. 1995; Valiela and Rietsma 1995). Also, saltwater intrusion in lower salinity coastal marshes disrupts system ecology, resulting in short-term vegetation dieback (Cahoon 2006). In contrast to the negative effects of hurricanes and storms, hurricane-generated sediment input can counterbalance relative sea-level rise and promote wetland sustainability (McKee and Cherry 2009). In addition, bare patches resulting from wrack deposition allow for the recruitment of other plant species, generating habitat heterogeneity and increasing plant diversity (Guntenspergen et al. 1995).

Mangrove forests in Florida have been periodically disturbed by hurricanes, such as Hurricane Donna in 1960 (Craighead and Gilbert 1962), Hurricane Andrew in 1992 (Smith et al. 1994; Baldwin et al. 1995), and Hurricane Wilma in 2006 (Whelan et al. 2009). These and other studies have documented the impacts of such storms on subsequent structure and function of mangrove forests, including defoliation and losses of branches or entire trees (Davis 1995; Baldwin et al. 2001; Davis et al. 2004; Milbrandt et al. 2006; Profitt et al. 2006; Ward et al. 2006; Smith et al. 2009; Whelan et al. 2009; Castaneda-Moya et al. 2010; Harris et al. 2010). Recovery of mangroves after hurricane disturbance is a function of a complex interaction of factors, including seedling recruitment and survival, resprouting capability, and colonization by herbaceous vegetation (Baldwin et al. 2001). Post-disturbance site productivity (Ross et al. 2006) and spatial variation in hurricane impact (Thaxton et al. 2007) further influence regeneration success. Recovery of ecological structure generally occurs, but it is a relatively slow process compared to herbaceous systems.

Hurricane damage to seagrass beds can be highly variable and depends on location as well as hurricane characteristics (Smith et al. 1994; Courtemanche et al. 1999; Paerl et al. 2001; Coles and Brown 2007; Cebrián et al. 2008; Anton et al. 2009). Hurricanes have been observed to cause widespread damage to seagrasses, but also to pass with little or no damage (Poiner et al. 1989; Hemminga and Duarte 2000). For example, a cyclone and its associated rainfall caused a loss of approximately 1,000 km^2 (247,000 ac) of seagrass in Hervey Bay, Australia (Preen et al. 1995). Other reports have shown hurricanes having only small impacts on seagrass beds and associated macrophytes (Fourqurean and Rutten 2004; Tilmant et al. 1994).

Hurricane Katrina was one of the most destructive storms in U.S. history, with winds over 264 km/h (164 mi/h) and a storm surge of 7.8 to 8.5 m (25.6 to 27.9 ft) in the western coast of Mississippi (Hsu et al. 2005). Anton et al. (2009) showed that this powerful hurricane had no major impact on seagrass density, biomass, or community structure (abundance of producers and consumers) in a seagrass bed located approximately 100 km (62 mi) to the east of the hurricane's landfall. Overall, this research showed that natural temporal changes in seagrass metabolism, recorded before the hurricane, were larger than any post-storm changes. Conversely, a combination of tropical storm activity and higher-than-average watershed discharge in the Big Bend area of Florida is believed to have caused severe reductions in seagrass distribution and abundance up to 2005; absence of storm activity and relatively low river discharges have resulted in significant expansion of seagrass cover, including into areas where seagrass species had not been observed for many years[3] (FDEP 2012b) Even when there are significant changes in the coastal wetland vegetation after hurricanes, faunal assemblages, especially mobile nekton, are resilient, and any immediate effects tend to be quickly reversed (Piazza and La Peyre 2009).

6.6.1.2   Floods and Drought

Flooding may elicit major changes in morphology and sediments of barrier beaches, salt marshes, seagrasses, and subtidal substrates, and result in significant (albeit temporary) losses of fauna associated with those habitats. Nearly freshwater conditions (freshets) may occur in areas normally classified as marine waters during extreme flooding events. Although best known for their damaging effects on oysters, freshets also can be responsible for mass mortalities among sessile epifauna (cnidarians such as *Renilla*) and many infaunal species. Motile epifauna (portunid crabs and penaeid shrimps) may move out of areas exposed to extreme reductions in salinity.

Adequate light levels have been identified as one of the most important factors influencing the presence of seagrasses (Hemminga and Duarte 2000). Water column turbidity can be influenced by a number of factors, both natural and anthropogenic. Regardless of source, the primary detrimental effect of turbidity is increased attenuation of light. Reduced light over a prolonged period can deplete seagrass carbon reserves, resulting in increased shoot mortality and ultimately the decline of whole meadows. In extreme cases, the lack of photosynthetically produced oxygen can lead to sediment anoxia and a more rapid rate of seagrass mortality (Ralph et al. 2006). Additionally, it has been suggested that elevations in turbidity may reduce irradiance to a point that stresses seagrasses, reducing their vitality and making them more vulnerable to disease (Giesen et al. 1990; Hemminga and Duarte 2000).

Drought conditions may favor expansion of drought-resistant vegetative community types. For example, mangrove populations survived while competing species, such as *Spartina alterniflora*, had high mortality caused by a regional drought in 2000 in coastal Louisiana (McKee et al. 2004). Hypersaline zones created in salt marshes during droughts may exhibit changes in vegetation composition, from species such as *Juncus roemerianus and Spartina patens* to salt-tolerant species such as *Sarcocornia pacifica* and *Distichlis spicata*. Bertness (1992) observed in a New England salt marsh that these salt-tolerant species shade the soil, reduce evaporation, and ameliorate salinity; as salinity decreases, other species (for example, *Juncus* and *Spartina*) can re-populate the area and outcompete earlier colonizers until they dominate the patches after 2 to 4 years. However, since 2000, numerous examples of drought-induced plant mortality have been documented along the Atlantic seaboard of the United States (Alber et al. 2008). These events have resulted in salt marsh dieback, often in the absence of significant recovery.

6.6.1.3   Subsidence and Sea-Level Rise

Natural subsidence in the northern GoM has resulted in loss of salt marsh habitat and its associated fauna, but expands habitats for other species. Subsidence in coastal Louisiana has converted large areas of marsh in Barataria Bay, Timbalier Bay, and other embayments to open-water habitat (Britsch and Dunbar 1993; Couvillion et al. 2011), resulting in reductions in primary productivity but increases in subtidal habitats and populations of infauna and epifauna. Work in Florida found that mangrove areas in the Ten Thousand Islands National

---

[3] Carl M. Way, Barry A. Vittor and Associates, Inc.

Wildlife Refuge had increased 35 % from 1927 to 2005; this increase was attributed to sea-level rise as well as factors such as subsidence, enhanced propagule dispersal via new waterways, and reduced freshwater delivery from overland flow (Krauss et al. 2011). Mangroves are also predicted to replace freshwater forests in the eastern Gulf (Doyle et al. 2010). Although some workers have examined rates of sediment accretion in expanding mangrove stands at their northern limits (Perry and Mendelssohn 2009), no information exists on the relative capacity of mangroves to keep up with sea-level rise compared to salt marsh habitat. Relative sea-level rise throughout the GoM is discussed in detail in Section 6.3.

6.6.1.4 Herbivory

Disturbances caused by herbivores may have important consequences for vegetated marine habitats. Grazers in wetlands include insects, crustaceans, snails, fish, waterfowl, and mammals (McKee and Baldwin 1999 and references therein). Arguably, two of the most damaging grazers in salt marshes of the GoM are mammals: nutria (*Myocaster coypus*) and muskrat (*Ondatra zibethicus*), nutria being introduced. Geese also are damaging grazers, but their impacts in the GoM are more localized. These animals can cause "eatouts" that result in denuded marsh surfaces that often recover slowly or not at all (Kerbes et al. 1990; Linscombe and Kinler 1997; Gough and Grace 1998; Randall and Foote 2005). Herbivore impacts from invertebrates (for example, snails in Louisiana and Georgia (Silliman et al. 2005) and crabs in New England (Holdredge et al. 2008)), causing denuding of marshes, also have been reported.

Mangrove forests can be damaged by wood-boring beetles (xylovores), which may cause death of individual trees or small groups of trees (Lugo and Patterson-Zucca 1977; Feller 1995; Smith et al. 1994). Such biotic agents of disturbance were unrecognized until Feller (1995) reported that up to 30 % of the canopy in a red mangrove forest in Belize was removed by the activities of wood-boring beetles. The larvae of the beetles are active in the phloem and outer xylem where they create extensive feeding galleries that are still evident in the standing dead wood and fallen litter for many years. Feller (1995) found that the activity of a single larva can ultimately girdle a branch or bole, resulting in a thinning of the canopy, and terminating in the creation of a light gap and standing dead wood that is secondarily invaded by other xylovores. Only a few estimates of herbivory rates have been made in mangrove forests in the northern GoM, mainly in Florida (Onuf et al. 1977; Erickson et al. 2004; Feller et al. 2007).

Herbivorous animals consume live seagrass blades, epiphytes, and macroalgae. Animals from several different taxonomic groups, such as gastropods, fish, sea urchins, waterfowl, sea turtles, and manatees can all be significant consumers of seagrasses. Sea urchins are primary invertebrate grazers in seagrass beds. In some instances, sea urchin grazing can be so intense that much of the seagrass primary production is consumed, occasionally resulting in the elimination of extensive seagrass patches (Larkum and West 1990; Hemminga and Duarte 2000). Such overgrazing is correlated with sea urchin population density, and can be common in some areas. As a result of these overgrazing events (and subsequent bare patches), young sea urchins are exposed to higher predation rates. This increase in predation results in a decline in the urchin population, which in turn results in a recovery of the seagrass vegetation. In this way, the grazing pressure exerted by sea urchins on seagrass systems may show an oscillating pattern over time, due to consecutive cycles of growth and decline of the sea urchin population (Heck and Valentine 1995; Hemminga and Duarte 2000). Herbivory by certain fish species can be intense under specific conditions; however, the proportion of seagrass production consumed by fish is generally low when compared to the intense grazing observed by sea urchins (Klumpp et al. 1993; Greenway 1995; Cebrián et al. 1996; Hemminga and Duarte 2000).

6.6.1.5 Other Natural Disturbances

Naturally occurring hypoxia has been reported in coastal waters for many years; as summer water temperatures and salinities rise and winds decrease, the potential for mass mortalities among motile fauna as well as benthic and sedentary fauna increases. While this phenomenon (known locally as a "jubilee") occurs mainly in estuarine waters, hypoxic events also occur in nearshore marine waters. In areas of the GoM that are frequently or chronically impacted by hypoxia, benthos tends to be dominated by short-lived, small deposit-feeding polychaetes, and long-lived infauna and epifauna are rare (Rabalais et al. 2002). Some common nekton species in the GoM, including penaeid shrimps, have a degree of tolerance for low oxygen conditions (Rosas et al. 1999), but most actively avoid hypoxic conditions (Renaud 1986). Localized and short-term cyclic hypoxic episodes may be common in tidal creeks of estuaries (Tyler et al. 2009), but nekton are capable of quickly emigrating from areas affected by localized hypoxia and repopulating the area when conditions improve (Hackney et al. 1976).

Mangroves are susceptible to damage by prolonged freeze conditions. Stevens et al. (2006) reported on the role of freezing on mangrove density and seedling establishment, and fluctuations in comparison to salt marsh species. Others have examined the effects of freezing on the structure of dwarf mangrove forests in Florida and how this may change mangrove diversity in the future with a warmer climate (Ross et al. 2009).

6.6.2 Human-Induced Disturbances

Human-induced stressors in coastal habitats are associated primarily with waste discharges, nutrient enrichment, navigation improvements, flood control measures, coastal development, and petro-chemical-related development. Unlike most natural stressors, human-induced disturbances are typically chronic and persistent. This section describes the major anthropogenic factors that affect coastal habitats in the GoM.

6.6.2.1 Nutrient Enrichment and Pollution

Worldwide population growth has led to an exponential increase in nutrient inputs into the coastal zone, primarily though massive use of fertilizers for agriculture (Nixon and Buckley 2002). Introduction of excess nutrients via the Mississippi River has caused a large area of GoM bottom waters (over 20,000 km^2 [7,722 mi^2] in 2000) to become hypoxic on an annual basis, resulting in die-offs of many epifauna and infauna (Rabalais et al. 2002). At the Louisiana Offshore Oil Port, densities of infauna decreased by over 80 % during persistent hypoxic conditions in the spring-summer seasons of 1990 to 1993; however, taxa richness only decreased by about 50 %, suggesting that some infauna are adapted to nearly anaerobic conditions (Barry A. Vittor and Associates, Inc. 1995). Chronic hypoxia also has been observed in Chandeleur Sound and western Mississippi Sound (Lopez et al. 2010a, b). Waste discharges into coastal waters also include municipal storm water, treated effluents from wastewater treatment plants, and industrial effluents. In general, ecological impacts are relatively minor due to the dilution effects of relatively large receiving waters, and due to improved levels of treatment for such discharges.

Although wetland primary productivity is nutrient limited, recent research has indicated that the input of nutrients, especially nitrogen, can accelerate the expansion of invasive genotypes of *Phragmites australis* into brackish marshes in New England (Bertness et al. 2002) and alter biomass allocation, possibly impacting the capacity of coastal wetlands to keep pace with rising sea levels, as reported for coastal Louisiana (Darby and Turner 2008). Increased inputs of nutrients to mangrove forests as a consequence of agriculture, urban sewage, and industrial effluents may have significant effects on forest structure and function by modifying net photosynthesis and other physiological processes (Lovelock et al. 2004), rates of herbivory (Feller and Chamberlain 2007), production-decomposition rates (Feller et al. 2007), organic matter accumulation and contribution to soil elevation maintenance (McKee et al. 2007a), or nutrient recycling (Feller et al. 1999; McKee et al. 2002; Whigham et al. 2009). Much of the foregoing work has been conducted outside the northern GoM but the findings are relevant to understanding the general responses of mangroves to nutrient enrichment.

Nutrient excess can also affect the competitive interactions between coastal wetland vegetation types. For example, McKee and Rooth (2008) examined effects of nitrogen addition on the competition between the mangrove, *Avicennia germinans*, and the salt marsh dominant, *Spartina alterniflora*, in coastal Louisiana. When grown in mixture in greenhouse mesocosms, nitrogen addition greatly favored *S. alterniflora*, which is a stronger competitor for nitrogen. Also, seedlings established in nitrogen-fertilized field plots in coastal Louisiana were more susceptible to crab herbivory than unfertilized seedlings (McKee and Rooth 2008). Consequently, nutrient enrichment in this plant community would be expected to modify the outcome of species interactions. Another study, conducted in the Florida Everglades (Castaneda-Moya et al. 2011), found that variation in the availability of phosphorus to below-ground roots could influence future response of mangroves in that system to sea-level rise through modifications of root production and organic contributions to soil volume.

Although eutrophication does not directly cause declines in seagrass because seagrasses generally benefit from higher nutrient levels, eutrophication results in a bottom-up effect in which other marine primary producers, such as phytoplankton and macroalgae, are allowed to proliferate (Hemminga and Duarte 2000). Nutrient requirements for seagrasses tend to be lower than those for phytoplankton and macroalgae (Duarte 1995). Therefore, at similar nutrient inputs, seagrasses are able to maintain primary production well above that of phytoplankton and macroalgae. Seagrasses also experience lower grazing than the other marine primary producers (Cebrián and Duarte 1994; Duarte 1995). For these reasons, seagrasses are able to maintain higher biomasses and outcompete other primary producers under nutrient-limited conditions (Hemminga and Duarte 2000).

As nutrient inputs increase, light becomes the limiting factor and the balance between primary producers is altered. Phytoplankton biomass in the water column increases and micro- and macroalgae proliferate, sometimes overgrowing the seagrasses. Carpets of epiphytes may cover seagrass leaves, further reducing light available for seagrass photosynthesis. This reduction of light is ultimately the most important factor responsible for the decline of seagrasses in eutrophied waters (Hemminga and Duarte 2000). Along with a reduction in light availability, eutrophication can also lead to decreased sediment oxidation status, anoxia of bottom waters, and, in some cases, concentrations of nutrients that reach toxic levels (Hemminga and Duarte 2000).

6.6.2.2   Fishery Activities

Coastal marine faunal communities are heavily exploited for commercial and recreational uses, especially as key components of fisheries. Penaeid shrimps, blue crab, stone crab, and bay scallop comprise vital fisheries in the GoM. Although infauna are not harvested for commercial use, they provide a major source of food for many fishery species, and environmental disturbances that decrease the abundance of infauna can also reduce populations of important epifauna.

Fishing practices, such as trawling and dredging, can disturb bottom sediments, damage shoots and rhizomes, or completely remove seagrasses from the substratum, possibly resulting in severe local reductions of seagrass cover (Hemminga and Duarte 2000). In areas where small boats are numerous, the cumulative effect of boat moorings and propeller scars can result in considerable loss of vegetation (Walker et al. 1989; Creed and Amado Filho; 1999; Hemminga and Duarte 2000), although losses due to these mechanical disturbances are likely small relative to damage caused by elevated water turbidity (Onuf 1994).

Rapid degradation and loss of coastal wetland habitats may be considered the greatest threat to many nekton species of special interest in the GoM (Thomas 1999; Contreras Espinosa and Warner 2004). However, nekton assemblages in both the northern (Gravitz 2008) and southern (Contreras-Espinosa and Warner 2004; Contreras-Espinosa and Castañeda-Lopez 2007; Ortiz and Lalana 2005) GoM are affected directly and indirectly by activities associated with commercial and recreational fisheries exploitation. The status of many managed fisheries in the GoM is poorly understood with respect to overharvest, but there are some well-known examples of depressed populations of formerly abundant inshore fish families such as the Sciaenidae, especially red drum (Tilmant et al. 1989) and reef fishes, whose juveniles commonly use coastal wetland habitats (Contreras-Espinosa and Warner 2004; Ortiz and Lalana 2005; Gravitz 2008). Removal of larger predatory fishes targeted in many recreational and commercial fisheries can have important implications for trophic dynamics within estuaries that result in shifts in associated nekton assemblages (Pauly et al. 1998; Hall et al. 2000).

Although it has been suggested that heavy fishing pressure following periods of high weather-related natural mortality events has hampered the recovery of some penaeid stocks within the GoM (Kutkuhn 1962), the penaeid shrimp fisheries of the GoM were long considered virtually impervious to overharvesting (Lindner and Anderson 1956). Unlike many larger, long-lived fish species targeted in recreational and commercial fisheries, penaeids are annual species that are reproductively prolific and population variation is largely controlled by environmental conditions, especially in estuarine wetland nursery habitats. In fact, because the fishery is considered fully exploited but not overharvested, it was proposed that trends in shrimp landings could be used as indicators of regional estuarine habitat quality (O'Connor and Matlock 2005). However, recent concerns have arisen about growth overfishing (i.e., decreases in the size of shrimp harvested) of some penaeid stocks in the GoM (Caillouet et al. 2008).

The unintended effects of fishing gear on nekton populations may have an even greater effect on nekton that are not specifically targeted by the principal fisheries in the GoM. Most mobile fishing gear disturbs habitats and potential prey resources for nekton populations (Watling and Norse 1998). Incidental taking of non-targeted species in by-catch is considered by many to be the most significant problem in fisheries management, with shrimp trawling producing the highest discard/catch ratio (Hall et al. 2000). The blue crab fishery, which uses largely stationary gear (i.e., crab traps), also has important indirect effects on many other nekton in the GoM. At least 23 species of fishes, including important recreational fisheries species among the Sciaenidae (drums), have been incidentally killed or injured in crab traps (Guillory et al. 2001). Diamondback terrapins (*Malaclemys terrapin*) are also among the coastal wetland nekton negatively impacted as by-catch of the crab fishery in the GoM (Butler and Heinrich 2007; Hart and Crowder 2011). Like many other long-lived species that are subjected to overharvesting, even unintentionally, it is the juvenile terrapins that are most susceptible to mortality in crab traps (Dorcas et al. 2007).

6.6.2.3 Invasive Species

Introduction of non-native plant species can alter diversity, productivity, and resilience of natural habitats, including mangrove forests. In south Florida, non-native mangrove species (from the Indo-Pacific) have become established in botanical gardens and these populations are reportedly expanding on the Atlantic coast (Fourqurean et al. 2010). However, the mangrove habitat is considered to be difficult to invade by non-mangrove species due to the fact that flooding and high salinity exclude most other plant species unadapted to these conditions (Lugo 1998). Nevertheless, some studies show negative effects of invasives, such as Brazilian pepper, through allelopathic effects of leached chemicals from the fruits on growth and leaf production of mangrove species (Donnelly et al. 2008). In salt marshes, invasive species are relatively few in number in the GoM, perhaps because few plant species have been able to evolutionarily adapt to salinity and flooding, and the niche space for invasion is limited. The European genotype of *Phragmites australis*, however, has become problematic along the mid-Atlantic and New England coasts of the United States, where it has replaced the native form of this species in many locations while expanding into brackish and saline marshes (Chambers et al. 1999). In the GoM, it also appears to have expanded its dominance, especially within the Birdfoot Delta of the Mississippi River (White et al. 2004a; Hauber et al. 2011) and coastal Mississippi (Peterson and Partyka 2006).

Invasions of seagrass systems by non-native producer species have not been reported to cause large-scale declines in seagrasses, although such a situation is not unlikely (Hemminga and Duarte 2000). Proliferation of *Caulerpa taxifolia* (a tropical green alga) was first noted in the Mediterranean in 1984 (Meinesz and Hesse 1991), and is considered potentially dangerous to that region's native seagrass meadows. These fears have been somewhat reduced by observations that algae appear to be successful colonizers only in polluted environments where seagrass vitality is poor, or in sparse meadows (DeVillèle and Verlaque 1995; Chisholm et al. 1997; Hemminga and Duarte 2000).

Unlike with non-native producer species, there are examples of unanticipated, indirect effects arising from the introduction into coastal marine systems of invasive consumer species. Arguably the most damaging invasive animal affecting coastal marshes in the northern GoM is the nutria, or marsh rat, *Myocaster coypus*. This rodent was introduced into the United States in the 1930s and 1940s, when it was brought to Louisiana to farm its fur (Wilsey et al. 1991). After escaping captivity, it spread throughout the northern GoM and elsewhere in the United States, with an estimated population size of 20 to 30 million (Byers 2009). Nutria not only consume marsh vegetation, they are also a cause of marsh loss due to their consumption of belowground plant parts and the resulting disturbance of the soil (McFalls et al. 2010). Invasive invertebrates in the GoM have also taken their toll. For example, the boring isopod, *Sphaeroma terebrans*, can cause substantial damage to red mangrove roots (Brooks and Bell 2002). The Asiatic green mussel, *Perna viridis*, has been found in estuaries of the eastern GoM, where it can occur in numbers that outcompete native species of mollusks, such as *Crassostrea virginica* (Baker et al. 2011). Other non-native introductions have led to some well-known examples of marine trophic cascades. The introduction of non-native green crabs (*Carcinus maenas*) from Europe to the Gulf of Maine reduced periwinkle (*Littorina littorea*) feeding and allowed ephemeral green algae (the periwinkle's preferred food) to dominate the substrate rather than less palatable brown and red algae (Lubchenko 1978; Vadas and Elner 1992). Other examples include the effects of the Asian clam (*Potamocorbula amurensis*) in San Francisco Bay (Carlton 1999), the non-native ctenophore (*Mnemiopsis leidyi*) in the Black Sea (Malyshev and Arkhipov 1992), and the invasive seastar (*Asterias amurensis*) in Australia (Buttermore et al. 1994). While such examples do not specifically refer to any situations occurring in the northern GoM, they serve to illustrate the ability of invasive predatory species to dramatically change food web structure of benthic ecosystems (Heck and Valentine 2007). The impacts of invasive species observed elsewhere also justify concerns about the potential effects in the GoM of recent exotic introductions of species such as lionfish (*Pterois* spp.) (Schofield 2010; Fogg et al. 2013) and penaeid shrimp species (e.g., Wakida-Kusunoki et al. 2011).

6.6.2.4 Navigation Improvements and Flood Control

Navigation improvements (channel dredging and dredged material disposal) can have localized impacts on coastal habitats, including marshes, seagrass beds, and subtidal soft bottoms. Construction of channels and basins involves deepening of existing open-water habitats, or converting vegetated areas to open-water. For example, dredging may have a direct mechanical impact on seagrass beds (Onuf 1994). Biota in such areas are destroyed or displaced by these activities. Excavated channels tend to accumulate fine-grained sediments which can become anaerobic; these areas support very few infaunal organisms and are generally avoided by motile epifauna and demersal fauna especially during warm months. Maintenance of navigation channels requires that re-dredging be performed periodically. Generally, sediments removed during maintenance dredging are placed in upland-confined disposal facilities or in open-water dredged material disposal sites. Biota in the latter sites are subject to periodic smothering or displacement, but recolonization is relatively rapid due to the presence of abundant fauna outside the affected areas, the ability of many burrowing infauna to survive smothering by reestablishing themselves in the new sediment layer, and the generally small spatial scale of disposal impacts.

Canal dredging to support oil and gas development has had a more significant impact on coastal habitats, especially salt marshes along the coasts of Louisiana and Texas. In Louisiana, which has approximately 40 % of the coastal wetlands in the conterminous United States, canal dredging for oil and gas well access has modified the coastal landscape and resulted in direct conversion of wetland to open water (Turner 1990). The resulting fragmentation of coastal marshes has been blamed in part for the rapid rate of marsh loss in Louisiana (estimated at 6,400 ha/year [Britsch and Dunbar 1993]), but the primary causes of marsh loss include regional subsidence and faulting and restriction of natural freshwater and sediment input to marshes due to construction of flood control levees along the Mississippi River. Furthermore, levees disrupt the natural delta cycle process that allows the Mississippi River to change course every 1,000–1,500 years, enabling new deltas to become established over time. The Mississippi River levee system, as well as construction of dams within the watershed, has reduced the amount of river-borne sediment that enters coastal marshes, resulting in an imbalance between subsidence and sedimentation. The resulting decrease in salt-marsh habitat has resulted in a decrease in primary productivity and detrital production that are key factors in fisheries production in the Gulf, and in the infaunal and epifaunal communities that inhabit marshes. Assemblages of organisms associated with open-water habitat, including species that have value to commercial and recreational fisheries, replace these communities.

6.6.2.5 Petroleum-Related Development

Oil and gas development in the GoM began in the 1920s and continues today. Aside from disturbances associated with canals and other access channels, petroleum-related activities have created ecosystem disturbances through installation of pipelines, discharges of drilling muds and fluids, discharges of produced water at well sites, installation of hard structures such as production platforms and wellheads, and incidental releases of oil. Oil pollution, *per se,* appears to have had little long-term impact on coastal vegetation, even though acute effects often occur during spill events (Pezeshki et al. 2000). Vegetation responses to petroleum hydrocarbons and vegetation capacity to recover are dependent on the toxicity of the oil, the volume of oil and extent of plant coverage, whether oil penetrates the soil, plant species impacted, oiling frequency, season during which the spill occurred, and cleanup methodologies employed (Lin and Mendelssohn 1996; Hester and Mendelssohn 2000; Pezeshki et al. 2000).

The literature describing the effects of petroleum hydrocarbons on coastal wetlands and their flora and fauna is vast and beyond the scope of this chapter. For published reviews on this topic, see, for example, Fang (1990), Baker et al. (1993), Catallo (1993), Proffitt (1998), Pezeshki et al. (2000), Ko and Day (2004), Michel and Rutherford (2013).

6.6.2.6 Marsh Burns

Besides natural lightning strikes, fire associated with managed burns is a relatively frequent disturbance in herbaceous wetlands such as salt marshes, although small burns do occur in mangroves, killing trees, but usually not initiating a large fire (McKee and Baldwin 1999). Fire creates disturbance patches that allow for subdominant species to gain dominance, often increasing diversity, until finally outcompeted by the surrounding climax vegetation (Nyman and Chabreck 1995). Fire in wetlands becomes particularly problematic when peat burns occur, reducing marsh elevation, increasing inundation, and stressing marsh vegetation.

6.7 SUMMARY

Vegetated marine habitats of the GoM provide a wealth of ecosystem services including food, employment, recreation, and natural system maintenance and regulation to the countries bordering the GoM: the United States, Mexico, and Cuba. The economic, ecological, and aesthetic values of these habitats benefit human well being as illustrated by the desire of humans to live on or near the coast. Ironically, the attraction of coastal shorelines and their habitats to people, along with associated demands for exploitation of natural resources, have led to environmental pressures that have taken their toll on many marine habitats. Nonetheless, coastal habitats of the GoM continue to represent vital components of the GoM ecosystem.

This chapter has reviewed the physical and biological processes that control habitat formation, change, and ecological structure and function. The goal has been to provide baseline information by which resource managers and decision makers can better manage these important natural resources. Those marine habitats that occur immediately adjacent to the GoM, including barrier islands and beaches, salt marshes and mangroves, seagrasses, intertidal and subtidal flats, and reed marshes at the mouth of the Mississippi River have been emphasized.

Although three distinct sedimentary provinces characterize the modern GoM basin, a wide variety of coastal depositional systems have evolved along the 6,077 km (3,776 mi) land–water interface in response to upland drainage; groundwater supply; sediment availability; wind, wave, and current processes; relative sea-level rise; and physiographic characteristics of margin deposits. Carbonate deposits dominate the Mexican States of Campeche (east of Laguna de Términos), Yucatán, and Quintana Roo, as well as the northwestern coast of Cuba and the southwestern coast of Florida. Terrigenous sediment is dominant in the northern GoM where 77 % of all fluvial flow entering the basin originates. Smaller fluvial watersheds along the Tamaulipas, Veracruz, and Tabasco coasts of Mexico contribute the remaining 23 % of flow to the Gulf, resulting in a mixture of fine-grained terrigenous clastics and carbonate sediment.

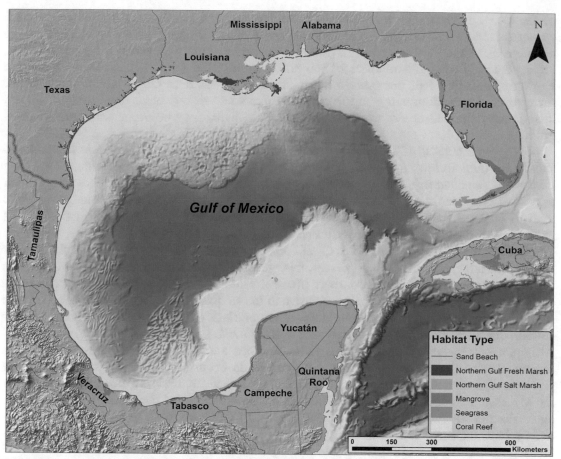

Figure 6.137. Summary distribution of coastal habitat type throughout the GoM (for data sources see Figures 6.80, 6.82–6.84, 6.86, 6.90, 6.92, 6.94, 6.96–6.101; basemap from French and Schenk, 2005).

Vegetated marine habitats are qualitatively similar throughout the GoM, though they vary in relative importance depending upon their location (Figure 6.137). Regional climate, geology, and riverine influence are key drivers of geographical habitat differences. For example, tropical and subtropical mangroves are more prevalent in the Southern GoM Ecoregion, as well as the South Florida/Bahamian Atlantic and Greater Antilles Ecoregions, compared with the Northern GoM Ecoregion, where temperate salt marshes dominate. Seagrasses occur throughout much of the GoM, but areal extent is less abundant in the northern GoM due to reduced water clarity and salinity associated with major riverine discharges of the Mississippi/Atchafalaya drainage basins. Also, arid environments resulting from low precipitation and high evapotranspiration in southern Texas-northwestern Mexico and the northern Yucatán generate hypersaline conditions and sedimentary habitats where rooted vegetation is stunted, absent, or replaced by algal assemblages. Such conditions stand in contrast to much of the remainder of the GoM, where high precipitation and lush vegetated marine habitats occur.

Mangroves in the GoM are dominated by four species: red mangrove (*Rhizophora mangle*), black mangrove (*Avicennia germinans*), white mangrove (*Laguncularia racemosa*), and button mangrove (*Conocarpus erectus*). The cold-tolerant black mangrove dominates in the northern extremes of the GoM, while all four species are important along the more southern shorelines of the GoM. Salt marshes also show distinct differences in species composition with smooth cordgrass (*Spartina alterniflora*) dominating many of the frequently flooded marshes of the northern GoM and black needle-rush (*Juncus roemerianus*) more prevalent in higher elevation, infrequently flooded salt marshes of the northeastern GoM (Mississippi, Alabama and the Florida panhandle). Salt marshes do occur in the more southern GoM, but generally where mangroves cannot dominate due to various stressors and disturbances (e.g., hypersaline and infrequently flooded areas, where herbaceous halophytes may co-occur with stunted mangroves).

Seagrasses consist of five major species in the GoM: turtlegrass (*Thalassia testudinum*), shoalgrass (*Halodule wrightii*), manatee grass (*Syringodium filiforme*), Engelman's seagrass (*Halophila engelmanni*), and widgeon grass (*Ruppia maritima*). Seagrass species composition varies regionally in the GoM. Across the northern Gulf, *Thalassia*, *Halophila*, and *Syringodium* tend to be much more common at lower latitudes in Florida and Texas, and these species gradually become less prevalent northward into Alabama, Mississippi, Louisiana, and north Texas. In this broad central GoM area, *Halodule wrightii* predominates and the three other species are largely absent. When present, they generally represent a minor component of the seagrass community unlike the southern areas of Florida and Texas where they occur abundantly. Seagrass beds are regionally extensive in the southern GoM, where many of the same species occur. Intertidal flats and subtidal soft bottoms lack rooted vascular vegetation but may contain marine macroalgae (seaweeds) such as *Avrainvillea*, *Caulerpa*, *Halimeda*, and *Udotea*. Although barrier islands and beaches are ubiquitous and many plant species are common throughout the GoM, unique species distributions do occur. For example, sea oats (*Uniola paniculata*) dominates much of the northern GoM, but is virtually absent along shorelines from the Mississippi River to northeastern Texas. In the tropical areas of the southern GoM, sea oats again disappears and species such as seagrape (*Coccoloba uvifera*), gullfeed (*Scaevola plumieri*), bay cedar (*Suriana maritima*), baybean (*Canavalia maritima*), sea rosemary (*Tournefortia gnaphaloides*), and others with more tropical affinities dominate.

GoM macroinvertebrates that live at or above the seafloor (epifauna) and on or within the substrate (infauna) are distributed primarily on the basis of sediment texture and quality, and vegetative cover type. Fewer species are adapted to the rigorous habitats provided by salt marshes and *Phragmites* marshes, despite the presence of abundant organic matter: epifauna such as the bivalves *Guekensia demissa* and *Polymesoda caroliniana* occur at the base of marsh plants, while fiddler crabs (*Uca* spp.) and mud crabs (*Rhithropanopeus harrisii*) forage across marsh mud flats. Marsh infauna have low diversity but the species that are present (e.g., the polychaetes *Capitella capitata*, *Neanthes succinea*, and *Laeonereis culveri*) can be relatively abundant. Diversity generally increases as the frequency and duration of inundation increases. Mangroves contain many more species than marshes, but include many of the same taxa, in addition to species such as penaeid shrimps (e.g., *Litopenaeus setiferus* and *Farfantepenaeus aztecus*), portunid crabs (especially blue crab, *Callinectes sapidus*), various sponges and tunicates, and spionid polychaetes. Beach habitats support higher numbers of epifaunal and infaunal taxa, including a wide variety of burrowing forms such as capitellid and nereid polychaetes, bivalves such as *Donax variabilis*, and crustaceans such as *Emerita talpoida*, *Lepidactylus triarticulatus*, and *Acanthohaustorius* spp. Intertidal flats and subtidal soft bottoms contain diverse and abundant faunal communities. These include many burrowing deposit feeders (especially polychaetes), as well as various bivalves, gastropods, echinoderms, and crustaceans. Epifauna and infauna are most diverse in seagrass meadows, which provide relatively stable habitat conditions, high productivity, and structure. Bay scallop (*Argopecten irradians*), green sea urchin (*Lytechinus variegates*), and sea star (*Echinaster serpentarius*) are associated primarily with seagrass beds, but may also occur on other soft bottoms.

Most of the numerically dominant epifaunal and infaunal taxa are found throughout the GoM, while others exhibit more limited geographic distributions. Species that are adapted to finer and organic-rich sediments characterize the Mississippi Estuarine and Texas Estuarine Ecoregions, while some species in the Eastern Gulf Neritic Ecoregion and South Florida/ Bahamian Atlantic Ecoregion are associated primarily with biogenic sediments on the West Florida Shelf and Campeche Banks in the Southern GoM Ecoregion. A faunal break has been described between the Eastern Gulf Neritic and Mississippi Estuarine Ecoregions (as defined by the DeSoto Canyon), but changes in species distributions are less abrupt in most other areas of the Gulf. Coastal habitat epifauna and infauna play an important role in the trophic dynamics of GoM ecosystems. They exhibit a wide range of feeding strategies and are critical to the conversion of vegetative detritus available to higher trophic levels. Few of these taxa are migratory; rather they are typically sedentary or have limited ranges of movement. As a result, their abundance and diversity serve as ideal indicators of habitat quality and perturbation.

Nekton are characterized by their mobility, and their assemblages in the region's vegetated marine habitats are a subset of the fishes, natant crustaceans, molluscs, marine reptiles, and marine mammals found along the beaches, bays, lagoons, and tidal channels of the GoM. It is difficult to describe a characteristic nekton assemblage for individual marine habitats because the habitat of many nekton species includes multiple types of coastal wetlands; species richness and abundance are often greatest at the boundaries (i.e., edges) between subtidal (e.g., embayments) and intertidal (e.g., salt marshes) wetland habitats. A few species like bay anchovy, *Anchoa mitchilli*, are ubiquitous in almost every coastal marine habitat within the GoM (Figure 6.108), while others like zostera shrimp, *Hippolyte zostericola*, and dwarf seahorse, *Hippocampus zostericola*, are closely associated with specific habitats such as subtidal seagrass beds (Figure 6.107). The relatively few species (e.g., fishes in the families *Fundulidae* and *Cyprinodontidae* and shrimps in genus *Palaemonetes*) that are abundant year-round residents of intertidal vegetated marine habitats in the northern GoM (Figure 6.106) are adapted to the wide range of environmental conditions typical of temperate intertidal estuaries. Many other species (e.g., penaeid shrimps, Gulf menhaden) are seasonally abundant as a result of life histories that involve the use of shallow coastal wetlands as nurseries by juvenile life stages (Figures 6.103 and 6.108) during seasons when environmental conditions are favorable for their survival and growth. Overall, nekton assemblages connect vegetated marine habitats across the coastal landscape of the GoM by facilitating significant energy transformations and production transfers among coastal wetland habitats and from estuaries to nearshore coastal marine environments *via* either diel, tidal, and ontogenetic migrations (e.g., penaeid shrimps, Gulf menhaden) or size-structured predator–prey interactions.

Greatest changes in coastal marine habitats occur in areas most susceptible to relative sea-level rise, tropical cyclones, and human disturbances. As such, the deltaic coast of Louisiana has experienced the greatest land and habitat changes in the GoM. Conversely, the more stable coasts of the Yucatán Peninsula, Cuba, and southwestern Florida have illustrated the least amount of change. Although vegetated marine habitats of the GoM are quite productive, human disturbances are recognized in areas of significant industrial activity and tourism. Human impacts are in large part tied to periodic and chronic stressors and disturbances associated with urban, agricultural, and industrial activities. The draining and filling of wetlands for human habitation, agricultural development, and industrial expansion have dramatically impacted coastal habitats throughout the GoM. Also, over-fishing and related activities have threatened important commercial fisheries in some areas of the Gulf. Other stressors such as nutrient enrichment, and resulting eutrophication and hypoxia, altered hydrology from multiple causes, invasive species, and chemical pollutants from agriculture and industry have challenged the health and sustainability of vegetated marine habitats. In addition, natural phenomena such as hurricanes, the underlying geology, and floods and drought exacerbate the human impacts. Information provided in this review is intended to help natural resource managers and policy makers better understand, manage, and restore these important natural ecosystems.

ACKNOWLEDGMENTS

BP sponsored the preparation of this chapter. This chapter was peer reviewed by anonymous and independent reviewers with substantial experience in the subject matter. We thank the peer reviewers, as well as others, including Joseph Baustian, who provided assistance with research and the compilation of information. In particular we thank Jennifer Berlinghoff who assisted with literature reviews, data extraction, figure preparation, and provided editorial and technical reviews of the document. Her contributions were invaluable.

A manuscript of this breadth required multiple authors to adequately address the physical and biological aspects of marine vegetated habitats. To that end, Irving A. Mendelssohn was responsible for documenting the vegetative characteristics of coastal habitats, Mark R. Byrnes was responsible for describing geological controls and physical processes affecting sedimentation and hydrodynamics in coastal habitats, Ronald T. Kneib summarized nekton population dynamics in and adjacent to coastal habitats, and Barry A. Vittor discussed benthic fauna and the distribution and ecology of seagrass habitat in the coastal GoM.

Maps throughout this chapter were created from World Imagery (credits: Esri, i-cubed, USDA, USGS, AEX, GeoEye, Getmapping, Aerogrid, IGN, IGP, and the GIS User Community) and Bing Maps Aerial (credits: © 2010 Microsoft Corporation and its data suppliers) aerial imagery web mapping services using ArcGIS® software by Esri. ArcGIS® and ArcMap™ are the intellectual property of Esri and are used herein under license. Copyright © Esri. All rights reserved. For more information about Esri® software, please visit www.esri.com. All original figures used in this document were reproduced with permission from copyright holders.

REFERENCES

Aagaard T, Greenwood B, Hughes M (2013) Sediment transport on dissipative, intermediate, and reflective beaches. Earth Sci Rev 124:32–50

Able KW, Hagan SM (2000) Effects of common reed (*Phragmites australis*) invasion on marsh surface macrofauna: Response of fishes and decapod crustaceans. Estuaries 23:633–646

Able KW, Hagan SM, Brown SA (2003) Mechanisms of marsh habitat alteration due to *Phragmites*: Response of young-of-the-year mummichog (*Fundulus heteroclitus*) to treatment for *Phragmites* removal. Estuaries 26:484–494

Abraham KF, Jefferies R, Rockwell RF (2005) Goose-induced changes in vegetation and land cover between 1976 and 1997 in an Arctic coastal marsh. Arct Antarct Alp Res 37:269–275

Absalonsen L, Dean RG (2010) Characteristics of shoreline change along the sandy beaches of the State of Florida: An atlas. Florida Sea Grant College Program R/C-S-35, University of Florida, Gainesville, FL, USA, 304 p. http://nsgl.gso.uri.edu/flsgp/flsgpm10001.pdf. Accessed 1 July 2013

Absalonsen L, Dean RG (2011) Characteristics of the shoreline change along Florida sandy beaches with an example for Palm Beach County. J Coast Res 27:16–26

Aburto-Oropeza O, Ezcurra E, Danemann G, Valdez V, Murray J, Sala E (2008) Mangroves in the Gulf of California increase fishery yields. Proc Natl Acad Sci USA 105:10456–10459

Adam P (1990) Salt marsh ecology. Cambridge University Press, Cambridge, MA, USA, 461 p

Akin S, Winemiller KO, Gelwick FP (2003) Seasonal and spatial variations in fish and macro-crustacean assemblage structure in Mad Island Marsh estuary, Texas. Estuar Coast Shelf Sci 57:269–282

Alber M, Swenson EM, Adamowicz SC, Mendelssohn IA (2008) Salt marsh dieback: An overview of recent events in the U.S. Estuar Coast Shelf Sci 80:1–11

Aleyev YG (1977) Nekton, Introduction: Nekton as an ecomorphological type of biont. Dr. W. Junk Publishers, The Hague, The Netherlands, pp 1–43

Alfaro AC (2006) Benthic macro-invertebrate community composition within a mangrove/seagrass estuary in northern New Zealand. Estuar Coast Shelf Sci 66:97–110

Alongi DM, Christofersсen P (1992) Benthic infauna and organism-sediment relations in a shallow, tropical coastal area: Influence of outwelled mangrove detritus and physical disturbance. Mar Ecol Prog Ser 81:229–245

Amante C, Eakins BW (2009) ETOPO1 1 Arc-Minute Global Relief Model: Procedures, Data Sources and Analysis. NOAA Technical Memorandum NESDIS NGDC-24. National Geophysical Data Center, NOAA, Boulder, CO, USA, 25 p

Andrew NL, Mapstone BD (1987) Sampling and the description of spatial pattern in marine ecology. Oceanogr Mar Biol Annu Rev 25:39–90

Angradi TR, Hagan SM, Able KW (2001) Vegetation type and the intertidal macroinvertebrate fauna of a brackish marsh: *Phragmites* vs *Spartina*. Wetlands 21:75–92

Anton A, Cebrián J, Duarte CM, Heck KL Jr, Goff J (2009) Low impact of Hurricane Katrina on seagrass community structure and functioning in the northern Gulf of Mexico. Bull Mar Sci 85:45–59

APG (Angiosperm Phylogeny Group) III (2009) An update of the Angiosperm Phylogeny Group classification for the orders and families of flowering plants: APG III. Bot J Linn Soc 161:105–121

Appendini CM, Paulo Salles E, Tonatiuh Mendoza E, López J, Torres-Freyermuth A (2012) Longshore sediment transport on the northern coast of the Yucatán Peninsula. J Coast Res 28:1404–1417

Arceo-Carranza D, Vega-Cendejas ME (2009) Spatial and temporal characterization of fish assemblages in a tropical coastal system influenced by freshwater inputs: Northwestern Yucatán Peninsula. Rev Biol Trop 57:89–103

Baird D, Ulanowicz RE (1989) The seasonal dynamics of the Chesapeake Bay ecosystem. Ecol Monogr 59:329–364

Baker JH, Jobe WD, Howard CL, Kimball KT, Janousek J, Chase PR (1981) Benthic biology, part 6. In: Bedinger CA Jr (ed) Ecological investigations of petroleum production platforms in the Central Gulf of Mexico. Southwest Research Institute, Houston, TX, USA, 209 p, plus appendices

Baker JM, Little DI, Owens EH (1993) A review of experimental shoreline oil spills. In: International oil spill conference (prevention, preparedness, response). American Petroleum Institute, Washington, DC, USA, pp 583–590

Baker P, Fajans JS, Baker SM (2011) Habitat dominance of a nonindigenous tropical bivalve, *Perna viridis* (Linnaeus, 1758), in a subtropical estuary in the Gulf of Mexico. J Molluscan Stud 78:28–33

Baldwin AH, Platt WJ, Gathen KL, Lessmann JM, Rauch TJ (1995) Hurricane damage and regeneration in fringe mangrove forests of southeastern Florida, USA. J Coast Res SI 21:169–183

Baldwin A, Egnotovich M, Ford M, Platt W (2001) Regeneration in fringe mangrove forests damaged by Hurricane Andrew. Plant Ecol 157:151–164

Balsillie JH, Donoghue JF (2011) Northern Gulf of Mexico sea-level history for the past 20,000 years. In: Buster NA, Holmes CW (eds) Gulf of Mexico: Origin, waters, and biota, vol 3, Geology. Texas A&M University Press, College Station, TX, USA, pp 53–69

Baltz DM, Rakocinski C, Fleeger JW (1993) Microhabitat use by marsh-edge fishes in a Louisiana estuary. Environ Biol Fish 36:109–126

Bao TQ (2011) Effect of mangrove forest structures on wave attenuation in coastal Vietnam. Oceanologia 53:807–818

Barba E, Raz-Guzman A, Sánchez AJ (2005) Distribution patterns of estuarine caridean shrimps in the southwestern Gulf of Mexico. Crustaceana 78:709–726

Barbier EB, Strand I (1998) Valuing mangrove-fishery linkages: A case study of Campeche, Mexico. Environ Resour Econ 12:151–166

Barbour MG (1992) Life at the leading edge: The beach plant syndrome. In: Seeliger U (ed) Coastal plant communities of Latin America. Academic Press, San Diego, CA, USA, pp 291–307

Barbour MG, De Jong TM, Pavlik BM (1985) Marine beach and dune plant communities. In: Chabot BF, Mooney HA (eds) Physiological ecology of North American plant communities. Chapman and Hall, New York, NY, USA, pp 296–318

Barbour MG, Rejmanek M, Johnson AF, Pavlik BM (1987) Beach vegetation and plant distribution patterns along the northern Gulf of Mexico. Phytocoenologia 15:201–233

Barry A. Vittor & Associates, Inc. (1986) Report on disposal site designation for the interim approved Port St. Joe and Panama City, Florida offshore dredged material disposal sites. Contract Report. USEPA (U.S. Environmental Protection Agency), Washington, DC, USA, 106 p

Barry A. Vittor & Associates, Inc (1991) Pensacola (FL) ocean dredged material disposal site benthic communities. Final report. USEPA, Washington, DC, USA, 34 p

Barry A. Vittor & Associates, Inc (1995) LOOP 14-year monitoring program synthesis report. Final report. LOOP, New Orleans, LA, USA, 74 p

Barry A. Vittor & Associates, Inc (1996) Benthic macroinfaunal analysis of open-water dredged material sites in Laguna Madre, Texas, May 1996 survey. Final report. Espy, Huston and Associates, Austin, TX, USA, 26 p

Barry A. Vittor & Associates, Inc (1999) Florida Keys National Marine Sanctuary Benthic Community Assessment. Final report. NOAA (National Oceanic and Atmospheric Administration), Silver Spring, MD, USA, 43 p

Barry A. Vittor & Associates, Inc (2004) Mapping of submerged aquatic vegetation in Mobile Bay and adjacent waters of Coastal Alabama. Final report. Mobile Bay National Estuary Program, Mobile, AL, USA, 30 p

Barry A. Vittor & Associates, Inc (2009) Submerged aquatic vegetation mapping in Mobile Bay and adjacent waters of coastal Alabama in 2008 and 2009. Final report. Mobile Bay National Estuary Program, Mobile, AL, USA, 16 p

Barry A. Vittor & Associates, Inc. (2010) Mapping of submerged aquatic vegetation in 2010: Mississippi Barrier Island Restoration Project. Report prepared for the Mobile District Corps of Engineers, Mobile, AL, USA, 6 p (excluding maps)

Barry A. Vittor & Associates, Inc (2011) Mississippi Sound Coastal Improvement Program (MsCIP): Mississippi Sound and the Gulf of Mexico Benthic Macroinfauna Community Assessment. Report. U.S. Army Corps of Engineers, Mobile District, Mobile, AL, USA

Battaglin WA, Aulenbach BT, Vecchia A, Buxton HT (2010) Changes in streamflow and the flux of nutrients in the Mississippi-Atchafalaya River Basin, USA, 1980-2007. USGS (U.S. Geological Survey) scientific investigations report 2009-5164. USGS, Reston, VA, USA, 47 p

Baum JK, Vincent ACJ (2005) Magnitude and inferred impacts of the seahorse trade in Latin America. Environ Conserv 32:305–319

Bearden BL, Hummell RL (1990) Geomorphology of coastal sand dunes, Morgan Peninsula, Baldwin County, Alabama. Alabama geological survey circular 150. Geological Survey of Alabama, Tuscaloosa, AL, USA, 71 p

Beck MW, Odaya M, Bachant JJ, Bergan J, Keller B, Martin R, Mathews R, Porter C, Ramseur G (2000) Identification of priority sites for conservation in the northern Gulf of Mexico: An ecoregional plan. The Nature Conservancy, Arlington, VA, USA, 48 p. http://sdms.cr. usgs.gov/pub/ngom/ngom.html. Vector digital data accessed 19 Jan 2015

Beck MW, Heck KL Jr, Able KW, Childers DL, Eggleston DB, Gillanders BM, Halpern B, Hays CG, Hoshino K, Minello TJ, Orth RJ, Sheridan PF, Weinstein MP (2001) The identification, conservation, and management of estuarine and marine nurseries for fish and invertebrates. BioScience 51:633–641

Becker JJ, Sandwell DT, Smith WHF, Braud J, Binder B, Depner J, Fabre D, Factor J, Ingalls S, Kim S-H, Ladner R, Marks K, Nelson S, Pharaoh A, Trimmer R, Von Rosenberg J, Wallace G, Weatherall P (2009) Global bathymetry and elevation data at 30 arc seconds resolution: SRTM30_PLUS. Mar Geod 32:355–371

Beddows PA, Smart PL, Whitaker FF, Smith SL (2007) Decoupled fresh-saline groundwater circulation of a coastal carbonate aquifer: Spatial patterns of temperature and specific electrical conductivity. J Hydrol 346:18–32

Beever JW III, Simberloff D, King LL (1979) Herbivory and predation by the mangrove tree crab Aratus pisonii. Oecologia 43:317–328

BEG (Bureau of Economic Geology) (1995) Shoreline types: Sabine Pass to Matagorda Peninsula, Environmental Sensitivity Index (ESI) Shoreline for the Upper Texas Coast. Vector digital data. The University of Texas at Austin, BEG, Austin, TX, USA. http://www.beg. utexas.edu/coastal/download.php. Accessed 19 Jan 2015

BEG (2000) Shoreline types: Matagorda Peninsula to the Rio Grande, Environmental Sensitivity Index (ESI) Shoreline for the Lower Texas Coast. Vector digital data. The University of Texas at Austin, BEG, Austin, TX, USA. http://www.beg.utexas.edu/coastal/download.php. Accessed 19 Jan 2015

Bell SS, Brooks RA, Robbins BD, Fonseca MS, Hall MO (2001) Faunal response to fragmentation in seagrass habitats: Implications for seagrass conservation. Biol Conserv 100:115–123

Bernard HA, Major CF Jr, Parrott BS, LeBlanc RJ Sr (1970) Recent sediments of Southeast Texas, a field guide to the Brazos Alluvial and Deltaic Plains and the Galveston Barrier Island Complex. The Bureau of Economic Geology guidebook 11. University of Texas at Austin, Austin, TX, USA, 47 p

Bertness MD (1984) Ribbed mussels and Spartina alterniflora production in a New England salt marsh. Ecology 65:1042–1055

Bertness MD (1985) Fiddler crab regulation of Spartina alterniflora production in a New England salt marsh. Ecology 66:1042–1055

Bertness MD (1988) Peat accumulation and the success of marsh plants. Ecology 69:703–713

Bertness MD (1992) The ecology of a New England salt marsh. Am Sci 80:260–268

Bertness MD, Ellison AM (1987) Determinants of pattern in a New England salt marsh plant community. Ecol Monogr 57:129–147

Bertness MD, Ewanchuk PJ, Silliman BR (2002) Anthropogenic modification of New England salt marsh landscapes. Proc Natl Acad Sci USA 99:1395–1398

Bethoux JP, Copin-Montégut G (1986) Biological fixation of atmospheric nitrogen in the Mediterranean Sea. Limnol Oceangr 31:1353–1358

Bielsa LM, Murdich WH, Labisky RF (1983) Species profiles: Life histories and environmental requirements of coastal fishes and invertebrates (south Florida)—pink shrimp. U.S. Fish and Wildlife Service, FWS/OBS-82/11.17. U.S. Army Corps of Engineers, TR EL-82-4. U.S. Fish and Wildlife Service, Washington, DC, USA, 21 p

Bird DE, Burke K, Hall SA, Casey JF (2011) Tectonic evolution of the Gulf of Mexico. In: Buster NA, Holmes CW (eds) Gulf of Mexico: Origin, waters, and biota, vol 3, Geology. Texas A&M University Press, College Station, TX, USA, pp 3–16

Blaha J, Sturges W (1981) Evidence for wind forced circulation in the Gulf of Mexico. J Mar Res 39:711–734

Blake ES, Rappaport EN, Landsea CW (2007) The deadliest, costliest, and most 34 intense United States tropical cyclones from 1851 to 2006. Technical Memorandum NWS 35 TPC-5. National Weather Service, National Hurricane Center, Miami, FL, USA, 47 p

Blum MD, Roberts HH (2009) Drowning of the Mississippi Delta due to insufficient sediment supply and global sea-level rise. Nat Geosci 2:488–491

Blundon JA, Kennedy VS (1982) Refuges for infaunal bivalves from blue crab, *Callinectes sapidum* (Rathbun), predation in Chesapeake Bay. J Exp Mar Biol Ecol 65:67–81

Boehm PD, Fiest DL (1982) Subsurface distributions of petroleum from an offshore well blowout: The Ixtoc I blowout, Bay of Campeche. Environ Sci Technol 16:67–74

BOEM (Bureau of Ocean Energy Management) (2011) Outer continental shelf oil and gas leasing program: 2012-2017. Draft environmental impact statement. U.S. Department of the Interior, Washington, DC, USA, 1492 p

Boesch DF (1972) Species diversity of marine macrobenthos in the Virginia area. Chesapeake Sci 13:206–211

Boesch DF (1977) A new look at the Zonation of Benthos along the estuarine gradient. In: Coull BC (ed) Ecology of marine benthos. University of South Carolina Press, Columbia, SC, USA, pp 245–266

Bologna PAX, Heck KL Jr (1999) Differential predation and growth rates of bay scallops within a seagrass habitat. J Exp Mar Biol Ecol 239:299–314

Bolongaro Crevenna Recaséns A (2012) Vulnerabilidad de Hábitats de Anidación de Tortugas Marinas por Efectos de Erosión Costera en el estado de Campeche en un Contexto de Cambio Climático. 2nd Congreso Nacional de Investigación en Cambio Climático. Academia Nacional de Investigación y Desarrollo

Borhidi A (1996) Phytogeography and vegetation ecology of Cuba. Akademiai Kiado, Budapest, Hungary

Boström C, Pittman SJ, Simenstad C, Kneib RT (2011) Seascape ecology of coastal biogenic habitats: Advances, gaps, and challenges. Mar Ecol Prog Ser 427:191–217

Bourget E, Messier D (1983) Macrobenthic density, biomass, and fauna of intertidal and subtidal sand in a Magdalen Islands lagoon, Gulf of St Lawrence Canadian. J Zool 61:2509–2518

Bowman RE, Stillwell CE, Michaels WL, Grosslein MD (2000) Food of Northwest Atlantic fishes and two common species of squid. NOAA Technical Memorandum NMFS NE 155. U.S. Department of Commerce, Washington, DC, USA, 137 p

Boyd R, Suter JR, Penland S (1989) Sequence stratigraphy of the Mississippi Delta. Trans Gulf Coast Geol Soc 39:331–340

Brenchley GA (1982) Mechanisms of spatial competition in marine soft-sediment communities. J Exp Mar Biol Ecol 60:17–33

Brinson MM (1977) Decomposition and nutrient exchange of litter in an alluvial swamp forest. Ecology 58:601–609

Britsch LD, Dunbar JB (1993) Land loss rates: Louisiana coastal plain. J Coast Res 9:324–338

Britton JC, Morton B (1989) Shore ecology of the Gulf of Mexico. University of Texas Press, Austin, TX, USA, 387 p

Brix H, Sorrell BK, Schierup H-H (1996) Gas fluxes achieved by in situ convective flow in *Phragmites australis*. Aquat Bot 54:151–163

Brooks DA (1984) Current and hydrographic variability in the Northwestern Gulf of Mexico. J Geophys Res 89:8022–8032

Brooks RA, Bell SS (2002) Mangrove response to attack by a root boring isopod: Root repair versus architectural modification. Mar Ecol Progr Ser 231:85–91

Brooks JM, Giammona CP (eds) (1991) Mississippi-Alabama Continental Shelf Ecosystem Study data summary and synthesis. U.S. Department of the Interior, MMS (Minerals Management Service), Gulf of Mexico OCS Region, New Orleans, LA, USA

Broome SW, Woodhouse WW Jr, Seneca ED (1975) The relationship of mineral nutrients to growth of Spartina alterniflora in North Carolina. II. The effects of N, P, and Fe fertilizers. Soil Sci Soc Am Proc 39:301–307

Browder JA, May LN Jr, Rosenthal A, Gosselink JG, Bauman RH (1989) Modeling future trends in wetland loss and brown shrimp production in Louisiana using thematic mapper imagery. Remote Sens Environ 28:45–59

Brown B (1982) Spatial and temporal distribution of a deposit-feeding polychaete on a heterogeneous tidal flat. J Exp Mar Biol Ecol 65:213–227

Bryant WR, Simmons GR, Grim PJ (1991) The morphology and evolution of basins on the continental slope, northwest Gulf of Mexico. Trans Gulf Coast Assoc Geol Soc 51:73–82

Buchsbaum RN, Catena J, Hutchins E, James-Pirri M-J (2006) Changes in salt marsh vegetation, *Phragmites australis*, and nekton in response to increased tidal flushing in a New England salt marsh. Wetlands 26:544–557

Burdick DM, Konisky RA (2003) Determinants of expansion for *Phragmites australis*, common reed, in natural and impacted coastal marshes. Estuaries 26(2B):407–416

Burdick DM, Buchsbaum R, Holt E (2001) Variation in soil salinity associated with expansion of *Phragmites australis* in salt marshes. Environ Exp Bot 46:247–261

Burfeind DD, Stunz GW (2006) The effects of boat propeller scarring intensity on nekton abundance in subtropical seagrass meadows. Mar Biol 148:953–962

Buster NA, Holmes CW (2011) Gulf of Mexico: Origin, waters, and biota, vol 3, Geology. Texas A&M University Press, College Station, TX, USA

Butler JA, Heinrich GL (2007) The effectiveness of bycatch reduction devices on crab pots at reducing capture and mortality of diamondback terrapins (*Malaclemys terrapin*) in Florida. Estuar Coast 30:179–185

Buttermore RE, Turner E, Morrige MG (1994) The introduced northern Pacific seastar *Asterias amurensis* in Tasmania. Mem Queensland Mus 36:21–25

Byers JE (2009) Invasive animals in marshes: Biological agents of change. In: Silliman BR, Grosholz ED, Bertness MD (eds) Human impacts on salt marshes: A global perspective, 1st edn. University of California Press, Oakland, CA, USA, pp 41–56

Byrne JV, Leroy DO, Riley CM (1959) The Chenier Plain and its stratigraphy, southwestern Louisiana. Trans Gulf Coast Assoc Geol Soc 9:237–259

Byrnes MR, McBride RA (2009) Coastal response to Hurricane Ike (2008): Southwest Louisiana and southeast Texas. Shore Beach 77:37–48

Byrnes MR, McBride RA, Tao Q, Davis L (1995) Historical shoreline dynamics along the Chenier Plain of southwestern Louisiana. Trans Gulf Coast Assoc Geol Soc 45:113–122

Byrnes MR, Hammer RM, Vittor BA, Ramsey JS, Snyder DB, Bosma KF, Wood JD, Thibaut TD, Phillips NW (1999) Environmental survey of identified sand resource areas offshore Alabama: Vol I: Main Text, Vol II: Appendices. OCS Report MMS 99-0052. U.S. Department of Interior, MMS, International Activities and Marine Minerals Division, Herndon, VA, USA, 326 p plus appendices

Byrnes MR, Griffee SF, Osler MS (2010) Channel dredging and geomorphic response at and adjacent to Mobile Pass, Alabama. Technical report ERDC/CHL TR-10-8. U.S. Army Engineer Research and Development Center, Vicksburg, MS, USA, 309 p

Byrnes MR, Rosati JD, Griffee SF, Berlinghoff JL (2012) Littoral sediment budget for the Mississippi Sound Barrier Islands. Technical report ERDC/CHL TR-12-9. U.S. Army Engineering Research and Development Center, Vicksburg, MS, USA, 180 p

Byrnes MR, Rosati JD, Griffee SF, Berlinghoff JL (2013) Historical sediment transport pathways and quantities for determining an operational sediment budget: Mississippi Sound barrier islands. In Brock JC, Barras JA, Williams SJ, eds, Understanding and predicting change in the coastal ecosystems of the Northern Gulf of Mexico. J Coast Res Special Issue 63:166–183

Cahoon DR (2006) A review of major storm impacts on coastal wetland elevations. Estuar Coast 29:889–898

Caillouet CW, Hart RA, Nance JM (2008) Growth overfishing in the brown shrimp fishery of Texas, Louisiana, and adjoining Gulf of Mexico EEZ. Fish Res 92:289–302

Cannicci S, Burrows D, Fratini S, Smith TJ, Offenberg J, Dahdouh-Guebas F (2008) Faunal impact on vegetation structure and ecosystem function in mangrove forests: A review. Aquat Bot 89:186–200

Canto-Maza WG, Vega-Cendejas ME (2007) Distribution, abundance and alimentary preferences of the fish *Opsanus phobetron* (Batrachoididae) at the Chelem coastal lagoon, Yucatán, México. Rev Biol Trop 55:979–988

Canto-Maza WG, Vega-Cendejas ME (2008) Feeding habits of the fish *Lagodon rhomboides* (Perciformes: Sparidae) at the coastal lagoon of Chelem, Yucatán, México. Rev Biol Trop 56:1837–1846

Carlton JT (1999) The scale and ecological consequences of biological invasions in the world's oceans. In: Sandlund O, Schei P, Viken A (eds) Invasive species and biodiversity management. Kluwer, Dordrecht, The Netherlands, pp 195–212

Carranza-Edwards A (2011) Mexican littoral of the Gulf of Mexico. In: Buster NA, Holmes CW (eds) Gulf of Mexico: Origin, waters, and biota, vol 3, Geology. Texas A&M University Press, College Station, TX, USA, pp 291–296

Carranza-Edwards A, Rosales-Hoz L, Chávez MC, Morales de la Garza E (2007) Environmental geology of the coastal zone. In: Withers K, Nipper M (eds) Environmental analysis of the Gulf of Mexico, Special publication series 1. Harte Research Institute for Gulf of Mexico Studies, Corpus Christi, TX, USA, pp 351–372

Carrera LM, Santander AP (1994) Los manglares de Cuba: Ecologia. In: Suman DO (ed) El ecosystema de manglar en America Latina y la cuenca del Caribe: su manejo y conservacion. University of Miami, Miami, FL, USA, pp 64–75

Carrillo L, Johns EM, Smith RH, Lamkin JT, Largier JL (2015) Pathways and hydrography in the Mesoamerican Barrier Reef System, Part 1: Circulation. Cont Shelf Res 109:164–176

Casey KS, Brandon TB, Cornillon P, Evans R (2010) The past, present and future of the AVHRR Pathfinder SST Program. In: Barale V, Gower JFR, Alberotanza L (eds) Oceanography from space: Revisited. Springer Science+Business Media, New York, NY, USA, pp 323–340

Castaneda-Moya E, Twilley RR, Rivera-Monroy VH, Zhang KQ, Davis SE, Ross M (2010) Sediment and nutrient deposition associated with Hurricane Wilma in mangroves of the Florida coastal Everglades. Estuar Coast 33:45–58

Castaneda-Moya E, Twilley RR, Rivera-Monroy VH, Marx BD, Coronado-Molina C, Ewe SML (2011) Patterns of root dynamics in mangrove forests along environmental gradients in the Florida coastal Everglades, USA. Ecosystems 14:1178–1195

Catallo WJ (1993) Ecotoxicology and wetland ecosystems: Current understanding and future needs. Environ Toxicol Chem 12:2209–2224

Caudill MC (2005) Nekton utilization of black mangrove (*Avicennia germinans*) and smooth cordgrass (*Spartina alterniflora*) sites in southwestern Caminada Bay, Louisiana. Master's Thesis, Louisiana State University, Baton Rouge, LA, USA, 71 p

Cavalieri AJ, Huang HA (1979) Evaluation of proline accumulation in the adaptation of diverse species of marsh halophytes to the saline environment. Am J Bot 66:307–312

Cebrián J (1999) Patterns in the fate of production in plant communities. Am Nat 154:449–468

Cebrián J (2002) Variability and control of carbon consumption, export and accumulation in marine communities. Limnol Oceanogr 47:11–22

Cebrián J, Duarte CM (1994) The dependence of herbivory on growth rate in natural plant communities. Funct Ecol 8:518–525

Cebrián J, Duarte CM, Marbà N, Enríquez S, Gallegos M, Olesen B (1996) Herbivory on *Posidonia oceanica*: Magnitude and variability in the Spanish Mediterranean. Mar Ecol Prog Ser 130:147–155

Cebrián J, Foster CD, Plutchak R, Sheehan KL, Miller MC, Anton A, Major K, Heck KL Jr, Powers SP (2008) The impact of Hurricane Ivan on the primary productivity and metabolism of marsh tidal creeks in the North Central Gulf of Mexico. Aquat Ecol 42:391–404

CEC (Commission for Environmental Cooperation) (2007) North America Elevation 1-Kilometer Resolution Map, 3rd edn. Collaborators include U.S. Department of the Interior, USGS, National Atlas of the U.S. CEC, Montréal, Québec, Canada. http://www.cec.org/Page.asp?PageID=924&ContentID=2841&SiteNodeID=497&BL_ExpandID=. Accessed 19 Jan 2015

CEC (2010) North American atlas-Basin Watersheds. Vector digital data. Collaborators include Natural Resources Canada, Instituto Nacional de Estadística y Geografía, USGS, Commission for Environmental Cooperation, Montréal, Québec, Canada. http://www.cec.org/Page.asp?PageID=924&ContentID=2866. Accessed 19 Jan 2015

CEC (2011a) North America climate-mean annual temperature. Raster digital data. Created by Museum of Vertebrate Zoology, University of California. Commission for Environmental Cooperation, Montréal, Québec, Canada. http://www.cec.org/Page.asp?PageID=122&ContentID=17628&SiteNodeID=659&AA_SiteLanguageID=1. Accessed 6 Feb 2013

CEC (2011b) North America climate-mean annual precipitation. Raster digital data. Created by Museum of Vertebrate Zoology, University of California. CEC, Montréal, Québec, Canada. http://www.cec.org/Page.asp?PageID=122&ContentID=17624&SiteNodeID=659&AA_SiteLanguageID=1. Accessed 19 Jan 2015

Chabreck RH (1972) Vegetation, water and soil characteristics of the Louisiana coastal region. Bulletin 664. Agricultural Experiment Station, Louisiana State University, Baton Rouge, LA, USA

Chambers RM, Meyerson LA, Saltonstall K (1999) Expansion of *Phragmites australis* into tidal wetlands of North America. Aquat Bot 64:261–273

Chang Y-L, Oey L-Y (2010) Eddy and wind-forced heat transports in the Gulf of Mexico. J Phys Oceanogr 40:2728–2742

Chapman VJ (1976) Coastal vegetation. Pergamon, Oxford, UK

Chauta J (2012) 10 Thousand Island Everglades FL. https://www.flickr.com/photos/juliochauta/7919077024/. Accessed 19 Feb 2017

Chesney EJ, Baltz DM, Thomas RG (2000) Louisiana estuarine and coastal fisheries and habitats: Perspectives from a fish's eye view. Ecol Appl 10:350–366

Chin-Leo G, Benner R (1991) Dynamics of bacterioplankton abundance and production in seagrass communities of a hypersaline lagoon. Mar Ecol Prog Ser 73:219–230

Chisholm JRM, Fernex FE, Mathieu D, Jaubert JM (1997) Wastewater discharge, seagrass decline and algal proliferation on the Côte d'Azur. Mar Pollut Bull 34:78–84

Christensen NL (2000) Vegetation of the southeastern coastal plain. In: Barbour MG, Billings WD (eds) North American Terrestrial Vegetation, 2nd edn. Cambridge University Press, Cambridge, UK, pp 398–448

Clark RR (1990) The Carbonate Beaches of Florida: An inventory of Monroe County Beaches. Beaches and Shores Technical and Design Memorandum 90-1. Florida Department of Natural Resources Division of Beaches and Shores, Tallahassee, FL, USA, 51 p

Cochrane JD, Kelly FJ (1986) Low-frequency circulation on the Texas-Louisiana Continental Shelf. J Geophys Res 91:10645–10659

Cocito S (2006) Bioconstruction and biodiversity: Their mutual influence. Sci Mar 68:137–144

Coen LD, Heck KL Jr (1991) The interacting effects of siphon nipping and habitat on bivalve (*Mercenaria mercenaria* (L.)) growth in a subtropical seagrass (*Halodule wrightii* Aschers.) meadow. J Exp Mar Biol Ecol 145:1–13

Coen LD, Heck KL Jr, Able LG (1981) Experiments on competition and predation among shrimps of seagrass meadows. Ecology 62:1484–1493

Coles SL, Brown EK (2007) Twenty-five years of change in coral coverage on a hurricane impacted reef in Hawaii: The importance of recruitment. Coral Reefs 26:705–717

Continental Shelf Associates, and Martel Laboratories (1985) Florida Big Bend seagrass habitat study narrative report. Contract 14-12-0001-30188. Mineral Management Services Metairie, LA, USA, 47 p + Appendices

Contreras-Espinosa F, Castañeda-Lopez O (2007) Coastal lagoons and estuaries of the Gulf of Mexico. In: Withers K, Nipper M (eds) Environmental analysis of the Gulf of Mexico. Harte Research Institute for Gulf of Mexico Studies, Houston, TX, USA, pp 230–261

Contreras-Espinosa F, Warner BG (2004) Ecosystem characteristics and management considerations for coastal wetlands in Mexico. Hydrobiologia 511:233–245

Costa CSB, Davy AJ (1992) Coastal salt marsh communities of Latin America. In: Seeliger U (ed) Coastal plant communities of Latin America. Academic Press, San Diego, CA, USA, pp 179–199

Costanza R, d'Arge R, de Groot R, Farber S, Grasso M, Hannon B, Limburg K, Naeem S, O'Neill RV, Paruelo J, Raskin RG, Suttonkk P, van den Belt M (1997) The value of the world's ecosystem services and natural capital. Nature 387:253–260

Courtemanche RP, Hester MW, Mendelssohn IA (1999) Recovery of a Louisiana barrier island marsh plant community following extensive hurricane-induced overwash. J Coast Res 15:872–883

Couvillion BR, Barras JA, Steyer GD, Sleavin-William W, Fisher M, Beck H, Trahan N, Griffin B, Heckman D (2011) Land area change in coastal Louisiana from 1932 to 2010. USGS Scientific Investigations Map 3164, scale 1:265,000. USGS, Reston, VA, USA, 12 p

Craighead FC, Gilbert VC (1962) The effects of Hurricane Donna on the vegetation of southern Florida. Q J Florida Acad Sci 25:1–28

Creed JC, Amado Filho GM (1999) Disturbance and recovery of the macroflora of a seagrass (*Halodule wrightii* Ascherson) meadow in the Abrolhos Marine National Park, Brazil: An experimental evaluation of anchor damage. J Exp Mar Biol Ecol 235:285–306

Crimaldi JP, Thompson JK, Rosman JH, Lowe RJ, Koseff JR (2002) Hydrodynamics of larval settlement: The influence of turbulent stress events at potential recruitment sites. Limnol Oceanogr 47:1137–1151

Croonen K, Froeling D, Marbus G, van Bemmel M (2006) Port of Altamira, Mexico. Master Project CT4061. Delft University of Technology, Delft, The Netherlands, 122 p

Currin CA, Wainright SC, Able KW, Weinstein MP, Fuller CM (2003) Determination of food web support and trophic position of the mummichog, *Fundulus heteroclitus*, in New Jersey smooth cordgrass (*Spartina alterniflora*), common reed (*Phragmites australis*), and restored salt marshes. Estuar Coast 26:495–510

Dahl BE, Fall BA, Lohse A, Appan SG (1974) Stabilization and reconstruction of Texas coastal foredunes with vegetation. Consortium, Gulf Universities Research, Galveston, TX, USA, 139 p

Daigle JJ, Griffith GE, Omernik JM, Faulkner PL, McCulloh RP, Handley LR, Smith LM, Chapman SS (2006) Ecoregions of Louisiana (color poster with map, descriptive text, summary tables, and photographs; map scale 1:1,000,000). USGS, Reston, VA, USA

Dames and Moore (1979) Mississippi, Alabama, Florida outer continental shelf baseline environmental survey; MAFLA, 1977/78. Program synthesis report. BLM/YM/ES-79/01-Vol-1-A. U.S. Department of the Interior, Bureau of Land Management, Washington, DC, USA, 278 p

Darby FA, Turner RE (2008) Below- and above-ground biomass of *Spartina alterniflora*: Response to nutrient addition in a Louisiana salt marsh. Estuar Coast 31:326–334

Dauby P (1989) The stable carbon isotope ratios in benthic food webs of the Gulf of Calvi, Corsica. Cont Shelf Res 9:181–195

Dauby P (1995) A δ^{13}C study of the feeding habits in four Mediterranean *Leptomysis* species (Crustacea: Mysidacea). PSZNI Mar Ecol 16:93–102

Davis DW (2010) Washed away? The invisible peoples of Louisiana's wetlands. University of Louisiana at Lafayette Press, Lafayette, LA, USA, 578 p

Davis RA (1995) Geologic impact of Hurricane Andrew on Everglades coast of southwest Florida. Environ Geol 25:143–148

Davis RA (1999) Complicated littoral drift systems in the Gulf Coast of peninsular Florida. In: Kraus NC (ed) Coastal sediments' 99. American Society of Civil Engineers, New York, NY, USA, pp 761–769

Davis RA (2011a) Sea-level change in the Gulf of Mexico. Texas A&M University Press, College Station, TX, USA, 172 p

Davis RA (2011b) Beaches, barrier islands, and inlets of the Florida Gulf Coast. In: Buster NA, Holmes CW (eds) Gulf of Mexico: Origin, waters, and biota, vol 3, Geology. Texas A&M University Press, College Station, TX, USA, pp 89–99

Davis RA, Yale KE, Pekala JM, Hamilton MV (2003) Barrier island stratigraphy and Holocene history of west-central Florida. Mar Geol 200:103–123

Davis SE, Cable JE, Childers DL, Coronado-Molina C, Day JW, Hittle CD, Madden CJ, Reyes E, Rudnick D, Sklar F (2004) Importance of storm events in controlling ecosystem structure and function in a Florida gulf coast estuary. J Coast Res 20:1198–1208

Dawes C, Siar K, Marlett D (1999) Mangrove structure, litter and macroalgal productivity in a northern-most forest of Florida. Mangroves Salt Marshes 3:259–267

Dawes CJ, Phillips RC, Morrison G (2004) Seagrass communities of the Gulf Coast of Florida: status and ecology. Florida Fish and Wildlife Conservation Commission, Fish and Wildlife Research Institute, and Tampa Bay Estuary Program, St. Petersburg, FL, USA, 74 p

Day JH, Field JG, Montgomery MP (1971) The use of numerical methods to determine the distribution of the benthic fauna across the continental shelf of North Carolina. J Anim Ecol 40:93–125

Day JW Jr, Hall CAS, Kemp WM, Yáñez-Arancibia A, Deegan LA (1989) Nekton, the free-swimming consumers. In: Day JW Jr, Hall CAS, Kemp WM, Yáñez-Arancibia A (eds) Estuarine ecology. John Wiley & Sons, New York, NY, USA, pp 377–437

de la Cruz A (1974) Primary productivity of coastal marshes in Mississippi. Gulf Res Rep 4:351–356

Dean RG, O'Brien MP (1987) Florida's West Coast inlets: Shoreline effects and recommended action. Florida Department of Natural Resources, Tallahassee, FL, USA, 100 p

Deegan LA (1993) Nutrient and energy transport between estuaries and coastal marine ecosystems by fish migration. Can J Fish Aquat Sci 50:74–79

Deegan LA, Day JW Jr, Gosselink JG, Yáñez-Arancibia A, Chávez GS, Sánchez-Gil P (1986) Relationships among physical characteristics, vegetation distribution and fisheries yield in Gulf of Mexico estuaries. In: Wolfe DA (ed) Estuarine variability. Academic Press, Orlando, FL, USA, pp 83–100

Deegan LA, Peterson BJ, Portier R (1990) Stable isotopes and cellulose activity as evidence for detritus as a food source for juvenile Gulf menhaden. Estuaries 13:14–19

Deegan LA, Hughes JE, Rountree RA (2000) Salt marsh ecosystem support of marine transient species. In: Weinstein MP, Kreeger DA (eds) Concepts and controversies in tidal marsh ecology. Kluwer Academic, Dordrecht, pp 333–365

Defant A (1960) Physical oceanography, vol 2. Pergamon Press, Oxford, UK, 598 p

Defenbaugh RE (1976) A study of the benthic macroinvertebrates of the continental shelf of northern Gulf of Mexico. PhD Dissertation, Texas A&M University, College Station, TX, USA, 476 p

DeLaune RD, Smith CJ, Patrick WH Jr (1983) Relationships of marsh elevation, redox potential, and sulfide to Spartina alterniflora productivity. Soil Sci Soc Am J 47:930–935

DeLaune RD, Smith CJ, Tolley MD (1984) The effect of sediment redox potential on nitrogen uptake, anaerobic root respiration and growth of *Spartina alterniflora* Loisel. Aquat Bot 18:223–230

den Hartog C (1970) The seagrasses of the world, vol 59. North-Holland Publishing, Amsterdam, The Netherlands

DeVillèle X, Verlaque M (1995) Changes and degradation in a *Posidonia aceanica* bed invaded by the introduced tropical alga *Caulerpa taxifolia* in the North Western Mediterranean. Bot Mar 38:79–87

Diaz RJ, Rosenberg R (1995) Marine benthic hypoxia: A review of its ecological effects and the behavioral responses of benthic macrofauna. Oceanogr Mar Biol Annu Rev 33:245–304

Dibajnia M, Sanchez C, Martinez M, Lara A, Nairn RB, Marván FG, Fournier CF, Risk M (2004) Why are Cancun beaches eroding? A question of integrated coastal zone management. Proceedings of XIII Congreso Panamericano de Ingenieria Oceanica y Costera. Mexico City, Mexico, 10 p

DiMarco SF, Reid RO (1998) Characterization of the principal tidal current constituents on the Texas-Louisiana Shelf. J Geophys Res 103:3092–3109

Dittel AI, Epifanio CE, Cifuentes LA, Kirchman DL (1997) Carbon and nitrogen sources for shrimp postlarvae fed natural diets from a tropical mangrove system. Estuar Coast Shelf Sci 45:629–637

Dittmann S (2000) Zonation of benthic communities in a tropical tidal flat of north-east Australia. J Sea Res 43:33–51

Dittmann S (2001) Abundance and distribution of small infauna in mangroves of Missionary Bay, North Queensland, Australia. Rev Biol Trop 49:535–544

Donnelly MJ, Green DM, Walters LJ (2008) Allelopathic effects of fruits of the Brazilian pepper *Schinus terebinthifolius* on growth, leaf production and biomass of seedlings of the red mangrove *Rhizophora mangle* and the black mangrove *Avicennia germinans*. J Exp Mar Biol Ecol 357:149–156

Donoghue JF, Demirpolat S, Tanner WF (1990) Recent shoreline changes, northwest Gulf of Mexico. In: Tanner WF (ed) Coastal sediments and processes. Proceedings, 9th Symposium on Coastal Sediments. Florida State University, Tallahassee, FL, USA, pp 51–66

Dorcas ME, Wilson JD, Gibbons JW (2007) Crab trapping causes population decline and demographic changes in diamondback terrapins over two decades. Biol Conserv 137:334–340

Dougherty KM, Mendelssohn IA, Monteferrafte FJ (1990) Effect of nitrogen, phosphorus and potassium additions on plant biomass and soil nutrient content of a swale barrier strand community in Louisiana. Ann Bot 66:265–271

Douglas BC (2005) Gulf of Mexico and Atlantic Coast sea level change. In: Sturges W, Lugo-Fernandez A (eds) Circulation in the Gulf of Mexico: Observations and models. American Geophysical Union, Washington, DC, USA, pp 111–121

Doyle LJ, Sharma DC, Hine AC, Pilkey OH Jr, Neal WJ, Pilkey OH Sr, Martin D, Belknap DF (1984) Living with the West Florida shore. Duke University Press, Durham, NC, USA, 222 p

Doyle TW, Krauss KW, Conner WH, From AS (2010) Predicting the retreat and migration of tidal forests along the northern Gulf of Mexico under sea-level rise. Forest Ecol Manage 259:770–777

Dring MJ (1982) The biology of marine plants. Thomson Litho, East Kilbride, Scotland

Duarte CM (1995) Submerged aquatic vegetation in relation to different nutrient regimes. Ophelia 41:87–112

Duarte CM, Agusti S (1998) The CO_2 balance of unproductive aquatic ecosystems. Science 281:234–236

Duarte CM, Cebrián J (1996) The fate of marine autotrophic production. Limnol Oceanogr 41:1758–1766

Duarte CM, Chiscano CL (1999) Seagrass biomass and production: A reassessment. Aquat Bot 65:159–174

Dugan P (1993) Wetlands in danger. Oxford University Press, New York, NY, USA, 187 p

Duke NC, Ball MC, Ellison JC (1998) Factors influencing biodiversity and distributional gradients in mangroves. Glob Ecol Biogeogr Lett 7:24–47

Dunne JA, Williams RJ, Martinez ND (2004) Network structure and robustness of marine food webs. Mar Ecol Prog Ser 273:291–302

Eckman JE (1987) The role of hydrodynamics in recruitment, growth, and survival of *Argopecten irradians* (L.) and *Anomia simplex* (D'Orbigny) within eelgrass meadows. J Exp Mar Biol Ecol 106:165–191

Edgar GJ (1990) Population regulation, population dynamics and competition amongst mobile epifauna associated with seagrass. J Exp Mar Biol Ecol 144:205–234

Edgar GJ, Shaw C (1995) The production and trophic ecology of shallow-water fish assemblages in southern Australia II. Diets of fishes and trophic relationships between fishes and benthos at Western Port, Australia. J Exp Mar Biol Ecol 194:83–106

Eggleston DB, Dahlgren CP, Johnson EG (2004) Fish density, diversity, and size-structure within multiple back reef habitats of Key West National Wildlife Refuge. Bull Mar Sci 75:175–204

Ela GM (2016) Vista de la playa de Balzapote en San Andrés Tuxtla. https://commons.wikime dia.org/wiki/File:Playa_de_Balzapote,_San_Andr%C3%A9s_Tuxtla.jpg. Accessed 19 Feb 2017

Eleuterius LN (1972) The marshes of Mississippi. Castanea 37:153–168

Elliott BA (1979) Anticyclonic rings and the energetics of the circulation of the Gulf of Mexico. PhD Dissertation, Texas A&M University, College Station, TX, USA

Elliott BA (1982) Anticyclonic rings in the Gulf of Mexico. J Phys Oceanogr 12:1292–1309

Ellis WL, Bell SS (2004) Conditional use of mangrove habitats by fishes: Depth as a cue to avoid predators. Estuaries 27:966–976

Ellis WL, Bell SS (2008) Tidal influence on a fringing mangrove intertidal fish community as observed by in situ video recording: Implications for studies of tidally migrating nekton. Mar Ecol Prog Ser 370:207–219

Ellis J, Stone GW (2006) Numerical simulation of net longshore sediment transport and granulometry of surficial sediments along Chandeleur Island, Louisiana, USA. Mar Geol 232:115–129

Ellis J, Nicholls P, Craggs R, Hofstra D, Hewitt J (2004) Effects of terrigenous sedimentation on mangrove physiology and associated macrobenthic communities. Mar Ecol Prog Ser 270:71–82

Ellison AM (1987) Effects of competition, disturbance, and herbivory on *Salicornia europaea*. Ecology 68:576–586

Ellison AM, Farnsworth EJ (1992) The ecology of Belizean mangrove-root fouling communities: Patterns of epibiont distribution and abundance, and effects on root growth. Hydrobiologia 247:87–98

Ellison AM, Farnsworth EJ (2001) Mangrove community ecology. In: Bertness MD, Gaines S, Hay ME (eds) Marine community ecology. Sinauer Press, Sunderland, MA, USA, pp 423–442

Enos P, Perkins RD (1979) Evolution of Florida Bay from island stratigraphy. Geol Soc Am Bull 90:59–83

Erickson AA, Saltis M, Bell SS, Dawes CJ (2003) Herbivore feeding preferences as measured by leaf damage and stomatal ingestion: A mangrove crab example. J Exp Mar Biol Ecol 289:123–138

Erickson AA, Bell SS, Dawes CJ (2004) Does mangrove leaf chemistry help explain crab herbivory patterns? Biotropica 36:333–343

Eriksson BK, Rubach A, Hillebrand H (2006) Biotic habitat complexity controls species diversity and nutrient effects on net biomass production. Ecology 87:246–254

Ernst CH, Lovich JE (2009) Turtles of the United States and Canada, 2nd edn. The Johns Hopkins University Press, Baltimore, MD, USA, pp 344–363

Escobar-Briones E, Winfield I (2003) Checklist of the benthic Gammaridea and Caprellidea (Crustacea: Peracarida: Amphipoda) from the Gulf of Mexico continental shelf and slope. Belg J Zool 133:37–44

Everitt JH, Judd FW, Escobar DE, Davis MR (1996) Integration of remote sensing and spatial information technologies for mapping black mangrove on the Texas gulf coast. J Coast Res 12:64–69

Ewel KC, Twilley RR, Ong JE (1998) Different kinds of mangrove forests provide different goods and services. Glob Ecol Biogeogr 7:83–94

Fang CS (1990) Petroleum drilling and production operations in the Gulf of Mexico. Estuaries 13:89–97

Fauchald K, Jumars P (1979) The diet of worms: A study of polychaete feeding guilds. Oceanogr Mar Biol Annu Rev 17:193–284

Faunce CH, Serafy JE (2006) Mangroves as fish habitat: 50 years of field studies. Mar Ecol Prog Ser 318:1–18

FDEP (Florida Department of Environmental Protection) (2008) Consultant's list of beaches controlled by placements of large amounts of sand. FDEP, Tallahassee, FL, USA. ftp://ftp.dep.state.fl.us/pub/water/beaches/HSSD/MHWfiles/. Accessed 19 Jan 2015

FDEP (2012a) Critically eroded beaches in Florida. Florida Department of Environmental Protection, Bureau of Beaches and Coastal Systems, Tallahassee, FL, USA, 76 p. Vector digital data. http://www.dep.state.fl.us/beaches/data/gis-data.htm. Accessed 19 Jan 2015

FDEP (2012b) Big bend seagrasses aquatic preserve management plan (draft). Florida Department of Environmental Protection, Florida Coastal Office, Tallahassee, FL, USA

Felder DK, Camp DL (eds) (2009) Gulf of Mexico—origins, waters and biota, vol 1, Biodiversity. Texas A&M University Press, College Station, TX, USA, 1392 p

Felder DL, Álvarez F, Goy JW, Lemaitre R (2009) Decapoda (Crustacea) of the Gulf of Mexico, with comments on the Amphionidacea. In: Felder DK, Camp DK (eds) Gulf of Mexico—origins, waters and biota, vol 1, Biodiversity. Texas A&M University Press, College Station, TX, USA, pp 1019–1104

Fell PE, Weissbach SP, Jones DA, Fallo MA, Zeppieri JA, Faison EK, Lennon KA, Newberry KJ, Reddington LK (1998) Does invasion of oligohaline tidal marshes by reed grass, *Phragmites australis* (Cav.) Trin. Ex. Steud., affect the availability of prey resources for the mummichog, *Fundulus heteroclitus* L.? J Exp Mar Biol Ecol 222:59–77

Feller IC (1995) Effects of nutrient enrichment on growth and herbivory of dwarf red mangrove (*Rhizophora mangle*). Ecol Monogr 65:477–505

Feller IC, Chamberlain A (2007) Herbivore responses to nutrient enrichment and landscape heterogeneity in a mangrove ecosystem. Oecologia 153:607–616

Feller IC, McKee KL (1999) Small gap creation in Belizean mangrove forests by a wood-boring insect. Biotropica 31:607–617

Feller IC, Whigham DF, O'Neill JP, McKee KL (1999) Effects of nutrient enrichment on within-stand cycling in a mangrove forest. Ecology 80:2193–2205

Feller IC, Lovelock CE, McKee KL (2007) Nutrient addition differentially affects ecological processes of *Avicennia germinans* in nitrogen versus phosphorus limited mangrove ecosystems. Ecosystems 10:347–359

Felley JD (1987) Nekton assemblages of three tributaries to the Calcasieu Estuary, Louisiana. Estuaries 10:321–329

Fenchel T (1969) The ecology of marine microbenthos, IV, Structure and function of the benthic ecosystem, its chemical and physical factors and the microfauna communities with special reference to the ciliated protozoa. Ophelia 6:1–182

FFWCC-FWRI (Florida Fish and Wildlife Conservation Commission-Fish and Wildlife Research Institute and Research Planning) (2003) Environmental Sensitivity Index shoreline classification lines Florida, 1st edn. Vector digital data, St. Petersburg, FL, USA. http://ocean.floridamarine.org/mrgis/Description_Layers_Marine.htm#benthic. Accessed 19 Jan 2015

Fleeger JW, Johnson DS, Galván KA, Deegan LA (2008) Top-down and bottom-up control of infauna varies across the salt marsh landscape. J Exp Mar Biol Ecol 357:20–34

Flint RW, Holland JS (1980) Benthic infaunal variability on a transect in the Gulf of Mexico. University of Texas Marine Science Institute Contribution 328. Academic Press, London, UK, 14 p

Flint RW, Rabalais NN (eds) (1980) Environmental studies, South Texas Outer Continental Shelf, 1975-1977. Final report. Bureau of Land Management, Washington, DC Contract AA551-CT8-51. University of Texas, Marine Science Institute, Austin, TX, USA

Flores-Verdugo F, Gonzalez-Farias F, Zamorano DS, Ramirez-Garcia P (1992) Mangrove ecosystems of the Pacific coast of Mexico: Distribution, structure, litterfall, and detritus dynamics. In: Seeliger U (ed) Coastal plant communities of Latin America. Academic, San Diego, CA, USA, pp 269–288

FNAI (Florida Natural Areas Inventory) (2010) Guide to natural communities of Florida, 2010 edn. Florida Natural Area Inventory, Tallahassee, FL, USA, 228 p

Fogg AQ, Hoffmayer ER, Driggers WB III, Campbell MD, Pellegrin GJ, Stein W (2013) Distribution and length frequency of invasive lionfish (Pterois sp.) in the northern Gulf of Mexico. Gulf Carib Res 25:111–115

Fonseca MS, Meyer DL, Hall MO (1996) Development of planted seagrass beds in Tampa Bay, Florida, USA. II. Faunal components. Mar Ecol Prog Ser 132:141–156

Fournier W, Hauber DP, White DA (1995) Evidence of infrequent sexual propagation of *Phragmites australis* throughout the Mississippi River delta. Am J Bot 82:71

Fourqurean JW, Rutten LM (2004) The impact of Hurricane Georges on soft-bottom, back reef communities: Site- and species-specific effects in south Florida seagrass beds. Bull Mar Sci 75:239–257

Fourqurean JW, Durako MJ, Hall MO, Hefty LN (2002) Seagrass distribution in south Florida: A multi-agency coordinated monitoring program. In: Porter JW, Porter KG (eds) The Everglades, Florida Bay, and the coral reefs of the Florida keys. CRC Press, Boca Raton, FL, USA, pp 497–522

Fourqurean JW, Smith TJ, Possley J, Collins TM, Lee D, Namoff S (2010) Are mangroves in the tropical Atlantic ripe for invasion? Exotic mangrove trees in the forests of South Florida. Biol Invasions 12:2509–2522

Franks JS, Christmas JY, Siler WL, Combs R, Waller R, Burns C (1972) A study of nektonic and benthic faunas of the shallow Gulf of Mexico off the state of Mississippi as related to some physical, chemical, and geological factors. Gulf Res Rep 4:1–148

Fredericq S, Cho TO, Earle SA, Gurgel CF, Krayesky DM, Mateo Cid LE, Mendoza-Gonzalez AC, Norris JN, Suarez AM (2009) Seaweeds of the Gulf of Mexico. In: Felder DL, Camp DK (eds) Gulf of Mexico origin, waters, and biota, vol 1, Biodiversity. Texas A&M University Press, College Station, TX, USA, pp 187–260

French CD, Schenk CJ (2005) Shaded relief image of the Gulf of Mexico (shadedrelief.jpg). USGS, Central Energy Resources Team, Reston, VA, USA. http://pubs.usgs.gov/of/1997/ofr-97-470/OF97-470L/graphic/data.htm. Accessed Jan 2015

Fritz HM, Blount C, Sokoloski R, Singleton J, Fuggle A, Mcadoo BG, Moore A, Grass C, Tate B (2007) Hurricane Katrina storm surge distribution and field observations on the Mississippi barrier islands. Estuar Coast Shelf Sci 74:12–20

Frolov SA, Sutyrin GG, Rowe GD, Rothstein LM (2004) Loop Current eddy interaction with the western boundary in the Gulf of Mexico. J Phys Oceanogr 34:2223–2237

Fry B (1984) $^{13}C/^{12}C$ ratios and the trophic importance of algae in Florida *Syringodium filiforme* seagrass meadows. Mar Biol 79:11–19

Gabrysch RK (1984) Ground-water withdrawals and land-surface subsidence in the Houston-Galveston Region, Texas, 1906-80. Report 287. Texas Department of Water Resources, Austin, TX, USA, 64 p

Galloway WE (2011) Pre-Holocene geological evolution of Northern Gulf of Mexico. In: Buster NA, Holmes CW (eds) Gulf of Mexico: Origin, waters, and biota, vol 3, Geology. Texas A&M University Press, College Station, TX, USA, pp 33–52

Gambrell RP, Patrick WJ Jr (1978) Chemical and microbiological properties of anaerobic soils and sediments. In: Hook DD, Crawford RMM (eds) Plant life in anaerobic environments. Ann Arbor Science Publishers, Ann Arbor, MI, USA, pp 375–423

Garrity CP, Soller DR (2009) Database of the geologic map of North America; adapted from the map by Reed JC, Jr. et al., 2005. U.S. Geological Survey data series 424. U.S. Geological Survey, Reston, VA, USA. http://pubs.usgs.gov/ds/424/. Accessed 19 Jan 2015

Gaston GR, Edds KA (1994) Long-term study of benthic communities on the continental shelf off Cameron, Louisiana: A review of brine effects and hypoxia. Gulf Res Rep 9:57–64

Gaston GR, Brown SS, Rakocinski CF, Heard RW, Summers JK (1995) Trophic structure of macrobenthic communities in northern Gulf of Mexico estuaries. Gulf Res Rep 9:111–116

Georgiou IY, Schindler J (2009) Numerical simulation of waves and sediment transport along a transgressive barrier island. Chapter H of Lavoie D (ed) Sand resources, regional geology, and coastal processes of the Chandeleur Islands Coastal System: An evaluation of the Breton National Wildlife Refuge. U.S. Geological Survey (USGS) Scientific Investigations Report 2009-5252. USGS, Arlington, VA, USA, pp 143–166

Georgiou IY, FitzGerald DM, Stone GW (2005) The impact of physical processes along the Louisiana coast. In Finkl CW, Khalil SM, eds, Saving America's wetland: Strategies for restoration of Louisiana's Coastal Wetlands and Barrier Islands. J Coast Res Special Issue 44:72–89

Georgiou I, McCorquodale JA, Meselhe E (2008) Hydrologic modeling and budget analysis of the Southwestern Louisiana Chenier Plain, part III: Sediment budget for the Chenier Plain. Louisiana Department of Natural Resources, Baton Rouge, LA, USA, 58 p

Gibson RN, Ansell AD, Robb L (1993) Seasonal and annual variations in abundance and species composition of fish and macrocrustacean communities on a Scottish sandy beach. Mar Ecol Prog Ser 98:89–105

Gibson RN, Robb L, Burrows MT, Ansell AD (1996) Tidal, diel and longer term changes in the distribution of fishes on a Scottish sandy beach. Mar Ecol Prog Ser 130:1–17

Giesen WBJT, Van Katwijk MM, den Hartog C (1990) Temperature, salinity, insolation and wasting disease of eelgrass (*Zostera marina* L.) in the Dutch Wadden Sea in the 1930's. Neth J Sea Res 25:395–404

Gilmore RG (1987) Subtropical-tropical seagrass communities of the southeastern United States: Fishes and fish communities. In: Durako MJ, Phillips RC, Lewis RR III (eds) Proceedings of a Symposium on Subtropical-Tropical Seagrasses of the Southeastern United States. Florida Department of Natural Resources, St. Petersburg, FL, USA, pp 117–138

Giraldo B (2005) Belowground productivity of mangroves in Southwest Florida. PhD Dissertation, Louisiana State University, Baton Rouge, LA, USA

Giri C, Ochieng E, Tieszen LL, Zhu Z, Singh A, Loveland T, Masek J, Duke N (2011a) Global Mangrove Distribution (USGS). Global Ecology and Biogeography 20:154–159. Vector digital data. UNEP-WCMC (United Nations Environment Programme—World Conservation Monitoring Centre), Cambridge, UK. http://data.unep-wcmc.org/datasets/21. Accessed 19 Jan 2015

Giri C, Long J, Tieszen L (2011b) Mapping and monitoring Louisiana's mangroves in the aftermath of the 2010 Gulf of Mexico oil spill. J Coast Res 27:1059–1064

Gleason ML, Zieman JC (1981) Influence of tidal inundation on internal oxygen supply of *Spartina alterniflora* and *Spartina patens*. Estuar Coast Shelf Sci 13:47–57

Gloeckner DR, Luczkovich JJ (2008) Experimental assessment of trophic impacts from a network model of a seagrass ecosystem: Direct and indirect effects of gulf flounder, spot and pinfish on benthic polychaetes. J Exp Mar Biol Ecol 357:109–120

González-Leija M, Mariño-Tapia I, Silva R, Enriquez C, Mendoza E, Escalante-Mancera E, Ruíz-Rentería, Uc-Sánchez E (2013) Morphodynamic evolution and sediment transport processes of Cancun beach. J Coast Res (published pre-print online). Coastal Education & Research Foundation, Coconut Creek, FL, USA

González-Sansón G, Aguilar-Betancourt C (2007) Marine ecosystems in the northwestern region of Cuba. In: Withers K, Nipper M (eds) Environmental analysis of the Gulf of Mexico, Special publication series 1. Harte Research Institute for Gulf of Mexico Studies, Corpus Christi, TX, USA, pp 373–390

Goodbred SLJ, Wright EE, Hine AC (1998) Sea-level change and storm-surge deposition in a Late-Holocene Florida salt marsh. J Sediment Res 68:240–252

Gough L, Grace JB (1998) Herbivore effects on plant species density at varying productivity levels. Ecology 79:1586–1594

Gould HR, McFarlan E Jr (1959) Geologic history of the Chenier Plain, southwestern Louisiana. Trans Gulf Coast Assoc Geol Soc 9:261–270

Grammer GL, Brown-Peterson NJ, Peterson MS, Comyns BH (2009) Life history of silver perch *Baridiella chrysoura* (Lecepèpe, 1803) in north-central Gulf of Mexico estuaries. Gulf Mexico Sci 27:62–73

Granados-Dieseldorff P, Baltz DM (2008) Habitat use by nekton along a stream-order gradient in a Louisiana estuary. Estuar Coast 31:572–583

Gravitz M (2008) The Gulf: From overfishing to healthy waters. Environment Texas Research and Policy Center, Austin, TX, USA, 11 p. www.environmenttexas.org. Accessed 19 Jan 2015

Greeley MS Jr, MacGregor R III (1983) Annual and semilunar reproductive cycles of the Gulf killifish, *Fundulus grandis*, on the Alabama coast. Copeia 1983:711–718

Green EP, Short FT (2003) World atlas of seagrasses, global distribution of seagrasses—polygons dataset, version 2 (2005). UNEP-WCMC (United Nations Environment Programme—World Conservation Monitoring Centre), Cambridge, UK. http://data.unep-wcmc.org/datasets/10. Accessed 17 Feb 2013

Greenway M (1995) Trophic relationships of macrofauna within a Jamaican seagrass meadow and the role of the echinoid *Lytechinus variegatus* (Lamarch). Bull Mar Sci 56:719–736

Griffith GE, Canfield DE Jr, Horsburgh CA, Omernik JM, Azevedo SH (1997) Lake regions of Florida (color poster with map, descriptive text, summary tables, and photographs; map scale 1:1,600,000). USEPA, Corvallis, OR, USA

Griffith GE, Omernik JM, Comstock JA, Lawrence S, Martin G, Goddard A, Hulcher VJ, Foster T (2001) Ecoregions of Alabama and Georgia (color poster with map, descriptive text, summary tables, and photographs; map scale 1:1,700,000). USGS, Reston, VA, USA

Griffith GE, Omernik JM, Pierson SM (2002) Level III and IV ecoregions of Florida. http://www.epa.gov/wed/pages/ecoregions/fl_eco.htm. Accessed 7 Feb 2013

Griffith GE, Bryce S, Omernik J, Rogers A (2007) Ecoregions of Texas. Texas Commission on Environmental Quality, Austin, TX, USA, 134 p

Grosslein MD (1976) Some results of fish surveys in the mid-Atlantic important for assessing environmental impacts. In: Gross G (ed) Middle Atlantic Continental Shelf and the New York Bight, Proceedings of the Symposium. American Museum of Natural History, New York City, November 3-5, 1975. American Society of Limnology and Oceanography, Allen Press, Lawrence, KS, USA, pp 312–328

Grothues TM, Able KW (2003) Response of juvenile fish assemblages in tidal salt marsh creeks treated for *Phragmites* removal. Estuar Coast 26:563–573

Guillory V, McMillen-Jackson A, Hartman L, Perry H, Floyd T, Wagner T, Graham G (2001) Blue crab traps and trap removal programs. Publication 88. Gulf States Marine Fisheries Commission, Ocean Springs, MS, USA, 13 p

Guntenspergen GR, Cahoon DR, Grace J, Steyer GD, Fournet S, Townson MA, Foote AL (1995) Disturbance and recovery of the Louisiana coastal marsh landscape from the impacts of Hurricane Andrew. J Coast Res Special Issue 21:324–339

Gutierrez-Espadas W (1983) Legendaria Llanura Sobre un Manto de Coral. Monografía Estatal, Secretaría de Educación Pública, Mexico City, Mexico, 271 p

Gutsell JS (1930) Natural history of the bay scallop. Bull US Bur Fish 46:569–632

Haas HL, Rose KA, Fry B, Minello TJ, Rozas LP (2004) Brown shrimp on the edge: Linking habitat to survival using an individual-based simulation model. Ecol Appl 14:1232–1247

Hackney CT, Hackney OP (1978) An improved, conceptually simple technique for estimation the productivity of marsh vascular flora. Gulf Res Rep 6:125–129

Hackney CT, Burbanck WD, Hackney OP (1976) Biological and physical dynamics of a Georgia tidal creek. Chesapeake Sci 17:271–280

Hackney CT, Stout JP, de la Cruz AA (1978) Standing crop and productivity of dominant marsh communities in the Alabama-Mississippi gulf coast. In: Brown LR (ed) Evaluation of the ecological role and techniques for the management of tidal marshes on the Mississippi and Alabama Gulf Coast. Mississippi-Alabama Sea Grant Publication MASGP-78-044. Ocean Springs, MS, USA, pp 1–29

Hagan SM, Brown SA, Able KW (2007) Production of mummichog (*Fundulus heteroclitus*): response in marshes treated for common reed (*Phragmites australis*) removal. Wetlands 27:54–67

Hale JA, Frazer TK, Tomasko DA, Hall MO (2004) Changes in the distribution of seagrass species along Florida's central Gulf Coast: Iverson and Bittaker revisited. Estuaries 27:36–43

Hall SJ (1994) Physical disturbance and marine benthic communities: Life in unconsolidated sediment. Oceanogr Mar Biol Annu Rev 32:179–239

Hall MO, Bell SS (1988) Response of small motile epifauna to complexity of epiphytic algae on seagrass blades. J Mar Res 116:613–630

Hall MO, Bell SS (1993) Meiofauna on the seagrass *Thalassia testudinum*: Population characteristic of harpacticoid copepods and associations with algal epiphytes. Mar Biol 116:137–146

Hall MA, Alverson DL, Metuzals KI (2000) By-catch: Problems and solutions. Mar Pollut Bull 41:204–219

Hamilton P (1992) Lower continental-slope cyclonic eddies in the central Gulf of Mexico. J Geophys Res 97:2185–2200

Hammerschlag N, Serafy JE (2009) Nocturnal fish utilization of a subtropical mangrove-seagrass ecotone. Mar Ecol 31:364–374

Hammerstrom KK, Kenworthy WJ, Fonseca MS, Whitfield PE (2006) Seed bank, biomass, and productivity of Halophila decipiens, a deep water seagrass on the west Florida continental shelf. Aquat Bot 84:110–120

Handley L, Altsman D, DeMay R (2007) Seagrass status and trends in the Northern Gulf of Mexico: 1940-2002. USGS Scientific Investigations Report 2006-5287 and USEPA 855-R-04-003. USGS, Reston, VA, USA, 267 p

Harper DE Jr (1991) Macroinfauna and macroepifauna. In: Brooks JM, Giamonna CP (eds) Mississippi-Alabama continental shelf ecosystem study data summary and synthesis, vol II: Technical narrative. U.S. Department of the Interior, Minerals Management Service. OCS Study MMS 91-0063. U.S. Department of the Interior, Gulf of Mexico OCS Region, New Orleans, LA, USA

Harrington RW Jr (1959) Delayed hatching in stranded eggs of marsh killifish, *Fundulus confluentus*. Ecology 40:430–437

Harrington RW Jr, Harrington ES (1961) Food selection among fishes invading a high subtropical salt marsh: From onset of flooding through the progress of a mosquito brood. Ecology 42:646–666

Harrington RW Jr, Harrington ES (1982) Effects on fishes and their forage organisms of impounding a Florida salt marsh to prevent breeding by salt marsh mosquitoes. Bull Mar Sci 32:523–531

Harris RJ, Milbrandt EC, Everham EM, Bovard BD (2010) The effects of reduced tidal flushing on mangrove structure and function across a disturbance gradient. Estuar Coast 33:1176–1185

Harrison PG (1989) Detrital processing in seagrass systems: A review of factors affecting decay rates, remineralization and detritivory. Aquat Bot 23:263–288

Hart KM, Crowder LB (2011) Mitigating by-catch of diamondback terrapins in crab pots. J Wildl Manage 75:264–272

Hart KM, McIvor CC (2008) Demography and ecology of mangrove diamondback terrapins in a wilderness area of Everglades National Park, Florida, USA. Copeia 1:200–208

Hauber DP, White DA, Powers SP, DeFrancesch FR (1991) Isozyme variation and correspondence with unusual infrared reflectance patterns in *Phragmites-australis* (Poaceae). Plant Syst Evol 178:1–8

Hauber DP, Saltonstall K, White DA, Hood CS (2011) Genetic variation in the common reed, *Phragmites australis*, in the Mississippi River Delta marshes: Evidence for multiple introductions. Estuar Coast 34:851–862

He R, Weisberg RH (2002) West Florida shelf circulation and temperature budget for the 1999 spring transition. Cont Shelf Res 22:719–748

Heard RW (1979) Guide to common tidal marsh invertebrates of the northeastern Gulf of Mexico. Mississippi-Alabama Sea Grant Consortium Publication MASGP-79-004. Mississippi-Alabama Sea Grant Consortium, Ocean Springs, MS, USA, 82 p

Heard RW (1982) Guide to Common Tidal Marsh Invertebrates of the Northeastern Gulf of Mexico. MASGP79004. Published by the Mississippi-Alabama Sea Grant Consortium. 87 p. https://www.usm.edu/gcrl/publications/docs/guide.to.common.tidal.marsh.invertebrates.heard.pdf. Accessed 2 Dec 2016

Heck KL Jr, Bryon D (2005) Post Hurricane Ivan damage assessment of seagrass resources of Coastal Alabama. Mobile Bay National Estuarine Program report, 14 p. http://www.mobi

lebaynep.com/images/uploads/library/Heck-and-Byron-ADCNR_SeagrassSurvey_finalre port.pdf. Accessed 7 Feb 2013

Heck KL Jr, Crowder LB (1991) Habitat structure and predator-prey interaction in vegetated aquatic systems. In: Bell SS, McCoy ED, Mushinsky HR (eds) Habitat complexity: The physical arrangement of objects in space. Chapman & Hall, New York, NY, USA, pp 281–299

Heck KL Jr, Orth RJ (1980) Seagrass habitats: The roles of habitat complexity, competition and predation in structuring associated fish and motile macroinvertebrate assemblages. In: Kennedy VS (ed) Estuarine perspectives. Academic Press, New York, NY, USA, pp 449–464

Heck KL Jr, Valentine JF (1995) Sea urchin herbivory: Evidence for long-lasting effects in subtropical seagrass meadows. J Exp Mar Biol Ecol 189:205–217

Heck KL Jr, Valentine JF (2006) Plant-herbivore interactions in seagrass meadows. J Exp Mar Biol Ecol 330:420–436

Heck KL Jr, Valentine JF (2007) The primacy of top-down effects in shallow benthic ecosystems. Estuar Coast 30:371–381

Heck KL Jr, Wetstone GS (1977) Habitat complexity and invertebrate species richness and abundance in tropical seagrass meadows. J Biogeogr 4:135–142

Heck KL Jr, Nadeau DA, Thomas R (1997) The nursery role of seagrass beds. Gulf Mexico Sci 1:50–54

Heck KL Jr, Coen LD, Morgan SG (2001) Pre- and post-settlement factors as determinants of juvenile blue crab *Callinectes sapidus* abundance: Results from the north-central Gulf of Mexico. Mar Ecol Prog Ser 222:163–176

Heck KL Jr, Hays G, Orth RJ (2003) Critical evaluation of the nursery role hypothesis for seagrass meadows. Mar Ecol Prog Ser 253:123–136

Heilman DJ, Edge BL (1996) Interaction of the Colorado River Project, Texas, with longshore sediment transport. In: Proceedings, 25th Coastal Engineering Conference. American Society of Civil Engineers (ASCE), Reston, VA, USA, pp 3309-3321

Heilman DJ, Kraus NC (1996) Beach fill functional design, town of South Padre Island, Texas. Final report. Conrad Blucher Institute for Surveying and Science, Texas A&M University, Corpus Christi, TX, USA, 70 p

Hemminga MA, Duarte DM (2000) Seagrass ecology. Cambridge University Press, Cambridge, UK, 298 p

Hemminga MA, Harrison PG, van Lent F (1991) The balance of nutrient losses and gains in seagrass meadows. Mar Ecol Prog Ser 71:85–96

Henriques PR (1980) Faunal community structure of eight soft shore, intertidal habitats in the Manukau Harbour. N Z J Ecol 3:97–103

Hentschel BT, Herrick BS (2005) Growth rates of interface-feeding spionid polychaetes in simulated tidal currents. J Mar Res 63:983–999

Herke WH (1995) Natural fisheries, marsh management, and mariculture: Complexity and conflict in Louisiana. Estuaries 18:10–17

Herke WH, Rogers BD (1984) Comprehensive estuarine nursery study completed. Fisheries 9:12–16

Herke WH, Wengert MW, LaGory ME (1987) Abundance of young brown shrimp in natural and semi-impounded marsh nursery areas: Relation to temperature and salinity. Northeast Gulf Sci 9:9–28

Hernández-Santana JR, Ortiz-Pérez MA, Mendez Linares AP, Gama-Campillo L (2008) Reconocimiento morfodinamico de la linea de costa del Estado de Tabasco, Mexico: Tendencias desde la segunda mitad del siglo XX hasta el presente. Investigaciones Geograficas, Boletin del Instituto de Geografia 65:7–21

Herrera-Silveira JA, Morales-Ojeda SM (2010) Subtropical Karstic Coastal Lagoon assessment, Southeast Mexico: The Yucatán Peninsula Case. In: Kennish MJ, Paert HW (eds) Coastal lagoons: Critical habitats of environmental change. CRC, Boca Raton, FL, USA, pp 307–333

Hester MW, Mendelssohn IA (1991) Expansion patterns and soil physicochemical characterization of three Louisiana populations of *Uniola paniculata* (sea oats). J Coast Res 7:387–401

Hester MW, Mendelssohn IA (2000) Long-term recovery of a Louisiana brackish marsh plant community from oil-spill impact: Vegetation response and mitigating effects of marsh surface elevation. Mar Environ Res 49:233–254

Hester MW, Wilsey BJ, Mendelssohn IA (1994) Grazing of Panicum amarum in a Louisiana barrier island dune plant community: Management implications for dune restoration projects. Ocean Coast Manage 23:213–224

Hester MW, Spalding EA, Franze CD (2005) Biological resources of the Louisiana coast: Part 1. An overview of coastal plant communities of the Louisiana Gulf shoreline. J Coast Res 44:134–145

Hik DS, Jefferies RL (1990) Increases in the net above-ground primary production of a salt-marsh forage grass: A test of the predictions of the herbivore-optimization model. J Ecol 78:180–185

Hillewaert H (2011) A variegated sea urchin at St. Lucie County Marine Center in Fort Pierce, St. Lucie County, Florida, USA. https://en.wikipedia.org/wiki/File:Lytechinus_variegatus. jpg. Accessed 19 Feb 2017

Hine AC (2009) Geology of Florida. Brooks/Cole Cengage Learning, Independence, KY, USA, 29 p

Hine AC, Locker SD (2011) Florida Gulf of Mexico continental shelf: Great contrasts and significant transitions. In: Buster NA, Holmes CW (eds) Gulf of Mexico: Origin, waters, and biota, vol 3, Geology. Texas A&M University Press, College Station, TX, USA, pp 101–127

Hine AC, Evans MW, Davis RA Jr, Belknap DF (1987) Depositional response to seagrass mortality along a low-energy, barrier-island coast: west-central Florida. J Sediment Petrol 57:431–439

Hine AC, Belknap DF, Hutton JG, Osking EB, Evans MW (1988) Recent geological history and modern sedimentary processes along an incipient, low-energy, epicontinental-sea coastline: Northwest Florida. J Sediment Petrol 58:567–579

Hoffman BA, Dawes CJ (1997) Vegetational and abiotic analysis of the salterns of mangals and salt marshes of the west coast of Florida. J Coast Res 13:147–154

Hogan MC (2013a) Petén-Veracruz moist forests. Encyclopedia of Earth. Environmental Information Coalition, National Council for Science and the Environment, Washington, DC, USA. http://www.eoearth.org/view/article/51cbf03e7896bb431f6a0dd1/. Accessed 5 Aug 2014

Hogan MC (2013b) Mesoamerican Gulf-Caribbean Mangroves. Encyclopedia of Earth. Environmental Information Coalition, National Council for Science and the Environment, Washington, DC, USA. http://www.eoearth.org/view/article/51cbfb967896bb431f6bfd52/. Accessed 5 Aug 014.

Holdredge C, Bertness MD, Altieri AH (2008) Role of crab herbivory in die-off on New England salt marshes. Conserv Biol 23:672–679

Hopkinson CS, Gosselink JG, Parrondo RT (1978) Aboveground production of seven marsh plant species in coastal Louisiana. Ecology 59:760–769

Howard RJ, Rafferty PS (2006) Clonal variation in response to salinity and flooding stress in four marsh macrophytes of the northern Gulf of Mexico, USA. Environ Exp Bot 56:301–313

Howard RJ, Travis SE, Sikes BA (2008) Rapid growth of a Eurasian haplotype of *Phragmites australis* in a restored brackish marsh in Louisiana, USA. Biol Invasions 10:369–379

Howe JC, Wallace RK, Rikard FS (1999) Habitat utilization by postlarval and juvenile penaeid shrimps in Mobile Bay, Alabama. Estuaries 22:971–979

Howes BL, Dacey JWH, Goehringer DD (1986) Factors controlling the growth form of *Spartina alterniflora*: Feedbacks between above-ground production, sediment oxidation, nitrogen and salinity. J Ecol 74:881–898

Hoyt JH (1969) Chenier versus barrier, genetic and stratigraphic definition. Am Assoc Pet Geol B 53:299–306

Hsu SA, Brand AD, Blanchard B (2005) Rapid estimation of maximum storm surges induced by Hurricanes Katrina and Rita in 2005. Coastal Studies Institute, Louisiana State University, Baton Rouge, LA, USA

Hughes AR, Williams SL, Duarte CM, Heck KL Jr, Waycott M (2009) Associations of concern: Declining seagrasses and threatened dependent species. Front Ecol Environ 7:242–246

Huh OK, Roberts HH, Rouse LJ, Rickman DA (1991) Fine grain sediment transport and deposition in the Atchafalaya and Chenier Plain sedimentary system. In: Kraus NC, Gingerich KJ, Kriebel DL (eds) Coastal Sediments '91. American Society of Civil Engineers (ASCE), New York, NY, USA, pp 817–830

Hummell RL, Parker SJ (1995) Holocene geologic history of Mobile Bay, Alabama. Alabama Geological Survey Circular 186, Tuscaloosa, AL, USA, 97 p

Humphrey WD (1979) Diversity, distribution and relative abundance of benthic fauna in a Mississippi tidal marsh. PhD Dissertation, Mississippi State University, Starkville, MS, USA, 93 p

Hunter KL, Fox DA, Brown LM, Able KW (2006) Responses of resident marsh fishes to stages of *Phragmites australis* invasion in three mid-Atlantic estuaries. Estuar Coast 29:487–498

IMaRS/USF (Institute for Marine Remote Sensing, University of South Florida); IRD, Centre de Nouméa (Institut de Recherché pour le Développement); NASA (National Aeronautics and Space Administration); UNEP-WCMC (United Nations Environment Programme—World Conservation Monitoring Centre) (2010) Global distribution of coral reefs. Vector digital data. UNEP-WCMC, Cambridge, UK. http://data.unep-wcmc.org/datasets/13. Accessed 7 Feb 2013

Irlandi EA (1994) Large-and small-scale effects of habitat structure on rates of predation: How percent coverage of seagrass affects rates of predation and siphon nipping on an infaunal bivalve. Oecologia 98:176–183

Irlandi EA (1996) The effects of seagrass patch size and energy regime on growth of a suspension-feeding bivalve. J Mar Res 54:161–185

Irlandi EA, Peterson CH (1991) Modification of animal habitat by large plants: Mechanisms by which seagrasses influence clam growth. Oecologia 87:307–318

Isphording WC (1975) The physical geology of Yucatán. Trans Gulf Coast Assoc Geol Soc 25:231–262

Isphording WC (1994) Erosion and deposition in northern Gulf of Mexico estuaries. Trans Gulf Coast Assoc Geol Soc 44:305–314

Isphording WC, Imsand FD, Flowers GC (1989) Physical characteristics and aging of gulf coast estuaries. Trans Gulf Coast Assoc Geol Soc 39:387–401

Iverson RL, Bittaker HF (1986) Seagrass distribution and abundance in eastern Gulf of Mexico coastal waters. Estuarine Coast Shelf Sci 22:577–602

Ivester MS (1978) Faunal dynamics. In: Brown LR, de la Cruz AA, Ivester MS, Stout JP (eds) Evaluation of the ecological role and techniques for the management of tidal marshes on the Mississippi and Alabama Gulf Coast. Mississippi-Alabama Sea Grant Consortium Publication MASGP-78-044. Mississippi-Alabama Sea Grant Consortium, Ocean Springs, MS, USA, pp 1–36

Jackson JBC, Kirby MX, Berger WH, Bjorndal KA, Botsford LW, Bourque BJ, Bradbury RH, Cooke R, Erlandson J, Estes JA, Hughes TP, Kidwell S, Lange CB, Lenihan HS, Pandolfi JM, Peterson CH, Steneck RS, Tegner MJ, Warner RR (2001) Historical overfishing and the recent collapse of coastal ecosystems. Science 293:629–638

Ji-Elle (2015) Mangrove sur l'île de Cayo Jutías (Cuba). http://wikivisually.com/wiki/File:Cayo_Jut%C3%ADas-Mangrove_(4).jpg. Accessed 19 Feb 2017

Jivoff PR, Able KW (2003) Blue crab, *Callinectes sapidus*, response to the invasive common reed, *Phragmites australis*: Abundance, size, sex ratio, and molting frequency. Estuaries 26:587–595

Jochens AE, DiMarco SF (2008) Physical oceanographic conditions in the deepwater Gulf of Mexico in summer 2000-2002. Deep-Sea Res Part II 55:2541–2554

Jochens AE, DiMarco SF, Nowlin WD Jr, Reid RO, Kennicutt MC II (2002) Northeastern Gulf of Mexico chemical oceanography and hydrography study. OCS Study 40 MMS 2002-055. MMS, Gulf of Mexico OCS Region, New Orleans, LA, USA

Johnson AF (1997) Rates of vegetation succession on a coastal dune system in northwest Florida. J Coast Res 13:373–384

Johnston IM (1924) Expedition of the California Academy of Sciences to the Gulf of California in 1921, The Botany (The Vascular Plants). Proc Calif Acad Sci 12:951–1118

Jones DL, Walter JF, Brooks EN, Serafy JE (2010) Connectivity through ontogeny: Fish population linkages among mangrove and coral reef habitats. Mar Ecol Prog Ser 401:245–258

Judd FW (2002) Tamaulipan Biotic Province. In: Tunnell JW Jr, Judd FW (eds) The Laguna Madre of Texas and Tamaulipas. Texas A&M University, College Station, TX, USA, pp 38–58

Judd FW, Lonard RI (2002) Species richness and diversity of brackish and salt marshes in the Rio Grande delta. J Coast Res 18:751–759

Judkins HL, Vecchione M, Roper CFE (2009) Cephalopoda (Mollusca) of the Gulf of Mexico. In: Felder DK, Camp DK (eds) Gulf of Mexico: Origins, waters and biota, vol 1, Biodiversity. Texas A&M University, College Station, TX, USA, pp 701–709

Kanouse S, La Peyre MK, Nyman JA (2006) Nekton use of *Ruppia maritima* and non-vegetated bottom habitat types within brackish marsh ponds. Mar Ecol Prog Ser 327:61–69

Kantha L (2005) Barotropic tides in the Gulf of Mexico. In: Sturges W, Lugo-Fernandez A (eds) Circulation in the Gulf of Mexico: Observations and models. American Geophysical Union, Washington, DC, USA, pp 159–163

Kao WY, Shih CN, Tsai TT (2004) Sensitivity to chilling temperatures and distribution differ in the mangrove species *Kandelia candel* and *Avicennia marina*. Tree Physiol 24:859–864

Kaput N, Koenis MPT, Nooij R, Sikkema T, Van der Waardt TP (2007) Erosion of the beach of historic Varadero, Cuba. Master Project Report CF 74. Delft University of Technology, Delft, The Netherlands, 271 p, plus appendices

Keddy PA (2010) Wetland ecology principles and conservation. Cambridge University Press, Cambridge, UK, 497 p

Keim BD, Muller RA, Stone GW (2007) Spatiotemporal patterns and return periods of tropical storm and hurricane strikes from Texas to Maine. J Climate 20:3498–3509

Kelly FJ (1991) Physical oceanography. In: Brooks JM, Giammona CP (eds) Mississippi-Alabama continental shelf ecosystem study data summary and synthesis. U.S. Department of the Interior, MMS, Gulf of Mexico OCS Region, New Orleans, LA, USA, pp 10–1 to 10–151

Kenworthy WJ, Zieman JC, Thayer GW (1982) Evidence for the influence of seagrasses on the benthic nitrogen cycle in a coastal plain estuary near Beaufort, North Carolina (USA). Oecologia 54:152–158

Kenworthy WJ, Currin C, Smith G, Thayer G (1987) The abundance, biomass and acetylene reduction activity of bacteria associated with decomposing rhizomes of two seagrasses, *Zostera marina* and *Thalassia testudinum*. Aquat Bot 27:97–119

Kerbes RH, Kotanen PM, Jefferies RL (1990) Destruction of wetland habitats by lesser snow geese: A keystone species on the West Coast of Hudson Bay. J Appl Ecol 27:242–258

Kikuchi T, Peres JM (1977) Consumer ecology of seagrass beds. In: McRoy CP, Helfferich C (eds) Seagrass ecosystems: A scientific perspective. Marcel Dekker, New York, NY, USA, pp 147–194

Kimball ME, Able KW, Grothues TM (2010) Evaluation of long-term response of intertidal creek nekton to *Phragmites australis* (common reed) removal in oligohaline Delaware Bay salt marshes. Restor Ecol 18:722–779

Kineke GC, Higgins EE, Hart K, Velasco D (2006) Fine-sediment transport associated with cold-front passages on the shallow shelf, Gulf of Mexico. Cont Shelf Res 26:2073–2091

King SP, Sheridan P (2006) Nekton of new seagrass habitats colonizing a subsided salt marsh in Galveston Bay, Texas. Estuaries 29:286–296

Kirby CJ, Gosselink J (1976) Primary production in a Louisiana Gulf Coast *Spartina alterniflora* marsh. Ecology 57:1052–1059

Kitting CL, Fry B, Morgan MD (1984) Detection of inconspicuous epiphytic algae supporting food webs in seagrass meadows. Oecologia 62:145–149

Kjerfve B (1986) Comparative oceanography of coastal lagoons. In: Wolfe DA (ed) Estuarine variability. Academic Press, New York, NY, USA, pp 63–81

Kjerfve B, Sneed JE (1984) Analysis and synthesis of oceanographic conditions in the Mississippi Sound Offshore Region. Final report. U.S. Army Corps of Engineers, Mobile District, Mobile, AL, USA, 252 p

Klumpp DW, Howard RK, Pollard DA (1989) Trophodynamics and nutritional ecology of seagrass communities. In: Larkum AWD, McComb AJ, Shepherd SA (eds) Biology of seagrasses. Elsevier, Amsterdam, The Netherlands, pp 394–457

Klumpp DW, Salita-Espinosa JT, Fortes MD (1993) Feeding ecology and trophic role of sea urchins in a tropical seagrass community. Aquat Bot 45:205–229

Kneib RT (1986) The role of *Fundulus heteroclitus* in salt marsh trophic dynamics. Integr Comp Biol 26:259–269

Kneib RT (1987) Predation risk and use of intertidal habitats by young fishes and shrimp. Ecology 68:379–386

Kneib RT (1997) The role of tidal marshes in the ecology of estuarine nekton. Oceanogr Mar Biol 35:163–220

Kneib RT (2000) Salt marsh ecoscapes and production transfers by estuarine nekton in the southeastern United States. In: Weinstein MP, Kreeger DA (eds) Concepts and controversies in tidal marsh ecology. Kluwer Academic Publishers, Dordrecht, The Netherlands, pp 267–291

Kneib RT (2003) Bioenergetic and landscape considerations for scaling expectations of nekton production from intertidal marshes. Mar Ecol Prog Ser 264:279–296

Knudsen EE, Paille RF, Rogers BD, Herke WH, Geaghan JP (1989) Effects of a fixed-crest weir on brown shrimp *Penaeus aztecus* growth, mortality, and emigration in a Louisiana coastal marsh. N Am J Fish Manage 9:411–419

Ko JY, Day JW (2004) A review of ecological impacts of oil and gas development on coastal ecosystems in the Mississippi Delta. Ocean Coast Manage 47:597–623

Koch MS, Mendelssohn IA, McKee KL (1990) Mechanism for the hydrogen sulfide-induced growth limitation in wetland plants. Limnol Oceanogr 35:399–408

Koch EW, Barbier EB, Silliman BR, Reed DJ, Perillo GME, Hacker SD, Granek EF, Primavera JH, Muthiga N, Polasky S, Halpern BS, Kennedy CJ, Kappel CV, Wolanski E (2009) Non-linearity in ecosystems services: Temporal and spatial variability in coastal protection. Front Ecol Environ 7:29–37

Kral R, Diamond AR, Ginzberg SL, Hansen CJ, Haynes RR, Keener BR, LeLong MG, Spaulding DD, Woods M (2011) Annotated checklist of the vascular plants of Alabama. Botanical Research Institute of Texas, Fort Worth, TX, USA, 110 p

Kraus NC, Heilman DJ (1997) Packery channel feasibility study: Inlet functional design and sand management. Technical report TAMU-CC-CBI-96-06. Texas A&M University, Corpus Christi, TX, USA, 106 p

Krauss KW, Lovelock CE, McKee KL, Lopez-Hoffman L, Ewe SML, Sous WP (2008) Environmental drivers in mangrove establishment and early development. Aquat Bot 89:105–127

Krauss KW, From AS, Doyle TW, Doyle TJ, Barry MJ (2011) Sea-level rise and landscape change influence mangrove encroachment onto marsh in the Ten Thousand Islands region of Florida, USA. J Coast Conserv 15:629–638

Krebs JM, Brame AB, McIvor CC (2007) Altered mangrove wetlands as habitat for estuarine nekton: Are dredged channels and tidal creeks equivalent? Bull Mar Sci 80:839–861

Kruczynski WL, Subrahmanyam CB, Drake SH (1978) Studies on the plant community of a north Florida salt-marsh. Part 1 Primary production. Bull Mar Sci 28:316–334

Krutak PR, Gío-Argáez R (1994) Ecology, taxonomy and distribution of dominant ostracode taxa in modern carbonate sediments, northeastern Yucatán shelf, Mexico. Revista Mexicana de Ciencias Geológicas 11:193–213

Kulp M, Penland S, Williams SJ, Jenkins C, Flocks J, Kindinger J (2005) Geological Framework, Evolution, and Sediment Resources for Restoration of the Louisiana Coastal Zone. In Finkl CW, Khalil SM, eds, Saving America's wetland: Strategies for restoration of Louisiana's Coastal Wetlands and Barrier Islands. J Coast Res 44:56–71

Kushlan JA, Mazotti F (1989) Historic and present distribution of the American crocodile in Florida. J Herpetol 23:1–7

Kutkuhn JH (1962) Gulf of Mexico commercial shrimp populations—trends and characteristics, 1956-59. Fish Bull Fish Wildl Ser 62:343–402

La Peyre MK, Gordon J (2012) Nekton density patterns and hurricane recovery in submerged aquatic vegetation, and along non-vegetated natural and created edge habitats. Estuar Coast Shelf Sci 98:108–118

Laliberté E, Turner BL, Costes T, Pearse SJ, Wyrwoll KH, Zemunik G, Lambers H (2012) Experimental assessment of nutrient limitation along a 2-million-year dune chronosequence in the south-western Australia biodiversity hotspot. J Ecol 100:631–642

Lambertini C, Mendelssohn IA, Gustafsson MHG, Olesen B, Riis T, Sorrell BK, Brix H (2012) Tracing the origin of gulf coast Phragmites (Poaceae): A story of long-distance dispersal and hybridization. Am J Bot 99:1–14

Lammerts EJ, Pegtel DM, Grootjans AP, van der Veen A (1999) Nutrient limitation and vegetation changes in a coastal dune slack. J Veg Sci 10:111–122

Lamotte RS (1952) Catalogue of the Cenozoic plants of North America through 1950. Geological Society of America Memoir 51. Geological Society of America, Boulder, CO, USA

Lance VA (2003) Alligator physiology and life history: The importance of temperature. Exp Gerontol 38:801–805

Lango-Reynoso F, Castañeda-Chávez MR, Landeros-Sánchez C, Galavíz-Villa I, Navarrete-Rodríguez G, Soto-Estrada A (2013) Cd, Cu, Hg and Pb, and organochlorines pesticides in commercially important benthic organisms coastal lagoons SW Gulf of Mexico. Agric Sci 1:63–79

Larkum AWD, West RJ (1990) Long-term changes of seagrass meadows in Botany Bay, Australia. Aquat Bot 37:55–70

Layman CA (2000) Fish assemblage structure of the shallow ocean surf-zone on the eastern shore of Virginia barrier islands. Estuar Coast Shelf Sci 51:201–213

Leber KM III (1985) The influence of predatory decapods, refuge, and microhabitat selection on seagrass communities. Ecology 66:1951–1964

Lee SY (2008) Mangrove macrobenthos: Assemblages, services, and linkages. J Sea Res 59:16–29

Lee K-S, Dunton KH (1999) Inorganic nitrogen acquisition in the seagrass Thalassia testudinum: Development of a whole-plant nitrogen budget. Limnol Oceanogr 44:1204–1215

Lepoint G, Nyssen F, Gobert S, Dauby P, Bouquegneau JM (2000) Relative impact of a seagrass bed and its adjacent epilithic algal community in consumer diets. Mar Biol 136:513–518

Levin LA, Talley TS, Thayer G (1996) Succession of macrobenthos in a created salt marsh. Mar Ecol Prog Ser 141:67–82

Levin LA, Talley TS, Hewitt J (1998) Macrobenthos of Spartina foliosa (Pacific cordgrass) salt marshes in southern California: Community structure and comparison to a Pacific mudflat and a Spartina alterniflora (Atlantic smooth cordgrass) marsh. Estuaries 21:129–144

Levinton J, Kelaher B (2004) Opposing organizing forces of deposit-feeding marine communities. J Exp Mar Biol Ecol 300:65–82

Lewis FG III (1984) Distribution of macrobenthic crustaceans associated with Thalassia, Halodule and bare sand substrata. Mar Ecol Prog Ser 19:101–113

Lewis FG III, Stoner AW (1981) An examination of methods for sampling macrobenthos in seagrass meadows. Bull Mar Sci 31:116–124

Lewis FG III, Stoner AM (1983) Distribution of macrofauna within seagrass beds: An explanation for patterns of abundance. Bull Mar Sci 33:296–304

Ley JA, Montague CL, McIvor CC (1999) Fishes in mangrove prop-root habitats of northeastern Florida Bay: Distinct assemblages across and estuarine gradient. Estuar Coast Shelf Sci 48:701–723

Li C (2012) WAVCIS—wave-current-surge information system. Coastal Studies Institute, Louisiana State University, Baton Rouge, LA, USA. http://wavcis.csi.lsu.edu/. Accessed 7 Feb 2013

Lin Q, Mendelssohn IA (1996) A comparison investigation of the effects of south Louisiana crude oil on the vegetation of fresh, brackish and salt marshes. Mar Pollut Bull 32:202–209

Lindner MJ, Anderson WW (1956) Growth, migrations, spawning and size distribution of shrimp Penaeus setiferus. Fish Bull Fish Wildl Ser 56:555–645

Linscombe G, Kinler N (1997) A survey of vegetative damage caused by nutria herbivory in the Barataria and Terrebonne Basins. Survey BTNEP-31. Barataria-Terrebonne National Estuary Program, Thibodaux, LA, USA

Litvin SY, Weinstein MP (2003) Life history strategies of estuarine nekton: The role of marsh macrophytes, benthic microalgae, and phytoplankton in the trophic spectrum. Estuar Coast 26:552–562

Livingston RJ (1982) Trophic organization of fishes in a coastal seagrass system. Mar Ecol Prog Ser 7:1–12

Livingston RJ (1984) The relationship of physical factors and biological response in coastal seagrass meadows. Estuaries 7:377–390

Livingston RJ, Kobylinski GJ, Lewis FG III, Sheridan PF (1976) Long-term fluctuations of epibenthic fish and invertebrate populations in Apalachicola Bay, Florida. Fish Bull 74:311–321

Llansó RJ, Bell SS, Vose FE (1998) Food habits of red drum and spotted seatrout in a restored mangrove impoundment. Estuaries 21:294–306

Loesch H (1960) Sporadic mass shoreward migrations of demersal fish and crustaceans in Mobile Bay, Alabama. Ecology 41:292–298

Loftus WF, Rehage JS (2005) Role of marsh-mangrove interface habitats as aquatic refuges for wetland fishes and other aquatic animals. Third annual report. U.S. Geological Survey—Florida Integrated Science, Everglades National Park Field Station, Homestead, FL, USA, 45 p

Loneragan NR, Bunn SE, Kellaway DM (1997) Are mangroves and seagrasses sources of organic carbon for penaeid prawns in a tropical Australian estuary? A multiple stable-isotope study. Mar Biol 130:289–300

Lopez J, Baker A, Boyd E (2010a) Water quality in Chandeleur Sound in 2008 and 2010. Lake Pontchartrain Basin Foundation, Metairie, LA, USA, 7 p

Lopez JD, Peterson MS, Lang ET, Charbonnet AM (2010b) Linking habitat and life history for conservation of the rare salt marsh topminnow *Fundulus jenkensi*: Morphometrics, reproduction, and trophic ecology. Endanger Species Res 12:141–155

Lopez JD, Peterson MS, Walker J, Grammer GL, Woodrey MS (2011) Distribution, abundance, and habitat characterization of the salt marsh topminnow, *Fundulus jenkinsi* (Evermann 1892). Estuar Coast 34:148–158

Lopez-Portillo J, Ezcurra E (1989) Zonation in mangrove and salt marsh vegetation at Laguna de Mecoacan, Mexico. Biotropica 21:107–114

Lovelock CE, Feller IC, McKee KL, Engelbrecht BMJ, Ball MC (2004) Experimental evidence for nutrient limitation of growth, photosynthesis, and hydraulic conductance of dwarf mangroves in Panama. Funct Ecol 18:25–33

Loza EL (1994) Los manglares de Mexico: Sinopsis general para su manejo. In: Suman DO (ed) El ecosystema de manglar en America Latina y la cuenca del Caribe: Su manejo y conservacion. University of Miami, Miami, pp 144–151

Lubchenko J (1978) Plant species diversity in a marine intertidal community: Importance of herbivore food preference and algal competitive abilities. Am Natl 12:23–39

Lugo AE (1998) Mangrove forests: A tough system to invade but an easy one to rehabilitate. Mar Pollut Bull 37:427–430

Lugo AE, Patterson-Zucca C (1977) The impact of low temperature stress on mangrove structure and growth. Trop Ecol 18:149–161

Lugo AE, Snedaker SC (1974) The ecology of mangroves. In: Johnston RF, Frank PW, Michener CD (eds) Annual review of ecology and systematics. Annual Reviews, Palo Alto, CA, USA, pp 39–64

Lugo-Fernandez A (2007) Is the Loop Current a chaotic oscillator? J Phys Oceanogr 37:1455–1469

Lugo-Fernandez A, Green RE (2011) Mapping the intricacies of the Gulf of Mexico's circulation. EOS 92:21–22

Luna LG, Montano AS, Castillo IM, Capote AJ (2009) Biodiversity, morphometry and diet of *Callinectes* crabs (Decapoda: Portunidae) in Santiago de Cuba. Rev Biol Trop 57:671–686

Lyon JG (1993) Practical handbook for wetland identification and delineation. CRC Press, Boca Raton, FL, USA, 157 p

Mairaro JL (2007) Disturbance effects on nekton communities of seagrasses and bare substrates in Biloxi Marsh, Louisiana. MS Thesis, Louisiana State University and Agricultural and Mechanical College, Baton Rouge, LA, USA, 72 p

Malyshev VI, Arkhipov AG (1992) The ctenophore *Mnemiopsis leidyi* in the western Black Sea. Hydro J 28:33–39

Marchand C, Baltzer F, Lallier-Verges E, Alberic P (2008) Pore-water chemistry in mangrove sediments: Relationship with species composition and developmental stages (French Guiana). Mar Geol 208:361–381

Markley JL, McMillan C, Thompson GA Jr (1982) Latitudinal differentiation in response to chilling temperatures among populations of three mangroves, *Avicennia germinans*, *Laguncularia racemosa*, and *Rhizophora mangle*, from the western tropical Atlantic and Pacific Panama. Can J Bot 60:2704–2715

Marmer HA (1954) Tides and sea level in the Gulf of Mexico. In: Galtsoff PS (ed) Gulf of Mexico, its origins, waters, and marine life. Fish Bulletin-NOAA, vol 55, pp 101–118

Martinez L, O'Brien S, Bethel M, Penland S, Kulp M (2009) Louisiana Barrier Island Comprehensive Monitoring Program (BICM), vol 2, Shoreline changes and barrier island land loss 1800's-2005. Pontchartrain Institute for Environmental Sciences, University of New Orleans, New Orleans, LA, USA. Main report, 32 p.; Appendices, 49 p

Martínez ML, Intralawan A, Vázquez G, Pérez-Maqueo O, Sutton P, Landgrave R (2007) The coasts of our world: Ecological, economic and social importance. Ecol Econ 63:254–272

Maas P (2006) Isla Contoy, Mexico. https://commons.wikimedia.org/wiki/File:IslaContoy-PeterMaas.JPG. Accessed 19 Feb 2017

Mateo MA, Cebrián J, Dunton K, Mutchler T (2006) Carbon flux in seagrass ecosystems. In: Larkum AWD, Orth RJ, Duarte CM (eds) Seagrasses: Biology, ecology and conservation. Springer, Dordrecht, The Netherlands, pp 159–192

Matheson RE Jr, Camp DK, Sogard SM, Bjorgo KA (1999) Changes in seagrass-associated fish and crustacean communities on Florida Bay mud banks: The effects of recent ecosystem changes? Estuaries 22:534–551

Mattson RA (2000) Seagrass ecosystem characteristics and research and management needs in the Florida Big Bend. In: Bortone SA (ed) Seagrasses: Monitoring, ecology, physiology, and management. CRC Press, Boca Raton, FL, USA, pp 259–277

Maun MA, Perumal J (1999) Zonation of vegetation on lacustrine coastal dunes: Effects of burial by sand. Ecol Lett 2:14–18

May EB (1973) Extensive oxygen depletion in Mobile Bay, Alabama. Limnol Oceanogr 18:353–366

May-Kú MA, Ordóñez-López U (2006) Spatial patterns of density and size structure of penaeid shrimps *Farfantepenaeus brasiliensis* and *Farfantepenaeus notialis* in a hypersaline lagoon in the Yucatán Peninsula, Mexico. Bull Mar Sci 79:259–271

Mazella L, Alberte RS (1986) Light adaptation and the role of autotrophic epiphytes in primary production of the temperate seagrass, *Zostera marina* L. J Exp Mar Biol Ecol 100:165–180

McBride RA, Taylor MJ, Byrnes MR (1992) Coastal morphodynamics and Chenier-Plain evolution in southwestern Louisiana, USA: A geomorphic model. Geomorphology 88:367–422

McBride RA, Taylor MJ, Byrnes MR (2007) Coastal morphodynamics and Chenier-Plain evolution in southwestern Louisiana, USA: A geomorphic model. Geomorphology 88:367–422

McEachran JD (2009) Fishes (Vertebrata: Pisces) of the Gulf of Mexico. In: Felder DK, Camp DK (eds) Gulf of Mexico—origins, waters and biota, vol 1, Biodiversity. Texas A&M University Press, College Station, TX, USA, pp 1223–1316

McFalls TB, Keddy PA, Campbell D, Shaffer G (2010) Hurricanes, floods, levees, and nutria: Vegetation responses to interacting disturbance and fertility regimes with implications for coastal wetland restoration. J Coast Res 26:901–911

McGowen JH, Garner LE, Wilkinson BH (1977) The Gulf shoreline of Texas: Processes, characteristics, and factors in use. Geological Circular 77-3. Bureau of Economic Geology (BEG), The University of Texas at Austin, Austin, TX, USA, 45 p

McIvor CC, Rozas LP (1996) Direct nekton use of intertidal salt marsh habitat and linkage with adjacent habitats: A review from the southeastern United States. In: Nordstrom KF, Roman CT (eds) Estuarine shores: Evolution, environments and human alterations. John Wiley & Sons, New York, NY, USA, pp 311–334

McKee KL (1995a) Mangrove species distribution and propagule predation in Belize: An exception to the dominance-predation hypothesis. Biotropica 27:334–345

McKee KL (1995b) Seedling recruitment patterns in a Belizean mangrove forest: Effects of establishment and physico-chemical factors. Oecologia 101:448–460

McKee KL (2012) Neotropical coastal wetlands. In: Batzer DP, Baldwin AH (eds) Wetland habitats of North America. University of California Press, Berkeley, CA, USA, pp 89–102

McKee KL, Baldwin AH (1999) Disturbance regimes in North American wetlands. In: Walke LR (ed) Ecosystems of disturbed ground. Elsevier, Amsterdam, The Netherlands, pp 331–363

McKee KL, Cherry JA (2009) Hurricane Katrina sediment slowed elevation loss in subsiding brackish marshes of the Mississippi River Delta. Wetland 29:2–15

McKee KL, Faulkner PL (2000) Restoration of biogeochemical function in mangrove forest. Restor Ecol 8:247–259

McKee KL, Rooth JE (2008) Where temperate meets tropical: Multifactorial effects of elevated CO_2, nitrogen enrichment, and competition on a mangrove-salt marsh community. Glob Change Biol 14:1–14

McKee KL, Mendelssohn IA, Hester MW (1988) Reexamination of pore water sulfide concentrations and redox potentials near the aerial roots of *Rhizophora mangle* and *Avicennia germinans*. Am J Bot 75:1352–1359

McKee KL, Feller IC, Popp M, Wanek W (2002) Mangrove isotopic fractionation across a nitrogen vs. phosphorous limitation gradient. Ecology 83:1065–1075

McKee KL, Mendelssohn IA, Materne MD (2004) Acute salt marsh dieback in the Mississippi River deltaic plain: A drought-induced phenomenon? Glob Ecol Biogeogr 13:65–73

McKee KL, Cahoon DR, Feller IC (2007a) Caribbean mangroves adjust to rising sea level through biotic controls on change in soil elevation. Glob Ecol Biogeogr 16:545–556

McKee KL, Rooth JE, Feller IC (2007b) Mangrove recruitment after forest disturbance is facilitated by herbaceous species in the Caribbean. Ecol Appl 17:1678–1693

McKenzie LJ, Yoshida RL, Coles RG (2006–2012) Seagrass-watch. www.seagrasswatch.org/seagrass.html. Accessed 7 Feb 2013

McLachlan A (1996) Physical factors in benthic ecology: Effects of changing sand grain size on beach fauna. Mar Ecol Prog Ser 131:205–217

McMichael RH Jr, Peters KM (1989) Early life history of spotted seatrout, *Cynoscion nebulosus* (Pisces: Sciaenidae), in Tampa Bay, Florida. Estuaries 12:98–110

McMillan C (1971) Environmental factors affecting seedling establishment of the black mangrove on the central Texas coast. Ecology 52:927–930

McMillan C, Sherrod CL (1986) The chilling tolerance of black mangrove, *Avicennia germinans*, from the Gulf of Mexico coast of Texas, Louisiana, and Florida. Contrib Mar Sci 29:9–16

Meinesz A, Hesse B (1991) Introduction et invasion de l'algue tropicale *Caulerpa taxifolia* en Méditerranée nord-occidentale. Oceanol Acta 14:415–426

Mendelssohn IA (1979) Nitrogen metabolism in the height forms of *Spartina alterniflora* in North Carolina. Ecology 60:574–584

Mendelssohn IA, Kuhn NL (2003) Sediment subsidy: Effects on soil-plant responses in a rapidly submerging coastal salt marsh. Ecol Eng 21:115–128

Mendelssohn IA, McKee KL (1988) *Spartina alterniflora* die-back in Louisiana: Time-course investigation of soil waterlogging effects. J Ecol 76:509–521

Mendelssohn IA, McKee KL (2000) Salt marshes and mangroves. In: Barbour MG, Billings WD (eds) North American vegetation. Cambridge University Press, New York, NY, USA, pp 501–536

Mendelssohn IA, Morris JT (2000) Eco-physiological controls on the productivity of *Spartina alterniflora* loisel. In: Weinstein MP, Kreeger DA (eds) Concepts and controversies in tidal marsh ecology. Kluwer Academic Publishers, Dordrecht, The Netherlands, pp 59–80

Mendelssohn IA, Postek MT (1982) Elemental analysis of deposits on the roots of *Spartina alterniflora*, Loisel. Am J Bot 69:904–912

Mendelssohn IA, Seneca ED (1980) The influence of soil drainage on the growth of salt marsh cordgrass *Spartina alterniflora* in North Carolina. Estuar Coast Mar Sci 2:27–40

Mendelssohn IA, McKee KL, Patrick WH Jr (1981) Oxygen deficiency in *Spartina alterniflora* roots: Metabolic adaptation to anoxia. Science 214:439–441

Mendelssohn IA, McKee KL, Postek MT (1982) Sublethal stresses controlling *Spartina alterniflora* productivity. In: Turner RE, Gopal B, Wetzel RG, Whigham DF (eds) Wetlands ecology and management. International Scientific Publications, Jaipur, India, pp 223–242

Mendelssohn IA, Jordan JW, Talbot F, Starkovich CJ (1983) Dune building and vegetative stabilization in a sand deficient barrier island environment. In: Magoon OT, Converse H (eds) Coastal Zone 83—third symposium on coastal and ocean management, vol 1. American Society of Civil Engineers, San Diego, CA, USA, pp 601–619

Mendoza-Carranza M, Hoeinghaus DJ, Garcia AM, Romero-Rodriguez Á (2010) Aquatic food webs in mangrove and seagrass habitats of Centla Wetland, a biosphere reserve in southeastern Mexico. Neotrop Ichthy 8:171–178

Metcalfe KN, Glasby CJ (2008) Diversity of Polychaeta (Annelida) and other worm taxa in mangrove habitats of Darwin Harbour, northern Australia. J Sea Res 59:70–82

Meyer DL, Johnson JM, Gill JW (2001) Comparison of nekton use of *Phragmites australis* and *Spartina alterniflora* marshes in the Chesapeake Bay, USA. Mar Ecol Prog Ser 209:71–84

Meyer-Arendt KJ (1993) Shoreline changes along the North Yucatán Coast. In: Laska S, Puffer A (eds) Coastlines of the Gulf of Mexico. American Society of Civil Engineers, New York, NY, USA, pp 103–117

Meyerson LA, Saltonstall K, Windham L, Kiviat E, Findlay S (2000) A comparison of *Phragmites australis* in freshwater and brackish marsh environments in North America. Wetl Ecol Manage 8:89–103

Michel J, Rutherford N (2013) Oil spills in marshes: Planning and response considerations September 2013. Report to the National Oceanographic and Atmospheric Administration and the American Petroleum Institute. American Petroleum Institute, Washington, DC, USA

Milbrandt EC, Greenawalt-Boswell JM, Sokoloff PD, Bortone SA (2006) Impact and response of southwest Florida mangroves to the 2004 hurricane season. Estuar Coast 29:979–984

Miller TL, Morton RA, Sallenger AH, Moore LJ (2004) The national assessment of shoreline change: A GIS compilation of vector shorelines and associated shoreline change data for the U.S. Gulf of Mexico. Open-file report 2004-1089. Vector digital data. U.S. Geological Survey, Coastal and Marine Geology Program, U.S. Geological Survey, Center for Coastal and Watershed Studies, St. Petersburg, FL, USA. http://pubs.usgs.gov/of/2004/1089/gis-data.html. Accessed 24 May 2013

Miller DL, Thetford M, Schneider M (2008) Distance from the Gulf influences survival and growth of three barrier Island dune plants. J Coast Res 24:261–266

Minchinton TE, Bertness MD (2003) Disturbance-mediated competition and the spread of Phragmites australis in a coastal marsh. Ecol Appl 13:1400–1416

Minello TJ (1999) Nekton densities in shallow estuarine habitats of Texas and Louisiana and the identification of essential fish habitat. In Benaka LR, ed, Fish habitat: Essential fish habitat and rehabilitation. Am Fish Soc Symp 22:43–75

Minello TJ, Rozas LP (2002) Nekton in Gulf coast wetlands: Fine-scale distributions, landscape patterns, and restoration implications. Ecol Appl 12:441–455

Minello TJ, Zimmerman RJ (1992) Utilization of natural and transplanted Texas salt marshes by fish and decapod crustaceans. Mar Ecol Prog Ser 90:273–285

Minello TJ, Zimmerman RJ, Medina R (1994) The importance of edge for natant macrofauna in a created salt marsh. Wetlands 14:184–198

Minello TJ, Matthews GA, Caldwell PA, Rozas LP (2008) Population and production estimates for decapod crustaceans in wetlands of Galveston Bay, Texas. Trans Am Fish Soc 137:129–146

Minello TJ, Rozas LP, Baker R (2012) Geographic variability in salt marsh flooding patterns may affect nursery value for fishery species. Estuar Coast 35:501–514

Miner MD, Kulp MA, FitzGerald DM, Flocks JG, Weathers HD (2009) Delta lobe degradation and hurricane impacts governing large-scale coastal behavior, South-central Louisiana, USA. Geo-Mar Lett 29:441–453

Mitsch WJ, Gosselink JG (1993) Wetlands, 2nd edn. Van Nostrand Reinhold, New York, NY, USA

MMS (2005) Hurricanes on the OCS: Powerful new lessons. MMS Ocean Sci 2:10–11

MMS (2007) Outer continental shelf oil and gas leasing program: 2007-2012. Final environmental impact statement. U.S. Department of the Interior MMS, Herndon, VA, USA, 506 p

Modde T, Ross ST (1980) Seasonality of fishes occupying a surf zone habitat in the northern Gulf of Mexico. Bulletin 78:911–922

Moksnes P, Heck KL Jr (2006) Relative importance of habitat selection and predation for the distribution of blue crab megalopae and young juveniles. Mar Ecol Prog Ser 308:165–181

Molina C, Rubinoff P, Carranza J (2001) Guidelines for low-impact tourism along the Coast of Quintana Roo, México. English ed. Coastal Resources Center, University of Rhode Island, Narragansett Bay, RI, USA, 106 p

Moncreiff CA, Sullivan MJ, Daehnick AE (1992) Primary production dynamics in seagrass beds of Mississippi Sound: The contributions of seagrass, epiphytic algae, sand microflora and phytoplankton. Mar Ecol Prog Ser 87:161–171

Monreal-Gomez MA, Salas-de-Leon AS, Velasco-Mendoza H (2004) The hydrodynamics of the Gulf of Mexico. In: Withers K, Nipper M (eds) Environmental analysis of the Gulf of Mexico, vol 1, Special publication series. Harte Research Institute for Gulf of Mexico Studies, Corpus Christi, TX, USA, pp 2–16

Montague CL, Odum HT (1997) The intertidal marshes of Florida's Gulf Coast. In: Coultas CL, Hsieh Y (eds) Ecology and management of tidal marshes: A model from the Gulf of Mexico. St. Lucie Press, Delroy, pp 1–9

Montague CL, Wiegert RG (1990) Salt marshes. In: Myers RL, Ewel JJ (eds) Ecosystems of Florida. University Presses of Florida, Gainesville, FL, USA, pp 481–516

Monteiro-Neto C, Cunha LPR, Musick JA (2003) Community structure of surf-zone fishes at Cassino Beach, Rio Grande do Sul, Brazil. J Coast Res 35:492–501

Montgomery JLM, Targett TE (1992) The nutritional role of seagrass in the diet of the omnivorous pinfish *Lagodon rhomboides* (L.). J Exp Mar Biol Ecol 158:37–57

Moody RM (2009) Trophic dynamics of salt marshes in the Northern Gulf of Mexico. PhD Dissertation, University of South Alabama, Mobile, AL, USA

Morán DK, Salles P, Sánchez JC, Espinal JC (2007) Beach nourishment evolution in the Cancún beach, Quintana Roo, México. In: Kraus NC, Rosati JD (eds) Proceedings, Coastal Sediments '07. American Society of Civil Engineers (ASCE), Reston, VA, USA, pp 2279–2291

Morang A, Waters JP, Khalil SM (2012) Gulf of Mexico Regional Sediment Budget. In Khalil SM, Parson LE, Waters JP, eds, Technical Framework for the Gulf Regional Sediment Management Master Plan(GRSMMP). J Coast Res 60:14–29

Moreno-Casasola P (1988) Patterns of plant-species distribution on coastal dunes along the Gulf of Mexico. J Biogeogr 15:787–806

Moreno-Casasola P (1993) Dry coastal ecosystems of the Atlantic coasts of Mexico and Central America. In: Maarel EV (ed) Ecosystems of the world 2B, dry coastal ecosystems Africa, America, Asia and Oceania. Elsevier, New York, NY, USA, pp 389–405

Moreno-Casasola P (2007) Beaches and Dunes of the Gulf of Mexico: A view of the current situation. In: Withers K, Nipper M (eds) Environmental analysis of the Gulf of Mexico. Harte Research Institute of the Gulf of Mexico, Houston, TX, USA, pp 302–313

Moreno-Casasola P, Espejel I (1986) Classification and ordination of coastal sand dune vegetation along the Gulf and Caribbean sea of Mexico. Vegetatio 66:147–182

Moretzsohn F, Sánchez-Chávez JA, Tunnell JW Jr (eds) (2012) GulfBase: Resource database for Gulf of Mexico Research. Texas A&M University, Corpus Christi, TX, USA. http://www.gulfbase.org. Accessed 7 Feb 2013

Morgan MD, Kitting CL (1984) Productivity and utilization of the seagrass *Halodule wrightii* and its attached epiphytes. Limnol Oceanogr 29:1066–1076

Morris JT (1984) Effects of oxygen and salinity on ammonium uptake by *Spartina alterniflora* Loisel. and Spartina patens (Aiton) Muhl. J Exp Mar Biol Ecol 78:87–98

Morrisey DJ, Howitt L, Underwood AJ, Stark JS (1992) Spatial variation in soft-sediment benthos. Mar Ecol Prog Ser 81:197–204

Morton RA, Barras JA (2011) Hurricane impacts on coastal wetlands: A half-century record of storm-generated features from southern Louisiana. J Coast Res 27:27–43

Morton RA, Pieper MJ (1977) Shoreline changes on Mustang Island and North Padre Island (Aransas Pass to Yarborough Pass): An analysis of historical changes of the Texas Gulf Shoreline. Geological circular 77-1. Bureau of Economic Geology (BEG), The University of Texas at Austin, Austin, TX, USA, 45 p

Mullin SJ (1995) Estuarine fish populations among red mangrove prop roots of small overwash islands. Wetlands 15:324–329

Mumby PJ (2006) Connectivity of reef fish between mangroves and coral reefs: Algorithms for the design of marine reserves as seascape scales. Biol Conserv 128:215–222

Musick JA, Harbin MM, Berkeley SA, Burgess GH, Eklund AM, Findley L, Gilmore RG, Golden JT, Ha DS, Huntsman GR, McGovern JC, Sedberry GR, Parker SJ, Poss SG, Sala E, Schmidt TW, Weeks H, Wright SG (2000) Marine, estuarine, and diadromous fish stocks at risk of extinction in North America (exclusive of Pacific salmonids). Fisheries 25:6–30

Nagelkerken I, van der Velde G (2004) Relative importance of interlinked mangroves and seagrass beds as feeding habitats for juvenile reef fish on a Caribbean island. Mar Ecol Prog Ser 274:153–159

Nagelkerken I, Kleijnen S, Klop T, van den Brand RACJ, Cocheret de la Moriniere E, van der Velde G (2001) Dependence of Caribbean reef fishes on mangroves and seagrass beds as nursery habitats: A comparison of fish faunas between bays with and without mangrove/seagrass beds. Mar Ecol Prog Ser 214:225–235

Nagelkerken I, Roberts CM, van der Velde G, Dorenbosch M, van Riel MC, Cocheret de la Moriniere E, Nienhuis PH (2002) How important are mangroves and seagrass beds for coral-reef fish? The nursery hypothesis tested on an island scale. Mar Ecol Prog Ser 244:299–305

Nagelkerken I, Blaber SJM, Bouillon S, Green P, Haywood M, Kirton LG, Meynecke J-O, Pawlik J, Penrose HM, Sasekumar A, Somerfield PJ (2008) The habitat function of mangroves for terrestrial and marine fauna: A review. Aquat Bot 89:155–185

Nagelkerken I, van der Velde G, Gorissen MW, Meijer GJ, van't Hof T, den Hartog C (2000) Importance of mangroves, seagrass beds and the shallow coral reef as a nursery for important coral reef fishes, using a visual census technique. Estuar Coast Shelf Sci 55:31–44

NDBC (National Data Buoy Center) (2012) National Oceanic and Atmospheric Administration. http://www.ndbc.noaa.gov/. Accessed 7 Feb 2013

Neckles HA, Neil C (1994) Hydrologic control of litter decomposition in seasonally flooded prairie marshes. Hydrobiolia 286:155–165

Negreiros-Fransozo ML, Barba E, Sanchez AJ, Fransozo A, Ráz-Guzmán A (1996) The species of *Hippolyte leach* (Crustacea, Caridea, Hippolytidae) from Términos Lagoon, southwestern Gulf of Mexico. Rev Bras Zool 13:539–551

Neill C, Turner RE (1987) Comparison of fish communities in open and plugged backfilled canals in Louisiana coastal marshes. N Am J Fish Manage 7:57–62

Nelson DM (ed) (1992) Distribution and abundance of fishes and invertebrates in Gulf of Mexico estuaries, vol 1: Data summaries. ELMR Report 10. NOAA/NOS (National Ocean Service) Strategic Environmental Assessments Division, Rockville, MD, USA, 273 p

Newell RC, Seiderer LJ, Hitchcock DR (1998) The impact of dredging works in coastal waters: A review of the sensitivity to disturbance and subsequent recovery of biological resources on the seabed. Oceanogr Mar Biol Annu Rev 36:127–178

NHC (National Hurricane Center) (2012) Tropical cyclone climatology. National Oceanic and Atmospheric Administration National Weather Service, Miami, FL, USA. http://www.nhc.noaa.gov/pastprofile.shtml. Accessed 7 Feb 2013

Niering WA, Warren RS, Weymouth CG (1977) Our dynamic tidal marshes: Vegetation changes as revealed by peat analysis. Connecticut Arboretum Bull 22:2–12

Nixon SW, Buckley BA (2002) A strikingly rich zone—nutrient enrichment and secondary production in coastal marine ecosystems. Estuaries 25:782–796

NMFS (National Marine Fisheries Service) (1988) Marine fisheries statistics of the United States. Department of Commerce, NOAA, Washington, DC, USA

NOAA (National Oceanic and Atmospheric Administration) (2004) Seagrass information for Alabama, Florida, Mississippi, and Texas. National Coastal Data Development Center. Coastal Habitats and Ecosystems. Washington, DC, USA. http://www.ncddc.noaa.gov/website/CHP/viewer.htm. Accessed 7 Feb 2013

NOAA (2011) The Gulf of Mexico at a glance: A second glance. U.S. Department of Commerce, Washington, DC, USA, 51 p

NOAA, NGS (National Geodetic Survey) (2013a) NOAA national shoreline: Shoreline data rescue project of ten thousand islands, Florida, FL2701. U.S. Department of Commerce, NOAA, National Ocean Service NGS, Silver Springs, MD, USA. http://www.ngs.noaa.gov/NSDE/. Accessed 1 July 2013

NOAA, NGS (2013b) NOAA national shoreline: Shoreline data rescue project of Everglades City to Venice, CM-7808. U.S. Department of Commerce, NOAA, National Ocean Service NGS, Silver Springs, MD, USA. http://www.ngs.noaa.gov/NSDE/. Accessed 1 July 2013

NOAA, NOS (National Ocean Service), ORR (Office of Response and Restoration), HMRD (Hazardous Material Response Division); MMS (Minerals Management Service); USFWS (U.S. Fish and Wildlife Service); LOSCO (The Louisiana Oil Spill Coordinator's Office); LDWF (The Louisiana Department of Wildlife and Fisheries); LDNR (Louisiana Department of Natural Resources) (2004) Louisiana ESI: ESI (Environmental Sensitivity Index Shoreline Types—Lines), 1st edn. Vector digital data. NOAA, Seattle, WA, USA. http://response.restoration.noaa.gov/esi. Accessed 5 April 2012

NOAA, NOS, ORR, ERD (Emergency Response Division) and ARD (Assessment and Restoration Division); DHS (Department of Homeland Security), USCG (U.S. Coast Guard), Office of Incident Management and Preparedness; USEPA, Region 4, Nonpoint Source and Wetlands Planning Section (2007) Alabama Environmental Sensitivity Index (ESI): ESI (Shoreline Types – Lines and Polygons), 2nd edn. Vector digital data, Seattle, WA, USA. http://response.restoration.noaa.gov/esi. Accessed 7 Feb 2013

NOAA, NOS, ORR, ERD, and DHS, USCG, Office of Incident Management and Preparedness (2009) Sensitivity of coastal environments and wildlife to spilled oil: Mississippi: ESI (Environmental Sensitivity Index Shoreline Types—Lines and Polygons), 2nd edn. Vector digital data, Seattle, WA, USA. http://response.restoration.noaa.gov/esi. Accessed 7 Feb 2013

Nowell ARM, Jumars PA (1984) Flow environments of aquatic benthos. Annu Rev Ecol Syst 15:303–328

Nowlin WD Jr, Jochens AE, Reid RO, DiMarco SF (1998) Texas-Louisiana shelf circulation and transport processes study: Synthesis report, vol I. Technical report, OCS Study MMS 98-0035. MMS, Gulf of Mexico OCS Region, New Orleans, LA, USA

Nowlin WD Jr, Jochens AF, DiMarco SF, Reid RO (2000) Physical oceanography. In: Continental Shelf Associates (ed) Deepwater Gulf of Mexico environmental and socioeconomic data search and literature synthesis, vol 1: Narrative report. U.S. Department of the Interior, MMS, Gulf of Mexico OCS Region, New Orleans, LA, USA, 340 p

NWCM (National Water Commission of Mexico) (2010) Statistics on water in Mexico, 2010 edn. Coyoacan, Mexico, 249 p

Nyman JA, Chabreck RH (1995) Fire in coastal marshes: History and recent concerns. In: Cerulean S, Engstrom RT (eds) Proceedings of the Tall Timbers Fire Ecology Conference, vol 19. Louisiana State University, Baton Rouge, LA, USA, pp 134–141

Nyman JA, DeLaune RD, Roberts HH, Patrick WH Jr (1993) Relationship between vegetation and soil formation in a rapidly submerging coastal marsh. Mar Ecol Prog Ser 96:269–279

O'Connor TP, Matlock GC (2005) Shrimp landing trends as indicators of estuarine habitat quality. Gulf Mexico Sci 2:192–196

O'Connor TP, Whitall D (2007) Linking hypoxia to shrimp catch in the northern Gulf of Mexico. Mar Pollut Bull 54:460–463

Odum WE (1988) Comparative ecology of tidal freshwater and salt marshes. Annu Rev Ecol Syst 19:147–176

Odum EP (2000) Tidal marshes as outwelling/pulsing systems. In: Weinstein MP, Kreeger DA (eds) Concepts and controversies in tidal marsh ecology. Kluwer Academic Publishers, Dordrecht, The Netherlands, pp 3–7

Odum EP, Fanning ME (1973) Comparison of the productivity of *Spartina alterniflora* and *Spartina cynosuroides* in Georgia coastal marshes. Bull Geo Acad Sci 31:1–12

Odum WE, Heald EJ (1972) Trophic analysis of an estuarine mangrove community. Bull Mar Sci 22:671–738

Odum WE, McIvor CC, Smith TJ III (1982) The ecology of the mangroves of South Florida: A community profile. Final report. FWS/OBS-81/24. U.S. Department of the Interior, MMS Gulf of Mexico OCS Office, Metairie, LA, USA, 144 p

Oey L-Y (1995) Eddy and wind-forced circulation. J Geophys Res 100:8621–8637

Oey L-Y, Ezer T, Lee HC (2005) Loop current, rings and related circulation in the Gulf of Mexico: A review of numerical models and future challenges. In: Sturges W, Lugo-Fernandez A (eds) Circulation in the Gulf of Mexico: Observations and models. American Geophysical Union, Washington, DC, USA, pp 31–56

Okazaki H, Stanley J-D, Wright EE (2001) Tecolutla and Nautla Deltas, Veracruz, Mexico: Texture to evaluate sediment entrapment on deltaic plains and bypassing onto the Gulf of Mexico margin. J Coast Res 17:755–761

Oliver JS, Slattery PN, Hulberg LW, Nybakken JW (1980) Relationship between wave disturbance and zonation of benthic invertebrate communities along a subtidal high-energy beach in Monterey Bay, California. Fish Bull 78:437–454

Olmi EJ III, Orth RJ (1995) Introduction to the Proceedings of the Blue Crab Recruitment Symposium. Bull Mar Sci 57:707–712

Olmsted I (1993) Wetlands of Mexico. In: Dykyjova D, Whigham DF, Hejny S (eds) Wetlands of the world I: Inventory, ecology and management. Kluwer Academic Publishers, Dordrecht, The Netherlands, pp 637–677

Olson DM, Dinerstein E, Wikramanayake ED, Burgess ND, Powell GVN, Underwood EC, D'Amico JA, Itoua I, Strand HE, Morrison JC, Loucks CJ, Allnutt TF, Ricketts TH, Kura Y, Lamoreux JF, Wettengel WW, Hedao P, Kassem KR (2001) Terrestrial ecoregions of the world: A new map of life on earth. BioScience 51:933–938

Onuf CP (1994) Seagrasses, dredging and light in Laguna Madre, Texas, U.S.A. Estuar Coast Shelf Sci 39:75–91

Onuf CP, Teal JM, Valiela I (1977) Interactions of nutrients, plant growth and herbivory in a mangrove ecosystem. Ecology 58:514–526

Onuf CP, Phillips RC, Moncreiff CA, Raz-Guzman A, Herrera-Silveira JA (2003) The seagrasses of the Gulf of Mexico. In: Green EP, Short FT (eds) World atlas of seagrasses. University of California Press, Berkeley, CA, USA, pp 224–233

Orth RJ (1977) The importance of sediment stability in seagrass communities. In: Coull BC (ed) Ecology of marine benthos. University of South Carolina Press, Columbia, SC, USA, pp 281–300

Orth RJ, van Montfrans J (1984) Epiphyte-seagrass relationships with an emphasis on the role of micrograzing: A review. Aquat Bot 18:43–69

Orth RJ, Heck KL Jr, van Montfrans J (1984) Faunal communities in seagrass beds: A review of the influence of plant structure and prey characteristics on predator-prey relationships. Estuaries 7:339–350

Orth RJ, van Montfrans J, Lipcius RN, Metcalf KS (1996) Utilization of seagrass habitat by the blue crab, *Callinectes sapidus* Rathbun, in Chesapeake Bay: A review. In: Kuo J, Phillips RC, Walker DI, Kirkman H (eds) Seagrass biology: Proceedings of an International Workshop. University of Western Australia, Perth, Australia, pp 213–224

Orth RJ, Carruthers TJB, Dennison WC, Duarte CM, Fourqurean JW, Heck KL Jr, Hughes AR, Kendrick GA, Kenworthy WJ, Olyarnik S, Short FT, Waycott M, Williams SL (2006) A global crisis for seagrass ecosystems. BioScience 56:987–996

Ortiz M, Lalana R (2005) Marine biodiversity of the Cuban archipelago: An overview. In: Miloslavich P, Klein E (eds) Caribbean marine biodiversity: The known and the unknown. DEStech Publications, Lancaster, PA, USA. http://cbm.usb.ve/CoMLCaribbean/pdf/I-03_Cuba_final.pdf. Accessed 7 Feb 2013

Ortiz-Pérez MA (1992) Retroceso y avance de la línea de costa del frente deltáico del río San Pedro. Campeche-Tabasco Investigaciones Geográficas Boletín del Instituto de Geografía 25:7–23

Ortiz-Pérez MA, Benítez J (1996) Elementos teóricos para el entendimiento de los problemas de impacto ambiental en las planicies delticas: La región de Tabasco y Campeche. In: Botello AV, Rojas-Galaviz JL, Benítez J, Zárate-Lomelí D, (eds) Golfo de México, Contaminación e Impacto Ambiental: Diagnóstico y Tendencias, EPOMEX, Serie científica 5. Universidad Autónoma de Campeche, Campeche, Mexico, pp 483–503

Ortiz-Pérez MA, Hernández-Santana JR, Figueroa Mah Eng JM, Gama CL (2010) Tasas del avance transgresivo y regresivo en el frente deltaico tabasqueno: En el periodo comprendido del ano 1995 al 2008. In: Botello AV, Villanueva-Fragoso S, Gutierrez J, Rojas-Galaviz JL (eds) Vulnerabilidad de las Zonas Costeras Mexicanas ante el Cambio Climatico. Semarnat-INE, UNAM-ICMyl, Universidad Autónoma de Campeche, San Francisco de Campeche, Campeche, Mexico, pp 305–324

Osgood DT, Yozzo DJ, Chambers RM, Jacobson D, Hoffman T, Wnek J (2003) Tidal hydrology and habitat utilization by resident nekton in *Phragmites* and non-*Phragmites* marshes. Estuaries 26:522–533

Otvos EG (1979) Barrier island evolution and history of migration, north central Gulf Coast. In: Leatherman SP (ed) Barrier Islands from the Gulf of St. Lawrence to the Gulf of Mexico. Academic Press, New York, NY, USA, pp 291–319

Otvos EG, Carter GA (2008) Hurricane degradation—barrier development cycles, Northeastern Gulf of Mexico: Landform evolution and island chain history. J Coast Res 24:463–478

Otvos EG, Giardino MJ (2004) Interlinked barrier chain and delta lobe development, northern Gulf of Mexico. Sediment Geol 169:47–73

Packham JR, Willis AJ (1997) Ecology of dunes, salt marsh and shingle. Chapman & Hall, London, UK, 335 p

Paerl HW, Bales JD, Ausley LW, Buzzelli CP, Crowder LB, Eby LA, Fear JM, Go M, Peierls BL, Richardson TL (2001) Ecosystem Impacts of Three Sequential Hurricanes (Dennis, Floyd, and Irene) on the United States' Largest Lagoonal Estuary, Pamlico Sound, NC. Proc Natl Acad Sci 98:5655–5660

Paine JG, Mathew S, Caudle T (2011) Texas Gulf shoreline change rates through 2007. Bureau of Economic Geology, University of Texas at Austin, Austin, TX, USA, 38 p

Parker SJ, Davies DJ, Smith WE (1997) Geological, economic, and environmental characterization of selected nearterm leasable offshore sand deposits and competing onshore sources for beach nourishment. Alabama geological survey circular 190. Geological Survey of Alabama, Tuscaloosa, AL, USA, 173 p

Parkinson RA (1989) Decelerating Holocene sea-level rise and its influence on Southwest Florida coastal evolution: A transgressive/regressive stratigraphy. J Sediment Petrol 50:960–972

ParksWatch-Mexico (2003) Park Profile-Mexico: Pantanos de Centla Biosphere Reserve. ParksWatch-Mexico, Delegación Benito Juárez, México, D.F., 20 p

Partyka ML, Peterson MS (2008) Habitat quality and salt-marsh species assemblages along an anthropogenic estuarine landscape. J Coast Res 24:1570–1581

Patterson CS, Mendelssohn IA (1991) A comparison of physicochemical variables across plant zones in a mangal/salt marsh community in Louisiana. Wetlands 11:139–161

Patterson CS, Mendelssohn IA, Swenson EM (1993) Growth and survival of *Avicennia germinans* seedlings in a mangal/salt marsh community in Louisiana, U.S.A. J Coast Res 9:801–810

Patterson S, McKee KL, Mendelssohn IA (1997) Effects of tidal inundation and predation on *Avicennia germinans* seedling establishment and survival in a sub-tropical mangal/salt marsh community. Mangrove Salt Marshes 1:103–111

Pauly D, Christensen V, Dalsgaard J, Froese R, Torres F Jr (1998) Fishing down marine food webs. Science 279:860–863

Penland S, Ramsey KE (1990) Relative sea-level rise in Louisiana and the Gulf of Mexico: 1908-1988. J Coast Res 6:323–342

Penland S, Suter JR (1989) The geomorphology of the Mississippi River Chenier Plain. Mar Geol 90:231–258

Penland S, Boyd R, Suter JR (1988) Transgressive depositional systems of the Mississippi delta plain: A model for barrier shoreline and shelf sand development. J Sediment Petrol 58:932–949

Penland S, Roberts HH, Williams SJ, Sallenger AH Jr, Cahoon DR, Davis DW, Groat CG (1990) Coastal land loss in Louisiana. Trans Gulf Coast Assoc Geol Soc 40:685–699

Penland S, Williams SJ, Davis DW, Sallenger AH Jr, Groat CG (1992) Barrier island erosion and wetland loss in Louisiana: Atlas of shoreline changes in Louisiana from 1985 to 1989. In: Williams SJ, Penland S, Sallenger AH (eds) Atlas of shoreline changes in Louisiana from 1853 to 1989. USGS Miscellaneous Investigations Series I-2150A. USGS, Reston, VA, USA, pp 2–7

Pennings SC, Grant MB, Bertness MD (2005) Plant zonation in low-latitude salt marshes: Disentangling the roles of flooding, salinity and competition. J Ecol 93:159–167

Peralta-Meixueiro MA, Vega-Cendejas ME (2010) Spatial and temporal evaluation of fish assemblages in the lagoon systems of Ria Lagartos, Mexico. In: Proceedings of the 63rd Gulf and Caribbean Fisheries Institute, Gulf and Caribbean Fisheries Institute, San Juan, Puerto Rico, pp 274–281

Peralta-Meixueiro MA, Vega-Cendejas ME (2011) Spatial and temporal structure of fish assemblages in a hyperhaline coastal system: Ría Lagartos, Mexico. Neotrop Ichthyol 9:3

Peresbarbosa-Rojas E (2005) Hacía un Diagnóstico de la Zona Costera de Veracruz y un Manejo Integral de la Zona Costera. Pronatura Veracruz and The Nature Conservancy, Xalapa, 91 p

Pérez-Castañeda R, Defeo O (2001) Population variability of four sympatric penaeid shrimps (*Farfantepenaeus* spp.) in a tropical coastal lagoon of Mexico. Estuar Coast Shelf Sci 52:631–641

Pérez-Hernández MA, Torres-Orozco RE (2000) Fish species richness evaluation in Mexican coastal lagoons: A case study in the Gulf of Mexico. Rev Biol Trop 48:425–438

Perry HM (ed) (1984) A profile of the blue crab fishery of the Gulf of Mexico. Gulf States Marine Fisheries Commission 9. Gulf States Marine Fisheries Commission, Ocean Springs, MS, USA, 80 p

Perry CL, Mendelssohn IA (2009) Ecosystem effects of expanding populations of *Avicennia germinans* in a Louisiana salt marsh. Wetlands 29:396–406

Peterson CGJ (1913) Valuation of the Sea II. The animal communities of the sea bottom and their importance for marine zoogeography. Report of the Danish Biological Station to the Board of Agriculture vol 21, pp 1–44

Peterson CH (1982) Clam predation by whelks (*Busycon spp.*): Experimental tests of the importance of prey size, prey density, and seagrass cover. Mar Biol 66:159–170

Peterson CH (1986) Enhancement of *Mercenaria mercenaria* densities in seagrass beds: Is pattern fixed during settlement season or altered by subsequent differential survival? Limnol Oceanogr 31:200–205

Peterson BJ, Heck KL Jr (2001) Positive interactions between suspension-feeding bivalves and seagrass—a facultative mutualism. Mar Ecol Prog Ser 213:143–155

Peterson MS, Partyka ML (2006) Baseline mapping of *Phragmites australis* (Common reed) in three coastal Mississippi estuarine basins. Southeast Nat 5:747–756

Peterson GW, Turner RE (1994) The value of salt marsh edge vs interior as a habitat for fish and decapod crustaceans in a Louisiana tidal marsh. Estuaries 17:235–262

Peterson CH, Able KW, DeJong CF, Piehler MF, Simenstad CA, Zedler JB (2008) Practical proxies for tidal marsh ecosystem services: Application to injury and restoration. Adv Mar Biol 54:221–266

Pezeshki SR, Hester MW, Lin Q, Nyman JA (2000) The effects of oil spill and clean-up on dominant U.S. Gulf coast marsh macrophytes: A review. Environ Pollut 108:129–139

Philipp KR, Field RT (2005) *Phragmites australis* expansion in Delaware Bay salt marshes. Ecol Eng 25:275–291

Phillips JD (1987) Shoreline processes and establishment of *Phragmites australis* in a coastal plain estuary. Vegetatio 71:139–144

Piazza BP, La Peyre MK (2009) The effect of Hurricane Katrina on nekton communities in the tidal freshwater marshes of Breton Sound, Louisiana, USA. Estuar Coast Shelf Sci 83:97–104

Pickens CN, Hester MW (2011) Temperature tolerance of early life history stages of black mangrove *Avicennia germinans*: Implications for range expansion. Estuar Coast Shelf Sci 34:824–830

Pielou EC, Routledge RD (1976) Salt-marsh vegetation—latitudinal gradients in zonation patterns. Oecologia 24:311–321

Pineda J (2000) Linking larval settlement to larval transport: Assumptions, potentials, and pitfalls. Oceanogr East Pac 1:84–105

Plunket JT (2003) A comparison of finfish assemblages on subtidal oyster shell (clutched oyster lease) and mud bottom in Barataria Bay, Louisiana. MS Thesis, Louisiana State University and Agricultural and Mechanical College, Baton Rouge, LA, USA, 84 p

Poggie JJ (1962) Coastal pioneer plants and habitat in the Tampico region, Mexico. Coastal Studies Institute, Louisiana State University, Baton Rouge, LA, USA

Poiner IR, Walker DI, Coles RG (1989) Regional studies—seagrasses of tropical Australia. In: Larkum AWD, McComb AJ, Shepherd SA (eds) Biology of seagrasses. A treatise on the biology of seagrasses with special reference to the Australian region. Elsevier, Amsterdam, The Netherlands, pp 279–303

Posey MH, Alphin TD, Meyer DL, Johnson JM (2003) Benthic communities of common reed *Phragmites australis* and marsh cordgrass *Spartina alterniflora* marshes in Chesapeake Bay. Mar Ecol Prog Ser 261:51–61

Preen AR, Long WJL, Coles RG (1995) Flood and cyclone related loss, and partial recovery, of more than 1000 km² of seagrass in Hervey-Bay, Queensland, Australia. Aquat Bot 52:3–17

Probert PK (1984) Disturbance, sediment stability, and trophic structure of soft-bottom communities. J Mar Res 42:893–921

Proffitt CE (ed) (1998) Effects and management of oil spills in marsh ecosystems. U.S. Department of the Interior, Minerals Management Service, Gulf of Mexico OCS Region, New Orleans, LA, USA

Proffitt CE, Devlin DJ (2005) Grazing by the intertidal gastropod Melampus coffeus greatly increases mangrove leaf litter degradation rates. Mar Ecol Prog Ser 296:209–218

Profitt CE, Milbrandt ED, Travis SE (2006) Red mangrove (*Rhizophora mangle*) reproduction and seedling colonization after Hurricane Charley: Comparisons of Charlotte Harbor and Tampa Bay. Estuar Coast 29:972–978

Psuty NP, Martínez ML, López-Portillo J (2008) Interaction of alongshore sediment transport and habitat conditions at Laguna La Mancha, Veracruz, Mexico. In: Proceedings of the International Pluridisciplinary Conference, Lille, France, 16–18 Jan 2008, 7 p

Psuty NP, Martínez ML, López-Portillo J, Silveira TM, Garcia-Franco JG, Rodriguez NA (2009) Interaction of alongshore sediment transport and habitat conditions at Laguna La Mancha, Veracruz, Mexico. J Coast Conserv 13(2–3):77–87

Pulich W (2001) Seagrass inventory of Christmas and Drum Bays: Webster, Texas. Special report. Galveston Bay National Estuary Program, Houston, TX, USA, 4 p

Pulich W, Blair C, White WA (1997) Current status and historical trends of seagrass in the Corpus Christi Bay National Estuary Program study area. Austin, Texas Natural Resource Conservation Commission, Publication CCBNEP-20, Austin, TX, USA, 131 p

Raabe EA, Streck AE, Stumpf RP (2004) Historic topographic sheets to satellite imagery: A methodology for evaluating coastal change in Florida's Big Bend Tidal Marsh. United States Geological Survey (USGS) Open File Report 02-211. USGS, Center for Coastal and Regional Marine Studies, St. Petersburg, FL, USA, 44 p

Rabalais NN, Turner RE, Wiseman WJ (2002) Gulf of Mexico, A.K.A. The Dead Zone. Annu Rev Ecol Evol Syst 33:235–263

Rabinowitz D (1978) Dispersal properties of mangrove propagules. Biotropica 10:47–57

Raichel DL, Able KW, Hartman JM (2003) The influence of *Phragmites* (common reed) on the distribution, abundance, and potential prey of a resident marsh fish in the Hackensack Meadowlands, New Jersey. Estuaries 26:511–521

Rakocinski CF, Drury D (2005) Early blue crab recruitment to alternative nursery habitats in Mississippi, USA. J Shellfish Res 24:253–259

Rakocinski CF, Heard RW, Simons T, Gledhill D (1991) Macroinvertebrate associations from beaches of selected barrier islands in the Northern Gulf of Mexico: Important environmental relationships. Bull Mar Sci 48:689–701

Rakocinski CF, Heard RW, LeCroy SE, McLelland HA, Simons T (1993) Seaward change and zonation of the sandy-shore macrofauna at Perdido Key, Florida, USA. Estuar Coast Shelf Sci 36:81–104

Rakocinski CF, LeCroy SE, McLelland JA, Heard RW (1998) Nested spatiotemporal scales of variation in sandy-shore macrobenthic community structure. Bull Mar Sci 63:343–362

Ralph PJ, Tomasko D, Moore K, Seddon S, MacGinnis-Ng CMO (2006) Human impacts on seagrasses: Eutrophication, sedimentation and contamination. In: Larkum WD, Orth RJ, Duarte CM (eds) Seagrasses: Biology, ecology and conservation. Springer, Dordrecht, The Netherlands, pp 567–593

Randall LAJ, Foote AL (2005) Effects of managed impoundments on herbivory and wetland plant production and stand structure. Wetlands 25:38–50

Raulerson GE (2004) Litter processing by macrodetritivores in natural and restored neotropical mangrove forests. PhD Dissertation, Louisiana State University, Baton Rouge, LA, USA

Raynie RC, Shaw RF (1994) Ichthyoplankton abundance along a recruitment corridor from offshore spawning to estuarine nursery ground. Estuar Coast Shelf Sci 39:421–450

Reese MM, Stunz GW, Bushon AM (2008) Recruitment of estuarine-dependent nekton through a new tidal inlet: The opening of Packery Channel in Corpus Christi, TX, USA. Estuar Coast 31:1143–1157

Rehm AE (1976) The effects of the wood-boring isopod *Sphaeroma terebrans* on the Mangrove communities of Florida. Environ Conserv 3:47–57

Reid RO, Whitaker RE (1981) Numerical model for astronomical tides in the Gulf of Mexico, vol 1: Theory and application. Coastal Engineering Research Center, U.S. Army Corps of Engineers, Vicksburg, MS, USA, 115 p

Renaud ML (1986) Detecting and avoiding oxygen deficient sea water by brown shrimp, *Penaeus aztecus* (Ives), and white shrimp *Penaeus setiferus* (Linnaeus). J Exp Biol Ecol 98:283–292

Reusch TBH, Chapman ARO (1995) Storm effects on eelgrass (*Zostera marina*) and blue mussel (*Mytilus edulis* L.) beds. J Exp Mar Biol Ecol 192:257–271

Revsbech NP, Madsen B, Jorgensen BB (1980) Oxygen in the sea bottom measured with a microelectrode. Science 207:1355–1356

Rhoads DC, Aller RC, Goldhaber MB (1977) The influence of colonizing benthos on physical properties and chemical diagenesis on the estuarine seafloor. In: Coull BC (ed) Ecology of marine benthos. University of South Carolina Press, Columbia, SC, USA, pp 113–138

Rivera-Arriaga E, Lara-Domínguez AL, Villalobos-Zapata G, Yáñez-Arancibia A (2003) Trophodynamic ecology of two critical habitats (seagrasses and mangroves) in Términos Lagoon, southern Gulf of Mexico. Fish Centre Res Rep 11:245–254

Robbins LL, Coble PG, Clayton TD, Cai WJ (2009) Ocean carbon and biogeochemistry scoping workshop on terrestrial and coastal carbon fluxes in the Gulf of Mexico, St. Petersburg, FL, USA, 6–8 May 2008. USGS Open-File Report 2009-1070. USGS, Reston, VA, USA, 46 p

Roberts HH (1997) Dynamic change of the Holocene Mississippi River delta plain: The delta cycle. J Coast Res 13:605–627

Robertson AI, Lenanton RCJ (1984) Fish community structure and food chain dynamics in the surf-zone of sandy beaches: The role of detached macrophyte detritus. J Exp Mar Biol Ecol 84:265–283

Robertson DR, Smith-Vaniz WF (2008) Rotenone: An essential but demonized tool for assessing marine fish diversity. Bioscience 58:165–170

Robertson ML, Mills AL, Zieman JC (1982) Microbial synthesis of detritus-like particulates from dissolved organic carbon released by tropical seagrasses. Mar Ecol Prog Ser 7:279–285

Rodríguez R (2010) Cuba. In: Bird ECF (ed) Encyclopedia of the world's coastal landforms. Springer, New York, NY, USA, pp 273–278

Rogers RM (1977) Trophic interrelationships of selected fishes on the continental shelf of the northern Gulf of Mexico. PhD Dissertation, Texas A&M University, College Station, TX, USA

Rogers AL, Ravens TM (2008) Measurement of longshore sediment transport rates in the surf zone on Galveston Island, Texas. J Coast Res 24:62–73

Rojas-Galaviz JL, Yáñez-Arancibia A, Day JW Jr, Vera-Herrera FR (1992) Estuarine primary producers: Laguna de Términos—a study case. In: Seeliger U (ed) Coastal plant communities of Latin America. Academic Press, San Diego, CA, USA, pp 141–154

Rooker JR, Holt SA, Holt GJ, Fuiman LA (1999) Spatial and temporal variability in growth, mortality, and recruitment potential of postsettlement red drum, *Sciaenops ocellatus*, in a subtropical estuary. Fish Bull 97:581–590

Rooth JE, Stevenson JC (2000) Sediment deposition patterns in *Phragmites australis* communities: implications for coastal areas threatened by rising sea-level. Wetl Ecol Manage 8:173–183

Rosado-Solórzano R, Guzmán del Próo SA (1998) Preliminary trophic structure model for Tampamachoco lagoon, Veracruz, Mexico. Ecol Model 109:141–154

Rosas C, Martinez E, Gaxiola G, Brito R, Sánchez A, Soto LA (1999) The effect of dissolved oxygen and salinity on oxygen consumption, ammonia excretion and osmotic pressure of *Penaeus setiferus* (Linnaeus) juveniles. J Exp Mar Biol Ecol 234:41–57

Rosati JD, Lawton C (2011) Channel shoaling with deepening of Houma Navigation Channel at Cat Island Pass, Louisiana. In: Roberts TM, Rosati JD, Wang P (eds) Proceedings, Symposium to Honor Dr. Nicholas Kraus. J Coast Res Special Issue 59:256–265. Coastal Education & Research Foundation, Inc. (CERF), West Palm Beach, FL, USA

Ross ST, Doherty TA (1994) Short-term persistence and stability of barrier island fish assemblages. Estuar Coast Shelf Sci 38:49–67

Ross ST, McMichael RH Jr, Ruple DL (1987) Seasonal and diel variation in the standing crop of fishes and macroinvertebrates from a Gulf of Mexico surf zone. Estuar Coast Shelf Sci 25:391–412

Ross MS, Ruiz PL, Sah JP, Reed DL, Walters J, Meeder JF (2006) Early post-hurricane stand development in fringe mangrove forests of contrasting productivity. Plant Ecol 185:283–297

Ross MS, Ruiz PL, Sah JP, Hanan EJ (2009) Chilling damage in a changing climate in coastal landscapes of the subtropical zone: A case study from south Florida. Glob Change Biol 15:1817–1832

Rosso PH, Cronin JT, Stevens RD (2008) Monitoring the invasion of *Phragmites australis* in coastal marshes of Louisiana, USA, using multi-source remote sensing data. Proc SPIE 7110:7110B-71100B-9. doi:10.1117/12.800269

Roth A-MF, Baltz DM (2009) Short-term effects of an oil spill on marsh-edge fishes and decapod crustaceans. Estuar Coast 32:565–572

Roth BM, Rose KA, Rozas LP, Minello TJ (2008) Relative influence of habitat fragmentation and inundation on brown shrimp *Farfantepenaeus aztecus* production in northern Gulf of Mexico salt marshes. Mar Ecol Prog Ser 359:185–202

Rozas LP (1992) A comparison of shallow-water and marsh-surface habitats associated with pipeline Marine Consortium for the U.S. Department of the Interior. OCS Study MMS 92-006. MMS, New Orleans, LA, USA, 25 p

Rozas LP (1993) Nekton use of salt marshes of the southeastern region of the United States. In: Magoon O, Wilson WS, Converse H, Tobin LT (eds) Proceedings of the 8th Symposium on

Coastal and Ocean Management. American Society of Civil Engineers, New York, NY, USA, pp 528–537

Rozas LP (1995) Hydroperiod and its influence on nekton use of the salt marsh: A pulsing ecosystems. Estuaries 18:579–590

Rozas LP, LaSalle MW (1990) A comparison of the diets of Gulf killifish, *Fundulus grandis* Baird and Girard, entering and leaving a Mississippi brackish marsh. Estuaries 13:332–336

Rozas LP, Minello TJ (2010) Nekton density patterns in tidal ponds and adjacent wetlands related to pond size and salinity. Estuar Coast 33:652–667

Rozas LP, Reed DJ (1993) Nekton use of marsh-surface habitats in Louisiana (USA) deltaic salt marshes undergoing submergence. Mar Ecol Prog Ser 96:147–157

Rozas LP, Reed DJ (1994) Comparing nekton assemblages of subtidal habitats in pipeline canals traversing brackish and saline marshes in coastal Louisiana. Wetlands 14:262–275

Rozas LP, Minello TJ, Dantin DD (2012) Use of shallow lagoon habitats by nekton of the northeastern Gulf of Mexico. Estuar Coast 35:572–586

Rucker JB, Snowden JO (1989) Relict progradational beach ridge complex on Cat Island in Mississippi Sound. Trans Gulf Coast Assoc Geol Soc 39:531–539

Ruiz GM, Hines AH, Posey MH (1993) Shallow water as a refuge habitat for fish and crustaceans in non-vegetated estuaries: An example from Chesapeake Bay. Mar Ecol Prog Ser 99:1–16

Ruple DL (1984) Occurrence of larval fishes in the surf zone of a northern Gulf of Mexico barrier island. Estuar Coast Shelf Sci 18:191–208

Rutherford ES, Schmidt TW, Tilmant JT (1989) Early life history of spotted seatrout (*Cynoscion nebulosus*) and gray snapper (Lutjanus griseus) in Florida Bay, Everglades National Park, Florida. Bull Mar Sci 44:49–64

Salvador A (1991a) Introduction. In: Salvador A (ed) The geology of North America, vol J, The Gulf of Mexico Basin. The Geological Society of America, Boulder, CO, USA, pp 1–12

Salvador A (1991b) Origin and development of the Gulf of Mexico basin. In: Salvador A (ed) The geology of North America, vol J, The Gulf of Mexico Basin. The Geological Society of America, Boulder, CO, USA, pp 389–444

Sanders HL (1958) Benthic studies of Buzzards Bay I animal-sediment relationships. Limnol Oceanogr 3:245–258

Sasser CE, Gosselink JG, Swenson EM, Evers DE (1995) Hydrologic, vegetation, and substrate characteristics of floating marshes in sediment-rich wetlands of the Mississippi river delta plain, Louisiana, USA. Wetl Ecol 3:171–187

Sasser CE, Visser JM, Mouton E, Linscombe J, Hartley SB (2008) Vegetation types in coastal Louisiana in 2007. 1 sheet, scale 1:550,000. USGS Open-File Report 2008-1224. USGS, Reston, VA, USA

Sastry AN (1978) Physiology and ecology of reproduction in marine invertebrates. In: Vernberg FJ (ed) Physiological ecology of estuarine organisms. University of South Carolina Press, Columbia, SC, USA, pp 279–299

Saucier MH, Baltz DM (1993) Spawning site selection by spotted seatrout, *Cynoscion nebulosus*, and black drum, *Pogonias cromis*, in Louisiana. Environ Biol Fish 36:257–272

Sauer J (1967) Geographic reconnaissance of the seashore vegetation along the Mexican Gulf coast. Louisiana State University, Baton Rouge, LA, USA

Schlacher TA, Schoeman DS, Dugan J, Lastra M, Jones A, Scapini F, Mclachlan A (2008) Sand beach ecosystems: Key features, sampling issues, management challenges and climate change impacts. Mar Ecol 29(suppl 1):70–90

Schmitz WJ Jr (2005) Cyclones and westward propagation in the shedding of anticyclonic rings from the Loop Current. In: Sturges W, Lugo-Fernandez A (eds) Circulation in the Gulf of

Mexico: Observations and models. American Geophysical Union, Washington, DC, USA, pp 241–261

Schmitz WJ Jr (2003) Notes on the circulation in and around the Gulf of Mexico. Volume I: A review of the deep water circulation. Texas A&M University Press, College Station, TX, USA

Schofield PJ (2010) Update on geographic spread of invasive lionfishes (Pterois volitans [Linnaeus, 1758] and P. miles [Bennett, 1828]) in the western North Atlantic Ocean, Caribbean Sea and Gulf of Mexico. Aquat Inv 5(suppl 1):S117–S122

Scott TM, Campbell KM, Rupert FR, Arthur JD, Green RC, Means GH, Missimer TM, Lloyd JM, Yon JW, Duncan JG (2001) Geologic map of the State of Florida. Map SERIES 146. Florida Geological Survey, Tallahassee, FL, USA. Vector digital data. http://www.dep.state.fl.us/geology/gisdatamaps/state_geo_map.htm. Accessed 15 July 2013

Seim HE, Kjerfve B, Sneed JE (1987) Tides of Mississippi Sound and the adjacent continental shelf. Estuar Coast Shelf Sci 25:143–156

Serafy JE, Faunce CH, Lorenz JJ (2003) Mangrove shoreline fishes of Biscayne Bay, Florida. Bull Mar Sci 72:161–180

Shaw JK, Johnson PG, Ewing RM, Comiskey CE, Brandt CC, Farmer TA (1982) Benthic macroinfauna community characterization in Mississippi Sound and adjacent waters. U.S. Army Corps of Engineers, Mobile District, Mobile, AL, USA, 442 p

Sheaves M (2005) Nature and consequences of biological connectivity in mangrove systems. Mar Ecol Prog Ser 302:293–305

Shepsis V, Bermudez HE, Carter JD, Feazel W (2010) Cameron Parish, Louisiana: 14.5 km beach nourishment project challenges and solutions. In: Proceedings, WEDA XXX Technical Conference and TAMU 41 Dredging Seminar, San Juan, Puerto Rico, USA, pp 106–120

Sheridan PF (1992) Comparative habitat utilization by estuarine macrofauna within the mangrove ecosystem of Rookery Bay, Florida. Bull Mar Sci 50:21–39

Sheridan P (1997) Benthos of adjacent mangrove, seagrass and non-vegetated habitats in Rookery Bay, Florida, USA. Estuar Coast Shelf Sci 44:455–469

Sheridan P, Hays C (2003) Are mangroves nursery habitat for transient fishes and decapods? Wetlands 23:449–458

Sheridan P, Minello TJ (2003) Nekton use of different habitat types in seagrass beds of lower Laguna Madre, Texas. Bull Mar Sci 72:37–61

Sheridan P, McMahan G, Conley G, Williams A, Thayer G (1997) Nekton use of macrophyte patches following mortality of turtlegrass, *Thalassia testudinum*, in shallow waters of Florida Bay (Florida, USA). Bull Mar Sci 61:801–820

Sherrod CL, McMillan C (1981) Black mangrove, *Avicennia germinans*, in Texas: Past and present distribution. Contrib Mar Sci 24:115–131

Sherrod CL, McMillan C (1985) The distributional history and ecology of mangrove vegetation along the northern Gulf of Mexico coastal region. Contrib Mar Sci 28:129–140

Sherrod CL, Hockaday DL, McMillan C (1986) Survival of red mangrove, *Rhizophora mangle*, on the Gulf of Mexico coast of Texas. Contrib Mar Sci 29:27–36

Shew DM, Baumann RH, Fritts TH, Dunn LS (1981) Texas barrier islands region ecological characterization: Environmental synthesis papers. Technical report FWS/OBS-81/32. Denver Wildlife Research Center, Tulane University Museum of Natural History, Belle Chasse, LA, USA

Silliman BR, van de Koppel J, Bertness MD, Stanton L, Mendelsohn I (2005) Drought, snails, and large-scale die-off of southern U.S. salt marshes. Science 310:1803–1806

Silvia C, Popma J, Moreno-Casasola P (1991) Coastal sand dune vegetation of Tabasco and Campeche, Mexico. J Veg Sci 2:73–88

Smart RM (1982) Distribution and environmental control of productivity and growth of *Spartina alterniflora* (Loisel.). In: Sen DN, Rajpurohit KS (eds) Contributions to the ecology of halophytes. Dr. W. Junk Publishers, The Hague, The Netherlands, pp 127–142

Smith DPB (2003) *Diadema antillarum* (black spiny Caribbean sea urchin). https://commons.wikimedia.org/wiki/File:Urchin003.jpg. Accessed 9 Feb 2017

Smith SV (1981) Marine macrophytes as a global carbon sink. Science 211:838–840

Smith EH (2002) Barrier Islands. In: Tunnell JW Jr, Judd FW (eds) The Laguna Madre of Texas and Tamaulipas. Texas A&M University Press, College Station, TX, USA, pp 127–136

Smith LC, Smith EH (2007) Final Report: Summary inventory of marine and fresh water fish of Padre Island National Seashore National Park Service. TAMU-CC-0703-CCS. U.S. Department of the Interior, Gulf Coast Network, Lafayette, LA, USA, 103 p

Smith TJ, Chan HT, McIvor CC, Robblee MB (1989) Comparisons of seed predation in tropical, tidal forests from three continents. Ecology 70:146–151

Smith TJ, Robblee MB, Wanless HR, Doyle TW (1994) Mangroves, hurricanes, and lightning strikes. BioScience 44:256–262

Smith TJ, Anderson GH, Balentine K, Tiling G, Ward GA, Whelan KRT (2009) Cumulative impacts of hurricanes on Florida mangrove ecosystems: Sediment deposition, storm surges, and vegetation. Wetlands 29:24–34

Snelgrove PVR, Butman CA (1994) Animal-sediment relationships revisited: Cause versus effect. Oceanogr Mar Biol Annu Rev 32:111–177

Snow AA, Vince SW (1984) Plant zonation in an Alaskan salt marsh. II. An experimental study of the role of edaphic conditions. J Ecol 72:669–684

Sogard SM, Powell GVN, Holmquist JG (1989) Utilization by fishes of shallow, seagrass-covered banks in Florida Bay: 2. Diel and tidal patterns. Environ Biol Fish 24:81–92

Solis-Marin FA, Herrero-Perezrul MD, Laguarda-Figueras A, Torres-Vega J (1993) Asteroideos y echinoideos de Mexico (Echinodermata). In: Salaza-Vallejo SI, Gonzalez NE (eds) Biodiversidad Marina y Costera de Mexico. Centre de Investigaciones de Quintana Roo (CIQRO), Mexico, pp 91–105

Sousa WP, Kennedy PG, Mitchell BJ, Ordonez BM (2007) Supply-side ecology in mangroves: Do propagule dispersal and seedling establishment explain forest structure? Ecol Monogr 77:53–76

Spalding MD, Fox HE, Allen GR, Davidson N, Ferdaña ZA, Finlayson M, Halpern BS, Jorge MA, Lombana A, Lourie SA, Martin KD, McManus E, Molnar J, Recchia CA, Robertson J (2007) Marine ecoregions of the world: A bioregionalization of coastal and shelf areas. BioScience 57:573–583

Spalding MD, Kainuma M, Collins L (2010) World atlas of mangroves. Earthscan, Oxford, UK

Stanton LE (2005) The establishment, expansion and ecosystem effects of *Phragmites australis*, an invasive species in coastal Louisiana. PhD Dissertation, The Department of Oceanography and Coastal Sciences, Louisiana State University, Baton Rouge, LA, USA, 182 p

Stapor FW Jr (1971) Origin of the Cabo Rojo beach-ridge plain, Veracruz, Mexico. Trans Gulf Coast Assoc Geol Soc 21:223–230

Stehlik L (1993) Diet of the brachyuran crabs *Cancer irroratus*, *C. borealis*, and *Ovalipes ocellatus* in the New York Bight. J Crust Biol 13:723–735

Stevens PF (2001) (onwards) Angiosperm phylogeny website. Version 9, June 2008. www.mobot.org/MOBOT/research/APweb. Accessed 11 Feb 2013

Stevens PW, Fox SL, Montague CL (2006) The interplay between mangroves and salt marshes at the transition between temperate and subtropical climate in Florida. Wetl Ecol Manage 14:435–444

Stevenson CST (2007). Enhancement of recruitment and nursery function by habitat creation in Pensacola Bay, Florida. MS Thesis, Department of Biology, The University of West Florida, Pensacola, FL, USA, 86 p

Stone GW, Zhang X (2001) A longshore sediment transport model for the Timbalier Islands and Isles Derniers. Louisiana Coastal Studies Institute, Louisiana State University, Baton Rouge, LA, USA, 26 p

Stone GW, Stapor FW, May JP, Morgan JP (1992) Multiple sediment sources and a cellular, non-integrated, longshore drift system: Northwest Florida and southeast Alabama coast, USA. Mar Geol 105:141–154

Stone GW, Liu B, Pepper DA, Wang P (2004) The importance of extratropical and tropical cyclones on the short-term evolution of barrier islands along the northern Gulf of Mexico, USA. Mar Geol 210:63–78

Stoner AW (1980a) The role of seagrass biomass in the organization of benthic macrofaunal assemblages. Bull Mar Sci 30:537–551

Stoner AW (1980b) Perception and choice of substratum by epifaunal amphipods associated with seagrasses. Mar Ecol Prog Ser 3:105–111

Stoner AW, Zimmerman RJ (1988) Food pathways associated with penaeid shrimps in a mangrove-fringed estuary. Fish Bull 86:543–551

Stout JP (1984) The ecology of irregularly flooded salt marshes of the Northeastern Gulf of Mexico: A community profile. U.S. Fish and Wildlife Service Department of the Interior, report 85(7.1), Washington, DC, USA, 98 p

Stuart SA, Choat B, Martin KC, Holbrook NM, Ball MC (2007) The role of freezing in setting the latitudinal limits of mangrove forests. New Phytol 173:576–583

Stumpf RP, Haines JW (1998) Variations in tidal level in the Gulf of Mexico and implications for tidal wetlands. Estuar Coast Shelf Sci 46:165–173

Sturges W (1993) The annual cycle of the western boundary current in the Gulf of Mexico. J Geophys Res 98:18053–18068

Sturges W, Blaha J (1975) A western boundary current in the Gulf of Mexico. Science 26:367–369

Sturges W, Kenyos KE (2008) Mean flow in the Gulf of Mexico. J Phys Oceanogr 38:1501–1514

Sturges W, Hoffmann NG, Leben RR (2010) A trigger mechanism for loop current ring separations. J Phys Oceanogr 40:900–913

Subrahmanyam CB, Coultas CL (1980) Studies on the animal communities in two north Florida salt marshes. Part III. Seasonal fluctuations of fish and macroinvertebrates. Bull Mar Sci 30:790–818

Subrahmanyam CB, Drake SH (1975) Studies on the animal communities in two north Florida salt marshes. Part I. Fish communities. Bull Mar Sci 25:445–465

Subrahmanyam CB, Kruczynski WL, Drake SH (1976) Studies on the animal communities in two north Florida salt marshes. Part II Macroinvertebrate communities. Bull Mar Sci 26:172–195

Suchanek TH, Williams SW, Ogden JC, Hubbard DK, Gill IP (1985) Utilization of shallow-water seagrass detritus by Caribbean deep-sea macrofauna: δ 13C evidence. Deep Sea Res 32:2201–2214

Sullivan CL (2009) Hurricanes of the Mississippi Gulf Coast: Three centuries of destruction. Mississippi Gulf Coast Community College Press, Perkinston, MS, USA, 174 p

Sullivan-Sealey K, Bustamante G (1999) Setting geographic priorities for Marine conservation in Latin America and the Caribbean. The Nature Conservancy, Arlington, VA, USA, 125 p

Summerson HC, Peterson CH (1984) Role of predation in organizing benthic communities of a temperate-zone seagrass bed. Mar Ecol Prog Ser 15:63–77

Swingle HA (1971) Biology of Alabama estuarine areas--cooperative Gulf of Mexico estuarine inventory. Bulletin 5. Alabama Marine Resources Laboratory, Dauphin Island, AL, USA, 140 p

Tagatz ME (1968) Biology of the blue crab, *Callinectes sapidus* Rathbun, in the St Johns River, Florida. Fish Bull 67:17–33

Tanner WF (1975a) Historical beach changes, Florida "Big Bend" coast. Trans Gulf Coast Assoc Geol Soc 25:379–382

Tanner WF (1975b) Symposium on beach erosion in middle America: Introduction. Trans Gulf Coast Assoc Geol Soc 25:365–368

Tanner WF, Stapor FW (1971) Tabasco beach-ridge plain: An eroding coast. Trans Gulf Coast Assoc Geol Soc 21:231–232

Taylor Engineering (2010) Analysis of rollover pass impacts to adjacent beaches in the littoral system. Texas General Land Office, Austin, TX, USA, 17 p

Teague WE, Jarosz E, Wang D, Mitchell D (2007) Observed oceanic response over the Upper Continental Slope and Outer Continental Shelf during Hurricane Ivan. J Phys Oceanogr 37:2181–2206

Teal JM (1962) Energy flow in the salt marsh ecosystem of Georgia. Ecology 43:614–624

Tenore KR (1985) Seasonal changes in soft bottom macrofauna of the U.S. South Atlantic Bight. In: Atkinson LP, Menzel DW, Bush KA (eds) Oceanography of the Southeastern U.S. Continental Shelf. American Geophysical Union, Washington, DC, USA, pp 130–140

Thaxton JM, Dewalt SJ, Platt WJ (2007) Spatial patterns of regeneration after Hurricane Andrew in two south Florida fringe mangrove forests. Florida Sci 70:148–156

Thayer GW, Bjorndal KA, Ogden JC, Williams SL, Zieman JC (1984) Role of larger herbivores in seagrass communities. Estuaries 7:351–376

Thayer GW, Colby DR, Hettler WF Jr (1987) Utilization of the red mangrove prop root habitat by fishes in south Florida. Mar Ecol-Prog Ser 35:25–38

Thibodeau PM, Gardner LR, Reeves HW (1998) The role of groundwater flow in controlling the spatial distribution of soil salinity and rooted macrophytes in a southeastern salt marsh, USA. Mangrove Salt Marshes 2:1–13

Thom BG (1967) Mangrove ecology and deltaic geomorphology: Tabasco, Mexico. J Ecol 55:301–343

Thom RM (1990) Spatial and temporal patterns in plant standing stock and primary production in a temperate seagrass system. Bot Mar 33:497–510

Thomas RG (1999) Fish habitat and coastal restoration in Louisiana. Am Fish Soc Symp 22:240–251

Thomas R, Dunkin L (2012) Erosion control and environmental restoration plan development, Matagorda County, Texas. Technical report ERDC/CHL TR-12-11. U.S. Army Engineering Research Center, Vicksburg, MS, USA, 102 p

Thomas JL, Zimmerman RJ, Minello TJ (1990) Abundance patterns of juvenile blue crabs (*Callinectes sapidus*) in nursery habitats of two Texas bays. Bull Mar Sci 46:115–125

Thompson RC, Wilson BJ, Tobin ML, Hill AS, Hawkins SJ (1996) Biologically generated habitat provision and diversity of rocky shore organisms at a hierarchy of spatial scales. J Exp Mar Biol Ecol 202:73–84

Thorson G (1957) Bottom communities (sublittoral or shallow shelf). In Hedgpeth JW, ed, Treatise on marine ecology and palaeoecology, volume I. Mem Geol Soc Am 67:461–534

Thrush SF, Whitlatch RB, Pridmore RD, Hewitt JE, Cummings VJ, Wilkinson MR (1996) Scale-dependent recolonization: The role of sediment stability in a dynamic sandflat habitat. Ecology 77:2472–2487

Thrush SF, Hewitt JE, Cummings VJ, Green MO, Funnell GA, Wilkinson MR (2000) The generality of field experiments: Interactions between local and broad-scale processes. Ecology 81:399–415

Thrush SF, Hewitt JE, Herman PM, Ysebaert T (2005) Multi-scale analysis of species-environment relationships. Mar Ecol Prog Ser 302:13–26

Tilmant JT, Rutherford ES, Thue EB (1989) Fishery harvest and population dynamics of red drum (*Sciaenops ocellatus*) from Florida Bay and adjacent waters. Bull Mar Sci 44:126–138

Tilmant JT, Curry RW, Jones R, Szmant A, Zieman JC, Flora M, Robblee MB, Smith D, Snow RW, Wanless H (1994) Hurricane Andrew's effects on marine resources. BioScience 44:230–237

Tiner RW (2003) Field guide to coastal wetland plants of the southeastern United States. University of Massachusetts Press, Amherst, MA, USA, 328 p

Tolley SG, Volety AK (2005) The role of oysters in habitat use of oyster reefs by resident fishes and decapod crustaceans. J Shellfish Res 24:1007–1012

Tomlinson PB (1994) The botany of mangroves. Cambridge University Press, Cambridge, UK

Törnqvist TE, Wallace DJ, Storms JEA, Wallinga J, Van Dam RL, Blaauw M, Derksen MS, Klerks CJW, Meijneken C, Snijders EMA (2008) Mississippi Delta subsidence primarily caused by compaction of Holocene strata. Nat Geosci 1:173–176

Torres-Rodríguez V, Márquez-García A, Bolongaro-Crevenna A, Chavarria-Hernández J, Expósito-Díaz G, Marquez-Garcia E (2010) Tasa de erosión y vulnerabilidad costera en el estado de Campeche debidos a efectos del cambio climático. In: Botello AV, Villanueva-Fragoso S, Gutierrez J, Rojas Galaviz JL (eds) Vulnerabilidad de las Zonas Costeras Mexicanas ante el Cambio Climático. Semarnat-INE, UNAM-ICMyL, Universidad Autónoma de Campeche, Campeche, Mexico, pp 325–344

Tunnell JW (2002a) The environment. In: Tunnel JW Jr, Judd FW (eds) The laguna madre of Texas and Tamaulipas. Texas A&M University Press, College Station, TX, USA, pp 73–84

Tunnell JW (2002b) Geography, climate, and hydrography. In: Tunnell Jr JW, Judd FW (eds) The Laguna Madre of Texas and Tamaulipas. Texas A&M University Press, Texas A&M University, College Station, TX, USA, pp 2–27

Turner RE (1976) Geographic variations in salt marsh macrophyte production: A review. Contrib Mar Sci 20:47–68

Turner RE (1977) Intertidal vegetation and commercial yields of penaeid shrimp. Am Fish Soc Trans 106:411–416

Turner RE (1990) Landscape development and coastal wetland losses in the northern Gulf of Mexico. Am Zool 30:89–105

Turner RE (1992) Coastal wetlands and penaeid shrimp habitat. In: Stroud RH (ed) Stemming the tide of coastal fish habitat loss. National Coalition for Marine Conservation, Savannah, GA, USA, pp 97–104

Turner RE, Gosselink JG (1975) A note on standing crops of *Spartina alterniflora* in Texas and Florida. Mar Sci 19:113–118

Turner RE, Swenson EM, Milan CS (2000) Organic and inorganic contributions to vertical accretion in salt marsh sediments. In: Weinstein NP, Kreeger DA (eds) Concepts and controversies in tidal marsh ecology. Kluwer Academic Publishers, Dordrecht / Boston / London, UK, pp 583–595

Turner RE, Baustian JJ, Swenson EM, Spicer JS (2006) Wetland sedimentation from Hurricanes Katrina and Rita. Science 314:449–452

Twichell DC, Andrews BD, Edmiston HL, Stevenson WR (2007) Geophysical mapping of Oyster habitats in a Shallow Estuary, Apalachicola Bay, Florida. USGS Open-File Report 2006-1381. USGS, Reston, VA, USA, 37 p

Twilley RR (1985) The exchange of organic carbon in basin mangrove forests in a southwest Florida estuary. Estuar Coast Shelf Sci 20:543–557

Twilley RR, Day JW (1999) The productivity and nutrient cycling of mangrove ecosystems. In: Yanez-Arancibia A, Lara-Dominguez AL (eds) Ecosistemas de Manglar en America Tropical. Instituto de Ecologia, A.C. Mexico, UICN/ORMA, Silver Spring, MD, USA, pp 127–152

Twilley RR, Lugo AE, Patterson-Zucca C (1986) Litter production and turnover in basin mangrove forests in southwest Florida. Ecology 67:670–683

Tyler RM, Brady DC, Targett TE (2009) Temporal and spatial dynamics of diel-cycling hypoxia in estuarine tributaries. Estuar Coast 32:123–145

U.S. Department of Commerce (1989) Galveston Bay: Issues, resources, status, and management. NOAA Estuarine Programs Office, Texas A&M University Sea Grant College Program, Washington, DC, USA, 116 p

Uebelacker JM, Johnson PG (eds) (1984) Taxonomic guide to the Polychaetes of the Northern Gulf of Mexico, vol I–VII. MMS, Gulf of Mexico OCS Region, Metarie, LA, USA

UNEP (United Nations Environment Program)/GPA (Global Programme of Action) (2003) Diagnosis of the erosion processes in the Caribbean Sandy Beaches. Report prepared by Environmental Agency, Ministry of Science, Technology and Environment, Government of Cuba. UNEP, The Netherlands, 74 p

USACE (U.S. Army Corps of Engineers) (2010) Louisiana Coastal Area (LCA), Louisiana beneficial use of dredged material program, January 2010 Final Programmatic Study Report and Programmatic Environmental Impact Statement. USACE, New Orleans District, New Orleans, LA, USA, 174 p

USACE (2012) Louisiana Coastal Area Barataria Basin barrier shoreline restoration final integrated construction report and final environmental impact statement. USACE—Mississippi Valley Division, New Orleans, LA, USA, 447 p

USACE (2014) U.S. Army Corps of Engineers Water Control Section Stage Date: Calcasieu River and Pass Near Cameron, LA. http://www2.mvn.usace.army.mil/cgi-bin/watercontrol.pl?73650. Accessed 8 June 2012

USDA (2002) Plant fact sheet: Common reed Phragmites australis. http://plants.usda.gov/factsheet/pdf/fs_phau7.pdf. Accessed 11 Feb 2013

USEPA (U.S. Environmental Protection Agency) (2011) Level IV ecoregions of the conterminous United States. U.S. EPA Office of Research & Development, National Health and Environmental Effects Research Laboratory, Corvallis, OR, USA. http://www.epa.gov/wed/pages/ecoregions/level_iii_iv.htm#Level%20IV. Accessed 7 Feb 2013

USGS (U.S. Geological Survey) (2012a) Water-resources data for the United States, water year 2011: U.S. Geological Survey water-data report WDR-US-2011, Site 02292900 Caloosahatchee River at S-79, near Olga, FL. USGS, Reston, VA, USA, 8 p

USGS (2012b) Water-resources data for the United States, water year 2011: USGS water-data report WDR-US-2011, Site 02330000 Ochlockonee River near Bloxham, FL. USGS, Reston, VA, USA, 5 p

USGS (2012c) Water-resources data for the United States, water year 2011: USGS water-data report WDR-US-2011, Site 02296750 Peace River at Arcadia, FL. USGS, Reston, VA, USA, 4 p

USGS (2012d) Water-resources data for the United States, water year 2011: USGS water-data report WDR-US-2011, Site 02323500 Suwannee River near Wilcox, FL. USGS, Reston, VA, USA, 5 p

USGS (2012e) Water-resources data for the United States, water year 2011: USGS water-data report WDR-US-2011, Site 02313000 Withlacoochee River near Holder, FL. USGS, Reston, VA, USA, 4 p

USGS (2012f) Water-resources data for the United States, water year 2011: USGS water-data report WDR-US-2011, Site 02359170 Apalachicola River near Sumatra, FL. USGS, Reston, VA, USA, 5 p

USGS (2012g) Water-resources data for the United States, water year 2011: USGS water-data report WDR-US-2011, Site 02366500 Choctawhatchee River near Bruce, FL. USGS, Reston, VA, USA, 5 p

USGS (2012h) Water-resources data for the United States, water year 2011: USGS water-data report WDR-US-2011, Site 02376033 Escambia River near Molino, FL. USGS, Reston, VA, USA, 5 p

USGS (2012i) Water-resources data for the United States, water year 2011: USGS water-data report WDR-US-2011, Site 02469761 Tombigbee River at Coffeeville Lock and Dam near Coffeeville, AL. USGS, Reston, VA, USA, 5 p

USGS (2012j) Water-resources data for the United States, Water Year 2011: U.S. Geological Survey Water-Data Report WDR-US-2011, Site 02428400 Alabama River at Claiborne Lock and Dam near Monroeville, AL. Reston, VA, USA, 5 p

USGS (2012k) Water-resources data for the United States, water year 2011: U.S. Geological Survey water-data report WDR-US-2011, Site 02479310 Pascagoula River at Graham Ferry, MS. Reston, VA, USA, 5 p

USGS (2012l) Water-resources data for the United States, Water Year 2011: U.S. Geological Survey water-data report WDR-US-2011, Site 02489500 Pearl River near Bogalusa, LA. Reston, VA, USA, 4 p

USGS (2012m) Water-resources data for the United States, water year 2011: USGS water-data report WDR-US-2011, Site 08015500 Calcasieu River near Kinder, LA. USGS, Reston, VA, USA, 4 p

USGS (2012n) Water-resources data for the United States, water year 2011: USGS water-data report WDR-US-2011, Site 08030500 Sabine River near Ruliff, TX. USGS, Reston, VA, USA, 3 p

USGS (2012o) Water-resources data for the United States, water year 2011: USGS water-data report WDR-US-2011, Site 08041000 Neches River at Everdale, TX. USGS, Reston, VA, USA, 3 p

USGS (2012p) Water-resources data for the United States, water year 2011: USGS water-data report WDR-US-2011, Site 08066500 Trinity River at Romayor, TX. USGS, Reston, VA, USA, 3 p

USGS (2012q) Water-resources data for the United States, water year 2011: USGS water-data report WDR-US-2011, Site 08116650 Brazos River near Rosharon, TX. USGS, Reston, VA, USA, 22 p

USGS (2012r) Water-resources data for the United States, water year 2011: USGS water-data report WDR-US-2011, Site 08162500 Colorado River near Bay City, TX. USGS, Reston, VA, USA, 3 p

USGS (2012s) Water-resources data for the United States, water year 2011: USGS water-data report WDR-US-2011, Site 08176500 Guadalupe River at Victoria, TX. USGS, Reston, VA, USA, 3 p

USGS (2012t) Water-resources data for the United States, water year 2011: USGS water-data report WDR-US-2011, Site 08188500 San Antonio River at Goliad, TX. USGS, Reston, VA, USA, 3 p

USGS (2012u) Water-resources data for the United States, water year 2011: USGS water-data report WDR-US-2011, Site 08211500 Nueces River at Calallen, TX. USGS, Reston, VA, USA, 3 p

Vadas RL, Elner RW (1992) Plant-animal interactions in the North-west Atlantic. In: John DM, Hawkins SJ, Price JH (eds) Plant-animal interactions in the marine benthos. Clarendon Press, Oxford, UK, pp 33–60

Valentine JF, Duffy JE (2006) The central role of grazing in seagrass ecology. In: Larkum AWD, Orth RJ, Duarte CM (eds) Seagrasses: Biology, ecology and conservation. Springer, Dordrecht, The Netherlands, pp 159–192

Valero A, Schipper J, Allnutt T (2014) Sierra de los Tuxtlas. World Wildlife Fund, Washington, DC, USA. http://www.worldwildlife.org/ecoregions/nt0161. Accessed 5 Aug 2014

Valiela I, Rietsma CS (1995) Disturbance of salt marsh vegetation by wrack mats in Great Sippewissett Marsh. Oecologia 102:106–112

Van der Valk AG (1974) Environmental factors controlling the distribution of forbs on coastal foredunes in Cape Hatteras National Seashore. Can J Bot 52:1057–1073

Van Hoey G, Guilini K, Rabaut M, Vinex M, Degraer S (2008) Ecological implications of the presence of the tube-building polychaete Lanice conchilega on soft-bottom benthic ecosystems. Mar Biol 154:1009–1019

Van Lieshout I (2007) Caya Levisa. https://www.flickr.com/photos/ingeborgvanlieshout/1580871364/in/photolist-dx3TKv-dx9o4N-dx3Uuxdx3Tdn-dx9nPL-3pGnYQ-dx3Une-dx3ThP-dx9kLY-3pGnUA-dx3TBX-dx9kCS-dx9mqs-dx3U6pdx3T78-dx9nGo-dx9mVY-dx9nWU-dx9kwf-dx3SdD-dx9ko3-dx9ndW-dx3S66-dx3RWa-66RXtQ-4TFrwH-ft1Pqg-ftgb6b-ftgaYf-ft1PDp-ftgayC-ftgarG-ftgaGW-ft1NZH-ftgb3j-ft1Pnp-ftgaN1-dx3TvF-66MEMP-66RX5f-66MExR. Accessed 19 Feb 2017

Vazquez De la Cerda AM (1993) Bay of Campeche cyclone. Ph.D. Thesis, Texas A&M University, College Station, TX, USA, 182 p

Vega-Cendejas ME, Arreguín-Sánchez F (2001) Energy fluxes in a mangrove ecosystem from a coastal lagoon in Yucatán Peninsula, Mexico. Ecol Mod 137:119–133

Vega-Cendejas ME, Hernández de Santillana M (2004) Fish community structure and dynamics in a coastal hypersaline lagoon: Rio Lagartos, Yucatán, Mexico. Estuar Coast Shelf Sci 60:285–299

Vince SW, Snow A (1984) Plant zonation in an Alaskan salt marsh. I. Distribution, abundance, and environmental factors. J Ecol 72:651–667

Virnstein RW, Mikkalsen PS, Kairns KD, Capone M (1983) Seagrass beds versus sand bottoms: The trophic importance of their associated benthic invertebrates. Florida Sci 45:491–509

Vismann B (1991) Sulfide tolerance: Physiological mechanisms and ecological implications. Ophelia 34:1–27

Visser JM, Sasser CE, Chabreck RH, Linscombe RG (1998) Marsh vegetation types of the Mississippi River Deltaic Plain. Estuaries 21:818–828

Visser JM, Sasser CE, Chabreck RH, Linscombe RG (2000) Marsh vegetation types of the Chenier Plain, Louisiana, USA. Estuaries 23:318–327

Vose FE, Bell SS (1994) Resident fishes and macrobenthos in mangrove-rimmed habitats: Evaluation of habitat restoration by hydrological modification. Estuar Coasts 17:585–596

Vukovich FM (2007) Climatology of ocean features in the Gulf of Mexico using satellite remote sensing data. J Phys Oceanogr 37:689–707

Wainright SC, Weinstein MP, Able KW, Currin CA (2000) Relative importance of benthic microalgae, phytoplankton and the detritus of smooth cordgrass *Spartina alterniflora* and the common reed *Phragmites australis* to brackish-marsh food webs. Mar Ecol Prog Ser 200:77–91

Wakida-Kusunoki AT, Amador-del Angel LE, Alejandro PC, Brahms CQ (2011) Presence of Pacific white shrimp Litopenaeus vannamei (Boone, 1931) in the southern Gulf of Mexico. Aquat Inv 6(suppl 1):S139–S142

Walker DI, Lukatelich RJ, Bastyan G, McComb AJ (1989) Effect of boat moorings on seagrass beds near Perth, Western Australia. Aquat Bot 36:69–77

Wanless HR, Vlaswinkel BM (2005) Coastal landscape and channel evolution affecting critical habitats at Cape Sable, Everglades National Park, Florida. Final report. Everglades National Park, National Park Service, U.S. Department of Interior, 196 p

Ward WC (2003) Introduction to Pleistocene geology of NE Quintana Roo. In: Ward WC (ed) Salt Water Intrusion & Coastal Aquifer Conference (SWICA) field trip to the Caribbean Coast of the Yucatán Peninsula (April), Yucatán, Mexico, pp 13–22

Ward GA, Smith TJ, Whelan KRT, Doyle TW (2006) Regional processes in mangrove ecosystems: Spatial scaling relationships, biomass, and turnover rates following catastrophic disturbance. Hydrobiologia 569:517–527

Warren RS, Fell PE, Grimsby JL, Buck EL, Rilling GC, Fertik RA (2001) Rates, patterns, and impacts of *Phragmites australis* expansion and effects of experimental *Phragmites* control on vegetation, macroinvertebrates, and fish within tidelands of the lower Connecticut River. Estuaries 24:90–107

Watling L, Norse EA (1998) Disturbance of the seabed by mobile fishing gear: A comparison to forest clearcutting. Conserv Biol 12:1180–1197

WAVCIS (Wave-Current-Surge Information System for Coastal Louisiana) (2012) Coastal Studies Institute, School of the Coast and Environment, Louisiana State University, Baton Rouge, LA, USA. http://wavcis.csi.lsu.edu/. Accessed 11 Feb 2013

Weaver JE, Holloway LF (1974) Community structure of fishes and macrocrustaceans in ponds of a Louisiana tidal marsh influenced by weirs. Contrib Mar Sci 18:57–69

Webb JW (1983) Soil water salinity variations and their effects on *Spartina alterniflora*. Contrib Mar Sci 26:1–13

Weinstein MP, Balletto JH (1999) Does the common reed, *Phragmites australis*, affect essential fish habitat? Estuaries 22:793–802

Weinstein MP, Litvin SY, Guida VG (2009) Essential fish habitat and wetland restoration success: A tier II approach to the biochemical condition of common mummichog *Fundulus heteroclitus* in common reed *Phragmites australis*—and smooth cordgrass *Spartina alterniflora*-dominated salt marshes. Estuar Coast 32:1011–1022

Weinstein MP, Litvin SY, Guida VG (2010) Stable isotope and biochemical composition of white perch in *Phragmites* dominated salt marsh and adjacent waters. Wetlands 30:1181–1191

Weis JS, Weis P (2003) Is the invasion of the common reed, *Phragmites australis*, into tidal marshes of the eastern U.S. an ecological disaster? Mar Pollut Bull 47:816–820

Wells JT, Coleman JM (1987) Wetland loss and the subdelta life cycle. Estuar Coast Shelf Sci 25:111–125

Wells JT, Peterson CH (1982) Restless ribbons of sand: Atlantic & Gulf Coast Barriers. Louisiana Sea Grant Program. U.S. Fish and Wildlife Service, Washington, DC, USA, 19 p

West RC (1977) Tidal salt-marsh and mangal formations of Middle and South America. In: Chapman VJ (ed) Ecosystems of the world, 1 Wet coastal ecosystems. Elsevier, Amsterdam, The Netherlands, pp 157–166

West DL, Williams AH (1986) Predation by *Callinectes sapidus* (Rathbun) within *Spartina alterniflora* (Loisel) marshes. J Exp Mar Biol Ecol 100:75–95

Weston DP (1988) Macrobenthos-sediment relationships on the continental shelf off Cape Hatteras, North Carolina. Cont Shelf Res 8:267–286

Whaley SD, Minello TJ (2002) The distribution of benthic infauna of a Texas salt marsh in relation to the marsh edge. Wetlands 22:753–766

Whelan KRT, Smith TJI, Anderson GH, Ouellette ML (2009) Hurricane Wilma's impact on overall soil elevation and zones within the soil profile in a mangrove forest. Wetlands 29:16–23

Whigham DF, Verhoeven JTA, Samarkin V, Megonigal PJ (2009) Responses of *Avicennia germinans* (black mangrove) and the soil microbial community to nitrogen addition in a hypersaline wetland. Estuar Coast 32:926–936

White M (2011). Trosclair Road, Near Creole, LA. http://www.panoramio.com/photo/50178598. Accessed 19 Feb 2017

White DA, Hauber DP, Hood CS (2004a) Clonal differences in *Phragmites australis* from the Mississippi River Delta. Southeast Nat 3:531–544

White WA, Paine JG (1992) Wetland plant communities, Galveston Bay System. Galveston Bay National Estuary Program, Houston, TX, USA, 124 p

White WA, Tremblay TA, Waldinger RL, Calnan TR (2002) Status and trends of wetland and aquatic habitats on Texas barrier islands, Matagorda Bay to San Antonio Bay. Bureau of Economic Geology, Austin, TX, USA, 66 p

White WA, Tremblay TA, Waldinger RL, Calnan TR (2004b) Status and trends of wetland and aquatic habitats on barrier islands, Upper Texas Coast, Galveston and Christmas Bays. Bureau of Economic Geology, Austin, TX, USA, 67 p

White WA, Tremblay TA, Waldinger RL, Thomas RC (2007) Status and trends of wetland and aquatic habitats on texas barriers: upper coast Strandplain-Chenier System and Southern Coast Padre Island National Seashore. Bureau of Economic Geology, Austin, TX, USA, 88 p

WHO (World Health Organization) (2005) Ecosystems and human well being—health synthesis. A report of the millennium ecosystem assessment. WHO, Geneva, Switzerland, 64 p

Wiken E, Nava FJ, Griffith G (2011) North American terrestrial ecoregions—level III. Commission for Environmental Cooperation, Montreal, Canada, 149 p

Wilbur DH, Clarke DG, Ray GL, Burlas M (2003) Response of surf zone fish to beach nourishment operations on the northern coast of New Jersey, USA. Mar Ecol Prog Ser 250:231–246

Wilkinson T, Wiken E, Bezaury-Creel J, Hourigan T, Agardy T, Herrmann H, Janishevski L, Madden C, Morgan L, Padilla M (2009) Marine ecoregions of North America. Commission for Environmental Cooperation, Montreal, Canada, 200 p

Willard DA, Holmes CW, Weimer LM (2001) The Florida Everglades Ecosystem: Climatic and anthropogenic impacts over the last two millennia. Bull Am Paleontol 361:41–56

Williams SL, Heck KL Jr (2001) Seagrass community ecology. In: Bertness MD, Gaines SD, Hay M (eds) Marine community ecology. Sinauer Associates, Sunderland, MA, USA, pp 317–337

Williams DD, Kraus NC, Anderson CM (2007) Morphological response to a new inlet, Packery Channel, Corpus Christi, Texas. In: Kraus NC, Rosati JD (eds) Proceedings, Coastal Sediments '07. American Society of Civil Engineers (ASCE), Reston, VA, USA, pp 1529–1542

Williams SJ, Kulp M, Penland S, Kindinger JL, Flocks JG (2011) Mississippi River delta plain, Louisiana coast, and inner shelf Holocene geologic framework, processes, and resources. In: Buster NA, Holmes CW (eds) Gulf of Mexico: Origin, waters, and biota, vol 3, Geology. Texas A&M University Press, College Station, TX, USA, pp 175–193

Wilsey BJ, Chabreck RH, Linscombe RG (1991) Variation in nutria diets in selected freshwater forested wetlands of Louisiana. Wetlands 11:263–278

Windham L (2001) Comparison of biomass production and decomposition between *Phragmites australis* (common reed) and *Spartina patens* (salt hay grass) in brackish tidal marshes of New Jersey, USA. Wetlands 21:179–188

Windham L, Lathrop RG Jr (1999) Effects of *Phragmites australis* (common reed) invasion on aboveground biomass and soil properties in brackish tidal marsh of the Mullica River, New Jersey. Estuaries 22:927–935

Wingard GL, Hudley JW, Holmes CW, Willard DA, Marot M (2007) Synthesis of Age data and chronology for Florida Bay and Biscayne Bay Cores collected for the ecosystem history of South Florida's estuaries projects. U.S. Geological Survey (USGS), Open File Report 2007-1203. USGS, Reston, VA, USA, 120 p

Woods CMC, Shiel DR (1997) Use of seagrass *Zostera novazelandica* (Setchell, 1993) as habitat and food by the crab *Macrophtalmus hirtepes* (Heller, 1862) (Brachyura: Ocypodidae) on rocky intertidal platforms in southern New Zealand. J Exp Mar Biol Ecol 214:49–65

Woodward-Clyde Consultants (1983) Southwest Florida shelf ecosystems study. MMS, Gulf of Mexico OCS Region, Metarie, LA, USA

Woodworth PL, Player R (2003) The permanent service for mean sea level: An update to the 21st century. J Coast Res 19:287–295

WWF (World Wildlife Fund) (2014a) Veracruz Moist Forests. Encyclopedia of Earth. Environmental Information Coalition, National Council for Science and the Environment. Washington, DC, USA. http://www.eoearth.org/view/article/156844/. Accessed Aug 5 2014

WWF (2014b) Veracruz dry forests. Encyclopedia of Earth. Environmental Information Coalition, National Council for Science and the Environment, Washington, DC, USA. http://editors.eol.org/eoearth/wiki/Veracruz_dry_forests. Accessed 5 Aug 2014

WWF (2014c) Alvarado mangroves. Encyclopedia of Earth. Environmental Information Coalition, National Council for Science and the Environment, Washington, DC, USA. http://editors.eol.org/eoearth/wiki/Alvarado_mangroves. Accessed 5 Aug 2014

WWF (2014d) Usumacinta mangroves. Encyclopedia of Earth. Environmental Information Coalition, National Council for Science and the Environment, Washington, DC, USA. http://editors.eol.org/eoearth/wiki/Usumacinta_mangroves. Accessed 5 Aug 2014

WWF (2014e) Cuban Cactus scrub. Encyclopedia of Earth. Environmental Information Coalition, National Council for Science and the Environment, Washington, DC, USA. http://editors.eol.org/eoearth/wiki/Cuban_cactus_scrub. Accessed 5 Aug 2014

WWF, Hogan MC (2013) Yucatán dry forests. Encyclopedia of Earth. Environmental Information Coalition, National Council for Science and the Environment, Washington, DC, USA. http://editors.eol.org/eoearth/wiki/Yucat%C3%A1n_dry_forests. Accessed 5 Aug 2014

WWF, Hogan MC (2014a) Petenes mangroves. Encyclopedia of Earth. Environmental Information Coalition, National Council for Science and the Environment, Washington, DC, USA. http://editors.eol.org/eoearth/wiki/Petenes_mangroves. Accessed 5 Aug 2014

WWF, Hogan MC (2014b) Bahamian mangroves. Encyclopedia of Earth. Environmental Information Coalition, National Council for Science and the Environment, Washington, DC, USA. http://editors.eol.org/eoearth/wiki/Bahamian_mangroves. Accessed 5 Aug 2014

Yamamuro M (1999) Importance of epiphytic cyanobacteria as food sources for heterotrophs in a tropical seagrass bed. Coral Reefs 18:263–271

Yáñez-Arancibia A, Day JW (2004) Environmental sub-regions in the Gulf of Mexico coastal zone: The ecosystem approach as an integrated management tool. Ocean Coast Manage 47:727–757

Yáñez-Arancibia A, Lara-Dominguez AL (1988) Ecology of three sea catfishes (Ariidae) in a tropical coastal ecosystem—southern Gulf of Mexico. Mar Ecol Prog Ser 49:215–230

Yáñez-Arancibia A, Lara-Dominguez AL, Day JW Jr (1993) Interactions between mangrove and seagrass habitats mediated by estuarine nekton assemblages: Coupling of primary and secondary production. Hydrobiologia 264:1–12

Yáñez-Arancibia A, Lara-Dominguez AL, Pauly D (1994) Coastal lagoons as fish habitat. Chapter 12. In: Kjerfve B (ed) Coastal lagoon processes. Elsevier, Amsterdam, The Netherlands, pp 363–376

Yáñez-Arancibia A, Ramírez-Gordillo J, Day JW, Yoskowitz W (2009) Environmental sustainability of economic trends in the Gulf of Mexico: What is the limit for Mexican coastal development? In: Tunnell JW Jr, Felder DL, Earle SA (eds) Gulf of Mexico: Origin, waters, and biota. Texas A&M University Press, College Station, TX, USA, pp 82–104

Yarbro LA, Carlson PR Jr (2011) Seagrass integrated mapping and monitoring for the State of Florida. Mapping and monitoring report 1. Florida Fish and Wildlife Conservation Commission, Tallahasscc, FL, USA, 202 p

Yoskowitz DW (2009) The productive value of the Gulf of Mexico. In: Cato J (ed) Gulf of Mexico: Origins, waters, and biota, vol 2, Ocean and coastal economy. Texas A&M University Press, College Station, TX, USA, pp 21–27

Yoskowitz D, Santos C, Allee B, Carollo C, Henderson J, Jordan S, Ritchie J (2010) Proceedings, Gulf of Mexico Ecosystem Services Workshop, Bay St. Louis, 16–18 June. Harte Research Institute for Gulf of Mexico Studies, Texas A&M University-Corpus Christi, TX, USA, 16 p

Yuhas CE, Hartman JM, Weis JS (2005) Benthic communities in Spartina alterniflora– and Phragmites australis–dominated salt marshes in the Hackensack Meadowlands, New Jersey. Urban Habitats Online Journal 3

Zajac RN, Whitlatch RB, Thrush SF (1998) Recolonization and succession in soft-sediment infaunal communities: The spatial scale of controlling factors. Hydrobiologia 375:227–240

Zajac RN, Lewis RS, Poppe LJ, Twichell DC, Vozarik J, DiGiacomo-Cohen ML (2003) Responses of infaunal populations to benthoscape structure and the potential importance of transition zones. Limnol Oceanogr 48:829–842

Zavala-Hidalgo J, Morey SL, O'Brien JJ (2003) Seasonal circulation on the Western Shelf of the Gulf of Mexico using a high-resolution numerical model. J Geophys Res 108:3389

Zeug SC, Shervette VA, Hoeinghaus DJ, Davis SE III (2007) Nekton assemblage structure in natural and created marsh-edge habitats of the Guadalupe Estuary, Texas, USA. Estuar Coast Shelf Sci 71:457–466

Ziegler S, Benner R (1999) Dissolved organic carbon cycling in a subtropical seagrass-dominated lagoon. Mar Ecol Prog Ser 180:149–160

Zieman J, Wetzel RG (1980) Productivity in seagrasses: Methods and rates. In: Phillips RC, McRoy CP (eds) Handbook of seagrass biology: An ecosystem perspective. Garland STPM Press, New York, NY, USA, pp 87–116

Zieman JC, Zieman RT (1989) The ecology of the seagrass meadows of the west coast of Florida: A community profile. U.S. Fish and Wildlife Service biological report 85(7.25). U.S. Fish and Wildlife Service, Washington, DC, USA, 155 p

Zimmerman RJ, Minello TJ (1984) Densities of *Penaeus aztecus*, *Penaeus setiferus*, and other natant macrofauna in a Texas salt marsh. Estuaries 7(4A):421–433

Zimmerman RJ, Minello TJ, Rozas LP (2000) Salt marsh linkages to productivity of penaeid shrimps and blue crabs in the northern Gulf of Mexico. In: Weinstein MP, Kreeger DA (eds) Concepts and controversies in tidal marsh ecology. Kluwer Academic Publishers, Dordrecht, The Netherlands, pp 293–314

Zomlefer WB, Judd WS, Giannasi DE (2006) Northernmost limit of *Rhizophora mangle* (red mangrove Rhizophoraceae) in St. Johns County, Florida. Castanea 71:239–244

CHAPTER 7

OFFSHORE PLANKTON AND BENTHOS OF THE GULF OF MEXICO

Gilbert T. Rowe[1]

[1]Texas A&M University—Galveston, Galveston, TX 77553, USA
roweg@tamug.edu

7.1 INTRODUCTION

This chapter summarizes baseline knowledge on the benthic communities of the seafloor and the plankton of the water column on the continental shelf, continental slope, and the abyssal plain of the Gulf of Mexico up through 2009 and prior to the Deepwater Horizon oil spill. As such, this review does not consider the higher components of a typical marine food web: fishes, turtles, mammals, and birds. An overview is provided of the general characteristics of benthos and plankton in terms of community structure—abundance, biomass, and biodiversity—in each habitat within the entire Gulf of Mexico large marine ecosystem (LME) [*sensu* Ken Sherman, National Oceanic and Atmospheric Administration (NOAA)]. This is followed by discussions of what is known about each unique or different assemblage's function within its habitat. In this context, function is defined as community dynamics in terms of elemental cycling or energetics of the organisms involved to the degree that this is known. Emphasis is principally on the seafloor, with some reference to the relationships between transient phytoplankton and zooplankton assemblages and their interactions with life on the bottom. The seafloor organisms or benthos are targeted because they are geographically static in space and time and thus can serve as better indicators of each habitat's characteristics and ostensibly its health. Plankton are included because they are the base of offshore food webs; all estimations of baseline conditions up a food web will reflect the nature or health of the phytoplankton and zooplankton. Variations in community structure—abundance, biomass, productivity, and diversity—from habitat to habitat and relationships to community function will be described from the literature reviewed when appropriate. The presumption is made that offshore life is, in general, food limited, and thus, sources of energy, carbon, and nitrogen, for example, become important in ultimately determining what species survive in each habitat—that is, food supplies determine community structure. Thus, where available, the relationships between community structure and function, in terms of food supplies, will be reviewed.

Summaries of the literature will consider each major habitat separately: (1) continental shelf benthos, (2) continental slope and abyssal plain level-bottom assemblages, (3) the biota and biological processes of methane seeps, and (4) corals and live bottoms. Peculiar features in each of these habitats will be mentioned but not treated exhaustively (for example, pinnacles and banks on the shelf and canyons on the slope). The general nature of offshore life in the Gulf of Mexico will be compared to other ocean basins, marginal seas, and continental margins. In addition to the natural assemblages of organisms in different habitats (1 through 4 above), some attention will be given to those areas of the Gulf in which human activities have altered or impacted natural processes significantly. The most salient of these are eutrophication and hypoxia associated with the Mississippi River plume, enrichment that is ostensibly derived from

© The Author(s) 2017

C.H. Ward (ed.), *Habitats and Biota of the Gulf of Mexico: Before the Deepwater Horizon Oil Spill*,
DOI 10.1007/978-1-4939-3447-8_7

offshore platforms and structures, and the impact of intensive bottom trawling on resident populations. Where possible, comparisons will be made between the stocks and diversities of major continental margin habitats. For example, numerous mesoscale surveys (10–100 kilometers [km] (6.2–62 miles [mi])) have been conducted across the entire northern continental shelf, but only a few comparisons of these have been attempted (Rabalais et al. 1999b). A singular goal of this chapter will be to identify gradients in ecosystem productivity, as represented by standing stocks, along with gradients in biodiversity (the relationships between biodiversity and productivity remain obscure, at best). Likewise, while there have been numerous disparate studies that together encompass the entire continental margin and deep basin of the Gulf of Mexico (Felder and Camp 2009; Fautin et al. 2010; Ellis et al. 2011), few ecological comparisons of them all have yet been attempted because methods have varied and finding original data is not always possible.

Some important generalizations have emerged from a review of the biota of the entire offshore Gulf of Mexico. In general, the open-ocean ecosystem—from the algal phytoplankton, the vertically migrating zooplankton and mesopelagic fishes, down to the level-bottom sediment-dwelling seafloor assemblages—is dependent on the physics of the ecosystem. That is, the water mass signature characteristics, along with contributions from the continental margin, ultimately control the biota and its food webs in ecological time scales of days to months. As a marginal basin, the ratio of coastline to Gulf of Mexico basin area (or volume) is high compared to major oceans, and thus, the surrounding land masses are more important to Gulf of Mexico processes than might be expected on the Atlantic, Pacific, and even Arctic margins of the United States. On the other hand, some of the most fascinating biotic assemblages in the deep Gulf of Mexico are the fossil hydrocarbon-based communities that are linked directly to the history of the Gulf over geologic time (centuries to millennia) and not to extant physics. The hermatypic corals living on banks and domes are able to exist on the tops of salt diapirs but are thus dependent on both year-to-year climate and almost day-to-day weather. Nevertheless, the coral assemblages could not exist without the salt extrusions on which they are perched. Likewise, deep-living cold-water mesophotic corals on the upper continental slope depend on sinking detritus from the surface for food but are anchored to hard authigenic carbonate substrates that are deposited as methane seeps age. Thus, the corals are dependent on both the present and the past conditions of the Gulf of Mexico. As those corals provide a living structure to thriving fish and invertebrate assemblages, so too do thousands of offshore platforms provide a hard substrate for thriving animal–plant communities that contribute to the high biodiversity within and along the margin of the Gulf of Mexico. The obvious similarities or links between parts of the system can be linked together in mass-balance models that illustrate the interdependence of the biotas of the different habitats of the offshore Gulf. Much is still unknown about life in the deep Gulf of Mexico and thus a penultimate section is devoted to these holes in our knowledge. Finally, an analysis of ostensibly vital ecosystem services of the offshore biota will be considered.

7.2   HISTORICAL PERSPECTIVES: EXPLORING THE DEEP GULF OF MEXICO

Exploration of the fauna living in the deep Gulf of Mexico began in the late nineteenth century aboard the steamer *Blake* (Milne-Edwards 1880; Geyer 1970; Roberts 1977) under the direction of Alexander Agassiz at Harvard's Museum of Comparative Zoology [for a thorough listing of these reports, along with descriptions of the fauna by taxon, see the compendium of Felder and Camp (2009)]. In the mid-twentieth century, the U.S. National Marine Fisheries Service (NMFS), using the U.S. Bureau of Commercial Fisheries' vessel *Oregon II*, sampled the deep Gulf of Mexico using large shrimp trawls along the upper continental slope. Although no new fisheries of economic importance were uncovered, the numerous large trawl samples continue to enhance taxonomic and zoogeographic knowledge of larger invertebrates (Wicksten and Packard 2005) and demersal fishes (McEachran and Fechhelm 1998, 2006) in the Gulf of Mexico and the Caribbean. In the early 1960s, Willis Pequegnat at Texas A&M University (TAMU) initiated studies of the deep Gulf of Mexico with support from the Office of Naval Research (ONR) using the R/V *Alaminos* and followed in the 1970s by work with the R/V *Gyre*. Pequegnat's group employed quantitative sampling for the first time in the deep Gulf of Mexico using a Campbell grab for the infauna and a skimmer to sample larger epifauna. The 2-meter (m) (6.6 feet [ft]) wide skimmer was armed with counter wheels that measured the distances over which this unique device traveled over the bottom surface. The results generated were included in numerous publications and theses by Pequegnat's associates and students, including an intricate scheme of bathymetric zonation (Roberts 1977; Pequegnat 1983; Pequegnat et al. 1990). In addition, they discovered a large area in the eastern Gulf of Mexico covered by ironstone (Pequegnat et al. 1972; Rowe and Kennicutt 2008) and deep bottom currents (Pequegnat 1972). The Woods Hole Oceanographic Institution (WHOI) also published contemporaneous quantitative data on the deep Gulf of Mexico in the 1970s. The rate of the decline in biomass with depth, they discovered, is log-normal and universal between ocean basins, but the intercept of the decline is a function of surface water primary production (PP) (Rowe and Menzel 1971; Rowe 1971; Rowe et al. 1974). The U.S. Department of Energy (DOE) supported this WHOI work, under the direction of John Ryther and David Menzel.

By the 1980s, the complexion of the investigations of the deep Gulf changed substantially. Prospects of offshore oil and gas resources led to intensified environmental studies supported by the Bureau of Land Management (BLM), which evolved, for the ocean, into the Minerals Management Service (MMS) of the U.S. Department of the Interior (DOI). This agency is now the Bureau of Ocean Energy Management (BOEM) of the DOI. All aspects of Gulf of Mexico processes have been investigated: physics, geology, chemistry, and biology. The environmental research has been conducted by competitive bidding by multi-institutional groups organized in response to requests for proposals published widely by the agency. Management of each project has been by a single academic institution or an independent consultancy. The earliest works in the 1970s dealt with the continental shelf (see separate section on Continental Shelf Studies); this was followed by several broad, rather general categories: physical oceanography; general, level-bottom seafloor ecology; methane seeps and their communities; and an experimental arena designed to determine the effects of oil and gas exploration and production in offshore waters. In addition, when special issues have been brought to the attention of the agency, such as potential response of Cetaceans or the possibility of mercury contamination, somewhat more narrow initiatives have been supported. Each of the many studies has had a distinctive name and acronym. This section of this chapter will deal only with those studies devoted to explication of deep-ocean faunal communities.

The world's view of the deep Gulf of Mexico changed abruptly again in the 1980s with the outstanding discovery of diverse communities of seafloor organisms that live apparently on oil and gas (Brooks et al. 1985; Kennicutt et al. 1985) rather than algal plankton. Alternatively, some of the foundation species of these seep communities use the sulfide produced by anaerobic bacteria as an energy source (Cordes et al. 2003). This profound discovery gave rise to almost three decades of invigorated surveys, sampling, and experimentation in the Gulf of Mexico to determine why and how organisms living on fossil hydrocarbons function and why they would appear so similar in structure to communities that survive in hydrothermal fluids rich in geothermally produced sulfide at spreading centers. These studies not only continue today in the Gulf, but also led to the realization that similar phenomena are being encountered on a yearly basis in the numerous depositional environments on continental margins (Levin and Sibuet 2012).

7.3 HABITAT DEFINITIONS

This section is a broad summary of the different physical habitats within the entire offshore ecosystem of the Gulf of Mexico. This classification is based for the most part on water depth, but also on other physical characteristics that are or can be important in determining what types of organisms live in that habitat. These categories are important because the abundance and diversity can vary widely between habitats, depending on the physical (chemical and geological) conditions. Each habitat and its biota will thus provide different ecosystem services.

7.3.1 Continental Shelf (Ken Sherman's Large Marine Ecosystem)

The most salient habitats of the northern Gulf of Mexico offshore are depicted in Figure 7.1 provided by the NOAA. This includes the northern continental shelf, which is mostly terrigenous mud west of the Mississippi Delta and carbonate material east of the delta. Note that the eastern shelf is interdigitated hard bottom and carbonate sands. The northern shelf in its entirety can be presumed to be temperate or Carolinian in composition (Engle and Summers 2000). Just west of the delta, the shelf water column becomes hypoxic due to stratification by freshwater and eutrophication from nutrient loading (Rabalais et al. 2002; Bianchi et al. 2010). The Carolinian biota transitions into tropical and semitropical species in lower Florida and about midway down the Mexican coast on the west side of the basin. The outer shelf of the northern Gulf of Mexico is characterized by banks and pinnacles whose foundations are carbonates in the eastern Gulf of Mexico or salt diapirs in the central Gulf of Mexico. The most notable is the Flower Garden Banks National Marine Sanctuary (FGBNMS), described in detail by Rezak et al. (1985). The most obvious feature of the southern Gulf of Mexico is the wide Campeche Bank and its numerous small coral islands, with some actually inhabited (West Triangles and Arrecife Alacranes).

7.3.2 Continental Slope and Abyssal Plain

It is difficult to provide a simple overview of the Gulf of Mexico continental slope because it contains so many complicated physiographic features, each being its own habitat with peculiar characteristics. Prominent among these are submarine canyons that cross isobaths. The largest—the Mississippi Trough—begins as a gouge in a narrow shelf just off the Mississippi Delta. Sediments pour out with the river plume and are deposited in the trough. Eventually the muds move offshore at unknown rates to unknown depths (Bianchi et al. 2006). This contrasts with the De Soto Canyon at the northeast corner of the Gulf; it is not off a river and thus does not actively transport material downslope that is known. Methane seeps and other fossil hydrocarbon assemblages are interspersed along the northwest slope at depths of less than 100 m (328 ft) to depths of at least 2,000 m (6,561 ft), emanating from fossil hydrocarbon deposits below kilometer-deep layers of pelagic sediments and salt. The overlying terrigenous and pelagic sediments are denser than the underlying salt. This forces the bathymetry to exhibit a varying array of diapirs (mounds) and intermediate basins between the mounds. In the north, and in similar sediments in the south, these salt and sediment deposits terminate in steep escarpments on the north (Sigsbee), south (Campeche), and east (Florida) margins of the basin. Each of these transitional physiographic features, as unique habitats, might be expected to harbor characteristic faunas. Below the steep escarpments lies the continental rise and Sigsbee Abyssal Plain (SAP). Below the Mississippi Canyon lies a thick, broad wedge of land-derived sediments—termed the Mississippi Sediment Cone—stretching down onto and bisecting the east from the west abyssal plain. The abyssal plain has been formed

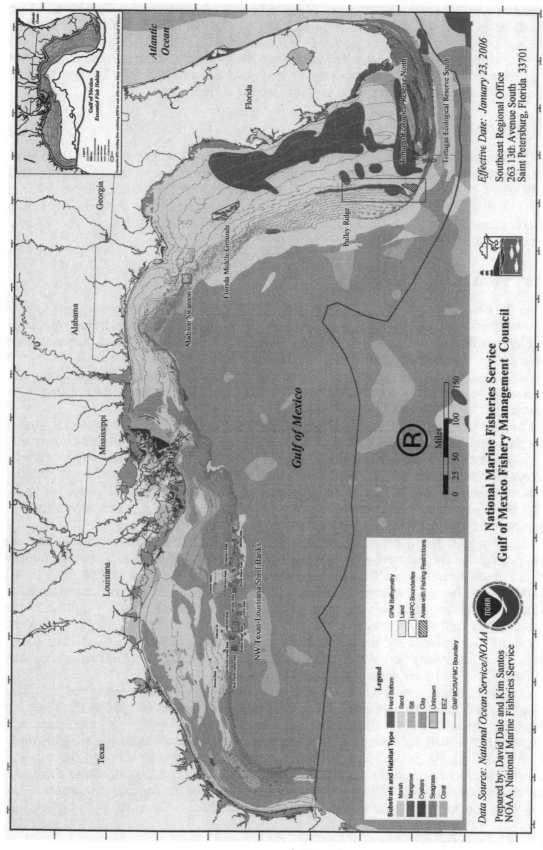

Figure 7.1. Habitats of the northern Gulf of Mexico (modified from GMFMC 2004, 2005).

by numerous intermittent turbidity flows from the margins. Its depths range from about 3.3 km (2.1 mi) down to about 3.7 km (2.3 mi) (most abyssal plains in the larger ocean basins have depths of 5–6 km [3.1–3.7 mi]). In general featureless, the SAP does contain small knolls that protrude up several hundred meters from the floor. An odd feature of the eastern boundary of the Mississippi Sediment Cone is an area of iron stone-like reddish crust that may be characterized by substantial bottom currents (Pequegnat et al. 1972).

7.4 PLANKTON

This section deals with the drifting plants and animals that occupy and drift through all habitats and depths of the open ocean. Generally small, these plants and animals together provide the food for most of the larger, often charismatic animals that make up higher levels of the food webs. The plankton are thus vital to a healthy ocean. This section treats the plankton in sections according to their function and taxonomic composition, as well as the different habitats in which they occur.

7.4.1 Functional Categories

At the base of open-ocean food webs is the plankton, defined as organisms that drift in currents. Plankton is composed of photosynthetic phytoplankton and heterotrophic zooplankton. Phytoplankton accomplishes the primary fixation of organic matter from carbon dioxide that supports the entire ecosystem biota. They are linked to higher trophic levels by the zooplankton, which is composed to a large degree of small crustaceans such as copepods. This section will describe the nature of each functional group in the Gulf of Mexico and what controls their distributions and productivity.

The plankton, or drifting organisms, is divided into two broad groups: the smaller phytoplankton, all small plant cells, and the larger zooplankton, all animals of various sizes. The phytoplankton (single-celled plants) synthesize organic matter from carbon dioxide, whereas the zooplankton (the animals) are the first step in the consumption of organic matter produced by the plants. The bulk of the biomass in all offshore ecosystems depends on this PP by the plants and the secondary (growth) production of the zooplankton, which then fall prey to larger species.

Phytoplankton is composed of single-celled organisms that are photosynthetic (use the energy of light to fix carbon dioxide into organic matter); they produce the bulk of the organic matter in aquatic ecosystems. They are divided into two general taxonomic groups: diatoms and dinoflagellates. In the open ocean, smaller nano- and pico-plankton are also important autotrophs, meaning they too are photosynthetic and produce organic matter. The growth of the phytoplankton depends on available light and inorganic nutrients such as nitrate, phosphate, and silicate to reproduce and thus produce new organic matter in the form of plant cells. The baseline characteristics of the phytoplankton outlined below are dependent on and vary directly as a function of these variables—light and inorganic nutrients.

Zooplankton is generally divided into categories based on taxonomic group and individual size of the animals. This determines the methods employed to sample them. The most frequently studied group is the *net* plankton (sometimes referred to as mesoplankton). This plankton is sampled with nets with a mesh size of about 100 micrometers (µm) up to just over 300 µm (1 µm = 3.9×10^{-5} inches [in.]). The dominant taxa are the copepod crustaceans. Nets of various sizes are held in a variety of frames, usually large rings, and these nets are hauled through the water column to filter out the drifting zooplankton. The mesh sizes are intended to be small enough to capture most zooplankton but large enough that they do not clog up with the smaller phytoplankton. Flow meters are placed in the net opening to determine the volume of water filtered during a tow. The resultant data are then presented in terms of water volume filtered, usually cubic meters (m^3). Often the total bulk of the sampled organisms is estimated as volume displacement per m^3, meaning the amount of water displaced by the organisms is considered an estimate of their total biomass. Thus, zooplankton biomass is often expressed as milliliter(s) per cubic meter (mL/m^3). The data also can be represented as number of species or number of a particular group per m^3. These quantitative estimations allow for comparisons among Gulf of Mexico habitats, offshore regions, and even other ocean basins. The baseline characteristics in Gulf of Mexico offshore habitats will thus be presented in these general quantitative terms, as presented in the available literature.

A second category of zooplankton is the macroplankton. Because they can swim and make large diurnal vertical migrations, they are sometimes referred to as the micronekton (the nekton being large swimming species). These larger animals are measured in centimeters (cm) rather than millimeters (mm). They are sampled with large nets that can be several meters across. The nets contain wider mesh than that for net plankton and are towed at several knots because these animals can be active swimmers and thus can avoid slow-moving nets. One dominant prey is the smaller abundant copepod crustacean in the net plankton. The macroplankton is a major source of food for large predators, including billfish, marine mammals, and squid.

An additional form of plankton is the neuston. It lives at the surface interface with the atmosphere. This suite of both plants and animals that drift within the surface boundary layer are sampled with floating nets that reach just above and below the interface. A major component of the neuston in the Gulf of Mexico are large windrows of floating *Sargassum* that act as protective nursery habitats for juvenile stages of large pelagic fish.

The zooplankton also can be defined in terms of their time in the plankton. The holoplankton are always planktonic throughout their entire life cycles. The meroplankton are residents of the plankton only as larval and juvenile stages. Their adult stages are either as benthic (seafloor) invertebrates or as freely swimming nektonic predators. This resume of the plankton baselines will treat each of these categories separately. A large section is devoted to the ichthyoplankton because they grow into important pelagic and benthic fishes. This form of meroplankton is sampled in the surface 200 m (656 ft) and in the neuston.

A further distinction within the plankton is between the neritic assemblages that live nearshore and the open-water groups that live offshore. Thus, this baseline survey will include this distinction because the species composition of the two areas is different, and in the Gulf of Mexico the studies of these two habitats have been very different in nature and results.

7.4.1.1   The Phytoplankton: Physical and Chemical Controls

The base of offshore food webs is the PP by photosynthesis of diatoms, dinoflagellates, prymnesiophytes, and others, the single-celled algae that float or drift in surface currents. Phytoplankton require light and inorganic nutrients (nitrogen compounds—nitrate, nitrite, nitrous oxide, free amino acids, ammonium, primary amines and phosphate and silicon), and the rate of PP by these one-celled microorganisms is proportional to the light and nutrients available. In marine systems, including the Gulf of Mexico, nitrate is considered to be the most important limiting nutrient, although phosphate may in some cases be limiting as well when there is an overabundance of nitrate. Direct measurements of PP are accomplished on discrete water samples from standard depths taken down through the water column within the photic (lighted) zone. The general method used since the 1950s is incubation of the water with radiolabelled bicarbonate. Carbon 14 (^{14}C) is incorporated into cells in a given volume of water over a given length of time under varying intensities of light and at varying concentrations of nutrients. At the end of the incubation, the water is filtered and the radiocarbon is then counted on a scintillation counter to determine carbon uptake rates. Alternative methods include measuring the photosynthetic pigment chlorophyll *a*, counting cell density per unit volume, or oxygen production over time. Species composition and cell densities (stock size and biomass) can be determined on the same discrete water samples. It would not be an exaggeration to say that hundreds of such measurements have been made all over the Gulf of Mexico in the last 50 years.

A less accurate but more comprehensive way to estimate PP is the use of satellite color images to estimate surface water photosynthetic pigments in cells. From this information, the total surface water phytoplankton standing stocks (biomass as mg C/m^3) can be estimated. Likewise PP can be estimated (in mg C/m^3/h [hour]) based on known relationships between photosynthetically active radiation (PAR) and pigment concentrations. The values of surface PP also can be entered into established first-order decay relationships between PP and delivery of particulate organic carbon (POC) at any depth. Surveys based on discrete samples and satellite-based maps will be used to provide an overview of present state of knowledge of the importance of phytoplankton offshore in the Gulf of Mexico.

The satellite information has been used to define ecoregions (Figures 7.2 and 7.3) that are characterized by specific levels of chlorophyll *a* concentrations based on satellite Sea-viewing Wide Field-of-view Sensor (SeaWiFS) images (Salmeron-Garcia et al. 2011). Each region also has a set of physical and chemical traits that give rise to that region's pigment concentrations. For example, the central Gulf of Mexico has very low pigments because it has no good source of nitrate. Regions 12 and 13 are bathed in Caribbean water but are characterized by upwelling

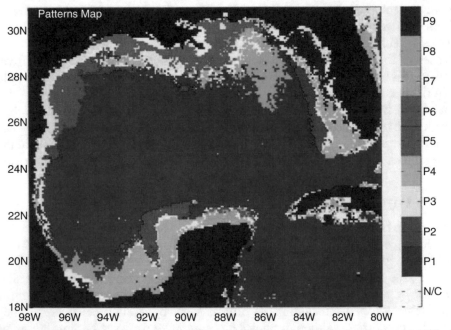

Figure 7.2. Ecoregion colors based on chlorophyll *a* concentrations assessed with SeaWiFS satellite images (from Figure 4 in Salmeron-Garcia et al. 2011; republished with kind permission from Springer Science+Business Media). The lowest levels of approximately 0.1–1 micrograms per liter (μg/L) are found in the *dark blue* (P1) area, whereas higher values are seen in the northeast (P6–P9) with values as high as 5–10 μg/L.

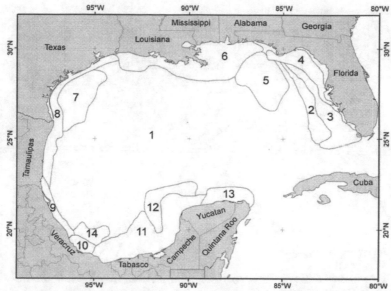

Figure 7.3. Ecoregions of surface water chlorophyll *a* pigments estimated from SeaWiFS satellite images (from Figure 5 in Salmeron-Garcia et al. 2011; republished with kind permission from Springer Science+Business Media). Each region corresponds to specific ranges of primary production (PP) and associated physical properties.

(12) and mixing (13). Mexican rivers influence regions 9, 10, 14, and 11. Region 6 is influenced by nitrate loading in the Mississippi River plume extending onto the continental shelf. Region 5 has high chlorophyll *a* concentrations because the water is pulled off of the shelf by eddies that break off from the loop current (LC). Each ecoregion, according to these authors, has its own seasonal variation patterns. The complicated set of three regions aligned with the Florida coast is a combination of upwelling and river flow.

The northeastern corner of the Gulf of Mexico is a healthy region of high PP (Figure 7.4) (Qian et al. 2003). Rate limiting nitrate is drawn offshore by warm eddies, but spatial distributions of algal biomass are controlled by riverine and estuarine input of nutrients. Both the Mississippi and the Apalachicola rivers are most important. On the other hand, the far western ecoregions of south Texas and northern Mexico are depleted of nutrients and support very low PP and algal biomass (Flint and Rabalais 1981). These two regions contrast markedly with the continental shelf just to the west of the Mississippi Delta, where hypoxia occurs during the spring, fall, and summer months when the water column is vertically stratified by freshwater (Wiseman and Sturges 1999; Rowe and Chapman 2002). The species composition of the phytoplankton in each ecoregion is also a function of the ratio of the nutrients (Dortch and Whitledge 1992). High nitrate input (greater than 100 micromoles per liter [μmol/L]) results in intense blooms that sink into and below the thermocline (Lohrenz et al. 1990). There the organic matter is respired and hypoxia ensues. As discussed in following sections, these processes have profound effects on the biota (see shelf benthos section).

To a large degree, the important role of circulation on open-ocean Gulf of Mexico productivity can be explained on the basis of sea-surface height (Figure 7.5). The best succinct description of the important processes related to the loop current system (LCS) is found in Jochens and DiMarco (2008). The water that flows into the Gulf of Mexico from the Caribbean is warm, devoid of nitrate at the surface, and has little plant biomass. It is the reddish water

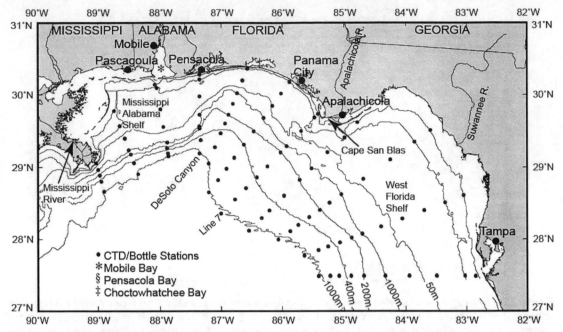

Figure 7.4. Phytoplankton study sites in the northeast Gulf of Mexico (from Figure 1 in Qian et al. 2003; reprinted with permission from Elsevier).

(Figure 7.5) flowing north between the Yucatán Peninsula and Cuba, often referred to as the loop current (LC) because it flows into the Gulf of Mexico and then abruptly curls around to the right (because it is a topographic high), returning around the Florida Keys to the Atlantic. When it penetrates deep into the north-central Gulf, it can spin off warm eddies, which are also areas of elevated sea-surface height that spin clockwise. With this flow pattern, the LC or the eddies can often pull shelf water east of the Mississippi River out into deep water, thus transferring productive water containing nitrate into deeper regions where it would normally be very oligotrophic (Maul 1974). The LC's warm eddies retain their original oligotrophic character as they move west across the entire Gulf of Mexico, degrading slowly and ending up against the continental shelf of Mexico, pictured as brown to orange blobs (Figure 7.5). The warm anticyclonic centers are topographic highs (Figure 7.5) and thus are less productive than their margins or the cool cyclonic regions adjacent to them (Biggs and Muller-Karger 1994; Biggs et al. 2008), which are topographic lows (blue in Figure 7.5). This variation all occurs in ecoregion 1 (Figures 7.2 and 7.3). The net PP in these offshore features varies between 100 and 200 mg C/m^2/day (El Sayed 1972).

An important comparison is the rate of new production between the various ecoregions of the Gulf of Mexico because this new organic matter is cycled up the food web at the surface or it is exported to the seafloor or down the water column to deep-living components. The highest PP rates on the continental shelf in the Mississippi River plume reach 3–5 g C/m^2/day (Lohrenz et al. 1990; Dagg and Breed 2003). However, narrow regions along all the coasts over much of the Gulf are substantially less—0.5 to 1.5 g C/m^2/day (Flint and Rabalais 1981; Qian et al. 2003)—and decrease offshore. The lowest rates in the central Gulf of Mexico are limited because of the depth of the nutricline at about 125 m (410 ft) (El Sayed 1972; Biggs et al. 2008); the phytoplankton in these waters produce 100–200 mg C/m^2/day at most (Bogdanov et al. 1969).

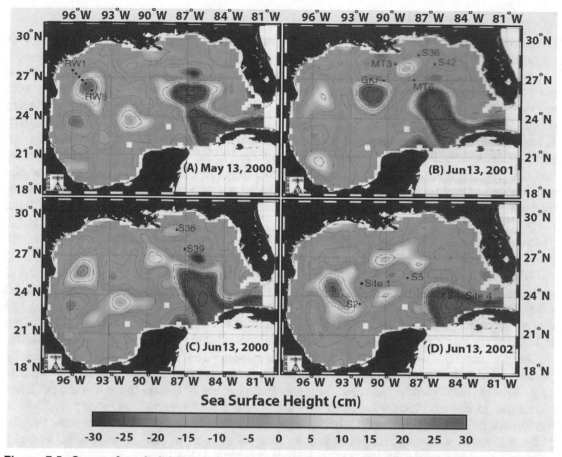

Figure 7.5. Sea-surface height showing warm eddies spun off the loop current (LC) (*reddish*) versus cold areas (*blue*) between warm eddies. Warm high areas spin clockwise; cold areas spin counterclockwise (from Plate 1 in Jochens and DiMarco 2008; reprinted with permission from Elsevier). The lowest phytoplankton production is in the *red areas*; the highest offshore is in the *blue*. However, these offshore sites are much lower than on the shelf; the highest are close to shore (see Figures 7.2 and 7.3).

The cyanobacteria, *Trichodesmium* spp., by fixing nitrogen, may play a significant role in the oligotrophic (nitrogen limited) central regions of the Gulf of Mexico (Carpenter and Roenneberg 1995; Letelier and Karl 1996). Referred to as *diazotrophs*, these organisms need energy such as light (the flat transparent surface of a calm ocean) or carbon compounds (as in the guts of termites) to transform unreactive dissolved nitrogen (N_2) into ammonium. When they are in a senescent stage, they are thought to release ammonium that could initiate a red tide bloom (see below). They could also be supplying limiting fixed nitrogen to phytoplankton in the warm oligotrophic eddies pictured in Figure 7.5. A bloom of *Trichodesmium* on the west Florida shelf may have been stimulated by iron fertilization from West African dust (Lenes et al. 2001).

Unfortunately, the phytoplankton can produce toxic blooms, often referred to as red tide. The west coast of Florida appears to be particularly susceptible to blooms of *Karenia brevis* and *Gymnodinium breve* (Chew 1956; Simon and Dauer 1972; Tester and Steidinger 1997; Gilbes et al. 1996). These can be poisonous to fish and invertebrates that consume them. The causes of such blooms remain obscure. It has been suggested that the blooms occur in the absence of adequate grazing by zooplankton to keep their densities in check.

7.4.1.2   The Zooplankton

Zooplankton are small heterotrophic organisms that also drift in currents (as opposed to swim against them). They are vital to a healthy productive ecosystem because they are the intermediary within the food web between primary producers and major consumers of economic importance—the pelagic fishes. Copepod crustaceans are the dominant taxon in both numbers and biomass in most coastal and open-ocean conditions, including the Gulf of Mexico (Bogdanov et al. 1969; Hopkins 1982; Dagg et al. 1988; Ortner et al. 1989; Elliott et al. 2012). A large fraction of the zooplankton is filter feeders that use phytoplankton cells directly, but some, such as arrow worms (chaetognaths, such as *Sagitta* spp.), are predators. The filter feeders, detritivores, and omnivores are all considered grazers of the algal standing stocks. Net zooplankton is quantified using opening and closing nets with mesh of 125–330 μm. The nets, towed at discrete depth intervals, have demonstrated that many species occupy specific depth ranges. Smaller microzooplankton are sampled with large-volume bottles and filtered. Large drifting zooplankton, such as jelly fish (Phylum Cnidaria), are important food for open-ocean turtle populations.

Most zooplankton migrate daily, swimming up to surficial waters (upper 50 m [164 ft]) at night and descending during daylight hours, and this is evident in the Gulf of Mexico (Hopkins 1982). However, as Hopkins notes, each species has its own pattern of migration, resulting in a mix of species at various depths over a 24-h cycle. Most of the migration occurs in the upper 100 m (328 ft), and it is all more or less confined to the top 1,000 m (3,281 ft) of the water column.

In continental shelf or neritic waters, the zooplankton plays a similar role—linking phytoplankton production to higher trophic levels. However, the species composition is markedly different and assemblages reach far higher biomass than offshore. In the Mississippi River plume the copepods, *Temora turbinate* and *Eucalanus pileatus*, can graze more than 50 % of the PP on a daily basis (Ortner et al. 1989; Dagg et al. 1996). The latter work documents the role of the grazers in removing lithogenic particles (suspended mud) as well as living cells. As major grazers, zooplankton can prevent toxic algal blooms before they occur.

Through frequent molting of their exoskeletons, crustacean zooplankton contribute considerable material (Dagg et al. 1988) to detrital food webs, especially offshore in deep water. Likewise zooplankton package the remains of the phytoplankton cells they graze into fecal pellets that sink far faster than the individual cells, thus adding a significant pathway for organic matter to reach great depths. Zooplankton, in sum, are a major functional group in clearing detrital organic matter out of surface layers and channeling it to food-starved deepwater biota; the slow rain of detrital particles is assumed to be a major source of food for much of the deep bottom fauna. This flux of fecal pellets, cell debris, and molts is often referred to as the *biological pump*.

The various habitats of the Gulf of Mexico neritic continental shelf contain largely the same dominant groups in the holoplankton, mostly copepod crustaceans (Ortner et al. 1989; Dagg 1995). However, the physical habitats themselves vary widely around the circumference of the Gulf of Mexico, as indicated in the above sections on phytoplankton. This variation in the physical nature of the habitats affects the species composition, diversity, productivity, and animal behavior of the assemblages. The most salient example of a modified, atypical environment is the seasonal hypoxia on the continental shelf off Louisiana. The net zooplankton between 2003 and 2008 were clustered into four assemblages dominated by calanoid copepod crustaceans (Elliott et al. 2012). Mean densities among the four groups they identified ranged from 23,000 individuals/m^3 down to 1,600/m^3. The groupings were related to temperature, salinity, and the vertical extent of hypoxic conditions, with severe restrictions (stress) in abundance below 2 mg of oxygen per liter of water (the upper limit of hypoxia) (Elliott et al. 2012). These authors suggest that the large fecal pellets of big copepods mediate vertical flux of organic matter and thus increase the extent of bottom water hypoxia. This reinforces the suggestion of Dagg et al. (2008) that a microbial food web intensifies the Louisiana shelf's bottom water hypoxia. It is evident that hypoxia reduces habitat size for aerobic metazoans and can reduce the mean individual size within planktonic assemblages (Kimmel et al. 2009).

To the west of the Louisiana hypoxia on the south Texas shelf, the PP is drastically reduced because of minimal river runoff (see above section of continental shelf phytoplankton). This is reflected in low densities of zooplankton. However, a near-bottom layer of particulate matter is an almost universal feature of the Texas continental shelf (Flint and Rabalais 1981). Thus, the zooplankton feeds predominantly in this near-bottom, 1–2 m thick *nepheloid* layer (Bird 1983), not near the surface. The exact origin of the nepheloid layer is unclear. It may be the westward extent of mud from the Mississippi River, and/or the resuspension of mud by trawlers, tidal currents, or by resident biota. This suspended particle layer is something that differentiates the shelf habitat west of the Mississippi Delta from the relatively transparent (particle free) water east of the Mississippi Delta on the Mississippi, Alabama, and Florida coastlines (Figure 7.1).

Further to the south, the typical zooplankton assemblages reflect a gradual change in habitat types within the zoogeographic temperate regime of the northern Gulf of Mexico to habitats in the semitropical/tropical regime of the southern Gulf of Mexico. This change occurs near Tampico, Mexico, at about 24° N latitude. Below this, the seasonality is more hospitable to coral reefs and the associated biota, including the plankton (De la Cruz 1972). Densities and biomass in the southern Gulf of Mexico are low but diversity is high. Biomass is low because phytoplankton production is limited by lack of inorganic nutrients, principally nitrate. Exceptions are the areas near the mouths of the rivers at the base of the Gulf of Campeche and the narrow zones of upwelling associated with the shallow but geographically extensive Campeche Bank.

The most productive region of the Gulf of Mexico shelf is east of the Mississippi Delta over to Florida, as indicated in the section above for phytoplankton (Figures 7.1, 7.2, 7.3, and 7.4). This is due to nutrient input from rivers, the complicated physical environment, and proximity to the LC. The complicated physics includes epipelagic nutrient enhancement due to wind-driven upwelling along the shelf edge. Additionally, the Mississippi River adds nutrients. These processes were first observed in the early studies of Riley (1937) in this region. Mesoscale eddies break off of the northern extension of the LC; this can draw nutrient-rich shelf water offshore, thus enhancing PP (Hamilton 1992; Sahl et al. 1997). This PP provides food for enlarged stocks of mesozooplankton (Ressler and Jochens 2003). Upwelling enhances production all along the outer west Florida continental shelf (Weisberg et al. 2000).

The broad carbonate platform that forms the west Florida continental shelf supports abundant and diverse zooplankton populations (see area in Figure 7.1). For example, zooplankton were aligned in three separate zones along shore: one composed of a nearshore high density assemblage of larvaceans, a second inshore zone of small copepods in low densities, and a third richer zone offshore of larger species of copepods (Kleppel et al. 1996; Sutton et al. 2001). Zooplankton grazing intensity may play a role in controlling toxic blooms of the dinoflagellate phytoplankton *Karenia brevis* that plagues the west Florida shelf and coastline (Milroy et al. 2008).

The most comprehensive investigation of the offshore holoplanktonic zooplankton concentrated on the vertical distribution of all size classes of animals at an offshore location in the eastern Gulf of Mexico (27° N × 86° W) (Hopkins 1982). The sizes—larger than 1 mm (0.04 in.)—are based on opening and closing net tows with a 162-μm mesh, whereas the metazoan animals—smaller than 1 mm (0.04 in.)—are based on large-volume bottle samples. The samples were taken at 25 m (82 ft) intervals down to a depth of 150 m (492 ft), and then at

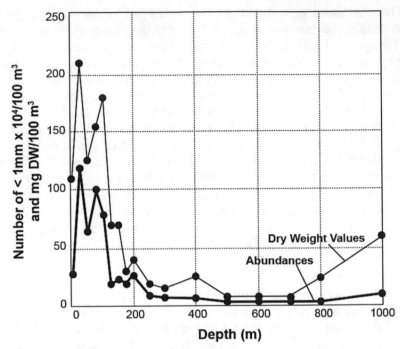

Figure 7.6. Vertical distribution of the standing stock of microzooplankton (less than 1 mm in length) at a deepwater location in the eastern Gulf of Mexico (modified from Hopkins 1982).

100 m (328 ft) intervals down to a maximum depth of 1 km (0.62 mi). The animals were sorted to species when possible and to major group, usually family or order, otherwise, for a total of 11 general categories. Of the totals, the copepod crustaceans were overwhelmingly dominant at all depths. The species composition was almost entirely different from those that dominated in the neritic habitats described above. Likewise, there was a distinct vertical partitioning of species. While much of this vertical zonation could be due to feeding habits, some of it may be related to sharp vertical gradients in temperature, according to Hopkins (1982). Hopkins' sampling was also taken during the day and at night to determine vertical migration behavior; his studies also suggested, however, that there was some net avoidance near the surface by larger motile species during daylight.

A distinct planktocline was observed in these samples at the 50–100 m (164–328 ft) depth (Figures 7.6 and 7.7). Most of the animals and the biomass were found at the surface at night and in the daytime, in spite of vertical migrations to avoid the light. The total biomass integrated over the 1 km (0.62 mi) water column that they sampled amounted to about 1.6 mg dry weight (dw)/m². This concentration near the surface was especially evident in the larger groups caught with the net (Figure 7.7). Of the total biomass, most was sampled in the larger size groups, amounting to about 1.2 g dw/m²; the smaller forms amounted to 0.4 g dw/m². The mean size of the larger than 1 mm (0.04 in.) group was about 26 micrograms (µg) dw per individual whereas the smallest group (smaller than 1 mm [0.04 in.]) averaged about 0.25 µg dw per individual. These would be equivalent to about 10.4 µg carbon and 0.1 µg carbon per individual, respectively.

The totals observed are comparable to other oligotrophic areas such as the central Sargasso Sea, according to Hopkins (1982), in agreement with observations of phytoplankton production and biomass discussed above. The principal predators on the zooplankton appeared to be mid-water fish populations such as the myctophids, which Hopkins was also able to assess

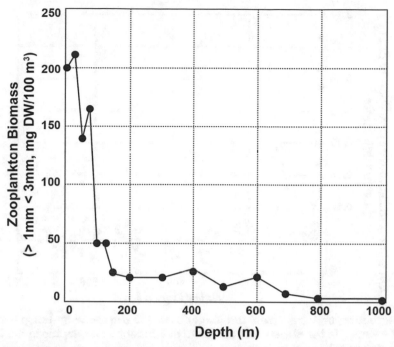

Figure 7.7. Vertical distribution of the biomass of net or mesozooplankton (larger than 1 mm, but smaller than 3 mm) sampled at night in the eastern Gulf of Mexico (modified from Hopkins 1982). Mesoplankton is traditionally sampled with a 330 μm mesh net.

with the study's opening–closing nets, but the greatest concentrations of fish were between depths of 50 and 100 m (164 and 328 ft), not near the surface.

The distribution of zooplankton is not uniform across the entire open Gulf of Mexico, as the above studies of the vertical distributions might imply. The flow from the Caribbean is the principal source of water and thus a source of plankton to the Gulf. This becomes the LC once it enters the Gulf, which pulses irregularly into the eastern gulf in an anticyclonic loop that enters through the Yucatán Channel and leaves through the Florida Straits (Hopkins 1982). On the northern boundary of the loop, warm eddies can spin off that move westward across the Gulf of Mexico (Figure 7.8), and these affect the distribution of both the phytoplankton and zooplankton. The warm anticyclonic eddies (turn clockwise and sea surface is elevated) are oligotrophic because the water comes from the Caribbean (Biggs 1992). However, small submesoscale cyclones (turn counter clockwise and are below mean sea level) can have enhanced nutrients and plankton concentrations, including mesoplankton and micronekton that can be assessed from acoustic backscatter (Ressler and Jochens 2003). Thus the open offshore Gulf of Mexico, while oligotrophic overall, is actually a patchwork of different concentrations of plankton that are controlled by physical circulation patterns on scales of tens to hundreds of kilometers (Jochens and DiMarco 2008).

The Southeast Area Monitoring and Assessment Program (SEAMAP) database contains extensive information on a wide variety of standing stocks, including zooplankton biomass distributions (Figure 7.8) in the upper 200 m (656 ft) of the water column (Rester 2011). A plot of more than 100 locations across a wide depth interval illustrates that the zooplankton baseline in general ranges from 0.025 to 0.075 mL/m^3 displacement volume over most of the Gulf of Mexico offshore (depths greater than 100 m [328 ft]), but that nearshore, the values can be much higher.

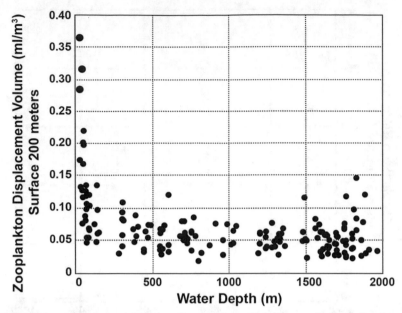

Figure 7.8. Distribution of net zooplankton (larger than 330 μm mesh net) displacement volume (a measure of biomass) in the surface 200 m (656 ft) at different water depths in the Gulf of Mexico (from SEAMAP database).

Zooplankton displacement volume (larger than 330 μm mesh net) nearshore is substantially higher than offshore (Figure 7.9). Note the parallels between the phytoplankton and zooplankton biomass levels by comparing Figure 7.9 with Figures 7.2 and 7.3: the highest are always close to shore and adjacent to river mouths.

7.4.1.3   Ichthyoplankton

A relatively small but vital component of the zooplankton in the upper 200 m (656 ft) of the water column are the ichthyoplankton, composed of fish eggs, larvae, and juveniles (SWFSC 2007). While fish eggs have their own food supply, fish larvae eat smaller plankton; both serve as an important prey base for marine invertebrates and fish. The distribution of ichthyoplankton is a function of the spawning locations of adult fish, currents, and sea-surface temperatures. Monitoring ichthyoplankton provides essential information on potential population sizes of adult fish since the survival rates of larval fish are assumed to contribute to recruitment success and year-class strength in adults (Houde 1997; Fuiman and Werner 2002; SWFSC 2007).

7.4.1.3.1   Baseline Ichthyoplankton Abundance and Distribution in the U.S. Gulf of Mexico

SEAMAP is a state/federal/university program for the collection, management, and dissemination of fishery-independent data obtained without the direct reliance on commercial or recreational fishermen (Rester 2011). A major goal of SEAMAP is to provide a large, standardized database for management agencies, industry, and scientists. The types of surveys conducted include plankton, reef fish, shrimp/groundfish, shrimp/bottomfish (trawl), and bottom longline, as well as occasional special surveys. Sampling is usually conducted at predetermined SEAMAP stations arranged in a fixed, systematic grid pattern, typically at approximately 56 km (34.8 mi) or 0.5° intervals, across the entire Gulf of Mexico (Rester 2011). All surveys are not conducted each year, and all stations and seasons are not sampled every

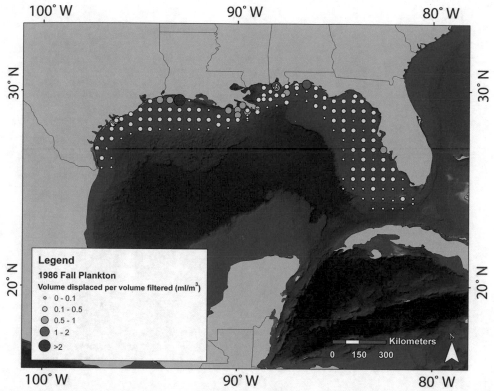

Figure 7.9. Zooplankton displacement volume in SEAMAP samples from fall sampling in the upper 200 m (656 ft) (larger than 330 μm mesh net).

year Gulf wide, with a particular deficiency in winter sampling (Lyczkowski-Shultz et al. 2004). The majority of SEAMAP plankton samples are collected using bongo nets and neuston nets. A 61 cm (24 in.) bongo net, fitted with 0.333 mm (0.013 in.) mesh netting, is fished in an oblique tow path from a maximum depth of 200 m (656 ft) or to 2–5 m (6.6–16.4 ft) off the bottom at depths less than 200 m (656 ft), and a mechanical flow meter is mounted off-center in the mouth of each bongo net to record the volume of water filtered (Rester 2011). A single or double 2 m × 1 m (6.6 ft × 3.3 ft) pipe frame neuston net, fitted with 0.937 mm (0.037 in.) mesh netting, is towed at the surface with the frame half submerged for 10 min (Rester 2011). Therefore, the two types of plankton nets used provide samples from distinct and separate segments of the water column: the neuston net samples the upper 0.5 m (1.6 ft) of the ocean surface, while the pair of bongo nets sample the entire water column from subsurface to near bottom, or to a maximum depth of 200 m (656 ft) (Lyczkowski-Shultz et al. 2004). Fish larvae are removed from the samples and identified to lowest possible taxon, typically to family.

A review of available SEAMAP data from 1982 through 2007 indicated that ichthyoplankton information collected during the spring and fall plankton surveys provided the most consistent results, both temporally and spatially; therefore, these data are summarized in the following sections.[1]

Spring plankton surveys typically cover the open Gulf of Mexico waters within the EEZ, as well as the Florida continental shelf on occasion (Figures 7.10 and 7.11), while fall plankton

---

[1] ENVIRON's Baseline Information Management System (BIMS) experts provided an interpreted SEAMAP dataset that contained ichthyoplankton data from 1982 through 2007.

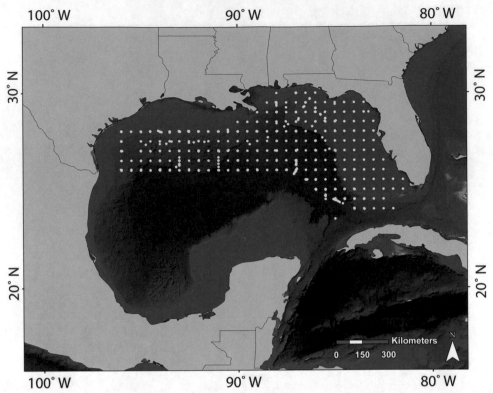

Figure 7.10. Generalized sampling locations of the SEAMAP spring plankton surveys from 1982 through 2007.

surveys typically sample the entire continental shelf of the U.S. Gulf of Mexico (Figures 7.12 and 7.13). Fish larvae in bongo net samples are expressed as number under 10 m^2 of sea surface, while larvae taken in neuston samples are expressed as number per 10-min tow. Note that the sampling sites in fall and spring were different (reason unknown).

Because of the large number of ichthyoplankton taxa collected from the U.S. Gulf of Mexico during the spring and fall from 1982 through 2007, summarizing the results for all taxa is not practical. Therefore, a small but representative number of fish taxa (11) were selected based on ecological and economic importance; baseline information for these taxa is summarized in the sections that follow. The selected taxa are listed below, and a description of the summarized SEAMAP data for each taxon is included in Tables 7.1 and 7.2.

- Family Carangidae: Jacks and pompanos
- Family Clupeidae: Herrings, shads, sardines, and menhadens
- Family Coryphaenidae: Dolphinfish
- Family Istiophoridae: Marlin and sailfish
- Family Lutjanidae: Snappers
- Family Mugilidae: Mullets
- Family Sciaenidae: Drums and croakers, includes redfish (*Sciaenops ocellatus*) and spotted seatrout (*Cynoscion nebulosus* and *Cynoscion regalis*)
- Family Scombridae: Mackerels, tunas, and bonitos (excluding *Thunnus*)

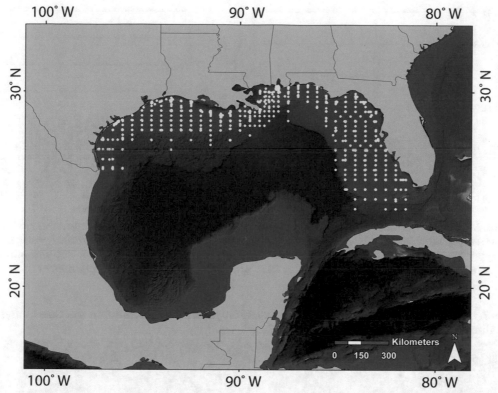

Figure 7.11. Generalized sampling locations of the SEAMAP fall plankton surveys from 1982 through 2007.

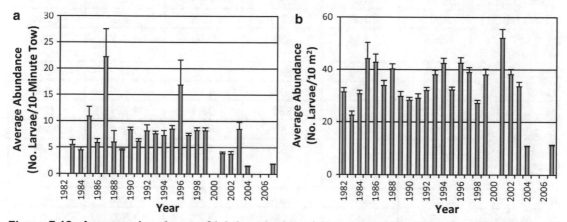

Figure 7.12. Average abundance of ichthyoplankton (all taxa combined) for neuston net (a) and bongo net (b) samples for the SEAMAP spring plankton surveys from 1982 through 2007. Spring plankton surveys were not conducted in 2000, 2005, or 2006, and only bongo net sampling was conducted during the spring plankton survey in 1982. Error bars = standard error.

- Genus *Thunnus*: Tuna (*Thunnus*), Atlantic bluefin tuna (*Thunnus thynnus*), blackfin tuna (*Thunnus atlanticus*), yellowfin tuna (*Thunnus albacares*), and bigeye tuna (*Thunnus obesus*)
- Family Serranidae: Seabasses and groupers
- Family Xiphiidae: Swordfish

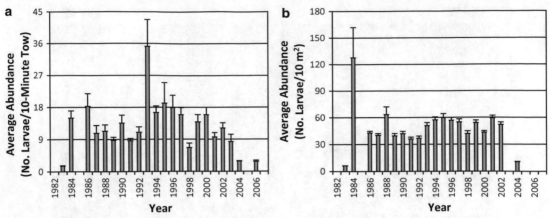

Figure 7.13. Average abundance of ichthyoplankton (all taxa combined) for neuston net (a) and bongo net (b) samples for the SEAMAP fall plankton surveys from 1982 through 2007. Fall plankton surveys were not conducted in 1982, 1985, 2005, or 2007, and only neuston net sampling was conducted during fall plankton surveys in 2003 and 2006. Error bars = standard error.

Table 7.1. Description of SEAMAP Data for the Spring and Fall Plankton Surveys Conducted from 1982 through 2007 for the 11 Selected Fish Taxa[a]

Family/ Genus	No. of Occurrences in Spring Plankton Surveys	Percent Occurrence in Spring Plankton Surveys	No. of Occurrences in Fall Plankton Surveys	Percent Occurrence in Fall Plankton Surveys	No. of Occurrences in Neuston Net Samples	No. of Occurrences in Bongo Net Samples
Carangidae	5,221	5.60	6,983	7.98	7,489	4,715
Clupeidae	896	0.96	3,717	4.25	2,429	2,184
Corypha enidae	1,983	2.13	436	0.50	2,072	347
Istiophoridae	456	0.49	286	0.33	644	98
Lutjanidae	577	0.62	3,608	4.12	1,320	2,865
Mugilidae	1,109	1.19	360	0.41	1,291	178
Sciaenidae	170	0.18	3,596	4.11	1,316	2,450
Scombridae (excluding *Thunnus*)	2,759	2.96	4,306	4.92	2,731	4,334
Thunnus	2,358	2.53	1,094	1.25	1,975	1,477
Serranidae	2,955	3.17	3,256	3.72	1,590	4,621
Xiphiidae	177	0.19	13	0.01	177	13

[a]Spring plankton surveys were not conducted in 2000, 2005, or 2006, and only bongo net sampling was conducted during the spring plankton survey in 1982; fall plankton surveys were not conducted in 1982, 1985, 2005, or 2007, and only neuston net sampling was conducted during fall plankton surveys in 2003 and 2006.

Family Carangidae

Jacks and pompanos are both ecologically important as predators and prey (Lyczkowski-Shultz et al. 2004). Some species are important in the commercial and recreational fisheries in the Gulf of Mexico and are highly regarded as food (e.g., pompano), game fish (e.g.,

Table 7.2. Description of SEAMAP Data for the 11 Selected Fish Taxa for the Spring and Fall Plankton Surveys Conducted from 1982 through 2007[a]

Taxa		No. of Occurrences in Spring Plankton Surveys	Percent Occurrence in Spring Plankton Surveys	No. of Occurrences in Fall Plankton Surveys	Percent Occurrence in Fall Plankton Surveys	No. of Occurrences in Neuston Net Samples	No. of Occurrences in Bongo Net Samples
Common Name	Scientific Name						
Carangidae							
Amberjacks	*Seriola*	766	14.67	283	4.05	930	119
Jacks and pompanos	*Carangidae*	745	14.27	638	9.14	497	886
Round scad	*Decapterus punctatus*	672	12.87	1,556	22.28	1,166	1,062
Blue runner	*Caranx crysos*	504	9.65	557	7.98	826	235
Pompanos	*Trachinotus*	281	5.38	127	1.82	403	5
Bigeye scad	*Selar crumenophthalmus*	271	5.19	738	10.57	483	526
Rough scad	*Trachurus lathami*	132	2.53	17	0.24	88	61
Rainbow runner	*Elagatis bipinnulata*	65	1.24	45	0.64	64	46
Lookdown	*Selene vomer*	39	0.75	365	5.23	151	253
Atlantic bumpers	*Chloroschombrus chrysurus*	31	0.59	2,051	29.37	1,074	1,008
Leatherjack	*Oligopolites saurus*	31	0.59	61	0.87	82	10
Mackerel scad	*Decapterus*	12	0.23	15	0.21	7	20
Pilot fish	*Naucrates doctor*	12	0.23	1	0.01	13	0
African pompano	*Alectis ciliaris*	9	0.17	17	0.24	24	2
Yellow jack	*Caranx bartholomai*	5	0.10	1	0.01	6	0
Crevall jack	*Caranx hippos*	4	0.08	2	0.03	1	5

(continued)

Table 7.2 (continued)

Taxa		No. of Occurrences in Spring Plankton Surveys	Percent Occurrence in Spring Plankton Surveys	No. of Occurrences in Fall Plankton Surveys	Percent Occurrence in Fall Plankton Surveys	No. of Occurrences in Neuston Net Samples	No. of Occurrences in Bongo Net Samples
Common Name	Scientific Name						
Palometa	*Trachinotus goodie*	3	0.06	0	0	3	0
Horse-eye jack	*Caranx latus*	2	0.04	0	0	2	0
Florida pompano	*Trachinotus carolinus*	2	0.04	1	0.01	3	0
Permit	*Trachinotus falcatus*	2	0.04	0	0	0	2
Jack mackerels	*Trachurus*	2	0.04	0	0	0	2
Threadfish	*Alectis*	1	0.02	0	0	0	1
Bumperfish	*Chloroscombrus*	1	0.02	4	0.06	2	3
Rainbow runner	*Elagatis*	1	0.02	0	0	1	0
Leatherjacks	*Oligoplites*	1	0.02	0	0	1	0
Lookdown	*Selene*	1	0.02	65	0.93	30	36
Atlantic moonfish	*Selene setapinnis*	1	0.02	10	0.14	8	3
Banded rudderfish	*Seriola zonata*	1	0.02	2	0.03	1	2
Bluntnose jack	*Hemicaranx amblyrhynchus*	0	0	1	0.01	1	0
Clupeidae							
Scaled sardine	*Harengula jaguana*	281	31.36	885	23.81	766	400
Red-eye round herring	*Etrumeus teres*	231	25.78	11	0.30	63	179
Gilt sardine	*Sardine Aurita*	198	22.10	920	24.75	568	550
Herrings, shads, and sardines	*Clupeidae*	105	11.72	630	16.95	438	297

(continued)

Table 7.2 (continued)

Taxa		No. of Occurrences in Spring Plankton Surveys	Percent Occurrence in Spring Plankton Surveys	No. of Occurrences in Fall Plankton Surveys	Percent Occurrence in Fall Plankton Surveys	No. of Occurrences in Neuston Net Samples	No. of Occurrences in Bongo Net Samples
Common Name	Scientific Name						
Atlantic thread herring	*Opisthonema oglinum*	70	7.81	1,253	33.71	578	745
Menhaden	*Brevoortia*	6	0.67	12	0.32	10	8
Atlantic menhaden	*Brevoortia tyrannus*	2	0.22	0	0	2	0
Gulf menhaden	*Brevoortia patronus*	1	0.11	0	0	1	0
Round herrings	*Etrumeus*	1	0.11	0	0	0	1
Sardines	*Sardinella*	1	0.11	2	0.05	0	3
Finescale menhaden	*Brevoortia gunteri*	0	0	1	0.03	0	1
Herrings	*Harengula*	0	0	1	0.03	1	0
Thread herrings	*Opisthonema*	0	0	2	0.05	2	0
Coryphaenidae							
Common dolphinfish	*Coryphaenidae hippurus*	930	46.90	265	60.78	1,087	108
Dolphinfishes	*Coryphaena*	669	33.74	104	23.85	562	211
Pompano dolphinfish	*Coryphaenidae equiselis*	364	18.36	51	11.70	407	8
Dolphinfishes	*Coryphaenidae*	20	1.01	16	3.67	16	20
Istiophoridae							
Marlins	*Istiophoridae*	267	58.55	167	58.39	382	52
Indo-Pacific sailfish	*Istiophorus platypterus*	169	37.06	102	35.66	229	42
Sailfish	*Istiophorus*	10	2.19	8	2.80	16	2
Blue marlin	*Makaira nigricans*	4	0.88	0	0.00	3	1

(continued)

Table 7.2 (continued)

Taxa		No. of Occurrences in Spring Plankton Surveys	Percent Occurrence in Spring Plankton Surveys	No. of Occurrences in Fall Plankton Surveys	Percent Occurrence in Fall Plankton Surveys	No. of Occurrences in Neuston Net Samples	No. of Occurrences in Bongo Net Samples
Common Name	Scientific Name						
Spearfish	*Tetrapturus*	4	0.88	9	3.15	13	0
Atlantic white marlin	*Tetrapturus albidus*	1	0.22	0	0	1	0
Lutjanidae							
Snappers	Lutjanidae	220	38.13	1,318	36.53	312	1,226
Vermillion snapper	*Rhomboplites aurorubens*	148	25.65	856	23.73	376	628
Wenchman	*Pristipomoides aquilonaris*	95	16.46	497	13.77	218	374
Snappers	*Lutjanus*	38	6.59	372	10.31	134	276
Red snapper	*Lutjanus campechanus*	36	6.24	425	11.78	208	253
Slopefishes	*Symphysanodon*	26	4.51	8	0.22	4	30
Mangrove snapper	*Lutjanus griseus*	13	2.25	108	2.99	56	65
Jobfish	*Pristipomoide*	1	0.17	4	0.11	4	1
Snappers	Etelinae	0	0	2	0.06	0	2
Queen snapper	*Etelis oculatus*	0	0	6	0.17	1	5
Mutton snapper	*Lutjanus analis*	0	0	1	0.03	1	0
Lane snapper	*Lutjanus synagris*	0	0	11	0.30	6	5
Mugilidae							
Mullets	*Mugil*	504	45.45	167	46.39	578	93
White mullet	*Mugil curema*	327	29.49	96	26.67	401	22
Mullets	Mugilidae	274	24.71	95	26.39	307	62
Flathead mullet	*Mugil cephalus*	4	0.36	2	0.56	5	1

(continued)

Table 7.2 (continued)

Taxa		No. of Occurrences in Spring Plankton Surveys	Percent Occurrence in Spring Plankton Surveys	No. of Occurrences in Fall Plankton Surveys	Percent Occurrence in Fall Plankton Surveys	No. of Occurrences in Neuston Net Samples	No. of Occurrences in Bongo Net Samples
Common Name	Scientific Name						
Drums and croakers	Sciaenidae	57	33.53	489	13.60	102	444
Kingfish	Menticirrhus	43	25.29	677	18.83	313	407
Banded drum	Larimus fasciatus	17	10.00	225	6.26	79	163
Sciaenidae							
Sand weakfish	Cynoscion arenarius	16	9.41	496	13.79	153	359
Atlantic croaker	Micropogonias undulates	10	5.88	256	7.12	83	183
Silver seatrout	Cynoscion nothus	5	2.94	369	10.26	149	225
Spot	Leiostomus xanthu	5	2.94	94	2.61	20	79
American silver perch	Bairdiella chrysoura	4	2.35	16	0.44	7	13
Drums	Cynoscion	3	1.76	159	4.42	59	103
Star drum	Stellifer lanceolatus	2	1.18	174	4.84	68	108
Weakfish	Cynoscion regalis	1	0.59	2	0.06	1	2
Black drum	Pogonias cromis	1	0.59	1	0.03	0	2
Redfish	Sciaenops ocella	1	0.59	609	16.94	272	338
Perch	Bairdiella spp.	0	0	1	0.03	0	1
Scombridae							
Skipjack tuna	Katsuwonus pelamis	961	34.83	248	5.76	509	700
Bullet tun and frigate tuna	Auxis	940	34.07	861	20.00	846	955

(continued)

Table 7.2 (continued)

Taxa			No. of Occurrences in Spring Plankton Surveys	Percent Occurrence in Spring Plankton Surveys	No. of Occurrences in Fall Plankton Surveys	Percent Occurrence in Fall Plankton Surveys	No. of Occurrences in Neuston Net Samples	No. of Occurrences in Bongo Net Samples
Common Name	Scientific Name							
Mackerels, tunas, and bonitos	Scombridae		388	14.06	673	15.63	179	882
Little tunny	Euthynnus alletteratus		295	10.69	1,016	23.59	544	767
King mackerel	Scomberomorus cavalla		50	1.81	901	20.92	348	603
Spanish mackerel	Scomberomorus macula		46	1.67	538	12.49	269	315
Wahoo	Acanthocybium solandri		34	1.23	19	0.44	12	41
Mackerels	Scomberomorus		12	0.43	38	0.88	8	42
Tuna	Euthynnus		7	0.25	0	0	0	7
Kingfish	Scomberomorus regalis		6	0.22	11	0.26	11	6
Atlantic mackerel	Scomber scombrus		4	0.14	0	0	0	4
Atlantic bonito	Sarda		2	0.07	1	0.02	2	1
Thunnus								
Tuna	Thunnus		1,512	64.12	962	87.93	1,451	1,023
Northern bluefin tuna	Thunnus thynnus		672	28.50	50	4.57	429	293
Blackfin tuna	Thunnus atlanticus		163	6.91	81	7.40	89	155
Yellowfin tuna	Thunnus albacares		10	0.42	0	0	5	5
Bigeyed tuna	Thunnus obesus		1	0.04	1	0.09	1	1

(continued)

Table 7.2 (continued)

Taxa		No. of Occurrences in Spring Plankton Surveys	Percent Occurrence in Spring Plankton Surveys	No. of Occurrences in Fall Plankton Surveys	Percent Occurrence in Fall Plankton Surveys	No. of Occurrences in Neuston Net Samples	No. of Occurrences in Bongo Net Samples
Common Name	Scientific Name						
Serranidae							
Seabasses and groupers	Serranidae	1,075	36.38	1,231	37.81	446	1,860
Combers	Serranus	463	15.67	87	2.67	221	329
Sand perch	Diplectrum	254	8.60	662	20.33	285	631
Reeffish, wreckfish, and jewelfish	Anthias	203	6.87	29	0.89	51	181
Reeffish, wreckfish, and jewelfish	Hemanthias	182	6.16	27	0.83	86	123
Yellowfin bass	Anthias nicholsi	148	5.01	67	2.06	31	184
Red barbier	Hemanthias vivanus	136	4.60	11	0.34	32	115
Seabasses	Centropristis	119	4.03	325	9.98	117	327
Longtail bass	Hemanthias leptus	102	3.45	20	0.61	22	100
Pygmy seabass	Serraniculus pumilio	33	1.12	503	15.45	151	385
Streamer bass	Hemanthias aureorubens	17	0.58	3	0.09	4	16
Basslets	Liopropoma	16	0.54	7	0.21	7	16
Groupers	Epinephelus	15	0.51	0	0	1	14
Reeffish, wreckfish, and jewelfish	Anthiinae	12	0.41	1	0.03	1	12

(continued)

Table 7.2 (continued)

Taxa		No. of Occurrences in Spring Plankton Surveys	Percent Occurrence in Spring Plankton Surveys	No. of Occurrences in Fall Plankton Surveys	Percent Occurrence in Fall Plankton Surveys	No. of Occurrences in Neuston Net Samples	No. of Occurrences in Bongo Net Samples
Common Name	Scientific Name						
Roughtongue bass	*Holanthias martinicensis*	12	0.41	0	0	3	9
Streamer bass	*Pronotogrammus aureorubens*	8	0.27	0	0	1	7
Reef bass	*Pseudogramma gregoryi*	7	0.24	7	0.21	5	9
Podges	*Pseudogramma*	7	0.24	4	0.12	0	11
Groupers	*Mycteroperca*	5	0.17	0	0	0	5
Black seabass	*Centropristis striata*	4	0.14	22	0.68	2	24
Seabasses	*Pronotogrammu*	4	0.14	0	0	0	4
Reeffish, wreckfish, and jewelfish	*Plectranthias*	3	0.10	0	0	2	1
Hamlets	*Hypoplectrus*	1	0.03	0	0	0	1
Seabass	*Serraninae*	1	0.03	0	0	1	0
Yellowtail bass	*Bathyanthias mexicanus*	0	0	1	0.03	1	0
Seabasses	*Serraniculus*	0	0	1	0.03	0	1
Xiphiidae							
Swordfish	*Xiphias gladius*	176	99.44	13	100	176	13
Swordfish	*Xiphias*	1	0.56	0	0	1	0

[a]Spring plankton surveys were not conducted in 2000, 2005, or 2006, and only bongo net sampling was conducted during the spring plankton survey in 1982; fall plankton surveys were not conducted in 1982, 1985, 2005, or 2007, and only neuston net sampling was conducted during fall plankton surveys in 2003 and 2006.

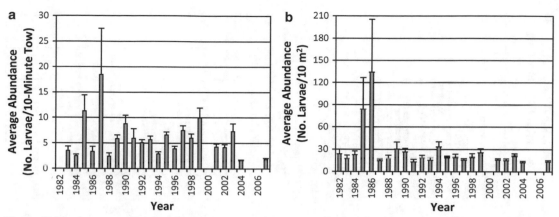

Figure 7.14. Average abundance of Carangidae for neuston net (a) and bongo net (b) samples for the SEAMAP spring plankton surveys from 1982 through 2007. Spring plankton surveys were not conducted in 2000, 2005, or 2006, and only bongo net sampling was conducted during the spring plankton survey in 1982. Error bars = standard error.

amberjack), or bait (e.g., blue runner, *Caranx crysos*) (Ditty et al. 2004; Lyczkowski-Shultz et al. 2004). Larval jacks and pompanos cannot be reliably identified to species; however, they can typically be identified to genus (Lyczkowski-Shultz et al. 2004).

Jacks and pompanos made up more of the total ichthyoplankton catch than any of the other groups selected for analysis (Table 7.1). More larval jacks and pompanos were captured during the fall along the continental shelf, as compared to spring in the open Gulf of Mexico and, while larvae were captured both at the surface and in the water column, the majority of larval jacks and pompanos were captured at the water surface in neuston nets (Table 7.1).

The average abundance of Carangidae larvae collected by neuston net during the spring ranged from 1.6 (2004) to 18.5 (1987) larvae per 10-min tow, and the average abundance of carangids collected by bongo net ranged from 12.3 (2004) to 134 (1986) larvae per 10 m^2 (Figure 7.14). Carangid average abundance during the fall along the continental shelf ranged from 1.4 (1983) to 23.1 (1999) larvae per 10-min tow for neuston net samples, while bongo net samples ranged from 6.9 (1983) to 98.4 (1999) larvae per 10 m^2 (Figure 7.15). In general, the average abundance of Carangidae larvae was typically higher along the continental shelf during the fall as compared to the spring in the open Gulf for both gear types. With the exception of 1985 and 1986, the average abundance of larvae for bongo net samples was within a similar range during the spring; however, average carangid larval abundances were highly variable from year to year during the spring for neuston samples and during the fall for both gear types from 1982 through 2007 (Figures 7.14 and 7.15).

During the spring, the majority of larvae captured were jacks (*Caranx*), while most of the jack and pompano larvae that were obtained during the fall were Atlantic bumper (*Chloroscombrus chrysurus*) and round scad (*Decapterus punctatus*) (Table 7.2). Jacks were distributed throughout the open Gulf of Mexico during the spring, as well as throughout most of the continental shelf during the fall (Figure 7.16). While Atlantic bumper larvae were distributed throughout the entire continental shelf during the fall, larvae were sparsely distributed throughout the Gulf during spring plankton surveys (Figure 7.17).

Family Clupeidae

As forage fish, herrings, shads, sardines, and menhadens are abundant coastal pelagic species that constitute an important, if not primary, food source for many predatory game and commercial fishes (Shaw and Drullinger 1990a). Most of the herring, shad, sardine, and

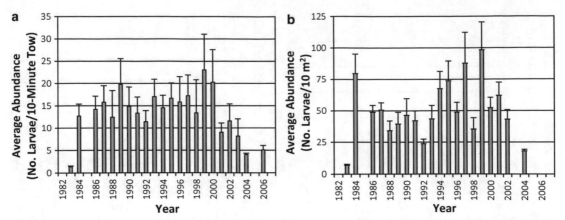

Figure 7.15. Average abundance of Carangidae for neuston net (a) and bongo net (b) samples for the SEAMAP fall plankton surveys from 1982 through 2007. Fall plankton surveys were not conducted in 1982, 1985, 2005, or 2007, and only neuston net sampling was conducted during fall plankton surveys in 2003 and 2006. Error bars = standard error.

menhaden larvae that were captured during the plankton surveys from 1982 through 2007 were captured during the fall along the continental shelf, and similar numbers were taken in both the neuston and bongo nets (Table 7.1). Thirteen taxa are included in this group, with the most larvae being scaled sardines (*Harengula jaguana*), round herring (*Etrumeus teres*), and Spanish sardine (*Sardinella aurita*) in the spring and Atlantic thread herring (*Opisthonema oglinum*), Spanish sardine, and scaled sardines in the fall (Table 7.2). While scaled sardines and Atlantic thread herring were distributed throughout the continental shelf during the fall, they were sparsely distributed throughout the open Gulf and Florida Shelf in the spring (Figures 7.18 and 7.19).

The Gulf menhaden (*Brevoortia patronus*) is one of the most abundant pelagic fishes in the northern coastal Gulf of Mexico; it is an exploited marine resource, the principal prey for many important commercial and recreational fish species, as well as marine birds and mammals. As both a planktivore and detritivore, Gulf menhaden are an integral and key component of the Gulf of Mexico ecosystem (Vaughan et al. 2011). However, because adults spawn primarily near the mouth of the Mississippi River during the winter and the plankton surveys were conducted in the spring and fall, only one Gulf menhaden juvenile was collected in the plankton surveys (Table 7.2).

From 1982 through 2007, average clupeid larval abundances were highly variable from year to year during the spring and fall for both gear types (Figures 7.20 and 7.21). For neuston net samples, the average abundance of Clupeidae larvae ranged from 0 (2004) to 68.5 (1983) larvae per 10-min tow in the spring in the open Gulf and from 1.3 (1983) to 97.4 (1993) larvae per 10-min tow during the fall along the continental shelf (Figures 7.20 and 7.21). Average clupeid larval abundance for bongo net samples ranged from 11.7 (1982) to 247.2 (1999) larvae per 10 m^2 and from 6.7 (1983) to 225.9 (1995) larvae per 10 m^2 during spring and fall plankton surveys, respectively (Figures 7.20 and 7.21). For both gear types, the average abundance of Clupeidae larvae was typically higher along the continental shelf during the fall, compared to spring in the open Gulf of Mexico.

Family Coryphaenidae

Dolphinfishes (sometimes referred to as mahi mahi or dorado) are an important commercial and recreational species distributed throughout the tropical and subtropical seas of the world and are highly prized for food (Ditty et al. 1994). They are often associated with

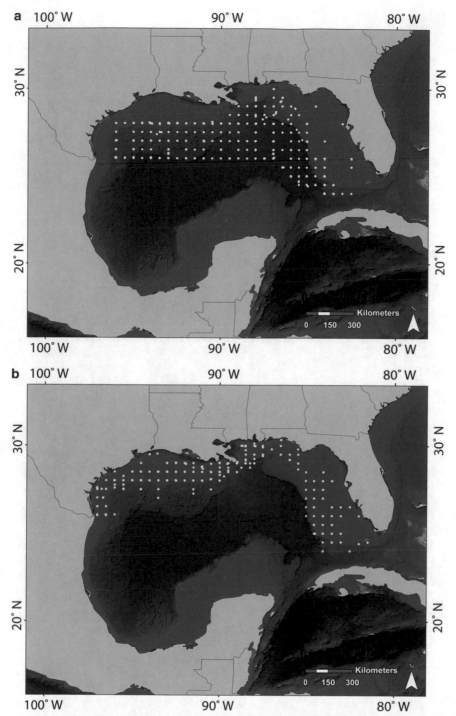

Figure 7.16. Distribution of jack (*Caranx*) larvae during the SEAMAP spring (a) and fall (b) plankton surveys from 1982 through 2007. Spring plankton surveys were not conducted in 2000, 2005, or 2006, and only bongo net sampling was conducted during the spring plankton survey in 1982. Fall plankton surveys were not conducted in 1982, 1985, 2005, or 2007, and only neuston net sampling was conducted during fall plankton surveys in 2003 and 2006.

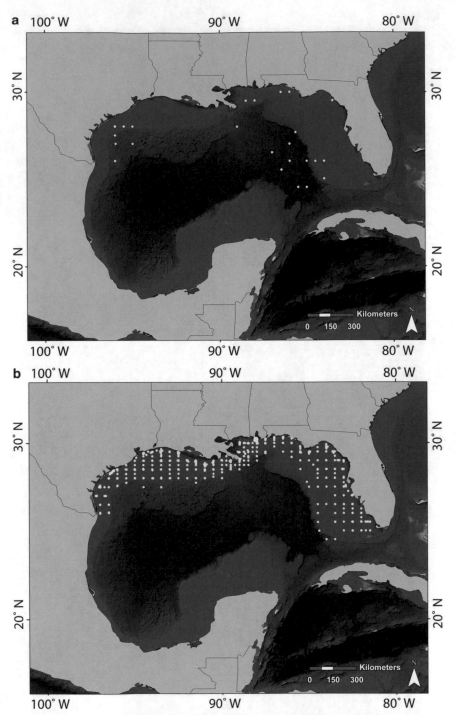

Figure 7.17. Distribution of Atlantic bumper (*Chloroscombrus chrysurus*) larvae during the SEAMAP spring (a) and fall (b) plankton surveys from 1982 through 2007. Spring plankton surveys were not conducted in 2000, 2005, or 2006, and only bongo net sampling was conducted during the spring plankton survey in 1982. Fall plankton surveys were not conducted in 1982, 1985, 2005, or 2007, and only neuston net sampling was conducted during fall plankton surveys in 2003 and 2006.

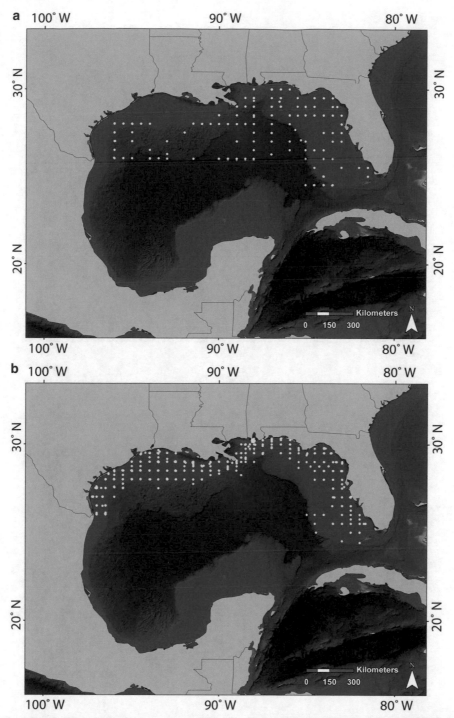

Figure 7.18. Distribution of scaled sardines (*Harengula jaguana*) larvae during the SEAMAP spring (a) and fall (b) plankton surveys from 1982 through 2007. Spring plankton surveys were not conducted in 2000, 2005, or 2006, and only bongo net sampling was conducted during the spring plankton survey in 1982. Fall plankton surveys were not conducted in 1982, 1985, 2005, or 2007, and only neuston net sampling was conducted during fall plankton surveys in 2003 and 2006.

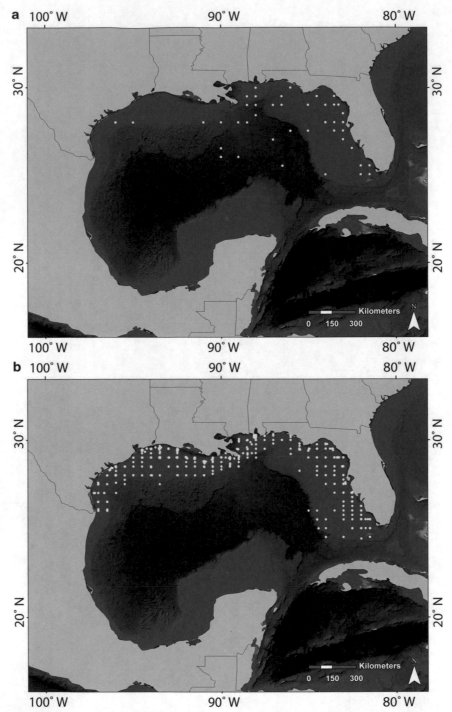

Figure 7.19. Distribution of Atlantic thread herring (*Opisthonema oglinum*) larvae during the SEA-MAP spring (a) and fall (b) plankton surveys from 1982 through 2007. Spring plankton surveys were not conducted in 2000, 2005, or 2006, and only bongo net sampling was conducted during the spring plankton survey in 1982. Fall plankton surveys were not conducted in 1982, 1985, 2005, or 2007, and only neuston net sampling was conducted during fall plankton surveys in 2003 and 2006.

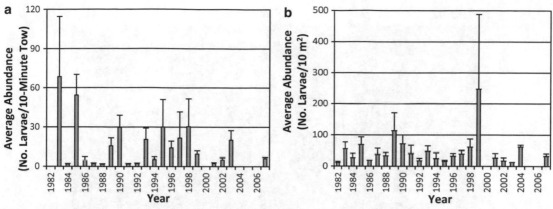

Figure 7.20. Average abundance of Clupeidae for neuston net (a) and bongo net (b) samples for the SEAMAP spring plankton surveys from 1982 through 2007. Spring plankton surveys were not conducted in 2000, 2005, or 2006, and only bongo net sampling was conducted during the spring plankton survey in 1982. Error bars = standard error.

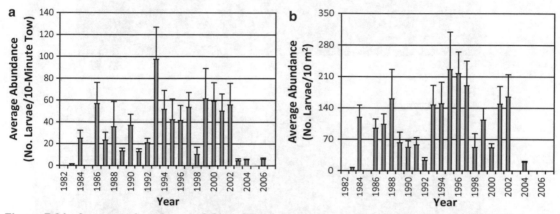

Figure 7.21. Average abundance of Clupeidae for neuston net (a) and bongo net (b) samples for the SEAMAP fall plankton surveys from 1982 through 2007. Fall plankton surveys were not conducted in 1982, 1985, 2005, or 2007, and only neuston net sampling was conducted during fall plankton surveys in 2003 and 2006. Error bars = standard error.

Sargassum spp. or other floating objects. One of the fastest growing species in the ocean, dolphinfish, serves as a primary food source for many pelagic predators (Palko et al. 1982). This group includes four taxa (Table 7.2). Most dolphinfish larvae were taken in the spring in the open Gulf of Mexico and at the water surface in neuston nets (Table 7.1). Dolphinfish were fairly well distributed throughout sampling stations during both spring and fall plankton surveys conducted from 1982 through 2007 (Figure 7.22). During the spring, as well as during the fall along the continental shelf, average abundances of dolphinfish larvae occurred at low densities and typically ranged from 1 to 3 larvae per 10-min neuston tow (Figures 7.23 and 7.24). For bongo net samples, the average abundance of coryphaenid larvae generally ranged from 5 to 9 larvae per 10 m² (Figures 7.23 and 7.24). In the spring in the open Gulf of Mexico, the highest average abundance of larval dolphinfish for samples collected by bongo net occurred in 2007, while the highest average abundance occurred in 1998 during the fall along the continental shelf (Figures 7.23 and 7.24).

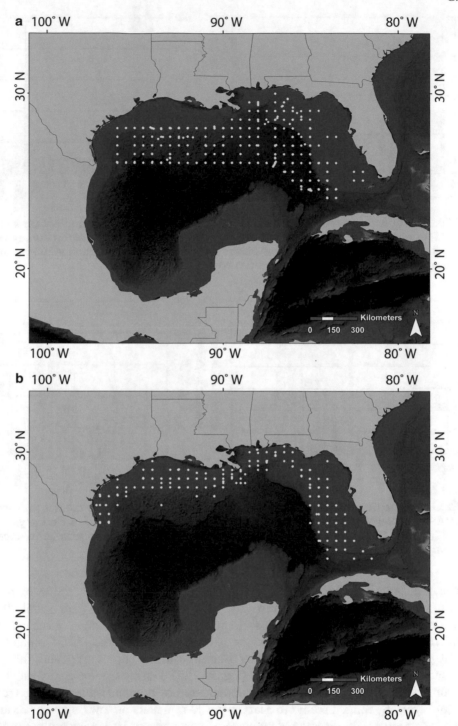

Figure 7.22. Distribution of dolphinfish (Coryphaenidae) larvae during the SEAMAP spring (a) and fall (b) plankton surveys from 1982 through 2007. Spring plankton surveys were not conducted in 2000, 2005, or 2006, and only bongo net sampling was conducted during the spring plankton survey in 1982. Fall plankton surveys were not conducted in 1982, 1985, 2005, or 2007, and only neuston net sampling was conducted during fall plankton surveys in 2003 and 2006.

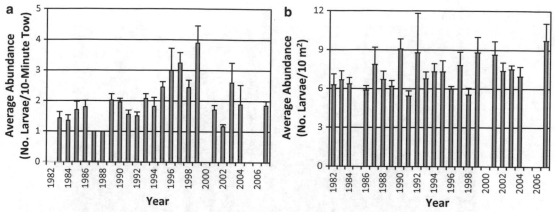

Figure 7.23. Average abundance of dolphinfish (Coryphaenidae) for neuston net (a) and bongo net (b) samples for the SEAMAP spring plankton surveys from 1982 through 2007. Spring plankton surveys were not conducted in 2000, 2005, or 2006, and only bongo net sampling was conducted during the spring plankton survey in 1982. Error bars = standard error.

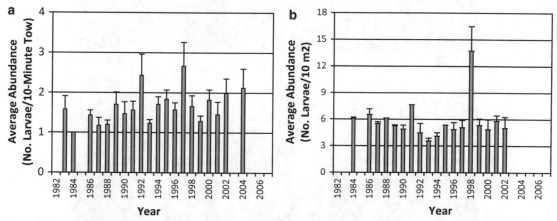

Figure 7.24. Average abundance of dolphinfish (Coryphaenidae) for neuston net (a) and bongo net (b) samples for the SEAMAP fall plankton surveys from 1982 through 2007. Fall plankton surveys were not conducted in 1982, 1985, 2005, or 2007, and only neuston net sampling was conducted during fall plankton surveys in 2003 and 2006. Error bars = standard error.

Family Istiophoridae

Billfish, marlin, and sailfish are highly migratory across vast expanses of open ocean; therefore, not much is known about their life histories, especially the larval stages (Tidwell et al. 2007). Billfish support a sport fishery worth hundreds of millions of dollars each year, and as top predators play a critical role in all pelagic ecosystems (Tidwell et al. 2007; Rooker et al. 2012).

Most larval marlin and sailfish were taken in the spring in the open Gulf and at the water's surface, and seven taxa were included in this group (Tables 7.1 and 7.2). Though they did not occur at all sampling stations, billfish were fairly well represented during spring and fall plankton surveys (Figure 7.25).

Average abundances of billfish larvae for neuston net samples typically ranged from 1 to 4 larvae per 10-min tow during both the spring and fall, indicating similar surface densities in both the open Gulf and continental shelf (Figures 7.26 and 7.27). For neuston net samples from

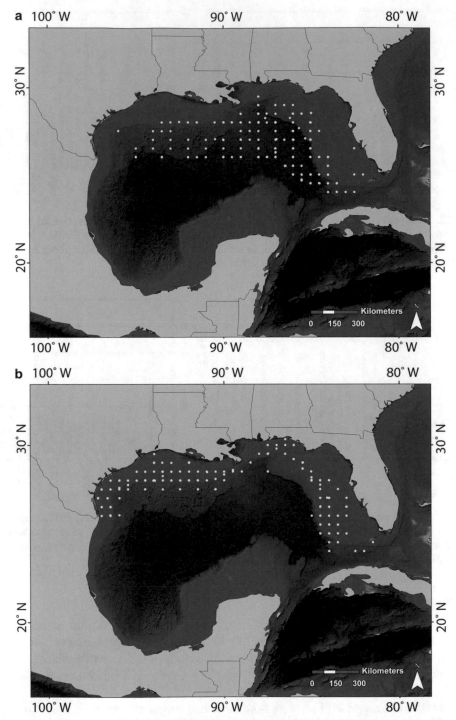

Figure 7.25. Distribution of billfish (Istiophoridae) larvae during the SEAMAP spring (a) and fall (b) plankton surveys from 1982 through 2007. Spring plankton surveys were not conducted in 2000, 2005, or 2006, and only bongo net sampling was conducted during the spring plankton survey in 1982. Fall plankton surveys were not conducted in 1982, 1985, 2005, or 2007, and only neuston net sampling was conducted during fall plankton surveys in 2003 and 2006.

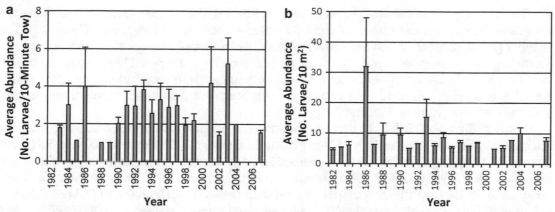

Figure 7.26. Average abundance of billfish (Istiophoridae) for neuston net (a) and bongo net (b) samples for the SEAMAP spring plankton surveys from 1982 through 2007. Spring plankton surveys were not conducted in 2000, 2005, or 2006, and only bongo net sampling was conducted during the spring plankton survey in 1982. Error bars = standard error.

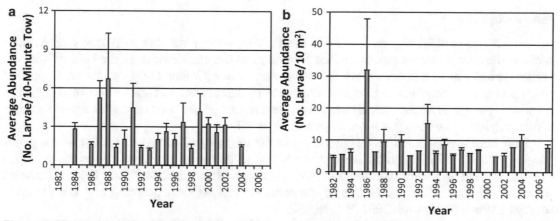

Figure 7.27. Average abundance of billfish (Istiophoridae) for neuston net (a) and bongo net (b) samples for the SEAMAP fall plankton surveys from 1982 through 2007. Fall plankton surveys were not conducted in 1982, 1985, 2005, or 2007, and only neuston net sampling was conducted during fall plankton surveys in 2003 and 2006. Error bars = standard error.

1982 through 2007, the highest average abundance of billfish larvae occurred in 2003 during spring surveys and in 1998 during fall surveys. The average abundance of billfish larvae was typically higher during the spring on the open Gulf as compared to along the continental shelf during the fall for bongo net samples (Figures 7.19 and 7.20). The highest abundance for all spring and fall bongo net samples, more than 30 larvae per 10 m^2, occurred in 1986.

Family Lutjanidae

The snapper family includes mostly reef-associated species, as well as several deepwater species; due to the excellent quality of its meat, snappers are of significant importance to the commercial and recreational fisheries in the Gulf of Mexico, and many species are overfished (Martinez-Andrade 2003). For example, red snapper (*Lutjanus campechanus*) is one of the most important food fishes in the Gulf of Mexico, and this fishery, which collapsed in the eastern Gulf of Mexico in the late 1980s, is the most controversial fishery in the U.S. Gulf of Mexico (Johnson et al. 2009; Cowan et al. 2010).

The majority of larval snapper were captured during the fall plankton surveys along the continental shelf, and most were taken in the water column in the bongo nets (Table 7.1). The snapper group includes 12 taxa, with most of the larvae captured during both the spring and fall consisting of the snapper family (Lutjanidae) and vermillion snapper (*Rhomboplites auroru-bens*) (Table 7.2). Both Lutjanidae and vermillion snapper were found along the entire continental shelf during fall plankton surveys, and they were not distributed widely during spring plankton surveys (Figures 7.28 and 7.29).

From 1982 through 2007, the average abundance of lutjanid larvae collected by neuston net during the spring in the open Gulf of Mexico ranged from 0 (1983, 1988, and 2004) to 6.8 (1990) larvae per 10-min tow, while the average abundance of snapper larvae collected by bongo net ranged from 4 (1982) to 22.7 (1986) larvae per 10 m^2 (Figure 7.30). Snapper larvae average abundance during the fall along the continental shelf ranged from 0 (2006) to 12.4 (1987) larvae per 10-min tow for neuston net samples, and bongo net samples ranged from 5.6 (1983) to 24.2 (2001) larvae per 10 m^2 (Figure 7.31). For both gear types, the average abundance of snapper larvae was typically higher along the continental shelf during the fall as compared to the spring in the open Gulf (Figures 7.30 and 7.31).

Family Mugilidae

Mullet are ecologically important in the flow of energy through estuarine communities because they are primary consumers that feed on plankton and detritus. In the Gulf of Mexico, mullet typically spawn many miles offshore in deep water (Collins 1985). Mullet are important prey species for many fish and are also important to the recreational and commercial fisheries. As silvery pelagic juveniles, mullet inhabit surface waters of the open ocean for several months before migrating inshore (Lyczkowski-Shultz et al. 2004).

Four taxa are included in this group, with the majority of larvae identified to the genus *Mugil* (Table 7.2). Most mullet larvae were taken in the spring plankton surveys in the open Gulf of Mexico and were captured in the neuston nets at the surface (Table 7.1). Mullet were found in the open Gulf, as well as in the continental shelf during spring and fall plankton surveys from 1982 through 2007 (Figure 7.32).

Average abundances of larval mullet were highly variable from year to year during spring and fall for both gear types from 1982 through 2007 (Figures 7.33 and 7.34). For neuston net samples, the average abundance ranged from 0.97 (1985) to 23.8 (1999) per 10-min tow in the spring in the open Gulf and from 0 (1983, 2003, and 2006) to 15.3 (1984) larvae per 10-min tow during the fall along the continental shelf (Figures 7.33 and 7.34). Average mugilid larval abundance for bongo net samples ranged from 2.8 (1983) to 24.1 (1986) per 10 m^2 and from 0 (1983, 1986, 1998, and 2002) to 18.6 (2004) larvae per 10 m^2 during spring and fall plankton surveys, respectively (Figures 7.33 and 7.34). The average abundance of mullet larvae was typically higher during the spring in the open Gulf, compared to fall along the continental shelf for both gear types.

Family Sciaenidae

Members of the Family Sciaenidae (drums and croakers) are an important sport and commercial fishery resource along the U.S. Gulf of Mexico and are perhaps the most prominent group of northern Gulf inshore fishes (Cowan and Shaw 1988). This group includes 15 taxa, with most of the larvae consisting of the drum and croaker family (Sciaenidae), with the kingfish genus (*Menticirrhus*) in the spring and the kingfish genus and redfish in the fall (Table 7.2). The vast majority of larval drum and croaker were found during fall plankton surveys, with the

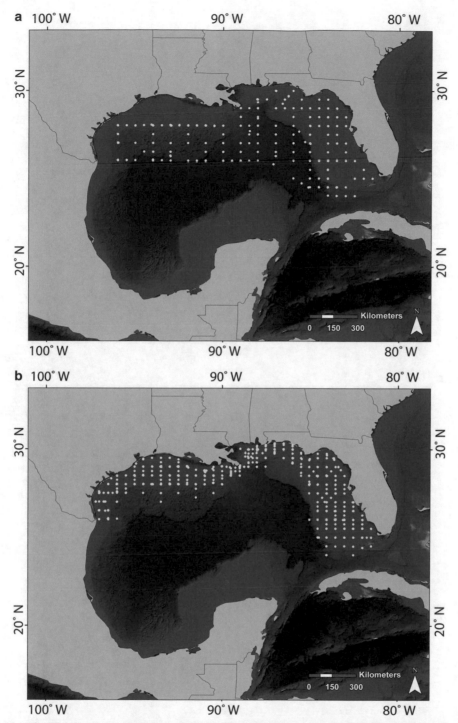

Figure 7.28. Distribution of snapper (Lutjanidae) larvae during the SEAMAP spring (a) and fall (b) plankton surveys from 1982 through 2007. Spring plankton surveys were not conducted in 2000, 2005, or 2006, and only bongo net sampling was conducted during the spring plankton survey in 1982. Fall plankton surveys were not conducted in 1982, 1985, 2005, or 2007, and only neuston net sampling was conducted during fall plankton surveys in 2003 and 2006.

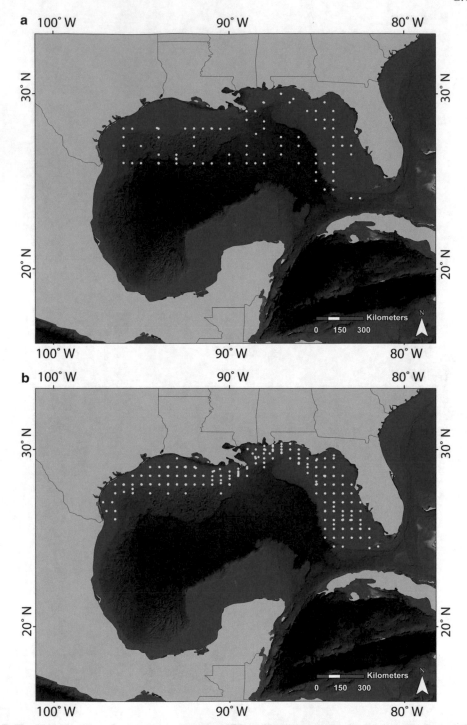

Figure 7.29. Distribution of vermilion snapper (*Rhomboplites aurorubens*) larvae during the SEA-MAP spring (a) and fall (b) plankton surveys from 1982 through 2007. Spring plankton surveys were not conducted in 2000, 2005, or 2006, and only bongo net sampling was conducted during the spring plankton survey in 1982. Fall plankton surveys were not conducted in 1982, 1985, 2005, or 2007, and only neuston net sampling was conducted during fall plankton surveys in 2003 and 2006.

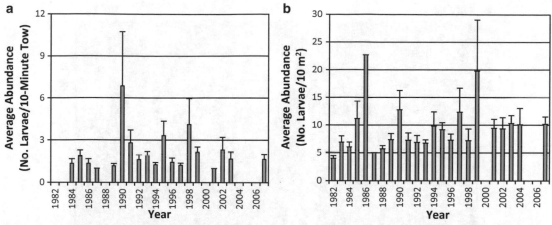

Figure 7.30. Average abundance of snapper (Lutjanidae) for neuston net (a) and bongo net (b) samples for the SEAMAP spring plankton surveys from 1982 through 2007. Spring plankton surveys were not conducted in 2000, 2005, or 2006, and only bongo net sampling was conducted during the spring plankton survey in 1982. Error bars = standard error.

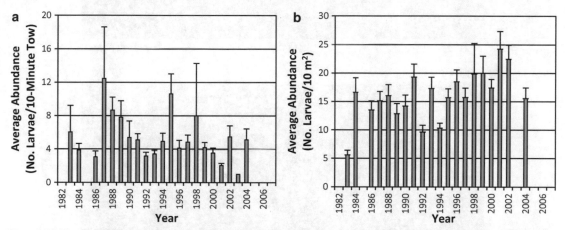

Figure 7.31. Average abundance of snapper (Lutjanidae) for neuston net (a) and bongo net (b) samples for the SEAMAP fall plankton surveys from 1982 through 2007. Fall plankton surveys were not conducted in 1982, 1985, 2005, or 2007, and only neuston net sampling was conducted during fall plankton surveys in 2003 and 2006. Error bars = standard error.

most in the water column in the bongo net samples (Table 7.1). The drum family (Sciaenidae) larvae were more extensive along the continental shelf during the fall compared to the spring (Figure 7.35); the seasonal distribution of this group was even more dramatic for the redfish (Figure 7.36).

From 1982 through 2007, with the exception of 1991 and 1995, the average abundance of sciaenid larvae collected by neuston net during the spring in the open Gulf of Mexico was fewer than 6 larvae per 10-min tow, and drum and croaker larval abundance averaged fewer than 20 larvae per 10 m^2 for bongo net samples during the spring, with the exception of 1986, when the average larval abundance was more than 120 larvae per 10 m^2 (Figure 7.36). Drum and croaker average abundance ranged from 1.3 (2006) to 32.2 (2000) larvae per 10-min tow for neuston net samples, and bongo net samples ranged from 4.8 (1983) to 208 (1988) larvae per 10 m^2 during the fall (Figure 7.38). For both gear types, the average abundance of sciaenid larvae was typically much higher along the continental shelf during the fall compared to the spring in the open Gulf (Figures 7.37 and 7.38).

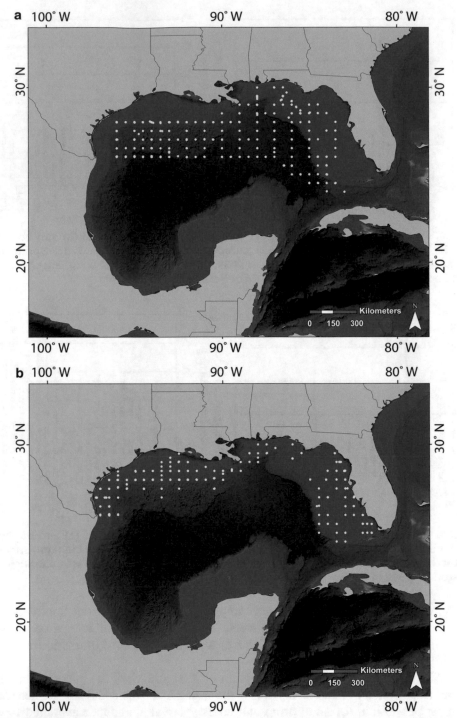

Figure 7.32. Distribution of mullet (Mugilidae) larvae during the SEAMAP spring (a) and fall (b) plankton surveys from 1982 through 2007. Spring plankton surveys were not conducted in 2000, 2005, or 2006, and only bongo net sampling was conducted during the spring plankton survey in 1982. Fall plankton surveys were not conducted in 1982, 1985, 2005, or 2007, and only neuston net sampling was conducted during fall plankton surveys in 2003 and 2006.

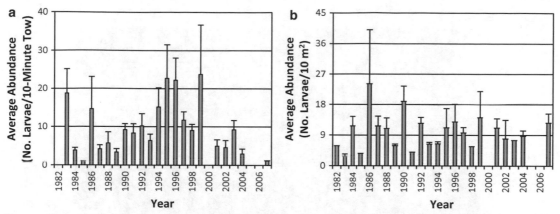

Figure 7.33. Average abundance of mullet (Mugilidae) for neuston net (a) and bongo net (b) samples for the SEAMAP spring plankton surveys from 1982 through 2007. Spring plankton surveys were not conducted in 2000, 2005, or 2006, and only bongo net sampling was conducted during the spring plankton survey in 1982. Error bars = standard error.

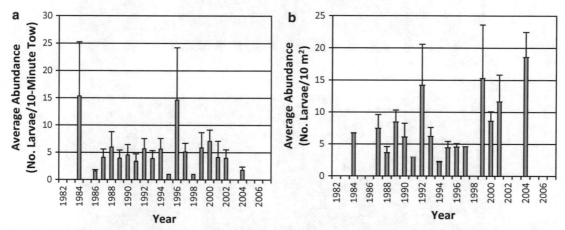

Figure 7.34. Average abundance of mullet (Mugilidae) for neuston net (a) and bongo net (b) samples for the SEAMAP fall plankton surveys from 1982 through 2007. Fall plankton surveys were not conducted in 1982, 1985, 2005, or 2007, and only neuston net sampling was conducted during fall plankton surveys in 2003 and 2006. Error bars = standard error.

Redfish larvae were concentrated higher in the water column during daylight hours than at night in the general area east of the Mississippi Delta and south of the Mississippi barrier island over the East Louisiana–Mississippi–Alabama shelf in September and October 1984 and 1985 (Lyczkowski-Shultz and Steen 1991). In addition, there was no clear relationship between vertical aggregation of red drum larvae and temperature or salinity profiles or microzooplankton prey distribution. Atlantic croaker (*Micropogonias undulatus*) larvae were found to be least abundant in surface waters at night, and the highest abundances at night were observed at the deepest depths sampled during an investigation conducted in inner-shelf waters off Mississippi during September and October 1984 and 1985 (Comyns and Lyczkowski-Schultz 2004). By midmorning, Atlantic croaker larvae had moved up the water column, and highest abundances were usually found at 5 m (16.4 ft); no consistent pattern was found in the vertical stratification of Atlantic croaker larvae during the midday or afternoon.

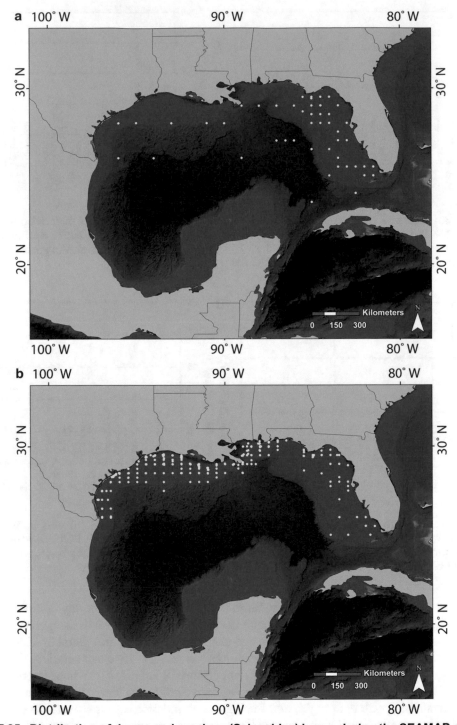

Figure 7.35. Distribution of drums and croakers (Sciaenidae) larvae during the SEAMAP spring (a) and fall (b) plankton surveys from 1982 through 2007. Spring plankton surveys were not conducted in 2000, 2005, or 2006, and only bongo net sampling was conducted during the spring plankton survey in 1982. Fall plankton surveys were not conducted in 1982, 1985, 2005, or 2007, and only neuston net sampling was conducted during fall plankton surveys in 2003 and 2006.

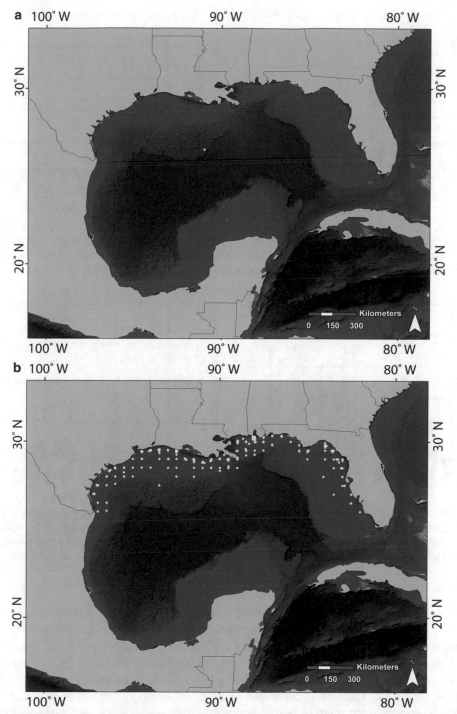

Figure 7.36. Distribution of redfish (*Sciaenops ocellatus*) larvae during the SEAMAP spring (a) and fall (b) plankton surveys from 1982 through 2007. Spring plankton surveys were not conducted in 2000, 2005, or 2006, and only bongo net sampling was conducted during the spring plankton survey in 1982. Fall plankton surveys were not conducted in 1982, 1985, 2005, or 2007, and only neuston net sampling was conducted during fall plankton surveys in 2003 and 2006.

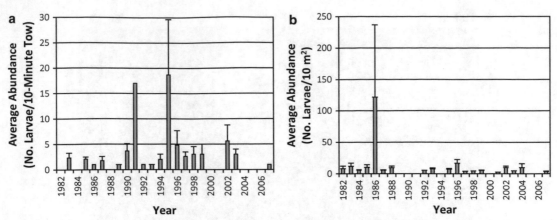

Figure 7.37. Average abundance of drums and croakers (Sciaenidae) for neuston net (a) and bongo net (b) samples for the SEAMAP spring plankton surveys from 1982 through 2007. Spring plankton surveys were not conducted in 2000, 2005, or 2006, and only bongo net sampling was conducted during the spring plankton survey in 1982. Error bars = standard error.

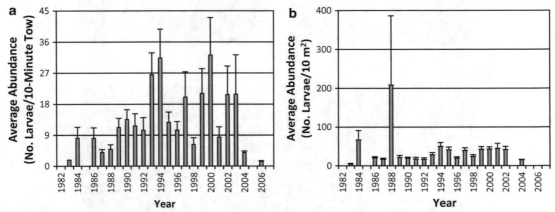

Figure 7.38. Average abundance of drums and croakers (Sciaenidae) for neuston net (a) and bongo net (b) samples for the SEAMAP fall plankton surveys from 1982 through 2007. Fall plankton surveys were not conducted in 1982, 1985, 2005, or 2007, and only neuston net sampling was conducted during fall plankton surveys in 2003 and 2006. Error bars = standard error.

Family Scombridae

Mackerels, tunas (with the exception of *Thunnus*, discussed in the section below), and bonitos are important recreational and commercial fish species. For example, king mackerel (*Scomberomorus cavalla*) and Spanish mackerel (*Scomberomorus maculatus*), which are abundant and highly migratory, are coastal members of the Scombridae family and support large commercial and recreational fisheries (De Vries et al. 1990). This group includes 13 taxa; the majority of larvae for this group consisted of skipjack tuna (*Katsuwonus pelamis*) and tuna (*Auxis*) in the spring in the open Gulf of Mexico and little tunny (*Euthynnus alletteratus*) in the fall along the continental shelf (Table 7.2). Mackerel, tuna, and bonito larvae were typically taken in the fall and in bongo net samples of the water column (Table 7.1). Skipjack tuna were distributed throughout the open Gulf of Mexico during the spring. During the fall, they occurred in locations along the near edge of the continental shelf (Figure 7.39). Little tunny were found at some locations during spring plankton surveys; however, they were densely distributed throughout the entire continental shelf during the fall (Figure 7.40).

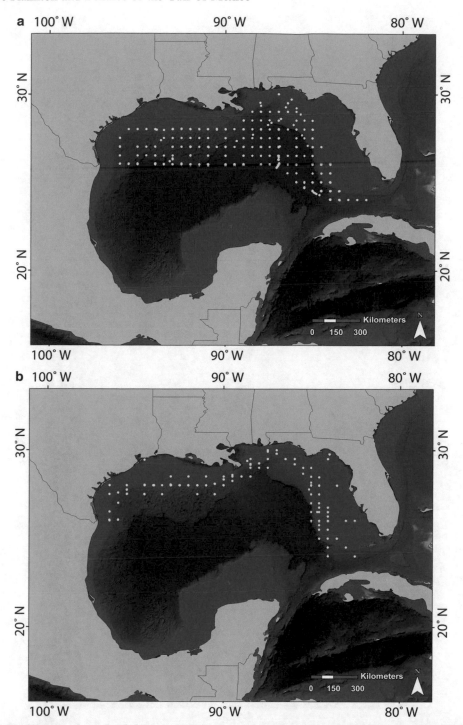

Figure 7.39. Distribution of skipjack tuna (*Katsuwonus pelamis*) larvae during the SEAMAP spring (a) and fall (b) plankton surveys from 1982 through 2007. Spring plankton surveys were not conducted in 2000, 2005, or 2006, and only bongo net sampling was conducted during the spring plankton survey in 1982. Fall plankton surveys were not conducted in 1982, 1985, 2005, or 2007, and only neuston net sampling was conducted during fall plankton surveys in 2003 and 2006.

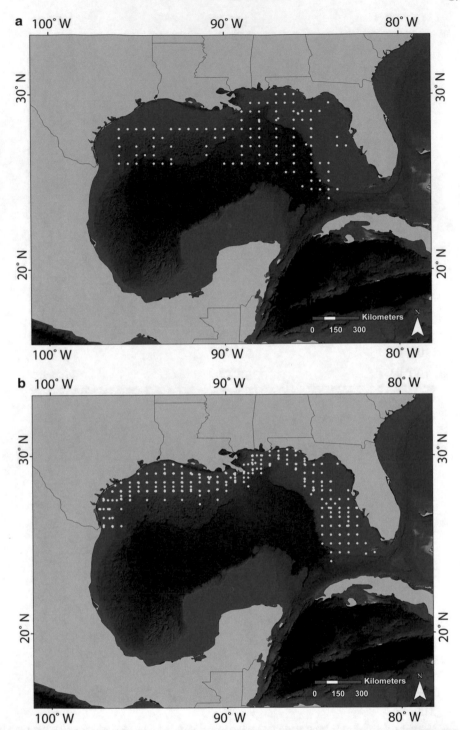

Figure 7.40. Distribution of little tunny (*Euthynnus alletteratus*) larvae during the SEAMAP spring (a) and fall (b) plankton surveys from 1982 through 2007. Spring plankton surveys were not conducted in 2000, 2005, or 2006, and only bongo net sampling was conducted during the spring plankton survey in 1982. Fall plankton surveys were not conducted in 1982, 1985, 2005, or 2007, and only neuston net sampling was conducted during fall plankton surveys in 2003 and 2006.

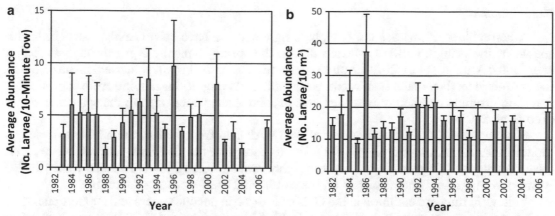

Figure 7.41. Average abundance of Scombridae for neuston net (a) and bongo net (b) samples for the SEAMAP spring plankton surveys from 1982 through 2007. Spring plankton surveys were not conducted in 2000, 2005, or 2006, and only bongo net sampling was conducted during the spring plankton survey in 1982. Error bars = standard error.

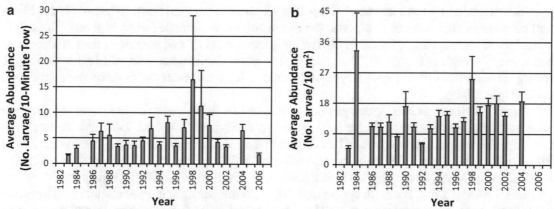

Figure 7.42. Average abundance of Scombridae for neuston net (a) and bongo net (b) samples for the SEAMAP fall plankton surveys from 1982 through 2007. Fall plankton surveys were not conducted in 1982, 1985, 2005, or 2007, and only neuston net sampling was conducted during fall plankton surveys in 2003 and 2006. Error bars = standard error.

Average scombrid larval abundances were highly variable from year to year during the spring and fall (Figures 7.41 and 7.42) for both gear types from 1982 through 2007. The average abundance of Scombridae larvae collected by neuston net during the spring ranged from 1.7 (1988) to 9.7 (1996) per 10-min tow, and the average abundance of scombrids collected by bongo net ranged from 8.7 (1985) to 37.4 (1986) larvae per 10 m^2 (Figure 7.41). During the fall, scombrid average abundance along the continental shelf ranged from 0 (2003) to 16.5 (1998) larvae per 10-min tow for neuston net samples, while bongo net samples ranged from 4.7 (1983) to 33.4 (1984) larvae per 10 m^2 (Figure 7.42). The mackerel, tuna, and bonito larvae average abundances were within a similar range during the spring in the open Gulf and during the fall along the continental shelf for both gear types (Figures 7.41 and 7.42).

Genus Thunnus

Atlantic bluefin tuna are large, highly migratory and have been heavily overfished. They spawn in the pelagic Gulf of Mexico during the spring, typically April through June (Teo et al. 2007; Muhling et al. 2010). Adult bluefin tuna have the broadest thermal niche of any of the Scombridae; they make fast, ocean basin-wide scale migrations ranging from cool subpolar foraging grounds to discrete breeding sites in subtropical waters during the spawning season (Teo et al. 2007).

Muhling et al. (2010) used a subset of SEAMAP data from 1982 through 2006 to develop a model of suitable Atlantic bluefin tuna larvae habitat in the northern Gulf of Mexico. The location and size of favorable habitat was highly variable among years. Habitats within the LC, warm-core rings, and cooler waters on the continental shelf were less favorable.

Yellowfin tuna are common in the Gulf of Mexico in pelagic waters and support one of the most valuable commercial fisheries in the Gulf of Mexico (Lang et al. 1994). Lang et al. (1994) determined that significant spawning of yellowfin tuna most likely occurred in the northern Gulf of Mexico in the vicinity of the Mississippi River discharge plume, when 801 larvae were collected during July and September 1987, and enhanced yellowfin tuna larval growth and survival occurred in the plume frontal waters.

Identification of tuna larvae of the genus *Thunnus* is very difficult (Richards et al. 1990), and because of this most of the larvae for this group were identified only to genus (Table 7.2). Tuna larvae were typically found in the spring in the open Gulf and usually at the surface in the neuston net samples (Table 7.1). They occurred throughout the open Gulf of Mexico during the spring. During the fall, they were typically found at locations near the edge of the continental shelf (Figure 7.43).

From 1982 through 2007, average abundances of tuna larvae were highly variable from year to year during spring and fall for both gear types (Figures 7.44 and 7.45). For neuston net samples, the annual abundances of larval tuna averaged fewer than 8 larvae per 10-min tow in the spring in the open Gulf, with the exception of 1985. During the fall along the continental shelf, average abundance ranged from 0 (2003 and 2006) to 16.6 (1987) larvae per 10-min tow (Figures 7.44 and 7.45). For both spring and fall plankton surveys, annual abundances of larval tuna for bongo net samples were within a similar range and typically averaged fewer than 25 larvae per 10 m^2 (Figures 7.44 and 7.45).

Family Serranidae

Twenty-eight taxa are included in this group of seabasses and groupers, with the majority of larvae identified to the seabass family (Table 7.2). Adult grouper are a commercially and recreationally important species that are highly susceptible to overfishing, largely due to their spawning behavior and slow growth (Marancik et al. 2012).

While fairly similar numbers of seabasses and groupers were captured during the spring in the open Gulf of Mexico and during the fall along the continental shelf, most larvae were collected from the water column using bongo nets (Table 7.1). In addition, seabasses and groupers were distributed throughout the open Gulf as well as the continental shelf during spring and fall plankton surveys from 1982 through 2007 (Figure 7.46).

With few exceptions (1987 and 1988 during the spring and 1986, 1990, and 1993 during the fall), annual abundances for larval serranids averaged fewer than 6 per 10-min tow for spring and fall neuston net samples from 1982 through 2007 (Figures 7.47 and 7.48). Average serranid larval abundance for bongo net samples ranged from 9.4 (2007) to 49.4 (1994) per 10 m^2 and from 7.1 (1983) to 32.4 (1984) per 10 m^2 during spring and fall plankton surveys, respectively

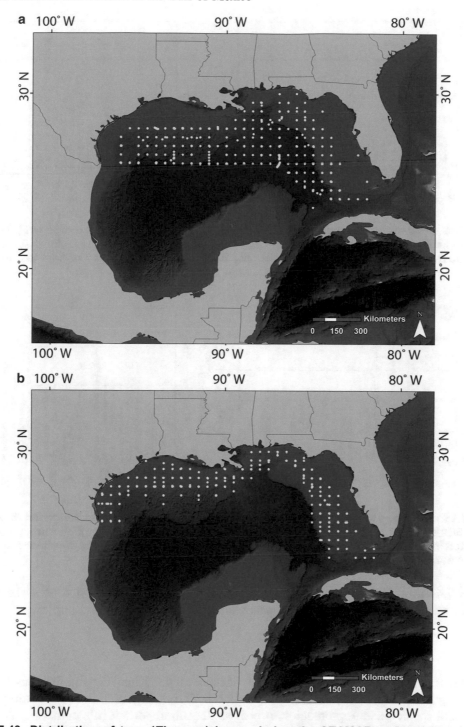

Figure 7.43. Distribution of tuna (*Thunnus*) larvae during the SEAMAP spring (a) and fall (b) plankton surveys from 1982 through 2007. Spring plankton surveys were not conducted in 2000, 2005, or 2006, and only bongo net sampling was conducted during the spring plankton survey in 1982. Fall plankton surveys were not conducted in 1982, 1985, 2005, or 2007, and only neuston net sampling was conducted during fall plankton surveys in 2003 and 2006.

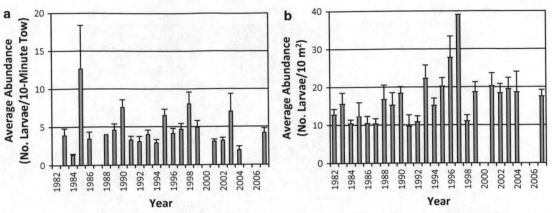

Figure 7.44. Average abundance of tuna (*Thunnus*) for neuston net (a) and bongo net (b) samples for the SEAMAP spring plankton surveys from 1982 through 2007. Spring plankton surveys were not conducted in 2000, 2005, or 2006, and only bongo net sampling was conducted during the spring plankton survey in 1982. Error bars = standard error.

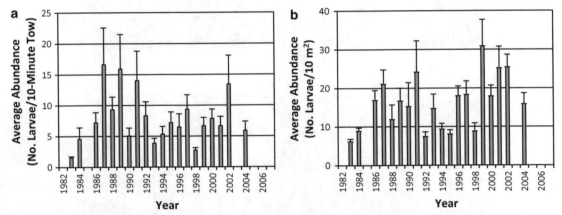

Figure 7.45. Average abundance of tuna (*Thunnus*) for neuston net (a) and bongo net (b) samples for the SEAMAP fall plankton surveys from 1982 through 2007. Fall plankton surveys were not conducted in 1982, 1985, 2005, or 2007, and only neuston net sampling was conducted during fall plankton surveys in 2003 and 2006. Error bars = standard error.

(Figures 7.47 and 7.48). For both spring and fall, the average abundance of seabasses and groupers was higher for bongo net samples than it was for neuston net samples.

Family Xiphiidae

Swordfish (*Xiphias gladius*) is the only species of the Xiphiidae family. This billfish is highly migratory and large, and while overfished, it has high value as a commercial and recreational species; as a top predator, swordfish play an important role in marine ecosystems (Rooker et al. 2012).

Swordfish larvae made up a very small percentage of the total ichthyoplankton catch; most were captured during the spring in the open Gulf at the water surface in neuston nets (Table 7.1). From 1982 through 2007, low numbers of swordfish larvae were collected by neuston net during the spring in the open Gulf, with average abundances ranging from 0 to 2.1 larvae per 10-min tow (Figure 7.49). Swordfish larvae were distributed sparsely throughout the open Gulf of

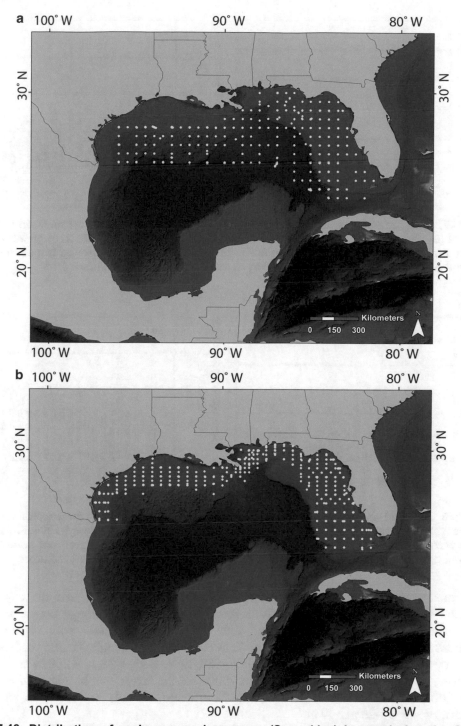

Figure 7.46. Distribution of seabasses and groupers (Serranidae) larvae during the SEAMAP spring (a) and fall (b) plankton surveys from 1982 through 2007. Spring plankton surveys were not conducted in 2000, 2005, or 2006, and only bongo net sampling was conducted during the spring plankton survey in 1982. Fall plankton surveys were not conducted in 1982, 1985, 2005, or 2007, and only neuston net sampling was conducted during fall plankton surveys in 2003 and 2006.

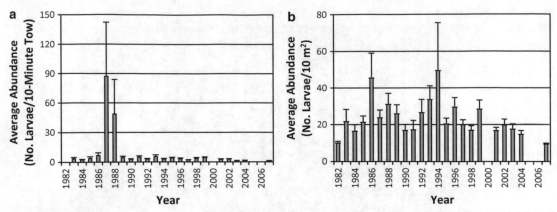

Figure 7.47. Average abundance of seabasses and groupers (Serranidae) for neuston net (a) and bongo net (b) samples for the SEAMAP spring plankton surveys from 1982 through 2007. Spring plankton surveys were not conducted in 2000, 2005, or 2006, and only bongo net sampling was conducted during the spring plankton survey in 1982. Error bars = standard error.

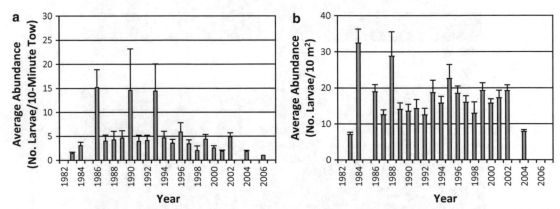

Figure 7.48. Average abundance of seabasses and groupers (Serranidae) for neuston net (a) and bongo net (b) samples for the SEAMAP fall plankton surveys from 1982 through 2007. Fall plankton surveys were not conducted in 1982, 1985, 2005, or 2007, and only neuston net sampling was conducted during fall plankton surveys in 2003 and 2006. Error bars = standard error.

Mexico during the spring. In the fall, they were occasionally found near the edge of the continental shelf (Figure 7.50). Swordfish larvae were collected by bongo net during spring plankton surveys in 1982, 1983, 1989, 1991, 1992, 1995, 1996, and 2004, with average annual abundances ranging from 0 to 7.6 larvae per 10 m^2. The average abundance of swordfish larvae collected by neuston net during the fall in 1986, 1988, 1989, 1995, 1998, and 2001 was 1 larva per 10-min tow, while the average abundance in 2000 was 2.3 larvae per 10-min tow. From 1982 through 2007, swordfish larvae were collected by bongo net only in 2001, with an average abundance of 4.9 larvae per 10 m^2.

7.4.1.3.2 Summary of SEAMAP Ichthyoplankton Database Information

The large SEAMAP database is intended to be a robust resource for fisheries stock assessments that could contribute to the management of Gulf of Mexico fisheries. It allows comparison of the distribution of larval and juvenile stages of a wide range of species from different habitats as adults. Surface-living juveniles from the neuston nets can be contrasted

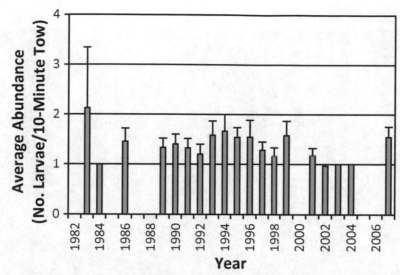

Figure 7.49. Average abundance of swordfish (Xiphiidae) for neuston net samples for the SEA-MAP spring plankton surveys from 1982 through 2007. Spring plankton surveys were not conducted in 2000, 2005, or 2006, and only bongo net sampling was conducted during the spring plankton survey in 1982. Error bars = standard error.

with those living throughout the water column caught with the bongo nets. Spawning season supposedly can be inferred from the season that a species appears in the ichthyoplankton. Yearly trends up or down can be inferred for each species and thus compared with variations in other species and with stock assessments of adults. However, the degree to which ichthyoplankton stock distributions are related to recruitment and adult stocks is a subject of considerable contentious debate (Haddon 2001).

The SEAMAP information does have problems. Determining trends in larval populations of the selected taxa from 1982 through 2007 is challenging because of the year-to-year variability in ichthyoplankton densities collected using both the neuston and bongo nets. In addition, comparing interannual variability is difficult because of the substantial differences in the temporal and spatial distribution of stations sampled each year under SEAMAP. The fall versus the spring sampling patterns are different for example, thus precluding seasonal comparisons. However, larval abundances appear to be stable or increasing for the majority of the selected taxa (e.g., Carangidae, Clupeidae, Coryphaenidae) that were summarized in the sections above. In addition, high densities of larvae occurred for many of the selected taxa (e.g., Carangidae, Clupeidae, Serranidae).

The SEAMAP sampling plan appears to have considered the entire EEZ as a monotypic habitat with little variation from place to place. That is, it is viewed as an LME. However, the habitats vary markedly over time and space, as reviewed in the initial section on habitat distributions. For example, what effect does the time-varying hypoxic zone off Louisiana have on ichthyoplankton distributions? How are ichthyoplankton partitioned between warm-core eddies and the cooler waters between them (Rooker et al. 2012)?

7.4.1.3.3 Baseline Ichthyoplankton Abundance and Distribution in Gulf of Mexico Regions

Various investigations have been conducted to determine the abundance and distribution of ichthyoplankton in specific regions of the Gulf of Mexico, and these are summarized below.

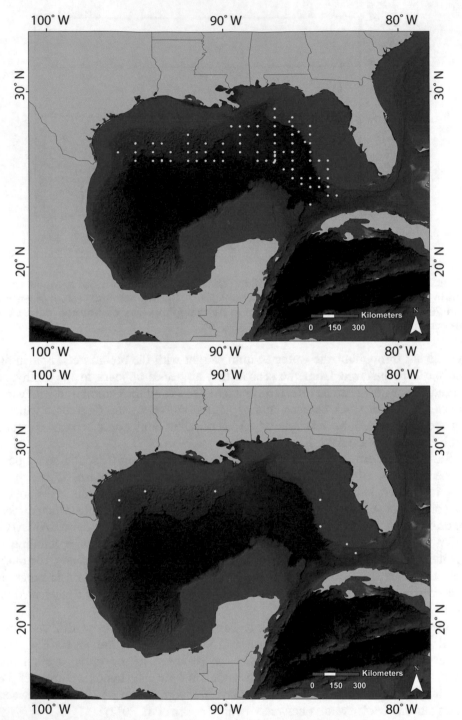

Figure 7.50. Distribution of swordfish (Xiphiidae) larvae during the SEAMAP spring (a) and fall (b) plankton surveys from 1982 through 2007. Spring plankton surveys were not conducted in 2000, 2005, or 2006, and only bongo net sampling was conducted during the spring plankton survey in 1982. Fall plankton surveys were not conducted in 1982, 1985, 2005, or 2007, and only neuston net sampling was conducted during fall plankton surveys in 2003 and 2006.

Northern Gulf of Mexico

The Gulf of Mexico continental shelf environment experiences seasonal changes in water temperature accompanied by discharges of low salinity, high nutrient water from rivers into the northern and eastern shelf areas. Using SEAMAP data, Muhling et al. (2012) characterized the spatial and temporal changes in abundances of larval fish assemblages on the northern Gulf of Mexico continental shelf from 1984 through 2008. Lanternfishes (Myctophidae) were the most common taxa collected and represented 14.65 % of the total collected ichthyoplankton, followed by codlets (Bregmacerotidae, 9.98 %) and gobies (Gobiidae, 9.29 %). Of the more than 500 taxa collected, the 20 most common fish families were evaluated. Larvae of some pelagic and mesopelagic families showed marked increases in abundance over the survey time period, while the abundances of some benthic fish families decreased (Muhling et al. 2012). Changes in fish assemblage structure were partially explained by changes in sea-surface temperature, as well as changes in the shrimp trawling effort. Interannual fish assemblage variability was also influenced by outflow from the Mississippi River. However, there was no explanation for spatial and temporal trends for many of the family groups (Muhling et al. 2012).

Carassou et al. (2012) investigated the spatial, seasonal, and depth-related structure of ichthyoplankton assemblages collected across a 77 km (47.8 mi) cross-shore gradient from March 2007 through December 2009 from highly productive estuarine waters to offshore oceanic waters on the Alabama shelf. A total of 350,766 larvae, in 17 orders and 70 families, were collected; the most common families were drums (Sciaenidae, approximately 42 % of total), followed by anchovies (Engraulidae, approximately 32 % of total). While the total density of fish larvae was significantly higher inshore, the number of families increased offshore. The total density of fish larvae also varied significantly among months, with the lowest values being observed in January and the highest in October and August. There were monthly variations in family richness, with minimum richness in December and maximum richness in May. Seven assemblages were associated with water masses characterized by distinct differences in temperature and salinity (Carassou et al. 2012). Families of larvae that were typically offshore included herrings, shads, sardines, and menhaden (Clupeidae); codlets (Bregmacerotidae); lizardfishes (Synodontidae); mackerels, tunas, and bonitos (Scombridae); and cusk-eels (Ophidiidae). Inshore families included anchovies (Engraulidae), gobies (Gobiidae), and clingfishes (Gobiesocidae). Larval fish assemblages varied seasonally and as a function of depth, but inshore and offshore assemblages remained clearly separated regardless of the season and depth considered; this strong and consistent structure was related to the combined effects of adult spawning behaviors and local oceanographic conditions, especially the influence of the Mobile River (Carassou et al. 2012).

Ichthyoplankton surveys were conducted in the northern Gulf of Mexico from 2006 through 2008 to determine the relative value of the region as early life habitat of sailfish (*Istiophorus platypterus*), blue marlin (*Makaira nigricans*), white marlin (*Kajikia albida*), and swordfish (*Xiphias gladius*) (Rooker et al. 2012). Sailfish were the dominant billfish collected in summer surveys, and larvae were present at 37.5 % of the stations sampled. Blue marlin and white marlin larvae were present at 25 % and 4.6 % of the stations sampled, respectively, and swordfish occurred at 17.2 % of the stations. Areas of peak production were detected and maximum density estimates for sailfish (22.09 larvae per 1,000 m^2) were significantly higher than the other species: blue marlin (9.62 larvae per 1,000 m^2), white marlin (5.44 larvae per 1000 m^2), and swordfish (4.67 larvae per 1,000 m^2) (Rooker et al. 2012). The distribution and abundance of billfish larvae varied spatially and temporally, and several environmental variables (sea-surface temperature, salinity, sea-surface height, distance to the LC, current velocity, water depth, and *Sargassum* biomass) were deemed to be influential variables. Densities of billfish were typically higher in frontal zones or areas proximal to the LC. Habitat suitability was strongly linked to physicochemical attributes of the water masses they inhabited, and observed abundance was higher in slope waters with lower sea-surface temperature and higher salinity. The study suggests that the northern Gulf of Mexico is very important in the early life habitat of billfishes (Rooker et al. 2012).

Tidwell et al. (2007) confirmed that the northern Gulf of Mexico provides important nursery habitat for billfish larvae. Ichthyoplankton surveys were conducted with neuston nets in the summers of 2005 and 2006 to identify areas in the northern Gulf of Mexico with high larval billfish densities. The mean density of larvae per sample ranged from 0 to 53.8 larvae per 1,000 m^2. The highest densities of billfish larvae were located at the fronts of anticyclonic eddies. The catch of 2,589 billfish larvae from 167 stations provides powerful support that the northern Gulf of Mexico is a billfish nursery.

Monthly samples of ichthyoplankton were collected from October 2004 through October 2006 from a site off the coast of Alabama in the northern Gulf of Mexico, about 18 km (11.2 mi) south of Dauphin Island, Alabama (Hernandez et al. 2010). Mean concentrations of total fish larvae peaked in August because of very high abundances of Atlantic bumper (290.6 larvae per 100 m^3) and sand seatrout (*Cynoscion arenarius*, 301.1 larvae per 100 m^3), while taxonomic diversity was generally higher from March through October. Taxonomic richness was generally highest during the late summer and early fall. Of the 58 different families of fish collected, the dominant groups included anchovies (Engraulidae), sand seatrout, Atlantic bumper, Atlantic croaker, Gulf menhaden, tonguefishes (*Symphurus* spp.), gobies (Gobiidae), drums (Sciaenidae), and cusk-eels (Ophidiidae) (Hernandez et al. 2010). Nearly all of the Atlantic bumpers (87 %) were collected in August, while sand seatrout were present throughout the year. The Atlantic croaker was the third most abundant taxon, with an October peak in abundance of 119.5 larvae per 100 m^3 (Hernandez et al. 2010). It is important to note that the SEAMAP data are in units of number of larvae per 10 m^2 or per 10-min tow, whereas the Rooker et al. data are in numbers per 1,000 m^2 and the Hernandez et al. data are in numbers per 100 m^3.

SEAMAP spring and fall surveys from 1982 through 2005 were analyzed to provide information on location and timing of spawning, larval distribution patterns, and interannual occurrence for groupers (Marancik et al. 2012). Shelf-edge habitat was determined to be important for spawning of many species of grouper. Spawning for some species may occur year round, but two peak seasons were evident: late winter and late summer through early fall. A shift in species dominance over the last three decades from spring-spawned species (most of the commercial species) to fall-spawned species also was documented.

The more than 4,000 oil and gas platforms in the Gulf of Mexico likely affect ichthyoplankton populations (Boswell et al. 2010). Lindquist et al. (2005) collected baseline information on vertical and horizontal distribution patterns of larval and juvenile fish from five offshore platforms off the Louisiana Coast from 1995 through 2000. Light traps and passively fished plankton nets were used at night to collect fish in surface and deep waters (15–23 m [49.2–75.4 ft] in depth) within the platform structure. Light traps were also used to collect fish from surface waters directly down-current of the platforms. Compared to light traps fished in deep water, light traps fished at the surface collected higher densities and diversities of ichthyoplankton. Herrings, shads, and sardines; anchovies; lizardfishes; and presettlement blennies were the most common in surface waters within the platforms, while postflexion mackerels and tunas and settlement-size blennies, damselfishes, and clownfishes were most common in surface waters down-current of the platforms. Deep plankton nets collected higher densities of non-herring/shad/sardine ichthyoplankton, while surface plankton nets collected higher numbers of taxa. The vertical distribution patterns described for dominant larval fish collected by plankton nets were generally consistent with those from other studies: herring/

shad/sardine, jack, drum, and mackerel/tuna larvae more abundant in surface waters at platforms and lizardfish, codlet, goby, and left-eye flounder larvae more abundant in deeper waters (Lindquist et al. 2005).

Ditty et al. (2004) reviewed SEAMAP data from bongo net samples collected from 1982 through 1986 to describe the distribution of carangid larvae in the northern Gulf of Mexico relative to areas of high zooplankton. Of the 29,000 larvae from 13 species or species complexes in 11 genera, Atlantic bumper and round scad accounted for 91.7 % of all larvae, which agrees with the summaries above. Atlantic bumper densities averaged 2.9, 20.5, and 42.8 larvae per 100 m^3 for the eastern, central, and western Gulf of Mexico, respectively, while densities of round scad averaged 6.7, 0.4, and 0.1 larvae per 100 m^3, respectively, for the same regions. Carangids, including Atlantic bumper and round scad, appeared to spawn at water mass boundaries (fronts) and/or along other hydrographic features that promote higher productivity (Ditty et al. 2004).

The seasonal occurrence, distribution, and abundance of dolphinfish larvae were determined primarily from 814 neuston net collections taken during SEAMAP ichthyoplankton surveys of the Gulf of Mexico between 1982 and 1984 (Ditty et al. 2004). Larval dolphinfish were collected during all months sampled, but small larvae and pompano dolphin were found primarily during warm months. Larvae of common dolphinfish were significantly more abundant than pompano dolphin. Larval dolphinfish of both species were widely distributed in neritic and oceanic waters and most were collected near the surface. Over 90 % of common dolphinfish and about 80 % of pompano dolphin occurred over the outer continental shelf and in oceanic waters; overall densities averaged 4.8 and 0.8 larvae per 10 neuston tows, respectively.

The distribution, abundance, and seasonality of four carangids (blue runner, Atlantic bumpers, round scad, and rough scad, *Trachurus lathami*) off the Louisiana coast were evaluated using SEAMAP data from 1982 and 1983 (Shaw and Drullinger 1990b). Maximum abundances of larval blue runner, Atlantic bumper, and round scad were found in July inside the 40 m (131.2 ft) isobath. Larval Atlantic bumpers were captured in June and July only; blue runner in May, June, and July; and round scad in all seasons. Atlantic bumper larvae, concentrated mostly off western Louisiana, were by far the most abundant carangid in 1982 and 1983. Larval blue runner was the second most abundant summer-spawned carangid in 1982 and 1983; however, their abundance and depth distribution varied considerably between years (Shaw and Drullinger 1990b). The relative abundance of larval round scad off Louisiana was low, and they were captured only west of the Mississippi Delta. Rough scad were winter/spring and outer-shelf spawners; while they ranked third in overall abundance, they were the most abundant carangid on the outer shelf (Shaw and Drullinger 1990b).

Shaw and Drullinger (1990a) evaluated the distribution, abundance, and seasonality of four coastal pelagic species from the Clupeidae family—round herring, scaled sardine, Atlantic thread herring, and Spanish sardine—in the northern Gulf of Mexico using SEAMAP data from 1982 to 1983. During the summer, larval Atlantic thread herring and scaled and Spanish sardines were abundant on the inner shelf (less than 40 m or 131.2 ft) but were rare or absent in deeper waters. Scaled sardine and thread herring were found in all sampled inner-shelf water locations, but Spanish sardines were rare in the north-central Gulf (Shaw and Drullinger 1990a). During 1982, larval Atlantic thread herring were the most abundant of the four clupeids, while Spanish sardines were the most abundant during 1983. On the West Florida shelf, Spanish sardines dominated larval clupeid populations both years. Scaled sardine larvae were the least abundant of the four species both years; however, they were still captured in 20 % of the inner-shelf bongo net collections. Round herring larvae were collected from February through early June and were abundant on the outer shelf, especially off Louisiana. Over the 2-year period, outer-shelf mean abundance for round herring was 40.2 larvae per 10 m², while inner-shelf mean abundance for scaled sardine, Atlantic thread herring, and Spanish sardine were 14.9, 39.2, and 41.9 larvae per 10 m², respectively (Shaw and Drullinger 1990a).

Ichthyoplankton cruises were conducted in continental shelf waters off west Louisiana from December 1981 through April 1982 to determine the distribution and abundance of larval drums and croakers (Cowan and Shaw 1988). The total sciaenid larval density was highest in April, and the high densities were associated with the coastal boundary layer, a horizontal density front caused by an intrusion of freshwater from the Atchafalaya River east of the study area. Sand seatrout larvae were the most abundant, followed by Atlantic croaker, spot (*Leiostomus xanthurus*), black drum (*Pogonias cromis*), southern kingfish (*Menticirrhus americanus*), and banded drum (*Larimus fasciatus*). Spawning by sand seatrout began in January. Both sand seatrout and Atlantic croaker larvae were captured at higher rates at night than during the day (Cowan and Shaw 1988). Sand seatrout larvae appeared to be somewhat surface oriented, while spot may undergo a vertical migration.

Sogard et al. (1987) collected ichthyoplankton at three inshore–offshore transects off Southwest Pass, Louisiana, Cape Sand Blas, Florida, and Galveston, Texas, from 1979 through 1981 to determine densities of larval Gulf menhaden, Atlantic croaker, and spot in the northern Gulf of Mexico. All species were more abundant at inshore than offshore stations. Gulf menhaden and Atlantic croaker were most abundant off Southwest Pass, Louisiana, a major outlet of the Mississippi River. Of the three species, only the Gulf menhaden demonstrated any consistent vertical distribution pattern. At inshore stations Gulf menhaden were concentrated near the surface at midday, while offshore and present at 70 m (229.7 ft), most were also caught near the surface (Sogard et al. 1987).

Southern Gulf of Mexico

Espinosa-Fuentes and Flores-Coto (2004) investigated the horizontal and vertical variation of ichthyoplankton assemblages in continental shelf waters of the southern Gulf of Mexico during each season in 1994 and 1995. A total of 21,814 ichthyoplankton, consisting of 25 families, 89 genera, and 92 species, was collected. Four assemblages were identified— coastal, inner neritic, outer neritic, and oceanic. Important members of the coastal assemblage in areas of the highest salinity fluctuations and in depths less than 30 m (98.4 ft) included estuarine-dependent species such as Atlantic bumper, sand weakfish, kingfishes (*Menticirrhus* spp.), croakers (*Micropogonias* spp.), and American stardrum (*Stellifer lanceolatus*). Abundant ichthyoplankton in the oceanic assemblage at depths of 50 and 100 m (164 and 328 ft) in areas with the least salinity fluctuations included pelagic species such as antenna cod (*Bregmaceros atlanticus*), lanternfishes (*Myctophum* spp.), pearly lanternfish (*Myctophum nitidulum*), large-finned lanternfish (*Hygophum macrochir*), and smallfin lanternfish (*Benthosema suborbital*) (Espinosa-Fuentes and Flores-Coto 2004). The main taxa in the inner neritic assemblage were hump-backed butterfish (*Selene setapinnis*), bigeye scad (*Selar crumenophthalmus*), shoal flounder (*Syacium gunteri*), eyed flounder (*Bothus ocellatus*), striped codlet (*Bregmaceros cantori*), and largehead hairtail (*Trichiurus lepturus*). The frequent and abundant species in the outer neritic assemblage of the outer-shelf stations and mid-depths were lanternfishes (*Diaphus* spp.), bristlemouths (*Cyclothone* spp.), fairy basslets (*Anthias* spp.), tunas (*Thunnus* spp.), bigeye scad, blue runner, rough scad (*Trachurus lathami*), bullet tuna (*Auxis rochei*), and striped codlet (Espinosa-Fuentes and Flores-Coto 2004).

Sanvicente-Añorve et al. (2000) evaluated the scales of the main physical and biological processes influencing the ichthyoplankton distribution in the southern Gulf of Mexico. These included the Bay of Campeche (spring 1983, winter 1984, and summer 1987), the littoral zone adjacent to Terminos Lagoon (bimonthly between July 1986 and May 1987), and the Carmen Inlet between the lagoon and the sea (monthly between April 1980 through January 1981). The main circulation patterns of the southern Gulf of Mexico, continental water discharges, mixing processes, and oceanic gyres were important processes affecting ichthyoplankton distribution patterns and community structure in the Bay of Campeche, and 81 families of ichthyoplankton, which included oceanic, neritic, and estuarine-dependent species, were collected (Sanvicente-Añorve et al. 2000). The neritic zone of the Bay of Campeche contained the highest densities of ichthyoplankton; highest densities (1,000–3,000 individuals per m^3) were found near the Grijalva-Usumacinta River delta in the summer, and the lowest densities (fewer than 300 larvae per m^3) were found in the winter. Distinct ichthyoplankton assemblages were identified and included a coastal assemblage characterized by Atlantic bumper (*Chloroscombrus chrysurus*, 5.4–209 larvae per m^3), Atlantic thread herring (*Opisthonema oglinum*, 152 larvae per m^3), sand weakfish (*Cynoscion arenarius*, 10.8–24 larvae per m^3), Atlantic croaker (2.1–4.8 larvae per m^3), and hogchoker (*Trinectes maculatus*, 3.9 larvae per m^3); a neritic assemblage characterized by tonguefishes (Cynoglossidae, 2.1–6.7 larvae per m^3), codlets (Bregmacerotidae, 5.8–63.6 larvae per m^3), and left-eye flounders (Bothidae, 0.5–7.8 larvae per m^3); and an oceanic assemblage dominated by lanternfishes (Myctophidae, 0.3–3.3 larvae per m^3) and bristlemouths (Gonostomatidae, 0.1–2.3 larvae per m^3). Twenty-three families of ichthyoplankton were collected from the littoral zone adjacent to Terminos Lagoon. Littoral currents, lagoon influence, spatial salinity variability, and meteorological conditions determined the structure and function of ichthyoplankton groups (Sanvicente-Añorve et al. 2000). In the littoral zone, a high abundance of ichthyoplankton occurred from May to September, followed by a strong decrease in January and March. While they changed in size, two groups occurred throughout the year; one group, which consisted of anchovies (46–197.6 larvae per m^3) and gobies (8.5–371.5 larvae per m^3), was typically located in the area adjacent to the Carmen Inlet. The second group, located near the Puerto Real inlet, was characterized by Atlantic thread herrings (49.4–56.4 larvae per m^3), Atlantic bumpers (44 larvae per m^3), and scaled sardines, 39.9 larvae per m^3, which dominated in May, July, and September. In the Carmen Inlet between the lagoon and the sea, 38 families of ichthyoplankton were collected. Tidal- and wind-induced currents, bottom topography, and salinity gradients were the major forces controlling ichthyoplankton distribution (Sanvicente-Añorve et al. 2000). In the inlet, greatest densities of ichthyoplankton were found in the central-western section and the deepest eastern channel, and strong vertical stratification was observed; 99 % of the total catch consisted of anchovies, gobies, herrings/shads/sardines, drums, and mojarras. Distinctive ichthyoplankton patterns were produced by the combination of the physical, biological, and oceanographic processes and the life history strategies of the fishes—the periods and spawning areas of the adults, larval stages, dispersal capabilities of larvae, and the larval stage duration (Sanvicente-Añorve et al. 2000).

7.4.1.4 Neuston and *Sargassum* spp.

The neuston are drifting organisms that inhabit the surface layer of the ocean (note above that the ichthyoplankton was sampled within this layer with a net designed to float at the surface); likewise numerous ichthyoplankton can be found in this narrow habitat. While a wide variety of organisms are encountered within this layer in general (Dooley 1972; Turner et al. 1979), the prolific assemblage is associated with the pelagic *Sargassum* algal mats (Parr 1939). These occur in the Gulf of Mexico in windrows measuring hundreds of meters long by tens of meters wide. The long, linear windrows are formed by Langmuir circulation.

In the North Atlantic and Gulf of Mexico, free-floating mats of *Sargassum*—pelagic brown algae—supplies a dynamic infrastructure for diverse assemblages of fishes, invertebrates, sea turtles, seabirds, and marine mammals (Casazza and Ross 2008). To date, a number of studies have documented ichthyofaunal assemblages associated with *Sargassum* in these waters, most notably those of two holopelagic species: *S. fluitans* and *S. natans* (Adams 1960; Parin 1970; Zaitsev 1971; Dooley 1972; Bortone et al. 1977; Fedoryako 1980, 1989; Gorelova and Fedoryako 1986; Settle 1993; Hoffmayer et al. 2002; Wells and Rooker 2004a, b; Casazza and Ross 2008). Pelagic *Sargassum* is ubiquitous throughout the surface waters of the northern Gulf of Mexico and waters adjacent to the southeastern coastal waters of the United States. (Hoffmayer et al. 2002; Wells and Rooker 2004a, b; Casazza and Ross 2008). In general, the pelagic zone of these waters is featureless apart from free-floating *Sargassum* mats, production platforms, flotsam, buoys, and fish aggregation devices (Wells and Rooker 2004a, b). Previous studies report that *Sargassum* mats function as an essential fish habitat (EFH), affording food sources and protection from predators to juvenile and adult fishes in what is otherwise a nutrient-poor, structure-free environment (Wells and Rooker 2004a, b; Rooker et al. 2006).

Conservation interests for commercially valuable fish species have encouraged efforts to gain a better scientific understanding of nursery habitats used by these and other species at early life stages (Wells and Rooker 2004a, b). Identification and understanding of *Sargassum* community structure as an EFH is necessary in building healthy and sustainable fisheries supported by effective management strategies (Wells and Rooker 2004b). The physical nature of the various forms of *Sargassum* habitat (e.g., individual clumps, small patches, large rafts, and weed lines) makes sampling these habitats extremely difficult and potentially inconsistent (Casazza and Ross 2008). Satellite observations suggest that the Gulf of Mexico is the source of windrows of *Sargassum* in the central north Atlantic (Gower and King 2011).

Wells and Rooker (2004b) examined the spatial and temporal patterns of habitat use and evaluated the role of *Sargassum* as nursery habitat for fishes in the northwestern Gulf of Mexico. Inshore and offshore comparisons were made; inshore waters were sampled from northern (Galveston) and southern (Port Aransas) Texas from May to August 2000 and offshore waters (15–70 nautical miles) off Galveston and Port Aransas, Texas. Replicate samples (3–5) were collected monthly from May to August 2000 in each zone. *Sargassum* mats were arbitrarily chosen during a period from 08:00 to 15:00 h using a larval purse seine (20 m [65.6 ft] long, 3.3 m [10.8 ft] deep, 1,000 μm mesh). Purse seines were used as the only collection material and deployed as the boat encircled a chosen mat. Once around the mat, the net was pursed. A total of 10,518 individuals representing 36 fish species from 17 families were collected using the purse seine method only. All taxa listed in the study were included in this review since all were identified to a species level. Dominant taxa included filefishes (Mon-acanthidae, 4,621), jacks (Carangidae, 1,827), triggerfishes (Balistidae, 1,604), pipefishes (Syng-nathidae, 1,096) and frogfishes (Antennariidae, 368), which accounted for 43.9 %, 17.4 %, 15.3 %, 10.4 %, and 3.5 % of the total capture, respectively. Hoffmayer et al. (2002) on the other hand sampled a total of 18,749 fishes representing 86 species in 138 collections with combined methods of neuston nets of two sizes and paired bongo nets. However, for the purposes of this study only 10,283 were considered due to a lack of family and species identification for much of the sampling; 19 taxa identification extended only to a family level. Surface tows with a neuston net supplied the greatest abundance and diversity of species collected (9,865 fishes; 79 species identified to species level). Oblique tow with paired bongo nets yielded far less abundance and diversity (418 fishes, 36 species identified to species level). Catches were dominated by flyingfishes (Exocoetidae, 3,876) and jacks (Carangidae, 1,521) and accounted for 37.7 % and 14.8 % of the total capture, respectively.

Species and individual counts were used to determine diversity and evenness of collections for each method and study. Species richness (S) was highest in the Casazza and Ross (2008) (76 species) and Hoffmayer et al. (2002) (86 species) studies. Higher fish diversity (H') was observed for methods of neuston net, nightlighting, bongo nets, and purse seine. Values of evenness (J') for species collections were noticeably greater for neuston net, nightlighting, bongo nets, and purse seine methods. When studies were compared, species diversity and evenness of distribution was higher in Hoffmayer et al.'s (2002) investigation; however, the number of individuals was lower (8,968) than those of the other studies.

The mean biomass, according to Robert Webster (Texas A&M University, personal communication), is about 140 mg dw/m^2 in the Gulf of Mexico. However, this can be extremely variable. Parr (1939) estimated values of 258 g dw/m^2, standard deviation = 174, for example, in the Gulf of Mexico. The gross and net productivity are higher in neritic waters than offshore due, it is presumed, to increased levels of inorganic nutrients (Lapoint 1995). Lapoint (1995) estimated doubling time at 20 days, although Robert Webster believes it could be as short as 10 days. This would equate to about 7 mg/m^2/day or 2.5 mg dw/m^2/year. About 40 % of the dry weight is carbon, meaning the contribution of *Sargassum* to total phytoplankton PP is rather small. Although *Sargassum* windrows are considered critical habitat because they serve as a refuge for fish larvae and juveniles, as indicated abundantly in the ichthyoplankton section above (Wells and Rooker 2004a, b), when it washes ashore, it becomes a nuisance. Using satellite images of windrow movements, Webster estimates that it takes about 60 days to move across the continental shelf onto the beaches of Texas.

Data for fish assemblages associated with *Sargassum* suggest the important natural function of *Sargassum* as an EFH. Samples of fishes taken in the north-central (Hoffmayer et al. 2002) and northwestern (Wells and Rooker 2004b) Gulf of Mexico showed similarities in species diversity and abundance. A small number of taxa dominate most of the collections. These include filefishes (Monacanthidae), jacks (Carangidae), triggerfishes (Balistidae), pipefishes (Syngnathidae), and frogfishes (Antennariidae), which accounted for 52.5 % (Hoffmayer et al. 2002), 87 % (Wells and Rooker 2004b), and 94 % (Casazza and Ross 2008). Similarly, these families represent a large proportion of the total catch in studies conducted in the western Atlantic (Dooley 1972) and eastern Gulf of Mexico (Bortone et al. 1977).

Fishes at larval and juvenile stages were predominately present across all three studies and all capture methods except hook-and-line (Wells and Rooker 2004b). The relationships between the quantity of *Sargassum* and species richness and abundance and biomass of fishes can be highly variable. Dooley (1972) and Fedoryako (1980) found no correlation between numbers of fishes and quantity of *Sargassum*, but significant positive correlations between fish abundances and quantity of algae have been catalogued in other studies (Moser et al. 1998; Wells and Rooker 2004b). The sampling methods chosen by the investigator may substantially influence these results. *Sargassum* habitat is a dynamic and difficult habitat to sample, and the structural complexity of this habitat strongly affects fish assemblages (Dooley 1972).

7.5 MESOPELAGIC (MID-WATER) FISHES AND PELAGIC MEGAFAUNAL INVERTEBRATES (MICRONEKTON OR MACROPLANKTON)

Mesopelagic (mid-water) fishes are relatively small species such as the Gonostomatidae and Myctophidae (lanternfish) that vertically migrate daily from depths somewhat less than 1,000 m (3,281 ft) up to the surface waters at night. They are sampled with an Isaacs-Kidd mid-water trawl, which is difficult to quantify, or a Tucker trawl (Hopkins et al. 1973), used as sets of opening and closing nets that sample vertical stratification. Mean weight of mid-water fishes in the Gulf of Mexico is about 16 g (0.04 pounds [lb]) wet weight (ww) per individual (Bangma and Haedrich 2008). Most mid-water fishes prey on net-sized zooplankton (Hopkins and Baird 1977; Hopkins et al. 1996), but are eaten by all sizes of large pelagic species (Sutton and Hopkins 1996). Beaked whales for example feed down to depths of 1 km (0.62 mi) preying on squid and mid-water fishes.

The Gulf of Mexico has been considered a distinct geographic region (Backus et al. 1977) on the basis of the lanternfish species distributions in the Sargasso Sea and the Caribbean. Out of about 209 species known to occur in the western Sargasso and Caribbean Sea complex (Gartner et al. 1988; Sutton and Hopkins 1996), approximately 140 have been sampled in the Gulf of Mexico. Bangma and Haedrich (2008) have suggested that the Gulf mid-water fishes be considered an *ecotone* or transition between the subtropical Atlantic and tropical faunas because the Gulf of Mexico has a mixture of species from both the north Atlantic and Caribbean. In any case, the mid-water fish play a significant role in the transfer of mass and energy up the food web to larger open-ocean pelagic species (Hopkins and Baird 1977; Hopkins et al. 1996). The deep Gulf of Mexico between about 1,500 m (4,921 ft) and 3,700 m (12,139 ft) is very poorly sampled to date. A biomass of 4.5 mg ww/m^3 (standard deviation = 1.9) between the 1,500 and 3,700 m (4,921 and 12,139 ft) depth can be estimated based on the work of Sutton et al. (2008) in the central Atlantic. This would be the equivalent of about 12 g ww/m^2 between 1.4 and 3.7 km (0.87 and 2.3 mi) in depth.

Much of our knowledge of deep macroplankton or micronekton is not quantitative in terms of numbers or biomass per volume. However, extensive information is available on the number of species (Gamma diversity) of the Gulf of Mexico and adjacent Caribbean. This is due to the exploratory fishing of the NMFS (now within NOAA) (Springer and Bullis 1956; Bullis and Thompson 1965). Summaries of catches of oplophorid shrimps (Decapoda: Caridea: Oplophoridae) by Pequegnat and Wicksten (2006) illustrate the wide geographic and depth distributions of this diverse group caught in mid-water trawls and bottom-trawled nets. Of the 25 species they reviewed, 21 were sampled in the water column.

Mesopelagic micronekton standing stocks (Hopkins and Lancraft 1984; Sutton and Hopkins 1996) and composition (Hopkins et al. 1989) assessments are available for the eastern Gulf of Mexico (Figure 7.51). The latter authors have constructed an energy budget for a typical mid-water fish species that defines their importance in consuming upper water column

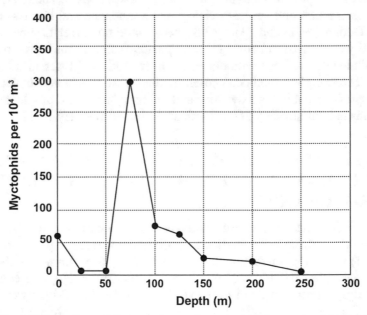

Figure 7.51. **Vertical distribution of the numbers of mesopelagic myctophid fishes in the eastern Gulf of Mexico (modified from Hopkins 1982).**

zooplankton, principally copepod crustaceans. They then estimate the potential production of these populations as potential prey for large terminal predators, such as billfish and beaked whales.

7.6 SEAFLOOR COMMUNITIES: THE BENTHOS

Level-bottom soft sediment (sand-, silt- and clay-sized particles) communities are composed of a wide range of size classes that are sampled by different methods. The sizes are also based on how they are sampled: the smaller the organism, the smaller the sampler (Table 7.3).

Each of these size groups will be considered separately, and a synthesis will be attempted that draws them together in a comparison and ultimately into a proposed food web. Three characteristics of biotic assemblages will be described, if adequate data are available:

1. **Densities** per unit area (or sediment volume), and associated **biomass** per unit area.

2. **Biodiversity** (a) within habitat diversity indices (Alpha diversity), (b) between habitats or species turnover or change along a gradient (Beta diversity or species turnover in space), and (c) species richness (Gamma diversity or total number of species samples).

3. **Species** composition in recurrent faunal groups or **zonation** as a function of depth (or some correlate with depth).

7.6.1 Continental Shelf Benthos

Numerous studies have been made of the biota and associated supporting habitat variables of the Gulf of Mexico. They encompass the entire Gulf periphery (Figure 7.52) (Rabalais et al. 1999b). Those studies on the northern coast (e.g., in U.S. waters) were funded by U.S. federal government agencies in anticipation of expanded offshore oil and gas exploration and production (BLM, MMS, and BOEM). Each study contains significant information that can be used to assess ecosystem processes that can be compared to each other and to other continental shelves. The databases were generated in order to establish baselines from which

Table 7.3. Level-Bottom Seafloor Assemblage Size Groupings

Size Class	Size	Sampling Device	References
Microbiota	<1 µm (bacteria and *Archaea*), and protists up to 40 µm	1–3 cm diameter subcorer	Deming and Carpenter (2008)
Meiofauna	>40 but <500 µm	3–6 cm diameter subcorer	Baguley et al. (2008)
Macrofauna	From 250 up to 500 µm, depending on location	GOMEX corer Spade corer Ekman grab Smith-McIntyre grab	Boland and Rowe (1991), Escobar-Briones et al. (2008a, b), Harper (1977), and multiple studies (see text)
Megafauna	>1 cm	Trawls, photos Skimmer, traps	Pequegnat et al. (1970) and Pequegnat (1983)
Demersal fishes	Trawl caught, 2.5 cm stretch mesh	Trawls, photos, skimmer, longline	Pequegnat et al. (1990)

cm centimeter, *µm* micrometer

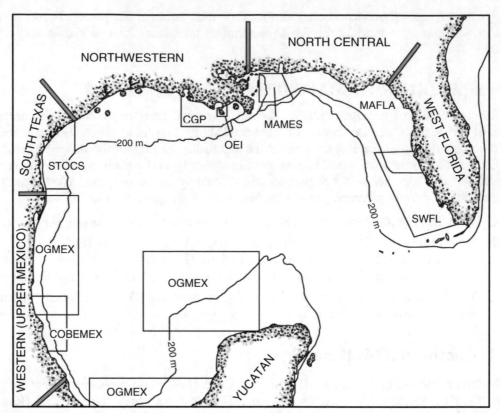

Figure 7.52. Regional studies of the continental shelf of the Gulf of Mexico (from Rabalais et al. 1999b).

damage or alterations could be assessed. In addition, extensive monitoring and associated experimental process measurements and numerical simulations have been made and are ongoing in the regional, seasonal hypoxic region that stretches west from the central Mississippi Delta to the border with Texas. NOAA (including Sea Grant), U.S. Geological Survey (USGS), U.S. Environmental Protection Agency (USEPA), and state agencies have supported the hypoxic area investigations. Studies of the biota in Mexican waters have been sponsored by the Consejo Nacional de Ciencia y Technología or National Council of Science and Technology (CNCYT) (the equivalent to the U.S. National Science Foundation). This section will attempt to summarize and compare the most salient features of the areas studied.

The faunas of the northern shelf are considered Carolinian or temperate, whereas the faunas of the southern shelf are semitropical to tropical (Engle and Summers 2000). The south Texas/northern Mexico shelf is composed of terrigenous sand, silt, and clay; the central hypoxic area of the north is mainly fluvial mud (silt and clay, with some sand), and the eastern Florida coast is hard bottom carbonate. Where the eastern Gulf of Mexico bottom off Florida is not hard carbonate (see Figure 7.1), carbonate sands replace it. The broad shelf of the Yucatán Peninsula is carbonate, but the narrow shelf at the southern end of the Bay of Campeche is terrigenous mud (silt, clay, and sand) that debouches from rivers. The biogeographic provinces and the sediment type play a big role in determining faunal composition in each area.

Quantitative seafloor samples and trawls were taken on the soft (sand, silt, and clay) substrates in each of the regions depicted in Figure 7.52 and Table 7.4 to estimate animal densities and species composition of the meiofauna, macrofauna, epibenthic megafauna, and

Table 7.4. Comparison of Macroinfaunal Assemblages, Continental Shelf (Northern Gulf of Mexico) (sample sizes varied, replication varied, all used 0.5 mm sieves) (nearshore are on the inner continental shelf at depths less than 50 m; offshore are in depths of greater than 50 m on the outer shelf)

Location/Area	Nearshore Densities	Offshore Densities	Total No. of Species
STOCS[a]	2,707 (1,561)	229 (62)	837
MAFLA	5,268 (3,533)	575 (342)	1,691
Hypoxic area[b]	3,741 (3,349)		185
Buccaneer field[c]	5,850 (2,902–10,937)		352
Bryan mound[d]	1,109 (709)		
CTGLF[e]	Range of 6–12,576		576
SWFES[f]	Range of 3,245–15,821		414

Values are arithmetic means of individuals per m^2 followed by standard deviation in parentheses; the last column is the total number of species in each study
[a]Values from Flint (1980), not Flint and Rabalais (1981)
[b]Nunnally et al. (2013)
[c]Harper (1977)
[d]September 1977 control site only—Don Harper data archived at TAMUG
[e]Bedinger (1981) (several locations subject to hypoxic conditions)
[f]Danek et al. (1985), soft-bottom locations only

demersal fishes. This information is embodied in numerous reports, government documents, and peer-refereed papers, as summarized in Table 7.4 and in the review by Rabalais et al. (1999b). Sampling locations were organized along the coast in transects that bisected the shelf, from depths as shallow as 6 m (19.7 ft) out to the edge of the shelf at depths approaching 200 m (656 ft). Recurrent groups or assemblages were determined among these sites, and maps were then used to illustrate the groupings. The entire northern Gulf of Mexico coastline exhibited some common features: (1) highest densities of macrofauna were encountered at the inshore locations, (2) lowest densities were at the outer-shelf margin, (3) macrofaunas were dominated by diverse assemblages of polychaete annelid worms followed by amphipod crustaceans and bivalve molluscs in lesser numbers, and (4) principal faunal groups were aligned parallel to the coastline within depth intervals in a predictable fashion. About 20 % of the dominant macrobenthos are shared between the three northern Gulf study areas—South Texas Outer Continental Shelf (STOCS), Mississippi Alabama Marine Ecosystem Study (MAMES), and the Mississippi, Alabama, Florida (MAFLA) ecosystem studies—and Rabalais et al. (1999b) suggest that there is regional endemism within the macrofaunal component of the benthic communities. However, that degree of overlap in similar species is substantially higher than might be expected, given the differences in the habitats (Figure 7.1).

The STOCS investigation on the south Texas shelf, summarized in Flint and Rabalais (1981), was designed to gain a quantitative understanding of how the shelf ecosystem food web functions relative to supplies of inorganic plant nutrients, phytoplankton productivity, stocks of zooplankton, and fate on the sea floor. The data clearly demonstrate that meager nutrient supply (nitrate) supports relatively low PP because chlorophyll a concentrations were consistently below 1 mg C/m^3 all year, with the exception of single modest spikes during brief spring and fall blooms. A carbon budget was created to illustrate how an estimated 103 g C/m^2/year of new production (a high value given the low chlorophyll a values) is cycled through the food web to the economically important brown shrimp (*Farfantepenaeus aztecus*) population. Modest gradients of ammonium (NH$_4$) at the seafloor suggested that benthic-pelagic coupling

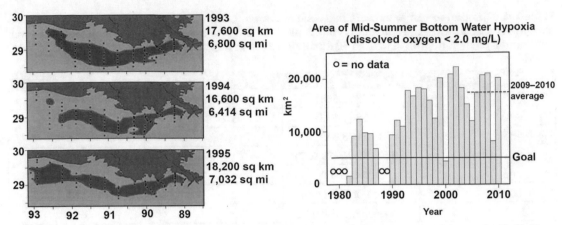

Figure 7.53. Area of continental shelf that habitually experiences seasonal hypoxia (*left*) (from Rabalais et al. 1999a); illustration of relative increase in size of hypoxic area over time (*right*) (from Rabalais and Turner (2011)); Goal refers to anticipated decrease in size if and when nitrate loading is reduced.

(regeneration by the sediment community) could be an important source of nitrogen to the water column. More recent advances in numerical modeling of food webs coupled to physical models should now be applied to this comprehensive set of shelf data.

The central and eastern Gulf of Mexico shelves are stark contrasts to the south Texas and Mexican shelves. The Louisiana shelf is bathed by freshwater from the Mississippi River and the Atchafalaya Bay diversion. This contributes high levels of inorganic nutrients (greater than 100 µmol/L nitrate concentration) that enhance PP. This is accompanied by freshwater that creates intense vertical stratification. This condition is seasonal, beginning in the late winter or early spring, and intensifying throughout the summer months of warming that contributes to the vertical stratification. The vertical stratification and surface water PP decline with water column mixing in the fall. The effect of this condition produces a large area (at times larger than approximately 20,000 km^2) of hypoxic (less than 2 mg O_2/L) bottom water that is stressful to most shelf biota (Figure 7.53). Motile swimmers escape; sessile organisms suffer. The region is often referred to in the public media as a *dead zone*. But this is a misnomer; it is not *dead*, although it supports a unique fauna (Gaston 1985; Rabalais et al. 2001; Baustian and Rabalais 2009; Baustian et al. 2009). The hypoxic fauna is dominated by polychaete (Rabalais et al. 2001) and nematode worms (Murrell and Fleeger 1989). A sulfur-oxidizing bacterium (*Beggiatoa* sp.) is often observed on the sediment surface under conditions approaching anoxia (Rowe et al. 2002). The diversity and abundance of the infauna is severely reduced by hypoxic conditions, and the longer hypoxic conditions persist without reoxygenation, the greater the decline in the surviving fauna (Baustian and Rabalais 2009). Recovery during the winter, when the bottom water is normoxic, is modest (Rabalais et al. 2001; Nunnally et al. 2013).

The causes, along with remedial strategies, are the subject of some debate. It has been advocated that agricultural runoff up the Mississippi River must decrease nitrogen loading from fertilizer in order to reduce the size and intensity of the hypoxia (Rabalais et al. 2002, 2007). Others question the overriding importance of fertilizer nitrogen as the cause. Dissolved organic matter (DOM) in the freshwater could contribute to the biological oxygen demand (Bianchi et al. 2010), and stratification prevents deepwater oxygenation (Rowe 2001). The plume of these discharges has been partitioned into zones in which different processes both cause and maintain hypoxia (Rowe and Chapman 2002). In the proximal zone near the river mouths (referred to as *brown*), the sediment loading prevents light penetration, and hypoxia is caused by enhanced sedimentation. The next zone (*green*) represents the now-classic paradigm in which high levels of nitrate cause eutrophication. The final zone (*blue*) is characterized by relatively clear water with low nitrate concentrations and PP is low, but hypoxia is maintained by vertical stratification of the water column. If too much freshwater and/or DOM are primary causes of hypoxia, then reducing the nutrient load up the river will have only a minor effect on the condition.

Benthic infaunal abundance reflects the overall productivity of a coastal ecosystem. Thus, within the LME there is a substantial difference between the areas. The relatively productive northeast has twice the macrofauna as south Texas, whereas the hypoxic area lies in between. It must be noted however that the hypoxic fauna is composed of an assemblage that is adapted to low oxygen stress. It lacks the numerous species of crustaceans and mollusks common to the other two areas.

7.6.2 Corals and Live-Bottom Assemblages

Extensive areas in the Gulf of Mexico are dominated by coral growth and hard carbonate bottoms (Figure 7.1). The entire Campeche Bank off the north extension of the Yucatán Peninsula is composed of carbonate that has been formed since the Triassic–Jurassic eras. The fauna is semitropical to tropical (Tunnell et al. 2007). Hermatypic (reef-building) species are common and extensive. The most salient big reef is Alacran in the middle of the bank, more or less (Kornicker et al. 1959). Lists of species are available for many groups (Rice and Kornicker 1962; Gonzalez-Gandara and Arias-Gonzalez 2001). It is also important to artisanal fishers (Bello et al. 2005). The northern Gulf of Mexico also has patchy areas of hermatypic corals but these are encountered on the tops of salt diapirs on the outer continental shelf or upper continental slope, the most prominent being the Flower Garden Banks, which now have been designated a national marine sanctuary—FGBNMS (Figure 7.54). The many similar banks on the outer continental shelf west of the Mississippi River are plotted on the NOAA habitat map (Figure 7.1). The fauna of these banks has been studied extensively. They are important habitats for shelf fishes, and thus, recreational fishermen and amateur scuba divers frequently visit them on charter boats. Recreational hook-and-line fishing is allowed in the FGBNMS but spearfishing is not. The most extensive descriptions of the many banks on the outer shelf can be found in Rezak et al. (1985).

Rezak et al. (1985) portray many of the banks in a similar fashion. The biodiversity of the fishes, corals, and associated invertebrates in the northern Gulf of Mexico is less than the Caribbean or the southern Gulf of Mexico because these structures are at the northern boundary of the corals' ranges. All the corals release their eggs and sperm simultaneously in late summer. This synchronous spawning is observed at specific tidal and lunar conditions in many coral reefs worldwide. All coral reefs in shallow water are dependent on clear water because they contain symbiotic photosynthetic zooxanthellae. Thus, they are threatened by eutrophication that increases planktonic algal growth.

Note the layer of particle-rich water at the deep margin of the bank in Figure 7.54; this nepheloid layer is a ubiquitous feature on the shelf and upper slope of the northern Gulf of Mexico west of the central Mississippi Delta region. This is the same feature referred to above in the zooplankton section. Zooplankton grazing occurs in this near-bottom layer rather than at the surface.

Live-bottom assemblages occur on hard carbonate bottoms on the Campeche Bank, as mentioned above, but also in extensive areas of the carbonate platform off west Florida (Figure 7.1). A large fraction of the eastern continental shelf hard bottom is thus substantially different from the fauna of the northwestern Gulf of Mexico fauna living primarily on soft sediments. The boundary between the two habitat types is more or less the De Soto Canyon to the north and the Florida Keys archipelago to the south. The outer margin of the southern half

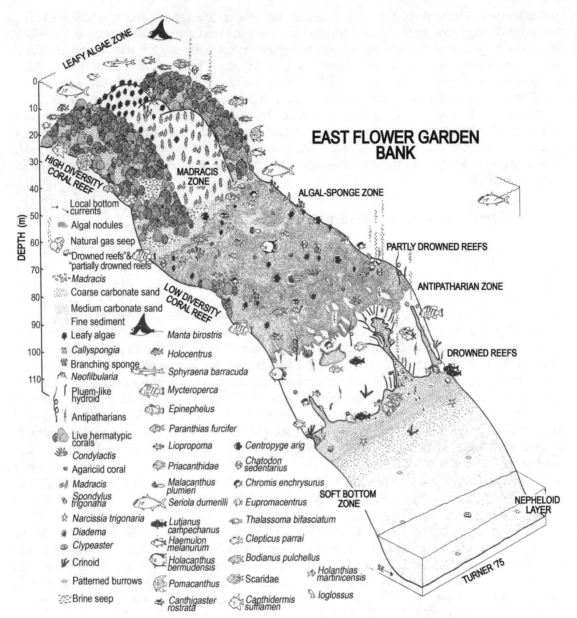

Figure 7.54. Diagram of faunal and floral zonation down the side of the East Flower Garden Bank coral reef on top of a salt diapir on the outer continental shelf off Texas. Note the salt pond and stream on the lower boundary and the bubbles appearing intermittently across the entire depth interval. Copied from Rezak et al. (1985) (republished with permission of John Wiley and Sons Inc.; permission conveyed through Copyright Clearance Center, Inc.) and based on Bright et al. (1984).

of these hard grounds is bathed by the loop current returning back south toward the Florida Straits (Figures 7.2 and 7.3). Sampling habitats sometimes referred to as *live bottoms* is far more difficult than soft bottoms of silt, clay, and sand. Scuba divers are often required to employ suction or pumping mechanisms (that sieve material through a mesh bag) or scrape off areas defined by a metal quadrat. Remotely operated vehicles (ROV) with still and video cameras have been used extensively for surveying hard bottoms. The foundation species that cover the bottom are sponges, attached algae, sea grasses such as *Zostera* and *Thalassia*, anemones, and individual corals. Mixed among them are a diverse assemblage of polychaete worms, crustaceans, and echinoderms. The diversity of the small forms living in among the foundation species is high because of the physical variety of the available space. The principal areas on the shelf are the Alabama Pinnacles, the Florida Middle Grounds (FMG), and the smaller Madison-Swanson Banks (Figure 7.1).

The FMG evolved about 20,000 years ago when sea level was lower. The FMG is a succession of ancient coral reefs covering about 1,193 km^2 (461 square miles [mi^2]) (Figure 7.1), 128.6 km (80 mi) to the northwest off the coast of Florida. The FMG is constructed of both high and low relief limestone ledges and pinnacles that exceed 15.2 m (50 ft) in some areas. The FMG is located about 150 km (93.2 mi) south of the Florida panhandle between 28° 10′ × 28° 45′ N and 084°00′ and 084°25′ W.

Several other live bottom areas off northwest Florida are being considered as potential sanctuaries to stimulate or at least preserve some important fish species that are popular game fish (Harder and David 2009). Their depths remain just beyond the accepted maximum depth for recreational scuba (e.g., 39.6 m [130 ft]), but they are fished commercially and by recreational fishermen.

The USGS study referred to as the Northeastern Gulf of Mexico-Coastal and Marine Ecosystem Program (NEGOM-CMEP) has to date conducted the most comprehensive recent study of the Alabama Pinnacles, but earlier studies have been extensive as well (Ludwick and Walton 1957; Brooks and Giammona 1990). The USGS surveyed both the shallow reef trend (65–80 m [213–262 ft]) and deep reef trend (85–110 m [279–361 ft]). Eight main reefs (five shallow, three deep) were selected for fish community structure and trophodynamics studies, all within the region designated in Figure 7.1. The combined sampling effort by the USGS study included 326 stations, apportioned into 112 angling, 63 trap, 22 bottom trawl, 58 ROV, 15 dredge/core/grab, and 37 plankton stations. The study collected over 6,000 specimens for food habits analyses, taxonomic verification and documentation, and subsequent life history analyses, plus photographs of 113 species. The ROV observations were quantified along transects with both video and still cameras positioned 1 m (3.28 ft) above bottom to provide known areas of coverage.

The FMG ecosystem has similarities to modern patch-reefs and supports a thriving complex assemblage of species that have affinities to temperate Carolinian and tropical Caribbean origins. The fish species are tropical, with megabenthic invertebrates characterized by stony coral, gorgonians, and large basket sponges. Recent surveys have tabulated 170 species of fish, 103 species of algae, approximately 40 sponges, 75 mollusks, 56 decapod crustaceans, 41 polychaetes, 23 echinoderms, and 23 species of stony corals (NOAA CCMA 2002).

Roughtongue Reef is a roughly elliptical (400 m [1,312 ft] major base diameter), high-profile, flat-top structure with steep vertical sides. Fishermen have historically called the general area containing this and the next two target reefs the "40 Fathom Fishing Ground." Roughtongue Reef belongs to the shallow pinnacle trend, with a base depth of 80 m (262 ft). The USGS-designated name refers to the common name for the small planktivorous serranid, *Pronotogrammus martinicensis*, the roughtongue bass, which was extremely abundant on this reef. Cat's Paw Reef is a group of six small, medium-to-high profile, flat-topped mounds arranged in the pattern of a cat's paw print, with a 5–10 m (16.4–32.8 ft) relief. This cluster of mounds lies about 1,000 m (3,281 ft) west of Roughtongue Reef in the 40 Fathom Fishing Ground. Individual reef formations within the feature have flat-top communities present with limited sediment cover and highly eroded and sculpted rock surfaces with vertical faces along edges of features. Small soft corals in the USGS study were abundant on horizontal surfaces; solitary coral colonies (including *R. manuelensis*), with spiral sea whips, antipatharians, and crinoids, were also common. Yellowtail Reef is a single, elliptical (200 m [656 ft] base diameter), high-profile, flat-top structure, that reaches the shallowest crest depth (60 m [197 ft]) of all study sites. This structure also belongs to the 40 Fathom Fishing Ground group. It forms the northwestern end of a reef arc with Cat's Paw Reef at the center and Roughtongue Reef lying at the southeastern end. Like other reef features in the group, an extensive flat-top area is present and is characterized by accumulated sediments and a dense invertebrate assemblage dominated by octocorals, antipatharians, sponges, and coralline algae. Rock outcrops characterize the northern extent of the feature, and sessile invertebrates and coralline algae are known to colonize these areas. The USGS-designated name refers to the yellowtail reef fish (*Chromis enchrysura*), which was particularly abundant on this reef.

Double Top Reef is a horseshoe shaped (100 m [328 ft] base diameter), high-profile structure that consists of multiple flat-top mounds with steep vertical sides. This area belongs to the shallow pinnacle trend in the northeastern Gulf of Mexico and also includes a similarly shaped series of mounds in the study area referred to as Triple Top Reef and an adjacent, low profile feature referred to as Pancake Reef. These features also have flat-top communities characterized by high sediment cover and dense invertebrate assemblages dominated by octocorals and antipatharians, with few solitary corals. Vertical rock walls and overhangs are dominated by *R. manuelensis* and other solitary corals. Alabama Alps is a long, narrow, north–south aligned, high-profile mound approximately 1,000 m (3,281 ft) in length. In previous studies, this same area was referred to as Lagniappe Delta Shallow and has historically been called the 36 Fathom Ridge by fishers. Alabama Alps forms the northwestern terminus of a long northwest-to-southeast-aligned ridge and pinnacle arc paralleling the shelf edge; it belongs to the shallow pinnacle trend of the northeastern Gulf. The top of this feature has sections of relatively flat terrain with scattered sections of sediment cover, particularly in the southern portion of the feature. Octocorals, antipatharians, and sponges dominate invertebrate assemblages on the flat sections. The sides of the feature range from vertical walls to large attached monoliths where the solitary coral *R. manuelensis* was the dominant sessile invertebrate with crinoids, antipatharians, coralline algae, sponges, and other solitary corals present. The USGS-designated name refers to the precipitous terrain, particularly the near-vertical west-face scarp of the structure and its position off the state of Alabama.

Ludwick and Walton Pinnacle 1 is the central member of a group of five medium- to high-profile, spire-top, shelf-edge structures with 10 m (32.8 ft) maximum relief and a base depth of 110 m (360.9 ft). This group belongs to the deep shelf-edge pinnacle trend in the northeastern Gulf. These pinnacles form a short east–west aligned arc on the shelf-slope break, bordering the northern edge of a massive shelf-edge slump of rubble. A fairly uniform coverage of debris surrounds the base with diminutive rocky reef outcrops and patch-reefs encrusted with *R. manuelensis*, octocorals, antipatharians, and crinoids. Emergent rocky features with vertical walls, rock ridges, and rock arches are distributed across the reef. Vertical rock faces had highly eroded surfaces and were densely covered with *R. manuelensis*, with low coverage of other solitary corals, octocorals, sponges, and antipatharians. Ludwick and Walton Pinnacle 2 is another of the deep shelf-edge pinnacle group. This structure, lying immediately to the east of Pinnacle 1, also was profiled and contoured by Ludwick and Walton (1957). Dense populations of *R. manuelensis*, other solitary corals, octocorals, crinoids, and basket stars colonized the elevated rocky features, while low relief hard bottom regions were characterized primarily by octocorals, antipatharians, and crinoids. Scamp Reef is a member of the Ludwick and Walton Pinnacles deep shelf-edge group with a precipitous southern reef face. This structure, lying immediately to the west of Pinnacle 1, also was profiled and contoured by Ludwick and Walton (1957). This feature has extensive vertical rock outcrops with profiles in excess of 5 m (16.4 ft). Spectacular arches, overhangs, and rugged topography occur along the southern face of the reef, with exposed rock colonized by *R. manuelensis*, antipatharians, crinoids, octocorals, and ahermatypic coral colonies. The USGS name Scamp Reef refers to the abundance of the scamp grouper (*Mycteroperca phenax*) that reside at this site.

Qualitative observations on the physical habitat and megafaunal invertebrates associated with particular biotopes and fish assemblages were made by USGS associates from the videotapes on the ROV. Fishes on flat-topped features were assigned to six biotopes: reef top, reef face, reef crest, reef base, reef talus around a reef base, and soft bottom. Reef top biotope invertebrate assemblages had high density and species richness and were dominated by erect sponges, octocorals (particularly sea fans such as *Nicella* sp.), antipatharians, gorgono-cephalid basket stars, bryozoans, comatulid crinoids, and coralline algae. Reef crest biotopes typically were characterized by extensive rocky outcrops, with small areas of sediment cover and low invertebrate densities. The USGS report distinguished the reef crest ecotone from the adjacent flat reef top and vertical reef face biotopes to identify the possible influence of currents on the reef fish community. Reef face biotopes were rugged, vertical rocky surfaces that were characterized by lower densities of epifauna than reef tops but had an abundance of ahermatypic corals, including *R. manuelensis*, *Madrepora* sp., and *Madracis/Oculina* sp., comatulid crinoids, octocoral fans, the antipatharians spiral whip *Stichopathes lutkeni*, coral-line algae (to a depth of about 75 m [246 ft]), and sea urchins. The reef base was an ecotone between the steep reef face and the talus zone, with the rugged rocky face sometimes undercut with small cave-like overhangs. It contained vertical faces with solitary corals and the coarse sediments. Reef talus biotopes (circum-reef sediment apron) were the flat areas of reef debris and coarse carbonate sediments extending out from the base of large, high relief mounds. Coarse sediments and debris appeared to have been produced by shell and rock fragments eroded from the main reef. Small rocky outcrops in this biotope were often encrusted with solitary corals, small octocoral fans, and antipatharians. The soft-bottom/sand-plain biotopes were flat and featureless but occasionally contoured by ripples, sand waves, and excavated burrows, pits, and mounds. Sessile invertebrates in this biotope were limited to small octocorals or antipatharians attached to rock surfaces. The intermittent soft-bottom sediments should be composed of polychaete worms, crustaceans, and bivalve molluscs similar to assemblages described above for the continental shelf.

Large corals and sponges are known to occur worldwide at the outer margins of continental shelves at depths of several hundred meters (Roberts and Hirschfield 2004). These complex structures occur on hard bottoms (Brook and Schroeder 2007) and serve as habitat for a complex assemblage of invertebrates and fishes (Baker and Wilson 2001; Sulak et al. 2007, 2008). In the northern Gulf of Mexico, the corals, *Lophelia pertusa* and *Madrepora oculata*, and the black coral, *Leiopathes* sp. (Prouty et al. 2011) (Figure 7.55) are known to occur along a narrow bathymetric zone of the upper continental slope from just east of the Mississippi Delta over to the east of the De Soto Canyon (CSA 2007).

The narrow distributions (Figure 7.55) of the deep-sea coral (DSC) worldwide indicate that they all have a common set of requirements. They live in the dark. Thus, they contain no photosynthetic zooxanthellae that are vital symbionts in shallow-water hermatypic (reef-building) coral species. As they do not rely on endosymbiont photosynthesis, they are thus heterotrophic and rely on a steady rain of organic detritus that rains down from the productive surface water (Duineveld et al. 2004; Davies et al. 2010; Mienis et al. 2012) or material that is exported from the adjacent continental shelf (Walsh et al. 1981). They occupy water that is relatively cold (less than 10 degree Celsius [°C]), probably substantially colder at high latitudes), below or more or less at the permanent thermocline. At these depths (200–1,000 m [656–3,281 ft]), they would not be subject to marked seasonal temperature variations. They require a hard substrate, and in the northern Gulf of Mexico, this is provided by authigenic carbonate deposition that precipitates as fossil hydrocarbon seeps age (Roberts et al. 2010) or

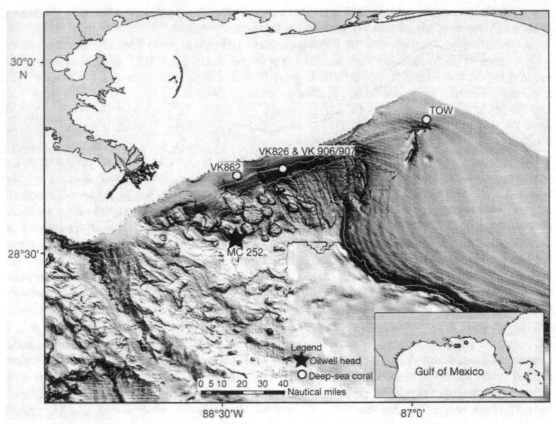

Figure 7.55. Locations of deep cold-water *Lophelia* reefs in the northeast Gulf of Mexico (from Prouty et al. 2011). The Alabama Pinnacles are located in shallower water north of the deep *Lophelia* complexes.

asphaltine solids (Williamson et al. 2008) that can support solitary sea pen and sea fan colonies. While the establishment of DSC assemblages requires these hard substrates (Hovland 1990), so far there is little evidence that the corals or sponges use the fossil organic matter as an energy or carbon source (e.g., food) (Becker et al. 2009). There is probably little to no predation on the foundation coral and sponge species themselves, but this is by inference, not actual observations. In life history models of the methane seep communities, an absence of predation is assumed because of the slow growth and long lives of the foundation species (Cordes et al. 2005a); it is thus reasonable to make this assumption—that they have no predators—with the corals as well. The DSC assemblages are considered biodiversity hot spots (Roberts et al. 2009).

West Florida *Lophelia* Lithoherms: This region consists of dozens and possibly hundreds of 5–15 m (16.4–49.2 ft) tall lithoherms (elongated carbonate mounds) off the southwest Florida shelf at depths of 500 m (1,640.4 ft), some of which are capped with thickets of live and dead *Lophelia*. The habitat extends more than 20 km (12.4 mi) along the shelf slope. In 2003, Reed et al. (2006) conducted a SEABEAM bathymetric survey over a small portion (1.85 × 1.85 km [1.15 × 1.15 mi]) of the region. They used Innovator ROV dives to ground-truth three features: a 36 m (118 ft) tall escarpment and two of the lithoherms. They examined a 36 m (118 ft) tall escarpment from 412 to 448 m (1,351.7–1,469.8 ft) at the eastern edge of the flat terrace that contained the lithoherms. The escarpment was nearly vertical and had very rugged topography with crevices, outcrops, and a series of narrow ledges. The dominant sessile fauna consisted of

Antipatharia (30 cm [11.8 in.] tall), numerous Octocorallia including Isididae (30–40 cm [11.8–15.8 in.]), and sponges, *Heterotella* spp., *Phakellia* spp., and Corallistidae. The SEA-BEAM bathymetry revealed dozens of lithoherms on a terrace west of the escarpment. Eight other lithoherms were reflected on the ROV's sonar within a 100 m (328 ft) radius. Estimated coral cover ranged from less than 5 % to greater than 50 % in some areas, with 1–20 % live. The dominant fauna was similar to the escarpment except for *Lophelia*, which was not observed on the escarpment. Common sessile benthic species included Cnidaria: Antipatharia (*Antipathes* spp. and *Cirrhipathes* spp.), *L. pertusa*, Octocorallia; and Porifera: *Heterotella* spp. and other hexactinellid vase sponges, and various plate and vase demospongiae (Pachastrellidae, Petro-siidae, Astrophorida). Common motile invertebrates included Mollusca, Holothuroidea, Cri-noidea, and decapod crustaceans (*Chaceon fenneri* and Galatheidae). Nine species of fish included Anthiinae, shortnose greeneye (*Chlorophthalmus agassizi*), conger eel (*Conger ocea-nicus*), blackbelly rosefish (*Helicolenus dactylopterus*), codling (*Laemonema melanurum*), beardfish (*Polymixia* spp.), and hake (*Urophycis* spp.). The high number of hard bottom lithoherms revealed by the limited SEABEAM mapping effort and few ROV dives led Reed et al. (2006) to believe that there was tremendous potential for unexplored coral and fish habitat in this region.

The narrow depth distribution of the deepwater corals in the eastern Gulf of Mexico is thought to require bottom currents in addition to specific temperatures (Davies et al. 2010; Mienis et al. 2012). The corals require particulate matter from the overlying phytoplankton as a food source, but particulate matter that is not useable as nutritional food could potentially also smother the corals. The authors provide evidence that bottom currents at these depths supply adequate nutritional material but also act to sweep the areas free of suspended matter that could be detrimental. Thus, in addition to a narrow temperature range and hard substrata, these species require currents that can supply adequate nutritional POC but eliminate inorganic, terrestrial, river-derived or resuspended particulate material that can smother them. Fluxes of particulate matter into a sediment trap moored above the corals indicated that supplies of POC would be adequate to support the coral metabolism and growth (Mienis et al. 2012). The intersection of requirements of temperatures of 5–10 °C, POC nutritional levels yet to be defined, persistent bottom currents and hard substrate in this habitat may explain why these species complexes are rare: the habitat is rare. This narrow intersection of requirements could explain why similar deep corals have not been encountered on knolls west of the Mississippi River where the persistent near-bottom nepheloid layer could smother them.

Slow growth is a common biological feature of all the species involved in the DSC assemblages; they live up to several hundred years or more (Prouty et al. 2011). This remarkable phenomenon is supported by age dating with ^{210}Pb and ^{14}C concentration gradients and observations of features in the skeletal material of the black coral, *Leiopathes* sp.

Two deeper habitats that need mention are the asphaltine assemblage that was discovered in association with the very deep (about 3.6 km [2.2 mi]) Sigsbee Knolls (MacDonald et al. 2004) and the iron stone crust that covers the sediment surface on the deep (greater than 2 km [1.2 mi]) eastern margin of the Mississippi sediment fan (Pequegnat et al. 1972; Rowe and Kennicutt 2008; Rowe et al. 2008a). The asphalt-like outcroppings appear to have formed from fossil hydrocarbon deposits (Williamson et al. 2008) and harbor sessile organisms such as sea fans and sea whips. The reddish iron stone crust is thought to have been formed on the surface of slump deposits that originated in shallow water (Santschi and Rowe 2008). Both occur at depths where the vital POC input is very limited, and thus, they both support minimal benthic biomass and diversity.

7.6.3 Cold Seep Communities

The first hint of methane expulsion from the sediments was the observation that acoustic records on echo-sounder recorders (ESRs) monitoring water depth and seafloor properties were occasionally, briefly, wiped out. Such wipeouts in sound were determined to be caused by gas bubbles in the water—evidently methane and other short-chained hydrocarbons bubbling out of the sediments. The sound was not transmitted through the bubbles; that is, it was wiped out on the ESR records. This gas was assumed to be coming from the dissolution of methane clathrates or ice-like material composed of sea water, clay, and short-chained hydrocarbons that together are known to form a solid (ice) at pressures of 30 to possibly greater than 100 atmosphere (atm) and temperatures of less than 10 °C As the ice warms up or as pressure diminishes, it turns to gas, thus forming bubbles. Clathrates and associated methane releases were first discovered in the Gulf of Mexico on the upper slope (Brooks et al. 1984). The methane released was then discovered to support seafloor communities that are reminiscent of hydrothermal vent communities (Kennicutt et al. 1985; Brooks et al. 1985). The ice or gas hydrates can break off and float, giving off bubbles in the process (MacDonald et al. 2003). Most information on gas expulsion has been developed during three substantial studies supported by the MMS (now BOEM). The investigations, CHEMO I (1991) and CHEMO II (1997), concentrated on locations at depths of less than 1 km (0.62 mi), whereas, the most recent project, CHEMO III, explored the deep GoM continental slope, with support from BOEM and NOAA. CHEMO III has been summarized in the special issue *Deep-Sea Res. II*, 57 (2010), Cold seeps are distributed extensively, reaching all over the northern and southern continental slopes where they are underlain by salt deposits (Figure 7.56).

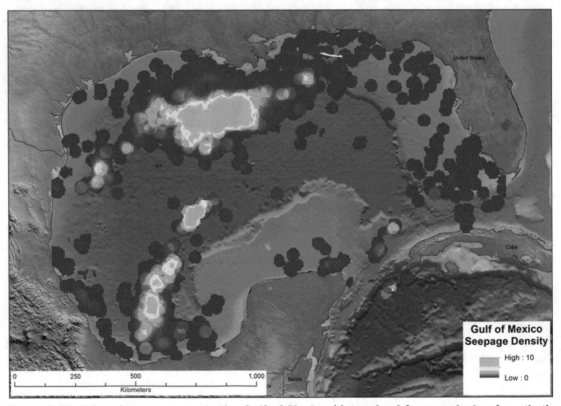

Figure 7.56. Oil and gas seepage in the Gulf of Mexico (determined from analysis of synthetic aperture radar, graphic provided by CGG's NPA Satellite Mapping, used with permission).

Finding or prospecting for seeps has taken many approaches. Sub-seafloor and seafloor surface three-dimensional seismic profiles, multibeam bathymetry, and side-scan sonar swaths are used to identify areas of potential fluid gas expulsion. Acoustic wipe out zones indicate bubbles near the seafloor. Sea surface slicks seen from satellites can be followed back to natural releases of oil and gas at the seafloor (MacDonald et al. 1993; De Beukelaer et al. 2003). With these three types of information, the next step is to confirm existence of seep communities using bottom photographic surveys, ROV observations, or deep submergence research vehicle sampling (Roberts et al. 2010). Although seafloor trawling provided some of the first clear confirmations that seep communities exist (Rosman et al. 1987), this is now frowned upon because of the damage it does to the structures. The Gulf of Mexico seeps are the most well known worldwide (Fisher et al. 2007).

The cold seep faunal assemblages occur in five categories: mussel beds, clam beds, vestimentiferan (tube worm) clumps, an epifauna of brachiopods and solitary corals, and gorgonian fields (Kennicutt et al. 1985; Rosman et al. 1987; MacDonald et al. 1989, 1990a, b, c). According to Roberts et al. (2010), of the thousands of seeps on the northern Gulf of Mexico slope, many surround the edges of the intraslope basins where shallow subsurface salt bodies give rise to bathymetry with faults that provide pathways for salt, gas, and oil to flow up to the seafloor.

Stable isotope measurements suggest that a principal energy source is hydrogen sulfide in addition to methane (Brooks et al. 1985; Demopoulos et al. 2010; Becker et al. 2010). Physiological studies have suggested that endosymbiotic relationships exist between mussels and methanotrophic bacteria, but clams and vestimentiferans contain sulfur-oxidizing bacteria (Cordes et al. 2005a). It is assumed that the seeping fossil hydrocarbons nourish sulfate-reducing bacteria that provide sulfide to the sulfide-oxidizing endosymbionts (Freytag et al. 2001). The bathymodiolids harbor at least four symbiotic functional groups: methanotrophs (consume methane as an energy and carbon source), methylotrophs (consume a methyl group at the end of a fatty acid), and two different thiotrophs (oxidize-reduced sulfur compounds for energy) (Dupperon et al. 2007). The community of organisms on the deep Florida Escarpment is not supported by fossil hydrocarbons (Paull et al. 1984).

The composition of the principal fauna associated with fluid expulsion varies over time as the seep matures. *Bathymodiolus* mussels with methanotrophic symbionts arrive first (Roberts et al. 1990; Bergquist et al. 2003). Prior to this, the sediments may need to be stabilized by carbonate precipitation that is a byproduct of the oxidation of the hydrocarbons (Aharon and Fu 2000; Joye et al. 2004; Luff et al. 2004). Vestimentiferan tubeworms follow after enough carbonate substrate is available (Cordes et al. 2003). The clumps of tubeworms and mussel beds are considered the foundation species of the seep communities (Cordes et al. 2010). The three species of mussels are *Bathymodiolus brooksi*, *B. childressi*, and *B. heckerae*. The tubeworms are known to be *Escarpia laminata* and *Lamellibrachia luymesi* (Miglietta et al. 2010) and *Seepiophila jonesi* (Gardiner et al. 2001), among others.

These foundation species serve as habitat for a speciose assemblage of smaller organisms, but the small organisms associated with the larger individual clumps are difficult to sample quantitatively (Bright et al. 2010; Cordes et al. 2010; Lessard-Pilon et al. 2010). Fauna associated with tubeworms appears to have a higher diversity than mussel beds on the upper slope (550 m [1,804 ft] depth), but at greater depths, this distinct difference is less obvious in rarefaction curves (Cordes et al. 2010). The mussel beds appear to have a mid-depth maximum (MDM) diversity but the fauna associated with the tubeworms did not. This is an interesting observation because MDMs have been observed on many continental margins, but their cause is equivocal at best. The nonseep macroinfauna of the Gulf of Mexico has a distinct MDM, but this was not apparent in the polychaete worms (Rowe and Kennicutt 2008). Many of the species associated with the foundation fauna are obviously seep-associated organisms such as the shrimp *Alvinocaris muricola*, the polychaete worm *Methanoaricia dendrobranchiata*, and the snail *Provanna sculpta*. There seems to be minimal overlap with nonseep fauna (Wei et al. 2010a). The α or within-habitat diversity in the mussel clumps and the wormtube bushes appears to be high: of 32 samples from the middle and deep slope (about 1–2.7 km [0.62–1.7 mi] depth), the mean of the expected number of species per 50 individuals (E (50)) was 6.5, $\sigma = 2.2$ (Cordes et al. 2010). These samples of the associated animals were obtained by washing the mussel clumps and the tubeworm bush samples with filtered seawater through a 1-mm sieve in the ship's laboratory (Cordes et al. 2010). The infauna from sediment samples in studies not associated with seeps was sieved through slightly finer sieves (generally 0.3 mm) (Wei et al. 2012a), making a comparison between the seep and nonseep faunas difficult. Had the seep fauna washings been done with a finer sieve, the diversity values might have been higher. Likewise, comparisons of biomass and densities are not possible because the seep foundation species, as habitats, are three dimensional, whereas the quantitative biomass and density estimates of faunas on level silt and clay sea floor away from seeps were all estimated as individuals or biomass/m^2.

The vestimentiferans, namely *Lamellibrachia luymesi*, form aggregates or bushes of up to thousands of individuals and they are estimated to live hundreds of years (Fisher et al. 1997; Julian et al. 1999; Bergquist et al. 2000), even though the individual worms are far smaller than those encountered at hydrothermal vents (Fisher et al. 1990).

An enigma in the Gulf of Mexico is the proximity of diverse, high biomass, and productive assemblages, supported by fossil hydrocarbon, to the more general, comparatively oligotrophic (low productivity and modest biomass) level-bottom assemblages away from seeps (Wei et al. 2012a). The possibility that the sites of fossil carbon expulsion and seepage are fertilizing wide areas from nearby nonseep fauna has not been supported by stable carbon and nitrogen isotope analyses in samples of fauna near seeps (Carney 1994, 2010). That is, the many different habitats that are characterized by fossil organic matter supporting high biomass and productivity on the seafloor have had very little influence on the organisms in the habitats away from the seeps. That said, the boundaries between the two (seep versus nonseep) remain poorly defined. Demopoulos et al. (2010), for example, found stable isotope evidence that a suite of free-living invertebrates in soft sediments associated with seep sites are feeding on the free-living sulfur-oxidizing white and pink *Beggiatoa*-like bacteria species living on sulfide diffusing out of the sediment.

7.6.4 Continental Slope and Abyssal Plain Assemblages

Groups of organisms also occur along the continental slope, as well as in the abyssal plain. These assemblages are described in the following paragraphs.

7.6.4.1 Microbiota (Heterotrophic Bacteria and Archaea)

Both the density and biomass of sediment microbes have been exhaustively documented by Deming and Carpenter (2008) in conjunction with the MMS study Deep Gulf of Mexico Benthos (DGoMB) (Rowe and Kennicutt 2008). Cross-slope sampling sites were spread from the western Gulf of Mexico off south Texas across the northern Gulf of Mexico to north Florida, at depths of about 200 m (656 ft) out across the SAP to depths of 3,650 m (11,975 ft). The top 15 cm (5.9 in.) of cores were counted at four sediment intervals using a combination of DAPI and Acridine Orange stains. Values ranged from 1.0×10^8 to 1.89×10^9 cells/cm^3, while depth-integrated biomass ranged from almost Log_{10} 0.5 g C/m^2 at the shallow sites down to Log_{10} 0.05 g C/m^2, with a consistent decline from the upper slope (less than 500 m [1,640 ft]) down to the low values at 3.7 km (2.5 mi) depth. Cell numbers declined with depth in the sediments. Cell densities followed no particular pattern as a function of water depth, but biomass ranged from 2.6 down to 1.0 g C/m^2 from the upper continental slope down to the low values on the abyssal plain. The reason for this difference in counts versus biomass is related to a general decrease in measured cell size with depth. The biomass of the microbiota was positively related to POC flux (Biggs et al. 2008) and negatively related to depth. Deming and Carpenter (2008) also measured whole-core respiration and microbial production on repressurized recovered cores, and these values have been used in seafloor food web models (Rowe et al. 2008b; Rowe and Deming 2011). No more detailed information is available on the specific types of bacteria and Archaea present in these counts or incubations, just that they are presumed to be heterotrophs that consume DOM.

7.6.4.2 Meiofauna: Foraminifera and Metazoa

The meiofauna are small (>40 μm) single-celled (Foraminifera) or multicelled (metazoan) organisms that consume detritus and smaller protists and bacteria living on or within the sediments. The most prevalent of the metazoans are nematodes (round worms), harpacticoid copepods (crustaceans), and kinorhynchs. Assessing the abundance of forams is difficult because the empty (dead) shells must be differentiated from living organisms (Bernhard et al. 2008). Forams have been investigated extensively because many species have calcium carbonate shells or agglutinated tests (volcanic glass shards) that are preserved as fossils, making them important sources of information on the history of Gulf of Mexico sediments (Parker 1954; Phleger and Parker 1951; Poag 1981; Reynolds 1982). Assemblages of forams are thought to be zoned with depth and associated with specific water masses (Denne and Sen Gupta 1991, 1993; Jones and Sen Gupta 1995). Some are associated with upwelling on the Florida slope (Sen Gupta et al. 1981), while others appear to occur in association with fossil hydrocarbon seeps (Sen Gupta and Aharon 1994) and bacterial mats (Sen Gupta et al. 1997). In samples of living forams across a wide depth interval, Bernhard et al. (2008) documented a mean density of 3.9×10^4 individuals/m^2, with a mean biomass of 31.5 mg C/m^2. The highest density (8.2×10^4 individuals/m^2) and biomass (98.1 mg C/m^2) were located at a known methane seep site (Bush Hill) at a depth of 548 m (1,798 ft) on the upper continental slope. Mean densities on the upper slope (4.0×10^4 individuals/m^2, $\sigma = 2.5$) were not different from those on the abyssal plain (4.6×10^4 individuals/m^2, $\sigma = 1.9$), but the biomass was higher on the slope (52 mg C/m^2, $\sigma = 34$) than on the abyssal plain (12.9 mg C/m^2, $\sigma = 6.6$). Smaller-sized forams on the abyssal plain explain the biomass difference. The mean size among all ten locations sampled by Bernhard et al. (2008) was 0.8 μg C per individual. Fifty-nine species were encountered at the ten sites sampled, but the fauna was dominated by *Saccorhiza ramosa* (51.7 % of the total individuals).

The metazoan meiofauna abundances and biomass were determined at all the same locations as the microbiota by Deming and Carpenter (2008, see above) during DGoMB (Baguley et al. 2008), making this survey of the northern Gulf of Mexico one of the most comprehensive available anywhere. In addition, this latter study measured grazing rates and estimated respiration based on temperature and animal size. Mean biomass was 43.4 mg C/m^2, with a high of 157 on the upper slope down to a low 3.5 mg C/m^2 on the abyssal plain. Nematode worms and harpacticoid copepods dominated the biomass at all depths. Densities and biomass declined with depth, as did the estimates of total respiration of this fraction of the fauna: from about 4.5 mg C/m^2/day respiration or production of carbon dioxide on the upper slope down to almost none on the abyssal plain. Harpacticoid copepod species composition and nematode genera have been used to define recurrent groups of meiofauna over this broad area of the slope and abyss (Baguley et al. 2006; Sharma et al. 2012). Groups of species were not aligned with depth (as is common in larger groups), but occurred in isolated patches that cross (rather than align with) depth intervals of hundreds of meters, probably due to their modes of reproduction and recruitment strategies. These estimates of respiration and biomass relative to depth are important because they are most likely controlled by food supply that is imported from the surface or exported to the seafloor from the adjacent continental shelf. Likewise, new or alien sources of organic matter, such as natural or accidentally spilled or leaked hydrocarbons, could affect them in either a positive or a negative manner.

7.6.4.3 Macrofauna

Quantitative investigations of the macrofauna were initiated in the mid-1960s (Rowe and Menzel 1971; Rowe 1971; Rowe et al. 1974; Pequegnat 1983). The published surveys used an anchor dredge or a van Veen grab to sample specific areas of the seafloor, followed by sediment sieving with a 0.42 mm mesh sieve. Since those early publications, the sieve size generally prescribed in studies supported by MMS in deep water has been reduced to 0.3 mm, meaning that total abundances of smaller organisms would have increased in the later studies (Recall that all the continental shelf studies used 0.5 mm sieves). These small changes, while affecting densities, probably have not affected biomass estimates (Rowe 1983). The most recent studies have used a GOMEX corer (Boland and Rowe 1991) or a spade corer (Escobar-Briones et al. 2008b, c), whereas some of the present ongoing sampling has gone to a multicorer (Barnett et al. 1984).

The Gulf of Mexico macrofauna biomass follows a log-normal relationship with depth, whether measured as wet weight, dry weight, or organic carbon (Rowe and Menzel 1971). The slope of the log-normal line appears to be the same regardless of which measure is used, but the slope of the densities can be less than that of the weight measures, indicating that abundances do not decline as fast as biomass; that is, animals in some ocean basins are getting smaller with depth. Recall that this was true of the microbiota and the meiofauna as well. It appears that the rate of decline of biomass with depth is a general feature on most continental margins, but the height of the line (the origin at shallow intercept on the shelf) above the x-axis is a function of the rate of PP in the surface water (Rowe 1971; Wei et al. 2010a). Thus, the biomass regression in the Gulf is steep but somewhat below most other ocean basins, a clear indication that the Gulf of Mexico is an oligotrophic ecosystem, with several exceptional habitats.

Most of the historical biomass measurements in the Gulf of Mexico (Figure 7.57) have been incorporated into a single database for the purpose of predicting macrofaunal biomass across large scales of depth and region (Wei et al. 2010b, 2012a). The densities and biomass are dominated by worms (Figure 7.58), either polychaetes or nematodes (Figure 7.59).

It is presumed that animal densities decline with depth because food becomes limiting (Rowe 1971, 1983). A log-normal relationship has been described for most of the world's oceans, including the Gulf of Mexico. The height of the line is related to the levels of PP in the surface water (Rowe 1971), as well as input from the margins (Walsh et al. 1981; Deming and Carpenter 2008; Santschi and Rowe 2008). Submarine canyons appear to concentrate organic matter, thus enhancing their biomass and animal abundances, especially in the Gulf of Mexico (Roberts 1977; Soliman and Rowe 2008; Escobar-Briones et al. 2008a; Rowe and Kennicutt 2008).

The biomass in the southern Gulf of Mexico is decidedly lower than that in the northern Gulf of Mexico, as illustrated in Figure 7.60 from Wei et al. (2012a), using data from Escobar-Briones et al. (2008a). This reflects the source of the water (the Caribbean via the Yucatán Strait) and the resulting low PP due to nitrate limitation. The high variance among the southern

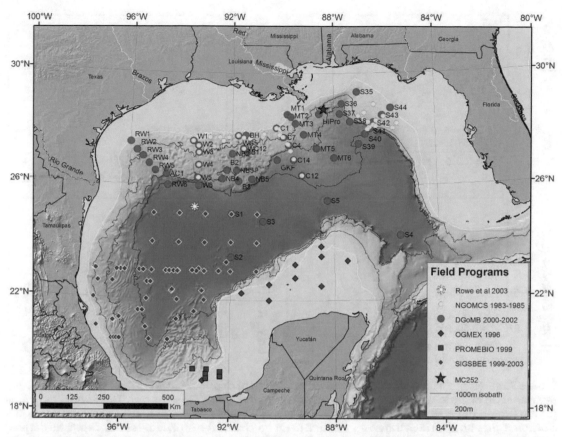

Figure 7.57. Distribution of offshore quantitative samples of macrofauna on which the biomass data are based (from Figure 1 in Wei et al. 2012a; reprinted with permission from Elsevier).

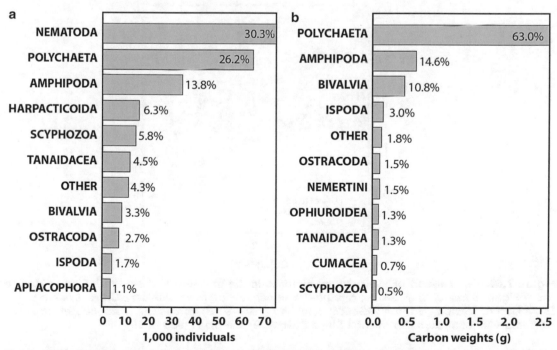

Figure 7.58. Distribution of macrofauna taxa within the samples used in the estimates of biomass (mg C per individual × total number of individuals at a location) and animal abundances (from Figure 2 in Wei et al. 2012a; reprinted with permission from Elsevier).

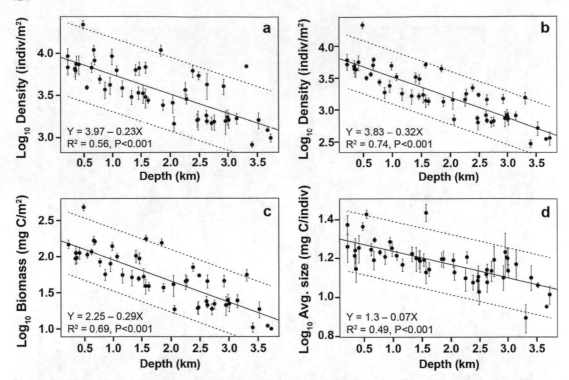

Figure 7.59. Regressions of macrofauna as a function of depth in the deep Gulf of Mexico. The *top left panel* (a) includes nematode worms and the *top right panel* (b) does not. The *bottom left panel* (c) illustrates the now classic log-normal decline in biomass as a function of depth, whereas the *bottom right panel* (d) illustrates the decline in mean size of the individuals with depth, as derived from biomass and abundance data (from Figure 3 in Wei et al. 2012a; reprinted with permission from Elsevier).

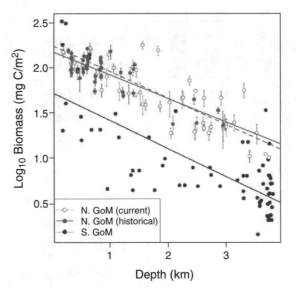

Figure 7.60. Comparison of macrofaunal biomass in the northern and southern Gulf of Mexico (from Figure 6 in Wei et al. 2012a; reprinted with permission from Elsevier). Current refers to the Deep Gulf of Mexico Benthos (DGoMB) sampling (2000–2003) versus the historical, which is the Northern Gulf of Mexico Continental Slope (NGoMCS) samples (1983–1985).

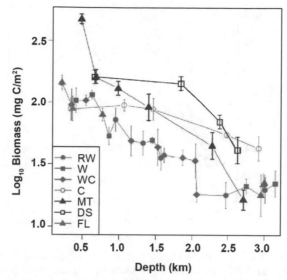

Figure 7.61. Comparison of biomass of macrofauna on transects in the MMS-sponsored DGoMB study across the northern Gulf of Mexico. The lines went from west (RW) to east (FL). While there is no apparent difference in these longitudinal extremes as illustrated, the highest values were in two large canyons (*DS* De Soto Canyon, *MT* Mississippi Trough) and the central transect, which was just west of the MT line (from Figure 5 in Wei et al. 2012a; reprinted with permission from Elsevier). Original data in or derived from Rowe and Kennicutt (2008).

Gulf of Mexico samples reflects their use of small subcores from a spade corer or a multicorer, which takes small samples (about 125 cm^2 versus 2,000 cm^2 in the GOMEX corer).

The transects across the northern margin of the Gulf of Mexico (Figure 7.61) illustrate that the highest macrofauna abundances and biomasses are found in the central locations of the Gulf, at all depths. All transects merge to very low values on the SAP. The highest numbers were encountered at a depth of 500 m (1,640 ft) in the head of the Mississippi Trough (Soliman and Rowe 2008). Much of this high density can at times be attributed to a single species of a small tube-dwelling amphipod crustacean (*Ampelisca mississippiana*) (Soliman and Rowe 2008). At mid-slope depths, however, the highest abundances and biomass were encountered in the De Soto Canyon.

Comparisons of biomass values between the Northern Gulf of Mexico Continental Slope (NGoMCS) (1983–1985) and DGoMB studies (2000–2003), about 20 years apart, revealed no significant differences (Wei et al. 2012a) (Figure 7.61). There was no indication that mid-slope basins, proximity to methane seeps or the base of steep escarpments affected the biomass or animal densities (Figure 7.62) in any of the previous studies (Wei et al. 2012a). Variations that could be attributed to season have not been tested adequately as yet, although Wei et al. (2012a) did try to estimate possible effects of what they termed *arrival time lag* of POC input to the seafloor. This is almost impossible because the settling rate of the surface-derived POC is unknown, and the rate at which newly arrived POM is incorporated into the biota is unknown and probably is a function of the different size or functional groups. Additionally, the horizontal contribution of material from the margins is thought to be important but is impossible to quantify (Bianchi et al. 2006; Rowe et al. 2008b; Santschi and Rowe 2008). Thus, the organic detritus has two sources—lateral transport from the margins and vertical transport from the surface—neither of which is well constrained or understood.

It is presumed that the severe decline in biomass and abundance of the fauna (all sizes) as a function of depth reflects the decline in POC input with depth (Figures 7.62 and 7.63). Thus, it

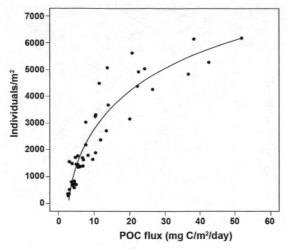

Figure 7.62. Density of macrofauna individuals in the Gulf of Mexico as a function of delivery of POC as estimated from sea surface—satellite estimated chlorophyll *a* concentration (modified from the data in Biggs et al. 2008 and Wei et al. 2012a).

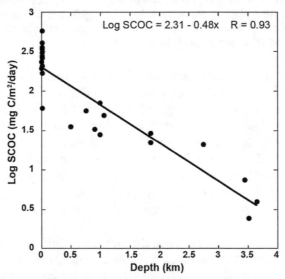

Figure 7.63. Sediment community oxygen consumption (SCOC) in the northern Gulf of Mexico (Rowe et al. 2008a). The deep samples on the slope and abyssal plain are from Rowe et al. (2003, 2008a), whereas the shallow (less than 100 m) data are from studies of the continental shelf, many of which were measured on sediments in the hypoxic area off Louisiana (Rowe et al. 2002).

would stand to reason that any additional input of labile (easily biodegradable) organic matter would enhance biomass locally. This is true for food falls or carcasses of fishes and marine mammals. That no effects could be discerned in the continental slope mesoscale basins (which could trap particulates), near methane seeps or at the base of escarpments indicates that the methods used cannot extract the effects from the highly variable database, or in fact these features do not enhance food resources.

Sediment community oxygen consumption (SCOC) (Figure 7.63) illustrates that the model-estimated POC flux and carbon turnover by the seafloor organisms are in good agreement. Both decline in significant log-normal fashion as a function of depth. However, the rate of decline in the SCOC is almost two times that of the biomass. This suggests that total community heterotrophic metabolic rates decline faster than biomass. It also demonstrates that the metabolic rate of the community as a function of biomass declines as a function of depth.

In the DGoMB samples (2000–2003), a total of about 957 different species were enumerated at the 43 designated locations. Taxonomic specialists at many different institutions generated the lists of these species. Type material is now archived in the benthic invertebrate collections at Texas A&M University—Galveston (TAMUG), whereas material collected in the earlier offshore MMS programs is housed at the Texas Cooperative Wildlife Collections Marine Invertebrate Collections at TAMU—College Station or has been deposited in the U.S. National Museum of Natural History (the Smithsonian). A large fraction of the macrofauna-sized material remains undescribed, although putative species designations have been given to each different species based on the judgment of the taxonomist in charge of a group.

A list of all the described species and the putative species with separate designations has been assembled into a single database. This database has been used to identify recurrent groups of organisms using measures of similarity (shared species) between each pair of samples across the entire northern Gulf of Mexico, excluding the continental shelf. Four major depth-related zones were apparent (Figure 7.64) (Wei et al. 2010a). The middle two were separated longitudinally as well. Each location in each demarcated group shared at least 20 % of its species with all the other locations in the zone. Wei et al. (2010a) concluded that the most important factor

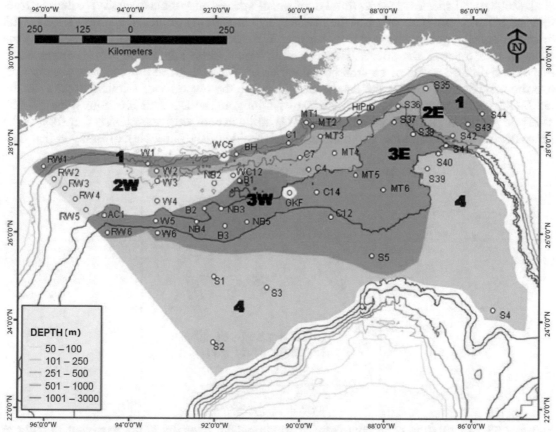

Figure 7.64. Zonation of macrofaunal species into four major depth-related zones based on percent species shared between locations, with the two intermediate zones divided between east versus west subzones (from Wei and Rowe 2006; Wei et al. 2010a).

giving rise to this pattern is the decline in POC input. That is, the variable that controls the sharp fall in biomass has also given rise to this alignment of groups of species along isobaths. The intermediate east versus west separations appeared to be a function of a difference in sediment grain sizes: a coarse sand fraction (composed of $CaCO_3$ pelagic foram tests) with a mean of 25 % in the east versus a coarse fraction of less than 5 % among the western locations. The western locations were dominated by terrigenous clays that were thought to dilute the pelagic carbonate fraction. It is not clear whether it was the sediment grain size or the mode of the pelagic input that was important. Roberts (1977) also describes four zones in the area of the De Soto Canyon based mostly on megafauna from skimmer samples (Pequegnat et al. 1970); likewise Powell et al. (2003) describes four zones of demersal fishes from the upper slope down to the shallow margin of the abyssal plain.

Biodiversity is often used as a measure of community, ecological, or environmental health. However, the causes of variations in diversity are numerous and inconclusive. The zonation referred to in Figure 7.64 is beta diversity, or the turnover or replacement of species along a physical gradient. Wilson (2008) described the *within-habitat* (alpha) diversity of isopod crustaceans in the macrofauna along the transects occupied by both NGoMCS (1983–1985) and DGoMB (2000–2002). This group, based on *expected species*, $E(s)$, displayed an MDM that occurred at the 1–1.5 km (0.62–0.75 mi) depth. To Wilson (2008), the distribution appeared to suggest that the deep Gulf of Mexico might have suffered some extinction events, and thus, the present-day deep fauna reflects invasions of shallow species from the margins. Haedrich et al. (2008) used species richness (total numbers of species or gamma diversity) to demonstrate that the MDM is not an artifact of the overlapping bathymetric ranges of multiple species with little ecological significance, but rather a significant nonrandom response to variations in the ecosystem. However, the species richness of different large taxonomic groups appeared to respond to different sets of environmental variables. Wei and Rowe (unpublished manuscript) use the macrofauna species list database to illustrate the response of within-habitat diversity [as E (100 individuals)] to POC flux estimates among all the DGoMB locations (Figure 7.65). This odd parabolic pattern could illustrate a relaxation in competitive exclusion that follows the sharp decline in POC input as depth increases (right side of the parabola); diversity in that data

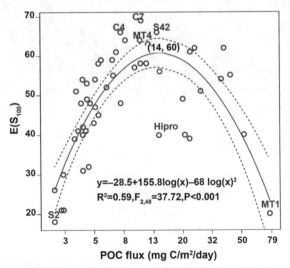

Figure 7.65. Macrofauna diversity (alpha or within-habitat diversity index Expected Number of Species per number of individuals), or rarefaction, (E (100) = number of species per 100 individuals) plotted as a function of estimated POC flux onto the seafloor (from Wei and Rowe, unpublished data, manuscript in preparation).

set attained a maximum at POC input values that are encountered on the mid- to upper slope at a depth of approximately 1.2–1.5 km (0.75–0.93 mi); then the E (100) declined again on the left arm of the parabola as POC input becomes more and more severely limiting on the abyssal plain. The "relaxation of competitive exclusion" hypothesis is just one of several possible explanations for the MDM and the increase as POC input declines offshore. An alternative is the MDM occurs in a region of intermediate levels of disturbance by physical and biological processes.

7.6.4.4 Megafaunal Invertebrates

Megafauna is defined in size as being identifiable in seafloor photographs, larger than 1 cm (0.4 in.) in diameter and caught in trawls with stretch mesh of about 2.5 cm (1 in.) (Table 7.4). It includes large sessile and motile invertebrates and in some instances authors have included bottom-living or demersal fishes as well. Here the demersal fishes have been treated separately (see below). Most of the invertebrate species encountered are documented in the monograph of Gulf of Mexico biota edited by Felder and Camp (2009); only a small fraction of this size group remains undescribed, compared to the macrofauna above, in which approximately 50 % remain undescribed.

An early goal of megafauna studies in the deep Gulf of Mexico was to document and describe patterns of bathymetric (depth) zonation (Roberts 1977; Pequegnat 1983; Pequegnat et al. 1990). The simplest approach has been to tabulate the depths with the most rapid change in species composition. This is done by observing the depth range of each species or the depths at which each species starts and then stops along the entire bathymetric gradient. Pequegnat (1983) and Pequegnat et al. (1990) used this approach and followed the overly intricate zonation nomenclature of Menzies et al. (1973) to describe Gulf of Mexico zonation patterns. Rather than looking at bathymetric starts and stops, Roberts (1977) and Pequegnat (1983) calculate percent similarities between individual skimmer samples. Roberts (1977) described four depth-related zones in the De Soto Canyon. As noted above, Wei et al. (2010a) used percent similarities to describe four zones that conformed to broad depths in the macrofauna across the entire northern Gulf of Mexico.

The compendium by Pequegnat (1983) is the most comprehensive account of Gulf of Mexico megafauna (Figure 7.64). It is a product of the environmental consultancy TerEco Corporation as a report of contract work for the MMS, but unfortunately, it was never published in the open literature either as a stand-alone book or as an individual or set of peer-refereed papers.[2] The groupings of species were determined using percent similarities, and then illustrated with a cluster diagram and a site-by-site foldout matrix illustration that is rarely used. An atlas-like section gives bathymetric distributions and quantitative abundances relative to depth of numerous species. These species distributions are presented as modified whisker plots. Each species has a dedicated page that includes the depth/abundance data, an illustration of the organism and a map of the sites where it was encountered in the Gulf of Mexico. Both fishes and large invertebrates captured with the skimmer are included. A peculiar feature of this survey was the lack of sampling in the prominent Mississippi Trough, which later studies found to be very important to deep Gulf processes and faunal groupings (see Rowe and Kennicutt 2008). It may be that Pequegnat was trying to describe the natural six zone zonation pattern with depth that Menzies et al. (1973) suggested was a worldwide feature, and Pequegnat suspected that a canyon fauna would be an exception to the rule. Or it may be

---

[2] It is however available online at http://www.data.boem.gov/PI/PDFImages/ESPIS/3/3898.pdf, thanks to the DOI's BOEM.

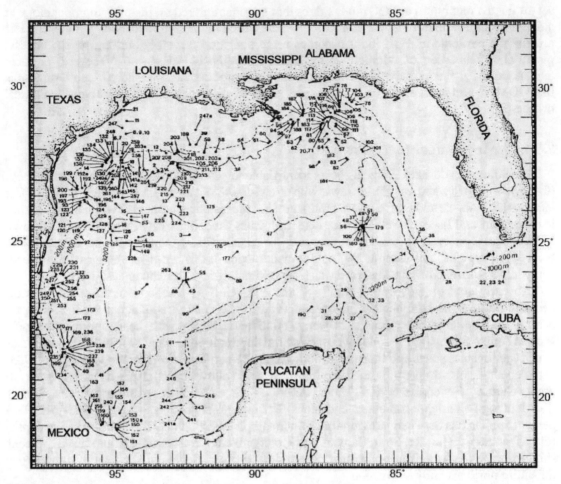

Figure 7.66. Distribution of sites sampled in the Gulf of Mexico for deepwater benthos by the R/V Alaminos, Office of Naval Research vessel operated by Texas A&M University (from Pequegnat 1983).

that the contractors (MMS) had advised Pequegnat against sampling there. It is interesting to note that the R/V *Alaminos* sampling (Figure 7.66) was not excluded from the Mexican EEZ as would be the case today without special permissions or participation with a Mexican institution on a Mexican research vessel.

An example of the illustrations in Pequegnat (1983) is the sea star, *Dytaster insignis*, with its broad depth distribution (Figure 7.67). The sea star is also common on the northwest Atlantic coast. The skimmer was particularly good at sampling the Echinodermata. Note that each major group within the echini has an MDM, as was observed in the macrofauna discussed above (Figure 7.68). However, note too that each major group's depth of maximum number of species is somewhat different. It is presumed that the megafauna prey on the macrofauna (Rowe et al. 2008b), but how this predation shapes or alters the variations in macrofauna diversity, as a function of depth is not known.

The megafauna are assumed to decline in numbers and biomass as a function of depth. They conform to the following equation:

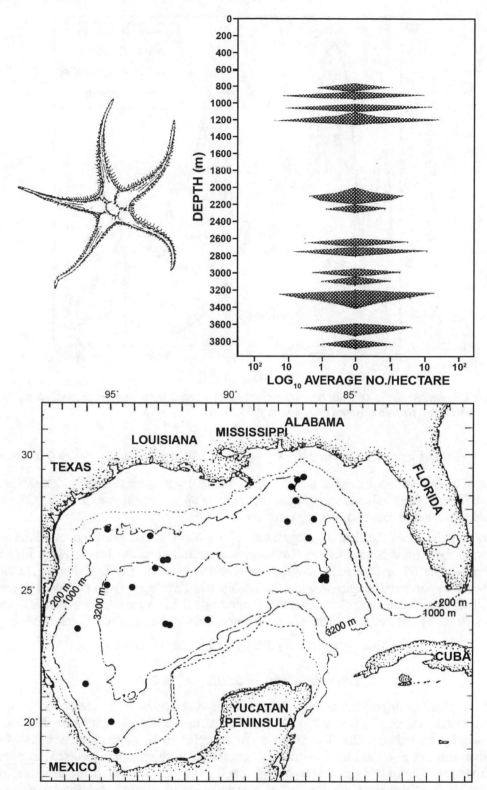

Figure 7.67. *Dytaster insignis*, a sea star, as an example of numerous illustrations of megafauna and fish distributions (from Pequegnat 1983).

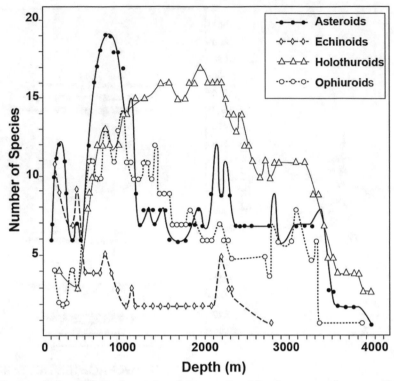

Figure 7.68. Bathymetric distribution of numbers of echinoderm species sampled by the R/V *Alaminos* using a skimmer (from Pequegnat 1983).

$$\text{Megafauna biomass}\ \left(\text{mg C/m}^2\right) = 12.1 - 2.36(\text{depth in km}),\quad r^2 = 0.02$$

But the relationship presented in Rowe et al. (2008b) was not statistically significant at the $P = 0.05$ level. Mean values ranged from about 12 mg up to a maximum of 55 mg C/m^2 on the upper slope down to less than 0.5 mg C/m^2 on the SAP.

Among the most fascinating megafauna of the Gulf is the giant isopod, *Bathynomus giganteus*, the largest isopod known (Briones-Fourzan and Lozano-Alvarez 1991). Individuals can be more than 35 cm (13.8 in.) in length. The largest weigh up to 1.4 kilograms (kg) (3.1 lb) wet weight. An exponential length–weight relationship was developed from collected specimens, and a linear relationship was found between body length (BL, cm) and body width (BW, cm), as demonstrated by the two equations below (Briones-Fourzan and Lozano-Alvarez 1991):

$$\text{Log Weight(kg)} = (-1.428 \log \text{BL}) + 2.957,\quad r^2 = 0.996$$

$$\text{BW} = 0.4338\,\text{BL} - 0.092,\quad r^2 = 0.982$$

These peculiar organisms occupy the upper slope at depths from about 200 to 1,000 m (656–3,281 ft). Although often taken in deep trawls, the most successful sampling has used large, steel wire baited traps. The animals are assumed to be general scavengers of small macrobenthos but also feed on slow-moving megafauna such as echinoderms. They appear to exhibit seasonal reproduction, although evidence for this is equivocal. Their age, respiration, and growth rates remain unknown, but it is reasonable to suggest that they play a role in cropping seafloor macrofauna. They can be kept alive in the laboratory for months and thus could be valid subjects of experimentation in the future (Mary Wicksten, 2012, Texas A&M University, personnel communication).

Solitary Cnidaria (sea fans, sea pens, anemones) are salient sessile members of the megafauna. Sea fans occur on small tar pillows on the Shenzi (oil and gas) field but not on soft mud nearby (Williamson et al. 2008). MacDonald et al. (2004) observed them associated with asphalt volcanism in the Campeche Knolls in the southern Gulf of Mexico. Trawl surveys in the northern margin of the deep Gulf of Mexico have noted that large anemones occur most frequently associated with submarine canyons and are especially common in the De Soto Canyon (Ammons and Daly 2008). It should be noted that these sessile organisms are all filter feeders that depend on a rain of detritus for nutrition, thus limiting their distributions to locations where a nutritional POC source is available. For food, they may also depend on horizontal or depth-contour controlled bottom currents to supply them with organic particulate material.

As illustrated in Figure 7.68, the echinoderms are an important component of the deep megafauna. Within this diverse phylum, the holothuroids (sea cucumbers) appear to be the most widely distributed in the deep Gulf of Mexico, with a prominent MDM. However, they are difficult to sample. This is suspected because of trawling and multishot seafloor photography in the same locations of the seafloor. For example, photographs of a species of *Peniagone* sp. illustrated that it maintained a density of about 160,000 individuals per hectare (10,000 m^2 or 107,640 ft^2). The mean length of this species in these photographs was about 2.75 cm, or just over an inch. But individuals of this species were never captured in the trawls at the same location and time. Many holothuroids are more or less neutrally buoyant and thus it was thought that these individuals were not captured because they were pushed or swept away by the trawl's bow wave. Many species of holothuroid are known to be able to swim or drift slowly over the deep seafloor. This information is contained in the *Northern Gulf of Mexico Continental Slope Study Annual Report, Year 3, Vol II, Technical Report* (Gallaway et al. 1988). This preliminary report is a font of knowledge that is not found in the final report (Pequegnat 1983) or a lone published summary of the work (Pequegnat et al. 1990).

According to Ziegler (2002), the invertebrate megafauna densities of the continental slope and abyssal plain of the Gulf of Mexico are one to two orders of magnitude less than equivalent depths in other studies at higher latitudes (Rowe and Menzies 1969; Ohta 1983; Lampitt et al. 1986; Mayer and Piepenburg 1996). This supports the suggestion that in general the Gulf of Mexico is oligotrophic (Smith and Hinga 1983), based on low densities and biomass of macrobenthos (see above section on macrofauna), but this generalization ignores the numerous slope assemblages supported by fossil hydrocarbons or the fauna in the Mississippi and De Soto canyons. While this generalization may apply to the open continental slope and the abyssal plain, it may not apply to exceptional habitats such as seeps and canyons where food supplies are enhanced.

7.6.4.5 Deepwater Demersal Fishes

The deep bottom-dwelling or demersal fish assemblages of the Gulf of Mexico are fairly well known (McEachran and Fechhelm 1998, 2006). Most species can be found in FishBase, where their worldwide distributions, age, maximum size, growth rates, and reproduction are documented. In the northern Gulf of Mexico, the species appear to occur in at least four somewhat overlapping depth-related assemblages (Roberts 1977; Pequegnat 1983; Pequegnat et al. 1990; Powell et al. 2003). Sampling deep-living species began in the 1950s by NMFS conducting exploratory fishing in the Gulf of Mexico and Caribbean aboard the *Oregon* II. Beginning in the 1960s, deep sampling on the R/V *Alaminos* by Pequegnat (1983) using a

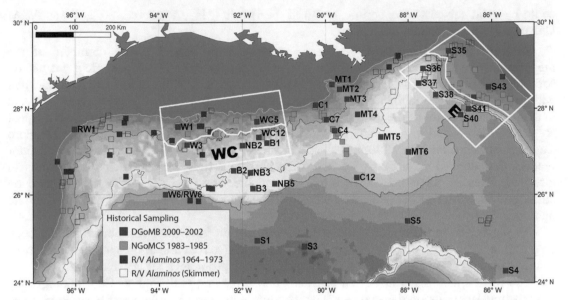

Figure 7.69. Epibenthic fish sampling in the deep northern Gulf of Mexico. The *solid symbols* are otter trawls versus *open symbols* for benthic skimmer. *Gray line* = 200 m isobath. The *black line* = 1,000 m isobath. The station names are used in the 2000–2002 sites (from Wei et al. 2012b).

skimmer resulted in a large dataset that sampled fish and invertebrates across a large spectrum of sizes. A creditable contribution of the latter was the semiquantitative estimates of fishes and megafauna that were based on the skimmer's bottom distance measuring odometer. Two other large surveys in 1983–1985 to 2000–2002 both used 40 ft semi-balloon or otter trawls (shrimp trawls) on a single warp to sample both megafauna and fishes.

Wei et al. (2012b) assembled all the demersal fish data from the above three *Alaminos* and *Gyre* surveys to determine if the fish faunal composition taken by the skimmer and the shrimp trawl were the same and also if there was any evidence that the fauna has changed during the 40 years over which the surveys were conducted (Figure 7.69). As indicated by the map, not all of the sampling was done at the same sites.

A cluster diagram of all the historical *Alaminos* samples (Figure 7.70, top) illustrates that there were four zones, according to all the data reviewed by Wei et al. (2012b). These are mapped across the area (Figure 7.70, bottom) and can be compared with the zones documented for smaller organisms above. They found no evidence that the skimmer data or time had affected the composition or the abundance of the fishes.

A cluster diagram of all the pooled data from 1964 through 2002 (Figure 7.70, top) was used to illustrate that the four zones were evident in both the old and the more recent data, according to Wei et al. (2012b). These are mapped across the area (Figure 7.71, bottom) and can be compared with the zones documented for smaller organisms above in the section on the megafauna and macrofauna. Wei et al. (2012b) found no evidence that the skimmer data or time had affected the composition or the abundance of the fishes with a 10 % similarity in species (the solid line); however, the cluster diagram does suggest that further structure exists within these large groups. This can be related to depth, but is likely a function of subtle differences in the habitats (Levin and Sibuet 2012).

Wei et al. (2012b) used violin diagrams to represent the depth and abundances of occurrences of the groups across the depth intervals using lumped data (Figure 7.72) and they separated the data as well into the most abundant species (Figure 7.73). These represent the range of the groups and the depths at which they are most abundant.

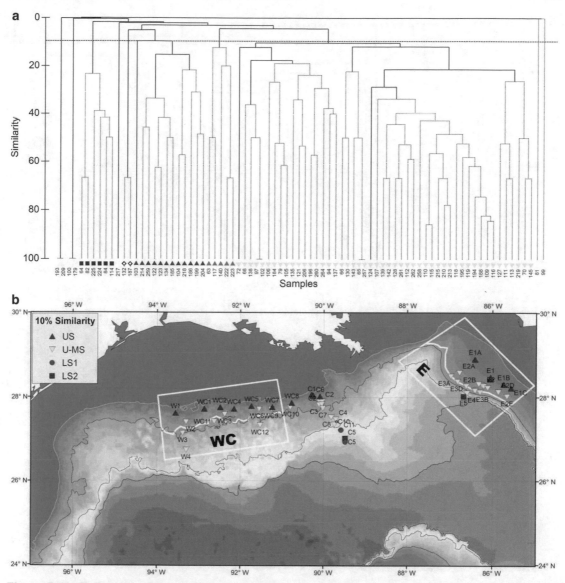

Figure 7.70. Epibenthic fish species composition and faunal zonation during the R/V *Alaminos* cruises from 1964 to 1973. (a) Group-average cluster analysis on intersample Sørensen's similarities. The *solid lines* indicate significant structure (SIMPROF test, *P* < 0.05). The *horizontal dashed line* shows 10 % similarity. (b) Distribution of the fish faunal zones with at least 10 % faunal similarity. *US* Upper-Slope Group, *U-MS* Upper-to-Mid-Slope Group, *LS1* Lower-Slope Group, *LS2* Lower-Slope-to-Abyssal Group. The *colors* on the cluster analysis dendrogram correspond to the locations of the *colors* on the map (from Wei et al. 2012b).

Multidimensional scaling (MDS) was used by Wei et al. (2012b) to view how environmental factors (as MDS1 and MDS2) affected the groups as a function of depth and as a function of the different cruises (Figure 7.74). The distances over the space in the figure are proportional to the similarity in species of the samples. That is, the shallow sites on the right were far different from the deep locations on the left. However, the red, yellow, green, and blue sites in the middle were different, but not by much (they hover close together in the center of space) in the top panel. On the other hand, the bottom panel illustrates that the colors representing cruises overlap a lot, indicating that the fauna was the same between them.

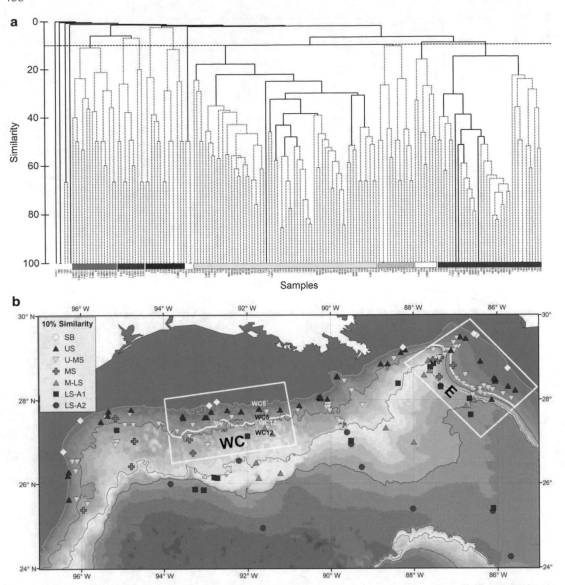

Figure 7.71. Epibenthic fish species composition and faunal zonation for the pooled data from 1964 to 2002. (a) Group-average cluster analysis on intersample Sørensen's similarities. The *solid lines* indicate significant structure (SIMPROF test, *P* < 0.05). The *horizontal dashed line* shows 10 % similarity. (b) Distribution of the fish faunal zones with at least 10 % of faunal similarity. *SB* Shelf-Break Group, *US* Upper-Slope Group, *U-MS* Upper-Slope-to-Mid-Slope Group, *MS* Mid-Slope Group, *M-LS* Mid-to-Lower-Slope Group, *LS-A1* Lower-Slope-to-Abyssal Group 1, *LS-A2* Lower-Slope-to-Abyssal Group 2. The *colors* on the cluster analysis dendrogram correspond to those on the map (from Wei et al. 2012b).

Wei et al. (2012b) plotted the fish similarities of MDS1 as a function of both depth and macrofauna biomass (Figure 7.75). The coherence of the dots indicates that depth is very important in determining where fish species live, and the right panel implies that this pattern exhibited by the fishes agrees with that of the biomass of the macrofauna. This could mean that they either depend on the macrofauna for food or that the same set of conditions that control the macrofauna also has a substantial influence on the distribution of the fishes, both of which seem logical.

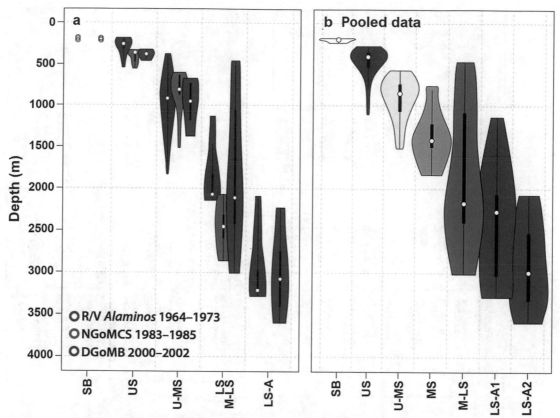

Figure 7.72. *Violin plots* of sampling depths for homogenous faunal groups in (a) R/V *Alaminos*, NGoMCS, and DGoMB studies, and (b) pooled data of all three surveys. A *violin plot* is a combination of *box plot* and kernel density plot (Wei et al. 2012b) that shows the probability of data at different values, the median and kernel density estimation. *SB* Shelf-Break Group, *US* Upper-Slope Group, *U-MS* Upper-Slope-to-Mid-Slope Group, *LS* Lower-Slope Group, *M-LS* Mid-to-Lower-Slope Group, *LS* Lower-Slope-to-Abyssal Group, *LS-A1* Lower-Slope-to-Abyssal Group 1, *LS-A2* Lower-Slope-to-Abyssal Group 2.

Biomass of the demersal fish assemblages declined with depth, according to Rowe et al. (2008b), in the following manner:

$$\text{Fish}\left(\text{mg C/m}^2\right) = 10.20\,e^{-0.93(\text{depth in km})}, \quad r^2 = 0.21$$

That is, demersal fish biomass declined exponentially down the continental margin to the abyssal plain. It can be surmised, therefore, that food supplies are increasingly limiting, and this lack of food supply is exacerbated as depth and distance from shore increase. However, it should be noted that while the above equation would predict about 10 mg C/m² at the shallow margin of the sampling, the shallow water data ranged from more than 40 mg down to about 2 mg C/m², suggesting that the upper margin is extremely variable. Also, there were hot spots of high biomass observed along the boundary between the shelf and the slope. Two of these were the De Soto Canyon and the Mississippi Trough in particular, according to Rowe et al. (2008b).

Commercial fisheries are extending down the continental slope in some regions of the world. However, this is unlikely in the Gulf of Mexico because the surface productivity is inadequate to support such a fishery, based on the data gathered to date.

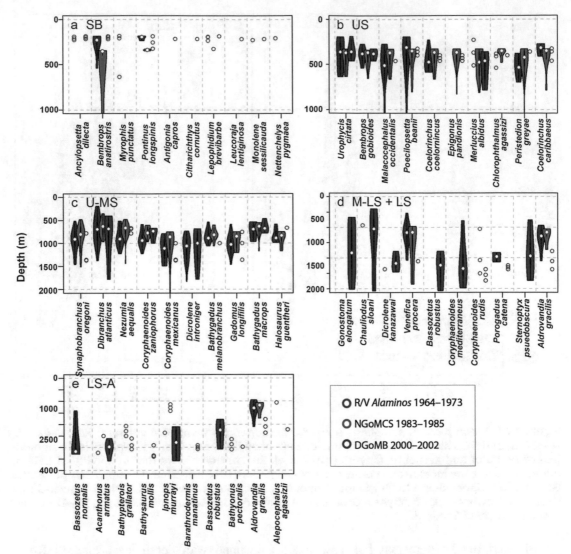

Figure 7.73. *Violin plots* of sampling depths for the top ten most common species (with highest occurrence) from (a) Shelf Break, (b) Upper Slope, (c) Upper-to-Mid-Slope, (d) Mid-to-Lower and Lower Slope, and (e) Lower-Slope-to-Abyssal Groups. *Colors* indicate different sampling times. The *violin plot* is a combination of *box plot* and kernel density plot (See Fig. 7.72). When the sampling depths were equal or fewer than three observations, the raw depth values are shown (from Wei et al. 2012b).

7.7   OFFSHORE COMMUNITY DYNAMICS, CARBON CYCLING, AND ECOSYSTEM SERVICES

Community function refers to the dynamics of the living components of assemblages of organisms. In the context of deep-ocean habitats this is considered to include such variables as growth, feeding, reproduction, recruitment, predation, mortality, respiration, and excretion (Figure 7.76). It can also include responses of the latter list to variables such as pollution, organic matter input, temperature, oxygen, and currents. In the deep ocean, these features of a

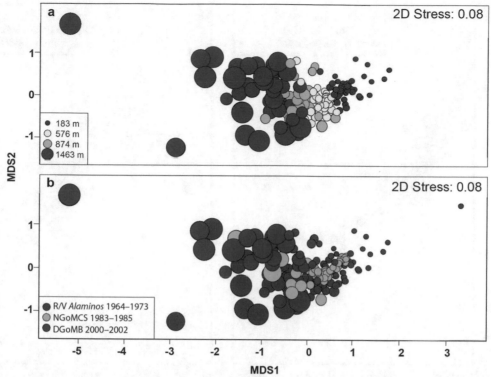

Figure 7.74. Nonmetric multidimensional scaling (MDS) on intersample Sørensen's similarities of pooled demersal fish data (from Wei et al. 2012b). The distances between samples represent dissimilarities in species composition. (a) Symbol sizes are relative water depth, with *small circles* being very shallow on the right and very deep on the left; *colors* indicate four depth intervals with equivalent numbers of samples. (b) Symbol sizes show relative depth, and *colors* indicate three studies of different sampling times.

community are substantially more difficult to assess than in shallow environments or compared to community structure characteristics (e.g., biomass, species composition, and diversity).

Methods for measuring community function include sediment traps to assess input of POC to the seafloor; use of natural and introduced radionuclides to define rates of change in time (Yeager et al. 2004; Santschi and Rowe 2008; Prouty et al. 2011); stable isotopes to infer food web structure; incubations in the laboratory or in situ to determine uptake rates of biologically active compounds such as oxygen, nitrate, and sulfide (Rowe et al. 2002, 2008a); and numerical simulations that solve for rates that are impossible to measure (Cordes et al. 2005b; Rowe et al. 2008b; Rowe and Deming 2011).

In the deep Gulf a number of studies have been undertaken to determine aspects of total level-bottom sediment community processes on the seafloor. Baguley et al. (2008) labeled sediment bacteria with ^{14}C and made them available to free-living nematode populations in small, repressurized incubation chambers. The results were inconclusive. There was little evidence that nematodes rely to any degree on bacterial cells as a food source. However, total microbial heterotrophic uptake of a ^{14}C labeled mixture of dissolved free amino acids was used to determine microbial uptake rates in combination with production of ^{14}C carbon dioxide and utilization of 3-H thymidine to determine respiration and growth rates simultaneously (Deming and Carpenter 2008). A free-falling benthic lander was used to implant incubation chambers on the seafloor to measure total SCOC (Figure 7.63). The secondary production of the

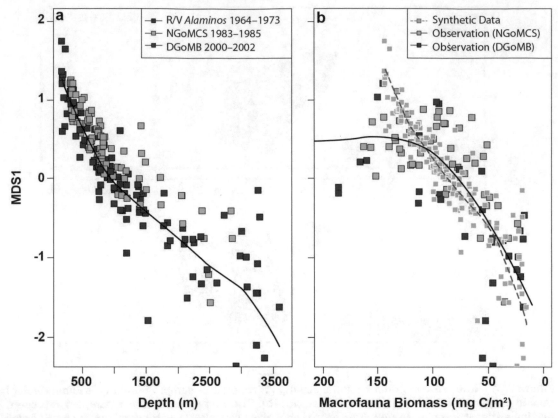

Figure 7.75. The *x*-axis of the nonmetric multidimensional scaling (MDS1) plotted against (a) depth and (b) total macrofaunal biomass, where MDS1 represents species composition of demersal fishes in multivariate space. The trend lines show the MDS1 as smooth spline functions of depth or macrofaunal biomass (from Wei et al. 2012b).

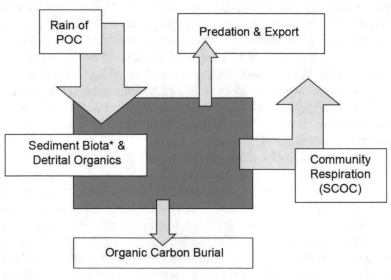

Rain of POC = Predation and Export + Community
Respiration (SCOC) + Organic Carbon Burial

Figure 7.76. Organic carbon budget for deep-sea bottom biota; * refers to "total living biomass" on and in the sea floor (microbes, meiofauna, macrofauna, and megafauna (from Rowe et al. 2008b; republished with permission of Elsevier Science and Technology Journals, permission conveyed through Copyright Clearance Center, Inc.).

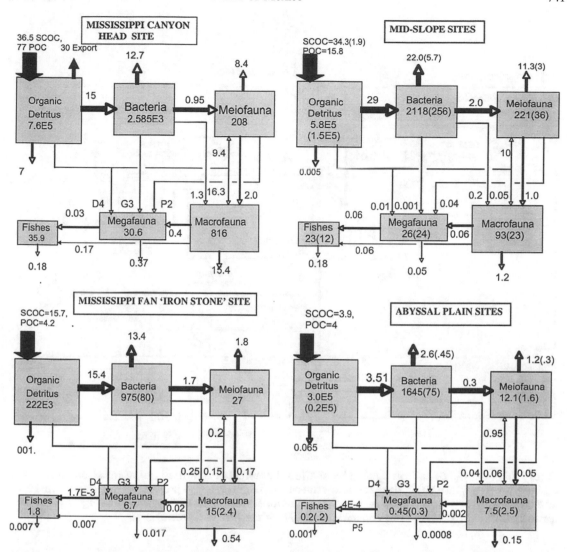

Figure 7.77. Four food web carbon budgets at depths 0.4 km (*upper left*), 1.5 km (*upper right*), 2.6 km (*lower left*), and 3.6 km (*lower right*) (from Rowe et al. 2008b; republished with permission of Elsevier Science and Technology Journals, permission conveyed through Copyright Clearance Center, Inc.), in mg C m^{-2} for the *boxes* and mg C m^{-2} day^{-1} for the *arrows*.

dominant amphipod *Ampelisca mississippiana* was estimated in the head of the Mississippi Trough using size frequencies in the population (Soliman and Rowe 2008), but it is rare that such rates can be measured in deep water because growth is slow, organisms are small, and numerous samples are required over time.

All of the stock and process data collected above during the DGoMB 2000–2002 survey have been incorporated into a model of presumed food webs at four deep locations: the Mississippi Trough head, at mid-slope depths, in the lower slope/abyssal iron stone region and on the abyssal plain (Rowe et al. 2008b) (Figure 7.77). Processes are driven by the input of POC as estimated from the SCOC regression equation (Figure 7.63) and model-estimated input inferred from satellite-determined surface chlorophyll *a* estimates (Biggs et al. 2008). This POC input to the organic carbon pool (Morse and Beazley 2008) is then divided up into five biological size categories (bacteria, meiofauna, macrofauna, megafauna, and fishes) using

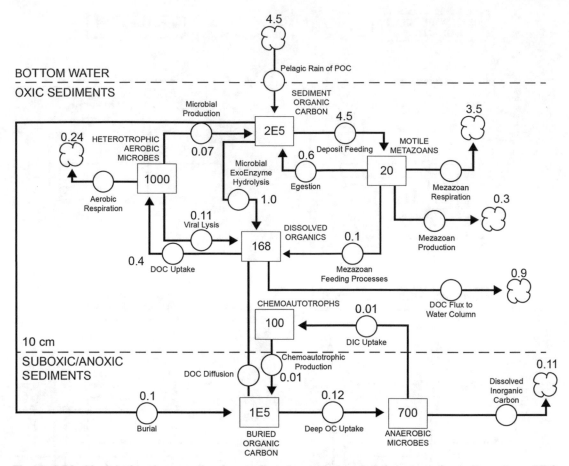

Figure 7.78. Model of carbon cycling by seafloor bacteria in relation to transformations from POC, by invertebrates, into DOC, thus reducing the role of bacteria in the processes (redrawn from Rowe and Deming 2011; reprinted by permission of Taylor & Francis Ltd.). The units are mg organic C/m^2 for the stocks (*boxes*) and mg organic C/m^2/day for the fluxes (*arrows*).

carbon as the basic model currency. The habitats at four depths are pictured: Mississippi Trough, mid-slope, iron stone area on the Mississippi Fan, and the abyssal plain, with standing stocks and total carbon flow decreasing exponentially as depth increases.

In the original rendition, most of the organic carbon was recycled by the bacteria, but a more recent assessment of the original rates in Deming and Carpenter (2008) led to a considerable downward revision of the microbial component (Rowe and Deming 2011) (Figure 7.78) because the microbes consume dissolved organic carbon (DOC) and not POC. The POC must be released into a dissolved form (DOC) before it is accessible to the bacteria. The authors suggest that this remobilization is done through "messy feeding" by motile invertebrates, viruses, or exoenzymes produced by the bacteria. How the bacterial assemblage as a whole would respond to an oil spill or free methane remains to be seen.

Table 7.5, accompanied by Figure 7.79, is a simplified summary of quantitative information on the major stocks and the fluxes or transfers between those stocks in the deep Gulf of Mexico as gleaned from the reviews in the above sections. This carbon cycle would require about 33 mg new N/m^2/day for the organic matter production by photosynthesis. The sources of this could be rain, dust, mixing up through the nutricline by storms, recycling from the zooplankton and

Table 7.5. Relationships in Carbon Biomass and Food Web Exchanges between Living Components of the Deep Offshore Water Column and Seafloor Generated from the Reviews in the Above Sections Taken from the Literature

Category	Biomass (mg C/m²)	Gains (mg C/m²/day)	Transfers (mg C/m²/day)
Phytoplankton[a]	1,000 (euphotic zone, 0–100 m)	100–200 (net primary production)	50–150 (grazing zooplankton, loss to DOC, or sinks)
Zooplankton and mid-water fishes[b]	500 (0–1,500 m)	50–100 (by grazing on phytoplankton)	15–30 (eaten by predators, wastes sink to deeper layers)
Pelagic predators	5–50	10–20 (predation on zooplankton and mid-water fishes)	1–3 (sinks as dead carcasses or feces)
Deepwater scavengers[c]	1,200 (poorly known, low concentrations but integrated over 2.7 km of water column)	12[d] (consumed over the deep water column, most lost to respiration)	3–5 (transferred as particulate matter or aggregates sinking to the bottom)
Seafloor communities[e]	1,660 (mostly inactive microbes), 3–3.7 km depth	3–5 (rain of particles from above)	0.2 (long-term burial)

The five listed stocks are represented in Fig. 7.79. Respiration is not explicit
[a]El Sayed (1972) and Biggs et al. (2008)
[b]Hopkins (1982) and Hopkins and Baird (1977)
[c]Estimated from Sutton et al. (2008) from the Atlantic Ridge
[d]Modified from Del Giorgio and Williams (2005)
[e]Rowe et al. (2008b)

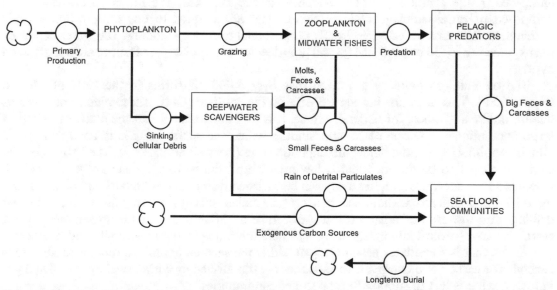

Figure 7.79. Simplified relationship between surface-produced organic matter and its routes to the deep ocean floor biota (modified from Rowe 2013).

microbiota, and nitrogen fixation by the species complex *Trichodesmium*. The major loss of organic matter from each heterotrophic stock is respiration, but that is not explicit in the budget. Even so, considerable carbon dioxide is produced over the deepwater column as the organic material that sinks into it is metabolized. The deep consumers in the water column are obscure deepwater scavengers. Although present in very low concentrations, this stock is integrated over a water column of about 2.5 km (1.6 mi). While this rendition represents the extreme deep abyssal plain of the Gulf of Mexico at a 3.2–3.7 km (2–2.3 mi) depth, at lesser depths up the continental slope, more particulate matter would reach the seafloor resulting in higher biomass, as is the case.

The effects of new or alien organic matter are not immediately apparent. Large plant detritus such as *Thalassia, Zostera* or *Sargassum* is probably of some importance. Carcasses may be as well. How fossil organics such as oil or gas would be incorporated into such a carbon budget is not as yet known.

7.8   STRESSORS AND ALTERED HABITATS

The Gulf of Mexico overall is an oligotrophic basin, in spite of its high margin-to-basin ratio and the input of nutrient-rich water from rivers, principally the Mississippi. The reason is that the largest source of water is the warm, nutrient-poor Caribbean. The nutricline is deep below the mixed layer and the euphotic zone. The result is that standing stocks of all levels of the complex, offshore food web are below comparable levels along the margins of other much larger ocean basins.

On the other hand, the Gulf suffers from a large region of hypoxia (less than 2 mg O_2/L) along the continental shelf of Louisiana. This is caused by nutrient loading and stratification resulting from the freshwater plume of the Mississippi River. The freshwater creates a vertical stratification that prevents mixing of oxygen-rich eutrophic surface water through the pycnocline into the bottom salty water. It is presumed that recreational and commercial fisheries are hampered by the condition, but the evidence for this is mixed. The fauna on the seafloor is composed of an assemblage of invertebrates that are adapted to low oxygen and organic enrichment. The area affected is directly proportional to spring flooding. The water depths of the hypoxia are 10–50 m (32.8–164 ft) and thus deepwater populations offshore are not affected by it.

The oil and gas industry at present has over 6,000 platforms in the Gulf of Mexico (Figure 7.80). This does not include PEMEX in the southern Gulf. These platforms serve as habitat unlike any other, for better or worse. It is well documented that the platforms support large populations of sessile plants and animals within the surface euphotic zone and sessile attached animals at depths below the euphotic zone (Boswell et al. 2010). The effects on the seafloor appear to be mixed. Right below a platform, the bottom fauna can be diminished, whereas a halo several kilometers away can be enriched with greater numbers and biomass than would be encountered without the platform. Detailed surveys in deep water indicate that drilling mud disposed of adjacent to a well can result in anoxic or reducing sediments for a restricted area (several kilometers at most) where the fauna is low in diversity and numbers.

A risk that is sometimes acknowledged is that the seep communities rely on fossil hydrocarbons for carbon and energy. If the supplies are diminished by withdrawal by the oil and gas industry, what is left to support these peculiar communities?

Platforms increase the primary and secondary productivity within an ecosystem by increasing the surface area on which plants can grow. This new PP is supplied with adequate nutrients by recycling of plant organic matter by the attached invertebrates and browsers within the complex of producers and consumers within the restricted habitat. Excess detritus that sinks

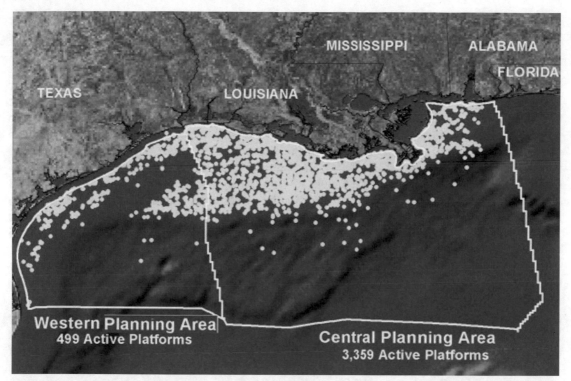

Figure 7.80. Offshore platforms in the Gulf of Mexico that serve as substrate for epibenthic organisms and habitat for numerous species of fishes popular in recreational fisheries and to commercial party boat patrons: map of the 3858 oil and gas platforms in the Gulf of Mexico in 2006. The size of the *dots* used to note platform locations is highly exaggerated and the density of platforms is low (http://oceanexplorer.noaa.gov/explorations/06mexico/background/oil/media/platform_600.html).

below the platform nourishes a deeper fauna. These processes can theoretically lead to low oxygen below a platform, but the rates of this input have not been established. The productivity of the continental shelf ecosystem would be substantially less without the platforms, but it is presently impossible to calculate this difference.

The numerous platforms are popular sites for recreational fishers and charter boat captains. Many fear that the removal of platforms after wells are no longer producing will remove and eliminate these important habitats. Some believe the removal of this widespread spatial rugosity will have severe effects on recreational fishing in the northern Gulf of Mexico (Joe Surovik, Coastal Safari Charters, personal communication).

The continental shelf of the Gulf has been subjected to shrimp trawling for almost a century (Watling and Norse 1998; Wells et al. 2008). Practically every square kilometer of surface is dragged over on a yearly basis. The exceptions are the sanctuaries such as the FGBNMS and where corals or platforms physically prevent bottom trawling. The effects of the trawling are not immediately apparent because the baseline prior to trawling is not known (Peterson et al. 2011).

It is widely believed that fishing pressure in general, worldwide, has led to an overall decrease in the mean size of the largest predatory species of finfish (Pauly et al. 1998). This is probably true for the Gulf of Mexico, but there is no historical baseline on which to verify this.

There is widespread support for regulating human activities in the upper continental slope zone because of the sensitive nature of the vulnerable DSC biotope (Rogers 1999). This assemblage is restricted to a limited set of environmental variables (rain of organic particulate matter, low and invariable temperatures, solid substrate, lack of predation), foundation species appear to be slow growing and old (decades to centuries), and the foundation species provide habitat structure to a wide variety of organisms, even though they are structurally delicate. These areas are termed vulnerable marine ecosystems (VMEs) by the International Council for the Exploration of the Sea (ICES). In addition to the corals, VMEs can contain large sponge aggregates (*Geodia* spp., *Pheronemia* spp.). Organizations supporting efforts to protect the upper continental slope VMEs are the Alaska Conservation Foundation, Earth Friends, The Rockefeller Brothers Fund, the Surdna Foundation, and the Pew Charitable Trusts, among others (Roberts and Hirschfield 2004), in addition to ICES. International agreements and national legislation to protect VMEs would be similar to marine protected areas (MPAs) and critical fisheries habitats (CFHs) in terms of regulating activities deemed harmful. According to Roberts (2002) the biggest threat to upper continental slope VMEs from human activity is bottom trawling, although oil and gas industry prospecting and production, anchoring, and some other forms of fishing might also pose some potential threats.

Overburdened sediments on the outer margin of the shelf and the upper continental slope can collapse, moving large masses of sediments downslope. These cataclysmic movements leave scars in the margin they left and hillocks where they come to rest. This process erodes away the shallower seafloor communities and then buries others, both potentially wiping out the biota of areas that are tens of kilometers in cross section. Altered or unexpected patterns in natural and bomb-produced radionuclides in the sediments are good after-the-fact evidence of where mass sediment slumping has occurred (Santschi and Rowe 2008). Such mass movements can also threaten oil and gas activities on the seafloor.

While the effects of actions near well heads on the biota on the shelf and offshore are fairly well documented (CSA 2006), the effects of massive blowouts and excessive oil, gas, and dispersant contamination remain unknown as yet. Human-derived trash is frequently encountered (Wei et al. 2012c), but deleterious effects of these alien materials have not been documented.

While hurricanes are known to have profound effects on coastlines, it should be recognized that they can resuspend sediments down to tens of centimeters out on the continental shelf to water depths of at least 50 m (164 ft). The wave action on the bottom is known to completely reorganize seafloor assemblages of organisms. For example, the well-adapted polychaete worm fauna that survives hypoxia off Louisiana was replaced by a more typical invertebrate assemblage after hurricane Katrina. Ironically, that new fauna turned out to be more susceptible to the stress of the following summer's hypoxia (Nunnally et al. 2013).

The effects of climate change on the offshore biota of the Gulf of Mexico are open to conjecture. More drought conditions will increase salinities nearshore and in isolated or closed embayments. Wet conditions will have the opposite effect: flooding will intensify or enlarge the hypoxic region off Louisiana. Increased water temperatures may be deleterious to organisms during the summer that are already near their upper limit of temperature tolerance. Lowered pH may make calcium carbonate deposition by organisms more difficult. A slight rise in seafloor temperatures in the areas of methane clathrate deposits may cause them to de-gas more intensively or even to break loose from the bottom.

7.9 REMAINING UNKNOWNS

Although the data to date suggest that seeps do not influence the nonseep fauna, it is difficult to accept that multiple seeps occurring in close proximity over an extended area of the bottom do not harbor their own associated sediment fauna. The hydrocarbon sources could fertilize adjacent fauna that is characteristic of a depth range or they could be supporting their own unique assemblage of species adapted for gassy or oily sediments. The continental shelf and slope are composed of layers of pelagic and terrigenous mixtures of sediment that overlay thick salt deposits. It is reasonable to assume that the salt is squeezed out horizontally when it reaches the steep escarpments that line much of the basin. If salt does squeeze out from the escarpments horizontally, it could be forming slow-moving rivers of dense salt that would have unknown effects on the biota (William Bryant, TAMU, personal communication). The deep basin of the central Gulf of Mexico is bordered on three sides by extremely steep escarpments. The fauna that lives on these unique formations is virtually unknown, except for small targeted areas (Paull et al. 1984; Reed et al. 2006). No consistent investigations to date have documented how the fauna might be changing offshore as a function of time. Such changes could be seasonal and a function of PP that responds to sunlight, nutrients, or mixing. The continental shelf hypoxia associated with the Mississippi River plume is an example of recurrent seasonality nearshore, but offshore the deepwater effects of the spring bloom have not been demonstrated in the Gulf of Mexico, although seasonality is widely recognized on other continental margins.

In the *Year 3* report on the NGoMCS investigation in 1982, a section prepared by Greg Boland illustrated that the small (2–5 cm [0.80–2 in.]) sea cucumber, *Peniagone* sp., was observed in great abundances (hundreds per mi^2), but they were not sampled by a trawl. It is not known if this was a function of gear or timing of the sampling. We can thus ask what other organisms have not been sampled because we have not had the means to capture them? At great depths (between about 2–3.7 km [1.2–2.3 mi]) in the water column, the resident fauna is relatively unknown; presumably the sparse fauna subsists on a meager rain of detrital particles from the surface, but that is just a presumption: no data is available on what lives in this large volume of water and what supplies this fauna with nutrition. While some information is available on this layer in some ocean basins (Sutton et al. 2008), we know almost nothing of this layer in the deep Gulf of Mexico. This is a huge volume of water, and its biota will undoubtedly prove to be sparse; quantifying it needs to be accomplished nonetheless.

7.10 SUMMARY

The purpose of establishing a baseline for the status of the plankton and benthos of the open Gulf of Mexico is because these broad categories of organisms support, as food sources, all the major groups of larger organisms of economic importance or charismatic megafauna (mammals, birds, turtles). The health of the benthos and plankton groups—defined by their abundance, biomass, diversity, and productivity—determines or controls the larger organisms in the food web. The terminal elements of a food web are not sustainable if their food supplies fail or if their food sources are altered significantly. This summary does not include finfish, commercially important invertebrates, mammals, turtles, or birds.

This summary addresses communities or assemblages of organisms, sometimes referred to as biotopes, in a variety of habitats. These assemblages of organisms can each be defined by their quantitative abundances and biomasses and their biodiversities within volumes of water or sea surface areas, usually per m^2. In addition, where useful and available, the several dominant organisms are listed by their common and scientific names. Species lists are not provided, although references in the literature that contain such lists are given. The Gulf of Mexico offshore ecosystem is divided up into salient habitats, and each contains its own suites of organisms (e.g., assemblages or biotopes). These include (1) continental shelves, (2) deep continental margins and adjacent abyssal plain, (3) methane seeps, and (4) live (hard) bottoms, partitioned according to water depths [hermatypic coral reefs in the Mexican EEZ, coral banks on diapirs (e.g., the FGBNMS, Alabama Pinnacles, FMG, Viosca Knolls, and Florida Litho-herms)]. In addition, some important exceptional habitats within those habitats are highlighted (shelf hypoxia off Louisiana, large submarine canyons [Mississippi, De Soto, Campeche], deep iron stone sediments, and asphaltine outcroppings).

The functional groups of organisms reviewed are (1) phytoplankton, separated into near-shore (neritic) and open-ocean assemblages, (2) zooplankton, again separated into neritic and offshore populations, with somewhat more extensive coverage of the ichthyoplankton because of its potential importance to fisheries, and (3) benthos, divided by habitat into level-bottom soft sediments, hard bottom coral-supporting sea floor and fossil hydrocarbon-supporting communities. In each case, some explanations are given about what biological processes or environmental characteristics of a particular habitat control the distributions of the organisms in question.

Several significant generalizations can be made based on the baseline information referred to above. In general, the low productivity and biomass of many of the larger habitats indicate that the Gulf of Mexico is oligotrophic compared to similar habitats at higher latitudes or continental margins characterized by tropical or equatorial upwelling. This generalization is based on geographically widespread assessments of phytoplankton, zooplankton, and benthic biomass. Deep benthos, regardless of size category, declines exponentially as a function of depth and delivery of detrital organic matter to the seafloor; the well-established statistical regressions of these declines tend to be below similar biomass estimates on other continental margins where such studies have been conducted. Likewise, the benthic biomass down across the continental margin of the northern Gulf of Mexico appears to be higher than that across the continental margin of the southern Gulf of Mexico. The deep zooplankton and the benthos species composition fall into depth-related zones along the continental margin of the northern Gulf of Mexico. That is, all groups of organisms appear to be zoned into discrete depth intervals, but with substantial overlap in species composition between zones.

Several important exceptions to oligotrophy are evident. The Louisiana continental shelf west of the Mississippi Delta is subjected to seasonal hypoxia because of excessive nitrate delivery in the river water and stratification caused by the freshwater. Ameliorating this harmful recurring condition is problematic; improving farming practices to reduce the nitrate loading and diverting the freshwater before it reaches the Gulf are possible helpful alternatives (Peterson et al. 2011). Much of the continental slope is characterized by patches of larger benthic organisms that are sustained by fossil hydrocarbons that seep up to the seafloor from deposits within the sediments. While many similar cold seep communities have now been discovered on continental margins worldwide, the Gulf of Mexico appears to support some of the most prolific that have been described to date. Clearly, what is known now about the species composition and the chemistry and physiological modes of existence of such communities is based on studies conducted in the Gulf of Mexico.

Another exceptional habitat type with high diversity and biomass are several large subma-rine canyons. It is presumed that they support high regional biomass by accumulating or focusing organic detritus. Likewise such habitats provide physical complexity that enhances species richness. Hard bottoms, sometimes referred to as live bottoms, are intermittently scattered across the entire Gulf of Mexico continental margin. They are inherently more difficult to evaluate because quantitative evaluations have to consider three dimensions in many cases. The hard bottom makes sampling difficult. Numerous sessile large benthic organisms, both animals and plants, attached to the seafloor in such habitats provide a diverse physical environment that provides niches for a long list of inhabitants, from small cryptic invertebrates to large finfishes. While diversity and species lists in such habitats have been evaluated with cameras and direct observations, quantifying biomass and rates of processes remains extremely difficult if not impossible; comparisons are relative between such habitats. The shallow banks on the continental shelf contain hermatypic corals that depend on light because the corals contain photosynthetic zooxanthellae. Many such banks are important to recreational fisheries, as are the many habitats formed by offshore platforms. Such complex structures are also fascinating destinations for scuba divers. An important example is the FGBNMS. At greater depths, such as the Alabama Pinnacles, hard bottoms on seafloor prominences have long provided popular fishing spots, although they are too deep for recreational scuba. We know little about what lives on the unexplored escarpments surrounding the deep Gulf of Mexico central basin.

A major shortcoming of a summary of the diversity, abundance, biomass, and productivity of the lower-level components of the various habitats of the Gulf is a general lack of valid long-term (centuries-long) baseline information. This is especially true for the continental shelves; they have been fished extensively for decades or more and what is now observed may not resemble the biota that existed prior to extensive exploitation. The continental slope of the northwest Gulf of Mexico is composed of alternating mesoscale basins and diapirs. Each basin might present a different habitat, depending on its underlying fossil hydrocarbon deposits and its relation to settling particulate matter. Virtually nothing is known about the fauna of the many individual basins and how they compare with each other or with the biota outside of a basin. In terms of food webs, the case has been made in the appropriate sections that the major supplies of energy and carbon that support the food webs of most habitats are either (1) PP in surface water that creates the slow rain of POC through the water column, and (2) seeps of naturally occurring fossil hydrocarbons that support extensive but patchy seep communities. However we know little of the relative importance of alternative sources such as carcasses or *Sargassum* and shallow water-attached plants. While it is widely acknowledged that the continental slopes of the Gulf are subject to slumps of sediments and that turbidity flows have formed the Mississippi Fan and adjacent abyssal plain, we know little of how such dynamic physical processes might affect the fauna.

This review of the plankton and benthos of the Gulf of Mexico demonstrates that the principal ecosystem components, at the lower end of the food web (phytoplankton, zooplankton, mid-water fishes, and seafloor organisms) in most habitats are characteristic of an oligotrophic ecosystem; that is, the biota is relatively low in numbers and biomass compared to other continental margins (e.g., upwelling regions, temperate and polar latitudes). The principal cause of this oligotrophy is the source water from the Caribbean depleted of nitrate in about the surface 125 m (410 ft). The penetration of the LC coming up through the Yucatán channel spins off warm anticyclonic (clockwise) eddies that travel west across the Gulf of Mexico. These features induce a counter flow in the opposite direction. Depending on location, this combination of complicated surface currents can draw nutrient-rich water off the continental shelf into deep water, and phytoplankton production can thus be marginally enhanced offshore. Upwelling zones along the west coast of the Yucatán Peninsula and Florida are also characterized by some intensification of PP. Most of the offshore regions of modestly enhanced productivity can be observed remotely by satellites.

Populations of plankton offshore represent a near-surface fauna that declines with depth in a biocline: the further from the surface, the more depauperate the biomass. This biocline occurs in the top 100–200 m (328–656 ft), and by a depth of 1 km (0.62 mi) the standing stocks are extremely limited. All size groups of multicellular organisms decline exponentially as a function of depth and distance from land, so that the abyssal plain supports only a few mg C/m^2 of total seafloor biota (fishes, zooplankton, mega-, macro-, and meiobenthos). Biodiversity of the macrobenthos, measured as alpha or within-habitat diversity, follows a different pattern as a function of depth, depending on the taxon studied. In general there is a mid-depth maximum (MDM) of the macrofauna alpha diversity at a depth of about 1.2 km (0.75 mi). Beta diversity (zonation or recurrent groups across a physical gradient) is clearly apparent in the macrofauna, megafauna, and fishes (Pequegnat 1983; Pequegnat et al. 1990; Powell et al. 2004; Wei et al. 2010a), and the steep decline of POC flux with depth has been suggested as a cause (Wei et al. 2010a). The oligotrophic (depauperate in biomass) conditions are reflected in low sediment mixing and biodegradation (Yeager et al. 2004; Santschi and Rowe 2008) and sediment community biomass and respiration (Rowe et al. 2008a, b).

The deep continental margin of the Gulf of Mexico has exceptionally complex layers of pelagic and terrigenous sediments overlying thick salt that is associated with fossil organic deposits (oil and gas). This oil and gas seeps up to the seafloor where it supports a peculiar fauna. The seep-supported assemblages are believed to live upwards of centuries, based on in situ growth rate experiments. Authigenic carbonate deposited at old seeps provides substrate for deep-living cold-water corals such as *Lophelia pertusa* that provide habitat for deep-living demersal fish, crustaceans, and echinoderms in a narrow depth band at the upper margin of the continental slope in the northeastern Gulf of Mexico (Sulak et al. 2008). Given that the open Gulf is relatively oligotrophic, these corals would not be expected to be as abundant in the Gulf of Mexico as they are in other more productive basins or at high latitudes.

Potential problems in sustaining the biota offshore include the possible effects of climate change, turbidity currents and slumps, eutrophication, oil and gas industry accidents, hypoxia, overfishing, trawling the bottom, and hurricanes. The luxuriant growths associated with pinnacles and salt diapirs are threatened by all the above, one way or the other. The establishment of areas such as the FGBNMS offers some protection from directly intrusive activities, but not from climate-induced changes that are more global. The thousands of oil and gas industry platforms in the Gulf seem to have had a positive effect on biodiversity and fishing, but there is no uniform acceptance of these relationships. Removal of platforms on the other hand is thought to be a threat to thriving recreational fishers and charter boat operators.

ACKNOWLEDGMENTS

BP sponsored the preparation of this chapter. This chapter has been peer reviewed by anonymous and independent reviewers with substantial experience in the subject matter. I thank the peer reviewers, as well as others, who provided assistance with research and the compilation of information. Completing this chapter would not have been possible without the tireless work of Kym Rouse Holzwart and Jonathan Ipock, ENVIRON International Corporation, in obtaining documents, compiling data and information, preparing text, maps, and graphs, and compiling references.

REFERENCES

Adams JA (1960) A contribution to the biology and postlarval development of the Sargassum fish, *Histrio histrio* (Linnaeus), with a discussion of the *Sargassum* complex. Bull Mar Sci 10:56–82

Aharon P, Fu B (2000) Microbial sulfate reduction rates and oxygen isotope fractionations at oil and gas seeps in deepwater Gulf of Mexico. Geochim Cosmochim Acta 62:233–246

Ammons A, Daly M (2008) Distribution, habitat use and ecology of deepwater anemones (Actiniaria) in the Gulf of Mexico. Deep-Sea Res II 55:2657–2666

Backus R, Craddock J, Haedrich R, Robison B (1977) Atlantic mesopelagic zoogeography. Mem Sears Found Mar Res 1(Part 7):266–286

Baguley J, Montagna P, Lee W, Hyde L, Rowe G (2006) Spatial and bathymetric trends in Harpacticoida (Copepoda) community structure in the Northern Gulf of Mexico deep sea. J Exp Mar Biol Ecol 320:327–341

Baguley J, Montagna L, Hyde L, Rowe G (2008) Metazoan meiofauna biomass and weight-dependent respiration in the northern Gulf of Mexico deep sea. Deep-Sea Res II 55:2607–2616

Baker M, Wilson C (2001) Use of bomb radiocarbon to validate otolith section ages of red snapper *Lutjanus campechanus* from the northern Gulf of Mexico. Limnol Oceanogr 46:1819–1824

Bangma J, Haedrich R (2008) Distinctiveness of the mesopelagic fish fauna in the Gulf of Mexico. Deep-Sea Res II 55:2594–2596

Barnett P, Watson J, Connelly D (1984) The multiple corer for taking virtually undisturbed samples from the shelf, bathyal and abyssal sediment. Oceanol Acta 7:399–408

Baustian M, Rabalais N (2009) Seasonal composition of benthic macrofauna exposed to hypoxia in the northern Gulf of Mexico. Estuar Coasts 32:975–983

Baustian MM, Craig JK, Rabalais NN (2009) Effects of summer 2003 hypoxia on macro-benthos and Atlantic croaker foraging selectivity in the northern Gulf of Mexico. J Exp Mar Biol Ecol 381:S3–S37

Becker E, Cordes E, Macko S, Fisher C (2009) Importance of seep primary production to *Lophelia pertusa* and associated fauna in the Gulf of Mexico. Deep-Sea Res I 56:786–800

Becker E, Lee R, Macko S, Faure B, Fisher C (2010) Stable carbon and nitrogen isotope compositions of hydrocarbon-seep bivalves on the Gulf of Mexico lower continental slope. Deep-Sea Res II 57:1957–1964

Bedinger C Jr (1981) Ecological investigations of petroleum production platforms in the central Gulf of Mexico, vol 1, Part 6, 7. U.S. Department of Commerce, National Technical Information Service, Alexandria, VA, USA, 527 p

Bello J, Rios V, Liceaga C, Zetina C, Cervera K, Arceo P, Hernandez H (2005) Incorporating spatial analysis of habitat into spiny lobster (*Panulirus argus*) stock assessment at Ala-cranes Reef, Yucatán, Mexico. Fish Res 73:37–47

Bergquist D, Williams F, Fisher C (2000) Longevity record for deep-sea invertebrate. Nature 403:499–500

Bergquist D, Ward T, Cordes EE, McNelis T, Howlett S, Kosoff R, Hourdez S, Carney R, Fisher CR (2003) Community structure of vestimentiferan-generated habitat islands from Gulf of Mexico cold seeps. J Exp Mar Biol Ecol 289:197–222

Bernhard J, Sen Gupta B, Baguley J (2008) Benthic foraminifera living in Gulf of Mexico bathyal and abyssal sediments: Community analysis and comparison to metazoan meio-faunal biomass. Deep-Sea Res II 55:2617–2626

Bianchi T, Allison M, Canuel E, Corbett D, McKee B, Sampere T, Wakeham S, Waterson E (2006) Rapid export of organic matter to the Mississippi Canyon. EOS Trans Am Geophys Union 87:565–573

Bianchi T, Dimarco S, Cowan R Jr, Hetland R, Chapman P, Day J, Allison M (2010) The science of hypoxia in the northern Gulf of Mexico: A review. Sci Total Environ 408:1471–1484

Biggs D (1992) Nutrients, plankton and productivity in a warm-core ring in the western Gulf of Mexico. J Geophys Res 97:2143–2154

Biggs D, Muller-Karger F (1994) Ship and satellite observations of chlorophyll stocks in interacting cyclone-anticyclone eddy pairs in the western Gulf of Mexico. J Geophys Res 99:7371–7384

Biggs D, Hu C, Muller-Karger F (2008) Remotely sensed sea-surface chlorophyll and POC flux at Deep Gulf of Mexico Benthos sampling stations. Deep-Sea Res II 55:2555–2562

Bird J (1983) Relationships between particle-grazing zooplankton and vertical phytoplankton distributions on the Texas continental shelf. Estuar Coast Shelf Sci 16:131–144

Bogdanov D, Sokolov V, Khromov N (1969) Regions of high geological and commercial productivity in the Gulf of Mexico and Caribbean Sea. Oceanology 8:371–381

Boland G, Rowe G (1991) Deep-sea benthic sampling with the GOMEX box corer. Limnol Oceanogr 36:1015–1020

Bortone S, Hastings P, Collard S (1977) The pelagic *Sargassum* ichthyofauna of the eastern Gulf of Mexico. Northeast Gulf Sci 1:60–67

Boswell K, Wells RJD, Cowan J Jr, Wilson C (2010) Biomass, density, and size distributions of fishes associated with a large-scale artificial reef complex in the Gulf of Mexico. Bull Mar Sci 86:879–889

Bright TJ, Kraemer GP, Minnery GA, Viada ST (1984) Hermatypes of the Flower Garden Banks, Northwestern Gulf of Mexico: A comparison to other western Atlantic reefs. Bull Mar Sci 34:461–476

Bright M, Plum C, Riavitz L, Nikolov N, Martinez-Arbizu P, Cordes E, Gollner S (2010) Epizooic metazoan meiobenthos associated with tubeworm and mussel aggregations from cold seeps of the northern Gulf of Mexico. Deep-Sea Res II 57:1982–1989

Briones-Fourzan P, Lozano-Alvarez E (1991) Aspects of the biology of the giant isopod *Bathynomus giganteus* A. Milne Edwards, 1879 (Flabellifera: Cirolanidae), off the Yucatán Peninsula. J Crust Biol 11:375–385

Brook S, Schroeder W (2007) State of deep coral ecosystems in the Gulf of Mexico region: Texas to the Florida Straits. In: Lumsden S, Hourigan T, Bruckner A (eds) The state of deep coral ecosystems of the United States. NOAA Technical Memorandum CRCP-3. NOAA Coral Reef Conservation Program, Silver Spring, MD, USA, pp 271–306

Brooks JM, Giammona CP (eds) (1990) Mississippi-Alabama marine ecosystem study annual report, year 2, vol 1, Technical narrative. OCS Study MMS 89-0095. Minerals Management Service, New Orleans, LA, USA. 348 p

Brooks JM, Kennicutt MC II, Fay RR, McDonald TJ, Sassen R (1984) Thermogenic gas hydrates in the Gulf of Mexico. Science 225:409–411

Brooks J, Kennicutt M II, Bidigare R, Fay R (1985) Hydrates, oil seepage, and chemosynthetic ecosystems on the Gulf of Mexico slope. EOS Trans Am Geophys Union 66:106

Bullis H, Thompson J (1965) Collections by the exploratory fishing vessels *Oregon*, *Silver Bay*, *Combat* and *Pelican* made during 1956-1960 in the southwestern North Atlantic, vol 510. Special scientific report-fisheries. U.S. Department of the Interior, Fish and Wildlife Service, Washington, DC, USA, 130 p

Carassou L, Hernandez F, Powers S, Graham W (2012) Cross-shore, seasonal, and depth-related structure of ichthyoplankton assemblages in coastal Alabama. Trans Am Fish Soc 141:1137–1150

Carney RS (1994) Consideration of the oasis analogy for chemosynthetic communities at Gulf of Mexico hydrocarbon vents. Geo-Mar Lett 14:149–159

Carney RS (2010) Stable isotope trophic patterns in echinoderm megafauna in close proximity to and remote from Gulf of Mexico lower slope hydrocarbon seeps. Deep-Sea Res II 57:1965–1971

Carpenter EJ, Roenneberg T (1995) The marine planktonic cyanobacteria *Trichodesmium* spp.: Photosynthetic rate measurements in the SW Atlantic Ocean. Mar Ecol Prog Ser 118:267–273

Casazza T, Ross S (2008) Fishes associated with pelagic *Sargassum* and open water lacking *Sargassum* in the Gulf Stream off North Carolina. Fish Bull 106:348–363

Chew F (1956) A tentative method for the prediction of the Florida red tide outbreaks. Bull Mar Sci Gulf Caribb 6:292–304

Collins M (1985) Species profiles: Life histories and environmental requirements of coastal fishes and invertebrates (South Florida)—Striped Mullet. U.S. Fish and Wildlife Service biological report 82 (11.34), U.S. Army Corps of Engineers, TR EL-82-4. Washington, DC, USA, 11 p

Comyns BH, Lyczkowski-Schultz J (2004) Diel vertical distribution of Atlantic croaker, *Micropogonas undulatus*, larvae in the Northcentral Gulf of Mexico with comparisons to red drum, *Sciaenops ocellatus*. Bull Mar Sci 74:69–80

Cordes E, Bergquist D, Shea K, Fisher C (2003) Hydrogen sulphide demand of long-lived vestimentiferan tube worm aggregations modifies the chemical environment at deep-sea hydrocarbon seeps. Ecol Lett 6:212–219

Cordes E, Hourdez S, Predmore B, Redding M, Fisher C (2005a) Succession of hydrocarbon seep communities associated with the long-lived foundation species *Lamellibrachia luymesi*. Mar Ecol Prog Ser 305:17–29

Cordes E, Arthur M, Shea K, Arvidson R, Fisher C (2005b) Modeling the mutualistic interactions between tubeworms and microbial consortia. PLoS Biol 3:0497–0506

Cordes E, Becker E, Hourdez S, Fisher C (2010) Influence of foundation species, depth and location on diversity and community composition at Gulf of Mexico lower-slope cold seeps. Deep-Sea Res II 57:1870–1881

Cowan J Jr, Shaw R (1988) The distribution, abundance, and transport of larval sciaenids collected during winter and early spring from the continental shelf waters off West Louisiana. Fish Bull 86:129–142

Cowan J Jr, Grimes C, Patterson W III, Walters C, Jones A, Lindberg W, Sheehy D, Pine W III, Powers J, Campbell M, Lindeman C, Diamon L, Hilborn R, Gibson H, Rose K (2010) Red snapper management in the Gulf of Mexico: Science- or faith-based? Rev Fish Biol Fish 21:187–204

CSA (Continental Shelf Associates) (2006) Effects of oil and gas exploration and development at selected continental slope sites in the Gulf of Mexico, vol 1, Executive summary. OCS Study MMS 2006-044. U.S. Department of the Interior, Minerals Management Service, Gulf of Mexico OCS Region, New Orleans, LA, USA, 45 p

CSA (2007) Characterization of northern Gulf of Mexico deepwater hard bottom communities with emphasis on *Lophelia* coral. U.S. Department of the Interior, Minerals Management Service, OCS Study MMS 2007-044. MMS, Gulf of Mexico OCS Region, New Orleans, LA, USA, 169 p + app

Dagg M (1995) Copepod grazing and the fate of phytoplankton in the northern Gulf of Mexico. Cont Shelf Res 15:1303–1317

Dagg MJ, Breed GA (2003) Biological effects of Mississippi River nitrogen on the northern Gulf of Mexico—a review and synthesis. J Mar Syst 43:133–152

Dagg M, Ortner P, Al-Yamini F (1988) Winter-time distribution and abundance of copepod nauplii in the northern Gulf of Mexico. Fish Bull 86:219–230

Dagg MJ, Green EP, McKee BA, Ortner PB (1996) Biological removal of fine-grained lithogenic particles from a large river plume. J Mar Res 54:149–160

Dagg M, Sato R, Liu H (2008) Microbial food web contributions to bottom water hypoxia in the northern Gulf of Mexico. Cont Shelf Res 28:1127–1137

Danek L, Tomlinson M, Tourtellotte G, Tucker W, Erickson K, Foster G, Lewbel G, Boland G, Baker J (1985) Southwest Florida Shelf benthic communities study year 4, annual report, vol 1, Executive summary. U.S. Department of the Interior, Minerals Management Service, Metairie, LA, USA, 49 p

Davies A, Duineveld GCA, van Weering TCE, Mienis F, Quattrini A, Seim H, Bane J, Ross S (2010) Short-term environmental variability in cold water coral habitat at Viosca Knoll, Gulf of Mexico. Deep-Sea Res I 57:199–212

De Beukelaer SM, MacDonald IR, Guinnasso NL, Murray JA (2003) Distinct side-scan sonar, RADARSAT SAR, and acoustic profiler signatures of gas and oil seeps on the Gulf of Mexico slope. Geo-Mar Lett 23:177–186

De la Cruz A (1972) Zooplankton de la region sureste del Golfo de Mexico. Ciencias 4:55

De Vries D, Grimes C, Lang K, White D (1990) Age and growth of king and Spanish mackerel larvae and juveniles from the Gulf of Mexico and U.S. South Atlantic Bight. Environ Biol Fish 29:135–143

Del Giorgio PA, Williams PJB (2005) Respiration in aquatic ecosystems. Oxford University Press, Oxford, UK, 315 p

Deming J, Carpenter S (2008) Factors influencing benthic bacterial abundance, biomass and activity on the northern continental margin and deep basin of the Gulf of Mexico. Deep-Sea Res II 55:2597–2606

Demopoulos A, Gualtieri D, Kovacs K (2010) Food-web structure of seep sediment macrobenthos from the Gulf of Mexico. Deep-Sea Res II 57:1972–1981

Denne R, Sen Gupta B (1991) Association of bathyal foraminifera with water masses in the northwestern Gulf of Mexico. Mar Micropaleontol 17:173–193

Denne R, Sen Gupta B (1993) Matching of benthic foraminiferal depth limits and water mass boundaries in the northwestern Gulf of Mexico: An investigation of species occurrences. J Foraminiferal Res 22:108–117

Ditty J, Shaw R, Grimes C (1994) Larval development, distribution, and abundance of common dolphin, *Coryphaena hippurus,* and pompano dolphin, *C. equiselis* (family: Coryphaenidae), in the northern Gulf of Mexico. Fish Bull 92:275–291

Ditty J, Shaw R, Cope J (2004) Distribution of carangid larvae (Teleostei: Carangidae) and concentrations of zooplankton in the northern Gulf of Mexico, with illustrations of early *Hemicaranx amblyrhynchus* and *Caranx* spp. larvae. Mar Biol 145:1001–1014

Dooley J (1972) Fishes associated with the pelagic *Sargassum* complex, with a discussion of the *Sargassum* community. Contrib Mar Sci 16:1–32

Dortch Q, Whitledge T (1992) Does nitrogen or silicon limit phytoplankton production in the Mississippi River plume and nearby regions? Cont Shelf Res 12:1293–1309

Duineveld G, Lavaleye M, Berghuis E (2004) Particle flux and food supply to a seamount cold-water coral community (Galicia Bank, NW Spain). Mar Ecol Prog Ser 277:13–23

Dupperon S, Sibuet M, MacGregor B, Kuypers M, Fisher C, Dubilier N (2007) Diversity, relative abundance and metabolic potential of bacterial endosymbionts in the *Bathymodiolus* mussel species from cold seeps in the Gulf of Mexico. Environ Microbiol 9:1423–1438

El Sayed S (1972) Primary productivity and standing crop of phytoplankton. In: Bushnell V (ed) Chemistry, primary productivity and benthic algae of the Gulf of Mexico, vol 22, Serial atlas of the marine environment (Folio 22). American Geographical Society, Brooklyn, NY, USA, pp 8–13

Elliott D, Pierson J, Roman M (2012) Relationship between environmental conditions and zooplankton community structure during summer hypoxia in the northern Gulf of Mexico. J Plankton Res 34:602–613

Ellis S, Incze L, Lawton P, Ojaveer H, MacKenzie B, Pitcher CR, Shirley TC, Ero M, Tunnell JW Jr, Doherty PJ, Zeller BM (2011) Four regional marine biodiversity studies: Approaches and contributions to ecosystem-based management. PLoS One 6:e18997. doi:10.1371/journal.pone.0018997

Engle V, Summers J (2000) Biogeography of benthic macroinvertebrates in estuaries along the Gulf of Mexico and western Atlantic coasts. Hydrobiologia 436:17–33

Escobar-Briones E, Estrada Santillan E, Legendre P (2008a) Macrofaunal density and biomass in the Campeche Canyon, Southwestern Gulf of Mexico. Deep-Sea Res II 55:2679–2685

Escobar-Briones E, Gaytan-Caballero A, Legendre P (2008b) Epibenthic megacrustaceans from the continental margin, slope and abyssal plain of the southwestern Gulf of Mexico: Factors responsible for variability in species composition and diversity. Deep-Sea Res II 55:2667–2678

Escobar-Briones E, Diaz C, Legendre P (2008c) Meiofaunal community structure of the deep-sea Gulf of Mexico: Variability due to sorting methods. Deep-Sea Res II 55:2627–2633

Espinosa-Fuentes M, Flores-Coto C (2004) Cross-shelf and vertical structure of ichthyoplankton assemblages in continental shelf waters of the southern Gulf of Mexico. Estuar Coast Mar Sci 59:333–352

Fautin D, Dalton P, Incze LS, Leong J-AC, Pautzke C, Rosenberg A, Sandifer P, Sedberry G, Tunnel JW Jr, Abbott I, Brainard RE, Brodeur M, Eldredge LG, Feldman M, Moretzshon F, Vroom PS, Wainstein M, Wolff N (2010) An overview of marine biodiversity in United States waters. PLoS One 5:e11914. doi:10.1371/journal.pone.0011914

Fedoryako B (1980) The ichthyofauna of the surface waters of the Sargasso Sea southwest of Bermuda. J Ichthyol 20:1–9

Fedoryako B (1989) A comparative characteristic of oceanic fish assemblages associated with floating debris. J Ichthyol 29:128–13

Felder D, Camp D (eds) (2009) Gulf of Mexico origin, waters and biota: Biodiversity. Texas A&M Press, College Station, TX, USA, 1393 p

Fisher C, Kennicutt M II, Brooks J (1990) Stable carbon isotopic evidence for carbon limitation in hydrothermal vent vestimentiferans. Science 247:1094–1096

Fisher C, Urcuyo I, Simkins M, Nix E (1997) Life in the slow lane: Growth and longevity of cold-seep vestimentiferans. PSZNI Mar Ecol 18:83–94

Fisher C, Roberts H, Cordes E, Bernard B (2007) Cold seeps and associated communities of the Gulf of Mexico. Oceanography 20:118–129

Flint W (ed) (1980) Environmental studies, South Texas outer continental shelf, 1975-1977, vol 3. University Texas Marine Science Institute, Port Aransas, TX, USA, 648 p

Flint W, Rabalais N (eds) (1981) Environmental studies of a marine ecosystem. University of Texas, Austin, TX, USA, 240 p

Freytag J, Girguis P, Bergquist D, Andras J, Childress J, Fisher CR (2001) A paradox resolved: Sulfide acquisition by roots of seep tubeworms sustains net chemoautotrophy. Proc Natl Acad Sci U S A 98:13408–13413

Fuiman L, Werner R (eds) (2002) Fishery science: The unique contributions of early life stages. Blackwell Scientific Publications, Oxford, UK, 336 p

Gallaway BJ, Martin, LR, Howard, RL (eds) (1988) Northern Gulf of Mexico continental slope study annual report year 3, vol 2. Technical report, 614 p

Gardiner S, McMullin E, Fisher C (2001) Seepiophila jonesi, a new genus and species of vestimentiferan tube worm (Annelida: Pogonophora) from hydrocarbon seep communities in the Gulf of Mexico. Proc Biol Soc Wash 114:694–707

Gartner J Jr, Hopkins T, Baird R, Milliken D (1988) The lanternfishes (Piscies: Myctophidae) of the eastern Gulf of Mexico. Fish Bull 85:81–98

Gaston G (1985) Effects of hypoxia on the macrobenthos of the inner shelf off Cameron, Louisiana. Estuar Coast Shelf Sci 20:603–613

Geyer R (1970) Preface. In: Pequegnat W, Chace F (eds) Contributions to the biology of the Gulf of Mexico, vol 1, Texas A&M University Oceanographic Studies. Gulf Publishing, Houston, TX, USA, pp v–xiv

Gilbes F, Tomas C, Walsh J, Muller-Karger F (1996) An episodic chlorophyll plume on the West Florida Shelf. Cont Shelf Res 16:1201–1224

GMFMC (Gulf of Mexico Fishery Management Council) (2004) Final environmental impact statement for the generic essential fish habitat amendment to the fishery management plans of the Gulf of Mexico. Gulf of Mexico Fishery Management Council, Tampa, FL, USA, 682 p

GMFMC (2005) Final generic amendment number 3 for addressing the essential fish habitat requirements to the fishery management plans of the Gulf of Mexico. Gulf of Mexico Fishery Management Council, Tampa, FL, USA, 106 p

Gonzalez-Gandara C, Arias-Gonzalez J (2001) Lista actualizada de los peces del arrecife Alacranes, Yucatán, Mexico. Anales del Instituto de Biologia, UNAM, Serie Zoologia 72:245–258

Gorelova T, Fedoryako B (1986) Topic and trophic relationships of fishes associated with drifting *Sargassum* algae. J Ichthyol 26:63–72

Gower J, King S (2011) Satellite images show the movement of floating *Sargassum* in the Gulf of Mexico. Int J Remote Sens 32:1917–1929

Haddon M (2001) Modelling and quantitative methods in fisheries. Chapman and Hall, New York, NY, USA, 406 p

Haedrich R, Devine J, Kendall V (2008) Predictors of species richness in the deep-benthic fauna of the northern Gulf of Mexico. Deep-Sea Res II 55:2650–2656

Hamilton P (1992) Lower continental slope cyclonic eddies in the central Gulf of Mexico. J Geophys Res Oceans 97:2185–2200

Harder S, David A (2009) Examination of proposed additional closed areas on the west Florida shelf. Report. Gulf of Mexico Fish Management Council, Tampa, FL, USA, 10 p

Harper D (1977) Distribution and abundance of macrobenthic and meiobenthic organisms in the vicinity of the Buccaneer oil/gas field. Final Report. National Marine Fisheries Service, Galveston, TX, USA

Hernandez F Jr, Powers SP, Graham WM (2010) Detailed examination of ichthyoplankton seasonality from a high-resolution time series in the northern Gulf of Mexico during 2004–2006. Trans Am Fish Soc 139:1511–1525

Hoffmayer E, Franks J, Comyns B, Hendon R, Waller R (2002) Larval and juvenile fishes associated with pelagic *Sargassum* in the Northcentral Gulf of Mexico. Gulf Caribb Fish Inst 56:261–269

Hopkins T (1982) The vertical distribution of zooplankton in the eastern Gulf of Mexico. Deep-Sea Res 29:1069–1083

Hopkins T, Baird R (1977) Aspects of the feeding ecology of oceanic midwater fishes. In: Andersen N, Zahuranec B (eds) Oceanic sound scattering prediction, vol 5, Marine Science. Plenum Press, New York, NY, USA, pp 325–360

Hopkins T, Lancraft T (1984) The composition and standing stock of mesopelagic micronekton at 27° N 86° W in the eastern Gulf of Mexico. University of Texas Marine Science Institute. Contrib Mar Sci 27:143–158

Hopkins T, Baird R, Milliken D (1973) A messenger-operated closing trawl. Limnol Oceanogr 18:488–490

Hopkins T, Gartner J Jr, Flock M (1989) The caridean shrimp (Decapoda; Natantia) assemblage in the mesopelagic zone of the eastern Gulf of Mexico. Bull Mar Sci 45:1–14

Hopkins T, Sutton T, Lancraft T (1996) The trophic structure and predation impact of low latitude midwater fish assemblage. Prog Oceanogr 38:205–239

Houde E (1997) Patterns and consequences of selective processes in teleost early life histories. In: Chambers R, Trippel E (eds) Early life history and recruitment in fish populations. Chapman and Hall, London, UK, pp 173–196

Hovland M (1990) Do carbonate reefs form due to fluid seepage? Terra Nova 2:8–18

Jochens A, DiMarco S (2008) Physical oceanographic conditions in the deepwater Gulf of Mexico in summer 2000-2002. Deep-Sea Res II 55:2541–2554

Johnson D, Perry H, Lyczkowski-Schultz J, Hanisko D (2009) Red snapper larval transport in the northern Gulf of Mexico. Trans Am Fish Soc 138:458–470

Jones M, Sen Gupta B (1995) Holocene benthic foraminiferal diversity and abundance variations in lower bathyal and abyssal environments, northwestern Gulf of Mexico. Gulf Coast Assoc Geol Soc Trans 45:304–311

Joye S, Boetius A, Orcutt B, Montoya J, Schulz H, Erickson M, Lugo S (2004) The anaerobic oxidation of methane and sulfate reduction in sediments from Gulf of Mexico cold seeps. Chem Geol 205:219–238

Julian D, Gaill F, Wood E, Arp A, Fisher C (1999) Roots as a site of hydrogen sulfide uptake in the hydrocarbon seep vestimentiferan *Lamellibranchia* sp. J Exp Biol 202:2245–2257

Kennicutt M II, Brooks J, Bidigare R, Fay R, Wade T, MacDonald T (1985) Vent-type taxa in a hydrocarbon seep region on the Louisiana slope. Nature 317:351

Kimmel D, Boicourt W, Pierson J, Roman M, Zhang X (2009) A comparison of the mesozooplankton response to hypoxia in Chesapeake Bay and the northern Gulf of Mexico using the biomass size spectrum. J Exp Mar Biol Ecol 381:S65–S73

Kleppel GS, Davis CS, Carter K (1996) Temperature and copepod growth in the sea: A comment on the temperature-dependent model of Huntley and Lopez. Am Nat 148:397–406

Kornicker L, Bonet F, Cann R, Hoskin C (1959) Alacran reef, Campeche Bank, Mexico. Inst Mar Sci 6:1–22

Lampitt R, Billett D, Rice A (1986) Biomass of the invertebrate megabenthos from 500-4100 m in the northeast Atlantic Ocean. Mar Biol 93:69–81

Lang KL, Grimes CB, Shaw RF (1994) Variations in the age and growth of yellowfin tuna larvae, *Thunnus albacores*, collected about the Mississippi River plume. Environ Biol Fish 39:259–270

Lapoint B (1995) A comparison of nutrient-limited productivity in *Sargassum natans* from neritic vs. oceanic waters of the western North Atlantic Ocean. Limnol Oceanogr 40:625–633

Lenes J, Darrow B, Cattrall C, Heil C, Callahan M, Vargo G, Byrne R, Prospero J, Bates D, Fanning K, Walsh J (2001) Iron fertilization and the *Trichodesmium* response on the West Florida shelf. Limnol Oceanogr 46:1261–1277

Lessard-Pilon S, Porter M, Cordes E, MacDonald I, Fisher C (2010) Community composition and temporal change in deep Gulf of Mexico cold seeps. Deep-Sea Res II 57:1891–1903

Letelier R, Karl D (1996) Role of *Trichodesmium* spp. in the productivity of the subtropical North Pacific Ocean. Mar Ecol Prog Ser 133:263–273

Levin L, Sibuet M (2012) Understanding continental margin biodiversity: A new imperative. Ann Rev Mar Sci 4:79–112

Lindquist D, Shaw R, Hernandez F Jr (2005) Distribution patterns of larval and juvenile fishes at offshore petroleum platforms in the north-central Gulf of Mexico. Estuar Coast Shelf Sci 62:655–665

Lohrenz R, Dagg M, Whitledge T (1990) Enhanced primary production at the plume/oceanic interface of the Mississippi River. Cont Shelf Res 19:639–664

Ludwick J, Walton W (1957) Shelf edge calcareous prominences in the northeastern Gulf of Mexico. AAPG Bull 41:2054–2101

Luff R, Wallmann K, Aloisi G (2004) Numerical modeling of carbonate crust formation at cold vent sites: Significance for fluid and methane budgets and chemosynthetic biological communities. Earth Planet Sci Lett 221:337–353

Lyczkowski-Shultz J, Steen JP Jr (1991) Diel vertical distribution of red drum *Sciaenops ocellatus* larvae in the northcentral Gulf of Mexico. Fish Bull 89:631–641

Lyczkowski-Shultz J, Hanisko D, Sulak K, Dennis G III (2004) Characterization of ichthyoplankton within the U.S. Geological Survey's Northeastern Gulf of Mexico study area—based on analysis of Southeast area monitoring and assessment program (SEAMAP) sampling surveys, 1982-1999. NEGOM ichthyoplankton synopsis final report. OCS report USGS-scientific investigation report SIR-2004-5059 and CEC NEGOM Program Investigation Report 2004-02. U.S. Department of the Interior, U.S. Geological Survey, Gainesville, FL, USA, 136 p

MacDonald I, Boland G, Baker J, Brooks J, Kennicutt M II, Bidigare R (1989) Gulf of Mexico hydrocarbon seep communities. II. Spatial distribution of seep organisms and hydrocarbons at Bush Hill. Mar Biol 101:235–247

MacDonald IR, Callender WR, Burke RA, McDonald SJ, Carney RS (1990a) Fine-scale distribution of methanotrophic mussels at a Louisiana cold seep. Prog Oceanogr 24:15–24

MacDonald IR, Guinasso NL, Reilly JF, Brooks JM, Callender WR, Gabrielle SG (1990b) Gulf of Mexico hydrocarbon seep communities. VI: Patterns in community structure and habitat. Geo-Mar Lett 10:244–252

MacDonald IR, Reilly JF, Guinasso NL, Brooks JM, Bryant WR (1990c) Chemosynthetic mussels at a brine-filled pockmark in the northern Gulf of Mexico. Science 248:1096–1099

MacDonald I, Guinasso N, Ackleson S, Amos J, Duckworth R, Sassen R, Brooks JM (1993) Natural oil slicks in the Gulf of Mexico are visible from space. J Geophys Res 98:16351–16364

MacDonald I, Sager W, Peccini M (2003) Gas hydrate and chemosynthetic biota in mounded bathymetry at mid-slope hydrocarbon seeps: Northern Gulf of Mexico. Mar Geol 198:133–158

MacDonald I, Bohrmann G, Escobar E, Abegg F, Blanchon P, Blinova B, Breckmann W, Drew M, Eisenhauer A, Han X, Heeschen K, Meier F, Mortera C, Naehr T, Orcutt B, Bernard B, Brooks J, de Farag M (2004) Asphalt volcanism and chemosynthetic life, Campeche Knolls, Gulf of Mexico. Science 304:999–1002

Marancik K, Richardson D, Lyczkowski-Shultz J, Cowen R, Konieczna M (2012) Spatial and temporal distribution of grouper larvae (Serranidae: Epinephelinae: Epinephelini) in the Gulf of Mexico and Straits of Florida. Fish Bull 110:1–20

Martinez-Andrade F (2003) A comparison of life histories and ecological aspects among snappers (Pisces: Lutjanidae). Louisiana State University, Baton Rouge, LA, USA, 194 p

Maul G (1974) The Gulf loop current. In: Smith R (ed) Proceedings of the marine environmental implications of offshore drilling in the eastern Gulf of Mexico. State University System of Florida, Institute of Oceanography, St. Petersburg, FL, USA, pp 87–97

Mayer M, Piepenburg D (1996) Epibenthic community patterns on the continental slope off east Greenland at 75° N. Mar Ecol Prog Ser 143:151–164

McEachran J, Fechhelm J (1998) Fishes of the Gulf of Mexico, vol 1. University of Texas Press, Austin, TX, USA, 1112 p

McEachran J, Fechhelm J (2006) Fishes of the Gulf of Mexico, vol 2. University of Texas Press, Austin, TX, USA, 1004 p

Menzies RJ, George RY, Rowe GT (1973) Abyssal environment and ecology of the world oceans. John Wiley and Sons, New York, NY, USA, 488 p

Mienis F, Duineveld G, Davies A, Ross S, Seim H, Bane J, van Weering T (2012) The influence of near-bed hydrodynamic conditions on cold-water corals in the Viosca Knoll area, Gulf of Mexico. Deep-Sea Res I 60:32–45

Miglietta M, Hourdez S, Cordes E, Fisher C (2010) Species boundaries of Gulf of Mexico vestimentiferans (Polychaeta, Siboglinidae) inferred from mitochondrial genes. Deep-Sea Res II 57:1916–1925

Milne-Edwards A (1880) Reports of the Results of Dredging, Under the Supervision of Alexander Agassiz, in the Gulf of Mexico and in the Caribbean Sea, 1877, '78, '79, by the United States Survey Steamer "Blake", Lieutenant Commander CD Sigsbee, USN, and Commander JR Bartlett, USN, Commanding. 8. Etudes Preliminaires sur les Crustaces. Bull Mus Comp Zool 8:1–68

Milroy S, Dieterle D, He R, Kirkpatrick G, Lester K, Steidinger K, Vargo G, Walsh J, Weisberg R (2008) A three-dimensional biophysical model of *Karenia brevis* dynamics on the west Florida shelf: A look at physical transport and potential zooplankton grazing controls. Cont Shelf Res 28:112–136

Morse J, Beazley M (2008) Organic matter in deepwater sediments of the northern Gulf of Mexico and its relationship to the distribution of benthic organisms. Deep-Sea Res II 55:2563–2571

Moser M, Auster P, Bichy J (1998) Effects of mat morphology on large *Sargassum* associated fishes: Observations from a remotely operated vehicle (ROV) and free-floating video camcorders. Environ Biol Fish 51:391–398

Muhling B, Lamkin J, Roffer M (2010) Predicting the occurrence of Atlantic bluefin tuna (*Thunnus thynnus*) larvae in the northern Gulf of Mexico: Building a classification model from archival data. Fish Oceanogr 19:526–539

Muhling B, Lamkin J, Richards J (2012) Decadal-scale responses of larval fish assemblages to multiple ecosystem processes in the northern Gulf of Mexico. Mar Ecol Prog Ser 450:37–53

Murrell M, Fleeger J (1989) Meiofauna abundance on the Gulf of Mexico continental shelf affected by hypoxia. Cont Shelf Res 9:1049–1062

NOAA, CCMA (National Oceanic and Atmospheric Administration, Center for Coastal Monitoring and Assessment) (2002) The state of coral reef ecosystems of the United States and Pacific Freely Associated States: 2002. National Centers for Coastal Ocean Science, Silver Spring, MD, USA, 265 p

Nunnally C, Rowe G, Thornton DCO, Quigg A (2013) Sedimentary oxygen consumption and nutrient regeneration in the Gulf of Mexico hypoxic zone. In: Brock JC, Barras JA, Williams SJ (eds) Understanding and predicting change in the coastal ecosystems of the northern Gulf of Mexico, Journal of Coastal Research, special issue no. 63, Coconut Creek, FL, USA, pp 84–96

Ohta S (1983) Photographic census of large-sized benthic organisms in the bathyal zone of Suruga Bay, central Japan. Bull Ocean Res Inst Univ Tokyo 15:1–244

Ortner P, Hill L, Cummings S (1989) Zooplankton community structure and copepod species composition in the northern Gulf of Mexico. Cont Shelf Res 9:387–402

Palko B, Beardsley G, Richards W (1982) Synopsis of the biological data on dolphin-fishes, *Coryphaena hippurus* Linnaeus and *Coryphaena equiselis* Linnaeus. NOAA technical report NMFS circular 443, Food and Agriculture Organization fisheries synopsis 130. National Marine Fisheries Service, Seattle, WA, USA, 28 p

Parin NV (1970) Ichthyofauna of the epipelagic zone. In: Mills H (ed) Israel program for Scientific Translations, Jerusalem. U.S. Department of Commerce, Springfield, VA, USA, 206 p

Parker F (1954) Distribution of the foraminifera in the northeastern Gulf of Mexico. Bull Mus Comp Zool 111:453–588

Parr A (1939) Quantitative observations on the pelagic *Sargassum* vegetation of the Western North Atlantic. Bull Bingham Oceanogr Collection 6:1–94

Paull C, Hecker B, Commeau R, Freeman-Lynde R, Neumann C, Corso W, Golubic S, Hook J, Sikes E, Curray J (1984) Biological communities at the Florida escarpment resemble hydrothermal vent taxa. Science 226:965–967

Pauly D, Christensen V, Dalsgaard J, Froese R, Torres F (1998) Fishing down marine food webs. Science 279:860–863

Pequegnat W (1972) A deep bottom current on the Mississippi Cone. In: Capurro L, Reid J (eds) Contributions on the physical oceanography of the Gulf of Mexico, vol 2, Texas A&M University Oceanographic Studies. Gulf Publishing Company, Houston, TX, USA, pp 65–87

Pequegnat W (1983) The ecological communities of the continental slope and adjacent regimes of the northern Gulf of Mexico. TerEco final report. U.S. Department of the Interior, Minerals Management Service, Gulf of Mexico OCS Office, Metarie, LA, USA, 40 p

Pequegnat L, Wicksten M (2006) Oplophorid shrimps (Decapoda: Caridea: Oplophoridae) in the Gulf of Mexico and Caribbean Sea from the collections of the research vessels Alaminos, Oregon and Oregon II. Crustacean Res 35:92–107

Pequegnat W, Bright T, James B (1970) The benthic skimmer, a new biological sampler for deep-sea studies. In: Pequegnat W, Chace F (eds) Contributions to the biology of the Gulf of Mexico, vol 1, Texas A&M University Oceanographic Studies. Gulf Publishing Company, Houston, TX, USA, pp 17–20

Pequegnat W, Bryant W, Fredricks A, McKee T, Spalding R (1972) Deep-sea ironstone deposits in the Gulf of Mexico. J Sediment Petrol 42:700–710

Pequegnat W, Galloway B, Pequegnat L (1990) Aspects of the ecology of the deepwater fauna of the Gulf of Mexico. Am Zool 30:45–64

Peterson C, Coleman F, Jackson JBC, Turner RE, Rowe G, Barber RT, Bjorndal KA, Carney RS, Cowen RK, Hoekstra JM, Holligaugh JT, Laska SB, Luettich RA Jr, Osenberg CW, Roady SE, Senner S, Teal JM, Wang P (2011) A once and future Gulf of Mexico ecosystem. The Pew Charitable Trusts, Philadelphia, PA, USA, 111 p

Phleger F, Parker F (1951) Ecology of foraminifera, northwest Gulf of Mexico. Geological Soc Am Mem 46 (Parts 1 and 11), 88 p + 64 p

Poag CW (1981) Ecologic atlas of benthic foraminifera of the Gulf of Mexico. Marine Science International, Woods Hole, MA, USA, 175 p

Powell S, Haedrich R, McEachran J (2003) The deep-sea demersal fish fauna of the northern Gulf of Mexico. J Northwest Atl Fish Sci 31:19–33

Powell E, Parsons-Hubbard K, Callender W, Staff G, Rowe G, Brett C, Walker S, Raymond A, Carlson D, White S, Heise E (2004) Taphonomy on the continental shelf and slope: Two-year trends—Gulf of Mexico and Bahamas. Palaeogeogr Palaeoclimatol Palaeoecol 184:1–35

Prouty N, Roark E, Buster N, Ross S (2011) Growth rate and age distribution of deep-sea black corals in the Gulf of Mexico. Mar Ecol Prog Ser 423:101–115

Qian Y, Jochens A, Kennicutt M II, Biggs D (2003) Spatial and temporal variability of phytoplankton biomass and community structure over the continental margin of the northeast Gulf of Mexico based on pigment analysis. Cont Shelf Res 23:1–17

Rabalais NN, Turner RE (2011) 2011 Forecast: Summer Hypoxic Zone Size, Northern Gulf of Mexico. http://www.gulfhypoxia.net/Research/Shelfwide%20Cruises/2011/HypoxiaForecast2011.pdf. Accessed 15 November 2016

Rabalais NN, Turner RE, Justic D, Dortch Q, Wiseman WJ Jr (1999a) Characterization of Hypoxia: Topic 1 Report for the Integrated Assessment on Hypoxia in the Gulf of Mexico. U.S. Department of Commerce, NOAA, National Ocean Service, Coastal Ocean Program. May. http://oceanservice.noaa.gov/products/hypox_t1final.pdf. Accessed 15 November 2016

Rabalais N, Carney R, Escobar-Briones E (1999b) Overview of continental shelf benthic communities of the Gulf of Mexico. In: Kumpf H, Sherman K (eds) The Gulf of Mexico large marine ecosystem: Assessment, sustainability and management. Blackwell Science, Malden, MA, USA, pp 211–240

Rabalais N, Smith L, Harper D, Justic D (2001) Effects of seasonal hypoxia on continental shelf benthos. In: Rabalais N, Turner R (eds) Coastal hypoxia: Consequences for living resources and ecosystems. American Geophysical Union, Washington, DC, USA, pp 211–240

Rabalais N, Turner RE, Scavia D (2002) Beyond science into policy: Gulf of Mexico hypoxia and the Mississippi River. Bioscience 52:129–142

Rabalais N, Turner R, Sen Gupta B, Boesch D, Chapman P, Murrell M (2007) Hypoxia in the northern Gulf of Mexico: Does the science support the plan to reduce, mitigate, and control hypoxia? Estuar Coast 30:753–772

Reed J, Weaver D, Pomponi S (2006) Habitat and fauna of deep-water Lophelia pertusa coral reefs off the southeastern U.S.: Blake Plateau, Straits of Florida, and Gulf of Mexico. Bull Mar Sci 78:343–375

Ressler P, Jochens A (2003) Hydrographic and acoustic evidence for enhanced plankton stocks in a small cyclone in the northeastern Gulf of Mexico. Cont Shelf Res 23:41–61

Rester JK (2011) SEAMAP environmental and biological atlas of the Gulf of Mexico, 2008. Gulf States Marine Fisheries Commission, report 191. http://www.gsmfc.org/pub lications/GSMFC%20Number%20191.pdf. Accessed 9 June 2013

Reynolds L (1982) Modern benthic foraminifera from the Gyre intraslope basin, northern Gulf of Mexico. Gulf Coast Assoc Geol Soc Trans 32:341–351

Rezak R, Bright T, McGrail D (1985) Reefs and banks of the northwestern Gulf of Mexico. Wiley, New York, NY, USA, 259 p

Rice W, Kornicker L (1962) Mollusks of Alacran reef, Campeche Bank, Mexico. Publ Inst Mar Sci 62:366–402

Richards W, Potthoff T, Kim J (1990) Problems identifying tuna larvae species (Pisces: Scombridae: Thunnus) from the Gulf of Mexico. Fish Bull 88:607–609

Riley GA (1937) The significance of the Mississippi River drainage for biological conditions in the northern Gulf of Mexico. J Mar Res 1:60–74

Roberts T (1977) An analysis of deep-sea benthic communities in the northeast Gulf of Mexico. Texas A&M University, College Station, TX, USA, 258 p

Roberts C (2002) Deep impacts: The rising toll of fishing in the deep-sea. Trends Ecol Evol 17:242–245

Roberts S, Hirschfield M (2004) Deep-sea corals: Out of sight, but no longer out of mind. Front Ecol Environ 2:123–130

Roberts HH, Aharon P, Carney R, Larkin J, Sassen R (1990) Sea floor responses to hydrocarbon seeps, Louisiana continental slope. Geo-Mar Lett 10:232–243

Roberts J, Wheeler A, Freiwald A, Cairns S (2009) Cold-water corals: the biology and geology of deep-sea coral habitats. Cambridge University Press, Cambridge, UK, 352 p

Roberts H, Shedd W, Hunt J Jr (2010) Dive site geology: DSV ALVIN (2006) and ROV JASON II (2007) dives to the middle-lower continental slope, northern Gulf of Mexico. Deep-Sea Res II 57:1837–1858

Rogers A (1999) The biology of *Lophelia pertusa* (Linnaeus 1758) and other deep-water reef-forming corals and impacts from human activities. Int Rev Hydrobiol 84:315–406

Rooker JR, Turner JP, Holt SA (2006) Trophic ecology of *Sargassum*-associated fishes in the Gulf of Mexico determined from stable isotopes and fatty acids. Mar Ecol Prog Ser 313:249–259

Rooker J, Simms J, Wells R, Holt S, Holt J, Graves J, Furey N (2012) Distribution and habitat associations of billfish and swordfish larvae across mesoscale features in the Gulf of Mexico. PLoS One 7(4):e34180

Rosman I, Boland G, Baker JS (1987) Epifaunal aggregations of Vesicomyidae on the continental slope off Louisiana. Deep-Sea Res 34:1811–1820

Rowe GT (1971) Benthic biomass and surface productivity. In: Costlow J (ed) Fertility of the sea, vol 2. Gordon and Breach, New York, NY, USA, pp 441–454

Rowe GT (1983) Biomass and production of the deep-sea macrobenthos. In: Rowe G (ed) Deep-sea biology, the sea, vol 8. Wiley, New York, NY, USA, pp 97–121

Rowe GT (2001) Seasonal hypoxia in the bottom water off the Mississippi River delta. J Environ Qual 30:281–290

Rowe GT (2013) Seasonality in deep-sea food webs—a tribute to the early works of Paul Tyler. Deep-Sea Res II 92:9–17

Rowe GT, Chapman P (2002) Hypoxia in the northern Gulf of Mexico: Some nagging questions. Gulf Mex Sci 20:153–160

Rowe GT, Deming J (2011) An alternative view of the role of heterotrophic microbes in the cycling of organic matter in deep-sea sediment. Mar Biol Res 7:629–636

Rowe GT, Kennicutt MC (2008) Introduction to the Deep Gulf of Mexico Benthos Program (DGoMB). Deep-Sea Res II 55:2536–2540

Rowe GT, Menzel D (1971) Quantitative benthic samples from the deep Gulf of Mexico with some comments on the measurement of deep-sea biomass. Bull Mar Sci 21:556–566

Rowe GT, Menzies R (1969) Zonation of large benthic invertebrates in the deep-sea off the Carolinas. Deep-Sea Res 16:531–537

Rowe GT, Polloni P, Hornor S (1974) Benthic biomass estimates from the northwestern Atlantic Ocean and the northern Gulf of Mexico. Deep-Sea Res 21:641–650

Rowe GT, Cruz-Kaegi M, Morse J, Boland G, Escobar-Briones E (2002) Sediment community metabolism associated with continental shelf hypoxia, northern Gulf of Mexico. Estuaries 25:1097–1116

Rowe GT, Lohse A, Hubbard GF, Boland G, Escobar-Briones E, Deming J (2003) Preliminary trophodynamic carbon budget for the Sigsbee Deep benthos, northern Gulf of Mexico. In: Stanley D, Scarborough-Bull A (eds) Fisheries, reefs and offshore development, vol 36, American Fisheries Society Symposium. American Fisheries Society, Bethesda, MD, USA, pp 225–238

Rowe GT, Morse J, Nunnally C, Boland G (2008a) Sediment community oxygen consumption in the deep Gulf of Mexico. Deep-Sea Res II 55:2686–2691

Rowe GT, Wei C, Nunnally C, Haedrich R, Montagna P, Baguley J, Bernhard JM, Wicksten M, Ammons A, Escobar-Briones E, Soliman Y, Deming JW (2008b) Comparative biomass structure and estimated carbon flow in food webs in the deep Gulf of Mexico. Deep-Sea Res II 55:2699–2711

Sahl LE, Weisenburg DE, Merrell WJ (1997) Interaction of mesoscale features with Texas shelf and slope waters. Cont Shelf Res 17:117–136

Salmeron-Garcia O, Zavala-Hidalgo J, Mateos-Jasso A, Romero-Centeno A (2011) Regionalization of the Gulf of Mexico from space-time chlorophyll-*a* concentration variability. Ocean Dyn 61:439–448

Santschi P, Rowe G (2008) Radiocarbon-derived sedimentation rates in the Gulf of Mexico. Deep-Sea Res II 55:2572–2576

Sanvicente-Añorve L, Flores-Coto C, Chiappa-Carrara X (2000) Temporal and spatial scales of ichthyoplankton distribution in the southern Gulf of Mexico. Estuar Coast Mar Sci 51:463–475

Sen Gupta B, Aharon P (1994) Benthic foraminifera of bathyal hydrocarbon vents of the Gulf of Mexico: Initial report on communities and stable isotopes. Geo-Mar Lett 14:88–96

Sen Gupta B, Lee R, May M III (1981) Upwelling and an unusual assemblage of benthic foraminifera on the northern Florida continental slope. J Paleontol 55:853–857

Sen Gupta B, Platon E, Bernhard J, Aharon P (1997) Foraminiferal colonization of hydrocarbon-seep bacterial mats and underlying sediment, Gulf of Mexico slope. J Foraminiferal Res 27:292–300

Settle L (1993) Spatial and temporal variability in the distribution and abundance of larval and juvenile fishes associated with pelagic *Sargassum*. University of North Carolina at Wilmington, Wilmington, NC, USA, 64 p

Sharma J, Baguley J, Montagna P, Rowe G (2012) Assessment of longitudinal gradients in nematode communities in the deep northern Gulf of Mexico and concordance with benthic taxa. Int J Oceanogr. Article ID 903018, 15 pages. doi: 10.1155/2012/903018

Shaw RF, Drullinger D (1990a) Early-life-history profiles, seasonal abundance, and distribution of four species of Clupeid larvae from the northern Gulf of Mexico, 1982 and 1983. NOAA technical report NMFS 88. U.S. Department of Commerce, Washington, DC, USA, 60 p

Shaw RF, Drullinger DL (1990b) Early-life-history profiles, seasonal abundance, and distribution of four species of Carangid larvae off Louisiana, 1982 and 1983. NOAA technical report NMFS 89. U.S. Department of Commerce, Washington, DC, USA, 37 p

Simon JL, Dauer DM (1972) A quantitative evaluation of red-tide induced mass mortalities of benthic invertebrates in Tampa Bay, Florida. Environ Lett 3:229–234

Smith K, Hinga K (1983) Sediment community respiration in the deep sea. In: Rowe G (ed) Deep-sea biology, the sea, vol 8. Wiley, New York, NY, USA, pp 331–370

Sogard S, Hoss D, Govoni J (1987) Density and depth distribution of larval Gulf menhaden, *Brevoortia patronus*, Atlantic croaker, *Micropogonias undulatus*, and spot, *Leiostomus xanthurus*, in the northern Gulf of Mexico. Fish Bull 85:601–609

Soliman Y, Rowe G (2008) Secondary production of *Ampelisca mississippiana* Soliman and Wicksten 2007 (Amphipoda, Crustacea) in the head of the Mississippi Canyon, northern Gulf of Mexico. Deep-Sea Res II 55:2692–2698

Springer S, Bullis HR (1956) Collections by the *Oregon* in the Gulf of Mexico. U.S. Fish and Wildlife Service special scientific report-fisheries no. 196. U.S. Fish and Wildlife Service, Washington, DC, USA, 134 p

Sulak K, Brooks R, Luke K, Norem D, Randall M, Quaid AJ, Yeargin GE, Miller JM, Harden WM, Caruso JH, Ross SW (2007) Demersal fishes associated with *Lophelia pertusa* coral and hard substrate biotopes on the continental slope, northern Gulf of Mexico. In: George R, Cairns S (eds) Conservation and adaptive management of seamount and deep-sea coral ecosystems. University of Miami, Miami, FL, USA, pp 65–92

Sulak K, Randall M, Luke K, Norem A, Miller J (eds) (2008) Characterization of northern Gulf of Mexico deep-water hard bottom communities with emphasis on *Lophelia* coral— *Lophelia* reef megafaunal community structure, biotopes, genetics, microbial ecology, and geology. USGS open-file report 2008-1148; OCS Study MMS 2008—15, 42 p. + DVDs. http://fl.biology.usgs.gov/coastaleco/OFR_2008-1148_MMS_2008-015/index.html. Accessed 26 Feb 2013

Sutton T, Hopkins T (1996) Species composition, abundance and vertical distribution of the stomiid (Pisces: Stomiiformes) fish assemblage of the Gulf of Mexico. Bull Mar Sci 59:530–542

Sutton TT, Hopkins TL, Remsen A, Burghart S (2001) Multisensor sampling of pelagic ecosystem variables in a coastal environment to estimate zooplankton grazing impact. Cont Shelf Res 21:69–87

Sutton T, Porteiro F, Heino M, Byrkjedal I, Langhelle G, Anderson C, Horne J, Soiland H, Falkenhaug T, Godo O, Bergstad O (2008) Vertical structure, biomass and topographic association of deep-pelagic fishes in relation to a mid-ocean ridge system. Deep-Sea Res II 55:161–184

SWFSC (Southwest Fisheries Science Center) (2007) What are ichthyoplankton? http://swfsc. noaa.gov/textblock.aspx?division=frd&id=6210. Accessed 26 Feb 2013

Teo S, Bustany A, Dewar H, Stokesbury M, Weng K, Beemer S, Seitz A, Farwell C, Prince E, Block B (2007) Annual migrations, diving behavior, and thermal biology of Atlantic bluefin tuna, *Thunnus thynnus*, on their Gulf of Mexico breeding grounds. Mar Biol 151:1–18

Tester PA, Steidinger KA (1997) *Gymnodinium breve* red tide blooms: Initiation, transport, and consequences of surface circulation. Limnol Oceanogr 42:1039–1051

Tidwell MT, Holt S, Rooker JR, Holt GJ (2007) The distribution and feeding ecology of larval billfish in the northern Gulf of Mexico. In: Proceedings, 60th Gulf and Caribbean Fisheries Institute annual meeting, Punta Cana, Dominican Republic, Gulf and Caribbean Fisheries Institute, Marathon, FL, USA, Nov 2007, pp 379–384

Tunnell W, Chavez E, Withers K (eds) (2007) Coral reefs of the southern Gulf of Mexico. Texas A&M University Press, College Station, TX, USA, 216 p

Turner J, Collard S, Wright J, Mitchell D, Steele P (1979) Summer distribution of pontellid copepods in the neuston of the eastern Gulf of Mexico continental shelf. Bull Mar Sci 29:287–297

Vaughan D, Govoni J, Shertzer W (2011) Relationship between Gulf menhaden recruitment and Mississippi River flow: Model development and potential application for management. Mar Coast Fish 3:344–352

Walsh J, Rowe G, Iverson R, McRoy C (1981) Biological export of shelf carbon is a sink of the global CO_2 cycle. Nature 291:196–201

Watling L, Norse E (1998) Disturbance of the seabed by mobile fishing gear: A comparison to forest clearcutting. Conserv Biol 12:1180–1197

Wei C-L, Rowe G (2006) The bathymetric zonation and community structure of deep-sea macrobenthos in the northern Gulf of Mexico. In: International Council for the Exploration of the Sea annual science conference, ICES CM 2006/D:0508, 77 p

Wei C-L, Rowe G, Hubbard GF, Scheltema AH, Wilson GDF, Petrescu I, Foster J, Wicksten M, Chen M, Davenport R, Soliman Y, Wang Y (2010a) Bathymetric zonation of deep-sea macrofauna in relation to export of surface phytoplankton production. Mar Ecol Prog Ser 399:1–14

Wei C-L, Rowe G, Escobar-Briones E, Boetius A, Soltwedel K, Caley J, Soliman Y, Huettmann F, Qu F, Yu Z, Pitcher CR, Haedrich R, Wicksten M, Rex M, Baguley J,

Sharma J, Danovaro R, MacDonald I, Nunnally C, Deming J, Montagna P, Levesque M, Weslawsk JM, Wlodarska-Kowalczuk M, Ingole B, Bett B, Billett D, Yool A, Bluhm B, Iken K, Narayanaswamy B (2010b) Global patterns and predictions of seafloor biomass using Random forests. PLoS One 5(12):e15323. doi:10.1371/journal.pone.0015323

Wei C-L, Rowe GT, Escobar-Briones E, Nunnally C, Soliman Y, Ellis N (2012a) Standing stocks and body size of deep-sea macrofauna: Predicting the baseline of 2010 *Deepwater Horizon* oil spill in the northern Gulf of Mexico. Deep-Sea Res I 69:82–99

Wei C-L, Rowe G, Haedrich R, Boland G (2012b) Long-term observations of epibenthic fish zonation in the northern Gulf of Mexico. PLoS One 7(10):e46707. doi:10.1371/journal.ponc.0046707

Wei C-L, Rowe G, Wicksten M, Nunnally C (2012c) Anthropogenic "litter" and macrophyte detritus in the deep northern Gulf of Mexico. Mar Pollut Bull 64:966–973

Weisberg R, Black B, Li Z (2000) An upwelling case study of Florida's west coast. J Geophys Res 105:11459–11469

Wells R, Rooker J (2004a) Distribution, age, and growth of young-of-the-year greater amberjack (*Seriola dumerili*) associated with pelagic *Sargassum*. Fish Bull 102:545–554

Wells R, Rooker J (2004b) Spatial and temporal patterns of habitat use by fishes associated with *Sargassum* mats in the northwestern Gulf of Mexico. Bull Mar Sci 74:81–99

Wells R, Cowan J Jr, Patterson W (2008) Habitat use and the effect of shrimp trawling on fish and invertebrate communities over the northern Gulf of Mexico continental shelf. ICES J Mar Sci 65:1610–1619

Wicksten M, Packard J (2005) A qualitative zoogeographic analysis of decapod crustaceans of the continental slopes and abyssal plain of the Gulf of Mexico. Deep-Sea Res I 52:1745–1765

Williamson S, Zois N, Hewitt A (2008) Integrated site investigation of seafloor features and associated fauna, Shenzi field, deepwater Gulf of Mexico. In: Proceedings, Offshore Technology Conference, BHP Billiton, Houston, TX, USA, May 2008, pp 1208–1216

Wilson DFG (2008) Local and regional species diversity of benthic Isopoda (Crustacea) in the deep Gulf of Mexico. Deep-Sea Res II 55:2634–2649

Wiseman W, Sturges W (1999) Physical oceanography of the Gulf of Mexico: Processes that regulate its biology. In: Kumpf H, Steidinger K, Sherman K (eds) The Gulf of Mexico large marine ecosystem. Blackwell Science, Malden, MA, USA, pp 77–92

Yeager KM, Santschi P, Rowe G (2004) Sediment accumulation and radionuclide inventories (^{239,240}Pu, ^{210}Pb and ^{234}Th) in the northern Gulf of Mexico, as influenced by organic matter and macrofaunal density. Mar Chem 91:1–14

Zaitsev Y (1971) Marine neustonology. Israel Program for Scientific Translations, Jerusalem, 207 p

Ziegler M (2002) The epibenthic megafauna of the northern Gulf of Mexico continental slope. Department of Oceanography, Texas A&M University, College Station, TX, USA, 93 p

APPENDIX A: WEBSITES SUPPORTED BY BOEM, NOAA, AND USGS WITH COMPREHENSIVE INFORMATION ON THE BIOTA OF THE DEEP GOM

MacDonald I, Schroeder W, Brooks J (1995) Chemosynthetic ecosystems study, final report. OCS Study MMS 95-0023. U.S. DOI, MMS, GoM OCS Region, New Orleans, LA, USA, 338 p. www.gomr.mms.gov/PI/PDFImages/ESPIS/3/3323.pdf

MacDonald I (ed) (2002) Stability and change in Gulf of Mexico chemosynthetic communities, vol II, Technical Rept. OCS Study MMS 2002-036. U.S. DOI, MMS, GoM OCS Region, New Orleans, LA, USA, 456 p.
www.gomr.mms.gov/PI/PDFImages/ESPIS/3/3072.pdf

Continental Shelf Associates, Inc (2004) Final report: Gulf of Mexico Comprehensive Synthetic Based Muds Monitoring Program, vol I, Technical.
www.gomr.mms.gov/PI/PDFImages/ESPIS/2/3049.pdf

Continental Shelf Associates, Inc (2008) Final report: Gulf of Mexico Synthetic Based Muds Monitoring Program, vol I, Technical.
www.gomr.mms.gov/PI/PDFImages/ESPIS/2/3050.pdf

Continental Shelf Associates, Inc (2008) Final report: Gulf of Mexico Synthetic Based Muds Monitoring Program, vol II, Technical.
www.gomr.mms.gov/PI/PDFImages/ESPIS/2/3051.pdf

Continental Shelf Associates, Inc (2007) Characterization of Northern Gulf of Mexico deep-water hard bottom communities with emphasis on *Lophelia* coral. OCS Study MMS 2007-044. U.S. DOI, MMS, GoM OCS Region, New Orleans, LA, USA, 169 p + app
www.gomr.mms.gov/PI/PDFImages/ESPIS/4/4264.pdf

Schroeder W (2007) Seafloor characteristics and distribution patterns of Lophelia pertusa and other sessile megafauna at two upper-slope sites in the Northwestern Gulf of Mexico. U.S. DOI, MMS, GoM OCS Region, New Orleans, LA, USA, 49 p.
OCS Study MMS 2007-035. www.gomr.mms.gov/PI/PDFImages/ESPIS/4/4264.pdf

Sulak K, Randall M, Luke KE, Norem AD, Miller JM (eds) (2008) Characterization of Northern Gulf of Mexico deepwater hard bottom communities with emphasis on Lophelia coral—Lophelia reef megafaunal community structure, biotopes, genetics, microbial ecology, and geology (2004-2006). USGS Open-File Report 2008-1148; OCS Study MMS 2008-015.

Brooks J, Fisher C, Roberts H, Bernard B, MacDonald I, Carney R, Joye S, Cordes E, Wolff G, Goehring E (2008) Investigations of chemosynthetic communities on the lower continental slope of the Gulf of Mexico: Interim reports 1 and 2. OCS Study MMS 2008-009. U.S. DOI, MMS, GoM OCS Region, New Orleans, LA, USA, 332 p. and 2009-046, 360 p.
www.gomr.mms.gov/PI/PDFImages/ESPIS/4/4320.pdf and 4/4877.pdf

NGOMCSS Study from 1988:
www.gomr.mms.gov/PI/PDFImages/ESPIS/3/3773.pdf
www.gomr.mms.gov/PI/PDFImages/ESPIS/3/3774.pdf
www.gomr.mms.gov/PI/PDFImages/ESPIS/3/3695.pdf
www.gomr.mms.gov/PI/PDFImages/ESPIS/3/3696.pdf

Previous *Lophelia* studies:

www.tdi-bi.com/Lophelia/Data/Loph_Cru1_Rpt-Final.pdf
Also Cru2_Rpt-Final.pdf

Ongoing *Lophelia* studies by BOEM and NOAA:
http://fl.biology.usgs.gov/coastaleco/OFR_2008-1148_MMS_2008-015/index.html.

CHAPTER 8

SHELLFISH OF THE GULF OF MEXICO

John W. Tunnell, Jr.

[1]Texas A&M University—Corpus Christi, Corpus Christi, TX 78412, USA
wes.tunnell@tamucc.edu

8.1   INTRODUCTION

Historically, the Gulf of Mexico and the Gulf of Maine produce the greatest amount of seafood by volume and value in the United States after Alaska (Upton 2011; NMFS 2011). During 2007–2009, the Gulf of Mexico annual average commercial landings by poundage was 1.4 billion pounds (635,000 metric tons) and value was $660 million (NOS 2011). Four of the top five species by poundage and value in the Gulf of Mexico were shellfish, or invertebrates (brown shrimp, white shrimp, blue crab, and Eastern oyster). Menhaden was the only species of finfish in the top five. Shrimp are the most valuable shellfish industry in the Gulf, followed by the Eastern oyster and blue crab. At least 49 species of shellfish arc taken as seafood in Gulf of Mexico waters from its three surrounding countries (the United States, Mexico, and Cuba). Penaeid shrimp (brown, white, and pink), Eastern oyster, and blue crab are widely taken and make up the main commercial industry species, whereas there are numerous additional local and artisanal shellfish fisheries for many other different species in Mexico and Cuba. The northern Gulf of Mexico is warm temperate, and the southern Gulf is subtropical to tropical. Iconic Caribbean species, like conch and spiny lobsters, have been taken commercially and recreationally in South Florida, Cuba, and Mexico. The purpose of this chapter is to summarize the primary commercial, marine and estuarine shellfish species of the northern Gulf of Mexico and their status and trends in the decades preceding the Deepwater Horizon oil spill of April 2010. However, since the waters and species of the Gulf do not recognize political boundaries, since many species range much wider than just the northern Gulf, and since the Gulf of Mexico is recognized as a Large Marine Ecosystem, an overview of all Gulf shellfish species is provided for better understanding of the species and their aquatic habitats. Readers interested in other aspects of fisheries should also review two other chapters presented in this series: Chapters 9 and 10 (in Volume 2). The latter focuses on the same five key species as this chapter, but its primary focus is on the economic value of the species and the fishing industry.

The term *shellfish* is usually used to refer to the edible species of invertebrates, as opposed to the term finfish, which is reserved for the true fishes. The largest shellfish groups include the mollusks (e.g., clams, oysters, conchs) and crustaceans (e.g., shrimps, crabs, lobsters), but shellfish also include invertebrate species without shells, such as sea cucumbers, octopus, squid, and jellyfish. The term *commercial invertebrates* is used to describe those invertebrate species that are captured for purposes other than food, such as ornamental shells and corals (Orensanz and Jaimieson 1998; Jennings et al. 2001). Within the Gulf of Mexico, as within all oceans of the world, the harvested species include only a very small percentage of the total biodiversity of any of the targeted fishery groups. For instance, a recent survey of all biota within the Gulf of Mexico revealed a total of 15,419 species (Felder and Camp 2009). Of that total 9,063 would be considered invertebrates, but only about 49, or 0.54 %, are taken as seafood. The primary commercial shellfish of the Gulf of Mexico are critically linked to estuaries for part (shrimp) or

© The Author(s) 2017
C.H. Ward (ed.), *Habitats and Biota of the Gulf of Mexico: Before the Deepwater Horizon Oil Spill*,
DOI 10.1007/978-1-4939-3447-8_8

all of their life cycles (blue crabs and oysters), and they are characterized by large numbers of eggs/young, rapid growth, and fairly short life cycles.

Historically, fishes and shellfishes have served as a source of food, commerce, and recreation for humans since ancient times (Ross 1997). Within the Gulf of Mexico, and considering our three primary shellfish groups, oysters were probably the first to be harvested by Native Americans due to their ease of capture near shorelines. Because of their popularity as a seafood delicacy and their importance as a significant commercial product, the Eastern oyster is perhaps the most studied marine species in the Gulf of Mexico (Berrigan et al. 1991). Although taken as seafood much earlier, the first commercial catches of shrimp and blue crabs were not recorded until the 1880s (Condrey and Fuller 1992; Steele and Perry 1990).

Today, finfish and shellfish are harvested intensively around the world as a source of protein in the family diet, for commercial or economic gain, or for recreational purposes (Ross 1997). Over 198 billion pounds (90 million metric tons) of finfish and shellfish are harvested annually around the world (www.fao.org, FAO 2011). In the United States in 2009 over 7.8 billion pounds (over 3.5 million metric tons) were harvested, and from the northern Gulf of Mexico, over 1.5 billion pounds (over 0.7 million metric tons) were taken (NMFS 2011). In the Gulf of Mexico, as in the rest of the United States, shellfish top the economic value list of species, indicating their continued importance to the fishery, and finfish top the weight or poundage list (NOS 2011). Using a 3-year average between 2007 and 2009, 78 % of all U.S. landings of shrimp came from the Gulf of Mexico with a 3-year average of 221 million pounds, and 62 % of all U.S. oyster landings came from the Gulf with a 3-year average of 22 million pounds of meats (NOS 2011).

Numerous techniques are utilized to evaluate the health or status of a selected fishery. This chapter will focus more on the biological populations of selected species through time, since the Commercial and Recreational Fisheries chapter will focus mainly on economic value and the fishing industry. Fishery scientists and managers use fishery-dependent and fishery-independent techniques to evaluate a given fishery, and both kinds of evaluation are critical in determining a fishery status through time. Fishery-dependent data are collected by state and federal fishery management agencies, so they can manage various species stocks, or populations, as well as the fishers who take them. Fisheries-independent data are collected by most Gulf States (e.g., Louisiana Department of Wildlife and Fisheries, [LDWF]), and these too are critical in fishery management work, because they reveal population trends that are independent of issues affecting the fishing industry. Within Federal waters (offshore) of the Gulf of Mexico fishery-independent data are collected by the Southeast Area Monitoring and Assessment Program (SEAMAP). SEAMAP is a cooperative state, federal, and academic program for the collection, management, and dissemination of fishery-independent data and information in the southeastern United States. Shrimp trawl surveys for SEAMAP occur in the summer and fall each year on the continental shelves of the northern Gulf. Fishery-dependent data are subject to numerous and varying conditions and circumstances, such as inaccurate "landings" (usually defined as pounds landed by the fishery, but not all landings are reported and not all reported landings are accurate), varying regulations (by state and federal agencies), and economic factors, such as fuel costs, market values, and foreign competition with some species (shrimp). Fishery-independent data are best for tracking time series of various species, but since this type of data is not universal, it will be used where available alongside fishery-dependent data in this document. For the purposes of fishery-independent data for this chapter, SEAMAP will be used for shrimp in the offshore areas. For inshore data on shrimp, oysters, and crabs, LDWF data will be used, as Louisiana has the highest catches, and it is not within the scope of this chapter to individually review all state fishery-independent programs.

Jurisdictional oversight and boundaries for the Gulf of Mexico shellfisheries vary in all three countries, but basically they are under federal jurisdiction in Cuba and Mexico with some province/state involvement, and in the United States both state and federal entities are involved. The five U.S. states have jurisdiction in the state territorial waters, or seas, which extend out into the Gulf of Mexico 3 nautical miles (5.6 km) off the shoreline of Louisiana, Mississippi, and Alabama, and 9 nautical miles off Texas and western Florida (NMFS 2010). Within each of these states their state fish and wildlife agency manages each fishery within its State Territorial Waters, and they all work together via the regional Gulf States Marine Fisheries Commission (GSMFC). In federal waters, the Exclusive Economic Zone (EEZ) extends from the seaward boundary of the State Territorial Waters out 200 nautical miles from the shoreline. The agency responsible for federal fishery management is the National Marine Fisheries Service (NMFS) of the National Oceanic and Atmospheric Administration (NOAA). Stakeholders within the Gulf fishery, along with NMFS, are engaged via the Gulf of Mexico Fisheries Management Council (GMFMC).

The Magnuson-Stevens Fishery Conservation and Management Act in the United States provides for the management and conservation of fishery resources within the U.S. EEZ (NMFS 2011). Under the Magnuson-Stevens Act, the GMFMC is charged with preparing fishery management plans for the fisheries needing management within the U.S. Gulf of Mexico. Of the three main shellfish groups covered in this document, only penaeid shrimp have a Gulf of Mexico Fishery Management Plan (FMP), since the other two main groups (blue crabs and oysters) are primarily inshore, or estuarine, species within State Territorial Waters. For blue crabs and oysters, each U.S. Gulf state may have its own rules and regulations regarding these two fisheries (Cody et al. 1992; Guillory 1996; Heath 1998; Perry et al. 1998; Quast et al. 1988; Steele and Bert 1998), but they all operate under a regional management plan prepared collectively by the GSMFC composed of members from each state (oysters, VanderKooy 2012; blue crabs, Guillory et al. 2001).

Understanding the biology, life cycles, and distribution of targeted species is central to understanding how they are affected by fishing and the environment. These aspects and environmental parameters are well known for Gulf of Mexico penaeid shrimp, blue crabs, and Eastern oysters. An examination of them along with their living space will help reveal the status and recent trends of their populations prior to the Deepwater Horizon oil spill in 2010. Healthy Gulf of Mexico waters and habitats are critical to the productivity and sustainability of Gulf shellfisheries. Diverse estuarine habitats provide nursery grounds (food, habitat, protection) for juvenile and young of many species. Seagrass beds, salt marshes, mangrove forests, and oyster reefs are particularly critical in the life cycles of many shellfish species. Indeed, these have all been labeled as Essential Fish Habitat (EFH) by both state and federal management authorities and warrant protection and conservation for future stocks of these species. Density, growth, and survival of the three primary Gulf taxa covered in this document are dependent upon the continued health of these key habitats and surrounding areas (Zimmerman et al. 2000; Beck et al. 2001, 2003; Minello et al. 2003; Jordan et al. 2009).

Substantial, long-term increases in shellfish harvests, along with increased human population levels and coupled coastal development, water quality issues, and habitat loss/degradation have all had impacts on fished stocks. Parasites and diseases, economic conditions, and management/regulatory decisions have also had effects and therefore can cause annual fluctuations in the fisheries (Lotze et al. 2006; Halpern et al. 2008; Jackson 2008).

In summary, regarding current status and historical trends of shrimp, oysters, and crabs in the northern Gulf of Mexico, it will be shown that there have been natural population fluctuations over the past several decades in all three groups. In addition, some fisheries, like shrimp, have been greatly affected in recent years by exogenous factors, such as rising fuel costs, market competition from imported shrimp, and fleet damage by hurricanes. Overall, shrimp populations seem to be flourishing, while the shrimp fishery is in decline. Oyster populations, which have been lost or degraded worldwide, appear to be fairly stable in the Gulf of Mexico, showing variable annual and multiyear fluctuations due mainly to environmental conditions but also sometimes due to economic/market conditions (VanderKooy 2012). In addition, there has been damage to some oyster reefs and oyster fisheries due to hurricanes, and a decadal decline in oyster stock assessment in Louisiana (LDWF 2010). There is considerable concern in the Gulf over the continued loss of oyster reef habitat. The blue crab fishery is quite variable from state to state with Louisiana showing a continued growth and the largest fishery in the northern Gulf over the past two decades, while Texas shows a decrease in not only the fishery, but species populations statewide during the same time frame. Gulf-wide there is agreement that healthy bays and estuaries lead to more productive fisheries, and therefore some habitats need to be conserved, while others need to be restored.

8.2 JURISDICTIONAL BOUNDARIES AND GOVERNING AGENCIES

It is important to know the jurisdictional boundaries of the three countries involved in Gulf of Mexico shellfisheries, as well as the agencies and organizations responsible for shellfish fisheries management, in order to better understand the status and trends of shellfish populations. In the United States, both state and federal jurisdiction and agency/organization management is important, but in Mexico most shellfisheries jurisdiction and management belongs to the federal government, although the States can make suggestions and become involved. In Cuba, jurisdiction and management is at the Federal level, but provincial fishing associations also have responsibilities. All three countries, of course, recognize the 200 nautical mile EEZ marked from the shoreline seaward.

8.2.1 The United States (Federal and State)

Unlike Mexico and Cuba where all Gulf waters are considered to be under federal jurisdiction, the United States allows Gulf States the sovereign right to govern and manage their own state territorial waters, or seas. These waters vary by state with Louisiana, Mississippi, and Alabama having 3 nautical miles out from the shoreline, and Texas and western Florida declaring 9 nautical miles. Shellfisheries in the northern Gulf States are managed by each state's fish and wildlife agency and include (east to west): Florida Fish and Wildlife Conservation Commission, Alabama Department of Conservation and Natural Resources, Mississippi Department of Marine Resources, LDWF, and Texas Parks and Wildlife Department. Each state, according to its own laws, can establish fishery management plans and regulations for species within its waters, or it can work collaboratively with other states under the fishery management plans of the GSMFC. This commission, established in 1949, maintains an active web site (www.gsmfc.org, GSMFC 2011) for programs, publications, regulations, databases of landings, and much more. Fishery management plans for the two major shellfishery groups discussed herein (blue crabs and oysters) can be found at this site, as well as news releases, regulations, and licenses and fees for all five states.

At the federal level, the Magnuson-Stevens Fishery Conservation and Management Act of 1976 established eight Regional Fishery Management Councils around the United States to assist in the stewardship of federal fishery resources that occur beyond state waters (ELI and CMDA 2011). Each council is charged with preparing and implementing fishery management plans for harvested stocks, which in turn must be reviewed and approved by the NMFS of

NOAA. Enforcement of fishery management plans is accomplished by the NOAA Office of Law Enforcement via shipboard observers and dockside enforcement, the U.S. Coast Guard through sea patrols, and the states via joint enforcement agreements with the Office of Law Enforcement. In addition, the fishery management plans must identify EFH, including ways to minimize adverse effects to EFH caused by fishing, as well as actions to conserve and enhance EFH (ELI and CMDA 2011).

The Gulf of Mexico Fishery Management Council (GMFMC) performs fishery management within the EEZ of the Gulf of Mexico, and it has developed three fishery management plans for Gulf shellfish species (shrimp, spiny lobster, and stone crab). The GMFMC maintains an extensive web site (www.gulfcouncil.org) with regulations, meetings, management plans, committees, and panels on Gulf fisheries and issues, news, and many other resources (library, stock assessments, scoping documents and proposed amendments, education information, and FAQs, to mention a few). Each fishery management plan on the web site has a current list of amendments documenting changes since the first fishery management plan was passed, so it is very easy to follow the history of adaptive management of each fishery through time. Appointed science and statistical committees within the Council are charged with the responsibility of ascertaining if a federally managed species is overfished, and if so, to decide on an appropriate rebuilding plan.

8.2.2 Mexico (Federal with State Input)

In Mexico, the majority of fishery development occurred in the late 1970s with the creation of the Departamento de Pesca (Fisheries Department) and a consequent, substantial investment in state-owned fishing fleets and industrial plants (Diaz-de-Leon et al. 2004). Mexican fisheries catches peaked in the early 1980s, followed by decreasing catches due to overexploitation and overcapitalization of most fisheries through the 1990s. During a 1999 public review of the Ley de Pesca 1992 (Fishery Law), the Instituto Nacional de Pesca (IPN, National Fishery Institute) proposed that the Carta Nacional Pesqueria (National Fisheries Charter) should inform Mexican fisheries by defining, inventorying, managing, regulating, and conserving the resources. The National Fisheries Charter subsequently became the regulatory instrument rather than just informing the fishery in the early to middle first decade of the 2000s (Diaz-de-Leon et al. 2004).

Today, the regulation of harvested species is established mainly by the Ley General de Pesca y Acuacultura Sustentable (General Law on Fisheries and Sustainable Aquaculture; ELI and CMDA 2011). This new law was published in 2007 with the main objective to promote and manage the exploitation of fisheries resources and aquaculture in a sustainable way. The main powers of the Federal government in regard to fisheries and aquaculture are implemented through the Comision Nacional de Acuacultura y Pesca (CONAPESCA, National Commission of Aquaculture and Fisheries), which is a decentralized component of the Secretaria de Agricultura, Ganaderia, Desarrollo Rural, Pesca, y Alimentacion (SAGARPA, Ministry of Agriculture, Livestock, Rural Development, Fisheries, and Foods). The General Law of Fisheries and Sustainable Aquaculture emphasizes the joint collaboration of CONAPESCA with the Secretaria de Medio Ambiente y Recursos Naturales (SEMARNAT, Ministry of the Environment and Natural Resources) concerning conservation and restoration of the environment to maintain healthy fisheries.

8.2.3 Cuba (Federal with Provincial Fishing Associations)

In Cuba, most Cuban fisheries were focused on the continental shelves of the island nation before 1960, using artisanal fishing gear and small boats (3–11 m; 9.9–36 ft), and most boats were without engines. Larger vessels (20–25 m; 72–89 ft) were restricted to the tuna fishery, along with a few shrimp trawlers (Claro et al. 2001, 2009). After the Cuban Revolution in 1959, growth of the fisheries industry was an important objective of the new government (Baisre 2006). Domestic catches dominated and increased throughout the 1960s, but increases in a significant long-distance fleet increased international catches during the 1970s to 1980s, which dominated as much as two-thirds of the Cuban catch (Baisre et al. 2003; Baisre 2006). However, as the Soviet Union assisted in the fishery fleet buildup in the 1960s and 1970s, its collapse and withdrawal from Cuba in the early 1990s severely curtailed the offshore fisheries (Adams et al. 2000). After this, the Cuban fishing industry changed dramatically with the emphasis shifting from high-volume, but low-value pelagic fisheries to high-value, coastal fin- and shell-fish species caught mainly in nearshore waters (Adams et al. 2000). The principal marine fisheries of Cuba today are lobster (most valuable), shrimp, small pelagics, demersal reef fishes, mullets, crabs, some mollusks, and sponges (Baisre 2006).

For statistical and data gathering purposes, the insular shelf of Cuba is divided into four sectors: Zone A—Southeast; Zone B—Southwest; Zone C—Northwest; and Zone D—Northeast (Claro et al. 2001; Baisre 2006). Zone C, the Northwest shelf and Gulf of Mexico portion of Cuba, is the smallest shelf area and therefore contributes the lowest amount to the catches (only 2–3 % of the total). A 60-year trend analysis by Baisre (2000) showed sustained increases in catches from the mid-1950s to 1970s, decreasing growth rate during the 1980s, and a revealing impact on overall fisheries during the 1990s. His study revealed that about 39 % of the fishery resources were in a senescent phase, about 49 % were in a mature phase with high exploitation, and only about 12 % were still in the developing phase with a possibility of increased catches. None of the fisheries remained underdeveloped.

Fisheries management in Cuba is under the control of the Ministerio de la Industria Pesquera (MIP, Ministry of Fishing Industries). Formerly it was more centralized, but today it is quite decentralized. The MIP is directly responsible for the national legal and administrative functions, but the production activities, control, and services have been delegated to Provisional Fishing Associations (PFAs) around the country. The PFAs are responsible for producing fin- and shellfish landings, which are in compliance with species-specific harvest plans. These harvest plans, which are developed by the individual PFAs, are then consulted on and approved by the Executive Board of MIP. The PFAs have legal and jurisdictional authority with independent control over the resources (e.g., vessels, fuel, supplies, ice, and labor) (Adams et al. 2000; Baisre et al. 2003).

The current regulatory framework for Cuban fisheries management is under Decreto Ley 164 (Decree Law 164, "Rules for Fisheries") passed in 1996. Other important laws passed in the late 1990s and early 2000s deal with protection and conservation of the environment, establishment of a national system of protected areas, and coastal zone management, all of which work together today to attempt to make Cuba's fisheries more sustainable (Baisre 2006).

8.3 SHELLFISH OF THE GULF OF MEXICO

At least 49 species of shellfish are fished within the Gulf of Mexico in the coastal and marine waters of Cuba, Mexico, and the United States (Table 8.1). For purposes of this chapter, shellfish of the Gulf of Mexico (Figure 8.1) are defined as those that live within Gulf marine habitats, coastal waters, and tidal wetlands as defined by Felder and Camp (2009). These include all the waters of Florida Bay and the Florida Keys west of a line from Key Largo to Punta Hicacos, Cuba, and north of a line between Cabo San Antonio, Cuba, and Cabo Catoche, Quintana Roo, Mexico. The northern Gulf of Mexico from Cabo Rojo, Veracruz, in the western Gulf of Mexico to Cape Romano, Florida, in the eastern Gulf is considered warm temperate, or

Table 8.1. Shellfish Fisheries of the Gulf of Mexico

Species[a]	Common Name[b] English (Spanish)	Country[c]/Remarks[d]/Citation[e]
Phylum Mollusca/Class Gastropoda/Family Strombidae		
1. *Aliger costatus* (syn. *Strombus costatus*)	Milk Conch (Caracol Blanco or Lanceta)	MX—fishery species in danger of extinction (Baqueiro 2004; Diario Oficial 2010)
2. *Eustrombus gigas* (syn. *Strombus gigas*)	Queen Conch (Caracol Rosado, de Abanico or Reina)	MX—fishery deteriorated in MX (Diaz-de-Leon et al. 2004); CU—populations controlled (Claro et al. 2001); US (South Florida)—overfished (CFMC 1996)
3. *Strombus pugilis*	West Indian Fighting Conch (Caracol Canelo or Lancetita)	MX—species with fishery potential (Baqueiro 2004; Diario Oficial 2010)
Family Fasciolariidae		
4. *Fasciolaria lilium*	Banded Tulip (Caracol Campechana)	MX—currently fished (Baqueiro 2004)
5. *Fasciolaria tulipa*	True Tulip (Caracol Campechana)	MX—currently fished (Baqueiro 2004; Diario Oficial 2010)
6. *Triplofusus giganteus* (syn. *Pleuroploca giganteus*)	Horse Conch (Caracol Rojo or Chac Pel)	MX—fishery species in danger of extinction (Baqueiro 2004; Diario Oficial 2010)
Family Melongenidae		
7. *Busycon perversum*	Knobbed Whelk (Sacabocados, Lix, or Caracol Trompillo)	MX—currently fished (Baqueiro 2004, listed as *B. carica*; Diario Oficial 2010)
8. *Melongena bispinosa*	Crown Conch (Caracol Negro or Moloncito)	MX—species with fishery potential (Baqueiro 2004, listed as *M. corona bispinosa*; Diario Oficial 2010)
9. *Melongena melongena*	Crown conch (Caracol Chivita or Chirita, or Molon)	MX—species with fishery potential (Baqueiro 2004; Diario Oficial 2010)
Family Turbinellidae		
10. *Turbinella angulata*	West Indian Chank (Caracol Tomburro or Negro)	MX—currently fished (Baqueiro 2004; Diario Oficial 2010)
Class Cephalopoda/Family Loliginidae		
11. *Doryteuthis pealeii*	Longfin Inshore Squid	All three species lumped in the US and MX statistics as "squids"; MX—incidental in shrimp trawls (Baqueiro 2004); US—primarily caught as bycatch in shrimp trawls (Patillo et al. 1997)

(continued)

Table 8.1. (continued)

Species[a]	Common Name[b] English (Spanish)	Country[c]/Remarks[d]/Citation[e]
12. *Doryteuthis plei*	Slender Inshore Squid	
13. *Lolliguncula brevis*	Atlantic Brief Squid (all = calamar)	
Family Octopodidae		
14. *Octopus maya*	Yucatán Octopus (Pulpo Rojo)	MX (Yucatán, Campeche)—current important fishery (Baqueiro 2004; Diario Oficial 2010); maximally exploited (Diaz-de-Leon et al. 2004)
15. *Octopus* cf. *vulgaris*	Common Octopus (Pulpo Paton)	MX—current fishery (Baqueiro 2004; Diario Oficial 2010); potential for exploitation (Diaz-de-Leon et al. 2004)
Class Bivalvia/Family Arcidae		
16. *Anadara transversa*	Transverse Ark (Arca Transversa)	MX—species with fishery potential (Baqueiro 2004)
Family Mytilidae		
17. *Geukensia granosissima*	Southern Ribbed Mussel (Mejillon Amarillo)	MX—species with fishery potential (Baqueiro 2004, listed as *G. demissa*)
18. *Modiolus americanus*	American Horse Mussel (Mejillon)	MX—species with fishery potential (Baqueiro 2004)
Family Ostreidae		
19. *Crassostrea rhizophorae*	Mangrove Oyster (Ostion de Mangle)	MX—current fishery (Diario Oficial 2010); CU—current fishery (Baisre 2000)
20. *Crassostrea virginica*	Eastern Oyster (Ostion Americano)	MX—current fishery (Baqueiro 2004; Diario Oficial 2010); US—major fishery (Berrigan et al. 1991)
Family Pinnidae		
21. *Atrina rigida*	Stiff Pen Shell (Callo de Hacha)	MX—current fishery (Baqueiro 2004)
Family Pectinidae		
22. *Argopecten irradians*	Bay Scallop (Almeja Abanico)	MX—species with fishery potential (Baqueiro 2004); US (Florida)—closed (Patillo et al. 1997)
Family Lucinidae		
23. *Codakia orbicularis*	Tiger Lucine (Almeja Rayada or Blanca)	MX—species with fishery potential (Baqueiro 2004; Diario Oficial 2010)
Family Corbiculidae		
24. *Polymesoda caroliniana*	Carolina Marsh Clam (Almeja negra)	MX—current fishery (Baqueiro 2004; Diario Oficial 2010)

(continued)

Table 8.1. (continued)

Species[a]	Common Name[b] English (Spanish)	Country[c]/Remarks[d]/Citation[e]
Family Veneridae		
25. *Chione elerata* (syn. *Chione cancellata*)	Florida Cross-barred Venus (Almeja China or Ronosa)	MX—species with fishery potential (Baqueiro 2004; listed as *C. cancellata*)
26. *Mercenaria campechensis*	Southern Quahog (Concha or Almeja Bola)	MX—species with fishery potential (Baqueiro 2004; Diario Oficial 2010)
Family Mactridae		
27. *Rangia cuneata*	Atlantic Rangia (Almeja Gallito)	MX—current fishery (Baqueiro 2004; Diario Oficial 2010)
28. *Rangia flexuosa*	Brown Rangia (Almeja Chira)	MX—species with fishery potential (Baqueiro 2004; Diario Oficial 2010)
Phylum Arthropoda/Subphylum Crustacea/Class Malacostraca/Order Decapoda/Family Penaeidae		
29. *Farfantepenaeus aztecus*	Brown Shrimp (Camaron Café)	MX (Tamaulipas, Veracruz, Tabasco)—current fishery (Diario Oficial 2010), maximal exploitation (Diaz-de-Leon et al. 2004); US (Texas, Louisiana)—current significant fishery (GMFMC 1981; Caillouet et al. 2008, 2011)
30. *Farfantepenaeus duorarum*	Pink Shrimp (Camaron Rosado)	MX (Campeche Bay) current fishery (Diario Oficial 2010), deteriorated fishery (Diaz-de-Leon et al. 2004); US (Southwest Florida)—current important fishery (GMFMC 1981; Hart 2008; Hart and Nance 2010)
31. *Litopenaeus setiferus*	White Shrimp (Camaron Blanco)	MX (Campeche Bay)—deteriorated fishery (Diaz-de-Leon et al. 2004); US (Northern Gulf of Mexico)—current significant fishery (GMFMC 1981; Nance et al. 2010)
32. *Xiphopenaeus kroyeri*	Atlantic Seabob (Camaron Siete Barbas)	MX (Tabasco)—potential for fishery development (Diaz-de-Leon et al. 2004), important coastal fishery (Wakida-Kusunoki 2005); US—important northern Gulf fishery (Gusmao et al. 2006)
Family Sicyonidae		
33. *Sicyonia brevirostis*	Rock Shrimp (Cameron de Roca)	MX (North Quintana Roo)—small fishery (Diario Oficial 2010); US—small deep-sea fishery (Stiles et al. 2007)

(continued)

Table 8.1. (continued)

Species[a]	Common Name[b] English (Spanish)	Country[c]/Remarks[d]/Citation[e]
Family Solenoceridae		
34. *Pleoticus robustus* (syn. *Hymenopenaeus robustus*)	Royal Red Shrimp	US (Deep Gulf)—small deep-sea fishery (Stiles et al. 2007)
Family Palinuridae		
35. *Panulirus argus*	Spiny Lobster (Langosta del Caribe)	CU (Northern Coast)—most valuable Cuban fishery (Claro et al. 2001); MX (Yucatán, Quintana Roo)—exploited to deteriorated fishery (Diaz-de-Leon et al. 2004), US (Florida Keys)—(GMSAFMC 1982)
36. *Panulirus guttatus*	Spotted Lobster (Langosta Pinta)	MX—small fishery, incidental with Spiny Lobster (Diario Oficial 2010)
Family Portunidae		
37. *Callinectes bocourti*	Bocourt Swimming Crab (Jaiba Roma)	MX—coastal zone fishery (Diario Oficial 2010)
38. *Callinectes danae*	Dana Swimming Crab (Jaiba Siri)	MX—coastal zone fishery (Diario Oficial 2010)
39. *Callinectes ornatus*	Swimming Crab (Jaiba)	MX—coastal zone fishery (Diario Oficial 2010)
40. *Callinectes rathbunae*	Sharptooth Swimming Crab (Jaiba Prieta)	MX—coastal zone fishery (Diario Oficial 2010)
41. *Callinectes sapidus*	Blue Crab (Jaiba Azul)	MX—coastal zone fishery (Diario Oficial 2010); US—significant fishery (Guillory et al. 2001); CU—current fishery (Baisre 2000)
42. *Callinectes similis*	Lesser Blue Crab (Jaiba Pequena Azul)	MX—coastal zone fishery (Diario Oficial 2010)
Family Menippidae		
43. *Menippe adina*	Gulf Stone Crab (Congrejo Moro or Congrejo de Piedra Negro)	MX (Tamaulipas)—coastal zone fishery, mainly incidental with Jaiba fishery (Diario Oficial 2010); US—small northern Gulf fishery (Patillo et al. 1997)
44. *Menippe mercenaria*	Florida Stone Crab (Congrejo Moro or Congrejo de Piedra Negro)	US (West Coast of Florida) southwest Florida fishery mainly (Patillo et al. 1997)
45. *Menippe nodifrans*	Cuban Stone Crab (Congrejo Moro or Congrejo de Piedra Negro)	US (Florida Keys)—Caribbean species fishery (Patillo et al. 1997); CU (Northwest Coast)—coastal fishery (Baisre 2000; Claro et al. 2001); MX (Southern Gulf)—coastal zone

(continued)

Table 8.1. (continued)

Species[a]	Common Name[b] English (Spanish)	Country[c]/Remarks[d]/Citation[e]
		fishery, mainly incidental with Jaiba fishery (listed as *M. mercenania* in Diario Oficial 2010)
Family Gecarcinidae		
46. *Cardiosoma guanhumi*	Blue Land Crab (Congrejo Azul or de Tierra)	MX—terrestrial crab fishery, up to 2 miles inland (Diario Oficial 2010); CU—small land crab fishery (Baisre 2000)
Phylum Echinodermata/Class Holothuroidea/Family Holothuridae		
47. *Holothuria floridana*	Sea Cucumber (Pepina del Mar)	MX (Yucatán)—local artisanal fishery at Progresso (Mexicano-Cintora et al. 2007)
Family Stichopodidae		
48. *Astichopus multifidus*	Sea Cucumber (Pepina del Mar)	MX (Yucatán)—local artisanal fishery at Progresso (Mexicano-Cintora et al. 2007)
49. *Isostichopus badionotus*	Sea Cucumber (Pepina del Mar)	MX (Yucatán)—local artisanal fishery at Progresso (Mexicano-Cintora et al. 2007)

[a]Species names and higher classification follow Felder and Camp (2009), except for Class Bivalvia, where Tunnell et al. (2010) is followed
[b]Common names are given in English and Spanish where appropriate and available; English common names follow the "official" common names used on the NOAA Fisheries (NOAA 2011a): Office of Science and Technology, Fisheries Statistics Division (ST1) data web site (http://www.st.nmfs.noaa.gov/st1/) or from McLaughlin et al. (2005) for crustaceans or Turgeon et al. (1998) for mollusks
[c]Country (and sometimes State): *CU* Cuba; *MX* Mexico; *US* United States. Country or State listing is when a fishery for that species is present, not just distribution of the species
[d]Remarks note the status or type of fishery in that country or state, such as large active and small artisanal only
[e]Citation—only key references are listed, either the Fishery Management Plan for that species, a recent paper with extensive citations or information, or a status of the species

Carolinian Province, by zoogeographers (Briggs 1974). From Cabo Rojo southward throughout the entire southern Gulf of Mexico, Yucatán Peninsula, northwest Cuba, and the Florida Keys up to Cape Romano is considered tropical, or Caribbean Province (Figure 8.1).

Of the 49 species of shellfish taken as fishery species within the Gulf of Mexico, 28 are mollusks, 18 are crustaceans, and three are echinoderms (Tables 8.1 and 8.2). Eleven of the species are warm temperate, or Carolinian Province, species that live in the northern Gulf of Mexico, and 47 are distributed in the tropical, or Caribbean Province, waters of the southern Gulf, including south Florida. Regarding the three countries that surround the Gulf of Mexico, 16 of the species are taken within the United States, 46 from Mexico, and six from Cuba (Tables 8.1 and 8.2). Some species, such as oysters, penaeid shrimp, and blue crabs overlap the biogeographic, as well as political boundaries, and are found in two or more provinces or countries, respectively. As is true in most places in the world, diversity (number) of species is higher in the tropical waters of the southern Gulf of Mexico and productivity of selected

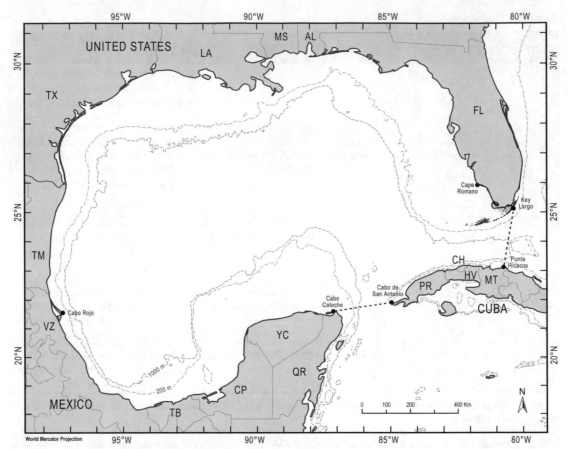

Figure 8.1. Gulf of Mexico, delimiting the geographic boundaries considered in this chapter. Abbreviations for the states (or provinces in Cuba; *counterclockwise*) from Florida: *FL* Florida, *AL* Alabama, *MS* Mississippi, *LA* Louisiana, *TX* Texas, *TM* Tamaulipas, *VZ* Veracruz, *TB* Tabasco, *CP* Campeche, *YC* Yucatán, *QR* Quintana Roo, *PR* Pinar del Rio, *CH* Ciudad de la Habana, *HV* La Habana, *MT* Matanzas (map by Fabio Moretzsohn (used with permission of the Harte Research Institute, Texas A&M University - Corpus Christi); adapted from Felder and Camp 2009).

fishery species is higher in northern Gulf temperate waters (Ekman 1953; Pianka 1966; Briggs 1974).

For purposes of presentation and discussion within this chapter, shellfisheries of the Gulf of Mexico have been divided into three categories according to the relative size and importance of each fishery: major, moderate, or minor (Table 8.3). A major, or primary, fishery is a large fishery of significant economic importance to one or more of the three countries. A moderate, but important, fishery is one which is either widely or regionally of modest economic importance, and a minor fishery is one which is only of local or artisanal importance. Major fishery species (five species) are widely distributed and abundant, moderate species (six species) are, or were, abundant in a specific region or habitat, and minor species (38 species) can be locally or regionally abundant, especially within specific habitats. The fisheries listed in this document have been recognized as such in the published literature or "federal register" (i.e., Diario Oficial 2010) of a given country. There are undoubtedly other shellfish species that are not listed that have been taken by local individuals either recreationally or commercially in various regions of the Gulf.

Table 8.2. Numbers of Gulf of Mexico Shellfish by Classification Category, Zoogeographic Distribution, and Country

Classification Category	
Taxonomic group	**Number of species**
Mollusks	**28**
Gastropods	10
Cephalopods	5
Bivalves	13
Crustaceans	**18**
Shrimp	6
Lobsters	2
Crabs	10
Echinoderms	**3**
Total	*49*
Zoogeographic distribution	**Number of species**
Warm temperature/Northern Gulf of Mexico (Carolinian Province)	11
Tropical/Southern Gulf of Mexico (Caribbean Province)	47
Country	**Number of species**
United States	16
Mexico	46
Cuba	6

8.3.1 Mollusks

Molluscan shellfish species include gastropods (snails, conchs, whelks, and others), cephalopods (squids, octopi), and bivalves (mussels, oysters, scallops, clams, and others). Of the ten gastropod species listed in Table 8.1, most are minor fishery species taken by local or small artisanal fisheries in Mexico, and only the queen conch is a moderate, but important species that has been widely, or commonly, taken in all three Gulf countries. Regarding the five species of cephalopods in Table 8.1, squid (three species) are minor species that are primarily taken as incidental bycatch in shrimp trawls in the United States and Mexico, and only the Yucatán octopus sustains a moderate, but important, regional fishery in the Yucatán Peninsula states of Campeche and Yucatán. Of the 13 species of bivalves listed, most are minor shellfish species taken in small local or artisanal fisheries of Mexico with only the Eastern oyster making a major, or primary, fishery in the United States and Mexico, so it will be covered in detail in Sections 8.4 and 8.5. The mangrove oyster is a moderate, but important, fishery in both Mexico and Cuba, and will therefore be covered below.

8.3.1.1 Queen Conch

The queen conch is a large and beautiful marine snail reaching over 30 cm (12 in.) in length and weighing up to 2.3 kg (5 lb). It has a flared outer lip with a bright pink or rosy aperture giving it the sometimes-used common name of pink conch (Figure 8.2) (Tunnell et al. 2010).

Table 8.3. Relative Size and Importance of Gulf of Mexico Shellfish Fisheries

Major fishery	Country
1. Eastern Oyster	US, MX
2. Brown Shrimp	US, MX
3. Pink Shrimp	US, MX
4. White Shrimp	US, MX
5. Blue Crab	US, MX
Moderate but important fishery	**Country**
1. Queen Conch	US, MX, CU
2. Yucatán Octopus	MX
3. Mangrove Oyster	MX, CU
4. Atlantic Seabob	US, MX
5. Spiny Lobster	US, MX, CU
6. Florida Stone Crab	US
Minor fishery	**Country**
1. Milk Conch	MX
2. West Indian Fighting Conch	MX
3. Banded Tulip	MX
4. True Tulip	MX
5. Horse Conch	MX
6. Knobbed Whelk	MX
7. Crown Conch	MX
8. West Indian Chank	MX
9. Squids (three species)	US, MX
10. Common Octopus	MX
11. Transverse Ark	MX
12. Southern Ribbed Mussel	MX
13. American Horse Mussel	MX
14. Stiff Pen Shell	MX
15. Bay Scallop	US, MX
16. Tiger Lucine	MX
17. Carolina Marsh Clam	MX
18. Florida Cross-barred Venus	MX
19. Southern Quahog	MX
20. Atlantic Rangia	MX
21. Brown Rangia	MX
22. Rock Shrimp	US, MX
23. Royal Red Shrimp	US
24. Spotted Lobster	MX
25. Swimming Crabs (six species)	MX
26. Gulf Stone Crab	US, MX
27. Cuban Stone Crab	US, MX, CU
28. Blue Land Crab	MX, CU
29. Sea Cucumbers (three species)	MX

US United States, *MX* Mexico, *CU* Cuba

Figure 8.2. Queen Conch (*Eustrombus gigas*) (photo by J.W. Tunnell).

It prefers sand, seagrass, and coral rubble habitats in warm, tropical seas of generally less than 21 m (70 ft). It is found throughout the Caribbean Sea and southern Gulf of Mexico, and it ranges as far north as Bermuda and as far south as Brazil (CFMC 1996).

Because the queen conch is prized for its meat and shell, population declines began throughout its range prior to the 1960s; however, most authorities and fishers did not acknowledge that overharvesting was occurring until the 1980s (Brownell and Stevely 1981; Iversen and Jory 1985; Appeldoorn and Meyers 1993; CFMC 1996). Conch fisheries in some localities, such as Florida Keys and Cuba, virtually collapsed due to overharvest (CFMC 1996). Once common on the Veracruz coral reefs, the queen conch essentially disappeared from that area in the 1980s (Tunnell et al. 2007). Likewise, it was common and being overfished during the 1980s on Alacran and other Campeche Bank reefs, but now it has low population levels due to over-harvesting (Figure 8.3) (Baqueiro 2004; Diaz-de-Leon et al. 2004; Tunnell et al. 2007).

Historically, the queen conch ranked second only to the spiny lobster in terms of export value of Caribbean-wide fishery products, and only second to a variety of finfish (mostly reef fish) in terms of local consumption (CFMC 1996). Even archaeological evidence strongly suggests its use and importance as a food source long before discovery of the New World (Stevely 1979). However, even though queen conch were once abundant and an important fishery resource throughout the wider Caribbean, today most localities no longer have a viable fishery due to overfishing (CFMC 1996).

In the United States, in the Florida Keys and surrounding area, commercial and sport conch fisheries (taken by hand while snorkeling or diving) had completely collapsed by the mid-1970s, primarily due to overharvest. Commercial harvest of queen conch was banned in the Florida Keys in 1975, and a ban on all commercial and recreational harvest was implemented in 1986 (CFMC 1996; SEDAR 2007). In Mexico, the queen conch fishery has deteriorated (Diaz-de-Leon et al. 2004), but some exploitation continues on Alacran and other Campeche Bank coral reefs. In Cuba, due to intense harvesting and overexploitation, takings have been prohibited since 1992, except small-quota catches permitted under special authorization in very selected areas (Claro et al. 2001, 2009). Interestingly, in Cuba the queen conch was historically, and even recently, taken as not only food but for bait. In all localities, the shells of the queen conch have historically been used in the shell-craft and handicraft trades.

Figure 8.3. Queen Conch fishing boat on Alacran Reef, Yucatán, Mexico (photo by J.W. Tunnell).

8.3.1.2 Yucatán Octopus

The Yucatán octopus fishery is one of the most important fishery resources in the southern Gulf of Mexico (Arreguin-Sanchez et al. 2000). The endemic Yucatán octopus makes up approximately 80 % of the catch, and the common octopus the remainder. The octopods are common in the nearshore limestone rocky bottom of the states of Campeche and Yucatán, with the latter having the largest part of the fishery. There are three fleets that participate in this fishery, two artisanal operating in shallow waters, and a mid-sized fleet of boats that operate in deeper waters. The fishing gear for all of these is locally known as the *jimba*, a long cane or bamboo pole extending from either end of a small boat with multiple lines on each (Figure 8.4) (Arreguin-Sanchez et al. 2000; Mexicano-Cintora et al. 2007). Small crabs (usually majids or portunids) are tied to the end of the line and allowed to drag at or near the bottom as the boat drifts with the current on the surface. When the octopus grabs the crab, the fisherman gently pulls the catch into the boat, kills it, and puts it into a storage box with ice. Fishermen operate out of nine small ports in Campeche and Yucatán, with many operating directly off the open Gulf beaches (Figure 8.5).

With upwards of 2,000 pangas (small fishing boats) and many more very small skiffs launched from the pangas, as well as about 500 mid-sized vessels with about ten skiffs each operating in the three fleets, catches in recent years have ranged from 11 to 22 million pounds, with a high of over 39 million pounds in 1997. The latter take is considered an overexploitation level, as noted by Arenas and de Leon (1999) and Arenas-Fuentes and Jimenez-Badillo (2004), who suggest a 22–26 million pounds sustainable level.

Figure 8.4. *Jimba* fishing rigs (cane or bamboo poles with multiple fishing lines) on Mexican fishing pangas at Chicxulub Puerto, Yucatán, Mexico (photo by J.W. Tunnell).

Figure 8.5. Mexican octopus fishermen on the beach at Chicxulub Puerto, Yucatán, Mexico (photo by J. W. Tunnell).

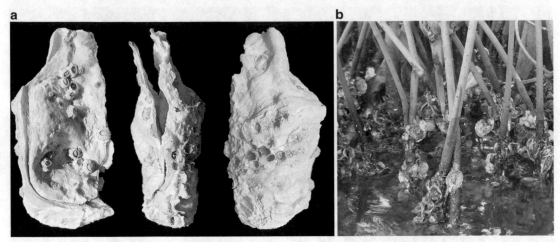

Figure 8.6. (a) Mangrove oysters (*Crassostrea rhizophorae*) and (b) growing on prop roots of red mangroves (photo 8.6a by John Wiley (used with permission of the Harte Research Institute, Texas A&M University - Corpus Christi) and 8.6b by Project Noah, http://www.projectnoah.org/).

8.3.1.3   Mangrove Oyster

Mangrove oysters, as implied by their common name, grow on the roots of mangrove trees (Figure 8.6) in estuarine conditions and are harvested commercially in both Mexico and Cuba (Baqueiro 2004; Baisre 2000). In Mexico, the largest catch of this species was taken along the mangrove-dominated coastline north of the town of Campeche (city), Campeche (state), in the western Yucatán Peninsula, but that population was apparently decimated by the Ixtoc I oil spill in 1979 and 1980 (Tunnell, personal communication with fishermen in 2010). The majority of mangrove oysters in Cuba are taken on the southern coast where extensive mangroves exist, but some are taken in the several small estuaries of the northwest coast facing the Gulf of Mexico.

8.3.2   Crustaceans

Crustacean shellfish species in the Gulf of Mexico include shrimp, lobster, and crab. Of the six shrimp species listed in Table 8.1, three are major shellfish species (brown, white, and pink shrimp), one is a moderate, but important, species (Atlantic seabob), and two are minor deep-sea species (rock shrimp and royal red shrimp (Stiles et al. 2007)). Two lobster fishery species are listed, one moderate (spiny lobster) and one minor (spotted lobster). Of the ten species of crab shellfishery species, only one is a major fishery (blue crab), one a moderate, but important, fishery (Florida stone crab), and all the rest are minor fishery species. In addition, there are two deep-sea crabs that are found within the Gulf of Mexico, golden crab (*Chaceon fenneri*) and deep-sea red crab (*C. quinquedens*), that are fished in the Atlantic and have unexploited potential in the Gulf (Waller et al. 1995; Trigg et al. 1997; Kilgour and Shirley 2008). Below are brief overviews of the moderate fishery species mentioned above: Atlantic seabob in the United States and Mexico; spiny lobster in the United States, Mexico, and Cuba; and, the Florida stone crab in the United States (Florida). The major fishery species mentioned above will be covered in detail in Sections 8.4 and 8.5 below.

8.3.2.1   Atlantic Seabob

The Atlantic seabob is a wide-ranging penaeid shrimp species that extends from North Carolina to southern Brazil and includes the entire Gulf of Mexico and Caribbean Sea

Figure 8.7. Seabob (*Xiphopenaeus kroyeri*). Photograph by Darryl L. Felder, University of Louisiana at Lafayette (all rights reserved by D.L. Felder).

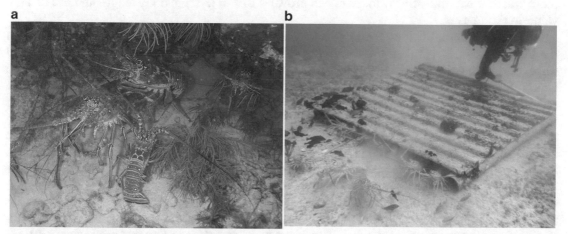

Figure 8.8. (a) Spiny lobster (*Panulirus argus*), Cayos Arcas coral reef, southern Gulf of Mexico (photograph by Dr. Jose Borges Souza, Centro Interdisciplinario de Ciencias Marinas, Instituto Politécnico Nacional, used with permission). (b) Lobster house or casita, which is commonly used in Mexico and Cuba (photograph from NOAA, http://www.habitat.noaa.gov/images/lobstertraps2.jpg).

(Figure 8.7). It is found on sandy and muddy bottoms of 1–70 m (3–230 ft) depth, but it seems to prefer water less than 27 m (88 ft) and near heavy freshwater outflows of estuaries and deltas. It is a very important shrimp fishery in two rather small specific areas within the Gulf of Mexico: In the United States between Pensacola, Florida, and Texas, and in Mexico off eastern Tabasco and western Campeche, specifically near Isla del Carmen and Laguna de Terminos in the Gulf. Annual landings of 3–4 million pounds of whole shrimp have been recorded in the distinctive seabob fishery near Ciudad del Carmen (Wakida-Kusunoki 2005).

8.3.2.2 Spiny Lobster

Like the queen conch, the spiny lobster is an iconic Caribbean species (Figure 8.8). It is widespread in shallow, warm tropical waters throughout the wider Caribbean and up to North Carolina on offshore banks and south to Brazil. Its preferred habitat is rocky bottom or coral

Figure 8.9. (a) Florida stone crab and (b) stone crab traps (image 8.9a and photo 8.9b by NOAA).

reefs where it can hide. Spiny lobsters can grow up to 1 m (3 ft) in body length and are of high value in the market.

In the continental United States, it is only taken commercially in South Florida, primarily in the Florida Keys. It is taken there commercially by diving or using wooden, plastic, or metal traps, and it constitutes the most valuable commercial fishery in Florida. Recreationally in the United States, it is primarily taken by diving. Spiny lobster is Florida's second most valuable recreational fishery (next to spotted sea trout), and overfishing is not occurring, according to the NOAA Fisheries Fish Watch program. The managed catch has averaged about 5.6 million pounds per year over the past decade. In Mexico, it has been maximally exploited on the Campeche Bank coral reefs and nearshore waters of Yucatán State (Diaz-de-Leon et al. 2004). In Cuba, spiny lobster is the most valuable fishery, but the majority of lobsters are taken along the southern portion of the country with only a small number/percentage being taken along the northwest coast facing the Gulf of Mexico (Claro et al. 2001). Landings in the mid-1990s were in the 21–27 million pounds range for Cuba (Baisre 2000), but most of those landings (60 %) are from the southwest shelf, and only 2–3 % are taken from the northwest (Claro et al. 2001).

8.3.2.3 Florida Stone Crab

The Florida stone crab (Figure 8.9) is one of three stone crab species within the Gulf of Mexico, but it is the only one with a targeted commercial fishery (Costello et al. 1979; Patillo et al. 1997). The Florida stone crab ranges from the Big Bend area of Florida near Apalachicola Bay and extends down the west coast around the tip of Florida and up the east coast to North Carolina (Williams and Felder 1986; Williams 1984). It also occurs in the Yucatán and Caribbean. Stone crab pots (wooden or plastic traps) are utilized off southwest Florida and in the Florida Keys for this distinctive regional fishery. Captured crabs have the large claw removed, and then they are replaced back into the environment, which makes this a uniquely sustainable fishery, since the crabs can regenerate the claw (Restrepo 1992). No overfishing is occurring in this fishery. There is a fishery management plan for stone crabs in the Gulf, but since their data are not separated by species the fishery is for "stone crabs" and not the three individual species (GMFMC 1979).

8.3.3 Echinoderms

Only three species of echinoderms are harvested commercially in the Gulf of Mexico, and all are minor shellfisheries located out of the Port of Progresso, Yucatán (Zetina et al. 2002; Mexicano-Cintora et al. 2007).

8.4 MAJOR SHELLFISH SPECIES

Since populations of key shellfish species vary naturally and greatly from year to year, it is important to understand the biology, ecology, and distribution of these species. General life cycles of each species, as well as affecting environmental parameters will be presented in this section. As the most valuable Gulf fishery, the penaeid shrimp (brown, pink, and white) will be discussed first, followed by Eastern oyster and blue crab.

8.4.1 Penaeid Shrimp

There are 20 species of shrimp in the Family Penaeidae in the Gulf of Mexico (Felder et al. 2009), but only three are of major importance as Gulf shellfisheries (brown, pink, white). These decapod (10 feet or legs) crustaceans are common to abundant in coastal estuaries and continental shelf waters of the Gulf. All three species have a similar life cycle, which includes spawning offshore with rapid development of eggs into larvae and juveniles that are carried inshore into extensive estuaries. These estuarine habitats serve as critical habitat and nursery grounds for the shrimp (Nelson et al. 1992; Patillo et al. 1997; Osborn et al. 1969). After 2–3 months of rapid growth, the shrimp approach maturity and migrate back offshore to complete their life cycle (Figure 8.10). The average life span of the three species is about 18 months, although they can live up to 3 years (Williams 1984). Regarding the shrimp fishery, these species are all considered to be an annual crop, but harvest time varies depending upon the species. The shrimp fishery is seasonal with most (about 80 %) of the catch taken between June and December each year. Historically, brown shrimp have been the largest fishery (usually over 50 %), followed by white shrimp and then pink shrimp, although white shrimp surpassed brown catches in 2005 and 2008 for the first time in 50 years (Nance 2011). The majority of shrimp by weight (about 80 %) are taken offshore and the remainder inshore (Osborn et al. 1969).

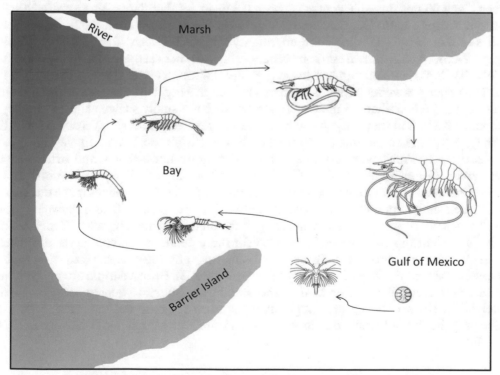

Figure 8.10. Typical life cycle of a penaeid shrimp in the Gulf of Mexico (drawing by J. W. Tunnell).

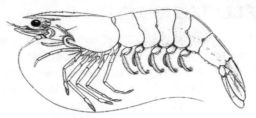

Figure 8.11. Brown shrimp (*Farfantepenaeus aztecus*) (drawing by NOAA).

8.4.1.1 Brown Shrimp

Brown shrimp (*Farfantepenaeus aztecus* Ives, 1891) (Figure 8.11) range farther north than any other U.S. penaeid shrimp, extending from Martha's Vineyard southward along the Atlantic coast around the tip of Florida, then around the Gulf to the northwestern Yucatán Peninsula (Williams 1984; Carpenter 2002). Adult brown shrimp females reach up to 236 mm (9.3 in.) in size; males reach 195 mm (7.7 in.) (Carpenter 2002) and have a brownish color. Ecologically, brown shrimp are an important food source for many species of finfish, with the type of fish varying with the size or life stage of the shrimp, and shrimp in turn feed upon a wide variety of food, depending on their life stage. Larval stages feed upon phytoplankton and zooplankton, and postlarvae feed on epiphytes, phytoplankton, and detritus. Juveniles and adults prey upon polychaetes, amphipods, and chironomid larvae, but they also feed upon algae and detritus (Cook and Lindner 1970; Patillo et al. 1997).

Habitat for brown shrimp ranges from offshore continental shelf waters for adults and eggs to shallow estuarine vegetated (preferred) and unvegetated bottoms for postlarvae and juveniles (Patillo et al. 1997). Salinity tolerance is generally wide-ranging, and optimal salinity depends on the life stage. Larvae tolerate salinities ranging between 24.1 and 36 ppt (Cook and Murphy 1966), and postlarvae have been collected in salinities from 0.1 to 69 ppt but grow best between 2 and 40 ppt. Juveniles range between 0 and 40 ppt, but they seem to prefer 10–20 ppt (Cook and Murphy 1966; Copeland and Bechtel 1974; Zimmerman et al. 1990). Adults tolerate salinities of 0.8–45 ppt, but their optimum salinity range is between 24 and 38.9 ppt (Cook and Murphy 1966). Adult brown shrimp generally spawn between depths of 46 and 91 m (151–299 ft), but they can range between 18 and 137 m (59–450 ft) (Renfro and Brusher 1982). The major spawning period is September through May, but it can occur throughout the year at depths greater than 46 m (150 ft). Brown shrimp usually spawn at night (Henley and Rauschuber 1981), and they may spawn more than once during a season (Perez-Farfante 1969). Generally, estuarine recruitment occurs when brown shrimp postlarvae move into estuaries from February to April with incoming tides and migrate into shallow and often vegetated nursery areas (Copeland and Truitt 1966; King 1971; Minello et al. 1989). In the northern Gulf, this recruitment can occur all year long (Baxter and Renfro 1966). Juveniles move out into open bays and then subadults migrate into coastal waters. Emigration to offshore spawning grounds occurs from May through August, coinciding with full moons and ebb tides (Copeland 1965).

The brown shrimp fishery (see Section 8.5 for more detail) is centered in the northwestern Gulf of Mexico, primarily off Texas and Louisiana, but there is a small fishery in the southeastern part of the Bay of Campeche (Carpenter 2002). Brown shrimp are most abundant from March to December but optimal catches are during March to September (Copeland and Bechtel 1974). Brown shrimp are caught at night when they are out and most active. They usually bury in the substrate during the day and are not caught by fishing gear (Osborn et al. 1969).

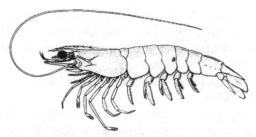

Figure 8.12. Pink shrimp (*Farfantepenaeus duorarum*) (drawing by NOAA).

8.4.1.2 Pink Shrimp

Pink shrimp (*Farfantepenaeus duorarum* Burkenroad, 1939) (Figure 8.12) range from the lower Chesapeake Bay area through the Straits of Florida and around the Gulf of Mexico to Cabo Catoche down to Isla Mujeres in the northeastern Yucatán Peninsula (Williams 1984; Carpenter 2002). The largest populations, and hence the largest catches, of pink shrimp are concentrated in two Gulf localities where the bottom is composed of calcareous muds and sands or a mixture of mud and sand (Hildebrand 1954, 1955; Springer and Bullis 1954): (1) off southwestern Florida, and (2) off the State of Campeche in the southeastern Bay of Campeche west of the Yucatán Peninsula. Adult pink shrimp females reach up to 280 mm (11 in.) in size and males reach 269 mm (10.6 in.), but are usually more in the 190 mm (7.5 in.) range, and color is quite variable from gray, blue gray, blue, or purplish in juveniles and young adults from estuaries and nearshore waters. Offshore adults from deeper waters often tend to be red, pinkish, blue gray, or nearly white (Williams 1984). Almost all are distinctly characterized by a dark spot of varying color at the juncture of the third and fourth abdominal segment. Ecologically, pink shrimp seem to prefer seagrasses in general and shoal grass (*Halodule wrightii*) in particular (Patillo et al. 1997). Large populations of juveniles appear to be important in supporting large populations of juvenile fish in these habitats. They also provide an important link in the estuarine food web by converting detritus to more available biomass for fish, birds, and other predators.

Habitat for pink shrimp eggs and planktonic larvae is pelagic, whereas postlarval and juvenile stages occur in oligohaline to euhaline estuarine waters and bays (Patillo et al. 1997). Adults occur in estuaries and nearshore waters to 64 m (210 ft), and mature pink shrimp in deep offshore waters but have highest concentrations between 9 and 44 m (30–144 ft). Largest numbers of pink shrimp are found where shallow bays and estuaries border a broad, shallow continental shelf (Perez-Farfante 1969; Costello and Allen 1970; Williams 1984), and where habitats have daily tidal flushing with marine water and large seagrass beds with high blade densities (Costello et al. 1986). Salinity requirements or preferences vary with shrimp size and geographic area (Costello and Allen 1970). Postlarval pink shrimp have been observed in salinities ranging from 12 to 43 ppt. Juveniles have been observed in waters less than 1–47 ppt, but they seem to prefer salinities greater than 20 ppt (Costello and Allen 1970; Copeland and Bechtel 1974). Adults are generally found in 25–45 ppt, although they have been found in salinities as high as 69 ppt (Patillo et al. 1997). Adult pink shrimp generally spawn in seawater depths of 4–48 m (13–158 ft) and probably deeper waters also (Perez-Farfante 1969). In the northern Gulf of Mexico, the two principal spawning grounds are the Sanibel grounds and Tortugas grounds in depths between 15 and 48 m (49–158 ft). The height of the spawning activity occurs from April through September in the Florida Bay region (Costello and Allen 1970; Williams 1984). Spawning occurs as water temperature rises, and maximum activity occurs between 27 and 30.8 °C (Rossler et al. 1969; Jones et al. 1970). Estuarine recruitment for pink shrimp into nursery grounds occurs during the summer months, and they remain there for 2–6 months (Costello and Allen 1970; Copeland and Bechtel 1974). Late juveniles and early adults migrate into deeper offshore waters. Although emigration occurs throughout the year, the main peak in activity occurs in the fall with a secondary peak in the spring. Decreasing water temperatures trigger the pink shrimp to move into deeper offshore waters (Costello and Allen 1970; Copeland and Bechtel 1974).

The pink shrimp fishery (see Section 8.5 for more detail) occurs almost continuously around the Gulf of Mexico, but concentrations are highest in the carbonate mud and sand areas of southwest Florida (Klima et al. 1986; Hart 2008) and southeastern Bay of Campeche. Pink shrimp, like brown shrimp, burrow in during the day and come out at night, which is when trawling activity is most intense for this species.

8.4.1.3   White Shrimp

White shrimp (*Litopenaeus setiferus* Linnaeus, 1767) (Figure 8.13) range from Fire Island, New York, to Saint Lucie Inlet, Florida; near the Dry Tortugas (rarely); and then around the Gulf of Mexico from the Ochlocknee River, Florida, to Campeche, Mexico. The centers of abundance in the Gulf occur off Louisiana, Texas, and Tabasco (Williams 1984; Klima et al. 1987), but the greatest densities occur off Louisiana (Klima et al. 1982). Adult white shrimp females reach up to 257 mm (10 in.) in size and males reach 175 mm (6.9 in.) (Carpenter 2002), and they have a translucent, bluish white body color with dusky bands and patches composed of scattered black specks (Williams 1984). Ecologically, white shrimp provide an important link in estuarine food webs by converting detritus and plankton into biomass available for fishes and other predators (Patillo et al. 1997). They are preyed upon by a large number of different estuarine and coastal finfish, and their postlarvae and juveniles tolerate lower salinities than other penaeid species. White shrimp also remain in estuaries longer and grow larger than brown shrimp (Christmas and Etzold 1977). White shrimp are omnivorous at all life stages, but they tend to rely more on plant matter than animal matter (McTigue and Zimmerman 1991). Larval stages of white shrimp are planktivorous, while adults and juveniles are scavengers. Adults combine predation with detrital feeding, including a wide variety of items such as detritus, insects, annelids, gastropods, copepods, bryozoans, sponges, corals, fish, filamentous algae, and vascular plant stems and roots (Darnell 1958; Perez-Farfante 1969; Christmas and Etzold 1977).

Habitat for white shrimp ranges from nearshore neritic to estuarine, and from pelagic to demersal, depending on life stage (Patillo et al. 1997). Eggs and early planktonic larval stages are most abundant in nearshore marine waters. Postlarve move into shallow water estuarine habitats of soft mud or clay bottoms (sometimes sand) high in organic detritus, or abundant marsh grass in oligohaline to euhaline salinities (Patillo et al. 1997; Carpenter 2002). White

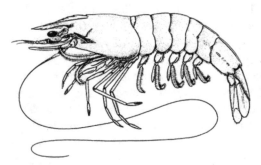

Figure 8.13. White shrimp (*Litopenaeus setiferus*) (drawing by NOAA).

shrimp are apparently more tolerant of lower salinities than brown shrimp (Gunter 1961). Postlarvae have been collected in salinities between 0.4 and 37.4 ppt, and juveniles seem to prefer salinities less than 10 ppt. Juveniles are frequently found in tidal rivers and tributaries throughout their range (Christmas and Etzold 1977). Collections of juveniles have occurred in salinities from 0.3 ppt in Florida to as high as 41.3 ppt in the Laguna Madre of Texas (Gunter 1961). Adults are generally found offshore in salinities greater than 27 ppt. Regarding depth, adults are usually found in Gulf waters less than 27 m (89 ft), and they are most abundant in waters less than 14 m (46 ft) (Perez-Farfante 1969; Renfro and Brusher 1982; Muncy 1984). Spawning takes place from spring through fall, but it peaks in summer (June–July) in offshore waters, where the eggs hatch and develop into larvae (Etzold and Christmas 1977; Klima et al. 1982). Like other penaeid shrimp, eggs are demersal and larval stages are planktonic. Postlarvae then migrate into estuarine nursery grounds through passes during May to November, with peaks in June, and a secondary one in September for the northwestern Gulf (Baxter and Renfro 1966). Juveniles migrate further up the estuary than brown or pink shrimp into less saline waters (Perez-Farfante 1969). As shrimp grow and mature, they leave the marsh habitat for open waters of the estuary and higher salinities. Emigration of juveniles and subadults from the estuaries into the open Gulf occurs in late August and September. Adults predominate in offshore, continental shelf waters during the fall and winter months and then move back nearshore in April and May (Patillo et al. 1997).

The white shrimp fishery (see Section 8.5 for more detail) is widely distributed throughout the nearshore Gulf of Mexico, but maximum catches occur along the Louisiana coast west of the Mississippi Delta (Christmas and Etzold 1977). White shrimp do not burrow into the bottom like brown and pink shrimp during the day, so the largest catches are predominantly made during daylight hours (Osborn et al. 1969). Most bays have a large bait shrimp fishery for white shrimp, and a wide variety of different kinds of nets are used for capture both commercially and recreationally (Patillo et al. 1997).

8.4.2 Eastern Oyster

The Eastern oyster (*Crassostrea virginica* Gmelin, 1791; also called American oyster) is by far the most important commercial mollusk landed in the Gulf of Mexico from Florida through Texas (Dugas et al. 1997), and it is perhaps the single most studied marine species in the entire Gulf of Mexico (see Galtsoff 1964; Berrigan et al. 1991, and VanderKooy 2012, for summaries) (Figure 8.14). Furthermore, oysters are considered to be a significantly important species in most estuaries along the Atlantic and Gulf of Mexico coasts, and self-sustaining populations

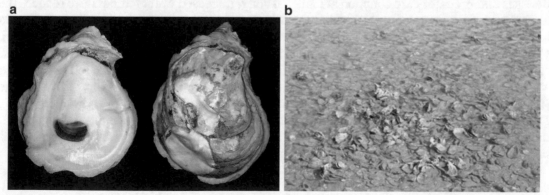

Figure 8.14. (a) Eastern oyster (*Crassostrea virginica*) and (b) an exposed intertidal oyster reef (photo 8.14a by John Wiley (used with permission of the Harte Research Institute, Texas A&M University - Corpus Christi) and 8.14b by J. W. Tunnell).

play an essential role in the ecology of these estuaries (NOAA 2007a). The Eastern oyster is easily recognized and distinguished from other species by the deep purple muscle scar, centrally located, on the interior of each valve.

Eastern oysters are bivalve mollusks in the family Ostreidae, and there are six total species found in this family within the Gulf of Mexico (Turgeon et al. 2009). The range of this species is from the Gulf of St. Lawrence in Canada through the Gulf of Mexico to the Yucatán Peninsula in Mexico (Galtsoff 1964) and perhaps further south (Carriker and Gaffney 1996). Gaffney later (2005 in NOAA 2007a) communicated that the Eastern oyster might only be confirmed genetically to the northern Yucatán Peninsula and that other distinct *Crassostrea* species may exist to the south. Although size and growth rate is highly dependent on salinity, temperature, food supply, and other environmental factors (Kennedy 1996; VanderKooy 2012), oysters generally grow rapidly during the first 6 months of life (up to 10 mm or 0.4 in. per month) and then slow down (Quast et al. 1988). Oysters may reach approximately 15 cm (5.9 in.) in 5 or 6 years (Hofstetter 1962; Berrigan et al. 1991), but a maximum size of 30 cm (11.8 in.) has been recorded in oysters living 25–30 years in Texas (Martin 1987). Harvest size (7.6–9.0 cm, 3–3.5 in.) is reached in the Gulf of Mexico within 18–24 months after setting (Hofstetter 1977; Berrigan et al. 1991). Oysters exposed to salinities that fluctuate within normal ranges (14–28 ppt; Quast et al. 1988; Shumway 1996) grow faster than those found in relatively constant salinity, but growth is stunted at 7.5 ppt and ceases below 5 ppt.

Ecologically, oysters are important in providing reef habitats that serve as areas of concentration for many other organisms (Wells 1961; Bahr and Lanier 1981), and they serve as a food source for a variety of estuarine fish and invertebrates (Burrell 1986; Eggleston 1990). Although oyster reefs have long been known as important ecological structures that participate in benthic-pelagic coupling via filtering vast quantities of water for feeding and then depositing rich organic material to the benthos, recent studies promote their importance as EFH with numerous important ecosystem services (Coen et al. 1999, 2007; Peterson et al. 2003; Grabowski and Peterson 2007). See more on this topic in Section 8.7.

Oysters are capable of surviving in a wide range of environmental conditions in coastal bays and estuaries (NOAA 2007a). However, their preferred or optimum habitat is on hard substrates in mid-salinity ranges (15–30 ppt) from intertidal to shallow subtidal. They prefer oyster shell for settlement but will settle on any available hard substrate, such as wooden pilings, concrete bulkheads, riprap shoreline, and boat hulls. In the Gulf of Mexico, depth ranges include 0.0–4.0 m (0.0–13 ft) (MacKenzie and Wakida-Kusunoki 1997; Dugas et al. 1997) and salinity optima of 10–27.5 ppt for larvae and about 5–40 ppt for adults (NOAA 2007a). Survival rate is better for adult oysters in the lower salinity range, as oyster diseases and predators are common at higher salinity. Increased water temperature reduces the ability of oysters to tolerate high salinities, while lower water temperatures allow oysters to tolerate lower salinity for longer periods (Berrigan 1988; Quast et al. 1988; Hofstetter 1990). However, prolonged exposure to freshwater during flood events, often referred to as *freshets*, can result in severe oyster mortalities (Galtsoff 1930; Hofstetter 1981; Marwitz and Bryan 1990; VanderKooy 2012). Temperature optima for oysters are 20.0–32.5 °C (68–90.5 °F) for larvae (Calabrese and Davis 1970) and 20–30 °C (68–86 °F) for adults (Stanley and Sellers 1986). Dissolved oxygen is 20–100 % saturation, but oysters can take low oxygen or no oxygen on a daily basis (NOAA 2007a; Berrigan et al. 1991). Water circulation is important for oysters for bringing in a constant food supply, but too much sedimentation is not good.

The Eastern oyster is a remarkably important and resilient organism within Gulf of Mexico estuaries. It is regarded as both a "colonizer" and an "ecosystem engineer" (NOAA 2007a). With favorable salinity and temperature regimes in the estuaries, successful reproduction and spawning of this highly fecund species provides widespread opportunity for settlement.

However, within the predominantly soft substrate estuaries of the Gulf of Mexico, available hard substrate habitat becomes the most limiting factor controlling oyster abundance (Berrigan et al. 1991). Where clean, hard substrate exists, oysters easily and abundantly colonize. As ecosystem engineers, they can even modify the physical environment and make it more suitable for their own long-term survival.

Eastern oysters are protandric, meaning that individuals first mature as males and then typically change to females later in life. Oysters may also change sex annually due to changes in environmental, nutritional, and physiological conditions. Although accurate fecundity is difficult to determine in oysters, estimates range from 2 to 115 million eggs per female, depending on size and geographic locality (Galtsoff 1964; NOAA 2007a). Initiation of spawning occurs with a combination of environmental factors including water temperature, salinity, and physiochemical interactions (Galtsoff 1964; Berrigan et al. 1991; NOAA 2007a). In Gulf waters, spawning occurs in all but the coldest months of the year. Generally, conditions for spawning include water temperature above 20 °C (68 °F) and salinity higher than 10 ppt.

Oysters develop through several free-swimming larval stages after fertilization, and then they attach to a suitable hard substrate and become sessile. The rate of development through the larval stages is variable and mainly dependent upon temperature (Shumway 1996). The process of settlement, metamorphosis (from veliger larva to spat with shell), and attachment normally occurs within 2–3 weeks of hatching, but it can be delayed for up to a month or more depending on environmental conditions (Kennedy 1996; NOAA 2007a).

Predation and disease is a significant factor to consider with oyster populations in the Gulf of Mexico. When oysters are young and their shells are thin, they are subject to predation by a variety of crabs and some fish species, particularly black drum, but as the oyster gets older it is more protected from many species. However, in more saline waters, the oyster drill (*Stramonita haemostoma*, a gastropod), the boring clionid sponge, and some boring polychaetes can inflict significant damage on oyster populations (Butler 1954; NOAA 2007a; VanderKooy 2012). Likewise, Dermo, which is a parasitic disease caused by the protozoan *Perkinsus marinus*, has caused extensive mortality to oyster populations in certain localities and in certain years (Ray 1987). It is most damaging in high salinities and high temperatures, particularly in times of drought (NOAA 2007a). Some harmful algal blooms are known to kill oysters (*Alexandrium monilatum*), yet others are known to only make them unfit for human consumption (red tide organism, *Karenia brevis*).

As filter feeders, oysters can bioaccumulate contaminants and microorganisms, including human pathogens and toxigenic microalgae, as noted above, when these organisms are present in the surrounding waters of oyster growing areas (Childress 1966; Calabrese et al. 1973; VanderKooy 2012). A number of commonly occurring bacteria, enterovirulents, parasites, and viruses can be contracted by eating raw or undercooked oysters. Since oysters are commonly consumed raw and whole, public health controls are now quite stringent to protect the consumer. Initially, the U.S. Public Health Service developed control measures through the National Shellfish Sanitation Program to reduce the risk of disease associated with the consumption of raw shellfish (oysters, clams, and mussels), and now many states also have similar programs (VanderKooy 2012).

The oyster fishery in the Gulf of Mexico (see Section 8.5 for more details on the fishery) has historically been a valuable fishery (Stanley and Sellers 1986). Although oyster production has been highly variable, Louisiana produces the most oysters in the commercial fishery and most of that is done via leases. Other Gulf states' oyster grounds are primarily public. Florida and Alabama allow tongs for harvesting oysters, while Mississippi, Louisiana, and Texas allow harvesting with dredges (NOAA 2007a).

Crab trap

Figure 8.15. (a) Blue crab (*Callinectes sapidus*) and (b) Blue crab trap (photos by South Carolina Department of Natural Resources).

8.4.3 Blue Crab

The blue crab (*Callinectes sapidus* Rathbun, 1896) supports one of the largest commercial and recreational fisheries in the Gulf of Mexico (Figure 8.15), and it is an abundant, environmentally tolerant, estuarine-dependent organism with year-round accessibility to the fishery (Guillory et al. 2001). Since the commercial harvest of blue crabs is primarily in state, rather than federal, territorial waters, the fisheries are managed by the various state resource management agencies in cooperation with the GSMFC (Patillo et al. 1997).

Blue crabs are swimming crabs in the family Portunidae, and 29 total species in this family are found within the Gulf of Mexico (Felder et al. 2009). In addition, there are eight species within the genus *Callinectes* within the Gulf, six of which are taken as fishery species within Mexico. The range of the blue crab is from Nova Scotia, Maine, and northern Massachusetts to northern Argentina, including Bermuda and the Antilles (Williams 1984). It has also been introduced into European waters and Japan (Carpenter 2002). Maximum size of adult blue crab is reported to be 246 mm (9.7 in.) in width including spines, and average is around 150 mm (5.9 in.) (Patillo et al. 1997). Blue crab begin to reach maturity as they go over 100 mm (3.9 in.), and they are almost all mature when they reach 130 mm (5.1 in.). Estimated life span of the blue crab is 3–4 years.

Ecologically, the blue crab performs a variety of functional roles in estuaries, and it plays an important role in trophic dynamics (Patillo et al. 1997). At different stages in its life cycle, the blue crab serves as predator or prey. Numerous species of fish, mammals, and birds prey upon the blue crab (Killam et al. 1992). In turn, the blue crab is an omnivore, scavenger, detritivore, predator, and cannibal that feeds on a wide variety of plants and animals, primarily selecting whatever is most available at the time and location where it is found (Menzel and Hopkins 1956; Darnell 1959; Costlow and Sastry 1966; Laughlin 1982).

Habitat for blue crab is in coastal waters on a variety of bottom types in freshwater, estuaries, and the shallow ocean from the water's edge to usually less than 35 m (115 ft) (Williams 1984). The biology of this species is better known than any of the others within this genus. Zoea larvae are usually found in pelagic waters and megalopa larvae may be found nearshore or in higher salinity estuarine areas. Megalopae settle into seagrass or other vegetated bottoms (Killam et al. 1992). Juveniles tend to be found in greatest numbers in low to intermediate salinities, which are characteristic of upper to middle estuaries (Steele and Perry

1990). They seem to prefer seagrass habitat as a nursery area, along with salt marshes (Thomas et al. 1990; Killam et al. 1992). Both juveniles and adults tend to be demersal. Adult males spend most of their time in low salinity areas, and females move from higher to lower salinity as they approach their terminal molt in order to mate (Patillo et al. 1997).

Environmental parameters that affect the growth, survival, and distribution of blue crab vary with life stages and sex (Killam et al. 1992). As might be expected, the eggs of blue crabs are the most sensitive to changing environmental conditions such as temperature and salinity, whereas juveniles and adults have greater tolerances. Since juveniles and adults are more motile, they can also avoid or leave when conditions are not right. Juvenile and adult blue crabs have been collected at temperatures ranging between 3 and 35 °C (37–95 °F), but they stop feeding at temperatures below about 11 °C (52 °F), and they burrow in the mud at 5 °C (41 °F). Juvenile blue crabs are usually found in lower salinity waters, typically between 2 and 21 ppt. Adult males seem to prefer salinities of less than 10 ppt, and egg-bearing females (sponge) in waters usually above about 20 ppt (Patillo et al. 1997). The blue crab is very sensitive to low dissolved oxygen.

Regarding reproduction, the sexes are separate in blue crabs, fertilization is internal, and the eggs develop oviparously (Williams 1984). Mating normally occurs in the low salinity waters in the upper estuaries. Females mate while they are in the soft-shell stage in the upper estuary, but they move out to higher salinity water near the mouths or inlets of estuaries, or into the Gulf of Mexico in preparation for spawning. Spawning may occur anytime within the 2–9 months after mating, but it usually occurs in the spring by females that mated the previous fall in August to September (Williams 1984). Two spawning peaks usually occur in the northern Gulf of Mexico, one in the late spring and the other during the late summer or early fall (Stuck and Perry 1981; Patillo et al. 1997). Fecundity estimates for blue crab range from 723,500 to 2,173,300 eggs per spawning (Truitt 1939), but usually the range is between 1,750,000 and 2,000,000 (Millikin and Williams 1984). Females may spawn more than once per year.

The blue crab fishery (see Section 8.5 for more details) is found within almost all estuaries of the northern Gulf coast. Catches are highest in areas with more freshwater inflow. Hard shell crabs predominate in the catch and almost all are taken in crab pots (traps) today, although high numbers were taken in the past via trotlines and drop nets. Recreational catches are important, making up 4–20 % depending on location within the Gulf.

8.4.4 Peak Spawning, Recruitment, and Migration

Although it is difficult to give exact times for major biological activities in these three main shellfish groups of the northern Gulf of Mexico due to environmental, temporal, and geographic variation, it is instructive to see their normal and peak times of spawning, recruitment, and migration to explain or reveal the complexities of their life cycles (Table 8.4). Because of the warm temperate nature of the environment of the northern Gulf, many species reproduce almost year round, except for the coldest months, but they do have peak times when their eggs, larvae, juveniles, or adults are most abundant. This composite table is a combination of information from many sources and gives a general picture of these biological activities. Nelson et al. (1992) and Patillo et al. (1997) present detailed information and hundreds of sources in two volumes on the distribution and abundance of fishes and invertebrates in northern Gulf of Mexico estuaries, and Guillory et al. (2001) and VanderKooy (2012) analyze and present decades of blue crab and oyster data, information, and literature, respectively.

Table 8.4. Composite Display of Peak Months/Seasons of Selected Biological Activity for Shrimp (Brown, Pink, White), Eastern Oyster, and Blue Crab in the Northern Gulf of Mexico

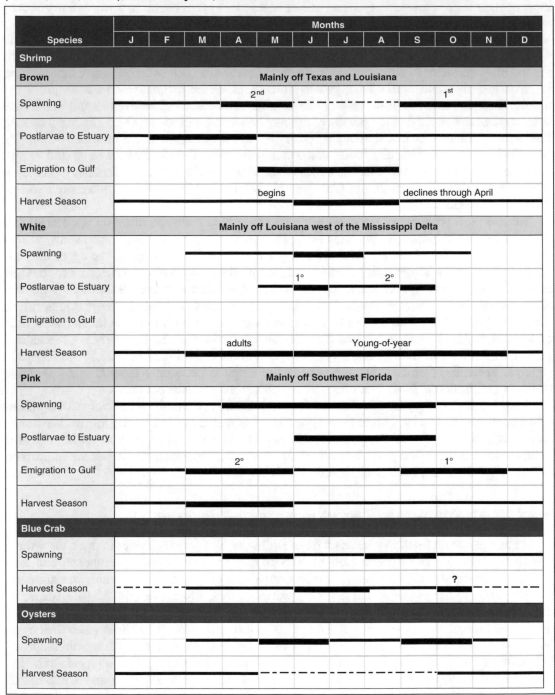

Compiled from: Williams (1984), Gauthier and Soniat (1989), Nelson et al. (1992), Patillo et al. (1997), Guillory et al. (2001), Hart and Nance (2010), and VanderKooy (2012).
Solid line means activity occurring, *bold line* means peak in activity, and *dashed line* means probable or possible activity occurring.

8.5 SHELLFISH SPECIES STATUS AND TRENDS

Current status (in 2009) and historical trends (1960–2009) of catches of the *seafood trinity* (shrimp, oysters, and blue crabs) for the northern Gulf of Mexico are presented in this section of this chapter. Both fishery-dependent and fishery-independent data are presented and graphed. As will be noted below, annual harvests vary considerably, primarily due to annual fluctuations in environmental conditions that variously affect the eggs, larvae, and juveniles of the various species. Some fluctuations in commercial harvest are also caused by management decisions (addition of turtle excluder devices (TEDs) and bycatch reduction devices (BRDs) on shrimp trawls, limited entry programs and closures), economic conditions (fuel costs, insurance costs), loss of critical or essential habitat (seagrass beds, coastal wetlands, oyster reefs), or other environmental problems (degraded water quality, hurricanes). The latter two issues will be dealt with in the following sections after this one (Sections 8.6 and 8.7).

8.5.1 Status and Trends of Shrimp

The penaeid shrimp fishery for brown, pink, and white shrimp is the most valuable fishery in the Gulf of Mexico. These three species are all very short-lived and highly fecund, making them inherently resilient to fishing pressure (MRAG 2010). Adult brown shrimp are typically caught in less than 55 m (180 ft) and white shrimp are generally caught in less than 37 m (120 ft), and both species favor muddy or peaty bottoms, often with sand, clay, or broken shells. Primary habitat for harvesting adult pink shrimp is sand, sand-shell, or carbonate mud bottoms from the intertidal zone out to 35–65 m (115–210 ft). Catch season varies sequentially during the year: brown shrimp during May through August; white shrimp during September through November; and pink shrimp during December through April (MRAG 2010). Brown shrimp make up the majority of the Gulf catch, followed by white and then pink shrimp.

The shrimp fishery in the Gulf of Mexico has a long history, with the white shrimp fishery being the oldest, starting in the areas around New Orleans and Biloxi (Condrey and Fuller 1992). Haul seines pulled by large rowboats fitted with sails in estuaries and bays were the primary means of harvest until the trawl was introduced into the Gulf in 1917. With the use of trawls, landings continued to increase as fishermen expanded their range and depth of fishing. In the late 1940s, there was a dramatic drop in the white shrimp fishery and sudden increase in abundance and catch of the brown shrimp (Condrey and Fuller 1992). This reduction in the white shrimp fishery initiated exploration for other shrimping grounds in the Gulf and led to the discovery of other brown shrimping grounds in the western and northwestern Gulf, as well as pink shrimping grounds off southwestern Florida and southeastern Bay of Campeche (Springer 1951).

The period from 1950 to 1976 was marked by continued growth and expansion in the Gulf of Mexico shrimping fleet, as well as maximal use of U.S. Gulf shrimping grounds, and continued expansion into foreign fishing grounds in Central and South America (Condrey and Fuller 1992). An overview of the entire Gulf of Mexico shrimp fishery by Osborn et al. (1969) provided the first and only Gulf-wide maps of shrimp catch distribution (Figure 8.16). In 1976, the Magnuson-Stevens Fishery Conservation and Management Act was passed, requiring fishery management plans for all significant fisheries, and the Gulf shrimp fishery entered into a new era. This new act established regional fishery management councils and focused attention on the newly recognized EEZ off the U.S. coastline. In 1981, the GMFMC implemented the FMP for the Shrimp Fishery of the Gulf of Mexico in U.S. waters (GMFMC 1981). The Shrimp FMP for the Gulf of Mexico states its management objectives:

Figure 8.16. Historic map of all shrimp catches from the entire Gulf of Mexico (from Osborn et al. 1969).

- To optimize the yield from shrimp recruited to the fishery
- To encourage habitat protection measures to prevent undue loss of shrimp habitat
- To coordinate the development of shrimp management measures by the GMFMC with the shrimp management programs of the several states, where feasible
- To promote consistency with the Endangered Species Act and the Marine Mammal Protection Act
- To minimize the incidental capture of finfish by shrimpers, when appropriate
- To minimize conflict between shrimp and stone crab fishermen
- To minimize adverse effects of obstructions to shrimp trawling
- To provide for a statistical reporting system

The Gulf Council has been very active in the past several decades updating and amending the shrimp management plans to protect shrimp stocks from overfishing, reduce turtle drowning, reduce finfish bycatch, and protect EFH (MRAG 2010). The Gulf Shrimp FMP has been amended 14 times, and a 15th amendment is under development and consideration. To limit effort in the fishery NMFS established a moratorium on issuing more fishing permits in 2005. All federally permitted commercial vessels must be fitted with certified TEDs and BRDs (MRAG 2010).

The Galveston Laboratory of NOAA's NMFS Southeast Science Center has been a focal point for shrimp research since the late 1950s (Klima 1981, 1989). Extensive research on the biology and distribution of various shrimp species, shrimp management issues, stock assessments, and critical habitat issues has been a hallmark of this laboratory. Decades of important shrimp research are credited to the scientists of that laboratory: C. W. Caillouet, R. A. Hart, E. F Klima, J. H. Kutkuhn, M. J. Lindner, T. J. Minello, J. M. Nance, L. P. Rozas, and R. J. Zimmerman to name a few.

Fishery-dependent data in the following sections come from the NOAA Fisheries Office of Science and Technology, Fisheries Statistics Division web site. Fishery-independent data comes from SEAMAP housed at the web site of the GSMFC. SEAMAP is a state/federal/university program for the collection, management, and dissemination of fishery-independent data (information collected without reliance on data reported by commercial or recreational fishermen) in U.S. waters of the Gulf of Mexico (Rester 2011). Annual reports, or SEAMAP Environmental and Biological Atlases, have been published annually since the data set began in 1983. A major objective of SEAMAP is to provide a large, standardized database needed by state and federal management agencies, industry, and scientists to make sound management decisions about Gulf fisheries (Rester 2011). SEAMAP data, as well as all NOAA Fisheries data are recorded in the Gulf of Mexico by shrimp statistical subareas (Nance 1992; Nance et al. 2006), which extend from the Florida Keys (subarea no. 1) to the Rio Grande in South Texas (no. 21). SEAMAP shrimp data comes from shrimp trawls collected on the continental shelves of the northern Gulf during summer and fall surveys. Inshore fisheries-independent data for shrimp within this section comes from the LDWF 16-foot trawl sampling program, which collects data year around.

Figure 8.17 shows the variable, long-term trend in total shrimp landings from fishery-dependent NOAA Fisheries data for the northern Gulf of Mexico during 1960 to 2009. The variability in the landings line reflects both annual fluctuations in fishery effort and natural population fluctuations of the three species through time governed by varying environmental conditions (Osborn et al. 1969; Condrey and Fuller 1992; MRAG 2010; Nance 2011). The gradual increase in landings and effort during the first three decades from 1960 to the late 1980s most

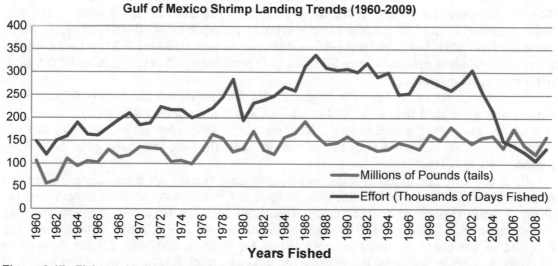

Figure 8.17. Fishery-dependent total Gulf of Mexico (United States) shrimp landing trends from 1960 to 2009 using NOAA Fisheries fishery-dependent data (from NOAA Fisheries).

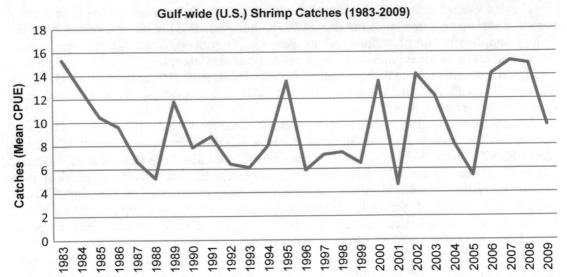

Figure 8.18. Fishery-independent mean catch per unit effort (*CPUE* mean count of shrimp caught per minute of fishing effort out of all sampled stations) in U.S. Gulf of Mexico shrimp trawl catches during 1983–2009 (from SEAMAP).

likely reflects the expanding shrimp fishery fleet and therefore increasing catch (Nance 2011). The precipitous drop in effort in the early 2000s primarily represents exogenous factors, such as rising fuel costs, competition from imported shrimp, damage to the fleet by recent hurricanes, and other issues (Caillouet et al. 2008; Hart 2008). Figure 8.18 shows the catch per unit effort (CPUE) of total offshore shrimp across the northern Gulf of Mexico as reported by SEAMAP using fishery-independent data, and it also clearly shows the natural fluctuations in annual populations mentioned above.

8.5.1.1 Brown Shrimp

The brown shrimp fishery is located primarily off Texas and Louisiana, but it extends from Texas to the westernmost part of Florida (shrimp statistical areas 10–21; Caillouet et al. 2008, 2011). Figures 8.19 and 8.20 reveal the trends in brown shrimp in the northern Gulf of Mexico by state and total catch from 1980 and 1983 to 2009, respectively, showing fishery-dependent and fishery-independent catches. Figure 8.19 clearly demonstrates the predominance of the catch off Texas and Louisiana.

Brown shrimp usually have the largest landings of northern Gulf shrimp (Figure 8.21). Brown shrimp reached an apex in 1990 at 103.4 million pounds (tails) followed by a low of 66.3 million pounds in 1997, a high of 96.8 million pounds in 2000, a low of 58.0 million pounds in 2005, and another high of 76.9 million pounds in 2009. The long-term average is 73.0 million pounds (Nance 2011).

Fishing effort (measured in thousands of 24-h days fished) for brown shrimp increased steadily from 1960 through 1989 but then dropped off in 1991 and remained almost level for about 7 years (Figure 8.22). Effort then fluctuated over the next several years, reaching 100 thousand days fished in 2004, which is similar to days fished in the 1970s. Effort then dropped to the upper 60 thousand days fished for 2005 to 2007 and further dropped to 61 thousand days in 2008, the lowest since the 1960s. In 2009, brown shrimp effort increased to 82 thousand days fished (Nance 2011).

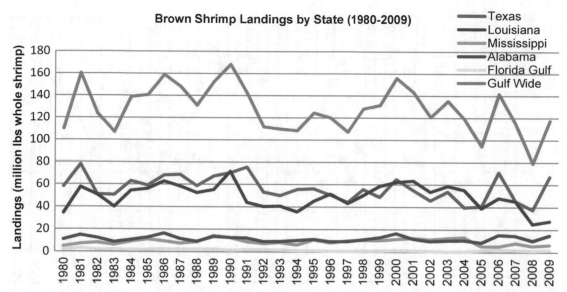

Figure 8.19. Fishery-dependent catches of brown shrimp in the northern Gulf of Mexico by state and Gulf-wide from 1980 to 2009 (from NOAA Fisheries).

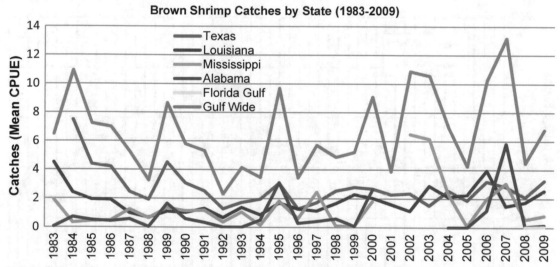

Figure 8.20. Fishery-independent mean catch per unit effort (*CPUE* mean count of shrimp caught per minute of fishing effort out of all sampled stations) of brown shrimp in the Northern Gulf of Mexico by state and Gulf-wide during 1983–2009 (from SEAMAP).

There are great fluctuations in CPUE fishery-dependent data, but generally there was a slow decline from 1960 to the late 1980s (Figure 8.23) (Nance 2011). Then, a general slow fluctuating increase was observed for 16–17 years. The brown shrimp CPUE value was 638 lb per day fished in 1998 and the best value since 1985. After fluctuation for several years, an upward trend in CPUE began in 2002 and reached an all-time record high value of 1,244 lb per day fished in 2006. In 2008, CPUE for brown shrimp dropped to 821 lb per day fished, but that is still above the long-term average of 643 lb per day fished. In 2009, the CPUE was 932 lb per day fished.

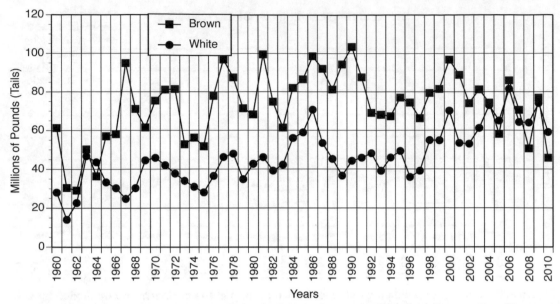

Figure 8.21. Annual catch data for northern Gulf of Mexico brown and white shrimp fisheries (from Nance 2011).

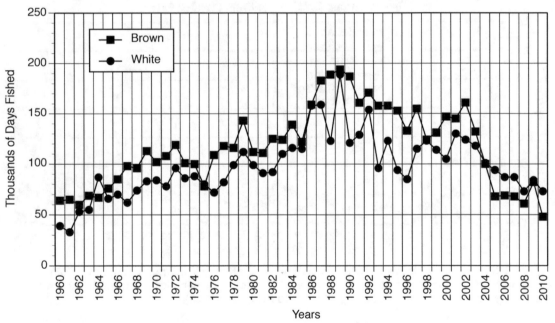

Figure 8.22. Annual effort data for northern Gulf of Mexico brown and white shrimp fisheries (from Nance 2011).

One important issue related to the brown shrimp fishery is the "Texas Closure." This became a tool implemented in 1981 as a primary objective of the Gulf of Mexico Shrimp FMP to increase the yield of brown shrimp harvested from Texas offshore waters (Jones et al. 1982; Nance 1996). This closure of the shrimp fishery from mid-May to mid-July each year allows the smaller shrimp to grow larger, thereby increasing the size of brown shrimp and subsequently getting a higher market value (Nance 1996).

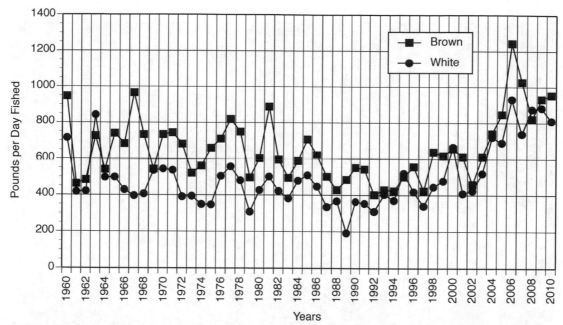

Figure 8.23. Annual CPUE data for the northern Gulf of Mexico brown and white shrimp fisheries (from Nance 2011).

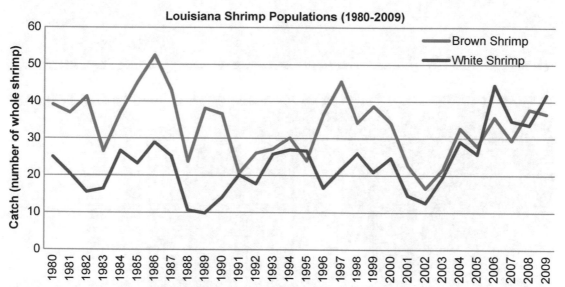

Figure 8.24. Annual fishery-independent CPUE (catch per 10-min 16-ft shrimp trawl) of brown and white shrimp caught in all Louisiana state waters during 1980–2009 (data from LDWF).

Examination of fishery-independent data gathered by Louisiana state biologists with the LDWF reveals natural population abundance variation and trends during 1980–2009 for brown and white shrimp in Louisiana state waters (Figure 8.24). Although demonstrated to fluctuate greatly over the past 30 years, these CPUE data (catch of whole shrimp per 10-min trawl) show a general upward trend in brown shrimp populations since 2002.

Historical and modern brown shrimp catch distribution is shown in Figure 8.25 (Osborn et al. 1969; NOAA 2011b).

a

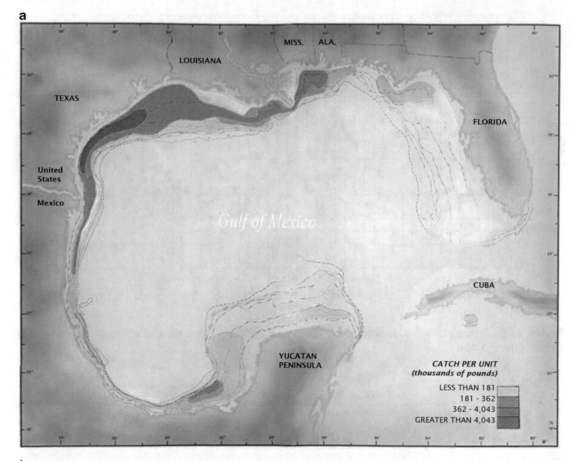

b

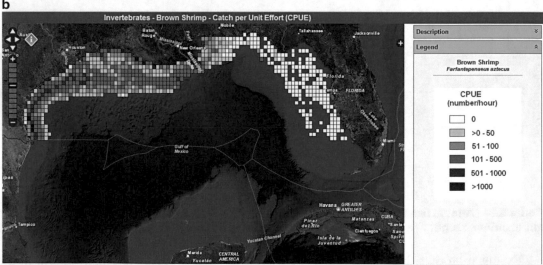

Figure 8.25. (a) Historical and (b) modern brown shrimp catch distribution in the northern Gulf of Mexico (images from (a) Osborn et al. (1969) and (b) NOAA Gulf of Mexico Data Atlas, http://gulfatlas.noaa.gov/).

8.5.1.2 Pink Shrimp

The pink shrimp fishery is primarily located off southwest Florida and secondarily off west Florida, but pink shrimp are also caught in all northern Gulf States (Figures 8.26 and 8.27). The main fishery encompasses statistical areas 1–9, with the Tortugas fishery in areas 1–3 and the west Florida area covering 4–9. Fishery-dependent data distinctly shows the predominance of the catch in Florida (Figure 8.26).

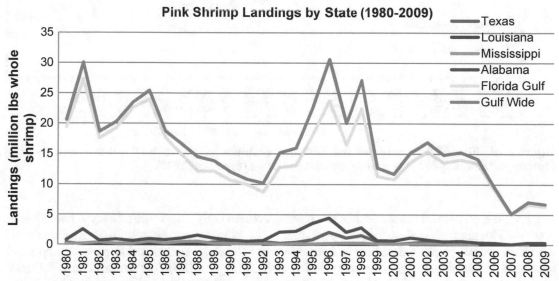

Figure 8.26. Fishery-dependent pink shrimp landings in the northern Gulf of Mexico by state and Gulf-wide during 1980–2009 (from NOAA Fisheries).

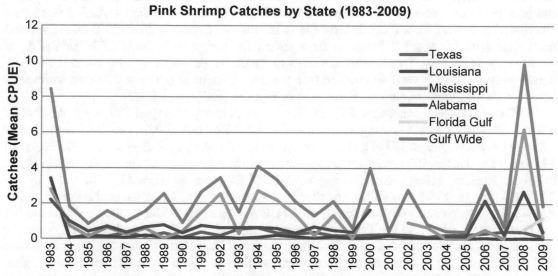

Figure 8.27. Fishery-independent mean catch per unit effort (*CPUE* mean count of shrimp caught per minute of fishing effort out of all sampled stations) of pink shrimp from the northern Gulf of Mexico by state and Gulf-wide during 1983–2009 (from SEAMAP).

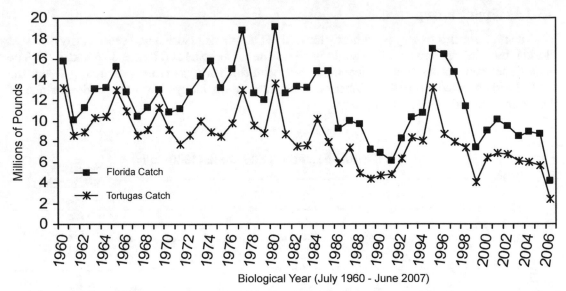

Figure 8.28. Pink shrimp catch on the Tortugas grounds (subareas 1–3) and the west coast of Florida (subareas 4–9) for biological years 1960–2006 (from Hart 2008).

Lacking an overall assessment of the entire northern Gulf pink shrimp fishery, focus herein is on the Florida pink shrimp fishery, since that is the main geographic region of the fishery and a recent overall biological review is available (Hart 2008). Annual Florida pink shrimp catch averaged 11.2 million pounds between 1960 and 2007 (Figure 8.28). Record numbers of Florida pink shrimp were landed in 1996 at 18.9 million pounds, but the catch subsequently declined and has remained near or below the long-term mean. Catches on the Tortugas grounds decreased considerably during 2005, 2006, and 2007 (Hart 2008).

Fishing effort in the Tortugas fishery was at a constant level from 1960 through the mid-1980s (average of 16.3 thousand days, Hart 2008) (Figure 8.29). Effort then dropped in the late 1980s and early 1990s but peaked in 1995 at 25 thousand days fished. Effort then fluctuated over the following years but began a continuous decline in 2002–2003 of 13 thousand days fished to 6.0 and 3.0 thousand days fished for the western coast of Florida and the Tortugas, respectively. These levels are most likely due to economic conditions in the fishery community, such as devastation caused by hurricanes Katrina and Rita in 2005, an increase in low-cost shrimp imports, and an increase in fuel prices (Haby et al. 2003).

CPUE for fishery-dependent data of Florida pink shrimp averaged 598 lb per day fished during 1960–1985 on the Tortugas grounds (Figure 8.30; Hart 2008). The CPUE was below average between 1986 and 1994 and then fluctuated a few years until 1999 when CPUE equaled 349 lb per day fished, the lowest value recorded over the entire data set on the Tortugas fishing grounds. The CPUE then began climbing to a high of 736 lb per day fished in 2005 and a drop in 2006 to 615 days, which was still one of the highest levels recorded over the past 20 years. So, as noted above, catch and effort declined, yet CPUE remained high. Thus, relative abundance of the Florida pink shrimp in the Tortugas fishery as measured by CPUE has been stable over the long-term data set for that area. This is an indication that the fishery is most likely not in decline and that the primary reason for the low harvest numbers is due to economic and not biological conditions (Hart 2008). A close examination of these latter trends and modeling efforts is provided by Hart and Nance (2010).

Historical and modern pink shrimp catch distribution is shown in Figure 8.31 (Osborn et al. 1969; NOAA 2011b).

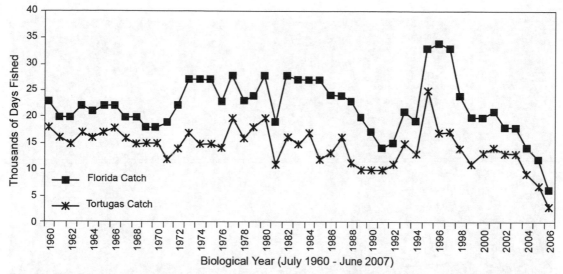

Figure 8.29. Pink shrimp fishing effort on the Tortugas grounds (subareas 1–3) and the west coast of Florida (subareas 4–9) for biological years 1960–2006 (from Hart 2008).

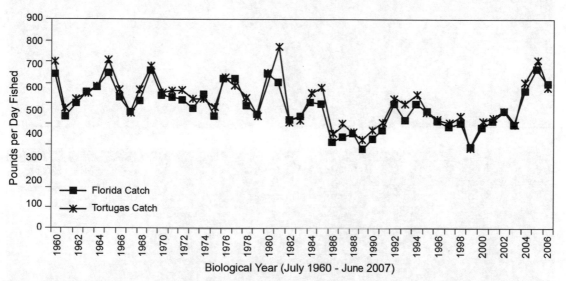

Figure 8.30. Pink shrimp catch per unit effort (CPUE) for the Tortugas grounds (subareas 1–3) and the west coast of Florida (subareas 4–9) for biological years 1960–2006 (from Hart 2008).

8.5.1.3 White Shrimp

The white shrimp fishery is located primarily off Louisiana and Texas but catches occur in all five northern Gulf States (Figures 8.32 and 8.33). Fishery-dependent data demonstrates that the catch for white shrimp is primarily off Louisiana and secondarily off Texas (Figure 8.32).

White shrimp landings in the northern Gulf of Mexico are second to brown shrimp, which is the largest catch of the three penaeid species (Nance 2011; Hart 2008) (see Figure 8.21). White shrimp reached its greatest harvest during 2006 at 81.5 million pounds (Nance 2011). Previous to that, 2004 was the highest (72.6 million pounds) followed by 1986 (70.7 million pounds). After the 1986 high catch, levels fluctuated around the long-term mean of 46.2 million pounds, but

a

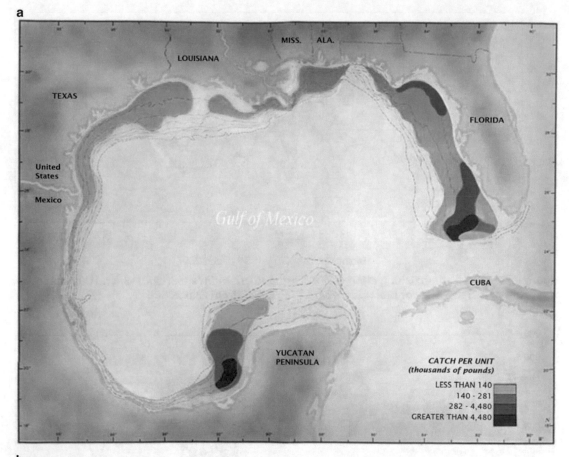

b

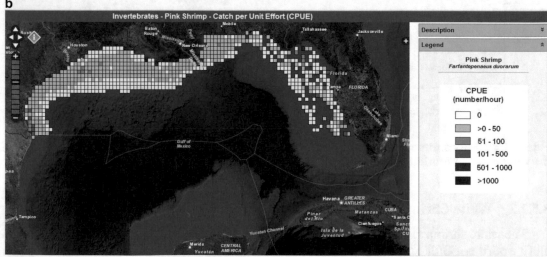

Figure 8.31. (a) Historical (image from Osborn et al. 1969) and (b) Modern pink shrimp catch distribution in the northern Gulf of Mexico (image from NOAA Gulf of Mexico Data Atlas, http://gulfatlas.noaa.gov/).

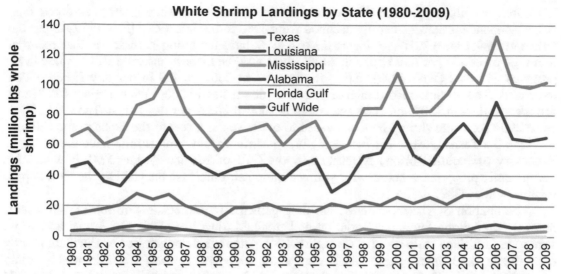

Figure 8.32. Fishery-dependent catches of white shrimp in the northern Gulf of Mexico by state and Gulf-wide from 1980 to 2009 (from NOAA Fisheries).

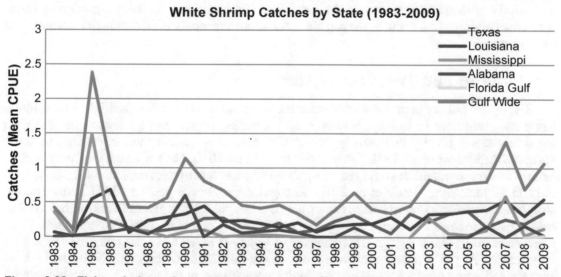

Figure 8.33. Fishery-independent mean catch per unit effort (*CPUE* mean count of shrimp caught per minute of fishing effort out of all sampled stations) of white shrimp in the northern Gulf of Mexico by state and Gulf-wide from 1983 to 2009 (from SEAMAP).

then in the late 1990s began increasing, but with yearly fluctuations all above the long-term mean. White shrimp landings were above brown shrimp landings for the first time in recent history (1960–2005) during 2005 and then again in 2008 (Nance 2011).

Fishing effort for white shrimp increased steadily from 1960 through 1989 (Nance 2011) (see Figure 8.22). From the 1989 high of almost 190,000 days fished, fishing effort had a fluctuating decrease to a low of 85,000 days fished in 1996, then a fluctuating increase to 130,000 days fished in 2001 when effort began declining again. Effort declined to a low of 73,000 days fished in 2008 and then had a slight increase in 2009 to 84,000 days (Nance 2011).

Catch per unit effort (CPUE; pounds per day fished) for white shrimp, as noted from fishery-dependent data, generally declined from 1960 to the late 1980s (low of 192 lb per day fished in 1989; Nance 2011) (see Figure 8.32). A slow, but fluctuating increase was then recorded to a high of 665 lb per day fished in 2000, which was the highest observed CPUE value in the previous 36 years (1964–1999). After another low in 2001 of 409 lb per day fished, CPUE increased to an all-time record high of 931 lb per day fished in 2006. White shrimp CPUE then dropped slightly in 2007 and 2008 and ended in 2009 at 882 lb per day fished. The CPUE levels since 2004 for white shrimp have increased to or above the levels of the 1960s (Nance 2011).

Like the Texas closure for increasing brown shrimp yield, pink shrimp have the Tortugas Sanctuary off south Florida, established by the Gulf of Mexico Shrimp FMP in 1981. This permanently protected sanctuary for young shrimp helps increase the yield of the Florida pink shrimp (Klima 1989).

Examination of fishery-independent data gathered by biologists with the LDWF reveals natural population abundance variation and trends during 1980–2009 for brown and white shrimp in Louisiana state waters (see Figure 8.24). Although demonstrated to fluctuate greatly over the past 30 years, these CPUE data show a general upward trend in white shrimp populations since 2002. Historical and modern white shrimp catch is shown in Figure 8.34 (Osborn et al. 1969; NOAA 2011b).

In summary, before 2009, the overall northern Gulf of Mexico shrimp stocks, as shown herein, appear to be flourishing, while the shrimp fishery appears to be in decline, primarily due to related economic and market conditions. Texas and Louisiana are the top-producing states for brown shrimp, Louisiana is the top state for white shrimp, and Florida is the top state for pink shrimp.

8.5.2 Status and Trends of Oysters

The oyster fishery in the Gulf of Mexico is the second most valuable shellfish fishery, and it has a long and diverse history. However, an evaluation of the current status and historical trends reveals a fishery in jeopardy on the U.S. east coast and beyond, according to some authors (Rothschild et al. 1994; Kirby 2004; Beck et al. 2011). Over a century of overfishing, habitat destruction, and degradation of water quality has left oyster reefs at risk globally with only 15 % remaining (Beck et al. 2011). Recognition of oyster reefs as EFH and estuarine structures with many important ecosystem services has placed significant focus on their critical role in estuaries and a need for widespread restoration (Coen et al. 1999; Peterson et al. 2003; Coen et al. 2007; Grabowski and Peterson 2007; Volety et al. 2009; Beck et al. 2011 to mention only a few). At the conclusion of their study in the mid-2000s (Beck et al. 2011), the Gulf of Mexico had some of the best remaining oyster populations in the world, and much attention began focusing on major restoration projects and programs. A more recent study analyzes changes in historic vs. present oyster habitat area (extent of coverage or distribution) and biomass (Zu Ermgassen et al. 2012). This new study suggests that biomass has declined, whereas the extent of habitat has been fairly stable in most areas.

Use of oysters as food and their shells as tools has been widely documented for prehistoric Native Americans in coastal areas of the Gulf of Mexico (Hester 1980; Ricklis 1996; Withers 2010). The first agency regulation of the oyster industry is found in the late 1800s, and many decades of oyster harvest data demonstrate the dramatic fluctuations in population levels and harvest (Berrigan et al. 1991; MacKenzie 1996; Dugas et al. 1997; NOAA 2007a).

The GSMFC published the first Regional Management Plan for the oyster fishery of the Gulf of Mexico in 1991 (Berrigan et al. 1991), and the second one was released in early 2012 (VanderKooy 2012). These comprehensive plans review all aspects of Gulf of Mexico oyster

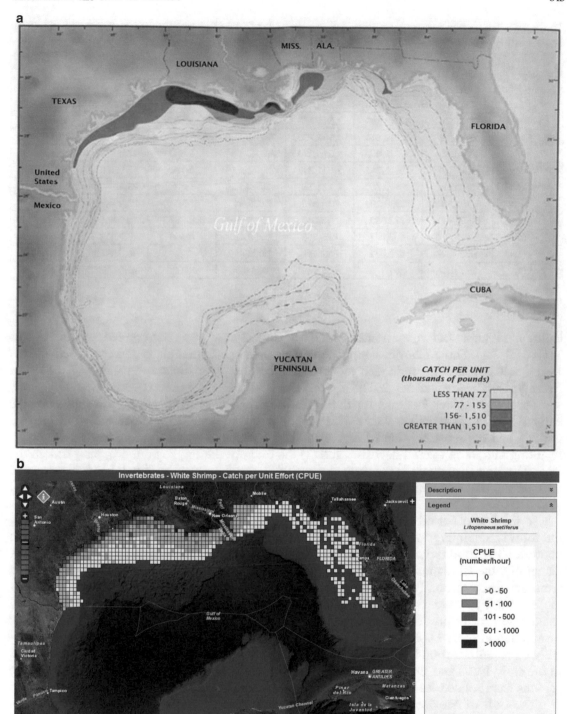

Figure 8.34. (a) Historical (image from Osborn et al. 1969) and (b) Modern white shrimp catch in the northern Gulf of Mexico (image from NOAA Gulf of Mexico Data Atlas, http://gulfatlas.noaa.gov/).

Table 8.5. Five-Year Average Landings (pounds of meats) of Eastern Oyster by Region 1950–2009 (from VanderKooy 2012)

Years	New England	South Atlantic	Mid Atlantic	Chesapeake	Pacific	Gulf	United States
1950–1954	2,135,820	3,751,800	16,036,900	34,500,400	19,920	12,545,120	68,989,960
1955–1959	437,400	3,030,760	6,396,360	36,639,000	12,440	13,166,120	59,682,080
1960–1964	378,478	4,063,460	1,548,720	22,983,980	9,360	20,139,800	49,123,798
1965–1969	283,628	3,139,440	1,144,700	22,610,780	13,340	20,917,340	48,109,228
1970–1974	267,280	1,766,900	2,526,980	24,943,560	8,580	17,206,040	46,719,340
1975–1979	620,220	1,940,041	2,941,240	21,152,660	2,776	18,978,066	45,635,003
1980–1984	1,245,660	2,438,736	2,228,180	17,184,700	462	23,357,919	46,455,657
1985–1989	1,162,178	1,580,296	370,520	9,030,011	32	20,294,850	32,437,887
1990–1994	5,624,089	773,492	845,210	2,356,109	2,287	15,902,540	25,503,727
1995–1999	2,465,268	507,927	825,208	1,969,435	8,408	22,760,376	28,536,622
2000–2004	433,476	588,632	832,557	1,000,412	725	25,516,329	28,372,131
2005–2009	337,167	801,178	601,069	604,004	43,020	21,017,328	23,340,168

biology, fishery, and management. All five Gulf States were represented on the Oyster Technical Task Force that developed these plans, and substantial plans and continued efforts were made to increase production yet protect the oyster populations and habitats (Arnold and Berrigan 2002; VanderKooy 2012).

Total U.S. oyster landings for the Eastern oyster have been declining steadily since the early 1950s with a peak in 1952 of 72.2 million pounds (Table 8.5) (VanderKooy 2012). Two periods had the most substantial declines: (1) New England region starting in the mid-1950s, resulting in a 32 % overall decrease, and (2) the Chesapeake Bay region, dropping first in the late 1950s and then again in the early to mid-1980s, resulting in an additional 37 % decrease in total production from the peak down to an average of 46.6 million pounds annually (Figure 8.35) (VanderKooy 2012).

In the 5-year period (2000–2004) just before the devastating hurricanes of the mid-2000s, the total U.S. landings of Eastern oyster had declined to only 28.3 million pounds, which was about a 60 % total reduction from the average harvest of the early 1950s (VanderKooy 2012). Generally, the Chesapeake Bay region was the nation's largest producer of all oyster species (four species: Eastern, Pacific, European flat, and Olympia) from the earliest landings records in 1880 until the mid-1970s (Figure 8.36) (VanderKooy 2012). The Gulf of Mexico generally ranked second in production, followed by the Pacific region. The remaining Eastern oyster production in other U.S. regions (South Atlantic, Mid-Atlantic, and New England) has historically represented around 10 % of the total domestic supply of oysters, with a few notable highs in the early 1950s and 1990s. However, since 2000, the combined landings for all three of these regions have totaled less than 7 % on average (VanderKooy 2012).

The Gulf of Mexico began dominating oyster production in the United States in the early 1980s when the northeastern areas began to decline. Despite the oyster reef-damaging hurricanes of 2004 and 2005, total Gulf production increased from the early 1980s to the present and has remained fairly stable (VanderKooy 2012). The Gulf of Mexico share of U.S. Eastern oyster production averaged about 40 % until 1980, but since then, it has increased from 50 % in the early 1980s to 60 % through the mid-1990s, and today represents 80–90 % of the U.S. total production (Table 8.6) (VanderKooy 2012).

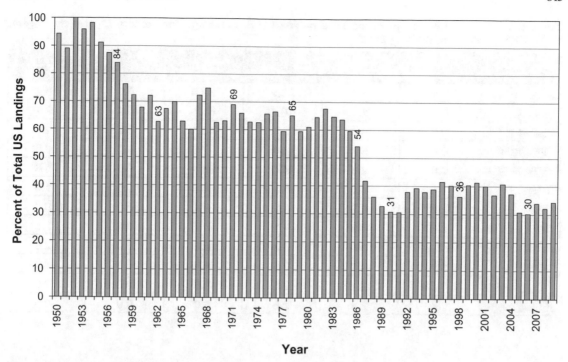

Figure 8.35. Percent decline from 1952 peak in total U.S. production of Eastern oysters from 1950 to 2009 (all regions combined). Peak production in this time period was 72.2 million pounds in 1952 (from VanderKooy 2012).

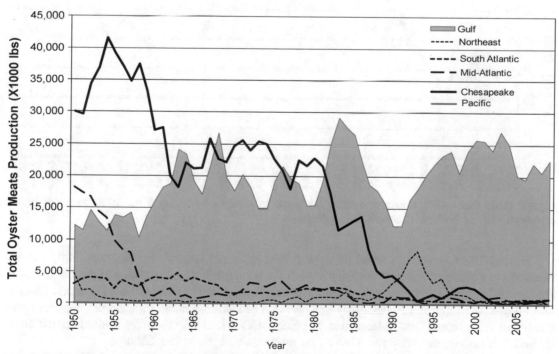

Figure 8.36. Total U.S. oyster landings for all four species (Eastern, Pacific, European flat, and Olympia) in pounds of meats by region from 1950 to 2009 (from VanderKooy 2012).

Table 8.6. Five-Year Average Percentage of Total U.S. Landings for Eastern Oyster by Region 1950–2009 (from VanderKooy 2012)

Years	New England	South Atlantic	Mid Atlantic	Chesapeake	Pacific	Gulf
1950–1954	3.1	5.4	23.2	50.0	0.0	18.2
1955–1959	0.7	5.1	10.7	61.4	0.0	22.1
1960–1964	0.8	8.3	3.2	46.8	0.0	41.0
1965–1969	0.6	6.5	2.4	47.0	0.0	43.5
1970–1974	0.6	3.8	5.4	53.4	0.0	36.8
1975–1979	1.4	4.3	6.4	46.4	0.0	41.6
1980–1984	2.7	5.2	4.8	37.0	0.0	50.3
1985–1989	3.6	4.9	1.1	27.8	0.0	62.6
1990–1994	22.1	3.0	3.3	9.2	0.0	62.4
1995–1999	8.6	1.8	2.9	6.9	0.0	79.8
2000–2004	1.5	2.1	2.9	3.5	0.0	89.9
2005–2009	1.4	3.4	2.6	2.6	0.0	90.0

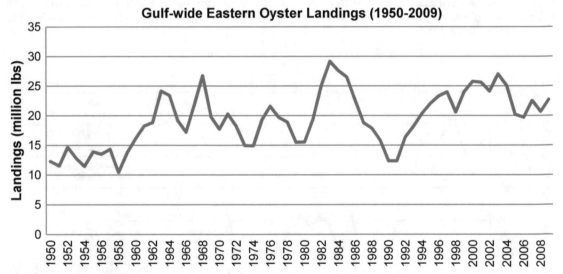

Figure 8.37. Gulf-wide Eastern oyster landings (pounds of meats) from 1950 to 2009 in the northern Gulf of Mexico (from NOAA Fisheries).

Figure 8.37 reveals the trend of oyster landings Gulf-wide from 1950 to 2009, and the fluctuating nature of catches is easily seen from this simple graph through time. Fluctuations are generally caused by changing environmental conditions (NOAA 2007a), but other species-related (diseases, parasites, harmful algal blooms) or fishery-related (market prices, fuel costs, etc.) issues can cause fluctuations also. See Sections 8.6 and 8.7 below for related environmental and habitat issues that govern oyster populations and the oyster fishery.

Louisiana is the top oyster-producing state in the northern Gulf of Mexico (Figure 8.38), as well as the entire United States, and most of its oysters are harvested from oyster leases.

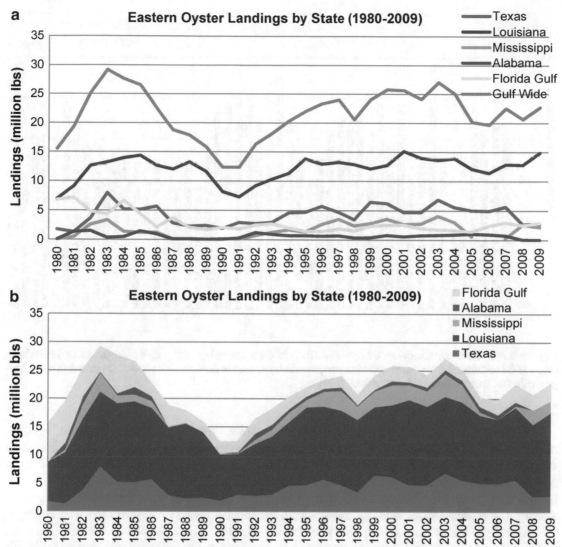

Figure 8.38. Eastern oyster landings by state in the northern Gulf of Mexico from 1980 to 2009 in two different formats: (a) line graph for ease of seeing Gulf-wide total and top-producing state Louisiana and (b) stacked graph for ease of comparing all five states together (from NOAA Fisheries).

Louisiana's average annual production of 11.9 million pounds represents nearly 60 % of the total Gulf of Mexico production during 1986–2005 (VanderKooy 2012). All other Gulf States primarily harvest oysters from public oyster grounds (NOAA 2007a). Florida and Alabama allow oysters to be harvested only with tongs on public oyster reefs. Mississippi, Louisiana, and Texas allow oysters to be harvested with dredges. Florida, Louisiana, and Texas market oysters year round, whereas Alabama and Mississippi follow seasonal harvest and marketing. Gulf of Mexico oyster reefs/resources are primarily subtidal and exhibit good sets and fast growth. In general, oyster landings increased gradually during the 1960s and 1970s then peaked in the early 1980s (NOAA 2007a). Oyster landings declined during the late 1980s due to a drought from 1986 to 1989, and a steady increase began after 1993. Confusion over the potential health risks associated with the consumption of raw oysters has eroded consumer confidence, and this may have caused an effect on oyster markets (NOAA 2007a).

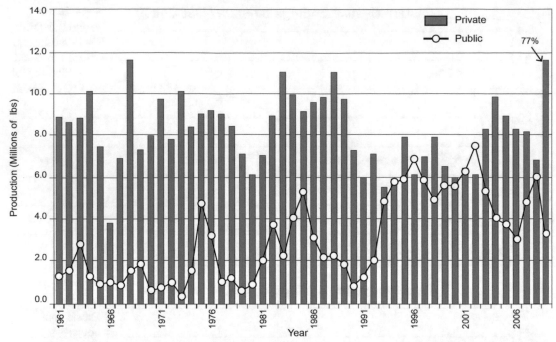

Figure 8.39. Total production of oysters from leases (private) and public grounds in Louisiana from 1961 to 2009. Long-term average for private landings is 8.007 million pounds and 3.065 million pounds for public (from VanderKooy 2012).

In Florida, oyster harvest mainly (90–95 %) comes from public oyster grounds, and the majority of that comes from Apalachicola Bay, which contains the state's most commercially valuable oyster reefs (NOAA 2007a). Alabama and Mississippi combined produce about 12 % of the Gulf of Mexico oyster landings. Both states suffered dramatic declines in oyster production from 1987 through 1992 (Dugas et al. 1997), but Alabama returned to long-term averages and Mississippi landings increased to the highest levels in 30 years (NOAA 2007a).

Although Louisiana oyster harvests are primarily from leased bottoms (Berrigan et al. 1991), public oyster grounds increased in production during the 1990s and early 2000s (LDWF 2005). As an example of the size and growth of the fishery, lease acreage expanded from less than 50,000 acres in 1960 to 130,000 acres in the early 1970s to 230,000 acres in the early 1980s and about 394,000 acres today (VanderKooy 2012). The CPUE data do not indicate a trend in the fishery, and fishing efforts remained stable from 1961 to 1986 (Berrigan et al. 1991). Public oyster grounds in Louisiana are used as seed areas for the leased areas and for harvest of market oysters. Harvest of market oysters from these public grounds has increased since 1992, and they even exceeded lease harvest of oysters in 1996 and 2002 (Figure 8.39; LDWF 2005; LDWF 2010; VanderKooy 2012). Fishery-independent trends in long-term population abundance data from public grounds show that Louisiana oyster stock was stable at relatively low levels from 1982 to the early 1990s then increased until 2001 and declined from 2002 to 2009 (LDWF 2005, 2010; VanderKooy 2012). These meter square counts of oysters to determine stock assessment on public grounds throughout Louisiana Coastal Study Areas (CSAs) clearly reveal the cyclical trends of natural oyster populations (Figure 8.40). The Louisiana Wildlife and Fisheries Commission uses oyster stock assessment data along with the Louisiana Oyster Task Force and LDWF, Marine Fisheries Division recommendations to set oyster harvest seasons. A

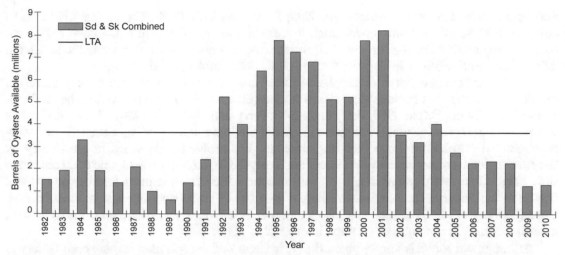

Figure 8.40. Historical estimated oyster stock size (*Sd* seed oysters; *Sk* sack or market-size oysters) on the public oyster areas of Louisiana. Meter square counts of natural populations (along with other information) are used to determine "barrels of oysters available" for the upcoming oyster season. *LTA* denotes the long-term average of 1982 to 2009 (from LDWF 2010).

lower stock availability of oysters generally results in a shorter Louisiana harvest season (NOAA 2007a).

Finally, in Texas, the Texas oyster fishery comprises two components: (1) a public reef fishery and (2) a leased bottom fishery (NOAA 2007a). Leases are only found in Galveston Bay, and they are used strictly as depuration areas for oysters transplanted from restricted waters. The lease harvest for oysters in Texas comprises between 20 and 25 % of the total commercial landings for the state. Long-term data indicate a general declining trend in oyster landings in Texas from 1956 to 1981, followed by an extremely large increase in 1982 and another decline in landings until 1987 (Quast et al. 1988). Since that time, landings have increased to more than 5.5 million pounds of meats harvested in 2004 (NOAA 2007a). More than half of Texas's public oyster reefs are found in Galveston Bay, and those account for 80 % or more of the Texas annual commercial oyster harvest.

In summary, through 2009, the northern Gulf of Mexico oyster fishery appears to be stable, but with observed annual, multiyear, or decadal fluctuations caused primarily by variable environmental conditions, but also at times by economic/market conditions. Louisiana is the top oyster-producing state in the northern Gulf, as well as in the United States.

8.5.3 Status and Trends of Blue Crabs

The blue crab fishery is the third most valuable shellfish fishery in the northern Gulf of Mexico, and it represents one of the largest commercial and recreational fisheries in the Gulf of Mexico (Guillory et al. 2001). Blue crabs are estuarine-dependent species that are highly productive, short-lived, and fast growing. All of these unique characteristics are important when considering the fishery and its management. Hard crabs are generally harvested almost exclusively in crab traps. During the 1990s (the last full decade of analysis), annual Gulf hard shell crab commercial landings averaged 61.6 million pounds, and the contribution of Gulf landings to the total U.S. landings ranged between 21.6 and 35.4 % (Guillory et al. 2001). Average contributions for each Gulf state included the following: Louisiana, 60.9 %; Florida, 17.7 %; Texas 14.3 %; Alabama, 4.9 %; and Mississippi, 1.9 % (Adkins 1972; Perry 1975; Guillory and Perret 1998; Hammerschmidt et al. 1998; Steele and Bert 1998). The recreational fishery equaled 4–20 % of the commercial catch in different areas of the Gulf (Guillory 1998), and there is a high-value fishery for soft-shell crabs, which averaged 188,000 lb annually during the 1990s (Perry and Malone 1985, 1989; Caffey et al. 1993; Guillory et al. 2001).

Significant changes have taken place in the Gulf of Mexico blue crab fishery since the publication of the first regional management plan (Steele and Perry 1990) and earlier descriptions of the fishery (Moss 1982; Perry et al. 1984; Perry and McIlwain 1986). Fishing effort has increased significantly, while harvests of blue crabs have stabilized or declined, and new management regulations have been implemented. The problems identified in the fishery by the first regional management plan (Steele and Perry 1990), including economic overcapitalization, habitat loss and/or degradation, as well as competition from imported crab products still persist in the fishery. The increase in count of number of crab fishermen and number of crab traps (Guillory et al. 1998, 2001) has also led to a decline in catch per fisherman, and a general overall increase in the number of traps in most Gulf states.

Although not much is known about the early history of the commercial blue crab fishery in the Gulf of Mexico, it is known that commercial landing statistics were first collected in the 1880s (Steele and Perry 1990). Long-handled dip nets were first used, and then drop nets and trotlines were employed. The first commercial fishery for blue crabs in the Gulf developed near New Orleans to supply the French Market and local restaurants (Perry et al. 1984). The first crab processing plant for Louisiana crabmeat was built at Morgan City in 1924, and others followed in Louisiana and other Gulf states. Hard crab fishing to be used for commercial processing did not become significant until World War II, and landings then increased gradually but erratically through the 1950s, 1960s, and 1970s, followed by a dramatic increase in the 1980s (Guillory et al. 2001). Although a very wide variety of fishing gears have been used to harvest blue crabs, today they are harvested almost exclusively with wire traps.

Figure 8.41 presents fishery-dependent blue crab catch Gulf-wide between 1950 and 2009, and Figure 8.42 presents fishery-dependent blue crab data for all five Gulf states from 1980 to 2009. Gulf-wide there is a fluctuating but continual increase in landings from 1950 until the late 1980s, followed by fluctuating but stable, or slightly declining, catches to 2009. Total reported

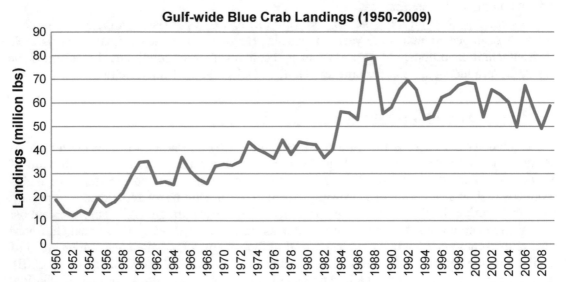

Figure 8.41. Fishery-dependent Gulf-wide blue crab landings in the northern Gulf of Mexico between 1950 and 2009 (from NOAA Fisheries).

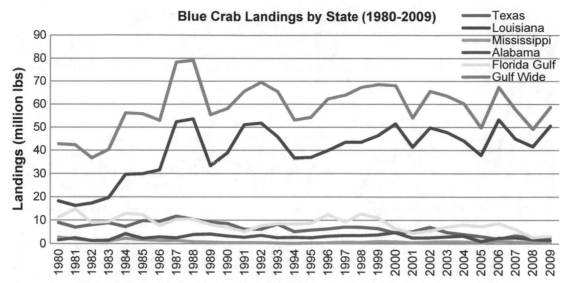

Figure 8.42. Fishery-dependent blue crab landings in the northern Gulf of Mexico by state between 1980 and 2009 (from NOAA Fisheries).

landings for blue crabs in the Gulf of Mexico increased from less than one million pounds in the late 1880s to approximately 18 million pounds before World War II. Landings then increased markedly in the late 1950s with the introduction of wire traps followed by increased processing capacity and market development (Guillory et al. 2001). Landings continued to rise in the 1980s with record landings of 78 and 79 million pounds occurring in 1987 and 1988, respectively. Landings of blue crab declined slightly after 1988 and then continued to fluctuate within the 50–70 million pound range. On the state graph (Figure 8.42), Louisiana clearly has the highest catch with a fluctuating but increasing trend since the mid-1980s. Florida and Texas both have a fluctuating but decreasing trend over the past three decades.

Stock assessment of Gulf of Mexico blue crab is limited by an absence of reliable fishery-dependent data (Guillory et al. 2001), and since there are no credible CPUE data available and no information on population age structure, many assumptions have to be made in modeling stock size. Fishery-independent data, however, gathered by the five Gulf States, does allow a better picture of blue crab status and trends.

Examination of Louisiana fishery-independent data gathered by biologists with the LDWF reveals natural population abundance variation and trends during 1980–2009 for blue crab in Louisiana state waters (Figure 8.43). Although demonstrated to fluctuate greatly over the past 30 years, these CPUE data show a general decline in crab populations with the trawl gear since the early 1990s extending to the present. Blue crab seine data (1986–2009) shows similar fluctuations over the past 25 years, and indicates a lower population level since the late 1990s.

In Texas, the blue crab fishery is shown to have matured as a fishery in the 1980s and moved into a senescent phase in the 1990s and 2000s (Sutton and Wagner 2007). Fishery-independent data show distinctive decreasing trends over the past three decades for bay bag seine, bay trawls, bay gill nets, and offshore trawls, and CPUE of Texas crab fishermen peaked in the mid-1980s with a continuing, but fluctuating, decrease since then (Sutton and Wagner 2007).

The blue crab fishery is characterized by seasonal, annual, and geographic fluctuations in landings (Guillory et al. 2001). Fluctuations have become more pronounced in recent years and

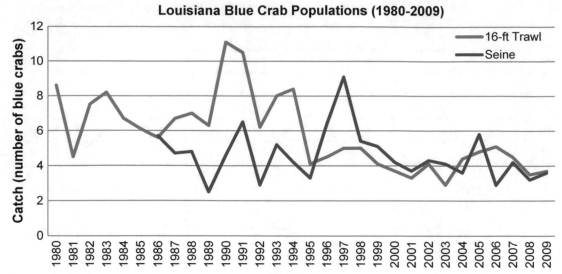

Figure 8.43. Fishery-independent annual CPUE (number of blue crabs per 10-min 16-ft trawl; average number of blue crabs per seine haul) in all Louisiana State waters during 1980/ 1986–2009 (data from LDWF).

include the following suggested causes: economic factors related to market demand and processing capacity (Moss 1982); economic interdependency with other kinds of fisheries (Steele and Perry 1990); changes in blue crab fishing effort (Guillory et al. 1996); and variability in year-class strength of blue crabs (Steele and Perry 1990).

In summary, through 2009, in the northern Gulf of Mexico, the blue crab fishery appears stable but is characterized by seasonal, annual, and geographic fluctuations in landings, which have become more pronounced in recent years due to various economic and other conditions. Louisiana is the top blue crab-producing state in the northern Gulf, whereas both fishery-dependent and fishery-independent data in Texas show long-term declines for blue crabs.

8.6   INFLUENCES ON SHELLFISH POPULATIONS AND THE FISHERY

Numerous factors are reported in the literature that influence or control natural populations of shellfish and their commercial landings. Environmental factors include natural variations in environmental parameters (salinity and temperature primarily), droughts, floods, hurricanes, anthropogenic impacts (e.g., degradation of water quality, habitat degradation, and loss) and outbreaks of diseases and parasites. Economic or market conditions include increased fuel costs, competition from imported shrimp causing reduced market prices for domestic shrimp, fishery overcapitalization, rising insurance costs, and loss of coastal habitat due to coastal development. After fishery management plans are developed for each fishery species, continued monitoring of the species and fishery, as well as periodic stock assessments, allow fishery managers to make management decisions about catch size, catch quota, or catch season, or in some cases, amend the fishery management plan. Important habitat issues are covered separately in Section 8.7 below.

8.6.1 Environmental Conditions

Natural fluctuations in environmental parameters, such as salinity and temperature, are known widely to cause fluctuations in animal populations and fisheries from year to year (Hofmann and Powell 1998). Many of the annual fluctuations seen on the graphs in Section 8.5 of this chapter are due to the annual variability in environmental conditions. Recent studies, however, have shown alarming negative changes and downward trends in marine species and habitats, especially in the coastal zone (Lotze et al. 2006; Halpern et al. 2008; Jackson 2008). Synergistic effects of habitat loss and destruction, overfishing, introduced species, global warming, ocean acidification, toxins and other pollutants, and massive runoff of nutrients are transforming once complex ecosystems, such as coral reefs and kelp forests, into monotonous level bottoms, transforming clear and productive coastal areas into anoxic dead zones, and transforming complex food webs formerly topped by large predators into simplified, microbial-dominated ecosystems with boom and bust cycles of toxic algal blooms, jellyfish, and disease (Jackson 2008). Globally, oysters were the first shellfish/invertebrate species to suffer extreme depletion, losing 85–90 % of populations in coastal bays and estuaries (Kirby 2004; Lotze et al. 2006; Jackson 2008; Beck et al. 2011). Many of the depleted areas are considered permanently depleted because of eutrophication, disease, and habitat loss not allowing recolonization (Jackson 2008). Fortunately, the Gulf of Mexico harbors some of the best remaining oyster reefs worldwide, and there is great opportunity in the Gulf for conservation and restoration (Beck et al. 2011).

Global climate change, including the issues of rising sea surface temperature, sea-level rise, and ocean acidification are all considered something like a large uncontrolled experiment with unknown, but potentially predictable, consequences. Some consider these to be the greatest challenge of humanity today (Jackson 2008). Some of the most obvious concerns for shellfish populations, which are dependent on estuarine conditions, are alterations to freshwater inflow and sea-level rise impacts to coastal marshes, the nursery grounds of shrimp and crabs (Montagna et al. 2007). Tolan (2007) demonstrated the impact of El Nino-Southern Oscillation translated to the watershed scale on salinity in all Texas estuaries along the Texas coast, showing significant correlations between global climate signals and local salinity patterns. These salinity patterns in turn affect the reproduction, recruitment, and survival of shellfish populations in this region of the northwestern Gulf of Mexico. Stenseth et al. (2002) and Hare and Able (2007) have likewise linked climate fluctuations or changes to ecological effects and fisheries.

Large-scale hypoxia in the northern Gulf of Mexico off Louisiana and Texas overlaps with the habitat and fishing grounds of brown and white shrimp (Zimmerman and Nance 2001). Adult brown shrimp are more affected than white shrimp due to their predominance further offshore where the hypoxic zone persists in the summer. When the hypoxic zone is widespread and persistent on the Louisiana shelf, the shrimp catch is always low (Zimmerman and Nance 2001). If the hypoxic zone blocks shrimp migration offshore, shrimp distributions and densities may be modified. White shrimp that are concentrated closer to shore and in bays and estuaries are not as affected by the hypoxic zone.

Hurricanes and tropical storms have the potential of affecting both the targeted shellfish species and the shellfish industry. In 2005, a large segment of the fishing and fishing-related businesses of the northern Gulf of Mexico and southwest Florida were devastated by Hurricanes Katrina, Rita, and Wilma. With the exception of oysters, an extensive study and report after these hurricanes revealed that Gulf coast living marine resources were not significantly impacted (NOAA 2007b). Commercial and recreational landings declined dramatically immediately after the storms, but they appeared to rebound to previous levels the following year.

However, the fishermen and fishing communities that were impacted by the storms were less resilient and did not rebound. Millions of tons of fishing-related debris were strewn across the land and shallow waters after the storms. Oysters are typically the shellfish species most severely impacted by hurricanes in the northern Gulf, as they are subject to direct physical damage or burial by mud and sand, or even hurricane-related debris, and freshets (Berrigan 1988; Haby et al. 2009). Fishery-independent surveys conducted by NOAA Fisheries after the storms indicated that none of the shrimp or crab stocks were significantly impacted and that most observed changes were within the normal, past interannual variation of CPUE (NOAA 2007b).

8.6.2 Parasites, Predators, and Diseases

Although parasites and diseases have generally not been a widespread issue in wild populations of shrimp, blue crabs are occasionally affected by diseases and parasites (Couch and Martin 1982; Davis and Sizemore 1982; Overstreet 1982; Overstreet and Rebarchik 1995; Messick 1998). Oysters, on the other hand, can be significantly impacted by parasites, pests, competitors, and diseases (White and Wilson 1996; NOAA 2007a). These impacts come from a wide variety of organisms including gastropods, crabs, flatworms, polychaetes, bivalves, sponges, bryozoans, and fish. Generally, young oyster spat is far more susceptible to predation than mature, market size oysters, as their thick calcium carbonate shell protects them against most intruders. Likewise, mortalities of oysters are much less in lower salinities (5–15 ppt) since most of the predators prefer higher salinities.

The main marine predator of adult oysters in the Gulf of Mexico is the oyster drill (*Stramonita haemastoma*, formerly *Thais haemostoma*), a gastropod with the ability to bore through the shell, kill, and then eat the oyster. Juvenile oysters are subject to predation by a number of crab species (stone crab, *Menippe* spp.; mud crab, *Panopeus herbstii*; blue crab, *Callinectes sapidus*). Polyclad flatworms in the genus *Stylochus* spp. can also predate oysters by slipping into the gaping shell and feeding on the oyster. Certain fishes are also known to feed upon oyster, such as sheepshead (*Archosargus probatocephalus*), skates (*Raja* spp.), and black drum (*Pagonias cromis*). Most fish feed only on thin-shelled juveniles, but the black drum can eat adult shells up to 8 cm (White and Wilson 1996).

Oyster pests detract from the health of the oyster by either weakening the shell or body, or by competing for food (White and Wilson 1996). Extensive damage rarely occurs, as pests are often small or uncommon. Well-known pests include the boring sponge (*Cliona* spp.), boring polychaetes (*Polydora* spp.), pea crabs (*Pinnotheres ostreum*), an ecoparasitic gastropod (*Boonea impressa*), and several bivalves that bore into the oyster shell (*Diplothyra smithi*, *Lithophaga bisulcata*, and *B. aristata*).

Oyster competitors can reduce the success of oyster populations by competing for food or space. Typical competitors include algae, arthropods, anemones, bryozoans, sponges, polychaetes, annelids, and mollusks.

There are two oyster diseases reported from Gulf of Mexico oysters: (1) Dermo and (2) MSX. Dermo is a parasitic disease caused by a protozoan *Perkinsus marinus* (Mackin et al. 1950). This is the first major oyster pathogen to be identified, and it was originally in the genus *Dermocystidium*, hence the shorten nickname Dermo. This parasite infects oysters during their first year of life and continues to proliferate causing up to 50 % mortalities in oysters living to their second year, and 80–90 % mortalities by the third year (NOAA 2007a). The parasite inhibits and affects the immune system of oysters, and it continues to cause significant mortalities along the Gulf coast. High salinity and high temperatures elevate the disease level in oysters, particularly during times of drought. MSX is also a protozoan disease, and it is caused by *Haplosporidium nelsoni*. This second lethal disease of oysters was first reported from Delaware Bay, where it was dubbed MSX for multinucleate sphere X (unknown) at first discovery during 1957 and 1958 with massive mortalities in oysters. Although MSX continues to be a problem on mid-Atlantic oyster reefs, it has not caused major mortalities in the Gulf (NOAA 2007a).

8.6.3 Economic Conditions

Economic conditions regarding shellfish fisheries in the Gulf of Mexico can cause fluctuations or trends in landings equivalent to major swings seen in environmental condition variations and subsequent animal population levels. Economic conditions affecting the shrimp industry include: rising costs (fuel, insurance, etc.), poor market prices for domestic shrimp, competition from imports, damage from hurricanes, overcapitalization of the fishery industry, erosion, and conversion of waterfront property in some areas from fishing industry use to tourism-based and alternative uses (Nance et al. 2006; NOAA 2007b; Caillouet et al. 2008). As an example, Caillouet et al. (2008) and Nance et al. (2010) showed that numbers and sizes of shrimp, as well as landings per unit effort, were increasing through 2006 and forecast that such trends would continue if shrimping effort continued to decline. Nance (2011) confirmed those forecasts. However, while shrimp stocks have been increasing, annual yields have been decreasing most recently because of a decline in shrimping effort driven by economic conditions. Figure 8.17 clearly shows the declining effort when plotted with landings. Economic conditions are only briefly mentioned here with a few examples as an influence on shellfish fisheries. For a fuller, more detailed coverage on the economics of the fishing industry, please see Chapter 10 in this series dealing with the economics of Commercial and Recreational Fisheries of the Gulf of Mexico.

8.6.4 Management Decisions

Management decisions made by state and federal fishery management agencies can also affect landings and trends in shellfish species within the Gulf of Mexico. As noted in Section 8.2 above, these agencies include the GMFMC and NOAA Fisheries for federal waters relating to shrimp, and all five state fish and wildlife, or natural resource agencies, along with the GSMFC, for oyster and blue crab, as well as inshore shrimp catches. The Gulf of Mexico FMP for Shrimp was established in 1981 (GMFMC 1981) and now has 14 amendments (and number 15 under consideration) adjusting the fishery according to sustainable management needs to keep the fishery and habitats healthy. Oyster and blue crab both have a regional management plan coordinated and published by the GSMFC in cooperation with all five states (oyster regional management plan, VanderKooy 2012; blue crab regional management plan, Guillory et al. 2001). These plans are usually updated about every 10 years or so, and the various state agencies make adjustments, as needed, within their state. Examples of fishery management tools that can affect landings within the fishery include: limits on catch size and number (or quota), establishment of fishing seasons, regulations on fishery gear (e.g., use of TEDs and BRDs in shrimp nets), and licensing or permits for fishery vessels.

8.7 IMPORTANCE OF ESTUARIES AND SELECTED HABITATS TO SHELLFISH SPECIES

Since healthy estuaries are critically important in the life cycles of all three primary groups covered in this white paper (shrimp, oyster, and blue crab), it is important to understand the status of several critical habitats on the populations of the commercial shellfish covered. EFH is now a critical element of all fishery management plans. Since two other chapters in this series deal with habitats (Chapter 6 and 7) and their importance to all species, only an overview is given here.

8.7.1 Estuaries and Delta/Coastal Marshes

Estuaries and coastal delta salt marshes develop at the mouths of rivers and coastal streams, and therefore are the sites where freshwater mixes with seawater and dilutes it. Both of these ecosystems exhibit very high levels of primary production and support multiple and important active food webs (Odum 1980; Teal 1986; Ross 1997). Of the commercially and recreationally important finfish and shellfish, over 80 % of species along the Atlantic coast, and nearly all species in the Gulf, are dependent on the estuaries and delta marshes at some stage of their life cycle (Arnett 1983; Gosselink 1984). The most valuable fisheries are dependent on functional coastal habitats for sustained productivity (Jordan et al. 2009).

As noted repeatedly in earlier portions of this chapter, shellfisheries are highly variable based on environmental fluctuations and population dynamics. Positive correlations are seen between fishery yields and intertidal vegetation (Turner 1977, 1986), yet negative correlations are seen between fish catch and nutrient enrichment (Deegan 2002) and habitat degradation, such as seagrass depletion and algal growth (Deegan et al. 2000; Deegan 2002). Humans have congregated in coastal areas from the beginnings of humanity, and the cumulative effects of exploitation, habitat destruction, and pollution are more severe in estuaries and the coastal zone than anywhere else in the ocean, except for coral reefs (Jackson 2008). Human impacts have depleted over 90 % of formerly important finfish and shellfish species, destroyed over 65 % of seagrass beds and wetland habitats, degraded water quality, and accelerated the invasion of multitudes of introduced species (Lotze et al. 2006). The sustainability of estuaries and delta/coastal marshes and their productivity are of concern with these conditions (Pauly 2010; Baltz and Yáñez-Arancibia 2011).

8.7.2 Seagrass Meadows, Mangrove Forests, Oyster Reefs, and Salt Marshes

Coastal ecosystems provide many vital ecological and economic services, including highly productive commercial and recreational fisheries (Beck et al. 2001, 2003). Key inshore habitats, such as seagrass beds, coastal marshes, and mangrove forests, are highly valued due to their very high productivity, which supports great abundances and diversity of fish and invertebrates, including shrimp, oyster, and blue crab. Because of the great abundance of juveniles found in these habitats, they are often referred to as nurseries where young organisms seek food and shelter (Beck et al. 2001, 2003). Strong evidence is available showing the linkages of penaeid shrimp, blue crabs, and other fishery species to these vital coastal habitats (Sheridan 1992; Heck et al. 1997; Zimmerman et al. 2000; Minello et al. 2003; Rozas and Minello 2010, and many more).

The reauthorization of the Magnuson-Stevens Fishery Conservation and Management Act in 1996 included a new explicit goal to protect, restore, and enhance all "essential fish habitats." The law defined EFH as "those waters and substrate necessary to fish for spawning, breeding, feeding, and or growth to maturity," and "fish" was defined to include all forms of fish, invertebrates, and plants but not birds and mammals. Most of the above discussed, vegetated habitats all fit within the definition of EFH. In addition, oyster reefs also fit within that definition, and great attention has recently focused on oyster reefs as EFH and their importance in providing various ecosystem services and their need for restoration (Coen et al. 1999, 2007;

Peterson et al. 2003; Grabowski and Peterson 2007; Volety et al. 2009; Beck et al. 2011). Indeed, the conservation, restoration, and management of healthy coastal ecosystems are the requirements for healthy and productive shellfisheries in the Gulf of Mexico.

8.8 SUMMARY

There are at least 49 recognized species of shellfish taken in commercial or artisanal fisheries in the Gulf of Mexico. Of these 49 species, 28 are mollusks, 18 are crustaceans, and three are echinoderms. The greatest diversity of species is found in the tropical waters of the southern Gulf of Mexico, but the largest abundances and value are found in the northern Gulf of Mexico. Regarding the three countries that surround the Gulf of Mexico, 16 species are taken within U.S. waters, 46 from Mexico, and six from Cuba. For purposes of this chapter shellfish species are broken into three categories: (1) major (five species); (2) moderate, but important (six species); and (3) minor (38 species) (see Table 8.3). Major species (brown, pink, and white shrimp, Eastern oyster, and blue crab) are the main focus of this document, providing biological, ecological, and fishery status and trends; moderate species are briefly covered in the text; minor species are not covered in the text but are listed in Table 8.1, along with all Gulf shellfish species by scientific name, common name (English and Spanish), distribution, remarks, and a key citation. Jurisdictional boundaries and governing agencies for fishery management in all three Gulf countries are presented and compared.

The current status and historical trends of penaeid shrimp, Eastern oyster, and blue crab all show natural population fluctuations over the past several decades due to varying environmental conditions. In addition, some fisheries, like shrimp, have been greatly affected in recent years by exogenous factors, such as rising fuel costs, market competition from imported shrimp, and fleet damage from hurricanes. Overall, shrimp populations seem to be flourishing, while the shrimp fishery is in decline due to these, and other, factors. Unlike other parts of the United States, Gulf oyster populations seem to be fairly stable, but they show variable annual and multiyear fluctuations, the result mainly of environmental conditions, but also sometimes due to economic/market conditions. As the lead oyster-producing state in the Gulf, Louisiana oyster populations have declined in the past decade when compared to the 1990s, but levels are still above what they were in the 1980s (LDWF 2010). There is considerable concern over the continued loss of oyster reef habitat. The blue crab fishery is quite variable from state to state with Louisiana showing a continued growth and the largest fishery over the past two decades, while Texas shows a decrease in not only the fishery but also species populations statewide during the same time frame. Gulf-wide there is agreement that healthy bays and estuaries lead to more productive fisheries, and therefore the need for conservation of some habitats and restoration of others.

REFERENCES

Adams C, Sanchez-Vega P, Garcia-Alvarez A (2000) An overview of the Cuban commercial fishing industry and recent changes in management structure and objectives. In: Proceedings, International Institute of Fisheries Economics and Trade, Corvallis, OR, USA, 10–14 July 2000. 7 p

Adkins G (1972) Study of the blue crab fishery in Louisiana. Technical Bulletin 3. Louisiana Wildlife and Fisheries Commission, Baton Rouge, LA, USA. 57 p

Appeldoorn RS, Meyers S (1993) Part Z: Puerto Rico and Hispaniola. Marine fishery resources of the Antilles, FAO fisheries technical paper 326 Rome. FAO (Food and Agriculture Organization of the United Nations), Rome, Italy, pp 99–158

Arenas F, Diaz de Leon CA (eds) (1999) Sustentabilidad y pesca responsable en Mexico: evaluacion y manejo 1997-1998. Instituto Nacional de la Pesca, SAGARPA, Delegación Benito Juárez, Mexico, Federal District. 1112 p

Arenas-Fuentes V, Jimenez-Badillo L (2004) Fishing in the Gulf of Mexico: Towards greater biomass in exploitation. In: Caso M, Pisanty I, Excurra E (eds) Environmental analysis of the Gulf of Mexico. Instituto Nacional de Ecologia (English translation edited by Withers K, Nipper M (2007) Harte Research Institute for Gulf of Mexico studies. Texas A&M University—Corpus Christi, TX, USA, pp 468–477

Arnett GR (1983) Introduction to the session—user group demands on the environment. In: Reintjes JW (ed) Improving multiple use of coastal and marine resources. American Fisheries Society, Bethesda, MD, USA, pp 11–13

Arnold WS, Berrigan ME (2002) A summary of the oyster (*Crassostrea virginica*) fishery in Florida. Florida Fish and Wildlife Conservation Commission, Tallahassee, FL, USA

Arreguin-Sanchez F, Solis-Ramirez MJ, Gonzalez de la Rosa ME (2000) Population dynamics and stock assessment for *Octopus maya* (Cephalopoda: Octopodidae) fishery in the Campeche Bank, Gulf of Mexico. Rev Biol Trop 48:323–331

Bahr M, Lanier WP (1981) The ecology of intertidal oyster reefs of the South Atlantic coast: A community profile. U.S. Fish and Wildlife Service Biological Report, FWS/OBS-81/15. U.S. Geological Survey, Reston, VA, USA. 105 p

Baisre JA (2000) Chronicle of Cuban marine fisheries (1935–1995): Trend analysis and fisheries potential. Fisheries Technical Paper 394. FAO, Rome, Italy. 26 p

Baisre JA (2006) Cuban Fisheries Management Regime: Current state and future prospects. The United Nations University-Fisheries Training Programme, Reykjavik, Iceland. 33 p

Baisre JA, Booth S, Zeller D (2003) Cuban fisheries catches within FAO area 31 (Western Central Atlantic): 1950-1999. Fish Cent Res Rep 11:133–139

Baltz DM, Yáñez-Arancibia A (2011) Ecosystem-based management of coastal fisheries in the Gulf of Mexico: Environmental and anthropogenic impacts and essential habitat protection (Chapter 19). In: Day JW, Yáñez-Arancibia A (eds) The Gulf of Mexico: Ecosystem-based management, vol 4, Harte Research Institute for Gulf of Mexico studies. Texas A&M University Press, College Station, TX, USA

Baqueiro ER (2004) Current state of molluscan resources of the Gulf of Mexico. In: Caso M, Pisanty I, Excurra E (eds) Instituto Nacional de Ecologia (English translation edited by Withers K, Nipper M (2007) Harte Research Institute for Gulf of Mexico studies. Texas A&M University–Corpus Christi, TX, USA, pp 195–220

Baxter KN, Renfro WC (1966) Seasonal distribution and size distribution of post larval brown and white shrimp near Galveston, Texas, with notes on species identification. U.S. Fish Bull 66:149–158

Beck MW, Heck KL Jr, Able K, Childers D, Eggleston D, Gillanders BM, Halpern B, Hays C, Hoshino K, Minello T, Orth R, Sheridan P, Weinstein M (2001) The identification, conservation, and management of estuarine and marine nurseries for fish and invertebrates. BioScience 51:633–641

Beck MW, Heck KL, Able KW, Childers DL, Eggleston DB, Gillanders BM, Halpern BS, Hays CG, Hoshino K, Minello TJ, Orth RJ, Sheridan PF, Weinstein MR (2003) The role of nearshore ecosystems as fish and shellfish nurseries. Issues Ecol 11:1–12

Beck WM, Brumbaugh RD, Airoldi L, Carranza A, Coen LD, Crawford C, Defeo O, Edgar GJ, Hancock B, Kay MC, Lenihan HS, Luckenbach MW, Toropova CL, Zhang G, Guo X (2011) Oyster reefs at risk and recommendations for conservation, restoration, and management. BioScience 61:107–116

Berrigan ME (1988) Management of oyster resources in Apalachicola Bay following Hurricane Elena. J Shellfish Res 7:281–288

Berrigan M, Candies T, Cirino J, Dugas R, Dyer C, Gray J, Herrington T, Keithly W, Leard R, Nelson JR, Van Hoose M (1991) The oyster fishery of the Gulf of Mexico, United States: A regional management plan. Publication 24, March. Gulf States Marine Fisheries Commission, Ocean Springs, MS, USA

Briggs JC (1974) Marine zoogeography. McGraw-Hill, New York, NY, USA

Brownell WN, Stevely JM (1981) The biology, fisheries, and management of the queen conch, *Strombus gigas*. Mar Fish Rev 43:1–12

Burrell VG Jr (1986) Species profiles: Life histories and environmental requirements of coastal fishes and invertebrates (south Atlantic)—American oyster. U.S. Fish and Wildlife Service Biological Report 82(11.57). U.S. Fish and Wildlife Service, Slidell, LA, USA

Butler PA (1954) The southern oyster drill. Proc Natl Shellfish Assoc 44:67–75

Caffey RH, Culley DD, Roberts KJ (1993) The Louisiana soft-shelled crab industry: A profile. Sea Grant Program Publication LSU-G-9-001. Louisiana State University, Baton Rouge, LA, USA. 46 p

Caillouet CW Jr, Hart RA, Nance JM (2008) Growth overfishing in the brown shrimp fishery of Texas, Louisiana, and adjoining Gulf of Mexico EEZ. Fish Res 92:289–302

Caillouet CW Jr, Hart RA, Nance JM (2011) Simulation of tail weight distribution in biological year 1986-2006 landings of brown shrimp, *Farfantepenaeus aztecus*, from the Northern Gulf of Mexico fishery. Mar Fish Rev 73:27–40

Calabrese A, Davis HC (1970) Tolerances and requirements of embryos and larvae of bivalve molluscs. Helgolander Meeresun 20:553–564

Calabrese A, Collier RS, Nelson DA, MacInnes JR (1973) The toxicity of heavy metals to embryos of the American oyster, *Crassostrea virginica*. Mar Biol 18:162–166

Carpenter KE (2002) The living marine resources of the Western Central Atlantic, vol 1, Introduction, molluscs, crustaceans, hagfishes, sharks, batoid fishes, and chimaeras. FAO species identification guide for fishery purposes and American Society of Ichthyologists and Herpetologists special publication 5. FAO, Rome, Italy. 600 p

Carriker MR, Gaffney PM (1996) A catalogue of selected species of living oysters (Ostreacea) of the world. In: Kenned VS, Newell RIE, Eble AF (eds) The eastern oyster *Crassostrea virginica*. MD Sea Grant, College Park, MD, USA, pp 1–18

CFMC (Caribbean Fishery Management Council) (1996) Fishery Management Plan, regulatory impact review, and final environmental impact statement for the queen conch resources of Puerto Rico and the United States Virgin Islands. Caribbean Fishery Management Council, San Juan, Puerto Rico. 63 p

Childress UR (1966) An investigation into levels of concentration, seasonal variations, and sources of pesticide toxicants in some species from selected bay areas. Project MP-R-2, Coastal Fisheries Project Technical Report. Texas Parks and Wildlife Department, Austin, TX, USA

Christmas JY, Etzold DJ (1977) The shrimp fishery of the Gulf of Mexico United States: A regional management plan. Gulf Coast Res Lab Tech Rep Ser 2(1). 125 p

Claro R, Lindeman KC, Parenti LR (2001) Ecology of the marine fishes of Cuba. Smithsonian Institution Press, Washington, DC, USA. 253 p

Claro R, Mitcheson YS, Lindeman KC, Garcia-Cagide AR (2009) Historical analysis of Cuban commercial fishing effort and the effects of management interventions on important reef fishes from 1960-2005. Fish Res 99:7–16

Cody TJ, Wagner T, Bryan CE, McEachron LW, Rayburn R, Bowling B, Mambretti J (1992) Texas blue crab fishery management plan, Fisheries management plan series 4. Texas Parks and Wildlife Department, Coastal Fisheries Branch, Austin, TX, USA

Coen LD, Luckenbach MW, Breitburg DL (1999) The role of oyster reefs as essential fish habitat: A review of current knowledge and some new perspectives. Am Fish Soc Symp 22:438–454

Coen LD, Brumbaugh RD, Bushek D, Grizzle R, Luckenbach MW, Posey MH, Powers SP, Tolley SG (2007) Ecosystem services related to oyster restoration. Mar Ecol Prog Ser 341:303–307

Condrey R, Fuller D (1992) The U.S. Gulf shrimp fishery. In: Giantz MH (ed) Climate variability, climate change and fisheries. Cambridge University Press, Cambridge, UK, pp 89–119

Cook HL, Lindner MJ (1970) Synopsis of biological data on the brown shrimp *Penaeus aztecus* Ives, 1891. FAO Fish Rep 57:1471–1497

Cook HL, Murphy MA (1966) Rearing penaeid shrimp from eggs to post larvae. Proc Conf SE Assoc Game Comm 19:283–288

Copeland BJ (1965) Fauna of the Aransas pass inlet, Texas. I. Emigration shown by tide trap collections. Inst Mar Sci Univ Tex Publ 10:9–21

Copeland BJ, Bechtel TJ (1974) Some environmental limits of six gulf coast estuarine organisms. Contrib Mar Sci 18:169–204

Copeland BJ, Truitt MV (1966) Fauna of Aransas pass inlet, Texas. II. Penaeid shrimp post larvae. Tex J Sci 18:65–74

Costello TJ, Allen DM (1970) Synopsis of biological data on the pink shrimp *Penaeus duorarum* Burkenroad, 1939. FAO Fish Rep 57:1499–1537

Costello TJ, Bert TM, Cartano DG, Davis G, Lyon G, Rockwood C, Stevely J, Tashiro J, Trent WL, Turgeon D, Zuboy J (1979) Fishery management plan for the stone crab fishery of the Gulf of Mexico. Gulf of Mexico Fishery Management Council, Tampa, FL, USA. 188 p

Costello TJ, Allen DM, Hudson JH (1986) Distribution, seasonal abundance, and ecology of juvenile northern pink shrimp, *Penaeus duorarum*. The Florida Bay area, NOAA Tech Memo NMFS-SEFC-161. National Oceanic and Atmospheric Administration, Silver Spring, MD, USA. 84 p

Costlow JD, Sastry AN (1966) Free amino acids in developing stages of two crabs, *Callinectes sapidus* and *Rhithropanopeus harrisii* Gould. Acta Embryol Morphol Exp 9:44–55

Couch JA, Martin S (1982) Protozoan symbionts and related diseases of the blue crab, *Callinectes sapidus* Rathbun from the Atlantic and gulf coasts of the United States. In: Perry HM, Van Engel WA (eds) Proceedings, blue crab colloquium, Publication 7. Gulf States Marine Fisheries Commission, Ocean Springs, MS, USA, pp 71–80

Darnell RM (1958) Food habits of fishes and larger invertebrates of Lake Pontchartrain, Louisiana, an estuarine community. Inst Mar Sci Univ Tex Publ 5:353–416

Darnell RM (1959) Studies of the life history of the blue crab *Callinectes sapidus* Rathbun in Louisiana waters. Trans Am Fish Soc 88:294–304

Davis JW, Sizemore RK (1982) Incidence of *Vibrio* species associated with blue crabs (*Callinectes sapidus*) collected from Galveston Bay, Texas. Appl Environ Microbiol 43:1092–1097

Deegan LA (2002) Lesson learned: The effects of nutrient enrichment on the support of nekton by seagrass and salt march ecosystems. Estuaries 25:727–742

Deegan LA, Hughes JE, Rountree RA (2000) Salt marsh ecosystem support of marine transient species. In: Weinstein MP, Kreeger DA (eds) Concepts and controversies in tidal marsh ecology. Kluwer, Dordrecht, The Netherlands, pp 333–365

Diario Oficial (2010) Carta Nacional Pequera. Segunda Seccion: Secretaria del agricultura, ganaderia, desarrollo rural, pesca y alimentacion. Mexico City, Mexico. 102 p

Diaz-de-Leon A, Fernandez JI, Alvarez-Torres P, Ramirez-Flores O, Lopez-Lumes LG (2004) The sustainability of the Gulf of Mexico's fishing grounds. In: Caso M, Pisanty I, Excurra E (eds) Instituto Nacional de Ecologia (English translation edited by Withers K, Nipper M (2007) Harte Research Institute for Gulf of Mexico Studies, Texas A&M University—Corpus Christi, TX, USA, pp 457–467

Dugas RJ, Joyce EA, Berrigan MA (1997) History and status of the oyster, *Crassostrea virginica*, and other molluscan fisheries of the Gulf of Mexico. In: MacKenzie CL Jr, Burrell VG Jr, Rosenfield A, Hobart WL (eds) The history, present condition, and future of the molluscan fisheries of North and Central America and Europe, NOAA Technical Report NMFS 127. NOAA, Silver Spring, MD, USA, pp 187–210

Eggleston DB (1990) Foraging behavior of the blue crab, *Callinectes sapidus*, on juvenile oysters, *Crassostrea virginica*: Effects of prey density and size. Bull Mar Sci 46:62–82

Ekman S (1953) Zoogeography of the sea. Sidgwick and Jackson, London, UK. 417 p

ELI and CMDA (Environmental Law Institute and Centro Mexicano de Derecho Ambiental) (2011) Gulf of Mexico habitat conservation and restoration: Comparing the Mexican and United States legal and institutional frameworks. Gulf of Mexico Foundation, Corpus Christi, TX, USA. 99 p

Etzold DJ, Christmas JY (1977) A comprehensive summary of the shrimp fishery for the Gulf of Mexico United States: A regional management plan. Gulf Coast Res Lab Tech Rep Ser 2(2). 20 p

FAO (2011) Food and Agriculture Organization of the United Nations. http://www.fao.org. Accessed 2 Feb 2013

Felder DL, Camp DK (2009) Gulf of Mexico origin, waters, and biota, vol 1, Biodiversity. Texas A&M University Press, College Station, TX, USA. 1393 p

Felder DL, Alvarez F, Goy JW, Lemaitre R (2009) Decapoda (Crustacea) of the Gulf of Mexico, with comments on the Amphionidacea. In: Felder DL, Camp DK (eds) Gulf of Mexico origin, waters, and biota, vol 1, Biodiversity. Texas A&M University Press, College Station, TX, USA, pp 1019–1104

Galtsoff PS (1930) Destruction of oyster bottoms in Mobile Bay by the flood of 1929. U.S. Bureau of Fisheries, Rep. Commission Fisheries for 1929 (Doc 1069). U.S. Government Printing Office, Washington, DC, USA, pp 741–758

Galtsoff PS (1964) The American oyster *Crassostrea virginica* Gmelin. Fish Bull 64:1–480

Gauthier JD, Soniat TM (1989) Changes in the gonadal state of Louisiana oysters during their autumn spawning season. J Shellfish Res 8:83–86

GMFMC (Gulf of Mexico Fishery Management Council) (1979) Fishery management plan for the stone crab fishery of the Gulf of Mexico. Gulf of Mexico Fishery Management Council, Tampa, FL, USA. 188 p

GMFMC (1981) Fishery management plan for the shrimp fishery of the Gulf of Mexico, United States waters. Gulf of Mexico Fishery Management Council, Tampa, FL, USA. 246 p

GMSAFMC (Gulf of Mexico and South Atlantic Fishery Management Councils) (1982) Fishery management plan environmental impact statement and regulatory review for spiny lobster in the Gulf of Mexico and South Atlantic. Gulf of Mexico Fishery Management Council, Tampa, FL, USA. 247 p

Gosselink JG (1984) The ecology of delta marshes of coastal Louisiana: A community profile. U.S. Fish and Wildlife Service FWS/OBS-84/09. U.S. Geological Survey, Reston, VA, USA

Grabowski JH, Peterson CH (2007) Restoring oyster reefs to recover ecosystem services. In: Cuddington K, Byers J, Wilson W, Hastings A (eds) Ecosystem engineers: Plants to protists. Academic Press, Burlington, MA, USA, pp 281–298

832   J.W. Tunnell, Jr.

GSMFC (Gulf States Marine Fisheries Commission) (2011) Gulf States Marine Fisheries Commission. http://www.gsmfc.org. Accessed 2 Feb 2013

Guillory V (1996) A management profile of Louisiana blue crab, *Callinectes sapidus*, Fisheries management plan series 8, Part 2. Louisiana Department of Wildlife and Fisheries, Baton Rouge, LA, USA. 34 p

Guillory V (1998) A survey of the recreational blue crab fishery in Terrebonne Parish, Louisiana. J Shellfish Res 17:4543–4550

Guillory V, Perret WE (1998) Management, history, and status and trends in the Louisiana blue crab fishery. J Shellfish Res 17:413–424

Guillory V, Bourgeois M, Prejean P, Burdon J, Merrell J (1996) A biological and fisheries profile of Louisiana blue crab, *Callinectes sapidus*, Part 1, Fisheries management plan series 5. Louisiana Department of Wildlife and Fisheries, Baton Rouge, LA, USA

Guillory V, Perry H, Steele P, Wagner T, Hammerschmidt P, Heath S, Moss C (1998) The Gulf of Mexico blue crab fishery: Historical trends, status, management, and recommendations. J Shellfish Res 17:395–403

Guillory V, Perry H, Steele P, Wagner T, Keithly W, Pellegrin B, Petterson J, Floyd T, Buckson B, Hartman L, Holder E, Moss C (2001) The blue crab fishery of the Gulf of Mexico, United States: A regional management plan, Publication 96. Gulf States Marine Fisheries Commission, Ocean Springs, MS, USA

Gunter G (1961) Habitat of juvenile shrimp (family Penaeidae). Ecology 42:598–600

Gusmao J, Lazoski C, Monteiro FA, Sole-Cava AM (2006) Cryptic species and population structuring of the Atlantic and Pacific seabob shrimp species, *Xiphopenaeus kroyeri* and *Xiphopenaeus riveti*. Mar Biol 149:491–502

Haby MG, Miget RJ, Falconer LL, Graham GL (2003) A review of current conditions in the Texas shrimp industry, an examination of contributing factors, and suggestions for remaining competitive in the global shrimp market, TAMU-SG-03-701. Texas Cooperative Extension Sea Grant College Program, College Station, TX, USA. 26 p

Haby MG, Miget MJ, Falconer LL (2009) Hurricane damage sustained by the oyster industry and the oyster reefs across the Galveston Bay system with recovery recommendations, TAMU-SG-09-201. TexasAgriLife Extension Service/Sea Grant Extension Program, College Station, TX, USA. 51 p

Halpern BS, Walbridge S, Selkoe KA et al (2008) A global map of human impact on marine ecosystems. Science 319:948–952

Hammerschmidt P, Wagner T, Lewis G (1998) Status and trends in the Texas blue crab (*Callinectes sapidus*) fishery. J Shellfish Res 17:405–412

Hare JA, Able KW (2007) Mechanistic links between climate and fisheries along the east coast of the United States: Explaining population outbursts of Atlantic croaker (*Micropogonias undulatus*). Fish Oceanogr 16:31–45

Hart RA (2008) A biological review of the Tortugas pink shrimp fishery 1960 through 2007. NOAA Technical Memorandum NMFS-SEFSC-573. NOAA, Silver Spring, MD, USA. 28 p

Hart, RA, Nance JM (2010) Gulf of Mexico pink shrimp assessment modeling update: From a static VPA to an integrated assessment model, stock synthesis. NOAA Technical Memorandum NMFS-SEFSC-604. NOAA, Silver Spring, MD, USA. 32 p

Heath SR (1998) The Alabama blue crab fishery: Historical trends, status, management, and the future. J Shellfish Res 17:435–439

Heck KL Jr, Nadeau DA, Thomas R (1997) The nursery role of seagrass beds. Gulf Mexico Sci 15:50–54

Henley DE, Rauschuber DG (1981) Freshwater needs of fish and wildlife resources in the Nueces-Corpus Christi Bay area, Texas: A literature synthesis. Biological Report FWS/OBS-80/10. U.S. Fish and Wildlife Service, Washington, DC, USA. 410 p

Hester TR (1980) Digging into South Texas prehistory: A guide for amateur archaeologists. Corona, San Antonio, TX, USA. 201 p

Hildebrand HH (1954) A study of the fauna of the brown shrimp (*Penaeus aztecus* Ives) grounds in the western Gulf of Mexico. Inst Mar Sci Univ Tex Publ 3:233–366

Hildebrand HH (1955) A study of the fauna of the pink shrimp (*Penaeus duorarum* Burkenroad) grounds in the Gulf of Campeche. Inst Mar Sci Univ Tex Publ 4:169–232

Hofmann EE, Powell TM (1998) Environmental variability effects on marine fisheries: Four case histories. Ecol Appl 8:S23–S32

Hofstetter RP (1962) Study of oyster growth and population structure of the public reefs in East Bay, Galveston and Trinity Bay. Coastal Fisheries Branch Project Reports. Texas Parks and Wildlife Department, Austin, TX, USA

Hofstetter RP (1977) Trends in population levels of the American oyster, *Crassostrea virginica* Gmelin on public reefs in Galveston Bay, Texas, Technical series 10. Texas Parks and Wildlife Department, Coastal Fisheries Branch, Austin, TX, USA. 90 p

Hofstetter RP (1981) Rehabilitation of public oyster reefs damaged or destroyed by a natural disaster, Management data series 21. Texas Parks and Wildlife Department, Coastal Fisheries Branch, Austin, TX, USA. 9 p

Hofstetter RP (1990) The Texas oyster fishery, 3rd Rev, Bulletin 40, TPWD-BK-3400-216-2/90. Texas Parks and Wildlife Department, Coastal Fisheries, Austin, TX, USA

Iversen ES, Jory DE (1985) Queen conch at the crossroads. Sea Front 31:151–159

Jackson JBC (2008) Ecological extinction and evolution in the brave new ocean. Proc Natl Acad Sci 105:11458–11465

Jennings S, Kaiser MJ, Reynolds JD (2001) Marine fisheries ecology. Blackwell Publishing, London, UK. 417 p

Jones AC, Dimitriou DE, Ewald JJ, Tweedy JH (1970) Distribution of early developmental stages of pink shrimp, *Penaeus duorarum*, in Florida waters. Bull Mar Sci 20:634–661

Jones A, Klima EF, Poffenberger J (1982) The effects of the 1981 closure on the Texas shrimp fishery. Mar Fish Rev 44:1–4

Jordan SJ, Smith LM, Nestlerode JA (2009) Cumulative effects of coastal habitat alterations on fishery resources: Toward prediction at regional scales. Ecol Soc 14:16, http://www.ecologyandsociety.org/vol14/iss1/art16/. Accessed 2 Feb 2013

Kennedy VS (1996) Biology of larvae and spat. In: Kennedy VS, Newell RIE, Eble AF (eds) The eastern oyster *Crassostrea virginica*. Maryland Sea Grant College, University of Maryland, College Park, MD, USA, pp 371–421

Kilgour MJ, Shirley TC (2008) Distribution of red deepsea crab (*Chaceon quiquedens*) by size and sex in the Gulf of Mexico. Fish Bull 106:317–320

Killam KA, Hochberg RJ, Rzemien EC (1992) Synthesis of basic life histories of Tampa Bay species, Technical publication 10-92. Tampa Bay National Estuary Program, St. Petersburg, FL, USA. 155 p

King BD III (1971) Study of migratory patterns of fish and shellfish through a natural pass, Technical series 9. Texas Parks and Wildlife Department, Austin, TX, USA. 54 p

Kirby MX (2004) Fishing down the coast: Historical expansion and collapse of oyster fisheries along continental margins. Proc Natl Acad Sci 101:13096–13099

Klima EF (1981) The National Marine Fisheries shrimp research program in the Gulf of Mexico. Kuwait Bull Mar Sci 2:185–207

Klima EF (1989) Approaches to research and management of U.S. Fisheries for penaeid shrimp in the Gulf of Mexico, 99. In: Caddy JF (ed) Marine invertebrate fisheries: Their assessment and management. FOA, Rome, Italy, pp 87–113

Klima EF, Baxter KN, Patella FJ (1982) A review of the offshore shrimp fishery and the 1981 Texas closure. Mar Fish Rev 44:16–30

Klima EF, Matthews GA, Patella FJ (1986) Synopsis of the Tortugas pink shrimp fishery, 1960-1983, and the impact of the Tortugas Sanctuary. N Am J Fish Manag 6:301–310

Klima EF, Castro Melendez RG, Baxter N, Patella FJ, Cody TJ, Sullivan LF (1987) MEXUS-Gulf shrimp research, 1978–84. Mar Fish Rev 49:21–30

Laughlin RA (1982) Feeding habits of the blue crab *Callinectes sapidus* Rathbun, in the Apalachicola Estuary, Florida. Bull Mar Sci 32:807–822

LDWF (Louisiana Department of Wildlife and Fisheries) (2005) Oyster stock assessment report of the public oyster areas in Louisiana, Oyster data report series 11. Baton Rouge, LA, USA

LDWF (2010) Oyster stock assessment report of the public oyster areas in Louisiana, Oyster data report series 16. LA, USA

Lotze HK, Lenihan HS, Bourque BJ, Bradbury RH, Cooke RG, Kay MC, Kidwell SM, Kirby MX, Peterson CH, Jackson JBC (2006) Depletion, degradation, and recovery potential of estuaries and coastal seas. Science 312:1806–1809

MacKenzie CL (1996) History of oystering in the United States and Canada, featuring North America's greatest oyster estuaries. Mar Fish Rev 58:1–78

MacKenzie CL Jr, Wakida-Kusunoki AT (1997) The oyster industry of Eastern Mexico. Mar Fish Rev 59:1–13

Mackin JG, Owen HM, Collier A (1950) Preliminary note on the occurrence of a new protistan parasite, *Dermocystidium marinum* n. sp. in *Crassostrea virginica* (Gmelin). Science 111:32–329

Martin N (1987) Raw deals. Tex Shores 20:4–8

Marwitz SR, Bryan CE (1990) Rehabilitation of public oyster reefs damaged by a natural disaster in San Antonio Bay, Management data series 32. Texas Parks and Wildlife Department, Austin, TX, USA. 10 p

McLaughlin PA, Camp DK, Angel MV, Bousfield EL, Brunel P et al (2005) Common and scientific names of aquatic invertebrates from the United States and Canada: Crustaceans, vol 31, American Fisheries Society special publication. American Fisheries Society, Bethesda, MD, USA. 533 p

McTigue TA, Zimmerman RJ (1991) Carnivory vs. herbivory in juvenile *Penaeus setiferus* (Linnaeus) and *Penaeus aztecus* (Ives). J Exp Mar Biol Ecol 151:1–16

Menzel RW, Hopkins SH (1956) Crabs as predators of oysters in Louisiana. Proc Natl Shellfish Assoc 46:177–184

Messick GA (1998) Diseases, parasites, and symbionts of blue crabs (*Callinectes sapidus*) dredged from Chesapeake Bay. J Crustac Biol 18:533–548

Mexicano-Cintora G, Leonce-Valencia C, Salas S, Vega-Cendejas ME (2007) Recursos pesqueros de Yucatán: Fichas tecnicas y referencias bibliograficas. Centro de Investigacion y Estudios Avanzados del I.P.N. (CINVESTAV) Unidad Merida. 1a. Edicion. Merida, Yucatán, Mexico. 150 p

Millikin MR, Williams AB (1984) Synopsis of biological data on the blue crab, *Callinectes sapidus*. FAO Fish Synop 138, NOAA Technical Report NMFS 1. NOAA/National Marine Fisheries Service, Silver Spring, MD, USA

Minello TJ, Zimmerman RJ, Martinez EX (1989) Mortality of young brown shrimp *Penaeus aztecus* in estuarine nurseries. Trans Am Fish Soc 118:693–708

Minello TJ, Able KW, Weinstein MP, Hayes CG (2003) Salt marshes as nurseries for nekton: Testing hypotheses on density, growth and survival through meta-analysis. Mar Ecol Prog Ser 246:39–59

Montagna PA, Gibeaut JC, Tunnell JW Jr (2007) South Texas climate 2100: Coastal impacts. In: Norwine J, John K (eds) The changing climate of South Texas 1900-2100: Problems and prospects, impacts and implications. Texas A&M University-Kingsville, Kingsville, TX, USA, pp 57–77

Moss CG (1982) The blue crab fishery of the Gulf of Mexico. In: Perry HM, VanEngel WA (eds) Proceedings of the blue crab colloquium, Gulf States Marine Fisheries Commission Publication 7, Ocean Springs, MS, USA, pp 93–104

MRAG (MRAG Americas Inc.) (2010) Pre-assessment of the Gulf of Mexico Shrimp fishery. Sustainable Fisheries Partnership, St. Petersburg, FL, USA. 7 p

Muncy RJ (1984) Species profiles: Life histories and environmental requirements of coastal fishes and invertebrates (Gulf of Mexico)—white shrimp. Biological Report FWS/OBS-82/11.20. U.S. Fish and Wildlife Services, Washington, DC, USA. 19 p

Nance JM (1992) Estimation of effort for the Gulf of Mexico shrimp fishery. NOAA Technical Memorandum NMFS-SEFSC-300. NOAA, Silver Spring, MD, USA

Nance JM (1996) Biological review of the 1995 Texas closure. NOAA Tech Mem NMFS-SEFSC-379. NOAA, Silver Spring, MD, USA. 7 p

Nance JM (2011) Stock assessment report 2010 Gulf of Mexico shrimp fishery. Gulf of Mexico Fishery Management Council, Tampa, FL, USA. 16 p

Nance JM, Keithly W Jr, Caillouet C Jr, Cole J, Gaidry W, Gallaway B, Griffin W, Hart R, Travis M (2006) Estimation of effort, maximum sustainable yield, and maximum economic yield in the shrimp fishery of the Gulf of Mexico. Gulf of Mexico Fishery Management Council, Tampa, FL, USA. 85 p

Nance JM, Caillouet CW Jr, Hart RA (2010) Size-composition of annual landings in the white shrimp fishery of the northern Gulf of Mexico, 1960–2006: Its trend and relationships with other fishery-dependent variables. Mar Fish Rev 72:1–13

Nelson DM, Monaco ME, Williams CD, Czapla TE, Patillo ME, Coston-Clements L, Settle LR, Irlandi EA (1992) Distribution and abundance of fishes and invertebrates in Gulf of Mexico estuaries, vol 1, Data summaries, ELMR Report 10. NOAA/NOS Strategic Environmental Assessments Division, Rockville, MD, USA. 273 p

NMFS (National Marine Fisheries Service) (2010) Fisheries of the United States 2009, Silver Spring, MD, USA. 103 p

NMFS (2011) Fisheries of the United States 2010, Silver Spring, MD, USA. 103 p

NOAA (National Oceanic and Atmospheric Administration) (2007a) Status review of the Eastern Oyster (*Crassostrea virginica*). NOAA Tech Mem NMFS-F/SPO-88, Silver Spring, MD, USA. 105 p

NOAA (2007b) Report to congress on the impacts of hurricanes Katrina, Rita, and Wilma on Alabama, Louisiana, Florida, Mississippi, and Texas fisheries, Silver Spring, MD, USA. 133 p

NOAA (2011a) NOAA Fisheries: Office of Service and Technology, Fisheries Statistics Division (ST1). http://www.st.nmfs.noaa.gov/st1/. Accessed 2 Feb 2013

NOAA (2011b) NOAA Gulf of Mexico Data Atlas. http://gulfatlas.noaa.gov/. Accessed 2 Feb 2013

NOS (National Ocean Service) (2011) The Gulf of Mexico at a glance: A second glance. U.S. Department of Commerce, Washington, DC, USA. 51 p

Odum EP (1980) The status of three ecosystem-level hypotheses regarding salt marsh estuaries: Tidal subsidy, outwelling, and detritus-based food chains. In: Kennedy VS (ed) Estuarine perspectives. Academic Press, New York, NY, USA, pp 485–495

Orensanz JM, Jaimieson GS (1998) The assessment and management of spatially structured stocks: An overview of the North Pacific symposium on invertebrate stock assessment and management. Can Spec Publ Fish Aquat Sci 125:441–459

Osborn KW, Maghan BW, Drummond SB (1969) Gulf of Mexico shrimp atlas. U.S. Department of the Interior, Bureau of Commercial Fisheries Circular 312. 20 p

Overstreet RM (1982) Metazoan symbionts of the blue crab. In: Perry HM, Van Engel WA (eds) Proceedings of the blue crab colloquium, Publication 7. Gulf States Marine Fisheries Commission, Ocean Springs, MS, USA, pp 129–136

Overstreet RM, Rebarchik D (1995) Assessment of infections of the blue crab in Pensacola Bay. Final Report. U.S. Environmental Protection Agency, Washington, DC, USA. 76 p

Patillo ME, Czapla TE, Nelson DM, Monaco ME (1997) Distribution and abundance of fishes and invertebrates in Gulf of Mexico estuaries, vol 2, Species life history summaries, ELMR Report 11. NOAA/NOS Strategic Environmental Assessments Division, Silver Spring, MD, USA. 377 p

Pauly D (2010) 5 easy pieces: The impact of fisheries on marine ecosystems (The State of the World's Oceans), Sea around us project. Island, Washington, DC, USA. 194 p

Perez-Farfante I (1969) Western Atlantic shrimps of the genus *Penaeus*. Fish Bull 67:461–591

Perry HM (1975) The blue crab fishery in Mississippi. Gulf Res Rep 5:39–57

Perry HM, Malone R (1985) Proceedings of the national symposium on the soft-shelled blue crab fishery. Gulf Coast Research Laboratory, Ocean Springs, MS, USA. 128 p

Perry HM, Malone R (1989) Blue crabs: Soft shell production. In: Cake EW Jr, Whicker LF, Ladner CM (eds) Mississippi Aquatic Ventures Center: Aquaculture profiles and opportunities in Mississippi. Mississippi Department of Wildlife, Fisheries and Parks, Bureau of Marine Resources, Jackson, MS, USA, pp 1–32

Perry HM, McIlwain TD (1986) Species profiles: Life histories and environmental requirements of coastal fishes and invertebrates (Gulf of Mexico)—blue crab. U.S. Fish Wildlife Services Biological Report 82 (11.55). U.S. Army Corp of Engineers, TR EL-82-4, Vicksburg, MS, USA. 21 p

Perry HM, Adkins G, Condrey R, Hammerschmidt PC, Heath S, Herring JR, Moss C, Perkins G, Steele P (1984) A profile of the blue crab fishery of the Gulf of Mexico, Publication 9. Gulf States Marine Fisheries Commission, Ocean Springs, MS, USA. 80 p

Perry HM, Warren J, Trigg C, Van Devender T (1998) The blue crab fishery of Mississippi. J Shellfish Res 17:425–433

Peterson CH, Grabowski JH, Powers SP (2003) Estimated enhancement of fish production resulting from restoring oyster reef habitat: Quantitative valuation. Mar Ecol Prog Ser 264:249–264

Pianka ER (1966) Latitudinal gradients in species diversity: A review of concepts. Am Nat 100:33–46

Quast WD, Johns MA, Pitts DE Jr, Matlock GC, Clark JE (1988) Texas oyster fishery management plan, Fishery management plan series number 1. Texas Parks and Wildlife Department, Coastal Fisheries Branch, Austin, TX, USA. 178 p

Ray SM (1987) Salinity requirements of the American oyster, *Crassostrea virginica*. In: Muller AJ, Matthews GA (eds) Freshwater inflow needs of the Matagorda Bay system with focus on the needs of Penaeid Shrimp, NOAA Technical Memorandum NMFS-SEFC:189. NOAA, Silver Spring, MD, USA, pp E1–E28

Renfro WC, Brusher HA (1982) Seasonal abundance, size distribution, and spawning of three shrimps (*Penaeus aztecus*, *P. setiferus*, and *P. duorarum*) in the northwestern Gulf of

Mexico, 1961-1962. NOAA Tech Memo NMFS-SEFC-94. NOAA, Silver Spring, MD, USA. 47 p

Rester JK (2011) SEAMAP environmental and biological atlas of the Gulf of Mexico, 2009. NOAA, Project NA11NMF4350028. NOAA, Silver Spring, MD, USA

Restrepo VR (1992) A mortality model for a population in which harvested individuals do not necessarily die: The stone crab. Fish Bull 90:412–416

Ricklis RA (1996) The Karankawa Indians of Texas: An ecological study of cultural tradition and change. University of Texas Press, Austin, TX, USA. 222 p

Ross MR (1997) Fisheries conservation and management. Prentice Hall, Upper Saddle River, NJ, USA. 374 p

Rossler MA, Jones AC, Munro JL (1969) Larval and postlarval pink shrimp *Penaeus duorarum* in south Florida. In: Mistakidis MN (ed) Proceedings of the world scientific conference on the biology and culture of shrimps and prawns, FAO Fish Rep 57, pp 859–866

Rothschild BJ, Ault JS, Goulletquer P, Heral M (1994) The decline of the Chesapeake Bay oyster population: A century of habitat destruction and overfishing. Mar Ecol 111:29–39

Rozas L, Minello T (2010) Nekton density patterns in tidal ponds and adjacent wetlands related to pond size and salinity. Estuar Coasts 33:652–667

SEDAR (Southeast Data, Assessment, and Review) (2007) SEDAR 14 stock assessment report: Caribbean Queen Conch. SEDAR Stock Assessment Report 3, North Charleston, SC, USA. 171 p

Sheridan PF (1992) Comparative habitat utilization by estuarine macro-fauna within the mangrove ecosystem of Rookery Bay, Florida. Bull Mar Sci 50:21–39

Shumway SE (1996) Natural environmental factors. In: Kennedy VS, Newell RIE, Eble AF (eds) The eastern oyster *Crassostrea virginica*. Maryland Sea Grant College, University of Maryland, College Park, MD, USA, pp 467–513

Springer S (1951) The *Oregon's* fishery explorations in the Gulf of Mexico, 1950. Commer Fish Rev 13:1–8

Springer S, Bullis HR (1954) Exploration shrimp fishing in the Gulf of Mexico. Summary report for 1952-54. Commer Fish Rev 16:1–16

Stanley JG, Sellers MA (1986) Species profiles: Life histories and environmental requirements of coastal fishes and invertebrates (Gulf of Mexico)—American oyster. U.S. Fish Wildlife Service Biological Report 82(11.64), U.S. Army Corps of Engineers Report TR EL-82-4. U.S. Army Corps of Engineers, Vicksburg, MS, USA. 25 p

Steele P, Bert T (1998) The Florida blue crab fishery. J Shellfish Res 17:441–450

Steele P, Perry HM (1990) The blue crab fishery of the Gulf of Mexico, United States: A regional management plan, Publication number 21. Gulf States Marine Fisheries Commission, Ocean Springs, MS, USA

Stenseth NC, Mysterud A, Ottersen G, Hurrell JW, Chan K-S, Lima M (2002) Ecological effects of climate fluctuations. Science 297:1292–1296

Stevely JM (1979) The biology and fishery of the queen conch (*Strombus gigas*); A review. In: Proceedings of the 4th annual tropical and subtropical fisheries technological conference of the Americas, TAMU-SG 80-101. Texas A&M University Sea Grant College Program, College Station, TX, USA pp 203–210

Stiles ML, Harrould-Kolieb E, Faure R, Ylitalo-Ward H, Hirshfield MF (2007) Deep sea trawl fisheries of the southeast U.S. and Gulf of Mexico: Rock shrimp, royal red shrimp, calico scallops. Oceana, New York, NY, USA. 13 p

Stuck KC, Perry HM (1981) Observations on the distributions and the seasonality of portunid megalopae in Mississippi coastal waters. Gulf Res Rep 7:93–95

Sutton G, Wagner T (2007) Stock assessment of blue crab (*Callinectes sapidus*) in Texas coastal waters, Management data series 249. Texas Parks and Wildlife Department, Austin, TX, USA. 42 p

Teal JM (1986) The ecology of regularly flooded salt marshes of New England: A community profile. FWS/OBS-81/01. U.S. Fish and Wildlife Service, Biological Service Program, Washington, DC, USA

Thomas JL, Zimmerman RJ, Minello TJ (1990) Abundance patterns of juvenile blue crabs (*Callinectes sapidus*) in nursery habitats of two Texas bays. Bull Mar Sci 46:115–125

Tolan JM (2007) El Nino-southern oscillation impacts translated to the watershed scale: estuarine salinity patterns along the Texas Gulf Coast, 1982 to 2004. Estuar Coast Shelf Sci 72:247–260

Trigg C, Perry H, Brehm W (1997) Size and weight relationships for the golden crab, *Chaceon fenneri*, and the red crab, *Chaceon quinquedens*, from the eastern Gulf of Mexico. Gulf Res Rep 9:339–343

Truitt RV (1939) The blue crab. In: Our water resources and their conservation, Contribution 27. Chesapeake Biological Lab, Solomons, MD, USA, pp 10–38

Tunnell JW Jr, Chavez RA, Withers K (2007) Coral reefs of the southern Gulf of Mexico. Texas A&M University Press, College Station, TX, USA. 194 p

Tunnell JW Jr, Andrews J, Barrera NC, Moretzsohn F (2010) Encyclopedia of Texas seashells: Identification, ecology, distribution, and history. Texas A&M University Press, College Station, TX, USA. 512 p

Turgeon DD, Quinn JF Jr, Bogan AE, Coan EV, Hockberg FG, Lyons WG, Mikkelsen PM, Neves RJ, Roper CFE, Rosenberg G, Roth B, Scheltema A, Thompson FG, Vecchione M, Williams JD (1998) Common and scientific names of aquatic invertebrates from the United States and Canada: Mollusks, 2nd edn, Special publication 26. American Fisheries Society, Bethesda, MD, USA. 509 p

Turgeon DD, Lyons WG, Mikkelsen P, Rosenberg G, Moretzsohn F (2009) Bivalvia (Mollusca) of the Gulf of Mexico. In: Felder DL, Camp DK (eds) Gulf of Mexico origin, waters, and biota, vol 1, Biodiversity. Texas A&M University Press, College Station, TX, USA, pp 711–744

Turner RE (1977) Intertidal vegetation and commercial yields of penaeid shrimp. Trans Am Fish Soc 106:411–416

Turner RE (1986) Relationship between coastal wetlands, climate, and penaeid shrimp yields. In: Proceedings of the FAOIIOC (IREP (OSLR) workshop on recruitment in coastal demersal stocks, Ciudad del Carmen, Campeche, 21–25 April 1986, IOC workshop report 44—supplement. FOA, Rome, Italy, pp 267–275

Upton HF (2011) The Deepwater Horizon oil spill and the Gulf of Mexico fishing industry. Service Report for Congress. Congressional Research Service, Washington, DC, USA. 14 p

VanderKooy SJ (2012) The oyster fishery of the Gulf of Mexico, United States: A fisheries management plan. Gulf States Marine Fisheries Commission, Ocean Springs, MS, USA

Volety AK, Savarese M, Tolley SG, Arnold WS, Sime P, Goodman P, Chamberlain RH, Doering PH (2009) Eastern oysters (*Crassostrea virginica*) as an indicator for restoration of everglades ecosystems. Ecol Indic 9:S120–S136

Wakida-Kusunoki AT (2005) Seabob shrimp small-scale fishery in southeastern of Mexico. Gulf Caribb Fish Inst 56:573–581

Waller R, Perry H, Trigg C, McBee J, Erdman R, Blake N (1995) Estimates of harvest potential and distribution of the deep sea red crab, *Chaceon quinquedens*, in the north central Gulf of Mexico. Gulf Res Rep 9:75–84

Wells HW (1961) The fauna of oyster beds, with special reference to the salinity factor. Ecol Monogr 31:239–266

White ME, Wilson EA (1996) Predators, pests, and competitors. In: Kennedy VS, Newell RIE, Eble AF (eds) The eastern oyster *Crassostrea virginica*. Maryland Sea Grant, College Park, MD, USA, pp 559–579

Williams AB (1984) Shrimps, lobsters and crabs of the Atlantic Coast. Smithsonian Institution Press, Washington, DC, USA. 550 p

Williams AB, Felder DL (1986) Analysis of stone crabs: *Menippe mercenaria* (Say), restricted, and a previously unrecognized species described (Decapoda: Xanthidae). Proc Biol Soc Wash 99:517–543

Withers K (2010) Shells in Texas coastal history. In: Tunnell JW Jr, Andrews J, Barrera NC, Moretzsohn F (eds) Encyclopedia of Texas seashells: Identification, ecology, distribution, and history. Texas A&M University Press, College Station, TX, USA, pp 5–20

Zetina MC, Rios V, Hernandez C, Guevara M, Ortiz E (2002) Catalogo de especies de pepino de mar comercializables del Estado de Yucatán. Ediciones de la Universidad Autonoma de Yucatán, Mexico

Zimmerman RJ, Nance JM (2001) Effects of hypoxia on the shrimp fishery of Texas and Louisiana. In: Rabalais NN, Turner RE (eds) Coastal hypoxia: Consequences for living resources and ecosystems, coastal and estuarine studies 58. American Geophysical Union, Washington, DC, USA, pp 293–310

Zimmerman RJ, Minello TJ, Castiglione MC, Smith DL (1990) Utilization of marsh and assorted habitats along a salinity gradient in Galveston Bay. NOAA Technical Memorandum, NMFS-SEFC-250. NOAA, Silver Spring, MD, USA. 68 p

Zimmerman RJ, Minello TJ, Rozas LP (2000) Salt marsh linkages to productivity of penaeid shrimps and blue crabs in the northern Gulf of Mexico. In: Weinstein MP, Kreeger DA (eds) Concepts and controversies in tidal marsh ecology. Kluwer Academic Publishers, Dordrecht, The Netherlands, pp 293–314

Zu Ermgassen PSE, Spalding MD, Blake B, Coen LD, Dumbauld B, Geiger S, Grabowski JH, Grizzle R, Luckenbach M, McGraw K, Rodney W, Ruesink JL, Powers SP, Brumbaugh RD (2012) Historical ecology with real numbers: Past and present extent and biomass of an imperiled estuarine habitat. In: Proceedings of the Royal Society B 279:3393–3400 (Biological Sciences), Published online 13 June 2012

APPENDIX A
LIST OF ACRONYMS, ABBREVIATIONS, AND SYMBOLS

°	Degree(s)	BMLD	Below mud-line depth
%	Percent	BOD	Biochemical oxygen demand
‰	Parts per thousand		
°C	Degree(s) Celsius	BOEM	Bureau of Ocean Energy Management
°F	Degree(s) Fahrenheit		
µg/g	Microgram(s) per gram	BOEMRE	Bureau of Ocean Energy Management, Regulation, and Enforcement
µg/L	Microgram(s) per liter		
µm	Micrometer(s)		
µmol	Micromole(s)		
2D	Two-dimensional	BP	BP Exploration & Production Inc.
3D	Three-dimensional		
AAM	American Academy of Microbiology	BRD	Bycatch reduction device
		BSR	Bottom simulating reflector(s)
ac	Acre		
AL	Alabama	BTEC	Barataria-Terrebonne Estuarine Complex
APG	Angiosperm Phylogeny Group		
		BTNEP	Barataria-Terrebonne National Estuary Program
APHA	American Public Health Association		
		BW	Body weight
atm	Atmosphere	cal yr BP	Calibration year Before Present
AUV	Autonomous underwater vehicle		
		CBBE	Coastal Bend Bays and Estuaries
bbl	Barrel(s)		
BEACH	Beaches Environmental Assessment, Closure, and Health Program	CBBEP	Coastal Bend Bays and Estuaries Program
		CCA	Chromated copper arsenate
BEG	Bureau of Economic Geology		
		CEC	Commission for Environmental Cooperation
BIMS	Baseline Information Management System		
BL	Body length	CENR	Committee on the Environment and Natural Resources
BLM	Bureau of Land Management		

C.H. Ward (ed.), *Habitats and Biota of the Gulf of Mexico: Before the Deepwater Horizon Oil Spill*, DOI 10.1007/978-1-4939-3447-8

CFH	Critical fisheries habitat	**DSC**	Deep sea coral
CFMC	Caribbean Fishery Management Council	**DW/dw**	Dry weight
		EEZ	Exclusive Economic Zone
cm	Centimeter(s)	**EFH**	Essential fish habitat
cm³	Cubic centimeters	**ELI**	Environmental Law Institute
CMDA	Centro Mexicano de Derecho Ambiental	**EMAP**	Environmental Monitoring and Assessment Program
CNCYT	Consejo Nacional de Ciencia y Technología (National Council of Science and Technology)	**EMAP-E**	Environmental Monitoring and Assessment Program Estuaries
CONAPESCA	Comisión Nacional de Acuacultura y Pesca (National Commission of Aquaculture and Fisheries)	**ERL**	Effects range low
		ERM	Effects range median
		ESA	Ecological Society of America
CPUE	catch per unit effort	**ESR**	Echo-sounder recorder
CSA	Coastal Study Area	**FAQ**	Frequently asked questions
CSA	Continental Shelf Associates	**FAO**	Food and Agriculture Organization
CU	Cuba		
cy	Cubic yards	**FB**	Florida Bay
DAPI	4′,6-Diamidino-2-phenylindole (fluorescent stain for DNA)	**FDEP**	Florida Department of Environmental Protection
DDT	Dichlorodiphenyltrichloroethane	**FFWCC-FWRI**	Florida Fish and Wildlife Conservation Commission—Fish and Wildlife Research Institute and Research Planning
DGoMB	Deep Gulf of Mexico Benthos		
DHS	U.S. Department of Homeland Security	**FGBNMS**	Flower Gardens Banks National Marine Sanctuary
DIN	Dissolved inorganic nitrogen		
DIP	Dissolved inorganic phosphorus	**FL**	Florida
		FMG	Florida Middle Grounds
DO	Dissolved oxygen	**FMP**	Fishery Management Plan
DOC	Dissolved organic carbon	**FNAI**	Florida Natural Areas Inventory
DoD	U.S. Department of Defense	**ft**	Foot/feet
DOE	U.S. Department of Energy	**ft³**	Cubic feet
		g	Gram(s)
DOI	U.S. Department of the Interior	**gal**	Gallon(s)
		GBEP	Galveston Bay Estuary Program
DOM	Dissolved organic matter		
dpm	Disintegrations per minute	**GCRL**	Gulf Coast Research Laboratory

GERG	Geochemical and Environmental Research Group	**LME**	Large Marine Ecosystem
		LSU	Louisiana State University
GMFMC	Gulf of Mexico Fishery Management Council	**LTA**	Long-term average
		m	Meter(s)
GMSAFMC	Gulf of Mexico and South Atlantic Fishery Management Councils	m^2	Square meter(s)
		m^3	Cubic meter(s)
		m^3/s	Cubic meter(s) per second
GoM/GOMEX	Gulf of Mexico	**MAFAC**	Marine Fisheries Advisory Committee
GoMRI	Gulf of Mexico Research Initiative	**MAFLA**	Mississippi-Alabama-Florida
GOOMEX	Gulf of Mexico Offshore Operations Monitoring Experiment	**MAMES**	Mississippi Alabama Marine Ecosystem Study
		MDM	Mid-depth maximum
GSMFC	Gulf States Marine Fisheries Commission	**MDS**	Multidimensional scaling
		mg	Milligram(s)
GT	Gigatonne(s)	**mg/L**	Milligram(s) per liter
h	Hour(s)	**mi**	Mile(s)
ha	Hectare(s)	mi^2	Square mile(s)
HAB	Harmful algal bloom	mi^3	Cubic mile(s)
HRI	Harte Research Institute for Gulf of Mexico Studies	**min**	Minute(s)
		MIP	Ministerio de la Industria Pesquera (Ministry of Fishing Industries)
ICES	International Commission for the Exploration of the Seas		
		mL/ml	Milliliter(s)
IMaRS/USF	Institute for Marine Remote Sensing, University of South Florida	**mm**	Millimeter(s)
		mM	millimolar
		MMS	Minerals Management Service
		MPA	Marine protected area
in.	Inch(es)	**MRFSS**	Marine Recreational Fisheries Statistics Survey
IPN	Instituto Nacional de Pesca (National Fishery Institute)	**MRIP**	Marine Recreational Information Program
kg	Kilogram(s)	**MS**	Mississippi
km	Kilometer(s)	**MsCIP**	Mississippi Coastal Improvements Program
km^2	Square kilometer(s)		
km^3	Cubic kilometer(s)	**mV**	Millivolt(s)
L/l	Liter(s)	**MX**	Mexico
LA	Louisiana	**NASA**	National Aeronautics and Space Administration
Lb/lb	Pound(s)		
LC	Loop current	**NCA**	National Coastal Assessment
LCS/lcs	Loop current system		
LDWF	Louisiana Department of Wildlife and Fisheries	**NCCR**	National Coastal Condition Report

NDBC	National Data Buoy Center		**PBS**	Public Broadcasting System
NEGOM-CME	Northeastern Gulf of Mexico-Coastal and Marine Ecosystem Program		**PCB**	Polychlorinated biphenyl
NEP	National Estuary Program		**PDB**	Pee Dee Belemnite
			PEMEX	Petróleos Mexicanos (Mexican Petroleums)
NEP CCR	National Estuary Program Coastal Condition Report		**PFA**	Provisional Fishing Associations
ng	Nanogram(s)		**POC**	Particulate organic carbon
NGoMCS	Northern Gulf of Mexico Continental Slope		**POTW**	Publicly owned treatment work
NHC	National Hurricane Center		**PP**	Primary production
			ppb	Part(s) per billion
NJ	New Jersey		**ppm**	Part(s) per million
NLFWA	National Listing of Fish and Wildlife Advisories		**ppt**	Part(s) per thousand
NMFS	National Marine Fisheries Service		**PRAWN**	Program Tracking, Advisories, Water Quality Standards, and Nutrients
NOAA	National Oceanic and Atmospheric Administration		**PSMSL**	Permanent Service for Mean Sea Level
NOAA/OER	NOAA Office of Ocean Exploration and Research		**psu**	Practical salinity unit
			REMAP	Regional Environmental Monitoring and Assessment Program
NOAA/CCMA	NOAA Center for Coastal Monitoring and Assessment		**RPD**	Redox potential discontinuity
NOS	National Ocean Service		**ROV**	Remotely operated (underwater) vehicle
NPDES	National Pollutant Discharge Elimination System		**s/sec**	Second(s)
			SAGARPA	Secretaria de Agricultura, Ganaderia, Desarrollo Rural, Pesca, y Alimentacion (Ministry of Agriculture, Livestock, Rural Development, Fisheries, and Foods)
NRC	National Research Council			
NS&T	National Status and Trends			
NWCM	National Water Commission of Mexico			
NWI	National Wetlands Inventory		**SAP**	Sigsbee Abyssal Plain
			SAV	Submerged aquatic vegetation
OCS	Outer continental shelf			
ONR	Office of Naval Research		**SBNEP**	Sarasota Bay National Estuary Program
PAH	Polycyclic aromatic hydrocarbon		**SCAR**	Scientific Committee on Antarctic Research
PAR	Photosynthetically active radiation		**SCOC**	Sediment community oxygen consumption

SCOPE	Scientific Committee on Problems of the Environment	**TCEQ**	Texas Commission for Environmental Quality
SCUBA	Self-contained underwater breathing apparatus	**TCM**	Trillion cubic meters
		TDS	Total dissolved solids
		TED	Turtle excluder device
SEAMAP	Southeast Area Monitoring and Assessment Program	**TOC**	Total organic carbon
		TSS	Total suspended solids
		TX	Texas
SeaWiFS	Sea-viewing wide field-of-view sensor	**UCM**	Unresolved complex mixture
SEDAR	Southeast Data, Assessment, and Review	**UNEP**	United Nations Environment Program
SEMARNAT	Secretary of Environment and Natural Resources, Mexico (Ministry of the Environment and Natural Resources)	**UNEP/GPA**	United Nations Environment Program/ Global Programme of Action
		U.S.	United States
		USA	United States of America
SEPM	Society for Sedimentology	**USACE**	U.S. Army Corps of Engineers
SERDP	Strategic Environmental Research and Development Program	**USDA**	U.S. Department of Agriculture
SIMB	Society of Industrial Microbiology and Biotechnology	**USEPA**	U.S. Environmental Protection Agency
		USF	University of South Florida
SIMPROF	Similarity profile analysis	**USFWS**	U.S. Fisheries and Wildlife Service
SJRWMD	St. Johns River Water Management District	**USGS**	U.S. Geological Survey
		USM	The University of Southern Mississippi
SLM	Southern Laguna Madre		
SOFLA	Southwest Florida Shelf	**VME**	Vulnerable Marine Ecosystem
STOCS	South Texas Outer Continental Shelf		
		VOC	Volatile organic compound
STP	Standard temperature and pressure	**VPDB**	Vienna PDB
SVOC	Semivolatile organic compound	**WAVCIS**	Wave-Current-Surge Information System for Coastal Louisiana
SWFSC	Southwest Fisheries Science Center	**WCI**	Water clarity indicator
TAMU	Texas A&M University	**WHO**	World Health Organization
TAMU-CC	Texas A&M University at Corpus Christi	**WHOI**	Woods Hole Oceanographic Institution
TAMUG	Texas A&M University at Galveston		
TB	Tampa Bay	**ww**	Wet weight
TBEP	Tampa Bay Estuary Program	**WWF**	World Wildlife Fund
		yr	Year

APPENDIX B
UNIT CONVERSION TABLE

Multiply	By	To Obtain
Acres	0.405	Hectares
Acres	1.56 E–3	Square miles (statute)
Centimeters	0.394	Inches
Cubic feet	0.028	Cubic meters
Cubic feet	7.48	Gallons (U.S. liquid)
Cubic feet	28.3	Liters
Cubic meter	35.3	Cubic feet
Cubic yard	0.76	Cubic meter
Feet	0.305	Meters
Gallons (U.S. liquid)	3.79	Liters
Hectares	2.47	Acres
Inches	2.54	Centimeters
Kilograms	2.20	Pounds (avoir)
Kilometers	0.62	Miles (statue)
Liters	0.035	Cubic feet
Liters	0.26	Gallons (U.S. liquid)
Meters	3.28	Feet
Metric ton(ne)s	1.102	U.S. short tons
Miles (statue)	1.61	Kilometers
Pounds (avoir)	0.45	Kilograms
Square feet	0.093	Square meters
Square kilometers	0.386	Square miles
Square miles	640	Acres
Square miles	2.59	Square kilometers

© The Author(s) 2017

C.H. Ward (ed.), *Habitats and Biota of the Gulf of Mexico: Before the Deepwater Horizon Oil Spill*,
DOI 10.1007/978-1-4939-3447-8

INDEX

A

Abyssal plain, 1, 13, 18, 22, 47, 48, 165, 168, 181, 183, 209, 331, 333, 344, 641, 644–646, 720–738, 741, 742, 744, 747, 749

Acanthocybium solandri, 666

Achiridae, 562, 571

Achirus lineatus, 523, 562, 571

Adinia xenica, 486, 523, 541

African pompano, 661

Agricultural runoff, 5, 61, 96, 219, 710

Alabama Pinnacles, 18, 20, 713, 716, 748, 749

Alaminos Canyon, 338

Alectis, 662

Alectis ciliaris, 661

Algae, 27, 61, 63, 64, 73, 152, 491, 555, 559, 580, 705, 712, 713, 790, 824

Algal bloom, 4, 5, 7, 34, 46, 62, 75, 131, 145, 146, 156, 157, 652, 823

Aliger costatus, 775

Alligator, 368, 453, 478, 489

Alpheus heterochaelis, 558

Amberjack, 26, 28, 49, 661, 669

Americamysis almyra, 558

American horse mussel, 23, 776, 782

American silver perch, 665

Americardia guppyi, 557

Ampelisca
   A. abdita, 473
   A. holmesi, 558
   A. mississippiana, 725, 741

Amphipod, 472, 473, 476, 506, 507, 521, 537, 539, 548, 557, 559, 560, 569, 570, 573, 709, 725, 741, 790

Anadara transversa, 776

Anchoa
   A. cayorum, 486, 541
   A. cubana, 486
   A. hepsetus, 486, 507, 541
   A. lamprotaenia, 486
   A. lyolepis, 486, 507
   A. mitchilli, 485, 486, 507, 523, 524, 541, 562, 572, 585

Anchovie, 484, 507, 540, 543, 544, 699, 700, 703

Anemone, 342, 713, 733, 824

Annelid, 472, 520, 521, 524, 538, 539, 570, 709, 792, 824

Annual rainfall, 56, 58, 368, 370, 470

Anthias, 667, 702

Anthias nicholsi, 667

Anthiinae, 667, 717

Apalachicola Bay, 196–197, 199, 225, 366, 383, 552, 788, 818

Apocorophium louisianum, 558

Arabella iricolor, 557

Aransas Bay, 91, 92, 204, 205

Archea, 276

Archosargus
   A. probatocephalus, 486, 494, 542, 824
   A. rhomboidalis, 486

Argopecten
   A. gibbus, 557
   A. irradians, 556, 560, 584, 776

Ariidae, 483, 484, 525, 562, 571

Ariopsis felis, 562

Armandia agilis, 557

Armases cinereum, 558

Arsenic (As), 9, 67, 97, 160, 219, 223, 239, 253, 258, 267, 268

Arthropod, 824

Artificial reef, 20

Asteroid, 473

Astichopus multifidus, 779

Astropecten, 473

Atchafalaya Bay, 87, 710

Atchafalaya drainage basin(s), 15, 583

Atherinidae, 484, 508, 522, 540, 541, 562

Atherinomorus stipes, 486, 541

Atlantic bluefin tuna, 26, 28, 49, 659, 692

Atlantic bonito, 666

Atlantic brief squid, 776

Atlantic bumper, 661, 669, 672, 700–703

Atlantic croaker, 573, 665, 685, 700, 702, 703

Atlantic mackerel, 666

Atlantic menhaden, 663

Atlantic moonfish, 662

Atlantic rangia, 24, 777, 782

Atlantic seabob, 23, 777, 782, 786–787

Atlantic thread herring, 26, 663, 670, 674, 701–703

© The Author(s) 2017
C.H. Ward (ed.), *Habitats and Biota of the Gulf of Mexico: Before the Deepwater Horizon Oil Spill*,
DOI 10.1007/978-1-4939-3447-8